To: Dr. Shin

My mentor and teacher,
and many thanks to NSF
for the research support that
enabled the work described
in this book.

Aug 6, 2009

Victor Giurgiutiu

Structural Health Monitoring

STRUCTURAL HEALTH MONITORING

WITH

PIEZOELECTRIC WAFER ACTIVE SENSORS

VICTOR GIURGIUTIU
University of South Carolina

AMSTERDAM • BOSTON • HEIDELBERG • LONDON
NEW YORK • OXFORD • PARIS • SAN DIEGO
SAN FRANCISCO • SINGAPORE • SYDNEY • TOKYO

Academic Press is an imprint of Elsevier

Cover photos © iStockphoto
Cover design by Lisa Adamitis

Academic Press is an imprint of Elsevier
30 Corporate Drive, Suite 400, Burlington, MA 01803, USA
525 B Street, Suite 1900, San Diego, California 92101-4495, USA
84 Theobald's Road, London WC1X 8RR, UK

This book is printed on acid-free paper. ♾

Library of Congress Cataloging-in-Publication Data
Giurgiutiu, Victor.
Structural health monitoring with piezoelectric wafer active sensors/Victor Giurgiutiu.
p. cm.
ISBN-13: 978-0-12-088760-6 (alk. paper)
1. Structural analysis (Engineering) 2. Piezoelectric devices. 3. Piezoelectric transducers.
4. Automatic data collection systems. I. Title
TA646.G55 2007
624.1′71–dc22
2007043697

British Library Cataloguing-in-Publication Data
A catalogue record for this book is available from the British Library.

ISBN: 978-0-12-088760-6

For information on all Academic Press publications
visit our Web site at www.books.elsevier.com

Printed in the United States of America
07 08 09 10 10 9 8 7 6 5 4 3 2 1

To My Loving and Understanding Family

CONTENTS

1

INTRODUCTION

1.1 STRUCTURAL HEALTH MONITORING PRINCIPLES AND CONCEPTS

Structural health monitoring (SHM) is an area of growing interest and worthy of new and innovative approaches. The United States spends more than $200 billion each year on the maintenance of plant, equipment, and facilities. Maintenance and repairs represents about a quarter of commercial aircraft operating costs. Out of approximately 576 600 bridges in the US national inventory, about a third are either 'structurally deficient' and in need of repairs, or 'functionally obsolete' and in need of replacement. The mounting costs associated with the aging infrastructure have become an on-going concern. Structural health monitoring systems installed on the aging infrastructure could ensure increased safety and reliability.

Structural health monitoring is an area of great technical and scientific interests. The increasing age of our existing infrastructure makes the cost of maintenance and repairs a growing concern. Structural health monitoring may alleviate this by replacing scheduled maintenance with as-needed maintenance, thus saving the cost of unnecessary maintenance, on one hand, and preventing unscheduled maintenance, on the other hand. For new structures, the inclusion of structural health monitoring sensors and systems from the design stage is likely to greatly reduce the life-cycle cost.

Structural health monitoring is an emerging research area with multiple applications. Structural health monitoring assesses the state of structural health and, through appropriate data processing and interpretation, may predict the remaining life of the structure. Many aerospace and civil infrastructure systems are at or beyond their design life; however, it is envisioned that they will remain in service for an extended period. SHM is one of the enabling technologies that will make this possible. It addresses the problem of aging structures, which is a major concern of the engineering community. SHM allows condition-based maintenance (CBM) inspection instead of schedule-driven inspections. Another potential SHM application is in new systems; that is, by embedding SHM sensors and associate sensory systems into a new structure, the design paradigm can be changed and considerable savings in weight, size, and cost can be achieved. A schematic representation of a generic SHM system is shown in Fig. 1.1.

Structural health monitoring can be performed in two main ways: (a) passive SHM; and (b) active SHM. *Passive SHM* is mainly concerned with measuring various operational parameters and then inferring the state of structural health from these parameters. For example, one could monitor the flight parameters of an aircraft (air speed, air turbulence,

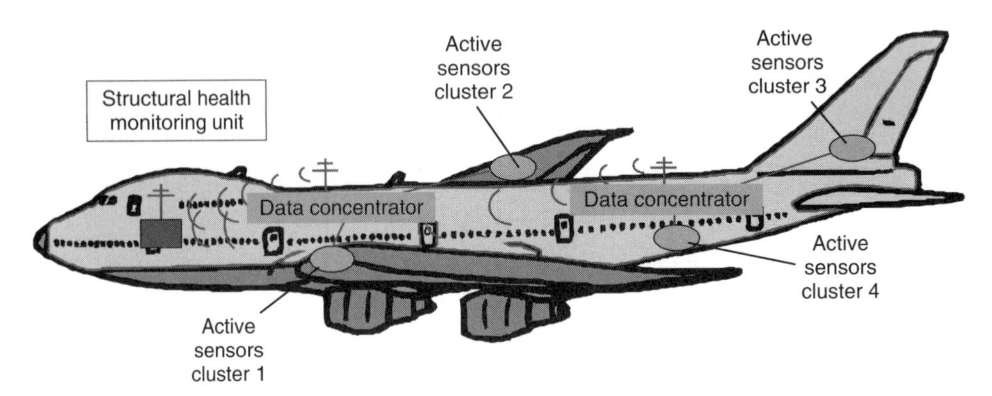

FIGURE 1.1 Schematic representation of a generic SHM systems consisting of active sensors, data concentrators, wireless communication, and SHM central unit.

g-factors, vibration levels, stresses in critical locations, etc.) and then use the aircraft design algorithms to infer how much of the aircraft useful life has been used up and how much is expected to remain. Passive SHM is useful, but it does not directly address the crux of the problem, i.e., it does not directly examine if the structure has been damaged or not. In contrast, *active SHM* is concerned with directly assessing the state of structural health by trying to detect the presence and extent of structural damage. In this respect, active SHM approach is similar with the approach taken by nondestructive evaluation (NDE) methodologies, only that active SHM takes it one-step further: active SHM attempts to develop damage detection sensors that can be permanently installed on the structure and monitoring methods that can provide on demand a structural health bulletin. Recently, damage detection through guided-wave NDE has gained extensive attraction. Guided waves (e.g., Lamb waves in plates) are elastic perturbations that can propagate for long distances in thin-wall structures with very little amplitude loss. In Lamb-wave NDE, the number of sensors required to monitor a structure can be significantly reduced. The potential also exist of using phased array techniques that use Lamb waves to scan large areas of the structure from a single location. However, one of the major limitations in the path of transitioning Lamb-wave NDE techniques into SHM methodologies has been the size and cost of the conventional NDE transducers, which are rather bulky and expensive. The permanent installation of conventional NDE transducers onto a structure is not feasible, especially when weight and cost are at a premium such as in the aerospace applications. Recently emerged *piezoelectric wafer active sensors* (PWAS) have the potential to improve significantly structural health monitoring, damage detection, and nondestructive evaluation. PWAS are small, lightweight, inexpensive, and can be produced in different geometries. PWAS can be bonded onto the structural surface, can be mounted inside built-up structures, and can be even embedded between the structural and nonstructural layers of a complete construction. Studies are also being performed to embed PWAS between the structural layers of composite materials, though the associated issues of durability and damage tolerance has still to be overcome.

Structural damage detection with PWAS can be performed using several methods: (a) *wave propagation*, (b) *frequency response transfer function*, or (c) *electromechanical (E/M) impedance*. Other methods of using PWAS for SHM are still emerging. However, the modeling and characterization of Lamb-wave generation and sensing using surface-bonded or embedded PWAS for SHM has still a long way to go. Also insufficiently advanced are reliable damage metrics that can assess the state of structural health with confidence and trust. The Lamb-wave-based damage detection techniques using structurally integrated PWAS for SHM is still in its formative years. When SHM systems are being developed, it is often found that little mathematical basis is provided for the choice of the various testing parameters involved such as transducer geometry, dimensions, location and materials, excitation frequency, bandwidth, etc.

Admittedly, the field of structural health monitoring is very vast. A variety of sensors, methods, and data reduction techniques can be used to achieve the common goal of asking the structure 'how it feels' and determining the state of its 'health', i.e., structural integrity, damage presence (if any), and remaining life. Attempting to give an encyclopedic coverage of all such sensors, methods, and techniques is not what this book intends to do. Rather, this book intends to present an integrated approach to SHM using as a case study the *PWAS* and then taking the reader through a step-by-step presentation of how these sensors can be used to detect and quantify the presence of damage in a given structure. In this process, the book goes from simple to complex, from the modeling and testing of simple laboratory specimens to evaluation of large, realistic structures. The book can be used as a textbook in the classroom, as a self-teaching text for technical specialists interested in entering this new field, or a reference monograph for practicing experts using active SHM methods.

1.2 STRUCTURAL FRACTURE AND FAILURE

1.2.1 REVIEW OF LINEAR ELASTIC FRACTURE MECHANICS PRINCIPLES

The stress intensity factor at a crack tip has the general expression

$$K(\sigma, a) = C\sigma\sqrt{\pi a} \tag{1}$$

where σ is the applied stress, a is the crack length, and C is a constant depending on the specimen geometry and loading distribution. It is remarkable that the stress intensity factor increases not only with the applied stress, σ, but also with the crack length, a. As the crack grows, the stress intensity factor also grows. If the crack grows too much, a critical state is achieved when the crack growth becomes rapid and uncontrollable. The value of K associated with rapid crack extension is called the *critical stress intensity factor* K_c. For a given material, the onset of rapid crack extension always occurs at the same stress intensity value, K_c. For different specimens, having different initial crack lengths and geometries, the stress level, σ, at which rapid crack extension occurs, may be different. However, the K_c value will always be the same. Therefore, K_c is a property of the material. Thus, the condition for fracture to occur is that the local stress intensity factor $K(\sigma, a)$ exceeds the value K_c, i.e.,

$$K(\sigma, a) \geq K_c \tag{2}$$

We see that K_c provides a single-parameter fracture criterion that allows the prediction of fracture. Although the detailed calculation of $K(\sigma, a)$ and determination of K_c may be difficult in some cases, the general concept of using K_c to predict brittle fracture remains nonetheless applicable. The K_c concept can also be extended to materials that posses some limited ductility, such as high-strength metals. In this case, the $K(\sigma, a)$ expression (1) is modified to account for a crack-tip plastic zone, r_Y, such that

$$K(\sigma, a) = C\sigma\sqrt{\pi(a + r_Y)} \tag{3}$$

where the maximum value of r_Y can be estimated as

$$r_{Y\sigma} = \frac{1}{2\pi}\sqrt{\frac{K_c}{Y}} \quad \text{(plane stress)} \tag{4}$$

$$r_{Y\sigma} = \frac{1}{6\pi}\sqrt{\frac{K_c}{Y}} \quad \text{(plane strain)} \tag{5}$$

In studying the material behavior, one finds that the plane strain conditions give the lowest value of K_c, whereas the plane stress conditions can give K_c values that may range from two to ten times higher. This effect is connected with the degree of constraint imposed upon the material. Materials with higher constraint effects have a lower K_c value. The plane strain condition is the condition with most constraint. The plane strain K_c is also called the *fracture toughness* K_{Ic} of the material. Standard test methods exist for determining the material fracture-toughness value. When used in design, fracture-toughness criteria gives a larger margin of safety than elastic–plastic fracture mechanics methods such as (a) crack opening displacement (COD) methods; (b) R-curve methods; (c) J-integral methods. However, the fracture toughness approach is more conservative: it is safer, but heavier. For a complete design analysis, the designer should consider, in most cases, both conditions: (a) the possibility of failure by brittle fracture; and (b) the possibility of failure by ductile yielding.

1.2.2 FRACTURE MECHANICS APPROACH TO CRACK PROPAGATION

The concepts of linear fracture mechanics can be employed to analyze a given structure and predict the crack size that will propagate spontaneously to failure under the specified loading. This critical crack size can be determined from the critical stress intensity factor as defined in Eq. (3). A fatigue crack that has been initiated by cyclic loading, or other damage mechanism, may be expected to grow under sustained cyclic loading until it reaches a critical size beyond which will propagate rapidly to catastrophic failure. Typically, the time taken by a given crack damage to grow to a critical size represents a significant portion of the operational life of the structure. In assessing the useful life of a structure, several things are needed such as:

- understanding of the crack-initiation mechanism
- definition of the critical crack size, beyond which the crack propagates catastrophically
- understanding the crack-growth mechanism that makes a subcritical crack propagate and expand to the critical crack size.

Experiments of crack length growth with number of cycles for various cyclic-load values have indicated that a high value of the cyclic load induces a much more rapid crack growth than a lower value (Collins, 1993). It has been found that crack growth phenomenon has several distinct regions (Fig. 1.2):

(i) An initial region in which the crack growth is very slow
(ii) A linear region in which the crack growth is proportional with the number of cycles
(iii) A nonlinear region in which the log of the crack growth rate is proportional with the log of the number of cycles.

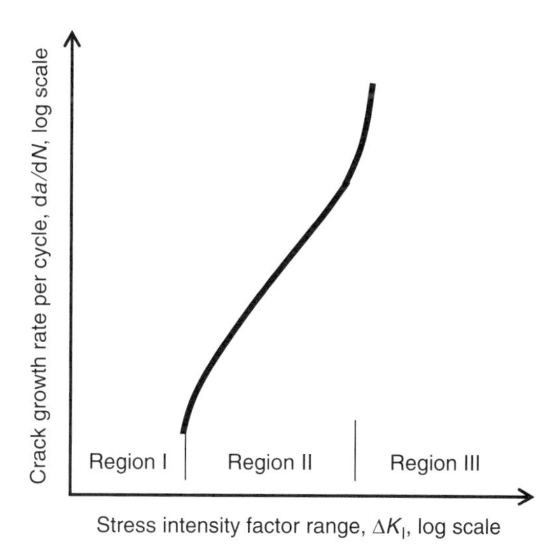

FIGURE 1.2 Schematic representation of fatigue crack growth in metallic materials.

In analyzing fatigue crack growth, Paris and Erdogan (1963) determined that the fatigue crack-growth rate depends on the alternating stress and crack length:

$$\frac{da}{dN} = f(\Delta\sigma, a, C) \tag{6}$$

where $\Delta\sigma$ is the peak-to-peak range of the cyclic stress, a is the crack length, and C is a parameter that depends on mean load, material properties, and other secondary variables.

In view of Eq. (1), it seems appropriate to assume that the crack-growth rate will depend on the cyclic stress intensity factor, ΔK, i.e.,

$$\frac{da}{dN} = g(\Delta K) \tag{7}$$

where ΔK is the peak-to-peak range of the cyclic stress intensity factor. Experiments have shown that, for various stress levels and various crack lengths, the data points seem to follow a common law when plotted as crack-growth rate versus stress intensity factor (Collins, 1993). This remarkable behavior came to be known as 'Paris law'; its representation corresponds to the middle portion of the curve shown in Fig. 1.2. Fatigue-crack growth-rate laws have been reported for a wide variety of engineering materials. As middle portion of the curve in Fig. 1.2 is linear on log-log scale, the corresponding Eq. (7) can be written as:

$$\frac{da}{dN} = C_{EP}(\Delta K)^n \tag{8}$$

where n is the slope of the log-log line, and C_{EP} is an empirical parameter that depends upon material properties, test frequency, mean load, and some secondary variables. If the parameter C_{EP} and n are known, then one can predict how much a crack has grown after N cycles, i.e.,

$$a(N) = a_0 + \int_1^N C_{EP}(\Delta K)^n dN \tag{9}$$

where a_0 is the initial crack length.

Paris law represents well the middle portion of the curve in Fig. 1.2. However, the complete crack-growth behavior has three separate phases:

(1) Crack nucleation
(2) Steady-state regime of linear crack growth on the log-log scale
(3) Transition to the unstable regime of rapid crack extension and fracture.

Such a situation is depicted in Fig. 1.2, where Region I corresponds to the crack nucleation phase, Region II to linear growth, and Region III to transition to the unstable regime. Threshold values for ΔK that delineate one region from the other seem to exist. As shown in Fig. 1.2, the locations of these regions in terms of stress intensity factor vary significantly from one material to another.

Paris law is widely used in engineering practice. Further studies have revealed several factors that also need to be considered when applying Paris law to engineering problems. Some of these factors are

- Influence of cyclic stress ratio on the threshold value of ΔK
- Difference between constant-amplitude tests and spectrum loading
- Effect of maximum stress on spectrum loading
- Retardation and acceleration effects due to overloads.

The influence of the stress ratio and threshold have been incorporated in the modified Paris law (Hartman and Schijve, 1970)

$$\frac{\mathrm{d}a}{\mathrm{d}N} = \frac{C_{\mathrm{HS}}(\Delta K - \Delta K_{\mathrm{TH}})^m}{(1-R)K_c - \Delta K} \tag{10}$$

where R is the stress ratio $\sigma_{\max}/\sigma_{\min}$, K_c is the fracture toughness for unstable crack growth under monotonic loading, ΔK_{TH} is the threshold cyclic stress intensity factor for fatigue propagation, and C_{HS} is an empirical parameter.

The difference between constant-amplitude loading and spectrum loading has been shown to depend on the maximum stress value. If the maximum stress is held at the same values in both constant-amplitude and spectrum loading, then the crack growth rates seem to follow the same law. However, if the maximum stress is allowed to vary, the spectrum loading results seem to depend strongly on the sequence in which the loading cycles are applied, with the overall crack growth being significantly higher for spectrum loading than for constant-amplitude loading (McMillan and Pelloux, 1967). The retardation effects due to overloads have been reported by several investigators as evidence of the *interaction effect* whereby fatigue damage and crack extension depend on preceding cyclic-load history. An interaction of considerable interest is the *retardation* of crack growth due to the application of occasional cycles of crack-opening overload. Retardation is characterized by a period of reduced crack-growth rate following the application of a peak load higher than the subsequent peak. The retardation has been explained by the inference that the overload will induce yield at the crack tip and will produce a zone of local plastic deformation in the crack-tip vicinity. When the overload is removed, the surrounding material forces the yielded zone into a state of residual compression that tends to inhibit the crack growth under the subsequent loads of lower value. The crack-growth rate will remain smaller until the growing crack has traversed the overload yield zone, when it returns to the normal value. Crack-growth *acceleration*,

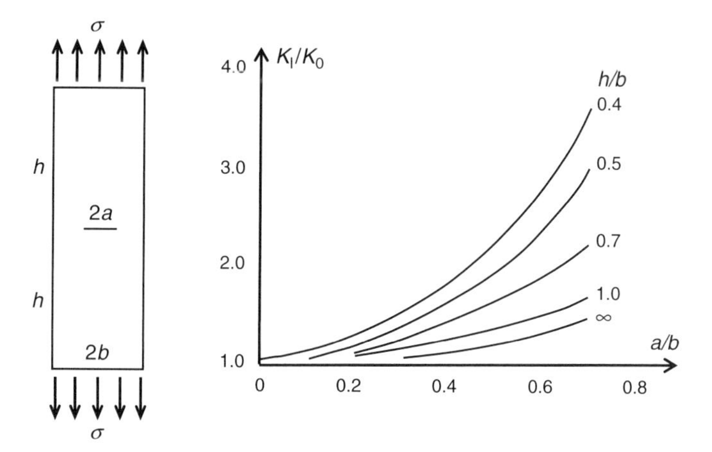

FIGURE 1.3 Plate (length $= 2h$, width $= 2b$), containing a central crack length of $2a$. Tensile stress σ acts in the longitudinal direction.

on the other hand, may occur after crack-closing overloads. In this case, the overload yield zone will produce residual tension stresses, which add to the subsequent loading and result in crack-growth acceleration.

For simple geometries, the stress intensity factor can be predicted analytically. Such predictions have been confirmed by extensive experimental testing; look-up tables and graphs have been made available for design usage. For example, a rectangular specimen with a crack in the middle has stress intensity factor for mode I cracking given by

$$K_I = \beta\sigma\sqrt{\pi a} \tag{11}$$

where σ is the applied tensile stress, a is half of the crack length, and $\beta = K_I/K_0$. The term K_0 represents the ideal stress intensity factor corresponding to an infinite plate with a single crack in the center. The parameter β represents the effect of having a plate of finite dimensions, i.e., the changes in the elastic field due to the plate boundaries not being infinitely far from the crack (Fig. 1.3). The value of the parameter β for a large variety of specimen geometries can be found in the literature.

1.3 IMPROVED DIAGNOSIS AND PROGNOSIS THROUGH STRUCTURAL HEALTH MONITORING

1.3.1 FRACTURE CONTROL THROUGH NDI/NDE

In-service inspection procedures play a major role in the fail-safe concept. Structural regions and elements are classified with respect to required nondestructive inspection (NDI) and NDE sensitivity. Inspection intervals are established on the basis of crack growth information assuming a specified initial flaw size and a 'detectable' crack size, a_{det}, the latter depending on the level of available NDI/NDE procedure and equipment. Cracks larger than a_{det} are presumed to be discovered and repaired. The inspection intervals must be such that an undetected flaw will not grow to critical size before the next inspection. The assumptions used in the establishment of inspection intervals are

- All critical points are checked at every inspection
- Cracks larger than a_{det} are all found during the inspection
- Inspections are performed on schedule
- Inspection techniques are truly nondamaging.

In practice, these assumptions are sometimes violated during infield operations, or are impossible to fulfill. For example, many inspections that require extensive disassembly for access may result in flaw nucleation induced by the disassembly/reassembly process. Some large aircraft can have as many as 22 000 critical fastener holes in the lower wing alone (Rich and Cartwright, 1977). Complete inspection of such a large number of sites is not only tedious and time consuming, but also subject to error born of the boredom of inspecting 20 000 holes with no serious problems, only to miss one hole with a serious crack (sometimes called the 'rogue' crack). Nonetheless, the use of NDI/NDE techniques and the establishment of appropriate inspection intervals have progressed considerably. Recent developments include automated scanning systems and pattern-recognition method that relive the operator of the attention consuming tedious decision making in routine situations and allow the human attention to be concentrated on truly difficult cases. Nevertheless, the current practice of scheduled NDI/NDE inspections leaves much to be desired.

1.3.2 DAMAGE TOLERANCE, FRACTURE CONTROL AND LIFE-CYCLE PROGNOSIS

A *damage tolerant* structure has a design configuration that minimizes the loss of aircraft due to the propagation of undetected flaws, cracks, and other damage. To produce a damage-tolerant structure, two design objectives must be met:

(1) Controlled safe flaw growth, or safe life with cracks
(2) Positive damage containment, i.e., a safe remaining (residual) strength.

These two objectives must be simultaneously met in a judicious combination that ensures effective fracture control. Damage-tolerant design and fracture control includes the following:

(i) Use of fracture-resistant materials and manufacturing processes
(ii) Design for inspectability
(iii) Use of damage-tolerant structural configurations such as multiple load paths or crack stoppers (Fig. 1.4).

In the application of fracture control principles, the basic assumption is that flaws do exist even in new structures and that they may go undetected. Hence, any member in the structure must have a *safe life* even when cracks are present. In addition, flight-critical components must be *fail-safe*. The concept of *safe life* implies the evaluation of the expected lifetime through margin-of-safety design and full-scale fatigue tests. The margin of safety is used to account for uncertainties and scatter. The concept of *fail-safe* assumes that flight-critical components cannot be allowed to fail, hence alternative load paths are

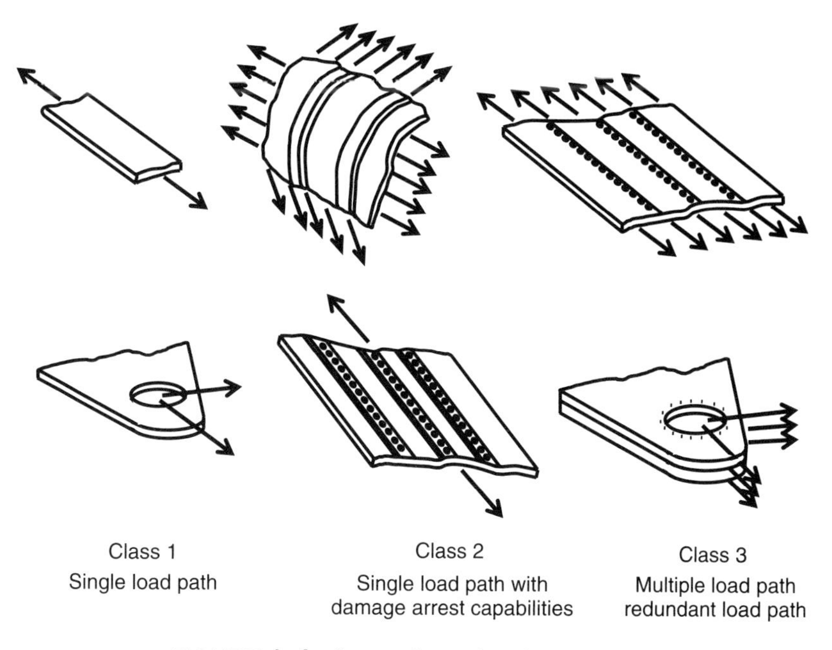

FIGURE 1.4 Structural types based on load path.

supplied through redundant components. These alternative load paths are assumed to be able to carry the load until the failure of the primary component is detected and a repair is made.

1.3.3 LIFE-CYCLE PROGNOSIS BASED ON FATIGUE TESTS

The estimated design life of an aircraft is based on full-scale fatigue testing of complete test articles under simulated fatigue loading. The benefits of full-scale fatigue testing include:

- Discover fatigue critical elements and design deficiencies
- Determine time intervals to detectable cracking
- Collect data on crack propagation
- Determine remaining safe life with cracks
- Determine residual strength
- Establish proper inspection intervals
- Develop repair methods.

The structural life proved through simulation test should be longer by a factor from two to four than the design life. Full-scale fatigue testing should be continued over the long term such that fatigue failures in the test article will stay ahead of the fleet experience by enough time to permit the redesign and installation of whatever modifications are required to prevent catastrophic fleet failures. However, full-scale fatigue testing of an article such as a newly designed aircraft is extremely expensive. In addition, the current aircraft in our fleets have exceeded the design fatigue life, and hence are no longer covered by the full-scale fatigue testing done several decades ago.

1.3.4 PERCEIVED SHM CONTRIBUTIONS THE STRUCTURAL DIAGNOSIS AND PROGNOSIS

Structural health monitoring could have a major contribution to the structural diagnosis and prognosis. Although NDE methods and practices have advanced remarkably in recent years, some of their inherent limitations still persist. NDI/NDE inspection sensitivity and reliability are driven by some very practical issues when dealing with actual airframes. Field inspection conditions may be quite different when compared with laboratory test standards.

Perhaps the major limitation of current NDI/NDE practices is the fact that NDI/NDE, as we know it, cannot provide a continuous assessment of the structural material state. This limitation is rooted in the way NDI/NDE inspections are performed: the aircraft has to be taken off line, stripped down to a certain extent, and scanned with NDI/NDE transducers. This process is time-consuming and expensive. This situation could be significantly improved through the implementation of a SHM system. Having the SHM transducers permanently attached to the structure (even inside closed compartments), would allow for structural interrogation (scanning) to be performed on demand, as often as needed. In addition, a consistent historical record can be accumulated because these on-demand interrogations are done always with the same transducers that are placed in exactly the same locations and interrogated in the same way.

Structural health monitoring could provide an advanced utilization of the existing sensing technologies to add progressive state-change information to a system reasoning process from which we can infer component capability and predict its future safe-use capacity (Cruse, 2004). Through monitoring the state of structural health, we can achieve a historical database and acquire change information to assist in the system reasoning process. Advanced signal processing methods can be used to detect characteristic changes in the material state and make that state-change information available to the prognosis reasoning system. The concept of change detection can be used to characterize the material state by identifying critical features that show changes with respect to a reference state that is stored in the information database and updated periodically. When this is performed in coordination with existing NDI/NDE practices, the structural health monitoring information performed in between current inspection intervals will provide supplementary data that would have a densifying effect on the historical information database.

Another advantage of implementing SHM systems is related to the nonlinear aspects of structural crack propagation. Most of the current life prognosis techniques are based on linear assumptions rooted in laboratory tests performed under well-defined conditions. However, actual operational conditions are far from ideal, and incorporate a number of unknown factors such as *constraint effects*, *load spectrum variation*, and *overloads*. These effects are in the realm of nonlinear fracture mechanics and make the prediction very difficult. However, the dense data that can be collected by an SHM system could be used as feedback information on, say, the crack-growth rate, and could allow the adjustment of the basic assumptions to improve the crack-growth prediction laws.

1.4 ABOUT THIS BOOK

The book is organized in 12 chapters. Chapter 1 presents an introduction to SHM, its motivation, and main approaches. Focus is brought on PWAS and their possible uses in the SHM process. Chapter 2 is dedicated to the description of active materials, which perform bidirectional transduction of electric or magnetic energy into mechanical vibration and wave energy. Active materials (piezoelectrics, electrostrictive, magnetostrictive, etc.)

are the essential ingredient in the construction of active sensors for SHM applications. Chapters 3 through 6 cover in some details the essential vibration and wave propagation theory needed to understand the active SHM approach. The presentation is done in a unified approach, with common notations spanning across these chapters. In writing these chapters, the author has insisted on presenting the fact that vibration and wave propagation phenomena have a common root, and thus deserve an unified treatment, which is not usually achieved in conventional textbooks. Chapters 7 through 11 address the various techniques that are employed to achieve structural health monitoring with PWAS. Thus, Chapter 7 describes the PWAS construction and their operation principles. Chapter 8 treats the methods used to achieve tuning between PWAS and the guided waves traveling in the structure such that single-mode excitation of multi-mode waves is achieved. Chapter 9 discusses standing-wave techniques in which PWAS are used as high-frequency modal sensors. In this method, the damage in the structure is detected from the changes observed in the high-frequency vibration spectrum measured with the E/M impedance method. Chapter 10 presents the wave propagation techniques in which PWAS are used as transmitters and receivers of guided waves and damage is detected through reflections, scatter, and modification of the wave signal. Chapter 11 presents the use of PWAS in phased arrays, which permits the creation of wave beams that are steered electronically such that a large structural area can be monitored from a single location. Chapter 12 presents the signal processing methods needed in performing structural health monitoring. A number of mathematical and elasticity prerequisites that are needed in understanding the book, but may be already known to some of the readers, are presented in the appendices.

This book is thought out as a textbook. This textbook can be used for both teaching and research. It not only provides students, engineers and other interested technical specialists with the foundational knowledge and necessary tools for understanding SHM transducers and systems, but also shows them how to employ this knowledge in actual-engineering situations. This textbook offers comprehensive teaching tools (workout examples, experiments, homework problems, and exercises). An extensive on-line instructor manual containing lecture plans and homework solutions that can be used at various instructional levels (undergraduate, Master and PhD) is posted on the publisher's website. The reader is encouraged to download the instructor's manual and use it for teaching, research, and/or self instruction.

2

ELECTROACTIVE AND MAGNETOACTIVE MATERIALS

2.1 INTRODUCTION

Electroactive and magnetoactive materials are materials that modify their shape in response to electric or magnetic stimuli. Such materials permit induced-strain actuation and strain sensing which are of considerable importance in SHM. Induced-strain actuation allows us to create motion at the micro scale without pistons, gears, or other mechanisms. Induced-strain actuation relies on the direct conversion of electric or magnetic energy into mechanical energy. It is a solid-state actuation, has much fewer parts than conventional actuation, and is much more reliable. It offers the opportunity for creating SHM systems that are miniaturized, effective, and efficient. On the other hand, strain sensing with electroactive and magnetoactive materials creates direct conversion of mechanical energy into electric and magnetic energy. With piezoelectric strain sensors, strong and clear voltage signals are obtained directly from the sensor without the need for intermediate gage bridges, signal conditioners, and signal amplifiers. These direct sensing properties are especially significant in dynamics, vibration, and audio applications in which alternating effects occur in rapid succession thus preventing charge leaking. Other applications of active materials are in sonic and ultrasonic transduction, in which the transducer acts as both sensor and actuator, first transmitting a sonic or ultrasonic pulse, and then detecting the echoes received from the defect or target.

In this chapter, we will discuss several types of active materials: piezoelectric ceramics, electrostrictive ceramics, piezoelectric polymers, and magnetostrictive compounds. Various formulations of these materials are currently available commercially. The names PZT (a piezoelectric ceramic), PMN (an electrostrictive ceramic), Terfenol-D (a magnetostrictive compound), and PVDF (a piezoelectric polymer) have become widely used. In this chapter, we will attempt a review of the principal active material types. We will treat each material type separately, will present their salient features, and introduce the modeling equations. In our discussion, we will start with a general perspective on the overall subject of piezoelectricity and ferroelectric ceramics, explaining some of the physical behavior underpinning their salient features, especially in relation to perovskite crystalline structures. We will continue by considering separately the piezoceramics and electrostrictive ceramics commonly used in current applications and commercially available to the interested user. The focus of the discussion is then switched toward piezoelectric polymers, such as PVDF, with their interesting properties, such as flexibility, resilience, and durability, which make them preferable to ferroelectric ceramics in certain applications. The discussion of magnetostrictive materials, such as Terfenol-D, concludes our review of the active materials spectrum. Thus, we will pave the way toward the next chapters, in which the use of active materials in the construction of induced-strain actuators and active sensors for SHM applications will be discussed.

2.2 PIEZOELECTRICITY

Piezoelectricity describes the phenomenon of generating an electric field when the material is subjected to a mechanical stress (direct effect), or, conversely, generating a mechanical strain in response to an applied electric field. The *direct piezoelectric effect* predicts how much electric field is generated by a given mechanical stress. This *sensing effect* is utilized in the development of piezoelectric sensors. The *converse piezoelectric effect* predicts how much mechanical strain is generated by a given electric field. This *actuation effect* is utilized in the development of piezoelectric induced-strain actuators. Piezoelectric properties occur naturally in some crystalline materials, e.g., quartz crystals (SiO_2) and Rochelle salt. The latter is a natural ferroelectric material, possessing an orientable domain structure that aligns under an external electric field and thus enhances its piezoelectric response. Piezoelectric response can also be induced by electrical poling certain polycrystalline materials, such as piezoceramics.

2.2.1 ACTUATION EQUATIONS

For linear piezoelectric materials, the interaction between the electrical and mechanical variables can be described by linear relations (ANSI/IEEE Standard 176-1987). A constitutive relation is established between mechanical and electrical variables in the tensorial form

$$S_{ij} = s^E_{ijkl} T_{kl} + d_{kij} E_k + \delta_{ij} \alpha^E_i \theta \tag{1}$$

$$D_i = d_{ikl} T_{kl} + \varepsilon^T_{ik} E_k + \tilde{D}_i \theta \tag{2}$$

where S_{ij} and T_{ij} are the strain and stress, E_k and D_i are the electric field and electric displacement, and θ is the temperature. The stress and strain variables are second-order tensors, whereas the electric field and the electric displacement are first-order tensors. The coefficient s_{ijkl} is the compliance, which signifies the strain per unit stress. The coefficients d_{ikl} and d_{kij} signify the coupling between the electrical and the mechanical variables, i.e., the charge per unit stress and the strain per unit electric field. The coefficient α_i is the coefficient of thermal expansion. The coefficient $\tilde{D}_i$ is the electric displacement temperature coefficient. Because thermal effects influence only the diagonal terms, the respective coefficients, α_i and $\tilde{D}_i$, have single subscripts. The term δ_{ij} is the Kroneker delta ($\delta_{ij} = 1$ if $i = j$; zero otherwise). The Einstein summation convention for repeated tensor indices (Knowles, 1997) is employed throughout. The superscripts T, D, E shown in these and other equations signify that the quantities are measured at zero stress ($T = 0$), zero electric displacement ($D = 0$), or zero electric field ($E = 0$), respectively. In practice, the zero electric displacement condition corresponds to open circuit (zero current across the electrodes), whereas the zero electric field corresponds to closed circuit (zero voltage across the electrodes). The strain is defined as

$$S_{ij} = \frac{1}{2}(u_{i,j} + u_{j,i}) \tag{3}$$

where u_i is the displacement, and the comma followed by an index signifies partial differentiation with respect to the space coordinate associated with that index.

Equation (1) is the *actuation equation*. It is used to predict how much strain will be created at a given stress, electric field, and temperature. The terms proportional with stress and temperature are common with the formulations of classical thermoelasticity. The term proportional with the electric field is specific to piezoelectricity and represents the *induced-strain actuation* (ISA), i.e.,

$$S_{ij}^{\text{ISA}} = d_{kij} E_k \tag{4}$$

For this reason, the coefficient d_{kij} can be interpreted as the *piezoelectric strain coefficient*.

Equation (2) is used to predict how much electric displacement, i.e., charge per unit area, is required to accommodate the simultaneous state of stress, electric field, and temperature. In particular, the term $d_{ikl}T_{kl}$ indicates how much charge is being produced by the application of the mechanical stress T_{kl}. For this reason, the coefficient d_{ikl} can be interpreted as the *piezoelectric charge coefficient*. Note that d_{kij} and d_{ikl} represent the same third-order tensor only that the indices have been named appropriately to the respective equations in which they are used.

2.2.2 SENSING EQUATIONS

So far, the piezoelectric equations have expressed the strain and electric displacement in terms of applied stress, electric field, and temperature using the constitutive tensorial Eqs. (1) and (2), and their matrix correspondents. However, these equations can be replaced by an equivalent set of equations that highlight the sensing effect, i.e., predict how much electric field will be generated by a given state of stress, electric displacement, and temperature. (As the electric voltage is directly related to the electric field, this arrangement is preferred for sensing applications.) Thus, Eqs. (1) and (2) can be expressed as

$$S_{ij} = s_{ijkl}^{D} T_{kl} + g_{kij} D_k + \delta_{ij} \alpha_i^{D} \theta \tag{5}$$

$$E_i = g_{ikl} T_{kl} + \beta_{ik}^{T} D_k + \tilde{E}_i \theta \tag{6}$$

Equation (6) predicts how much electric field, i.e., voltage per unit thickness, is generated by 'squeezing' the piezoelectric material, i.e., represents the direct piezoelectric effect. This formulation is useful in piezoelectric sensor design. Equation (6) is called the sensor equation. The coefficient g_{ikl} is the *piezoelectric voltage coefficient* and represents how much electric field is induced per unit stress. The coefficient $\tilde{E}_i$ is the *pyroelectric voltage coefficient* and represents how much electric field is induced per unit temperature change.

2.2.3 STRESS EQUATIONS

The piezoelectric constitutive equations can also be expressed in such a way as to reveal stress and electric displacement in terms of strain and electric field. This formulation is especially useful for defining the piezoelectric constitutive equations in stress and strength analyses. The stress formulation of the piezoelectric constitutive equations are

$$T_{ij} = c^E_{ijkl} S_{kl} - e_{kij} E_k - c^E_{ijkl} \delta_{kl} \alpha^E_k \theta \tag{7}$$

$$D_i = e_{ikl} S_{kl} + \varepsilon^T_{ik} E_k + \tilde{D}_i \theta \tag{8}$$

where c^E_{ijkl} is the stiffness tensor, and e_{kij} is the piezoelectric stress constant. The term $c^E_{ijkl}\delta_{kl}\alpha^E_k\theta$ represents the stress induced in a piezoelectric material by temperature changes when the strain is forced to be zero. For example, the material being fully constraint against deformation. Such stresses, which are induced by temperature effects, are also known as *residual thermal stresses*. They are very important in calculating the strength of piezoelectric materials, especially when they are processed at elevated temperatures.

2.2.4 ACTUATOR EQUATIONS IN TERMS OF POLARIZATION

In practical piezoelectric sensor and actuator design, the use of electric field, E_i, and electric displacement, D_i, is more convenient, as these variables relate directly to the voltage and current that can be experimentally measured. However, theoretical explanations of the observed phenomena using solid-state physics are more direct when the polarization P_i is used instead of the electric displacement D_i. The polarization, electric displacement, and electric field are related by

$$D_i = \varepsilon_0 E_i + P_i \tag{9}$$

where ε_0 is the free-space dielectric permittivity. On the other hand, the electric field and electric displacement are related by

$$D_i = \varepsilon_{ik} E_k \tag{10}$$

Here ε_{ik} is the effective dielectric permittivity of the material. Thus, the polarization can be related to the electric field in the form

$$P_i = (\varepsilon_{ik} - \delta_{ik}\varepsilon_0) E_k = \kappa_{ik} E_k \tag{11}$$

In terms of polarization P_i and coefficient $\kappa_{ik} = \varepsilon_{ik} - \delta_{ik}\varepsilon_0$, Eqs. (1) and (2) can be expressed in the form

$$S_{ij} = s^E_{ijkl} T_{kl} + d_{kij} E_k + \delta_{ij} \alpha_i \theta \tag{12}$$

$$P_i = d_{ikl} T_{kl} + \kappa^T_{ik} E_k + \tilde{P}_i \theta \tag{13}$$

where $\tilde{P}_i$ is the coefficient of pyroelectric polarization. One notes that, in Eq. (13), the coefficient d_{ikl} signifies the induced polarization per unit stress, hence it can be viewed as *polarization coefficient*.

2.2.5 COMPRESSED MATRIX NOTATIONS

To write the elastic and piezoelectric tensors in matrix form, a compressed matrix notation is introduced to replace the tensor notation (Voigt notations). This compressed matrix notation consists of replacing ij or kl by p or q, where $i, j, k, l = 1, 2, 3$ and $p, q = 1, 2, 3, 4, 5, 6$ according to Table 2.1.

TABLE 2.1 Conversion from tensor to matrix indices for the Voigt notations

ij or *kl*	*p* or *q*
11	1
22	2
33	3
23 or 32	4
31 or 13	5
12 or 21	6

Thus, the 3×3 stress and strain tensors, T_{ij} and S_{ij}, are replaced by 6–element long column matrices of elements T_p and S_p. The $3 \times 3 \times 3 \times 3$ fourth order stiffness and compliance tensors c_{ijkl}^E and s_{ijkl}^E are replaced by 6×6 stiffness and compliance matrices of elements c_{pq}^E and s_{pq}^E. Similarly, c_{ijkl}^D and s_{ijkl}^D are replaced by c_{pq}^D and s_{pq}^D. The $3 \times 3 \times 3$ piezoelectric tensors, d_{ikl}, e_{ikl}, g_{ikl}, and h_{ikl}, are replaced by 3×6 piezoelectric matrices of elements d_{ip}, e_{ip}, g_{ip}, h_{ip}. The following rules apply

$$T_p = T_{ij}, \quad p = 1, 2, 3, 4, 5, 6 \quad \text{whereas } i, j = 1, 2, 3 \quad \text{(Stress)} \tag{14}$$

$$\begin{matrix} S_p = S_{ij}, & i = j, & p = 1, 2, 3 \\ S_p = 2S_{ij}, & i \neq j, & p = 4, 5, 6 \end{matrix} \quad \text{whereas } i, j = 1, 2, 3 \quad \text{(Strain)} \tag{15}$$

The factor of two in the strain equation is related to a factor of two in the definition of shear strains in the tensor and matrix formulation.

$$c_{pq}^E = c_{ijkl}^E, \quad c_{pq}^D = c_{ijkl}^D \quad p = 1, 2, 3, 4, 5, 6 \quad \text{(Stiffness coefficients)} \tag{16}$$

$$\begin{cases} s_{pq}^E = s_{ijkl}^E, & i = j \text{ and } k = l, \quad p, q = 1, 2, 3 \\ s_{pq}^E = 2s_{ijkl}^E, & i = j \text{ and } k \neq l, \quad p = 1, 2, 3 \quad q = 4, 5, 6 \\ s_{pq}^E = 4s_{ijkl}^E, & i \neq j \text{ and } k \neq l, \quad p, q = 4, 5, 6 \end{cases} \quad \text{(Compliance coefficients)} \tag{17}$$

Similar expressions can be derived for s_{pq}^D. The factors of 2 and 4 are associated with the factor of 2 from the strain equations.

$$e_{ip} = e_{ikl}, \quad h_{ip} = h_{ikl} \quad \text{(Piezoelectric stress constants)} \tag{18}$$

$$\begin{cases} d_{iq} = d_{ikl}, & k = l, \quad q = 1, 2, 3 \\ d_{iq} = 2d_{ikl}, & k \neq l, \quad q = 4, 5, 6 \end{cases} \quad \text{(Piezoelectric strain constants)} \tag{19}$$

$$\begin{cases} g_{iq} = g_{ikl}, & k = l, \quad q = 1, 2, 3 \\ g_{iq} = 2g_{ikl}, & k \neq l, \quad q = 4, 5, 6 \end{cases} \quad \text{(Piezoelectric voltage constants)} \tag{20}$$

The compressed matrix notations have the advantage of brevity. They are commonly used in engineering applications. The values of the elastic and piezoelectric constants given by the active material manufacturers in their product specifications are given in compressed matrix notations.

2.2.6 PIEZOELECTRIC EQUATIONS IN COMPRESSED MATRIX NOTATIONS

In engineering practice, the tensor Eqs. (1) and (2) can be rearranged in matrix form using the compressed matrix notations (Voigt notations), in which the stress and strain tensors are arranged as 6-component vectors, with the first three components representing *direct* stress and strain, whereas the last three components representing *shear* stress and strain. Thus,

$$\begin{Bmatrix} S_{11} \\ S_{22} \\ S_{33} \\ S_{23} \\ S_{31} \\ S_{12} \end{Bmatrix} \Longrightarrow \begin{Bmatrix} S_1 \\ S_2 \\ S_3 \\ S_4 \\ S_5 \\ \frac{1}{2}S_6 \end{Bmatrix}, \quad \begin{Bmatrix} T_{11} \\ T_{22} \\ T_{33} \\ T_{23} \\ T_{31} \\ T_{12} \end{Bmatrix} \Longrightarrow \begin{Bmatrix} T_1 \\ T_2 \\ T_3 \\ T_4 \\ T_5 \\ T_6 \end{Bmatrix} \tag{21}$$

Hence, the constitutive Eqs. (1) and (2) take the matrix form

$$\begin{Bmatrix} S_1 \\ S_2 \\ S_3 \\ S_4 \\ S_5 \\ S_6 \end{Bmatrix} = \begin{bmatrix} s_{11} & s_{12} & s_{13} & 0 & 0 & 0 \\ s_{21} & s_{22} & s_{23} & 0 & 0 & 0 \\ s_{31} & s_{32} & s_{33} & 0 & 0 & 0 \\ 0 & 0 & 0 & s_{44} & 0 & 0 \\ 0 & 0 & 0 & 0 & s_{55} & 0 \\ 0 & 0 & 0 & 0 & 0 & s_{66} \end{bmatrix} \begin{Bmatrix} T_1 \\ T_2 \\ T_3 \\ T_4 \\ T_5 \\ T_6 \end{Bmatrix} + \begin{bmatrix} d_{11} & d_{21} & d_{31} \\ d_{12} & d_{22} & d_{32} \\ d_{13} & d_{23} & d_{33} \\ d_{14} & d_{24} & d_{34} \\ d_{15} & d_{25} & d_{35} \\ d_{16} & d_{26} & d_{36} \end{bmatrix} \begin{Bmatrix} E_1 \\ E_2 \\ E_3 \end{Bmatrix} + \begin{Bmatrix} \alpha_1 \\ \alpha_2 \\ \alpha_3 \\ 0 \\ 0 \\ 0 \end{Bmatrix} \theta \tag{22}$$

$$\begin{Bmatrix} D_1 \\ D_2 \\ D_3 \end{Bmatrix} = \begin{bmatrix} d_{11} & d_{12} & d_{13} & d_{14} & d_{15} & d_{16} \\ d_{21} & d_{22} & d_{23} & d_{24} & d_{25} & d_{26} \\ d_{31} & d_{32} & d_{33} & d_{34} & d_{35} & d_{36} \end{bmatrix} \begin{Bmatrix} T_1 \\ T_2 \\ T_3 \\ T_4 \\ T_5 \\ T_6 \end{Bmatrix} + \begin{bmatrix} \varepsilon_{11} & \varepsilon_{12} & \varepsilon_{13} \\ \varepsilon_{21} & \varepsilon_{22} & \varepsilon_{23} \\ \varepsilon_{31} & \varepsilon_{32} & \varepsilon_{33} \end{bmatrix} \begin{Bmatrix} E_1 \\ E_2 \\ E_3 \end{Bmatrix} + \begin{Bmatrix} \tilde{D}_1 \\ \tilde{D}_2 \\ \tilde{D}_3 \end{Bmatrix} \theta \tag{23}$$

Please note that the piezoelectric matrix in Eq. (22) is the transpose of the piezoelectric matrix in Eq. (23). When written in compact form, Eqs. (22) and (23) become

$$S_p = s^E_{pq} T_q + d_{kp} E_k + \delta_{pq} \alpha^E_q \theta, \qquad p, q = 1, \ldots, 6; \qquad k = 1, 2, 3 \tag{24}$$

$$D_i = d_{iq} T_q + \varepsilon^T_{ik} E_k + \tilde{D}_i \theta, \qquad q = 1, \ldots, 6; \qquad i, k = 1, 2, 3 \tag{25}$$

Equations (22) and (23) can also be written in matrix format, i.e.,

$$\{\mathbf{S}\} = [\mathbf{s}]\{\mathbf{T}\} + [\mathbf{d}]^t\{\mathbf{E}\} + \{\boldsymbol{\alpha}\}\theta \tag{26}$$

$$\{\mathbf{D}\} = [\mathbf{d}]\{\mathbf{T}\} + [\boldsymbol{\varepsilon}]\{\mathbf{E}\} + \{\tilde{\mathbf{D}}\}\theta \tag{27}$$

Compressed matrix (Voigt) expressions similar to Eqs. (22) through (27) can be derived for the other constitutive equations such as Eqs. (5)–(8), (12)–(13), etc.

The values of the piezoelectric coupling coefficients differ from material to material. Most piezoelectric materials of interest are crystalline solids. These can be single crystals (either natural or synthetic) or polycrystalline materials like ferroelectric ceramics. In certain crystalline piezoelectric materials, the piezoelectric coefficient, d_{ji} ($i = 1, \ldots, 6$;

$j = 1, 2, 3$) may be enhanced or diminished through preferred crystal-cut orientation. The piezoceramics are polycrystalline materials, with randomly polarized microscopic properties. As fabricated, piezoceramics do not display macroscopic piezoelectricity due to the random microscopic polarization. This situation is overcome through *poling*. The poling process, which consists of applying a strong electric field at elevated temperatures, confers polycrystalline piezoceramic materials macroscopic piezoelectric properties similar to those observed in piezoelectric single crystals.

In practical applications, many of the piezoelectric coefficients, d_{ji}, have negligible values as the piezoelectric materials respond preferentially along certain directions depending on their intrinsic (spontaneous) polarization. For example, consider the situation of piezoelectric wafer as depicted in Fig. 2.1. To illustrate the d_{33} and d_{31} effects, assume that the applied electric field, E_3, is parallel to the spontaneous polarization, P_s (Fig. 2.1a). If the spontaneous polarization, P_s, is aligned with the x_3 axis, then such a situation can be achieved by creating a vertical electric field, E_3, through the application of a voltage V between the bottom and top electrodes depicted by the grey shading in Fig. 2.1a. The application of such an electric field that is parallel to the direction of spontaneous polarization ($E_3||P_s$) results in a vertical (thickness-wise) expansion $\varepsilon_3 = d_{33}E_3$ and a lateral (in plane) contractions $\varepsilon_1 = d_{31}E_3$ and $\varepsilon_2 = d_{32}E_3$ (the lateral strains are contracted as the coefficient d_{31} and d_{32} have opposite sign to d_{33}). So far, the strains experienced by the piezoelectric wafer have been direct strains. Such an arrangement can be used to produce thickness-wise and in-plane vibrations of the wafer.

However, if the electric field is applied perpendicular to the direction of spontaneous polarization, then the resulting strain will be shear. This can be obtained by electroding the lateral faces of the piezoelectric wafer. The application of a voltage to the lateral electrodes shown in Fig. 2.1b results in an in-plane electric field, E_1 that is perpendicular to the spontaneous polarization, ($E_1 \perp P_s$). This produces an induced shear strain $\varepsilon_5 = d_{15}E_1$. Similarly, if the electrodes were applied to the front and back faces, the resulting electric field would be E_2 and the resulting strain would be $\varepsilon_4 = d_{24}E_2$. The shear–strain arrangements discussed here can be used to induce shear vibrations in the piezoelectric

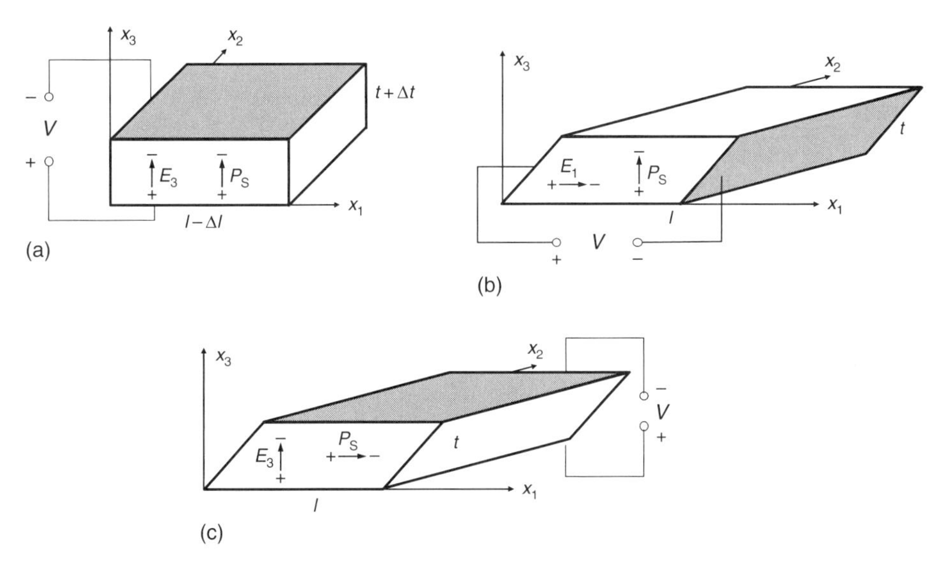

FIGURE 2.1 Basic induced-strain responses of piezoelectric materials: (a) direct strains $\varepsilon_3 = d_{33}E_3$ (thickness), $\varepsilon_1 = d_{31}E_3$, $\varepsilon_2 = d_{32}E_3$ (in plane); (b) shear strain $\varepsilon_5 = d_{15}E_1$; (c) shear strain $\varepsilon_5 = d_{35}E_3$ (*Note*: grey shading depicts the electrodes).

wafer. The use of lateral electrodes may not be feasible in the case of a thin wafer. In this case, top and bottom electrodes can be used again, but the spontaneous polarization of the wafer must be aligned with an in-plane direction. This latter situation is depicted in Fig. 2.1c, where the spontaneous polarization is shown in the x_1 direction, whereas the electric field is applied in the x_3 direction. The shear strain induced by this arrangement would be $\varepsilon_5 = d_{35}E_3$.

For piezoelectric materials with transverse isotropy, $d_{32} = d_{31}$, $d_{24} = d_{15}$, $\varepsilon_{22} = \varepsilon_{11}$. Hence, for piezoelectric materials with transverse isotropy, such as common piezoceramics, the constitutive piezoelectric equations become

$$\begin{Bmatrix} S_1 \\ S_2 \\ S_3 \\ S_4 \\ S_5 \\ S_6 \end{Bmatrix} = \begin{bmatrix} s_{11} & s_{12} & s_{13} & 0 & 0 & 0 \\ s_{12} & s_{22} & s_{23} & 0 & 0 & 0 \\ s_{13} & s_{23} & s_{33} & 0 & 0 & 0 \\ 0 & 0 & 0 & s_{44} & 0 & 0 \\ 0 & 0 & 0 & 0 & s_{55} & 0 \\ 0 & 0 & 0 & 0 & 0 & s_{66} \end{bmatrix} \begin{Bmatrix} T_1 \\ T_2 \\ T_3 \\ T_4 \\ T_5 \\ T_6 \end{Bmatrix} + \begin{bmatrix} 0 & 0 & d_{31} \\ 0 & 0 & d_{32} \\ 0 & 0 & d_{33} \\ 0 & d_{15} & 0 \\ d_{15} & 0 & 0 \\ 0 & 0 & 0 \end{bmatrix} \begin{Bmatrix} E_1 \\ E_2 \\ E_3 \end{Bmatrix} + \begin{Bmatrix} \alpha_1 \\ \alpha_2 \\ \alpha_3 \\ 0 \\ 0 \\ 0 \end{Bmatrix} \theta \tag{28}$$

$$\begin{Bmatrix} D_1 \\ D_2 \\ D_3 \end{Bmatrix} = \begin{bmatrix} 0 & 0 & 0 & 0 & d_{15} & 0 \\ 0 & 0 & 0 & d_{15} & 0 & 0 \\ d_{31} & d_{32} & d_{33} & 0 & 0 & 0 \end{bmatrix} \begin{Bmatrix} T_1 \\ T_2 \\ T_3 \\ T_4 \\ T_5 \\ T_6 \end{Bmatrix} + \begin{bmatrix} \varepsilon_{11} & 0 & 0 \\ 0 & \varepsilon_{11} & 0 \\ 0 & 0 & \varepsilon_{33} \end{bmatrix} \begin{Bmatrix} E_1 \\ E_2 \\ E_3 \end{Bmatrix} + \begin{Bmatrix} \tilde{D}_1 \\ \tilde{D}_2 \\ \tilde{D}_3 \end{Bmatrix} \theta \tag{29}$$

Compressed matrix (Voigt) expressions similar to Eqs. (28) and (29) can be derived for the other constitutive equations such as Eqs. (5)–(8), (12)–(13), etc.

2.2.7 RELATIONS BETWEEN THE CONSTANTS

The constants that appear in the equations described in the previous sections can be related to each other. For example, the stiffness tensor, c_{ijkl}, is the inverse of the strain tensor, s_{ijkl}. Similar relations can be established for the other constants and coefficients. In writing these relations, we use the compressed matrix notation with $i, j, k, l = 1, 2, 3$ and $p, q, r = 1, 2, 3, 4, 5, 6$. We also use the 3×3 unitmatrix δ_{ij} and the 6×6 unitmatrix δ_{pq}. As before, Einstein convention of implied summation over the repeated indices applies.

$$c^E_{pr} s^E_{qr} = \delta_{pq}, \quad c^D_{pr} s^D_{qr} = \delta_{pq} \quad \text{(Stiffness–compliance relations)} \tag{30}$$

$$\varepsilon^S_{ik} \beta^S_{jk} = \delta_{ij}, \quad \beta^T_{ik} \varepsilon^T_{jk} = \delta_{ij} \quad \text{(Permittivity–impermittivity relations)} \tag{31}$$

$$c^D_{pq} = c^E_{pq} + e_{kp} h_{kq}, \quad s^D_{pq} = s^E_{pq} - d_{lp} g_{kq} \quad \text{(Close circuit–open circuit effects on elastic constants)} \tag{32}$$

$$\varepsilon^T_{ij} = \varepsilon^S_{ij} + d_{iq} e_{jq}, \quad \beta^T_{ij} = \beta^S_{ij} - g_{iq} h_{jq} \quad \text{(Stress–strain effects on dielectric constants)} \tag{33}$$

$$\begin{cases} e_{ip} = d_{iq} c^E_{qp}, & d_{ij} = \varepsilon^T_{ik} g_{kp} \\ g_{ip} = \beta^T_{ik} d_{kq}, & h_{ip} = g_{iq} \varepsilon^D_{qp} \end{cases} \quad \text{(Relations between piezoelectric constants)} \tag{34}$$

2.2.8 ELECTROMECHANICAL COUPLING COEFFICIENT

Electromechanical coupling coefficient is defined as the square root of the ratio between the mechanical energy stored and the electrical energy applied to a piezoelectric material

$$k = \sqrt{\frac{\text{Mechanical energy stored}}{\text{Electrical energy applied}}} \tag{35}$$

For direct actuation, we have $k_{33} = \frac{d_{33}}{\sqrt{s_{33}\varepsilon_{33}}}$, for transverse actuation, $k_{31} = \frac{|d_{31}|}{\sqrt{s_{11}\varepsilon_{33}}}$, and for shear actuation, $k_{15} = \frac{d_{15}}{\sqrt{s_{55}\varepsilon_{11}}}$. For uniform inplane actuation, we obtain the planar coupling coefficient, $\kappa_p = \kappa_{13}\sqrt{\frac{2}{1-\nu}}$, where ν is the Poisson ratio.

2.2.9 HIGHER ORDER MODELS OF THE ELECTROACTIVE RESPONSE

Higher order models of the electroactive ceramics contain both linear and quadratic terms. The linear terms are associated with the conventional *piezoelectric response*. The quadratic terms are associated with the *electrostrictive response*, whereas the application of electric field in one direction induces constriction (squeezing) of the material. The electrostrictive effect is not limited to piezoelectric materials, and is present in all materials, though with different amplitudes. The electrostrictive response is quadratic in electric field. Hence, the direction of the electrostriction does not switch as the polarity of the electric field is switched. The constitutive equations that incorporate both piezoelectric and electrostrictive response have the form

$$S_{ij} = s^E_{klij}T_{kl} + d_{kij}E_k + M_{klij}E_kE_l \tag{36}$$

Note that the first two terms are the same as for piezoelectric materials. The third term is due to electrostriction. The coefficients M_{klij} are the electrostrictive coefficients.

2.3 PIEZOELECTRIC PHENOMENA

Polarization is a phenomenon observed in dielectrics and it consists in the separation of positive and negative electric charges at different ends of the dielectric material on the application of an external electric field. A typical example is the polarization of the dielectric material inside a capacitor on the application of an electric voltage across the capacitor plates. Polarization is the explanation for the fact that the dielectric capacitor can hold much more charge than the vacuum capacitor, since

$$D = \varepsilon_0 E + P \tag{37}$$

where D, the electric displacement, represents charge per unit area; E, the electric field, represents voltage divided by the distance between the capacitor plates; and ε_0 is the electric permittivity of the vacuum. It is apparent from Eq. (37) that the polarization P represents the additional charge stored in a dielectric capacitor as compared with a vacuum capacitor.

Spontaneous polarization is the phenomenon by which polarization appears without the application of an external electric field. Spontaneous polarization has been observed in certain crystals in which the centers of positive and negative charges do not coincide. Crystals are classified into 32-point groups according to their crystallographic symmetry (international and Schonflies crystallographic symbols). These 32-point groups can be divided into two large classes, one containing point groups that have a center of symmetry, the other containing point groups that do not have a center of symmetry, and hence display some spontaneous polarization. Of the 21-point groups that do not display a center of symmetry, 20 contain crystals that may display spontaneous polarization. Spontaneous polarization can occur more easily in perovskite crystal structures.

Permanent polarization is the phenomenon by which the polarization is retained even in the absence of an external electric field. The process through which permanent polarization is induced in a material is known as *poling*.

Paraelectric materials do not display permanent polarization, i.e., they have zero polarization in the absence of an external electric field. When an external field is applied, their polarization is roughly proportional with the applied electric field. It increases when the electric field is increased, and decreases back to zero when the field is reduced. If the field is reversed, the polarization also reverses (Fig. 2.2a). Paraelectric behavior represents the behavior of common dielectrics.

Ferroelectric materials have permanent polarization that can be altered by the application of an external electric field. The term 'ferroelectric materials' was derived by analogy with the term 'ferromagnetic materials,' in which the permanent magnetization is altered by the application of an external magnetic field. Figure 2.2b describes graphically the ferroelectric behavior during the cyclic application of an electric field. As the electric field is increased beyond the critical value, called *coercive field*, E_c, the polarization suddenly increases to a high value. This value is roughly maintained when the electric field is decreased, such that at zero electric field the ferroelectric material retains a permanent spontaneous polarization P_s. When the electric field is further reduced beyond the negative value $-E_c$, the polarization suddenly switches to a large negative value, which is roughly maintained as the electric field is decreased. At zero electric field, the permanent spontaneous polarization is now $-P_s$. As the electric field is again increased into the positive range, the polarization is again switched to a positive value, as the field increased beyond E_c. Characteristic of this behavior is the high hysteresis of the loop traveled during a cycle. The ferroelectric behavior can be explained through the existence of aligned internal dipoles that have their direction switched when the electric field is sufficiently strong. The slight horizontal slopes observed in Fig. 2.2b are attributable to the paraelectric component of the total polarization.

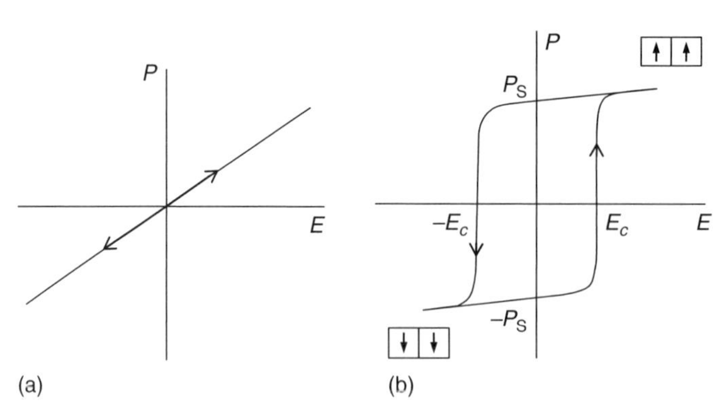

FIGURE 2.2 Polarization vs. applied electric field for three types of materials: (a) paraelectric behavior; (b) ferroelectric behavior.

Piezoelectricity[1] is the property of a material to display electric charge on its surface under the application of an external mechanical stress. In other words, a piezoelectric material changes its polarization under stress. Piezoelectricity is related to permanent polarization, and can be attributed to the permanent polarization being changed when the material undergoes mechanical deformation due to the applied stress. Conversely, the change in permanent polarization produces a mechanical deformation, i.e., strain.

Pyroelectricity is the property of a material to display electric charge on its surface due to changes in temperature. In other words, a pyroelectric material changes its polarization when the temperature changes. Pyroelectricity is related to spontaneous polarization, and can be attributed to spontaneous polarization being changed when the material undergoes geometric changes due to changes in temperature. If the material is also piezoelectric, and if its boundaries are constraint, change in temperature produces thermal stresses that result in high polarization through the piezoelectric effect.

Rochelle salt was one of the first observed ferroelectric materials. Most ferroelectric materials are piezoelectric and pyroelectric. Remarkable about the Rochelle salt was that its piezoelectric coefficient was much larger than that of quartz. However, quartz is much more stable and rugged.

2.4 PEROVSKITE CERAMICS

Perovskites are a large family of crystalline oxides with the metal to oxygen ratio 2:3. Perovskites derive their name from a specific mineral known as perovskite. The simplest perovskite lattice has the expression, $X_m Y_n$, in which the X atoms are rectangular close packed and the Y atoms occupy the octahedral interstices. The rectangular close packed X atoms may be a combination of various species, X^1, X^2, X^3, etc. For example, in the barium titanate perovskite, $BaTiO_3$, we have $X^1 = Ba^{2+}$ and $X^2 = Ti^{4+}$, whereas $Y = O^{2-}$ (Fig. 2.3). In the lattice structure, the Ba^{2+} divalent metallic cations are at the corners, the Ti^{4+} tetravalent metallic cation is in the center, whereas the O^{2-} anions are on the faces. The Ba^{2+} cations are larger, whereas the Ti^{4+} cations are smaller. The size of the Ba^{2+} cation affects the overall size of the lattice structure. Perovskite arrangements like in $BaTiO_3$ are generically designated ABO_3. Their main commonality is that they have

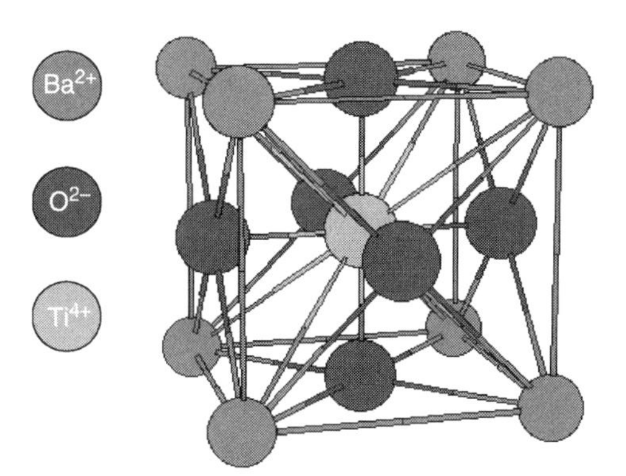

FIGURE 2.3 Crystal structure of a typical perovskite, $BaTiO_3$: the Ba^{2+} cations are at the cube corners, the Ti^{4+} cation is in the cube center, and the O^{2-} anions on the cube faces.

[1] The prefixes 'piezo' and 'pyro' are derived from the Greek words for 'force' and 'fire', respectively.

a small, tetravalent metal ion, e.g., titanium or zirconium, in a lattice of larger, divalent metal ions, e.g., lead or barium, and oxygen ions (Fig. 2.3). Under conditions that confer tetragonal or rhombohedral symmetry, each crystal has a dipole moment.

2.4.1 POLARIZATION OF THE PEROVSKITE STRUCTURE

At elevated temperatures, the primitive perovskite arrangement is symmetric faced-centered cubic (FCC) and does not display electric polarity (Fig. 2.4a). This symmetric lattice arrangement forms the *paraelectric phase* of the perovskite, which exist at elevated temperatures. As the temperature decreases, the lattice shrinks and the symmetric arrangement is no longer stable. For example, in barium titanate, the Ti^{4+} cation snaps from the cube center to other minimum-energy locations situated off center. This is accompanied by corresponding motion of the O^{2-} anions. Shifting of the Ti^{4+} and O^{2-} ions causes the structure to be altered, creating strain and electric dipoles. The crystal lattice becomes distorted, i.e., slightly elongated in one direction, i.e., tetragonal (Fig. 2.4b). In barium titanate, the distortion ratio is $c/a = 1.01$, corresponding to 1% strain in the c-direction with respect to the a-direction. This change in dimensions along the c-axis is called *spontaneous strain*, S_s. The orthorhombic tetragonal structure has polarity because the centers of the positive and negative charges no longer coincide, yielding a net electric dipole. This polar lattice arrangement forms the *ferroelectric phase* of the perovskite, which exists at lower temperatures. The transition from one phase into the other takes place at the phase transition temperature, commonly called the *Curie temperature*. In barium titanate, $BaTiO_3$, the phase transition temperature is around 130° C. As the perovskite is cooled below the transition temperature, T_c, the paraelectric phase changes into the ferroelectric phase, and the material displays spontaneous strain, S_s, and spontaneous polarization, P_s. Alternatively, when the perovskite is heated above the transition temperature, the ferroelectric phase changes into the paraelectric phase, and the spontaneous strain and spontaneous polarization are no longer present.

2.4.1.1 Temperature Dependence of Spontaneous Polarization, Spontaneous Strain, and Dielectric Permittivity

In the ferroelectric phase, the perovskite displays spontaneous polarization, which decreases with temperature. In the same time, the dielectric permittivity increases with temperature. At the transition temperature, i.e., at the Curie point, T_c, the polarization

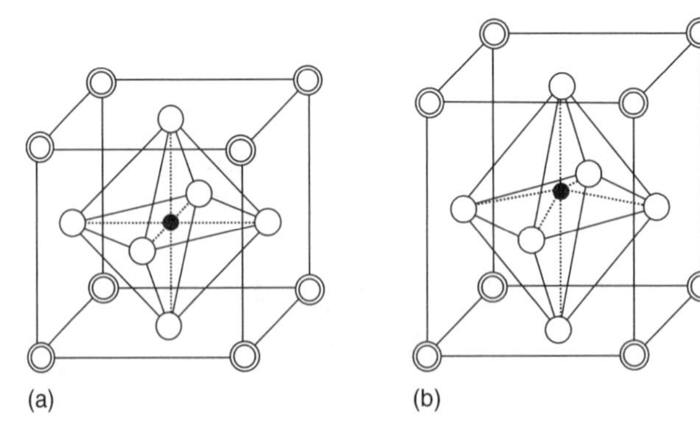

FIGURE 2.4 Spontaneous strain and polarization in a perovskite structure: (a) above the Curie point, the crystal has cubic lattice, displaying a symmetric arrangement of positive and negative charges and no polarization (paraelectric phase); (b) below the Curie point, the crystal has tetragonal lattice, with asymmetrically placed central atom, thus displaying polarization (ferroelectric phase).

vanishes, and the permittivity suddenly jumps to a very large value. These phenomena are associated with the transition from the ferroelectric phases into the paraelectric phase, i.e., from the distorted orthotropic lattice to the symmetric FCC lattice. As the temperature further increases, the permittivity decreases drastically with temperature following a $1/T$ rule. The spontaneous strain, S_s, which only exists in the ferroelectric phase, decreases with temperature. As the Curie temperature, T_c, is crossed, transition from ferroelectric phase into paraelectric phase takes place, and the spontaneous strain vanishes. The piezoelectric strain coefficient, d, increases with temperature up to T_c, and then vanishes.

2.4.1.2 Induced Strain and Induced Polarization

When the perovskite is in the ferroelectric phase, strain and polarization can be also induced by the application of an electric field. When applied in the direction of spontaneous polarization, the electric field increases the net polarization, and increases the lattice distortion. The additional strain and polarization are called *induced strain* and *induced polarization*. At first, the induced strain and induced polarization increase linearly with the applied electric field. However, as the field increases to higher values, saturation-induced nonlinear effects set in. The maximum displacement allowed by the crystal structure is termed the "polarization saturation".

When the applied electric field is contrary to the direction of spontaneous polarization (i.e., *reverse field*), the induced strain and induced polarization add algebraically to the spontaneous strain and spontaneous polarization already existing in the ferroelectric perovskite, and the net strain and polarization decrease. At high reverse fields, the polarization and strain may suddenly increase in the direction of the applied reverse field. This spontaneous switching of polarization and strain occurs as the central atom suddenly jumps into an opposite off-center location, more appropriate to the direction of the externally applied electric field. The sudden jump reverses the relative positions of the asymmetric ions in the crystal lattice and aligns the spontaneous polarization with the applied electric field. This phenomenon, called *polarization reversal*, is the main characteristics of ferroelectric materials. The value of the electric field at which polarization reversal takes place is called *coercive field*. Polarization reversal is also accompanied by large strains. However, it results in a large hysteresis loop. Due to its drastic character, repeated application of polarization reversal subjects the crystal lattice to considerable internal stresses, increases the lattice fatigue, and shortens the life of the ferroelectric materials.

When the perovskite is in the paraelectric phase, linear piezoelectricity is absent. However, strain can be still induced through the *electrostrictive effect*, which is quadratic in the applied field.

2.4.2 POLING OF PEROVSKITE CERAMICS

During fabrication, perovskite ceramics undergo phase transformation from *paraelectric* state to *ferroelectric* state. This transformation takes place as the material cools below the *Curie temperature*, T_c. The resulting ferroelectric ceramic has a polycrystalline structure (grains) with randomly oriented ferroelectric domains (Fig. 2.5a). If the grains are large, ferroelectric domains can exist even inside each grain. Due to the random orientation of the electric domains, individual polarizations cancel each other, and the net polarization of the virgin ferroelectric ceramic is zero.

This random orientation can be transformed into a preferred orientation through **poling**. Poling aligns the dipole domains and gives the piezoceramic material a net polarization. A poled ferroelectric ceramics behaves more or less like a single crystal. Poling of

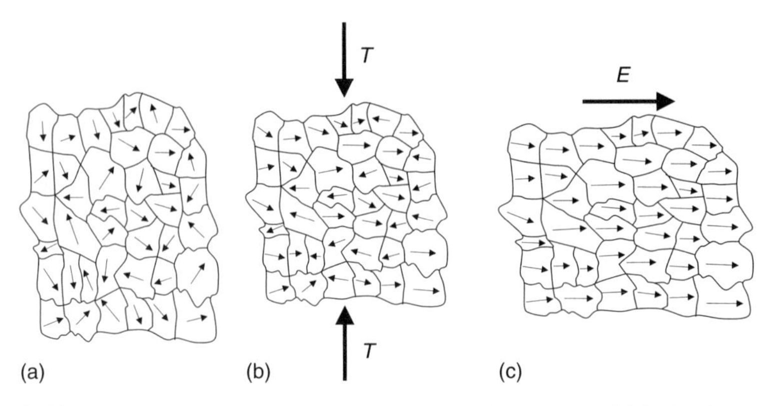

FIGURE 2.5 Piezoelectric effect in polycrystalline perovskite ceramics: (a) in the absence of stress and electric field, the electric domains are randomly oriented; (b) application of stress produces orientation of the electric domains perpendicular to the loading direction. The oriented electric domains yield a net polarization; (c) application of an electric field orients the electric domains along the field lines and produces induced strain.

piezoceramics is attained at elevated temperatures in the presence of a high electric field. The application of a high electric field at elevated temperatures results in the alignment of the crystalline domains. This alignment is locked in place when the piezoceramic is cooled (permanent polarization). During poling, the orientation of the piezoelectric domains also produces a mechanical deformation. When the piezoceramic is cooled, this deformation is locked in place (permanent strain). Poling is performed in silicon oil bath at elevated temperature under a d.c. electric field of 1–3 kV/mm.

A poled ferroelectric ceramic responds to the application of an applied electric field or mechanical stress with typical piezoelectric behavior. The subsequent application of an electric field or a mechanical strain affects this state and changes the interplay between mechanical deformation and polarization. When a mechanical strain is applied, the polarization is changed and the *direct piezoelectric effect* is obtained. When an electric field is applied, the mechanical strain is changed, and the *converse piezoelectric effect*, i.e., the *induced-strain actuation*, results.

In a poled ferroelectric ceramic, electric domains exist mainly in two varieties:

(1) Electric domains that became more or less aligned with the direction of the electric field during the poling operation, and are now more or less parallel to the direction of spontaneous polarization
(2) 90° electric domains that did not orient themselves during poling and are left perpendicular to the direction of spontaneous polarization.

If an external electric field is applied in the direction of spontaneous polarization, the induced-strain actuation process takes place in three major steps. First, through the *intrinsic effect*, the strain of the piezoelectric domains increases under the influence of the applied electric field (curves *a* and *b* in Fig. 2.6). This effect is rather linear, and related to conventional piezoelectricity. The induced strain adds to the already existing spontaneous strain, which was created during the poling process (approximately 0.275% in Fig. 2.6).

During the intrinsic response, the electric domains that are better aligned with the electric field deform more, whereas those that are less aligned deform less, since the field strength is projected through the individual orientation angle of each domain. In addition,

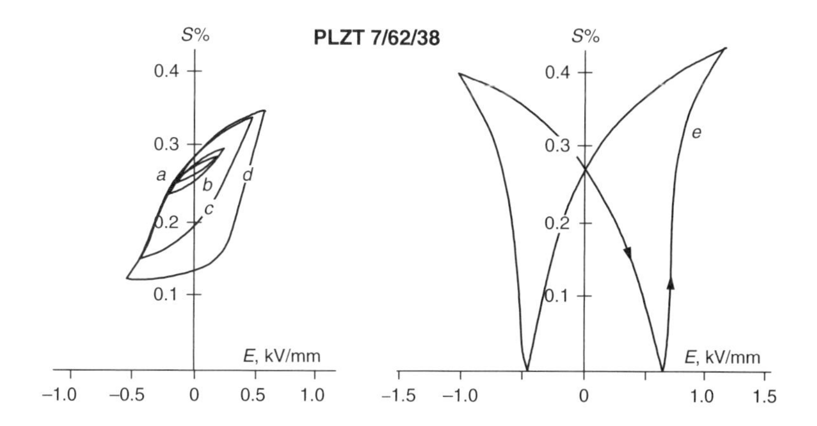

FIGURE 2.6 Induced-strain curves in PLZT 7/62/38 for various levels of electric field.

the resulting strain in each domain has to be projected back onto the main strain direction, which in tetragonal lattices coincides with the electric field direction. Thus, the overall effect is strongly affected by the percentage orientation of the electric domains with respect to the direction of overall polarization. Not surprisingly, the overall piezoelectric strain coefficient of a ferroelectric ceramic is less than that of an equivalent single crystal of the same formulation.

Further increase of the electric field triggers the *extrinsic effect* during which the domains undergo rotation (Fig. 2.7). Ferroelectric domains, which initially were not oriented with the applied field, now tend to orient themselves with the applied field. The most dramatic reorientation that can take place is that of the 90° domains. During this process, high strains can be produced as the rotation of the 90° domains adds the full strength of the lattice spontaneous strain existing in the 90° domains. (For example, $BaTiO_3$ perovskite has a spontaneous strain of 1%. The rotation of a 90° domain in this compound will produce a local strain of 1%.) Of course, the local strains will have to combine through elastic interactions with the strains of the adjacent domains, but the total effect can be quite significant. The extrinsic effect is shown in curves *c*, *d*, and *e* of Fig. 2.6. This extrinsic effect produces very spectacular results during the upswing of the induced-strain actuation process, and induced strains of up to 0.15% are shown in Fig. 2.6. However, during the downswing, as the electric field is reduced, considerable permanent strain remains, as the ex-90° domains do not rotate back. The extrinsic effect is believed to be the cause of nonlinearity and hysteresis losses in piezoceramics.

Thirdly, the material undergoes electrostriction. This is a volumetric effect, which is proportional to the square of the electric field. In conventional piezoceramics, the

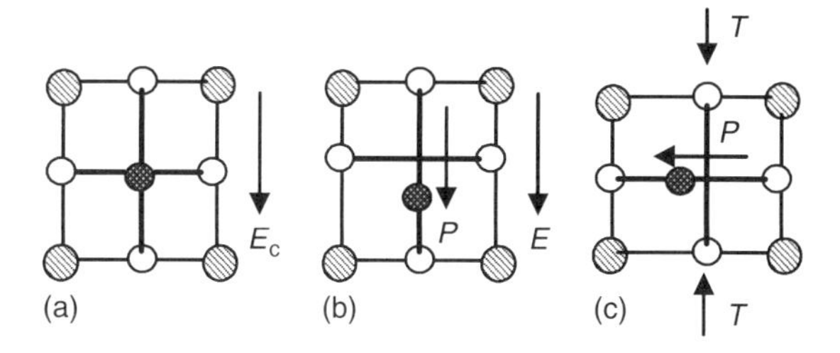

FIGURE 2.7 (a) Depolarization under coercive field E_C; (b) 180° domain switch for $E < E$; (c) 90° domain switch for stresses higher than the coercive stress.

electrostrictive effect is negligible. However, it becomes quite significant in electrostrictive ceramics, as discussed in Section 2.4.5 of this chapter.

If, after being decreased to zero, the electric field is increased in the opposite direction, the strain decreases at first, until all the polarization-induced spontaneous strain is cancelled, and the overall strain is zero. Further increase of the reverse electric field beyond this point produces the phenomenon of domain switching, whereby the crystal lattice snaps into a new position, which is now aligned with the applied reversed field. At this point, the reverse coercive field, $-E_C$, was attained. If the reverse field is further increased, small additional strains may still be obtained through the intrinsic effect and through electrostriction, but these are diminishing returns. If now the electric field is brought back to zero, the strain decreases, but significant permanent strain still remains. Increasing the field in the positive direction will, at first, induce further reduction of the strain because the electric domains are almost all of reverse polarity. As the field is further increased, domain switching will again occur as the value of forward-coercive field, $+E_C$, is attained. The result is the "butterfly curve" (curve *e* in Fig. 2.6).

The high-field nonlinear behavior is frequency dependent. The typical butterfly curve (curve *e* of Fig. 2.6) can only be attained under quasi-static application of the electric field, in which case sufficient time is given to the electric domain switching to propagate through the full body of the piezoceramic. If the frequency is increased, the domain reorientation and domain switching cannot fully develop before the electric field is reversed. As a result, as the frequency increases, the maximum attainable strains under maximum field tend to diminish. However, hysteresis also decreases.

The linear piezoelectric equation can only describe a fraction of the full operating range of the materials. Outside the linear range, advanced theories considering the material micromechanics must be employed. Within the linear range, piezoelectric ceramics produce strains that are more or less proportional to the applied electric field or voltage. Induced strains around 0.1% (1000 μstrain) are encountered.

Due to nonlinear behavior, the dielectric permittivity, piezoelectric coefficient, loss factor and coupling coefficients vary with the applied electric field and mechanical stress. For example, the fact that the dielectric permittivity is strongly dependent on stress and electric field is correlated with the observation that the capacitance of PZT wafers driven at different frequencies and electric fields display an almost proportional increase with voltage. This capacitance variation further affects the power requirements of a piezoelectric actuator.

2.4.3 COMMON PEROVSKITE CERAMICS

Perovskite arrangements such as in $BaTiO_3$ are generically designated $A^{2+}B^{4+}O_3^{2-}$. In $BaTiO_3$, the sites A^{2+} are occupied by Ba^{2+} cations, whereas the sites B^{4+} are occupied by Ti^{4+} cations. However, the A^{2+} and B^{4+} sites can be also taken by other similar cations, of similar sizes. For example the A^{2+} site can be occupied by other large-size divalent metallic cations such as $A^{2+} = Ba^{2+}, Sr^{2+}, Pb^{2+}, Sn^{2+}$, etc. The B^{4+} site can be occupied by other small-size tetravalent metallic cations like $B^{4+} = Ti^{4+}, Zr^{4+}$, etc. Through such replacements, other perovskite compounds are obtained.

Besides the basic replacement of the cations in the formulation, $A^{2+}B^{4+}O_3^{2-}$, it is also possible to have mixtures of similar cations forming solid solution alloys of various proportions. For example, the solid solution alloying of the B site can yield combination of the form $A^{2+}(B_{1-x}^{4+}B_x'^{4+})O_3^{2-}$. In solid solutions, the distribution of the B and B′ cations can be *ordered* or *disordered*. Solid solution perovskites may have several ferroelectric phases. The alloying proportions of the B and B′ components influence the type of phases present in the solid solution. Commercial piezoelectric ceramics are typically made of simple perovskites and solid solution perovskite alloys. Typical examples of simple perovskites are

- Barium titanate (BT) with chemical formula $BaTiO_3$
- Lead titanate (PT) with chemical formula $PbTiO_3$.

Typical examples of solid-solution perovskite alloys are

- Lead zirconate titanate (PZT) with chemical formula $Pb(Zr, Ti)O_3$
- Lead lanthanum zirconate titanate (PLZT) with chemical formula $(Pb, La)(Zr, Ti)O_3$
- Ternary ceramics, e.g., $BaO\text{-}TiO_2\text{-}R_2O_3$, where R is a rare earth.

2.4.4 PIEZOELECTRIC CERAMICS

Piezoelectric ceramics are perovskite varieties in which the linear piezoelectric response dominates. At low electric fields, piezoelectric ceramics are well described by the linear piezoelectric equations. When piezoceramics are classified according to their coercive field during field-induced strain actuation, two main categories emerge. If the coercive field is large, say greater than 1 kV/mm, then the piezoceramic is 'hard'. A hard piezoceramic shows an extensive linear drive region, but a relatively small strain magnitude. If the coercive field is moderate, say between 0.1 and 1 kV/mm, then the piezoceramic is classified as 'soft'. A soft piezoceramic shows a large field induced strain, but relatively large hysteresis. The stress-induced 90° domains switching for hard PZT compositions (acceptor doped) occurs at higher stresses than in the case of soft PZT compositions (donor doped). The 'hard' and 'soft' behavior is also related to the Curie temperature. Hard piezoceramics tend to have a higher Curie point, $250°\,C < T_c$, whereas soft piezoceramics have a moderate Curie point, $150°\,C < T_c < 250°\,C$. The DOD-STD-1376A(SH) standard defines six piezoelectric ceramic types also known as Navy Type I–VI. This standard is used by piezoelectric ceramics manufacturers and suppliers as a minimum quality requirement for their products.

A solid solution ferroelectric perovskite with wide applications is the *lead zirconate titanate*, $Pb(Zr_{1-x}Ti_x)O_3$, commonly known as **PZT**. In the PZT perovskite unit cell, lead, Pb^{2+}, occupies the corners, oxygen, O^{2-}, the faces, and titanium/zirconium, Zr^{4+}/Ti^{4+}, the octahedral voids. To date, many PZT formulations exist, the main differentiation being between 'soft' (e.g., PZT 5-H) and 'hard' (e.g., PZT 8). PZT attains the highest piezoelectric coupling and the maximum electric permittivity near the morphotropic phase boundary (MPB). This corresponds to the change in the crystal structure from the tetragonal phase to the rhombohedral phase, which occurs when the Zr/Ti ratio is approximately 53/47. The explanation for this phenomenon is as follows. Above the Curie temperature, PZT has a cubic lattice and is paraelectric. The Curie temperature varies with the alloying proportion, from $\sim 250°\,C$ for pure $PbZrO_3$ to $\sim 500°\,C$ for pure $PbTiO_3$. Below the Curie temperature, PZT is ferroelectric; but its lattice can be either tetragonal or rhombohedral, according to the alloying proportion. On the phase diagram, the line separating the two phases is called the MPB. The tetragonal lattice has six distortion variants, i.e., the central cation can be displaced in any one of the six possible positions parallel to the three lattice axes. The rhombohedral lattice has eight distortion variants, i.e., the central cation can be displaced in any one of the eight possible positions parallel to the four diagonals. On the line separating the two phases in the phase diagram, i.e., on the MPB, both the tetragonal phase and the rhombohedral phase may exist. Hence, the total number of distortion variants on MPB is 14, which is the cumulative effect of both phases. Having more distortion variants increases the material responsiveness,

TABLE 2.2 Properties of APC piezoelectric ceramic (www.americanpiezo.com)

Property	APC 840	APC 841	APC 850	APC 855	APC 856	APC 880
$\rho(\mathrm{kg/m^3})$	7600	7600	7700	7500	7500	7600
$d_{33}(10^{-12}\,\mathrm{m/V})$	290	275	400	580	620	215
$d_{31}(10^{-12}\,\mathrm{m/V})$	−125	109	−175	270	260	−95
$d_{15}(10^{-12}\,\mathrm{m/V})$	480	450	590	720	710	330
$g_{33}(10^{-3}\,\mathrm{Vm/N})$	26.5	25.5	26	19.5	18.5	25
$g_{31}(10^{-3}\,\mathrm{Vm/N})$	−11	10.5	−12.4	8.8	8.1	−10
$g_{15}(10^{-3}\,\mathrm{Vm/N})$	38	35	36	27	25	28
$s_{11}{}^{E}(10^{-12}\,\mathrm{m^2/N})$	11.8	11.7	15.3	14.8	15.0	10.8
$s_{33}{}^{E}(10^{-12}\,\mathrm{m^2/N})$	17.4	17.3	17.3	16.7	17.0	15.0
$\varepsilon_{33}^{T}/\varepsilon_0$	1250	1350	1750	3250	4100	1000
k_p	0.59	0.60	0.63	0.65	0.65	0.50
k_{33}	0.72	0.68	0.72	0.74	0.73	0.62
k_{31}	0.35	0.33	0.36	0.38	0.36	0.30
k_{15}	0.70	0.67	0.68	0.66	0.65	0.55
Poisson ratio, σ	0.30	0.40	0.35	0.32	0.39	0.28
Young modulus $Y_{11}{}^{E}$ (GPa)	80	76	63	61	58	90
Young modulus $Y_{33}{}^{E}$ (GPa)	68	63	54	48	45	72
Curie temperature (°C)	325	320	360	195	150	310
Dissipation factor, $\tan\delta$ (%)	0.4	0.35	1.4	2	2.7	0.35
Mechanical Q_M	500	1400	80	75	72	1000

Note: $\varepsilon_0 = 8.85 \times 10^{-12}$ Farad/m. Poisson ratio is calculated from the formula, $k_p^2 = \frac{1}{1-\sigma}k_{31}^2$.

as the material has more options to deform under the action of external factors, e.g., electric field or mechanical pressure. At room temperature, the MPB is placed around the 47/53 alloying ration. Several PZT formulations are commercially available. For example, Table 2.2 gives the properties of piezoelectric ceramic wafers available from American Piezo Ceramics, Inc.

2.4.5 ELECTROSTRICTIVE CERAMICS

Electrostrictive ceramics are perovskite materials in which the electrostrictive response is dominant. If the coercive field is less than 0.1 kV/mm, the material is rather an electrostrictor, which displays an approximately quadratic dependence of strain on electric field. The perovskites that display a large electrostrictive response are the disordered complex perovskites, which have high electrostrictive coefficient with respect to electric field and a diffuse transition temperature (diffuse Curie point).

2.4.5.1 Relaxor Ferroelectrics

Electrostrictive ceramics are also called *relaxor ferroelectric* because they display large dielectric relaxation, i.e., frequency dependence of the dielectric permittivity. In a relaxor ferroelectric, the permittivity decreases as the test frequency increases. In addition, the value of temperature at which the permittivity peaks shifts upward. This behavior is in contrast with that of conventional ferroelectrics, for which the temperature at which the permittivity peaks hardly changes with frequency. The dielectric relaxation phenomenon can be attributed to the presence of microdomains in the crystal structure. Mulvihill et al. (1995) subjected a <111> single crystal of lead zirconate niobate, $Pb(Zr_{1/3}Nb_{2/3})O_3$, to dielectric constant measurements in two states: (1) unpoled; (2) poled. The unpoled state, which has only microdomains, exhibited the dielectric relaxation phenomenon.

The poled state, which has macrodomains induced by the applied electric field, did not exhibit dielectric relaxation; it behaved more like a conventional dielectric material. In relaxor materials, the transition between piezoelectric behavior and loss of piezoelectric capability does not occur at a specific temperature (Curie point), but instead occurs over a temperature range (Curie range), that can be formulated to be lower than the room temperature. Thus, electrostrictive ceramics have a rather diffused phase transition that spans a temperature range around the transition temperature. Hence, the temperature dependence of electrostrictive ceramics around the transition temperature is markedly less than that of normal perovskite solid solutions.

Lead magnesium niobate, lead magnesium niobate/lanthanum formulations, and lead nickel niobate are currently among the most studied relaxor materials. The electrostrictive ferroelectrics have very high dielectric permittivity and polarization. Due to their very large dielectric permittivity, the electrostrictive ferroelectrics have found wide applications in the construction of compact-chip capacitors. The coercive field of electrostrictive ceramics is much smaller than that of piezoelectric ceramics. A common electrostrictive ceramics is lead magnesium niobate, $Pb(Mg_{1/3}Nb_{2/3})O_3$, also known as PMN. Another commonly used electrostrictive ceramic is lead titanate, $PbTiO_3$, also known as PT. Combination of these two formulations are also common, under the designation PMN-PT (Fig. 2.8). Another electrostrictive ceramic is lead larthanium zirconium titarate $(Pb, La)(Zr, Ti)O_3$, also known as PLZT. Other ferroelectric ceramic systems that have been formulated to display strong electrostrictive behavior include lead barium zirconate titanate, $(Pb, Ba)(Zr, Ti)O_3$, a.k.a., PBZT and barium stannate titanate, $Ba(Sn, Ti)O_3$, a.k.a., BST. To obtain large (apparent) electrostriction, it is essential that ferroelectric microdomains in the ceramic structure are generated. Various methods, such as the doping with ions of a different valence or ionic radius, or the creation of vacancies, which introduce spatial microscopic inhomogeneity, are used.

2.4.5.2 Constitutive Equations of Electrostrictive Ceramics

The strain-field curves of electrostrictive ceramics display a typical quadratic behavior. On such curves, a positive mechanical strain is obtained for both positive and negative electric fields. However, the strain field curve is strongly nonlinear, as appropriate to quadratic behavior. What is remarkable about electrostrictive ceramics is their very low hysteresis. The constitutive equations of electrostrictive ceramics are similar to those for piezoelectric ceramics, only that it also incorporates significant second-order terms are included, e.g.,

$$S_{ij} = s^E_{ijkl} T_{kl} + d_{kij} E_k + m_{klij} E_k E_l \tag{38}$$

$$D_m = d_{mkl} T_{kl} + 2m_{mnij} E_n T_{ij} + \varepsilon^T_{mn} E_n \tag{39}$$

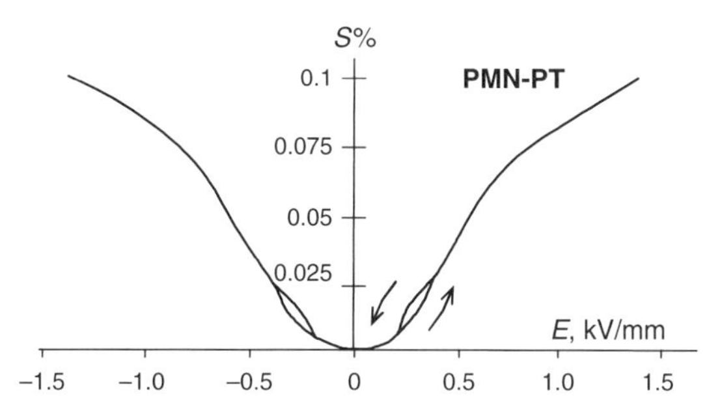

FIGURE 2.8 Field-induced strain in 90-10 PMN-PT electrostrictive ceramic.

In this equation, the first two terms are the same as those used to describe the piezoelectric constitutive law, i.e., Hooke's law and the converse piezoelectric effect. The third term represents the electrostriction effect. The components of m_{klij} are the electrostrictive coefficients. Equation (38) indicates that electrostriction appears as quadratic addition to the linear piezoelectric effect. In fact, the two effects are separable because the piezoelectric effect is possible only in noncentrosymmetric materials, whereas the electrostrictive effects are not limited by symmetry and are present in all materials. In addition to the direct electrostrictive effect, the converse electrostrictive effect also exists.

Commercially available PMN formulations are internally biased and optimized to give quasi-linear behavior. In this situation, they display much less nonlinearity than the conventional quadratic electrostriction, and resemble more the conventional linear piezoelectricity. The linearized electrostrictive ceramics retain the very low hysteresis of quadratic electrostrictive ceramics. Thus, from this standpoint, they are superior to conventional piezoelectric ceramics. However, linearized electrostrictive ceramics do not accept field reversal. After linearization, the constitutive equations of electrostrictive ceramics resemble those of conventional piezoceramics, i.e.,

$$S_{ij} = s^{E}_{ijkl} T_{kl} + \tilde{d}_{ijk} E_k \tag{40}$$

$$D_m = \tilde{d}_{mkl} T_{kl} + \varepsilon^{T}_{mn} E_n \tag{41}$$

The symbol $\sim$ indicates that the piezoelectric constants, $\tilde{d}_{ijk}$, of Eqs. (40) and (41) are different from the corresponding constants d_{ijk} in the original Eqs. (38) and (39). This is due to the linearization process. In Eqs. (38) and (39), the d_{ijk} constants were quite small, because the main effect was due to the quadratic effects represented by the m_{klij} constants. In Eqs. (40) and (41), the $\tilde{d}_{ijk}$ constants are quite significant, as they represent the effect of the linearization of Eqs. (38) and (39).

2.5 PIEZOPOLYMERS

Piezoelectric polymers are polymers that display piezoelectric properties similar to those of quartz and piezo ceramics. Piezoelectric polymers are supplied in the form of thin films. It is flexible and shows large compliance. Piezoelectric polymers are cheaper and easier to fabricate than piezoceramics. The flexibility of piezoelectric polymers overcomes some of the drawbacks associated with the piezoelectric ceramics brittleness. A typical piezoelectric polymer is the polyvinylidene fluoride, abbreviated PVDF or PVF_2. This polymer has strong piezoelectric and pyroelectric properties. Its chemical formulation is $(–CH_2–CF_2–)_n$. This polymer displays a crystallinity of 40–50%. The PVDF crystal is dimorphic, the two types being designated I (or β) and II (or α). In the β phase (i.e., type I), PVDF is polar and piezoelectric. When in the α phase, PVDF is not polar and is commonly used as electrical insulator, among other applications. To impart the piezoelectric properties, the α phase is converted to the β phase and then polarized. Stretching α-phase material produces the β phase.

2.5.1 PIEZOPOLYMER PROPERTIES AND CONSTITUTIVE EQUATIONS

The piezoelectric properties of piezopolymers are comparable to those of piezoceramics (Table 2.3). However, its modulus of elasticity is much lower. Remarkable about piezopolymers is their large pyroelectric constant, which makes them good candidates for infrared sensor applications. Another beneficial property is that, unlike piezoceramics,

TABLE 2.3 Comparison of PVDF properties with those of piezoelectric ceramics

Property	Units	PVDF film	PZT ($PbZrTiO_3$)	$BaTiO_3$
Density	kg/m^3	1780	7500	5700
Relative permittivity	$\varepsilon/\varepsilon_0$	12	1200	1700
d_{31}	10^{-12} C/N	23	110	78
g_{31}	10^{-3} Vm/N	216	10	5
k_{31}	at 1 kHz	0.12	0.30	0.21
Young modulus	GPa	~3	~60	~110
Acoustic impedance	10^6 kg/m²-s	2.7	30	30

they can be operated at high strain levels. The use at strain of up to 0.2% has been reported. The constitutive relations for PVDF can be described as

$$\begin{aligned} S_{ij} &= s^E_{ijkl} T_{kl} + d_{kij} E_k + \alpha^E_i \theta \\ D_j &= d_{jkl} T_{kl} + \varepsilon^T_{jk} E_k + \tilde{D}_m \theta \end{aligned} \tag{42}$$

where S_{ij} is the mechanical strain, T_{ij} is the mechanical stress, E_i is the electrical field, D_i is the electrical displacement (charge per unit area). The coefficient s^E_{ijkl} is the mechanical compliance of the material measured at zero electric field ($E = 0$). The coefficient ε^E_{jk} is the dielectric constant measured at zero mechanical stress ($T = 0$). The coefficient d_{ijk} is the *piezoelectric strain constant* (also known as the *piezoelectric charge constant*), which couples the electrical and mechanical variables and expresses how much strain is obtained per unit applied electric field, θ is the absolute temperature, α^E_i is the coefficient of thermal expansion under constant electric field; $\tilde{D}_i$ is the temperature coefficient of the electric displacement.

2.5.2 TYPICAL PIEZOPOLYMER APPLICATIONS

The PVDF material is flexible, and not brittle like piezoelectric ceramics. This property is especially important for applications involving complicated shapes and/or significant structural strains. Its easy formability, along with this property, makes it superior to ceramics as a sensor. As a sensor, PVDF provides higher voltage/electric field in response to mechanical stress than piezoceramics. The piezoelectric g-constant (i.e., the voltage generated per unit mechanical stress) is typically 10 to 20 times larger than for piezoceramics. PVDF film also produces an electric voltage in response to infrared light due to its strong pyroelectric coefficient. Hence, they have found wide applications as sensors.

Piezopolymers are often used for sensing. PVDF can be formed in thin films and bonded to many surfaces. Uniaxial films, which are electrically poled in one direction, can measure stresses along one axis, whereas biaxial films can measure stresses in a plane. A PVDF sensor can be used like a strain gage; however, it does not require a conditioning power supply. The output signal is also comparable to that of an amplified strain gage signal. This high sensitivity is due to the low thickness of the typical PVDF film (25 μm). Because of its good sensor properties (i.e., the high g constant), light weight, flexibility; toughness, PVDF is used in numerous sensor applications. When used as an actuator, PVDF gives a much lower force than piezoceramics, due to is much lower modulus (Table 2.3). Hence, it is best use in the actuation of compliant microstructures, with a low inherent stiffness. It is inappropriate for structural control applications involving conventional structural materials.

2.6 MAGNETOSTRICTIVE MATERIALS

In simple terms, **magnetostriction** is the material property that causes certain materials to change shape when an external magnetic field is applied. Magnetostrictive materials expand in the presence of a magnetic field, as their magnetic domains align with the field lines. Magnetostriction was initially observed in nickel, cobalt, iron, and their alloys but the values were small ($<50\,\mu$strain). Large strains ($\sim 10\,000\,\mu$strain) were observed in the rare-earth elements terbium (Tb) and dysprosium (Dy) at cryogenic temperatures (i.e., below 180° K). Very large magnetostriction exists at room temperature in the terbium-iron alloy $TbFe_2$. The binary alloy Terfenol-D ($Tb_{0.3}Dy_{0.7}Fe_{1.9}$), developed at Ames Laboratory and the Naval Ordnance Laboratory (now Naval Surface Weapons Center), displays magnetostriction of up to $2000\,\mu$strain at room temperature and up to 80° C and higher. Current Terfenol-D binary alloy formulations are of the form $Tb_{1-x}Dy_xFe_{1.9-2}$ where x is the relative proportion of dysprosium, whereas the proportion of iron can vary between 1.9 and 2. In the foregoing discussion, we will use the generic value 2, while understanding that the actual value may be between 1.9 and 2, according to the detail formulation of the particular Terfenol-D alloy.

The magnetostrictive constitutive equations contain both linear and quadratic terms

$$S_{ij} = s^E_{ijkl} T_{kl} + d_{kij} H_k + m_{klij} H_k H_l \tag{43}$$

$$B_j = d_{jkl} T_{kl} + \mu^T_{jk} H_k \tag{44}$$

where, in addition to the already defined variables, H_k is the magnetic field intensity, B_j is the magnetic flux density, and μ^T_{jk} is the magnetic permeability under constant stress. The coefficients d_{kij} and m_{klij} are defined in terms of magnetic units. The magnetic field intensity, H, in a rod surrounded by a coil with n turns per unit length is related to the current, I, through the relation

$$H = nI \tag{45}$$

2.6.1 LINEARIZED EQUATIONS OF PIEZOMAGNETISM

Magnetostrictive material response is basically quadratic in magnetic field, i.e., the magnetostrictive response does not change sign when the magnetic field is reversed. However, the nonlinear magnetostrictive behavior can be linearized about an operating point through the application of a bias magnetic field. In this case, piezomagnetic behavior, in which response reversal accompanies field reversal, can be obtained. The equations of linear piezomagnetism in compact matrix (Voigt) notations are

$$S_p = s^H_{pq} T_q + d_{kp} H_k, \qquad p, q = 1, \ldots, 6; \quad k = 1, 2, 3 \tag{46}$$

$$B_i = d_{iq} T_q + \mu^T_{ik} H_k, \qquad q = 1, \ldots, 6; \quad i, k = 1, 2, 3 \tag{47}$$

where, S_p is the mechanical strain, T_q is the mechanical stress, H_k is the magnetic field intensity, B_i is the magnetic flux density, and μ^T_{ik} is the magnetic permeability. The coefficient s^H_{pq} is the mechanical compliance of the material measured at zero magnetic field ($M = 0$). The coefficient μ^T_{ik} is the magnetic permeability measured at zero mechanical stress ($T = 0$). The coefficient d_{ik} is the *piezomagnetic constant*, which couples the magnetic and mechanical variables and expresses how much strain is obtained per unit magnetic field. For common magnetoactive materials, Eqs. (46) and (47) take the long-hand form

$$\begin{Bmatrix} S_1 \\ S_2 \\ S_3 \\ S_4 \\ S_5 \\ S_6 \end{Bmatrix} = \begin{bmatrix} s_{11} & s_{12} & s_{13} & 0 & 0 & 0 \\ s_{12} & s_{11} & s_{13} & 0 & 0 & 0 \\ s_{13} & s_{13} & s_{33} & 0 & 0 & 0 \\ 0 & 0 & 0 & s_{44} & 0 & 0 \\ 0 & 0 & 0 & 0 & s_{55} & 0 \\ 0 & 0 & 0 & 0 & 0 & s_{66} \end{bmatrix} \begin{Bmatrix} T_1 \\ T_2 \\ T_3 \\ T_4 \\ T_5 \\ T_6 \end{Bmatrix} + \begin{bmatrix} 0 & 0 & d_{31} \\ 0 & 0 & d_{31} \\ 0 & 0 & d_{33} \\ 0 & d_{15} & 0 \\ d_{15} & 0 & 0 \\ 0 & 0 & 0 \end{bmatrix} \begin{Bmatrix} H_1 \\ H_2 \\ H_3 \end{Bmatrix} \tag{48}$$

$$\begin{Bmatrix} B_1 \\ B_2 \\ B_3 \end{Bmatrix} = \begin{bmatrix} 0 & 0 & 0 & 0 & d_{15} & 0 \\ 0 & 0 & 0 & d_{15} & 0 & 0 \\ d_{31} & d_{31} & d_{33} & 0 & 0 & 0 \end{bmatrix} \begin{Bmatrix} T_1 \\ T_2 \\ T_3 \\ T_4 \\ T_5 \\ T_6 \end{Bmatrix} + \begin{bmatrix} \mu_{11}^T & 0 & 0 \\ 0 & \mu_{11}^T & 0 \\ 0 & 0 & \mu_{33}^T \end{bmatrix} \begin{Bmatrix} H_1 \\ H_2 \\ H_3 \end{Bmatrix} \tag{49}$$

The magnetomechanical coupling coefficient, k, is defined as the ratio of the magnetoelastic energy to the geometric mean of the elastic and magnetic energies, i.e.,

$$k = \frac{U_{\mathrm{me}}}{\sqrt{U_{\mathrm{e}} U_{\mathrm{m}}}} \tag{50}$$

where U_{e} is the elastic energy, U_{m} is the magnetic energy, and U_{me} is the magnetoelastic energy in the material.

Typical physical properties of Terfenol-D material are given in Table 2.4.

TABLE 2.4 Physical properties of Terfenol-D (http://etrema-usa.com/terfenol)

Nominal composition	$Tb_{0.3}Dy_{0.7}Fe_{1.92}$
Mechanical properties	
Young modulus	25–35 GPA
Sound speed	1640–1940 m/s
Tensile strength	28 MPa
Compressive strength	700 MPa
Thermal properties	
Coefficient of thermal expansion	12 ppm/°C
Specific heat	0.35 kJ/kg-K
Thermal conductivity	13.5 W/m-k
Electrical properties	
Resistivity	58×10^{-8} O-m
Curie temperature	380° C
Magnetostrictive properties	
Strain (estimated linear)	800–1200 ppm
Energy density	14–25 kJ/m^2
Magnetomechanical properties	
Relative permeability	3–10
Coupling factor	0.75

2.7 SUMMARY AND CONCLUSIONS

This chapter has reviewed and briefly discussed the basic types of electroactive and magnetoactive materials. Electroactive and magnetoactive materials are materials that modify their shape in response to electric or magnetic stimuli. Such materials permit induced-strain actuation and strain sensing which are of considerable importance in mini mechatronics.

On one hand, induced-strain actuators are based on active materials that display dimensional changes when energized by electric, magnetic, or thermal fields. Piezoelectric, electrostrictive, and magnetostrictive materials have been presented and analyzed. Of these, piezoelectric (PZT), electrostrictive (PMN), and magnetostrictive (Terfenol-D) materials are commonly used. On the other hand, strain sensing with electroactive and magnetoactive materials creates direct conversion of mechanical energy into electric and magnetic energy. With piezoelectric strain sensors, strong and clear voltage signals are obtained directly from the sensor without the need for intermediate gage bridges, signal conditioners, and signal amplifiers.

Figure 2.9 compares induced-strain response of some commercially available piezoelectric, electrostrictive, and magnetostrictive actuation materials. It can be seen that the electrostrictive materials have less hysteresis, but more nonlinearity. The little hysteresis of electrostrictive ceramics can be an important plus in certain applications, especially at high frequencies. However, one should be aware that this low hysteresis is strongly temperature dependent. As the temperature decreases, the hysteresis of electrostrictive ceramics increases, such that, below a certain temperature, the hysteresis of electrostrictive ceramics may exceed that of piezoelectric ceramics. In general, because the beneficial behavior of the electrostrictive ceramics is related to the diffuse phase transition in the relaxor range, their properties degrade as the operation temperature gets outside the relaxor phase-transition range.

In summary, one can conclude that the potential of active materials for sensing and actuation applications has been demonstrated in several successful applications.

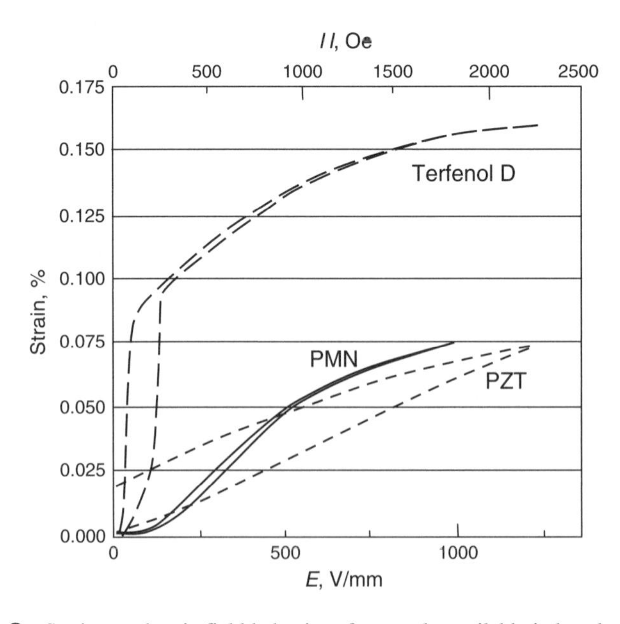

FIGURE 2.9 Strain vs. electric field behavior of currently available induced-strain materials.

However, this field is still in its infancy and further research and development is being undertaken to establish active materials as reliable, durable, and cost-effective options for large-scale engineering applications.

2.8 PROBLEMS AND EXERCISES

1. Explain the difference between tensor notations and Voigt matrix notations in the writing of the compliance and stiffness matrices
2. Explain the following difference in subscripts usage: the (1, 3) term in the compliance matrix is denoted s_{13}, whereas the (1, 3) term in the piezoelectric coefficient matrix is denoted d_{31}
3. Calculate the spontaneous strain, S_s, and the spontaneous polarization, P_s, for the barium titanate lattice shown in Fig. 2.10.

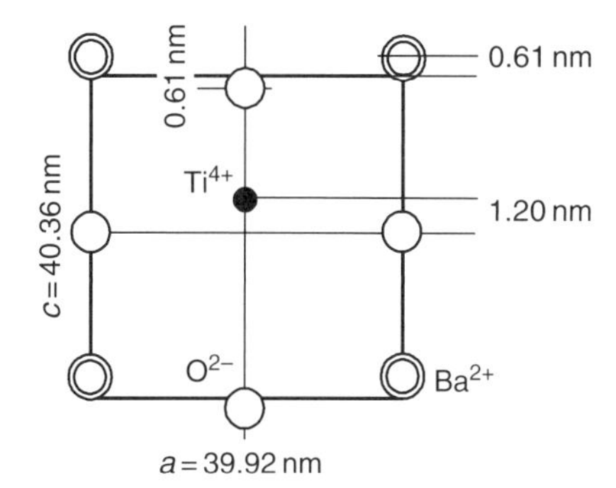

FIGURE 2.10 Ionic shifts inducing spontaneous strain and spontaneous polarization in barium titanate.

3

Vibration of Solids and Structures

3.1 Introduction

This chapter offers a brief introduction to vibration theory. This introduction is necessary because many of the SHM methods to be discussed in later chapters will utilize concepts and formulae from vibration theory.

The chapter starts with the theory of vibration of a single degree of freedom (1-dof) system, the *particle vibration*. This simple system will be used as a springboard for the analysis of more complicated system later in the chapter. The 1-dof particle vibration will be used to introduce fundamental basic concepts such as the differential equation of motion, harmonic solutions, free vs. forced vibrations, and damped vs. undamped vibrations. Energy methods approach to vibration analysis will also be discussed.

The second part of the chapter covers the vibration of continuous systems. Partial differential equations (PDE) in space and time will govern this type of vibrations. Assuming harmonic behavior in time, the equation of motion is reduced to an ordinary differential equation (ODE) in the space domain. This is a boundary value problem, which yields eigenvalues and eigenmodes, and the associate *natural frequencies* and *mode shapes*. The axial vibration of bars, flexural vibration of beams, and torsional vibration of shafts will be considered. In each case, the study of free vibrations is followed by the study of forced vibrations.

The chapter ends with a set of problems and exercises that will assist the student in consolidating the understanding of the basic concepts and in applying the theory to practical situations.

3.2 Single Degree of Freedom Vibration Analysis

Consider a 1-dof vibration system consisting of a particle of mass m supported by an elastic spring of stiffness k. Initially, the particle is at equilibrium (Fig. 3.1a). In its

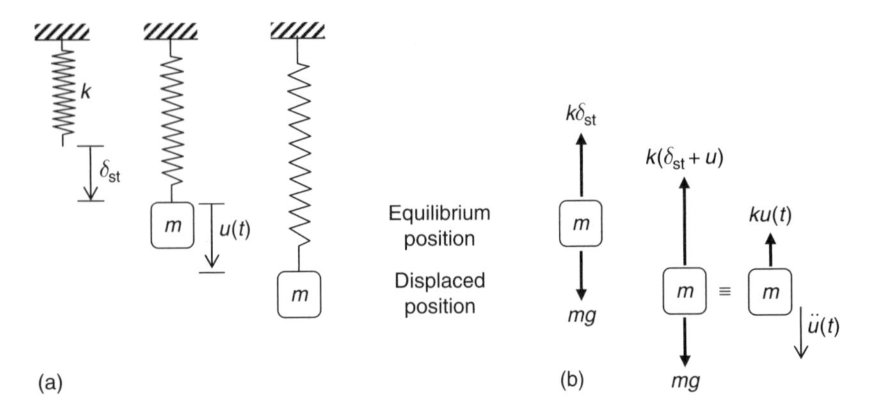

FIGURE 3.1 Particle during free vibration: (a) physical mechanism; (b) free-body diagram.

equilibrium state, the weight of the particle, W, is balanced by a force in the spring, $k\delta_{\text{st}}$, where δ_{st} is the *static displacement* of the spring under the weight, $W = mg$, i.e.,

$$\delta_{\text{st}} = \frac{mg}{k} \tag{1}$$

If the particle is displaced from this equilibrium state and then let free, it will oscillate up and down about the equilibrium position, i.e., it will experience a state of free vibration with time-dependent displacement $u(t)$ (Fig. 3.1b). Due to friction, the vibration will decrease in amplitude and die out after some time. This is the case of *damped free vibration*, in which the initial energy of vibration is being gradually dissipated through friction or another damping mechanics. In the ideal case when friction is absent, the vibration will continue indefinitely with constant amplitude. This is the case of *undamped free vibration*. Undamped free vibrations are easier to analyze, but do not usually happen in practice.

Damped vibrations involve a more complicated analysis, but the results are more of practical application. When the vibration is not developing freely, but due to the excitation of an external oscillatory force, it is called *forced vibration*. Depending on the presence or absence of energy dissipation mechanisms, we may have *damped forced vibration* or *undamped forced vibration*. Again, the undamped forced vibration is easier to analyze, but the damped forced vibration is more representative of actual phenomena.

3.2.1 FREE VIBRATION OF A PARTICLE

3.2.1.1 Oscillatory Motion

An oscillatory motion can be defined by the formula

$$u(t) = C\cos(\omega t + \psi) \tag{2}$$

where C is the amplitude measured in length units, ω is the *angular frequency* measured in radians per second (rad/s), and ψ is the *initial phase*, measured in radians.

The *frequency*, f, which is measured in cycles per second (c/s) or Hz, is related to the angular frequency by the formula

$$f = \frac{1}{2\pi}\omega \tag{3}$$

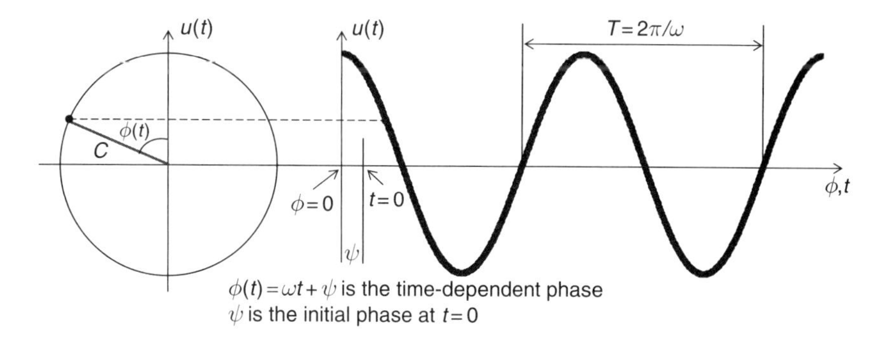

FIGURE 3.2 Schematic representation of an oscillatory motion of amplitude C, angular frequency ω, and initial phase ψ.

The *period*, τ, measured in seconds is related to the frequency, f, by the formula

$$\tau = \frac{1}{f} \tag{4}$$

Phasor representation of oscillatory motion

In phasor notation, the vibrational motion is represented by the phasor $C\angle\psi$, where C is the magnitude and ψ is the phase angle.

Complex representation of oscillatory motion

Recall Euler identity (Kreyszig, 1999)

$$e^{i\alpha} = \cos\alpha + i\sin\alpha, \quad \alpha \in \mathbb{R} \tag{5}$$

Using Euler identity given by Eq. (5), we can view the cosine function as the real part of the complex exponential function, i.e.,

$$\cos\alpha = \operatorname{Re} e^{i\alpha} \tag{6}$$

Hence, the vibrational motion described by Eq. (2) can be viewed as the real part of a complex exponential representation, i.e.,

$$u(t) = C\cos(\omega t + \psi) = C \operatorname{Re} e^{i(\omega t + \psi)} \tag{7}$$

Therefore, we can simply deal with a complex function $\tilde{u}(t)$ remembering the convention that the actual physical motion is only its real part, i.e., $u(t) = \operatorname{Re}\tilde{u}(t)$. Hence

$$\tilde{u}(t) = Ce^{i\psi}e^{i\omega t} = \tilde{C}e^{i\omega t} \tag{8}$$

where

$$\tilde{C} = Ce^{i\psi} \tag{9}$$

is the complex amplitude. The constants C and ψ are the magnitude and phase of the complex amplitude, i.e.,

$$C = \left|\tilde{C}\right| \quad \text{and} \quad \psi = \arg\tilde{C} \tag{10}$$

3.2.1.2 Undamped Free Vibration

Consider the particle displaced from the equilibrium position by a time-dependent displacement $u(t)$. The additional spring force due to this displacement is $ku(t)$, whereas the acceleration is $\ddot{u}(t)$, as shown in Fig. 3.3. Newton law of motion and Eq. (1) gives

$$m\ddot{u}(t) = -ku(t) \tag{11}$$

Equation (11) can be rearranged in the form of a homogenous linear differential equation of the form

$$m\ddot{u}(t) + ku(t) = 0 \tag{12}$$

The solution of a homogenous linear ODE (Kreyszig, 1999) is sought in the form $e^{\lambda t}$. Substitution in Eq. (12) yields the characteristic equation

$$m\lambda^2 + k = 0 \tag{13}$$

which has the complex solutions

$$\lambda_{1,2} = \pm i\sqrt{\frac{k}{m}} \tag{14}$$

The square root of the ratio between stiffness and mass is usually denoted by ω_{n}, where the subscript n stands for *natural*, and the quantity ω_{n} is called the *natural angular frequency* of vibration

$$\omega_{\mathrm{n}} = \sqrt{\frac{k}{m}}, \quad \text{or} \quad \omega_{\mathrm{n}}^2 = \frac{k}{m} \tag{15}$$

Note that division of Eq. (16) by the mass m and utilization of the natural angular frequency ω_{n} defined by Eq. (15) yields the *normalized form* of Eq. (12), i.e.,

$$\ddot{u}(t) + \omega_{\mathrm{n}}^2 u(t) = 0 \tag{16}$$

General solution of undamped free vibration

The general solution of Eq. (12) is expressed in the form of complex exponentials, i.e.,

$$u(t) = C_1 e^{i\omega_{\mathrm{n}} t} + C_2 e^{-i\omega_{\mathrm{n}} t} \tag{17}$$

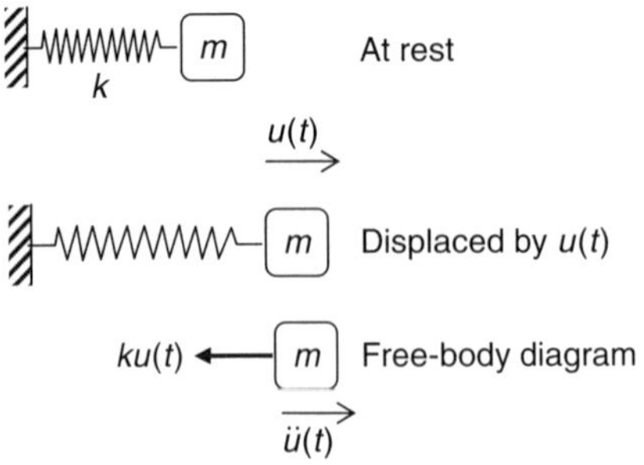

FIGURE 3.3 Free-body diagram of a particle during free horizontal vibration.

Using Euler identity given by Eq. (5), we can conveniently express Eq. (17) in the trigonometric form

$$u(t) = A\cos\omega_n t + B\sin\omega_n t \tag{18}$$

where the constants A, B and C_1, C_2 are directly related to each other through simple trigonometry. Equation (18) can be rewritten in the form

$$u(t) = C\cos(\omega_n t + \psi) \tag{19}$$

where

$$A = C\cos\psi, \quad B = -C\sin\psi \tag{20}$$

and

$$C = \sqrt{A^2 + B^2}, \quad \psi = \tan^{-1}\left(\frac{B}{A}\right) \tag{21}$$

In complex notations, the undamped free vibration solution is given by

$$\tilde{u}(t) = Ce^{i\psi}e^{i\omega_n t} = \tilde{C}e^{i\omega_n t} \tag{22}$$

where

$$\tilde{C} = Ce^{i\psi} \tag{23}$$

is the complex amplitude. The constants C and ψ are the magnitude and phase of the complex amplitude, i.e.,

$$C = |\tilde{C}| \quad \text{and} \quad \psi = \arg\tilde{C} \tag{24}$$

Expansion of Eq. (23) through the Euler identity given by Eq. (5) and comparison with Eq. (20) yields

$$A = \operatorname{Re}\tilde{C}, \quad B = -\operatorname{Im}\tilde{C} \tag{25}$$

General solution for given initial displacement and initial velocity

Assume that the initial displacement, u_0, and the initial velocity, $\dot{u}_0$, are known. Using Eq. (18), we write

$$\begin{aligned} u_0 = u(0) &= A\cos\omega_n t + B\sin\omega_n t|_{t=0} = A \\ \dot{u}_0 = \dot{u}(0) &= -\omega_n A\sin\omega_n t + \omega_n B\cos\omega_n t|_{t=0} = \omega_n B \end{aligned} \tag{26}$$

Solving Eq. (26) for A and B yields

$$\begin{aligned} A &= u_0 \\ B &= \frac{\dot{u}_0}{\omega_n} \end{aligned} \tag{27}$$

Substitution of Eq. (27) into Eq. (18) gives the general solution of undamped free vibrations with initial displacement, u_0, and initial velocity, $\dot{u}_0$, in the form

$$u(t) = u_0 \cos \omega_n t + \frac{\dot{u}_0}{\omega_n} \sin \omega_n t \tag{28}$$

3.2.1.3 Damped Free Vibration

Consider the 1-dof damped system consisting of a spring, k, mass, m, and dashpot damper, c, as presented in Fig. 3.4a. The equation of motion for damped free vibration is given by the linear ODE

$$m\ddot{u}(t) + c\dot{u}(t) + ku(t) = 0 \tag{29}$$

It is convenient to study the normalized form of Eq. (29), i.e.,

$$\ddot{u}(t) + 2\zeta\omega_n\dot{u}(t) + \omega_n^2 u(t) = 0 \tag{30}$$

where ω_n is the natural angular frequency already defined by Eq. (15), whereas ζ is the *damping ratio*, defined by

$$\zeta = c/c_{cr}, \quad c_{cr} = 2\omega_n m = 2\sqrt{mk} \tag{31}$$

For reasons that will become apparent shortly, the value c_{cr} is called *critical damping*.

The solution of a linear ODE (Kreyszig, 1999) is sought in the form $e^{\lambda t}$. Substitution into Eq. (29) yields the characteristic equation

$$\lambda^2 + 2\zeta\omega_n\lambda + \omega_n^2 = 0 \tag{32}$$

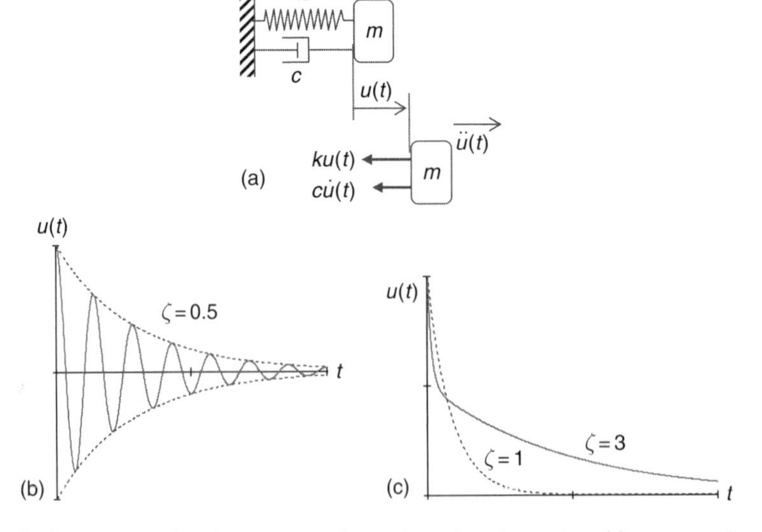

FIGURE 3.4 Damped vibration response for various damping ratios: (a) system schematic and free-body diagram; (b) underdamped ($\zeta < 1$); (c) critically damped ($\zeta = 1$), overdamped ($\zeta > 1$).

which has the complex solutions

$$\lambda_{1,2} = -\zeta\omega_n \pm i\omega_n\sqrt{1-\zeta^2} \tag{33}$$

Equation (33) can be expressed as

$$\lambda_{1,2} = \sigma \pm i\omega_d \tag{34}$$

where σ is the real exponent given by

$$\sigma = -\zeta\omega_n \tag{35}$$

whereas ω_d is the *damped natural frequency* given by

$$\omega_d = \omega_n\sqrt{1-\zeta^2} \tag{36}$$

General solution of damped free vibration

Using Eqs. (33) and (35), the general solution of Eq. (29) is expressed in the form of complex exponentials, i.e.,

$$u(t) = C_1 e^{\left(-\zeta\omega_n + i\omega_n\sqrt{1-\zeta^2}\right)t} + C_2 e^{\left(-\zeta\omega_n - i\omega_n\sqrt{1-\zeta^2}\right)t} \tag{37}$$

Using the damped natural frequency, ω_d, given by Eq. (36), we can rewrite Eq. (37) in the form

$$u(t) = C_1 e^{(-\zeta\omega_n + i\omega_d)t} + C_2 e^{(-\zeta\omega_n - i\omega_d)t} \tag{38}$$

Using Euler identity (Kreyszig, 1999) and trigonometric manipulations, Eq. (37) can be rewritten in the form

$$u(t) = Ce^{-\zeta t}\cos(\omega_d t + \psi) \tag{39}$$

The constants C and ψ, just as C_1 and C_2, depend on the initial conditions. In complex notations, the damped free vibration solution can be expressed as

$$\tilde{u}(t) = \tilde{C}e^{-\zeta t}e^{i\omega_d t} \tag{40}$$

where

$$\tilde{C} = Ce^{i\psi}. \tag{41}$$

is the complex amplitude.

Effect of damping on vibration response

Depending on the value of the damping ratio, ζ, the vibration response can be categorized as follows:

- underdamped response, $\zeta < 1$, i.e., $c < c_{cr}$
- critically damped response, $\zeta = 1$, i.e., $c = c_{cr}$
- overdamped response, $\zeta > 1$, i.e., $c > c_{cr}$

The underdamped response consists of a decreasing–amplitude oscillation (Fig. 3.4b). The amplitude decrease is due to the exponential decay factor in Eq. (40). For structural applications, the damping ratio is relatively small ($\zeta < 5\%$). For such *lightly damped* structures, the damped natural frequency, $\omega_{\mathrm{d}} = \omega_{\mathrm{n}}\sqrt{1-\zeta^2}$ is not much different from the undamped natural frequency, ω_{n}. Hence, the damped response is similar to the undamped response, only that the amplitude displays the exponential decay.

As the damping increases, the difference between the damped natural frequency and the undamped natural frequency increases, and the damped response depart more and more from the undamped response. When damping exceeds the critical damping (overdamped case, $\zeta > 1$), the damped response is no longer oscillatory (Fig. 3.4c, $\zeta = 3$). In fact, analysis of Eqs. (36) and (37) indicates that for the overdamped case ($\zeta > 1$), the overdamped response is composed of two decaying exponentials

$$u(t) = C_1 e^{-\left(\zeta-\sqrt{\zeta^2-1}\right)\omega_{\mathrm{n}} t} + C_2 e^{-\left(\zeta+\sqrt{\zeta^2-1}\right)\omega_{\mathrm{n}} t} \tag{42}$$

When damping equals the critical damping ($\zeta = 1$), the two roots of the characteristic Eq. (32) coalesce, $\lambda_1 = \lambda_2 = -\zeta\omega_{\mathrm{n}}$, and the solution takes the form

$$u(t) = (C_1 + C_2 t)e^{-\omega_{\mathrm{n}} t} \tag{43}$$

The plot of Eq. (43) is shown in Fig. 3.4c for $\zeta = 1$, which somehow resembles the overdamped plot for $\zeta = 3$, only that the mathematical relation that generated this curve contains only one decaying exponential instead of two. It is apparent from Fig. 3.4c that the overdamped response ($\zeta = 3$) decays more rapidly at first, but then it takes longer to settle down. The critically damped response ($\zeta = 1$) decays a little slower at first, but then it settles down more rapidly.

Logarithmic decrement, δ

It is frequently desirable to determine the damping through experimental methods. Several methods exist. For example, the determination of damping through the *logarithmic decrement method* consists of measuring two consecutive peaks on the free oscillatory response and then using a formula to determine the damping ratio ζ. The *logarithmic decrement*, δ, is defined as the logarithm of the ratio between two consecutive peaks in the oscillatory free-decay response, i.e.,

$$\delta = \ln\left(\frac{u_1}{u_2}\right) \tag{44}$$

To illustrate how the logarithmic decrement works, assume (Fig. 3.5) that we measure two consecutive oscillatory peaks, u_1 and u_2, occurring at times t_1 and $t_2 = t_1 + \tau_{\mathrm{d}}$, respectively, where τ_{d} is the damped period,

$$\tau_{\mathrm{d}} = \frac{2\pi}{\omega_{\mathrm{d}}} \tag{45}$$

Using Eq. (40), we write the ratio of the two amplitudes as

$$\frac{u_1}{u_2} = \frac{e^{-\zeta\omega_{\mathrm{n}} t}}{e^{-\zeta\omega_{\mathrm{n}}(t+\tau)}} = e^{\zeta\omega_{\mathrm{n}}\tau} \tag{46}$$

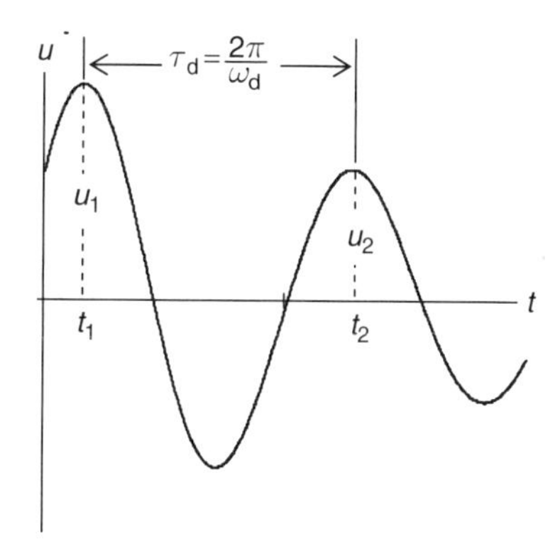

FIGURE 3.5 Determination of the logarithmic decrement.

where the oscillatory part cancelled out in view of Eq. (45). Substitution of Eq. (46) into Eq. (44) yields

$$\delta = \zeta \omega_n \tau_d = \frac{2\pi}{\sqrt{1-\zeta^2}}\zeta \tag{47}$$

Upon solution,

$$\zeta = \frac{\delta}{\sqrt{(2\pi)^2 + \delta^2}} \tag{48}$$

For lightly damped system, $\zeta << 1$ and Eq. (47) simplifies to

$$\delta \cong 2\pi\zeta \quad \text{for} \quad \zeta << 1 \tag{49}$$

Hence, for lightly damped systems, the relation between the damping ratio and logarithmic decrement simplifies to the widely used formula

$$\zeta = \frac{\delta}{2\pi} \quad \text{for} \quad \zeta << 1 \tag{50}$$

These concepts can be easily extended to the case when more than one period of oscillation is taken into account between t_1 and t_2, as would be the need for very lightly damped systems.

3.2.2 FORCED VIBRATION OF A PARTICLE

In this section, we will analyze the response of a 1-dof system to external excitation. In particular, we will start with considering an external *excitation force*, $F(t)$. To simplify the analysis, we will assume the excitation to be harmonic, i.e.,

$$F(t) = \hat{F}\cos(\omega t) \tag{51}$$

where ω is the *excitation frequency*, sometimes referred to as the *driving frequency*. If the need arises, the analysis of the response to harmonic single-frequency excitation can always be extended to more complicated time-dependant excitations through *Fourier Analysis* (Kreyszig, 1999). We will start with the analysis of a simple undamped 1-dof system and then proceed to damped 1-dof systems.

3.2.2.1 Undamped Forced Vibration

Consider the differential equation for undamped forced vibration under harmonic excitation

$$m\ddot{u}(t) + ku(t) = \hat{F}\cos(\omega t) \tag{52}$$

Upon normalization by mass m, we get

$$\ddot{u}(t) + \omega_{\mathrm{n}}^2 u(t) = \hat{f}\cos(\omega t) \tag{53}$$

where

$$\hat{f} = \frac{\hat{\hat{F}}}{m} \tag{54}$$

and

$$f(t) = \hat{f}\cos(\omega t) \tag{55}$$

is the *forcing function.*

Equation (53) is an inhomogeneous linear ODE. The solution of this type of equations (Kreyszig, 1999) consists of the superposition of the *complementary solution* that satisfies the homogeneous ODE and a *particular solution* that satisfies to inhomogeneous (right hand side) part of Eq. (52). The complimentary solution is in fact the solution already discussed in previous section as given by Eq. (17) or (19). The particular solution is sought of the same form as the forcing function and is found to be

$$u_p(t) = \frac{1}{-\omega^2 + \omega_{\mathrm{n}}^2}\hat{f}\cos(\omega t) \tag{56}$$

Superposing the complementary solution given by Eq. (19) with the particular solution given by Eq. (56) yields the complete solution for undamped forced vibration

$$u(t) = C\cos(\omega_{\mathrm{n}} t + \psi) + \frac{1}{-\omega^2 + \omega_{\mathrm{n}}^2}\hat{f}\cos(\omega t) \tag{57}$$

where the constants C and ψ are to be determined from the imposition of the initial conditions.

Dynamic amplification factor

When a load is suddenly applied to a vibrating system, the stresses and strains in the system are larger than in the case of a gradually applied load. In other words, the *dynamic loading is more severe than the static loading of same overall value.* This is known as the *dynamic amplification factor.* In the case of a simple vibrating system, the dynamic amplification factor has value of 2, i.e., the dynamic loading is twice as severe as the static loading. For more complicated vibrating systems, the value of the dynamic amplification factor may differ. A simple analysis of the dynamic amplification factor for a simple vibrating system is shown next.

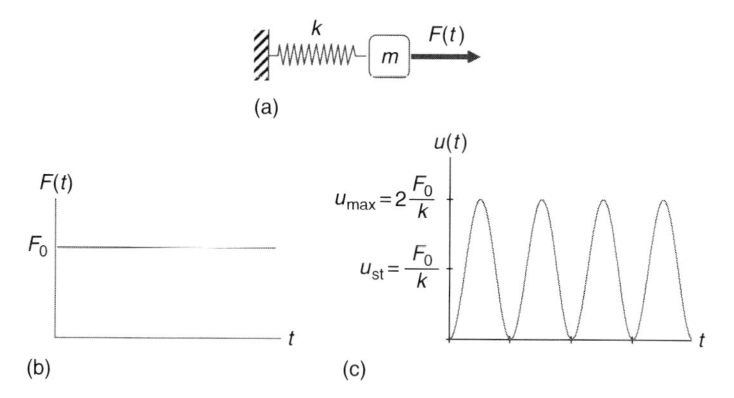

FIGURE 3.6 Dynamic amplification factor: (a) simple vibrating system; (b) suddenly applied load; (c) time response of the system.

Consider the simple undamped vibrating system of Fig. 3.6a. At $t = 0$, the system is at rest ($u_0 = 0$, $\dot{u}_0 = 0$). A force F_0 is suddenly applied at $t = 0$. Free-body diagram yields, for $t > 0$, the differential equation

$$m\ddot{u}(t) + ku(t) = F_0, \quad t > 0 \tag{58}$$

The general solution of Eq. (58) consists of a complementary solution, $u_c(t)$, as well as a particular solution, $u_p(t)$, i.e.,

$$u(t) = u_c(t) + u_p(t) \tag{59}$$

The complementary solution has the form of Eq. (18). The simplest form of the particular solution is

$$u_p(t) = \frac{F_0}{k}, \quad t > 0 \tag{60}$$

It can be easily verified that the particular solution of Eq. (60) verifies the differential equation (58). Substituting Eqs. (18) and (60) into Eq. (59) yields the general solution

$$u(t) = A\cos\omega_n t + B\sin\omega_n t + \frac{F_0}{k}, \quad t > 0 \tag{61}$$

Applying the 'at rest' initial conditions, we write

$$\begin{aligned} u(0) &= A\cos\omega_n t + B\sin\omega_n t + \frac{F_0}{k}\bigg|_{t=0} = A + \frac{F_0}{k} = 0 \\ \dot{u}(0) &= -\omega_n A\sin\omega_n t + \omega_n B\cos\omega_n t|_{t=0} = \omega_n B = 0 \end{aligned} \tag{62}$$

Upon solution of Eq. (62), we get

$$A = -\frac{F_0}{k}, \quad B = 0 \tag{63}$$

Hence, the general solution (61) becomes

$$u(t) = \frac{F_0}{k}(1 - \cos\omega_n t), \quad t > 0 \tag{64}$$

A plot of Eq. (64) is shown in Fig. 3.6c. It is apparent that the displacement starts from zero, and then climbs to a maximum value, then returns to zero, and continues as an oscillatory motion. It is also apparent that the maximum displacement is

$$u_{\max} = 2\frac{F_0}{k} \tag{65}$$

The maximum displacement is obtained for values of t at which the cosine function takes the value -1. Equation (65) illustrates the assertion that *the dynamic displacement is twice the static displacement*, as the static displacement under load F_0 would simply be $u_{st} = F_0/k$. In the case of an undamped vibrating system as considered here, *the dynamic amplification factor* is equal to 2. If damping were applied to the system, the dynamic amplification factor will take a lower value, as some vibration decay will take place before the point of maximum amplitude is reached.

3.2.2.2 Damped Forced Vibration

Consider the 1-dof system of Fig. 3.7 consisting of a mass, m, a spring, k, and a damper, c. When excited by force $F(t)$, the system satisfies the differential equation

$$m\ddot{u}(t) + c\dot{u}(t) + ku(t) = F(t) \tag{66}$$

If the exciting force is harmonic, $F(t) = \hat{F}e^{i\omega t}$, we obtain the differential equation for damped forced vibration under harmonic excitation in the form

$$m\ddot{u}(t) + c\dot{u}(t) + ku(t) = \hat{F}\cos(\omega t) \tag{67}$$

However, the case of damped forced vibration is treated more conveniently in complex notations. Apply the convention $\cos(\omega t) = \mathrm{Re}\, e^{i\omega t}$ and write Eq. (67) in the form

$$m\ddot{u}(t) + c\dot{u}(t) + ku(t) = \hat{F}e^{i\omega t} \tag{68}$$

Upon normalization by m, we get

$$\ddot{u}(t) + 2\zeta\omega_n\dot{u}(t) + \omega_n^2 u(t) = \hat{f}e^{i\omega t} \tag{69}$$

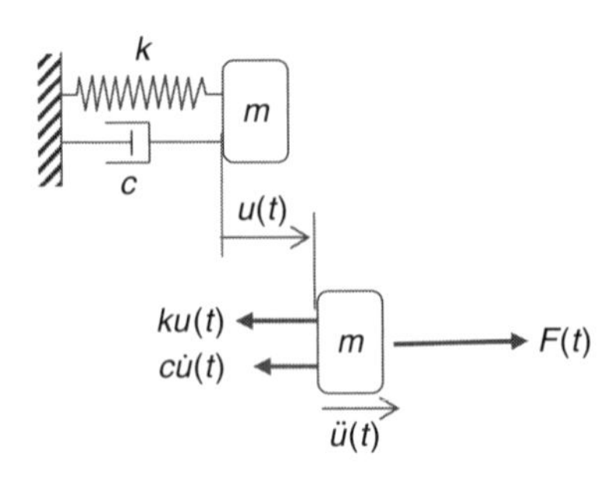

FIGURE 3.7 Damped 1-dof system under force excitation.

where $\hat{f} = \hat{F}/m$ as defined by Eq. (54). The function

$$\tilde{f}(t) = \hat{f}e^{i\omega t} \tag{70}$$

is the *complex forcing function*. Equation (69) is a homogenous linear ODE with solution consisting of the superposition of (a) the *complementary solution* given by Eqs. (40) and (41) and (b) a *particular solution*. We seek the particular solution in the same form as the forcing function, i.e.,

$$u_P(t) = \hat{u}_P e^{i\omega t} \tag{71}$$

The first and second derivatives of Eq. (71) are

$$\begin{aligned} \dot{u}_P(t) &= \hat{u}_P i\omega e^{i\omega t} = i\omega u_P(t) \\ \ddot{u}_P(t) &= \hat{u}_P (i\omega)^2 e^{i\omega t} = -\omega^2 u_P(t) \end{aligned} \tag{72}$$

Substitution of Eqs. (71) and (72) into Eq. (69) yields

$$u_p(t) = \frac{1}{-\omega^2 + i2\zeta\omega_n\omega + \omega_n^2}\hat{f}e^{i\omega t} \tag{73}$$

Superposing the complementary solution given by Eq. (40) with the particular solution given by Eq. (73) yields the complete solution for damped forced vibration

$$u(t) = Ce^{-\zeta t}e^{i(\omega_d t + \psi)} + \frac{1}{-\omega^2 + i2\zeta\omega_n\omega + \omega_n^2}\hat{f}e^{i\omega t} \tag{74}$$

where the constants C and ψ are to be determined from the initial conditions.

Equation (74) can be extended to the case of a generic excitation function, $\tilde{f}(t)$, using the Fourier expansion:

$$\tilde{f}(t) = \sum_{-\infty}^{+\infty} \tilde{f}_k e^{i\omega_k} \tag{75}$$

We can write Eq. (69) in the generic way

$$\ddot{u}(t) + 2\zeta\omega_n\dot{u}(t) + \omega_n^2 u(t) = \tilde{f}(t) \tag{76}$$

Then use the Fourier expansion (75) to reduce it to the form of Eq. (69). Subsequently, one would calculate the response for each harmonic excitation of frequency ω_k and amplitude $\tilde{f}_k$, and then reassemble the total response through Fourier summation.

Steady-state damped forced vibration solution

Examination of Eq. (74) reveals that complementary solution $Ce^{-\zeta t}e^{i(\omega_d t+\psi)}$ is only of interest at the beginning of the forced vibration, as it soon dies out due to damping. For this reason, the complementary solution is known as the *transient solution*. After some time, the damped forced vibration will settle into a *steady-state motion* that only contains the particular solution. Most forced vibration studies are only interested in the steady-state solution, and not in the transient solution. Hence, we will henceforth only analyze the steady-state solution

$$u(t) = \frac{1}{-\omega^2 + i2\zeta\omega_n\omega + \omega_n^2}\hat{f}e^{i\omega t} \tag{77}$$

which has the compact representation

$$u(t) = \hat{u}e^{i\omega t} \tag{78}$$

where $\hat{u}(\omega)$ is the frequency-dependent displacement amplitude given by

$$\hat{u}(\omega) = \frac{1}{-\omega^2 + i2\zeta\omega_n\omega + \omega_n^2}\hat{f} \tag{79}$$

It should be noted that the amplitude $\hat{u}(\omega)$ given by Eq. (79) is a frequency-dependent complex function that could be expressed using the magnitude-phase representation

$$\hat{u}(\omega) = |\hat{u}(\omega)|\, e^{i\phi(\omega)} \tag{80}$$

The phase angle, $\phi(\omega) = \arg \hat{u}(\omega)$, represents the phase difference between the response and the forcing function.

Dynamic stiffness and mechanical impedance

Substitution of Eqs. (15) and (54) into Eq. (79) and summary manipulation yields the steady-state damped forced vibration amplitude in the form

$$\hat{u}(\omega) = \frac{\hat{F}}{-\omega^2 m + ic\omega + k} \tag{81}$$

Equation (81) can bc vicwcd as the ratio between the force, F, and a frequency-dependent *dynamic stiffness*, $k_{\text{dyn}}(\omega)$, i.e.,

$$\hat{u}(\omega) = \frac{\hat{F}}{k_{\text{dyn}}(\omega)} \tag{82}$$

where

$$k_{\text{dyn}}(\omega) = -\omega^2 m + ic\omega + k \tag{83}$$

is the frequency-dependent dynamic stiffness of the 1-dof system.

Similarly, we can develop an expression for the *mechanical impedance* that is defined as ratio between the excitation force and velocity response. Differentiation of Eq. (78) yields the velocity

$$\dot{u}(t) = i\omega u(t) \tag{84}$$

The amplitude of Eq. (84) is $\hat{\dot{u}} = i\omega\hat{u}$. Hence, Eq. (81) can be rearranged in the form

$$\hat{\dot{u}} = i\omega\frac{\hat{F}}{-\omega^2 m + ic\omega + k} = \frac{\hat{F}}{i\omega m + c + \frac{k}{i\omega}} \tag{85}$$

Equation (85) can be rewritten in the form

$$\hat{u}(\omega) = \frac{\hat{F}}{Z(\omega)} \tag{86}$$

where $Z(\omega)$ is the *mechanical impedance* of the 1-dof system given by

$$Z(\omega) = i\omega m + c + \frac{k}{i\omega} \tag{87}$$

Frequency response function

Equation (81) can be rearranged in the form

$$\hat{u}(\omega) = \frac{\hat{F}}{k} \frac{1}{-\left(\frac{\omega}{\omega_n}\right)^2 + i2\zeta\frac{\omega}{\omega_n} + 1} \tag{88}$$

or

$$\hat{u}(p) = u_{st} \frac{1}{-p^2 + i2\zeta p + 1} \tag{89}$$

where

$$u_{st} = \frac{\hat{F}}{k} \tag{90}$$

is the *static deflection*, i.e., the deflection that the spring would display if the force amplitude $\hat{F}$ were statically applied, whereas

$$p = \frac{\omega}{\omega_n} \tag{91}$$

is the *normalized frequency*. The frequency-dependent part of Eq. (89) is called the *frequency response function* (FRF) of the 1-dof system and is given by

$$H(p) = \frac{1}{-p^2 + i2\zeta p + 1} \tag{92}$$

Plots of the frequency response function vs. the normalized frequency for a range of damping values are shown in Fig. 3.8. Substitution of Eq. (92) into Eq. (89) yields

$$\hat{u}(p) = u_{st} H(p) \tag{93}$$

The magnitude of the frequency response function is also known as *magnification factor*, M, given by

$$M(p) = |H(p)| = \frac{1}{\sqrt{(1-p^2)^2 + 4\zeta^2 p^2}} \tag{94}$$

Figure 3.8a presents a plot of the magnitude vs. normalized frequency. The plot indicates two distinct regions, one to the left, and the other to the right of the line $p = 1$. The $p = 1$

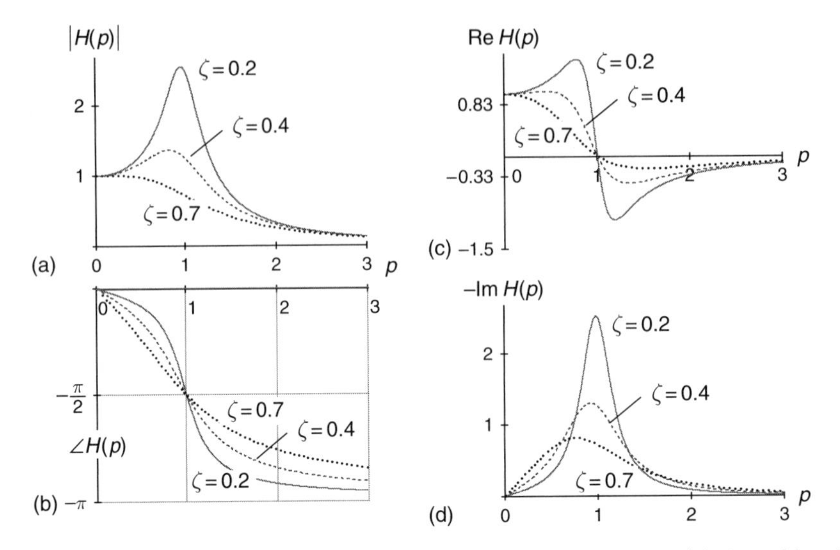

FIGURE 3.8 Frequency response function for a 1-dof system: (a) magnitude; (b) phase; (c) real part; (d) imaginary part.

point, where $\omega = \omega_n$, corresponds approximately to a peak in the response amplitude. This situation of maximum response is commonly referred to as *mechanical resonance*. Figure 3.8a indicates that the amplitude at resonance increases as the damping decreases. In the theoretical case of zero damping (undamped forced vibrations), examination of Eq. (56) indicates that at resonance the denominator goes through zero, i.e., the response at resonance becomes infinite. However, practical systems always have some damping, hence an "infinite" response is not usually found in practice. However, the resonance response of lightly damped systems can be very large and can endanger the safe operation of a dynamic system if not properly controlled.

Regarding the exact location of the resonance point, examination of Fig. 3.8a and Eq. (92) reveals that point of maximum response is at $p = 1$ location only for the case of zero damping. For nonzero damping, the point of maximum response differs from the $p = 1$ location; the exact location of the *resonance frequency* is given by

$$p_r = \sqrt{1 - 2\zeta^2} \tag{95}$$

In physical terms, Eq. (95) can be expressed as

$$\omega_r = \omega_n \sqrt{1 - 2\zeta^2} \tag{96}$$

Note that the damped resonance frequency given by Eq. (96) is different from the damped natural frequency given by Eq. (36). At low damping ratios, the damped resonance frequency, ω_r, and the damped natural frequency, ω_d, are only marginally different from the undamped natural frequency, ω_n. However, as the damping increases, these three frequencies start to differ considerably.

The phase of the frequency response function of Eq. (92) can be calculated as

$$\phi(p) = \arg H(p) = \arg\left(\frac{1}{-p^2 + i2\zeta p + 1}\right) \tag{97}$$

The phase angle $\phi(p)$ represents the phase difference between response and excitation. The plot of phase vs. frequency (Fig. 3.8b) indicates that the phase angle is always negative, i.e., the response lags behind excitation. Figure 3.8b also indicates that the phase angle increases with frequency, i.e., the lag between excitation and response increases with frequency, which is consistent with common experience. As the excitation frequency sweeps the entire range from 0 to ∞, the phase angle goes through a 180° (π radians) change. It is also apparent from Fig. 3.8b that the phase angle equals $-90°$ ($-\pi/2$ radians) at the $p=1$ point. This indicates that when the excitation frequency, ω, matches the undamped natural frequency, ω_{n}, the response is in *quadrature* with the excitation (i.e., it lags by 90° behind the excitation). This is quite apparent from a cursory evaluation of Eq. (92) as, for $p=1$, the frequency response function becomes purely imaginary, i.e.,

$$H(1) = \left.\frac{1}{-p^2 + i2\zeta p + 1}\right|_{p=1} = \frac{1}{i2\zeta} \tag{98}$$

which corresponds to a phase angle of $-90°$. The magnitude of the response at $p=1$ point is easily obtained by taking the magnitude of Eq. (98), i.e., it takes the simple form

$$|H(1)| = M_{90} = \frac{1}{2\zeta} \tag{99}$$

where M_{90} signifies the value of the magnification factor at the quadrature point, which also corresponds to the point when the excitation frequency, ω, equals the natural undamped frequency, ω_{n}, i.e., at the $p=1$ point. Using Eq. (99) in conjunction with Eqs. (15), (31), (90), (91), (93) yields the *response amplitude at the quadrature point* as

$$|\hat{u}_{90}| = \frac{\hat{F}}{c\omega} \tag{100}$$

As already pointed out, the difference between the undamped resonance $p=1$ and the actual resonance p_{r} is only slight for lightly damped systems. In this case, the frequency at which the phase is $-90°$ could acceptably approximate the location of the actual resonance. This fact is especially useful in experimental work on lightly damped systems, because it permits the estimation of the resonance frequency from a plot of the phase response vs. frequency. However, for higher damped systems, this approach does no longer apply.

Estimation of system damping from the frequency response function

When measured experimentally, the magnification plot vs. frequency can be used to estimate the systems damping. Several methods can be used; among the more common ones, we cite: (a) the quadrature (90° phase) method; (b) the resonance peak method; (c) the quality factor method.

Quadrature phase method for damping estimation

The quadrature phase method relies on measuring the response magnitude at the frequency for which the response is in quadrature with the excitation ($-90°$ phase). Recalling Eq. (99), we write

$$M_{90} = \frac{1}{2\zeta} \tag{101}$$

Solution of Eq. (101) yields the damping ratio

$$\zeta = \frac{1}{2M_{90}} \tag{102}$$

Resonance peak method for damping estimation

The resonance peak method utilizes the fact that, at resonance, the resonance peak has the expression

$$M_r = M(p_r) = \left. \frac{1}{\sqrt{(1-p_r^2)^2 + 4\zeta^2 p_r^2}} \right|_{p_r^2 = 1-2\zeta^2} = \frac{1}{2\zeta\sqrt{1-\zeta^2}} \tag{103}$$

For lightly damped systems ($\zeta \ll 1$), Eq. (103) simplifies to

$$\zeta \cong \frac{1}{2M_r} \tag{104}$$

Equation (104) is similar to Eq. (102) used in the quadrature method. However, as damping increases, the simplified expression of Eq. (104) does no longer apply, and the exact solution of the nonlinear Eq. (103) must be used. In addition, the resonance peak becomes flatter and flatter, as illustrated by the $\zeta > 0.5$ curve in Fig. 3.8a. For these reasons, the resonance peak method for damping estimation is most useful for lightly damped systems only.

Quality factor method for damping estimation

Examination of the magnification plot in Fig. 3.8a indicates that the width of the resonance peak varies with damping values. (The smaller the damping, the taller and slimmer the resonance peak is!) It seems appropriate to try to use a measure of the resonance peak width in order to determine the system damping. Such a measure is offered by the *quality factor*, Q. Quality factor, Q, is a term originating in electrical engineering where it is usually used as a figure of merit for evaluating narrow-band pass filters, such as used in tuning a radio receiver. A good highly-selective band-pass filter must have a strong response at the tuning frequency, and a fast decrease of the response to the left and the right of the tuning frequency.

The simplest band-pass filters are second-order resonant circuits with the frequency response very similar to that of the damped vibration systems analyzed here. Hence, the electrical engineering terminology used in describing the quality of second-order band-pass filters has been found useful in the analysis of lightly damped vibration systems. The narrowness of the band-pass filter frequency response around the tuning frequency is characterized by the *frequency bandwidth* defined as

$$\Delta\omega = \omega_U - \omega_L \tag{105}$$

where ω_U and ω_L are the upper and lower *half-power frequencies* (3 dB *points*) located to the right and the left of the circuit resonance frequency. Because the power is proportional to amplitude squared, the half-power points correspond to points where the amplitude has decreased by a factor $\sqrt{2}$, i.e., by 3 dB. For lightly damped systems, the bandwidth takes the simple expression

$$\omega_U - \omega_L \cong 2\zeta\omega_n \tag{106}$$

In electrical engineering textbooks (e.g., Johnson et al., 1995), the quality factor of a band-pass filter is defined as the ratio of the resonance frequency, ω_r, to the frequency bandwidth, $\Delta\omega$, i.e.,

$$Q = \frac{\omega_r}{\Delta\omega} = \frac{\omega_r}{\omega_U - \omega_L} \tag{107}$$

For lightly damped systems, the resonance frequency is well approximated by the undamped frequency, i.e., $\omega_r \simeq \omega_n$. On substitution, in Eq. (107), the quality factor becomes

$$Q = \frac{\omega_n}{2\zeta\omega_n} = \frac{1}{2\zeta} \tag{108}$$

Equation (108) can be used to estimate the system damping, i.e.,

$$\zeta \cong \frac{1}{2}\frac{\omega_n}{\omega_U - \omega_L} \tag{109}$$

Another important feature of the quality factor, which can be useful in applications, is that the quality factor is 2π times the ratio between the maximum energy stored during a cycle and the total energy dissipated per cycle (Lindner, 1999), i.e.,

$$Q = 2\pi \frac{\text{maximum energy stored during a cycle}}{\text{total energy dissipated per cycle}} \tag{110}$$

Mechanical–electrical equivalents

An important analogy exists between electrical circuits and mechanical spring-mass-damper systems. Figure 3.9 shows a series circuit consisting of an a.c. voltage source, $\mathrm{v}(t)$, of angular frequency, ω, and the typical circuit elements such as an electrical resistance, R, an inductance, L, and a capacitance, C. The electrical current flowing through the circuit is $i(t)$, and the electrical charge accumulated in the circuit is $q(t) = \int i(t)\mathrm{d}t$. Basic electrical engineering analysis (Fitzgerald et al., 1967) gives the following differential equation between the time-dependent charge, $q(t)$, the applied voltage $\mathrm{v}(t)$, and the circuit elements R, L, C,

$$L\ddot{q}(t) + R\dot{q}(t) + \frac{1}{C}q(t) = \mathrm{v}(t) \tag{111}$$

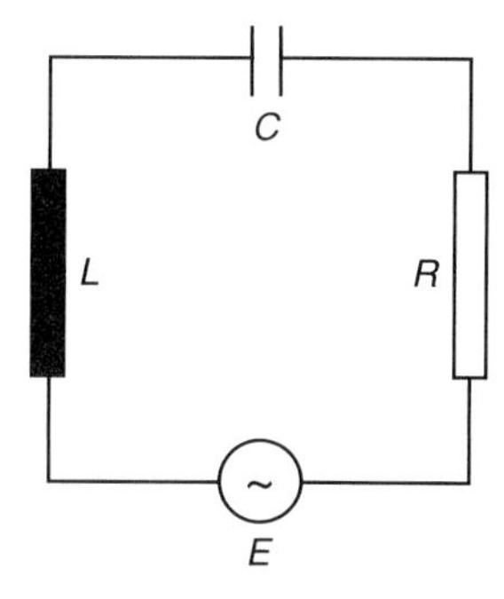

FIGURE 3.9 Typical a.c. electric circuit containing resistance, R, inductance, L, capacitance, C, and voltage source, E.

TABLE 3.1 Mechanical–electrical equivalents

Mechanical			Electrical		
Name	Symbol	Units	Name	Symbol	Units
Displacement	u	m (meter)	Charge	q	C (coulomb)
Velocity	$\dot{u}$	m/s (meter/second)	Current	i	A (ampere)
Force	F	N (newton)	Voltage	v	V (volts)
Mass	m	kg (kilogram)	Inductance	L	H (Henry)
Stiffness	k	N/m (newton/meter)	1/Capacitance	$1/C$	F (Farad)
Damping	c	Ns/m (newton-second/meter)	Resistance	R	Ω (ohm)

This equation has the same form as Eq. (68), the forced damped vibration equation for the 1-dof mechanical system. Thus, by a simple interchange of symbols (Table 3.1), knowledge about the behavior of the electrical circuit can be used to infer knowledge about the behavior of the mechanical system. This fact allows us to analyze the behavior of mechanical system by using analytical models and software predictors developed for electrical system. It also allows us to experimentally simulate the behavior of mechanical system using electrical circuits built using the mechanical–electrical equivalents. Such circuits form the basis of *analog computers*.

If the applied voltage is harmonic, $\mathrm{v}(t) = \hat{V}e^{i\omega t}$, then the response is also harmonic, $q(t) = \hat{Q}e^{i\omega t}$, and Eq. (111) becomes

$$\left(-\omega^2 L + i\omega R + \frac{1}{C}\right)\hat{Q} = \hat{V} \tag{112}$$

Recall that the current, i, is the time derivative of electric charge, i.e.,

$$i(t) = \dot{q}(t) = \mathrm{i}\omega i(t) \tag{113}$$

Then, Eq. (112) becomes

$$\left(i\omega L + R + \frac{1}{i\omega C}\right)\hat{I} = \hat{V} \tag{114}$$

Equation (114) can be written in the usual form

$$Z(\omega)\hat{I} = \hat{V} \tag{115}$$

where $Z(\omega)$ is the *electrical impedance* given by

$$Z(\omega) = i\omega L + R + \frac{1}{i\omega C} \tag{116}$$

Comparison of Eq. (116) and (87) reveals the correspondence between the electrical impedance and the mechanical impedance.

3.2.3 ENERGY METHODS IN 1-DOF VIBRATION ANALYSIS

In Sections 3.2.1 and 3.2.2, we derived the equation of motion of the vibrating particle by applying Newton second law to the particle isolated in a free-body diagram. In this approach, we figured out the action of all forces acting on the particle and equated it with the product between mass and acceleration. An alternate way of deriving the equation of motion is by *energy methods*. We will illustrate the energy methods approach with a few simple examples based on the 1-dof motion of the particle. However, the true advantage of the energy methods' approach becomes apparent when applied to complicated mechanical systems which, even if essentially 1-dof, are difficult to analyze in detail.

3.2.3.1 Undamped 1-dof Vibration Analysis by Energy Methods

Consider the undamped 1-dof horizontal vibration of a particle attached to a spring, as illustrated in Fig. 3.10a. The particle undergoes horizontal displacement, $u(t)$, around the equilibrium position. The kinetic energy, T, of the particle is given by

$$T(t) = \frac{1}{2}m\dot{u}^2(t) \tag{117}$$

whereas the elastic energy, V, consisting of the energy stored in the displaced spring, is given by

$$V(t) = \frac{1}{2}ku^2(t) \tag{118}$$

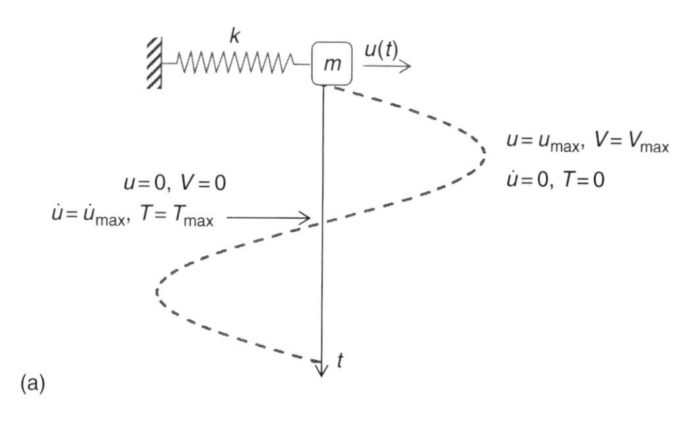

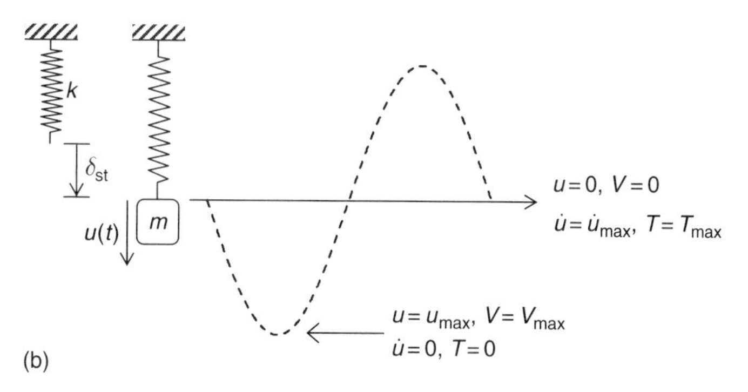

FIGURE 3.10 Energy analysis of the undamped vibration of a particle: (a) horizontal motion; (b) vertical motion.

The total energy of the system is

$$E = T(t) + V(t) = \frac{1}{2}m\dot{u}^2(t) + \frac{1}{2}ku^2(t) \tag{119}$$

Substituting $u(t) = \hat{u}\cos\omega t$, we write

$$E = T(t) + V(t) = \frac{1}{2}m\omega^2\hat{u}^2\sin^2\omega t + \frac{1}{2}k\hat{u}^2\cos^2\omega t \tag{120}$$

Recalling Eq. (15), i.e., $k = \omega_n^2 m$, we can write Eq. (120) as

$$E = T(t) + V(t) = \frac{1}{2}m\hat{u}^2\left(\omega^2\sin^2\omega t + \omega_n^2\cos^2\omega t\right) \tag{121}$$

Derivation of the equation of motion by energy methods

The principle of energy conservation stipulates that the energy is conserved (i.e., it is stationary), which gives:

$$E = E_0 = \text{const} \tag{122}$$

Hence, the derivative of energy with respect to time should be zero. Taking the time derivative of Eq. (119) and equating it to zero gives

$$\frac{d}{dt}[T(t) + V(t)] = \frac{d}{dt}\left[\frac{1}{2}m\dot{u}^2(t) + \frac{1}{2}ku^2(t)\right] = 0 \tag{123}$$

Upon performing the differentiation, we get

$$m\ddot{u}(t)\dot{u}(t) + k\dot{u}(t)u(t) = 0 \tag{124}$$

Simplification by $\dot{u}(t)$ yields the equation of motion

$$m\ddot{u}(t) + ku(t) = 0 \tag{125}$$

Equation (125) is identical with Eq. (12) derived from Newton law of motion.

Estimation of the natural frequency by energy methods

Energy methods can be expediently used to estimate the natural frequency of the vibrating system. As we analyze the oscillatory motion, we noticed that there are two *salient points*: (1) the extreme position and (2) the equilibrium position. Assume the 1-dof system oscillates freely, i.e., at its natural frequency. Recall the expression of the oscillatory motion, $u(t) = C\cos\omega_n t$, and hence $\dot{u}(t) = -\omega_n C\sin\omega_n t$. When the particle passes through the extreme position, the displacement amplitude reaches the maximum, u_{max}, while the velocity vanishes, because the sine function vanishes when the cosine function reaches an extreme (± 1). Hence the kinetic energy at the extreme position reaches the maximum, whereas the elastic energy is zero. The law of energy conservation (121) gives

$$E|_{u=0} = V_{max} = \frac{1}{2}ku_{max}^2 = \frac{1}{2}k\hat{u}^2 = E_0 \tag{126}$$

When the particle passes through the equilibrium position (i.e., through the center), its displacement is zero, hence the spring is undeformed and stores no energy. This means that the elastic energy is zero at the equilibrium position. At the same time, the particle passing through the equilibrium position has maximum velocity, because the sine function reaches an extreme (± 1) when the cosine vanishes. Hence the kinetic energy reaches a maximum as the particle passes through the equilibrium position. The law of energy conservation gives

$$E|_{u=0} = T_{\max} = \frac{1}{2} m \dot{u}_{\max}^2 = \frac{1}{2} m \omega_n^2 \hat{u}^2 = E_0 \tag{127}$$

Comparison of Eqs. (126) and (127) yields

$$V_{\max} = T_{\max}, \quad \text{i.e.,} \ \frac{1}{2} m \omega_n^2 \hat{u}^2 = \frac{1}{2} k \hat{u}^2 \tag{128}$$

Upon simplification, Eq. (128) yields

$$\omega_n^2 = \frac{k}{m} \quad \text{or} \quad \omega_n = \sqrt{\frac{k}{m}} \tag{129}$$

Equation (129) has established a relationship between the natural frequency and the maximum values of the potential and kinetic energies. This concept leads to the *Rayleigh quotient* and the *Rayleigh–Ritz method* used in the estimation of the fundamental frequency of more complicated vibration systems (Meirovitch, 1986).

Effect of gravitational field on energy methods formulation of vibration analysis

So far, we considered the case of horizontal vibration in which the gravitational field does not intervene. Now, we consider the case of vertical vibration, which takes place in the same direction as the gravitational field (Fig. 3.10b). We assume, that the vibrational motion takes place about a static equilibrium position δ_{st}. With this datum, the potential energy associated with the vibrational motion takes the form

$$V(t) = \frac{1}{2} k [\delta_{st} + u(t)]^2 - \frac{1}{2} k \delta_{st}^2 - mgu(t) \tag{130}$$

where the potential energy stored in the spring under static conditions, $\frac{1}{2} k \delta_{st}^2(t)$, was subtracted from the potential energy of the spring deformed during the vibrational motion, $\frac{1}{2} k [\delta_{st} + u(t)]^2$. The quantity $-mgu(t)$ is the gravitational potential. Upon expansion, Eq. (130) becomes

$$V(t) = \frac{1}{2} k \delta_{st}^2 + \frac{1}{2} 2k \delta_{st} u(t) + \frac{1}{2} k u(t)^2 - \frac{1}{2} k \delta_{st}^2 - mgu(t) \tag{131}$$

Recalling the equilibrium condition $k\delta_{st} = mg$, Eq. (130) simplifies to the form

$$V(t) = \frac{1}{2} k u^2(t) \tag{132}$$

which is identical with Eq. (118) derived for horizontal motion. Thus, it is apparent that the gravitational field can be eliminated from the energy methods formulation of the vibration analysis by taking the vibration datum at the static equilibrium position.

3.2.3.2 Damped 1-dof Vibration Analysis by Energy Methods

Energy methods can also be used for the analysis of damped systems. In this case, the total energy of the system is not conserved, as energy is dissipated by the damping forces. Two cases will be analyzed here: (1) the derivation of the damped 1-dof equation of motion via energy methods, and (2) the energy and power associated with the damped 1-dof response to forced excitation.

Derivation of the damped 1-dof equation by energy methods

We use the first law of thermodynamics to express the fact that the incremental change in total energy equals the incremental work performed by the external forces acting on the system, i.e.,

$$\delta E = \delta W \tag{133}$$

The incremental change in the total energy is expressed by the incremental change in the kinetic and potential energies. Recalling Eq. (119), we get

$$\delta E = \delta T(t) + \delta V(t) = m\dot{u}(\ddot{u}\delta t) + ku(\dot{u}\delta t) \tag{134}$$

The incremental work δW is estimated by the product between the dissipation forces and the incremental displacement, i.e.,

$$\delta W = (-c\dot{u})\delta u = (-c\dot{u})\dot{u}\delta t \tag{135}$$

Combination of Eqs. (133) through (135) yields

$$m\dot{u}(\ddot{u}\delta t) + ku(\dot{u}\delta t) = (-c\dot{u})\dot{u}\delta t \tag{136}$$

Upon simplification and rearrangement, Eq. (136) yields Eq. (29) of damped free vibration, i.e.,

$$m\ddot{u}(t) + c\dot{u}(t) + ku(t) = 0 \tag{137}$$

Power and energy associated with damped 1-dof response to harmonic excitation

Recall Eq. (77) describing the steady-state vibration response and Eq. (84) expressing the particle velocity in terms of particle motion and excitation frequency, i.e.,

$$u(t) = \hat{u}(\omega)e^{i\omega t}, \quad \dot{u}(t) = i\omega u(t) \tag{138}$$

Recalling the convention associated with the use of complex notation, we retain only the real parts and rewrite Eq. (138) as

$$u(t) = |\hat{u}(\omega)|\cos(\omega t + \psi), \quad \dot{u}(t) = -\omega\,|\hat{u}(\omega)|\sin(\omega t + \psi) \tag{139}$$

The instantaneous *power input* is defined as the product between force and velocity, i.e.,

$$P(t) = F(t)\dot{u}(t) = \hat{F}\cos(\omega t)\,[-\omega\,|\hat{u}(\omega)|\sin(\omega t + \psi)] \tag{140}$$

Upon simplification,

$$P(t) = -\omega\hat{F}\,|\hat{u}(\omega)|\cos(\omega t)\sin(\omega t + \psi) \tag{141}$$

Applying trigonometric formulae, we can express Eq. (141) as

$$P(t) = -\frac{1}{2}\omega\hat{F}\,|\hat{u}(\omega)|\,[\sin\psi + \sin(2\omega t + \psi)] \tag{142}$$

The *energy input* to the system per cycle of oscillation is obtained by integrating the power over a period of oscillation $\tau = 2\pi/\omega$, i.e.,

$$\Delta E_{\text{cyc}} = \int_0^{2\pi/\omega} P(t)\mathrm{d}t = -\omega\hat{F}\,|\hat{u}(\omega)|\int_0^{2\pi/\omega}\cos(\omega t)\sin(\omega t + \psi)\mathrm{d}t = -\pi\hat{F}\,|\hat{u}(\omega)|\sin\psi \tag{143}$$

Substituting Eq. (81) into Eq. (143) yields, upon manipulation,

$$\Delta E_{\text{cyc}}(\omega) = \pi c\omega\,|\hat{u}(\omega)|^2 \tag{144}$$

Equation (144) indicates that the required energy input per cycle (i.e., the *energy dissipated* by the damped system) increases linearly with frequency and quadratic with response amplitude. The *average power* is obtained by dividing the energy per cycle by the cycle duration, $\tau = 2\pi/\omega$, i.e.,

$$P_{\text{av}}(\omega) = \frac{1}{2}c\omega^2\,|\hat{u}(\omega)|^2 \tag{145}$$

It is noted that power is proportional to the square of the vibration amplitude. It is also proportional to the system damping and the frequency squared. The response amplitude reaches the maximum at resonance. Hence, the power intake to the system also reaches a local maximum at resonance, i.e.,

$$P_{\text{max}} = \frac{1}{2}c\omega_{\text{r}}^2\,|\hat{u}_{\text{r}}|^2 \tag{146}$$

For lightly damped systems, the resonance point can be sufficiently well approximated by the quadrature point, i.e., $\omega_{\text{r}} \cong \omega_{90} = \omega_{\text{n}}$, $\hat{u}_{\text{r}} \cong \hat{u}_{90}$, and

$$P_{\text{max}} = \frac{1}{2}c\omega_{\text{n}}^2\,|\hat{u}_{90}|^2 \tag{147}$$

where the response amplitude at quadrature, $\hat{u}_{90}$, is given by Eq. (100). Upon substitution into Eq. (145), the power at resonance is found to be

$$P_{\text{max}} = \frac{1}{2}\frac{\hat{F}^2}{c} \tag{148}$$

At resonance, where the response amplitude reaches a maximum, the energy per cycle reaches a maximum too.

3.3 VIBRATION OF CONTINUOUS SYSTEMS

3.3.1 AXIAL VIBRATION OF A BAR

Consider a uniform bar of length l, axial stiffness EA, and mass per unit length m, as shown in Fig. 3.11a. Assume motion in the longitudinal direction, $u(t)$. For compactness, use the notations

$$\frac{\partial}{\partial x}(\,) = (\,)' \quad \text{and} \quad \frac{\partial}{\partial t}(\,) = (\dot{\,}) \tag{149}$$

3.3.1.1 Free Axial Vibration of a Bar

Free-body analysis of the infinitesimal element dx shown in Fig. 3.11b yields

$$N(x,t) + N'(x,t)\mathrm{d}x - N(x,t) = m\ddot{u}(x,t) \tag{150}$$

where $N(x, t)$ is the axial stress resultant. Upon simplification, we get

$$N'(x,t)\mathrm{d}x = m\ddot{u}(x,t) \tag{151}$$

The N stress resultant is evaluated by integration across the cross-sectional area of the direct stress shown in Fig. 3.11c, i.e.,

$$N(x,t) = \int_A \sigma(x,z,t)\mathrm{d}A \tag{152}$$

Recall the strain-displacement relation

$$\varepsilon = u' \tag{153}$$

and the stress–strain constitutive relation

$$\sigma = E\varepsilon \tag{154}$$

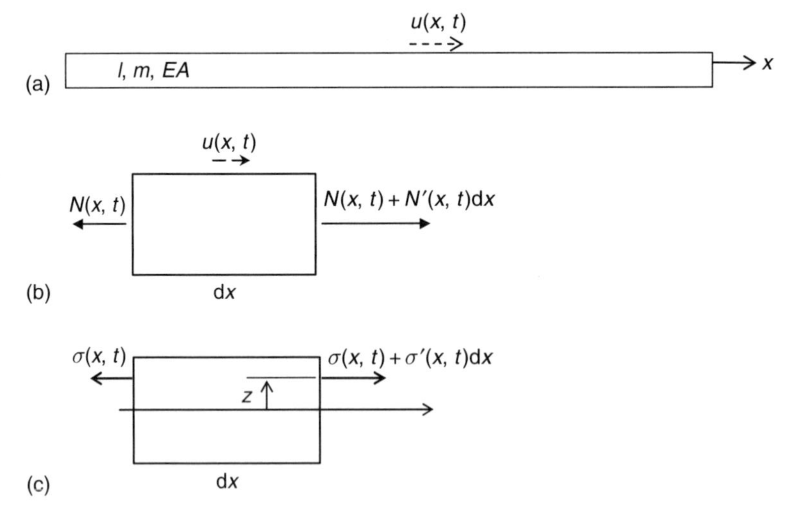

FIGURE 3.11 Uniform bar undergoing axial vibration: (a) general schematic; (b) infinitesimal axial element; (c) thickness-wise stress distribution.

where E is Young modulus of elasticity. Substitution of Eqs. (153) and (154) into Eq. (152) yields

$$N(x,t) = \int_A Eu'(x,t)\mathrm{d}A = EA\,u'(x,t) \tag{155}$$

where EA is the *axial stiffness*. Differentiation of Eq. (155) w.r.t. x and substitution into Eq. (151) yields the *equation of motion for axial vibration*, i.e.,

$$EA\,u'' = m\ddot{u} \tag{156}$$

Upon division by m, we get the equation of motion for axial vibration of a bar in the form of the *wave equation*, i.e.,

$$c^2 u'' = \ddot{u} \tag{157}$$

where the constant c^2 is given by

$$c^2 = \frac{EA}{m} \quad \text{or} \quad c = \sqrt{\frac{EA}{m}} \tag{158}$$

As the bar is uniform, the mass per unit length is the product between density and cross-sectional area, $m = \rho A$, and Eq. (158) takes the alternate form

$$c^2 = \frac{E}{\rho} \quad \text{or} \quad c = \sqrt{\frac{E}{\rho}} \tag{159}$$

Natural frequencies and mode shapes for fixed-fixed boundary conditions

Equation (357) is a PDE in space, x, and time, t. One way of seeking the solution of Eq. (357) is through the method of the separation of variables, i.e., assuming

$$u(x,t) = \hat{u}(x)e^{i\omega t} \tag{160}$$

where $\hat{u}(x)$ depends only on x, whereas $e^{i\omega t}$ depends only on t. This assumption implies synchronous motion within the elastic body, i.e., the general vibration shape of the body deformation is preserved during a vibration cycle, only that the amplitude of motion varies cyclically. Stating it differently, all the points in the body execute the same cycling motion in time, passing through the equilibrium position and then through the maxima and minima simultaneously. This condition seems intuitive, at least for non-dissipative systems. Upon substitution of Eq. (160) into Eq. (357), we obtain a second-order ODE in the form

$$c^2\hat{u}'' + \omega^2\hat{u} = 0 \tag{161}$$

Introducing the notation

$$\gamma = \frac{\omega}{c} \tag{162}$$

we write Eq. (161) in the form

$$\hat{u}'' + \gamma^2\hat{u} = 0 \tag{163}$$

Equation (163) admits the general solution

$$\hat{u}(x) = C_1 \sin \gamma x + C_2 \cos \gamma x \tag{164}$$

We notice that the vibration shape given by Eq. (164) is harmonic, i.e., it depends on sine and cosine functions in the space variable x. The shape of such deformation is wave-like (Fig. 3.12); hence, the constant γ is called the *wavenumber*. An associate constant is the *wavelength* λ.

The constants C_1 and C_2 are to be determined from the boundary conditions. For illustration, assume the *fixed-fixed boundary conditions*, i.e.,

$$\begin{aligned} u(0, t) &= 0 \\ u(l, t) &= 0 \end{aligned} \tag{165}$$

Since conditions of Eq. (165) must be satisfied at any time t, then they are to be satisfied by the space-dependent amplitude $\hat{u}(x)$, i.e.,

$$\begin{aligned} \hat{u}(0) &= 0 \\ \hat{u}(l) &= 0 \end{aligned} \tag{166}$$

Substitution of Eq. (164) into Eq. (165) yields

$$\begin{cases} C_1 \sin \gamma x + C_2 \cos \gamma x|_{x=0} = 0 \\ C_1 \sin \gamma x + C_2 \cos \gamma x|_{x=l} = 0 \end{cases} \Rightarrow \begin{cases} C_2 = 0 \\ C_1 \sin \gamma l + C_2 \cos \gamma l = 0 \end{cases} \tag{167}$$

Since the first condition yields $C_2 = 0$, the second condition yields

$$C_1 \sin \gamma l = 0 \tag{168}$$

Equation (168) has a nontrivial solution $C_1 \neq 0$ only if $\sin \gamma l = 0$. Under these conditions, the general solution (164) takes the form

$$\hat{u}(x) = C_1 \sin \gamma x \tag{169}$$

The conditions under which Eq. (168) accepts nontrivial solutions are obtained by finding solutions of the equation

$$\sin \gamma l = 0 \tag{170}$$

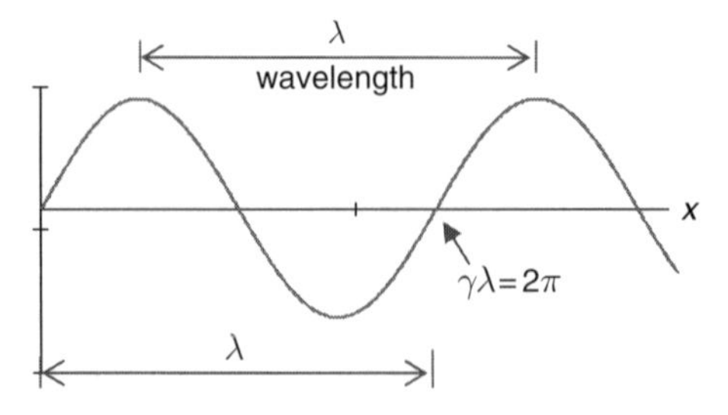

FIGURE 3.12 Typical vibration shape illustrating the relation between wavelength and wavenumber.

Equation (170), called the *characteristic equation*, defines the values of γ for which nontrivial solutions of Eq. (168) exist. These are the *characteristic values* or *eigenvalues* of the system. Solution of Eq. (170) yields

$$\gamma l = \pi, 2\pi, 3\pi \ldots \quad \text{or} \quad \gamma_j = j\frac{\pi}{l}, \quad j = 1, 2, 3, \ldots \tag{171}$$

Combining Eq. (171) with Eqs. (158) and (162) yields the *natural angular frequencies*

$$\omega_j = j\frac{\pi}{l}\sqrt{\frac{EA}{m}}, \quad j = 1, 2, 3, \ldots \tag{172}$$

Since $\omega = 2\pi f$, the corresponding circular frequencies, f_j, are given by the formula

$$f_j = j\frac{1}{2l}\sqrt{\frac{EA}{m}}, \quad j = 1, 2, 3, \ldots \tag{173}$$

We can express Eq. (173) in a more compact form using Eq. (158). Hence, we write

$$f_j = j\frac{c}{2l}, \quad j = 1, 2, 3, \ldots \tag{174}$$

We note that Eqs. (171) and (172) predict a denumerable infinite set of eigenvalues and natural frequencies, which is typical of the vibration of continuous system. The first frequency, ω_1, is called the *fundamental frequency*, whereas the other frequencies ($\omega_j, j = 2, 3, \ldots$) are called *overtones*. The overtones are integral multiples of the fundamental frequency. The fundamental frequency is also referred to as the *fundamental harmonic*, whereas the overtones are also referred to as *higher harmonics*.

Mode shapes

For each eigenvalue and natural frequency, Eq. (169) defines an *eigen function* or *natural mode* of vibration (*mode shape*), i.e.,

$$U_j(x) = C_j \sin \gamma_j l, \quad j = 1, 2, 3, \ldots \tag{175}$$

It should be noted that all the mode shapes $U_j(x)$ satisfy the differential Eq. (163) and the boundary conditions (166). As mentioned earlier, the constants C_j cannot be determined from the differential equation and the boundary conditions. This is quite alright, as it can be easily verified that if $U_j(x)$ is a mode shape that satisfies the differential equation and the boundary condition, then any scaled version of it, $\alpha U_j(x)$, $\alpha \in \mathbb{R}$, also satisfies them. Hence, the value of the constants C_j can be arbitrarily chosen. For example, one can chose $C_j = 1$, which might have some advantage in mode-shape plotting.

Orthogonality of mode shapes

Mode shapes orthogonality is an important property that allows one to verify if some independently derived functions qualify to be considered as mode shapes. Recall that the mode shapes (175) satisfy Eq. (161), which can be conveniently expressed explicitly in m, EA, and ω_j, i.e.,

$$EA\,U_j'' + \omega_j^2 m U_j = 0, \quad j = 1, 2, 3, \ldots \tag{176}$$

Consider two separate mode shapes, $U_p(x)$ and $U_q(x)$. They satisfy Eq. (176), i.e.,

$$\begin{aligned} EA\,U_p'' &= -\omega_p^2 m U_p \\ EA\,U_q'' &= -\omega_q^2 m U_q \end{aligned} \tag{177}$$

To analyze *orthogonality with respect to mass*, consider the mass-weighted integral

$$\int_0^l mU_p(x)U_q(x)\mathrm{d}x \tag{178}$$

Substitution of Eq. (175) into Eq. (178) yields

$$\int_0^l m\left(C_p \sin\gamma_p x\right)\left(C_q \sin\gamma_q x\right)\mathrm{d}x \tag{179}$$

Using standard trigonometric formulae, we write

$$\int_0^l \left(\sin\gamma_p x\right)\left(\sin\gamma_q x\right)\mathrm{d}x = \frac{1}{2}\int_0^l \left[\cos(\gamma_p-\gamma_q)x - \cos(\gamma_p+\gamma_q)x\right]\mathrm{d}x = \frac{l}{2}\delta_{pq} \tag{180}$$

Equation (171) was used to evaluate γ_p, γ_q. The symbol δ_{pq} is the Kronecker delta ($\delta_{pq}=1$ for $p=q$, and $\delta_{pq}=0$ for $p\neq q$). Using Eqs. (179) and (180) and assuming $p\neq q$, Eq. (178) yields the *orthogonality condition with respect to mass* in the form

$$\int_0^l mU_p(x)U_q(x)\mathrm{d}x = 0, \quad p\neq q \tag{181}$$

It should be noted that the mass orthogonality condition expressed by Eq. (181) is not restricted to the simple mode shapes expressed by Eq. (175). To prove this, recall Eq. (177), multiply the first line by U_q and the second line by U_p, and integrate each line over the length of the bar to obtain

$$\begin{aligned} \int_0^l EA\,U_p''(x)U_q(x)\mathrm{d}x &= -\omega_p^2\int_0^l mU_p(x)U_q(x) \\ \int_0^l EA\,U_q''(x)U_p(x)\mathrm{d}x &= -\omega_q^2\int_0^l mU_q(x)U_p(x) \end{aligned} \tag{182}$$

Integration by parts yields

$$\begin{aligned} -\int_0^l EA\,U_p'(x)U_q'(x)\mathrm{d}x + EA\,U_p''(x)U_q(x)\big|_0^l &= -\omega_p^2\int_0^l mU_p(x)U_q(x) \\ -\int_0^l EA\,U_q'(x)U_p'(x)\mathrm{d}x + EA\,U_q''(x)U_p(x)\big|_0^l &= -\omega_q^2\int_0^l mU_q(x)U_p(x) \end{aligned} \tag{183}$$

The left hand side of Eq. (183) contains two parts: (a) an integral term containing the first derivative of the mode shapes, U_p', U_q', and (b) a boundary-evaluated term expressed in terms of the mode shapes, U_p, U_q, and their second derivate, U_p'', U_q''. Recalling that the mode shapes, U_p, U_q satisfy the boundary conditions (166), it becomes apparent that, in our case, the boundary-evaluated terms in Eq. (190) vanish. Hence, Eq. (190) takes the simpler form

$$\begin{aligned} \int_0^l EA\,U_p'(x)U_q'(x)\mathrm{d}x &= \omega_p^2\int_0^l mU_p(x)U_q(x) \\ \int_0^l EA\,U_q'(x)U_p'(x)\mathrm{d}x &= \omega_q^2\int_0^l mU_q(x)U_p(x) \end{aligned} \tag{184}$$

Subtraction of the second line of Eq. (184) from the first line yields

$$\left(\omega_p^2 - \omega_q^2\right) \int_0^l m U_p(x) U_q(x) \mathrm{d}x = 0 \tag{185}$$

For distinct mode numbers, $p \neq q$, the frequencies are also distinct, $\omega_p^2 \neq \omega_q^2$, and hence Eq. (185) implies

$$\int_0^l m U_p(x) U_q(x) \mathrm{d}x = 0, \quad p \neq q \tag{186}$$

which is exactly the mass orthogonalization condition derived earlier as Eq. (181). Notice that the above derivation is quite general, if the boundary conditions are such that the boundary-evaluated terms in Eq. (183) vanish.

To analyze *orthogonality with respect to stiffness*, consider the integral

$$\int_0^l \left[EA\, U_p''(x)\right] U_q(x) \mathrm{d}x \tag{187}$$

Recalling Eq. (182), we write Eq. (187) as

$$\int_0^l \left[EA\, U_p''(x)\right] U_q(x) \mathrm{d}x = -\int_0^l m\omega_p^2 U_p(x) U_q(x) \mathrm{d}x \tag{188}$$

Using Eq. (181), Eq. (188) becomes

$$\int_0^l \left[EA\, U_p''(x)\right] U_q(x) \mathrm{d}x = 0, \quad p \neq q \tag{189}$$

Equation (189) is the *orthogonality condition with respect to stiffness*. Integration by parts of Eq. (189) yields an alternate expression of the orthogonality condition with respect to stiffness, i.e.,

$$\int_0^l EA\, U_p'(x) U_q'(x) \mathrm{d}x - \left[EA U_p''(x) U_q(x)\right]_0^l = 0, \quad p \neq q \tag{190}$$

Equation (190) contains two parts: (a) an integral term containing the first derivative of the mode shapes, U_p', U_q', and (b) a boundary-evaluated term expressed in terms of one of the mode shapes, U_q, and the second derivate of the other mode shape, U_p''. Recalling that the mode shapes U_q satisfy the boundary conditions (165), it becomes apparent that the second term in Eq. (190) vanishes. Hence, the orthogonality expression with respect to stiffness can be alternately expressed as

$$\int_0^l EA\, U_p'(x) U_q'(x) \mathrm{d}x = 0, \quad p \neq q \tag{191}$$

if the boundary conditions are such that the second term in Eq. (190) is automatically zero. In practical applications, where independently derived mode shape candidates are tested for stiffness orthogonality, the formulation (190) may be preferred to formulation (189) since first derivatives are easier to evaluate with reasonable accuracy than second derivatives.

Normalization of mode shapes: normal modes

Recalling the mode shapes orthogonality analysis, we notice that, for $p = q = j$, Eq. (179) can be resolved using Eq. (180) to yield

$$C_j = \sqrt{\frac{2}{ml}} \tag{192}$$

Substitution of Eq. (192) into Eq. (175) yields

$$U_j(x) = \sqrt{\frac{2}{ml}} \sin \gamma_j l, \quad j = 1, 2, 3, \ldots \tag{193}$$

The mode shapes of Eq. (193) are *mass normalized*, i.e., they satisfy the mass-weighted integral condition

$$\int_0^l mU_j^2(x)\mathrm{d}x = 1, \quad j = 1, 2, 3, \ldots \tag{194}$$

Mode shapes, such as those given by Eq. (193), that satisfy Eq. (194) are called *normalized mode shapes* or *normal modes*. Plots of the mode shapes and the corresponding natural frequencies for a fixed-fixed elastic bar are given in Table 3.2.

TABLE 3.2 Mode shapes of a fixed-fixed elastic bar

Mode #	Eigenvalue	Resonant frequency	Mode shape	Wavelength multiplicity
1	$\gamma l = \pi$	$f_1 = \frac{c}{2l}$	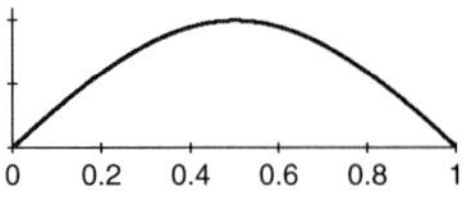 $U_1 = \sin \pi \frac{x}{l}$	$l_1 = \frac{\lambda}{2}$
2	$\gamma l = 2\pi$	$f_2 = 2\frac{c}{2l}$	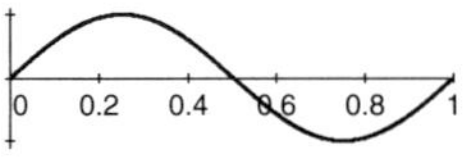 $U_2 = \sin 2\pi \frac{x}{l}$	$l_2 = 2\frac{\lambda}{2}$
3	$\gamma l = 3\pi$	$f_3 = 3\frac{c}{2l}$	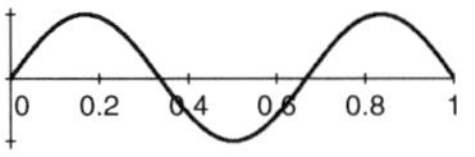 $U_3 = \sin 3\pi \frac{x}{l}$	$l_3 = 3\frac{\lambda}{2}$
4	$\gamma l = 4\pi$	$f_4 = 4\frac{c}{2l}$	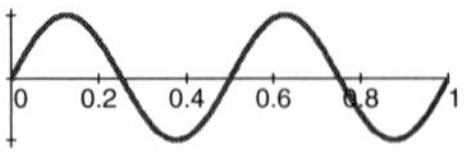 $U_4 = \sin 4\pi \frac{x}{l}$	$l_4 = 4\frac{\lambda}{2}$
5	$\gamma l = 5\pi$	$f_5 = 5\frac{c}{2l}$	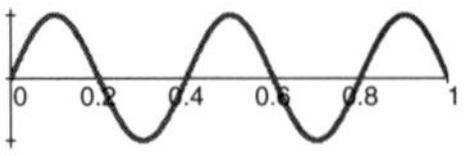 $U_5 = \sin 5\pi \frac{x}{l}$	$l_5 = 5\frac{\lambda}{2}$
6	$\gamma l = 6\pi$	$f_6 = 6\frac{c}{2l}$	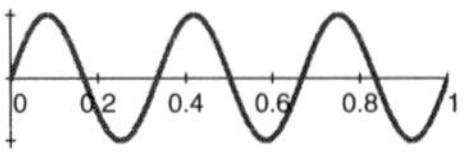 $U_6 = \sin 6\pi \frac{x}{l}$	$l_6 = 6\frac{\lambda}{2}$

To analyze *normalization with respect to stiffness*, we recall Eq. (184) and impose $p = q = j$ to get

$$\int_0^l EA\,U_p'(x)U_q'(x)\mathrm{d}x = \omega_p^2 \int_0^l mU_p(x)U_q(x)\bigg|_{p=q=j}, \quad j = 1, 2, 3, \ldots \tag{195}$$

or

$$\int_0^l EA\,U_j'^2(x)\mathrm{d}x = \omega_j^2 \int_0^l mU_j^2(x), \quad j = 1, 2, 3, \ldots \tag{196}$$

Using Eq. (194), we simplify Eq. (196) to express the *normalization condition with respect to stiffness* in the form

$$\int_0^l EA\,U_j'^2(x)\mathrm{d}x = \omega_j^2, \quad j = 1, 2, 3, \ldots \tag{197}$$

It is also useful to note that imposing $p = q = j$ in Eq. (182) yields, via Eq. (194),

$$\int_0^l EA\,U_j''(x)U_j(x)\mathrm{d}x = -\omega_j^2 \tag{198}$$

Orthonormal modes

We have established so far the conditions for mode shape orthogonality and mode shapes normalization. Mode shapes that are simultaneously normal and orthogonal are called *orthonormal modes*; they satisfy the *orthonormality condition*:

$$\begin{aligned} &\int_0^l mU_p(x)U_q(x)\mathrm{d}x = \delta_{pq} = \begin{cases} 1 & \text{for } p = q \\ 0 & \text{otherwise} \end{cases} \\ &\int_0^l EA\,U_p'(x)U_q'(x)\mathrm{d}x = \omega_p^2\delta_{pq} = \begin{cases} \omega_p^2 & \text{for } p = q \\ 0 & \text{otherwise} \end{cases} \end{aligned} \tag{199}$$

It is also useful to note that

$$\int_0^l EA\,U_p''(x)U_q(x)\mathrm{d}x = -\omega_p^2\delta_{pq} = \begin{cases} -\omega_p^2 & \text{for } p = q \\ 0 & \text{otherwise} \end{cases} \tag{200}$$

Modal mass and stiffness: modal coefficients

If the mode shapes are normalized according to Eq. (192), then their weighted integral with respect to mass evaluates to unity, as indicated by Eq. (194). However, mode shapes may not necessarily be normalized in accordance with Eq. (192); in fact, we have already shown that the mode shape amplitudes, which are the solution of a homogenous system, have one degree of indeterminacy and hence can be generally scaled by any arbitrary factor. Hence, for a generic mode shape $U_j(x)$, Eq. (194) would be

$$\int_0^l mU_j^2(x)\mathrm{d}x = m_j, \quad j = 1, 2, 3, \ldots \tag{201}$$

where m_j is the *modal mass*. By a similar argument, Eq. (197) would become

$$\int_0^l EA\,U_j'^2(x)\mathrm{d}x = k_j, \quad j = 1, 2, 3, \ldots \tag{202}$$

where k_j is the *modal stiffness*. Using Eq. (196), we write

$$k_j = \omega_j^2 m_j, \quad j = 1, 2, 3, \ldots \tag{203}$$

The modal mass, m_j, and the modal stiffness, k_j, are the *modal coefficients*. If damping were present, then an additional modal parameter, the *modal damping*, c_j, would be similarly derived. For mass-normalized mode shapes, the modal coefficients would be the *modal frequency*, ω_j, and the modal damping ratio, ζ_j.

It is also useful to note that, in virtue of Eq. (182),

$$\int_0^l EA\, U_j''(x) U_j(x) \mathrm{d}x = -k_j^2 \tag{204}$$

Rayleigh quotient

Equation (203) can be used to express the frequency of the jth natural mode of vibration in terms of the modal stiffness and modal mass, i.e.,

$$\omega_j^2 = \frac{k_j}{m_j}, \quad j = 1, 2, 3, \ldots \tag{205}$$

Using Eqs. (201) and (202), we rewrite Eq. (206) in the form

$$\omega_j^2 = \frac{\int_0^l EA\, U_j'^2(x) \mathrm{d}x}{\int_0^l m U_j^2(x) \mathrm{d}x}, \quad j = 1, 2, 3, \ldots \tag{206}$$

Equation (206) can be used to calculate *estimates of the natural frequencies* using *tests functions*, $X(x)$, that are not necessarily actual mode shapes but resemble (approximate) the mode shapes. In this case, formula (206) becomes the *Rayleigh quotient* given by

$$\omega_j^2 \cong \frac{\int_0^l EA\, X_j'^2(x) \mathrm{d}x}{\int_0^l m X_j^2(x) \mathrm{d}x}, \quad X_j(x) \cong U_j(x), \quad j = 1, 2, 3, \ldots \tag{207}$$

where $X_j(x)$ approximates *the jth mode shape*, $U_j(x)$. The most common use of the Rayleigh quotient is the approximation of the first (fundamental) natural frequency, ω_1.

3.3.1.2 Other Boundary Conditions

Free-free bar

Consider the free-free boundary conditions in the form

$$\begin{aligned} N(0, t) &= 0 \\ N(l, t) &= 0 \end{aligned} \tag{208}$$

Substitution of Eq. (155) into Eq. (208) yields the free-free boundary conditions in terms of displacements, i.e.,

$$\begin{aligned} \hat{u}'(0) &= 0 \\ \hat{u}'(l) &= 0 \end{aligned} \tag{209}$$

Recall the general solution (164), i.e.,

$$\hat{u}(x) = C_1 \sin \gamma x + C_2 \cos \gamma x \tag{210}$$

Differentiation and substitution of Eq. (210) into Eq. (209) yields

$$\begin{cases} C_1 \gamma \cos \gamma x - C_2 \gamma \sin \gamma x|_{x=0} = 0 \\ C_1 \gamma \cos \gamma x - C_2 \gamma \sin \gamma x|_{x=l} = 0 \end{cases} \Rightarrow \begin{cases} C_1 = 0 \\ C_1 \cos \gamma l - C_2 \sin \gamma l = 0 \end{cases} \tag{211}$$

Since the first condition yields $C_1 = 0$, the second condition yields

$$C_2 \sin \gamma l = 0 \tag{212}$$

Equation (212) has a nontrivial solution $C_2 \neq 0$ only if $\sin \gamma l = 0$. This condition yields the same eigenvalues and natural frequencies as previously found for the fixed-fixed case, i.e.,

$$\gamma_j = j\frac{\pi}{l}, \quad \omega_j = j\frac{\pi}{l}\sqrt{\frac{EA}{m}}, \quad j = 1, 2, 3, \ldots \tag{213}$$

Under these conditions, the general solution (210) yields the mode shapes

$$U_j(x) = C_j \cos \gamma_j x, \quad j = 1, 2, 3, \ldots \tag{214}$$

Mode shape normalization yields, as before, $C_j = \sqrt{2/ml}$, and hence

$$U_j(x) = \sqrt{\frac{2}{ml}} \cos \gamma_j x, \quad j = 1, 2, 3, \ldots \tag{215}$$

Plots of the mode shapes and the corresponding natural frequencies for a free-free elastic bar are given in Table 3.3. We note that the displacements reach a maximum at the ends. The *odd modes*, corresponding to $j = 1, 3, \ldots$, are *antisymmetric*, i.e., the end displacements are in opposite directions, whereas the middle displacement is zero. The *even modes*, corresponding to $j = 2, 4, \ldots$, are *symmetric*, i.e., the end displacements are in the same directions, whereas the middle displacement is nonzero.

Fixed-free bar

Consider the fixed-free boundary conditions in the form

$$\begin{aligned} u(0, t) &= 0 \\ N(l, t) &= 0 \end{aligned} \tag{216}$$

Substitution of Eqs. (155) and (153) into Eq. (216) yields the fixed-free boundary conditions in the form

$$\begin{aligned} \hat{u}(0) &= 0 \\ \hat{u}'(l) &= 0 \end{aligned} \tag{217}$$

Recall the general solution (164), i.e.,

$$\hat{u}(x) = C_1 \sin \gamma x + C_2 \cos \gamma x \tag{218}$$

TABLE 3.3 Mode shapes of a free-free elastic bar

Mode #	Eigenvalue	Resonant frequency	Mode shape		Wavelength multiplicity
1	$\gamma l = \pi$	$f_1 = \frac{c}{2l}$	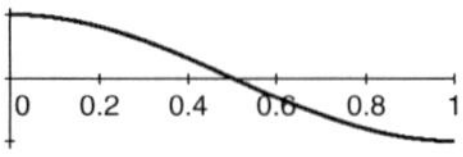	$U_1 = \cos \pi \frac{x}{l}$	$l_1 = \frac{\lambda}{2}$
2	$\gamma l = 2\pi$	$f_2 = 2\frac{c}{2l}$	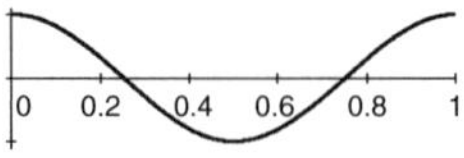	$U_2 = \cos 2\pi \frac{x}{l}$	$l_2 = 2\frac{\lambda}{2}$
3	$\gamma l = 3\pi$	$f_3 = 3\frac{c}{2l}$	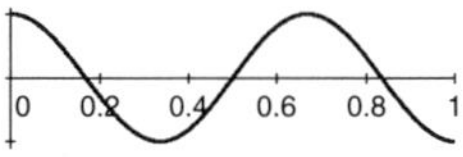	$U_3 = \cos 3\pi \frac{x}{l}$	$l_3 = 3\frac{\lambda}{2}$
4	$\gamma l = 4\pi$	$f_4 = 4\frac{c}{2l}$	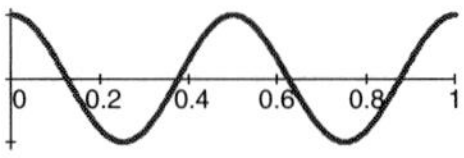	$U_4 = \cos 4\pi \frac{x}{l}$	$l_4 = 4\frac{\lambda}{2}$
5	$\gamma l = 5\pi$	$f_5 = 5\frac{c}{2l}$	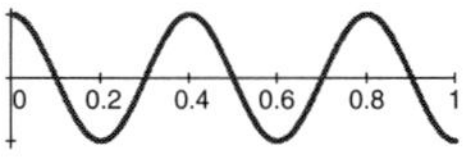	$U_5 = \cos 5\pi \frac{x}{l}$	$l_5 = 5\frac{\lambda}{2}$
6	$\gamma l = 6\pi$	$f_6 = 6\frac{c}{2l}$	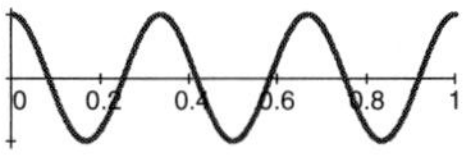	$U_6 = \cos 6\pi \frac{x}{l}$	$l_6 = 6\frac{\lambda}{2}$

Substitution of Eq. (218) into Eq. (217) yields

$$\begin{cases} C_1 \sin \gamma x + C_2 \cos \gamma x|_{x=0} = 0 \\ C_1 \gamma \cos \gamma x - C_2 \gamma \sin \gamma x|_{x=l} = 0 \end{cases} \Rightarrow \begin{cases} C_2 = 0 \\ C_1 \cos \gamma l - C_2 \sin \gamma l = 0 \end{cases} \tag{219}$$

Since the first condition yields $C_2 = 0$, the second condition yields

$$C_1 \cos \gamma l = 0 \tag{220}$$

Equation (220) has a nontrivial solution $C_1 \neq 0$ only if

$$\cos \gamma l = 0 \tag{221}$$

This condition yields the eigenvalues and natural frequencies, i.e.,

$$\gamma_j = (2j-1)\frac{\pi}{2l}, \quad \omega_j = (2j-1)\frac{\pi}{2}\frac{1}{l}\sqrt{\frac{EA}{m}}, \quad j = 1, 2, 3, \ldots \tag{222}$$

We notice that the eigenvalues and natural frequencies of the fixed-free case correspond to odd multiples of $\pi/2$, whereas those of the fixed-fixed and free-free cases corresponded to even multiples of $\pi/2$. Under these conditions, the general solution (210) yields the mode shapes

$$U_j(x) = C_j \sin \gamma_j x, \quad j = 1, 2, 3, \ldots \tag{223}$$

TABLE 3.4 Mode shapes of a fixed-free elastic bar

Mode #	Eigenvalue	Resonant frequency	Mode shape	Wavelength multiplicity
1	$\gamma l = \frac{\pi}{2}$	$f_1 = \frac{c}{4l}$	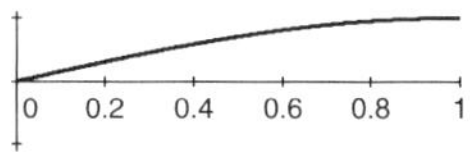 $U_1 = \sin\frac{\pi}{2}\frac{x}{l}$	$l_1 = \frac{\lambda}{4}$
2	$\gamma l = 3\frac{\pi}{2}$	$f_2 = 3\frac{c}{4l}$	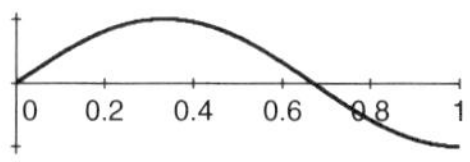 $U_2 = \sin 3\frac{\pi}{2}\frac{x}{l}$	$l_2 = 3\frac{\lambda}{4}$
3	$\gamma l = 5\frac{\pi}{2}$	$f_3 = 5\frac{c}{4l}$	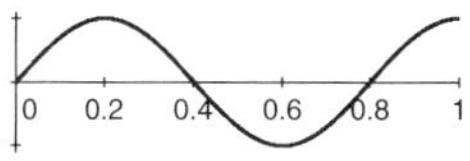 $U_3 = \sin 5\frac{\pi}{2}\frac{x}{l}$	$l_3 = 5\frac{\lambda}{4}$
4	$\gamma l = 7\frac{\pi}{2}$	$f_4 = 7\frac{c}{4l}$	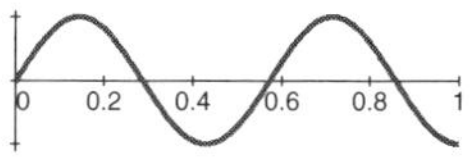 $U_4 = \sin 7\frac{\pi}{2}\frac{x}{l}$	$l_4 = 7\frac{\lambda}{4}$
5	$\gamma l = 9\frac{\pi}{2}$	$f_5 = 9\frac{c}{4l}$	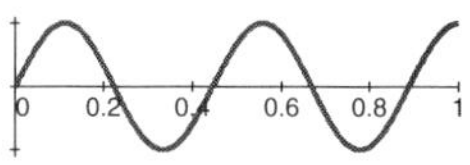 $U_5 = \sin 9\frac{\pi}{2}\frac{x}{l}$	$l_5 = 9\frac{\lambda}{4}$
6	$\gamma l = 11\frac{\pi}{2}$	$f_6 = 11\frac{c}{4l}$	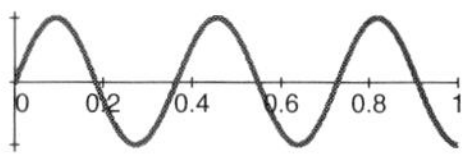 $U_6 = \sin 11\frac{\pi}{2}\frac{x}{l}$	$l_6 = 11\frac{\lambda}{4}$

with γ_j given by Eq. (222). Mode shape normalization yields, as before, $C_j = \sqrt{2/ml}$, and hence

$$U_j(x) = \sqrt{\frac{2}{ml}} \sin \gamma_j x, \quad j = 1, 2, 3, \ldots \tag{224}$$

Plots of the mode shapes and the corresponding natural frequencies for a fixed-free elastic bar are given in Table 3.4.

3.3.1.3 Forced Axial Vibration of a Bar

Assume an uniform bar undergoing axial vibration under the excitation of an externally applied time-dependent axial force per unit length, $f(x, t)$ as shown in Fig. 3.13. Considering the combined effect of internal stress resultant, $N(x, t)$, and external excitation, $f(x, t)$, we perform the free-body analysis of an infinitesimal bar element dx to get

$$N'(x, t)\mathrm{d}x + f(x, t) = m\ddot{u}(x, t) \tag{225}$$

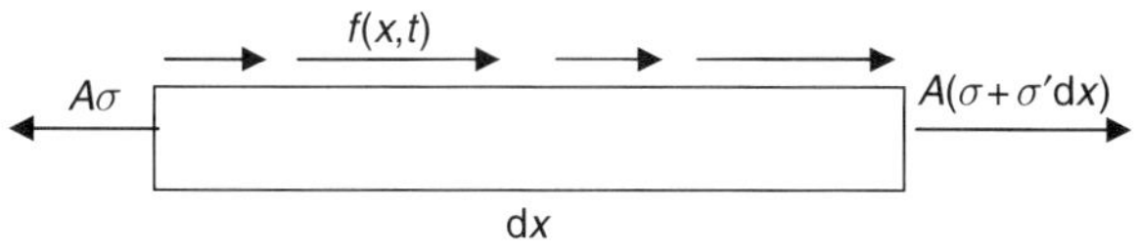

FIGURE 3.13 Infinitesimal element for the analysis of forced axial vibrations.

Relating, as before, the stress resultant $N(x, t)$ to the displacement $u(x, t)$ yields the equation of motion for forced axial vibration

$$m\,\ddot{u}(x,t) - EA \times u''(x,t) = f(x,t) \tag{226}$$

Without loss of generality, we assume the external excitation to be harmonic in the form

$$f(x,t) = \hat{f}(x)e^{i\omega t} \tag{227}$$

Then, we obtain

$$m\,\ddot{u}(x,t) - EA \times u''(x,t) = \hat{f}(x)e^{i\omega t} \tag{228}$$

Modal expansion theorem

Assume the modal expansion

$$u\,(x,t) = \sum_{j=1}^{\infty} C_j U_j(x) e^{i\omega t} \tag{229}$$

where $U_j(x)$ are natural modes satisfying the free-vibration equation of motion (176) and the orthogonality conditions (186) and (189), whereas C_j are the *modal participation factors*. Equation (229) is harmonic in the excitation frequency, which implies that we are only concerned with the steady-state response. Substitution of Eq. (229) into Eq. (228) and division by $e^{i\omega t}$ yields

$$-m\omega^2 \sum_{p=1}^{\infty} C_p U_p(x) - EA \sum_{p=1}^{\infty} C_p U_p''(x) + = \hat{f}(x) \tag{230}$$

where the index p was used instead of j for convenience. Multiplication of Eq. (230) by $U_q(x)$ and integration over the length of the beam yields

$$\begin{aligned} -\omega^2 \sum_{p=1}^{\infty} C_p \int_0^0 m\,U_q(x)U_p(x)\mathrm{d}x - \sum_{p=1}^{\infty} C_p \int_0^l EA\,U_p''(x)U_q(x)\mathrm{d}x \\ = \int_0^l \hat{f}(x)U_q(x)\mathrm{d}x, \quad q = 1,2,3,\ldots \end{aligned} \tag{231}$$

Using the orthogonality conditions given by Eqs. (186) and (189), and the definition of modal mass and stiffness given by Eqs. (201) and (204), we recast Eq. (231) in the form of a set of decoupled linear algebraic equations, i.e.,

$$C_j\left(-\omega^2 m_j + k_j\right) = f_j, \quad j = 1,2,3,\ldots \tag{232}$$

where f_j is the *modal excitation* given by

$$f_j = \int_0^l \hat{f}(x)U_j(x)\mathrm{d}x, \quad j = 1,2,3,\ldots \tag{233}$$

If the modes are orthonormal, then substituting the orthonormality conditions (199) and (200) into Eq. (231) yields the simpler form

$$C_j\left(-\omega^2 + \omega_j^2\right) = f_j, \quad j = 1,2,3,\ldots \tag{234}$$

Response by modal analysis

Solution of Eq. (235) yields the modal participation factors in the form

$$C_j = \frac{f_j}{-\omega^2 + \omega_j^2}, \quad j = 1, 2, 3, \ldots \tag{235}$$

Equation (235) corresponds to the amplitude the expression in Eq. (56) derived for 1-dof systems. Substitution of Eq. (235) into Eq. (229) yields

$$u(x, t) = \sum_{j=1}^{\infty} \frac{f_j}{-\omega^2 + \omega_j^2} U_j(x) e^{i\omega t}, \tag{236}$$

So far, the vibration has been considered undamped. However, real phenomena are always associated with some damping ζ_j. The modal participation factors for a damped system are given by

$$C_j = \frac{f_j}{-\omega^2 + 2i\zeta_j \omega_j \omega + \omega_j^2}, \quad j = 1, 2, 3, \ldots \tag{237}$$

The modal response defined by Eq. (237) resemble the response of the 1-dof vibrating system, Eq. (79). In fact, as the excitation frequency passes through a natural frequency, ω_j, the continuous system will *resonate* and will vibrate in the natural mode $U_j(x)$. Around a natural frequency, the continuous system behaves more or less like a 1-dof system vibrating in the corresponding natural mode. If the excitation frequency covers a large frequency span, several resonances can be encountered corresponding to several natural frequencies and normal modes.

Substitution of Eq. (237) into Eq. (229) yields the frequency response of the damped axial vibration system in the form

$$u(x, t) = \sum_{j=1}^{\infty} \frac{f_j}{-\omega^2 + 2i\zeta_j \omega_j \omega + \omega_j^2} U_j(x) e^{i\omega t} \tag{238}$$

Equation (238) represents a superposition of a number of terms, each term corresponding to a natural frequency and normal mode of vibration. Equation (238) allows us to determine the response of the continuous structure to harmonic excitation of variable frequency. This leads to the *frequency response function* (FRF) concept. The FRF could be calculated by letting the excitation function take the form of *unit excitation*. Several unit excitation functions could be envisages, depending on the choice of spatial variation. As the excitation frequency sweeps through natural frequency values, ω_j, the corresponding term in the series becomes very large, the structure passes through a structural resonance, and the response displays a resonance peak. Over a frequency interval, the response can display several peaks, corresponding to several resonances, as illustrated in Fig. 3.14.

Generalized coordinates and modal equations

The modal expansion method is not limited to the case of harmonic excitation illustrated in previous section. In fact, we can assume modal expansion in the general form

$$u(x, t) = \sum_{j=1}^{\infty} \eta_j(t) U_j(x), \tag{239}$$

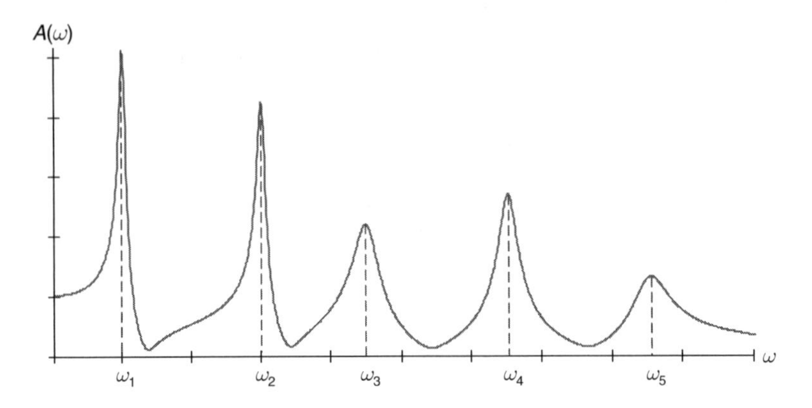

FIGURE 3.14 Multi-peak frequency response plot for a continuous system.

where the generic functions $\eta_j(t)$ are time-dependent *modal* (*normal*) *coordinates*. Substitution of modal expansion (239) into the equation of motion (226) yields

$$EA\sum_{j=1}^{\infty}\eta_j(t)U_j''(x)+m\sum_{j=1}^{\infty}\ddot{\eta}_j(t)U_j(x)=-f(x,t) \tag{240}$$

As before, multiply Eq. (240) by each of the mode shapes in turn, use the orthonormality conditions (199), (200) and get, upon rearrangement, a set of decoupled equations in the form

$$m_j\ddot{\eta}_j(t)\omega_j^2+k_j\eta_j(t)=f_j(t), \quad j=1,2,3,\ldots \tag{241}$$

where $f_j(t)$ is the *time-dependent modal excitation* given by

$$f_j(t)=\int_0^l f(x,t)U_j(x)\mathrm{d}x, \quad j=1,2,3,\ldots \tag{242}$$

Introducing the modal damping ratio, ζ_j, to account for the inherent dissipative losses encountered in practice, we obtain the modal equations for damped vibrations of a continuous system in the form

$$\ddot{\eta}_j(t)+2\zeta\omega_j\dot{\eta}_j(t)+\eta_j(t)\omega_j^2=f_j(t), \quad j=1,2,3,\ldots \tag{243}$$

Each of the modal equations (243) resembles the 1-dof Eq. (76). Since the modal equations are uncoupled, each of them can be solved independently, and then reassembled through modal summation to obtain the total response.

3.3.1.4 Axial Vibration Energy in a Vibrating Bar

We have seen in the study of the 1-dof vibration of a particle that the vibration energy, E, contains a kinetic energy, T, and an elastic energy (potential energy), V, i.e.,

$$E=T(t)+V(t)=\frac{1}{2}m\dot{u}^2(t)+\frac{1}{2}ku^2(t) \tag{244}$$

To investigate these principles in the context of an elastic bar we will consider the kinetic and elastic energies stored in the bar, i.e.,

$$\begin{aligned} T(t) &= \int_\Omega \frac{1}{2}\rho \dot{u}^2(x,t)\mathrm{d}\Omega = \int_0^l \frac{1}{2} m\,\dot{u}^2(x,t)\mathrm{d}x \\ V(t) &= \int_\Omega \frac{1}{2}\sigma(x,t)\varepsilon(x,t)\mathrm{d}\Omega = \int_0^l \frac{1}{2}EA\,u'^2(x,t)\mathrm{d}x \end{aligned} \quad \text{(axial vibration of a bar)} \quad (245)$$

where Ω is the total volume of the bar. Using the modal expansion of Eq. (229), we write

$$\begin{aligned} T(t) = \int_0^l \frac{1}{2} m &\left[\sum_{p=1}^{\infty} C_p U_p(x)\ (-\omega \sin \omega t)\right] \\ &\left[\sum_{q=1}^{\infty} C_q U_q(x)\ (-\omega \sin \omega t)\right] \mathrm{d}x \end{aligned} \quad \text{(kinetic energy)} \quad (246)$$

$$\begin{aligned} V(t) = \int_0^l \frac{1}{2} EA &\left[\sum_{p=1}^{\infty} C_p U_p'(x)\ \cos \omega t\right] \\ &\left[\sum_{p=1}^{\infty} C_p U_p'(x)\ \cos \omega t\right] \mathrm{d}x \end{aligned} \quad \text{(elastic energy)} \quad (247)$$

Processing of Eqs. (246) and (247) through the use of the orthogonality conditions (186), (191) and of the modal mass and stiffness definitions (201), (202) yields

$$T(t) = \omega^2 \sin^2 \omega t \sum_{p=1}^{\infty} \frac{1}{2} C_j^2 m_j \quad \text{(kinetic energy)} \quad (248)$$

$$V(t) = \cos^2 \omega t \sum_{p=1}^{\infty} \frac{1}{2} k_j C_j^2 = \cos^2 \omega t \sum_{p=1}^{\infty} \frac{1}{2} m_j \omega_j^2 C_j^2 \quad \text{(elastic energy)} \quad (249)$$

Hence, the total energy is given by

$$\begin{aligned} E &= T(t) + V(t) \\ &= \omega^2 \sin^2 \omega t \sum_{p=1}^{\infty} \frac{1}{2} C_j^2 m_j + \cos^2 \omega t \sum_{p=1}^{\infty} \frac{1}{2} C_j^2 k_j \\ &= \sum_{p=1}^{\infty} \frac{1}{2} m_j C_j^2 \left(\omega^2 \sin^2 \omega t + \omega_j^2 \cos^2 \omega t\right) \end{aligned} \quad (250)$$

If the system oscillates at one of the resonant frequencies, say $\omega = \omega_j$, then the kinetic and elastic energies are concentrated into just one mode of vibration, and resemble 1-dof behavior, i.e.,

$$T_j(t) = \sin^2 \omega t \frac{1}{2} m_j \omega_j^2 C_j^2 \quad \text{(modal kinetic energy)} \quad (251)$$

$$V_j(t) = \cos^2 \omega t \frac{1}{2} k_j C_j^2 = \cos^2 \omega t \frac{1}{2} m_j \omega_j^2 C_j^2 \quad \text{(modal elastic energy)} \quad (252)$$

The total energy of the system vibrating in the jth natural mode is given by the *modal vibration energy*:

$$E_j = T_j(t) + V_j(t) = \frac{1}{2}m_j\omega_j^2 C_j^2 \sin^2\omega_j t + \frac{1}{2}m_j\omega_j^2 C_j^2 \cos^2\omega_j t = \frac{1}{2}m_j\omega_j^2 C_j^2 \quad \text{(modal vibration energy)} \tag{253}$$

During the vibration cycle, the total vibration energy, E_j, is constant, but the contributions of the kinetic and elastic parts vary. At the point of maximum displacement the elastic energy reaches a maximum while the kinetic energy is zero, whereas at the point of zero displacement (but maximum velocity) the elastic energy is zero while the kinetic energy reaches a maximum. Recall that the assumption of natural modes of vibrations is that the complete body undergoes the vibration in phase, i.e., all the points in the body reach the maximum amplitude and then the zero amplitude, etc., in the same time.

3.3.2 FLEXURAL VIBRATION OF A BEAM

3.3.2.1 Free Flexural Vibration of a Beam

Consider a uniform beam of length l, mass per unit length m, bending stiffness EI, undergoing flexural vibration of displacement $w(x, t)$ as shown in Fig. 3.15a. Centroidal axes are assumed. An infinitesimal beam element of length dx is subjected to the action of bending moments, $M(x, t)$, $M(x, t) + M'(x, t)dx$, axial forces, $N(x, t) + N'(x, t)dx$, and shear forces $V(x, t)$, $V(x, t) + V'(x, t)dx$ (Fig. 3.15b).

We will briefly review the Euler–Bernoulli theory of bending. (Shear deformation and rotary inertia effects are ignored.) Free-body analysis of the infinitesimal element of Fig. 3.15b yields

$$N'(x, t) = 0 \tag{254}$$

$$V'(x, t) = m\ddot{w}(x, t) \tag{255}$$

$$M'(x, t) + V(x, t) = 0 \tag{256}$$

The N and M stress resultants are evaluated by integration of the direct stress across the cross-sectional area shown in Fig. 3.15c, i.e.,

$$N(x, t) = \int_A \sigma(x, z, t)dA \tag{257}$$

$$M(x, t) = -\int_A \sigma(x, z, t)z dA \tag{258}$$

Using the stress–strain constitutive relation of Eq. (154), the axial force and moment stress resultants (257) and (258) can be expressed as

$$N(x, t) = E\int_A \varepsilon(x, z, t)dA \tag{259}$$

$$M(x, t) = -E\int_A \varepsilon(x, z, t)z dA \tag{260}$$

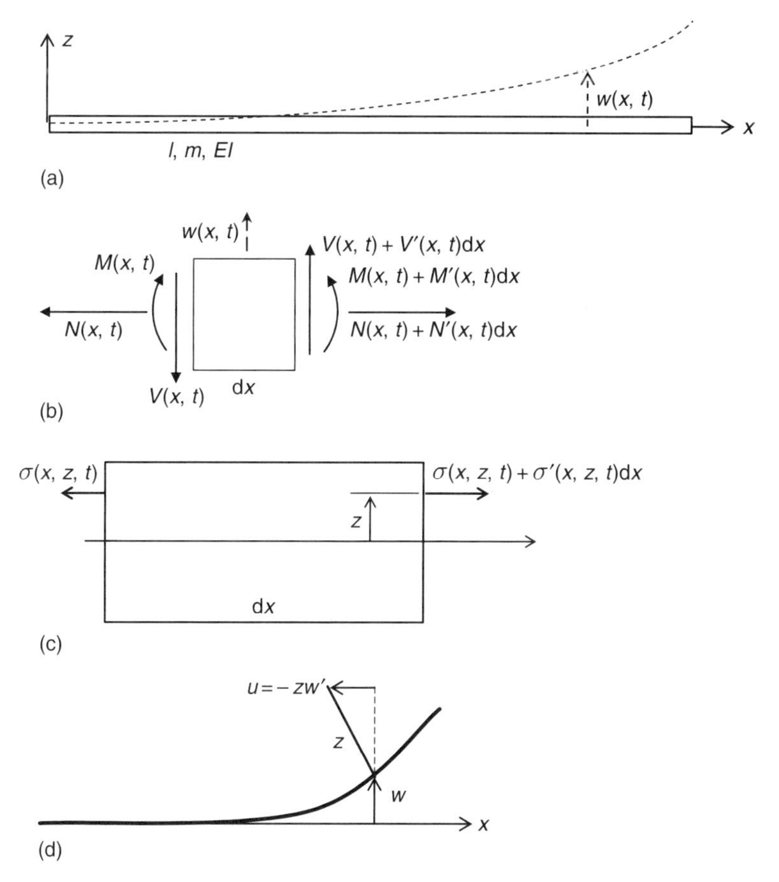

FIGURE 3.15 Beam undergoing flexural vibration: (a) general layout; (b) free-body diagram of an infinitesimal beam element; (c) stress distribution across the thickness; (d) horizontal displacement induced by flexure.

Kinematic analysis (Fig. 3.15) yields the direct strain $\varepsilon(x, z, t)$ in terms of the flexural motion $w(x, t)$ and the thickness-wise location z, i.e.,

$$\begin{aligned} u(x, z, t) &= -zw'(x, t) \\ \varepsilon(x, z, t) &= u'(x, z, t) = -zw''(x, t) \end{aligned} \tag{261}$$

Substitution of Eq. (261) into Eq. (260) and integration over the area yields

$$N(x, t) = -Ew''(x, t)\int_A z\mathrm{d}A = 0 \tag{262}$$

$$M(x, t) = Ew''(x, t)\int_A z^2\mathrm{d}A \tag{263}$$

Note that Eq. (262) indicates that the axial stress resultant is zero, i.e., $N(x, t) = 0$, since centroidal axes were assumed. On the other hand, Eq. (263) yields

$$M(x, t) = EI\ w''(x, t) \tag{264}$$

Substitution of Eq. (264) into Eq. (256) yields

$$V(x,t) = -EI\ w'''(x,t) \tag{265}$$

Differentiation of Eq. (265) with respect to x, and substitution into Eq. (256) yields the *equation of motion for free flexural vibration* of a beam, i.e.,

$$EI\ w'''(x,t) + m\ \ddot{w}(x,t) = 0 \tag{266}$$

Upon division by m and rearrangement, we get

$$a^4 w'''' + \ddot{w} = 0 \tag{267}$$

where the constant a^4 is given by

$$a^4 = \frac{EI}{m} \quad \text{or} \quad a = \left(\frac{EI}{m}\right)^{1/4} \tag{268}$$

Natural frequencies and mode shapes for pin-pin boundary conditions

The discussion of natural frequencies and mode shapes for flexural vibration follows the general pattern developed for the axial vibration. Eq. (267) is a PDE in time, t, and space, x. As before, assume separation of variables and write

$$w(x,t) = \hat{w}(x)e^{i\omega t} \tag{269}$$

Upon substitution of Eq. (269) into Eq. (266), we obtain a fourth-order ODE in the form

$$a^4\hat{w}'''' - \omega^2\hat{w} = 0 \tag{270}$$

Upon division by a^4, Eq. (270) becomes

$$w'''' - \gamma^4\hat{w} = 0 \tag{271}$$

where

$$\gamma^4 = \frac{\omega^2}{a^4} = \frac{m}{EI}\omega^2 \quad \text{or} \quad \gamma = \left(\frac{m}{EI}\right)^{1/4}\sqrt{\omega} \tag{272}$$

Equation (271) admits solutions of the form

$$\hat{u}(x) = Ce^{\lambda x} \tag{273}$$

Substitution of Eq. (273) into Eq. (271) yields the *characteristic equation*

$$\lambda^4 - \gamma^4 = 0 \tag{274}$$

which has the roots

$$\lambda_1 = i\gamma, \quad \lambda_2 = -i\gamma, \quad \lambda_3 = \gamma, \quad \lambda_4 = -\gamma \tag{275}$$

Correspondingly, we have the general solution

$$\hat{w}(x) = A_1 e^{i\gamma x} + A_2 e^{-i\gamma x} + A_3 e^{\gamma x} + A_4 e^{-\gamma x} \tag{276}$$

Equation (276) can be written in the alternate form

$$\hat{w}(x) = C_1 \sin \gamma x + C_2 \cos \gamma x + C_3 \sinh \gamma x + C_4 \cosh \gamma x \tag{277}$$

The constants C_1, C_2, C_3, C_4 are to be found from the boundary conditions. For illustration, assume the **pin-pin boundary conditions**, i.e.,

$$\begin{aligned} w(0,t) &= 0 \quad M(0,t) = 0 \\ w(l,t) &= 0 \quad M(l,t) = 0 \end{aligned} \tag{278}$$

The use of Eqs. (264) and (269) transforms Eq. (278) into

$$\begin{aligned} \hat{w}(0) &= 0 \quad \hat{w}''(0) = 0 \\ \hat{w}(l) &= 0 \quad \hat{w}''(l) = 0 \end{aligned} \tag{279}$$

Substitution of Eq. (277) into Eq. (279) yields

$$\begin{aligned} C_1 \sin \gamma x + C_2 \cos \gamma x + C_3 \sinh \gamma x + C_4 \cosh \gamma x|_{x=0} &= 0 \\ \gamma^2 [-C_1 \sin \gamma x - C_2 \cos \gamma x + C_3 \sinh \gamma x + C_4 \cosh \gamma x]\big|_{x=0} &= 0 \\ C_1 \sin \gamma x + C_2 \cos \gamma x + C_3 \sinh \gamma x + C_4 \cosh \gamma x|_{x=l} &= 0 \\ \gamma^2 [-C_1 \sin \gamma x - C_2 \cos \gamma x + C_3 \sinh \gamma x + C_4 \cosh \gamma x]\big|_{x=l} &= 0 \end{aligned} \tag{280}$$

Upon simplification, we get

$$\begin{aligned} C_2 + C_4 &= 0 \\ -C_2 + C_4 &= 0 \\ C_1 \sin \gamma l + C_2 \cos \gamma l + C_3 \sinh \gamma l + C_4 \cosh \gamma l &= 0 \\ -C_1 \sin \gamma l - C_2 \cos \gamma l + C_3 \sinh \gamma l + C_4 \cosh \gamma l &= 0 \end{aligned} \tag{281}$$

Further simplification yields

$$\begin{aligned} C_2 = C_4 &= 0 \\ C_1 \sin \gamma l + C_3 \sinh \gamma l &= 0 \\ -C_1 \sin \gamma l + C_3 \sinh \gamma l &= 0 \end{aligned} \tag{282}$$

or

$$\begin{aligned} C_2 = C_3 = C_4 &= 0 \\ C_1 \sin \gamma l &= 0 \end{aligned} \tag{283}$$

since $\sinh \gamma l \neq 0$ for $\gamma l \neq 0$. We note that a nontrivial solution is only attained if

$$\sin \gamma l = 0 \tag{284}$$

Equation (284) is the *characteristic equation* that defines the *eigenvalues* γ for which nontrivial solutions exist. Solution of Eq. (284) yields

$$\gamma l = \pi, 2\pi, 3\pi, \ldots \quad \text{or} \quad \gamma_j = j\frac{\pi}{l}, \quad j = 1, 2, 3, \ldots \tag{285}$$

Combining Eq. (285) with Eq. (272) yields the *natural frequencies*

$$\omega_j = \gamma_j^2 \sqrt{\frac{EI}{m}} \quad \text{or} \quad \omega_j = \left(j\frac{\pi}{l}\right)^2 \sqrt{\frac{EI}{m}} = (j\pi)^2 \sqrt{\frac{EI}{ml^4}}, \quad j = 1, 2, 3, \ldots \tag{286}$$

Mode shapes

For each eigenvalue and natural frequency, we have an *eigenfunction* or *natural mode* of vibration (*mode shape*), i.e.,

$$W_j(x) = C_j \sin \gamma_j l, \quad j = 1, 2, 3, \ldots \tag{287}$$

It should be noted that all the mode shapes $W_j(x)$ satisfy Eq. (271) and the boundary conditions (279). As mentioned earlier, the constants C_j cannot be determined from the differential equation and the boundary conditions. This is quite alright, since it can be easily verified that if $W_j(x)$ is a mode shape that satisfies the differential equation and the boundary conditions, then any scaled version of it, $\alpha W_j(x)$, $\alpha \in \mathbb{R}$, also satisfies them. Hence, the value of the constants C_j can be arbitrarily chosen. For example, one can chose $C_j = 1$, which might have some advantage in mode-shape plotting.

Orthogonality of mode shapes

Recall that the mode shapes $W_j(x)$ satisfy Eq. (271), which can be conveniently expressed explicitly in m, EI, and ω_j, i.e.,

$$EI\, W_j'''' - \omega_j^2 m\, W_j = 0, \quad j = 1, 2, 3, \ldots \tag{288}$$

Consider two separate mode shapes, $W_p(x)$ and $W_q(x)$. They satisfy the differential Eq. (288), i.e.,

$$\begin{aligned} EI\, W_p'''' &= \omega_p^2 m\, W_p \\ EA\, W_q'''' &= \omega_q^2 m\, W_q \end{aligned} \tag{289}$$

To analyze *orthogonality with respect to mass*, consider the mass-weighted integral

$$\int_0^l m\, W_p(x) W_q(x) \mathrm{d}x \tag{290}$$

Substitution of Eq. (287) into Eq. (290) yields

$$\int_0^l m \left(C_p \sin \gamma_p x\right) \left(C_q \sin \gamma_q x\right) \mathrm{d}x \tag{291}$$

Using standard trigonometric formulae, we write

$$\int_0^l \left(\sin \gamma_p x\right) \left(\sin \gamma_q x\right) \mathrm{d}x = \frac{1}{2} \int_0^l \left[\cos(\gamma_p - \gamma_q)x - \cos(\gamma_p + \gamma_q)x\right] \mathrm{d}x = \frac{l}{2}\delta_{pq} \tag{292}$$

Equation (285) was used to evaluate γ_p, γ_q. As before, δ_{pq} is the Kronecker delta ($\delta_{pq}=1$ for $p=q$, and $\delta_{pq}=0$ for $p\neq q$). Using Eqs. (291), (292) and assuming $p\neq q$, Eq. (290) yields the *orthogonality condition with respect to mass* in the form

$$\int_0^l m\,W_p(x)W_q(x)\mathrm{d}x=0,\quad p\neq q \tag{293}$$

The mass orthogonality condition expressed by Eq. (293) is not restricted to the simple mode shapes expressed by Eq. (287). To prove this, recall Eq. (289), multiply the first line by W_q and the second line by W_p, and integrate each line over the length of the bar to obtain

$$\begin{aligned}\int_0^l EI\,W_p''''(x)W_q(x)\mathrm{d}x&=\omega_p^2\int_0^l mW_p(x)W_q(x)\\ \int_0^l EI\,W_q''''(x)W_p(x)\mathrm{d}x&=\omega_q^2\int_0^l mW_q(x)W_p(x)\end{aligned} \tag{294}$$

Integration by parts yields

$$\begin{aligned}&\int_0^l EI\,W_p''(x)W_q''(x)\mathrm{d}x+EI\,W_p'''(x)W_q(x)\big|_0^l-EI\,W_p''(x)W_q'(x)\big|_0^l\\ &=\omega_p^2\int_0^l mW_p(x)W_q(x)\\ &\int_0^l EI\,W_p''(x)W_q''(x)\mathrm{d}x+EI\,W_q'''(x)W_p(x)\big|_0^l-EI\,W_q''(x)W_p'(x)\big|_0^l\\ &=\omega_q^2\int_0^l mW_p(x)W_q(x)\end{aligned} \tag{295}$$

The left hand side of Eq. (295) contains two parts: (a) an integral containing the second derivative of the mode shapes, and (b) two boundary-evaluated terms expressed in terms of the mode shapes, and their first, second, and third derivates. Recalling that the mode shapes satisfy the boundary conditions (279), it becomes apparent that, in our case, the boundary-evaluated terms in Eq. (295) vanish. Hence, Eq. (295) takes the simpler form

$$\begin{aligned}\int_0^l EI\,W_p''(x)W_q''(x)\mathrm{d}x&=\omega_p^2\int_0^l m\,W_p(x)W_q(x)\\ \int_0^l EI\,W_q''(x)W_p''(x)\mathrm{d}x&=\omega_q^2\int_0^l m\,W_q(x)W_p(x)\end{aligned} \tag{296}$$

Subtraction of the second line of Eq. (296) from the first line yields

$$\left(\omega_p^2-\omega_q^2\right)\int_0^l m\,W_p(x)W_q(x)\mathrm{d}x=0 \tag{297}$$

For distinct mode numbers, $p\neq q$, the frequencies are also distinct, $\omega_p^2\neq\omega_q^2$, and hence Eq. (297) implies

$$\int_0^l mW_p(x)W_q(x)\mathrm{d}x=0,\quad p\neq q \tag{298}$$

which is exactly the mass orthogonalization condition derived earlier as Eq. (293). Notice that the above derivation is quite general, if the boundary conditions are such that the boundary-evaluated terms in Eq. (295) vanish.

To analyze *orthogonality with respect to stiffness*, consider the integral

$$\int_0^l \left[EI\, W_p''''(x)\right] W_q(x)\mathrm{d}x, \quad p \neq q \tag{299}$$

Recalling Eq. (294), we write Eq. (299) as

$$\int_0^l EI\, W_p''''(x) W_q(x)\mathrm{d}x = \omega_p^2 \int_0^l m\, W_p(x) W_q(x) \tag{300}$$

Recalling Eq. (298), Eq. (300) becomes

$$\int_0^l \left[EI\, W_p''''(x)\right] W_q(x)\mathrm{d}x = 0, \quad p \neq q \tag{301}$$

Equation (301) is the *orthogonality condition with respect to stiffness*. Integration by parts of Eq. (301) yields an alternate expression of the orthogonality condition with respect to stiffness, i.e.,

$$\int_0^l EI\, W_p''(x) W_q''(x)\ \mathrm{d}x + EI\, W_p'''(x) W_q(x)\Big|_0^l - EI\, W_p''(x) W_q'(x)\Big|_0^l = 0, \quad p \neq q \tag{302}$$

Recalling that the mode shapes satisfy the boundary conditions (279), it becomes apparent that the second term in Eq. (302) vanishes. Hence, the orthogonality expression with respect to stiffness can be alternately expressed as

$$\int_0^l EI\, W_p''(x) W_q''(x)\mathrm{d}x = 0, \quad p \neq q \tag{303}$$

if the boundary conditions are such that boundary-evaluated terms in Eq. (302) is automatically zero. In practical applications, where independently derived mode shape candidates are tested for stiffness orthogonality, the formulation (303) may be preferred since second derivatives are easier to evaluate with reasonable accuracy than fourth derivatives.

Normalization of mode shapes: normal modes

Recalling the mode shapes orthogonality analysis, we notice that, for $p = q = j$, Eq. (291) can be resolved using Eq. (292) to yield

$$C_j = \sqrt{\frac{2}{ml}} \tag{304}$$

Substitution of Eq. (304) into Eq. (287) yields the *mass-normalized mode shapes* (*normal modes*)

$$W_j(x) = \sqrt{\frac{2}{ml}} \sin \gamma_j l, \quad j = 1, 2, 3, \ldots \tag{305}$$

The mass-normalized mode shapes satisfy the *normalization condition with respect to mass*, i.e.,

$$\int_0^l m W_j^2(x)\mathrm{d}x = 1, \quad j = 1, 2, 3, \ldots \tag{306}$$

The corresponding *normalization condition with respect to stiffness* is

$$\int_0^l EI\, W_j''^2(x)\mathrm{d}x = \omega_j^2, \quad j = 1, 2, 3, \ldots \tag{307}$$

and also

$$\int_0^l EI\, W_j''''(x)W_j(x)\mathrm{d}x = \omega_j^2, \quad p \neq q \tag{308}$$

Orthonormal modes

Mode shapes that are simultaneously normal and orthogonal are called *orthonormal modes*; they satisfy the *orthonormality condition*:

$$\begin{aligned} \int_0^l m\, W_p(x)W_q(x)\mathrm{d}x = \delta_{pq} &= \begin{cases} 1 & \text{for } p = q \\ 0 & \text{otherwise} \end{cases} \\ \int_0^l EI\, W_p''(x)W_q''(x)\mathrm{d}x = \omega_p^2\delta_{pq} &= \begin{cases} \omega_p^2 & \text{for } p = q \\ 0 & \text{otherwise} \end{cases} \end{aligned} \tag{309}$$

and also

$$\int_0^l EI\, W_p''''(x)W_q(x)\mathrm{d}x = \omega_p^2\delta_{pq} = \begin{cases} \omega_p^2 & \text{for } p = q \\ 0 & \text{otherwise} \end{cases} \tag{310}$$

Modal mass and stiffness: modal coefficients

If the mode shapes are mass normalized in accordance with Eq. (304), then their weighted integral with respect to mass evaluates to unity. However, mode shapes may not necessarily be normalized in accordance with Eq. (304); in fact, we have already shown that the mode shape amplitudes, which are the solution of a homogenous system, have one degree of indeterminacy and hence can be generally scaled by any arbitrary factor. Hence, for a generic mode shape $W_j(x)$, Eq. (306) would be

$$\int_0^l mW_j^2(x)\mathrm{d}x = m_j, \quad j = 1, 2, 3, \ldots \tag{311}$$

where m_j is the *modal mass*. By a similar argument, Eq. (307) would become

$$\int_0^l EI\, W''^2_j(x)\mathrm{d}x = k_j^2, \quad j = 1, 2, 3, \ldots \tag{312}$$

where k_j is the *modal stiffness*, and

$$k_j = \omega_j^2 m_j, \quad j = 1, 2, 3, \ldots \tag{313}$$

The modal mass, m_j, and the modal stiffness, k_j, are the *modal coefficients*. If damping were present, then an additional modal parameter, the *modal damping*, c_j, would be similarly derived. For mass-normalized mode shapes, the modal coefficients would be the *modal frequency*, ω_j, and the modal damping ratio, ζ_j.

It is also useful to note that, in virtue of Eq. (294),

$$\int_0^l EI\, W_j''''W_j(x)\mathrm{d}x = k_j^2 \tag{314}$$

3.3.2.2 Other Boundary Conditions

Free-free beam

Consider the free-free boundary conditions in the form

$$\begin{aligned} M(0,t) &= 0 \quad V(0,t) = 0 \\ M(l,t) &= 0 \quad V(l,t) = 0 \end{aligned} \tag{315}$$

Using Eqs. (264), (265), and (269) into Eq. (315) yields the boundary conditions in terms of displacement and its derivatives, i.e.,

$$\begin{aligned} \hat{w}''(0) &= 0 \quad \hat{w}'''(0) = 0 \\ \hat{w}''(l) &= 0 \quad \hat{w}'''(l) = 0 \end{aligned} \tag{316}$$

Substitution of the general solution (277) into the boundary conditions (316) yields

$$\begin{aligned} &\gamma^2 [-C_1 \sin \gamma x - C_2 \cos \gamma x + C_3 \sinh \gamma x + C_4 \cosh \gamma x]\big|_{x=0} = 0 \\ &\gamma^3 [-C_1 \cos \gamma x + C_2 \sin \gamma x + C_3 \cosh \gamma x + C_4 \sinh \gamma x]\big|_{x=0} = 0 \\ &\gamma^2 [-C_1 \sin \gamma x - C_2 \cos \gamma x + C_3 \sinh \gamma x + C_4 \cosh \gamma x]\big|_{x=l} = 0 \\ &\gamma^3 [-C_1 \cos \gamma x + C_2 \sin \gamma x + C_3 \cosh \gamma x + C_4 \sinh \gamma x]\big|_{x=l} = 0 \end{aligned} \tag{317}$$

Upon simplification, we get

$$\begin{aligned} -C_2 + C_4 &= 0 \\ -C_1 + C_3 &= 0 \\ -C_1 \sin \gamma l - C_2 \cos \gamma l + C_3 \sinh \gamma l + C_4 \cosh \gamma l &= 0 \\ -C_1 \cos \gamma l + C_2 \sin \gamma l + C_3 \cosh \gamma l + C_4 \sinh \gamma l &= 0 \end{aligned} \tag{318}$$

Further simplification yields

$$\begin{aligned} C_4 &= C_2 \\ C_3 &= C_1 \\ C_1 (-\sin \gamma l + \sinh \gamma l) + C_2 (-\cos \gamma l + \cosh \gamma l) &= 0 \\ C_1 (-\cos \gamma l + \cosh \gamma l) + C_2 (\sin \gamma l + \sinh \gamma l) &= 0 \end{aligned} \tag{319}$$

The last two lines in Eq. (319) form a homogenous linear algebraic system in C_1 and C_2

$$\begin{cases} (-\sin \gamma l + \sinh \gamma l)\, C_1 + (-\cos \gamma l + \cosh \gamma l)\, C_2 = 0 \\ (-\cos \gamma l + \cosh \gamma l)\, C_1 + (\sin \gamma l + \sinh \gamma l)\, C_2 = 0 \end{cases} \tag{320}$$

The homogenous algebraic system (320) accepts nontrivial solutions only if its determinant has value zero, i.e.,

$$\begin{vmatrix} -\sin \gamma l + \sinh \gamma l & -\cos \gamma l + \cosh \gamma l \\ -\cos \gamma l + \cosh \gamma l & \sin \gamma l + \sinh \gamma l \end{vmatrix} = 0 \tag{321}$$

Expansion and simplification of Eq. (321) yields the *characteristic equation*

$$\cos \gamma l \cosh \gamma l - 1 = 0 \tag{322}$$

Equation (322) is a transcendental equation that yields the eigenvalues γl corresponding to the free flexural vibration of a free-free beam. Numerical values of the first five eigenvalues γl are as shown in Table 3.5. As γl becomes large, the numerical values approach a rational sequence. From the sixth eigenvalue onwards, they can be approximated to reasonable accuracy by

$$(\gamma l)_j = (2j+1)\frac{\pi}{2}, \quad j = 6, 7, 8, \ldots \tag{323}$$

Recalling Eq. (272), we calculate the corresponding natural frequencies as

$$\omega_j = (\gamma l)_j^2 \sqrt{\frac{EI}{ml^4}}, \quad f_j = \frac{1}{2\pi}(\gamma l)_j^2 \sqrt{\frac{EI}{ml^4}} \quad j = 1, 2, 3, \ldots \tag{324}$$

where the values $(\gamma l)_j$ are given in Table 3.5. Also given in Table 3.5 are the corresponding values for ω_j and f_j.

Solution of the algebraic system (320) and substitution into the general solution (277) yields the *mode shapes for flexural vibration of a free-free beam*, i.e.,

$$W_j(x) = A_j\left[(\cosh \gamma_j x + \cos \gamma_j x) - \beta_j(\sinh \gamma x + \sin \gamma x)\right], \quad j = 1, 2, 3, \ldots \tag{325}$$

where β_j is a modal parameter given by the solution of either the first or the second of Eq. (320), i.e.,

$$\beta_j = -\left(\frac{C_1}{C_2}\right)_j = \frac{\cosh \gamma_j l - \cos \gamma_j l}{\sinh \gamma_j l - \sin \gamma_j l} = \frac{\sinh \gamma_j l + \sin \gamma_j l}{\cosh \gamma_j l - \cos \gamma_j l} \tag{326}$$

TABLE 3.5 Eigenvalues, γl, natural frequencies, ω, and the modal parameter, β, for the flexural vibration of a free-free beam

j	$(\gamma l)_j$	ω_j	f_j	β_j
1	4.73004074	$22.373287\sqrt{\frac{EI}{ml^4}}$	$3.5608190\sqrt{\frac{EI}{ml^4}}$	0.982502215
2	7.85320462	$61.673823\sqrt{\frac{EI}{ml^4}}$	$9.8155346\sqrt{\frac{EI}{ml^4}}$	1.000777312
3	10.9956078	$120.903391\sqrt{\frac{EI}{ml^4}}$	$19.2423723\sqrt{\frac{EI}{ml^4}}$	0.999966450
4	14.1371655	$199.859448\sqrt{\frac{EI}{ml^4}}$	$19.2421376\sqrt{\frac{EI}{ml^4}}$	1.000001450
5	17.2787597	$298.55554\sqrt{\frac{EI}{ml^4}}$	$31.808619\sqrt{\frac{EI}{ml^4}}$	0.999999937
6,7,8, ...	$(2j+1)\frac{\pi}{2}$	$\left[(2j+1)\frac{\pi}{2}\right]^2\sqrt{\frac{EI}{ml^4}}$	$[(2j+1)]^2\frac{\pi}{8}\sqrt{\frac{EI}{ml^4}}$	1.00000000

The scale factor A_j in Eq. (325) is to be determined from the normalization condition. For free-free boundary conditions, a closed-form solution for the normalization factor is not readily available, unlike the previous situation of pin-pin boundary conditions where a closed-form solution was readily obtained. For free-free boundary conditions, the scale factor should be determined by imposing the normalization condition with A_j factored out, and then solving the resulting equation to determine the actual value of A_j. For example, if normalization with respect to mass is sought, Eq. (306) yields

$$A_j = \frac{1}{\sqrt{\int_0^l m W_j^2(x)\,\mathrm{d}x}} \tag{327}$$

Cantilever beam

A cantilever beam is fixed at one end and free at the other end; it has *fixed-free boundary conditions* described as

$$\begin{aligned} w(0,t) &= 0 \quad w'(0,t) = 0 \\ M(l,t) &= 0 \quad V(l,t) = 0 \end{aligned} \tag{328}$$

Using Eqs. (264), (265), and (269) into Eq. (328) yields the boundary conditions in terms of displacement and its derivatives, i.e.,

$$\begin{aligned} \hat{w}(0) &= 0 \quad \hat{w}'(0) = 0 \\ \hat{w}''(l) &= 0 \quad \hat{w}'''(l) = 0 \end{aligned} \tag{329}$$

Substitution of the general solution (277) into the boundary conditions (329) yields

$$\begin{aligned} C_1 \sin\gamma x + C_2 \cos\gamma x + C_3 \sinh\gamma x + C_4 \cosh\gamma x|_{x=0} &= 0 \\ \gamma\,[C_1 \cos\gamma x - C_2 \sin\gamma x + C_3 \cosh\gamma x + C_4 \sinh\gamma x]|_{x=0} &= 0 \\ \gamma^2\,[-C_1 \sin\gamma x - C_2 \cos\gamma x + C_3 \sinh\gamma x + C_4 \cosh\gamma x]|_{x=l} &= 0 \\ \gamma^3\,[-C_1 \cos\gamma x + C_2 \sin\gamma x + C_3 \cosh\gamma x + C_4 \sinh\gamma x]|_{x=l} &= 0 \end{aligned} \tag{330}$$

Upon simplification, we get

$$\begin{aligned} C_2 + C_4 &= 0 \\ C_1 + C_3 &= 0 \\ -C_1 \sin\gamma l - C_2 \cos\gamma l + C_3 \sinh\gamma l + C_4 \cosh\gamma l &= 0 \\ -C_1 \cos\gamma l + C_2 \sin\gamma l + C_3 \cosh\gamma l + C_4 \sinh\gamma l &= 0 \end{aligned} \tag{331}$$

Further simplification yields

$$\begin{aligned} C_4 &= -C_2 \\ C_3 &= -C_1 \\ C_1(\sin\gamma l + \sinh\gamma l) + C_2(\cos\gamma l + \cosh\gamma l) &= 0 \\ C_1(\cos\gamma l + \cosh\gamma l) + C_2(-\sin\gamma l + \sinh\gamma l) &= 0 \end{aligned} \tag{332}$$

The last two lines in Eq. (332) form a homogenous linear algebraic system in C_1 and C_2

$$\begin{aligned}(\sin \gamma l + \sinh \gamma l)C_1 + (\cos \gamma l + \cosh \gamma l)C_2 = 0 \\ (\cos \gamma l + \cosh \gamma l)C_1 + (-\sin \gamma l + \sinh \gamma l)C_2 = 0\end{aligned} \tag{333}$$

The homogenous algebraic system (333) accepts nontrivial solutions only if its determinant has value zero, i.e.,

$$\begin{vmatrix} \sin \gamma l + \sinh \gamma l & \cos \gamma l + \cosh \gamma l \\ \cos \gamma l + \cosh \gamma & -\sin \gamma l + \sinh \gamma l \end{vmatrix} = 0 \tag{334}$$

Expansion and simplification of Eq. (334) yields the *characteristic equation*

$$\cos \gamma l \cosh \gamma l + 1 = 0 \tag{335}$$

Equation (335) is a transcendental equation that yields the eigenvalues γl corresponding to the free flexural vibration of a cantilever beam. Numerical values of the first five eigenvalues γl are shown in Table 3.6. As γl becomes large, the numerical values approach a rational sequence. From the sixth eigenvalue onwards, they can be approximated to reasonable accuracy by

$$(\gamma l)_j = (2j-1)\frac{\pi}{2}, \quad j = 6, 7, 8, \ldots \tag{336}$$

Recalling Eq. (272), we calculate the corresponding natural frequencies as

$$\omega_j = (\gamma l)_j^2 \sqrt{\frac{EI}{ml^4}}, \quad f_j = \frac{1}{2\pi}(\gamma l)_j^2 \sqrt{\frac{EI}{ml^4}} \quad j = 1, 2, 3, \ldots \tag{337}$$

where the values $(\gamma l)_j$ are given in Table 3.6. Also given in Table 3.6 are the corresponding values for ω_j and f_j.

TABLE 3.6 Eigenvalues, γl, natural frequencies, ω, and the modal parameter, β, for the flexural vibration of a cantilever beam

j	$(\gamma l)_j$	ω_j, rad/s	f_j, Hz	β_j
1	1.87510407	$3.51601527\sqrt{\frac{EI}{ml^4}}$	$0.55959121\sqrt{\frac{EI}{ml^4}}$	0.73409551
2	4.6940911	$22.0344916\sqrt{\frac{EI}{ml^4}}$	$3.5068983\sqrt{\frac{EI}{ml^4}}$	1.01846732
3	7.8547574	$61.697214\sqrt{\frac{EI}{ml^4}}$	$9.8194166\sqrt{\frac{EI}{ml^4}}$	0.99922450
4	10.9955407	$120.901916\sqrt{\frac{EI}{ml^4}}$	$19.2421376\sqrt{\frac{EI}{ml^4}}$	1.00003355
5	14.1371684	$199.859530\sqrt{\frac{EI}{ml^4}}$	$31.808632\sqrt{\frac{EI}{ml^4}}$	0.99999855
6,7,8, ...	$(2j-1)\frac{\pi}{2}$	$\left[(2j-1)\frac{\pi}{2}\right]^2\sqrt{\frac{EI}{ml^4}}$	$[(2j-1)]^2\frac{\pi}{8}\sqrt{\frac{EI}{ml^4}}$	1.00000000

Solution of the algebraic system (333) and substitution into the general solution (277) yields the *mode shapes for flexural vibration of a cantilever beam*, i.e.,

$$W_j(x) = A_j\left[(\cosh\gamma_j x - \cos\gamma_j x) - \beta_j\,(\sinh\gamma x - \sin\gamma x)\right], \quad j = 1, 2, 3, \ldots \tag{338}$$

where β_j is a modal parameter given by the solution of either the first or the second line in Eq. (333), i.e.,

$$\beta_j = -\left(\frac{C_1}{C_2}\right)_j = \frac{\cosh\gamma_j l + \cos\gamma_j l}{\sinh\gamma_j l + \sin\gamma_j l} = \frac{\sinh\gamma_j l - \sin\gamma_j l}{\cosh\gamma_j l + \cos\gamma_j l} \tag{339}$$

3.3.2.3 Forced Flexural Vibration of a Beam

Assume and uniform beam undergoing flexural vibration under the excitation of a time-dependant distributed excitation $f(x, t)$, as shown in Fig. 3.16. The units of $f(x, t)$ are force per length. Free-body analysis of an infinitesimal element similar to that presented in Fig. 3.15b reveals that, due to the distributed excitation $f(x, t)$, Eq. (255) becomes

$$V'(x, t) + f(x, t) = m\ddot{w}(x, t) \tag{340}$$

Pursuing the argument outlined in Eqs. (257) through (265) yields the *equation of motion for forced flexural vibration* of a beam, i.e.,

$$m\ddot{w}(x, t) + EI\ w''''(x, t) = f(x, t) \tag{341}$$

Without loss of generality, we assume the external excitation to be harmonic in the form

$$f(x, t) = \hat{f}(x)e^{i\omega t} \tag{342}$$

and obtain

$$m\ddot{w}(x, t) + EI\ w''''(x, t) = \hat{f}(x)e^{i\omega t} \tag{343}$$

Modal expansion theorem

Assume the modal expansion

$$w(x, t) = \sum_{j=1}^{\infty} C_j W_j(x) e^{i\omega t}, \tag{344}$$

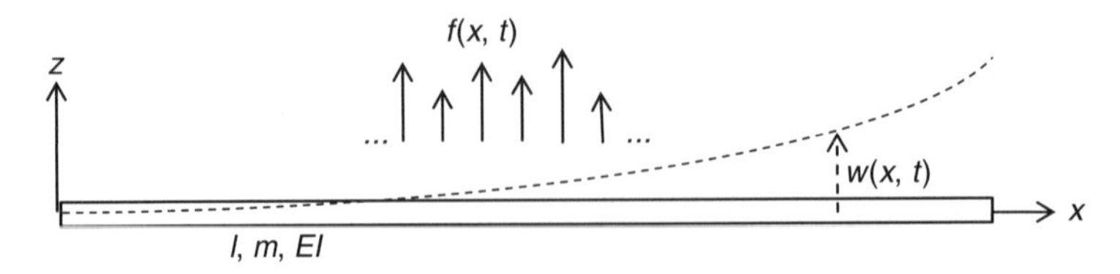

FIGURE 3.16 Uniform beam undergoing flexural vibrations under time-dependant distributed excitation $f(x, t)$.

where $W_j(x)$ are natural modes satisfying the free-vibration equation of motion (288) and the orthonormality conditions (298) and (301), whereas C_j are the *modal participation factors*. Equation (344) is harmonic in the excitation frequency, which implies that we are only concerned with the steady-state response. Substitution of Eq. (344) into Eq. (343) and division by $e^{i\omega t}$ yields

$$-m\omega^2 \sum_{p=1}^{\infty} C_p W_p(x) + EI \sum_{p=1}^{\infty} C_p W_p''''(x) = \hat{f}(x) \tag{345}$$

where the index p was used instead of j for convenience. Multiplication of Eq. (345) by $W_q(x)$ and integration over the length of the beam yields

$$\begin{aligned} &-\omega^2 \sum_{p=1}^{\infty} C_p \int_0^0 m W_q(x) W_p(x) \mathrm{d}x + \sum_{p=1}^{\infty} C_p \int_0^l EI\, W_p''''(x) W_q(x) \mathrm{d}x \\ &= \int_0^l \hat{f}(x) W_q(x) \mathrm{d}x, \quad q = 1, 2, 3, \ldots \end{aligned} \tag{346}$$

Using the orthogonality conditions (298), (301), and the definition of modal mass and stiffness given by Eqs. (311), (314), we can recast Eq. (346) in the form of a set of decoupled linear algebraic equations, i.e.,

$$C_j \left(-m_j \omega^2 + k_j\right) = f_j, \quad j = 1, 2, 3, \ldots \tag{347}$$

where f_j is the *modal excitation* given by

$$f_j = \int_0^l \hat{f}(x) W_j(x) \mathrm{d}x, \quad j = 1, 2, 3, \ldots \tag{348}$$

If the modes are orthonormal, then substituting the orthonormality conditions (309) and (310) into Eq. (346) yields the simpler form

$$C_j \left(-\omega^2 + \omega_j^2\right) = f_j, \quad j = 1, 2, 3, \ldots \tag{349}$$

Response by modal analysis

Solution of Eq. (349) yields the modal participation factors in the form

$$C_j = \frac{f_j}{-\omega^2 + \omega_j^2}, \quad j = 1, 2, 3, \ldots \tag{350}$$

Equation (350) corresponds to the amplitude of Eq. (56) derived for the 1-dof systems. Substitution of Eq. (350) into Eq. (344) yields

$$w(x, t) = \sum_{j=1}^{\infty} \frac{f_j}{-\omega^2 + \omega_j^2} W_j(x)\, e^{i\omega t} \tag{351}$$

So far, the vibration has been considered undamped. However, real phenomena are always associated with some damping ζ_j. In that case, the modal participation factors take the form

$$C_j = \frac{f_j}{-\omega^2 + 2i\zeta_j \omega_j \omega + \omega_j^2}, \quad j = 1, 2, 3, \ldots \tag{352}$$

The modal response defined by Eq. (352) resembles the response of the 1-dof vibrating system, Eq. (79). In fact, as the excitation frequency passes through a natural frequency, ω_j, the continuous structure will *resonate* and will vibrate in the natural mode $W_j(x)$. Around a natural frequency, the continuous system behaves more or less like a 1-dof system vibrating in the corresponding natural mode. If the excitation frequency covers a large frequency span, several resonances can be encountered corresponding to several natural frequencies and normal modes.

Substitution of Eq. (237) into Eq. (229) yields the frequency response of the damped flexural vibration system in the form

$$w(x,t) = \sum_{j=1}^{\infty} \frac{f_j}{-\omega^2 + 2i\zeta_j\omega_j\omega + \omega_j^2} W_j(x)\, e^{i\omega t} \tag{353}$$

Equation (238) represents a superposition of a number of terms, each term corresponding to a natural frequency and normal mode of vibration. Equation (238) allows us to determine the response of the continuous structure to harmonic excitation of variable frequency. This leads to the *frequency response function* (FRF) concept. The FRF could be calculated by letting the excitation function take the form of *unit excitation*. Several unit excitation functions could be envisages, depending on the choice of spatial variation. As the excitation frequency sweeps through natural frequency values, ω_j, the corresponding term in the series becomes very large. As the structure passes through a structural resonance, ω_j, the response displays a resonance peak. Over a frequency interval, the response can display several peaks, corresponding to several resonances (Fig. 3.14).

3.3.3 TORSIONAL VIBRATION OF A SHAFT

3.3.3.1 Free Torsional Vibration of a Shaft

Consider a uniform shaft of length l, mass torsional inertia per unit length ρI_p, torsional stiffness GJ, undergoing torsional vibration of displacement $\phi(x,t)$ as shown in Fig. 3.17a. An infinitesimal shaft element of length dx is subjected to the action of torsional moments (twisting moments a.k.a. torques), $T(x,t)$ and $T(x,t) + T'(x,t)\mathrm{d}x$ as shown in Fig. 3.17b. Free-body analysis of the infinitesimal element yields

$$T'(x,t) = \rho I_p \ddot{\phi}(x,t) \tag{354}$$

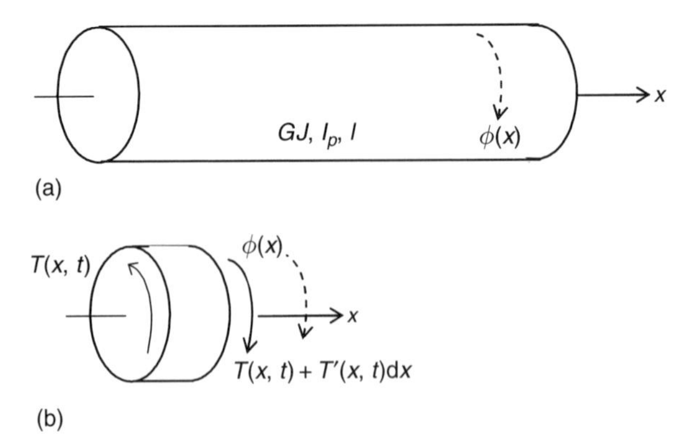

FIGURE 3.17 Uniform shaft undergoing torsional vibration: (a) general layout; (b) free-body diagram of an infinitesimal shaft element.

Simple torsion analysis relates the torsional stiffness to the twist, which is the space-derivative of the torsional displacement, i.e.,

$$T(x, t) = GJ\phi'(x, t) \tag{355}$$

Substitution of Eq. (355) into Eq. (354) yields the *equation of motion for torsional vibration*, i.e.,

$$GJ\phi''(x, t) = \rho I_p \ddot{\phi}(x, t) \tag{356}$$

Upon division by ρI_p, we get

$$c^2 \phi'' = \ddot{\phi} \tag{357}$$

where the constant c^2 is given by

$$c^2 = \frac{GJ}{\rho I_p} \quad \text{or} \quad c = \sqrt{\frac{GJ}{\rho I_p}} \tag{358}$$

Note that Eq. (357) is the same as the Eq. (157) derived for analysis of axial vibration. Hence, we will be able to simply map the axial vibration results into torsional vibration results through simple substitution of the appropriate terms, thus avoiding repetitious derivations.

Natural frequencies and mode shapes modes for fixed-fixed boundary conditions

Following the axial vibration analysis, we assume

$$\phi(x, t) = \hat{\phi}(x)\, e^{i\omega t} \tag{359}$$

and hence

$$\hat{\phi}(x) = C_1 \sin \gamma x + C_2 \cos \gamma x \tag{360}$$

where

$$\gamma = \frac{\omega}{c} \tag{361}$$

For **fixed-fixed boundary conditions** of the form

$$\begin{aligned} \phi(0, t) &= 0 \\ \phi(l, t) &= 0 \end{aligned} \tag{362}$$

we obtain the eigenvalues and natural frequencies

$$\gamma_j = j\frac{\pi}{l}, \quad \omega_j = j\frac{\pi}{l}\sqrt{\frac{GJ}{\rho I_p}}, \quad j = 1, 2, 3, \ldots \tag{363}$$

and the orthonormal mode shapes

$$\Phi_j(x) = \sqrt{\frac{2}{\rho I_p l}} \sin \gamma_j l, \quad j = 1, 2, 3, \ldots \tag{364}$$

The mode shapes of Eq. (364) are orthonormal with respect to mass torsional inertia.

Other boundary conditions

Consider *free-free* boundary conditions of the form

$$
\begin{aligned}
T(0, t) &= 0 \\
T(l, t) &= 0
\end{aligned}
\tag{365}
$$

Substitution of Eq. (355) into Eq. (365) yields the free-free boundary conditions in terms of displacements, i.e.,

$$
\begin{aligned}
\hat{\phi}'(0) &= 0 \\
\hat{\phi}'(l) &= 0
\end{aligned}
\tag{366}
$$

For these boundary conditions, the eigenvalues and natural frequencies are given by

$$
\gamma_j = j\frac{\pi}{l}, \quad \omega_j = j\frac{\pi}{l}\sqrt{\frac{GJ}{\rho I_p}}, \quad j = 1, 2, 3, \ldots \tag{367}
$$

The corresponding mode shapes are given by

$$
\Phi_j(x) = \sqrt{\frac{2}{\rho I_p l}} \cos \gamma_j l, \quad j = 1, 2, 3, \ldots \tag{368}
$$

Plots of the mode shapes and the corresponding natural frequencies for the vibration of a free-free elastic shaft are the same as those given for the axial vibration of an elastic bar in Table 3.3.

Consider *fixed-free* boundary conditions of the form

$$
\begin{aligned}
\phi(0, t) &= 0 \\
T(l, t) &= 0
\end{aligned}
\tag{369}
$$

Substitution of Eq. (355) into Eq. (369) yields the fixed-free boundary conditions in terms of displacements, i.e.,

$$
\begin{aligned}
\hat{\phi}(0) &= 0 \\
\hat{\phi}'(l) &= 0
\end{aligned}
\tag{370}
$$

For these boundary conditions, the eigenvalues and natural frequencies are given by

$$
\gamma_j = (2j-1)\frac{\pi}{2l}, \quad \omega_j = (2j-1)\frac{\pi}{2}\frac{1}{l}\sqrt{\frac{GJ}{\rho I_p}}, \quad j = 1, 2, 3, \ldots \tag{371}
$$

The corresponding mode shapes are given by

$$
\Phi_j(x) = \sqrt{\frac{2}{\rho I_p l}} \sin \gamma_j l, \quad j = 1, 2, 3, \ldots \tag{372}
$$

Plots of the mode shapes and the corresponding natural frequencies for a fixed-free elastic shaft are the same as those given for the axial vibration of an elastic bar in Table 3.4.

3.3.3.2 Forced Torsional Vibration of a Shaft

Assume an uniform shaft under externally applied time-dependent torsional moment per unit length $f(x, t)$. Considering the combined effect of internal stress resultant, $T(x, t)$, and external excitation, $f(x, t)$, we perform the free-body analysis of an infinitesimal shaft element $\mathrm{d}x$ to get

$$T'(x, t)\mathrm{d}x + f(x, t) = \rho I_p \ddot{\phi}(x, t) \tag{373}$$

Relating, as before, the stress resultant $T(x, t)$ to the displacement $\phi(x, t)$ and assuming, without loss of generality, the external excitation to be harmonic in the form

$$f(x, t) = \hat{f}(x)e^{i\omega t} \tag{374}$$

we obtain

$$\rho I_p \ddot{\phi}(x, t) - GJ\phi''(x, t) = \hat{f}(x)e^{i\omega t} \tag{375}$$

Modal expansion theorem

Assume modal expansion

$$\phi(x, t) = \sum_{j=1}^{\infty} C_j \Phi_j(x)e^{i\omega t}, \tag{376}$$

where $\Phi_j(x)$ are natural mode and C_j are the modal participation factors. Similar to the analysis of axial vibrations, substitution of Eq. (376) into Eq. (375) and application of the orthogonality conditions yields the modal equations

$$C_j\left(-m_j\omega_j^2 + k_j\right) = f_j, \quad j = 1, 2, 3, \ldots \tag{377}$$

where f_j is the modal excitation given by

$$f_j = \int_0^l \hat{f}(x)\Phi_j(x)\mathrm{d}x, \quad j = 1, 2, 3, \ldots \tag{378}$$

If the modes are normalized with respect to mass moment of inertia, then Eq. (377) takes the simpler form

$$C_j\left(-\omega^2 + \omega_j^2\right) = f_j, \quad j = 1, 2, 3, \ldots \tag{379}$$

Response by modal analysis

Solution of Eq. (379) yields the modal participation factors in the form

$$C_j = \frac{f_j}{-\omega^2 + \omega_j^2}, \quad j = 1, 2, 3, \ldots \tag{380}$$

Substitution of Eq. (380) into Eq. (376) gives

$$u(x, t) = \sum_{j=1}^{\infty} \frac{f_j}{-\omega^2 + \omega_j^2}\Phi_j(x)e^{i\omega t} \tag{381}$$

For damped vibrations, we add modal damping ζ_j to get

$$u(x,t) = \sum_{j=1}^{\infty} \frac{f_j}{-\omega^2 + 2i\zeta_j\omega_j\omega + \omega_j^2} \Phi_j(x)\, e^{i\omega t} \tag{382}$$

The general discussion of frequency sweeps and structural resonances performed for forced axial vibration also applies here for torsional vibration and will not be repeated.

3.4 SUMMARY AND CONCLUSIONS

This chapter has offered a brief introduction to vibration theory. This introduction is necessary because many of the SHM methods to be discussed in later chapters will utilize concepts and formulae from vibration theory.

The chapter started with the vibration analysis of a 1-dof system of a *particle*. This simple system was used as a springboard for the analysis of more complicated system later in the chapter. The 1-dof particle vibration was used to introduce fundamental basic concepts such as the differential equation of motion, harmonic solutions, free vs. forced vibrations and damped vs. undamped vibrations. Energy methods approach to vibration analysis were also discussed.

The second part of the chapter has covered the vibration of continuous systems. PDE in space and time govern this type of vibrations. Assuming harmonic behavior in time, the equation of motion was reduced to an ODE in the space domain. This boundary value problem was solved to yield the eigenvalues and eigenmodes, and the associate *natural frequencies* and *mode shapes*. The axial vibration of bars, flexural vibration of beams, and torsional vibration of shafts was considered. In each case, the study of free vibrations was followed by the study of forced vibrations.

3.5 PROBLEMS AND EXERCISES

1. Prove that $u(t) = A\cos\omega_n t + B\sin\omega_n t$ can be also expressed as $u(t) = C\cos(\omega_n t + \psi)$, and find the relationship between A, B, C, and ψ
2. Prove that $m\ddot{u}(t) + c\dot{u}(t) + ku(t) = 0$ can be also expressed as $\ddot{u}(t) + 2\zeta\omega_n\dot{u}(t) + \omega_n^2 u(t) = 0$ and derive the relations between the constants in the two equations
3. Prove that $u(t) = C_1 e^{(-\zeta\omega_n + i\omega_d)t} + C_2 e^{(-\zeta\omega_n - i\omega_d)t}$ can be rewritten as $u(t) = Ce^{-\zeta t}\cos(\omega_d t + \psi)$ and derive the relations between the constants in the two equations
4. Prove that when damping equals critical damping ($\zeta = 1$), the solution of $\ddot{u}(t) + 2\zeta\omega_n\dot{u}(t) + \omega_n^2 u(t) = 0$ is $u(t) = (C_1 + C_2 t)e^{-\omega_n t}$
5. Prove that the particular solution of $\ddot{u}(t) + \omega_n^2 u(t) = \hat{f}\cos(\omega t)$ is $u_p(t) = \frac{1}{-\omega^2 + \omega_n^2}\hat{f}\cos(\omega t)$
6. Prove that using Eq. (99) in conjunction with Eqs. (15), (31), (90), (91), and (93) yields the *response amplitude at the quadrature point* as $|\hat{u}_{90}| = \hat{F}/c\omega_n$
7. Prove that, for lightly damped systems, the bandwidth of the frequency response function $H(p) = \frac{1}{-p^2 + i2\zeta p + 1}$ takes the simple expression $\omega_2 - \omega_1 \cong 2\zeta\omega_n$.
8. Prove that the power at resonance of a lightly-damped 1-dof system is given by $P_{max} = \frac{1}{2}\frac{\hat{F}^2}{c}$
9. Find the first, second, and third natural frequencies of in-plane axial vibration of a steel beam of thickness $h_1 = 2.6\,\text{mm}$, width $b_1 = 8\,\text{mm}$, length $l = 100\,\text{mm}$, modulus

$E = 200\,\text{GPa}$, and density $\rho = 7.750\,\text{g/cm}^3$. The beam is in free-free boundary conditions. Then consider double the thickness ($h_2 = 5.2\,\text{mm}$), wider width ($b_2 = 19.6\,\text{mm}$), and then both. Recalculate the three frequencies for these other combinations of thickness and width. Discuss your results.

10. Find all the natural frequencies in the interval 1 kHz to 30 kHz of in-plane axial vibration of a steel beam of thickness $h_1 = 2.6\,\text{mm}$, width $b_1 = 8\,\text{mm}$, length $l = 100\,\text{mm}$, modulus $E = 200\,\text{GPa}$, and density $\rho = 7.750\,\text{g/cm}^3$. The beam is in free-free boundary conditions. Then consider double the thickness ($h_2 = 5.2\,\text{mm}$), wider width ($b_2 = 19.6\,\text{mm}$), and then both. Recalculate the frequencies for these other combinations of thickness and width. Discuss your results
11. Find the first, second, and third natural frequencies of out-of-plane flexural vibration of a steel beam of thickness $h_1 = 2.6\,\text{mm}$, width $b_1 = 8\,\text{mm}$, length $l = 100\,\text{mm}$, modulus $E = 200\,\text{GPa}$, and density $\rho = 7.750\,\text{g/cm}^3$. The beam is in free-free boundary conditions. Then consider double the thickness ($h_2 = 5.2\,\text{mm}$), wider width ($b_2 = 19.6\,\text{mm}$), and then both. Recalculate the three frequencies for these other combinations of thickness and width. Discuss your results.
12. Find all the natural frequencies in the interval 1 kHz to 30 kHz of out-of-plane flexural vibration of a steel beam of thickness $h_1 = 2.6\,\text{mm}$, width $b_1 = 8\,\text{mm}$, length $l = 100\,\text{mm}$, modulus $E = 200\,\text{GPa}$, and density $\rho = 7.750\,\text{g/cm}^3$. The beam is in free-free boundary conditions. Then consider double the thickness ($h_2 = 5.2\,\text{mm}$), wider width ($b_2 = 19.6\,\text{mm}$), and then both. Recalculate the three frequencies for these other combinations of thickness and width. Discuss your results.

4

VIBRATION OF PLATES

4.1 ELASTICITY EQUATIONS FOR PLATE VIBRATION

In this chapter, we will analyze plate vibration. After reviewing the general plate equations, we will consider two situations separately: (a) axial vibration of plates; and (b) flexural vibration of plates.

Because the plate surface is free, the z-direction stress is assumed to be zero ($\sigma_{zz} = 0$). Also zero are assumed to be the surface shear stresses $\sigma_{yz} = 0$ and $\sigma_{zx} = 0$. Hence, the 3-D elasticity relations given in Appendix B reduce to

$$\begin{aligned} \varepsilon_{xx} &= \frac{1}{E}\sigma_{xx} + \frac{-\nu}{E}\sigma_{yy} \\ \varepsilon_{yy} &= \frac{-\nu}{E}\sigma_{xx} + \frac{1}{E}\sigma_{yy} \qquad \text{and} \qquad \varepsilon_{xy} = \frac{1}{2G}\sigma_{xy} \\ \varepsilon_{zz} &= \frac{-\nu}{E}\sigma_{xx} + \frac{-\nu}{E}\sigma_{yy} \end{aligned} \tag{1}$$

In our analysis, we are only interested in the strains ε_{xx}, ε_{yy}, ε_{xy}. Solution of Eq. (1) yields

$$\begin{aligned} \sigma_{xx} &= \frac{E}{1-\nu^2}\left(\varepsilon_{xx} + \nu\varepsilon_{yy}\right) \\ & \qquad\qquad\qquad \sigma_{xy} = 2G\varepsilon_{xy} \\ \sigma_{yy} &= \frac{E}{1-\nu^2}\left(\nu\varepsilon_{xx} + \varepsilon_{yy}\right) \end{aligned} \tag{2}$$

4.2 AXIAL VIBRATION OF RECTANGULAR PLATES

Axial vibrations are related to extension/compression motion in the plate, i.e., the motion is *in-plane polarized.*

4.2.1 GENERAL EQUATIONS FOR AXIAL VIBRATION OF RECTANGULAR PLATES

Assume in-plane displacements $u(x, y, t)$ and $v(x, y, t)$ which are uniform across the plate thickness. The strains of interest are

$$\varepsilon_{xx} = \frac{\partial u}{\partial x} \quad \varepsilon_{yy} = \frac{\partial v}{\partial y} \quad \varepsilon_{yx} = \frac{1}{2}\left(\frac{\partial u}{\partial y} + \frac{\partial v}{\partial x}\right) \tag{3}$$

The strains are constant across the plate thickness. Substitution of Eq. (3) into Eq. (2) yields

$$\sigma_{xx} = \frac{E}{1-\nu^2}\left(\frac{\partial u}{\partial x} + \nu\frac{\partial v}{\partial y}\right) \qquad \sigma_{xy} = G\left(\frac{\partial u}{\partial y} + \frac{\partial v}{\partial x}\right) \qquad (4)$$
$$\sigma_{yy} = \frac{E}{1-\nu^2}\left(\nu\frac{\partial u}{\partial x} + \frac{\partial v}{\partial y}\right)$$

Note that the stresses are also constant across the plate thickness. Integration of stresses across the plate thickness gives the stress resultants (forces per unit width) N_x, N_y, and N_{xy} shown in Fig. 4.1, i.e.,

$$N_x = \int_{-h/2}^{h/2} \sigma_{xx}\,\mathrm{d}z = \frac{Eh}{1-\nu^2}\left(\frac{\partial u}{\partial x} + \nu\frac{\partial v}{\partial y}\right)$$
$$N_y = \int_{-h/2}^{h/2} \sigma_{yy}\,\mathrm{d}z = \frac{Eh}{1-\nu^2}\left(\nu\frac{\partial u}{\partial x} + \frac{\partial v}{\partial y}\right) \qquad (5)$$
$$N_{xy} = N_{yx} = \int_{-h/2}^{h/2} \sigma_{xy}\,\mathrm{d}z = Gh\left(\frac{\partial u}{\partial y} + \frac{\partial v}{\partial x}\right)$$

Newton second law applied to the infinitesimal element of Fig. 4.1 yields

$$\frac{\partial N_x}{\partial x} + \frac{\partial N_{yx}}{\partial y} = \rho h\ddot{u}$$
$$\frac{\partial N_y}{\partial y} + \frac{\partial N_{xy}}{\partial x} = \rho h\ddot{v} \qquad (6)$$

Equation (6) corresponds to forces in the x and y directions. The moment equation is not needed as in-plane rotation inertia is ignored. Substitution of Eq. (5) into Eq. (6) yields a system of coupled second-order PDE in u and v depending on the space and time variables. This system has a general solution representing plate vibration simultaneously in the x and y directions. The solution of this coupled system of differential equations is not immediate; this case will not be treated here.

When certain conditions are imposed on the vibration pattern, the system (6) simplifies and yields simple uncoupled solutions, which we will discuss next.

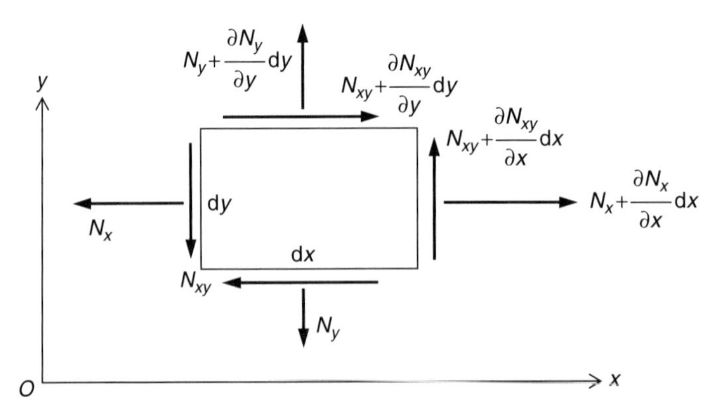

FIGURE 4.1 Infinitesimal plate element in Cartesian coordinates for the analysis of in-plane vibration.

4.2.2 STRAIGHT-CRESTED AXIAL VIBRATION OF RECTANGULAR PLATES

Let's consider a simplified case in which the particle motion is self-similar along any line parallel to the y-axis. If we view plate vibration as a system of standing waves in the plate, then this case can be considered as a system of standing *straight-crested axial waves* with the wave crest along the y-axis (Fig. 4.2). The problem is y-invariant and depends only on x, i.e.,

$$\begin{aligned} u(x, y, t) &\rightarrow u(x, t) \\ v(x, y, t) &\equiv 0 \end{aligned} \qquad \text{(straight-crested axial plate vibration)} \tag{7}$$

The particle motion in Eq. (7) was assumed parallel to the x-axis (longitudinally polarized). Because the problem is y-invariant, derivatives with respect to y are zero. Substitution of Eq. (7) into Eq. (5) and imposition of y-invariant conditions yields

$$\begin{aligned} N_x &= \frac{Eh}{1-\nu^2}\frac{\partial u}{\partial x} \\ N_y &= \frac{Eh}{1-\nu^2}\nu\frac{\partial u}{\partial x} \\ N_{xy} &= N_{yx} = 0 \end{aligned} \tag{8}$$

Substitution of Eq. (8) into the equation of motion (6) gives

$$\frac{\partial N_x}{\partial x} + \frac{\partial N_{yx}}{\partial y} = \frac{Eh}{1-\nu^2}\frac{\partial^2 u}{\partial x^2} = \rho h \ddot{u} \tag{9}$$

Denote with c_L is *longitudinal wave speed in the plate* given by

$$c_L^2 = \frac{1}{1-\nu^2}\frac{E}{\rho}, \qquad c_L = \sqrt{\frac{1}{1-\nu^2}\frac{E}{\rho}} \tag{10}$$

Hence, Eq. (9) yields the wave equation

$$c_L^2 \frac{\partial^2 u}{\partial x^2} = \ddot{u} \tag{11}$$

From here onward, the analysis duplicates the analysis of axial vibrations in a bar, with the only difference of using c_L, as defined by Eq. (10), instead of $c = \sqrt{E/\rho}$.

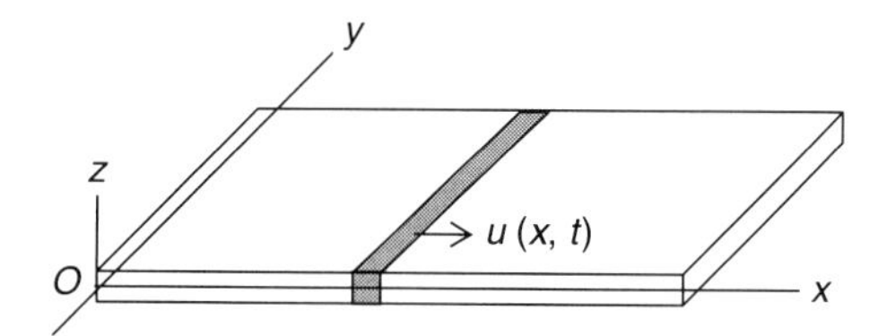

FIGURE 4.2 Straight-crested axial vibration in a plate.

4.3 AXIAL VIBRATION OF CIRCULAR PLATES

This section will consider the axial vibration of circular plates. To analyze them, we change from Cartesian coordinates (x, y) to polar coordinates (r, θ). Similarly the displacements (u, v) are replaced by displacements (u_r, u_θ).

4.3.1 GENERAL EQUATIONS FOR AXIAL VIBRATION OF CIRCULAR PLATES

Recall the stress displacement relations in polar coordinates given in Appendix B, i.e.,

$$\begin{aligned}\sigma_r &= \frac{E}{1-\nu^2}\left(\frac{\partial u_r}{\partial r} + \nu\frac{u_r}{r} + \nu\frac{1}{r}\frac{\partial u_\theta}{\partial \theta}\right)\\ \sigma_\theta &= \frac{E}{1-\nu^2}\left(\nu\frac{\partial u_r}{\partial r} + \frac{u_r}{r} + \frac{1}{r}\frac{\partial u_\theta}{\partial \theta}\right)\\ \sigma_{r\theta} &= G\left(\frac{1}{r}\frac{\partial u_r}{\partial \theta} + \frac{\partial u_\theta}{\partial r} - \frac{u_\theta}{r}\right)\end{aligned} \tag{12}$$

Note that stresses are constant across the plate thickness. Consider the infinitesimal plate element in polar coordinates shown in Fig. 4.3. Integration of stresses across the thickness gives the stress resultants N_r, N_θ, and $N_{r\theta}$ (forces per unit width) shown in Fig. 4.3, i.e.,

$$\begin{aligned}N_r &= \int_{-h/2}^{h/2} \sigma_r\, \mathrm{d}z = \frac{Eh}{1-\nu^2}\left(\frac{\partial u_r}{\partial r} + \nu\frac{u_r}{r} + \nu\frac{1}{r}\frac{\partial u_\theta}{\partial \theta}\right)\\ N_\theta &= \int_{-h/2}^{h/2} \sigma_\theta\, \mathrm{d}z = \frac{Eh}{1-\nu^2}\left(\nu\frac{\partial u_r}{\partial r} + \frac{u_r}{r} + \frac{1}{r}\frac{\partial u_\theta}{\partial \theta}\right)\\ N_{r\theta} &= \int_{-h/2}^{h/2} \sigma_{r\theta}\, \mathrm{d}z = Gh\left(\frac{1}{r}\frac{\partial u_r}{\partial \theta} + \frac{\partial u_\theta}{\partial r} - \frac{u_\theta}{r}\right)\end{aligned} \tag{13}$$

Newton law of motion applied to the infinitesimal element of Fig. 4.3 is applied by summing up the forces in the r and θ directions. Let's start with the r-direction equation of motion. We equal the sum of forces in the r-direction with the product between mass and acceleration in the same direction and obtain

$$r \text{ direction: } \left(N_r\, \mathrm{d}r\, \mathrm{d}\theta + r\, \mathrm{d}\theta \frac{\partial N_r}{\partial r}\mathrm{d}r\right) + \left(\frac{\partial N_{\theta r}}{\partial \theta}\mathrm{d}\theta\right)\mathrm{d}r - N_\theta\, \mathrm{d}r\, \mathrm{d}\theta = (\rho h\, r\, \mathrm{d}r\, \mathrm{d}\theta)\, \ddot{u}_r \tag{14}$$

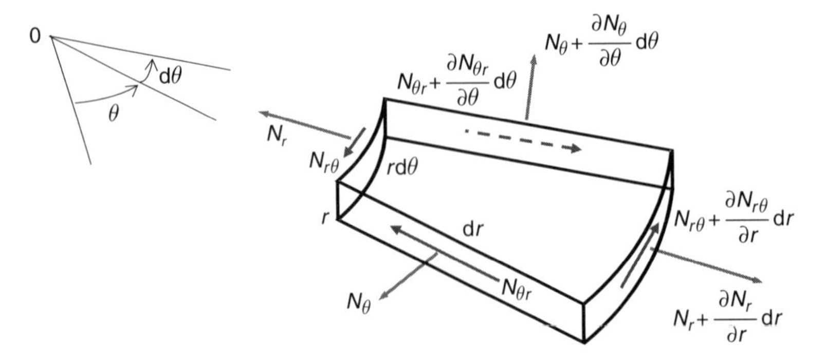

FIGURE 4.3 Infinitesimal plate element for the analysis of in-plane plate vibration in polar coordinates.

The terms in Eq. (14) deserve a little explanation. Let's start with the parenthesis: the first term in the parenthesis is due to the increase in arc length, whereas the second term in the parenthesis is due to increase in N_r. Now, let's consider the middle term: this term is due to the increase in $N_{\theta r}$. Finally, the last term is due to the inclination of the N_θ stress resultants which creates a projection along the r direction. Dividing through by $r\,\mathrm{d}r\,\mathrm{d}\theta$ and recalling $N_{\theta r} = N_{r\theta}$, Eq. (14) becomes

$$\frac{\partial N_r}{\partial r} + \frac{1}{r}\frac{\partial N_{\theta r}}{\partial \theta} + \frac{N_r - N_\theta}{r} = \rho h \ddot{u}_r \tag{15}$$

The θ-direction calculations follow a similar process. We equal the sum of forces in the θ-direction with the product between mass and acceleration in the same direction and obtain

$$\theta\text{-direction} \quad \left(N_{r\theta}\,\mathrm{d}r\,\mathrm{d}\theta + r\,\mathrm{d}\theta\frac{\partial N_{r\theta}}{\partial r}\mathrm{d}r\right) + \left(\frac{\partial N_\theta}{\partial \theta}\mathrm{d}\theta\right)\mathrm{d}r + N_{\theta r}\,\mathrm{d}r\,\mathrm{d}\theta = (\rho h r\,\mathrm{d}r\,\mathrm{d}\theta)\,\ddot{u}_\theta \tag{16}$$

The terms in Eq. (16) deserve a little explanation. Let's start with the parenthesis: the first term in the parenthesis is due to the increase in arc length, whereas the second term in the parenthesis is due to increase in $N_{r\theta}$. Now, let's consider the middle term: this term is due to the increase in N_θ. Finally, the last term is due to the inclination of the $N_{\theta r}$ stress resultants which creates a projection along the r-direction. Dividing through by $r\,\mathrm{d}r\,\mathrm{d}\theta$ and recalling $N_{\theta r} = N_{r\theta}$, Eq. (14) becomes

$$\frac{\partial N_{r\theta}}{\partial r} + \frac{1}{r}\frac{\partial N_\theta}{\partial \theta} + \frac{2}{r}N_{r\theta} = \rho h \ddot{u}_\theta \tag{17}$$

If we substitute the stress resultant expression (13) into Eqs. (15) and (17), we obtain a set of coupled second order partial differential equations in space and time involving the independent variables r, θ, t and the dependent variables u_r and u_θ. The solution of this coupled system of differential equations represents plate vibration simultaneously involving u_r and u_θ motions. The solution of this coupled system of differential equations is not immediate; this case will not be treated here. When certain conditions are imposed on the vibration pattern, the system of coupled partial differential equations simplifies and yields simple uncoupled solutions. One such case is the axisymmetric axial vibration of circular plates, which we will discuss next.

4.3.2 AXISYMMETRIC AXIAL VIBRATION OF CIRCULAR PLATES

In this section, we will restrict our attention to the axisymmetric in-plane (axial) vibration of circular plates. In this case, the θ dependence vanishes. The axisymmetric assumption implies that the motion is entirely radial, i.e., $u_\theta = 0$. In virtue of axial symmetry we also have $\partial u_r/\partial\theta = 0$. Eqs. (12) become

$$\begin{aligned}
\sigma_r &= \frac{E}{1-\nu^2}\left(\frac{\partial u_r}{\partial r} + \nu\frac{u_r}{r}\right) \\
\sigma_\theta &= \frac{E}{1-\nu^2}\left(\nu\frac{\partial u_r}{\partial r} + \frac{u_r}{r}\right) \\
\sigma_{r\theta} &= 0
\end{aligned} \tag{18}$$

Hence, Eq. (13) becomes

$$
\begin{aligned}
N_r &= \int_{-h/2}^{h/2} \sigma_r \, dz = \frac{Eh}{1-\nu^2}\left(\frac{\partial u_r}{\partial r} + \nu \frac{u_r}{r}\right) \\
N_\theta &= \int_{-h/2}^{h/2} \sigma_\theta \, dz = \frac{Eh}{1-\nu^2}\left(\nu \frac{\partial u_r}{\partial r} + \frac{u_r}{r}\right) \\
N_{r\theta} &= 0
\end{aligned}
\tag{19}
$$

4.3.2.1 Equation of Motion for Axisymmetric Vibration of Circular Plates

Under the axisymmetric assumption, Newton law of motion applied to the infinitesimal element of Fig. 4.3 yields

$$
\frac{\partial N_r}{\partial r} + \frac{N_r - N_\theta}{r} = \rho h \ddot{u}_r \tag{20}
$$

Substitution of Eqs. (19) into Eq. (20) yields, upon simplification,

$$
\frac{E}{1-\nu^2}\left(\frac{\partial^2 u_r}{\partial r^2} + \frac{1}{r}\frac{\partial u_r}{\partial r} - \frac{u_r}{r^2}\right) = \rho \ddot{u}_r \tag{21}
$$

Upon rearrangement, we get the wave equation in polar coordinates

$$
c_L^2\left(\frac{\partial^2 u_r}{\partial r^2} + \frac{1}{r}\frac{\partial u_r}{\partial r} - \frac{u_r}{r^2}\right) = \ddot{u}_r \tag{22}
$$

where c_L is the *longitudinal wave speed in a plate* previously defined by Eq. (10), i.e.,

$$
c_L^2 = \rho \frac{E}{(1-\nu^2)} \tag{23}
$$

Axisymmetric axial vibration of a circular plate can be understood in terms of standing circular-crested waves that propagate in a concentric circular pattern from the center of the plate and reflect at the plate circumference. Equation indicates that circular-crested axial waves in a plate propagate with the same wave speed c_L as the straight-crested axial waves. Assume the motion to be harmonic, i.e.,

$$
u_r(r, t) = \hat{u}(r) e^{i\omega t} \tag{24}
$$

Substitution of Eq. (24) into Eq. (22) and rearrangement yields

$$
c_L^2\left(\frac{\partial^2 \hat{u}_r}{\partial r^2} + \frac{1}{r}\frac{\partial \hat{u}_r}{\partial r} - \frac{\hat{u}_r}{r^2}\right) + \omega^2 \hat{u}_r = 0 \tag{25}
$$

Division by c_L^2, multiplication by r, and the use of the wavenumber definition $\gamma = \omega / c_L$ gives

$$
\left(\frac{\partial^2 \hat{u}_r}{\partial r^2} + \frac{1}{r}\frac{\partial \hat{u}_r}{\partial r} - \frac{\hat{u}_r}{r^2}\right) + \gamma^2 \hat{u}_r = 0 \tag{26}
$$

Perform the change of variable

$$x = \gamma r \tag{27}$$

Substitution of Eq. (27) into Eq. (26) yields the Bessel equation

$$x^2 \frac{\partial^2 \hat{u}}{\partial x^2} + x \frac{\partial \hat{u}}{\partial x} + (x^2 - 1)\hat{u} = 0 \tag{28}$$

where $x = \gamma r$. The solution of Eq. (28) is

$$\hat{u} = AJ_1(x) + BY_1(x) \tag{29}$$

where $J_1(x)$ is the Bessel function of the first kind and order 1, whereas $Y_1(x)$ is the Bessel function of the second kind and order 1. The arbitrary constants A and B are to be determined from boundary conditions. Substituting $x = \gamma r$ into Eq. (29) yields

$$\hat{u}(r) = AJ_1(\gamma r) + BY_1(\gamma r) \tag{30}$$

However, the Bessel function of the second kind $Y_1(\gamma r)$ has infinite value at $r = 0$ and has to be discarded (unless the plate has a hole around $r = 0$, which is not the case considered here). Eq. (30) becomes

$$\hat{u}_r(r) = AJ_1(\gamma r) \tag{31}$$

Hence, the *general solution for the axisymmetric vibration of circular plates* is given by

$$u_r(r, t) = AJ_1(\gamma r)e^{i\omega t} \tag{32}$$

The constant A, the frequency ω, and the wavenumber γ are determined from the boundary conditions.

4.3.2.2 Frequencies and Mode shapes of Free Circular Plates in Axisymmetric Axial Vibration

Consider a free circular plate of radius a undergoing axisymmetric axial vibration (Fig. 4.4). The boundary conditions correspond to tractions-free conditions at $r = a$, i.e.,

$$N_r(a) = 0 \tag{33}$$

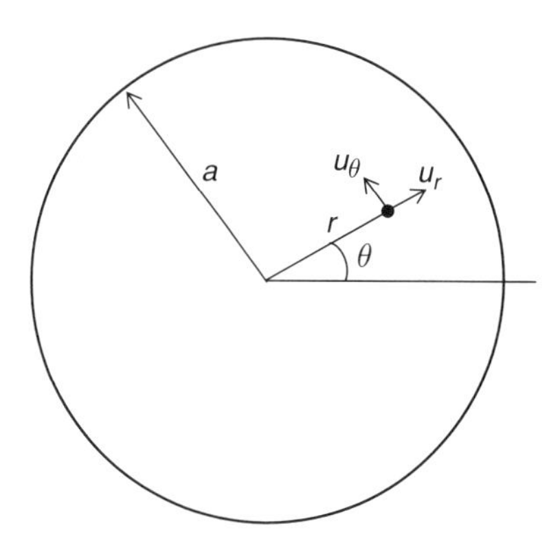

FIGURE 4.4 Coordinates' definition for in-plane axial vibration of a free circular plate.

Substituting Eq. (19) into Eq. (33) yields

$$\left(\frac{\partial u_r}{\partial r} + \nu \frac{u_r}{r}\right)\bigg|_{r=a} = 0 \tag{34}$$

Substituting Eq. (31) into Eq. (34) yields

$$A\gamma J_1'(\gamma a) + \frac{\nu}{a} A J_1(\gamma a) = 0 \tag{35}$$

Recall the following identity between Bessel functions

$$J_1'(z) = J_0(z) - \frac{1}{z} J_1(z) \tag{36}$$

Equation (35) becomes

$$A\left(\gamma J_0(\gamma a) - \frac{1}{a} J_1(\gamma a) + \frac{\nu}{a} J_1(\gamma a)\right) = 0 \tag{37}$$

Equation (37) should give the value of A. However, this is a homogenous equation, and the trivial solution $A = 0$ applies unless the parenthesis vanishes. Hence, the condition for Eq. (37) to yield nonzero solutions for A is that the parenthesis should vanish, i.e.,

$$\gamma J_0(\gamma a) - \frac{1}{a} J_1(\gamma a) + \frac{\nu}{a} J_1(\gamma a) = 0 \tag{38}$$

Equation (38) is the *characteristic equation*. The solutions of Eq. (38) are the *eigenvalues* for the axisymmetric vibration of free circular plates. Denoting $z = \gamma a$, we rewrite Eq. (38) as

$$z J_0(z) - (1 - \nu) J_1(z) = 0 \tag{39}$$

Equation (39) can be rearranged as

$$\frac{z J_0(z)}{J_1(z)} = (1 - \nu) \tag{40}$$

The left hand side of Eq. (40) is also known as the **modified quotient of Bessel functions** defined as

$$\tilde{J}_1(z) = \frac{z J_0(z)}{J_1(z)} \tag{41}$$

Note that Eq. (40) depends on the Poisson ratio, ν. Eq. (41) indicates that nonzero values of A are only possible at particular values of z, the **eigenvalues** of Eq. (39). Eq. (40) is transcendental in z and does not accept closed-form solution. Numerical solution of Eq. (40) for $\nu = 0.30$ yields

$$z = (\gamma a) = 2.048652; \quad 5.389361; \quad 8.571860; \quad 11.731771\ldots \tag{42}$$

It should be noted that, as these eigenvalues depend on ν, the ratio between successive eigenvalues and the fundamental eigenvalues could be used to determine the Poisson ratio experimentally through a curve-fitting process.

For each eigenvalue, $(\gamma a)_n$, we can determine the corresponding **resonance frequency** with the formula

$$\omega_n = \frac{c}{a}(\gamma a)_n \text{ and } f_n = \frac{1}{2\pi}\frac{c}{a}(\gamma a)_n, \quad n = 1, 2, 3 \tag{43}$$

where $c = c_L$ given by Eq. (23). The **mode shapes** are calculated with Eq. (31) using the wavenumber γ corresponding to each eigenvalue. For the nth eigenvalue, $(\gamma a)_n$, the nth wavenumber is calculated with the formula

$$\gamma_n = \frac{1}{a}(\gamma a)_n \tag{44}$$

Hence, the nth mode shape is given by

$$R_n(r) = A_n J_1(z_n r/a) \tag{45}$$

The constant A_n is determined through modes normalization, and depends on the normalization procedure used. A common normalization procedure, based on equal modal energy, yields

$$A_n = \sqrt{J_1^2(z_n) - J_0(z_n)J_2(z_n)} \tag{46}$$

An alternate normalization methods is to simply take $A_n = 1$. A graphical representation of the mode shapes is given in Table 4.1.

TABLE 4.1 Resonance mode shapes for axial vibration of a free elastic disc of radius $a = 10\,\text{mm}$

Mode	Eigenvalue	Resonant frequency	Mode shape	
R_1	$z_1 = \gamma_1 a = 2.048652$	$f_2 = \frac{1}{2\pi} z_1 \frac{c}{a}$	0 1 2 3 4 5 6 7 8 9 10	$R_1 = J_1(z_1 r/a)$
R_2	$z_2 = \gamma_2 a = 5.389361$	$f_2 = \frac{1}{2\pi} z_2 \frac{c}{a}$	0 1 2 3 4 5 6 7 8 9 10	$R_2 = J_1(z_2 r/a)$
R_3	$z_3 = \gamma_3 a = 8.571860$	$f_3 = \frac{1}{2\pi} z_3 \frac{c}{a}$	0 1 2 3 4 5 6 7 8 9 10	$R_3 = J_1(z_3 r/a)$

4.4 FLEXURAL VIBRATION OF RECTANGULAR PLATES

Flexural vibration appears from bending action. The Love-Kirchhoff theory of plate bending assumes that straight normals to the mid-surface remain straight after deformation. This implies a linear distribution of axial displacement across the thickness.

4.4.1 GENERAL EQUATIONS FOR FLEXURAL VIBRATION OF RECTANGULAR PLATES

For a displacement w of the plate mid surface, the displacements at any location z in the plate thickness are given by

$$u = -z\frac{\partial w}{\partial x}$$
$$v = -z\frac{\partial w}{\partial y} \tag{47}$$
$$w = w$$

The strains of interest are

$$\varepsilon_{xx} = \frac{\partial u}{\partial x} = -z\frac{\partial^2 w}{\partial x^2} \quad \varepsilon_{yy} = \frac{\partial v}{\partial y} = -z\frac{\partial^2 w}{\partial y^2} \quad \varepsilon_{xy} = \frac{1}{2}\left(\frac{\partial u}{\partial y} + \frac{\partial v}{\partial x}\right) = -z\frac{\partial^2 w}{\partial x \partial y} \tag{48}$$

Substitution of Eq. (48) into Eq. (2) yields

$$\sigma_{xx} = -z\frac{E}{1-\nu^2}\left(\frac{\partial^2 w}{\partial x^2} + \nu\frac{\partial^2 w}{\partial y^2}\right) \quad \sigma_{yy} = -z\frac{E}{1-\nu^2}\left(\nu\frac{\partial^2 w}{\partial x^2} + \frac{\partial^2 w}{\partial y^2}\right) \quad \sigma_{xy} = -2zG\frac{\partial^2 w}{\partial x \partial y} \tag{49}$$

Note that the stresses vary linearly across the plate thickness. Integration of stresses across the thickness gives the stress resultants M_x, M_y, and M_{xy} (moments per unit width) shown in Fig. 4.5, i.e.,

$$M_x = \int_{-h/2}^{h/2} \sigma_{xx} z\,\mathrm{d}z = -D\left(\frac{\partial^2 w}{\partial x^2} + \nu\frac{\partial^2 w}{\partial y^2}\right)$$
$$M_y = \int_{-h/2}^{h/2} \sigma_{yy} z\,\mathrm{d}z = -D\left(\nu\frac{\partial^2 w}{\partial x^2} + \frac{\partial^2 w}{\partial y^2}\right) \tag{50}$$
$$M_{xy} = -M_{yx} = \int_{-h/2}^{h/2} (-\sigma_{xy}) z\,\mathrm{d}z = (1-\nu)D\frac{\partial^2 w}{\partial x \partial y}$$

where D is the *flexural plate stiffness* defined as

$$D = \frac{Eh^3}{12(1-\nu^2)} \tag{51}$$

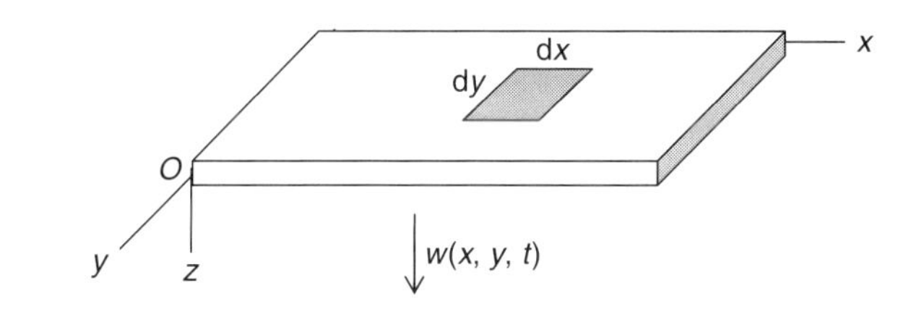

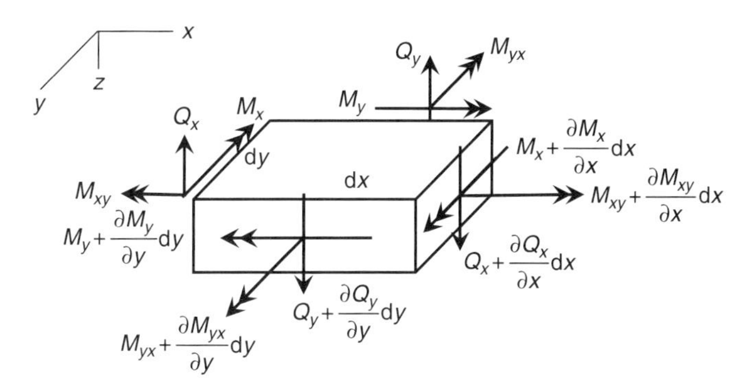

FIGURE 4.5 Infinitesimal plate element in Cartesian coordinates for the analysis of flexural vibration of rectangular plates.

Newton second law applied to the infinitesimal element of Fig. 4.5 yields

$$\begin{aligned}
&\text{vertical forces:} \quad \frac{\partial Q_x}{\partial x} + \frac{\partial Q_y}{\partial y} = \rho h \ddot{w} \\
&M_x \text{ moments:} \quad \frac{\partial M_x}{\partial x} + \frac{\partial M_{yx}}{\partial y} - Q_x = 0 \\
&M_y \text{ moments:} \quad \frac{\partial M_y}{\partial y} - \frac{\partial M_{xy}}{\partial x} - Q_y = 0
\end{aligned} \tag{52}$$

Simplification of Eqs. (52) yields

$$\frac{\partial^2 M_x}{\partial x^2} - \frac{\partial M_{xy}}{\partial x \partial y} + \frac{\partial^2 M_{yx}}{\partial x \partial y} + \frac{\partial^2 M_y}{\partial y^2} = \rho h \ddot{w} \tag{53}$$

Substitution of Eq. (50) into Eq. (53) yields the general equation for flexural vibration of a plate

$$D\left(\frac{\partial^4 w}{\partial x^4} + 2\frac{\partial^4 w}{\partial x^2 \partial y^2} + \frac{\partial^4 w}{\partial y^4}\right) + \rho h \ddot{w} = 0 \tag{54}$$

Equation (54) can be expressed in terms of the biharmonic operator for z-invariant motion $\nabla^4 = \nabla^2 \nabla^2 = \frac{\partial^4}{\partial x^4} + 2\frac{\partial^4}{\partial x^2 \partial y^2} + \frac{\partial^4}{\partial y^4}$ given in Appendix B, i.e.,

$$D\nabla^4 w + \rho h \ddot{w} = 0 \tag{55}$$

Assume harmonic motion in the form

$$w(x, y, t) = \hat{w}(x, y)\, e^{i\omega t} \tag{56}$$

Equation (55) becomes

$$D\nabla^4 w - \rho h \omega^2 \hat{w} = 0 \tag{57}$$

Introduce the parameter

$$\gamma^4 = \frac{\rho h}{D}\omega^2 \tag{58}$$

Hence, Eq. (57) becomes

$$\left(\nabla^4 - \gamma^4\right)\hat{w} = 0 \tag{59}$$

It is sometimes convenient to factor Eq. (59) into the form

$$\left(\nabla^2 + \gamma^2\right)\left(\nabla^2 - \gamma^2\right)\hat{w} = 0 \tag{60}$$

By the theory of differential equations, the complete solution to Eq. (60) can be obtained by superposition of the solutions to the following system of lower-order differential equations

$$\begin{cases} \left(\nabla^2 + \gamma^2\right)\hat{w}_1 = 0 \\ \left(\nabla^2 - \gamma^2\right)\hat{w}_2 = 0 \end{cases} \tag{61}$$

The plate bending and twisting moments can be related to the displacements through the relations of Eq. (50), i.e.,

$$\begin{aligned} M_x &= -D\left(\frac{\partial^2 w}{\partial x^2} + \nu\frac{\partial^2 w}{\partial y^2}\right) \\ M_y &= -D\left(\nu\frac{\partial^2 w}{\partial x^2} + \frac{\partial^2 w}{\partial y^2}\right) \\ M_{xy} &= -D(1-\nu)\frac{\partial^2 w}{\partial x \partial y} \end{aligned} \tag{62}$$

Transverse shear forces in the plate are given by

$$\begin{aligned} Q_x &= -D\frac{\partial}{\partial x}\left(\nabla^2 w\right) \\ Q_y &= -D\frac{\partial}{\partial y}\left(\nabla^2 w\right) \end{aligned} \tag{63}$$

The Kelvin–Kirchhoff edge reactions are given by

$$\begin{aligned} V_x &= Q_x + \frac{\partial M_{xy}}{\partial y} \\ V_y &= Q_y + \frac{\partial M_{xy}}{\partial x} \end{aligned} \tag{64}$$

The bending and twisting strain energy of a rectangular plate undergoing flexural vibration is given by

$$U = \frac{1}{2}D\int_A \left\{\left(\frac{\partial^2 w}{\partial x^2} + \nu\frac{\partial^2 w}{\partial y^2}\right)^2 - 2(1-\nu)\left[\frac{\partial^2 w}{\partial x^2}\frac{\partial^2 w}{\partial y^2} - \left(\frac{\partial^2 w}{\partial x\,\partial y}\right)^2\right]\right\} \mathrm{d}A \tag{65}$$

4.4.2 STRAIGHT-CRESTED FLEXURAL VIBRATION OF RECTANGULAR PLATES

Equation (54) has a general solution representing flexural vibration taking place simultaneously in the x and y directions. However, when certain conditions are imposed, the problem simplifies and yields closed-form solutions, as discussed next.

Consider *straight-crested flexural plate vibrations*. If we view plate vibration as system of standing waves in the plate, then this case can be considered a system of standing straight-crested flexural waves with the wave crest along the y-axis (Fig. 4.6). Taking the y-axis along the wave crest yields a y-invariant problem that depends only on x, i.e.,

$$w(x, y, t) \rightarrow w(x, t) \quad \text{(straight-crested flexural plate wave)} \tag{66}$$

Equation (66) implies $\partial w/\partial y = 0$, etc. Substitution of Eq. (66) into Eq. (54) yields

$$Dw'''' + \rho h \ddot{w} = 0 \tag{67}$$

where, as before, $(\ )'$ represents derivative with respect to the space variable x. Equation (67) has the same form as the equation for flexural vibration of beams presented in Chapter 3 ($D \leftrightarrow E$, $\rho h \leftrightarrow m$); hence Eq. (67) has the same general solution

$$w(x, t) = A_1 e^{i(\gamma_F x + \omega t)} + A_2 e^{-i(\gamma_F x - \omega t)} + A_3 e^{\gamma_F x} e^{i\omega t} + A_4 e^{-\gamma_F x} e^{i\omega t} \tag{68}$$

where $\gamma_F = \omega / c_F$, whereas c_F is the *flexural wave speed in plates* given by

$$c_F = \left(\frac{D}{\rho h} \right)^{1/4} \sqrt{\omega} = \left(\frac{Eh^2}{12\rho(1-\nu^2)} \right)^{1/4} \sqrt{\omega} \tag{69}$$

Using the half-thickness $d = h/2$, Eq. (69) becomes

$$c_F = \left(\frac{Ed^2}{3\rho(1-\nu^2)} \right)^{1/4} \sqrt{\omega} \tag{70}$$

Comparing Eq. (69) to the corresponding equation for beams presented in a previous chapter indicates that the only difference between the flexural wave speed in beams and plates is the presence of the correction term $(1-\nu^2)$, which accounts for the additional constraint (plain strain state) imposed by the plate geometry.

Straight-crested flexural vibrations of a plate are easy to derive mathematically, but they are not as easy to excite experimentally. For rectangular plates of finite aspect ratio,

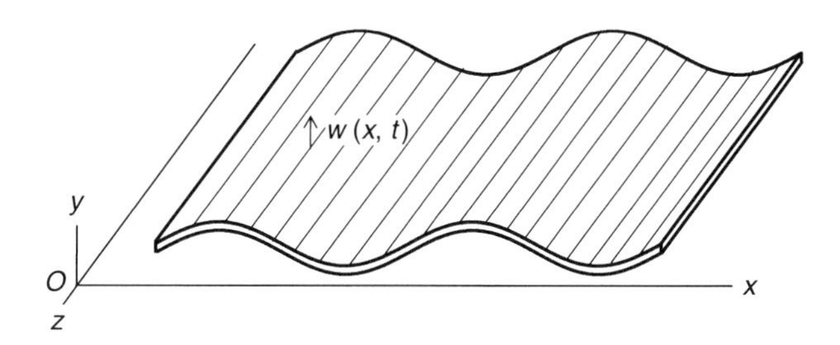

FIGURE 4.6 Straight-crested flexural vibration of a plate.

a special excitation system should be devised that imposes a plan-parallel motion across the width such that the standing wave motion propagates only along the length. However, certain boundary conditions and plate aspect ratio are conductive of being analyzed under the straight-crested assumption discussed here. If the aspect ratio has an extreme value that clearly favors one of the dimension (e.g., a length to width ratio of 10 or greater), then the straight-crested analysis would predict acceptable results, at least for the lower modes, which will manifest a beam-like behavior. For example, a free strip of plate material would have the lower frequencies and mode shapes quite acceptably predicted by an appropriately modified beam theory.

4.4.3 GENERAL FLEXURAL VIBRATION OF RECTANGULAR PLATES

The general flexural vibration of a rectangular plate (Fig. 4.7) is obtained from the general solution of Eq. (59) subject to specific boundary conditions. Here, we will consider two specific boundary conditions:

(a) simply supported rectangular plate
(b) free rectangular plate.

4.4.3.1 Flexural Vibration of Simply Supported Rectangular Plates

The problem of flexural vibration of a rectangular plate with all sides simply supported is quite easy to resolve. The simply supported boundary conditions are

$$\begin{aligned} w = 0, M_x = 0 \quad &\text{for} \quad x = 0 \text{ and } x = a \\ w = 0, M_y = 0 \quad &\text{for} \quad y = 0 \text{ and } y = b \end{aligned} \tag{71}$$

Expressing these boundary conditions with the use of Eqs. (62) yields

$$\begin{aligned} \frac{\partial^2 w}{\partial x^2} + \nu \frac{\partial^2 w}{\partial y^2} = 0 \quad &\text{for} \quad x = 0 \text{ and } x = a \\ \nu \frac{\partial^2 w}{\partial x^2} + \frac{\partial^2 w}{\partial y^2} = 0 \quad &\text{for} \quad y = 0 \text{ and } y = b \end{aligned} \tag{72}$$

Assuming time-harmonic vibrations allows us to separate the space and time dependencies and express the displacement in the form

$$w(x, y, t) = \hat{w}(x, y)e^{i\omega t} \tag{73}$$

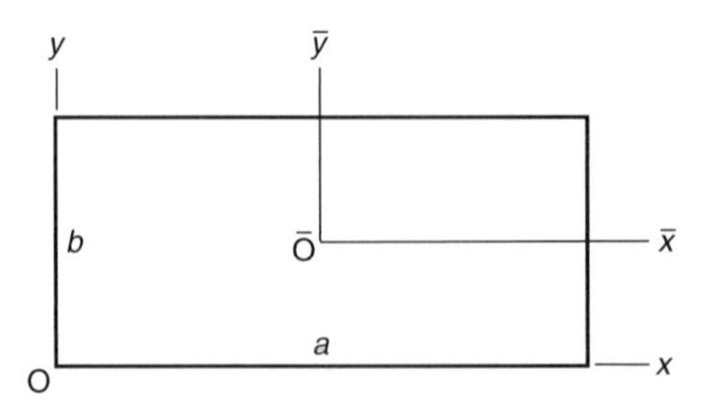

FIGURE 4.7 Coordinates definition for flexural vibration analysis of a free rectangular plate.

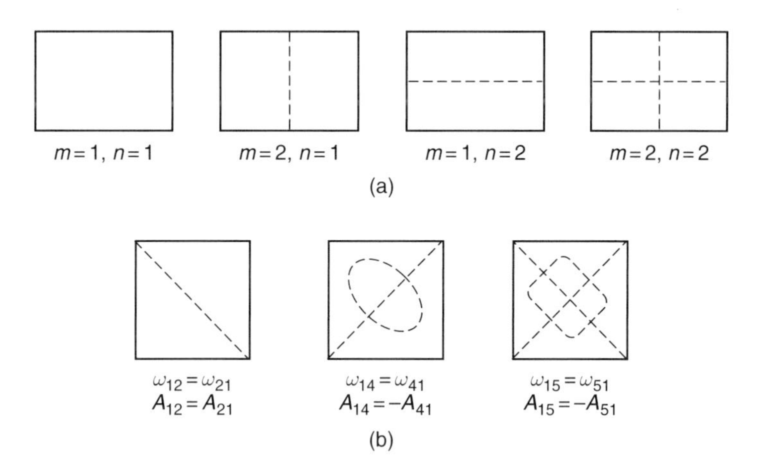

FIGURE 4.8 Nodal lines of flexural vibration mode shapes for simply supported rectangular plates: (a) arbitrary rectangular plate with a > b showing well-separated mode shapes; (b) coincidental frequencies and associated mode shapes appearing in square plates.

The remaining problem is to find the mode shape $\hat{w}(x, y)$. Examination of Eqs. (71) and (72) indicates that both would be satisfied by all harmonic functions that have nodes at the boundaries. Hence, the general solution for flexural vibration of simply supported plates takes the form

$$\hat{w}_{mn} = A_{mn} \sin \frac{m\pi x}{a} \sin \frac{n\pi y}{b} \tag{74}$$

where A_{mn} is the mode shape amplitude. Substitution of the general solution (74) into Eq. (57) yields

$$D\left[\left(\frac{m\pi}{a}\right)^4 + 2\left(\frac{m\pi}{a}\right)^2\left(\frac{n\pi}{b}\right)^2 + \left(\frac{n\pi}{b}\right)^4\right]\hat{w} - \rho h\omega^2\hat{w} = 0 \tag{75}$$

Hence, we can write the frequency equation

$$\omega_{mn} = \sqrt{\frac{D}{\rho h}}\left[\left(\frac{m\pi}{a}\right)^2 + \left(\frac{n\pi}{b}\right)^2\right] \tag{76}$$

For each natural frequency, ω_{mn}, one can use Eq. (74) to write the corresponding mode shape, where the amplitude A_{mn} would be determined through a normalization process. Schematics of a few flexural mode shapes of a rectangular plate are given in Fig. 4.8a. In the case of square plate vibration, for which the dimensions in x and y directions are identical, certain frequencies will coincide, resulting in combination mode shapes (Fig. 4.8b).

4.4.3.2 Flexural Vibration of Free Rectangular Plates

The problem of flexural vibration of a free rectangular plate is more difficult to resolve and does not yield closed-form solutions. The problem has received attention over the years, and has stirred the interest of Rayleigh (1873) and Ritz (1909). Here, we will present the solution due to Iguchi (1953) as cited by Leissa (1969).

Consider again the rectangular plate of Fig. 4.7. The free boundary conditions are expressed as

$$\begin{aligned} &M_x = 0, V_x = 0 \quad \text{for} \quad \bar{x} = \pm\frac{1}{2}a \\ &M_y = 0, V_y = 0 \quad \text{for} \quad \bar{y} = \pm\frac{1}{2}b \\ &M_{xy} = 0 \quad \text{(at corners)} \end{aligned} \tag{77}$$

Expressing these boundary conditions with the use of Eqs. (62) and (64) yields

$$\begin{aligned} &\frac{\partial^2 w}{\partial \bar{x}^2} + \nu\frac{\partial^2 w}{\partial \bar{y}^2} = 0 \quad \text{for} \quad \bar{x} = \pm\frac{1}{2}a \\ &\nu\frac{\partial^2 w}{\partial \bar{x}^2} + \frac{\partial^2 w}{\partial \bar{y}^2} = 0 \quad \text{for} \quad \bar{y} = \pm\frac{1}{2}b \\ &\frac{\partial^2 w}{\partial \bar{x}\partial \bar{y}} = 0 \quad \text{(at corners)} \end{aligned} \tag{78}$$

$$\begin{aligned} &\frac{\partial}{\partial \bar{x}}\left[\frac{\partial^2 w}{\partial \bar{x}^2} + (2-\nu)\frac{\partial^2 w}{\partial \bar{y}^2}\right] = 0 \quad \text{for} \quad \bar{x} = \pm\frac{1}{2}a \\ &\frac{\partial}{\partial \bar{y}}\left[(2-\nu)\frac{\partial^2 w}{\partial \bar{x}^2} + \frac{\partial^2 w}{\partial \bar{y}^2}\right] = 0 \quad \text{for} \quad \bar{y} = \pm\frac{1}{2}b \end{aligned} \tag{79}$$

Assuming time-harmonic vibrations allows us to separate the space and time dependencies and express the displacement in the form

$$w(\bar{x}, \bar{y}, t) = W(\bar{x}, \bar{y})e^{i\omega t} \tag{80}$$

Iguchi (1953) formulated the solution in the form of the series

$$W(\bar{x}, \bar{y}) = \sum_{n=0}^{\infty} X_n \cos n\pi\left[\frac{1}{2} + \eta\right] + \sum_{m=0}^{\infty} Y_m \cos m\pi\left[\frac{1}{2} + \xi\right] \tag{81}$$

where $\xi = \bar{x}/a$ and $\eta = \bar{y}/b$. The functions X_n and Y_m are expressed in terms of hyperbolic sines and cosines, i.e.,

$$\begin{aligned} X_n &= A_n\frac{\cosh \pi\lambda_{\alpha n}\xi}{\sinh \frac{\pi}{2}\lambda_{\alpha n}} + A_n^*\frac{\cosh \pi\lambda_{\alpha n}^*\xi}{\sinh \frac{\pi}{2}\lambda_{\alpha n}^*} + A_n^{**}\frac{\cosh \pi\lambda_{\alpha n}\xi}{\sinh \frac{\pi}{2}\lambda_{\alpha n}} + A_n^{***}\frac{\cosh \pi\lambda_{\alpha n}^*\xi}{\sinh \frac{\pi}{2}\lambda_{\alpha n}^*} \\ Y_m &= B_m\frac{\cosh \pi\lambda_{\beta m}\eta}{\sinh \frac{\pi}{2}\lambda_{\beta m}} + B_m^*\frac{\cosh \pi\lambda_{\beta m}^*\eta}{\sinh \frac{\pi}{2}\lambda_{\beta m}^*} + B_m^{**}\frac{\cosh \pi\lambda_{\beta m}\eta}{\sinh \frac{\pi}{2}\lambda_{\beta m}} + B_m^{***}\frac{\cosh \pi\lambda_{\beta m}^*\eta}{\sinh \frac{\pi}{2}\lambda_{\beta m}^*} \end{aligned} \tag{82}$$

where

$$\begin{aligned} &\lambda_{\alpha n}, \lambda_{\alpha n}^* = \sqrt{\alpha^2 n^2 \pm \mu}, \qquad \mu = \frac{\omega a^2}{\pi^2}\sqrt{\frac{\rho h}{D}}, \qquad \alpha = \frac{a}{b} \\ &\lambda_{\beta m}, \lambda_{\beta m}^* = \sqrt{\beta^2 m^2 \pm \mu^*}, \quad \mu^* = \frac{\omega b^2}{\pi^2}\sqrt{\frac{\rho h}{D}}, \quad \beta = \frac{b}{a} \end{aligned} \tag{83}$$

Further details of the method can be found in Iguchi (1953) or Leissa (1969). The application of the method yields an infinite characteristic determinant that can be truncated

Mode shape nodal pattern	$\omega a^2\sqrt{\rho/D}$	n	α_n	λ_n	λ_n^*
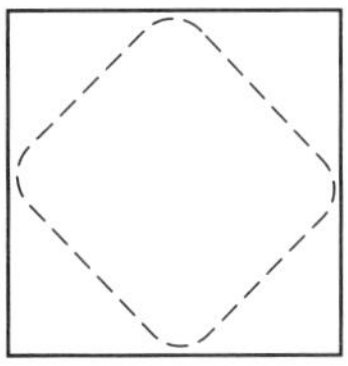		0	8.51935	–	–
		2	1.00000	2.54147	1.24133
	24.2702	4	0.04225	4.29641	3.67990
		6	0.01173	6.20154	5.79145
		8	0.00494	8.15225	7.84480
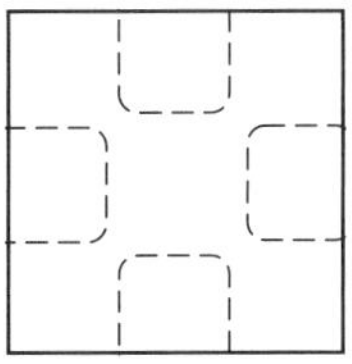		0	–0.11966	–	–
		2	1.00000	3.23309	1.56615i
	63.6870	4	0.03422	4.73844	3.08985
		6	0.01065	6.51558	5.43573
		8	0.00473	8.39362	7.58598
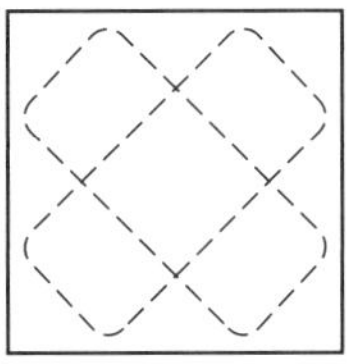		0	–8.81714	–	–
		2	1.00000	4.05046	2.89935i
	122.4449	4	–1.19356	5.32975	1.89572
		6	–0.08213	6.95746	4.85734
		8	–0.02402	8.74107	7.18288
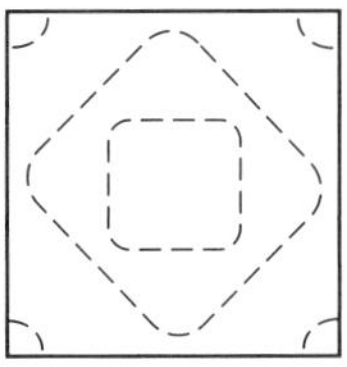		0	–0.07482	–	–
		2	1.00000	4.59037	3.61545i
	168.4888	4	0.44885	5.75078	1.03513i
		6	0.03590	7.28502	4.35069
		8	0.01347	9.00397	6.85044
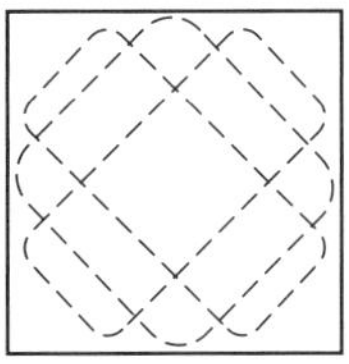		0	–8.90424	–	–
		2	1.00000	5.86426	5.13707i
	299.9325	4	–0.59521	6.81099	3.79335i
		6	–1.39192	8.14998	2.36864
		8	–0.13703	9.71543	5.79745

FIGURE 4.9 Mode shapes and frequencies of free rectangular plates for the case symmetric about coordinate axes and symmetric about diagonals (Iguchi, 1953).

to a finite order of terms. The eigenvalues of the truncated determinant were found to converge rapidly with the increasing order of the determinant. Frequencies, modal patterns, and numerical constants for the mode shapes are presented in Figs. 4.9–4.11.

4.5 FLEXURAL VIBRATION OF CIRCULAR PLATES

To analyze the flexural vibration of circular plates, we change from Cartesian coordinates (x, y, z) to cylindrical coordinates (r, θ, z), as shown in Fig. 4.12. Consider the circular plate element shown in Fig. 4.13. In cylindrical coordinates, the volume of the

Mode shape nodal pattern	$\omega a^2\sqrt{\rho/D}$	n	α_n	λ_n	λ_n^*
	19.5961	0 2 4 6 8	−19.46060 1.00000 0.00264 −0.00487 −0.00290	– 2.44653 4.24093 6.16324 8.12315	– 1.41933 3.74359 5.83219 7.87493
	65.3680	0 2 4 6 8	3.93698 1.00000 −0.09935 −0.01507 −0.00451	– 3.25932 4.75638 6.53864 8.40376	– 1.61926i 3.06216 5.43004 7.57475
	117.1093	0 2 4 6 8	3.84826 1.00000 −0.48091 −0.02845 −0.00453	– 3.98317 5.27879 6.91850 8.71009	– 2.80458i 2.03331 4.91267 7.22041
	161.5049	0 2 4 6 8	−0.02833 1.00000 −0.24428 −0.01363 −0.00297	– 4.51264 5.68893 7.23629 8.96459	– 3.51623i 0.60322i 4.43127 6.90189
	293.7190	0 2 4 6 8	5.79354 1.00000 0.66331 −0.61699 −0.05732	– 5.81033 6.76461 8.10925 9.68297	– 5.07543i 3.70944i 2.49801 5.85150

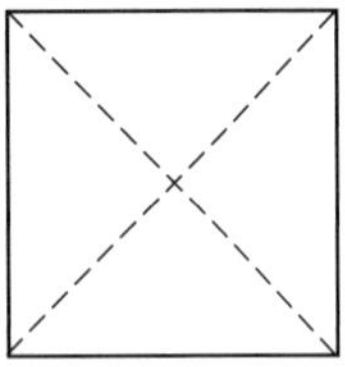

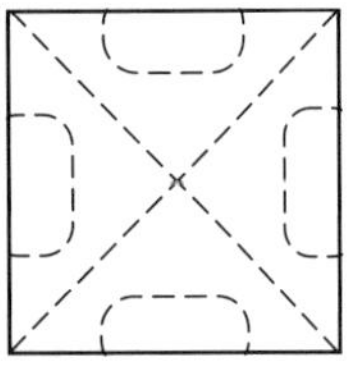

FIGURE 4.10 Mode shapes and frequencies of free square plates for the case symmetric about coordinate axes and antisymmetric about diagonals (Iguchi, 1953).

differential element is $dV = h_{\text{plate}} r\,dr\,d\theta$. The equation of motion is obtained by using the forces and moments relative to the chosen axis.

4.5.1 FORCE SUMMATION

Newton law of motion in the vertical z direction neglecting second-order terms and division by $r\,dr\,d\theta$, yields the equation relating the transverse forces with the flexural acceleration $\ddot{w}$, i.e.,

$$Q_r + r\frac{\partial Q_r}{\partial r} + \frac{\partial Q_\theta}{\partial \theta} = \rho h r \ddot{w} \tag{84}$$

Mode shape nodal pattern	$\omega a^2\sqrt{\rho/D}$	n	α_n	λ_n	λ_n^*
(a) Modes antisymmetric about coordinate axes and symmetric about diagonals					
	13.4728	1 3 5 7	1.00000 0.00766 0.00100 0.00041	1.53788 3.21949 5.13469 7.09684	0.06042i 2.76314 4.86158 6.90181
	77.5897	1 3 5 7	1.00000 0.23339 0.00888 0.00178	2.97685 4.10632 5.73251 7.54066	2.61947i 1.06694 4.14985 6.41392
	156.2387	1 3 5 7	1.00000 −4.56065 −0.05491 −0.01457	4.10247 4.98299 6.38986 8.05176	3.85101i 2.61348i 3.02815 5.75931
(b) Modes antisymmetric about coordinate axes and diagonals					
	69.5020	1 3 5 7	1.00000 −0.12837 −0.00557 −0.00101	2.833585 4.00525 5.66057 7.48612	2.45805i 1.29928 4.23769 6.47750
	173.6954	1 3 5 7	1.00000 2.68336 −0.13566 −0.02103	4.31266 5.15742 6.52679 8.16082	4.07419i 2.93241i 2.72047 5.60366
	204.6527	1 3 5 7	1.00000 0.15411 −0.13841 −0.01080	4.66215 5.45304 6.76282 8.35079	4.44248i 3.42573i 2.06503 5.31642

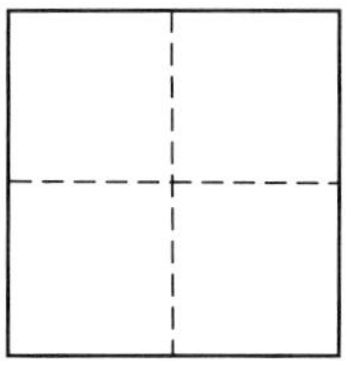

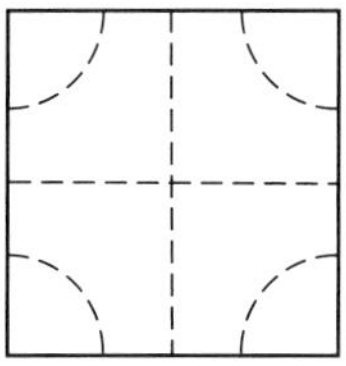

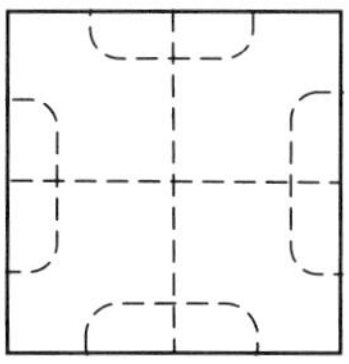

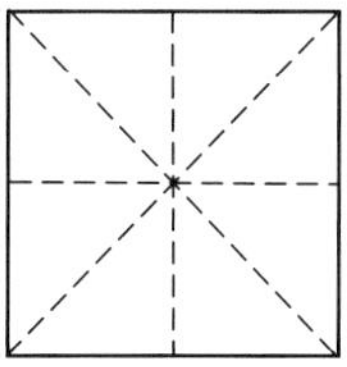

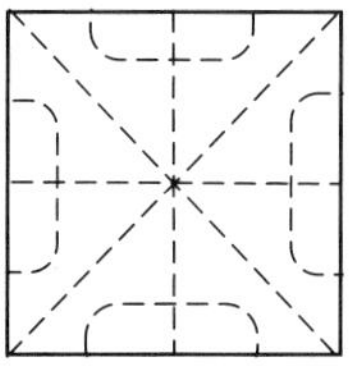

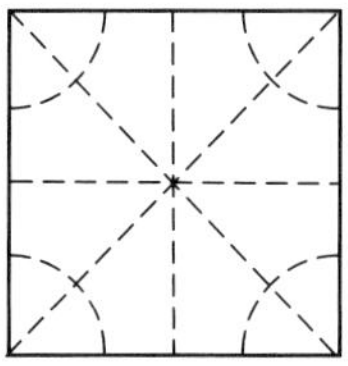

FIGURE 4.11 Mode shapes and frequencies of free square plates: (a) antisymmetric about coordinate axes and symmetric about diagonals; (b) antisymmetric about coordinate axes and diagonals (Iguchi, 1953).

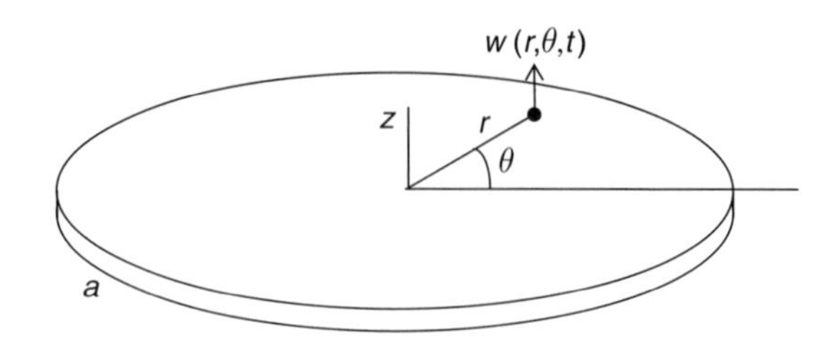

FIGURE 4.12 Definition of coordinates for flexural vibration analysis of a free circular plate.

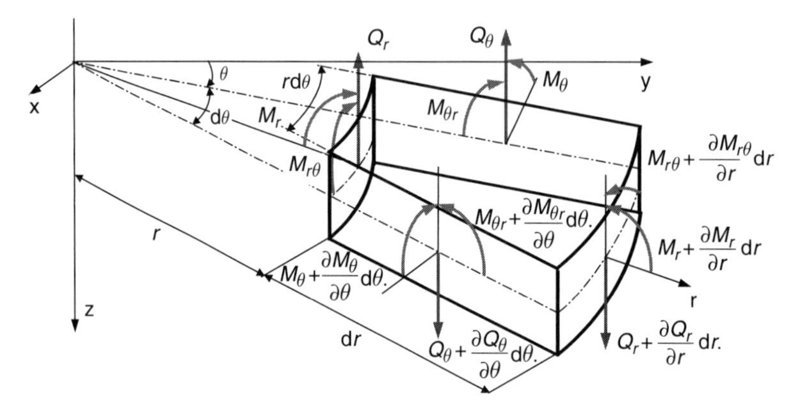

FIGURE 4.13 Infinitesimal element of a circular plate undergoing flexural vibrations.

4.5.2 MOMENT SUMMATION

Summating the moments, neglecting second- and third-order terms, and dividing by $r\,\mathrm{d}r\,\mathrm{d}\theta$ gives the moment equation for bending of circular plates

$$\frac{\partial M_r}{\partial r} + \frac{1}{r}\frac{\partial M_{\theta r}}{\partial \theta} + \frac{M_r - M_\theta}{r} - Q_r = 0 \tag{85}$$

Classical plate theory (e.g., Ugural, 1999, p. 107) relates the bending and twisting moments, M_r, M_θ, $M_{r\theta}$, to the flexural displacement, w, in the form

$$\begin{aligned} M_r &= -D\left[\frac{\partial^2 w}{\partial r^2} + \nu\left(\frac{1}{r}\frac{\partial w}{\partial r} + \frac{1}{r^2}\frac{\partial^2 w}{\partial \theta^2}\right)\right] \\ M_\theta &= -D\left[\frac{1}{r}\frac{\partial w}{\partial r} + \frac{1}{r^2}\frac{\partial^2 w}{\partial \theta^2} + \nu\frac{\partial^2 w}{\partial r^2}\right] \\ M_{r\theta} &= -(1-\nu)\,D\left[\frac{1}{r}\frac{\partial^2 w}{\partial r\partial\theta} - \frac{1}{r^2}\frac{\partial w}{\partial \theta}\right] \end{aligned} \tag{86}$$

where $D = Eh^3/12\left(1-\nu^2\right)$ is the flexural plate stiffness, as defined by Eq. (51). Similarly, the shear forces Q_r, Q_θ are related to the flexural displacement, w, in the form

$$\begin{aligned} Q_r &= -D\frac{\partial}{\partial r}\left(\nabla^2 w\right) \\ Q_\theta &= -D\frac{1}{r}\frac{\partial}{\partial \theta}\left(\nabla^2 w\right) \end{aligned} \tag{87}$$

where the Laplace operator has the form appropriate to polar coordinates as given in the Appendix B, i.e.,

$$\nabla^2 = \frac{\partial^2}{\partial r^2} + \frac{1}{r}\frac{\partial}{\partial r} + \frac{1}{r^2}\frac{\partial^2}{\partial \theta^2} \quad \text{(Laplace operator)} \tag{88}$$

The Kelvin–Kirchhoff edge reactions are given by

$$\begin{aligned} V_r &= Q_r + \frac{1}{r}\frac{\partial M_{r\theta}}{\partial \theta} \\ V_\theta &= Q_\theta + \frac{\partial M_{r\theta}}{\partial r} \end{aligned} \tag{89}$$

The bending and twisting strain energy of a circular plate undergoing flexural vibration is given by

$$\begin{aligned} U = \frac{1}{2}D\int_A \Bigg\{ & \left(\frac{\partial^2 w}{\partial r^2} + \frac{1}{r}\frac{\partial w}{\partial r} + \frac{1}{r^2}\frac{\partial^2 w}{\partial \theta^2}\right)^2 \\ & -2(1-\nu)\left[\frac{\partial^2 w}{\partial r^2}\left(\frac{1}{r}\frac{\partial w}{\partial r} + \frac{1}{r^2}\frac{\partial^2 w}{\partial \theta^2}\right) - \left(\frac{\partial}{\partial r}\left(\frac{1}{r}\frac{\partial w}{\partial \theta}\right)\right)^2\right]\Bigg\} \mathrm{d}A \end{aligned} \tag{90}$$

where $\mathrm{d}A = r\,\mathrm{d}r\,\mathrm{d}\theta$. Substitution of Eqs. (87) into Eq. (84) yields the equation of motion (55) in the general form

$$D\nabla^4 w + \rho h \ddot{w} = 0 \tag{91}$$

where the biharmonic operator has the form appropriate to polar coordinates, as given in the Appendix B, i.e.,

$$\begin{aligned} \nabla^4 &= \nabla^2\nabla^2 \\ &= \left(\frac{\partial^2}{\partial r^2} + \frac{1}{r}\frac{\partial}{\partial r} + \frac{1}{r^2}\frac{\partial^2}{\partial \theta^2}\right)\left(\frac{\partial^2}{\partial r^2} + \frac{1}{r}\frac{\partial}{\partial r} + \frac{1}{r^2}\frac{\partial^2}{\partial \theta^2}\right) \quad \text{(biharmonic operator)} \end{aligned} \tag{92}$$

Assuming time-harmonic vibration, we separate the space and time dependencies and express the displacement in the form

$$w(r, \theta, t) = \hat{w}(r, \theta)\, e^{i\omega t} \tag{93}$$

The remaining problem is to find the space-dependant solution $\hat{w}(r, \theta)$ such that it satisfies the differential equation (91) and the boundary conditions.

Equation (91) becomes

$$D\nabla^4 w - \rho h \omega^2 \hat{w} = 0 \tag{94}$$

Introduce the parameter

$$\gamma^4 = \frac{\rho h}{D}\omega^2 \tag{95}$$

Hence, Eq. (94) becomes

$$\left(\nabla^4 - \gamma^4\right)\hat{w} = 0 \tag{96}$$

It is convenient to factor Eq. (96) into the form

$$\left(\nabla^2 + \gamma^2\right)\left(\nabla^2 - \gamma^2\right)\hat{w} = 0 \tag{97}$$

By the theory of differential equations, the complete solution to Eq. (97) can be obtained by superposition of the solutions of the following system of two lower-order differential equations

$$\begin{cases} \left(\nabla^2 + \gamma^2\right)\hat{w}_1 = 0 \\ \left(\nabla^2 - \gamma^2\right)\hat{w}_2 = 0 \end{cases} \tag{98}$$

We seek the solution of Eq. (91) in the general form

$$\hat{w}(r, \theta) = \sum_{n=0}^{\infty} W_n(r)\cos n\theta + \sum_{n=0}^{\infty} W_n^*(r)\sin n\theta \tag{99}$$

Substitution of Eq. (99) into Eq. (98) yields

$$\begin{aligned} \frac{d^2 W_{1_n}}{dr^2} + \frac{1}{r}\frac{dW_{1_n}}{dr} - \left(\frac{n^2}{r^2} - \gamma^2\right) W_{1_n} = 0 \\ \frac{d^2 W_{2_n}}{dr^2} + \frac{1}{r}\frac{dW_{2_n}}{dr} - \left(\frac{n^2}{r^2} + \gamma^2\right) W_{2_n} = 0 \end{aligned} \tag{100}$$

and two other similar equations for W_n^*. Eqs. (100) are recognized as forms of the Bessel equation (McLachlan, 1948) having solutions

$$\begin{aligned} W_{1_n} = A_n J_n(\gamma r) + B_n Y_n(\gamma r) \\ W_{2_n} = C_n I_n(\gamma r) + D_n K_n(\gamma r) \end{aligned} \tag{101}$$

where J_n and Y_n are the Bessel functions of the first and second kinds, respectively, and where I_n and K_n are the modified Bessel functions of the first and second kinds, respectively. The coefficients A_n, B_n, C_n, D_n are found by the imposition of the boundary and initial conditions. Similar argument applies for W_n^*. Thus, the general solution of Eq. (96) is

$$\begin{aligned} \hat{w}(r, \theta) = & \sum_{n=0}^{\infty} \left[A_n J_n(\gamma r) + B_n Y_n(\gamma r) + C_n I_n(\gamma r) + D_n K_n(\gamma r)\right]\cos n\theta \\ & + \sum_{n=0}^{\infty} \left[A_n^* J_n(\gamma r) + B_n^* {}_n(\gamma r) + C_n^* I_n(\gamma r) + D_n^* K_n(\gamma r)\right]\sin n\theta \end{aligned} \tag{102}$$

However, the Bessel functions $Y_n(\gamma r)$ and $K_n(\gamma r)$ have infinite values at $r = 0$ and are discarded (unless the plate has a hole around $r = 0$, which is not the case considered here). For solid plates without a central hole, the terms of Eq. (102) involving $Y_n(\gamma r)$ and $K_n(\gamma r)$ are discarded because they become unbounded at $r = 0$. In addition, if the boundary conditions have some symmetry with respect to at least one diameter, then the terms in $\sin n\theta$ are not needed. When these assumptions are employed, Eq. (102) simplifies, and a typical mode shape has the expression:

$$W_n = \left[A_n J_n(\gamma r) + C_n I_n(\gamma r)\right]\cos n\theta \tag{103}$$

where $n = 0, 1, \ldots$ represents the number of nodal diameters.

4.5.3 FLEXURAL VIBRATION OF FREE CIRCULAR PLATES

The study of the flexural vibration of free circular plates has a rich history. Poisson (1829) presented the first study on the vibration of circular plates and calculated the ratio of the radii of the nodal circles to the radius of the plate when the vibrating plate had no nodal diameter and only one or two nodal circles. Three boundary conditions were considered: (a) fixed; (b) simply supported; (c) free. Kirchhoff (1850) extended Poisson's work for the vibration of free circular plates and calculated six ratios of the radii when the plate vibrates with one, two, or three nodal diameters. Airy (1911) presented a comprehensive treatment of the problem and its relation to the Bessel functions. An extensive treatment of the natural frequencies and mode shapes was done by Colwell and Hardy (1937).

The boundary conditions for a free circular plate of outer radius a are

$$\begin{aligned} M_r(a) &= 0 \\ V_r(a) &= 0 \end{aligned} \tag{104}$$

Substitution of boundary conditions (104) into Eqs. (86) and (89) with the use of Eqs. (87) yields the characteristic equation (Airy, 1911)

$$\frac{\lambda^2 J_n(\lambda) + (1-\nu)\left[\lambda^2 J_n'(\lambda) - n^2 J_n(\lambda)\right]}{\lambda^2 I_n(\lambda) + (1-\nu)\left[\lambda^2 I_n'(\lambda) - n^2 I_n(\lambda)\right]} = \frac{\lambda^3 J_n'(\lambda) + (1-\nu)n^2\left[\lambda J_n'(\lambda) - J_n(\lambda)\right]}{\lambda^3 I_n'(\lambda) + (1-\nu)n^2\left[\lambda I'(\lambda) - I_n(\lambda)\right]} \tag{105}$$

where $\lambda = \gamma a$. Itao and Crandall (1979) performed a comprehensive numerical solution of eigenvalue roots of the characteristic equation (105) and of the associated mode shapes. The eigenvalues of Eq. (105) were presented in the form $\lambda_{j,p}$, where $p = 0, 1, \ldots$ is the number of nodal diameters and $j = 1, 2, \ldots$ is the number of nodal circles. (The case $j = 0$ yields a triple root at $\lambda = 0$ that corresponds to three rigid-body motion modes of a free plate.) The mode shapes were presented in the form

$$W_{j,p}(r,\theta) = A_{j,p}\left[J_p(\lambda_{j,p} r/a) + C_{j,p} I_p(\lambda_{j,p} r/a)\right]\cos p\theta \tag{106}$$

A sample of values for the eigenvalue $\lambda_{j,p}$, the mode shape parameter $C_{j,p}$, and amplitude $A_{j,p}$ are given in Table 4.2. Further numerical values can be found in Itao and Crandall (1979).

TABLE 4.2 Eigenvalue $\lambda_{j,p}$, the mode shape parameter $C_{j,p}$, and amplitude $A_{j,p}$ for calculating the flexural vibration of a free circular plate (Itao and Crandall, 1979)

	$j=0$			$j=1$		
p	$\lambda_{0,p}$	$C_{0,p}$	$A_{0,p}$	$\lambda_{1,p}$	$C_{1,p}$	$A_{1,p}$
0	–	–	–	3.01146	−0.83810E−01	0.21979E+01
1	–	–	–	4.52914	−0.19007E−01	0.38359E+01
2	2.29391	0.22191E+00	0.36597E+01	5.93654	−0.55654E−02	0.44282E+01
3	3.49913	0.96357E−01	0.45349E+01	7.27468	−0.18637E−02	0.49565E+01
4	4.63974	0.45825E−01	0.53282E+01	8.56611	−0.67890E−03	0.54458E+01
5	5.74994	0.22796E−01	0.60587E+01	9.82382	−0.26225E−03	0.58981E+01
6	6.84169	0.11661E−01	0.67422E+01	11.05592	−0.10584E−03	0.63291E+01
7	7.92082	0.60806E−02	0.73889E+01	12.26783	−0.44198E−04	0.67401E+01
8	8.99069	0.32162E−02	0.80058E+01	13.46335	−0.18972E−04	0.71345E+01
9	10.05343	0.17200E−02	0.85981E+01	14.64527	−0.83316E−05	0.75145E+01
10	11.11048	0.92790E−03	0.91694E+01	15.81570	−0.37296E−05	0.78820E+01

The natural frequencies associated with each eigenvalue and mode shape are calculated with the formula

$$\omega_{j,p} = \left(\frac{D}{\rho h a^4}\right)^{1/2} \lambda_{j,p}^2 \tag{107}$$

The mode shape amplitudes, $A_{j,p}$, were mass-normalized with the formula

$$\int_0^{2\pi} \int_0^a \rho h \, W_{j,p}^2(r,\theta) r \, \mathrm{d}r \, \mathrm{d}\theta = \rho \pi a^2 h = m \tag{108}$$

where m is the total mass of the plate.

The mode shapes described by Eq. (106) are orthogonal in the sense that

$$\int_0^{2\pi} \int_0^a \rho h \, W_{j,p} W_{i,q} r \, \mathrm{d}r \, \mathrm{d}\theta = m \delta_{i,j} \delta_{p,q} \tag{109}$$

where $\delta_{i,j}$ is the Kronecker delta ($\delta_{i,j} = 1$ for $i = j$; and $\delta_{i,j} = 0$ otherwise). Explicit expressions for $A_{j,p}$ are obtained by substituting Eq. (106) into Eq. (108). Itao and Crandall (1979) give

$$A_{j,p}^{-2} = \begin{cases} 2\left\{\left[J_0(\lambda_{j,0}) + C_{j,0} I_0(\lambda_{j,0})\right]^2 + \left[J_0'(\lambda_{j,0})\right]^2 - \left[C_{j,0} I_0'(\lambda_{j,0})\right]^2\right\} & \text{for } p = 0 \\ 2\left\{\begin{array}{l}\left[J_p(\lambda_{j,p}) + C_{j,p} I_p(\lambda_{j,p})\right]^2 - \dfrac{p^2}{\lambda_{j,p}^2}\left[J_p^2(\lambda_{j,p}) + C_{j,p}^2 I_p^2(\lambda_{j,p})\right] \\ -\left[J_p'(\lambda_{j,p})\right]^2 - \left[C_{j,p} I_p'(\lambda_{j,p})\right]^2\end{array}\right\} & \text{for } p \neq 0 \end{cases} \tag{110}$$

An alternative form of Eq. (110) could be

$$A_{j,p} = \frac{1}{\sqrt{2}}\left\{\begin{array}{l}\left[J_p(\lambda_{j,p}) + C_{j,p} I_p(\lambda_{j,p})\right]^2 - \dfrac{p^2}{\lambda_{j,p}^2}\left[J_p^2(\lambda_{j,p}) + C_{j,p}^2 I_p^2(\lambda_{j,p})\right] \\ -\left[J_p'(\lambda_{j,p})\right]^2 - \left[C_{j,p} I_p'(\lambda_{j,p})\right]^2\end{array}\right\}^{-1/2} \tag{111}$$

4.5.4 AXISYMMETRIC FLEXURAL VIBRATION OF CIRCULAR PLATES

Axisymmetric flexural vibration of a circular plate can be understood in terms of standing circular-crested waves that propagate in a concentric circular pattern from the center of the plate and reflect at the plate circumference. The problem is θ-invariant and the biharmonic operator simplifies to the form

$$\begin{aligned}\nabla^4 = \nabla^2\nabla^2 &= \left(\frac{\partial^2}{\partial r^2} + \frac{1}{r}\frac{\partial}{\partial r}\right)\left(\frac{\partial^2}{\partial r^2} + \frac{1}{r}\frac{\partial}{\partial r}\right) \quad \text{(biharmonic operator for axial symmetry)} \\ &= \frac{\partial^4}{\partial r^4} + \frac{2}{r}\frac{\partial^3}{\partial r^3} - \frac{1}{r^2}\frac{\partial^2}{\partial r^2} + \frac{1}{r^3}\frac{\partial}{\partial r}\end{aligned} \tag{112}$$

Hence, Eq. (91) takes the form

$$D\left(\frac{\partial^4 w}{\partial r^4} + \frac{2}{r}\frac{\partial^3 w}{\partial r^3} - \frac{1}{r^2}\frac{\partial^2 w}{\partial r^2} + \frac{1}{r^3}\frac{\partial w}{\partial r}\right) + \rho h \ddot{w} = 0 \tag{113}$$

Assume harmonic motion with frequency ω, i.e.,

$$w(r, t) = \hat{w}(r)e^{i\omega t} \tag{114}$$

Substitution of Eq. (114) into Eq. (91) yields

$$D\nabla^4 \hat{w} - \omega^2 \rho h \hat{w} = 0 \tag{115}$$

Define the constant

$$\gamma^4 = \frac{\rho h}{D}\omega^2 \quad \text{or} \quad \gamma = \left(\frac{\rho h}{D}\right)^{1/4} \sqrt{\omega} \tag{116}$$

Substitution of Eq. (116) into Eq. (115) gives

$$\left(\nabla^4 - \gamma^4\right) \hat{w} = 0 \tag{117}$$

This can be also expressed as

$$\left(\nabla^2 - \gamma^2\right)\left(\nabla^2 + \gamma^2\right) \hat{w} = 0 \tag{118}$$

As the order in which the differentiation is done does not matter, Eq. (118) will be satisfied when either of the following expression is satisfied:

$$\text{either} \quad \left(\nabla^2 + \gamma^2\right) \hat{w} = 0 \quad \text{or} \quad \left(\nabla^2 - \gamma^2\right) \hat{w} = 0 \tag{119}$$

i.e.,

$$\text{either} \quad \left(\frac{\partial^2}{\partial r^2} + \frac{1}{r}\frac{\partial}{\partial r} + \gamma^2\right) \hat{w} = 0 \quad \text{or} \quad \left(\frac{\partial^2}{\partial r^2} + \frac{1}{r}\frac{\partial}{\partial r} - \gamma^2\right) \hat{w} = 0 \tag{120}$$

Define the substitution

$$x = \begin{cases} \gamma r & \text{for } +\gamma^2 \\ i\gamma r & \text{for } -\gamma^2 \end{cases} \tag{121}$$

Upon substitution of Eq. (121) into Eq. (120) and observing that the partial derivatives are actually simple derivatives as $\hat{w}$ is function only of the space variable, we get the Bessel equation of order zero, i.e.,

$$x^2 \frac{d^2\hat{w}}{dx^2} + x\frac{d\hat{w}}{dx} + x^2\hat{w} = 0 \tag{122}$$

The general solution of Eq. (122) is obtained in terms of Bessel functions, i.e.,

$$\hat{w} = DJ_0(x) + EY_0(x) + FI_0(x) + GK_0(x) \tag{123}$$

where $J_0(x)$ and $Y_0(x)$ are Bessel functions of first and second kind and order zero, whereas $I_0(\gamma r)$ and $K_0(\gamma r)$ are the modified Bessel functions of first and second kind and order zero. The constants D, E, F, G are to be determined from the initial and boundary conditions. Substituting $x = \gamma r$ in Eq. (123) yields

$$\hat{w}\ (r) = DJ_0(\gamma r) + EY_0(\gamma r) + FI_0(\gamma r) + GK_0(\gamma r) \tag{124}$$

However, the Bessel functions $Y_0(\gamma r)$ and $K_0(\gamma r)$ have infinite values at $r = 0$ and have to be discarded (unless the plate has a hole around $r = 0$, which is not the case considered here). Hence, Eq. (124) becomes

$$\hat{w}\,(r) = DJ_0(\gamma r) + FI_0(\gamma r) \tag{125}$$

Using Eq. (125) into Eq. (114) yields the *general solution for axisymmetric flexural vibration of circular plates* in the form

$$w\,(r,t) = [DJ_0(\gamma r) + FI_0(\gamma r)]\,e^{i\omega t} \tag{126}$$

The boundary conditions for a free circular plate of outer radius a are given by

$$\begin{aligned} M_r(a) &= 0 \\ V_r(a) &= 0 \end{aligned} \tag{127}$$

For axisymmetric vibration, Eqs. (86), (87), (89) loose their θ dependence and simply become

$$\begin{aligned} M_r &= -D\left[\frac{\partial^2 w}{\partial r^2} + \nu\frac{1}{r}\frac{\partial w}{\partial r}\right] \\ M_\theta &= -D\left[\frac{1}{r}\frac{\partial w}{\partial r} + \nu\frac{\partial^2 w}{\partial r^2}\right] \\ M_{r\theta} &= 0 \end{aligned} \tag{128}$$

$$\begin{aligned} Q_r &= -D\frac{\partial}{\partial r}\left(\nabla^2 w\right) \\ Q_\theta &= 0 \end{aligned} \tag{129}$$

where the Laplace operator in polar coordinates for axisymmetric motion has the form

$$\nabla^2 = \frac{\partial^2}{\partial r^2} + \frac{1}{r}\frac{\partial}{\partial r} \quad \text{(axisymmetric Laplace operator)} \tag{130}$$

For axisymmetric vibration, the Kelvin–Kirchhoff edge reactions are simply given by

$$\begin{aligned} V_r &= Q_r = -D\frac{\partial}{\partial r}\left(\nabla^2 w\right) \\ V_\theta &= Q_\theta = 0 \end{aligned} \tag{131}$$

Substitution of boundary conditions (127) into Eqs. (128), (130) and use of Eqs. (125), (126) yields a homogenous algebraic system for calculating the constants D and F. Nontrivial solutions of the homogenous algebraic system exist if the determinant is zero. Hence, one gets the characteristic equation

$$\frac{\lambda^2 J_0(\lambda) + (1-\nu)\lambda^2 J_0'(\lambda)}{\lambda^2 I_0(\lambda) + (1-\nu)\lambda^2 I_0'(\lambda)} = \frac{\lambda^3 J_0'(\lambda)}{\lambda^3 I_0'(\lambda)} \tag{132}$$

where $\lambda = \gamma a$. Eq. (132) is a transcendental equation that can be solved numerically. Upon numerical solution, one finds

$$\lambda = 3.01146,\ 6.20540,\ 9.37084, \ldots \tag{133}$$

The natural frequencies associated with each eigenvalue and mode shape are calculated with the formula

$$\omega_j = \left(\frac{D}{\rho h a^4}\right)^{1/2} \lambda_j^2 \tag{134}$$

For each eigenvalue, λ, one finds the corresponding mode shape by calculating the constants D and F in Eq. (125). However, only the ratio of the constants D and F are determined through this process; their exact value is determined through a normalization process. Hence, we write the general expression of the mode shape using an amplitude A_j and a mode shape parameter C_j, i.e.,

$$W_j(r) = A_j\left[J_0(\lambda_j r/a) + C_j I_0(\lambda_j r/a)\right] \tag{135}$$

The mode shape amplitudes A_j are obtained through a mass-normalization process, i.e.,

$$\int_0^{2\pi}\int_0^a \rho h\ W_j^2(r)\ r\,\mathrm{d}r\,\mathrm{d}\theta = \rho\pi a^2 h = m \tag{136}$$

where m is the total mass of the plate.

The mode shapes described by Eq. (135) are orthogonal in the sense that

$$\int_0^{2\pi}\int_0^a \rho h\ W_j W_i r\,\mathrm{d}r\,\mathrm{d}\theta = m\delta_{i,j} \tag{137}$$

where $\delta_{i,j}$ is the Kronecker delta ($\delta_{i,j} = 1$ for $i = j$; and $\delta_{i,j} = 0$ otherwise). Explicit expressions for A_j are obtained by substituting Eq. (135) into Eq. (136), i.e.,

$$A_j = \frac{1}{\sqrt{2}}\left\{\left[J_0(\lambda_j) + C_j I_0(\lambda_j)\right]^2 - \left[J_0'(\lambda_j)\right]^2 - \left[C_j I_0'(\lambda_j)\right]^2\right\}^{-1/2} \tag{138}$$

Values of the eigenvalue λ_j, mode shape parameter C_j, and mode shape amplitude A_j are given in Table 4.3. Representative plots of the mode shapes are given in Fig. 4.14.

TABLE 4.3 Eigenvalue λ_j, the mode shape parameter C_j, and amplitude A_j for calculating the axisymmetric flexural vibration of a free circular plate

j	λ_j	C_j	A_j
1	3.01146	−0.83810E − 01	2.1979
2	6.20540	0.31191E − 02	3.1389
3	9.37084	−0.12770E − 03	3.8468
4	12.52518	0.53684E − 05	4.4425
5	15.67466	−0.22815E − 06	4.9671
6	18.82161	0.97495e − 08	5.4413
7	21.96708	−0.41794E − 09	5.8778
8	25.11160	0.17952E − 10	6.2831

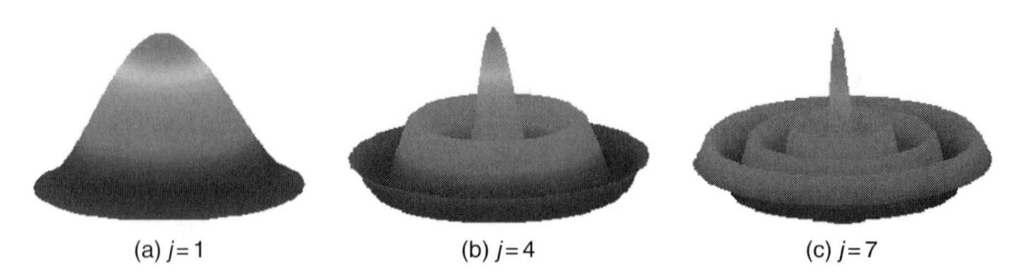

FIGURE 4.14 Flexural mode shapes of free circular plate undergoing axisymmetric vibration.

4.6 PROBLEMS AND EXERCISES

1. Find the first, second, and third natural frequencies of in-plane axial vibration of a circular aluminum plate of thickness 0.8 mm, diameter 100 mm, modulus E = 70 GPa, Poisson ratio $\nu = 0.33$ and density $\rho = 2.7\,\text{g/cm}^3$.
2. Find all the natural frequencies in the interval 10 kHz to 40 kHz of in-plane axial vibration of a circular aluminum plate of thickness 0.8 mm, diameter 100 mm, modulus E = 70 GPa, Poisson ratio $\nu = 0.33$, and density $\rho = 2.7\,\text{g/cm}^3$.
3. Find the first, second, and third natural frequencies of out-of-plane flexural vibration of a circular plate aluminum of thickness 0.8 mm, diameter 100 mm, modulus E = 70 GPa, Poisson ratio $\nu = 0.33$, and density $\rho = 2.7\,\text{g/cm}^3$.
4. Find all the natural frequencies in the interval 10 kHz to 40 kHz of out-of-plane flexural vibration of a circular aluminum plate of thickness 0.8 mm, diameter 100 mm, modulus E = 70 GPa, Poisson ratio $\nu = 0.33$, and density $\rho = 2.7\,\text{g/cm}^3$.

5

ELASTIC WAVES IN SOLIDS AND STRUCTURES

5.1 INTRODUCTION

5.1.1 CHAPTER OVERVIEW

This chapter presents a review of elastic wave propagation in elastic media. SHM methods based on elastic waves propagation are very diverse, and a number of approaches exist. However, a good understanding of SHM wave propagation methods cannot be achieved before a fundamental grasp of the basic principles that lay at the foundation of wave generation and propagation in solid media.

The chapter adopts an incremental step-by-step approach to the description and presentation of the wave propagation problem, which can become quite complicated in some cases. The chapter starts with the discussion of the simplest case of wave propagation – the study of the axial waves propagating in a straight bar. This simple physical example is used to develop fundamental principles of wave propagation, such as wave equation and wave speed; d'Alembert (generic) solution and separation of variables (harmonic) solution; the contrast between wave speed and particle velocity; acoustic impedance of the medium, and the wave propagation at material interfaces. The concept of standing waves is introduced, and the correspondence between standing waves and structural vibration is established. The power and energy associated with wave propagation in a simple bar are introduced and discussed.

After studying the propagation of simple axial waves in bars, the more complicated problem of flexural wave propagation in beams is introduced and discussed. The equation of motion for flexural waves is derived and discussed. The general solution in terms of propagating and evanescent waves is derived. The dispersive nature of flexural waves is identified and studied. In this context, the concept of *group velocity* as different from the *wave speed* (a.k.a., *phase velocity*) is introduced, and its implications on wave propagation are studied. Power and energy of flexural waves and the energy velocity of dispersive waves are discussed.

The discussion of wave propagation is next extended to plate waves. Two types are considered: (1) axial waves in plates and (2) flexural waves in plates. Each case is discussed separately. The general equations of wave motion are derived. Solutions are derived for certain selected cases such as straight-crested axial waves; straight-crested shear waves; circular-crested axial waves; straight-crested flexural waves; circular-crested flexural waves, etc. The dispersive nature of flexural waves in plates and its connection to the dispersive nature of flexural waves in beams are identified and discussed.

The last part of this chapter is allotted to the discussion of 3-D waves that appear in unbounded solids. The general equations of 3-D wave propagation in unbounded solid media are developed from first principles. The eigenvalues and eigenvectors of the wave equation are identified. The two corresponding basic wave types, *pressure waves* (a.k.a., longitudinal waves) and *shear waves* (a.k.a., transverse waves), are discussed. Dilatational, rotational, irrotational, and equivolume waves are identified and discussed. The case of z-invariant wave propagation (plane strain) is presented.

5.1.2 OVERVIEW OF WAVE PROPAGATION THEORY

Waves are disturbances that travel, or propagate, from one region of space to another. Different types of waves must be studied under the underlying phenomenon. Our objective is to mode the wave propagation for different types of ultrasonic waves. Table 5.1 shows some of the waves types possible in elastic solid media.

TABLE 5.1 Waves in elastic solids

Wave type	Particle motion, main assumptions
Pressure (a.k.a. longitudinal; compressional; dilatational; P-waves, axial waves)	Parallel to the direction of wave propagation
Shear (a.k.a., transverse waves; distortional waves; S-waves)	Perpendicular to the direction of wave propagation
Flexural (a.k.a., bending waves)	Elliptical, plane sections remain plane
Rayleigh (a.k.a., surface acoustic waves, SAW)	Elliptical, amplitude decays quickly with depth
Lamb (a.k.a., guided plate waves)	Elliptical, free-surface conditions satisfied at the upper and lower plate surface

To achieve better understanding of these waves, visualization of the waveforms was done. Putting the wave equations into mathematics software, the particle displacement as function of space and time was calculated. The resultant displacement can be shown in an array of vectors, which is like a snapshot of the particles at that moment. By showing the snapshots at different times, the propagation of the wave can be animated. The wave figures in the following sections are part of the result of this effort. Animations of waves were posted on the Internet at http://www.me.sc.edu/research/lamss/default.htm under the research section.

5.2 AXIAL WAVES IN BARS

Axial waves in bars are the simplest conceptualization of elastic wave motion. Assume a uniform bar of axial stiffness EA and mass per unit length $m = \rho A$, where ρ is the mass density and A is the cross-sectional area (Fig. 5.1a). The length of the bar and the boundary conditions are, so far, unspecified. However, the bar is considered to be long and slender. The bar undergoes time-varying axial displacement $u(x, t)$. Consider

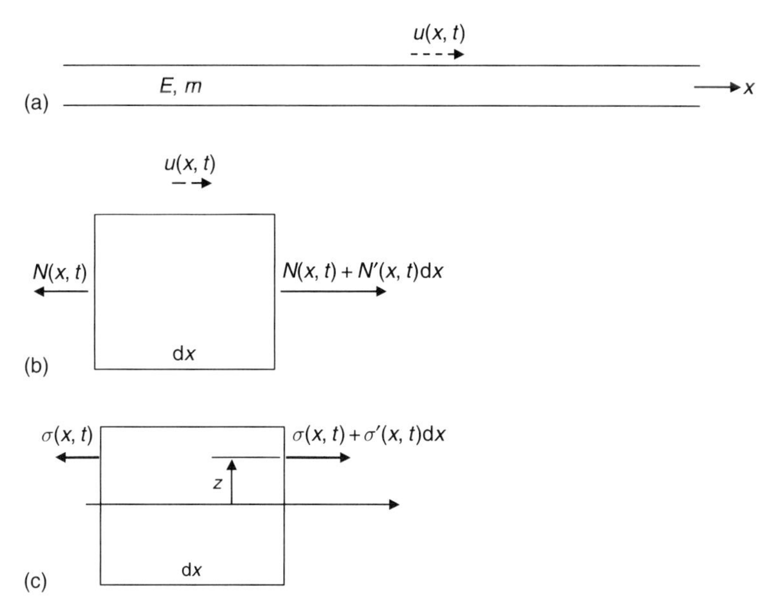

FIGURE 5.1 Uniform bar undergoing axial vibration: (a) general schematic; (b) infinitesimal axial element; (c) thickness-wise stress distribution.

an infinitesimal element, dx, cut out of the bar as shown in Fig. 5.1b. Assume that the stresses are uniformly distributed across the cross-section A.

5.2.1 WAVE EQUATION

Free body analysis of the infinitesimal element dx shown in Fig. 5.1b yields

$$N(x, t) + N'(x, t)\mathrm{d}x - N(x, t) = m\ddot{u}(x, t) \tag{1}$$

where $N(x, t)$ is the axial load. Upon simplification, we get

$$N'(x, t)\mathrm{d}x = m\ddot{u}(x, t) \tag{2}$$

The N stress resultant is evaluated by integration across the cross-sectional area of the direct stress shown in Fig. 5.1c, i.e.,

$$N(x, t) = \int_A \sigma(x, z, t)\mathrm{d}A \tag{3}$$

Recall the strain–displacement relation

$$\varepsilon = u' \tag{4}$$

and the stress–strain constitutive equation

$$\sigma = E\varepsilon \tag{5}$$

where E is Young modulus of elasticity. Substitution of Eqs. (4) and (5) into Eq. (3) yields

$$N(x,t) = \int_A Eu'(x,t)\mathrm{d}A = EAu'(x,t) \tag{6}$$

where EA is the *axial stiffness*. Differentiation of Eq. (6) with respect to x and substitution into Eq. (2) yields the *equation of motion for axial waves in a bar*, i.e.,

$$EAu'' = m\ddot{u} \tag{7}$$

Upon division by m, we get the *wave equation*, i.e.,

$$c^2u'' = \ddot{u} \tag{8}$$

where c is the *wave speed in the bar* given by

$$c = \sqrt{\frac{EA}{m}} \quad \text{or} \quad c^2 = \frac{EA}{m} \tag{9}$$

For a uniform bar, we have $m = \rho A$. Hence, Eq. (9) simplifies to

$$c = \sqrt{\frac{E}{\rho}} \quad \text{or} \quad c^2 = \frac{E}{\rho} \tag{10}$$

Typical values of the wave speed c in various materials are given in Table 5.2.

5.2.2 D'ALEMBERT SOLUTION TO THE WAVE EQUATION

The propagation aspects of the wave motion are highlighted by the d'Alembert solution. Recall Eq. (8) in the explicit differentiation format

$$c^2\frac{\partial^2 u}{\partial x^2} = \frac{\partial^2 u}{\partial t^2} \tag{11}$$

TABLE 5.2 Wave speed in a bar for several materials

Material	10^{-3} m/s	10^{-4} in./s
Aluminum	5.23	20.6
Brass	3.43	13.5
Cadmium	2.39	9.4
Copper	3.58	14.1
Gold	2.03	8.0
Iron	5.18	20.4
Lead	1.14	4.5
Magnesium	4.90	19.3
Nickel	4.75	18.7
Silver	2.64	10.4
Steel	5.06	19.9
Tin	2.72	10.7
Tungsten	4.29	16.9
Zinc	3.81	15.0

Assume that change of independent variables

$$\xi = x - ct \qquad \eta = x + ct \tag{12}$$

Upon differentiation,

$$\begin{aligned} \frac{\partial \xi}{\partial x} &= 1 \qquad \frac{\partial \eta}{\partial x} = 1 \\ \frac{\partial \xi}{\partial t} &= -c \qquad \frac{\partial \eta}{\partial t} = c \end{aligned} \tag{13}$$

Using chain differentiation and the results of Eq. (13), we write

$$\begin{aligned} \frac{\partial u}{\partial x} &= \frac{\partial u}{\partial \xi}\frac{\partial \xi}{\partial x} + \frac{\partial u}{\partial \eta}\frac{\partial \eta}{\partial x} = \frac{\partial u}{\partial \xi} + \frac{\partial u}{\partial \eta} \\ \frac{\partial u}{\partial t} &= \frac{\partial u}{\partial \xi}\frac{\partial \xi}{\partial t} + \frac{\partial u}{\partial \eta}\frac{\partial \eta}{\partial t} = -c\frac{\partial u}{\partial \xi} + c\frac{\partial u}{\partial \eta} \end{aligned} \tag{14}$$

Further differentiation yields, upon simplification,

$$\begin{aligned} \frac{\partial^2 u}{\partial x^2} &= \frac{\partial^2 u}{\partial \xi^2} + 2\frac{\partial^2 u}{\partial \xi\,\partial \eta} + \frac{\partial^2 u}{\partial \eta^2} \\ \frac{\partial^2 u}{\partial t^2} &= c^2\frac{\partial^2 u}{\partial \xi^2} - 2c^2\frac{\partial^2 u}{\partial \xi\,\partial \eta} + c^2\frac{\partial^2 u}{\partial \eta^2} \end{aligned} \tag{15}$$

Substitution of Eq. (15) into Eq. (11) yields, upon simplification,

$$\frac{\partial^2 u}{\partial \xi\,\partial \eta} = 0 \tag{16}$$

Equation (16) is integrated with respect to η and then with respect to ξ. Integration with respect to η yields

$$\frac{\partial u}{\partial \xi} = F(\xi) \tag{17}$$

where $F(\xi)$ is a function of ξ only. Then, integration of Eq. (17) with respect to ξ yields

$$u = \int F(\xi)\mathrm{d}\xi + g(\eta) \tag{18}$$

Denoting $f(\xi) = \int F(\xi)\mathrm{d}\xi$, Eq. (18) is recast in the form

$$u(\xi, \eta) = f(\xi) + g(\eta) \tag{19}$$

Using Eq. (12) to change back to the original variables (x, t) gives Eq. (19) the form

$$u(x, t) = g(x + ct) + f(x - ct) \tag{20}$$

The functions $f(x - ct)$ and $g(x + ct)$ in Eq. (20) are single-argument functions of two variables, x and t. To facilitate understanding, the single-argument character of these

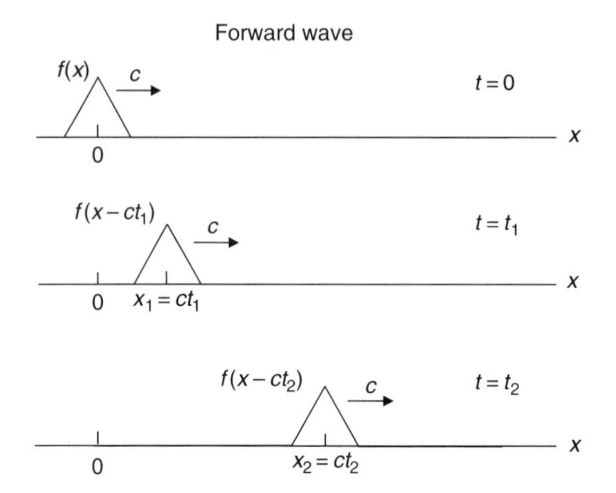

FIGURE 5.2 Propagation of a forward wave pulse $f(x-ct)$.

functions can be highlighted using a generic arbitrary argument z, and then examine the functions $f(z)$ and $g(z)$ to understand their shape and form. The actual two-variable dependence of these functions is recovered by replacing the generic z with the appropriate expressions of x and t, i.e., $x-ct$ to get $f(x-ct)$ or $x+ct$ to get $g(x+ct)$.

Equation (20) highlights the propagating-wave character of the solution to the wave equation. The function $f(x-ct)$ represents a *forward propagating wave*, whereas the function $g(x+ct)$ represents a *backward propagating wave*. This can be easily verified. Consider, for example, a wave pulse $f(x-ct)$ propagating forward as depicted in Fig. 5.2. Consider three time instances, $0 < t_1 < t_2$, and examine the function $f(x-ct)$ at these three instances. At $t=0$, the function sits at the origin. This is the *initial wave*. At $t=t_1$, the peak of the initial wave has moved to the position $x_1=ct_1$ because for this value, the argument of the function $f(z)$ is zero, i.e., $z_1=x_1-ct_1=0$. For $t=t_2$, the wave has moved even further to the right to position $x_2=ct_2$, and so on. This proves that $f(x-ct)$ *propagates in the positive x direction with speed c*. For the backward wave $g(x+ct)$, the same argument applies, only that the wave motion occurs backwards, i.e., to the left.

5.2.3 THE INITIAL VALUE PROBLEM FOR WAVE PROPAGATION

The initial value problem (Cauchy problem) for wave propagation consists in finding the solution of the wave equation given in Eq. (8) subject to the *initial conditions*

$$\begin{aligned} u(x,0) &= u_0(x), \quad \text{initial deformed shape of the bar} \\ \dot{u}(x,0) &= v_0(x), \quad \text{initial velocity of the points along the bar} \end{aligned} \tag{21}$$

The general solution (20) will be substituted into the initial conditions (21). Before doing so, we first notice that

$$\dot{u}(x,t) = \frac{\partial}{\partial t}[f(x-ct)+g(x-ct)] = -cf'(x-ct)-cg'(x-ct) \tag{22}$$

Then, substitution of Eqs. (20) and (22) into Eq. (21) yields

$$u(x,0) = f(x)+g(x) = u_0(x) \tag{23}$$

and

$$\dot{u}(x,0) = -cf'(x) + cg'(x) = v_0(x) \tag{24}$$

Integration of Eq. (24) with respect to x gives, upon rearrangement,

$$f(x) - g(x) = -\frac{1}{c}\int_b^x v_0(x^*)\mathrm{d}x^* \tag{25}$$

where b is an arbitrary auxiliary constant and x^* is an auxiliary integration variable. Combining Eqs. (23) with (25) and changing from x to z yields the expressions of the single-variable functions $f(z)$ and $g(z)$, i.e.,

$$\begin{aligned} f(z) &= \frac{1}{2}u_0(z) - \frac{1}{2c}\int_b^z v_0(z^*)\mathrm{d}z^* \\ g(z) &= \frac{1}{2}u_0(z) + \frac{1}{2c}\int_b^z v_0(z^*)\mathrm{d}z^* \end{aligned} \tag{26}$$

where z^* is an auxiliary integration variable. Substitution of Eq. (26) into Eq. (20) yields

$$u(x,t) = \frac{1}{2}\left[u_0(x-ct) + u_0(x+ct)\right] + \frac{1}{2c}\int_{x-ct}^{x+ct} v_0(z^*)\mathrm{d}z^* \tag{27}$$

In arriving at Eq. (27), we used the integration rule

$$-\int_b^{x-ct} + \int_b^{x+ct} = \int_{x-ct}^{b} + \int_b^{x+ct} = \int_{x-ct}^{x+ct} \tag{28}$$

5.2.4 STRAIN WAVES AND STRESS WAVES

So far, we have discussed the wave phenomenon in terms of $u(x, t)$, which has the physical meaning of a displacement disturbance propagating along the medium. However, associated with the same wave phenomenon are strain and stress disturbances. Therefore, it is beneficial to consider a description of the wave phenomenon in terms of strain $\varepsilon(x, t)$ and stress $\sigma(x, t)$.

Recall Eq. (8), i.e.,

$$c^2 u'' = \ddot{u} \tag{29}$$

Differentiation with respect to x yields

$$c^2 u''' = \ddot{u}' \tag{30}$$

Recall Eq. (4) defining the strain $\varepsilon = u'$; upon substitution into Eq. (30), we get the *strain wave equation*

$$c^2 \varepsilon'' = \ddot{\varepsilon} \tag{31}$$

Multiplication of Eq. (31) by E on both sides yields

$$c^2 E\varepsilon'' = E\ddot{\varepsilon} \tag{32}$$

Recalling the strain–strain relation (5), we obtain the *stress wave equation*

$$c^2\sigma'' = \ddot{\sigma} \tag{33}$$

We note that both the strain wave equation (31) and the stress wave equation (33) have exactly the same form as the displacement wave equation (8). The results obtained for the displacement wave formulation can be directly transferred to the strain wave formulation and the stress wave formulation. For example, it is apparent from Eqs. (8), (31), and (33) that displacement waves, strain waves, and stress waves travel at exactly the same speed, c. However, this should not be of any surprise as all three waves are just different facets of the same physical phenomenon, i.e., a disturbance traveling as an elastic wave through the medium.

5.2.5 PARTICLE VELOCITY VS. WAVE VELOCITY

We have seen in Table 5.2 that the speed with which the wave disturbance can propagate through the elastic media is quite large, of the order of several kilometers or miles per second. However, the question arises about the speed with which the actual medium moves; the answer to this question is provided by the study of the *particle velocity*. Recall that, in the displacement-wave formulation, the displacement $u(x, t)$ signifies the physical motion of the particles inside the medium. Then, the speed with which these particles move, or the *particle velocity*, is given by $\dot{u}(x, t) = \frac{\partial}{\partial t}u(x, t)$. Assume, for the sake of argument, a generic forward wave of the form

$$u(x, t) = f(x - ct) \tag{34}$$

Differentiation with respect to t yields the particle velocity

$$\dot{u}(x, t) = -cf'(x - ct) \tag{35}$$

where f' is the derivative of the single-argument function f. On the other hand, the stress produced by this wave is given by

$$\sigma(x, t) = Eu'(x, t) = E\frac{\partial}{\partial x}f(x - ct) = Ef'(x - ct) \tag{36}$$

Combining Eqs. (35) and (36) yields the *particle velocity in terms of the stress in the wave*, i.e.,

$$\dot{u}(x, t) = -\frac{c}{E}\sigma(x, t) \quad \text{(forward wave)} \tag{37}$$

Equation (37) indicates that the particle velocity for a forward wave has sign opposite to the wave stress, i.e., a compression wave would impart a forward particle velocity, which is to be expected. For a backward wave of the form $u(x, t) = g(x + ct)$, a similar analysis yields an expression resembling Eq. (37) but with an opposite sign, i.e.,

$$\dot{u}(x, t) = \frac{c}{E}\sigma(x, t) \quad \text{(backward wave)} \tag{38}$$

Example

Given: Consider a steel medium with $E = 207\,\text{GPa}$ and wave speed $c = 5.1\,\text{km/s}$.
Find: Calculate the particle velocity amplitude for a compression wave of amplitude $\sigma_{\max} = -100\,\text{MPa}$.

Answer: Use Eq. (37) and write

$$\dot{u}_{\max} = \frac{c}{E}\sigma_{\max} = \frac{5.1 \times 10^3}{207 \times 10^9} 100 \times 10^6 \cong 2.5\,\text{m/s}$$

It is apparent from this result that although the wave disturbance travels with a mighty 5100 m/s, the velocity of each particle involved in the wave phenomenon does not exceed a mere 2.5 m/s! This numerical example highlights the difference between the wave speed (representing the speed with which wave information and wave energy propagate) and the particle velocity (representing the actual motion of the particles involved in the wave phenomenon). The two physical quantities can be different by orders of magnitude. Waves can travel at high speeds, while particles in the medium move rather slowly.

5.2.6 ACOUSTIC IMPEDANCE OF THE MEDIUM

The relation between wave stress $\sigma(x, t)$ and particle velocity $\dot{u}(x, t)$ can be described in terms of the *acoustic impedance*. Consider again a forward wave $u(x, t) = f(x - ct)$ and recall Eq. (37), which indicates that the particle velocity amplitude depends not only on the stress amplitude but also on the ratio c/E. Recalling the definition of the wave speed given by Eq. (10), we write

$$\frac{E}{c} = \frac{\rho c^2}{c} = \rho c \tag{39}$$

Hence, Eqs. (37) and (38) can be written as

$$\begin{aligned} \sigma(x, t) &= -\rho c \cdot \dot{u}(x, t) \quad \text{(forward wave)} \\ \sigma(x, t) &= \rho c \cdot \dot{u}(x, t) \quad \text{(backward wave)} \end{aligned} \tag{40}$$

In terms of the relation between the stress and particle velocity amplitudes, both forms of Eq. (40) give

$$\hat{\sigma} = \rho c \cdot \hat{\dot{u}} \tag{41}$$

The quantity ρc is a characteristic property of the medium called the *acoustic impedance*. It signifies how much stress is needed to impart a prescribed velocity to the medium particles. An often used notation for the acoustic impedance is Z. Alternative forms of the acoustic impedance that might be encountered are

$$Z = \rho c = \sqrt{\rho E} = \frac{E}{c} \quad \text{(acoustic impedance of the medium)} \tag{42}$$

Hence, the relation between stress and particle velocity amplitudes can be written as

$$\hat{\sigma} = Z\hat{\dot{u}} \tag{43}$$

5.2.7 WAVE PROPAGATION AT INTERFACES

If the medium is not infinite, but bounded by interfaces with other media, the problem arises of how the wave phenomenon interacts with the interface. For exemplification, consider a wave pulse propagating forward in a bar consisting of two different materials, (E_1, ρ_1) and (E_2, ρ_2), interfaced at $x = 0$, as shown in Fig. 5.3a.

Consideration of the situation at the interface leads as to assume that there will be three waves present in the bar (Fig. 5.3b):

1. incident forward wave, u_i
2. reflected backward wave, u_r
3. transmitted forward wave, u_t.

Express these waves in terms of $f(ct - x)$ and $g(ct + x)$ functions, using subscripts appropriate to the medium in which they travel, i.e.,

$$\begin{aligned} u_i(x,t) &= f_1(c_1 t - x) \\ u_r(x,t) &= g_1(c_1 t + x) \\ u_t(x,t) &= f_2(c_2 t - x) \end{aligned} \tag{44}$$

Note that, for convenience but without loss of generality, the expression of the f and g functions in Eq. (44) is slightly different than that in Eq. (20). Using the strain–displacement relation (4) we write

$$\begin{aligned} \varepsilon_i(x,t) &= u_i'(x,t) = -f_1'(c_1 t - x) \\ \varepsilon_r(x,t) &= u_r'(x,t) = g_1'(c_1 t + x) \\ \varepsilon_t(x,t) &= u_t'(x,t) = -f_2'(c_2 t - x) \end{aligned} \tag{45}$$

The total motion results from the superposition of the individual motions, i.e.,

$$\begin{aligned} u_1(x,t) &= u_i(x,t) + u_r(x,t) = f_1(c_1 t - x) + g_1(c_1 t + x) \\ u_2(x,t) &= u_t(x,t) = f_2(c_2 t - x) \end{aligned} \tag{46}$$

Similarly, the total strains are

$$\begin{aligned} \varepsilon_1(x,t) &= \varepsilon_i(x,t) + \varepsilon_r(x,t) = -f_1'(c_1 t - x) + g_1'(c_1 t + x) \\ \varepsilon_2(x,t) &= \varepsilon_t(x,t) = -f_2'(c_2 t - x) \end{aligned} \tag{47}$$

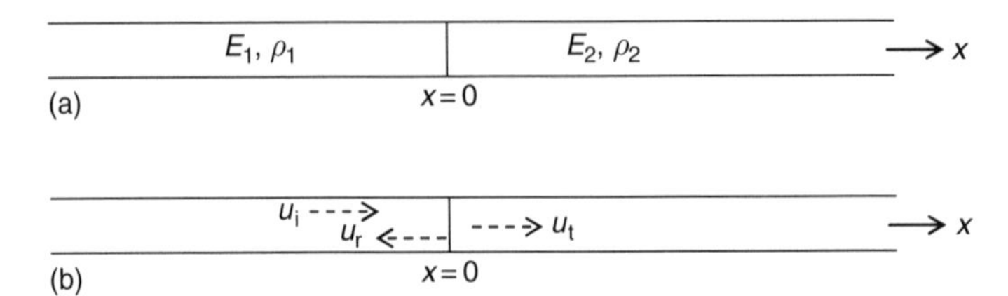

FIGURE 5.3 Uniform bar consisting of two materials interfacing at $x = 0$: (a) geometric and material properties; (b) transmitted and reflected waves.

At the interface, equilibrium and compatibility conditions must be satisfied, i.e.,

$$\sigma_1(0, t) = \sigma_2(0, t) \quad \text{(equilibrium)} \tag{48}$$

$$u_1(0, t) = u_2(0, t) \quad \text{(compatibility)} \tag{49}$$

Using the stress–strain constitutive relation (5) and Eq. (47), we write

$$\begin{aligned} \sigma_1(x, t) &= E_1\varepsilon_1(x, t) = E_1\left[-f_1'(c_1t - x) + g_1'(c_1t + x)\right] \\ \sigma_2(x, t) &= E_2\varepsilon_2(x, t) = -E_2 f_2'(c_2t - x) \end{aligned} \tag{50}$$

Substitution of Eqs. (46) and (50) into Eqs. (48) and (49) yields the interface conditions in the form

$$-E_1 f_1'(c_1t - x) + E_1 g_1'(c_1t + x)|_{x=0} = -E_2 f_2'(c_2t - x)|_{x=0} \quad \text{(equilibrium)} \tag{51}$$

$$f_1(c_1t - x) + g_1(c_1t + x)|_{x=0} = f_2(c_2t - x)|_{x=0} \quad \text{(compatibility)} \tag{52}$$

Imposing the $x = 0$ condition explicitly and differentiating Eq. (52) with respect to t yields the linear system

$$\begin{aligned} -E_1 f_1'(c_1t) + E_1 g_1'(c_1t) &= -E_2 f_2'(c_2t) \\ c_1 f_1'(c_1t) + c_1 g_1'(c_1t) &= c_2 f_2'(c_2t) \end{aligned} \tag{53}$$

The linear algebraic system (53) is used to express the unknown single-argument functions, g_1' and f_2' representing the reflected and transmitted waves with of the known function, f_1' representing the incident wave. Rearrange Eq. (53) in the form

$$\begin{aligned} E_1 g_1' + E_2 f_2' &= E_1 f_1' \\ c_1 g_1' - c_2 f_2' &= -c_1 f_1' \end{aligned} \tag{54}$$

Upon solution, we get

$$\begin{aligned} g_1' &= \frac{E_1c_2 - E_2c_1}{E_1c_2 + E_2c_1} f_1' \\ f_2' &= \frac{2E_1c_1}{E_1c_2 + E_2c_1} f_1' \end{aligned} \tag{55}$$

Using the wave-speed definitions $c_1^2 = E_1/\rho_1$ and $c_2^2 = E_2/\rho_2$, we can write $E_1 = c_1^2\rho_1$ and $E_2 = c_2^2\rho_2$. Substitution into Eq. (55) and simplification by c_1c_2 yields

$$\begin{aligned} g_1' &= \frac{\rho_1c_1 - \rho_2c_2}{\rho_1c_1 + \rho_2c_2} f_1' \\ f_2' &= \frac{c_1}{c_2}\frac{2\rho_1c_1}{\rho_1c_1 + \rho_2c_2} f_1' \end{aligned} \tag{56}$$

Recalling Eq. (45), we write the relation between the strains at the interface $x = 0$ as

$$\begin{aligned} \varepsilon_r &= -\frac{\rho_1c_1 - \rho_2c_2}{\rho_1c_1 + \rho_2c_2}\varepsilon_i \\ \varepsilon_t &= \frac{c_1}{c_2}\frac{2\rho_1c_1}{\rho_1c_1 + \rho_2c_2}\varepsilon_i \end{aligned} \tag{57}$$

As the waves are assumed to propagate self-similar without dispersion and dissipation, Eq. (57) also applies between the peak values elsewhere in the bar. Using the stress–strain relations

$$\begin{aligned}\sigma_i &= E_1\varepsilon_i = \rho_1 c_1^2 \varepsilon_i \\ \sigma_r &= E_1\varepsilon_r = \rho_1 c_1^2 \varepsilon_r \\ \sigma_t &= E_2\varepsilon_t = \rho_2 c_2^2 \varepsilon_t\end{aligned} \tag{58}$$

we write Eq. (57) in terms of stresses, i.e.,

$$\begin{aligned}\sigma_r &= -\frac{\rho_1 c_1 - \rho_2 c_2}{\rho_1 c_1 + \rho_2 c_2}\sigma_i \\ \sigma_t &= \frac{2\rho_2 c_2}{\rho_1 c_1 + \rho_2 c_2}\sigma_i\end{aligned} \tag{59}$$

Note that the second term in Eq. (59), i.e., the expression of the transmitted stress, is substantially different from the corresponding term in Eq. (57). Equation (59) can be expressed using the acoustic impedance notation $Z = \rho c$, i.e.,

$$\begin{aligned}\sigma_r &= -\frac{Z_1 - Z_2}{Z_1 + Z_2}\sigma_i \\ \sigma_t &= \frac{2Z_2}{Z_1 + Z_2}\sigma_i\end{aligned} \tag{60}$$

Special cases that might arise at the interface are (a) acoustically matched interface; (b) soft interface and free end; (c) stiff interface and built-in end (a.k.a., fixed end). These cases are discussed briefly next.

(a) *Acoustically matched interface*. If $Z_2 = Z_1$, then Eq. (60) predicts that the reflected wave becomes zero and the transmitted wave equals the incident wave. This situation is known as *acoustically matched interface*, in which case the interface is 'invisible' to the incoming wave and does not affect it at all.
(b) *Soft interface*. For $Z_2 \ll Z_1$, Eq. (60) predicts that the transmitted wave is much weaker than the reflected wave, $\sigma_t \ll \sigma_r$. In the extreme case of $Z_2 \to 0$, we encounter the case of a *free end*, in which case the second medium disappears altogether. At a free end, the reflected wave is equal and of opposite sign to the incident wave, i.e., an incoming compression wave is reflected as a tension wave. If the material is brittle, this tension wave produced by the reflection may produce spalling of the free end. Another characteristic of the free-end reflection (which can be easily proved from the above analysis) is that the particle velocity at the free-end amplitude is doubled as compared to the particle velocity in the rest of the bar.
(c) *Stiff interface*. For $Z_2 \gg Z_1$, Eq. (60) predicts that the transmitted wave is much stronger than the reflected wave. In the extreme case $Z_2 \to \infty$, we encounter the case of a *rigid built-in end*, in which case the reflected stress wave is equal and of the same sign with the incident wave, and hence the total stress in the bar has twice the amplitude of the incoming wave. This would explain why most failures would happen at a rigid built-in end. The transmitted wave is also double the amplitude of the incoming wave, with corresponding effects on the built-in support material.

5.2.8 SEPARATION OF VARIABLES SOLUTION TO THE WAVE EQUATION: HARMONIC WAVES

Recall the wave equation (8), i.e.,

$$c^2 u''(x, t) = \ddot{u}(x, t) \tag{61}$$

Equation (61) is a PDE in space, x, and time, t. One way of seeking the solution of Eq. (61) is through the method of separation of the variables, which means to express the solution, $u(x, t)$, in terms of a function $X(x)$ that depends only on x, and a function $T(t)$ that depends only on t, i.e.,

$$u(x, t) = X(x) \cdot T(t) \tag{62}$$

Upon differentiation and substitution in the wave Eq. (61), we get

$$c^2 X''(x) T(t) = X(x) \ddot{T}(t) \tag{63}$$

Now, assume that the time behavior is a persistent harmonic oscillation with angular frequency ω, i.e.,

$$T(t) = e^{i\omega t} \tag{64}$$

Such persistent harmonic oscillation in the time domain corresponds to the concept of continuous waves (CW), i.e., harmonic waves that continuously propagate in the medium under a steady-state condition.

Substitution of Eq. (64) into Eq. (63) yields

$$c^2 X''(x) e^{i\omega t} = -\omega^2 X(x) e^{i\omega t} \tag{65}$$

Upon simplification by $e^{i\omega t}$ and rearrangement, we get a simple ordinary differential equation (ODE) in x, i.e.,

$$c^2 X''(x) + \omega^2 X(x) = 0 \tag{66}$$

Denote

$$\gamma = \frac{\omega}{c} \quad \text{(wavenumber)} \tag{67}$$

Substitution of Eq. (67) into Eq. (66) yields

$$X''(x) + \gamma^2 X(x) = 0 \tag{68}$$

Equation (68) has the harmonic solution

$$X(x) = A e^{i\gamma x} + B e^{-i\gamma x} \tag{69}$$

A representation of space-wise harmonic solution given by Eq. (69) is shown in Fig. 5.4, which illustrates the meaning of the wavenumber, γ, and its relation to the *wavelength*, λ, i.e.,

$$\lambda = \frac{2\pi}{\gamma} \quad \text{(wavelength–wavenumber relationship)} \tag{70}$$

Figure 5.4 illustrates the intuitive definition of the wavelength, λ, as the length scale after which the wave repeats itself. This general definition supersedes the concept of

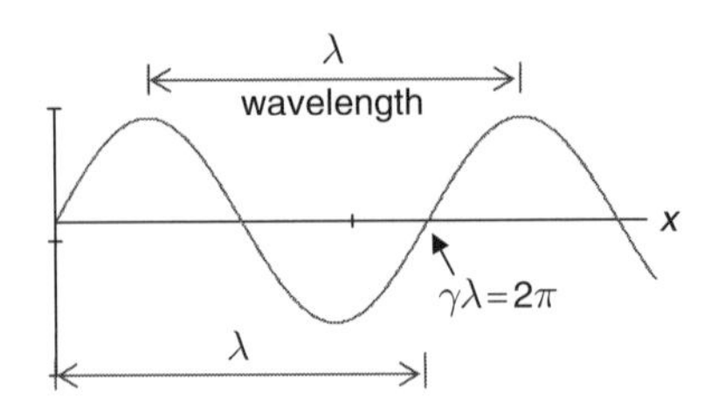

FIGURE 5.4 Relationship between wavenumber and wavelength in a harmonic wave.

harmonic waves and can be applied to any periodic wave. However, Eq. (70) only applies to harmonic waves. An alternative definition of the wavelength is as the distance traveled by the wave in a time period, i.e.,

$$\lambda = cT \quad \text{(wavelength–period relationship)} \tag{71}$$

As the time period, T, is the inverse of frequency, f, the wavelength can be also expressed in terms of frequency, i.e.,

$$\lambda = \frac{c}{f} \quad \text{(wavelength–frequency relationship)} \tag{72}$$

where $f = \omega/2\pi$. Application of Eq. (67) yields Eq. (70).

Substitution of Eqs. (64) and (69) into Eq. (63) yields the general *harmonic wave solution*

$$u(x, t) = Ae^{i(\gamma x+\omega t)} + Be^{-i(\gamma x-\omega t)} \tag{73}$$

where the constants A and B have to be determined specific to each case under consideration.

The wavenumber can be also written in terms of the wavelength, $\lambda = cT$, where T is the oscillation period given by $T = 1/f = 2\pi/\omega$. Alternative ways of writing a forward harmonic wave are

$$u_{\text{forward}}(x, t) = \hat{u}e^{i\omega\left(\frac{x}{c}-t\right)} = \hat{u}e^{i2\pi\left(\frac{x}{\lambda}-\frac{t}{T}\right)} \tag{74}$$

The corresponding expressions for the backward harmonic wave would be

$$u_{\text{backward}}(x, t) = \hat{u}e^{i\omega\left(\frac{x}{c}+t\right)} = \hat{u}e^{i2\pi\left(\frac{x}{\lambda}+\frac{t}{T}\right)} \tag{75}$$

The harmonic wave solution (73) is a subcase of the general propagating wave solution (20) developed through the d'Alembert method. In fact, the first term in Eq. (73) represents a backward wave, whereas the second term represents a forward wave. This assertion becomes even more apparent if we use Eq. (67) to cast Eq. (73) in the form

$$u(x, t) = Ae^{i\gamma(x+ct)} + Be^{-i\gamma(x-ct)} \tag{76}$$

The persistent harmonic solution obtained through the separation-of-variables method is a subcase of the more general periodic waves. The persistent harmonic solution is useful when applying harmonic analysis to analyze more complicated wave forms. However, this solution is rather restrictive. It is a continuous harmonic wave, whereas the general wave solution does not need to be either persistent or harmonic. Nevertheless, the persistent harmonic solution forms the core of wave propagation studies.

5.2.9 STANDING WAVES

The concept of *standing waves* bridges the gap between wave analysis and vibration analysis. We have already noticed that the differential equation of motion for a vibrating bar has the same form as the wave equation. Similarly, the general solution for a vibrating bar is the same as that of a continuous harmonic wave propagating in the bar. The difference between vibration and wave propagation resides mainly in the presence or absence of boundaries. In an unbounded medium, a disturbance source performing continuous oscillations will produce sustained harmonic waves continuously propagating outward toward infinity. However, if we have a bounded medium, then waves will be reflected by the boundary back toward the source and will interfere with the waves emanating from the source, resulting in a complicated process; under certain conditions, this process may yield *standing waves*, or structural vibration. A typical example would be that of a tuning fork. For illustration, consider two identical continuous harmonic waves propagating in opposite directions in an infinite bar, i.e.,

$$u_1(x,t) = \frac{1}{2}e^{-i(\gamma x-\omega t)} \quad \text{(forward wave)} \tag{77}$$

$$u_2(x,t) = \frac{1}{2}e^{i(\gamma x+\omega t)} \quad \text{(backward wave)} \tag{78}$$

The total motion is obtained via the superposition of these two waves, i.e.,

$$\begin{aligned} u(x,t) &= u_1(x,t)+u_2(x,t) = \frac{1}{2}e^{-i(\gamma x-\omega t)}+\frac{1}{2}e^{i(\gamma x+\omega t)} \\ &= \frac{1}{2}\left(e^{i\gamma x}+e^{-i\gamma x}\right)e^{i\omega t} \end{aligned} \tag{79}$$

Recall Euler formula $\cos\alpha = \frac{1}{2}(e^{i\alpha}+e^{-i\alpha})$; hence, Eq. (79) becomes

$$u(x,t) = (\cos\gamma x)e^{i\omega t} \quad \text{(standing wave)} \tag{80}$$

Equation (80) represents a *standing wave* of space-wise shape $\cos\gamma x$ and harmonic time-varying amplitude $e^{i\omega t}$. The standing wave has specific *nodes* and *anti-nodes* (*bows*) that have space-wise fixed positions (Fig. 5.5a). As time progresses, the amplitude of the bows increases and decreases but the nodes always remain fixed. The location of the nodes, x_n, is given by the simple requirement that γx_n is an odd multiple of $\pi/2$ such that the cosine function vanishes. Hence, the spacing of the nodes is π/γ. In view of Eq. (70), this condition becomes

$$\Delta x = \frac{\lambda}{2} \quad \text{(nodes spacing)} \tag{81}$$

Equation (81) indicates that the nodes are spaced at half-wavelength intervals.

Standing waves can be created in finite-length bars under certain conditions. Consider, for example, a bar of length L fixed at both ends. Clearly, the boundary conditions are those of zero displacement at the end. Hence, the ends of the bar must be nodes of the standing wave. This means that a standing wave could exist in this bar if the wave properties are such that the length is a multiple of the half wavelength, i.e.,

$$L = j\frac{\lambda}{2}, \quad j = 1, 2, 3, \ldots \tag{82}$$

An illustration of Eq. (72) is shown in Fig. 5.5b. Half wavelengths may be accommodated with the bar length. Several possibilities exist: one, two, three, etc. Each case corresponds to a different standing wave. These standing waves are actually the natural modes of

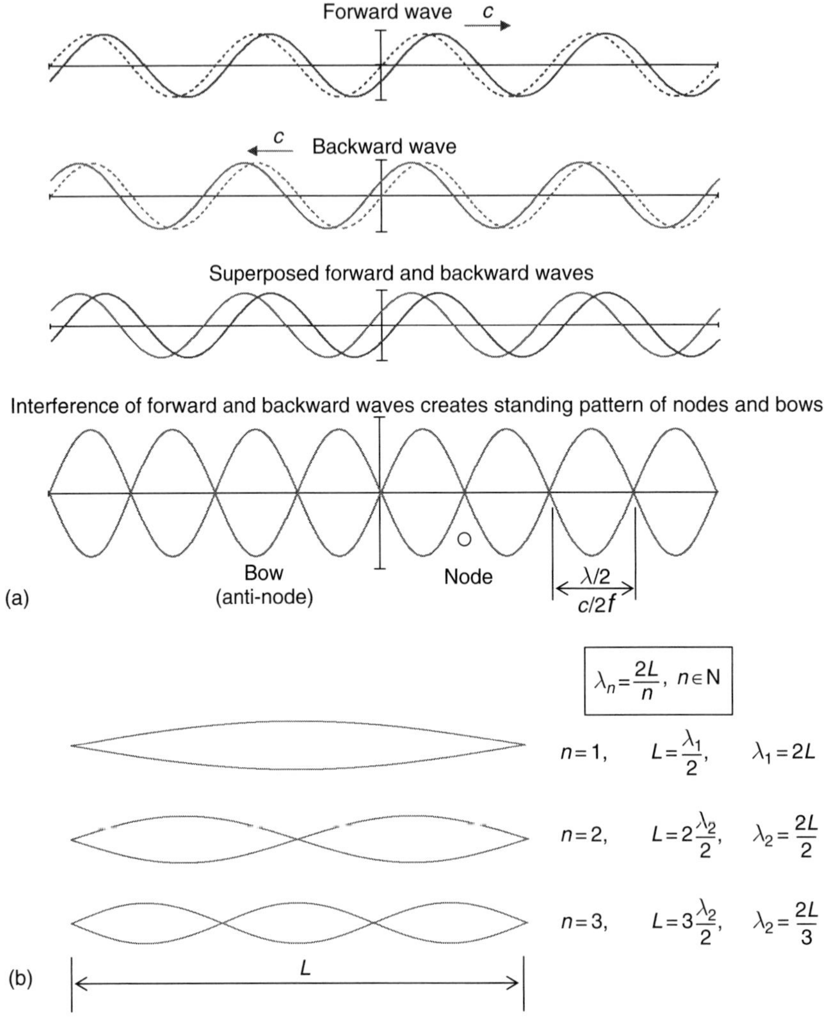

FIGURE 5.5 Standing waves: (a) interference of forward and backward waves results in standing waves; (b) multiple resonances in a finite bar of length L fixed at both ends.

vibration of the bar discussed in Chapter 3. Thus, the connection has been made between standing waves and natural vibration modes of a bounded object. If the bar length and the wavelength of the wave are such that condition (82) is satisfied, then the bar resonates. Using Eq. (72), the condition (82) can be rewritten as

$$L = j\frac{c}{2f}, \quad j = 1, 2, 3, \ldots \tag{83}$$

Equation (83) allows one to either design a bar that would resonate at certain frequencies, or, by inversion, predict the frequencies at which the bar will resonate. Using Eqs. (10) and (83), we can write the resonant axial frequencies of this fixed-fixed bar as

$$f = j\frac{1}{2L}\sqrt{\frac{E}{\rho}}, \quad j = 1, 2, 3, \ldots \tag{84}$$

It is apparent that Eq. (84) is identical with the corresponding equation from Chapter 3.

For bars with other boundary conditions, the same kind of argument applies, only that the appropriate conditions must be imposed at the boundary.

5.2.10 POWER AND ENERGY OF AXIAL WAVES IN A BAR

In this section, we study the energy and power associated with the propagation of axial waves in a bar.

5.2.10.1 Axial Wave Energy

By definition, the kinetic and elastic energy densities per unit volume are given by

$$k = \frac{1}{2}\rho \dot{u}^2 \quad \text{(kinetic energy per unit volume)} \tag{85}$$

$$v = \frac{1}{2}\sigma\varepsilon \quad \text{(elastic energy per unit volume)} \tag{86}$$

To obtain the energy distribution per unit length of bar, we integrate Eqs. (85) and (86) across the cross-section. The particle motion $u(x, t)$ associated with the axial waves propagating in a bar is uniform over the bar cross-section (Fig. 5.6a). Hence,

$$k(x, t) = \int_A \frac{1}{2}\rho \dot{u}^2(x, t)\mathrm{d}A = \frac{1}{2}m\dot{u}^2(x, t) \quad \text{(kinetic energy per unit length)} \tag{87}$$

$$v(x, t) = \int_A \frac{1}{2}\sigma(x, t)\varepsilon(x, t)\mathrm{d}A = \frac{1}{2}(EA)u'^2(x, t) \quad \text{(elastic energy per unit length)} \tag{88}$$

The total energy is calculated by adding the kinetic and elastic energies, i.e.,

$$e(x, t) = k(x, t) + v(x, t) \quad \text{(total energy per unit length)} \tag{89}$$

Consider a forward propagating wave, i.e.,

$$u(x, t) = f(x - ct) \quad \text{(forward wave)} \tag{90}$$

Substitution of Eq. (90) into Eqs. (87) and (88) yields

$$k(x, t) = \frac{1}{2}m(-c)^2 f'^2(x - ct) = \frac{1}{2}m\rho c^2 f'^2(x - ct) \tag{91}$$

$$v(x, t) = \frac{1}{2}EAf'^2(x - ct) = \frac{1}{2}mc^2 f'^2(x - ct) \tag{92}$$

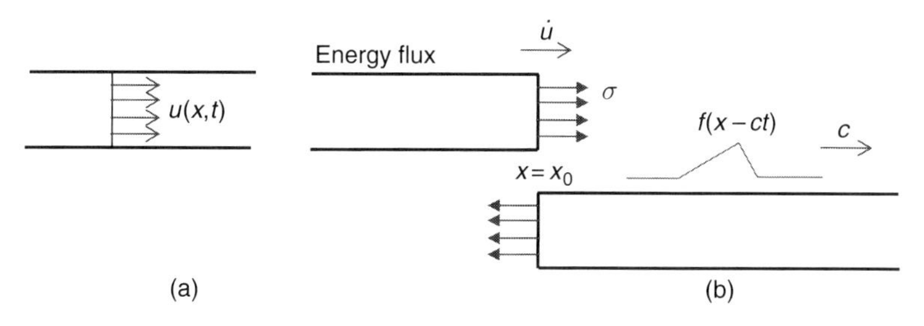

FIGURE 5.6 Power and energy of a propagating wave: (a) displacement distribution over the cross-section; (b) energy flux through a cross-section.

where the relation $EA = c^2 m$ of Eq. (9) was invoked. It is apparent that the kinetic energy expression (91) is equal to the elastic energy expression (92), i.e.,

$$k(x,t) = v(x,t) = \frac{1}{2} mc^2 f'^2(x - ct) \tag{93}$$

Equation (93) indicates that a propagating wave has its energy equally partitioned into kinetic and elastic components. Substitution into Eq. (89) yields the *total energy density per unit length of a propagating axial wave in a bar* as

$$e(x,t) = k(x,t) + v(x,t) = mc^2 f'^2(x - ct) \tag{94}$$

It is apparent that, for this simple waves, the energy propagates with the same wave speed, c, as the wave itself. For a backward wave, the same analysis would apply.

5.2.10.2 Axial Wave Power

The *wave power* is seen as the rate with which energy is transferred past a certain point during wave propagation. Assume that we cut the bar at a given point $x = x_0$ and study the energy flux across the interface (Fig. 5.6b). Consider the right hand part of the cut bar, i.e., $x_0 \le x$, and replace the removed part of the bar with the interface tractions $\vec{t} = -\sigma \vec{e}_x$. The work done by the tractions onto the interface represents the energy input into the bar. The time rate of this interface work is the *energy flux* into the bar or the *power input* into the bar. We write that power is the product of tractions and particle velocity, $\dot{\vec{u}} = \dot{u}\vec{e}_x$, i.e.,

$$P = \int_A \vec{t} \cdot \dot{\vec{u}} \, \mathrm{d}A = -\sigma \dot{u} A = -EAu'\dot{u} = -mc^2 u'\dot{u} \tag{95}$$

where Eqs. (4), (5), and (9) have been invoked. For the forward propagating wave (90), we have $u'(x,t) = f'(x - ct)$ and $\dot{u}(x,t) = -cf'(x - ct)$; hence, Eq. (95) becomes

$$P(x,t) = -mc^2 f'(x - ct)[-cf'(x - ct)] = mc^3 f'^2(x - ct) \tag{96}$$

Comparison of Eqs. (94) and (96) reveals that *power is the product of energy and wave speed*, i.e.,

$$P(x,t) = ce(x,t) \tag{97}$$

The same conclusion can be reached via the definition of power as the rate of energy transferred from the left-hand part of the bar, i.e.,

$$P(x_0,t) = -\frac{\partial}{\partial t} \int_0^{x_0} e(x,t)\mathrm{d}x = ce(x_0,t) \tag{98}$$

The proof is immediate, since Eq. (96) suggests that the energy is a single-argument function of the form $e(x - ct)$ and hence

$$\begin{aligned} -\frac{\partial}{\partial t} \int_0^{x_0} e(x - ct)\mathrm{d}x &= -\int_0^{x_0} \frac{\partial}{\partial t} e(x - ct)\mathrm{d}x = -\int_0^{x_0} (-c)e'(x - ct)\mathrm{d}x \\ &= c\int_0^{x_0} e'(x - ct)\mathrm{d}x = c \cdot e(x_0 - ct) \end{aligned} \tag{99}$$

where $e'(z)$ is the derivative of the single-argument function $e(z)$.

Power and energy of harmonic waves

The case of the harmonic waves is a particular case of the general propagating waves, and hence the previous results hold. Recall Eq. (76) and assume a forward harmonic wave of the form

$$u(x,t) = \mathrm{Re}\left(\hat{u}e^{-i(\gamma x-\omega t)}\right) = \hat{u}\cos(\gamma x-\omega t) \tag{100}$$

To obtain the *energy*, substitution of Eq. (100) into Eqs. (87) and (88) gives

$$k(x,t) = \frac{1}{2}m\dot{u}^2(x,t) = \frac{1}{2}m\omega^2\hat{u}^2\sin^2(\gamma x-\omega t) \quad \text{(kinetic energy density)} \tag{101}$$

$$v(x,t) = \frac{1}{2}EAu'^2(x,t) = \frac{1}{2}EA\gamma^2\hat{u}^2\sin^2(\gamma x-\omega t) \quad \text{(elastic energy density)} \tag{102}$$

Application of Eqs. (9) and (67) into Eqs. (101) and (102) verifies that the equality of the kinetic and elastic energies, i.e.,

$$\begin{aligned} v(x,t) &= \frac{1}{2}EA\gamma^2\hat{u}^2\sin^2(\gamma x-\omega t) = \frac{1}{2}mc^2\frac{\omega^2}{c^2}\hat{u}^2\sin^2(\gamma x-\omega t) \\ &= \frac{1}{2}m\omega^2\hat{u}^2\sin^2(\gamma x-\omega t) = k(x,t) \end{aligned} \tag{103}$$

The total energy is

$$e(x,t) = k(x,t)+v(x,t) = m\omega^2u^2(x,t) = m\omega^2\hat{u}^2\sin^2(\gamma x-\omega t) \tag{104}$$

The amplitude of wave energy is given by

$$\hat{e} = m\omega^2\hat{u}^2 \tag{105}$$

Equation (105) indicates that, for a harmonic wave, the energy amplitude is proportional to the frequency squared.

The *time-averaged energy* can be calculated by taking the average of Eq. (104) over a period T given by $T = 2\pi/\omega$, i.e.,

$$<e> = \frac{1}{T}\int_0^T e(x,t)\mathrm{d}t = \frac{1}{T}\int_0^T m\omega^2u^2(x,t)\mathrm{d}t \tag{106}$$

The integration in Eq. (106) can be performed using the explicit expression of Eq. (104). However, a more immediate way is to use the results of Appendix A, which state that the time-average product of two harmonic variables is one half the real part of the product of one variable times the conjugate of the other, i.e., $<VI> = \mathrm{Re}\left(\frac{1}{2}V\bar{I}\right) = \mathrm{Re}\left(\frac{1}{2}\bar{V}I\right)$. Applying these results to Eq. (106) yields the time-averaged energy as

$$<e> = \frac{1}{2}m\omega^2\,\mathrm{Re}(u\cdot\bar{u}) = \frac{1}{2}m\omega^2\hat{u}^2 \tag{107}$$

Equation (107) indicates that the time-averaged energy equals one half of the amplitude of the energy function given by Eq. (105) and is proportional to the frequency squared.

To obtain the *power*, substitute Eq. (104) into Eq. (97) and get

$$P(x,t) = ce(x,t) = mc\omega^2\hat{u}^2\sin^2(\gamma x-\omega t) \tag{108}$$

The power amplitude is

$$\hat{P} = mc\omega^2\hat{u}^2 \tag{109}$$

The time-averaged power is

$$<P> = \frac{1}{2}mc\omega^2\hat{u}^2 \tag{110}$$

As before, the time-averaged power is the time-averaged energy times the wave velocity, i.e.,

$$<P> = c\left(\frac{1}{2}m\omega^2\hat{u}^2\right) = c<e> \tag{111}$$

5.3 FLEXURAL WAVES IN BEAMS

Flexural waves appear from bending action. In beams, the Euler–Bernoulli theory of bending assumes that plane sections remain plane after the bending deformation is applied. This implies a linear distribution of axial displacement across the thickness. Shear deformation and rotary inertia effects are ignored.

5.3.1 EQUATION OF MOTION FOR FLEXURAL WAVES

Consider a uniform beam of mass per unit length m and bending stiffness EI, undergoing flexural wave motion with vertical displacement $w(x, t)$ as shown in Fig. 5.7a. Centroidal axes are assumed. An infinitesimal beam element of length dx is subjected to the action of bending moments, $M(x, t), M(x, t) + M'(x, t)dx$, and shear forces $V(x, t), V(x, t) + V'(x, t)dx$ (Fig. 5.7b).

Free-body analysis of the infinitesimal element of Fig. 5.7b yields

$$N'(x, t) = 0 \tag{112}$$

$$V'(x, t) = m\ddot{w}(x, t) \tag{113}$$

$$M'(x, t) + V(x, t) = 0 \tag{114}$$

The N and M stress resultants are evaluated by integration across the cross-sectional area of the direct stress shown in Fig. 5.7c, i.e.,

$$N(x, t) = \int_A \sigma(x, z, t)dA \tag{115}$$

$$M(x, t) = -\int_A \sigma(x, z, t)z\, dA \tag{116}$$

Using the stress–strain constitutive relations of Eq. (5), the axial force and moment stress resultants (115) and (116) can be expressed as

$$N(x, t) = E\int_A \varepsilon(x, z, t)dA \tag{117}$$

$$M(x, t) = -E\int_A \varepsilon(x, z, t)z\, dA \tag{118}$$

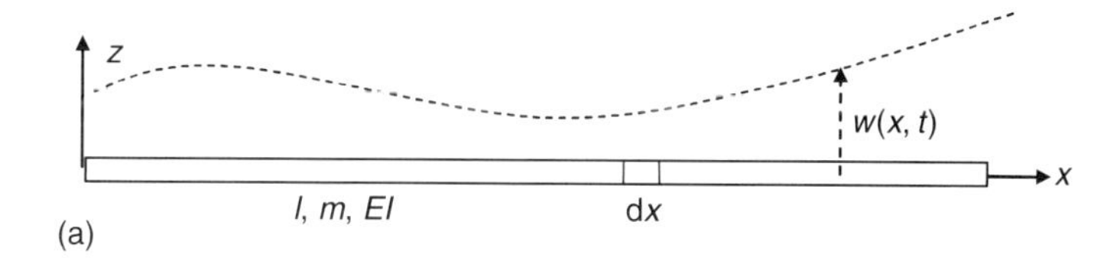

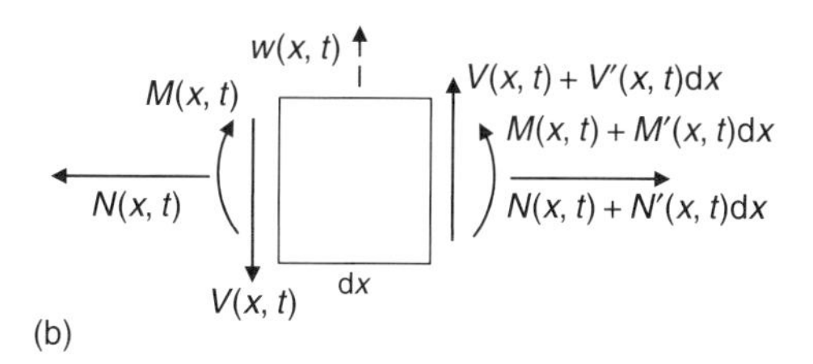

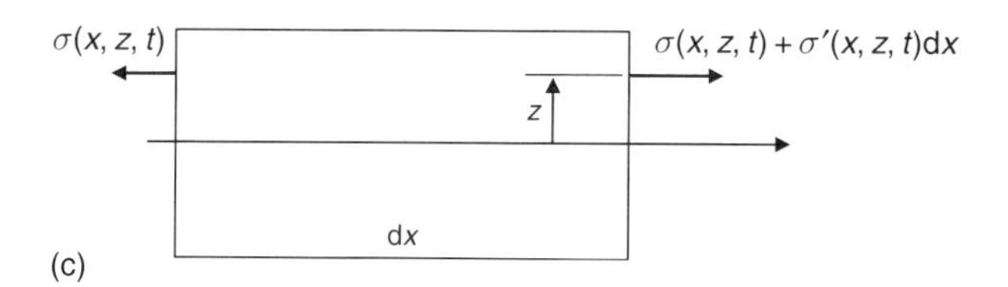

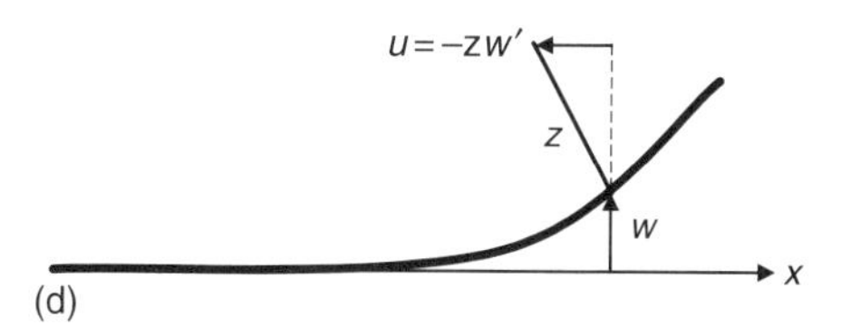

FIGURE 5.7 Beam undergoing flexural vibration: (a) general layout; (b) free-body diagram of an infinitesimal beam element; (c) stress distribution across the thickness; (d) horizontal displacement induced by flexure.

Kinematic analysis (Fig. 5.7) yields the direct strain $\varepsilon(x, z, t)$ in terms of the flexural motion $w(x, t)$ and the thickness-wise location z, i.e.,

$$\begin{aligned} u(x, z, t) &= -zw'(x, t) \\ \varepsilon(x, z, t) &= u'(x, z, t) = -zw''(x, t) \end{aligned} \tag{119}$$

Substitution of Eq. (119) into Eq. (118) and integration over the area yields

$$N(x, t) = -Ew''(x, t) \int_A z \, \mathrm{d}A = 0 \tag{120}$$

$$M(x, t) = Ew''(x, t) \int_A z^2 \, \mathrm{d}A \tag{121}$$

Note that Eq. (120) indicates that the axial stress resultant is zero, i.e., $N(x, t) = 0$, as centroidal axes were assumed. On the other hand, Eq. (121) yields

$$M(x, t) = EI \, w''(x, t) \tag{122}$$

Substitution of Eq. (122) into Eq. (114) yields

$$V(x, t) = -EI \, w'''(x, t) \tag{123}$$

Differentiation of Eq. (123) with respect to x and substitution into Eq. (112) yields the *equation of motion*, i.e.,

$$EI\, w''''(x,t) + m\ddot{w}(x,t) = 0 \tag{124}$$

Upon division by m and rearrangement, we get

$$a^2 w'''' + \ddot{w} = 0 \tag{125}$$

where the constant a^4 is given by

$$a^4 = \frac{EI}{m} \quad \text{or} \quad a = \left(\frac{EI}{m}\right)^{1/4} \tag{126}$$

Note that, for a beam of rectangular cross-section with height h and width b, the mass per unit length is $m = \rho bh$, and the area moment of inertia is $I = bh^3/12$. Hence, Eq. (126) becomes

$$a^4 = \frac{Eh^2}{12\rho} \quad \text{or} \quad a = \left(\frac{Eh^2}{12\rho}\right)^{1/4} \tag{127}$$

Assume separation of variables and write

$$w(x,t) = X(x)e^{i\omega t} \tag{128}$$

Upon substitution of Eq. (128) into Eq. (124), we obtain a fourth-order ordinary differential equation (ODE) in the form

$$a^4 X'''' - \omega^2 X = 0 \tag{129}$$

Upon division by a^2, Eq. (129) becomes

$$X'''' - \gamma^4 X = 0 \tag{130}$$

where

$$\gamma^4 = \frac{\omega^2}{a^4} = \frac{m}{EI}\omega^2 \quad \text{or} \quad \gamma = \left(\frac{m}{EI}\right)^{1/4}\sqrt{\omega} = \frac{\sqrt{\omega}}{a} \tag{131}$$

The characteristic equation of Eq. (130) is

$$\lambda^4 - \gamma^4 = 0 \tag{132}$$

Equation (132) has four roots, two imaginary roots $\lambda_{1,2} = \pm i\gamma$ and two real roots $\lambda_{3,4} = \pm\gamma$. Hence, Eq. (130) accepts a solution of the form

$$X(x) = Ae^{i\gamma x} + Be^{-i\gamma x} + Ce^{\gamma x} + De^{-\gamma x} \tag{133}$$

Substitution of Eq. (133) into Eq. (128) yields the general solution

$$w(x,t) = Ae^{i(\gamma x+\omega t)} + Be^{-i(\gamma x-\omega t)} + Ce^{\gamma x}e^{i\omega t} + De^{-\gamma x}e^{i\omega t} \tag{134}$$

We notice that Eq. (134) contains two types of terms, two propagating terms, $Ae^{i(\gamma x+\omega t)}$ and $Be^{-i(\gamma x-\omega t)}$, and two nonpropagating terms (evanescent), $Ce^{\gamma x}e^{i\omega t}$ and $De^{-\gamma x}e^{i\omega t}$. The propagating terms are carrying energy away through propagating elastic waves (energy radiation), whereas the non-propagating terms store energy in local vibrations confined to a small area. The choice of constants A, B, C, D is dictated by the *radiation condition*, which requires finite amplitudes at large distances. Thus, for forward waves ($0 < x$), we retain the constants B and D, whereas for backward waves ($x < 0$), we retain the constants A and C, i.e.,

$$w(x, t) = Be^{-i(\gamma x-\omega t)} + De^{-\gamma x}e^{i\omega t} \quad \text{(forward wave)} \tag{135}$$

$$w(x, t) = Ae^{i(\gamma x+\omega t)} + Ce^{\gamma x}e^{i\omega t} \quad \text{(backward wave)} \tag{136}$$

At sufficiently large distances from the source, the evanescent wave becomes insignificant, and only the propagating wave matters, i.e.,

$$w(x, t) \simeq Be^{-i(\gamma x-\omega t)} \qquad \text{(forward wave far from the source)} \tag{137}$$

$$w(x, t) = Ae^{i(\gamma x+\omega t)} \qquad \text{(backward wave far from the source)} \tag{138}$$

The *wave speed* of a flexural wave can be determined from the relation (67) between wavenumber and frequency, i.e.,

$$\gamma = \frac{\omega}{c} \tag{139}$$

Hence, the *flexural wave speed*, a.k.a. *flexural phase velocity*, is given by

$$c_{\mathrm{F}} = \frac{\omega}{\gamma} = \frac{\omega}{\frac{\sqrt{\omega}}{a}} = a\sqrt{\omega} = \left(\frac{EI}{m}\right)^{1/4}\sqrt{\omega} \tag{140}$$

For a beam of rectangular cross-section of height h, Eq. (140) becomes

$$c_{\mathrm{F}} = \left(\frac{Eh^2}{12\rho}\right)^{1/4}\sqrt{\omega} \tag{141}$$

It is apparent from Eq. (140) that the wave speed varies with frequency. A plot of Eq. (140) is given in Fig. 5.8. It is noticed that the flexural wave speed is zero when the frequency is zero. As frequency increases, the flexural wave speed also increases,

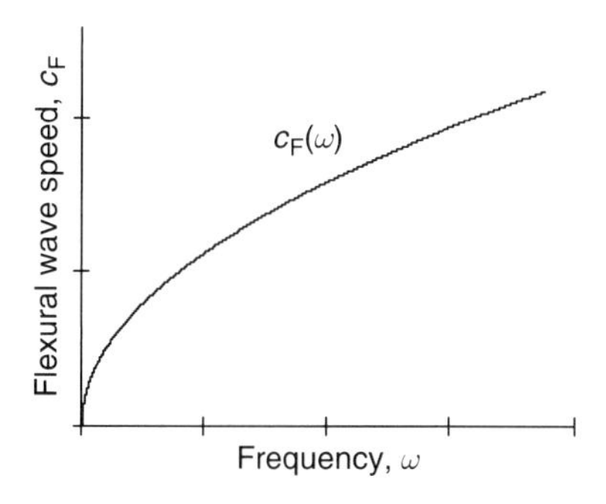

FIGURE 5.8 Wave speed vs. frequency curve for flexural waves.

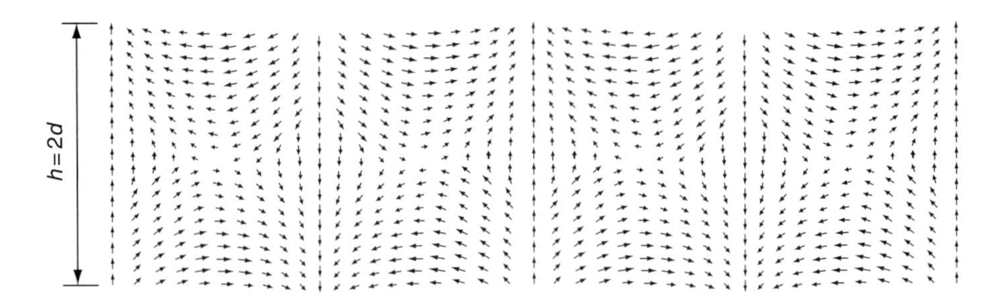

FIGURE 5.9 Simulation of flexural waves.

following the $\sqrt{\omega}$ rule. When the wave speed changes with frequency, the waves are called *dispersive waves*. The curve of Fig. 5.8 is called the *wave-speed dispersion curve*.

Assuming that vertical displacements follow the forward-speed harmonic expression

$$w(x, y, t) = \hat{w} \cdot e^{i(\gamma x - \omega t)} \tag{142}$$

yields the in-plane displacement field across the thickness of a flexing beam in the form

$$u(x, z, t) = z\frac{\partial}{\partial x}\hat{w}e^{i(\gamma x - \omega t)} = i\gamma z\hat{w}e^{i(\gamma x - \omega t)} \tag{143}$$

The particle motion described by Eqs. (142) and (143) is shown in Fig. 5.9. It is apparent that the particle motion is elliptical, with the ellipse aspect ratio varying from maximum at the surface to zero in the middle of the transverse thickness.

5.3.2 DISPERSION OF FLEXURAL WAVES

The wave dispersion phenomenon is best illustrated through the study of the propagation of single-frequency wave packets (a.k.a., tone bursts). Such a tone burst consists of a single-frequency carrier wave of carrier frequency, ω_c, whose amplitude is modulated such as to generate a burst-like behavior. (The modulation can be achieved with one of the common window functions such as Hanning window, Gaussian window, Hamming window).

Figure 5.10 shows comparatively the propagation of such a tone burst in two cases: (a) non-dispersive waves and (b) dispersive waves. Non-dispersive propagation occurs

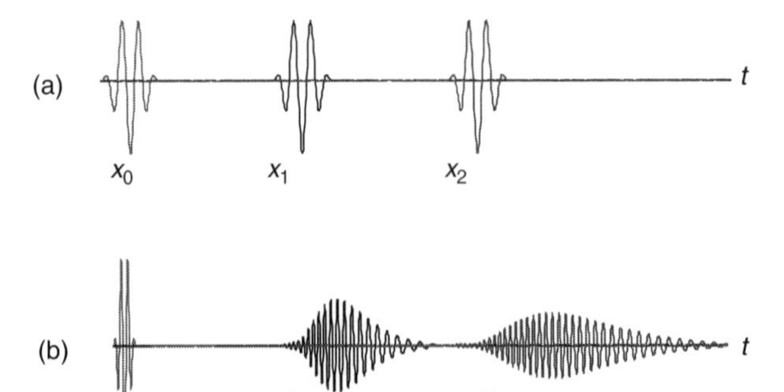

FIGURE 5.10 Propagation of tone bursts: (a) non-dispersive constant-wave speed, e.g., axial waves; (b) dispersive frequency-dependent wave speed, e.g., flexural waves.

when the wave speed is constant, e.g., in the case of axial waves. In this case, the wave speed is constant ($c = \text{const}$) and the wave packet at $x = x_1$ and $x = x_2$ has the same shape as it had initially at $x = x_0$ (Fig. 5.10a). However, if the wave speed varies with frequency, $c = c(\omega)$, then the shape of the wave packet changes as it propagates along. For this reason, the shape of the wave packet at $x = x_1$ and $x = x_2$ is significantly different from the initial shape it had at $x = x_0$. The packet has stretched out and elongated. In other words, *the waves inside the packet have dispersed.* The explanation for this phenomenon lies in the fact that the tone-burst, though of a single carrier frequency ω_c, in fact contains a multitude of frequencies introduced by the windowing operation. Figure 5.11 shows how the application of the windowing function generates a bell-shaped spectrum centered on the carrier frequency, ω_c. This aspect is further explained in Fig. 5.12. The presence of more than one frequency in the wave packet generates the dispersion phenomenon, because each frequency in the wave packet propagates with a different wave speed, as dictated by Fig. 5.8. Hence, an initially well-behaved wave packet spreads out and disperses after propagating through a dispersive medium.

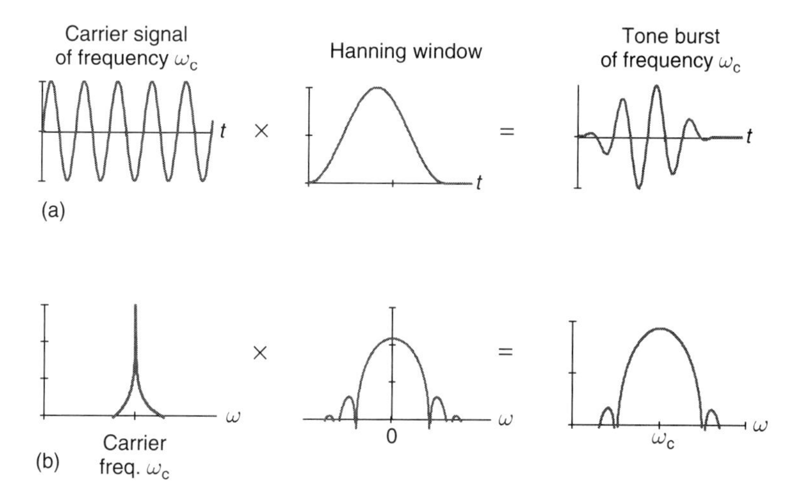

FIGURE 5.11 Modulation of a carrier wave by a time window: (a) time–domain multiplication of the two components; (b) frequency domain shift of the window spectrum to be centered on the carrier frequency, ω_c.

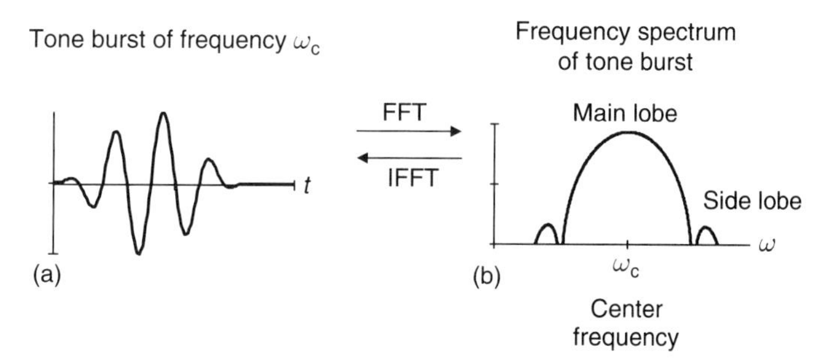

FIGURE 5.12 Frequency contents of a single-frequency wave packet (tone burst): (a) time–domain plot showing the windowed carrier wave of frequency ω_c; (b) frequency domain plot showing the bell-shaped window spectrum centered on the carrier frequency ω_c.

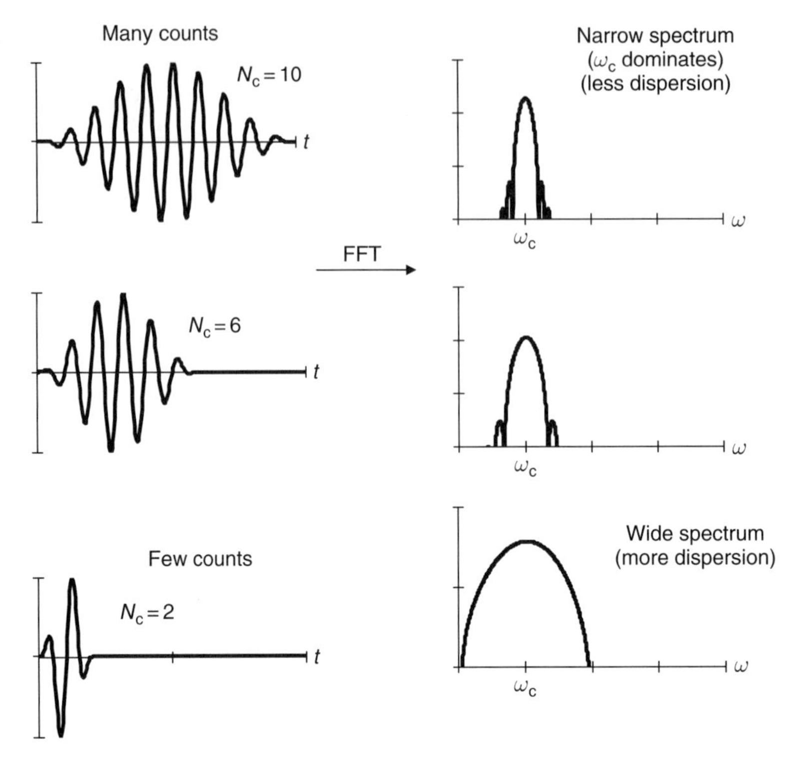

FIGURE 5.13 Frequency spectrum of tone bursts: (a) a long many-counts tone burst has a longer time–domain presence, but its frequency spectrum is narrower around the carrier frequency (b) a short tone burst of just a few counts has a short time–domain presence, but its frequency spectrum is wide covering many frequencies.

The intensity of dispersion depends on the wave-packet length. The relation between the tone-burst length and the frequency-spectrum width is illustrated in Fig. 5.13. If the tone burst has a long length (i.e., many wave cycles or 'counts' in the wave packet), then the frequency spectrum is narrow and resembles more closely the single-frequency spike of a continuous harmonic wave. The propagation of such a wave will be weakly dispersive. However, if the tone burst is rather short, i.e., it has just a few counts, then its frequency spectrum is spread out over many frequencies and the tone burst is highly dispersive. It is apparent that the time–domain duration, Δt, and the frequency spread, $\Delta\omega$, are in inverse relation, i.e., their product is constant,

$$\Delta t \cdot \Delta\omega = \text{const.} \tag{144}$$

Equation (144) is an expression of the Heisenberg uncertainty principle.

5.3.3 GROUP VELOCITY

When the wave propagation is dispersive, the wave packet travels at a velocity different from the speed of the individual waves in the packet. As the wave packet is actually a group of individual waves, the velocity with which the wave packet travels is called the *group velocity*, $c_g(\omega)$. In contrast, the wave speed of the individual waves, $c(\omega)$, is called the *phase velocity* because it directly affects the wave phase, $\psi(x, t) = \gamma(x - ct)$. The group velocity, $c_g(\omega)$, and phase velocity, $c(\omega)$, are related, and various interrelation formula will be given further down in this section.

5.3.3.1 Definition of Group Velocity

A commonly used definition of group velocity is

$$c_g = \frac{d\omega}{d\gamma} \tag{145}$$

To prove the formula of Eq. (145), consider a wave packet consisting of just two waves of same amplitude and slightly different frequencies, ω_1 and ω_2, i.e.,

$$u(x, t) = \frac{1}{2}\left[\cos(\gamma_1 x + \omega_1 t) + \cos(\gamma_2 x + \omega_2 t)\right] \tag{146}$$

where $\Delta\omega = \omega_2 - \omega_1 \ll 1$. Using the relation between wave speed, c, and frequency, ω, e.g., Fig. 5.8, we write the corresponding wavenumbers, i.e.,

$$\gamma_1 = \frac{\omega_1}{c_1}, \quad c_1 = c(\omega_1) \tag{147}$$

$$\gamma_2 = \frac{\omega_2}{c_2}, \quad c_2 = c(\omega_2) \tag{148}$$

The wavenumbers are, of course, also very close, i.e., $\Delta\gamma = \gamma_2 - \gamma_1 \ll 1$. Using the trigonometric identity

$$\cos\alpha + \cos\beta = 2\cos\frac{\beta-\alpha}{2}\cos\frac{\beta+\alpha}{2} \tag{149}$$

we write Eq. (146) as

$$u(x, t) = \cos\left(\frac{\Delta\gamma}{2}x - \frac{\Delta\omega}{2}t\right) \cdot \cos\left(\gamma_{av}x + \omega_{av}t\right) \tag{150}$$

Equation (150) represents a typical beats phenomenon (Fig. 5.14), where the fast-varying average frequency $\omega_{av} = (\omega_1 + \omega_2)/2$ represents a *carrier wave*, whereas $\Delta\omega/2$ represents a slow-varying *modulation wave*. The carrier wave propagates with the average phase velocity

$$c_{av} = \frac{\omega_{av}}{\gamma_{av}} \cong c(\omega_{av}) \tag{151}$$

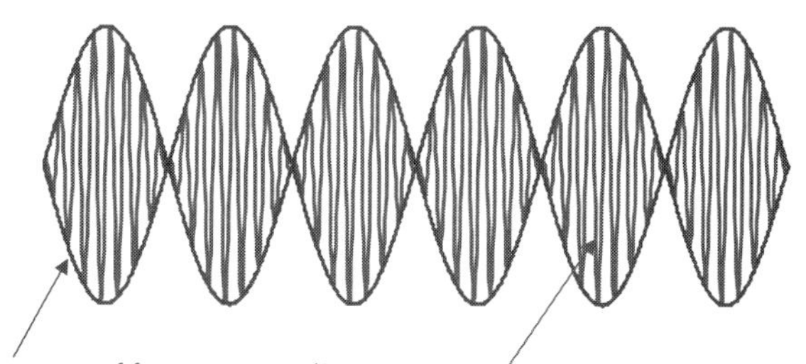
Modulation wave of frequency $\Delta\omega/2$: the wave group travels with speed
$c_g = \frac{\Delta\omega/2}{\Delta\gamma/2} = \frac{\Delta\omega}{\Delta\gamma}$
Carrier wave of frequency ω_{av} travels with speed
$c_{av} = \frac{\omega_{av}}{\gamma_{av}} \approx c(\omega_{av})$

FIGURE 5.14 Beats phenomenon created by two waves of slightly different frequencies.

The wave bursts generated by the modulation wave propagate with a different speed is given by

$$c_{\mathrm{g}} = \frac{\left(\dfrac{\Delta\omega}{2}\right)}{\left(\dfrac{\Delta\gamma}{2}\right)} = \frac{\Delta\omega}{\Delta\gamma} \tag{152}$$

In the limit, as $\dfrac{\Delta\omega}{\Delta\gamma} \underset{\substack{\Delta\omega\to 0\\ \Delta\gamma\to 0}}{\longrightarrow} \dfrac{\mathrm{d}\omega}{\mathrm{d}\gamma}$, Eq. (152) becomes

$$c_{\mathrm{g}} = \frac{\mathrm{d}\omega}{\mathrm{d}\gamma} \tag{153}$$

Alternative formulae for the calculation of group velocity are

$$c_{\mathrm{g}} = c + \gamma\frac{\mathrm{d}c}{\mathrm{d}\gamma} \tag{154}$$

$$c_{\mathrm{g}} = \frac{c^2}{c - \omega\frac{\mathrm{d}c}{\mathrm{d}\omega}} \tag{155}$$

5.3.3.2 Group Velocity of Flexural Waves

In the case of flexural waves, Eq. (140) gives the wave speed as

$$c_{\mathrm{F}} = \frac{\omega}{\gamma} = a\sqrt{\omega} \tag{156}$$

One can calculate the group velocity as follows: Resolve Eq. (156) to give

$$\gamma = \frac{1}{a}\sqrt{\omega} \tag{157}$$

Write Eq. (157) as

$$\omega = a^2\gamma^2 \tag{158}$$

Differentiate Eq. (158) with respect to γ and get

$$\frac{\mathrm{d}\omega}{\mathrm{d}\gamma} = a^2(2\gamma) \tag{159}$$

Use Eqs. (153) and (159) to get

$$c_{\mathrm{gF}} = \frac{\mathrm{d}\omega}{\mathrm{d}\gamma} = a^2(2\gamma) \tag{160}$$

Use Eq. (157) into Eq. (160) and obtain

$$c_{\mathrm{gF}} = 2a\sqrt{\omega} \tag{161}$$

Comparison of Eqs. (156) and (161) indicates that, for flexural waves, the group velocity is twice the wave speed, i.e.,

$$c_{\mathrm{gF}}(\omega) = 2c_{\mathrm{F}}(\omega) \tag{162}$$

A comparative plot of the group velocity $c_{\mathrm{gF}}(\omega)$ and wave speed $c(\omega)$ is given in Fig. 5.15.

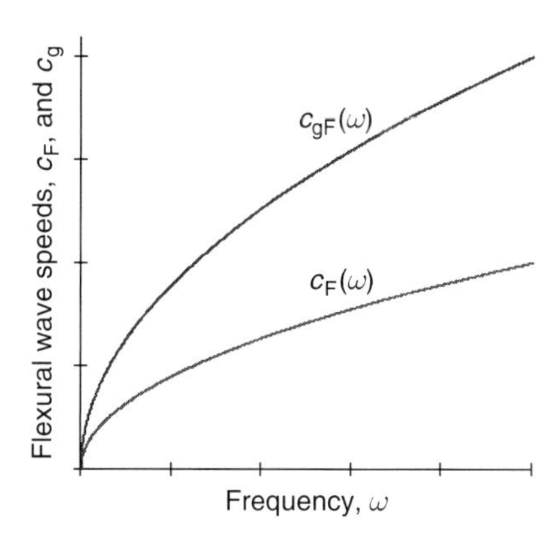

FIGURE 5.15 Phase velocity and group velocity for flexural waves.

5.3.4 ENERGY VELOCITY

The term 'energy velocity' signifies the velocity with which energy is transported by the wave motion. In case of nondispersive waves in which the wave speed and group velocity are equal, the wave packet propagates with the unique wave speed, and the packet energy is transported with the same unique speed. It will be shown that, in the case of dispersive waves, where wave packet propagates with the group velocity whereas individual waves propagate with the wave speed, the energy of the wave packet is transported with the group velocity. The proof is as follows.

Consider the same example of two slightly different waves as employed for the deduction of group velocity, i.e.,

$$u(x,t) = \cos\left(\frac{\Delta\gamma}{2}x - \frac{\Delta\omega}{2}t\right)\cdot\cos(\gamma x + \omega t) \tag{163}$$

Recall that the wave energy, e, is given by the sum of the kinetic energy, k, and elastic energy, v. Also recall that, for propagating waves, the kinetic and elastic energies are equal, i.e.,

$$e = k + v, \quad k = \frac{1}{2}\rho\dot{u}^2, \quad v = \frac{1}{2}\sigma\varepsilon, \quad k = v = \frac{e}{2} \tag{164}$$

In Eq. (164), the energies were expressed per unit volume. Using Eq. (164), we can express the wave energy as simply

$$e = \rho\dot{u}^2 \tag{165}$$

Differentiation of Eq. (163) with respect to t yields

$$\dot{u}(x,t) = \cos\left(\frac{\Delta\gamma}{2}x - \frac{\Delta\omega}{2}t\right)\cdot(-\omega)[-\sin(\gamma x + \omega t)] + \left(-\frac{\Delta\omega}{2}\right)\left[-\sin\left(\frac{\Delta\gamma}{2}x - \frac{\Delta\omega}{2}t\right)\right] \tag{166}$$

Substitution of Eq. (166) into Eq. (165) yields

$$e(x,t) = \rho\omega^2\cos^2\left(\frac{\Delta\gamma}{2}x - \frac{\Delta\omega}{2}t\right)\cdot\sin^2(\gamma x + \omega t) + O\left(\Delta\omega^2\right) \tag{167}$$

where $O(\Delta\omega^2)$ signifies negligible higher order terms. Equation (167) indicates that the energy has both a fast variation with carrier frequency, ω, and a slow variation with modulation frequency, $\Delta\omega/2$. In order to eliminate the fast variation, we take the time average over a period $T = 2\pi/\omega$ and get

$$\frac{1}{T}\int_0^T \sin^2(\gamma x + \omega t) = \frac{1}{2} \tag{168}$$

As $\Delta\omega \ll \omega$, the slow-varying term in Eq. (167) would have hardly changed while the integration over the fast period was performed. Hence, the time-averaged wave energy is given by

$$e(x,t) \cong \frac{1}{2}\rho\omega^2 \cos^2\left(\frac{\Delta\gamma}{2}x - \frac{\Delta\omega}{2}t\right) \tag{169}$$

Examination of Eq. (169) reveals that the time-averaged energy propagates with velocity

$$c_e = \frac{\Delta\omega/2}{\Delta\gamma/2} = \frac{\Delta\omega}{\Delta\gamma} = c_g \tag{170}$$

Equation (170) proves that, for dispersive waves, the energy velocity and the group velocity are equal, i.e., *energy propagates with group velocity.*

5.3.5 POWER AND ENERGY OF FLEXURAL WAVES IN A BEAM

5.3.5.1 Flexural Wave Energy

Recall the kinetic and elastic energy densities per unit volume given by

$$k = \frac{1}{2}\rho(\dot{u}^2 + \dot{w}^2) \quad \text{(kinetic energy per unit volume)} \tag{171}$$

$$v = \frac{1}{2}(\sigma_{xx}\varepsilon_{xx} + \sigma_{xz}\varepsilon_{xz}) \quad \text{(elastic energy per unit volume)} \tag{172}$$

To obtain the energy density per unit beam length, we integrate Eqs. (171) and (172) across the cross-section.

$$k(x,t) = \int_A \frac{1}{2}\rho\left(\dot{u}^2(x,z,t) + \dot{w}^2(x,t)\right) dA \quad \text{(kinetic energy per unit length)} \tag{173}$$

$$v(x,t) = \int_A \frac{1}{2}(\sigma_{xx}(x,z,t)\varepsilon_{xx}(x,z,t) + \sigma_{xz}(x,z,t)\varepsilon_{xz}(x,z,t))dA \quad \text{(elastic energy per unit length)} \tag{174}$$

Assume harmonic wave motion is of the form given by Eqs. (142) and (143), i.e.,

$$\begin{aligned} w &= \mathrm{Re}\left(\hat{w}e^{i(\gamma x - \omega t)}\right) \\ u &= -zw' = \mathrm{Re}\left(-i\gamma z\hat{w}e^{i(\gamma x - \omega t)}\right) = \mathrm{Re}(-i\gamma zw) \end{aligned} \tag{175}$$

Under the Euler-Bernoulli bending assumptions employed in this section, the shear deformation and rotary inertia are ignored, and the kinetic and elastic energy expressions

(173) and (174) only retain the terms in $\dot{w}$ and $\sigma_{xx}\varepsilon_{xx}$, respectively. The time-averaged expressions of the kinetic energy and elastic energy are

$$<k> \simeq \frac{1}{4} m\omega^2 \hat{w}^2 \tag{176}$$

$$<v> = \frac{1}{4} EI\gamma^4 \hat{w}^2 \tag{177}$$

Equations (139) and (140) imply that $EI\gamma^4 = m\omega^2$. Hence, Eqs. (176) and (177) have the same value, i.e.,

$$<k> = <v> = \frac{1}{4} m\omega^2 \hat{w}^2 \tag{178}$$

Subsequently, the time-averaged total energy is written as

$$<e> = <k> + <v> = \frac{1}{2} m\omega^2 \hat{w}^2 \tag{179}$$

The proof of Eqs. (176) and (177) is as follows. Differentiate Eq. (175) with respect to t and get

$$\begin{aligned} \dot{w} &= -i\omega\hat{w}e^{i(\gamma x - \omega t)} = -i\omega w \\ \dot{u} &= (-i\omega)(-i\gamma z w) = -\gamma\omega z w \end{aligned} \tag{180}$$

where the real-part operator Re is omitted but implied. Substitute Eq. (180) into Eq. (173), and take the time average over a period $T = 2\pi/\omega$. Using the results of Appendix, i.e., $<VI> = \mathrm{Re}\left(\frac{1}{2}V\bar{I}\right) = \mathrm{Re}\left(\frac{1}{2}\bar{V}I\right)$, get the time-averaged kinetic energy as

$$<k> = \frac{1}{2}\int_A \frac{1}{2}\rho(\dot{u}\bar{\dot{u}} + \dot{w}\bar{\dot{w}})\mathrm{d}A = \frac{1}{4}\rho I\gamma^2\omega^2\hat{w}^2 + \frac{1}{4}\rho A\omega^2\hat{w}^2 \tag{181}$$

The first term of Eq. (181) is associated with the rotary inertia of the cross-section, whereas the second term is associated with the translation inertia. For low-frequency long-wavelength motion, the wavenumber approaches zero ($\gamma \to 0$) and the first term in Eq. (181), i.e., the rotary inertia term, can be ignored. Hence, the time-averaged kinetic energy can be approximated by its translational term alone, i.e.,

$$<k> \simeq \frac{1}{4} m\omega^2 \hat{w}^2 \tag{182}$$

where $m = \rho A$.

To calculate the elastic energy given by Eq. (174), evaluate the strains in accordance with the elasticity relations of Appendix B, i.e.,

$$\varepsilon_{xx} = \frac{\partial u_x}{\partial x} \quad \text{and} \quad \varepsilon_{zx} = \frac{1}{2}\left(\frac{\partial u_z}{\partial x} + \frac{\partial u_x}{\partial z}\right) \tag{183}$$

where $u_x = u$ and $u_z = w$ of Eq. (175). Substitution of Eq. (175) into Eq. (183) yields

$$\varepsilon_{xx} = -(i\gamma)^2 z w = \gamma^2 z w \quad \text{and} \quad \varepsilon_{zx} = \frac{1}{2}(i\gamma w - i\gamma w) = 0 \tag{184}$$

Substituting Eq. (184) into Eq. (174) and taking the time average yields the time-averaged elastic energy

$$\langle v\rangle = \frac{1}{2}\int_A \frac{1}{2} E\varepsilon_{xx}\bar{\varepsilon}_{xx}\mathrm{d}A = \frac{1}{4}EI\gamma^4\hat{w}^2 \tag{185}$$

Recalling Eq. (140), we note that

$$EI\gamma^4 = m\omega^2 \tag{186}$$

Using Eq. (186) into Eq. (185) yields

$$\langle v\rangle = \frac{1}{4}m\omega^2\hat{w}^2 = \langle k\rangle \tag{187}$$

Hence, the time-averaged total energy is given by

$$\langle e\rangle = \langle k\rangle + \langle v\rangle = \frac{1}{2}m\omega^2\hat{w}^2 \tag{188}$$

However, the apparent simplicity of Eqs. (182) and (185) hides a fundamental flaw: the Euler-Bernoulli theory of bending (which is only an approximate theory) does not satisfy all the elasticity equations. For example, the equation of motion of the infinitesimal element in the x direction (see Appendix) is given by

$$\frac{\partial\sigma_{xx}}{\partial x} + \cancel{\frac{\partial\sigma_{xy}}{\partial y}} + \frac{\partial\sigma_{xz}}{\partial z} = \rho\ddot{u} \tag{189}$$

The shear stress σ_{xy} in Eq. (3.5.1) is obviously zero as there is no motion and no constraints in the y direction. But the stress σ_{xz} is not zero because $\partial\sigma_{xx}/\partial x \neq 0$ and $\rho\ddot{u}_x \neq 0$, as it can be easily verified. Hence, according to Eq. (3.5.1), $\partial\sigma_{xz}/\partial z = \rho\ddot{u}_x - \partial\sigma_{xx}/\partial x \neq 0$, which implies that $\sigma_{xz} \neq 0$ too. For low-frequency, long-wavelength motion, the shear deformation effects are negligible, and the approximate Eq. (177) holds. However, as the frequency and wavenumber increase, these approximations are no longer appropriate. An improved theory for the analysis of flexural waves in beams is the Timoshenko beam theory, which, by taking into account the shear deformation and rotary inertia, approaches much better the exact results obtained via the theory of elasticity.

5.3.5.2 Flexural Wave Power

The power flow associated with flexural waves is calculated by taking the integral over the cross-section of the product between the tractions vector, $\vec{t} = -\sigma_{xx}\vec{e}_x - \sigma_{xz}\vec{e}_z$, and the velocity vector, $\dot{\vec{u}} = \dot{u}_x\vec{e}_x + \dot{u}_z\vec{e}_z$, i.e.,

$$P = \int_A (\vec{t}\cdot\dot{\vec{u}})\mathrm{d}A = -\int_A (\sigma_{xx}\dot{u}_x + \sigma_{xz}\dot{u}_z)\,\mathrm{d}A \tag{190}$$

where $u_x = u$ and $u_z = w$ of Eq. (175). The negative sign is implied by the tractions sign convention. Recalling Eq. (184) and the stress–strain relations, we write

$$\sigma_{xx} = E\varepsilon_{xx} = E\gamma^2 zw \quad \text{and} \quad \sigma_{zx} = 2G\varepsilon_{zx} = 0 \tag{191}$$

where the real-part operator, Re, is omitted but implied. Substitution of Eqs. (191) into Eq. (190) and taking the time average yields time-averaged flexural wave power as

$$<P> = -\frac{1}{2}\int_A \bar{\sigma}_{xx}\dot{u}_x \, \mathrm{d}A = -\frac{1}{2}\int_A \left(E\gamma^2 z\bar{w}\right)(-\gamma\omega z w)\, \mathrm{d}A = \frac{1}{2}EI\gamma^3\omega\hat{w}^2 \tag{192}$$

Recalling Eq. (186), we note that

$$EI\gamma^3\omega = c_{\mathrm{F}} m\omega^2 \tag{193}$$

Substitution of Eq. (193) into Eq. (192) yields

$$<P> = \frac{1}{2}c_{\mathrm{F}} m\omega^2\hat{w}^2 \tag{194}$$

Comparison of Eqs. (179) and (194) yields

$$<P> = c_{\mathrm{F}}<e> \tag{195}$$

Equation (195) indicates that the power of flexural waves is the product between energy and flexural wave speed. This conclusion is similar with that of Eq. (111).

We note that Eq. (195) is consistent with our earlier conclusion that the group velocity is also the energy velocity, i.e., the velocity with which the energy of a wave packet travels through the dispersive medium.

5.4 TORSIONAL WAVES IN SHAFTS

Consider a uniform shaft of length l, mass torsional inertia per unit length ρI_p, torsional stiffness GJ, undergoing torsional vibration of displacement $\phi(x, t)$ as shown in Fig. 5.16a. An infinitesimal shaft element of length $\mathrm{d}x$ is subjected to the action of torsional moments (twisting moments or torques), $T(x, t)$ and $T(x, t) + T'(x, t)\mathrm{d}x$, as shown in Fig. 5.16b. Free-body analysis of the infinitesimal element yields

$$T'(x, t) = \rho I_p \ddot{\phi}(x, t) \tag{196}$$

Simple torsion analysis relates the torsional stiffness to the twist, which is the space derivative of the torsional displacement, i.e.,

$$T(x, t) = GJ\phi'(x, t) \tag{197}$$

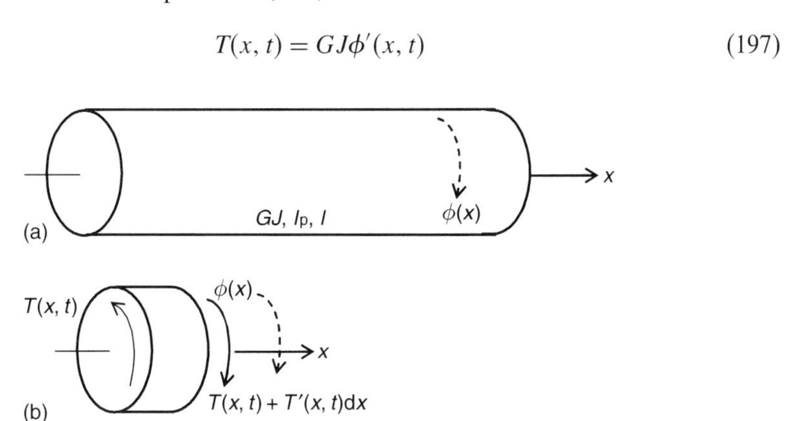

FIGURE 5.16 Uniform shaft undergoing torsional vibration: (a) general layout; (b) free-body diagram of an infinitesimal shaft element.

Substitution of Eq. (197) into Eq. (196) yields the *equation of motion for torsional waves*, i.e.,

$$GJ\phi''(x,t) = \rho I_p \ddot{\phi}(x,t) \tag{198}$$

Upon division by ρI_p, we get

$$c^2\phi'' = \ddot{\phi} \tag{199}$$

where the torsional wave speed c is given by

$$c = \sqrt{\frac{GJ}{\rho I_{\mathrm{p}}}} \quad \text{or} \quad c^2 = \frac{GJ}{\rho I_{\mathrm{p}}} \tag{200}$$

Note that Eq. (199) is the same as the Eq. (8) derived for analysis of axial waves in a bar. Hence, one could simply map the axial wave results into torsional wave results through simple substitution of the appropriate terms, thus avoiding repetitious derivations. For sake of brevity, this exercise is left to the reader.

5.5 PLATE WAVES

In this section, we will deal with plate waves. After reviewing the general plate equations, we will consider two situations separately: (a) axial plate waves and (b) flexural plate waves.

As the plate surface is free, the z direction stress is assumed zero ($\sigma_{zz} = 0$). Hence, the 3-D elasticity relations given in the Appendix B reduce to

$$\begin{aligned} \varepsilon_{xx} &= \frac{1}{E}\sigma_{xx} + \frac{-\nu}{E}\sigma_{yy} & \qquad \varepsilon_{xy} &= \frac{1}{2G}\sigma_{xy} \\ \varepsilon_{yy} &= \frac{-\nu}{E}\sigma_{xx} + \frac{1}{E}\sigma_{yy} & \text{and} \qquad \varepsilon_{yz} &= \frac{1}{2G}\sigma_{yz} \\ \varepsilon_{zz} &= \frac{-\nu}{E}\sigma_{xx} + \frac{-\nu}{E}\sigma_{yy} & \qquad \varepsilon_{zx} &= \frac{1}{2G}\sigma_{zx} \end{aligned} \tag{201}$$

Symmetry of the shear strains applies ($\varepsilon_{yx} = \varepsilon_{xy}$, $\varepsilon_{zy} = \varepsilon_{yz}$, $\varepsilon_{zx} = \varepsilon_{xz}$). In our analysis, we are only interested in ε_{xx}, ε_{yy}, ε_{xy}. Solution of Eq. (201) yields

$$\begin{aligned} \sigma_{xx} &= \frac{E}{1-\nu^2}\left(\varepsilon_{xx} + \nu\varepsilon_{yy}\right) \\ \sigma_{yy} &= \frac{E}{1-\nu^2}\left(\nu\varepsilon_{xx} + \varepsilon_{yy}\right) \end{aligned} \qquad \sigma_{xy} = 2G\varepsilon_{xy} \tag{202}$$

Symmetry of the shear stresses also applies ($\sigma_{yx} = \sigma_{xy}$).

5.5.1 AXIAL WAVES IN PLATES

Axial waves appear from extension/compression motion in the plate. Assume in-plane displacements $u(x, y, t)$ and $v(x, y, t)$ that are uniform across the plate thickness.

5.5.1.1 Equation of Motion for Axial Waves in Plates

The strains of interest are

$$\varepsilon_{xx} = \frac{\partial u}{\partial x} \quad \varepsilon_{yy} = \frac{\partial v}{\partial y} \quad \varepsilon_{yx} = \frac{1}{2}\left(\frac{\partial u}{\partial y} + \frac{\partial v}{\partial x}\right) \tag{203}$$

The strains are constant across the plate thickness. Substitution of Eq. (203) into Eq. (202) yields

$$\begin{aligned} \sigma_{xx} &= \frac{E}{1-\nu^2}\left(\frac{\partial u}{\partial x} + \nu\frac{\partial v}{\partial y}\right) \\ \sigma_{yy} &= \frac{E}{1-\nu^2}\left(\nu\frac{\partial u}{\partial x} + \frac{\partial v}{\partial y}\right) \end{aligned} \qquad \sigma_{xy} = G\left(\frac{\partial u}{\partial y} + \frac{\partial v}{\partial x}\right) \tag{204}$$

Note that the stresses are constant across the plate thickness. Integration of stresses across the thickness gives the stress resultants N_x, N_y, N_{xy} (forces per unit width) shown in Fig. 5.17, i.e.,

$$\begin{aligned} N_x &= \int_{-h/2}^{h/2} \sigma_{xx}\,\mathrm{d}z = \frac{Eh}{1-\nu^2}\left(\frac{\partial u}{\partial x} + \nu\frac{\partial v}{\partial y}\right) \\ N_y &= \int_{-h/2}^{h/2} \sigma_{yy}\,\mathrm{d}z = \frac{Eh}{1-\nu^2}\left(\nu\frac{\partial u}{\partial x} + \frac{\partial v}{\partial y}\right) \\ N_{xy} = N_{yx} &= \int_{-h/2}^{h/2} \sigma_{xy}\,\mathrm{d}z = Gh\left(\frac{\partial u}{\partial y} + \frac{\partial v}{\partial x}\right) \end{aligned} \tag{205}$$

Newton second law applied to the infinitesimal element of Fig. 5.17 yields

$$\begin{aligned} \frac{\partial N_x}{\partial x} + \frac{\partial N_{yx}}{\partial y} &= \rho h\ddot{u} \\ \frac{\partial N_y}{\partial y} + \frac{\partial N_{xy}}{\partial x} &= \rho h\ddot{v} \end{aligned} \tag{206}$$

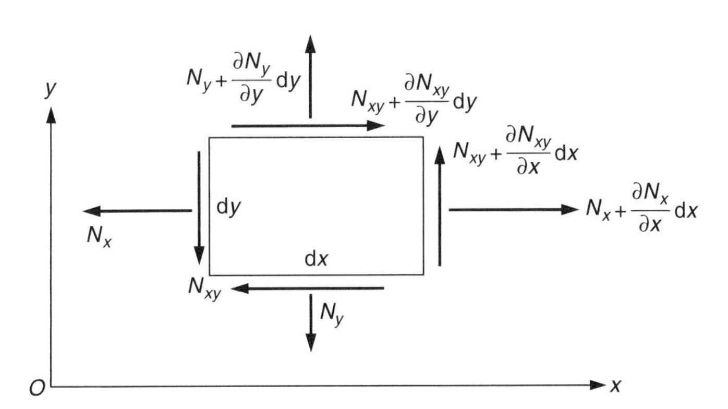

FIGURE 5.17 Infinitesimal plate element in Cartesian coordinates for the analysis of in-plane waves.

Equation (206) corresponds to the forces in the x and y directions. Note that the moment equation is identical zero, since in-plane rotation inertia is ignored. Substitution of Eq. (205) into Eq. (206) yields a system of two second-order partial differential equations in u and v depending on the space and time variables. This system has a general solution representing waves moving simultaneously in the x and y directions. However, when certain conditions are imposed on the wave-propagation pattern, the system simplifies and yields simple uncoupled solutions, which we will discuss next.

5.5.1.2 Straight-Crested Axial Waves in Plates

Consider *straight-crested axial plate waves* (Fig. 5.18a). The analysis is restricted to the study of waves having particle motion in the direction of wave propagation, i.e., *longitudinally (pressure) polarized.* Taking the y-axis along the wave crest yields a y-invariant problem that depends on x only, i.e.,

$$\begin{aligned} u(x, y, t) &\to u(x, t) \\ v(x, y, t) &\equiv 0 \end{aligned} \quad \text{(straight-crested axial plate wave)} \tag{207}$$

where the particle motion was assumed to be parallel to the x-axis. Substitution of Eq. (207) into Eqs. (205) and (206) yields

$$\begin{aligned} N_x &= \frac{Eh}{1-\nu^2}\frac{\partial u}{\partial x} \\ N_y &= \frac{Eh}{1-\nu^2}\nu\frac{\partial u}{\partial x} \\ N_{xy} &= N_{yx} = 0 \end{aligned} \tag{208}$$

and

$$\frac{\partial N_x}{\partial x} + \frac{\partial N_{yx}}{\partial y} = \frac{Eh}{1-\nu^2}\frac{\partial^2 u}{\partial x^2} = \rho h \ddot{u} \tag{209}$$

Upon simplification, we get the wave equation

$$c_L^2 \frac{\partial^2 u}{\partial x^2} = \ddot{u} \tag{210}$$

where c_L is the *longitudinal wave speed in a plate* given by

$$c_L = \sqrt{\frac{1}{1-\nu^2}\frac{E}{\rho}}, \quad c_L^2 = \frac{1}{1-\nu^2}\frac{E}{\rho} \tag{211}$$

From here onward, the analysis duplicates the analysis of axial waves in a bar, with the only difference of using c_L, as defined by Eq. (211), instead of c, as defined by Eq. (9).

5.5.1.3 Straight-Crested Shear Waves in Plates

Let us now consider another kind of straight-crested plate waves, in particular *straight-crested shear waves* (Fig. 5.18b). The analysis is restricted to the study of particle motion parallel to the wave front, i.e., perpendicular to the direction of wave propagation. This

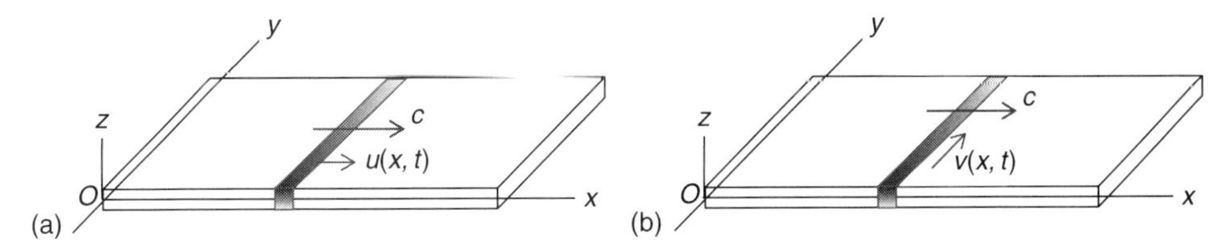

FIGURE 5.18 Straight-crested in-plane waves in a plate: (a) straight-crested axial waves; (b) straight-crested shear waves.

means that the wave is *transverse (shear) polarized*. Taking the y-axis along the wave crest yields a y-invariant problem that depends only on x, i.e.,

$$\begin{aligned} u(x, y, t) &\equiv 0 \\ v(x, y, t) &\to v(x, t) \end{aligned} \quad \text{(straight-crested shear wave)} \tag{212}$$

where the particle motion was assumed to be parallel to the y-axis. Substitution of Eq. (212) into Eqs. (205) and (206) yields

$$\begin{aligned} N_x = 0, \quad N_y = 0 \\ N_{xy} = N_{yx} = Gh\frac{\partial v}{\partial x} \end{aligned} \tag{213}$$

and

$$\frac{\partial N_{xy}}{\partial x} = Gh\frac{\partial^2 v}{\partial x^2} = \rho h\ddot{v} \tag{214}$$

Upon simplification, we get the wave equation

$$c_S^2\frac{\partial^2 v}{\partial x^2} = \ddot{v} \tag{215}$$

where c_S is the *shear wave speed* given by

$$c_S = \sqrt{\frac{G}{\rho}}, \quad c_S^2 = \frac{G}{\rho} \tag{216}$$

5.5.1.4 Circular-Crested Axial Waves Plate in Plates

Circular-crested waves arrive from a point source that will generate waves propagating in a concentric circular pattern. To analyze circular-crested waves, we change from Cartesian coordinates (x, y) to polar coordinates (r, θ). Similarly, the displacements (u, v) are replaced by displacements (u_r, u_θ). We assume that the motion is radial, $u_\theta = 0$. In virtue of axial symmetry, we also have $\partial u_r/\partial\theta = 0$. Recalling the strain–displacement and stress–strain relations in polar coordinates presented in Appendix B and applying the axial symmetry condition, $u_\theta = 0$, we get

$$\begin{aligned} \sigma_r &= \frac{E}{1-\nu^2}\left(\frac{\partial u_r}{\partial r} + \nu\frac{u_r}{r}\right) \\ \sigma_\theta &= \frac{E}{1-\nu^2}\left(\nu\frac{\partial u_r}{\partial r} + \frac{u_r}{r}\right) \end{aligned} \qquad \sigma_{r\theta} = 0 \tag{217}$$

Note that the stresses are constant across the plate thickness.

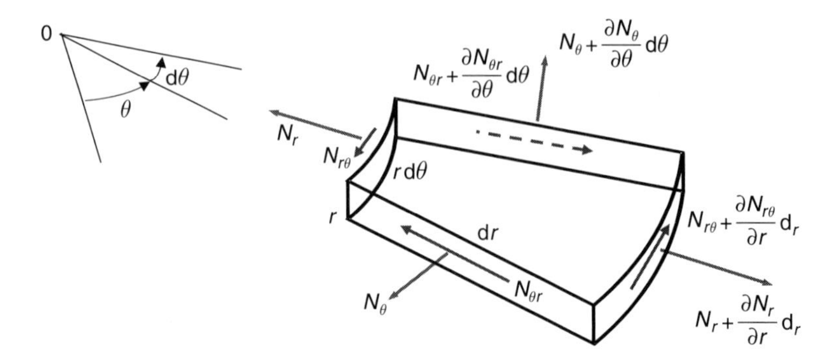

FIGURE 5.19 Infinitesimal plate element in polar coordinates for the analysis of in-plane waves.

Consider the infinitesimal plate element in polar coordinates shown in Fig. 5.19. Integration of stresses across the thickness gives the stress resultants N_r, N_θ, $N_{r\theta}$ (forces per unit width), i.e.,

$$N_r = \int_{-h/2}^{h/2} \sigma_r \, dz = \frac{Eh}{1-\nu^2}\left(\frac{\partial u_r}{\partial r} + \nu \frac{u_r}{r}\right) \qquad N_{r\theta} = N_{\theta r} = 0 \tag{218}$$
$$N_\theta = \int_{-h/2}^{h/2} \sigma_\theta \, dz = \frac{Eh}{1-\nu^2}\left(\nu \frac{\partial u_r}{\partial r} + \frac{u_r}{r}\right)$$

Newton second law applied to the infinitesimal element of Fig. 5.19 yields

$$\frac{\partial N_r}{\partial r} + \frac{1}{r}\frac{\partial N_{\theta r}}{\partial \theta} + \frac{N_r - N_\theta}{r} = \rho h \ddot{u} \tag{219}$$

Substitution of Eq. (218) into Eq. (219) yields, upon simplification,

$$\frac{E}{1-\nu^2}\left(\frac{\partial^2 u_r}{\partial r^2} + \frac{1}{r}\frac{\partial u_r}{\partial r} - \frac{u_r}{r^2}\right) = \rho \ddot{u}_r \tag{220}$$

Upon rearrangement, we get the wave equation in polar coordinates

$$c_L^2\left(\frac{\partial^2 u_r}{\partial r^2} + \frac{1}{r}\frac{\partial u_r}{\partial r} - \frac{u_r}{r^2}\right) = \ddot{u}_r \tag{221}$$

where c_L is the *longitudinal wave speed in a plate* given by Eq. (211). Equation (221) indicates that circular-crested axial waves in a plate propagate with the same wave speed, c_L, as the straight-crested axial wave described by Eq. (210).

Solution of Eq. (221) can be obtained in terms of Bessel functions if the motion is assumed to be harmonic, i.e.,

$$u_r(r,t) = \hat{u}(r)\, e^{i\omega t} \tag{222}$$

Substitution of Eq. (222) into Eq. (221), division by c_L^2, multiplication by r, use of the wavenumber definition $\gamma = \omega / c_L$, and the change of variable $x = \gamma r$ yields the Bessel equation

$$x^2 \frac{\partial^2 \hat{u}}{\partial x^2} + x \frac{\partial \hat{u}}{\partial x} + (x^2 - 1)\hat{u} = 0 \tag{223}$$

The solution of Eq. (223) is

$$\hat{u} = AJ_1(x) + BY_1(x) \tag{224}$$

where $J_1(x)$ is the Bessel function of the first kind and order 1, whereas $Y_1(x)$ is the Bessel function of the second kind and order 1. The arbitrary constants A and B are to be determined from initial and boundary conditions. Substituting $x = \gamma r$ in Eq. (224) yields

$$\hat{u}(r) = AJ_1(\gamma r) + BY_1(\gamma r) \tag{225}$$

However, the Bessel function $Y_1(\gamma r)$ has infinite values at $r = 0$ and has to be discarded (unless the plate has a hole around $r = 0$, which is not the case considered here). Hence, the *general solution for circular-crested axial plate waves* emanating from a source at the origin is given by

$$u_r(r, t) = AJ_1(\gamma r)e^{i\omega t} \tag{226}$$

A similar analysis can be performed for *circular-crested shear waves* in a plate by assuming $u_r = 0$ and $u_\theta \neq 0$. The circular-crested shear waves will be found to propagate with the shear wave speed, c_S, given by Eq. (216). The motion of circular-crested shear waves can be visualized as torsional waves created by the rotation of a cylinder oscillating at the origin.

5.5.2 FLEXURAL WAVES IN PLATES

Flexural waves appear from bending action. The Love-Kirchhoff plate theory assumes that straight normals to the mid-surface remain straight after deformation. This implies a linear distribution of axial displacement across the thickness.

5.5.2.1 Equation of Motion for Flexural Waves in Plates

For a displacement w of the plate mid-surface, the displacements at any location z in the plate thickness are given by

$$\begin{aligned} u &= -z\frac{\partial w}{\partial x} \\ v &= -z\frac{\partial w}{\partial y} \\ w &= w \end{aligned} \tag{227}$$

The strains of interest are

$$\varepsilon_{xx} = \frac{\partial u}{\partial x} = -z\frac{\partial^2 w}{\partial x^2}, \quad \varepsilon_{yy} = \frac{\partial v}{\partial y} = -z\frac{\partial^2 w}{\partial y^2}, \quad \varepsilon_{xy} = \frac{1}{2}\left(\frac{\partial u}{\partial y} + \frac{\partial v}{\partial x}\right) = -z\frac{\partial^2 w}{\partial x\,\partial y} \tag{228}$$

Substitution of Eq. (228) into Eq. (202) yields

$$\begin{aligned} \sigma_{xx} &= -z\frac{E}{1-\nu^2}\left(\frac{\partial^2 w}{\partial x^2} + \nu\frac{\partial^2 w}{\partial y^2}\right) \\ \sigma_{yy} &= -z\frac{E}{1-\nu^2}\left(\nu\frac{\partial^2 w}{\partial x^2} + \frac{\partial^2 w}{\partial y^2}\right) \end{aligned} \qquad \sigma_{xy} = -2zG\frac{\partial^2 w}{\partial x\,\partial y} \tag{229}$$

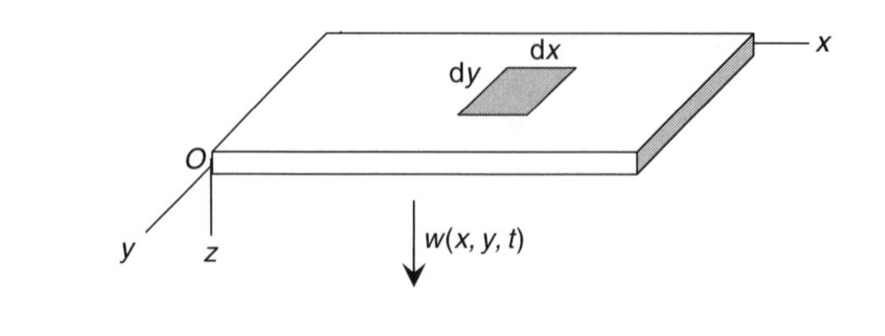

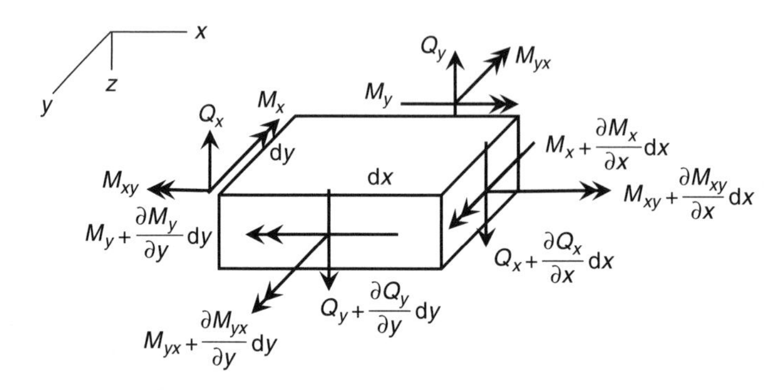

FIGURE 5.20 Infinitesimal plate element in Cartesian coordinates for the analysis of flexural waves.

Note that the stresses vary linearly across the plate thickness. Integration of stresses across the thickness gives the stress resultants M_x, M_y, M_{xy} (moments per unit width) shown in Fig. 5.20, i.e.,

$$
\begin{aligned}
M_x &= \int_{-h/2}^{h/2} \sigma_{xx} z \, \mathrm{d}z = -D\left(\frac{\partial^2 w}{\partial x^2} + \nu \frac{\partial^2 w}{\partial y^2}\right) \\
M_y &= \int_{-h/2}^{h/2} \sigma_{yy} z \, \mathrm{d}z = -D\left(\nu \frac{\partial^2 w}{\partial x^2} + \frac{\partial^2 w}{\partial y^2}\right) \\
M_{xy} = -M_{yx} &= \int_{-h/2}^{h/2} (-\sigma_{xy}) z \, \mathrm{d}z = (1-\nu) D \frac{\partial^2 w}{\partial x \, \partial y}
\end{aligned}
\tag{230}
$$

where D is the *plate flexural stiffness* defined as

$$
D = \frac{Eh^3}{12(1-\nu^2)} \tag{231}
$$

Newton second law applied to the infinitesimal element of Fig. 5.20 yields

$$
\begin{aligned}
&\text{vertical forces:} \quad \frac{\partial Q_x}{\partial x} + \frac{\partial Q_y}{\partial y} = \rho h \ddot{w} \\
&M_x \text{ moments:} \quad \frac{\partial M_x}{\partial x} + \frac{\partial M_{yx}}{\partial y} - Q_x = 0 \\
&M_y \text{ moments:} \quad \frac{\partial M_y}{\partial y} - \frac{\partial M_{xy}}{\partial x} - Q_y = 0
\end{aligned}
\tag{232}
$$

Simplification of Eqs. (232) yields

$$\frac{\partial^2 M_x}{\partial x^2} - \frac{\partial M_{xy}}{\partial x\,\partial y} + \frac{\partial^2 M_{yx}}{\partial x\,\partial y} + \frac{\partial^2 M_y}{\partial y^2} = \rho h \ddot{w} \tag{233}$$

Substitution of Eq. (230) into Eq. (233) yields

$$D\left(\frac{\partial^4 w}{\partial x^4} + 2\frac{\partial^4 w}{\partial x^2\,\partial y^2} + \frac{\partial^4 w}{\partial y^4}\right) + \rho h \ddot{w} = 0 \tag{234}$$

Equation (234) can be expressed in terms of the biharmonic operator for z-invariant motion $\nabla^4 = \nabla^2\nabla^2 = \frac{\partial^4}{\partial x^4} + 2\frac{\partial^4}{\partial x^2 \partial y^2} + \frac{\partial^4}{\partial y^4}$ (see Appendix B), i.e.,

$$D\nabla^4 w + \rho h \ddot{w} = 0 \tag{235}$$

Equation (234) has a general solution representing flexural plate waves propagating simultaneously in the x and y directions. However, when certain conditions are imposed on the wave propagation pattern, the system simplifies and yields simple uncoupled solutions, which we will discuss next.

5.5.2.2 Straight-Crested Flexural Waves in Plates

Consider *straight-crested flexural plate waves* (Fig. 5.21). Taking the y-axis along the wave crest yields a y-invariant problem that depends on x only, i.e.,

$$w(x, y, t) \to w(x, t) \quad \text{(straight-crested flexural plate wave)} \tag{236}$$

Equation (236) implies $\partial w/\partial y = 0$, etc. Substitution of Eq. (236) into Eq. (234) yields

$$Dw'''' + \rho h \ddot{w} = 0 \tag{237}$$

where, as before, $(\ )'$ represents the derivative with respect to the space variable x. Equation (237) has the same form as Eq. (124) for flexural waves in beams ($D \leftrightarrow E$, $\rho h \leftrightarrow m$) and hence has the same general solution

$$w(x, t) = A_1 e^{i(\gamma_F x + \omega t)} + A_2 e^{-i(\gamma_F x - \omega t)} + A_3 e^{\gamma_F x} e^{i\omega t} + A_4 e^{-\gamma_F x} e^{i\omega t} \tag{238}$$

where $\gamma_F = \omega / c_F$ is the flexural wavenumber and c_F is the *flexural wave speed in plates* given by

$$c_F = \left(\frac{D}{\rho h}\right)^{1/4} \sqrt{\omega} = \left(\frac{Eh^2}{12\rho(1-\nu^2)}\right)^{1/4} \sqrt{\omega} \tag{239}$$

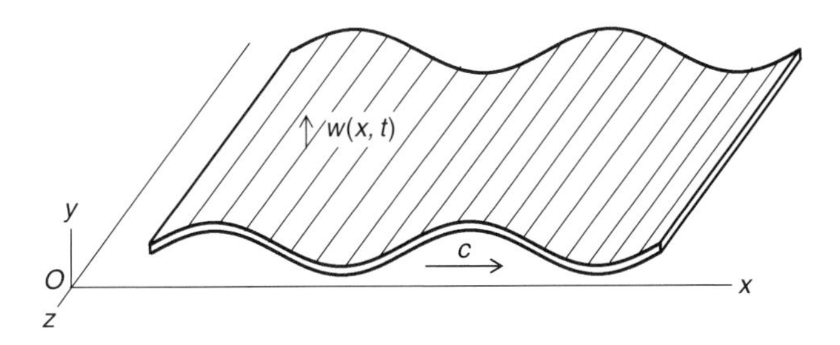

FIGURE 5.21 Straight-crested flexural wave in a plate.

Equation (239) can be further simplified to yield

$$c_{\mathrm{F}} = \frac{h}{2}\left(\frac{Ed^2}{3\rho(1-\nu^2)}\right)^{1/4}\sqrt{\omega} \tag{240}$$

where $d = h/2$.

Comparing Eq. (239) to the corresponding Eq. (141) for beams indicates that the only difference between the flexural wave speed in beams and plates is the presence of the correction term $(1-\nu^2)$, which accounts for the additional constraint (plain strain state) imposed by the plate geometry. Just as flexural waves in beams, the flexural waves in plates are highly dispersive, and the previous discussion on this subject applies.

5.5.2.3 Circular-Crested Flexural Waves in Plates

Circular-crested waves arrive from a point source that will generate waves propagating in a concentric circular pattern. To analyze circular-crested flexural waves, we change from Cartesian coordinates (x, y, z) to cylindrical coordinates (r, θ, z). The equation of motion (235) preserves its general form

$$D\nabla^4 w + \rho h\ddot{w} = 0 \tag{241}$$

only that the biharmonic operator has the appropriate form for polar coordinates as given in Appendix B i.e.,

$$\nabla^4 = \nabla^2\nabla^2 = \\ = \left(\frac{\partial^2}{\partial r^2} + \frac{1}{r}\frac{\partial}{\partial r} + \frac{1}{r^2}\frac{\partial^2}{\partial\theta^2}\right)\left(\frac{\partial^2}{\partial r^2} + \frac{1}{r}\frac{\partial}{\partial r} + \frac{1}{r^2}\frac{\partial^2}{\partial\theta^2}\right) \quad \text{(biharmonic operator)} \tag{242}$$

For the case of circular-crested waves considered here, the problem is θ-invariant and the biharmonic operator simplifies to the form

$$\nabla^4 = \nabla^2\nabla^2 = \left(\frac{\partial^2}{\partial r^2} + \frac{1}{r}\frac{\partial}{\partial r}\right)\left(\frac{\partial^2}{\partial r^2} + \frac{1}{r}\frac{\partial}{\partial r}\right) \\ = \frac{\partial^4}{\partial r^4} + \frac{2}{r}\frac{\partial^3}{\partial r^3} + \frac{1}{r^2}\frac{\partial^2}{\partial r^2} + \frac{1}{r^3}\frac{\partial}{\partial r} \quad \text{(biharmonic operator for axial symmetry)} \tag{243}$$

Hence, Eq. (241) takes the form

$$D\left(\frac{\partial^4 w}{\partial r^4} + \frac{2}{r}\frac{\partial^3 w}{\partial r^3} + \frac{1}{r^2}\frac{\partial^2 w}{\partial r^2} + \frac{1}{r^3}\frac{\partial w}{\partial r}\right) + \rho h\ddot{w} = 0 \tag{244}$$

Assume harmonic motion with frequency ω, i.e.,

$$w(r, t) = \hat{w}(r)e^{i\omega t} \tag{245}$$

Substitution of Eq. (245) into Eq. (241) yields

$$D\nabla^4\hat{w} - \omega^2\rho h\hat{w} = 0 \tag{246}$$

Define the wavenumber

$$\gamma = \left(\frac{\rho h}{D}\right)^{1/4}\sqrt{\omega} \quad \text{or} \quad \gamma^4 = \frac{\rho h}{D}\omega^2 \tag{247}$$

Substitution of Eq. (247) into Eq. (246) gives

$$\left(\nabla^4 - \gamma^4\right)\hat{w} = 0 \tag{248}$$

which can also be expressed as

$$\left(\nabla^2 - \gamma^2\right)\left(\nabla^2 + \gamma^2\right)\hat{w} = 0 \tag{249}$$

As the order in which the differentiation is done does not matter, Eq. (249) will be satisfied when either of the following expressions is satisfied:

$$\text{either } \left(\nabla^2 + \gamma^2\right)\hat{w} = 0 \quad \text{or} \quad \left(\nabla^2 - \gamma^2\right)\hat{w} = 0 \tag{250}$$

Equation (250) can be also expressed as

$$\text{either } \left(\frac{\partial^2}{\partial r^2} + \frac{1}{r}\frac{\partial}{\partial r} + \gamma^2\right)\hat{w} = 0 \quad \text{or} \quad \left(\frac{\partial^2}{\partial r^2} + \frac{1}{r}\frac{\partial}{\partial r} - \gamma^2\right)\hat{w} = 0 \tag{251}$$

Define the substitution

$$x = \begin{cases} \gamma r & \text{for } +\gamma^2 \\ i\gamma r & \text{for } -\gamma^2 \end{cases} \tag{252}$$

Upon substitution of Eq. (252) into Eq. (251) and observing that the partial derivatives are actually simple derivatives as $\hat{w}$ is only function of the space variable, we get the Bessel equation of order zero, i.e.,

$$x^2\frac{d^2\hat{w}}{dx^2} + x\frac{d\hat{w}}{dx} + x^2\hat{w} = 0 \tag{253}$$

The general solution of Eq. (253) is obtained in terms of Bessel functions, i.e.,

$$\hat{w} = CJ_0(x) + DI_0(x) + EY_0(x) + FK_0(x) \tag{254}$$

where $J_0(x)$ and $Y_0(x)$ are Bessel functions of first and second kind and order zero, whereas $I_0(\gamma r)$ and $K_0(\gamma r)$ are the modified Bessel functions. The constants C, D, E, F are to be determined from the initial and boundary conditions. Substituting $x = \gamma r$ in Eq. (254) yields

$$\hat{w}(r) = CJ_0(\gamma r) + DI_0(\gamma r) + EY_0(\gamma r) + FK_0(\gamma r) \tag{255}$$

However, the Bessel functions $Y_0(\gamma r)$ and $K_0(\gamma r)$ have infinite values at $r = 0$ and have to be discarded (unless the plate has a hole around $r = 0$, which is not the case considered here). Similarly, we discard $I_0(\gamma r)$ because it becomes unbounded as $r \to \infty$ (radiation condition). Hence, the *general solution for circular-crested flexural plate waves* emanating from a source at the origin is given by

$$w(r, t) = CJ_0(\gamma r)\, e^{i\omega t} \tag{256}$$

5.6 3-D WAVES

In an unbound elastic body, waves propagate freely in all directions. To analyze waves in an unbound elastic body, we will use the 3-D elasticity theory. Assume a generic plane wave propagating forward along a direction $\vec{n} = n_1\vec{e}_1 + n_2\vec{e}_2 + n_3\vec{e}_3$, i.e.,

$$\vec{u} = \vec{A} f(\vec{n} \cdot \vec{r} - ct) \tag{257}$$

where $\vec{A}$ is the wave amplitude vector given by

$$\vec{A} = A_1\vec{e}_1 + A_2\vec{e}_2 + A_3\vec{e}_3 \tag{258}$$

Recall the Navier equations defined in the Appendix A, i.e.,

$$\begin{aligned}
(\lambda+\mu)\left(\frac{\partial^2 u_x}{\partial x^2} + \frac{\partial^2 u_y}{\partial x\,\partial y} + \frac{\partial^2 u_z}{\partial x\,\partial z}\right) + \mu\left(\frac{\partial^2 u_x}{\partial x^2} + \frac{\partial^2 u_x}{\partial y^2} + \frac{\partial^2 u_x}{\partial z^2}\right) &= \rho \ddot{u}_x \\
(\lambda+\mu)\left(\frac{\partial^2 u_x}{\partial y\,\partial x} + \frac{\partial^2 u_y}{\partial y^2} + \frac{\partial^2 u_z}{\partial y\,\partial z}\right) + \mu\left(\frac{\partial^2 u_y}{\partial x^2} + \frac{\partial^2 u_y}{\partial y^2} + \frac{\partial^2 u_y}{\partial z^2}\right) &= \rho \ddot{u}_y \\
(\lambda+\mu)\left(\frac{\partial^2 u_x}{\partial z\,\partial x} + \frac{\partial^2 u_y}{\partial z\,\partial y} + \frac{\partial^2 u_z}{\partial z^2}\right) + \mu\left(\frac{\partial^2 u_z}{\partial x^2} + \frac{\partial^2 u_z}{\partial y^2} + \frac{\partial^2 u_z}{\partial z^2}\right) &= \rho \ddot{u}_z
\end{aligned} \tag{259}$$

To assist with the evaluation of Eq. (259), we note that the argument of Eq. (257), i.e., the phase of the wave front, is a function of time and space defined as

$$\vec{n} \cdot \vec{r} - ct = n_1x_1 + n_2x_2 + n_3x_3 - ct \tag{260}$$

Hence

$$\frac{\partial(\vec{n} \cdot \vec{r} - ct)}{\partial x_i} = n_i \quad \text{and} \quad \frac{\partial(\vec{n} \cdot \vec{r} - ct)}{\partial t} = -c \tag{261}$$

Using Eqs. (257) and (261), we write

$$\begin{aligned}
\frac{\partial \vec{u}}{\partial x_i} &= \vec{A} f'(\vec{n} \cdot \vec{r} - ct) \frac{\partial(\vec{n} \cdot \vec{r} - ct)}{\partial x_i} = \vec{A}\, n_i f'(\vec{n} \cdot \vec{r} - ct) \\
\frac{\partial \vec{u}}{\partial t} &= \vec{A} f'(\vec{n} \cdot \vec{r} - ct) \frac{\partial(\vec{n} \cdot \vec{r} - ct)}{\partial t} = \vec{A}(-c) f'(\vec{n} \cdot \vec{r} - ct)
\end{aligned} \quad \text{, etc.} \tag{262}$$

Substitution of Eq. (262) and its higher derivatives into Eq. (259) and cancellation of f'' throughout yields the expression

$$\begin{aligned}
(\lambda+\mu)\left(A_1n_1^2 + A_2n_2n_1 + A_3n_3n_1\right) + \mu A_1 &= \rho c^2 A_1 \\
(\lambda+\mu)\left(A_1n_1n_2 + A_2n_2^2 + A_3n_3n_2\right) + \mu A_2 &= \rho c^2 A_2 \\
(\lambda+\mu)\left(A_1n_1n_3 + A_2n_2n_3 + A_3n_3^3\right) + \mu A_3 &= \rho c^2 A_3
\end{aligned} \tag{263}$$

Upon rearrangement, Eq. (263) becomes

$$\begin{bmatrix} (\lambda+\mu)n_1^2 + (\mu - \rho c^2) & n_2n_1 & n_3n_1 \\ n_1n_2 & (\lambda+\mu)n_2^2 + (\mu - \rho c^2) & n_3n_2 \\ n_1n_3 & n_2n_3 & (\lambda+\mu)n_3^2 + (\mu - \rho c^2) \end{bmatrix} \begin{bmatrix} A_1 \\ A_2 \\ A_3 \end{bmatrix} = \begin{bmatrix} 0 \\ 0 \\ 0 \end{bmatrix} \tag{264}$$

The homogenous linear algebraic system of Eq. (265) accepts a nontrivial (i.e., nonzero) solution only if its determinant is zero, i.e.,

$$\begin{vmatrix} (\lambda+\mu)n_1^2+(\mu-\rho c^2) & n_2 n_1 & n_3 n_1 \\ n_1 n_2 & (\lambda+\mu)n_2^2+(\mu-\rho c^2) & n_3 n_2 \\ n_1 n_3 & n_2 n_3 & (\lambda+\mu)n_3^2+(\mu-\rho c^2) \end{vmatrix} = 0 \quad (265)$$

5.6.1 EIGENVALUES OF THE WAVE EQUATION AND WAVE SPEEDS

Upon expansion and simplification, Eq. (264) becomes

$$\left[(\lambda+2\mu)-\rho c^2\right](\mu-\rho c^2)^2 = 0 \quad (266)$$

It is apparent that Eq. (266) accepts three solutions (eigenvalues), of which two are identical, i.e.,

$$\begin{aligned} c_1 &= c_P \\ c_2 &= c_3 = c_S \end{aligned} \quad (267)$$

where

$$c_P = \sqrt{\frac{\lambda+2\mu}{\rho}}, \quad c_P^2 = \frac{\lambda+2\mu}{\rho} \quad \text{(pressure wave speed)} \quad (268)$$

$$c_S = \sqrt{\frac{\mu}{\rho}}, \quad c_S^2 = \frac{\mu}{\rho} \quad \text{(shear wave speed)} \quad (269)$$

The pressure wave speed and the shear wave speed can be also expressed in engineering notations, E, G, ν, i.e.,

$$c_P = \sqrt{\frac{1-\nu}{(1+\nu)(1-2\nu)}\frac{E}{\rho}}, \quad c_P^2 = \frac{1-\nu}{(1+\nu)(1-2\nu)}\frac{E}{\rho} \quad \text{(pressure wave speed)} \quad (270)$$

$$c_S = \sqrt{\frac{G}{\rho}} = \sqrt{\frac{1}{2(1+\nu)}\frac{E}{\rho}}, \quad c_S^2 = \frac{G}{\rho} = \frac{1}{2(1+\nu)}\frac{E}{\rho} \quad \text{(shear wave speed)} \quad (271)$$

5.6.2 EIGENVECTORS OF THE WAVE EQUATIONS

Substitution of the eigenvalues of Eq. (267) into Eq. (264) yields the eigenvectors of the algebraic linear system (264). It can be shown that the first eigenvalue, $c_1 = c_P$, is associated with a *pressure wave*, for which the particle motion is parallel with the direction of propagation, $\vec{n}$, whereas the second and third eigenvalues, $c_2 = c_S$ and $c_3 = c_S$, are associated with *shear waves*, for which the particle motion is perpendicular to the direction of propagation, $\vec{n}$. In addition, the two shear waves, though traveling with the same wave speed, c_S, will have different polarizations, in particular orthogonal to each other. To prove these assertions, assume, without lack of generality, that the wave front propagates parallel to the x-axis, i.e.,

$$\vec{n} = \vec{e}_1 \quad (272)$$

which means that $n_1 = 1, n_2 = 0, n_3 = 0$. In this case, Eq. (257) becomes

$$\vec{u} = \vec{A} f(x_1 - ct) \tag{273}$$

whereas the algebraic system (264) becomes

$$\begin{bmatrix} (\lambda + 2\mu) - \rho c^2 & 0 & 0 \\ 0 & \mu - \rho c^2 & 0 \\ 0 & 0 & \mu - \rho c^2 \end{bmatrix} \begin{bmatrix} A_1 \\ A_2 \\ A_3 \end{bmatrix} = \begin{bmatrix} 0 \\ 0 \\ 0 \end{bmatrix} \tag{274}$$

It is apparent that the eigenvalues of Eq. (274) are, as before, $c_1 = c_P$, $c_2 = c_S$, and $c_3 = c_S$. It is also apparent that the corresponding eigenvectors are $\{A_1, 0, 0\}^T$, $\{0, A_2, 0\}^T$, $\{0, 0, A_3\}^T$. Hence, the three possible waves are

$$\vec{u}_P = A_1 \vec{e}_1 f(x_1 - c_P t) \quad \text{(pressure wave, i.e., P-wave)} \tag{275}$$

$$\vec{u}_{SH} = A_2 \vec{e}_2 f(x_1 - c_S t) \quad \text{(shear-horizontal wave, i.e., SH-wave)} \tag{276}$$

$$\vec{u}_{SV} = A_3 \vec{e}_3 f(x_1 - c_S t) \quad \text{(shear-vertical wave, i.e., SV-wave)} \tag{277}$$

The *total wave solution*, $\vec{u}$, results from the superposition of the individual solutions, i.e.,

$$\vec{u} = \vec{u}_P + \vec{u}_{SH} + \vec{u}_{SV} \quad \text{(total wave)} \tag{278}$$

where the coefficients A_1, A_2, A_3 are to be determined from the initial and boundary conditions. For a backward propagating wave, the same approach can be applied.

Pressure waves are also called *compressional, axial, dilatational, longitudinal,* or simply *P-waves*. The main characteristic of the pressure waves is that the *particle motion is parallel to direction of wave propagation*. Shear waves are also called *transverse waves, distortional waves,* or *S-wave*. The main characteristic of the shear waves is that the *particle motion is perpendicular to the direction of wave propagation*.

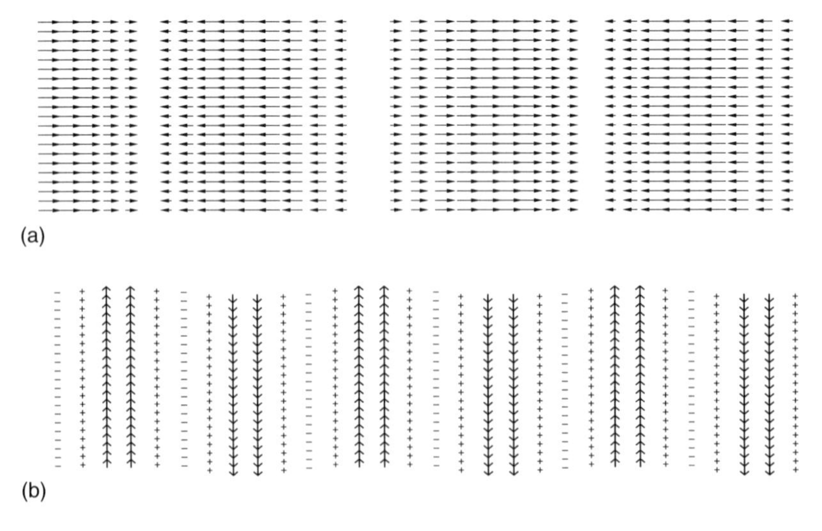

FIGURE 5.22 Simulation of plane waves: (a) pressure wave; (b) shear wave.

In the case of *harmonic waves* of frequency ω, the pressure and shear waves take the form

$$\vec{u}_{\mathrm{P}} = A_x \vec{i} e^{i(\gamma_{\mathrm{P}} x - \omega t)} \quad \text{(P-wave)} \tag{279}$$

$$\vec{u}_{\mathrm{SH}} = A_y \vec{j} e^{i(\gamma_{\mathrm{S}} x - \omega t)} \quad \text{(SH-wave)} \tag{280}$$

$$\vec{u}_{\mathrm{VH}} = A_z \vec{k} e^{i(\gamma_{\mathrm{S}} x - \omega t)} \quad \text{(SV-wave)} \tag{281}$$

where we have reversed to the x, y, z and $\vec{i}, \vec{j}, \vec{k}$ notations, for future convenience. The corresponding wavenumbers, γ_{P}, γ_{S}, are given by

$$\gamma_{\mathrm{P}} = \frac{\omega}{c_{\mathrm{P}}} \quad \gamma_{\mathrm{S}} = \frac{\omega}{c_{\mathrm{S}}} \tag{282}$$

Graphical representations of the *P*- and *S*-waves, i.e., plots of Eqs. (279) and (281), are given in Fig. 5.22.

5.6.3 WAVE POTENTIALS

Wave potentials are useful and convenient in several wave theory derivations. Assume that the displacement $\vec{u}$ can be expressed in terms of two potential functions, a *scalar potential* Φ and a *vector potential* $\vec{H} = H_x \vec{i} + H_y \vec{j} + H_z \vec{k}$, i.e.,

$$\vec{u} = \vec{\nabla}\Phi + \vec{\nabla} \times \vec{H} \quad \text{(potentials representation of displacements field)} \tag{283}$$

Equation (283) is known as the Helmholtz solution. Equation (283) is complemented by the *uniqueness condition*

$$\vec{\nabla} \cdot \vec{H} = 0 \quad \text{(uniqueness condition)} \tag{284}$$

Recall the Navier equations of Appendix B in the vector form

$$(\lambda + \mu)\vec{\nabla}\left(\vec{\nabla} \cdot \vec{u}\right) + \mu \nabla^2 \vec{u} = \rho \ddot{\vec{u}} \tag{285}$$

Using Eq. (283), we write the components of Eq. (285) as

$$\vec{\nabla} \cdot \vec{u} = \vec{\nabla} \cdot (\vec{\nabla}\Phi + \vec{\nabla} \times \vec{H}) = (\vec{\nabla} \cdot \vec{\nabla})\Phi + \cancel{\vec{\nabla} \cdot (\vec{\nabla} \times \vec{H})} = \nabla^2 \Phi \tag{286}$$

$$\nabla^2 \vec{u} = \nabla^2 (\vec{\nabla}\Phi + \vec{\nabla} \times \vec{H}) = \nabla^2 \vec{\nabla}\Phi + \nabla^2 \vec{\nabla} \times \vec{H} \tag{287}$$

$$\ddot{\vec{u}} = \vec{\nabla}\ddot{\Phi} + \vec{\nabla} \times \ddot{\vec{H}} \tag{288}$$

Substituting Eqs. (286), (287), (288) into Eq. (285) yields

$$(\lambda + \mu)\vec{\nabla}\left(\nabla^2 \Phi\right) + \mu\left(\nabla^2 \vec{\nabla}\Phi + \nabla^2 \vec{\nabla} \times \vec{H}\right) = \rho\left(\vec{\nabla}\ddot{\Phi} + \vec{\nabla} \times \ddot{\vec{H}}\right) \tag{289}$$

Noticing that $\vec{\nabla}\nabla^2 = \nabla^2 \vec{\nabla}$ (differentiation commutativity), Eq. (289) yields, upon rearrangement,

$$\vec{\nabla}\left((\lambda + 2\mu)\nabla^2 \Phi - \rho \vec{\nabla}\ddot{\Phi}\right) + \vec{\nabla} \times \left(\mu \nabla^2 \vec{H} - \rho \ddot{\vec{H}}\right) = \vec{0} \tag{290}$$

For Eq. (290) to hold at any place and any time, the components in parentheses must be independently zero, i.e.,

$$(\lambda+2\mu)\nabla^2\Phi-\rho\vec{\nabla}\ddot{\Phi}=0 \tag{291}$$

$$\mu\nabla^2\vec{H}-\rho\ddot{\vec{H}}=\vec{0} \tag{292}$$

Upon division by ρ and rearrangement, Eqs. (291) and (292) become the *wave equations* for the scalar potential Φ and the vector potential $\vec{H}$, i.e.,

$$c_{\mathrm{P}}^2\nabla^2\Phi=\vec{\nabla}\ddot{\Phi} \tag{293}$$

$$c_{\mathrm{S}}^2\nabla^2\vec{H}=\ddot{\vec{H}} \tag{294}$$

Equation (293) indicates that the scalar potential, Φ, propagates with the pressure wave speed, c_P, whereas Eq. (294) indicates that the vector potential, $\vec{H}$, propagates with the shear wave speed, c_S.

5.6.4 DILATATIONAL AND ROTATIONAL WAVES

To calculate the dilatational waves, recall the Navier equations in vector form as given in Appendix B, i.e.,

$$(\lambda+2\mu)\vec{\nabla}\Delta-2\mu\vec{\nabla}\times\vec{\omega}=\rho\ddot{\vec{u}} \tag{295}$$

Dot premultiply Eq. (295) by $\vec{\nabla}\cdot$ and get

$$(\lambda+2\mu)\vec{\nabla}\cdot\vec{\nabla}\Delta-\cancel{2\mu\vec{\nabla}\cdot(\vec{\nabla}\times\vec{\omega})}=\rho\vec{\nabla}\cdot\ddot{\vec{u}} \tag{296}$$

Upon simplification, Eq. (296) yields the *dilatation wave equation*, i.e.,

$$(\lambda+2\mu)\nabla^2\Delta=\rho\ddot{\Delta} \tag{297}$$

The simplification of Eq. (296) to yield Eq. (297) was made possible by the general property $\vec{a}\cdot(\vec{a}\times\vec{b})=0$ and by the definitions $\nabla^2=\vec{\nabla}\cdot\vec{\nabla}$ and $\Delta=\vec{\nabla}\cdot\vec{u}$ presented in Appendix B. The inter-commutativity of space and time derivatives was also employed. Division of Eq. (297) by ρ and utilization of definition (268) yields

$$c_{\mathrm{P}}^2\nabla^2\Delta=\ddot{\Delta} \tag{298}$$

which indicates that dilatational waves propagate with the pressure wave speed, c_{P}.

To calculate the rotational waves, recall the Navier equations in vector form as given in Appendix B, i.e.,

$$(\lambda+\mu)\vec{\nabla}\Delta+\mu\nabla^2\vec{u}=\rho\ddot{\vec{u}} \tag{299}$$

Cross premultiply Eq. (299) by $\vec{\nabla}\times$ and get

$$(\lambda+2\mu)\cancel{\vec{\nabla}\times\vec{\nabla}\Delta}-2\mu\nabla^2\vec{\nabla}\times\vec{u}=\rho\vec{\nabla}\times\ddot{\vec{u}} \tag{300}$$

Upon simplification, Eq. (300) yields the *rotational wave equation*, i.e.,

$$(\lambda + 2\mu)\nabla^2 \vec{\omega} = \rho \ddot{\vec{\omega}} \tag{301}$$

The simplification of Eq. (300) to yield Eq. (301) was made possible by the fact that $\vec{a} \times \vec{a} = \vec{0}$ and by the definition $\vec{\omega} = \frac{1}{2}\vec{\nabla} \times \vec{u}$ presented in Appendix B. The inter-commutativity of space and time derivatives was also employed. Division of Eq. (297) by ρ and utilization of definition (269) yields

$$c_S^2 \nabla^2 \vec{\omega} = \ddot{\vec{\omega}} \tag{302}$$

which indicates that rotational waves propagate with the shear wave speed, c_S.

The scalar Eq. (297) and the vector Eq. (301) indicate that both dilatation and rotation obey a wave equation and propagate as waves into the elastic body. Depending on the initial and boundary conditions, the dilation and rotational waves can either exist alone or coexist in the elastic body.

5.6.5 IRROTATIONAL AND EQUIVOLUME WAVES

It can be shown that the pressure waves are *irrotational waves*, i.e., have zero rotation ($\vec{\omega} = 0$), whereas the shear waves are *equivolume waves*, i.e., they have zero dilatation ($\Delta = \vec{\nabla} \cdot \vec{u} = 0$). The proof is as follows.

Assume a pressure wave in the form of Eq. (275), i.e.,

$$\vec{u} = A_1 \vec{e}_1 f(x_1 - c_P t) \quad \text{(pressure wave)} \tag{303}$$

Calculate the rotation and show that it is zero, i.e.,

$$\vec{\omega} = \frac{1}{2}\vec{\nabla} \times \vec{u} = \frac{1}{2}\vec{\nabla} \times A_1 \vec{e}_1 f(x_1 - c_P t) = \frac{1}{2} A_1 \left(\frac{\partial}{\partial x_3}\vec{e}_2 - \frac{\partial}{\partial x_2}\vec{e}_3 \right) f(x_1 - c_P t) = \vec{0} \tag{304}$$

This proves that *pressure waves are irrotational waves*, i.e., it has zero rotation.

The dilatation is calculated as

$$\Delta = \vec{\nabla} \cdot \vec{u} = \vec{\nabla} \cdot A_1 \vec{e}_1 f(x_1 - c_S t) = A_1 \frac{\partial}{\partial x_1} f(x_1 - c_S t) = A_1 f'(x_1 - c_S t) \neq 0 \tag{305}$$

This proves that the *pressure waves are dilatational waves*. However, the shear strains are nonzero, e.g.,

$$\varepsilon_{12} = \frac{1}{2}\left(\frac{\partial u_1}{\partial x_2} + \frac{\partial u_2}{\partial x_1} \right) = \frac{1}{2} A_1 f'(x_1 - ct) \neq 0 \tag{306}$$

This means that the pressure waves are *not* non-distortional.

In terms of wave potentials formulation, we note that waves based only on the scalar potential, Φ, are irrotational waves. To prove this, assume

$$\vec{u} = \vec{\nabla}\Phi \tag{307}$$

and calculate the rotation

$$\vec{\omega} = \frac{1}{2}\vec{\nabla} \times \vec{u} = \frac{1}{2}(\vec{\nabla} \times \vec{\nabla})\Phi = \vec{0} \tag{308}$$

Equation (308) is zero in virtue of the general property $\vec{a} \times \vec{a} = \vec{0}$. On the other hand, the dilation is given by

$$\Delta = \vec{\nabla} \cdot \vec{u} = \vec{\nabla} \cdot \vec{\nabla}\Phi = \nabla^2\Phi \neq 0 \tag{309}$$

This implies that the scalar potential, Φ, can be also seen as the *dilatation potential.*

Assume a shear wave in the form of, say, Eq. (276), i.e.,

$$\vec{u} = A_2\vec{e}_2 f(x_1 - c_S t) \quad \text{(shear wave)} \tag{310}$$

Calculate the dilatation and show that it is zero, i.e.,

$$\Delta = \vec{\nabla} \cdot \vec{u} = \vec{\nabla} \cdot A_2\vec{e}_2 f(x_1 - c_S t) = A_2 \frac{\partial}{\partial x_2} f(x_1 - c_S t) = 0 \tag{311}$$

This proves that the shear waves are *equivolume waves*, as the dilatation is a direct measure of volumetric expansion. The only deformation that the medium undergoes is distortion. This implies that the equivolume waves are also *distortional waves.*

In terms of wave potentials formulation, we notice that waves based on only the vector potential, $\vec{H}$, are equivolume waves. To prove this, assume

$$\vec{u} = \vec{\nabla} \times \vec{H} \tag{312}$$

and calculate the dilatation

$$\Delta = \vec{\nabla} \cdot \vec{u} = \vec{\nabla} \cdot (\vec{\nabla} \times \vec{H}) = 0 \tag{313}$$

Equation (313) is zero in virtue of the general property $\vec{a} \cdot (\vec{a} \times \vec{b}) = 0$. On the other hand, the rotation is given by

$$\vec{\omega} = \frac{1}{2}\vec{\nabla} \times \vec{u} = \frac{1}{2}\vec{\nabla} \times (\vec{\nabla} \times \vec{H}) \neq \vec{0} \tag{314}$$

This implies that the vector potential, $\vec{H}$, can be also seen as the *rotation potential.* As shown earlier, equivolume waves are also distortional waves. This implies that the vector potential, $\vec{H}$, can be also seen as a *distortional potential.*

An interesting note to be made is that while shear waves are distortional and non-dilatational, the dilatational (pressure) waves are not necessarily non-distortional. In fact, dilatational waves contain both bulk expansion and distortion. A simple example is provided by the case of spherical waves emanating from a point blast. By virtue of spherical symmetry, the motion is clearly irrotational. However, examination of an infinitesimal element indicates that its deformation is not solely volumetric. In fact, in order to achieve radial expansion, both bulk expansion and distortion (shear strains) are necessary. This observation if substantiated numerically by the relation between the constant $\lambda + 2\mu$, on one hand, and the bulk modulus, B, and shear modulus, μ, on the other hand. Using the definitions of Appendix B, we write

$$\lambda + 2\mu = B + \frac{4}{3}\mu \tag{315}$$

It is apparent that the constant $\lambda + 2\mu$, which defines the pressure wave speed, c_P, via Eq. (268), depends on both the bulk modulus, B, and the shear modulus, μ. Irrotational waves involve both bulk expansion and distortion.

5.6.6 Z-INVARIANT 3-D WAVES

Of particular interest are 3-D plane waves that are invariant in one direction along the wave front. This situation is encountered, for example, in straight-crested waves, with the wave crest parallel to the z direction. Traditionally (see, for example, Graff, 1975), the invariant direction is taken as the z-axis, and for this reason these waves are called *z-invariant*. It will be shown that under the z-invariant conditions, certain simplifications apply that are particularly useful in ultrasonic SHM practice.

Assume, as shown in Fig. 5.23, that the wave front is parallel to the z axis and that the wave disturbance is invariant along the z axis. This means that all the functions involved in the analysis will not depend on z, and their derivatives with respect to z will be zero. In addition, the wave front normal, $\vec{n}$, will be perpendicular to the z axis, i.e., $\vec{n} \perp \vec{k}$, where $\vec{k}$ is the unit vector of the z axis.

Recall Eq. (283), i.e.,

$$\vec{u} = \vec{\nabla}\Phi + \vec{\nabla} \times \vec{H} \tag{316}$$

As the problem is z-invariant, we have

$$\frac{\partial}{\partial z} \equiv 0 \quad \text{and} \quad \vec{\nabla} = \vec{i}\frac{\partial}{\partial x} + \vec{j}\frac{\partial}{\partial y} \tag{317}$$

Substitution of Eq. (317) into Eq. (316) yields, upon expansion,

$$\vec{u} = \left(\frac{\partial \Phi}{\partial x} + \frac{\partial H_z}{\partial y}\right)\vec{i} + \left(\frac{\partial \Phi}{\partial y} - \frac{\partial H_z}{\partial x}\right)\vec{j} + \left(\frac{\partial H_y}{\partial x} - \frac{\partial H_x}{\partial y}\right)\vec{k} \tag{318}$$

Equation (318) indicates that, though the motion is z-invariant, the displacement has components in all three directions (x, y, and z). The potentials Φ, H_x, H_y, H_z satisfy the wave equations and the uniqueness condition, i.e.,

$$c_P^2 \nabla^2 \Phi = \ddot{\Phi} \begin{cases} c_S^2 \nabla^2 H_x = \ddot{H}_x \\ c_S^2 \nabla^2 H_y = \ddot{H}_y \\ c_S^2 \nabla^2 H_z = \ddot{H}_z \end{cases} \quad \frac{\partial H_x}{\partial x} + \frac{\partial H_y}{\partial y} + \frac{\partial H_z}{\partial z} = 0 \tag{319}$$

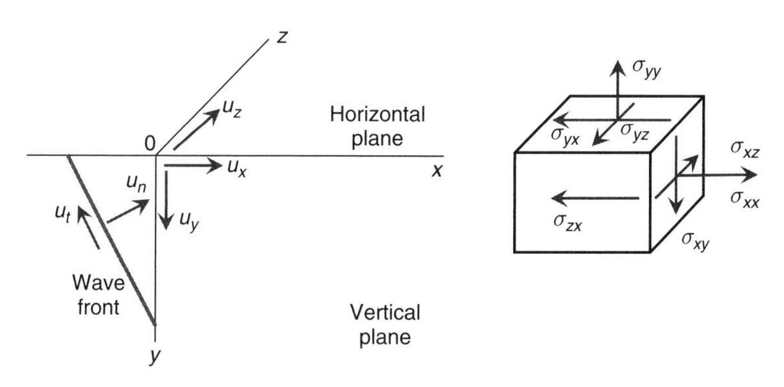

FIGURE 5.23 The general setup for the study of z-invariant plane waves.

The components of the displacement vector are

$$
\begin{aligned}
u_x &= \frac{\partial \Phi}{\partial x} + \frac{\partial H_z}{\partial y} \\
u_y &= \frac{\partial \Phi}{\partial y} - \frac{\partial H_z}{\partial x} \\
u_z &= \frac{\partial H_y}{\partial x} - \frac{\partial H_x}{\partial y}
\end{aligned}
\tag{320}
$$

Examinations of Eq. (320) indicates that it is possible to partition the solution into two parts: (a) a solution in terms of only two potentials, H_x and H_y, i.e., u_z and (b) a solution in terms of the other two potentials, Φ and H_z, i.e., u_x and u_y. The first solution, which accepts only the u_z displacement, will be a shear motion polarized in the horizontal plane Oxz, i.e., an SH wave. This motion is described in terms of the two potentials, H_x and H_y. The second solution, which accepts u_x and u_y displacements, will be the combination of a pressure wave, P (represented by the potential Φ) and a shear wave (represented by the potential H_z). Because this second solution is constrained to the vertical plane, the associated shear wave will be an SV wave. Consequently, the second solution is denoted P+SV. The two solutions are hence treated separately:

SH wave solution. Motion is contained in the horizontal plane, and the relevant potentials are H_x and H_y.

$$
u_x = u_y = 0, \quad u_z \neq 0, \quad \frac{\partial}{\partial z} = 0 \quad H_x \text{ and } H_y \text{ only} \tag{321}
$$

Recall the stress–displacement relations of Appendix B, i.e.,

$$
\begin{aligned}
\sigma_{xx} &= (\lambda + 2\mu)\frac{\partial u_x}{\partial x} + \lambda \frac{\partial u_y}{\partial y} + \lambda \frac{\partial u_z}{\partial z} & \qquad \sigma_{xy} &= \mu\left(\frac{\partial u_x}{\partial y} + \frac{\partial u_y}{\partial x}\right) \\
\sigma_{yy} &= \lambda \frac{\partial u_x}{\partial x} + (\lambda + 2\mu)\frac{\partial u_y}{\partial y} + \lambda \frac{\partial u_z}{\partial z} \quad \text{and} & \qquad \sigma_{yz} &= \mu\left(\frac{\partial u_y}{\partial z} + \frac{\partial u_z}{\partial y}\right) \\
\sigma_{zz} &= \lambda \frac{\partial u_x}{\partial x} + \lambda \frac{\partial u_y}{\partial y} + (\lambda + 2\mu)\frac{\partial u_z}{\partial z} & \qquad \sigma_{zx} &= \mu\left(\frac{\partial u_z}{\partial x} + \frac{\partial u_x}{\partial z}\right)
\end{aligned}
\tag{322}
$$

Substitution of Eqs. (320) and (321) into the stress–displacement relations (322) yields, upon rearrangement,

$$
\sigma_{yz} = \mu\left(-\frac{\partial^2 H_z}{\partial y^2} + \frac{\partial^2 H_z}{\partial x\,\partial y}\right) \quad \text{(SH waves)} \tag{323}
$$

$$
\sigma_{zx} = \mu\left(-\frac{\partial^2 H_z}{\partial x\,\partial y} + \frac{\partial^2 H_z}{\partial x^2}\right) \quad \text{(SH waves)} \tag{324}
$$

P+SV wave solution. Motion is contained in the vertical plane, and the relevant potentials are Φ and H_z.

$$
u_x \neq 0, \quad u_y \neq 0, \quad u_z = 0, \quad \frac{\partial}{\partial z} = 0, \quad \Phi \text{ and } H_z \text{ only} \tag{325}
$$

Substitution of Eqs. (320) and (325) into the stress–displacement relations (322) yields, upon rearrangement,

$$\begin{aligned}\sigma_{xx} &= (\lambda+2\mu)\left(\frac{\partial^2\Phi}{\partial x^2}+\frac{\partial^2\Phi}{\partial y^2}\right)-2\mu\left(\frac{\partial^2\Phi}{\partial y^2}-\frac{\partial^2 H_z}{\partial x\,\partial y}\right)\\ &= \lambda\left(\frac{\partial^2\Phi}{\partial x^2}+\frac{\partial^2\Phi}{\partial y^2}\right)+2\mu\left(\frac{\partial^2\Phi}{\partial x^2}+\frac{\partial^2 H_z}{\partial x\,\partial y}\right)\end{aligned} \quad \text{(P+SV waves)} \tag{326}$$

$$\begin{aligned}\sigma_{yy} &= (\lambda+2\mu)\left(\frac{\partial^2\Phi}{\partial x^2}+\frac{\partial^2\Phi}{\partial y^2}\right)-2\mu\left(\frac{\partial^2\Phi}{\partial x^2}+\frac{\partial^2 H_z}{\partial x\,\partial y}\right)\\ &= \lambda\left(\frac{\partial^2\Phi}{\partial x^2}+\frac{\partial^2\Phi}{\partial y^2}\right)+2\mu\left(\frac{\partial^2\Phi}{\partial y^2}-\frac{\partial^2 H_z}{\partial x\,\partial y}\right)\end{aligned} \quad \text{(P+SV waves)} \tag{327}$$

$$\sigma_{zz} = \lambda\left(\frac{\partial^2\Phi}{\partial x^2}+\frac{\partial^2\Phi}{\partial y^2}\right) \quad \text{(P+SV waves)} \tag{328}$$

$$\sigma_{xy} = \mu\left(2\frac{\partial^2\Phi}{\partial x\,\partial y}-\frac{\partial^2 H_z}{\partial x^2}+\frac{\partial^2 H_z}{\partial y^2}\right) \quad \text{(P+SV waves)} \tag{329}$$

5.7 SUMMARY AND CONCLUSIONS

This chapter has presented a review of elastic wave propagation in elastic media. Because the description and presentation of the wave propagation problem can become quite complicated in some cases, this chapter has adopted an incremental step-by-step approach. The chapter started with the discussion of the simplest case of wave propagation – the study of the axial waves propagating in a straight bar. This simple physical example was used to develop the fundamental principles of wave propagation, such as wave equation and wave speed; d'Alembert (generic) solution and separation of variables (harmonic) solution; the contrast between wave speed and particle velocity; acoustic impedance of the medium, and the wave propagation at material interfaces. The concept of standing waves was introduced, and the correspondence between standing waves and structural vibration was established. The power and energy associated with wave propagation in a simple bar were introduced and discussed.

After studying the propagation of simple axial waves in bars, the slightly more complicated problem of flexural wave propagation in beams was introduced and discussed. The equation of motion for flexural waves was developed. The general solution in terms of propagating and evanescent waves was derived. The dispersive nature of flexural waves was identified and studied. In this context, the concept of *group velocity* as different from the *phase velocity* (a.k.a., *wave speed*) was introduced, and its implications on wave propagation were studied. Power and energy of flexural waves and the energy velocity of dispersive waves were discussed.

The discussion of wave propagation was then extended to plate waves. Two types of plate waves were considered: (1) axial waves in plates and (2) flexural waves in plates. Each case was discussed separately. The general equations of wave motion were derived. Solutions were derived for certain selected cases such as straight-crested axial waves; straight-crested shear waves; circular-crested axial waves; straight-crested flexural waves; and circular-crested circular waves. The dispersive nature of flexural waves in plates and its connection to the dispersive nature of flexural waves in beams was identified and discussed.

The last part of the chapter was allotted to the discussion of 3-D waves that appear in unbounded solids. The general equations of 3-D wave propagation in unbounded solid media were developed from first principles. The eigenvalues and eigenvectors of the wave equation were identified. The two corresponding basic wave types, pressure waves and shear waves, were discussed. Dilatational, rotational, irrotational, and equivolume waves were identified and discussed. The case of z-invariant wave propagation was presented. The latter serves as a bridge toward the study of guided wave, such as Lamb waves, which makes the object of the next chapter.

5.8 PROBLEMS AND EXERCISES

1. Consider the 1-D wave equation $c^2u'' = \ddot{u}$ and recall the d'Alembert solution. (a) Verify by direct substitution that the functions $f(x-ct)$ and $g(x+ct)$ of the d'Alembert solution satisfy the wave equation. (b) Find the expressions of f and g for the following initial conditions at $t=0$, $u(x,0)=u_0(x)$ and $\dot{u}(x,0)=0$. (c) Sketch the behavior of $f(x-ct)$ and $g(x+ct)$ for various times $t>0$ and explain which is the forward wave and which is the backward wave.
2. For each material given in Table 5.3, find: (a) the wave speed. How long does it take the wave to travel 10 m? (b) the pressure wave amplitude (expressed in absolute value and as percentage of yield stress) needed to achieve a maximum strain of 5000 $\mu\varepsilon$.[1] Is this feasible? (c) the acoustic impedance; (d) the particle velocity for a pressure wave with amplitude $p_{\max} = \frac{1}{2}Y$; comment on the result; (e) the pressure wave amplitude (expressed in absolute value and as percentage of yield stress) needed to achieve a maximum particle velocity of 20 m/s. Is this feasible? (f) the displacement wave amplitude for a 100 kHz harmonic pressure wave with amplitude $p_{\max} = \frac{1}{2}Y$.

TABLE 5.3 Typical material properties of aluminum and steel

	Aluminum (7075 T6)	Steel (AISI 4340 normalized)
Modulus, E	70 GPa	200 GPa
Poisson ratio	−0.33	0.3
Density, ρ	2700 kg/m^3	7750 kg/m^3
Yield stress, Y	500 MPa	860 MPa

3. Consider a semi-infinite slender bar subjected to end displacement excitation $u_0(t)$ of the form: $u(t,0) = u_0(t) = bt\exp(-t/\tau), 0 < t$, where $\tau = 50\,\text{ms}$, $b = 10^{-4}c$, and c is the wave velocity in the bar. The bar is made of aluminum (Table 5.3). The cross-sectional area is $A = 25\,\text{mm}^2$. (a) Plot the value of u_0 at 1 ms intervals up to $t_{\max} = 10\tau$. (b) Predict the value of time when the maximum value of u_0 occurs and verify this value on the plot. (c) Find the solution $u(t,x)$. (d) Follow the wave propagation up to $x_{\max} = 10\tau c$. Sketch the solution at times $t = 0.1\tau,\ 0.5\tau,\ 2\tau,\ 3\tau,\ 4\tau,\ 6\tau,\ 8\tau$. Present results in the form of subsequent plots, one below the other. Describe what you see.

[1] $1\,\mu\varepsilon = 1$ micro-strain $= 10^{-6}$ units of strain

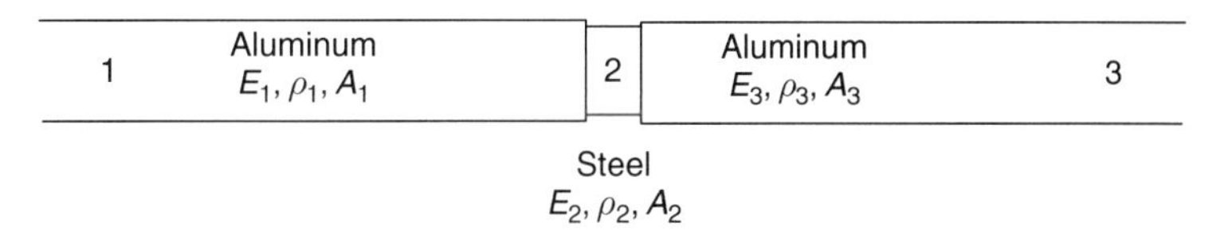

FIGURE 5.24 Split Hopkinson bar.

4. Consider the split Hopkinson bar. A long and slender aluminum bar ($r_1 = r_3 = 10\,\mathrm{mm}$) is split at the center and a steel piece ($r_2 = 10\,\mathrm{mm}$) is inserted in between as shown in Fig. 5.24. The contact at the two interfaces is assumed perfect. Assume that an incident compressive stress wave pulse of 400 MPa is traveling forward in bar 1. At the interface with bar 2 (the steel piece), some of the wave will be transmitted and some will be reflected. The transmitted wave will hit the second interface. Again, part of the incident wave will be reflected and part will be transmitted.
   Find: (a) the amplitude of the stress waves transmitted into bars 2 and 3. Comment on the stress values in comparison with the yield stress of the materials and explain what this means. (b) The amplitude of the stress waves reflected at the 1–2 and 2–3 interfaces. Comment on negative stress values (if they appear) and explain what this means. (c) What radius should the steel piece have such that no reflection occurs at the two interfaces, and the amplitude of the wave transmitted in bar 3 is the same as that of the incident wave in bar 1. After finding the value, verify that indeed no reflection takes place.
5. Consider a semi-infinite bar starting at $x = 0$. A rectangular pressure pulse is applied at the $x = 0$ end, $p(t) = p_0$ for $0 < t < T$; $p(t) = 0$ otherwise. The bar has a cross-sectional area $A = 10\,\mathrm{mm}^2$ and the material properties given in Table 5.3. The pressure pulse has $p_0 = Y/2,\ T = 10\,\mathrm{ms}$. Find (a) stress wave expression and its maximum value; (b) particle velocity expression and its maximum value; (c) kinetic, elastic, and total energy density per unit length of the bar (expression and maximum value); (d) power flow in the bar (expression and maximum value).
6. Consider a semi-infinite bar under 100-kHz harmonic pressure excitation at one end. The bar has a cross-sectional area $A = 10\,\mathrm{mm}^2$ and the material properties given in Table 5.3. The pressure amplitude is $p_0 = Y/2$. Find (a) displacement wave expressions and its amplitude; (b) stress wave expression and its amplitude; phase relation between stress and displacement; (c) particle velocity expression and its amplitude; phase relation between stress and particle velocity; (d) kinetic, elastic, and total energy densities per unit length of the bar (expression, amplitude, and time-averaged value); (e) power flow in the bar (expression amplitude, and time-averaged value).
7. Consider elastic waves in a free aluminum bar of length $L = 100\,\mathrm{mm}$ as generated by harmonic pressure excitation with amplitude $p_0 = Y/2$ applied at the left-hand end. (a) Analyze the wave reflection process in the bar (boundary condition at $x = 0$; the reflection at $x = L$). (b) Analyze the resonance condition and calculate the eigenvalues, and eigen frequencies; (c) calculate the eigen lengths for an excitation frequency of 50 kHz. (d) for the stress wave solution at resonance; sketch first, second, and third mode shapes; (e) for the displacement wave solution at resonance; sketch first, second, and third mode shapes.
8. Consider harmonic elastic waves in an infinite slender bar. (a) Derive the expression for standing waves considering the superposition of two identical harmonic waves traveling in opposite directions; accompany your derivation with the appropriate

sketches. (b) Determine the expression for the bar length L, such that it will permit the generation of standing waves of a given frequency, f; consider the following cases: fundamental; overtone; and second overtone. (c) Determine an expression for the frequency that will generate standing waves in a bar of a given length, L; consider the following cases: fundamental; overtone; and second overtone.

9. Consider the properties of aluminum and steel shown in Table 5.3. (a) Write the expressions for the elastic constants λ, μ, and bulk modulus, k; calculate their values. (b) Calculate the following wave-speed values: $c_P = \sqrt{(\lambda+2\mu)/\rho}$ (3-D pressure wave); $c_S = \sqrt{\mu/\rho}$ shear wave); $c = \sqrt{E/\rho}$ (1-D pressure wave); $c_L = \sqrt{E/\rho(1-\nu^2)}$ (axial waves in plates). Comment on why c and c_P have different values, while both are "pressure" wavespeeds .
10. Consider a 3-D plane wave traveling with speed c along an arbitrary direction $\vec{n}$. State the general expression of the particle motion, $\vec{u}$. Substitute it into the Navier equations and deduce the characteristic equation for the wave speed c. Solve the characteristic equation. Explain how many types of waves can travel in the 3-D material. Give the appropriate wave speeds. Sketch the particle motion for each wave type.
11. Consider the z-invariant 3-D conditions. Explain the meaning of the 'z-invariant' condition and write out all the related conditions that apply. Derive the expressions of strains in terms of displacements for the z-invariant case starting from the general 3-D strain–displacement relations. Derive the stress–strain expressions for the z-invariant case starting from the general stress–strain expressions (use expressions in terms of Lame constants). Using these results, derive the expressions of stress in terms of displacements for the z-invariant case.
12. Consider 3-D plane wave traveling in a direction $\vec{n} \perp Oz$. Assume the motion z-invariant. State the general solution of wave propagation in a 3-D medium using Φ and $\vec{H}$ potentials. Particularize the general solution to the case of z-invariant plane waves. State what type of plane waves can travel in the 3-D material under the z-invariance assumption. Give the expressions for the particle motion and stresses for each of these wave types in terms of the Φ and $\vec{H}$ potentials.
13. Consider the definition of group velocity in the form $c_g = \mathrm{d}\omega/\mathrm{d}\gamma$. Prove the following equivalent formulae for calculating group velocity: (a) $c_g = c + \gamma\frac{\mathrm{d}c}{\mathrm{d}\gamma}$; (b) $c_g = c^2/(c-\omega\frac{\mathrm{d}c}{\mathrm{d}\omega})$ and $c_g = c^2/(c-f\frac{\mathrm{d}c}{\mathrm{d}f})$; (c) $c_g = c^2/(c-(fd)\frac{\mathrm{d}c}{\mathrm{d}(fd)})$.
14. Consider a generic spherical wave, $\Phi(r,t)$, of energy E_0 emanating from a point source. At $r = r_0$, the wave amplitude is $A(r_0) = A_0$. Starting with the generic wave equation $\ddot{\Phi} = c^2\nabla^2\Phi$, deduce (a) the form of the wave equation that applies to this situation; (b) the d'Alembert solution for $\Phi(r,t)$; (c) the wave amplitude expression, $A(r)$, as function of r and A_0; and (d) the wave energy density expression, $e(r)$, as function of r and E_0.
15. Consider a generic circular wave, $\Phi(r,t)$, of energy E_0 emanating from a point source. At $r = r_0$, the wave amplitude is $A(r_0) = A_0$. Consider the problem to be 2-D. Starting with the generic wave equation $\ddot{\Phi} = c^2\nabla^2\Phi$, deduce (a) the form of the wave equation that applies to this situation; (b) the d'Alembert solution for $\Phi(r,t)$; (c) the wave amplitude expression, $A(r)$, as function of r and A_0; and (d) the wave energy density expression, $e(r)$, as function of r and E_0.

6

GUIDED WAVES

6.1 INTRODUCTION

This chapter will deal with an important class of waves that have widespread applications in structural health monitoring (SHM) – the class of *guided waves*. Guided waves are especially important for structural health monitoring because they can travel at large distances in structures with only little energy loss. Thus, they enable the SHM of large areas from a single location. Guided waves have the important property that they remain confined inside the walls of a thin-wall structure, and hence can travel over large distances. In addition, guided waves can also travel inside curved walls. These properties make them well suited for the ultrasonic inspection of aircraft, missiles, pressure vessels, oil tanks, pipelines, etc.

The chapter will start with the discussion of *Rayleigh waves*, a.k.a., *surface acoustic waves* (SAW). Rayleigh waves are found in solids that contain a free surface. The Rayleigh waves travel close to the free surface with very little penetration in the depth of the solid. For this reason, Rayleigh waves are also known as *surface-guided waves*.

The chapter will continue with a description of guided waves in plates. In flat plates, ultrasonic-guided waves travel as Lamb waves and as shear horizontal (SH) waves. Lamb waves are vertically polarized, whereas SH waves are horizontally polarized. Ultrasonic guided waves in flat plates were first described by Lamb (1917). A comprehensive analysis of Lamb wave was given by Viktorov (1967), Achenbach (1973), Graff (1975), Rose (1999), and Royer and Dieulesaint (2000).

A simple form of guided plate waves are the SH waves. The particle motion of SH waves is polarized parallel to the plate surface and perpendicular to the direction of wave propagation. The SH waves can be symmetric and antisymmetric. With the exception of the very fundamental mode, the SH wave modes are all dispersive.

Lamb waves are more complicated guided plate waves. Lamb waves are of two basic varieties, *symmetric Lamb-waves modes* ($S_0, S_1, S_2, \ldots$) and *antisymmetric Lamb-waves modes* ($A_0, A_1, A_2, \ldots$). Both Lamb wave types are quite dispersive. At any given value of the frequency–thickness product *fd*, a multitude of symmetric and antisymmetric Lamb waves may exist. The higher the *fd* value, the larger the number of Lamb-wave modes that can simultaneously exist. For relatively small values of the *fd* product, only the basic symmetric and antisymmetric Lamb-wave modes (S_0 and A_0) exist. As the *fd* product approaches zero, the S_0 and A_0 modes degenerate in the basic axial and flexural plate modes discussed in chapter 5. At the other extreme, as $fd \to \infty$, the S_0 and A_0 Lamb-wave modes degenerate into Rayleigh waves confined to the plate surface.

Guided waves also exist in other thin-wall structures, such as rods, tubes, and shells. Though the underlying physical principles of guided-wave propagation still apply, their study is more elaborate. For illustration, this chapter will cover a brief description of guided waves in cylindrical shells. It will be shown that dispersive longitudinal (axial), flexural, and torsional multi-mode guided waves can simultaneously exist in such structures.

The last part of this chapter will be devoted to the discussion of guided waves in composite materials. This analysis is rather elaborate, as wave reflection and transmission at each layer interface has to be considered across the composite thickness. However, a solution that uses the transfer matrix method will be illustrated.

6.2 RAYLEIGH WAVES

Rayleigh waves, a.k.a., SAW, have the property of propagating close to the body surface, with the motion amplitude decreasing rapidly with depth (Fig. 6.1). The polarization of Rayleigh waves lies in a plane perpendicular to the surface. The effective depth of penetration is less than a wavelength.

6.2.1 RAYLEIGH WAVES EQUATIONS

We analyze Rayleigh waves under the z-invariant assumption, i.e., $\frac{\partial}{\partial z} = 0$. The Rayleigh-wave particle motion is contained in the vertical plane, hence the P + SV wave solution of the previous section applies, i.e.,

$$u_x \neq 0, \quad u_y \neq 0, \quad u_z = 0 \tag{1}$$

According to the 3-D waves theory of Chapter 5, only the Φ and H_z potential are required to describe this motion. Recall that the potentials satisfy the wave equation, i.e.,

$$c_{\rm P}^2 \nabla^2 \Phi = \ddot{\Phi} \quad c_{\rm S}^2 \nabla^2 H_z = \ddot{H}_z \tag{2}$$

Assume that the Rayleigh wave motion takes place in the x direction with speed c and wavenumber ξ. Hence, the potentials Φ and H_z can be written as

$$\Phi(x, y, t) = f(y)\, e^{i(\xi x - \omega t)} \quad H_z(x, y, t) = h_z(y)\, e^{i(\xi x - \omega t)} \tag{3}$$

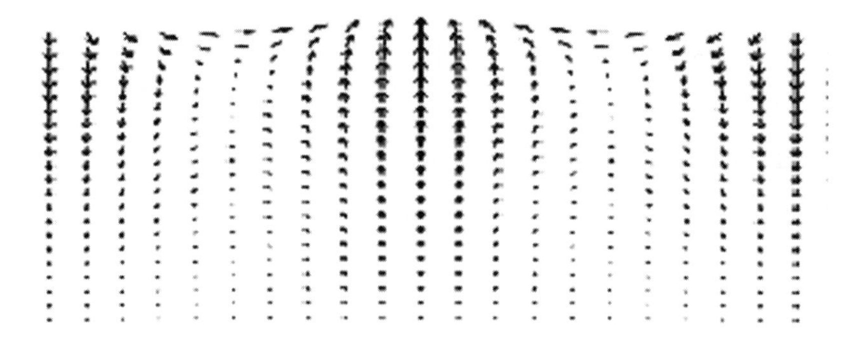

FIGURE 6.1 Simulation of Rayleigh wave in a semi-infinite medium.

Substitution of Eqs. (3) into Eqs. (2) yields, after division by $e^{i(\xi x-\omega t)}$,

$$c_P^2\left[(i\xi)^2 f(y)+f''(y)\right]=(-i\omega)^2 f(y) \qquad c_S^2\left[(i\xi)^2 h(y)+h''(y)\right]=(-i\omega)^2 h(y) \tag{4}$$

Upon rearrangement, we get

$$f''(y)-\left(\xi^2-\frac{\omega^2}{c_P^2}\right)f(y)=0 \qquad h''(y)-\left(\xi^2-\frac{\omega^2}{c_S^2}\right)h(y)=0 \tag{5}$$

Denote

$$\alpha^2=\xi^2-\frac{\omega^2}{c_P^2} \qquad \beta^2=\xi^2-\frac{\omega^2}{c_S^2} \tag{6}$$

Substitution of Eqs. (6) into Eqs. (5) yields the ordinary differential equations

$$f''(y)-\alpha^2 f(y)=0 \qquad h''(y)-\beta^2 h(y)=0 \tag{7}$$

Solutions of Eqs. (7) are

$$f(y)=A_1e^{\alpha y}+A_2e^{-\alpha y} \qquad h(y)=B_1e^{\beta y}+B_2e^{-\beta y} \tag{8}$$

The functions given by Eqs. (8) contain both increasing and decreasing exponentials in the space variable y. However, the increasing exponentials do not satisfy the radiation condition, because they would become unbound as $y\to\infty$. Hence, we only retain the decreasing exponentials and write the general solution as

$$f(y)=Ae^{-\alpha y} \qquad h(y)=Be^{-\beta y} \tag{9}$$

Substitution of Eqs. (9) into the general expressions (3) yields

$$\Phi(x,y,t)=Ae^{-\alpha y}e^{i(\xi x-\omega t)} \qquad H_z(x,y,t)=Be^{-\beta y}e^{i(\xi x-\omega t)} \tag{10}$$

6.2.2 BOUNDARY CONDITIONS

The boundary conditions for Rayleigh waves correspond to a tractions-free half-plane surface, i.e.,

$$\sigma_{yy}\big|_{y=0}=0 \qquad \sigma_{xy}\big|_{y=0}=0 \tag{11}$$

Recall the corresponding equations of Chapter 5, i.e.,

$$\sigma_{yy}=(\lambda+2\mu)\left(\frac{\partial^2\Phi}{\partial x^2}+\frac{\partial^2\Phi}{\partial y^2}\right)-2\mu\left(\frac{\partial^2\Phi}{\partial x^2}+\frac{\partial^2 H_z}{\partial x\,\partial y}\right) \tag{12}$$

$$\sigma_{xy}=\mu\left(2\frac{\partial^2\Phi}{\partial x\,\partial y}-\frac{\partial^2 H_z}{\partial x^2}+\frac{\partial^2 H_z}{\partial y^2}\right) \tag{13}$$

Substitution of the general expression (10) into Eqs. (12) and (13) yields

$$\sigma_{yy}=\left[(\lambda+2\mu)\left((i\xi)^2+(-\alpha)^2\right)Ae^{-\alpha y}-2\mu\left((i\xi)^2Ae^{-\alpha y}+i\xi(-\beta)Be^{-\beta y}\right)\right]e^{i(\xi x-\omega t)} \tag{14}$$

$$\sigma_{xy}=\mu\left[2i\xi(-\alpha)Ae^{-\alpha y}+\left(-(i\xi)^2+(-\beta)^2\right)Be^{-\beta y}\right]e^{i(\xi x-\omega t)} \tag{15}$$

Upon rearrangement, Eqs. (14) and (15) become

$$\sigma_{yy} = \left[(\lambda+2\mu)\left(-\xi^2+\alpha^2\right)Ae^{-\alpha y}+2\mu\xi^2Ae^{-\alpha y}+2\mu i\xi\beta Be^{-\beta y}\right]e^{i(\xi x-\omega t)} \tag{16}$$

$$\sigma_{xy} = \mu\left[-2i\alpha\xi Ae^{-\alpha y}+\left(\xi^2+\beta^2\right)Be^{-\beta y}\right]e^{i(\xi x-\omega t)} \tag{17}$$

Eq. (16) can be further simplified. Recall the definition of pressure-wave and shear-wave speeds $c_{\mathrm{P}}^2=(\lambda+2\mu)/\rho$, $c_{\mathrm{S}}^2=\mu/\rho$; upon substitution into Eqs. (6), we obtain

$$(\lambda+2\mu)\left(-\xi^2+\alpha^2\right)=-\rho\omega^2 \quad \text{and} \quad \mu\beta^2=-\mu\xi^2-\rho\omega^2 \tag{18}$$

Substitution of Eq. (18) into Eq. (16) yields

$$\begin{aligned}\sigma_{yy} &= \left[\left(-\rho\omega^2+2\mu\xi^2\right)Ae^{-\alpha y}+2\mu i\xi\beta Be^{-\beta y}\right]e^{i(\xi x-\omega t)}\\ &= \mu\left[\left(\beta^2+\xi^2\right)Ae^{-\alpha y}+2i\xi\beta Be^{-\beta y}\right]e^{i(\xi x-\omega t)}\end{aligned} \tag{19}$$

Substitution of Eqs. (19) and (18) into the boundary conditions (11) yields the homogeneous linear algebraic system

$$\begin{aligned}\left(\beta^2+\xi^2\right)A+2i\xi\beta B&=0\\ -2i\alpha\xi A+\left(\beta^2+\xi^2\right)B&=0\end{aligned} \tag{20}$$

6.2.3 WAVE SPEED OF THE RAYLEIGH SURFACE WAVE

Nontrivial solutions of (20) exists only when the determinant vanishes, i.e.,

$$\begin{vmatrix}\beta^2+\xi^2 & 2i\xi\beta\\ -2i\alpha\xi & \beta^2+\xi^2\end{vmatrix}=0 \tag{21}$$

which leads to the characteristic equation

$$\left(\beta^2+\xi^2\right)^2-4\alpha\beta\xi^2=0 \tag{22}$$

where α and β depend on the frequency ω and the wavenumber ξ as given by Eqs. (6). At a given frequency, numerical solution of Eq. (22) yields the corresponding wavenumbers. Hence, the definition $c=\omega/\xi$ yields the wave speed. In principle, Eq. (22) has three double roots. However, for practical situations, only one is real, corresponding to *wave speed of the Rayleigh surface wave.*

Equation (22) can be further processed to yield an equation in c. Recall Eqs. (6); using the definition $\omega=\xi c$, we write

$$\alpha^2=\xi^2\left(1-\frac{c^2}{c_{\mathrm{P}}^2}\right) \quad \beta^2=\xi^2\left(1-\frac{c^2}{c_{\mathrm{S}}^2}\right) \tag{23}$$

Substitution of Eq. (23) into Eq. (22) and division by ξ^4 yields

$$\left(2-\frac{c^2}{c_{\mathrm{S}}^2}\right)^2=4\left(1-\frac{c^2}{c_{\mathrm{P}}^2}\right)^{1/2}\left(1-\frac{c^2}{c_{\mathrm{S}}^2}\right)^{1/2} \tag{24}$$

Rationalization of Eq. (24) and division by $(c/c_{\mathrm{S}})^2$ yields

$$\left(\frac{c^2}{c_{\mathrm{S}}^2}\right)^3-8\left(\frac{c^2}{c_{\mathrm{S}}^2}\right)^2+24\left(\frac{c^2}{c_{\mathrm{S}}^2}\right)-16\left(\frac{c^2}{c_{\mathrm{P}}^2}\right)-16+\left(\frac{c_{\mathrm{P}}^2}{c_{\mathrm{S}}^2}\right)=0 \tag{25}$$

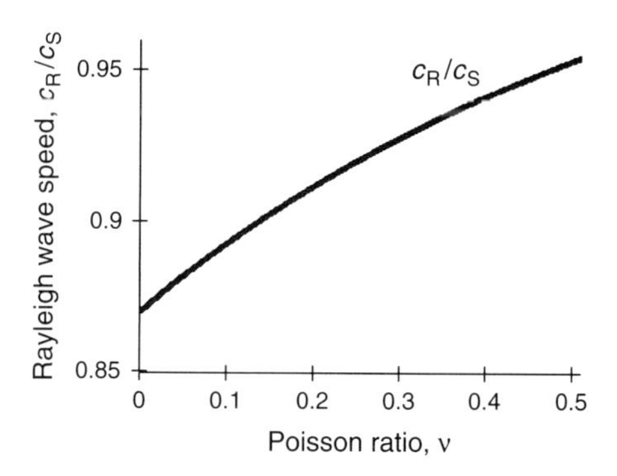

FIGURE 6.2 Plot of Rayleigh wave speed vs. Poisson ratio.

Note that the denominators of Eq. (25) contain both c_P and c_S. Define the ratio of the wave speeds in the material as

$$k^2 = \frac{c_P^2}{c_S^2} = \frac{\lambda + 2\mu}{\mu} = \frac{2(1-\nu)}{1-2\nu} \tag{26}$$

Substitution of Eq. (26) into Eq. (25) yields

$$\left(\frac{c^2}{c_S^2}\right)^3 - 8\left(\frac{c^2}{c_S^2}\right)^2 + \left(24 - 16k^{-2}\right)\left(\frac{c^2}{c_S^2}\right) - 16\left(1 - k^{-2}\right) = 0 \tag{27}$$

where k is a function of ν, as indicated by Eq. (26). Equation (27) is a cubic equation in $(c/c_S)^2$. For practical condition, Eq. (27) only accepts one real root, *the Rayleigh wave speed*, c_R, that depends on the shear wave speed, c_S, and the Poisson ratio, ν. A common approximation of the wave speed of the Rayleigh surface wave is given in the form

$$c_R(\nu) = c_S\left(\frac{0.87 + 1.12\nu}{1+\nu}\right) \tag{28}$$

Once the Rayleigh wave speed, c_R, is known, one can calculate the Rayleigh wavenumber

$$\xi_R = \frac{\omega}{c_R} \tag{29}$$

A plot of the Rayleigh wave velocity, $c_R(\nu)$, vs. the Poisson ratio, ν, is shown in Fig. 6.2. Notice that, for common Poisson ratio values, the Rayleigh wave speed, c_R, takes values close to and just below the shear wave speed, c_S.

6.2.4 PARTICLE MOTION OF THE RAYLEIGH SURFACE WAVE

To calculate Rayleigh-wave particle motion, recall the expression of the particle motion in terms of potentials, i.e.,

$$\begin{aligned} u_x &= \frac{\partial \Phi}{\partial x} + \frac{\partial H_z}{\partial y} \\ u_y &= \frac{\partial \Phi}{\partial y} - \frac{\partial H_z}{\partial x} \end{aligned} \tag{30}$$

Substitution of Eq. (10) into Eq. (30) yields

$$
\begin{aligned}
u_x(x, y, t) &= \left(i\xi A e^{-\alpha y} - \beta B e^{-\beta y}\right) e^{i(\xi x - \omega t)} = \hat{u}_x(y) e^{i(\xi x - \omega t)} \\
u_y(x, y, t) &= \left(-\alpha A e^{-\alpha y} - i\xi B e^{-\beta y}\right) e^{i(\xi x - \omega t)} = \hat{u}_y(y) e^{i(\xi x - \omega t)}
\end{aligned} \tag{31}
$$

where $\xi = \xi_R$. Eq. (20) can be used to express B in terms of A, *i.e.*,

$$
B = -\frac{\beta^2 + \xi^2}{2i\beta\xi} A = \frac{2i\alpha\xi}{\beta^2 + \xi^2} A \tag{32}
$$

Substitution of Eq. (32) into Eq. (31) yields the Rayleigh wave mode shape

$$
\begin{aligned}
\hat{u}_x(y) &= Ai\left(\xi e^{-\alpha y} - \frac{\beta^2 + \xi^2}{2\xi} e^{-\beta y}\right) \\
\hat{u}_y(y) &= A\left(-\alpha e^{-\alpha y} + i\frac{\beta^2 + \xi^2}{2\beta} e^{-\beta y}\right)
\end{aligned} \tag{33}
$$

Simulation of particle motion for a Rayleigh wave is shown in Fig. 6.1.

6.3 SH PLATE WAVES

Shear horizontal waves have a shear-type particle motion contained in the horizontal plane. The axes definition is shown in Fig. 6.3. Note that y axis is placed vertically, whereas the x axis and the z axis define the horizontal plane. The x axis is placed along the wave propagation direction, whereas the z axis is perpendicular to it.

6.3.1 GENERAL EQUATION OF SH WAVES

An SH wave has the particle motion along the z axis, whereas the wave propagation takes place along the x-axis. The particle motion has only the u_z component, which is given by

$$
u_z(x, y, t) = h(y)\, e^{i(\xi x - \omega t)} \tag{34}
$$

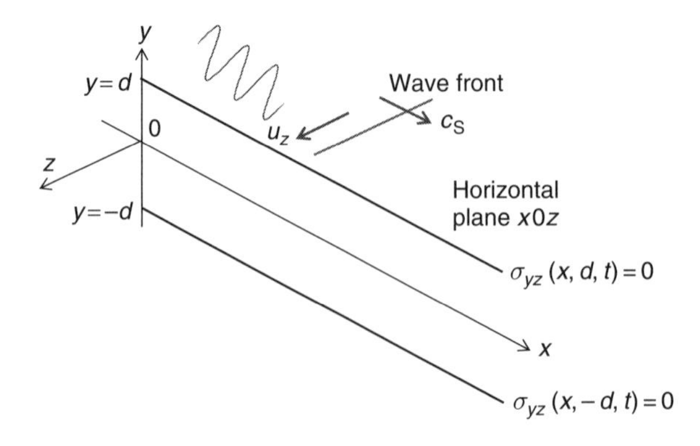

FIGURE 6.3 Axes definition and particle motion for SH waves.

The first part of Eq. (34) represents a standing wave $h(y)$ across the plate thickness. The second part, $e^{i(\xi x-\omega t)}$, represents a wave propagating in the x direction. As before, the problem is assumed z-invariant, i.e.,

$$\frac{\partial(\cdot)}{\partial z} \equiv 0 \quad (z\text{-invariant}) \tag{35}$$

The particle displacement satisfies the wave equation, i.e.,

$$\nabla^2 u_z = \frac{1}{c_S^2}\ddot{u}_z \tag{36}$$

Recalling the definition of the Laplacian ∇^2 and remembering that the motion is z-invariant, Eq. (36) becomes

$$\left(\frac{\partial^2}{\partial x^2}+\frac{\partial^2}{\partial y^2}\right)u_z = \frac{1}{c_S^2}\ddot{u}_z \tag{37}$$

Differentiation of Eq. (34) with respect to x, y, t yields, respectively,

$$\begin{aligned}
\frac{\partial^2 u_z}{\partial x^2} &= (i\xi)^2 h(y)\, e^{i(\xi x-\omega t)} \\
\frac{\partial^2 u_z}{\partial y^2} &= h''(y)\, e^{i(\xi x-\omega t)} \\
\ddot{u}_z &= (-i\omega)^2\, h(y)\, e^{i(\xi x-\omega t)}
\end{aligned} \tag{38}$$

Substitution of Eq. (38) into Eq. (37) and division of both sides by $e^{i(\xi x-\omega t)}$ yields

$$(i\xi)^2 + h''(y) = \frac{(-i\omega)^2}{c_S^2}h(y) \tag{39}$$

Denote

$$\eta^2 = \frac{\omega^2}{c_S^2} - \xi^2 \tag{40}$$

Using Eq. (40), we rewrite Eq. (39) in the form

$$h''(y) + \eta^2 h(y) = 0 \tag{41}$$

The solution of Eq. (41) has the general form

$$h(y) = C_1 \sin\eta y + C_2 \cos\eta y \tag{42}$$

Substitution of Eq. (42) into Eq. (34) yields the general solution

$$u_z(x, y, t) = (C_1 \sin\eta y + C_2 \cos\eta y)e^{i(\xi x-\omega t)} \tag{43}$$

6.3.2 BOUNDARY CONDITIONS

The tractions-free boundary conditions at the upper and lower plate surfaces yield

$$\sigma_{yz}(x, \pm d, t) = 0 \tag{44}$$

Recall the shear stress definition for z-invariant problems

$$\sigma_{yz} = \mu \frac{\partial u_z}{\partial y} \tag{45}$$

Substitution of Eq. (43) into Eq. (45) yields

$$\sigma_{yz} = \mu\eta(C_1 \cos \eta y - C_2 \sin \eta y)\, e^{i(\xi x - \omega t)} \tag{46}$$

Substitute Eq. (46) into boundary conditions (44) and get

$$\begin{aligned} C_1 \cos \eta(+d) - C_2 \sin \eta(+d) = 0 \\ C_1 \cos \eta(-d) - C_2 \sin \eta(-d) = 0 \end{aligned} \tag{47}$$

or

$$\begin{aligned} C_1 \cos \eta d - C_2 \sin \eta d = 0 \\ C_1 \cos \eta d + C_2 \sin \eta d = 0 \end{aligned} \tag{48}$$

The system of linear homogenous equations (48) accepts a nontrivial solution if the determinant of the coefficients is null, i.e.,

$$\begin{vmatrix} \cos \eta d & -\sin \eta d \\ \cos \eta d & \sin \eta d \end{vmatrix} = 0 \tag{49}$$

The determinant (49) yields the characteristic equation

$$\sin \eta d \cos \eta d = 0 \tag{50}$$

Equation (50) is zero when either $\sin \eta d = 0$ or $\cos \eta d = 0$. For reasons that will become apparent shortly, the solutions of $\sin \eta d = 0$ will lead to *symmetric modes* (S-modes) of the SH waves, whereas the solutions of $\cos \eta d = 0$ will lead to *antisymmetric modes* (A-modes) of the SH waves. Let us analyze these two situations separately.

6.3.2.1 Symmetric SH Modes

Consider the characteristic equation corresponding to the symmetric modes, i.e.,

$$\sin \eta d = 0 \quad \text{symmetric SH waves (S-modes)} \tag{51}$$

The roots of Eq. (51) are

$$\eta^S d = 0, \pi, 2\pi, \ldots, (2n)\frac{\pi}{2}, \quad n = 0, 1, \ldots \tag{52}$$

The values $\eta^S d$ given by Eq. (52) are the *eigenvalues for symmetric motion*. For every eigenvalue $\eta^S d$ value, we have

$$\sin \eta^S d = 0 \quad \text{and} \quad \cos \eta^S d = \pm 1 \tag{53}$$

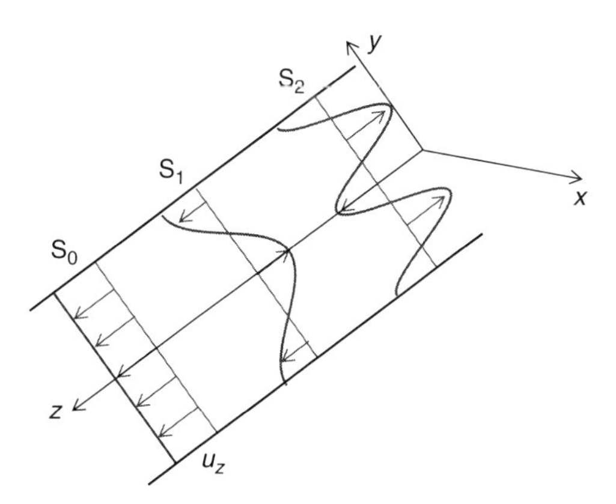

FIGURE 6.4 Plot of the first, second, and third symmetric modes of the SH waves.

Substitution of Eq. (53) into the linear system (48) yields $C_1 = 0$. Hence, the general solution (43) becomes

$$u_z^S(x, y, t) = C_2 \cos \eta y \, e^{i(\xi x - \omega t)} \qquad \text{symmetric SH waves (S-modes)} \tag{54}$$

where the sign $\pm$ was absorbed into the constant C_2. As expected, the function (54) is symmetric in the y variable, i.e., it represents a motion symmetric about the plate midplane. A sketch of the first, second, and third symmetric SH modes (S_0, S_1, and S_2, respectively) is given in Fig. 6.4.

6.3.2.2 Antisymmetric SH Modes

Consider the characteristic equation corresponding to the antisymmetric modes, i.e.,

$$\cos \eta d = 0 \qquad \text{antisymmetric SH waves (A-modes)} \tag{55}$$

The roots of Eq. (55) are

$$\eta^A d = \frac{\pi}{2}, 3\frac{\pi}{2}, 5\frac{\pi}{2}, \ldots, (2n+1)\frac{\pi}{2}, \quad n = 0, 1, \ldots \tag{56}$$

The values $\eta^A d$ given by Eq. (56) are the *eigenvalues for antisymmetric motion*. For every eigenvalue, $\eta^A d$, we have

$$\cos \eta^A d = 0 \quad \text{and} \quad \sin \eta^A d = \pm 1 \tag{57}$$

Substitution of Eq. (57) into the linear system (48) yields $C_2 = 0$. Hence, the general solution (43) becomes

$$u_z^A(x, y, t) = C_1 \sin \eta y \, e^{i(\xi x - \omega t)} \qquad \text{antisymmetric SH waves (A-modes)} \tag{58}$$

where the sign $\pm$ was absorbed into the constant C_1. As expected, the function (58) is antisymmetric in the y-variable, i.e., it represents a motion antisymmetric about the plate midplane. A sketch of the first, second, and third antisymmetric SH modes (A_0, A_1, and A_2, respectively) is given in Fig. 6.5.

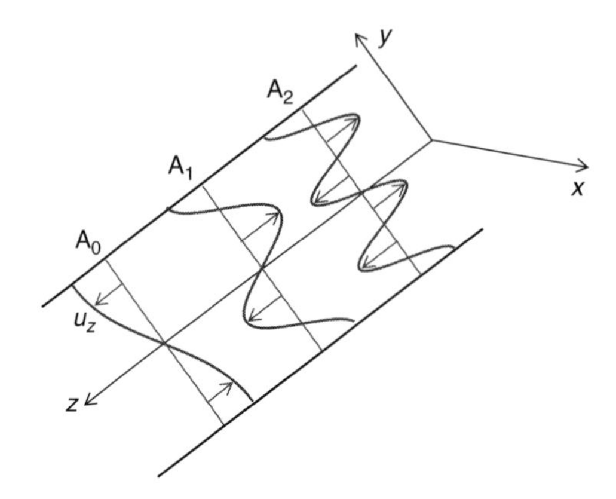

FIGURE 6.5 Plot of the first, second, and third antisymmetric modes of the SH waves.

6.3.3 DISPERSION OF SH WAVES

In this section, we will examine the dispersion behavior of the SH waves. We will start with the study of the dispersion of the wave speed (a.k.a., phase velocity). Then, we will study the dispersion of the group velocity.

6.3.3.1 Wave Speed Dispersion Curves

Recall Eq. (40), i.e.,

$$\eta^2 = \frac{\omega^2}{c_S^2} - \xi^2 \tag{59}$$

Recall $\xi = \omega/c$, where c is the wave speed. Upon substitution, Eq. (59) becomes

$$\eta^2 = \frac{\omega^2}{c_S^2} - \frac{\omega^2}{c^2} \tag{60}$$

Equation (60) can be solved for c, to give

$$c^2(\omega) = \frac{\omega^2}{\dfrac{\omega^2}{c_S^2} - \eta^2}, \qquad \text{i.e.,}\ c(\omega) = \frac{\omega}{\sqrt{\dfrac{\omega^2}{c_S^2} - \eta^2}} \tag{61}$$

Equation (61) can be expressed in terms of the eigenvalues (52) and (56), i.e.,

$$c(\omega) = \frac{c_S}{\sqrt{1 - (\eta d)^2 \left(\dfrac{c_S}{\omega d}\right)^2}} \tag{62}$$

By substituting the appropriate eigenvalue, one gets an analytical expression for the wave-speed dispersion curve of each SH wave mode. Table 6.1 gives the expressions for the wave-speed dispersion formulae for the first three symmetric and antisymmetric SH wave modes. Plots of these formulae are given in Fig. 6.6a. It should be noted that

TABLE 6.1 Dispersion functions of the SH waves

Mode	Eigenvalue ηd	Wave speed dispersion formula	ω_{cr}	Group velocity dispersion formula
S_0	0	$c^{S_0}(\omega) = c_S$	0	
A_0	$\frac{\pi}{2}$	$c^{A_0}(\omega) = \dfrac{c_S}{\sqrt{1-\left(\frac{\pi}{2}\right)^2\left(\frac{c_S}{\omega d}\right)^2}}$	$\frac{\pi}{2}\frac{c_S}{d}$	$c_g^{A_0}(\omega) = c_S\sqrt{1-\left(\frac{\pi}{2}\right)^2\left(\frac{c_S}{\omega d}\right)^2}$
S_1	$2\frac{\pi}{2}$	$c^{S_1}(\omega) = \dfrac{c_S}{\sqrt{1-\pi^2\left(\frac{c_S}{\omega d}\right)^2}}$	$\pi\frac{c_S}{d}$	$c_g^{S_1}(\omega) = c_S\sqrt{1-\pi^2\left(\frac{c_S}{\omega d}\right)^2}$
A_1	$3\frac{\pi}{2}$	$c^{A_1}(\omega) = \dfrac{c_S}{\sqrt{1-\left(3\frac{\pi}{2}\right)^2\left(\frac{c_S}{\omega d}\right)^2}}$	$3\frac{\pi}{2}\frac{c_S}{d}$	$c_g^{A_1}(\omega) = c_S\sqrt{1-\left(3\frac{\pi}{2}\right)^2\left(\frac{c_S}{\omega d}\right)^2}$
S_2	$4\frac{\pi}{2}$	$c^{S_2}(\omega) = \dfrac{c_S}{\sqrt{1-(2\pi)^2\left(\frac{c_S}{\omega d}\right)^2}}$	$2\pi\frac{c_S}{d}$	$c_g^{S_2}(\omega) = c_S\sqrt{1-(2\pi)^2\left(\frac{c_S}{\omega d}\right)^2}$
A_2	$5\frac{\pi}{2}$	$c^{A_2}(\omega) = \dfrac{c_S}{\sqrt{1-\left(5\frac{\pi}{2}\right)^2\left(\frac{c_S}{\omega d}\right)^2}}$	$5\frac{\pi}{2}\frac{c_S}{d}$	$c_g^{A_2}(\omega) = c_S\sqrt{1-\left(5\frac{\pi}{2}\right)^2\left(\frac{c_S}{\omega d}\right)^2}$
...	...	...	...	...

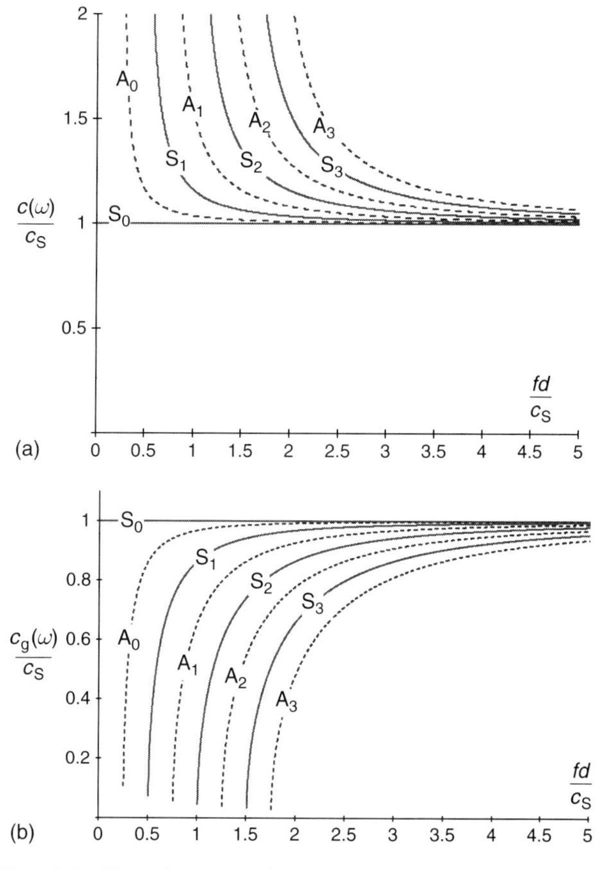

FIGURE 6.6 Plot of the dispersion curves for various SH wave modes: (a) dispersion of the wave speed (a.k.a., phase velocity); (b) dispersion of the group velocity.

the first symmetric SH wave mode, S_0, is not dispersive because its eigenvalue is zero ($\eta_0^S d = 0$), and hence Eq. (62) yields $c^{S_0}(\omega) = c_S$.

Figure 6.6a also indicates the asymptotic behavior of the SH waves wave speed. If $\omega \to \infty$, then $c \to c_S$. This behavior is also apparent from Eq. (62), as the second term under the square root sign vanishes for $\omega \to \infty$.

6.3.3.2 Cut-off Frequencies

At low frequencies, not all the SH modes may exist. Examination of Eq. (62) reveals that, as ω decreases, the term under the square root sign may become negative, and hence the wave speed would turn 'imaginary'. The condition for this to happen is

$$1-(\eta d)^2\left(\frac{c_S}{\omega d}\right)^2 = 0, \text{ i.e., } \omega_{cr} = \frac{c_S}{d}(\eta d) \tag{63}$$

The values of ω_{cr} are also given in Table 6.1. The critical frequency given by Eq. (63) represents the minimum frequency value at which a certain mode exists. Because the first mode, S_0, has zero eigenvalue ($\eta_0^S d = 0$), the corresponding critical frequency given by Eq. (63) will also zero ($\omega_{cr}^{S_0} = 0$). This means that the S_0 mode of the SH waves exists at all frequencies. However, this is not true for the other modes, which only appear at frequencies above their critical frequencies. At frequencies below their critical frequencies, these modes do not exist as propagating waves.

6.3.3.3 Significance of Imaginary WaveNumbers: Evanescent Waves

The previous section has shown that each SH mode has a critical frequency, ω_{cr}, below which the wave speed formula would yield imaginary values. In this section, we will examine what meaning can be attached to such imaginary wave speed value. Recall Eq. (59), i.e.,

$$\eta^2 = \frac{\omega^2}{c_S^2} - \xi^2 \tag{64}$$

Rearrangement of Eq. (64) yields the wavenumber expression

$$\xi = \sqrt{\frac{\omega^2}{c_S^2} - \eta^2} = \frac{\omega}{c_S}\sqrt{1-(\eta d)^2\left(\frac{c_S}{\omega d}\right)^2} \tag{65}$$

Using Eq. (63), we rewrite Eq. (65) as

$$\xi = \frac{\omega}{c_S}\sqrt{1-\left(\frac{\omega_{cr}}{\omega}\right)^2} \tag{66}$$

It is apparent from Eq. (66) that, for $\omega < \omega_{cr}$, the quantity under the square root sign becomes negative and the wavenumber becomes imaginary, i.e.,

$$\xi = i\xi^* \quad \text{for } \omega < \omega_{cr} \tag{67}$$

where ξ^* is a real number. To understand the meaning of Eq. (67), let us substitute it into the general expression (34); we get

$$u_z(x, y, t) = h(y)e^{i(\xi x - \omega t)} = h(y)e^{i(i\xi^* x)}e^{-i\omega t} \tag{68}$$

i.e.,

$$u_z(x, y, t) = \left[h(y)e^{-\xi^* x}\right]e^{-i\omega t} \tag{69}$$

Equation (69) describes an *evanescent*[1] *wave*, which has a decaying amplitude and only exists close to the excitation source. We have previously met evanescent waves in connection with flexural waves in beams. The space-wise behavior of an evanescent wave is non-harmonic, because it has a decaying exponential in the x-variable. On the other hand, the propagating waves, as described by Eq. (34), are harmonic in the x-variable, i.e., they are space-wise harmonic. A complete description of the wave response to a generic excitation should consider both the propagating (i.e., space-wise harmonic) an evanescent (i.e., space-wise non-harmonic) wave components.

6.3.3.4 Group Velocity Dispersion Curves

Recall the definition of group velocity as the derivative of frequency with respect to wavenumber, i.e.,

$$c_g = \frac{d\omega}{d\xi} \tag{70}$$

Hence, we will attempt to calculate this derivative. Recall Eq. (40), i.e.,

$$\eta^2 = \frac{\omega^2}{c_S^2} - \xi^2 \tag{71}$$

Upon rearrangement, we write

$$\omega^2 = c_S^2(\xi^2 + \eta^2) \tag{72}$$

Differentiation of Eq. (72) yields

$$2\omega\frac{d\omega}{d\xi} = c_S^2(2\xi) \tag{73}$$

Upon simplification and use of the definition $\xi = \omega/c$ one gets

$$\frac{d\omega}{d\xi} = \frac{c_S^2\xi}{\omega} = \frac{c_S^2}{c} \tag{74}$$

Substitution of Eq. (74) into Eq. (70) yields the group velocity of SH waves as

$$c_g = \frac{c_S^2}{c} \tag{75}$$

[1] Vanishing or likely to vanish like vapor. Synonym with transient.

We notice that the group velocity of SH waves is inversely proportional to the wave speed. Substituting the wave speed expression (62) into Eq. (75) yields

$$c_g(\omega) = c_S\sqrt{1-(\eta d)^2\left(\frac{c_S}{\omega d}\right)^2} \tag{76}$$

By substituting the appropriate eigenvalue, one gets an analytical expression for the group-velocity dispersion curve of each SH wave mode. Table 6.1 gives the expressions for the group-velocity dispersion formulae for the first three symmetric and antisymmetric SH wave modes. Plots of these formulae are given in Fig. 6.6b. It should be noted that the first symmetric SH wave mode, S_0, is not dispersive, because its eigenvalue is zero ($\eta_0^S d = 0$), and hence Eq. (76) yields $c_g^{S_0}(\omega) = c_S$.

Figure 6.6b also indicates the asymptotic behavior of the SH group velocity. If $\omega \to \infty$, then $c_g \to c_S$. This behavior is also apparent from Eq. (76), since the second term vanishes for $\omega \to \infty$.

6.4 LAMB WAVES

Lamb waves, a.k.a., guided plate waves, are a type of ultrasonic waves that are guided between two parallel free surfaces, such as the upper and lower surfaces of a plate. Lamb waves can exist in two basic types, symmetric and antisymmetric. It will be shown in this section that for each propagation type there exist a number of modes corresponding to the solutions of the Rayleigh–Lamb equation. The symmetric modes are designated S_0, S_1, S_2, . . . , whereas the antisymmetric are designated A_0, A_1, A_2, ... The symmetric Lamb waves resemble the axial waves (Fig. 6.7a), whereas the antisymmetric Lamb waves resemble the flexural waves (Fig. 6.7b). In fact, it can be proven that, at low frequencies, the symmetric Lamb waves approach the behavior of the axial plate waves, whereas the

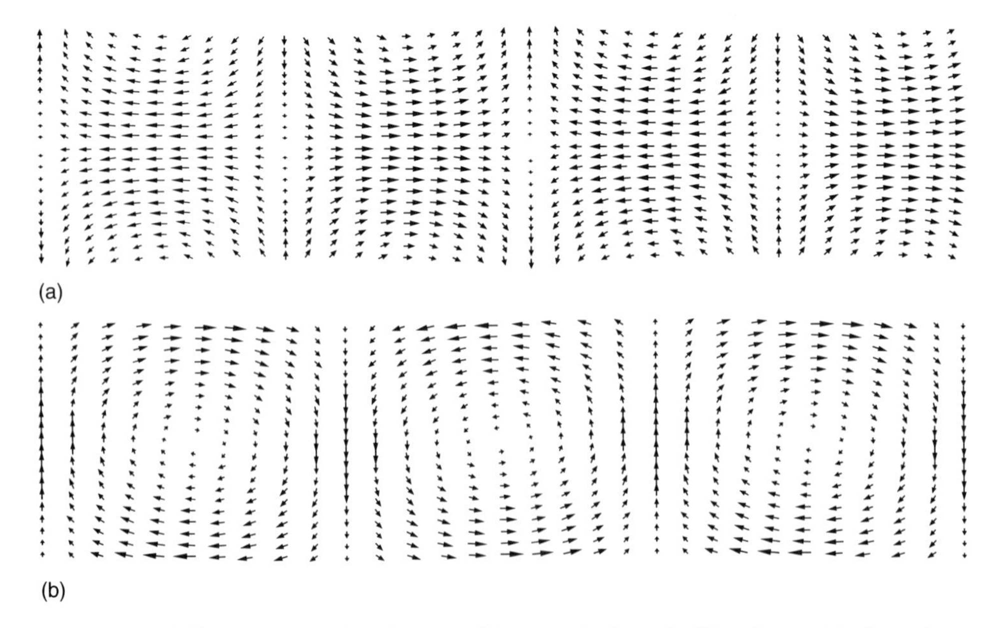

FIGURE 6.7 Simulation of Lamb waves: (a) symmetric S_0 mode; (b) antisymmetric A_0 mode.

antisymmetric Lamb waves approach the behavior of the flexural plate waves. Lamb waves are highly dispersive, and their speed depends on the product between frequency and the plate thickness. The wave-speed dispersion curves are obtained from the solution of the Rayleigh–Lamb equation. At a given frequency–thickness product, for each solution of the Rayleigh–Lamb equation, one finds a corresponding Lamb-wave speed and a corresponding Lamb-wave mode. Details of the derivation of the Lamb-waves equations and the corresponding solutions are given next.

6.4.1 THE DERIVATION OF LAMB-WAVES EQUATIONS

The derivation of the Lamb-wave equations is provided in this section. In this derivation, we will consider straight-crested Lamb waves propagating in a plate of thickness $h = 2d$ (Fig. 6.8). The assumption of straight-crested waves makes the problem z-invariant. In Chapter 5, we have shown that the z-invariant problem can be split into two separate cases, (1) SH waves, and (2) P + SV waves. The case of SH waves was treated in the previous section. Here we will treat the case of P + SV waves in a plate, i.e., we will assume that P-waves and SV-waves simultaneously exist in the plate. Through multiple reflections on the plate's lower and upper surfaces, and through constructive and destructive interference, the P-waves and SV-waves give rise to the Lamb waves, which consist of a pattern of standing waves in the thickness y direction (Lamb-wave modes) behaving like traveling waves in the x direction.

Recall that in Chapter 5 we have established that the P + SV waves can be described in terms of two scalar potentials, Φ and H_z. The potentials satisfy the wave equation, i.e.,

$$
\begin{aligned}
c_P^2 \nabla^2 \Phi &= \ddot{\Phi} \\
c_S^2 \nabla^2 H_z &= \ddot{H}_z
\end{aligned}
\tag{77}
$$

where $c_P^2 = (\lambda + 2\mu)/\rho$ and $c_S^2 = \mu/\rho$ are the pressure (longitudinal) and shear (transverse) wave speeds, whereas λ and μ are Lame constants, ρ is the mass density. To simplify the notations, we will denote the two potential functions, Φ and H_z by ϕ and ψ. Substitution in Eq. (77) and assumption of harmonic motion, $e^{-i\omega t}$, yield

$$
\begin{aligned}
\frac{\partial^2 \phi}{\partial x^2} + \frac{\partial^2 \phi}{\partial y^2} + \frac{\omega^2}{c_P^2}\phi &= 0 \\
\frac{\partial^2 \psi}{\partial x^2} + \frac{\partial^2 \psi}{\partial y^2} + \frac{\omega^2}{c_S^2}\psi &= 0
\end{aligned}
\tag{78}
$$

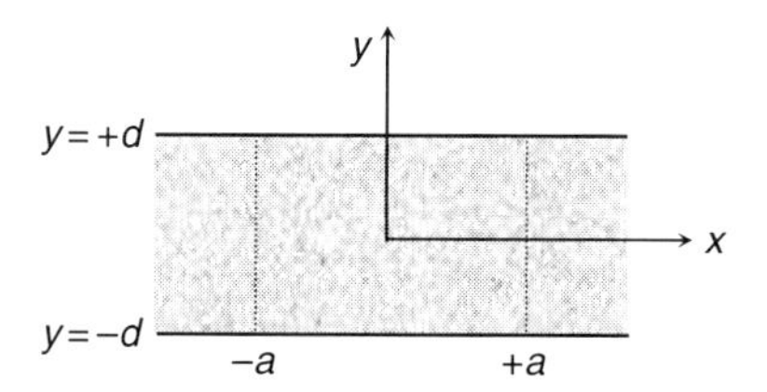

FIGURE 6.8 Plate of thickness $2d$, with a PWAS of width $2a$, under harmonic loading on the top surface.

Recall the displacement and stress expressions in terms of potentials, as derived in Chapter 5, i.e.,

$$
\begin{aligned}
&u_x = \frac{\partial \phi}{\partial x} + \frac{\partial \psi}{\partial y}, \quad u_y = \frac{\partial \phi}{\partial y} - \frac{\partial \psi}{\partial x}, \quad \varepsilon_x = \frac{\partial u_x}{\partial x} \\
&\tau_{yx} = \mu\left(2\frac{\partial^2 \phi}{\partial x\,\partial y} - \frac{\partial^2 \psi}{\partial x^2} + \frac{\partial^2 \psi}{\partial y^2}\right), \quad \tau_{yy} = \lambda\left(\frac{\partial^2 \phi}{\partial x^2} + \frac{\partial^2 \phi}{\partial y^2}\right) + 2\mu\left(\frac{\partial^2 \phi}{\partial y^2} - \frac{\partial^2 \psi}{\partial x\,\partial y}\right)
\end{aligned} \tag{79}
$$

For harmonic-wave propagation in the x direction, $e^{i(\xi x - \omega t)}$, the equations take the form

$$
\begin{aligned}
&\frac{\mathrm{d}^2 \phi}{\mathrm{d}y^2} + \left(\frac{\omega^2}{c_\mathrm{P}^2} - \xi^2\right)\phi = 0 \\
&\frac{\mathrm{d}^2 \psi}{\mathrm{d}y^2} + \left(\frac{\omega^2}{c_\mathrm{S}^2} - \xi^2\right)\psi = 0
\end{aligned} \tag{80}
$$

$$
\begin{aligned}
&u_x = i\xi\phi + \frac{\mathrm{d}\psi}{\mathrm{d}y}, \quad u_y = \frac{\mathrm{d}\phi}{\mathrm{d}y} - i\xi\psi, \quad \varepsilon_x = i\xi u_x \\
&\tau_{yx} = \mu\left(2i\xi\frac{\mathrm{d}\phi}{\mathrm{d}y} + \xi^2\psi + \frac{\partial^2 \psi}{\partial y^2}\right), \quad \tau_{yy} = \lambda\left(-\xi^2\phi + \frac{\mathrm{d}^2 \phi}{\mathrm{d}y^2}\right) + 2\mu\left(\frac{\mathrm{d}^2 \phi}{\mathrm{d}y^2} - i\xi\frac{\mathrm{d}\psi}{\mathrm{d}y}\right)
\end{aligned} \tag{81}
$$

Introduce the notations

$$
p^2 = \frac{\omega^2}{c_\mathrm{P}^2} - \xi^2, \quad q^2 = \frac{\omega^2}{c_\mathrm{S}^2} - \xi^2 \tag{82}
$$

Equation (80) becomes

$$
\begin{aligned}
&\frac{\mathrm{d}^2 \phi}{\mathrm{d}y^2} + p^2\phi = 0 \\
&\frac{\mathrm{d}^2 \psi}{\mathrm{d}y^2} + q^2\psi = 0
\end{aligned} \tag{83}
$$

Equation (83) accepts the general solution

$$
\begin{aligned}
&\phi = A_1 \sin py + A_2 \cos py \\
&\psi = B_1 \sin qy + B_2 \cos qy
\end{aligned} \tag{84}
$$

Hence, the displacements are

$$
\begin{aligned}
&u_x = i\xi(A_1 \sin py + A_2 \cos py) + q(B_1 \cos qy - B_2 \sin qy) \\
&u_y = p(A_1 \cos py - A_2 \sin py) - i\xi(B_1 \sin qy + B_2 \cos qy)
\end{aligned} \tag{85}
$$

These can be grouped into symmetric and antisymmetric components.

$$
\begin{aligned}
&u_x = (A_2 \mathrm{i}\xi \cos py + B_1 q \cos qy) + (A_1 \mathrm{i}\xi \sin py - B_2 q \sin qy) \\
&u_y = -(A_2 p \sin py + B_1 \mathrm{i}\xi \sin qy) + (A_1 p \cos py - B_2 \mathrm{i}\xi \cos qy)
\end{aligned} \tag{86}
$$

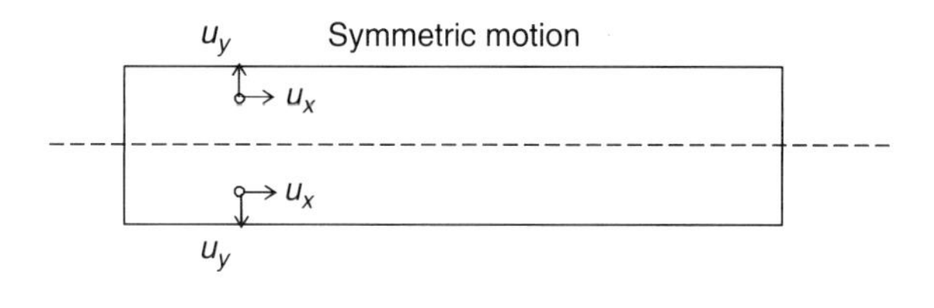

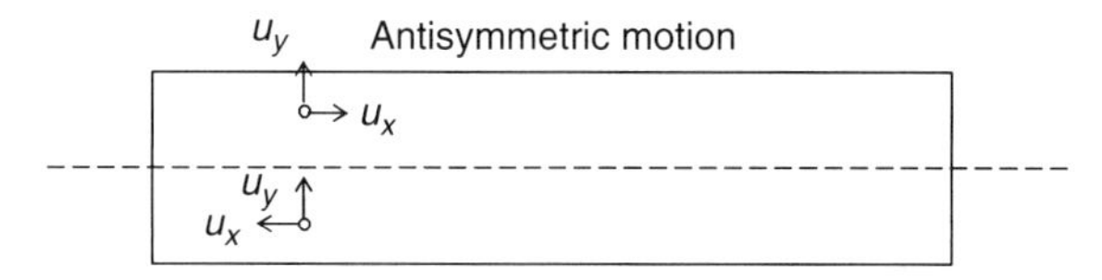

FIGURE 6.9 Symmetric and antisymmetric particle motion across the plate thickness.

The four integration constants are to be found from the boundary conditions. At this stage, it is advantageous to split the problem into symmetric and antisymmetric behavior (Fig. 6.9).

The derivatives of the potentials with respect to y are

$$\begin{aligned} &\frac{\partial \phi}{\partial y} = A_1 p \cos py - A_2 p \sin py \qquad \frac{\partial^2 \phi}{\partial y^2} = -A_1 p^2 \sin py - A_2 p^2 \cos py = -p^2 \phi \\ &\frac{\partial \psi}{\partial y} = B_1 q \cos qy - B_2 q \sin qy' \qquad \frac{\partial^2 \psi}{\partial y^2} = -B_1 q^2 \sin qy - B_2 q^2 \cos qy = -q^2 \psi \end{aligned} \tag{87}$$

The stresses are given by

$$\begin{aligned} \tau_{yx} =& \mu\left(2i\xi\frac{\mathrm{d}\phi}{\mathrm{d}y} + \xi^2\psi + \frac{\mathrm{d}^2\psi}{\mathrm{d}y^2}\right) \\ =& \mu\big[2i\xi(A_1 p\cos py - A_2 p \sin py) + \xi^2 (B_1 \sin qy + B_2 \cos qy) \\ & -q^2 (B_1 \sin qy + B_2 \cos qy)\big] \\ =& \mu\big[(A_1 2i\xi p \cos py - A_2 2i\xi p \sin py) + \left(B_1 \xi^2 \sin qy + B_2 \xi^2 \cos qy\right) \\ & + \left(-B_1 q^2 \sin qy - B_2 q^2 \cos qy\right)\big] \end{aligned} \tag{88}$$

$$\begin{aligned} \tau_{yx} =& \mu\big[-A_2 2i\xi p \sin py + B_1 \left(\xi^2 - q^2\right) \sin qy + A_1 2i\xi p \cos py \\ & + B_2\left(\xi^2 - q^2\right) \cos qy\big] \end{aligned} \tag{89}$$

$$\begin{aligned} \tau_{yy} =& \lambda\left(-\xi^2\phi + \frac{\mathrm{d}^2\phi}{\mathrm{d}y^2}\right) + 2\mu\left(\frac{\mathrm{d}^2\phi}{\mathrm{d}y^2} - i\xi\frac{\mathrm{d}\psi}{\mathrm{d}y}\right) \\ =& \lambda\big[-\xi^2 (A_1 \sin py + A_2 \cos py) + \left(-A_1 p^2 \sin py - A_2 p^2 \cos py\right)\big] \\ & + 2\mu\big[-p^2 (A_1 \sin py + A_2 \cos py) - i\xi(B_1 q \cos qy - B_2 q \sin qy)\big] \end{aligned} \tag{90}$$

$$\begin{aligned} \tau_{yy} =& \lambda\big[\left(-A_1 \xi^2 \sin py - A_2 \xi^2 \cos py\right) + \left(-A_1 p^2 \sin py - A_2 p^2 \cos py\right)\big] \\ & + 2\mu\big[\left(-A_1 p^2 \sin py - A_2 p^2 \cos py\right) + (-iB_1 \xi q \cos qy + B_2 i\xi q \sin qy)\big] \\ =& \lambda\big[-A_1\left(\xi^2 \sin py + p^2 \sin py\right) - A_2\left(\xi^2 \cos py + p^2 \cos py\right)\big] \\ & + 2\mu\big[-A_1 p^2 \sin py - A_2 p^2 \cos py\big] + 2\mu(-iB_1 \xi q \cos qy + B_2 i\xi q \sin qy) \end{aligned} \tag{91}$$

$$\begin{aligned}\tau_{yy} =& - A_1\lambda(\xi^2 \sin py + p^2 \sin py) - A_2\lambda(\xi^2 \cos py + p^2 \cos py)\\ &- A_1 2\mu p^2 \sin py - A_2 2\mu p^2 \cos py + 2\mu(-iB_1\xi q \cos qy + B_2 i\xi q \sin qy)\\ =& - A_1(\lambda\xi^2 \sin py + 2\mu p^2 \sin py + \lambda p^2 \sin py)\\ &- A_2(\lambda\xi^2 \cos py + 2\mu p^2 \cos py + \lambda p^2 \cos py)\\ &+ 2\mu(-iB_1\xi q \cos qy + B_2 i\xi q \sin qy)\end{aligned} \tag{92}$$

$$\begin{aligned}\tau_{yy} =& -A_1(\lambda\xi^2 + 2\mu p^2 + \lambda p^2)\sin py - A_2(\lambda\xi^2 + 2\mu p^2 + \lambda p^2)\cos py\\ &+ 2\mu(-iB_1\xi q \cos qy + B_2 i\xi q \sin qy)\end{aligned} \tag{93}$$

But

$$\begin{aligned}\lambda\xi^2 + 2\mu p^2 + \lambda p^2 &= \lambda\xi^2 + (\lambda + 2\mu)\,p^2 = \lambda\xi^2 + (\lambda + 2\mu)\left(\frac{\omega^2}{c_{\mathrm{L}}^2} - \xi^2\right)\\ &= \lambda\xi^2 - \xi^2(\lambda + 2\mu) + (\lambda + 2\mu)\left(\frac{\omega^2}{c_{\mathrm{L}}^2}\right) = -2\xi^2\mu + \omega^2\rho = \mu\left(-2\xi^2 + \frac{\omega^2}{c_{\mathrm{T}}^2}\right)\\ &= \mu\left[\left(\frac{\omega^2}{c_{\mathrm{T}}^2} - \xi^2\right) - \xi^2\right] = \mu(q^2 - \xi^2)\end{aligned} \tag{94}$$

Hence,

$$\begin{aligned}\tau_{yy} &= -A_1\mu(q^2 - \xi^2)\sin py - A_2\mu(q^2 - \xi^2)\cos py + 2\mu(-iB_1\xi q \cos qy + B_2 i\xi q \sin qy)\\ \tau_{yy} &= \mu\left[A_2(\xi^2 - q^2)\cos py - B_1 2i\xi q \cos qy + A_1(\xi^2 - q^2)\sin py + B_2 2i\xi q \sin qy\right]\end{aligned} \tag{95}$$

6.4.1.1 Symmetric Solution

The symmetric solution of the Lamb-waves equations assumes symmetry of displacements and stresses about the midplane. It can be easily observed that this amounts to

$$\begin{aligned}u_x(x, -d) &= u_x(x, d) \qquad & \tau_{yx}(x, -d) &= -\tau_{yx}(x, d)\\ u_y(x, -d) &= -u_y(x, d) \qquad & \tau_{yy}(x, -d) &= \tau_{yy}(x, d)\end{aligned} \tag{96}$$

Note that positive shear stresses have opposite directions on the upper and lower surfaces, hence the negative sign on τ_{yx} in Eq. (96). Therefore, the displacements and potentials for symmetric motion are

$$\begin{aligned}u_x &= A_2 i\xi \cos py + B_1 q \cos qy \qquad & \phi &= A_2 \cos py\\ u_y &= -A_2 p \sin py - B_1 i\xi \sin qy \qquad & \psi &= B_1 \sin qy\end{aligned} \tag{97}$$

Upon substitution in Eq. (95), we obtain

$$\begin{aligned}\tau_{yx} &= \mu\left[-A_2 2i\xi p \sin py + B_1(\xi^2 - q^2)\sin qy\right]\\ \tau_{yy} &= \mu\left[A_2(\xi^2 - q^2)\cos py - B_1 2i\xi q \cos qy\right]\end{aligned} \tag{98}$$

The symmetric boundary conditions are

$$\begin{aligned}\tau_{yx}(x, -d) &= -\tau_{yx}(x, d) = 0\\ \tau_{yy}(x, -d) &= \tau_{yy}(x, d) = 0\end{aligned} \tag{99}$$

Upon substitution, we obtain the linear system

$$\begin{aligned} -A_2 2i\xi p \sin pd + B_1(\xi^2 - q^2)\sin qd &= 0 \\ A_2(\xi^2 - q^2)\cos pd - B_1 2i\xi q \cos qd &= 0 \end{aligned} \tag{100}$$

Solution of this homogenous system of linear equation is only possible if the determinant is zero

$$D_S = (\xi^2 - q^2)^2 \cos pd \sin qd + 4\xi^2 pq \sin pd \cos q\mathrm{d} = 0 \tag{101}$$

This is the Rayleigh–Lamb equation for symmetric modes. An often used form of the symmetric Rayleigh–Lamb equation is

$$\frac{\tan pd}{\tan qd} = -\frac{(\xi^2 - q^2)^2}{4\xi^2 pq} \tag{102}$$

The solution of this transcendental equation is not easy, because p and q also depend on ξ. Numerical solution of Eq. (102) yields the symmetric (S) eigenvalues, $\xi_0^S, \xi_1^S, \xi_2^S, \ldots$ The relationship $c = \omega/\xi$ yields the *dispersive wave speed*, which is a function of the product *fd* between the frequency, $f = \omega/2\pi$, and the half thickness, d. At a given *fd* product, several Lamb modes may exist. At low *fd* values, only the lowest Lamb mode S_0 exists. A plot of the wave-speed dispersion curves for the symmetric Lamb modes is given in Fig. 6.10.

Substitution of the eigenvalues into the homogeneous linear system of Eq. (100) yields the eigen coefficients (A_2, B_1) in the form

$$\begin{aligned} A_2 &= N_{A_2} \quad N_{A_2} = 2i\xi q \cos qd \\ B_1 &= N_{B_1} \quad N_{B_1} = (\xi^2 - q^2)\cos pd \end{aligned} \tag{103}$$

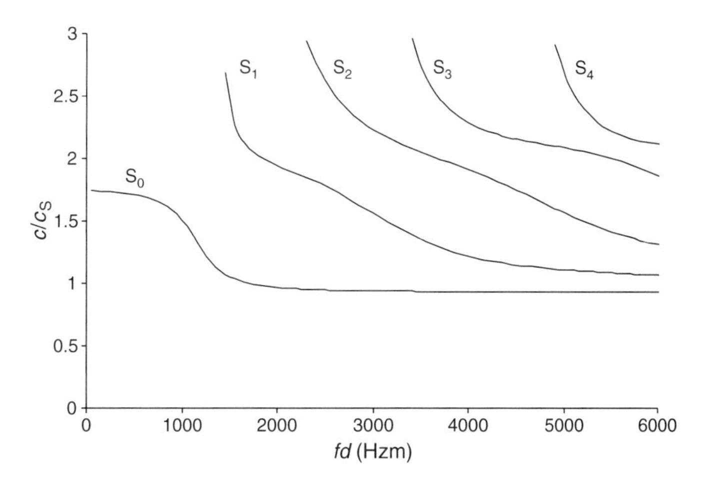

FIGURE 6.10 Wave speed dispersion curves for symmetric Lamb waves in an aluminum plate (c_S = shear wave speed, d = half thickness of the plate).

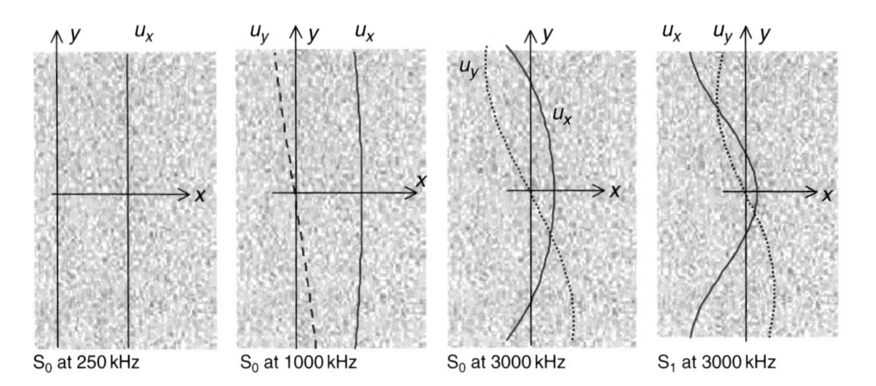

FIGURE 6.11 Across-the-thickness displacement fields in a 1-mm thick aluminum plate for various S_0 Lamb modes at various frequencies.

Substitution of the coefficients (A_2, B_1) into Eq. (97) yields the symmetric Lamb modes

$$\begin{aligned} u_x &= -2\xi^2 q \cos qd \cos py + q(\xi^2 - q^2)\cos pd \cos qy \\ u_y &= -2i\xi pq \cos qd \sin py - i\xi(\xi^2 - q^2)\cos pd \sin qy \end{aligned} \tag{104}$$

Simulations of the symmetric Lamb-wave modes at various frequencies and mode numbers are shown in Fig. 6.11. As $fd \to 0$, the S_0 modes resemble the across-the-thickness variation of conventional axial plate waves, i.e., constant across the thickness.

6.4.1.2 Antisymmetric Solution

The antisymmetric solution of the Lamb wave equations assumes that displacements and stresses are antisymmetric about the midplane

$$\begin{aligned} u_x(x,-d) &= -u_x(x,d) \qquad \tau_{yx}(x,-h) = \tau_{yx}(x,h) \\ u_y(x,-d) &= u_y(x,d) \qquad \tau_{yy}(x,-h) = -\tau_{yy}(x,h) \end{aligned} \tag{105}$$

Note that positive shear stresses have opposite directions on the upper and lower surfaces. They are inherently antisymmetric. Hence, the displacements and potentials for antisymmetric motion are

$$\begin{aligned} u_x &= (A_1 i\xi \sin py - B_2 q \sin qy) \qquad \phi = A_1 \sin py \\ u_y &= (A_1 p \cos py - B_2 i\xi \cos qy) \qquad \psi = B_2 \cos qy \end{aligned} \tag{106}$$

Equation (95) gives

$$\begin{aligned} \tau_{yx} &= \mu\left[A_1 2i\xi p \cos py + B_2(\xi^2 - q^2)\cos qy\right] \\ \tau_{yy} &= \mu\left[A_1(\xi^2 - q^2)\sin py + B_2 2i\xi q \sin qy\right] \end{aligned} \tag{107}$$

The antisymmetric boundary conditions are

$$\begin{aligned} \tau_{yx}(x,-d) &= \tau_{yx}(x,d) = 0 \\ \tau_{yy}(x,-d) &= -\tau_{yy}(x,d) = 0 \end{aligned} \tag{108}$$

Upon substitution, we obtain the linear system

$$\begin{aligned} A_1 2i\xi p \cos pd + B_2(\xi^2 - q^2)\cos qd = 0 \\ A_1(\xi^2 - q^2)\sin pd + B_2 2i\xi q \sin qd = 0 \end{aligned} \tag{109}$$

Solution of this homogenous system of linear equations is only possible if the determinant is zero

$$D_A = (\xi^2 - q^2)^2 \sin pd \cos qd + 4\xi^2 pq \cos pd \sin qd = 0 \tag{110}$$

This is the Rayleigh–Lamb equation for antisymmetric modes. An often used form of the antisymmetric Rayleigh–Lamb equation is

$$\frac{\tan pd}{\tan qd} = -\frac{4\xi^2 pq}{(\xi^2 - q^2)^2} \tag{111}$$

The solution of the Rayleigh–Lamb transcendental equation is not straightforward, because p and q also depend on ξ. Numerical solution of Eq. (111) yields the antisymmetric (A) eigenvalues, $\xi_0^A, \xi_1^A, \xi_2^A, \ldots$. The relationship $c = \omega/\xi$ yields the dispersive wave speed, which is a function of the product *fd* between the frequency, $f = \omega/2\pi$, and the half thickness, d. At a given *fd* product, several Lamb modes may exist. At low *fd* values, only the lowest Lamb mode A_0 exists. A plot of the wave-speed dispersion curves for the antisymmetric Lamb modes is given in Fig. 6.12.

Substitution of the eigenvalues into the homogeneous linear system of Eq. (109) yields the eigen coefficients A_1, B_2 in the form

$$\begin{aligned} A_1 = N_{A_1} \quad N_{A_1} = 2i\xi q \sin qd \\ B_2 = N_{B_2} \quad N_{B_2} = -(\xi^2 - q^2)\sin pd \end{aligned} \tag{112}$$

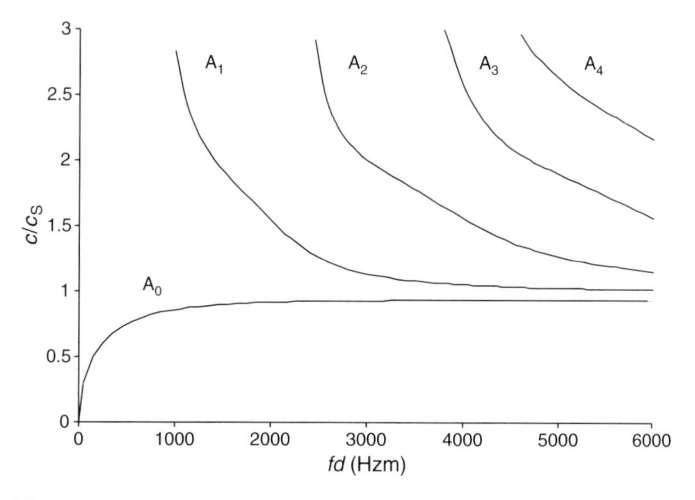

FIGURE 6.12 Wave speed dispersion curves for antisymmetric Lamb waves (c_S = shear wave speed, d = half thickness of the plate).

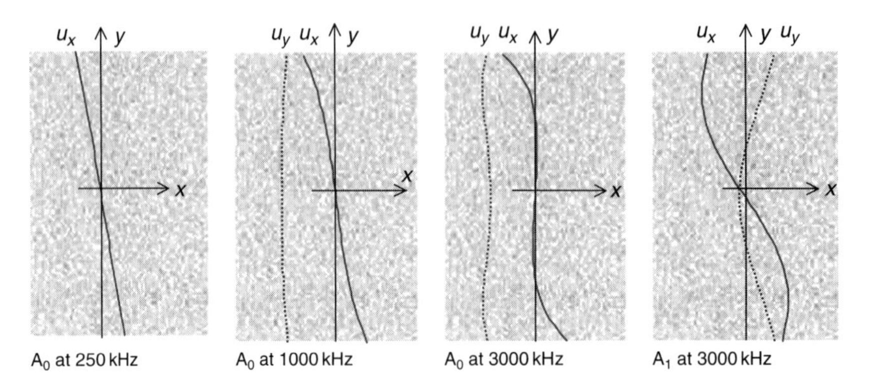

FIGURE 6.13 Displacement fields across the thickness for various A_0 Lamb modes at various frequencies in an aluminum plate (c_S = shear wave speed, d = half thickness of the plate).

Substitution of the coefficients (A_1, B_2) into Eq. (105) yields the antisymmetric Lamb modes

$$\begin{aligned} u_x &= -2\xi^2 q \sin qd \sin py + q(\xi^2 - q^2) \sin pd \sin qy \\ u_y &= 2i\xi pq \sin qd \cos py + i\xi(\xi^2 - q^2) \sin pd \cos qy \end{aligned} \tag{113}$$

Simulations of the antisymmetric Lamb-wave modes displacement fields at various frequencies and mode numbers are shown in Fig. 6.13. As $fd \to 0$, the A_0 modes resemble the across-the-thickness variation of conventional flexural plate waves, i.e., linear across the thickness.

6.4.1.3 Group Velocity Dispersion Curves

Another important property of the Lamb waves is the group-velocity dispersion curves. The group velocity of Lamb waves is important when examining the traveling of Lamb-wave packets, as in the experiments described later in this book. The group velocity, c_g, can be derived from the phase velocity, c, through the relation

$$c_g = c - \lambda \frac{\partial c}{\partial \lambda} \tag{114}$$

To reduce the programming efforts, some manipulation is applied to Eq. (114). Using the wavelength definition $\lambda = c/f$, we write

$$\frac{\partial c}{\partial \lambda} = \partial c \Big/ \partial \frac{c}{f} = \partial c \Big/ \left(\frac{1}{f} \partial c - \frac{c}{f^2} \partial f \right) = \frac{f^2(\partial c)}{f(\partial c) - c(\partial f)} \tag{115}$$

Hence,

$$\begin{aligned} c_g &= c - \frac{c}{f} \frac{f^2(\partial c)}{f(\partial c) - c(\partial f)} = c\left(1 - \frac{f(\partial c)}{f(\partial c) - c(\partial f)}\right) = c\left(-\frac{f(\partial c) - c(\partial f)}{c(\partial f)}\right)^{-1} \\ &= c^2 \left(c - fd \frac{\partial c}{\partial (fd)} \right)^{-1} \end{aligned} \tag{116}$$

Therefore,

$$c_g = c^2 \left(c - fd \frac{\partial c}{\partial (fd)} \right)^{-1} \tag{117}$$

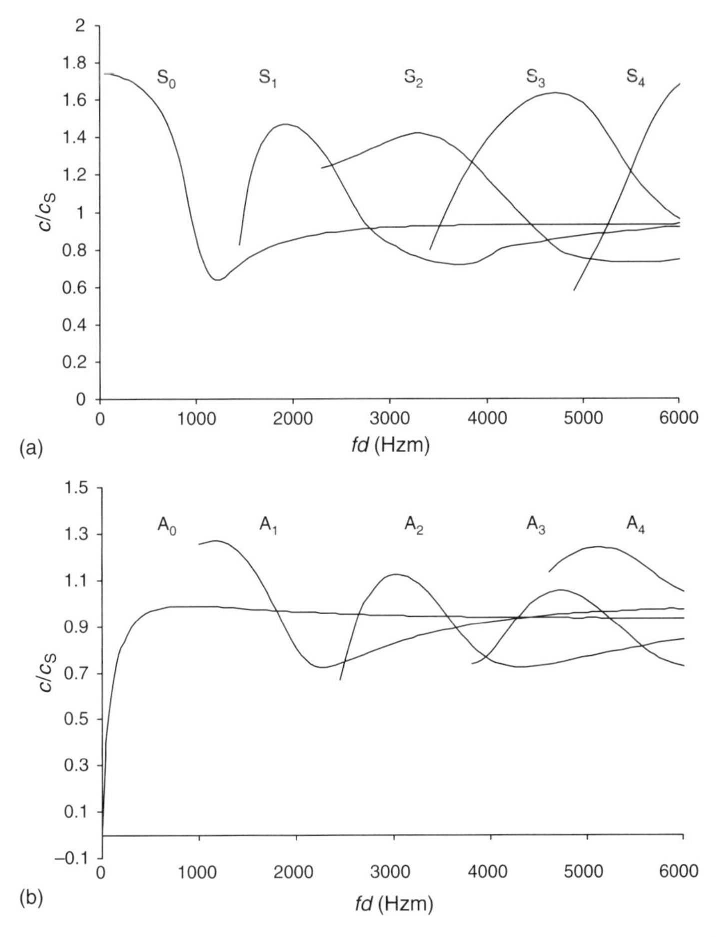

FIGURE 6.14 Group velocity dispersion curves: (a) symmetric Lamb-wave modes; (b) antisymmetric Lamb-wave modes.

Equation (117) uses the derivative of c with respect to the frequency–thickness product fd. This derivative is calculated from the phase-velocity dispersion curve. The numerical derivation can be done by the finite difference formula

$$\frac{\partial c}{\partial (fd)} \cong \frac{\Delta c}{\Delta (fd)} \tag{118}$$

Plots of the group-velocity dispersion curves for the symmetric and antisymmetric Lamb modes are given in Fig. 6.14.

6.4.1.4 Section Summary

Conventional axial and flexural waves in thin plates assume a certain displacement field across the plate thickness (constant for axial waves, linear for flexural waves). However, such simplified fields violate the stress-free boundary conditions on the plate's lower and upper surfaces. When the stress-free boundary conditions are imposed and the exact analysis is performed, one obtains Lamb waves. The Lamb-wave modes represent standing waves across the plate thickness and traveling waves along-the-plate length.

Several Lamb-wave modes exist, corresponding to the real roots of the transcendental Rayleigh–Lamb equation. Each wave mode has a different speed, a different wavelength, and a different standing wave pattern across the plate thickness. The Lamb modes pattern across the thickness varies with frequency. At low frequencies, the fundamental S_0 and A_0 Lamb modes approach the conventional axial and flexural plate waves. Lamb waves are dispersive, i.e., the wave speed varies with frequency, as shown by Fig. 6.10 and Fig. 6.12.

Figure 6.15 shows the wave-speed dispersion curves for S_0 and A_0 Lamb waves, axial plate waves, flexural plate waves, and Rayleigh waves in a 1-mm aluminum plate. Figure 6.15 shows the similarity, at low frequency, between the flexural waves and the A_0 Lamb waves. However, as the frequency increases, they become separated. This indicates that the conventional flexural waves are just a low-frequency approximation of the A_0 Lamb-wave mode. Similarly, the conventional axial waves are a low-frequency approximation of the S_0 Lamb-wave mode. At low frequency, the wave speeds of the axial waves and the S_0 Lamb-wave mode are very close. However, at higher frequency, they differ substantially. At the other end of the spectrum, Rayleigh waves are a high-frequency approximation of the S_0 and A_0 Lamb waves. As indicated in Fig. 6.15, as the frequency becomes very high, the S_0 and the A_0 wave speeds coalesce, and both have the same value. This value is exactly the Rayleigh wave speed. At high frequency, the particle motion of Lamb waves becomes restricted to the proximity of the free surfaces, and thus resembles that of the Rayleigh waves.

Lamb waves can have 1-D propagation (straight-crested Lamb waves) or 2-D propagation (circular-crested Lamb waves). Conventional ultrasonic transducers generate straight-crested Lamb waves. This is done through either an oblique impingement of the plate with a tone-burst through a coupling wedge, or through a patterned tapping through a 'comb' coupler. Circular-crested Lamb waves are not easily excited with conventional transducers. However, circular-crested Lamb waves can be easily excited with PWAS transducers, which have omnidirectional effects and generate 2-D Lamb waves propagating in a circular pattern. This section has presented the straight-crested Lamb waves. The circular-crested Lamb waves are discussed in the next section.

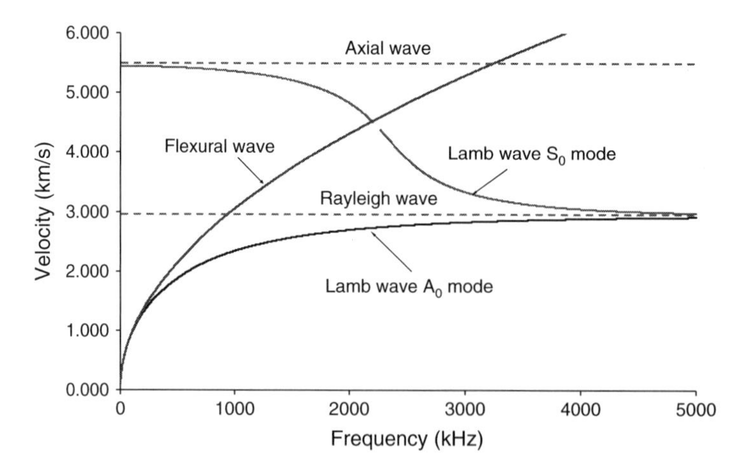

FIGURE 6.15 Calculation of frequency dependence of wave speed for axial, flexural, Lamb, and Rayleigh waves in 1-mm thick aluminum plate. Note that axial and flexural waves are only low-frequency approximations of the Lamb wave S_0 and A_0 modes. The Rayleigh wave is the high-frequency asymptote of the Lamb wave S_0 and A_0 modes.

6.4.2 CIRCULAR CRESTED LAMB WAVES

Circular-crested Lamb waves are developed in axisymmetric cylindrical coordinates. The wave equations are expressed in terms of the dilatation Δ and the vertical displacement, u_z,

$$(\lambda+\mu)\frac{\partial\Delta}{\partial r}+\mu\nabla^2 u_r-\mu\frac{u_r}{r^2}=\rho\ddot{u}_r \quad (\lambda+\mu)\frac{\partial\Delta}{\partial z}+\mu\nabla^2 u_z=\rho\ddot{u}_z \tag{119}$$

The displacements and the dilatation are assumed to follow Bessel functions in r, and harmonic functions in t, i.e.,

$$\begin{aligned} u_r &= f_1(z)J_1(\xi r)e^{-i\omega t} \\ u_z &= f_2(z)J_0(\xi r)e^{-i\omega t} \end{aligned} \qquad \Delta = A\frac{\mu}{\lambda+\mu}J_0(\xi r)e^{-i\omega t}g(\alpha z),$$

$$g(\alpha z)=\begin{cases}\cosh(\alpha z) & \text{(symmetric)}\\ \sinh(\alpha z) & \text{(antisymmetric)}\end{cases} \tag{120}$$

Imposition of the stress-free boundary conditions yields the same Rayleigh–Lamb characteristic equations as for straight-crested waves. After manipulation and utilization of several properties of the Bessel functions, we express the general solution in the form

$$\begin{cases} u_r = A^*\left[-2\xi^2 q\cos qd\cos pz+q\left(\xi^2-q^2\right)\cos pd\cos qz\right]J_1(\xi r)e^{-i\omega t} \\ u_z = A^*\left[2\xi pq\cos qd\sin pz+\xi\left(\xi^2-q^2\right)\cos pd\sin qz\right]J_0(\xi r)e^{-i\omega t} \end{cases} \text{(Symmetric motion)} \tag{121}$$

$$\begin{cases} u_r = A^*\left[-2\xi^2 q\sin qd\sin pz+q\left(\xi^2-q^2\right)\sin pd\sin qz\right]J_1(\xi r)e^{-i\omega t} \\ u_z = A^*\left[2\xi pq\sin qd\cos pz+\xi\left(\xi^2-q^2\right)\sin pd\cos qz\right]J_0(\xi r)e^{-i\omega t} \end{cases} \text{(Antisymmetric motion)} \tag{122}$$

Circular-crested Lamb waves have the same characteristic equation and the same across-the-thickness Lamb modes as the straight-crested Lamb waves. The main difference between straight-crested and circular-crested Lamb waves lies in their space dependence, which is through harmonic functions in the first case and through Bessel functions in the second case. A detailed derivation of circular-crested Lamb waves is given next.

6.4.2.1 Equations and Derivations

The problem at hand is that of a plate with traction-free surfaces under axisymmetric loading. Hence, the development is performed in cylindrical coordinates. Consider a plate with thickness $2d$, and a circular wave front propagating in the r direction (Fig. 6.16).

The traction-free surface condition can be described as $\sigma_{zz}=\sigma_{zr}=\sigma_{z\theta}=0$ at $z=\pm d$. The axisymmetric condition implies $\frac{\partial}{\partial\theta}=0$. In Cartesian coordinates, plane

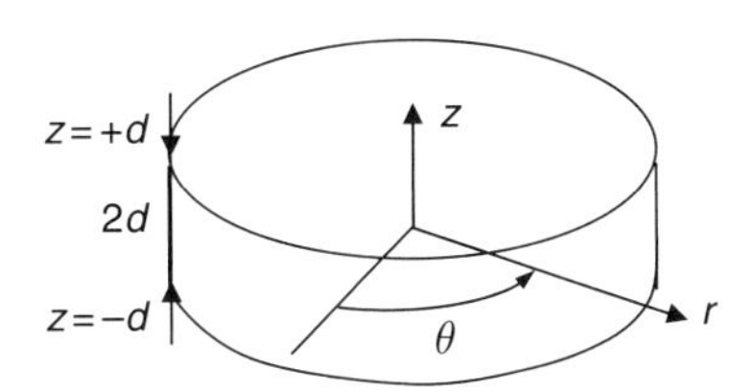

FIGURE 6.16 A plate with thickness $2d$, extends infinitely in the r direction.

strain condition can be described as z-invariant. In cylindrical coordinates, this condition can be described as θ-invariant. The governing equations in terms of displacement are the Navier equations. As all the $\frac{\partial}{\partial\theta}$ terms vanish, only two of these equations are nontrivial. These two equations are

$$(\lambda+\mu)\frac{\partial\Delta}{\partial r}+\mu\nabla^2 u_r-\mu\frac{u_r}{r^2}=\rho\ddot{u}_r$$
$$(\lambda+\mu)\frac{\partial\Delta}{\partial z}+\mu\nabla^2 u_z=\rho\ddot{u}_z \tag{123}$$

The dilatation Δ is assumed in the form

$$\Delta=A\frac{\mu}{\lambda+\mu}J_0(\xi r)e^{i\omega t}\begin{cases}\cosh(\alpha z) & \text{(symmetric motion)}\\ \sinh(\alpha z) & \text{(antisymmetric motion)}\end{cases} \tag{124}$$

The $\cosh(\alpha z)$ form corresponds to symmetric waves, in which the plate midsurface is stretched and compressed, whereas the motion across the plate thickness is symmetric with respect to the midsurface. The $\sinh\alpha z$ form corresponds to antisymmetric waves, in which the midsurface is unstretched, whereas the motion across the plate thickness is antisymmetric with respect to the midsurface. The symmetric waves resemble 'axial waves,' whereas the antisymmetric waves resemble 'flexural waves.'

In addition, we assume the displacements u_r and u_z to have the forms

$$u_r=f_1(z)J_1(\xi r)e^{i\omega t}$$
$$u_z=f_2(z)J_0(\xi r)e^{i\omega t} \tag{125}$$

Their first derivatives are

$$\frac{\partial}{\partial r}u_r=f_1(z)\xi\left(J_0(\xi r)-\frac{1}{\xi r}J_1(\xi r)\right)e^{i\omega t}$$
$$\frac{\partial}{\partial z}u_r=f_1'(z)\,J_1(\xi r)e^{i\omega t}$$
$$\frac{\partial}{\partial r}u_z=-f_2(z)\xi J_1(\xi r)\,e^{i\omega t} \tag{126}$$
$$\frac{\partial}{\partial z}u_z=f_2'(z)\,J_0(\xi r)e^{i\omega t}$$

The second derivatives are

$$\frac{\partial^2}{\partial r^2}u_r=f_1(z)\xi\left(-\xi J_1(\xi r)+\frac{2}{\xi r^2}J_1(\xi r)-\frac{1}{r}J_0(\xi r)\right)e^{i\omega t}$$
$$\frac{\partial^2}{\partial z^2}u_r=f_1''(z)\,J_1(\xi r)e^{i\omega t}$$
$$\frac{\partial^2}{\partial z^2}u_z=f_2''(z)\,J_0(\xi r)e^{i\omega t} \tag{127}$$
$$\frac{\partial^2}{\partial r^2}u_z=-f_2(z)\xi^2\left(J_0(\xi r)-\frac{1}{\xi r}J_1(\xi r)\right)e^{i\omega t}$$

In this derivation, we used the standard results

$$\begin{aligned} J_0'(z) &= -J_1(z) \\ nJ_n(z) + zJ_n'(z) &= zJ_{n-1}(z) \\ J_1(z) + zJ_1'(z) &= zJ_0(z) \\ \Rightarrow J_1'(z) &= J_0(z) - \frac{1}{z}J_1(z) \end{aligned} \tag{128}$$

Hence,

$$\begin{aligned} \frac{\partial}{\partial r}J_0(\xi r) &= \frac{\partial(\xi r)}{\partial r}\frac{\partial}{\partial(\xi r)}J_0(\xi r) = -\xi J_1(\xi r) \\ \frac{\partial}{\partial r}J_1(\xi r) &= \frac{\partial(\xi r)}{\partial r}\frac{\partial}{\partial(\xi r)}J_1(\xi r) = \xi\left(J_0(\xi r) - \frac{1}{\xi r}J_1(\xi r)\right) \end{aligned} \tag{129}$$

The dilatation in cylindrical coordinates is defined as

$$\Delta = \frac{\partial u_r}{\partial r} + \frac{1}{r}\frac{\partial u_\theta}{\partial \theta} + \frac{u_r}{r} + \frac{\partial u_z}{\partial z} \tag{130}$$

Substituting Eqs. (125) and (126) into Eq. (130) yields

$$\begin{aligned} \Delta &= \frac{\partial u_r}{\partial r} + \frac{u_r}{r} + \frac{\partial u_z}{\partial z} \\ &= f_1(z)\xi\left(J_0(\xi r) - \frac{1}{\xi r}J_1(\xi r)\right)e^{i\omega t} + \frac{1}{r}f_1(z)J_1(\xi r)e^{i\omega t} + f_2'(z)\,J_0(\xi r)e^{i\omega t} \\ &= [\xi f_1(z) + f_2'(z)]\,J_0(\xi r)\,e^{i\omega t} \end{aligned} \tag{131}$$

Equation (131) indicates that, in order to satisfy Eq. (124), it is necessary to have

$$\xi f_1(z) + f_2'(z) = A\frac{\mu}{\lambda+\mu}\begin{cases} \cosh(\alpha z) & \text{(symmetric)} \\ \sinh(\alpha z) & \text{(antisymmetric)} \end{cases} \tag{132}$$

Using Eqs. (124) and (125), the terms of the Navier equations can be derived as follows. The Laplacian in cylindrical coordinates is defined as

$$\nabla^2\phi = \frac{\partial^2\phi}{\partial r^2} + \frac{1}{r}\frac{\partial\phi}{\partial r} + \frac{1}{r^2}\frac{\partial^2\phi}{\partial\theta^2} + \frac{\partial^2\phi}{\partial z^2} \tag{133}$$

The Laplacian of u_r is

$$\begin{aligned} \nabla^2 u_r &= \frac{\partial^2 u_r}{\partial r^2} + \frac{1}{r}\frac{\partial u_r}{\partial r} + \frac{1}{r^2}\frac{\partial^2 u_r}{\partial\theta^2} + \frac{\partial^2 u_r}{\partial z^2} \\ &= \frac{\partial^2 u_r}{\partial r^2} + \frac{1}{r}\frac{\partial u_r}{\partial r} + \frac{\partial^2 u_r}{\partial z^2} \\ &= f_1(z)\xi\left(-\xi J_1(\xi r) + \frac{2}{\xi r^2}J_1(\xi r) - \frac{1}{r}J_0(\xi r)\right)e^{i\omega t} \\ &\quad + \frac{1}{r}f_1(z)\xi\left(J_0(\xi r) - \frac{1}{\xi r}J_1(\xi r)\right) + f_1''(z)\,J_1(\xi r)e^{i\omega t} \\ &= \left[f_1''(z) - f_1(z)\left(\xi^2 - \frac{1}{r^2}\right)\right]J_1(\xi r)\,e^{i\omega t} \end{aligned} \tag{134}$$

The Laplacian of u_z is

$$\begin{aligned}\nabla^2 u_z &= \frac{\partial^2 u_z}{\partial r^2} + \frac{1}{r}\frac{\partial u_z}{\partial r} + \frac{1}{r^2}\frac{\partial^2 u_z}{\partial \theta^2} + \frac{\partial^2 u_z}{\partial z^2} \\ &= \left[-f_2(z)\xi^2 J_0(\xi r) + f_2''(z)\, J_0(\xi r)\right] e^{i\omega t} \\ &= J_0(\xi r) e^{i\omega t}\left[-f_2(z)\,\xi^2 + f_2''(z)\right]\end{aligned} \tag{135}$$

The first derivatives of the dilatation Δ are

$$\begin{aligned}\frac{\partial}{\partial r}\Delta &= -\xi J_1(\xi r) A \frac{\mu}{\lambda+\mu} g\,(\alpha z)\, e^{i\omega t} \\ \frac{\partial}{\partial z}\Delta &= \alpha J_0(\xi r) A \frac{\mu}{\lambda+\mu} g'\,(\alpha z)\, e^{i\omega t}\end{aligned} \tag{136}$$

where

$$g\,(\alpha z) = \begin{cases}\cosh(\alpha z) & \text{(symmetric)} \\ \sinh(\alpha z) & \text{(antisymmetric)}\end{cases} \tag{137}$$

Substitution of Eqs. (125), (134), (135), (136) into Navier Eq. (123) yields

$$\begin{aligned}&(\lambda+\mu)\frac{\partial \Delta}{\partial r} + \mu \nabla^2 u_r - \mu \frac{u_r}{r^2} = \rho \ddot{u}_r \\ &-\lambda \xi J_1(\xi r) A g\,(\alpha z)\, e^{i\omega t} + \mu J_1(\xi r)\, e^{i\omega t}\left(f_1''(z) - \xi^2 f_1(z)\right) = \rho\left(-\omega^2\right) f_1(z) J_1(\xi r)\, e^{i\omega t} \\ &f_1''(z) - f_1(z)\left(\xi^2 - \frac{\rho}{\mu}\omega^2\right) = A\xi g\,(\alpha z)\end{aligned} \tag{138}$$

$$\begin{aligned}&(\lambda+\mu)\frac{\partial \Delta}{\partial z} + \mu \nabla^2 u_z = \rho \ddot{u}_z \\ &\alpha A \mu g'\,(\alpha z)\, J_0(\xi r) e^{i\omega t} + \mu\left[f_2''(z) - f_2(z)\,\xi^2\right] J_0(\xi r) e^{i\omega t} = \rho\left(-\omega^2\right) f_2(z) J_0(\xi r)\, e^{i\omega t} \\ &f_2''(z) - f_2(z)\left(\xi^2 - \frac{\rho}{\mu}\omega^2\right) = -A\alpha g'\,(\alpha z)\end{aligned} \tag{139}$$

Introduce

$$\beta^2 = \xi^2 - \frac{\rho}{\mu}\omega^2 = \xi^2 - \frac{\omega^2}{c_S^2} \tag{140}$$

$$\alpha^2 = \xi^2 - \frac{\rho}{\lambda+2\mu}\omega^2 = \xi^2 - \frac{\omega^2}{c_P^2} \tag{141}$$

Hence, Eqs. (138) and (139) can be written as

$$\begin{aligned}f_1''(z) - f_1(z)\,\beta^2 &= A\xi g\,(\alpha z) \\ f_2''(z) - f_2(z)\,\beta^2 &= -A\alpha g'\,(\alpha z)\end{aligned} \tag{142}$$

By solving Eq. (142) and Eq. (132), the two functions, $f_1(z)$ and $f_2(z)$, can be determined. Hence, the displacement functions can be also determined. For the symmetric case,

$$f_1''(z) - f_1(z)\,\beta^2 = A\xi \cosh(\alpha z) \tag{143}$$

$$f_2''(z) - f_2(z)\,\beta^2 = -A\alpha \sinh(\alpha z) \tag{144}$$

The Eq. (143) is solved as follows. The general solution of the homogeneous differential equation

$$f_1''(z) + \beta^2 f_1(z) = 0 \tag{145}$$

is

$$f_{1c}(z) = \overline{C}_1 e^{\beta z} + \overline{C}_2 e^{-\beta z} = C_1 \sinh(\beta z) + C_2 \cosh(\beta z) \tag{146}$$

The particular solution is assumed in the form

$$f_{1p}(z) = D \cosh(\alpha z) \tag{147}$$

Then

$$f_{1p}''(z) = \alpha^2 f_{1p}(z) \tag{148}$$

Upon substitution into Eq. (143),

$$f_{1p}(z) = A \frac{\xi}{\alpha^2 - \beta^2} \cosh(\alpha z) \tag{149}$$

Hence,

$$f_1(z) = f_{1c}(z) + f_{1p}(z) = C_1 \sinh(\beta z) + C_2 \cosh(\beta z) + A \frac{\xi}{\alpha^2 - \beta^2} \cosh(\alpha z) \tag{150}$$

To satisfy the symmetric motion condition, we must have $f_1(z) = f_1(-z)$. This yields $C_1 = 0$; hence the solution for $f_1(z)$ is

$$f_1(z) = C_2 \cosh(\beta z) + A \frac{\xi}{\alpha^2 - \beta^2} \cosh(\alpha z) \tag{151}$$

Equation (144) is solved in a similar way. The general solution of

$$f_2''(z) + \beta^2 f_2(z) = 0 \tag{152}$$

is

$$f_{2c}(z) = \overline{E}_1 e^{\beta z} + \overline{E}_2 e^{-\beta z} = E_1 \sinh(\beta z) + E_2 \cosh(\beta z) \tag{153}$$

Assume the particular solution

$$f_{2p}(z) = F \sinh(\alpha z) \tag{154}$$

Then

$$f_{2p}''(z) = \alpha^2 f_{2p}(z) \tag{155}$$

Upon substitution into Eq. (144),

$$f_{2p}(z) = -A \frac{\alpha}{\alpha^2 - \beta^2} \sinh(\alpha z) \tag{156}$$

Hence,

$$f_2(z) = f_{2c}(z) + f_{2p}(z) = E_1 \sinh(\beta z) + E_2 \cosh(\beta z) - A\frac{\alpha}{\alpha^2 - \beta^2}\sinh(\alpha z) \tag{157}$$

To satisfy the symmetric motion condition, we must have $f_2(z) = -f_2(-z)$. This yields $E_2 = 0$. Hence, the solution for $f_2(z)$ is

$$f_2(z) = E_1 \sinh(\beta z) - A\frac{\alpha}{\alpha^2 - \beta^2}\sinh(\alpha z) \tag{158}$$

To obtain the constants C_2 and E_1, substitute Eqs. (151) and (158) into Eq. (132). This gives

$$\begin{aligned}
&\xi f_1(z) + f_2'(z)\\
&= \xi C_2 \cosh(\beta z) + A\frac{\xi^2}{\alpha^2 - \beta^2}\cosh(\alpha z) + \beta E_1 \cosh(\beta z) - \alpha A\frac{\alpha}{\alpha^2 - \beta^2}\cosh(\alpha z)\\
&= (\xi C_2 + \beta E_1)\cosh(\beta z) + A\frac{\xi^2 - \alpha^2}{\alpha^2 - \beta^2}\cosh(\alpha z)\\
&= A\frac{\mu}{\lambda + \mu}\cosh(\alpha z)
\end{aligned} \tag{159}$$

Note that

$$\frac{\xi^2 - \alpha^2}{\alpha^2 - \beta^2} = \frac{\xi^2 - \left(\xi^2 - \dfrac{\omega^2 \rho}{\lambda + 2\mu}\right)}{\left(\xi^2 - \dfrac{\omega^2 \rho}{\lambda + 2\mu}\right) - \left(\xi^2 - \dfrac{\omega^2 \rho}{\mu}\right)} = \frac{\mu}{\lambda + \mu} \tag{160}$$

Equation (159) yields

$$\xi C_2 + \beta E_1 = 0 \tag{161}$$

i.e.,

$$C_2 = -\frac{\beta}{\xi} E_1 \tag{162}$$

Hence, the displacements in the symmetric case are

$$\begin{aligned}
u_r &= J_1(\xi r)e^{i\omega t}\left[-B\frac{\beta}{\xi}\cosh(\beta z) + A\frac{\xi}{\alpha^2 - \beta^2}\cosh(\alpha z)\right]\\
u_z &= J_0(\xi r)e^{i\omega t}\left[B\sinh(\beta z) - A\frac{\alpha}{\alpha^2 - \beta^2}\sinh(\alpha z)\right]
\end{aligned} \tag{163}$$

Similarly, in the antisymmetric case,

$$\begin{aligned}
u_r &= J_1(\xi r)e^{i\omega t}\left[-B\frac{\beta}{\xi}\sinh(\beta z) + A\frac{\xi}{\alpha^2 - \beta^2}\sinh(\alpha z)\right]\\
u_z &= J_0(\xi r)e^{i\omega t}\left[B\cosh(\beta z) - A\frac{\alpha}{\alpha^2 - \beta^2}\cosh(\alpha z)\right]
\end{aligned} \tag{164}$$

Using

$$p^2 = \frac{\omega^2}{c_P^2} - \xi^2 = -\alpha^2, \quad \alpha^2 = -p^2, \quad \alpha = ip \tag{165}$$

$$q^2 = \frac{\omega^2}{c_S^2} - \xi^2 = -\beta^2, \quad \beta^2 = -q^2, \quad \beta = iq \tag{166}$$

and the standard results $\sinh(ix) = i \sin x$, $\cosh(ix) = \cos x$, Eqs. (31), (32) become

$$\begin{aligned} u_r &= J_1(\xi r) e^{i\omega t} \left[-B\frac{iq}{\xi} \cos qz + A\frac{\xi}{q^2 - p^2} \cos pz \right] \\ u_z &= J_0(\xi r) e^{i\omega t} \left[Bi \sin qz - A\frac{ip}{q^2 - p^2} i \sin pz \right] \\ &= J_0(\xi r) e^{i\omega t} \left[Bi \sin qz + A\frac{p}{q^2 - p^2} \sin pz \right] \end{aligned} \tag{167}$$

$$\begin{aligned} u_r &= J_1(\xi r) e^{i\omega t} \left[-B\frac{iq}{\xi} i \sin qz + Ai\frac{\xi}{q^2 - p^2} \sin pz \right] \\ &= J_1(\xi r) e^{i\omega t} \left[B\frac{q}{\xi} \sin qz + Ai\frac{\xi}{q^2 - p^2} \sin pz \right] \\ u_z &= J_0(\xi r) e^{i\omega t} \left[B \cos qz - A\frac{ip}{q^2 - p^2} \cos pz \right] \end{aligned} \tag{168}$$

Symmetric modes of the circular-crested Lamb waves

Once the displacement functions are known, the strains and stress can be derived. Recall the symmetric part of Eq. (124)

$$\Delta = A\frac{\mu}{\lambda + \mu} J_0(\xi r)\, e^{i\omega t} \cosh(\alpha z) \tag{169}$$

Substitution of Eq. (163) into the definition of strains yields

$$\begin{aligned} \varepsilon_{rr} &= \frac{\partial u_r}{\partial r} = \left[-B\frac{\beta}{\xi} \cosh(\beta z) + A\frac{\xi}{\alpha^2 - \beta^2} \cosh(\alpha z) \right] \left(\xi J_0(\xi r) - \frac{1}{r} J_1(\xi r) \right) e^{i\omega t} \\ \varepsilon_{\theta\theta} &= \frac{1}{r}\frac{\partial u_\theta}{\partial \theta} + \frac{u_r}{r} = \frac{1}{r} J_1(\xi r) e^{i\omega t} \left[-B\frac{\beta}{\xi} \cosh(\beta z) + A\frac{\xi}{\alpha^2 - \beta^2} \cosh(\alpha z) \right] \\ \varepsilon_{zz} &= \frac{\partial u_z}{\partial z} = J_0(\xi r) e^{i\omega t} \left[\beta B \cosh(\beta z) - A\frac{\alpha^2}{\alpha^2 - \beta^2} \cosh(\alpha z) \right] \\ \varepsilon_{r\theta} &= \frac{1}{2}\left(\frac{1}{r}\frac{\partial u_r}{\partial \theta} + \frac{\partial u_\theta}{\partial r} - \frac{u_\theta}{r} \right) = 0 \\ \varepsilon_{rz} &= \frac{1}{2}\left(\frac{\partial u_z}{\partial r} + \frac{\partial u_r}{\partial z} \right) = \frac{1}{2} J_1(\xi r) e^{i\omega t} \left[-B\frac{\xi^2 + \beta^2}{\xi} \sinh(\beta z) + A\frac{2\alpha\xi}{\alpha^2 - \beta^2} \sinh(\alpha z) \right] \\ \varepsilon_{\theta z} &= \frac{1}{2}\left(\frac{\partial u_\theta}{\partial z} + \frac{1}{r}\frac{\partial u_z}{\partial \theta} \right) = 0 \end{aligned} \tag{170}$$

Using Eqs. (169) and (171), the stresses are derived as

$$\begin{aligned}\sigma_z &= \lambda\Delta + 2\mu\varepsilon_{zz}\\ &= \left[\lambda\left(A\frac{\xi^2-\alpha^2}{\alpha^2-\beta^2}\cosh(\alpha z)\right)+2\mu\left(\beta B\cosh(\beta z)-A\frac{\alpha^2}{\alpha^2-\beta^2}\cosh(\alpha z)\right)\right]J_0(\xi r)\,e^{i\omega t}\\ &= \mu\left(2\beta B\cosh(\beta z)-A\frac{2\alpha^2}{\alpha^2-\beta^2}\cosh(\alpha z)+A\frac{\frac{\lambda}{\mu}(\xi^2-\alpha^2)}{\alpha^2-\beta^2}\cosh(\alpha z)\right)J_0(\xi r)\,e^{i\omega t}\\ &= \mu\left(2\beta B\cosh(\beta z)-A\frac{\xi^2+\beta^2}{\alpha^2-\beta^2}\cosh(\alpha z)\right)J_0(\xi r)\,e^{i\omega t}\end{aligned} \tag{171}$$

$$\begin{aligned}\sigma_{rz} &= 2\mu\varepsilon_{rz}\\ &= \mu\left[-B\frac{\xi^2+\beta^2}{\xi}\sinh(\beta z)+A\frac{2\alpha\xi}{\alpha^2-\beta^2}\sinh(\alpha z)\right]J_1(\xi r)e^{i\omega t}\end{aligned} \tag{172}$$

Now, impose the free-surface boundary conditions on these two stresses,

$$\begin{aligned}\sigma_z|_{z=\pm d} &= \mu\left(2\beta B\cosh(\beta d)-A\frac{\xi^2+\beta^2}{\alpha^2-\beta^2}\cosh(\alpha d)\right)J_0(\xi r)\,e^{i\omega t}=0\\ \sigma_{rz}|_{z=\pm d} &= \mu\left[-B\frac{\xi^2+\beta^2}{\xi}\sinh(\beta d)+A\frac{2\alpha\xi}{\alpha^2-\beta^2}\sinh(\alpha d)\right]J_1(\xi r)e^{i\omega t}=0\end{aligned} \tag{173}$$

For non-zero solutions, the determinant of the coefficients of A and B in Eq. (172) must be equal to zero, i.e.,

$$\begin{vmatrix} 2\beta\cosh(\beta d) & -\dfrac{\xi^2+\beta^2}{\alpha^2-\beta^2}\cosh(\alpha d)\\ -\dfrac{\xi^2+\beta^2}{\xi}\sinh(\beta d) & \dfrac{2\alpha\xi}{\alpha^2-\beta^2}\sinh(\alpha d)\end{vmatrix}=0 \tag{174}$$

This yields the characteristic equation

$$\frac{\tanh(\beta d)}{\tanh(\alpha d)}=\frac{4\alpha\beta\xi^2}{(\xi^2+\beta^2)^2} \tag{175}$$

This is the familiar Rayleigh–Lamb frequency equation for the symmetric Lamb-wave modes. At the same time, the ratio of A and B is determined as

$$\frac{B}{A}=\frac{1}{2\beta}\cdot\frac{\xi^2+\beta^2}{\alpha^2-\beta^2}\cdot\frac{\cosh(\alpha d)}{\cosh(\beta d)}=\frac{2\alpha\xi^2}{(\alpha^2-\beta^2)(\xi^2+\beta^2)}\cdot\frac{\sinh(\alpha d)}{\sinh(\beta d)} \tag{176}$$

Thus, the displacements for symmetric circular-crested Lamb waves can be expressed in the form

$$\begin{aligned}u_r &= A\left(\frac{\xi}{\alpha^2-\beta^2}\right)J_1(\xi r)\left[-\frac{\xi^2+\beta^2}{2\xi^2}\frac{\cosh(\alpha d)}{\cosh(\beta d)}\cosh(\beta z)+\cosh(\alpha z)\right]e^{i\omega t}\\ u_z &= -A\left(\frac{\alpha}{\alpha^2-\beta^2}\right)J_0(\xi r)\left[-\frac{2\xi^2}{\xi^2+\beta^2}\frac{\sinh(\alpha d)}{\sinh(\beta d)}\sinh(\beta z)+\sinh(\alpha z)\right]e^{i\omega t}\end{aligned} \tag{177}$$

Recall the notations of Eqs. (165) and (166), i.e.,

$$p^2 = \frac{\omega^2}{c_P^2} - \xi^2 = -\alpha^2, \quad \alpha^2 = -p^2, \quad \alpha = ip \tag{178}$$

$$q^2 = \frac{\omega^2}{c_S^2} - \xi^2 = -\beta^2 g \quad \beta^2 = -q^2, \quad \beta = iq \tag{179}$$

Upon substitution, and recalling the standard results $\sinh(ix) = i \sin x$, $\cosh(ix) = \cos x$, yields

$$\frac{B}{A} = \frac{1}{2iq} \cdot \frac{\xi^2 - q^2}{q^2 - p^2} \cdot \frac{\cos pd}{\cos qd} = -i \frac{1}{2q} \cdot \frac{\xi^2 - q^2}{q^2 - p^2} \cdot \frac{\cos pd}{\cos qd} \tag{180}$$

Recall Eq. (163) in the form

$$\begin{aligned} u_r &= AJ_1(\xi r)e^{i\omega t}\left[-\frac{B}{A}\frac{iq}{\xi}\cos qz + \frac{\xi}{q^2 - p^2}\cos pz\right] \\ u_z &= AJ_0(\xi r)e^{i\omega t}\left[\frac{B}{A} i \sin qz + \frac{p}{q^2 - p^2}\sin pz\right] \end{aligned} \tag{181}$$

Upon substitution of Eq. (180), we get

$$\begin{aligned} u_r &= AJ_1(\xi r)e^{i\omega t}\frac{2\xi^2 q \cos qd \cos pz - q\left(\xi^2 - q^2\right)\cos pd \cos qz}{2\xi q\left(q^2 - p^2\right)\cos qd} \\ u_z &= AJ_0(\xi r)e^{i\omega t}\frac{2\xi qp \cos qd \sin pz + \xi\left(\xi^2 - q^2\right)\cos pd \sin qz}{2\xi q\left(q^2 - p^2\right)\cos qd} \end{aligned} \tag{182}$$

or

$$\begin{aligned} u_r &= -AJ_1(\xi r)e^{i\omega t}\frac{-2\xi^2 q \cos qd \cos pz + q\left(\xi^2 - q^2\right)\cos pd \cos qz}{2\xi q\left(q^2 - p^2\right)\cos qd} \\ u_z &= -AJ_0(\xi r)e^{i\omega t}\frac{-2\xi qp \cos qd \sin pz - \xi\left(\xi^2 - q^2\right)\cos pd \sin qz}{2\xi q\left(q^2 - p^2\right)\cos qd} \end{aligned} \tag{183}$$

i.e.,

$$\begin{aligned} u_r &= A^* J_1(\xi r)e^{i\omega t}\left[-2\xi^2 q \cos qd \cos pz + q\left(\xi^2 - q^2\right)\cos pd \cos qz\right] \\ u_z &= A^* J_0(\xi r)e^{i\omega t}\left[-2\xi pq \cos qd \sin pz - \xi\left(\xi^2 - q^2\right)\cos pd \sin qz\right] \end{aligned} \tag{184}$$

where the denominator and the minus sign were absorbed into A^*. The expressions in the brackets are identical to the expression for straight-crested Lamb waves. The factor of i in u_z, which appears in the straight-crested waves, does not need to appear here because the Bessel functions J_0 and J_1 are intrinsically in quadrature. Thus,

$$\begin{aligned} u_r &= A^*\left[2\xi^2 q \cos qd \cos pz - q\left(\xi^2 - q^2\right)\cos pd \cos qz\right] J_1(\xi r)e^{i\omega t} \\ u_z &= A^*\left[2\xi pq \cos qd \sin pz + \xi\left(\xi^2 - q^2\right)\cos pd \sin qz\right] J_0(\xi r)e^{i\omega t} \end{aligned} \tag{185}$$

Antisymmetric modes of the circular-crested Lamb waves

Recall the antisymmetric part of Eq. (124), i.e.,

$$\Delta = A\frac{\mu}{\lambda+\mu}J_0(\xi r)\, e^{i\omega t}\sinh(\alpha z) \tag{186}$$

Substitution of Eq. (164) into the definition of strains yields

$$\begin{aligned}
\varepsilon_{rr} &= \frac{\partial u_r}{\partial r} = \left[-B\frac{\beta}{\xi}\sinh(\beta z) + A\frac{\xi}{\alpha^2-\beta^2}\sinh(\alpha z)\right]\left(\xi J_0(\xi r) - \frac{1}{r}J_1(\xi r)\right)e^{i\omega t} \\
\varepsilon_{\theta\theta} &= \frac{1}{r}\frac{\partial u_\theta}{\partial\theta} + \frac{u_r}{r} = \frac{1}{r}J_1(\xi r)e^{i\omega t}\left[-B\frac{\beta}{\xi}\sinh(\beta z) + A\frac{\xi}{\alpha^2-\beta^2}\sinh(\alpha z)\right] \\
\varepsilon_{zz} &= \frac{\partial u_z}{\partial z} = J_0(\xi r)e^{i\omega t}\left[\beta B\sinh(\beta z) - A\frac{\alpha^2}{\alpha^2-\beta^2}\sinh(\alpha z)\right] \\
\varepsilon_{r\theta} &= \frac{1}{2}\left(\frac{1}{r}\frac{\partial u_r}{\partial\theta} + \frac{\partial u_\theta}{\partial r} - \frac{u_\theta}{r}\right) = 0 \\
\varepsilon_{rz} &= \frac{1}{2}\left(\frac{\partial u_z}{\partial r} + \frac{\partial u_r}{\partial z}\right) = \frac{1}{2}J_1(\xi r)e^{i\omega t}\left[-B\frac{\xi^2+\beta^2}{\xi}\cosh(\beta z) + A\frac{2\alpha\xi}{\alpha^2-\beta^2}\cosh(\alpha z)\right] \\
\varepsilon_{\theta z} &= \frac{1}{2}\left(\frac{\partial u_\theta}{\partial z} + \frac{1}{r}\frac{\partial u_z}{\partial\theta}\right) = 0
\end{aligned} \tag{187}$$

Using Eq. (186) and (188), the stresses are derived as

$$\begin{aligned}
\sigma_z &= \lambda\Delta + 2\mu\varepsilon_{zz} \\
&= \left[\lambda\left(A\frac{\xi^2-\alpha^2}{\alpha^2-\beta^2}\sinh(\alpha z)\right) + 2\mu\left(\beta B\sinh(\beta z) - A\frac{\alpha^2}{\alpha^2-\beta^2}\sinh(\alpha z)\right)\right]J_0(\xi r)\, e^{i\omega t} \\
&= \mu\left(2\beta B\sinh(\beta z) - A\frac{2\alpha^2}{\alpha^2-\beta^2}\sinh(\alpha z) + A\frac{\frac{\lambda}{\mu}(\xi^2-\alpha^2)}{\alpha^2-\beta^2}\sinh(\alpha z)\right)J_0(\xi r)\, e^{i\omega t} \\
&= \mu\left(2\beta B\sinh(\beta z) - A\frac{\xi^2+\beta^2}{\alpha^2-\beta^2}\sinh(\alpha z)\right)J_0(\xi r)\, e^{i\omega t}
\end{aligned} \tag{188}$$

$$\begin{aligned}
\sigma_{rz} &= 2\mu\varepsilon_{rz} \\
&= \mu\left[-B\frac{\xi^2+\beta^2}{\xi}\cosh(\beta z) + A\frac{2\alpha\xi}{\alpha^2-\beta^2}\cosh(\alpha z)\right]J_1(\xi r)e^{i\omega t}
\end{aligned} \tag{189}$$

Imposing the free-surface boundary conditions at the upper and lower surfaces, sets these two stresses to zero, i.e.,

$$\begin{aligned}
\sigma_z|_{z=\pm d} &= \mu\left(2\beta B\sinh(\beta d) - A\frac{\xi^2+\beta^2}{\alpha^2-\beta^2}\sinh(\alpha d)\right)J_0(\xi r)\, e^{i\omega t} = 0 \\
\sigma_{rz}|_{z=\pm d} &= \mu\left[-B\frac{\xi^2+\beta^2}{\xi}\cosh(\beta d) + A\frac{2\alpha\xi}{\alpha^2-\beta^2}\cosh(\alpha d)\right]J_1(\xi r)e^{i\omega t} = 0
\end{aligned} \tag{190}$$

Setting the determinant of coefficients of A and B in Eq. (172) to zero,

$$\begin{vmatrix} 2\beta\sinh(\beta d) & -\dfrac{\xi^2+\beta^2}{\alpha^2-\beta^2}\sinh(\alpha d) \\ -\dfrac{\xi^2+\beta^2}{\xi}\cosh(\beta d) & \dfrac{2\alpha\xi}{\alpha^2-\beta^2}\cosh(\alpha d) \end{vmatrix} = 0 \tag{191}$$

yields the characteristic equation

$$\frac{\tanh(\beta d)}{\tanh(\alpha d)} = \frac{(\xi^2+\beta^2)^2}{4\alpha\beta\xi^2} \tag{192}$$

This is the familiar Rayleigh–Lamb frequency equation for the antisymmetric Lamb-wave modes. At the same time the ratio of A and B is determined as

$$\frac{B}{A} = \frac{1}{2\beta} \cdot \frac{\xi^2+\beta^2}{\alpha^2-\beta^2} \cdot \frac{\sinh(\alpha d)}{\sinh(\beta d)} = \frac{2\alpha\xi^2}{(\alpha^2-\beta^2)(\xi^2+\beta^2)} \cdot \frac{\cosh(\alpha d)}{\cosh(\beta d)} \tag{193}$$

Thus, the displacements for antisymmetric circular-crested Lamb waves can be expressed in the form

$$\begin{aligned} u_r &= A\left(\frac{\xi}{\alpha^2-\beta^2}\right) J_1(\xi r)\left[-\frac{\xi^2+\beta^2}{2\xi^2}\frac{\sinh(\alpha d)}{\sinh(\beta d)}\sinh(\beta z)+\sinh(\alpha z)\right]e^{i\omega t} \\ u_z &= -A\left(\frac{\alpha}{\alpha^2-\beta^2}\right) J_0(\xi r)\left[-\frac{2\xi^2}{\xi^2+\beta^2}\frac{\cosh(\alpha d)}{\cosh(\beta d)}\cosh(\beta z)+\cosh(\alpha z)\right]e^{i\omega t} \end{aligned} \tag{194}$$

Recall the notations of Eqs. (165) and (166), i.e.,

$$p^2 = \frac{\omega^2}{c_{\mathrm{P}}^2} - \xi^2 = -\alpha^2, \quad \alpha^2 = -p^2, \quad \alpha = ip \tag{195}$$

$$q^2 = \frac{\omega^2}{c_{\mathrm{S}}^2} - \xi^2 = -\beta^2, \quad \beta^2 = -q^2, \quad \beta = iq \tag{196}$$

Upon substitution, and recalling the standard results $\sinh(ix) = i\sin x$, $\cosh(ix) = \cos x$, yields

$$\frac{B}{A} = \frac{1}{2iq} \cdot \frac{\xi^2-q^2}{q^2-p^2} \cdot \frac{\sin pd}{\sin qd} = -i\frac{1}{2q} \cdot \frac{\xi^2-q^2}{q^2-p^2} \cdot \frac{\sin pd}{\sin qd} \tag{197}$$

Recall Eq. (164)

$$\begin{aligned} u_r &= AJ_1(\xi r)e^{i\omega t}\left[i\frac{\xi}{q^2-p^2}\sin pz + \frac{B}{A}\frac{q}{\xi}\sin qz\right] \\ u_z &= AJ_0(\xi r)e^{i\omega t}\left[-\frac{ip}{q^2-p^2}\cos pz + \frac{B}{A}\cos qz\right] \end{aligned} \tag{198}$$

Upon substitution of Eq. (197) into Eq. (198), we get

$$\begin{aligned} u_r &= -AJ_1(\xi r)e^{i\omega t}i\frac{-2\xi^2 q\sin qd\sin pz + q(\xi^2-q^2)\sin pd\sin qz}{2\xi q\sin qd(q^2-p^2)} \\ u_z &= -AJ_0(\xi r)e^{i\omega t}i\frac{2\xi pq\sin qd\cos pz + \xi(\xi^2-q^2)\sin pd\cos qz}{2\xi q\sin qd(q^2-p^2)} \end{aligned} \tag{199}$$

or

$$\begin{aligned} u_r &= A^* J_1(\xi r)e^{i\omega t}\left[-2\xi^2 q\sin qd\sin pz + q\left(\xi^2-q^2\right)\sin pd\sin qz\right] \\ u_z &= A^* J_0(\xi r)e^{i\omega t}\left[2\xi pq\sin qd\cos pz + \xi\left(\xi^2-q^2\right)\sin pd\cos qz\right] \end{aligned} \tag{200}$$

where the denominator and the minus sign were absorbed into A^*. The expressions in the brackets are identical to the expression for straight-crested Lamb waves. The factor i in u_z, which appears in the straight-crested waves, does not need to appear here because the Bessel functions J_0 and J_1 are intrinsically in quadrature. Thus,

$$\begin{aligned} u_r &= A^*\left[-2\xi^2 q \sin qd \sin pz + q\left(\xi^2 - q^2\right)\sin pd \sin qz\right] J_1(\xi r)e^{i\omega t} \\ u_z &= A^*\left[2\xi pq \sin qd \cos pz + \xi\left(\xi^2 - q^2\right)\sin pd \cos qz\right] J_0(\xi r)e^{i\omega t} \end{aligned} \tag{201}$$

6.4.2.2 The Asymptotic Behavior at Large Radial Distance from the Origin

Different from the straight-crested Lamb waves, which follow the trigonometric functions, the circular-crested Lamb waves follow the Bessel functions. However, for large value of r, the Bessel functions converge to the asymptotic expression

$$J_v(z) \underset{z\to\infty}{\to} \sqrt{2/(\pi z)}\cos\left(z - \frac{1}{2}v\pi - \frac{1}{4}\pi\right) \tag{202}$$

In our case,

$$\begin{aligned} J_0(\xi r)|_{\xi r\to\infty} &\to \frac{\sqrt{2}\cos(\xi r - \pi/4)}{\sqrt{\pi\xi r}} = \frac{\cos\xi r + \sin\xi r}{\sqrt{\pi\xi r}} \\ J_1(\xi r)|_{\xi r\to\infty} &\to \frac{\sqrt{2}\cos(\xi r - 3\pi/4)}{\sqrt{\pi\xi r}} = \frac{-\cos\xi r + \sin\xi r}{\sqrt{\pi\xi r}} \end{aligned} \tag{203}$$

This means that for large values of ξr the displacements pattern becomes periodic at large radial distances from the origin. In this limit, the displacements reduce to the following forms.

For symmetric motion,

$$\begin{aligned} u_r &= A\left(\frac{\xi}{\alpha^2-\beta^2}\right)\frac{\sin\xi r - \cos\xi r}{\sqrt{\pi\xi r}}\left[\cosh(\alpha z) - \frac{\xi^2+\beta^2}{2\xi^2}\frac{\cosh(\alpha b)}{\cosh(\beta b)}\cosh(\beta z)\right]e^{i\omega t} \\ u_z &= -A\left(\frac{\alpha}{\alpha^2-\beta^2}\right)\frac{\sin\xi r + \cos\xi r}{\sqrt{\pi\xi r}}\left[\sinh(\alpha z) - \frac{2\xi^2}{\xi^2+\beta^2}\frac{\sinh(\alpha b)}{\sinh(\beta b)}\sinh(\beta z)\right]e^{i\omega t} \end{aligned} \tag{204}$$

For antisymmetric motion,

$$\begin{aligned} u_r &= A\left(\frac{\xi}{\alpha^2-\beta^2}\right)\frac{\sin\xi r - \cos\xi r}{\sqrt{\pi\xi r}}\left[\sinh(\alpha z) - \frac{\xi^2+\beta^2}{2\xi^2}\frac{\sinh(\alpha b)}{\sinh(\beta b)}\sinh(\beta z)\right]e^{i\omega t} \\ u_z &= -A\left(\frac{\alpha}{\alpha^2-\beta^2}\right)\frac{\sin\xi r + \cos\xi r}{\sqrt{\pi\xi r}}\left[\cosh(\alpha z) - \frac{2\xi^2}{\xi^2+\beta^2}\frac{\cosh(\alpha b)}{\cosh(\beta b)}\cosh(\beta z)\right]e^{i\omega t} \end{aligned} \tag{205}$$

These two sets of equations are periodic in ξr. The displacement patterns across the thickness are controlled by the terms in the bracket, which are of the same form as for the straight-crested Lamb waves. If we define 'half wave length' $\lambda'/2$ as the distance between adjacent zeros of the radial displacement, u_r, the limiting wave length λ'_∞ is given by the wavelength of $\sin(\xi r)$ and $\cos(\xi r)$, i.e.,

$$\lim_{r/d\to\infty}\lambda' = \lambda'_\infty = \frac{2\pi}{\xi} \tag{206}$$

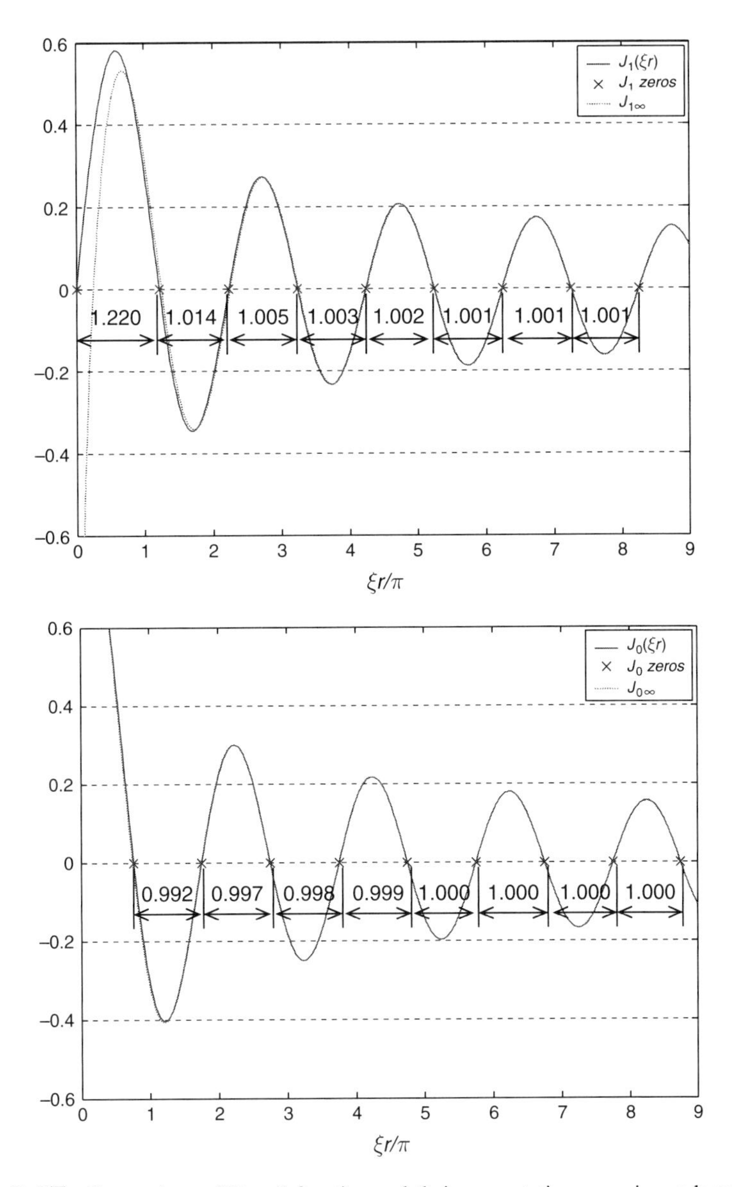

FIGURE 6.17 Comparison of Bessel function and their asymptotic expression at large ξr values. The numbers indicate the local ratio between apparent wavelength and the asymptotic wavelength.

Figure 6.17 shows the comparison between J_0 and J_1 and their asymptotic expressions. Note that the ratio between the apparent wavelength and the asymptotic wavelength, λ'_{∞}, approaches unity after four oscillations.

6.4.2.3 Section Summary

The particle displacements of circular-crested Lamb waves are governed by Bessel functions, whereas those of the straight-crested Lamb waves are governed by harmonic (sine and cosine) functions. In spite of this difference, the Rayleigh–Lamb frequency equation, which was developed for straight-crested Lamb waves, also applies to the circular-crested Lamb waves. Hence, the wave speed and group velocity of circular-crested Lamb waves are the same as those of straight-crested Lamb waves.

Close to the origin, the circular-crested Lamb waves have a shape that varies rapidly with the radial distance. However, beyond three wavelengths' distance from the origin, the behavior of circular-crested Lamb waves has almost stabilized, and the wavelength has come to within 0.1% of that of straight-crested Lamb waves. At large distances from the origin, the behavior of circular-crested Lamb waves approaches asymptotically that of straight-crested Lamb waves.

6.5 GENERAL FORMULATION OF GUIDED WAVES IN PLATES

In flat plates, ultrasonic guided waves propagate as Lamb waves and as SH waves. Lamb waves are vertically polarized, whereas SH waves are horizontally polarized. Both Lamb waves and SH waves can be symmetric or antisymmetric with respect to the plate midplane. In previous sections, we have analyzed the SH waves and the Lamb waves separately. Here, we will present briefly an unified approach which reveals both the Lamb waves and the SH waves can originate from a single set of equations. Our work will follow the notations introduced in an earlier chapter for straight-crested z-invariant 3-D waves.

Consider a plate with stress-free upper and lower surfaces situated at $y = \pm d$ (Fig. 6.18). The equation of motion for an isotropic elastic medium is given by

$$\mu \nabla^2 \mathbf{u} + (\lambda + \mu) \nabla \nabla \cdot \mathbf{u} = \rho \frac{\partial^2 \mathbf{u}}{\partial t^2} \tag{207}$$

where λ and μ are the Lame constants, ρ is the mass density, and $\mathbf{u}$ is the displacement vector. Assume straight-crested waves. Hence the problem becomes z-invariant and that displacement vector is given by

$$\vec{u} = \nabla \Phi + \nabla \times \vec{H} \tag{208}$$

where Φ and $\vec{H}$ are the scalar and vector potential functions given by

$$\Phi = f(y) e^{i(\xi x - \omega t)}, \quad \vec{H} = \left(h_x(y)\vec{i} + h_y(y)\vec{j} + h_z(y)\vec{k} \right) e^{i(\xi x - \omega t)} \tag{209}$$

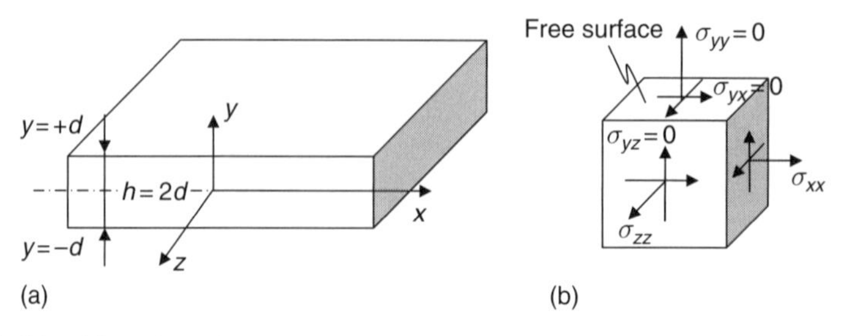

FIGURE 6.18 (a) A plate with thickness $2d$, extending infinitely in x and z direction; (b) free-body diagram of an infinitesimal area extracted from the plate.

where ω is the circular frequency, ξ is the wavenumber, and c is the wave speed, $c = \omega/\xi$. As shown in Chapter 5, the potentials satisfy the governing equations

$$\nabla^2\Phi = \frac{1}{c_P^2}\frac{\partial^2\Phi}{\partial t^2}, \quad \nabla^2\vec{H} = \frac{1}{c_S^2}\frac{\partial^2\vec{H}}{\partial t^2}, \quad \text{and} \quad \vec{\nabla}\cdot\vec{H} = 0 \tag{210}$$

Substitution of Eq. (209) into Eq. (210) yields

$$\begin{aligned} f'' - \xi^2 f &= -\omega^2 f/c_P^2 \\ h_x'' - \xi^2 h_x &= -\omega^2 h_x/c_S^2 \\ h_y'' - \xi^2 h_y &= -\omega^2 h_y/c_S^2 \\ h_z'' - \xi^2 h_z &= -\omega^2 h_z/c_S^2 \end{aligned} \tag{211}$$

where $c_P^2 = (\lambda + 2\mu)/\rho$ and $c_S^2 = \mu/\rho$ are the pressure (longitudinal) and shear (transverse) wave speeds. Solution of Eq. (211) is in the form

$$\begin{aligned} \Phi &= (A\cos\alpha y + B\sin\alpha y)e^{i(\xi x - \omega t)} \\ H_x &= (C\cos\beta y + D\sin\beta y)e^{i(\xi x - \omega t)} \\ H_y &= (E\cos\beta y + F\sin\beta y)e^{i(\xi x - \omega t)} \\ H_z &= (G\cos\beta y + H\sin\beta y)e^{i(\xi x - \omega t)} \end{aligned} \tag{212}$$

where $\alpha^2 = (\omega^2/c_P^2) - \xi^2$, and $\beta^2 = (\omega^2/c_S^2) - \xi^2$. The constants A through H are determined from the stress-free boundary conditions at the plate's upper and lower surfaces, which yield

$$\begin{bmatrix} -c_3\sin\alpha d & c_4\sin\beta d & 0 & 0 & 0 & 0 & 0 & 0 \\ c_1\cos\alpha d & c_2\cos\beta d & 0 & 0 & 0 & 0 & 0 & 0 \\ 0 & 0 & c_1\sin\alpha d & -c_2\sin\beta d & 0 & 0 & 0 & 0 \\ 0 & 0 & c_3\cos\alpha d & c_4\cos\beta d & 0 & 0 & 0 & 0 \\ 0 & 0 & 0 & 0 & -c_5\sin\beta d & \beta^2\sin\beta d & 0 & 0 \\ 0 & 0 & 0 & 0 & -\beta\sin\beta d & i\xi\sin\beta d & 0 & 0 \\ 0 & 0 & 0 & 0 & 0 & 0 & \beta^2\cos\beta d & c_5\cos\beta d \\ 0 & 0 & 0 & 0 & 0 & 0 & i\xi\cos\beta d & \beta\cos\beta d \end{bmatrix} \begin{pmatrix} A \\ H \\ B \\ G \\ E \\ D \\ C \\ F \end{pmatrix} = 0 \tag{213}$$

with $c_1 = (\lambda + 2\mu)\alpha^2 + \lambda\xi^2$, $c_2 = 2i\mu\xi\beta$, $c_3 = 2i\xi\alpha$, $c_4 = \xi^2 - \beta^2$, $c_5 = i\xi\beta$. Nontrivial solutions of the homogeneous Eq. (213) are obtained when the determinant of the coefficients matrix is zero, which yields the *characteristic equation*. Examination of Eq. (213) reveals that the determinant can be broken up into the product of four smaller determinants, corresponding to the coefficient of pairs law (A, H), (B, G), (E, D), (C, F). The first two pairs correspond to the symmetric and antisymmetric Lamb waves, whereas the last two pairs correspond to symmetric and antisymmetric SH waves. The characteristic equations for the symmetric and antisymmetric Lamb waves are also known as the *Rayleigh–Lamb equations*. These implicit transcendental equations are solved numerically to determine the permissible guided-wave solutions. For each solution of the characteristic equation, one determines a specific value of the wavenumber, ξ, and hence the wave speed, c. The variation of wave speed with frequency is shown in Fig. 6.19a,b for Lamb waves and in Fig. 6.19c for SH waves. At lower frequency–thickness products, only two Lamb wave types exist: S_0, which is a symmetrical Lamb wave resembling the longitudinal

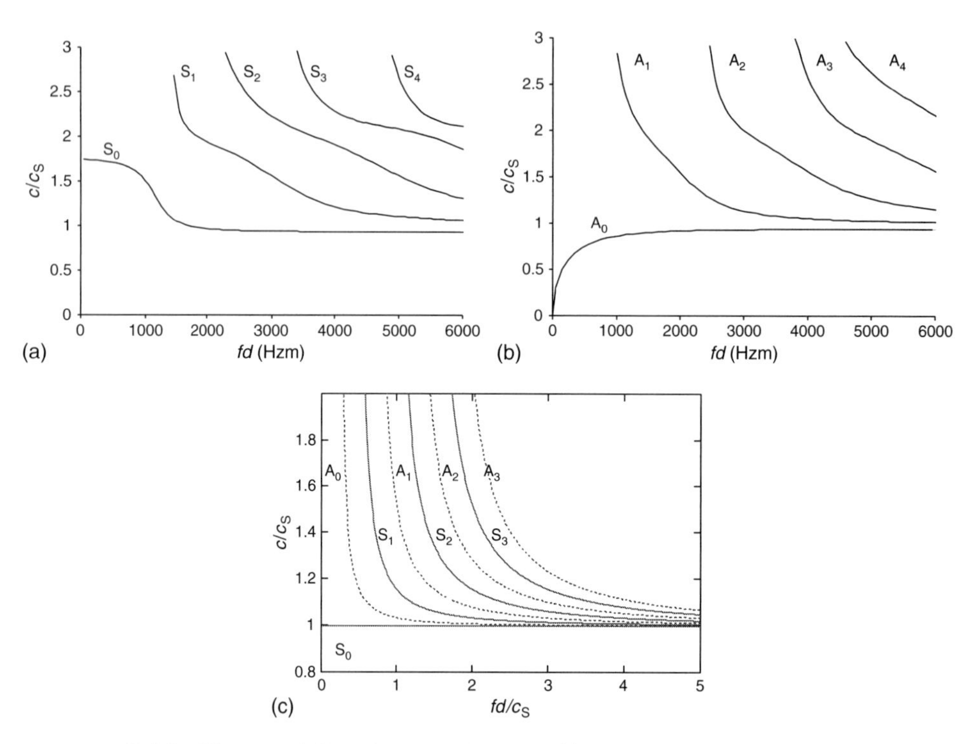

FIGURE 6.19 Wave speed dispersion curves in plates: (a) symmetric Lamb waves, S_n; (b) antisymmetric Lamb waves, A_n; (c) SH waves: solid lines are for symmetric SH modes (S_n), dotted lines are for antisymmetric SH modes (A_n) (c_S = shear wave speed, d = half thickness of the plate).

waves; and A_0, which is an antisymmetric Lamb wave resembling the flexural waves. At higher frequency–thickness values, a number of Lamb waves are present, S_n and A_n, where $n = 0, 1, 2$, etc. . . . At very high frequencies, the S_0 and A_0 Lamb waves coalesce into the Rayleigh waves, which are confined to the plate's upper and lower surfaces. The Lamb waves are highly dispersive (wave speed varies with frequency); however, S_0 waves at low frequency–thickness values show very little dispersion. The SH waves are also dispersive, with the exception of the first mode, SH_0, which does not show any dispersion at all. For circular-crested waves, a similar analysis can be conducted in polar coordinates.

6.6 GUIDED WAVES IN TUBES AND SHELLS

Guided waves have the important property that they stay confined inside the walls of a thin-wall structure, and hence can travel over large distances. In addition, the guided waves can also travel inside curved walls; this property makes them ideal for applications in the ultrasonic inspection of aircraft, missiles, pressure vessel, oil tanks, pipelines, etc. The study of guided-wave propagation in cylindrical shells can be considered a limiting case of the study of guided-wave propagation in hollow cylinders. As the wall-thickness of the hollow cylinder decreases with respect to its radius, the hollow cylinder approaches the case of a thin-wall cylindrical shell. Several investigators have considered the propagation of waves in solid and hollow cylinders. Love (1944) studied wave propagation in an isotropic solid cylinder and showed that three types of solutions are possible: (1) longitudinal; (2) flexural; and (3) torsional. At high frequencies, each of these solutions is multimodal and dispersive. Meitzler (1961) showed that, under certain conditions, mode coupling could exist between various wave types propagating in solid cylinders such as wires. Extensive numerical simulation and experimental testing of these phenomena was done by Zemenek (1972).

Comprehensive work on wave propagation in hollow circular cylinders was done by Gazis (1959). A comprehensive analytical investigation was complemented by numerical studies. The nonlinear algebraic equations and the corresponding numerical solutions of the wave-speed dispersion curves were obtained. These results found important applications in the ultrasonic NDE of tubing and pipes. Silk and Bainton (1979) found equivalences between the ultrasonic in hollow cylinders and the Lamb waves in flat plates and used them to detect cracks in heat exchanger tubing. Rose et al. (1994) used guided pipe waves to find cracks in nuclear steam generator tubing. Alleyne et al. (2001) used guided waves to detect cracks and corrosion in chemical plant pipe work.

6.6.1 DERIVATION OF EQUATIONS FOR CYLINDRICAL-SHELL GUIDED WAVES

A brief review of the mathematical modeling of guided waves in cylindrical shells are given next (Gazis, 1959; Rose et al., 1994; Silk and Bainton, 1979). The coordinates and characteristic dimensions are shown in Fig. 6.20; a and b are the inner and outer radii of the tube, and h is the tube thickness. The variables r, θ, and z are the radial, circumferential, and longitudinal coordinates. The modeling starts from the equation of motion for an isotropic elastic medium, in invariant form:

$$\mu\Delta\mathbf{u}+(\lambda+\mu)\nabla\nabla\cdot\mathbf{u}=\rho(\partial^2\mathbf{u}/\partial t^2) \tag{214}$$

where $\mathbf{u}$ is the displacement vector, ρ is the density, λ and μ are Lame's constants, and Δ is the three-dimensional Laplace operator. The vector $\mathbf{u}$ is expressed in terms of the dilation scalar potential ϕ and the equivolume vector potential $\mathbf{H}$ according to

$$\mathbf{u}=\nabla\phi+\nabla\times\mathbf{H} \tag{215}$$

$$\nabla\cdot\mathbf{H}=F(\mathbf{r},t) \tag{216}$$

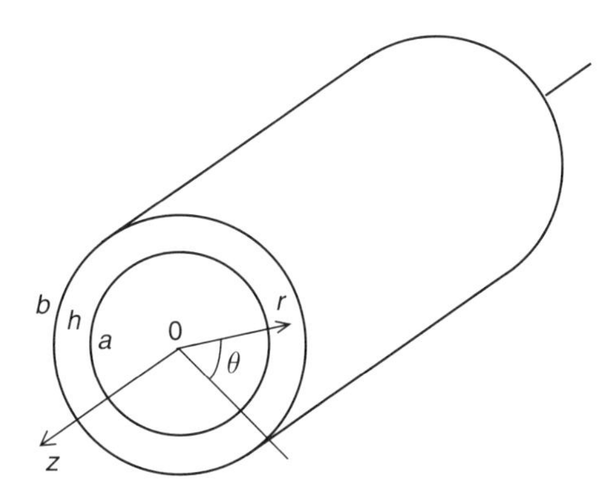

FIGURE 6.20 Reference coordinates and characteristic dimensions of a hollow cylinder in which the propagation of guided waves is studied.

For free motion, the displacement equations of motion are satisfied if the potentials ϕ and $\mathbf{H}$ satisfy the wave equations

$$c_P^2 \Delta \phi = \partial^2 \phi / \partial t^2 \tag{217}$$

$$c_S^2 \Delta \mathbf{H} = \partial^2 \mathbf{H} / \partial t^2 \tag{218}$$

where $c_P^2 = (\lambda + 2\mu)/\rho$ and $c_S^2 = \mu/\rho$ are the pressure and shear wave speeds, respectively. We express the potentials and the wave equations in cylindrical coordinates as

$$\begin{aligned} \phi &= f(r) \cos n\theta \cos(\omega t + \xi z) \\ H_r &= g_r(r) \sin n\theta \sin(\omega t + \xi z) \\ H_\theta &= g_\theta(r) \cos n\theta \sin(\omega t + \xi z) \\ H_z &= g_z(r) \sin n\theta \cos(\omega t + \xi z) \end{aligned} \tag{219}$$

where wave motion with wavenumber ξ along the z axis is assumed. Substitution of Eq. (219) into Eqs. (217) and (218) yields

$$\begin{aligned} (\Delta + \omega^2/c_P^2)\phi &= 0 \\ (\Delta + \omega^2/c_S^2)H_z &= 0 \\ (\Delta - 1/r^2 + \omega^2/c_S^2)H_r - (2/r^2)(\partial H_\theta/\partial \theta) &= 0 \\ (\Delta - 1/r^2 + \omega^2/c_S^2)H_\theta + (2/r^2)(\partial H_r/\partial \theta) &= 0 \end{aligned} \tag{220}$$

where $\Delta = \nabla^2$ is the Laplace operator. The following notations are introduced:

$$\alpha^2 = \omega^2/c_P^2 - \xi^2, \quad \beta^2 = \omega^2/c_S^2 - \xi^2 \tag{221}$$

6.6.2 GENERAL SOLUTION FOR CYLINDRICAL-SHELL GUIDED WAVES

The general solution is expressed in terms of Bessel functions J and Y or the modified Bessel functions I and K of arguments $\alpha_1 r = |\alpha r|$ and $\beta_1 r = |\beta r|$. As determined by Eq. (221), α and β can be either real or imaginary. The general solution is of the form

$$\begin{aligned} f &= A \cdot Z_n(\alpha_1 r) + B \cdot W_n(\alpha_1 r) \\ g_3 &= A_3 \cdot Z_n(\beta_1 r) + B_3 \cdot W_n(\beta_1 r) \\ g_1 &= \frac{1}{2}(g_r - g_\theta) = A_1 \cdot Z_{n+1}(\beta_1 r) + B_1 \cdot W_{n+1}(\beta_1 r) \\ g_2 &= \frac{1}{2}(g_r + g_\theta) = A_2 \cdot Z_{n-1}(\beta_1 r) + B_2 \cdot W_{n-1}(\beta_1 r) \end{aligned} \tag{222}$$

where, for brevity, Z denotes a J or I Bessel function, and W denotes a Y or a K Bessel function, as appropriate. Thus, the potentials are expressed in terms of the unknowns $A, B, A_1, B_1, A_2, B_2, A_3, B_3$. Two of these unknowns are eliminated using the gauge invariance property of the equivolume potentials. The strain–displacement and stress–strain relations are used to express the stresses in terms of the potential functions. Then, we impose the free-motion boundary conditions

$$\sigma_{rr} = \sigma_{rz} = \sigma_{r\theta} = 0 \quad \text{at} \quad r = a \quad \text{and} \quad r = b \tag{223}$$

Thus, one arrives at a linear system of six homogeneous equations in six unknowns. For nontrivial solution, the system determinant must vanish, i.e.,

$$\left|c_{ij}\right| = 0, i, j = 1, 2, \ldots 6 \tag{224}$$

The coefficients c_{ij} in Eq. (224) have complicated algebraic expressions that are not reproduced here for sake of brevity. We retain, however, that a characteristic equation exists in the form

$$\Omega_n(a, b, \lambda, \mu, fd, c) = 0 \tag{225}$$

where a and b represent the inner and outer radii of the tube, whereas λ and μ represent the Lame constants. This implicit transcendental equation is solved numerically to determine the permissible guided-wave solutions.

6.6.3 GUIDED-WAVE MODES IN CYLINDRICAL SHELLS

As shown by Gazis (1959), three basic families of guided waves may exist in hollow cylinders

(i) Longitudinal axially symmetric modes, $L(0, m), m = 1, 2, 3, \ldots$
(ii) Torsional axially symmetric modes, $T(0, m), m = 1, 2, 3, \ldots$
(iii) Flexural non-axially symmetric modes, $F(n, m), n = 1, 2, 3, \ldots \quad m = 1, 2, 3, \ldots.$

Within each family, an infinite number of modes exist such that their phase velocities, c, for a given frequency–thickness product, fd, represent permissible solutions of the implicit transcendental Eq. (225). The index m represents the number of the mode shape across the wall of the tube. The index n determines the manner in which the fields generated by the guided-wave modes vary with angular coordinate θ in the cross-section of the cylinder. For the F modes of family n, each field component can be considered to vary as either $\sin(n\theta)$ or $\cos(n\theta)$. It is also observed that the index n represents the mode shape of flexing of the tube as a whole.

The longitudinal and torsional modes are also referred to as the axial symmetric or $n = 0$ modes. These axial symmetric modes are preferred for defect detection in long pipes as the pipe circumference is uniformly insonified. The longitudinal modes are easier to generate with conventional ultrasonic transducers. They are good for locating circumferential cracks. However, for locating axial cracks and corrosion, the torsional modes, though more difficult to generate with conventional ultrasonic transducers, are recommended.

Thus, the existence in thin-wall cylinders of guided waves similar to the Lamb waves present in flat plates was identified. An examination of the differential equations dependence on the ratios h/r and h/λ indicates that, for shallow shells (h/r and $h/\lambda \ll 1$), the longitudinal modes approach the Lamb-wave modes, whereas the torsional modes approach the SH modes. In fact, it can be shown that the $L(0, 1)$ mode corresponds to the A_0 Lamb mode, whereas the $L(0, 2)$ mode corresponds to the S_0 Lamb mode, etc. The flexural modes remain a mode type specific to tube waves and without equivalence in flat plate waves.

6.6.4 SECTION SUMMARY

This section has discussed guided waves in thin-wall cylindrical shells. The mathematical model for the analysis of wave propagation in hollow cylinders was reviewed and its general solution was presented. It was shown that the general solution accepts three wave types:

(i) Longitudinal axially symmetric modes, $L(0, m), m = 1, 2, 3, \ldots$
(ii) Torsional axially symmetric modes, $T(0, m), m = 1, 2, 3, \ldots$
(iii) Flexural non-axially symmetric modes, $F(n, m), n = 1, 2, 3, \ldots, m = 1, 2, 3, \ldots$.

An examination of the differential equations dependence on the ratios h/r and h/λ indicates that, for shallow shells (h/r and $h/\lambda \ll 1$), the longitudinal modes approach the Lamb-wave modes, whereas the torsional modes approach the SH modes. In fact, it can be shown that the $L(0, 1)$ mode corresponds to the A_0 Lamb mode, whereas the $L(0, 2)$ mode corresponds to the S_0 Lamb mode, etc. The flexural modes remain a mode type specific to tube waves and without equivalence in flat plate waves.

6.7 GUIDED WAVES IN COMPOSITE PLATES

6.7.1 STATE OF THE ART IN MODELING GUIDED WAVE PROPAGATION IN COMPOSITES

Tang and Henneke (1989) used an approximate method for the derivation of Lamb-wave phase velocities in composite laminates. The method is useful to obtain the lowest symmetric S_0 mode and antisymmetric A_0 mode. In this case the approximation of the elementary shear-deformation plate theory is applicable because the frequencies are low and hence the wavelength is long.

Damage detection techniques for composite structures are based on the study of complicated wave mechanisms; they rely strongly on the use of predictive modeling tools. The response method and the modal method are the two most used inspection techniques. In the response method, the reflection and transmission characteristics of the plate are examined. In the modal method, the standing-wave properties of the system are evaluated. Both methods make use of the matrix formulation, which describes elastic waves in layered media with arbitrary number of layers. Lowe (1995) gives a review of the matrix technique for modeling ultrasonic waves in multilayered media, where the layers are isotropic. Wave propagation in multilayered media with an arbitrary number of flat layers was derived by Thomson (1950) and improved by Haskell (1953). In the Thomson–Haskell formulation, the displacements and the stresses at the bottom of a layer are described in terms of those at the top of the layer through a transfer matrix (TM). After propagating the TM process through all the layers, one obtains a single matrix that represents the behavior of the complete system. However, this process may show numerical instability for large layer thickness and high-excitation frequencies. This problem was caused by the poor numerical conditioning of the TM process that may be due to a combination of several factors involving the presence of decaying evanescent waves at the interface.

In the TM method, the field equations for the displacements and stresses in a flat isotropic elastic solid layer are expressed as the superposition of the fields of four bulk waves within the layer. The approach therefore is to derive the field equations for bulk waves, which are solutions of the wave equation in an infinite medium, and then to introduce the boundary conditions at an interface between two layers (Snell's law), so defining the rules for coupling between layers and for the superposition of the bulk waves. The analysis of the layers is restricted to 2-D, with the imposition of plane strain and in-plane motion conditions. At each interface eight waves are assumed: longitudinal and shear waves arriving from 'above' the interface and leaving 'below' the interface and (L+, S+), similarly, longitudinal and shear waves arriving from below the interface and leaving above the interface (L−, S−). There are thus four waves in each layer of the multilayered plate. Snell's law requires that for interaction of the waves, they must all share the same frequency and spatial properties in the x_1-direction at each interface. It follows that all displacement and stress equations have the same ω and k_1 wavenumber component, where k_1 is the projection of the bulk wavenumber onto the interface. The angles of incidence, transmission, and reflection of homogeneous bulk waves in the layers are then constrained.

An alternative to the Thomson–Haskell formulation is the global matrix (GM) formulation proposed by Knopoff (1964). In the GM formulation, all the equations of all the layers of the structure are considered. The system matrix consist of $4(n-1)$ equations, where n is the total number of layers. There is no a priori assumption on the interdependence between the sets of equations for each interface. The solution is carried out on the full matrix, addressing all the equation concurrently

$$\begin{bmatrix} [D_{1b}^-] & [-D_{2t}] & & & \\ & [D_{2b}] & [-D_{3t}] & & \\ & & [D_{3b}] & [-D_{4t}] & \\ & & & [D_{4b}] & [-D_{5b}^+] \end{bmatrix} \begin{bmatrix} A_1 \\ A_2 \\ A_3 \\ A_4 \\ A_5 \end{bmatrix} = 0 \tag{226}$$

where $[D]$ is a 4×4 field matrix that relates the amplitude of partial waves to the displacement and stress fields in a layer and A_i is the amplitude of the partial waves in layer i. This technique is robust but slow to compute for many layers because the matrix is rather large.

Both methods share a common characteristic: a solution of the characteristic function does not strictly prove the existence of a modal solution, but only that the system matrix is singular. Furthermore, the calculation of the determinant for the modal solution needs the use of a good algorithm because the aim of the problem is to find the zero of the determinant whereas the matrix is frequently close to being singular.

Nayfeh (1991, 1995) extended the Thomson–Haskell formulation to the case of anisotropic materials and composites of anisotropic layers. The wave is allowed to propagate on an arbitrary incidence angle θ, measured with respect to the direction normal to the plate (Fig. 6.21) and along any azimuthal angle, ϕ, measured in the plane of the plate Fig. 6.22.

Consider a layered plate of n layer made of anisotropic material. Let the thickness of the kth layer be d_k (Fig. 6.22). Each layer has a local coordinate system, $(x_i')_k$, such that

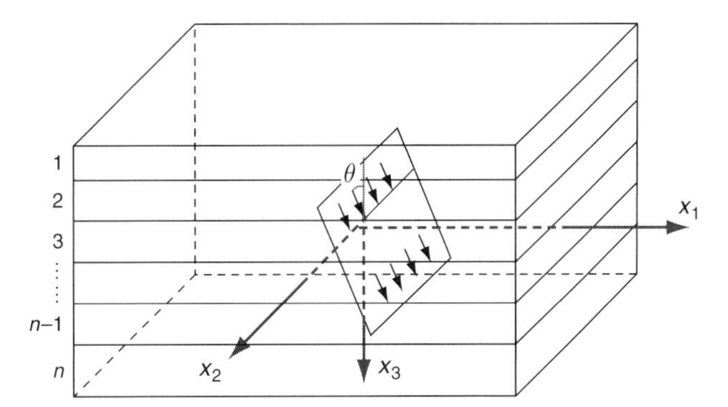

FIGURE 6.21 Plate of an arbitrary number of layer rigidly bonded with a plane wave propagating in the x_1–x_3 direction.

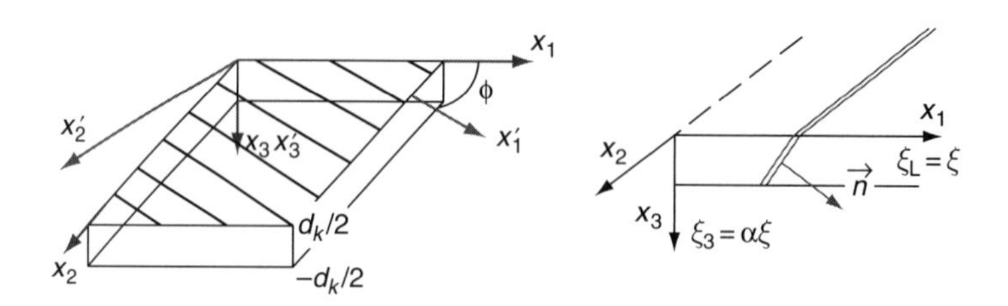

FIGURE 6.22 *k*th layer of the layered plate.

its origin is located in the upper plane of the layer. The *k*th layer occupies the region $0 \leq (x_3')_k \leq d^{(k)}$. For each layer, one can write the formal solution

$$(u_1, u_2, u_3) = (U_1, U_2, U_3)\, e^{i\xi(x_1 \sin\theta + \alpha x_3 - vt)} \tag{227}$$

Substituting the formal solution in the wave equation and imposing the existence of a nontrivial solution leads to the formal solution for the displacements and stresses in the expanded 6×6 matrix form. Through the matrix it is possible to relate the displacements and the stresses of the bottom to those of the top surface

$$\begin{Bmatrix} \{u_k^+\} \\ \{\sigma_k^+\} \end{Bmatrix} = [A_k] \begin{Bmatrix} \{u_k^-\} \\ \{\sigma_k^-\} \end{Bmatrix} \tag{228}$$

The matrix A_k is the transfer matrix for the monoclinic layer *k*. Invoking the continuity of displacement and stress components at the layer interfaces, it is possible to relate the displacements and stresses of the top of the layered plate to those at its bottom

$$A = \prod_{k=1}^{N} A_k \tag{229}$$

To find the dispersion curves of a layered plate, it is sufficient to set $\theta = 0$ and solve the determinant of the 3×3 submatrix

$$\begin{vmatrix} A_{41} & A_{42} & A_{43} \\ A_{51} & A_{52} & A_{53} \\ A_{61} & A_{62} & A_{63} \end{vmatrix} \tag{230}$$

In a bulk material there are three types of waves: longitudinal, shear vertical, and shear horizontal. In the case of a isotropic thin plate, the longitudinal and the shear vertical waves are coupled and their interaction gives two different family of mode propagation of the Lamb waves: symmetric and antisymmetric modes. The shear horizontal wave is decoupled from the other two and can be treated separately. In a generic anisotropic plate this is no longer true. The three waves are coupled and the solution of the 3×3 matrix determinant gives three modes of propagation. Even for orthotropic or higher than orthotropic material symmetry the three waves are coupled. This happens if the wave propagates in direction other than that of material symmetry. For wave propagation in the direction of material symmetry, the shear horizontal is decoupled and the TM formulation is much simplified.

Kausel (1986), Wang and Rokhlin (2001), Rokhlin and Wang (2002) described methods to resolve the numerical instability of the TM by introducing the layer stiffness matrix (SM) and using an efficient recursive algorithm to calculate the global stiffness matrix for an arbitrary anisotropic layered structure. In this method, a layer SM is used to replace the layer TM. The SM relates the stresses at the top and the bottom of the layer with the displacements at the top and bottom layer; the terms in the matrix have only exponentially decaying terms and the matrix had the same dimension and simplicity as the TM. This method is unconditionally stable and is slightly more computationally efficient than the TM method. For each layer, the local coordinate origin is settled at the top of the jth layer for waves propagating along the $-z$ direction and at the bottom of the jth layer for waves propagating along the $+z$ direction. This selection of coordinate system is very important for eliminating the numerical overflow of the exponential terms when the waves become nonhomogeneous (large fd values). In this way, the exponential terms are normalized and the nonhomogeneous exponentials are equal to one at the interface and decay toward the opposite surface of the layer.

The SM is defined as a matrix that relates the stresses at the top and bottom of its layer to the displacements at the top and bottom,

$$\begin{bmatrix} \sigma_{j-1} \\ \sigma_j \end{bmatrix} = \mathbf{K}_j \begin{bmatrix} u_{j-1} \\ u_j \end{bmatrix} \tag{231}$$

where $\mathbf{K}$ is the stiffness matrix, The TM of each layer has the principal diagonal terms depending on $e^{ik_z^{+1}h_j}$ that for large fd goes to zero and make the TM singular. The SM principal diagonal does not have these terms and it is independent of layer thickness, thus the SM is always regular. Through a recursive algorithm it is possible to calculate the total SM for a layered system.

Wang et al. (2002) studied wave propagation in orthotropic laminated spherical shells with arbitrary thickness through elastodynamic solution. Wang solves the elastodynamic equation for each separate orthotropic spherical shell by means of finite Hankel and Laplace transforms. By the interface continuity conditions between layers and the boundary conditions at the internal and external surfaces, he determines the unknown constants. As stated by Wang et al. (2002), the method can be useful in order to determine the lamination material and lay-up pattern in the design of laminated spherical-shell structures subjected to shock loading.

6.7.2 GUIDED WAVES DISPERSION CURVES IN COMPOSITE PLATES

In the case of a isotropic plate, the dispersion curves can be derived from the solution of the Rayleigh–Lamb equation. In the case of composite material it is not possible to find a close form solution of the dispersion curves; as described in the previous section, there are different methods (TM, GM, SM, etc.) that can be used to solve the problem. Here, we illustrate the use of the TM method applied to anisotropic laminated composite plates (Nayfeh, 1995), which is stable for low fd product values.

Consider a composite plate (Fig. 6.23) made up of orthotropic layers (e.g., unidirectional fibers composite plies).

Assume that each layer is made up of an orthotropic material with known properties, as defined by the layer stiffness matrix

$$C' = \begin{bmatrix} c'_{11} & c'_{12} & c'_{13} & 0 & 0 & 0 \\ c'_{12} & c'_{22} & c'_{23} & 0 & 0 & 0 \\ c'_{13} & c'_{23} & c'_{33} & 0 & 0 & 0 \\ 0 & 0 & 0 & c'_{44} & 0 & 0 \\ 0 & 0 & 0 & 0 & c'_{55} & 0 \\ 0 & 0 & 0 & 0 & 0 & c'_{66} \end{bmatrix} \tag{232}$$

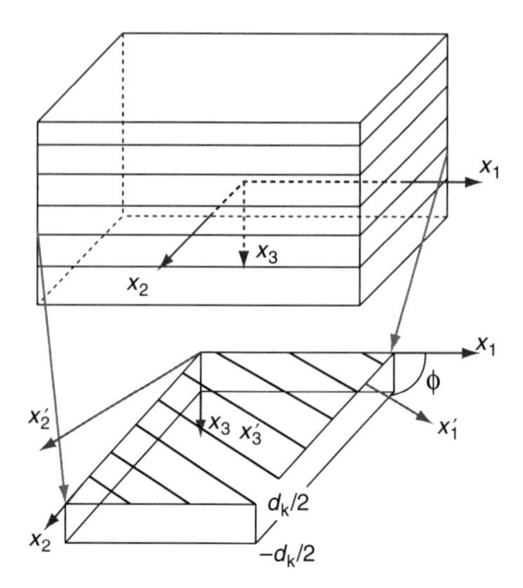

FIGURE 6.23 Laminated composite plate model showing the kth layer with orthotropic properties (e.g., an unidirectional composite ply).

Or, explicitly,

$$C' = \begin{bmatrix} \frac{(\nu_{23}^2 - 1)E_1^2}{\Delta} & -\frac{\nu_{12}E_1E_2}{\Delta} & -\frac{\nu_{12}E_1E_2}{\Delta} & 0 & 0 & 0 \\ -\frac{\nu_{12}E_1E_2}{\Delta} & \frac{(\nu_{12}^2E_2 - E_1)\Gamma}{\Delta} & -\frac{(\nu_{12}^2E_2 + \nu_{23}E_1)\Gamma}{\Delta} & 0 & 0 & 0 \\ -\frac{\nu_{12}E_1E_2}{\Delta} & -\frac{(\nu_{12}^2E_2 + \nu_{23}E_1)\Gamma}{\Delta} & \frac{(\nu_{12}^2E_2 - E_1)\Gamma}{\Delta} & 0 & 0 & 0 \\ 0 & 0 & 0 & G_4 & 0 & 0 \\ 0 & 0 & 0 & 0 & G_5 & 0 \\ 0 & 0 & 0 & 0 & 0 & G_6 \end{bmatrix} \tag{233}$$

where

$$\Delta = E_1(\nu_{23} - 1) + 2\nu_{12}^2E_2 \qquad \Gamma = \frac{E_2}{(1 + \nu_{23})} \tag{234}$$

The unidirectional layer in the matrix can have any orientation. The stiffness matrix in the global coordinate system is

$$C = T_1^{-1}C'T_2 = \begin{bmatrix} c_{11} & c_{12} & c_{13} & 0 & 0 & c_{16} \\ c_{12} & c_{11} & c_{13} & 0 & 0 & c_{16} \\ c_{13} & c_{13} & c_{33} & 0 & 0 & c_{36} \\ 0 & 0 & 0 & c_{44} & c_{45} & 0 \\ 0 & 0 & 0 & c_{45} & c_{44} & 0 \\ c_{16} & c_{16} & c_{36} & 0 & 0 & c_{66} \end{bmatrix} \tag{235}$$

where T_1 and T_2 are transformation matrices. Assume the wave propagation direction set at $\theta = 90°$. Then, the solution can be written as

$$(u_1, u_2, u_3) = (U_1, U_2, U_3)\, e^{i\zeta(x_1 + \alpha x_3 - vt)} \tag{236}$$

where ζ is the wavenumber, $v = \omega/\zeta$ is the phase velocity, ω is the circular frequency, α is an unknown parameter, and U_i is the displacement amplitude. The equation of motion is

$$\begin{aligned}
&C_{11}\frac{\partial^2 u_1}{\partial x_1^2} + C_{66}\frac{\partial^2 u_1}{\partial x_2^2} + C_{55}\frac{\partial^2 u_1}{\partial x_3^2} + 2C_{16}\frac{\partial^2 u_1}{\partial x_1 \partial x_2} + C_{16}\frac{\partial^2 u_2}{\partial x_1^2} + C_{26}\frac{\partial^2 u_2}{\partial x_2^2} + C_{45}\frac{\partial^2 u_2}{\partial x_3^2} \\
&\quad + (C_{12}+C_{66})\frac{\partial^2 u_2}{\partial x_1 \partial x_2} + (C_{13}+C_{55})\frac{\partial^2 u_3}{\partial x_1 \partial x_3} + (C_{36}+C_{45})\frac{\partial^2 u_3}{\partial x_2 \partial x_3} = \rho\frac{\partial^2 u_1}{\partial t^2} \\
&C_{16}\frac{\partial^2 u_1}{\partial x_1^2} + C_{26}\frac{\partial^2 u_1}{\partial x_2^2} + C_{45}\frac{\partial^2 u_1}{\partial x_3^2} + (C_{12}+C_{66})\frac{\partial^2 u_1}{\partial x_1 \partial x_2} + C_{66}\frac{\partial^2 u_2}{\partial x_1^2} + C_{22}\frac{\partial^2 u_2}{\partial x_2^2} + C_{44}\frac{\partial^2 u_2}{\partial x_3^2} \\
&\quad + 2C_{26}\frac{\partial^2 u_2}{\partial x_1 \partial x_2} + (C_{36}+C_{45})\frac{\partial^2 u_3}{\partial x_1 \partial x_3} + (C_{23}+C_{44})\frac{\partial^2 u_3}{\partial x_2 \partial x_3} = \rho\frac{\partial^2 u_2}{\partial t^2} \\
&(C_{13}+C_{55})\frac{\partial^2 u_1}{\partial x_1 \partial x_3} + (C_{36}+C_{45})\frac{\partial^2 u_1}{\partial x_2 \partial x_3} + (C_{36}+C_{45})\frac{\partial^2 u_2}{\partial x_1 \partial x_3} + (C_{23}+C_{44})\frac{\partial^2 u_2}{\partial x_2 \partial x_3} \\
&\quad + C_{55}\frac{\partial^2 u_3}{\partial x_1^2} + C_{44}\frac{\partial^2 u_3}{\partial x_2^2} + C_{33}\frac{\partial^2 u_3}{\partial x_3^2} + 2C_{45}\frac{\partial^2 u_2}{\partial x_1 \partial x_2} = \rho\frac{\partial^2 u_3}{\partial t^2}
\end{aligned} \tag{237}$$

Substituting Eq. (236) into Eq. (237), we obtain:

$$\begin{aligned}
&\left(C_{11} + C_{55}\alpha^2 - \rho v^2\right) u_1 + \left(C_{16} + C_{45}\alpha^2\right) u_2 + (C_{13} + C_{55})\,\alpha u_3 = 0 \\
&\left(C_{16} + C_{45}\alpha^2\right) u_1 + \left(C_{66} + C_{44}\alpha^2 - \rho v^2\right) u_2 + (C_{36} + C_{45})\,\alpha u_3 = 0 \\
&(C_{13} + C_{55})\,\alpha u_1 + (C_{36} + C_{45})\,\alpha u_2 + \left(C_{55} + C_{33}\alpha^2 - \rho v^2\right) u_3 = 0
\end{aligned} \tag{238}$$

Note that, if the material coordinate and the global coordinate systems coincide, then direction 2 is not coupled with the other two directions. This means that the SH wave would be decoupled from the other two modes of propagation and the mathematical formulation is simpler.

The above system accepts nontrivial solution if

$$\alpha^6 + A_1\alpha^4 + A_2\alpha^2 + A_3 = 0 \tag{239}$$

Following Nayfeh (1995) we obtain the following formulation

$$\begin{bmatrix} u_1 \\ u_2 \\ u_3 \\ \sigma_{33}^* \\ \sigma_{13}^* \\ \sigma_{23}^* \end{bmatrix} = \begin{bmatrix} 1 & 1 & 1 & 1 & 1 & 1 \\ V_1 & V_1 & V_3 & V_3 & V_5 & V_5 \\ W_1 & -W_1 & W_3 & -W_3 & W_5 & -W_5 \\ D_{11} & D_{11} & D_{13} & D_{13} & D_{15} & D_{15} \\ D_{21} & -D_{21} & D_{23} & -D_{23} & D_{25} & -D_{25} \\ D_{31} & -D_{31} & D_{31} & -D_{31} & D_{35} & -D_{35} \end{bmatrix} \begin{bmatrix} U_{11}e^{i\xi(x_1\sin\theta+\alpha_1 x_3 - vt)} \\ U_{11}e^{i\xi(x_1\sin\theta-\alpha_1 x_3 - vt)} \\ U_{13}e^{i\xi(x_1\sin\theta+\alpha_3 x_3 - vt)} \\ U_{13}e^{i\xi(x_1\sin\theta-\alpha_3 x_3 - vt)} \\ U_{15}e^{i\xi(x_1\sin\theta+\alpha_5 x_3 - vt)} \\ U_{15}e^{i\xi(x_1\sin\theta-\alpha_5 x_3 - vt)} \end{bmatrix} \tag{240}$$

where

$$\begin{aligned}
D_{1q} &= C_{13} + C_{36}V_q + C_{33}\alpha_q W_q \\
D_{2q} &= C_{55}\left(\alpha_q + W_q\right) + C_{45}\alpha_q V_q \\
D_{3q} &= C_{45}\left(\alpha_q + W_q\right) + C_{44}\alpha_q V_q
\end{aligned} \tag{241}$$

$$V_q = \frac{K_{11}(\alpha_q)K_{23}(\alpha_q) - K_{13}(\alpha_q)K_{12}(\alpha_q)}{K_{13}(\alpha_q)K_{22}(\alpha_q) - K_{12}(\alpha_q)K_{23}(\alpha_q)} \tag{242}$$

$$W_q = \frac{K_{11}(\alpha_q)K_{23}(\alpha_q) - K_{13}(\alpha_q)K_{12}(\alpha_q)}{K_{12}(\alpha_q)K_{33}(\alpha_q) - K_{23}(\alpha_q)K_{13}(\alpha_q)} \tag{243}$$

Denote by P the left hand side of Eq. (240), i.e., the vector that contains Eq. (244) P, the displacements and stresses. Then, denote by X the 6×6 matrix, by U the vector containing the U_{1i} elements, and by D the diagonal matrix with elements $e^{i\zeta(x_1+\alpha_i x_3 - vt)}$. Thus Eq. (240) can be written as

$$P_k = X_k D_k U_k \tag{244}$$

Through this equation it is possible to link the displacements and the stresses of the bottom layer with those of the top layer. For the upper and lower layer of the kth layer we have, respectively

$$P^- = X_k D_k^- U_k \tag{245}$$

$$P^+ = X_k D_k^+ U_k \tag{246}$$

Where D_k^- is specialized for the case of $x_3 = 0$ (in this case $D_k^- = I$) and D_k^+ for $x_3 = d_k$. We can write

$$P_k^+ = A_k P_k^- \tag{247}$$

where

$$A_k = X_k D_k X_k^{-1} \tag{248}$$

Appling the above procedure for each layer, it is possible to relate the displacements and the stresses at the upper surface of the layered plate to those of the lower surface via the transfer matrix multiplication

$$A = A_n A_{n-1}, \ldots, A_1 \tag{249}$$

The total transfer matrix expression is

$$\begin{Bmatrix} \{u^+\} \\ \{\sigma^+\} \end{Bmatrix} = \begin{bmatrix} [A_{uu}] & [A_{u\sigma}] \\ [A_{u\sigma}] & [A_{\sigma\sigma}] \end{bmatrix} \begin{Bmatrix} \{u^-\} \\ \{\sigma^-\} \end{Bmatrix} \tag{250}$$

To obtain the dispersion curved we must impose stress-free conditions at the upper and lower surfaces, this leads to the characteristic equation

$$|A_{u\sigma}| = 0 \tag{251}$$

This algorithm has been coded in a MATLAB program, which allows one to enter a vector of the layer orientation. Based on this input, the program calculates the transfer matrix of each layer in the global coordinate system and then the total transfer matrix (249) for different value of velocity and frequency. The program determines the velocity at which, for a given frequency, the matrix determinant changes sign and then finds the solution by bisection method.

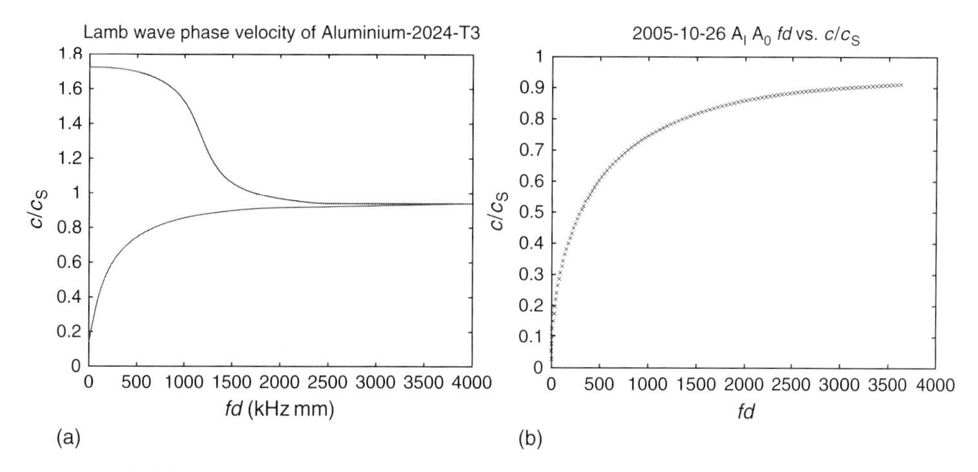

FIGURE 6.24 Dispersion curves for a aluminum plate 2024-T3 solved with (a) the Rayleigh–Lamb equation; (b) the transfer matrix method.

To verify that the algorithm works, we first test it on an isotropic plate 2024-T7; the dispersion curves obtained are in good agreement with those obtained from the solution of the Rayleigh–Lamb equation (Fig. 6.24).

Next, we considered a composite plate made up of one layer of unidirectional fibers for wave propagating in different orientation. The layer properties material is A534/AF252, which has stiffness properties with T300/5208 (Table 6.2)

The algorithm gives the results in terms of wavenumber–thickness product versus the phase velocity of the plate. For $\phi = 0/90$ layup, the SH wave is decoupled from the other two modes (Fig. 6.25a), whereas in all other cases the three waves are coupled. The value of the S_0 Lamb-wave mode at low frequencies decreases with the increase in fiber angle, whereas the value of the SH-wave increases and is no longer constant.

One drawback of the TM method is that it looses its numerical stability at high frequency–thickness products; when numerical stability is lost, one obtains plots like in Fig. 6.26.

We also studied the case of a quasi-isotropic plate $[(0/45/90/-45)_{2s}]$. The obtained dispersion curves are represented in Fig. 6.27.

TABLE 6.2 Ply material properties (Herakovich, 1998)

	T300/5208
Density (g/cm^3)	1.54
Axial Modulus E_1 (GPa)	132
Transverse Modulus E_2 (GPa)	10.8
Poisson's ratio ν_{12}	0.24
Poisson's ratio ν_{23}	0.59
Shear Modulus G_{12} (GPa)	5.65
Shear Modulus G_{23} (GPa)	3.38
Modulus ratio E_1/E_2	12.3
Axial tensile strength X_T (MPa)	1513
Transverse tensile strength Y_T (MPa)	43.4
Strength ratio X_T/Y_T	35

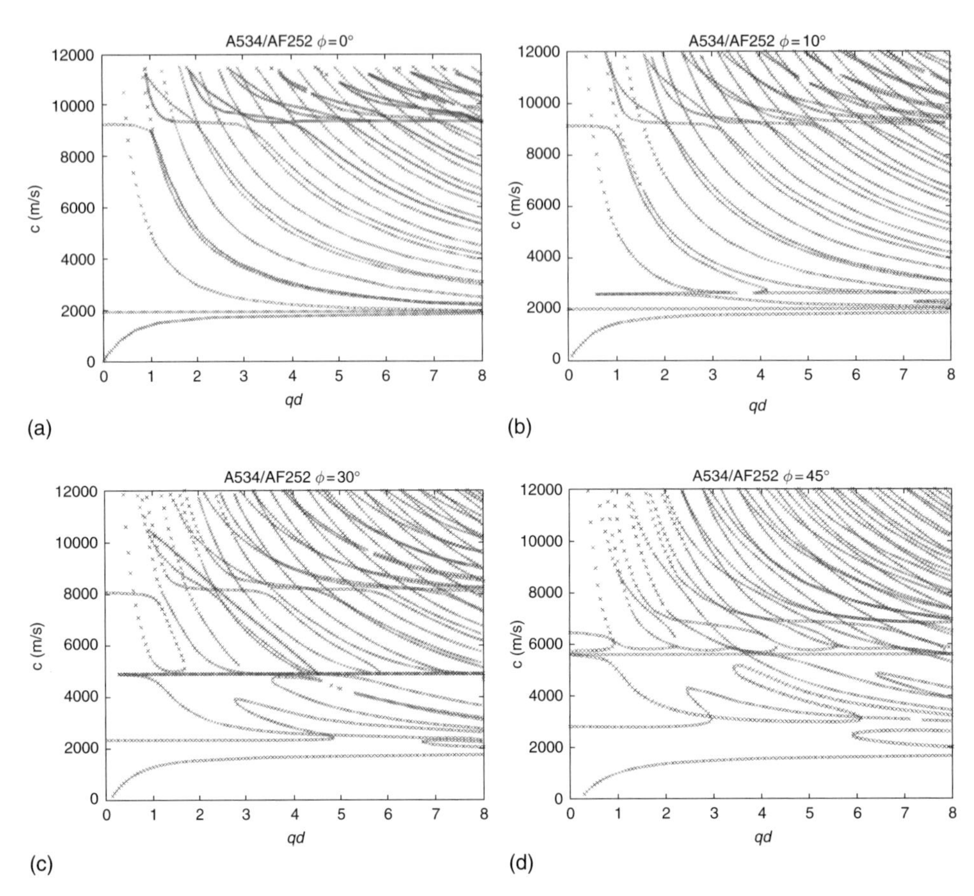

FIGURE 6.25 Dispersion curves for a layered plate A534/F252 of unidirectional fibers for different wave propagation orientations: (a) $\phi = 0°$; (b) $\phi = 10°$; (c) $\phi = 30°$; (d) $\phi = 45°$.

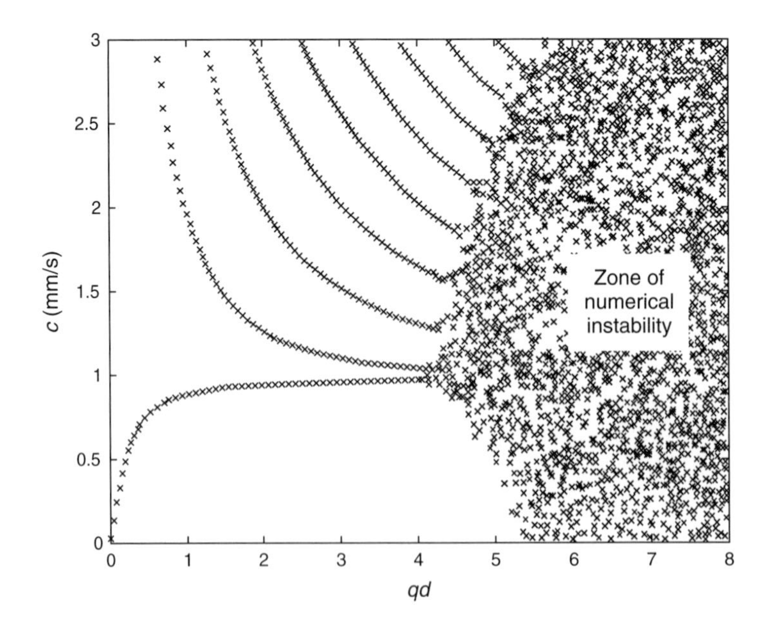

FIGURE 6.26 Transfer matrix instability for high values of the frequency–thickness product.

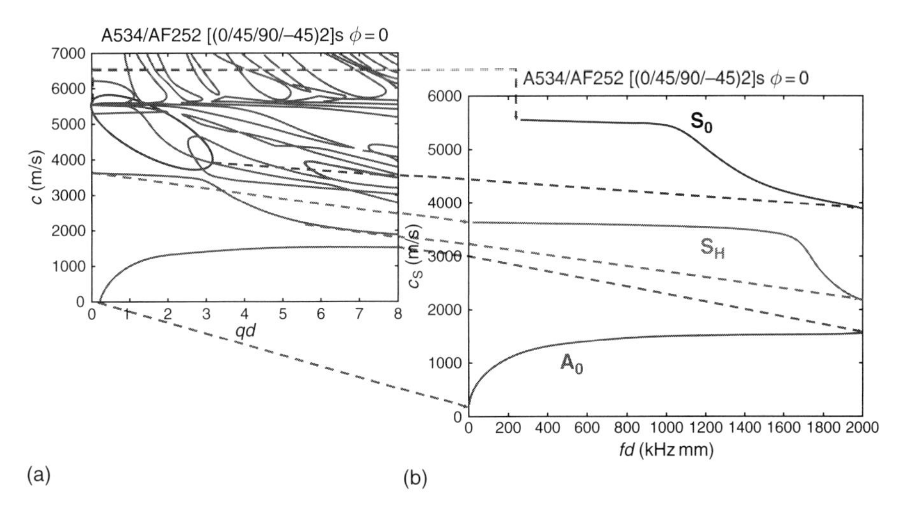

FIGURE 6.27 Dispersion curves for a quasi-isotropic plate $[(0/45/90/-45)_{2s}]$: (a) output from the program; (b) zoom in on the S_0, SH, A_0 modes.

6.8 SUMMARY AND CONCLUSIONS

This chapter has dealt with an important class of waves that have widespread applications in SHM – the class of *guided waves*. Guided waves are especially important for SHM because they can travel at large distances in structures with only little energy loss. Thus, they enable the SHM of large areas from a single location. Guided waves have the important property that they remain confined inside the walls of a thin-wall structure, and hence can travel over large distances. In addition, guided waves can also travel inside curved walls. These properties make them well suited for the ultrasonic inspection of aircraft, missiles, pressure vessel, oil tanks, pipelines, etc.

The chapter started with the discussion of *Rayleigh waves*, a.k.a., SAW. Rayleigh waves are found in solids that contain a free surface. The Rayleigh waves travel close to the free surface with very little penetration in the depth of the solid. For this reason, Rayleigh waves are also known as *surface-guided waves*.

The chapter continued with a description of guided waves in plates. In flat plates, ultrasonic guided waves travel as Lamb waves and as SH waves. Lamb waves are vertically polarized, whereas SH waves are horizontally polarized.

The discussion started with the simpler form of guided plate waves, i.e., with the SH waves. The particle motion of SH waves is polarized parallel to the plate surface and perpendicular to the direction of wave propagation. The SH waves can be symmetric and antisymmetric. With the exception of the very fundamental mode, the SH wave modes are all dispersive.

The discussion continued with the *Lamb waves*, which are more complicated guided plate waves. Lamb waves are of two basic varieties, *symmetric Lamb waves modes* $(S_0, S_1, S_2, \ldots)$ and *antisymmetric Lamb waves modes* $(A_0, A_1, A_2, \ldots)$. Both Lamb-wave types are quite dispersive. At any given value of the frequency–thickness product *fd*, a multitude of symmetric and antisymmetric Lamb waves may exist. The higher the *fd* value, the larger the number of Lamb-wave modes that can simultaneously exist. For relatively small values of the *fd* product, only the basic symmetric and antisymmetric Lamb-wave modes (S_0 and A_0) exist. As the *fd* product approaches zero, the S_0 and A_0 modes degenerate in the basic axial and flexural plate modes discussed in

Chapter 5. At the other extreme, as $fd \to \infty$, the S_0 and A_0 Lamb-wave modes degenerate into Rayleigh waves confined to the plate surface.

Guided waves also exist in other thin-wall structures, such as rods, tubes, and shells. Though the underlying physical principles of guided-wave propagation still apply, their study is more elaborate. For illustration, this chapter has covered a brief description of guided waves in cylindrical shells. It was shown that dispersive longitudinal (axial), flexural, and torsional multi-mode guided waves can simultaneously exist in such structures.

The last part of this chapter was devoted to the discussion of guided waves in composite materials. This analysis is rather elaborate, as wave reflection and transmission at each layer interface has to be considered across the composite thickness. However, a solution that uses the transfer matrix method was illustrated.

Because of guided-waves' multi-modal behavior, their use in SHM is quite complicated. Thus, their capability of traveling at far distances from a single location, which is beneficial for SHM applications, is somehow countered by the necessity of specialized excitation, more elaborate signal processing, and complex interpretation to compensate for their multi-modal and dispersive character. However, these aspects will be covered in later chapters.

6.9 PROBLEMS AND EXERCISES

1. Explain with sketches how the S_0 and A_0 Lamb waves may be similar or different from the axial and flexural waves in plates.
2. The wave speed (phase velocity) for flexural waves in a plate is given by $c_F = a\sqrt{\omega}$ where $a = [Ed^2/3\rho(1-\nu^2)]^{1/4}$ – Show that the group velocity is twice the wave speed, i.e., $c_g = 2c_F$. Calculate the energy velocity.
3. The guided waves in a plate can be Lamb waves and SH waves. Each of these wave types can be multimodal. Which wave mode is always nondispersive?

7

PIEZOELECTRIC WAFER ACTIVE SENSORS

7.1 INTRODUCTION

Piezoelectric wafer active sensors (PWAS) are inexpensive transducers that operate on the piezoelectric principle. Initially, PWAS were used for vibrations control, as pioneered by Crawley et al. (1987, 1990) and Fuller et al. (1990). Tzou and Tseng (1990) and Lester and Lefebvre (1993) modeled the piezoelectric sensor/actuator design for dynamic measurement/control. For damage detection, Banks et al. (1996) used PWAS to excite a structure and then sense the free decay response. The use of PWAS for structural health monitoring has followed three main paths: (a) modal analysis and transfer function; (b) electromechanical E/M impedance; (c) wave propagation. The use of PWAS for damage detection with Lamb-wave propagation was pioneered by Chang and his coworkers (Chang, 1995, 1998, 2001; Wang and Chang, 1999; Ihn and Chang, 2002). They have studied the use of PWAS for generation and reception of elastic waves in composite materials. Passive reception of elastic waves was used for impact detection. Pitch-catch transmission–reception of low-frequency Lamb waves was used for damage detection. PWAS wave propagation was also studied by Lakshmanan and Pines (1997), Culshaw et al. (1998), Lin and Yuan (2001), Dupont et al. (2000), Osmont et al. (2000), Diamanti, Hodgkinson, and Soutis (2002). The use of PWAS for high-frequency local modal sensing with the E/M impedance method was pursued by Liang et al. (1994), Sun et al. (1994), Chaudhry et al. (1994), Ayres et al. (1996), Park et al. (2001), Giurgiutiu et al. (1997–2002), and others.

PWAS couple the electrical and mechanical effects (mechanical strain S_{ij}, mechanical stress T_{kl}, electrical field E_k, and electrical displacement D_j) through the tensorial piezoelectric constitutive equations

$$
\begin{aligned}
S_{ij} &= s^E_{ijkl} T_{kl} + d_{kij} E_k, \\
D_j &= d_{jkl} T_{kl} + \varepsilon^T_{jk} E_k
\end{aligned}
\tag{1}
$$

where s^E_{ijkl} is the mechanical compliance of the material measured at zero electric field ($E = 0$), ε^T_{jk} is the dielectric permittivity measured at zero mechanical stress ($T = 0$), and d_{kij} represents the piezoelectric coupling effect. As apparent in Fig. 7.1, PWAS are small and unobtrusive. PWAS utilize the d_{31} coupling between in-plane strain and transverse

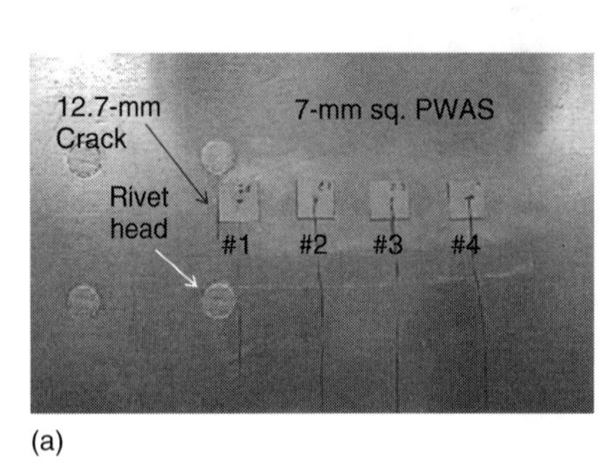

(a)

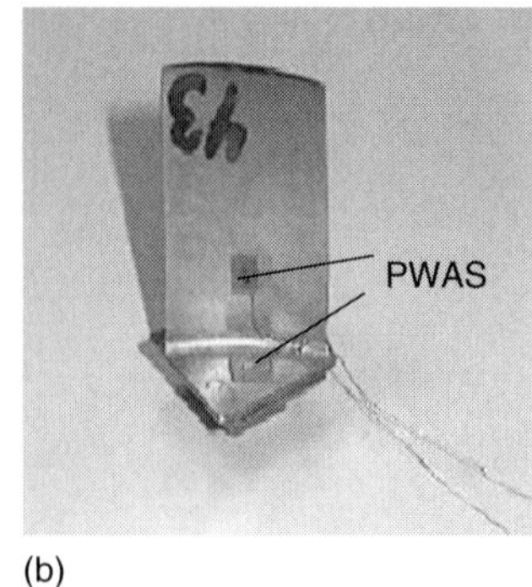

(b)

FIGURE 7.1 Piezoelectric wafer active sensors mounted on various structures: (a) array of 7-mm square PWAS on an aircraft panel; (b) PWAS on a turbine blade.

electric field. A 7-mm diameter PWAS, 0.2-mm thin, weighs a bare 78g. At less than \$10 each, PWAS are no more expensive than conventional high-quality resistance strain gages. However, the PWAS performance exceeds by far that of conventional resistance strain gages because PWAS are active devices that can interrogate the structure at will, whereas strain gages are passive devices that can only listen to the structure. Besides, PWAS can address high-frequency applications at hundreds of kHz and beyond. There are several ways in which PWAS can be used, as shown next.

As *high-bandwidth strain sensors*, PWAS convert directly mechanical energy to electrical energy. The conversion constant is linearly dependent on the signal frequency. In the kHz range, signals of the order of hundreds of mV are easily obtained. No conditioning amplifiers are needed; the PWAS can be directly connected to a high-impedance measuring instrument, such as a digitizing oscilloscope.

As *high-bandwidth strain exciters*, PWAS convert directly the electrical energy into mechanical energy. Thus, it can easily induce vibrations and waves in the substrate material. PWAS act very well as an embedded generator of waves and vibration. High-frequency waves and vibrations are easily excited with input signals as low as 10 V. This dual sensing and excitation characteristics of PWAS justifies their name of 'active sensors'.

As *resonators*, PWAS can perform resonant mechanical vibration under direct electrical excitation. Thus, very precise frequency standards can be created with a simple setup consisting of the PWAS and the signal generator. The resonant frequencies depend only on the wave speed (which is a material constant) and the geometric dimensions. Precise frequency values can be obtained through precise machining of the PWAS geometry.

As *embedded modal sensors*, PWAS can directly measure the high-frequency modal spectrum of a support structure. This is achieved with the E/M impedance method, which reflects the mechanical impedance of the support structure into the real part of the E/M impedance measured at PWAS terminals. The high-frequency characteristics of this method, which has been proven to operate at hundreds of kHz and beyond, cannot be achieved with conventional modal testing techniques. Thus, PWAS are the sensors of choice for high-frequency modal measurement and analysis.

The above PWAS functionalities will be described in several chapters from now onwards. We will start with the basic resonator equations and progress through the other functionalities such as Lamb-wave transmitters and receivers (i.e., embedded ultrasonic transducers) and high-frequency modal sensors. The theoretical developments are performed in a step-by-step manner, with full presentation of the intermediate steps.

This allows the readers the opportunity to duplicate the theoretical developments and extend them to fit their particular applications. To substantiate the theoretical developments and verify their predictions, several experimental tests are presented. Full description of the experimental setup and the performed measurements are provided. Again, the intention is to give readers the opportunity to duplicate the findings and to extend them to suit their particular application needs.

7.2 PWAS RESONATORS

This section addresses the behavior of a free PWAS. When excited by an alternating electric voltage, the free PWAS acts as an electromechanical resonator. The modeling of a free piezoelectric sensor is useful for (a) understanding the electromechanical coupling between the mechanical vibration response and the complex electrical response of the sensor; and (b) sensor screening and quality control prior to installation on the monitored structure.

Consider a piezoelectric wafer of length l, width b, and thickness t, undergoing piezoelectric expansion induced by the thickness polarization electric field, E_3 (Fig. 7.2). The electric field is produced by the application of a harmonic voltage $V(t) = \hat{V}e^{i\omega t}$ between the top and bottom surface electrodes. The resulting electric field, $E_3 = V/t$, is assumed uniform over the piezoelectric wafer.

Assume that the length, width, and thickness have widely different values ($t \ll b \ll l$) such that the length, width, and thickness motions are practically uncoupled. The motion, u_1, in the longitudinal direction, x_1, will be considered to be predominant (1-D assumption). Because the electric field is uniform over the piezoelectric wafer, its derivative is zero, i.e., $\partial E_3/\partial x_1 = 0$. Because the voltage excitation is harmonic, the electric field has the expression $E_3 = \hat{E}_3 e^{i\omega t}$, and the response is also harmonic, i.e., $u = \hat{u}e^{i\omega t}$, where $\hat{u}_1(x)$ is the x-dependent complex amplitude that incorporates any phase difference between the excitation and response. For compactness, we use the notations

$$\frac{\partial}{\partial x_1}(\ ) = (\ )' \quad \text{and} \quad \frac{\partial}{\partial t}(\ ) = (\dot{\ })$$

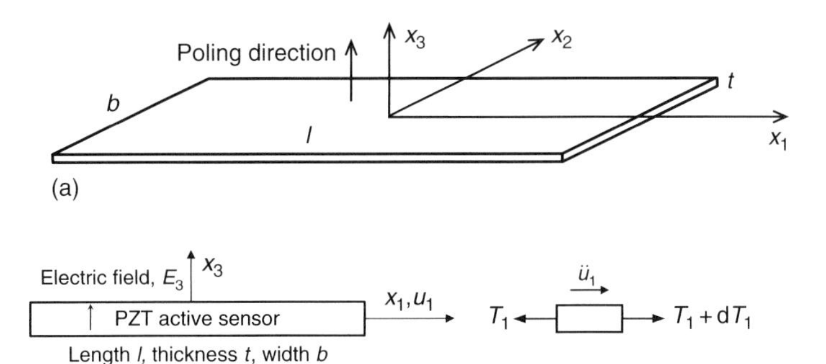

FIGURE 7.2 Schematic of a piezoelectric active sensor and infinitesimal axial element.

TABLE 7.1 Piezoelectric wafer properties (APC-850)

Property	Symbol	Value
Compliance, in plane	s_{11}^E	$15.30 \cdot 10^{-12}\,\text{Pa}^{-1}$
Compliance, thickness wise	s_{33}^E	$17.30 \cdot 10^{-12}\,\text{Pa}^{-1}$
Dielectric constant	ε_{33}^T	$\varepsilon_{33}^T = 1750\varepsilon_0$
Thickness wise induced-strain coefficient	d_{33}	$400 \cdot 10^{-12}\,\text{m/V}$
In-plane induced-strain coefficient	d_{31}	$-175 \cdot 10^{-12}\,\text{m/V}$
Coupling factor, parallel to electric field	k_{33}	0.72
Coupling factor, transverse to electric field	k_{31}	0.36
Coupling factor, transverse to electric field, polar motion	k_p	0.63
Poisson ratio	ν	0.35
Density	ρ	$7700\,\text{kg/m}^3$
Sound speed	c	2900 m/s

Note: $\varepsilon_0 = 8.85 \cdot 10^{-12}\,\text{F/m}$.

Under the 1-D assumption, the general constitutive equations reduce to the simpler expressions

$$S_1 = s_{11}^E T_1 + d_{31} E_3 \quad \text{(strain)} \tag{1}$$

$$D_3 = d_{31} T_1 + \varepsilon_{33}^T E_3 \quad \text{(electric displacement)} \tag{2}$$

where S_1 is the strain, T_1 is the stress, D_3 is the electrical displacement (charge per unit area), s_{11}^E is the mechanical compliance at zero field, ε_{33}^T is the dielectric constant at zero stress, and d_{31} is the induced strain coefficient, i.e., mechanical strain per unit electric field. Typical values of these constants are given in Table 7.1.

Recall Newton law of motion and the strain-displacement relation

$$T_1' = \rho \ddot{u}_1 \quad \text{(Newton law of motion)} \tag{3}$$

$$S_1 = u_1' \quad \text{(strain-displacement relation)} \tag{4}$$

Differentiate Eq. (1) with respect to x; because E_3 is constant (i.e., $E_3' = 0$), the strain rate becomes

$$S_1' = s_{11}^E \cdot T_1' \tag{5}$$

Substitution of Eqs. (3) and (4) into Eq. (5) gives

$$u_1'' = s_{11}^E \rho \ddot{u}_1 \tag{6}$$

Introduce the longitudinal wave speed of the material

$$c^2 = \frac{1}{\rho s_{11}^E} \quad \text{(wave speed)} \tag{7}$$

Substitution of Eq. (7) into Eq. (6) yields the 1-D wave equation:

$$\ddot{u}_1 = c^2 u_1'' \tag{8}$$

The general solution of the wave equation is

$$u_1(x, t) = \hat{u}(x) e^{i\omega t} \tag{9}$$

where

$$\hat{u}_1(x) = C_1 \sin \gamma x + C_2 \cos \gamma x \tag{10}$$

and

$$\gamma = \frac{\omega}{c} \tag{11}$$

The quantity γ is the *wavenumber*. An associated quantity is the *wavelength*, $\lambda = cT = c/f$, where $f = \omega/2\pi$. The constants C_1 and C_2 are determined from the boundary conditions.

7.2.1 MECHANICAL RESPONSE

For a piezoelectric wafer of length l with the origin at its center, the interval of interest is $x \in \left(-\frac{1}{2}l, \frac{1}{2}l\right)$. Stress-free boundary conditions imply $T_1\left(-\frac{1}{2}l\right) = T_1\left(+\frac{1}{2}l\right) = 0$. Substitution into Eq. (1) yields

$$S_1\left(-\frac{1}{2}l\right) = d_{31}E_3 \tag{12}$$

$$S_1\left(+\frac{1}{2}l\right) = d_{31}E_3 \tag{13}$$

Using the strain-displacement relation of Eq. (4) gives

$$\hat{u}_1'\left(-\frac{1}{2}l\right) = \gamma\left(C_1 \cos \gamma\frac{1}{2}l + C_2 \sin \gamma\frac{1}{2}l\right) = d_{31}\hat{E}_3 \tag{14}$$

$$\hat{u}_1'\left(\frac{1}{2}l\right) = \gamma\left(C_1 \cos \gamma\frac{1}{2}l - C_2 \sin \gamma\frac{1}{2}l\right) = d_{31}\hat{E}_3 \tag{15}$$

Addition of the two equations yields an equation in only C_1, i.e.,

$$\gamma C_1 \cos \gamma\frac{1}{2}l = d_{31}\hat{E}_3 \tag{16}$$

Hence,

$$C_1 = \frac{d_{31}\hat{E}_3}{\gamma \cos \frac{1}{2}\gamma l} \tag{17}$$

Subtraction of the two equations yields an equation in only C_2, i.e.,

$$\gamma C_2 \sin \gamma\frac{1}{2}l = 0 \tag{18}$$

Assuming $\sin \frac{1}{2}\gamma l \neq 0$, we get

$$C_2 = 0 \tag{19}$$

An alternative derivation of C_1 and C_2 can be obtained using Cramer's rule on the algebraic system

$$\begin{cases} C_1 \gamma \cos \frac{1}{2} \gamma l + C_2 \gamma \sin \frac{1}{2} \gamma l = d_{31} \hat{E}_3 \\ C_1 \gamma \cos \frac{1}{2} \gamma l - C_2 \gamma \sin \frac{1}{2} \gamma l = d_{31} \hat{E}_3 \end{cases} \tag{20}$$

Hence,

$$C_1 = \frac{\begin{vmatrix} d_{31}\hat{E}_3 & \gamma \sin \frac{1}{2}\gamma l \\ d_{31}\hat{E}_3 & -\gamma \sin \frac{1}{2}\gamma l \end{vmatrix}}{\begin{vmatrix} \gamma \cos \frac{1}{2}\gamma l & \gamma \sin \frac{1}{2}\gamma l \\ \gamma \cos \frac{1}{2}\gamma l & -\gamma \sin \frac{1}{2}\gamma l \end{vmatrix}} = \frac{d_{31}\hat{E}_3(-2\gamma \sin \frac{1}{2}\gamma l)}{-\gamma^2 2 \sin \frac{1}{2}\gamma l \cos \frac{1}{2}\gamma l} = \frac{d_{31}\hat{E}_3}{\gamma \cos \frac{1}{2}\gamma l} \tag{21}$$

and

$$C_2 = \frac{\begin{vmatrix} \gamma \cos \frac{1}{2}\gamma l & d_{31}\hat{E}_3 \\ \gamma \cos \frac{1}{2}\gamma l & d_{31}\hat{E}_3 \end{vmatrix}}{\begin{vmatrix} \gamma \cos \frac{1}{2}\gamma l & \gamma \sin \frac{1}{2}\gamma l \\ \gamma \cos \frac{1}{2}\gamma l & -\gamma \sin \frac{1}{2}\gamma l \end{vmatrix}} = \frac{0}{-\gamma^2 2 \sin \frac{1}{2}\gamma l \cos \frac{1}{2}\gamma l} = 0 \tag{22}$$

In this derivation, we have assumed that the determinant appearing at the denominator is nonzero, i.e., $\Delta \neq 0$. Because $\Delta = -\gamma^2 2 \sin \frac{1}{2}\gamma l \cos \frac{1}{2}\gamma l = -\gamma^2 \sin \gamma l$, this condition implies that

$$\sin \gamma l \neq 0 \tag{23}$$

Substitution of C_1 and C_2 into Eq. (10) yields

$$\hat{u}_1(x) = \frac{d_{31}\hat{E}_3}{\gamma} \cdot \frac{\sin \gamma x}{\cos \frac{1}{2}\gamma l} \tag{24}$$

Using Eq. (4), we get the strain as

$$\hat{S}_1(x) = d_{31}\hat{E}_3 \frac{\cos \gamma x}{\cos \frac{1}{2}\gamma l} \tag{25}$$

7.2.1.1 Solution in Terms of the Induced Strain, S_{ISA}, and Induced Displacement, u_{ISA}

Introduce the notations

$$S_{ISA} = d_{31}\hat{E}_3 \quad \text{(induced strain)} \tag{26}$$

$$u_{ISA} = S_{ISA} l = (d_{31}\hat{E}_3) l \quad \text{(induced displacement)} \tag{27}$$

where ISA signifies 'induced strain actuation'. Hence, Eqs. (24)–(25) can be written as

$$\hat{u}(x) = \frac{1}{2} \frac{u_{ISA}}{\frac{1}{2}\gamma l} \frac{\sin \gamma x}{\cos \frac{1}{2}\gamma l} \tag{28}$$

$$\hat{S}_1(x) = S_{ISA} \frac{\cos \gamma x}{\cos \frac{1}{2}\gamma l} \tag{29}$$

Equations (28)–(29) shows that, under dynamic conditions, the strain along the piezoelectric wafer is not constant. This is in contrast to the static case, for which the strain is uniform along the length of the actuator. The maximum strain amplitude is observed in the middle and its value is

$$S_{\max} = \frac{S_{\mathrm{ISA}}}{\cos \frac{1}{2}\gamma l} \tag{30}$$

Introducing the notation $\phi = \frac{1}{2}\gamma l$, the displacement and strain equations can be rewritten as

$$\hat{u}(x) = \frac{1}{2}\frac{u_{\mathrm{ISA}}}{\phi}\frac{\sin \gamma x}{\cos \phi} \tag{31}$$

$$\hat{S}_1(x) = S_{\mathrm{ISA}}\frac{\cos \gamma x}{\cos \phi} \tag{32}$$

7.2.1.2 Tip Strain and Displacement

At the wafer tips, $x = \pm\frac{1}{2}l$, the tip strain and displacement are

$$\hat{u}\left(\pm\frac{1}{2}l\right) = \pm\frac{1}{2}\frac{u_{\mathrm{ISA}}}{\frac{1}{2}\gamma l}\frac{\sin\frac{1}{2}\gamma l}{\cos\frac{1}{2}\gamma l} = \pm\frac{1}{2}\frac{u_{\mathrm{ISA}}}{\frac{1}{2}\gamma l}\tan\frac{1}{2}\gamma l \tag{33}$$

$$\hat{S}_1\left(\pm\frac{1}{2}\gamma l\right) = S_{\mathrm{ISA}}\frac{\cos\frac{1}{2}\gamma l}{\cos\frac{1}{2}\gamma l} = S_{\mathrm{ISA}} \tag{34}$$

We notice that, at the tip, the strain takes exactly the value S_{ISA}.

7.2.2 ELECTRICAL RESPONSE

Consider a 1-D PWAS under electric excitation (Fig. 7.3). Recall Eq. (2) representing the electrical displacement

$$D_3 = d_{31}T_1 + \varepsilon_{33}^T E_3 \tag{35}$$

Equation (1) yields the stress as function of strain and electric field, i.e.,

$$T_1 = \frac{1}{s_{11}}(S_1 - d_{31}E_3) \tag{36}$$

Hence, the electric displacement can be expressed as

$$D_3 = \frac{d_{31}}{s_{11}}(S_1 - d_{31}E_3) + \varepsilon_{33}^T E_3 \tag{37}$$

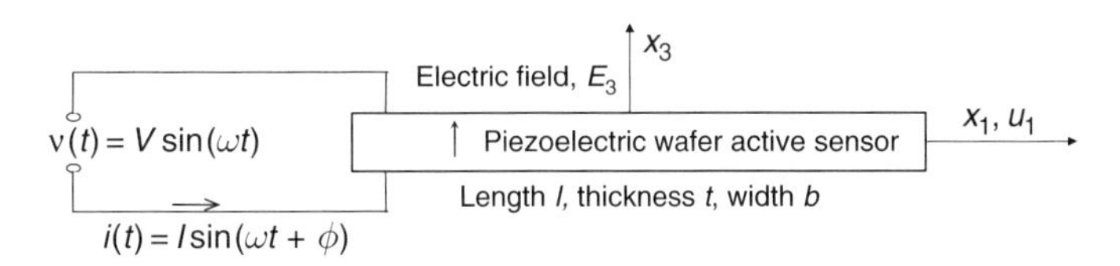

FIGURE 7.3 Schematic of a 1-D PWAS under electric excitation.

Upon substitution of the strain-displacement relation of Eq. (4), we get

$$D_3 = \frac{d_{31}}{s_{11}} \cdot u_1' - \frac{d_{31}^2}{s_{11}} \cdot E_3 + \varepsilon_{33}^T E_3 \tag{38}$$

i.e.,

$$D_3 = \varepsilon_{33}^T E_3 \left[1 - k_{31}^2 \left(1 - \frac{u_1'}{d_{31} E_3} \right) \right] \tag{39}$$

where $k_{31}^2 = d_{31}^2 / (s_{11} \varepsilon_{33})$ is the electromechanical coupling coefficient. Integration of Eq. (39) over the electrodes area $A = bl$ yields the total charge

$$Q = \int_A D_3 \, \mathrm{d}x_1 \, \mathrm{d}x_2 = \int_{-l/2}^{l/2} \int_0^b D_3 \, \mathrm{d}x_1 \, \mathrm{d}x_2 = \varepsilon_{33}^T E_3 b \left[l - k_{31}^2 \left(l - \frac{1}{d_{31} E_3} \cdot u_1 \Big|_{-\frac{1}{2}l}^{\frac{1}{2}l} \right) \right] \tag{40}$$

Assuming harmonic time dependence, $Q = \hat{Q} e^{i\omega t}$, yields $\hat{Q}$ in the form

$$\hat{Q} = \varepsilon_{33}^T \hat{E}_3 b \left[l - k_{31}^2 \left(l - \frac{1}{d_{31} \hat{E}_3} \cdot \left[\hat{u}_1 \left(\frac{1}{2} l \right) - \hat{u}_1 \left(-\frac{1}{2} l \right) \right] \right) \right] \tag{41}$$

Rearranging,

$$\hat{Q} = \varepsilon_{33}^T bl \left[1 - k_{31}^2 + k_{31}^2 \left(\frac{\hat{u}_1(\frac{1}{2}l) - \hat{u}_1(-\frac{1}{2}l)}{d_{31} \hat{E}_3 \cdot l} \right) \right] \tag{42}$$

Using the definitions $C = \varepsilon_{33}^T (A/h) \, A = bl$, $u_{\mathrm{ISA}} = d_{31} \hat{E}_3 \cdot l$, $\hat{V} = \hat{E}_3 / h$, we obtain the total charge as

$$\hat{Q} = C\hat{V} \left[1 - k_{31}^2 + k_{31}^2 \left(\frac{\hat{u}_1(\frac{1}{2}l) - \hat{u}_1(-\frac{1}{2}l)}{u_{\mathrm{ISA}}} \right) \right] \tag{43}$$

The electric current is obtained as the time derivative of the electric charge, i.e.,

$$I = \dot{Q} = i\omega Q \tag{44}$$

Hence,

$$\hat{I} = i\omega C\hat{V} \left[1 - k_{31}^2 + k_{31}^2 \left(\frac{\hat{u}_1(\frac{1}{2}l) - \hat{u}_1(-\frac{1}{2}l)}{u_{\mathrm{ISA}}} \right) \right] \tag{45}$$

Alternatively, the electric current could have been obtained by performing first the derivative with respect to time, and then the integration with respect to space, i.e.,

$$I = \int_A \frac{\mathrm{d}D_3}{\mathrm{d}t} \mathrm{d}A = i\omega \int_A D_3 \, \mathrm{d}A \tag{46}$$

The integral in Eq. (46) is the same as the integral in Eq. (40).

The admittance, Y, is defined as the ratio between the current and voltage, i.e.,

$$Y = \frac{\hat{I}}{\hat{V}} = i\omega C\left[1 - k_{31}^2 + k_{31}^2\left(\frac{\hat{u}_1(\frac{1}{2}l) - \hat{u}_1(-\frac{1}{2}l)}{u_{\mathrm{ISA}}}\right)\right] \tag{47}$$

Recall Eq. (28) giving the mechanical displacement solution

$$\hat{u}_1(x) = \frac{d_{31}\hat{E}_3}{\gamma} \cdot \frac{\sin\gamma x}{\cos\frac{1}{2}\gamma l}$$

Hence,

$$\frac{\hat{u}_1(\frac{1}{2}l) - \hat{u}_1(-\frac{1}{2}l)}{u_{\mathrm{ISA}}} = \frac{1}{2}\frac{\sin\frac{1}{2}\gamma l - (-\sin\frac{1}{2}\gamma l)}{\frac{1}{2}\gamma l\cos\frac{1}{2}\gamma l} = \frac{1}{2}2\frac{\sin\frac{1}{2}\gamma l}{\frac{1}{2}\gamma l\cos\frac{1}{2}\gamma l} = \frac{\tan\frac{1}{2}\gamma l}{\frac{1}{2}\gamma l} \tag{48}$$

Substitution in Eq. (47) yields

$$Y = i\omega C\left[1 - k_{31}^2\left(1 - \frac{\tan\frac{1}{2}\gamma\ell}{\frac{1}{2}\gamma\ell}\right)\right] \tag{49}$$

This result agrees with Ikeda (1996). However, it may be more convenient, at times, to write it as

$$Y = i\omega C\left[1 - k_{31}^2\left(1 - \frac{1}{\frac{1}{2}\gamma l\cdot\cot\frac{1}{2}\gamma l}\right)\right] \tag{50}$$

Note that the admittance is purely imaginary and consists of the capacitive admittance, $i\omega C$, modified by the effect of piezoelectric coupling between mechanical and electrical fields. This effect is apparent in the term containing the electromechanical coupling coefficient, k_{31}^2. The impedance, Z, is obtained as the ratio between the voltage and current, i.e.,

$$Z = \frac{\hat{V}}{\hat{I}} = Y^{-1} \tag{51}$$

Hence,

$$Z = \frac{1}{i\omega C}\left[1 - k_{31}^2\left(1 - \frac{\tan\frac{1}{2}\gamma\ell}{\frac{1}{2}\gamma\ell}\right)\right]^{-1} \tag{52}$$

Introducing the notation $\phi = \frac{1}{2}\gamma l$, we can write

$$Y = i\omega C\left[1 - k_{31}^2\left(1 - \frac{\tan\phi}{\phi}\right)\right] \tag{53}$$

$$Z = \frac{1}{i\omega C}\left[1 - k_{31}^2\left(1 - \frac{\tan\phi}{\phi}\right)\right]^{-1} \tag{54}$$

7.2.3 RESONANCES

In previous derivations, we made the assumption that $\sin \gamma l \neq 0$. This assumption implies that resonances do not happen. If resonances happened, they could be of two types:

1. Electromechanical resonances
2. Mechanical resonances.

Mechanical resonances take place in the same conditions as in a conventional elastic bar. They happen under mechanical excitation, which produces a mechanical response in the form of mechanical vibration. Electromechanical resonances are specific to piezoelectric materials. They reflect the coupling between the mechanical and electrical fields. Electromechanical resonances happen under electric excitation, which produces an electromechanical response, i.e., both a mechanical vibration and a change in the electric admittance and impedance. We will consider these two situations separately.

7.2.3.1 Mechanical Resonances

If a PWAS is excited mechanically with a frequency sweep, certain frequencies will appear at which the response is very large, i.e., the PWAS resonates. To study the mechanical resonances, assume that the material is not piezoelectric, i.e., $d_{31} = 0$. Hence, we develop the results for the classical mechanical resonances of an elastic bar. The stress-free boundary conditions become

$$\begin{cases} C_1 \gamma \cos \frac{1}{2}\gamma l + C_2 \gamma \sin \frac{1}{2}\gamma l = 0 \\ C_1 \gamma \cos \frac{1}{2}\gamma l - C_2 \gamma \sin \frac{1}{2}\gamma l = 0 \end{cases} \tag{55}$$

This homogenous system of equations accepts nontrivial solutions when the determinant of the coefficients is zero, i.e.,

$$\Delta = \begin{vmatrix} \gamma \cos \frac{1}{2}\gamma l & \gamma \sin \frac{1}{2}\gamma l \\ \gamma \cos \frac{1}{2}\gamma l & -\gamma \sin \frac{1}{2}\gamma l \end{vmatrix} = 0 \tag{56}$$

Upon expansion,

$$\Delta = -2 \sin \frac{1}{2}kl \cos \frac{1}{2}kl = 0 \tag{57}$$

Equation (57) yields the *characteristic equation*

$$\sin \frac{1}{2}\gamma l \cos \frac{1}{2}\gamma l = 0 \tag{58}$$

Recall the notation $\phi = \frac{1}{2}\gamma l$. Hence, the characteristic equation can be written as

$$\sin \phi \cos \phi = 0 \tag{59}$$

This equation admits two sets of solutions, one corresponding to $\cos \phi = 0$ and the other corresponding to $\sin \phi = 0$. For reasons that will become apparent, the first condition leads to *antisymmetric resonances*, the second to *symmetric resonances*.

Antisymmetric resonances ($\cos\phi = 0$)

Antisymmetric resonances happen when $\cos\phi = 0$. This happens when ϕ is an odd multiple of $\pi/2$, i.e.,

$$\cos\phi = 0 \rightarrow \phi = (2n-1)\frac{\pi}{2} \quad (60)$$

This means that, at antisymmetric resonances, the variable ϕ can take one of the following values:

$$\phi^A = \frac{\pi}{2}, \frac{3\pi}{2}, \frac{5\pi}{2}, \ldots \quad (61)$$

These values are the *antisymmetric eigenvalues* of the system. The superscript A denotes antisymmetric resonances. Recall the notations $\phi = \frac{1}{2}\gamma l$, $\gamma = \omega/c$, and $\omega = 2\pi f$. Hence,

$$\gamma l = \frac{\omega l}{c} = \frac{2\pi f l}{c} = (2n-1)\pi \quad (62)$$

Therefore, the *antisymmetric resonance frequencies* are given by the formula

$$f_n^A = (2n-1)\frac{c}{2l} \quad (63)$$

For each resonance frequency, Eq. (55) accepts nontrivial values for the coefficients C_1 and C_2. However, they can only be determined up to a scaling factor. Using the notation $\phi = \frac{1}{2}\gamma l$, we write Eq. (55) as

$$\begin{cases} C_1\cos\phi + C_2\sin\phi = 0 \\ C_1\cos\phi - C_2\sin\phi = 0 \end{cases} \quad (64)$$

Examination of Eq. (64) reveals that, for $\cos\phi = 0$, the constant C_2 vanishes, whereas the constant C_1 is indeterminate. We remove the indeterminacy by choosing $C_1 = 1$. Substituting $C_1 = 1$ and $C_2 = 0$ into Eq. (10) yields the response at antisymmetric resonances, i.e., the *antisymmetric mode shapes*

$$U_n^A = \sin\gamma_n x \quad (65)$$

Using Eq. (62), we write the mode shapes in the convenient form

$$U_n^A = \sin(2n-1)\pi\frac{x}{l} \quad (66)$$

For a graphical representation of these mode shapes, see the modes $A_1, A_2, A_3, \ldots$ in Table 7.2.

Another useful interpretation of these results is through the wavelength λ. Recall that the wavelength is defined as the distance traveled by the wave in a time period, i.e., $\lambda = cT$. Because the time period is the inverse of frequency, the wavelength is $\lambda = c/f$. Hence, Eq. (62) gives

$$l_n^A = (2n-1)\frac{\lambda}{2} \quad (67)$$

Equation (67) indicates that antisymmetric resonances happen when the PWAS length is an odd multiple of the half wavelength of the elastic waves traveling in the PWAS. This relation is used in constructing PWAS that have to resonate at a given frequency.

TABLE 7.2 1-D resonance mode shapes of piezoelectric elastic bar of length l

Mode	Eigenvalue	Resonant frequency	Mode shape	Half wavelength multiplicity	Note
A_1	$\phi_1^A = \frac{\pi}{2}$	$f_1^A = \frac{c}{2l}$	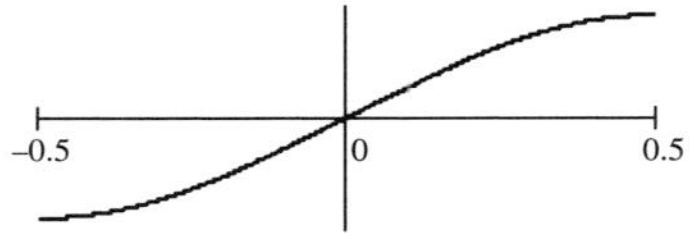 $U_1^A = \sin \pi \frac{x}{l}$	$l_1^A = \frac{\lambda}{2}$	EM
S_1	$\phi_1^S = 2\frac{\pi}{2}$	$f_1^S = 2\frac{c}{2l}$	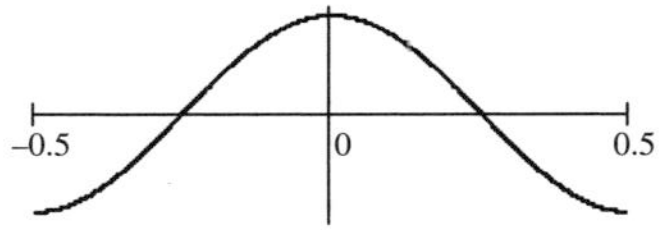 $U_1^S = \cos 2\pi \frac{x}{l}$	$l_1^S = 2\frac{\lambda}{2}$	
A_2	$\phi_2^A = 3\frac{\pi}{2}$	$f_2^A = 3\frac{c}{2l}$	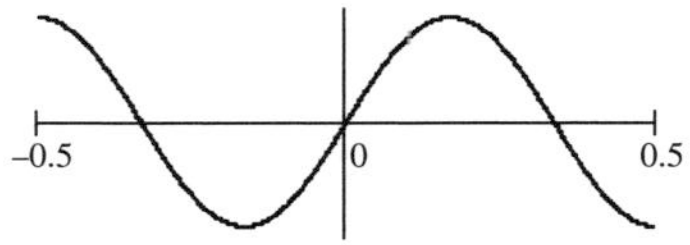 $U_2^A = \sin 3\pi \frac{x}{l}$	$l_2^A = 3\frac{\lambda}{2}$	EM
S_2	$\phi_2^S = 4\frac{\pi}{2}$	$f_2^S = 4\frac{c}{2l}$	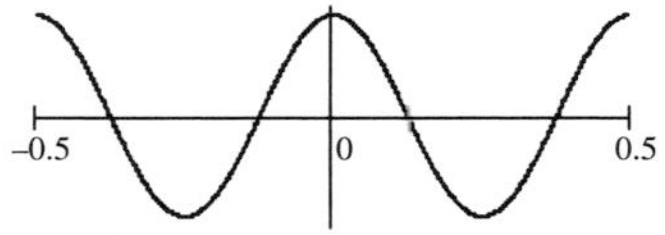 $U_2^S = \cos 4\pi \frac{x}{l}$	$l_2^S = 4\frac{\lambda}{2}$	
A_3	$\phi_3^A = 5\frac{\pi}{2}$	$f_3^A = 5\frac{c}{2l}$	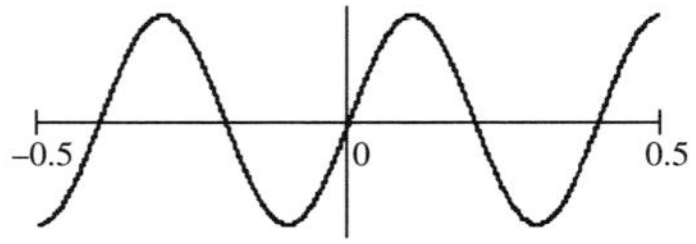 $U_3^A = \sin 5\pi \frac{x}{l}$	$l_3^A = 5\frac{\lambda}{2}$	EM
S_3	$\phi_3^S = 6\frac{\pi}{2}$	$f_3^S = 6\frac{c}{2l}$	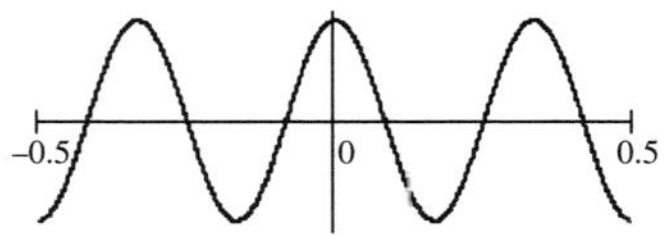 $U_3^S = \cos 6\pi \frac{x}{l}$	$l_3^S = 6\frac{\lambda}{2}$	

Symmetric resonances ($\sin\phi = 0$)

Symmetric resonances happen when $\sin\phi = 0$. This happens when ϕ is an even multiple of $\pi/2$, i.e.,

$$\sin\phi = 0 \rightarrow \phi = 2n\frac{\pi}{2} \tag{68}$$

This means that, at a symmetric resonance, the variable ϕ can take one of the following values:

$$\phi^S = 2\frac{\pi}{2}, 4\frac{\pi}{2}, 6\frac{\pi}{2}, \ldots \tag{69}$$

These values are the *symmetric eigenvalues* of the system. The superscript S denotes symmetric resonances. Recall the notations $\phi = \frac{1}{2}\gamma l$, $\gamma = \omega/c$, and $\omega = 2\pi f$. Hence,

$$\gamma l = \frac{\omega l}{c} = \frac{2\pi f l}{c} = 2n\pi \tag{70}$$

Therefore, the *symmetric resonance frequencies* are given by the formula

$$f_n^S = 2n\frac{c}{2l} \tag{71}$$

Recall Eq. (64)

$$\begin{cases} C_1\cos\phi + C_2\sin\phi = 0 \\ C_1\cos\phi - C_2\sin\phi = 0 \end{cases} \tag{72}$$

For each resonance frequency, Eq. (55) accepts nontrivial values for the coefficients C_1 and C_2. However, they can only be determined up to a scaling factor. Examination of this equation reveals that, for $\sin\phi = 0$, the constant C_1 vanishes, whereas the constant C_2 is indeterminate. We remove the indeterminacy by choosing $C_2 = 1$. Substituting $C_1 = 0$ and $C_2 = 1$ into Eq. (10) yields the response at symmetric resonances, i.e., the *symmetric mode shapes.*

$$U_n^S = \cos\gamma_n x \tag{73}$$

Using Eq. (70), we write the mode shapes in the convenient form

$$U_n^S = \cos 2n\pi\frac{x}{l} \tag{74}$$

Another useful interpretation of these results is through the wavelength $\lambda = c/f$. Hence, Eq. (70) becomes

$$l_n^S = 2n\frac{\lambda}{2} \tag{75}$$

Equation (75) indicates that symmetric resonances will happen when the PWAS length is an even multiple of half wavelengths of the elastic waves traveling in the PWAS. This relation is used in constructing PWAS that have to resonate at a given frequency.

7.2.3.2 Electromechanical Resonances

Recall the expressions for admittance and impedance given by Eqs. (53)–(54). These expressions can be rearranged as

$$Y = i\omega C\left[1 - k_{31}^2\left(1 - \frac{\tan\phi}{\phi}\right)\right] = i\omega C\left[\left(1 - k_{31}^2\right) + k_{31}^2\frac{\tan\phi}{\phi}\right] \tag{76}$$

$$Z = \frac{1}{i\omega C}\left[1 - k_{31}^2\left(1 - \frac{\tan\phi}{\phi}\right)\right]^{-1} = \frac{1}{i\omega C}\left[\left(1 - k_{31}^2\right) + k_{31}^2\frac{\tan\phi}{\phi}\right]^{-1} \tag{77}$$

The following conditions are considered:

- *Resonance*, when $Y \to \infty$, i.e., $Z = 0$
- *Anti-resonance*, when $Y = 0$, i.e., $Z \to \infty$.

Electrical resonance is associated with the situation in which a device is drawing very large currents when excited harmonically with a constant voltage at a given frequency. At resonance, the admittance becomes very large, whereas the impedance goes to zero. As the admittance becomes very large, the current drawn under constant-voltage excitation also becomes very large as $I = Y \cdot V$. In piezoelectric devices, the mechanical response at electrical resonance also becomes very large. This happens because the electromechanical coupling of the piezoelectric material transfers energy from the electrical input into the mechanical response. For these reasons, the resonance of an electrically driven piezoelectric device must be seen as an *electromechanical resonance.* A piezoelectric wafer driven at electrical resonance may undergo mechanical deterioration and even break up.

Electrical anti-resonance is associated with the situation in which a device under constant-voltage excitation draws almost no current. At anti-resonance, the admittance goes to zero, whereas the impedance becomes very large. Under constant-voltage excitation, this condition results in very small current being drawn from the source. In a piezoelectric device, the mechanical response at electrical anti-resonance is also very small. A piezoelectric wafer driven at the electrical anti-resonance hardly moves at all.

The *condition for electromechanical resonance* is obtained by studying the poles of Y, i.e., the values of ϕ that make $Y \to \infty$. Equation (76) reveals that $Y \to \infty$ as $\tan\phi \to \infty$. This happens when $\cos\phi \to 0$, i.e., when ϕ is an odd multiple of $\pi/2$

$$\cos\phi = 0 \to \phi = (2n-1)\frac{\pi}{2} \tag{78}$$

The angle ϕ can take the following values

$$\phi^{\mathrm{EM}} = \frac{\pi}{2}, \frac{3\pi}{2}, \frac{5\pi}{2}, \ldots \tag{79}$$

These values are the electromechanical eigenvalues. They are marked by superscript EM. Because $\phi = \frac{1}{2}\gamma l$, Eq. (79) implies that

$$\gamma l = \pi, 3\pi, 5\pi \tag{80}$$

Recall the definitions $\gamma = \omega/c$ and $\omega = 2\pi f$. Hence,

$$\gamma l = \frac{\omega l}{c} = \frac{2\pi f l}{c} = (2n-1)\pi \tag{81}$$

Therefore, the *electromechanical resonance frequencies* are given by the formula

$$f_n^{\mathrm{EM}} = (2n-1)\frac{c}{2l} \tag{82}$$

It is remarkable that the electromechanical resonance frequencies do not depend on any of the electric or piezoelectric properties. They depend entirely on the speed of sound in the material and on the geometric dimensions. In fact, they coincide with the antisymmetric resonance frequencies of Table 7.2. For each resonance frequency, f_n^{EM}, the angle ϕ_n is given by the formula

$$\phi_n^{\mathrm{EM}} = \frac{\pi l}{c} f_n \tag{83}$$

The vibration modes corresponding to the first, second, and third electromechanical resonances are marked EM in Table 7.2.

The explanation of electromechanical resonance

The electromechanical resonance can be explained by analyzing the electromechanical response given by Eq. (24). This equation states that

$$\hat{u}(x) = \frac{d_{31}\hat{E}_3}{\gamma} \cdot \frac{\sin \gamma x}{\cos \frac{1}{2}\gamma l} \tag{84}$$

The poles of Eq. (84) correspond to frequency values where the mechanical response to electrical excitation becomes unbounded, i.e., when electromechanical resonance happens. The function $\cos \frac{1}{2}\gamma l$ in the denominator goes to zero when the argument $\phi = \frac{1}{2}\gamma l$ is an odd multiple of $\pi/2$, i.e.,

$$\phi^{\mathrm{EM}} = \frac{\pi}{2}, \frac{3\pi}{2}, \frac{5\pi}{2}, \ldots \tag{85}$$

As $\cos \frac{1}{2}\gamma l$ goes to zero, the response becomes very large and resonance happens. Comparing Eqs. (85) and (61), we notice that these equations are the same, which means that the conditions for electromechanical resonance are the same as the conditions for antisymmetric mechanical resonance. This implies that electromechanical resonance is associated with the antisymmetric modes of mechanical resonance.

Explanation of why these particular modes are excited electrically resides in the electromechanical coupling that takes place in the piezoelectric material. As indicated by the constitutive Eq. (1), the electric field, E_3, couples directly with the strain, S_1, through the piezoelectric strain coefficient, d_{31}. The applied electric field is uniformly distributed over the piezoelectric wafer. A uniform distribution is a symmetric pattern. Because the strain is the space derivative of displacement, $S_1 = u_1'$, the strain distribution for an antisymmetric mode is also a symmetric pattern (Table 7.3). Because both the electric field and the strain have symmetric patterns, coupling between strain and electric field is promoted, and resonances are excited. For this reason, the antisymmetric modes of mechanical resonance are also modes of electromechanical resonance.

By the same token, we can explain why the electromechanical resonance does not happen at the frequencies corresponding to the symmetric modes of mechanical resonance. The symmetrical modes have an antisymmetric strain pattern, whereas the electric field continues to have a symmetric pattern. When a symmetric electric field pattern tries to

TABLE 7.3 Strain distributions corresponding to the first, second, and third antisymmetric modes of mechanical resonance. Note that the strain follows a symmetric pattern

Mode	Eigenvalue	Resonant frequency	Strain distribution		Half wavelength multiplicity
A_1	$\phi_1^A = \frac{\pi}{2}$	$f_1^A = \frac{c}{2l}$	–0.5 0 0.5	$S_1^A = \cos \pi \frac{x}{l}$	$l_1^A = \frac{\lambda}{2}$
A_2	$\phi_2^A = 3\frac{\pi}{2}$	$f_2^A = 3\frac{c}{2l}$	–0.5 0 0.5	$S_2^A = \cos 3\pi \frac{x}{l}$	$l_2^A = 3\frac{\lambda}{2}$
A_3	$\phi_3^A = 5\frac{\pi}{2}$	$f_3^A = 5\frac{c}{2l}$	–0.5 0 0.5	$S_3^A = \cos 5\pi \frac{x}{l}$	$l_3^A = 5\frac{\lambda}{2}$

excite an antisymmetric strain pattern, the net result is zero, because the effects on the left and right sides have opposite signs and cancel each other.

In conclusion, a piezoelectric wafer undergoing longitudinal vibrations, the electromechanical resonances happen at the poles of Y (zeros of Z) and correspond to the antisymmetric modes of mechanical resonances.

The *condition for electromechanical anti-resonance* is obtained by studying the zeros of Y, i.e., the values of ϕ that make $Y = 0$. Because electromechanical anti-resonances correspond to the zeros of the admittance (i.e., poles of the impedance), the current at anti-resonance is zero, $I = 0$. Equation (76) indicates that $Y = 0$ happens when

$$\left(1 - k_{31}^2\right) + k_{31}^2 \frac{\tan\phi}{\phi} = 0 \rightarrow \frac{\tan\phi}{\phi} = -\frac{1 - k_{31}^2}{k_{31}^2} \quad \text{(anti-resonance)} \qquad (86)$$

This equation is a transcendental equation that does not have closed-form solutions. Its solutions are found numerically.

In practice, the numerical difference between the values of the electromechanical resonances and electromechanical anti-resonances diminishes as we go to higher mode numbers. Table 7.4 shows the admittance and impedance poles calculated for $k_{31} = 0.36$, which is a common value in piezoelectric ceramics. The admittance poles, φ_Y, and impedance poles, φ_Z, differ significantly only for the first mode. By the fourth mode, the difference between them drops below 0.1%. This is important when trying to determine the resonance and anti-resonance frequencies experimentally.

Equations (76)–(77) can be used to predict the frequency response of the admittance and impedance functions. To this purpose, we note that $\varphi = \frac{1}{2}\omega l/c$. As the excitation frequency varies and resonance and anti-resonance frequencies are encountered, the admittance and impedance go through $+\infty$ to $-\infty$ transitions. Outside resonance, the admittance follows the linear function $i\omega C$, whereas the impedance follows the inverse function $1/(i\omega C)$.

TABLE 7.4 Admittance and impedance poles for $k_{31} = 0.36$

Admittance poles, φ_Y (electromechanical resonances)	$\pi/2$	$3\pi/2$	$5\pi/2$	$7\pi/2$	$9\pi/2$	$11\pi/2$	...
Impedance poles, φ_Z (electromechanical resonances)	$1.0565\pi/2$	$3.021\pi/2$	$5.005\pi/2$	$7\pi/2$	$9\pi/2$	$11\pi/2$	...
Ratio φ_Z/φ_Y	1.0565	1.0066	1.0024	1.0012	1.0007	1.0005	...
Difference between φ_Z and φ_Y	5.6%	0.66%	0.24%	0.12%	0.07%	0.05%	

7.2.3.3 Effect of Internal Damping

Materials under dynamic operation display internal heating due to several loss mechanisms. Such losses can be incorporated in the mathematical model of the PWAS by assuming complex compliance and dielectric constant

$$\bar{s}_{11} = s_{11}(1 - \mathrm{i}\eta), \quad \bar{\varepsilon}_{33} = \varepsilon_{33}(1 - \mathrm{i}\delta) \tag{87}$$

where a bar over a variable signfies a complex quantity. The values of η and δ vary with the piezoceramic formulation but are usually small (η, $\delta < 5\%$). The admittance and impedance become complex expressions

$$\overline{Y} = \mathrm{i}\omega \cdot \overline{C}\left[1 - \bar{k}_{31}^2\left(1 - \frac{1}{\bar{\varphi}\cot\bar{\varphi}}\right)\right], \quad \overline{Z} = \frac{1}{\mathrm{i}\omega \cdot \overline{C}}\left[1 - \bar{k}_{31}^2\left(1 - \frac{1}{\bar{\varphi}\cot\bar{\varphi}}\right)\right]^{-1} \tag{88}$$

where $\bar{k}_{31}^2 = d_{31}^2/(\bar{s}_{11}\bar{\varepsilon}_{33})$ is the complex coupling factor, $\overline{C} = (1 - \mathrm{i}\delta)C$, and $\bar{\varphi} = \varphi\sqrt{1 - \mathrm{i}\eta}$. Similar expressions can be derived for the width and thickness vibrations, with appropriate use of indices.

7.2.3.4 Admittance and Impedance Plots

Frequency plots of admittance and impedance are useful for the graphical determination of the resonance and anti-resonance frequencies. Figure 7.4 presents the numerical simulation of admittance and impedance response for a piezoelectric active sensor ($l = 7$ mm, $b = 1.68$ mm, $t = 0.2$ mm, APC-850 piezoceramic). Light damping, $\delta = \varepsilon = 1\%$ was assumed. As shown in Fig. 7.4a, the admittance and impedance essentially behave like $Y = i\omega C$ and $Z = 1/i\omega C$ outside resonances and anti-resonances. For example, the imaginary part of the admittance follows a straight-line pattern outside resonances, whereas the real part is practically zero. At resonances and anti-resonances, these basic patterns are modified by the addition of a pattern of behavior specific to resonance and anti-resonance. These patterns include zigzags of the imaginary part and a sharp peak of the real part. The admittance shows zigzags of the imaginary part and peaks of the real part around the resonance frequencies (Fig. 7.4a, left). The impedance shows the same behavior around the anti-resonance frequencies (Fig. 7.4a, right).

Figure 7.4b shows log-scale plots of the real parts of admittance and impedance. The log-scale plots are better for graphically identifying the resonance and anti-resonance frequencies. Table 7.5 lists, to four-digit accuracy, the frequencies read from these plots. Because the damping considered in this simulation was light, the values are given

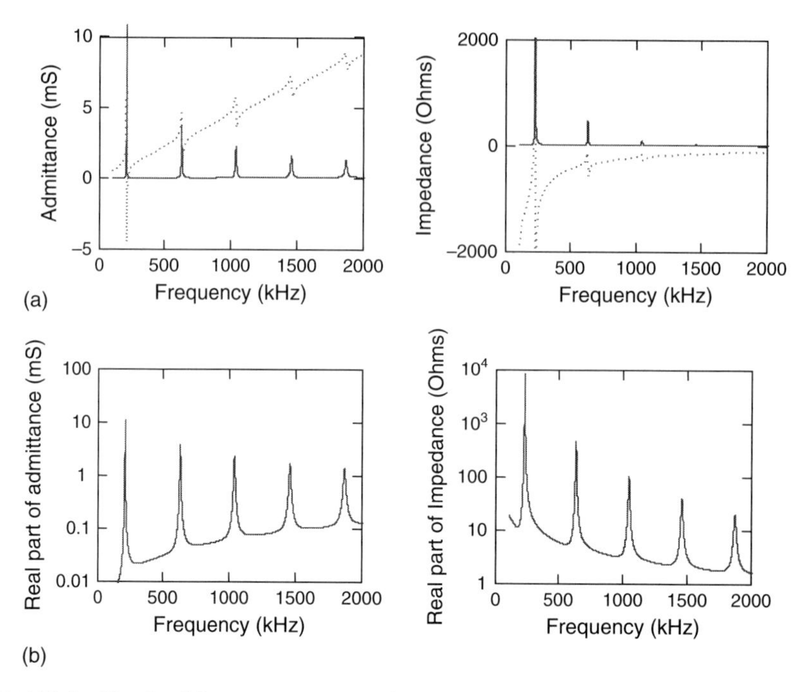

FIGURE 7.4 Simulated frequency response of admittance and impedance of a piezoelectric active sensor ($l = 7\,\text{mm},\ b = 1.68\,\text{mm},\ t = 0.2\,\text{mm}$, APC-850 piezoceramic, $\delta = \varepsilon = 1\%$): (a) complete plots showing both real (full line) and imaginary (dashed line) parts; (b) plots of real part only, log scale.

TABLE 7.5 Admittance and impedance poles determined from the numerical simulation of a piezoelectric wafer active sensor ($l = 7\,\text{mm},\ b = 1.68\,\text{mm},\ t = 0.2\,\text{mm}$, APC-850 piezoceramic, $\delta = \varepsilon = 1\%$)

Mode no.	1	2	3	4	5	6	...
Resonance frequencies (peaks of admittance real part spectrum), f_Y, kHz	207.17	621.57	1035.80	1450.20	1864.50	2278.90	...
Anti-resonance frequencies (peaks of impedance real part spectrum), f_Z, kHz	218.94	625.70	1038.30	1452.00	1866.00	2280.00	...
f_Z/f_Y	1.0568	1.0066	1.0024	1.0012	1.0008	1.0005	...
f_{Yn}/f_{Y1}	1.000	3.000	5.000	7.000	9.000	11.000	

in Table 7.5 compare well with the undamped data of Table 7.4. In fact, to three-digit accuracy, the ratio of the resonance frequencies for the lightly damped case is the same as for the undamped case (ratios 1, 3, 5,... corresponding to eigenvalues $\pi/2, 3\pi/2, 5\pi/2, \ldots$). This confirms that the peaks of the admittance and impedance real part spectra can be successfully used to measure the resonance and anti-resonance frequencies. The same information could be also extracted from the plots of the imaginary part, but this approach would be less practical. In the imaginary part plots, the resonance and anti-resonance patterns are masked by the intrinsic $Y = i\omega C$ and $Z = 1/i\omega C$ behaviors. In addition, the imaginary parts may undergo sign changes at resonances and anti-resonances; this would not allow log-scale plots to be applied, thus making the readings less precise.

7.2.4 EXPERIMENTAL RESULTS

A batch of 25 APC-850 small PWAS (7 mm sq., 0.2 mm thick, silver electrodes on both sides), from American Piezo Ceramics, Inc., was acquired and subjected to experimental measurements and statistical evaluation. We started with the mechanical and electrical properties declared by the vendor and then conducted our own measurements (Fig. 7.5). The material properties of the basic PZT material are given in Table 7.1.

The mechanical tolerances of these wafers, as presented by the vendor on their website, are given in Table 7.6. Other tolerances declared by the vendor were ±5% for resonance frequency, ±20% for capacitance, and ±20% for the d_{33} constant. For in-process quality assurance, we selected the following indicators:

1. geometrical dimensions
2. electrical capacitance
3. electromechanical impedance and admittance spectra

The geometric measurements were the initial indicators that, when showing an acceptable tolerance and a narrow spread, would build up the investigator's confidence. The electrical capacitance was used to verify electrical consistency of the fabrication process. It was found to be an important indicator but also somehow elusive. The E/M impedance and admittance spectra, though labor-intensive, were found to be the most comprehensive and rewarding.

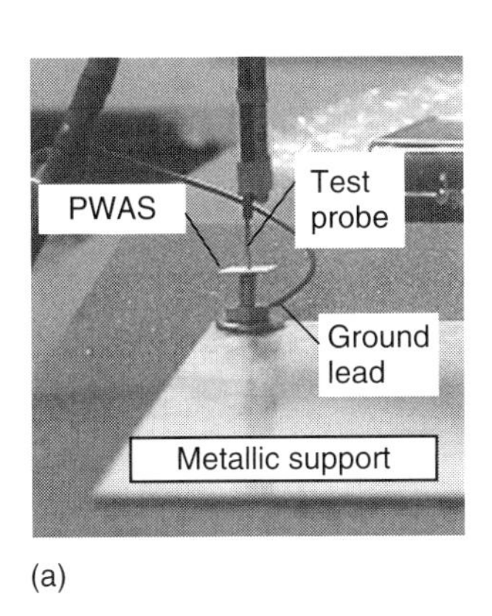

(a)

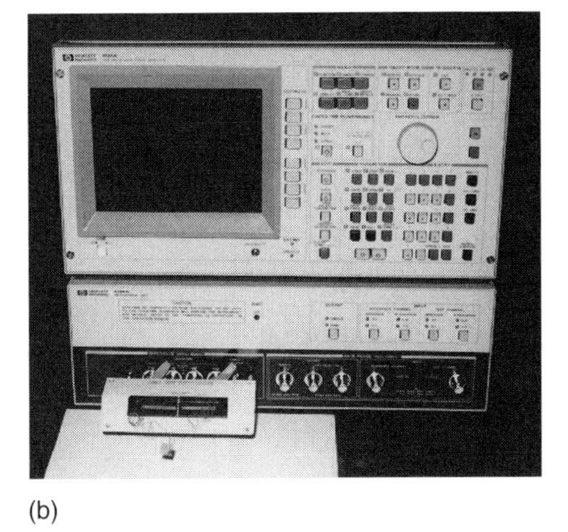

(b)

FIGURE 7.5 Dynamic measurement of PWAS frequency response: (a) test jig ensuring unrestraint support; (b) HP 4194A impedance phase-gain analyzer.

TABLE 7.6 Manufacturing tolerances for small piezoelectric wafers from the vendor APC, Inc. (www.americanpiezo.com)

Dimension	Units	Standard tolerance
Length or width of plates	< 13 mm	±0.13 mm
Thickness of all parts	0.20–0.49 mm	±0.025 mm

7.2.4.1 Geometric Measurements

Twenty-five nominally identical APC-850 wafers were measured with precision instrumentation consisting of a Mitutoyo Corp. CD 6″ CS digital caliper (0.01-mm precision) for length and width, and a Mitutoyo Corp. MCD 1″ CE digital micrometer (0.001-mm precision) for thickness. Statistical analysis of the data obtained from these measurements showed good agreement with the normal (Gaussian) distribution. Mean and standard deviation values for length/width and thickness were 6.9478 mm $\pm 0.5\%$ and 0.2239 mm $\pm 1.4\%$, respectively. The nominal values were 7 mm and 0.2 mm, respectively.

7.2.4.2 Electrical Capacitance Measurements

Electrical capacitance was measured with a BK Precision® Tool Kit™ 27040 digital meter with a resolution of 1 pF. PWAS capacitance was measured directly by putting the PZT square on a flat metallic support plate. The negative probe was connected to the plate, whereas the positive probe was placed to the top of the PWAS. Readings were taken when the tester readings had converged to a stable value. At least six readings were recorded and the average was taken. The process was iteratively improved until consistent results were obtained. The statistical analysis of the direct capacitance test results gave $C = 3.276\,\text{nF} \pm 3.8\%$.

7.2.4.3 Intrinsic E/M Impedance and Admittance of PWAS Resonators

The measurement of the intrinsic E/M impedance and admittance was done with an HP 4194A impedance phase-gain analyzer (Fig. 7.5). The test fixture for measuring the intrinsic E/M impedance/admittance of the PWAS is shown in Fig. 7.5a. We used a metallic plate with a lead connected at one corner. The PWAS was centered on the bolt head and held in place with the probe tip. Thus, the PWAS could vibrate freely. The PWAS samples were tested in the 100 Hz–12 MHz frequency range. Typical impedance spectra are shown in Fig. 7.6. The peaks appearing in the frequency range up to 3000 kHz are associated with the *anti-resonances of the in-plane modes*. These peaks are progressively smaller, with the fundamental resonance being the strongest. This is consistent with the fact that higher modes need more energy to get excited. Under constant energy excitation, higher modes would have lower amplitudes. At around 11 MHz, a new solitary strong peak appears. This is associated with the *fundamental resonance of the thickness mode*.

The *resonances of the in-plane modes* were identified from the E/M admittance spectra. Figure 7.7 shows the spectra of the E/M admittance real and imaginary parts in the frequency range up to 1200 kHz. The first, second, and third in-plane resonance frequencies are clearly observed. Also recorded were the values of the corresponding resonance peaks. Histograms of the statistical distribution of the resonance frequencies and resonance amplitudes for a batch of 25 PWAS specimens are given in Fig. 7.8. The PWAS were made of APC-850 piezoceramic with nominal dimensions $7\,\text{mm} \times 7\,\text{mm} \times 0.2\,\text{mm}$. The resonance frequency was found to have the value 251 kHz $\pm 1.2\%$. The admittance amplitude at resonance was found to be 67.152 mS $\pm 21\%$. These results indicate a narrow-band dispersion of the resonance frequency but a wider dispersion of the amplitude at resonance (Fig. 7.8).

7.2.4.4 Comparison between Measured and Calculated E/M admittance Spectra

The capability of Eqs. (76)–(77) to predict the E/M impedance and admittance response and subsequently identify resonance frequencies was investigated. From the beginning, it was observed that a rectangular PWAS with an aspect ratio of 1:1 could not be well

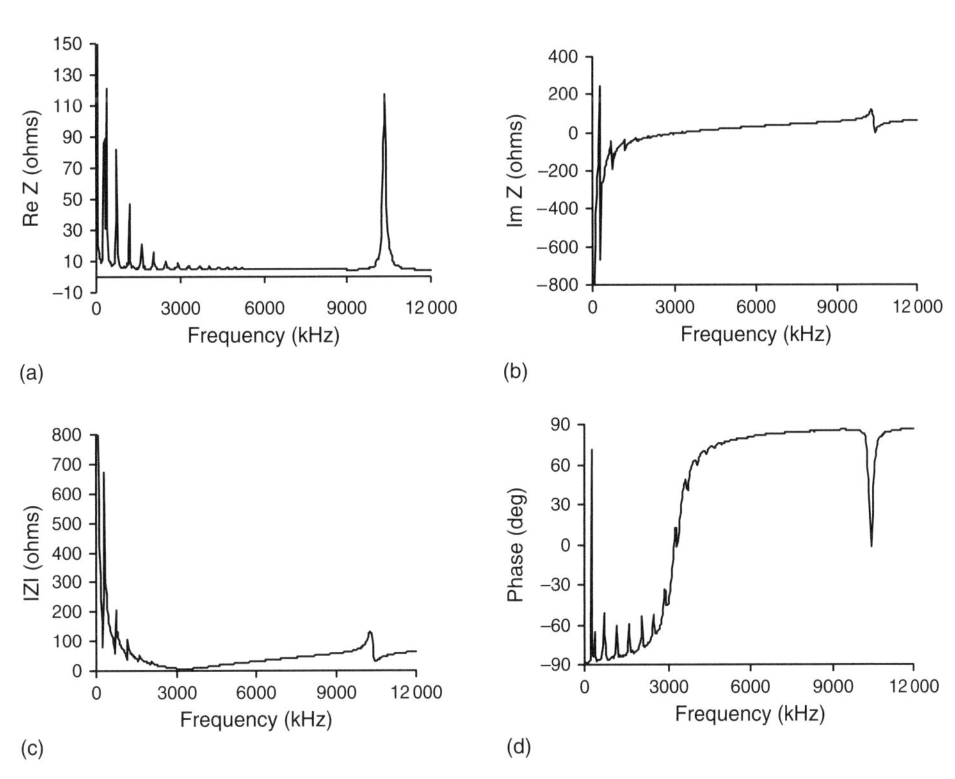

FIGURE 7.6 Intrinsic E/M impedance of a piezoelectric wafer active sensors up to 12 MHz (7 mm^2, 0.2 mm thick, APC-850 piezoceramic): (a) real part; (b) imaginary part; (c) amplitude; (d) phase.

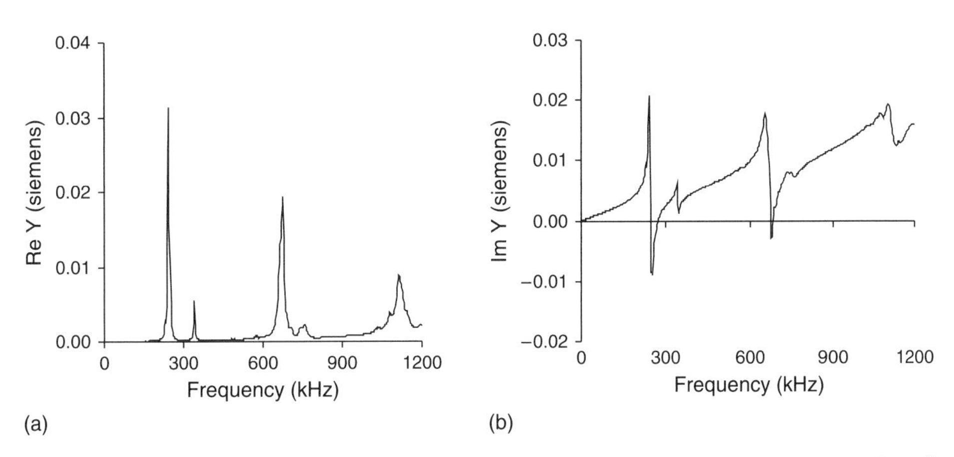

FIGURE 7.7 Intrinsic E/M admittance of a piezoelectric wafer active sensors up to 1200 kHz (7 mm^2, 0.2 mm thick, APC-850 piezoceramic): (a) real part; (b) imaginary part.

modeled by the 1-D analysis of Eqs. (76)–(77), and a 2-D analysis would be required. However, for higher aspect ratios, a better agreement between the 1-D analysis and the experimental results would be expected. Hence, it was decided to progressively modify the aspect ratio and observe this effect on the comparison between the theoretical and experimental results. It was expected that as the aspect ratio increased, the experimental results would converge on the 1-D predictions. To achieve this, we fabricated PWAS

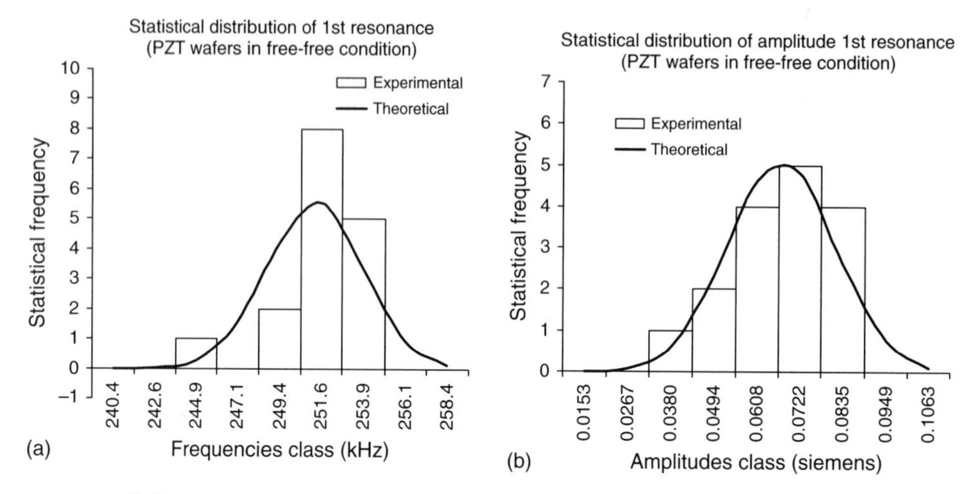

FIGURE 7.8 Statistical histograms of 1st resonance frequency and amplitude resulting from the testing of 25 PWAS specimens (7 mm^2, 0.2 mm thick, APC-850 piezoceramic): (a) 1st resonance frequencies; (b) admittance amplitude at the 1st resonance frequency.

specimens of aspect ratios increasing from 1:1 to 2:1 and 4:1. These specimens were fabricated from a 7-mm^2 square wafer, by cutting the width in half and then in half again. The designations 'square', 'half-width' and 'quarter-width' were used. Superposed plots of the measured and calculated E/M admittance real parts are given in Fig. 7.9 through Fig. 7.11 for the frequency range up to 1500 kHz. The in-plane resonance frequencies for each specimen were measured from the admittance plots. Also measured was the thickness resonance frequency obtained from impedance plots. The measured and calculated results for various aspect ratios are given in Table 7.7, where L = length modes, W = width modes, and T = thickness modes.

The results for the square PWAS are discussed first. The admittance real-part spectra are given in Fig. 7.9, whereas the corresponding frequency values are given in the first major row of Table 7.7. The actual length and width values of this nominally identical specimen were 6.99 mm and 6.56 mm, respectively. Because the length and the width were nearly identical, the corresponding lengthwise and widthwise resonance

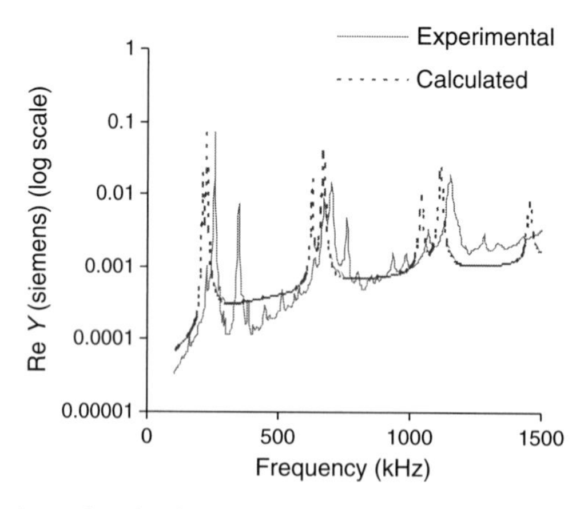

FIGURE 7.9 Experimental and calculated admittance spectra for square PWAS ($l = 6.99$ mm, $b = 6.56$ mm, $t = 0.215$ mm, $\varepsilon_{33}^{T} = 15.470 \cdot 10^{9}$ F/m, $s_{11}^{E} = 15.3 \cdot 10^{-12}$ Pa^{-1}, $d_{31} = -175 \cdot 10^{-12}$ m/V, $k_{31} = 0.36$).

TABLE 7.7 Results of the dynamic characterization of three rectangular PWAS of APC-850 piezoceramic (L = in-plane length vibration; W = in-plane width vibration; T = out-of-plane thickness vibration)

		Frequency, kHz								
Square wafer	Exp.	257	352	670	702	1070	1150	1572	1608	10 565
6.99-mm ×	Calc.	207.6	221.6	623	665	1038	1108	1453	1551	10 488
6.56-mm ×		(1L)	(1W)	(2L)	(2W)	(3L)	(3W)	(4L)	(4W)	(1T)
0.215-mm	Error	23.8%	58.8%	7.5%	5.6%	3.1%	3.8%	8.2%	3.4%	0.7%
½ width wafer	Exp.	208	432	597	670	821	1153	1307	1491	10 567
6.99-mm ×	Calc.	207.6	439	621			1038	1318	1451	10 488
3.53-mm ×		(1L)	(1W)	(2L)			(3L)	(2W)	(4L)	(1T)
0.215-mm	Error	0.2%	−1.6%	−4.0%			11.1%	−0.8%	2.8%	0.7%
¼ width wafer	Exp.	212		597	950	1020			1496	10 905
6.99-mm ×	Calc.	207.6		621	940	1038			1451	10 488
1.64-mm ×		(1L)		(2L)	(1W)	(3L)			(4L)	(1T)
0.215-mm	Error	2.1%		−4.0%	1.1%	−1.8%			3.1%	1.0%

Note: APC-850 piezoceramic has $\varepsilon_{33}^T = 15.470 \cdot 10^{-9}\,\mathrm{F/m}$, $s_{11}^E = 15.3 \cdot 10^{-12}\,\mathrm{m/N}$, $d_{31} = -175 \cdot 10^{-12}\,\mathrm{m/V}$, $k_{31} = 0.36$.

frequencies were close together, forming twin-peaks in the admittance plots. The 1-D analysis predicted that, in the 0–1500 kHz frequency band, seven peaks would exist (1L, 1W, 2L, 2W, 3L, 3W, 4L). The corresponding resonance frequencies (Table 7.7) were 207.6 kHz (1L), 221.6 kHz (1W), 623 kHz (2L), 665 kHz (2W), 1038 kHz (3L), 1108 kHz (3W) and 1453 kHz (4L). It should be noted that the calculated frequencies are in the harmonic ratio 1:3:5:7, as predicted by Table 7.3. The 1L and 1W experimental results were found to be significantly different from the theoretical predictions, showing a high positive error (19% and 37%, respectively). The high positive errors are indicative of the 2-D stiffening effect, typical of in-plane vibrations of low-aspect ratio plates. These 2-D stiffening effects could not be captured by the 1-D theory. At higher modes, the 2-D stiffening effect diminishes, and the agreement between theory and experiment improves (7.3% and 5.6% for the 2L and 2W modes; 2.3% and 3.6% for the 3L and 3W modes). We conclude that, with the exception of fundamental modes, 1-D theory gives a reasonable approximation even at aspect ratios as low as 1:1. The agreement for the out-of-plane 1T thickness frequency (10,565 kHz, 0.7% error) was also good, because this mode is very little affected by the in-plane shape of the PWAS.

The results for the half-width PWAS with a nominal aspect ratio 2:1 are discussed next. The actual length and width values were 6.99 mm and 3.53 mm, respectively. The admittance real-part spectra are given in Fig. 7.10, whereas the corresponding frequency values are given in the second major row of Table 7.7. In the 0–1500 kHz frequency band, the theoretical curve displays six peaks. The corresponding resonance frequencies are 207.6 kHz (1L), 621.6 kHz (2L), 1038 kHz (3L), 1451 kHz (4L), for lengthwise vibration, and 439 kHz (1W), 1318 kHz (2W), for widthwise vibration. The experimental results are 208 kHz (1L), 597 kHz (2L), 1153 kHz (3L), 1491 kHz (4L), for lengthwise vibration, and 432 kHz (1W) and 1307 kHz (2W), for widthwise vibration. The experimental results agree very well for the 1L and 1W modes (<2% error), a little less for the 2L and 2W modes (4–5% error), and reasonably well for the 3L mode (11% error). In addition to these clearly identifiable modes, several other peaks are present in the experimental curve. These modes are attributed to the edge roughness generated during the manufacturing process. This edge roughness produces secondary vibration effects. Overall, we conclude that a clear trend toward mode separation and a definite improvement in the first-mode

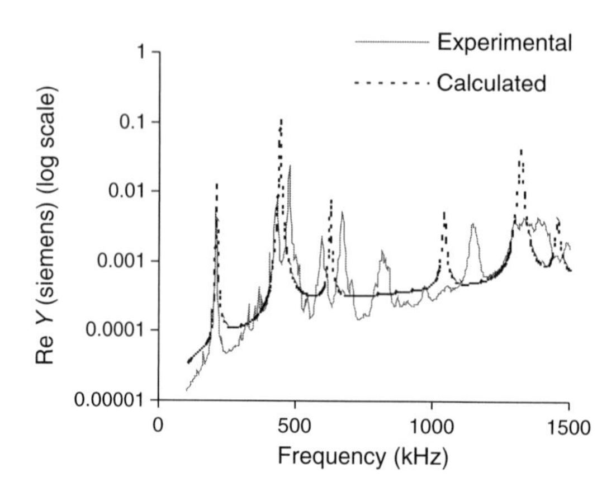

FIGURE 7.10 Experimental and calculated admittance spectra for half-width PWAS (l = 6.99 mm, $b = 3.53$ mm, $t = 0.215$ mm, $\varepsilon_{33}^T = 15.470 \cdot 10^9$ F/m, $s_{11}^E = 15.3 \cdot 10^{-12}$ Pa^{-1}, $d_{31} = -175 \cdot 10^{-12}$ m/V, $k_{31} = 0.36$).

prediction accuracy can be observed, although the aspect ratio (2:1) of the half-width PWAS is still far from a proper 1-D situation.

Finally, we discuss the results for the quarter-width PWAS with a nominal aspect ratio of 4:1. The actual length and width values were 6.99 mm and 1.64 mm, respectively. The admittance real-part spectra are given in Fig. 7.11, whereas the corresponding frequency values are given in the last major row of Table 7.7. In the 0–1500 kHz frequency band, the theoretical analysis predicts five resonance frequencies, as given in Table 7.7: 207.6 kHz (1L), 621.6 kHz (2L), 1038 kHz (3L), 1451 kHz (4L), and 940 kHz (1W). The corresponding theoretical curve displays four peaks, because the 1W and 3L peaks, having very close frequencies, have coalesced into a twin peak. The experimental results, as presented in Fig. 7.11 and Table 7.7, are in fairly good agreement with the theory, especially for the 1L, 2L, 3L, and 1W resonances (<4% error). The experimental curve shows an additional peak at 1167 kHz, which is not predicted by the 1-D analysis, but may be due to edge roughness. The overall conclusion is that, for 4:1 aspect ratio, good agreement between experiment and 1-D theory is obtained.

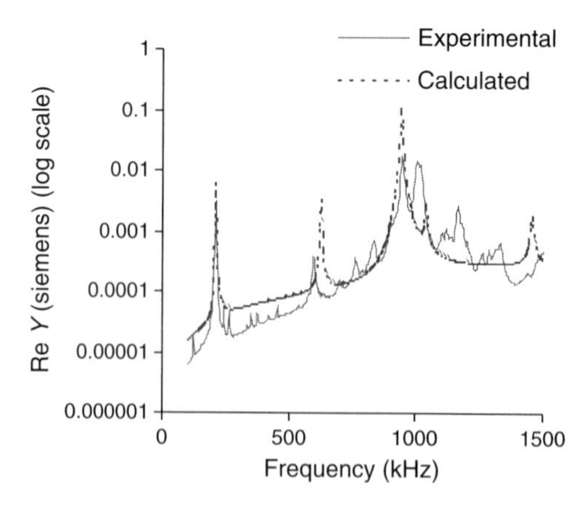

FIGURE 7.11 Experimental and calculated admittance spectra for quarter-width PWAS (l = 6.99 mm, b = 1.65 mm, t = 0.215 mm, ε_{33}^T = 15.470 $\cdot 10^9$ F/m, s_{11}^E = 15.3 $\cdot 10^{-12}$ Pa^{-1}, d_{31} = $-175 \cdot 10^{-12}$ m/V, $k_{31} = 0.36$).

7.3 CIRCULAR PWAS RESONATORS

7.3.1 MODELING OF A CIRCULAR PIEZOELECTRIC WAFER ACTIVE SENSOR

This section addresses the behavior of circular PWAS resonators, i.e., of free PWAS of circular form. Consider a circular piezoelectric wafer of radius r, and thickness t and width b, excited by the thickness polarization electric field, E_3 (Fig. 7.12).

The electric field is produced by the application of a harmonic voltage $Ve^{i\omega t}$ between the top and bottom surface electrodes. The resulting electric field, $E_3 = V/h$, is assumed uniform over the electrodes area. Because the electric field is uniform, the response will be assumed to be axially symmetric. This means that the circular piezoelectric wafer undergoes uniform radial and circumferential expansion. Hence, $u_\theta = 0$, and the only displacement is u_r and u_θ. The term 'uniform' implies that the derivative with respect to θ is zero, i.e., $\frac{\partial}{\partial\theta}$, and that the circumferential displacement is also zero, i.e., $u_\theta = 0$. Hence, the strain-displacement relations in polar coordinates reduce to

$$
\begin{aligned}
S_{rr} &= \frac{\partial u_r}{\partial r} & & & S_{rr} &= \frac{\mathrm{d}u_r}{\mathrm{d}r} \\
S_{\theta\theta} &= \frac{1}{r}\cancel{\frac{\partial u_\theta}{\partial\theta}} + \frac{u_r}{r} & &\rightarrow & S_{\theta\theta} &= \frac{u_r}{r} \\
S_{r\theta} &= \frac{1}{2}\left(\frac{1}{r}\cancel{\frac{\partial u_r}{\partial\theta}} + \cancel{\frac{\partial u_\theta}{\partial r}} - \cancel{\frac{u_\theta}{r}}\right) & & & S_{r\theta} &= 0
\end{aligned}
\tag{89}
$$

Recall the strain constitutive equation in polar coordinates

$$
\begin{aligned}
S_{rr} &= s_{rr}^E T_{rr} + s_{r\theta}^E T_{\theta\theta} + d_{3r} E_3 \\
S_{\theta\theta} &= s_{r\theta}^E T_{rr} + s_{\theta\theta}^E T_{\theta\theta} + d_{3\theta} E_3
\end{aligned}
\tag{90}
$$

Because the piezoelectric wafer is assumed in-plane isotropic, we will replace the s_{rr}^E, $s_{\theta\theta}^E$, $s_{r\theta}^E$ by s_{11}^E, s_{11}^E, s_{12}^E, respectively, such that

$$
\begin{aligned}
S_{rr} &= s_{11}^E T_{rr} + s_{12}^E T_{\theta\theta} + d_{31} E_3 \\
S_{\theta\theta} &= s_{12}^E T_{rr} + s_{11}^E T_{\theta\theta} + d_{31} E_3
\end{aligned}
\tag{91}
$$

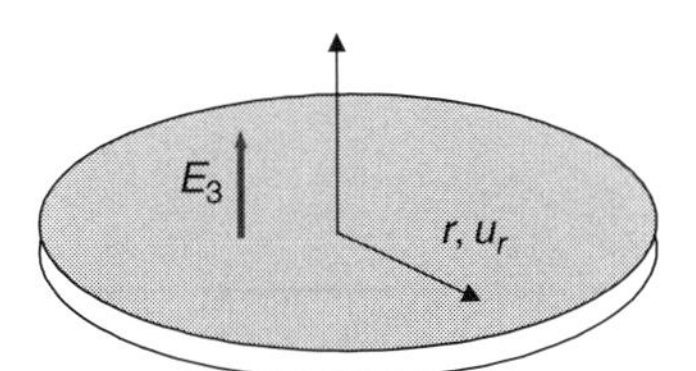

FIGURE 7.12 Circular PWAS.

Upon inversion,

$$\begin{aligned} T_{rr} &= \frac{1}{s_{11}^{E}(1-\nu^2)}\left[(S_{rr}+\nu S_{\theta\theta})-(1+\nu)\,d_{31}E_3\right] \\ T_{\theta\theta} &= \frac{1}{s_{11}^{E}(1-\nu^2)}\left[(\nu S_{rr}+S_{\theta\theta})-(1+\nu)\,d_{31}E_3\right] \end{aligned} \tag{92}$$

where ν is the Poisson ratio defined as

$$\nu = -\frac{s_{12}^{E}}{s_{11}^{E}} \quad \text{(Poisson ratio)} \tag{93}$$

Substituting the strain-displacements relations of Eq. (89) yields the stress expressed in terms of displacements, i.e.,

$$\begin{aligned} T_{rr} &= \frac{1}{s_{11}^{E}(1-\nu^2)}\left[\left(\frac{\mathrm{d}u_r}{\mathrm{d}r}+\nu\frac{u_r}{r}\right)-(1+\nu)\,d_{31}E_3\right] \\ T_{\theta\theta} &= \frac{1}{s_{11}^{E}(1-\nu^2)}\left[\left(\nu\frac{\mathrm{d}u_r}{\mathrm{d}r}+\frac{u_r}{r}\right)-(1+\nu)\,d_{31}E_3\right] \end{aligned} \tag{94}$$

Newton law of motion applied to the infinitesimal element of Fig. 7.13 yields

$$\frac{\mathrm{d}T_{rr}}{\mathrm{d}r}+\frac{T_{rr}-T_{\theta\theta}}{r} = -\omega^2\rho u_r \tag{95}$$

Substitution of the stresses T_{rr} and $T_{\theta\theta}$ yields

$$\begin{aligned} \frac{\mathrm{d}T_{rr}}{\mathrm{d}r} &= \frac{1}{s_{11}^{E}(1-\nu^2)}\frac{\mathrm{d}}{\mathrm{d}r}\left\{\left[\left(\frac{\mathrm{d}u_r}{\mathrm{d}r}+\nu\frac{u_r}{r}\right)-(1+\nu)\,d_{31}E_3\right]\right\} \\ &= \frac{1}{s_{11}^{E}(1-\nu^2)}\left[\frac{\mathrm{d}^2u_r}{\mathrm{d}r^2}+\nu\left(\frac{1}{r}\frac{\mathrm{d}u_r}{\mathrm{d}r}-\frac{u_r}{r^2}\right)\right] \end{aligned} \tag{96}$$

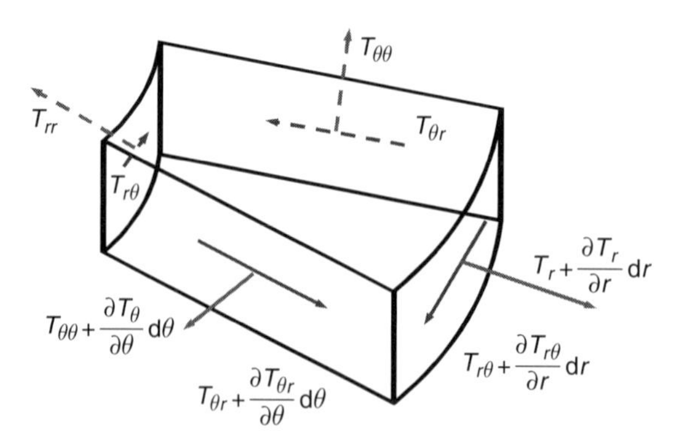

FIGURE 7.13 Stresses acting on an infinitesimal element of a circular piezoelectric wafer.

and

$$\frac{T_{rr} - T_{\theta\theta}}{r} = \frac{1}{s_{11}^E(1-\nu^2)} \frac{1}{r}\left(\frac{du_r}{dr} + \nu\frac{u_r}{r} - \frac{u_r}{r} - \nu\frac{du_r}{dr}\right)$$
$$= \frac{1}{s_{11}^E(1-\nu^2)}(1-\nu)\left(\frac{1}{r}\frac{du_r}{dr} - \frac{u_r}{r^2}\right) \tag{97}$$

Upon addition,

$$\frac{dT_{rr}}{dr} + \frac{T_{rr} - T_{\theta\theta}}{r} = \frac{1}{s_{11}^E(1-\nu^2)}\left(\frac{d^2u_r}{dr^2} + \frac{1}{r}\frac{du_r}{dr} - \frac{u_r}{r^2}\right) \tag{98}$$

Substitution into Eq. (94) yields the equation of motion

$$\frac{1}{s_{11}^E(1-\nu^2)}\left(\frac{d^2u_r}{dr^2} + \frac{1}{r}\frac{du_r}{dr} - \frac{u_r}{r^2}\right) = -\omega^2\rho u_r \tag{99}$$

Introduce the wave speed of axial waves in the piezoelectric wafer

$$c_P = \sqrt{\frac{1}{\rho s_{11}^E(1-\nu^2)}} \quad \text{(wave speed)} \tag{100}$$

Hence, the equation of motion becomes the wave equation

$$\frac{d^2u_r}{dr^2} + \frac{1}{r}\frac{du_r}{dr} - \frac{u_r}{r^2} = -\frac{\omega^2}{c_p^2}u_r \tag{101}$$

Introducing the wavenumber, $\gamma = \frac{\omega}{c_P}$, and rearranging yields

$$r^2\frac{d^2u_r}{dr^2} + r\frac{du_r}{dr} + \left(r^2\gamma^2 - 1\right)u_r = 0 \tag{102}$$

Using the change of variables

$$z = \gamma r \rightarrow r = \frac{1}{\gamma}z, \quad \frac{d}{dr} = \gamma\frac{d}{dz} \tag{103}$$

Equation (99) becomes

$$z^2\frac{d^2u_r}{dz^2} + z\frac{du_r}{dz} + \left(z^2 - 1\right)u_r = 0 \tag{104}$$

which is exactly the Bessel differential equation of order 1, accepting as solutions the Bessel function $J_1(z)$. Hence, the general solution of Eq. (99) is

$$u_r(r) = A \cdot J_1(\gamma r) \tag{105}$$

The constant A is determined from the boundary conditions. For stress-free boundary conditions at $r = a$, we write

$$T_{rr}(a) = 0 \tag{106}$$

Using Eq. (93), the boundary condition becomes

$$T_{rr}(a) = \frac{1}{s_{11}^E(1-\nu^2)}\left[\left(\frac{\mathrm{d}u_r}{\mathrm{d}r} + \nu\frac{u_r}{r}\right) - (1+\nu)\, d_{31}E_3\right] = 0 \tag{107}$$

i.e.,

$$\frac{\mathrm{d}u_r}{\mathrm{d}r} + \nu\frac{u_r}{r} - (1+\nu)\, d_{31}E_3 = 0 \tag{108}$$

Substitution of Eq. (105) into Eq. (107) yields

$$A\gamma J_1'(\gamma a) + \frac{\nu}{a} A J_1(\gamma a) = (1+\nu)\, d_{31}E_3 \tag{109}$$

Recall the Bessel functions identity

$$J_1'(z) = J_0(z) - \frac{1}{z} J_1(z) \tag{110}$$

Equation (109) becomes

$$A\left(\gamma J_0(\gamma a) - \frac{1}{a} J_1(\gamma a) + \frac{\nu}{a} J_1(\gamma a)\right) = (1+\nu)\, d_{31}E_3 \tag{111}$$

Hence,

$$A = \frac{(1+\nu)\, a d_{31}E_3}{(\gamma a)\, J_0(\gamma a) - (1-\nu)\, J_1(\gamma a)} \tag{112}$$

Substitution into the general solution (17) yields the displacement response

$$u_r(r) = d_{31}E_3 a \frac{(1+\nu)\, J_1(\gamma r)}{(\gamma a)\, J_0(\gamma a) - (1-\nu)\, J_1(\gamma a)} \tag{113}$$

Introducing the notation

$$u_{\mathrm{ISA}} = \left(d_{31}\hat{E}_3\right) a \quad \text{(Induced displacement)} \tag{114}$$

where ISA signifies 'induced strain actuation', we can express Eq. (113) as

$$u_r(r) = \frac{(1+\nu)\, J_1(\gamma r)}{(\gamma a)\, J_0(\gamma a) - (1-\nu)\, J_1(\gamma a)} u_{\mathrm{ISA}} \tag{115}$$

Introducing $z = \gamma a$, we rewrite Eq. (115) as

$$u_r(r) = \frac{(1+\nu)\, J_1(\gamma r)}{z J_0(z) - (1-\nu)\, J_1(z)} u_{\mathrm{ISA}} \tag{116}$$

7.3.2 ELECTRICAL RESPONSE

Consider a circular PWAS under electric excitation (Fig. 7.14). Recall the equation representing the electrical displacement as a function of stress and electric field

$$D_3 = d_{31}T_{rr} + d_{31}T_{\theta\theta} + \varepsilon_{33}^T E_3 \tag{117}$$

This equation can be integrated to determine the electric charge, and hence the current. Before doing so, we need to express the stress as a function of applied electric field, such that Eq. (116) relates the electric displacement to the applied electric field only. To achieve this, we will use the displacement solution of Eq. (113) and the stress-displacement relation of Eq. (94). First, note that the stresses in Eq. (116) can be grouped as

$$D_3 = d_{31}(T_{rr} + T_{\theta\theta}) + \varepsilon_{33}^T E_3 \tag{118}$$

Then, recall Eq. (94)

$$\begin{aligned} T_{rr} &= \frac{1}{s_{11}^E(1-\nu^2)}\left[\left(\frac{du_r}{dr} + \nu\frac{u_r}{r}\right) - (1+\nu)\,d_{31}E_3\right] \\ T_{\theta\theta} &= \frac{1}{s_{11}^E(1-\nu^2)}\left[\left(\nu\frac{du_r}{dr} + \frac{u_r}{r}\right) - (1+\nu)\,d_{31}E_3\right] \end{aligned} \tag{119}$$

Upon addition,

$$T_{rr} + T_{\theta\theta} = \frac{1}{s_{11}^E(1-\nu^2)}(1+\nu)\left[\frac{du_r}{dr} + \nu\frac{u_r}{r} - 2d_{31}E_3\right] \tag{120}$$

The displacement terms in Eq. (119) can be rewritten as

$$\frac{du}{dr} + \frac{u_r}{r} = \frac{1}{r}\left(r\frac{du_r}{dr} + u_r\right) = \frac{1}{r}\frac{d}{dr}(ru_r) \tag{121}$$

Hence,

$$\begin{aligned} T_{rr} + T_{\theta\theta} &= \frac{1}{s_{11}^E(1-\nu^2)}(1+\nu)\left[\frac{1}{r}\frac{d}{dr}(ru_r) - 2d_{31}E_3\right] \\ &= \frac{1}{s_{11}^E(1-\nu)}\left[\frac{1}{r}\frac{d}{dr}(ru_r) - 2d_{31}E_3\right] \end{aligned} \tag{122}$$

Substitution of Eq. (121) into Eq. (117) yields

$$D_3 = d_{31}\frac{1}{S_{11}^E(1-\nu)}\left[\frac{1}{r}\frac{d}{dr}(ru_r) - 2d_{31}E_3\right] + \varepsilon_{33}^T E_3 \tag{123}$$

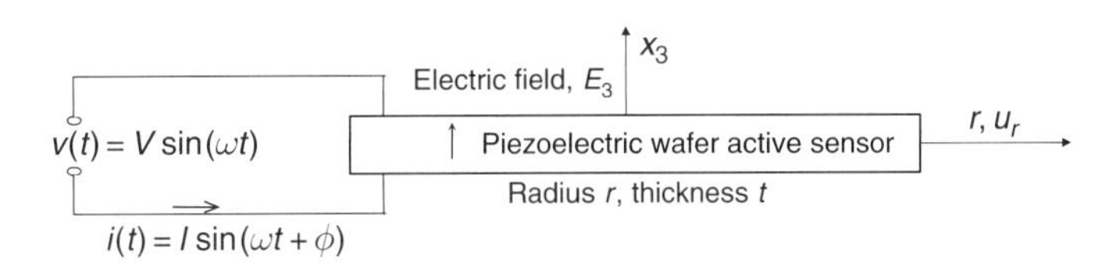

FIGURE 7.14 Schematic of a circular PWAS under electric excitation.

Upon rearrangement,

$$D_3 = \varepsilon_{33}^T E_3 \left\{ 1 - \frac{2}{(1-\nu)} \frac{d_{31}^2}{S_{11}^E \varepsilon_{33}^T} \left[1 - \frac{1}{2d_{31}E_3} \frac{1}{r} \frac{\mathrm{d}}{\mathrm{d}r}(ru_r) \right] \right\} \tag{124}$$

Introduce the *planar coupling coefficient*, k_p, is defined as

$$k_p^2 = \frac{2}{(1-\nu)} \frac{d_{31}^2}{s_{11}^E \varepsilon_{33}^T} \tag{125}$$

Note that

$$k_p^2 = \frac{2}{(1-\nu)} k_{31}^2 \tag{126}$$

Hence,

$$D_3 = \varepsilon_{33}^T E_3 \left\{ \left(1 - k_p^2\right) + k_p^2 \left[\frac{1}{2d_{31}E_3} \frac{1}{r} \frac{\mathrm{d}}{\mathrm{d}r}(ru_r) \right] \right\} \tag{127}$$

The total charge Q is obtained by integration of the electric displacement D_3 over the electrodes area $A = \pi a^2$, i.e.,

$$Q = \int_A D_3 r \,\mathrm{d}r \,\mathrm{d}\theta = 2\pi \int_0^a D_3 r \,\mathrm{d}r \tag{128}$$

Substitution of Eq. (126) into Eq. (127) yields

$$Q = 2\pi \int_0^a \varepsilon_{33}^T E_3 \left\{ \left(1 - k_p^2\right) + k_p^2 \left[\frac{1}{2d_{31}E_3} \frac{1}{r} \frac{\mathrm{d}}{\mathrm{d}r}(ru_r) \right] \right\} r \,\mathrm{d}r \tag{129}$$

Hence,

$$\begin{aligned} Q &= \pi\varepsilon_{33}^T E_3 \left\{ \left(1 - k_p^2\right) \int_0^a 2r \,\mathrm{d}r + k_p^2 \frac{1}{d_{31}E_3} \int_0^a \left[\frac{1}{\cancel{r}} \frac{\mathrm{d}}{\mathrm{d}r}(ru_r) \right] \cancel{r} \,\mathrm{d}r \right\} \\ &= \pi\varepsilon_{33}^T E_3 \left\{ \left(1 - k_p^2\right) r^2 \Big|_0^a + k_p^2 \frac{1}{d_{31}E_3} (ru_r) \Big|_0^a \right\} \\ &= \pi\varepsilon_{33}^T E_3 \left\{ \left(1 - k_p^2\right) a^2 + k_p^2 \frac{1}{d_{31}E_3} a u_r(a) \right\} \end{aligned} \tag{130}$$

i.e.,

$$Q = \pi a^2 \varepsilon_{33}^T E_3 \left\{ \left(1 - k_p^2\right) - k_p^2 \frac{u_r(a)}{d_{31}E_3 a} \right\} \tag{131}$$

Using the definitions $C = \varepsilon_{33}^T \left(A/h\right)$, $A = \pi a^2$, $u_{\mathrm{ISA}} = d_{31}E_3 a$, $V = E_3/h$, we obtain

$$Q = CV \left[\left(1 - k_p^2\right) + k_p^2 \frac{u_r(a)}{u_{\mathrm{ISA}}} \right] \tag{132}$$

The electric current is obtained as the time derivative of the electric charge, i.e.,

$$I = \dot{Q} = i\omega Q \tag{133}$$

Hence,

$$Q = i\omega CV\left[\left(1-k_p^2\right)+k_p^2\frac{u_r(a)}{u_{\text{ISA}}}\right] \tag{134}$$

The admittance, Y, is defined as the ratio between the current and voltage, i.e.,

$$Y = \frac{I}{V} = i\omega C\left[\left(1-k_p^2\right)+k_p^2\frac{u_r(a)}{u_{\text{ISA}}}\right] \tag{135}$$

Recall Eq. (115) giving the mechanical displacement solution

$$u_r(r) = u_{\text{ISA}}\frac{(1+v)\,J_1(\gamma r)}{(\gamma a)\,J_0(\gamma a)-(1-v)\,J_1(\gamma a)} \tag{136}$$

Hence,

$$Y = i\omega C\left[\left(1-k_p^2\right)-k_p^2\frac{(1+v)\,J_1(\gamma a)}{(\gamma a)\,J_0(\gamma a)-(1-v)\,J_1(\gamma a)}\right] \tag{137}$$

Recalling the notation $z=\gamma a$, we rewrite Eq. (137) as

$$Y = i\omega C\left[\left(1-k_p^2\right)-k_p^2\frac{(1+v)\,J_1(z)}{zJ_0(z)-(1-v)\,J_1(z)}\right] \tag{138}$$

This result agrees with that of Ikeda (1996). However, it may be more convenient at times to write it as

$$Y = i\omega C\left[1-k_p^2\left(1-\frac{(1+v)\,J_1(\gamma a)}{(\gamma a)\,J_0(\gamma a)-(1-v)\,J_1(\gamma a)}\right)\right] \tag{139}$$

Note that the admittance is purely imaginary and consists of the capacitive admittance, $i\omega C$, modified by the effect of piezoelectric coupling between mechanical and electrical fields. This effect is apparent in the term containing the electromechanical coupling coefficient, k_{31}^2. The impedance, Z, is obtained as the ratio between the voltage and current, i.e.,

$$Z = \frac{\hat{V}}{\hat{I}} = Y^{-1} \tag{140}$$

Hence,

$$Z = \frac{1}{i\omega C}\left[1-k_p^2\left(1-\frac{(1+v)\,J_1(\gamma a)}{(\gamma a)\,J_0(\gamma a)-(1-v)\,J_1(\gamma a)}\right)\right]^{-1} \tag{141}$$

Recalling the notation $z=\gamma a$, we rewrite

$$Y = i\omega C\left[1-k_p^2\left(1-\frac{(1+v)\,J_1(z)}{zJ_0(z)-(1-v)\,J_1(z)}\right)\right] \tag{142}$$

$$Z = \frac{1}{i\omega C}\left[1-k_p^2\left(1-\frac{(1+v)\,J_1(z)}{zJ_0(z)-(1-v)\,J_1(z)}\right)\right]^{-1} \tag{143}$$

7.3.3 RESONANCES

If resonances happen, they could be of two types:

(1) Electromechanical resonances
(2) Mechanical resonances

Mechanical resonances take place in the same conditions as in a conventional elastic disc. They happen under mechanical excitation, which produces a mechanical response in the form of mechanical vibrations. Electromechanical resonances are specific to piezoelectric materials. They reflect the coupling between the mechanical and electrical fields. Electromechanical resonances happen under electric excitation, which produces an electromechanical response, i.e., both a mechanical vibration and a change in the electric admittance and impedance. We will consider these two situations separately.

7.3.3.1 Mechanical Resonances

If a PWAS is excited mechanically with a frequency sweep, certain frequencies will be encountered at which the response is very large, i.e., the PWAS resonates. To study the mechanical resonances, assume that the material is not piezoelectric, i.e., $d_{31} = 0$. Hence, we develop the analysis of the classical mechanical resonances of an elastic disc. Our discussion will be restricted to axisymmetric in-plane vibrations. Recall Eq. (101) representing the wave equation for axisymmetric in-plane vibrations of an elastic disc

$$\frac{\mathrm{d}^2 u_r}{\mathrm{d}r^2} + \frac{1}{r}\frac{\mathrm{d}u_r}{\mathrm{d}r} - \frac{u_r}{r^2} = -\frac{\omega^2}{c_P^2} u_r \tag{144}$$

We have already shown that introducing the wavenumber, $\gamma = \frac{\omega}{c_P}$, and the substitution $z = \gamma r$ recovers the Bessel differential equation of order 1 as given by Eq. (104), i.e.,

$$z^2 \frac{\mathrm{d}^2 u_r}{\mathrm{d}z^2} + z\frac{\mathrm{d}u_r}{\mathrm{d}z} + (z^2 - 1)\, u_r = 0 \tag{145}$$

Hence, the general solution is of the form given by Eq. (105)

$$u_r(r) = A \cdot J_1(\gamma r) \tag{146}$$

where the constant A must be determined from the boundary conditions. For stress-free boundary conditions at $r = a$, we have $T_{rr}(a) = 0$. Using Eq. (93) and imposing non piezoelectric behavior, we obtain the stress-free boundary condition in the form

$$T_{rr}(a) = \frac{1}{s_{11}^E (1 - \nu^2)} \left(\frac{\mathrm{d}u_r}{\mathrm{d}r} + \nu \frac{u_r}{r} \right) = 0 \tag{147}$$

Substitution of the solution $u_r = A \cdot J_1(\gamma r)$ yields

$$A\gamma J_1'(\gamma a) + \frac{\nu}{a} A J_1(\gamma a) = 0 \tag{148}$$

Using the identity $J_1'(z) = J_0(z) - \frac{1}{z}J_1(z)$, dividing through by A and multiplying by a, we obtain the condition

$$(\gamma a)\, J_0(\gamma a) - (1 - \nu)\, J_1(\gamma a) = 0 \tag{149}$$

Equation (149) can be rearranged as

$$\frac{(\gamma a) J_0(\gamma a)}{J_1(\gamma a)} = (1 - \nu) \tag{150}$$

The left hand side of Eq. (148) is also known as the *modified quotient of Bessel functions* defined as

$$\tilde{J}_1(z) = \frac{z J_0(z)}{J_1(z)} \tag{151}$$

Note that Eq. (149) depends on the Poisson ratio, ν. Equation (149) indicates that nonzero values of A are only possible at particular values of (γa), which are the elastic system *eigenvalues*. Equation (149) is transcendental and does not accept closed-form solution. Numerical solution of Eq. (149) for $\nu = 0.30$ yields

$$z = (\gamma a) = 2.048652; \quad 5.389361; \quad 8.571860; \quad 11.731771\ldots \tag{152}$$

It should be noted that because these eigenvalues depend on ν, the ratio between successive eigenvalues and the fundamental eigenvalues could be used for determining the Poisson ratio experimentally through a curve-fitting process.

For each eigenvalue, (γa), we can determine the corresponding resonance frequency with the formula

$$f_n = \frac{1}{2\pi} \frac{c}{a} (\gamma a)_n \tag{153}$$

where $n = 1, 2, 3, \ldots$ The *mode shapes* are calculated with Eq. (105) by putting the wavenumber γ corresponding to each eigenvalue. Thus, for the nth eigenvalue, $(\gamma a)_n$, the nth wavenumber is calculated with the formula

$$\gamma_n = \frac{1}{a} (\gamma a)_n \tag{154}$$

Thus, the nth mode shape is

$$R_n(r) = A_n \cdot J_1(z_n r/a) \tag{155}$$

The constant A_n is determined through modes normalization and depends on the normalization procedure used. A common normalization procedure, based on equal modal energy, yields

$$A_n = \sqrt{J_1^2(z_n) - J_0(z_n) J_2(z_n)} \tag{156}$$

Other normalization methods simply take $A_n = 1$. A graphical representation of the mode shapes is given in Table 7.8.

TABLE 7.8 Axisymmetric mode shapes of piezoelectric elastic disc of radius $a = 10\,\text{mm}$

Mode	Eigenvalue	Resonant frequency	Mode shape
R_1	$z_1 = \gamma_1 a = 2.048652$	$f_2 = \frac{1}{2\pi} z_1 \frac{c}{a}$	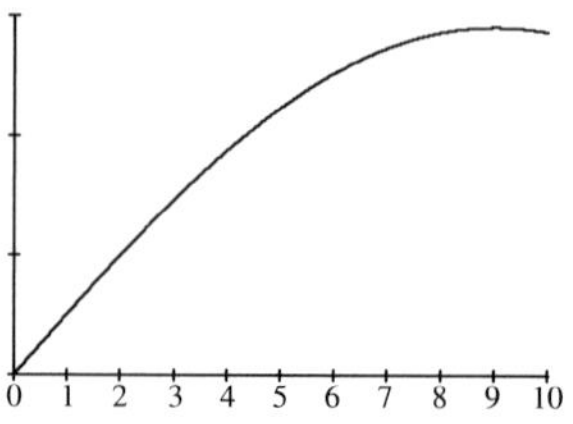 $R_1 = J_1(z_1 r/a)$
R_2	$z_2 = \gamma_2 a = 5.389361$	$f_2 = \frac{1}{2\pi} z_2 \frac{c}{a}$	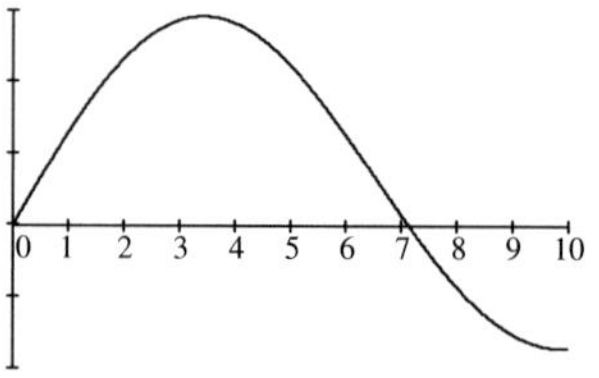 $R_2 = J_1(z_2 r/a)$
R_3	$z_3 = \gamma_3 a = 8.571860$	$f_3 = \frac{1}{2\pi} z_3 \frac{c}{a}$	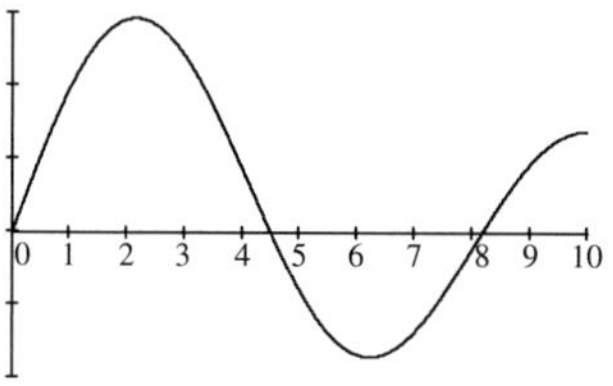 $R_3 = J_1(z_3 r/a)$

7.3.3.2 Electromechanical Resonances

Recall the expressions for admittance and impedance given by Eqs. (142)–(143). These expressions can be rearranged as

$$Y = i\omega C\left[1 - k_p^2\left(1 - \frac{(1+\nu)\,J_1(z)}{zJ_0(z) - (1-\nu)\,J_1(z)}\right)\right] \tag{157}$$

$$Z = \frac{1}{i\omega C}\left[1 - k_p^2\left(1 - \frac{(1+\nu)\,J_1(z)}{zJ_0(z) - (1-\nu)\,J_1(z)}\right)\right]^{-1} \tag{158}$$

The following conditions are considered:

- *Resonance*, when $Y \to \infty$, i.e., $Z = 0$
- *Anti-resonance*, when $Y = 0$, i.e., $Z \to \infty$.

Electrical resonance is associated with the situation in which a device is drawing very large currents when excited harmonically with a constant voltage at a given frequency. At resonance, the admittance becomes very large, whereas the impedance goes to zero. As the admittance becomes very large, the current drawn under constant-voltage excitation also becomes very large as $I = YV$. In piezoelectric devices, the mechanical response at electrical resonance also becomes very large. This happens because the electromechanical coupling of the piezoelectric material transfers energy from the electrical input into the mechanical response. For these reasons, the resonance of an electrically driven piezoelectric device must be seen as an *electromechanical resonance.* A piezoelectric wafer driven at resonance with high amplitudes may undergo mechanical deterioration and even break up.

Electrical anti-resonance is associated with the situation in which a device under constant-voltage excitation draws almost no current. At anti-resonance, the admittance goes to zero, whereas the impedance becomes very large. Under constant-voltage excitation, this condition results in very small current being drawn from the source. In a piezoelectric device, the mechanical response at electrical anti-resonance is also very small. A piezoelectric wafer driven at the electrical anti-resonance hardly moves at all. The resonance of an electrically driven piezoelectric device must be also seen as an *electromechanical anti-resonance.*

The *condition for electromechanical resonance* is obtained by studying the poles of Y, i.e., the values of z that make $Y \to \infty$. The poles of Y are roots of the denominator. These are obtained by solving the equation

$$zJ_0(z) - (1-\nu)\,J_1(z) = 0 \quad \text{(resonance)} \tag{159}$$

This equation is the same as the equation used to determine the mechanical resonances. This is not surprising, because in our analysis of mechanical resonances we only considered the axisymmetric modes, which couple well with a uniform electric field excitation. Hence, the frequencies of electromechanical resonance correspond identically to the frequencies for axisymmetric mechanical resonance given in Table 7.8.

The *condition for electromechanical anti-resonance* is obtained by studying the zeros of Y, i.e., the values of z which make $Y = 0$. Because electromechanical anti-resonances correspond to zeros of the admittance (i.e., poles of the impedance), the current at anti-resonance is zero, $I = 0$. Equation (157) indicates that $Y = 0$ happens when

$$1 - k_p^2\left[1 - \frac{(1+\nu)\,J_1(z)}{zJ_0(z) - (1-\nu)\,J_1(z)}\right] = 0 \tag{160}$$

Upon rearrangement,

$$\begin{gathered} 1 = k_p^2\left[1 - \frac{(1+\nu)\,J_1(z)}{zJ_0(z) - (1-\nu)\,J_1(z)}\right] \\ zJ_0(z) - (1-\nu)\,J_1(z) = k_p^2\,[zJ_0(z) - (1-\nu)\,J_1(z) - (1+\nu)\,J_1(z)] \end{gathered} \tag{161}$$

or

$$\begin{gathered} 1 = k_p^2\left[1 - \frac{(1+\nu)\,J_1(z)}{zJ_0(z) - (1-\nu)\,J_1(z)}\right] \\ zJ_0(z) - (1-\nu)\,J_1(z) = k_p^2\,[zJ_0(z) - (1-\nu)\,J_1(z) - (1+\nu)\,J_1(z)] \end{gathered} \tag{162}$$

Hence, the anti-resonance condition is

$$\frac{zJ_0(z)}{J_1(z)} = \frac{1 - \nu - 2k_p^2}{(1 - k_p^2)} \quad \text{(anti-resonance)} \tag{163}$$

This equation is also transcendental and does not accept closed-form solutions. Its solutions are found numerically.

7.3.4 EXPERIMENTAL RESULTS

Measured results and calculated predictions for circular PWAS are given in Table 7.9. A superposed plot of the measured and calculated E/M admittance spectra is given in Fig. 7.15. The E/M admittance of a circular PWAS undergoing axisymmetric in-plane radial vibrations was modeled with Eq. (158). Figure 7.15 shows the predicted and measured results superposed on the same plot. Three resonance peaks are clearly visible. The corresponding frequencies (300 kHz, 784 kHz, and 1247 kHz, as indicated in Table 7.9) correspond to the first three in-plane radial modes. The forth in-plane frequency (1697 kHz), that lies outside the 0–1500 kHz plotting range, was not plotted. However, its value appears in Table 7.9. During our experiments, we also noticed a very high frequency peak in the E/M impedance real-part response at 10 895 kHz. This value can be identified with the out-of-plane thickness vibration. Comparison of measured and calculated results listed in Table 7.9 for the circular disk case indicate very good agreement between theory and experiments (2.1% maximum error).

TABLE 7.9 Results of the dynamic characterization of a circular PWAS ($d = 6.98\,\text{mm}$, $t = 0.216\,\text{mm}$, $\varepsilon_{33}^T = 15.470 \cdot 10^9\,\text{F/m}$, $s_{11}^E = 18 \cdot 10^{-12} \cdot \text{Pa}^{-1}$, $d_{31} = -175 \cdot 10^{-12}\,\text{m/V}$, $\text{k}_\text{p} = 0.63$). (R = axisymmetric radial vibration, T = out-of-plane thickness vibration)

	Frequency (kHz)				
Experimental	300 (1R)	784 (2R)	1,247 (3R)	1,697 (4R)	10 895 (1T)
Calculated	303 (1R)	796 (2R)	1,267 (3R)	1,733 (4R)	10 690 (1T)
Error	−1.0%	−1.5%	−1.6%	−2.1%	1.9%

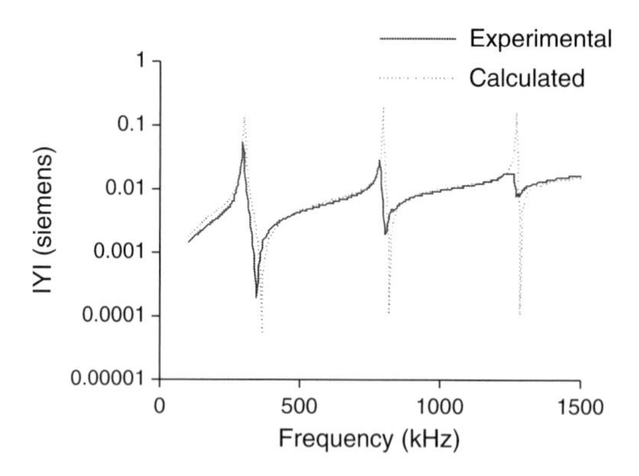

FIGURE 7.15 Experimental and calculated admittance spectra for circular PWAS ($d = 6.98\,\text{mm}$, $t = 0.216\,\text{mm}$, $\varepsilon_{33}^T = 15.470 \cdot 10^9\,\text{F/m}$, $s_{11}^E = 18 \cdot 10^{-12}\,\text{Pa}^{-1}$, $d_{31} = -175 \cdot 10^{-12}\,\text{m/V}$, $k_p = 0.63$).

7.4 COUPLED-FIELD ANALYSIS OF PWAS RESONATORS

In previous sections, we used solely analytical models to predict the electromechanical behavior of PWAS resonators and to predict their E/M admittance and impedance response during a frequency sweep. However, our results were not always perfect, as for example in the approximate analysis of a square PWAS resonator using 1-D analysis. In the engineering community, an alternative to analytical methods has developed in the form of the finite element method (FEM), which can perform the numerical analysis of complicated structures using small-element discretization. The FEM approach has gained wide-ranging acceptance especially for its ease of use in modeling complicated structures. Though the coupled-field FEM approach was initially developed for thermo-elasto-dynamic analysis, current developments in the FEM theory have produced advanced codes that are capable of handling more complex coupled-field (multi-physics) problems, such as the interaction of elastic, dynamic, and electrical fields that takes place in piezoelectric devices. Commercially available FEM packages have started to offer the option of coupled-field elements (see, for example, ANSYS, ABAQUS, etc.) In this section, we will explore how the coupled-field FEM approach can be applied to predict the E/M impedance spectrum of PWAS resonators and compare such results with analytical predictions and experimental measurements. Free PWAS will be modeled with the coupled-field FEM approach. Voltage constraints will be applied to the PWAS electrodes. A time harmonic analysis will be performed on the FEM models and the variation of the electrical charge with frequency will be monitored. From the electrical charge data, we can calculate the electric current and then the E/M impedance of the free PWAS. The PWAS E/M impedance $Z(\omega)$ is calculated at each frequency value as the ratio V/I, where V is the voltage and I is the electric current. In this calculation, the voltage V was the actual voltage constrain applied to the PWAS surface electrodes. The current I was calculated form the electric charge dynamically accumulated on the surface electrodes of the PWAS. Comparison of the simulated E/M impedance results with measured experimental data will be provided.

To perform the coupled-field analysis of a PWAS resonator, we used coupled-field elements, which could deal with both mechanical and electrical fields. For the coupled-field piezoelectric analysis, the stress field and the electric field are coupled to each other such that change in one field will induce change in the other field. The ABAQUS coupled-field finite element used in our analysis is the 3-D brick element that has eight nodes with up to six degrees of freedom (dof's) at each node. When used for piezoelectric analysis, it could have an additional dof, the electrical voltage. This electrical dof comes in addition to the displacement dof's. For each dof of a node, there is a reaction force. Reaction force FX, FY, FZ corresponds to the X, Y, Z displacement dof's, respectively. The electric charge Q is the electrical reaction corresponding to the voltage dof. We will utilize this charge Q to calculate the impedance data. An alternating electric voltage V is applied to all the coupled-field elements in the form of prescribed electrical dof's; the electrical charge will accumulate as electrical reaction Q on the PWAS surface electrodes. The impedance Z is then calculated as V/I, where I is the current value and V is the applied potential voltage. The current I comes from the charge accumulated on the PWAS surface electrodes and is calculated as $I = j\omega\Sigma Q_i$, where ω is the operating frequency, j is $\sqrt{-1}$, and ΣQ_i is the summed nodal charge (electrical reaction load at the nodes). When excited by an alternating electric voltage, the free PWAS acts as an electromechanical resonator. The modeling of a free PWAS is useful for understanding the electromechanical coupling between the mechanical vibration response and the complex electrical response of the sensor.

Harmonic coupled-field FEM analysis was performed on a square PWAS and on a circular PWAS (Table 7.10) The frequency response of the E/M impedance was obtained. In the harmonic analysis, excitation of sinusoidal voltage was applied on the nodes at the top and bottom surface. The excitation signal swept a certain frequency range similar to the way the impedance analyzer equipment measures the PWAS impedance. When the

TABLE 7.10 Free PWAS model used in coupled-field FEM analysis

			FEM model			
Shape	Thickness	Dimension	Element length	Number of elements	Nodes	Materials
Square	0.2 mm	7 mm × 7 mm	0.25 mm	392	675	APC 850
Circular	0.2 mm	7 mm diameter	0.25 mm	441	507	APC 850

PWAS is excited with alternating voltage, the corresponding current flows through the nodes placed on the PWAS electrodes. Then, the impedance Z at sensor terminals is calculated as V/I, where I is the current and V is the applied potential.

7.4.1 COUPLED-FIELD FEM ANALYSIS OF RECTANGULAR PWAS RESONATORS

The square PWAS considered in this analysis is 7 mm long, 7 mm wide, and 0.2 mm thick, APC 850 piezoceramic. This PWAS is similar to that considered in an earlier section (Fig. 7.5). The FEM mesh for this square PWAS is shown in Fig. 7.16a. The simulation of PWAS is simplified by using the symmetry conditions. Thus, considerable computer time can be saved. Only one quarter of the PWAS was modeled. The PWAS was modeled using 3-D eight-node coupled-field elements that has four dof's at each node (three displacements dof's and one voltage dof). This brick coupled-field element is capable of modeling piezoelectric materials when its VOLTAGE dof is activated. The nodes at the top and bottom surfaces had their VOLTAGE dof coupled to a common master node in order to simulate the existence of electrodes on these PWAS surfaces. This approach simplifies the solution process and yields a faster solution.

When a free PWAS is excited either electrically or mechanically, resonance may happen when the response is very large. These resonances can be of two types:

- Electromechanical resonances
- Mechanical resonances.

Mechanical resonances take place in the same condition as in a conventional elastic structure whereas electromechanical resonances are specific to piezoelectric materials.

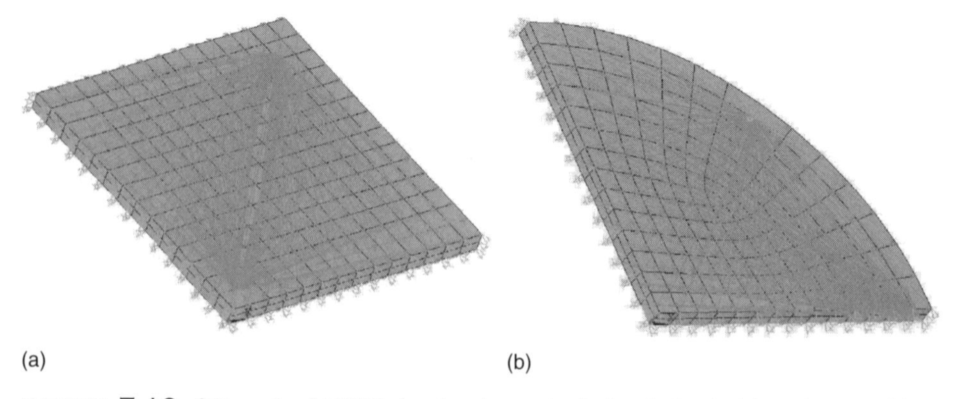

(a) (b)

FIGURE 7.16 3-D mesh of PWAS showing the mechanical and electrical boundary conditions. All the electrical dof's are joined together to a common voltage potential, V: (a) rectangular PWAS; (b) circular PWAS (taking advantage of symmetry, only one quarter of the PWAS is meshed).

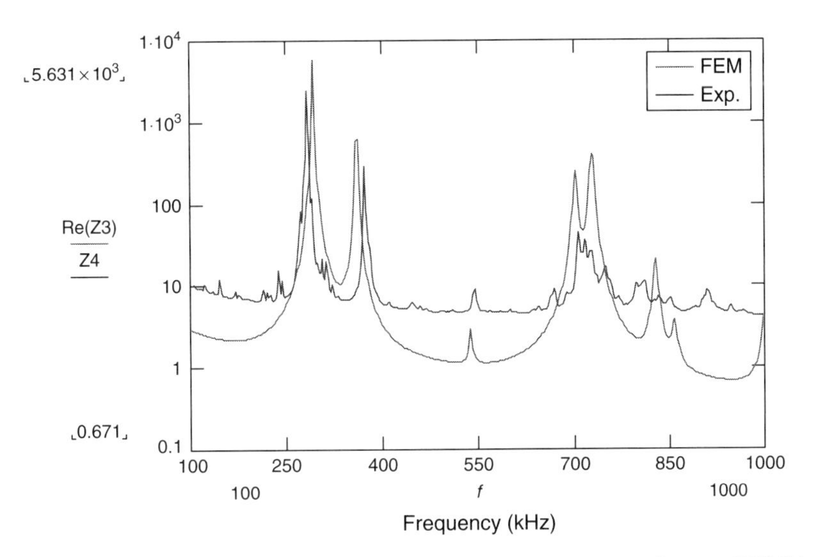

FIGURE 7.17 Comparison of real part frequency response of impedance of square PWAS between simulated results, Z3, and experimental data, Z4.

Electromechanical resonances reflect the coupling between the mechanical and electrical variables, they happen under electric excitation, which produces electromechanical response (i.e., both a mechanical vibration and a change in the electric admittance and impedance). When a PWAS is excited harmonically with a constant voltage at a given frequency, electrical resonance is associated with the situation in which a device is drawing very large currents. At resonance, the admittance becomes very large whereas the impedance goes to zero. As the admittance becomes very large, the current drawn under constant-voltage excitation also becomes very large because $I = YV$. In piezoelectric devices, the mechanical response at electrical resonance also becomes very large. This happens because the electromechanical coupling of the piezoelectric materials transfers energy from the electrical input into the mechanical response.

Figure 7.17 shows the log-scale plot of the real part of the electromechanical impedance calculated with the couple-field FEM analysis in the 100 kHz–1 MHz frequency range. E/M resonance peaks are well distinguished.

To verify these E/M impedance prediction results, we acquired experimental data on an actual PWAS with the HP 4194A Impedance Analyzer. The measured frequency response of the PWAS impedance is also plotted in Fig. 7.17. It is apparent that the simulation results compare very well with the experimental measurements, especially in the detection of the first four resonances. At higher frequencies, the results are not as well in agreement. The experimental curve shows an additional peak at around 920 kHz, which may be due to the edge roughness generated during the manufacturing process.

7.4.2 COUPLED-FIELD FEM ANALYSIS OF CIRCULAR PWAS RESONATORS

We continue our analysis with the consideration of a circular PWAS of 7-mm diameter and 0.2-mm thickness. The FEM mesh for this circular PWAS is shown in Fig. 7.16b. The PWAS simulation is again simplified by using the symmetry conditions. The impedance plot for this circular PWAS is given in Fig. 7.18. The frequency range from 100 kHz to 2 MHz was considered. Four resonance peaks are clearly visible in Fig. 7.18; these peaks corresponded to the first four in-plane radial modes.

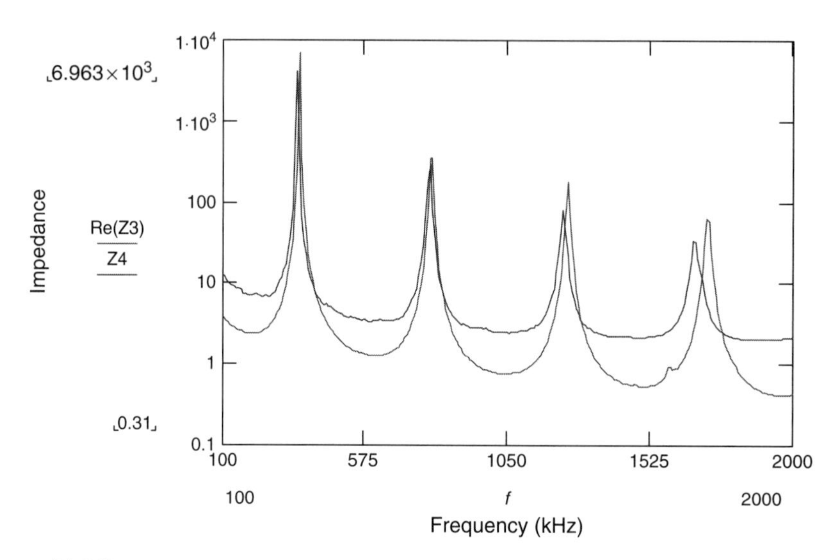

FIGURE 7.18 Comparison of real part frequency response of impedance of circular PWAS between simulated results, Z3, and experimental data, Z4.

The simulated impedance results were compared with impedance measurements performed on the actual circular PWAS using the HP 4194A Impedance Analyzer. Excellent agreement between finite element simulation and experimental measurements is indicated. We can also see that in high-frequency range (1500 kHz to 2000 kHz), the simulated impedance peak is shifted a little with respect to the experimental measurements. This may be explained by the fact that the finite element size and the time step of the harmonic analysis may not be sufficiently fine for such high frequencies and thus can yield corrupted results. The effect of element size and time step at high frequencies should always be explored when doing FEM analysis.

These two examples, the square PWAS and the circular PWAS, indicate the overall conclusion that very good agreement can be found between the coupled-field FEM simulation and experimental data of E/M impedance of free PWAS resonators. The coupled-field FEM results seem better than the analytical results, especially in predicting the response of rectangular PWAS (compare Fig. 7.17 with Fig. 7.9).

7.5 CONSTRAINED PWAS

The analysis of a constraint PWAS is essential for the analysis of the PWAS as a structural modal sensor, to be described in Chapter 9. When affixed to a structure, the PWAS is constrained by the structure and its dynamic behavior is essentially modified. In this section, we will consider that the structure constraining the PWAS is represented by an unspecified dynamic structural stiffness, k_{str}. Because this dynamic structural stiffness is frequency dependent, the way it interacts with the PWAS will also be frequency dependent and can significantly alter the PWAS resonances. As it will be shown in Chapter 9, the structural dynamics can overpower the inherent PWAS dynamics. In this case, the PWAS E/M impedance will closely follow the dynamics of the structure and the PWAS becomes a sensor of the dynamical modal behavior of the structure.

7.5.1 ONE-DIMENSIONAL ANALYSIS OF A CONSTRAINED PWAS

Consider a PWAS of length l_a, thickness t_a, and width b_a, undergoing longitudinal expansion, u_1, induced by the thickness polarization electric field, E_3. The electric field is produced by the application of a harmonic voltage $V(t) = \hat{V}e^{i\omega t}$ between the top and bottom surface electrodes. The resulting electric field, $E = V/t$, is assumed uniform with respect to $x_1 (\partial E/\partial x_1 = 0)$. The length, width, and thickness are assumed to have widely separated values ($t_a \ll b_a \ll l_a$) such that the length, width, and thickness motions are practically uncoupled.

The constitutive equations of the piezoelectric material are

$$S_1 = s_{11}^E T_1 + d_{31} E_3 \tag{164}$$

$$D_3 = d_{31} T_1 + \varepsilon_{33}^T E_3 \tag{165}$$

where S_1 is the strain, T_1 is the stress, D_3 is the electrical displacement (charge per unit area), s_{11}^E is the mechanical compliance at zero field, ε_3^T is the dielectric constant at zero stress, d_{31} is the induced strain coefficient, i.e., mechanical strain per unit electric field.

When the PWAS is bonded to the structure, the structure will constrain the PWAS motion with a structural stiffness, k_{str}. Hence, the present analysis will be concerned with the study of an elastically constrained PWAS, as shown in Fig. 7.19. In our model, the overall structural stiffness applied to the PWAS has been split into two equal components applied to the PWAS ends. The values of these components are $2k_{\text{str}}$ each, such that

$$k_{\text{total}} = \left[(2k_{\text{str}})^{-1} + (2k_{\text{str}})^{-1}\right]^{-1} = k_{\text{str}} \tag{166}$$

The boundary conditions applied at the PWAS ends balance the stress resultant, $T_1 b_a t_a$, with the spring reaction force, $2k_{\text{str}} u_1$, i.e.,

$$\begin{aligned} T_1\left(\frac{1}{2}l_a\right) b_a t_a &= -2k_{\text{str}} u_1\left(\frac{1}{2}l_a\right) \\ T_1\left(-\frac{1}{2}l_a\right) b_a t_a &= 2k_{\text{str}} u_1\left(-\frac{1}{2}l_a\right) \end{aligned} \tag{167}$$

Note that the + and − signs in Eq. (167) are chosen such as to be consistent with the sign convention. Recall the strain-displacement relation

$$S_1 = u_1' \tag{168}$$

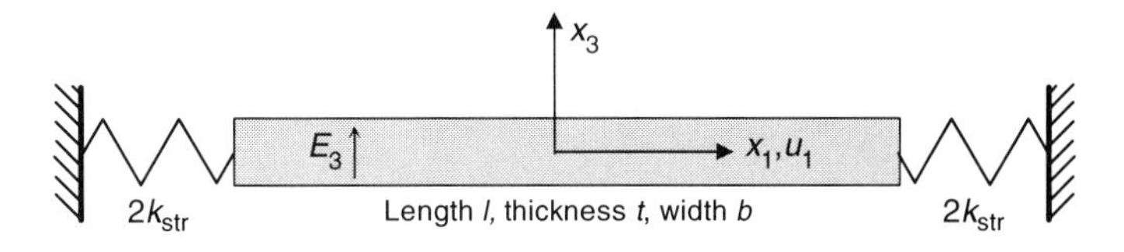

FIGURE 7.19 PWAS constrained by structural stiffness, k_{str}.

Substitution of Eqs. (164) and (168) into Eq. (167) gives

$$\begin{aligned} u_1'\left(\frac{1}{2}l_a\right)b_a t_a &= -2k_{\text{str}}\frac{s_{11}^E}{bt}u_1\left(\frac{1}{2}l_a\right)+d_{31}E_3 \\ u_1'\left(-\frac{1}{2}l_a\right)b_a t_a &= 2k_{\text{str}}\frac{s_{11}^E}{bt}u_1\left(-\frac{1}{2}l_a\right)+d_{31}E_3 \end{aligned} \tag{169}$$

Introducing the quasi-static PWAS stiffness,

$$k_{\text{PWAS}} = \frac{A_a}{s_{11}^E l_a}, \tag{170}$$

and the stiffness ratio

$$r = \frac{k_{\text{str}}}{k_{\text{PWAS}}}, \tag{171}$$

allows us to rearrange Eq. (169) in the form

$$\begin{aligned} u_1'\left(+\frac{1}{2}l_a\right)b_a t_a + \frac{r}{\frac{1}{2}l_a}u_1\left(\frac{1}{2}l_a\right) &= d_{31}E_3 \\ u_1'\left(-\frac{1}{2}l_a\right)b_a t_a - \frac{r}{\frac{1}{2}l_a}u_1\left(\frac{1}{2}l_a\right) &= d_{31}E_3 \end{aligned} \tag{172}$$

7.5.1.1 Mechanical Response

Substitution of Newton law of motion, $T_1' = \rho\ddot{u}_1$, and of the strain-displacement relation, $S_1 = u_1'$, into Eq. (164) yields the axial waves equation

$$\ddot{u}_1 = c_a^2 u_1'' \tag{173}$$

where $\dot{u} = \partial u/\partial t$, $u' = \partial u/\partial x$, and $c_a^2 = 1/\rho s_{11}^E$ is the piezoelectric material wave speed. The general solution of Eq. (173) is

$$u_1(x,t) = \hat{u}_1(x)e^{i\omega t} \tag{174}$$

where

$$\hat{u}_1(x) = (C_1\sin\gamma x + C_2\cos\gamma x) \tag{175}$$

The variable $\gamma = \omega/c_a$ is the wavenumber. The constants C_1 and C_2 are to be determined from the boundary conditions. Substitution of the general solution Eq. (175) into the boundary conditions Eq. (172) yields the following linear system in C_1 and C_2

$$\begin{aligned} \frac{1}{2}l\cdot\gamma\left(C_1\cos\frac{1}{2}\gamma l - C_2\sin\frac{1}{2}\gamma l\right) + r\left(C_1\sin\frac{1}{2}\gamma l + C_2\cos\frac{1}{2}\gamma l\right) &= \frac{1}{2}l\cdot d_{31}\hat{E}_3 \\ \frac{1}{2}l\cdot\gamma\left(C_1\cos\frac{1}{2}\gamma l + C_2\sin\frac{1}{2}\gamma l\right) - r\left(-C_1\sin\frac{1}{2}\gamma l + C_2\cos\frac{1}{2}\gamma l\right) &= \frac{1}{2}l\cdot d_{31}\hat{E}_3 \end{aligned} \tag{176}$$

Rearranging, we get

$$\begin{aligned}
&\frac{1}{2}\gamma l\left(\cos\frac{1}{2}\gamma l + r\sin\frac{1}{2}\gamma l\right)C_1 + \left(-\frac{1}{2}\gamma l\sin\frac{1}{2}\gamma l + r\cos\frac{1}{2}\gamma l\right)C_2 = \frac{1}{2}l\cdot d_{31}\hat{E}_3\\
&\frac{1}{2}\gamma l\left(\cos\frac{1}{2}\gamma l + r\sin\frac{1}{2}\gamma l\right)C_1 + \left(+\frac{1}{2}\gamma l\sin\frac{1}{2}\gamma l - r\cos\frac{1}{2}\gamma l\right)C_2 = \frac{1}{2}l\cdot d_{31}\hat{E}_3
\end{aligned} \tag{177}$$

Recall the notations $u_{\text{ISA}} = d_{31}\hat{E}_3 l$ and $\phi = \frac{1}{2}\gamma l$. Upon substitution, we get the following linear system in C_1 and C_2

$$\begin{aligned}
(\varphi\cos\varphi + r\sin\varphi)C_1 - (\varphi\sin\varphi - r\cos\varphi)C_2 &= \frac{1}{2}u_{\text{ISA}}\\
(\varphi\cos\varphi + r\sin\varphi)C_1 + (\varphi\sin\varphi - r\cos\varphi)C_2 &= \frac{1}{2}u_{\text{ISA}}
\end{aligned} \tag{178}$$

Assume the system determinant is nonzero, i.e., $\Delta \neq 0$. The solution can be obtained as follows. Subtraction of the first equation from the second equation yields

$$2(\varphi\sin\varphi - r\cos\varphi)C_2 = 0 \tag{179}$$

Hence, $C_2 = 0$. Now, add the two equations. The result is

$$2(\varphi\cos\varphi + r\sin\varphi)C_1 = 2\frac{1}{2}u_{\text{ISA}} \tag{180}$$

Hence,

$$C_1 = \frac{1}{2}u_{\text{ISA}}\frac{1}{(\varphi\cos\varphi + r\sin\varphi)}, \quad C_2 = 0 \tag{181}$$

Substituting C_1 and C_2 into Eq. (175) yields the solution

$$\hat{u}_1(x) = \frac{1}{2}u_{\text{ISA}}\frac{\sin\gamma x}{\varphi\cos\varphi + r\sin\varphi} \tag{182}$$

Substituting $\varphi = \frac{1}{2}\gamma l$, we write

$$\hat{u}_1(x) = \frac{1}{2}u_{\text{ISA}}\frac{\sin\gamma x}{\frac{1}{2}\gamma l\cos\frac{1}{2}\gamma l + r\sin\frac{1}{2}\gamma l} \tag{183}$$

7.5.1.2 Electrical Response

Consider the constrained PWAS under harmonic electric excitation as shown in Fig. 7.20. Recall Eq. (165) representing the electrical displacement

$$D_3 = d_{31}T_1 + \varepsilon_{33}^T E_3 \tag{184}$$

Equation (164) yields the stress as function of strain and electric field, i.e.,

$$T_1 = \frac{1}{s_{11}^E}(S_1 - d_{31}E_3) \tag{185}$$

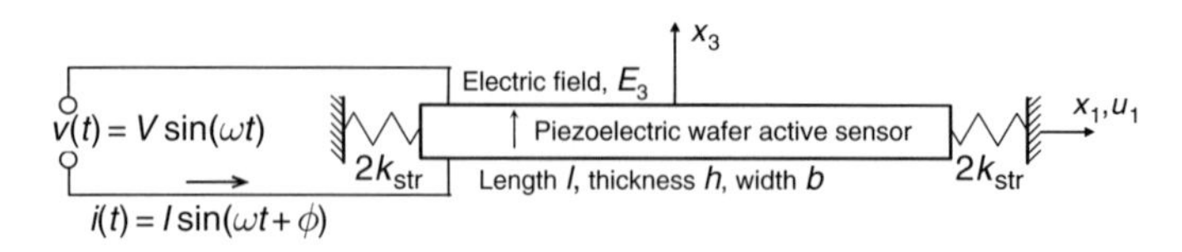

FIGURE 7.20 Schematic for the electrical response analysis of a constraint PWAS under harmonic electric excitation.

Hence, the electric displacement can be expressed as

$$D_3 = \frac{d_{31}}{s_{11}}(S_1 - d_{31}E_3) + \varepsilon_{33}^T E_3 \tag{186}$$

Upon substitution of the strain-displacement relation of Eq. (168), we get

$$D_3 = \frac{d_{31}}{s_{11}^E}\cdot u_1' - \frac{d_{31}^2}{s_{11}^E}\cdot E_3 + \varepsilon_{33}^T E_3 \tag{187}$$

i.e.,

$$D_3 = \varepsilon_{33}^T E_3\left[1 - k_{31}^2\left(1 - \frac{u_1'}{d_{31}E_3}\right)\right] \tag{188}$$

where $k_{13}^2 = d_{31}^2/(s_{11}^E\varepsilon_{33}^T)$ is the electromechanical coupling coefficient. Integration of Eq. (188) over the electrodes area $A = bl$ yields the total charge.

Equation (165) can be re-expressed as

$$D_3 = \frac{d_{31}}{s_{11}^E}(u_1' - d_{31}E_3) + \varepsilon_{33}^T E_3 = \varepsilon_{33}^T E_3\left[1 + k_{31}^2\left(\frac{u_1'}{d_{31}E_3} - 1\right)\right] \tag{189}$$

where $k_{31}^2 = d_{31}^2/(s_{11}^E\varepsilon_{33}^T)$ is the electromechanical coupling coefficient. Integration of Eq. (189) over the area of the piezoelectric wafer yields the total charge

$$Q = \int_{-\frac{l_a}{2}}^{+\frac{l_a}{2}}\int_{-\frac{b_a}{2}}^{+\frac{b_a}{2}} D_3\,\mathrm{d}x\,\mathrm{d}y = \varepsilon_{33}^T\frac{b_a l_a}{t_a}V\left[1 + k_{31}^2\left(\frac{1}{l_a}\frac{1}{d_{31}E_3}u_1\Big|_{-\frac{1}{2}l}^{\frac{1}{2}l} - 1\right)\right] \tag{190}$$

Assuming harmonic time dependence ($Q = \hat{Q}e^{i\omega t}$, etc.), we write

$$\hat{Q} = C\hat{E}_3\left\{1 + k_{31}^2\left(\frac{1}{l_a}\frac{1}{d_{31}\hat{E}_3}\left[\hat{u}_1\left(\frac{1}{2}l\right) - \hat{u}_1\left(-\frac{1}{2}l\right)\right] - 1\right)\right\} \tag{191}$$

where C is the conventional stress-free capacitance of the PWAS given by

$$C = \varepsilon_{33}^T\frac{b_a l_a}{t_a} \tag{192}$$

Using the definitions, $u_{ISA} = d_{31}\hat{E}_3 l_a$, $\hat{V} = \hat{E}_3/t_a$, we obtain

$$\hat{Q} = C\hat{V}\left[1 - k_{31}^2 + k_{31}^2\left(\frac{\hat{u}_1(\frac{1}{2}l) - \hat{u}_1(-\frac{1}{2}l)}{u_{ISA}}\right)\right] \tag{193}$$

The electric current is obtained as the time derivative of the electric charge, i.e.,

$$I = \dot{Q} = i\omega Q \tag{194}$$

Hence,

$$\hat{I} = i\omega C\hat{V}\left[1 - k_{31}^2 + k_{31}^2\left(\frac{\hat{u}_1(\frac{1}{2}l) - \hat{u}_1(-\frac{1}{2}l)}{u_{\mathrm{ISA}}}\right)\right] \tag{195}$$

The admittance, Y, is defined as the ratio between the current and voltage, i.e.,

$$Y = \frac{\hat{I}}{\hat{V}} = i\omega C\left[1 - k_{31}^2 + k_{31}^2\left(\frac{\hat{u}_1(\frac{1}{2}l) - \hat{u}_1(-\frac{1}{2}l)}{u_{\mathrm{ISA}}}\right)\right] \tag{196}$$

Recall Eq. (182) that defines the displacement solution

$$\hat{u}_1(x) = \frac{1}{2}u_{\mathrm{ISA}}\frac{\sin\gamma x}{\varphi\cos\varphi + r\sin\varphi} \tag{197}$$

Hence, the term of Eq. (196) containing $\hat{u}_1$ becomes

$$\hat{u}_1\left(\frac{1}{2}l\right) - \hat{u}_1\left(-\frac{1}{2}l\right) = \frac{1}{2}\frac{\sin\frac{1}{2}\gamma l - (-\sin\frac{1}{2}\gamma l)}{\phi\cos\phi + r\sin\phi} = \frac{1}{2}2\frac{\sin\phi}{\phi\cos\phi + r\sin\phi} = \frac{1}{r + \phi\cot\phi} \tag{198}$$

where the notation $\phi = \frac{1}{2}\gamma l$ was invoked. Upon substitution, we get

$$Y = \frac{\hat{I}}{\hat{V}} = i\omega C\left(1 - k_{31}^2 + k_{31}^2\frac{1}{r + \phi\cot\phi}\right) \tag{199}$$

or

$$Y = \frac{\hat{I}}{\hat{V}} = i\omega C\left[1 - k_{31}^2\left(1 - \frac{1}{r + \phi\cot\phi}\right)\right] \tag{200}$$

Note that the admittance is purely imaginary and consists of a capacitive admittance, $i\omega C$, modified by the effect of piezoelectric coupling between mechanical and electrical variables. The impedance, Z, is obtained as the ratio between the voltage and current, i.e.,

$$Z = \frac{\hat{V}}{\hat{I}} = \frac{1}{i\omega C}\left[1 - k_{31}^2\left(1 - \frac{1}{r + \phi\cot\phi}\right)\right]^{-1} \tag{201}$$

Thus, we have arrived at the admittance and impedance expressions for a PWAS constrained by the structural substrate with an equivalent stiffness ratio r.

In Eqs. (200) and (201), the structural stiffness ratio, r, is additive to the PWAS resonance term, $\varphi\cot\varphi$. When the PWAS is used in a frequency sweep, the apparent structural stiffness, k_{str}, will vary with frequency, going through zero at structural resonances and through extreme values at structural anti-resonances. Equations (200) and (201) imply that both structural resonances and PWAS resonances will be reflected in the admittance and impedance frequency spectra.

7.5.1.3 Asymptotic Behavior

Analysis of the asymptotic behavior of Eq. (201) allows us to recover situations for which the results are either known from previous investigations or can be easily determined through simple analysis. The asymptotic conditions to be considered here are

- Free PWAS, i.e., $r = 0$
- Fully constrained (blocked) PWAS, i.e., $r \to \infty$
- Constrained PWAS under quasi-static conditions, i.e., $\phi = 0$.

These asymptotic conditions will be considered in turn.

Free PWAS

A free piezoelectric wafer corresponds to $k_{str} = 0$, i.e., $r = 0$. In this case, the admittance and impedance expressions for a free piezoelectric wafer are recovered. Indeed, as the r term in the denominator of Eqs. (200) and (201) vanishes, the following expressions are obtained

$$\begin{aligned} Y_{free} &= i\omega C\left[1 - k_{31}^2\left(1 - \frac{1}{\phi \cot \phi}\right)\right] \\ Z_{free} &= \frac{1}{i\omega C}\left[1 - k_{31}^2\left(1 - \frac{1}{\phi \cot \phi}\right)\right]^{-1} \end{aligned} \tag{202}$$

These are exactly the previously determined expressions for the admittance and impedance expressions of a free piezoelectric wafer.

Fully constrained (blocked) PWAS

A fully constrained piezoelectric wafer has $k_{str} \to \infty$, i.e., $r \to \infty$. In this case, the fraction containing r at the denominator vanishes all together, and the admittance and impedance expressions become

$$\begin{aligned} Y_{blocked} &= i\omega C\left(1 - k_{31}^2\right) \\ Z_{blocked} &= \frac{1}{i\omega C}\left(1 - k_{31}^2\right)^{-1} \end{aligned} \tag{203}$$

These are indeed the expressions for a 'blocked' piezoelectric resonator cited in the specialized literature (Ikeda, 1996).

Constrained PWAS under quasi-static conditions

Quasi-static conditions are met when the frequency of oscillation is so low that the dynamic effects inside the piezoelectric wafer are negligible. This implies that the driving frequency is well below the first natural frequency of the piezoelectric wafer. Another way of looking at this is to say that the wavelength associated with this frequency is much larger than the wafer length, i.e., $\lambda \gg l$. In other words, the length of wafer is so small that the elastic waves travel very quickly from one end to the other, and no stress and strain gradients due to the dynamic effects are present. In terms of the wavenumber, γ, this assumption reduces to saying that $\gamma l \to 0$, because $\gamma = \frac{\omega}{c} = 2\pi\frac{1}{\lambda}$ and $\gamma l = \frac{\omega}{c} l = 2\pi \frac{l}{\lambda} \xrightarrow[\lambda \gg l]{} 0$. For $\phi \to 0$, Eqs. (200) and (201) becomes

$$\begin{aligned} Y &= i\omega C\left(1 - k_{31}^2 \frac{r}{1+r}\right) \\ Z &= \frac{1}{i\omega C}\left(1 - k_{31}^2 \frac{r}{1+r}\right)^{-1} \end{aligned} \tag{204}$$

This analysis of the asymptotic behavior illustrated how general the admittance and impedance expressions of Eqs. (200) and (201) are. The simpler expressions contained in Eqs. (204) can be used for low-frequency structure-focused analysis, in which the PWAS behavior can be considered quasi-static. However, for high-frequency analysis in which the PWAS resonances are also apparent, the complete expressions contained in Eqs. (200) and (201) must be used. These comprehensive expressions cover the complete frequency spectrum and encompass both structure and PWAS dynamics.

7.5.1.4 Damping Effects

The damping effects can be associated either with the piezoelectric material or with the elastic constrained. The damping effects in the piezoelectric material are covered through the adoption of complex compliance and dielectric constant expressions

$$\bar{s}_{11} = s_{11}(1 - i\eta), \quad \bar{\varepsilon}_{33} = \varepsilon_{33}(1 - i\delta) \tag{205}$$

The values of η and δ vary with the piezoceramic formulation but are usually small ($\eta, \delta < 5\%$).

The damping in the elastic constraint is similarly accounted for by assuming a complex stiffness expression, $\bar{k}_{str}$. As a result, the stiffness ratio will also take complex values, $\bar{r} = \bar{k}_{str}/\bar{k}_{PZT}$. This frequency-dependent complex stiffness ratio reflects both the elastic constraint damping and the sensor dissipation mechanisms. Therefore, the admittance and impedance expressions of Eq. (201) takes the complex notation form

$$\begin{aligned} \bar{Y} &= i\omega \bar{C}\left[1 - \bar{k}_{31}^2\left(1 - \frac{1}{\bar{\varphi}\cot\bar{\varphi} + \bar{r}}\right)\right] \\ \bar{Z} &= \frac{1}{i\omega \bar{C}}\left[1 - \bar{k}_{31}^2\left(1 - \frac{1}{\bar{\varphi}\cot\bar{\varphi} + \bar{r}}\right)\right]^{-1} \end{aligned} \tag{206}$$

It is worth noting that the elastic constraint can actually have a dynamic behavior of its own, e.g., of the form

$$\bar{k}_{str}(\omega) = \left[k(\omega) - \omega^2 m(\omega)\right] - i\omega c(\omega) \tag{207}$$

where $k(\omega)$, $m(\omega)$, and $c(\omega)$ are some frequency-dependent stiffness, mass, and damping coefficients.

7.5.1.5 Resonances

In order to determine resonances, we analyze the behavior of the determinant of system (179), i.e.,

$$\Delta = \begin{vmatrix} (\varphi \cos\varphi + r\sin\varphi) & -(\varphi\sin\varphi - r\cos\varphi) \\ (\varphi\cos\varphi + r\sin\varphi) & (\varphi\sin\varphi - r\cos\varphi) \end{vmatrix} \tag{208}$$

or

$$\Delta = 2(\varphi\cos\varphi + r\sin\varphi)(\varphi\sin\varphi - r\cos\varphi) \tag{209}$$

This determinant Δ is zero when either the first parenthesis or the second parenthesis is zero. When the first parenthesis in Δ is zero, the denominator of Eq. (182) vanishes and the response of the system to electrical excitation increases indefinitely. We identify this situation with an *electromechanical resonance*. When the second parenthesis in Δ vanishes, the denominator of the Eq. (182) does not vanish, and the electromechanical response does not increase indefinitely. We identify this situation with a purely *mechanical resonance*, which cannot be excited electrically under the constant-field distribution considered here. Therefore, the following two resonance conditions are identified.

$$\varphi\cos\varphi + r\sin\varphi = 0 \quad \text{(electromechanical resonance)} \tag{210}$$

$$\varphi\sin\varphi + r\cos\varphi = 0 \quad \text{(mechanical resonance only)} \tag{211}$$

The condition that Δ is zero can be also expressed as follows:

$$\Delta = (\phi\cos\phi + r\sin\phi)(\phi\sin\phi - r\cos\phi) = 0 \tag{212}$$

$$\phi^2\cos\phi\sin\phi + r\phi(\sin^2\phi - \cos^2\phi) - r^2\sin\phi\cos\phi = 0 \tag{213}$$

$$(\phi^2 - r^2)\sin 2\phi - r\phi\cos 2\phi = 0 \tag{214}$$

$$\tan 2\phi = \frac{r\phi}{\phi^2 - r^2} \tag{215}$$

Solution of Eq. (215) gives all the mechanical resonances, including those that are only electromechanical resonances.

7.5.2 TWO-DIMENSIONAL ANALYSIS OF A CONSTRAINED CIRCULAR PWAS

To model the circular-shaped PWAS, we start from the piezoelectric constitutive equations in cylindrical coordinates

$$\begin{aligned} S_{rr} &= s_{11}^E T_{rr} + s_{12}^E T_{\theta\theta} + d_{31}E_z \\ S_{\theta\theta} &= s_{12}^E T_{rr} + s_{11}^E T_{\theta\theta} + d_{31}E_z \\ D_z &= d_{31}(T_{rr} + T_{\theta\theta}) + \varepsilon_{33}^T E_z \end{aligned} \tag{216}$$

S_{rr} and $S_{\theta\theta}$ are the mechanical strains, T_{rr} and $T_{\theta\theta}$ the mechanical stresses, E_z the electrical field, D_z the electrical displacement, s_{11}^E and s_{12}^E the mechanical compliances at zero electric field ($E = 0$), ε_{33}^T the dielectric permittivity at zero mechanical stress ($T = 0$), and d_{31} the piezoelectric coupling between the electrical and mechanical variables. For axisymmetric motion, the problem is θ-independent and the space variation is in r only.

Hence, $S_{rr} = \partial u_r / \partial r$ and $S_{\theta\theta} = u_r / r$. Applying Newton law of motion, one recovers, upon substitution, the wave equation in polar coordinates

$$\frac{\partial^2 u_r}{\partial r^2} + \frac{1}{r}\frac{\partial u_r}{\partial r} - \frac{u_r}{r^2} = \frac{1}{c_P^2}\frac{\partial^2 u_r}{\partial t^2} \tag{217}$$

where

$$c_P = \sqrt{\frac{1}{\rho s_{11}^E (1 - \nu_a^2)}} \quad \text{(wave speed)} \tag{218}$$

is the wave speed in the PWAS for axially symmetric radial motion, with ν_a the Poisson's ratio of the piezoelectric material ($\nu_a = -s_{12}^E / s_{11}^E$). Equation (217) admits a general solution in terms of Bessel functions of the first kind, J_1, in the form

$$u_r(r, t) = A J_1\left(\frac{\omega r}{c}\right) e^{i\omega t} \tag{219}$$

where the coefficient A is determined from the boundary conditions.

When the PWAS is mounted on the structure, its circumference is elastically constrained by the dynamic structural stiffness (Fig. 7.21). At the boundary $r = r_a$, we have the boundary condition

$$T_{rr}(r_a) \cdot t_a = k_{\text{str}}(\omega) \cdot u_r(r_a) \tag{220}$$

where t_a is the PWAS thickness. Hence, the radial stress can be expressed as

$$T_{rr}(r_a) = \frac{k_{\text{str}}(\omega) u_r(r_a)}{t_a} \tag{221}$$

Substitution of Eq. (221) into Eq. (216) gives, upon rearrangement,

$$\frac{\partial u_r(r_a)}{\partial r} = \chi(\omega) \cdot (1 + \nu_a) \frac{u_r(r_a)}{r_a} - \nu_a \frac{u_r(r_a)}{r_a} + (1 + \nu_a) d_{31} E_z \tag{222}$$

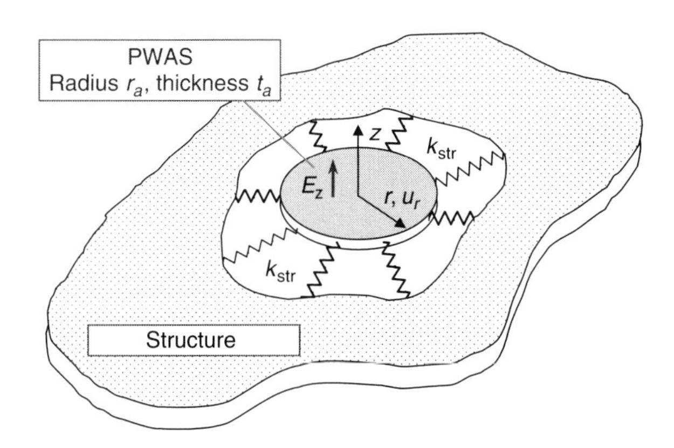

FIGURE 7.21 Circular PWAS constrained by structural stiffness, $k_{\text{str}}(\omega)$.

where

$$\chi(\omega) = \frac{k_{\rm str}(\omega)}{k_{\rm PWAS}} \tag{223}$$

is the dynamic stiffness ratio and

$$k_{\rm PWAS} = \frac{t_a}{\left[r_a s_{11}^E (1 - \nu_a)\right]} \tag{224}$$

is the static stiffness of the circular PWAS. Substitution of Eq. (219) into Eq. (222) yields

$$A = \frac{(1 + \nu_a)\, d_{31} E_0}{\frac{\omega}{c} J_0\left(\frac{\omega r_a}{c}\right) - \frac{[1 - \nu_a + \chi(\omega)(1 + \nu_a)]}{r_a} J_1\left(\frac{\omega r_a}{c}\right)} \tag{225}$$

The electrical admittance is calculated as the ratio between the current and the voltage amplitudes, i.e., $Y = \hat{I}/\hat{V}$. The current is calculated by integrating the electric displacement D_3 over the PWAS area to obtain the total charge, and then differentiating the result with respect to time, whereas the voltage is calculated by multiplying the electric field by the PWAS thickness. Hence, the electrical admittance is expressed as

$$\begin{aligned} Y(\omega) &= i\omega C\left(1 - k_p^2\right) \\ &\times \left[1 + \frac{k_p^2}{1 - k_p^2} \frac{(1 + \nu_a)\, J_1(\varphi_a)}{\varphi_a J_0(\varphi_a) - (1 - \nu_a)\, J_1(\varphi_a) - \chi(\omega)(1 + \nu_a)\, J_1(\varphi_a)}\right] \end{aligned} \tag{226}$$

where $\varphi_a = \omega r_a / c$ and $k_p = \sqrt{2 d_{31}^2 / \left[s_{11}^E (1 - \nu_a) \varepsilon_{33}^T\right]}$ is the planar coupling factor. Then, the inverse relationship between impedance and admittance, $Z(\omega) = 1/Y(\omega)$, yields

$$\begin{aligned} Z(\omega) = \Bigg\{ & i\omega C\left(1 - k_p^2\right) \\ & \times \left[1 + \frac{k_p^2}{1 - k_p^2} \frac{(1 + \nu_a)\, J_1(\varphi_a)}{\varphi_a J_0(\varphi_a) - (1 - \nu_a)\, J_1(\varphi_a) - \chi(\omega)(1 + \nu_a)\, J_1(\varphi_a)}\right]\Bigg\}^{-1} \end{aligned} \tag{227}$$

Equation (227) predicts the E/M impedance spectrum as it would be measured by the impedance analyzer at the embedded PWAS terminals during an SHM process. It allows for direct comparison between calculated predictions and experimental results. The structural dynamics is reflected in Eq. (227) through the dynamic stiffness ratio, $\chi(\omega) = k_{\rm str}(\omega)/k_{\rm PWAS}$, which contains the dynamic stiffness of the structure, $k_{\rm str}(\omega)$. This latter quantity results from the analysis of the circular plate dynamics.

7.6 PWAS ULTRASONIC TRANSDUCERS

For embedded NDE applications, PWAS can be used as embedded ultrasonic transducers. PWAS act as both exciters and detectors of elastic waves. PWAS couple their in-plane motion with the particle motion generated on the material surface by the elastic waves. The in-plane PWAS motion is excited by the applied oscillatory voltage through the d_{31} piezoelectric coupling. The PWAS action as ultrasonic transducers is fundamentally different from that of conventional ultrasonic transducers. Conventional ultrasonic

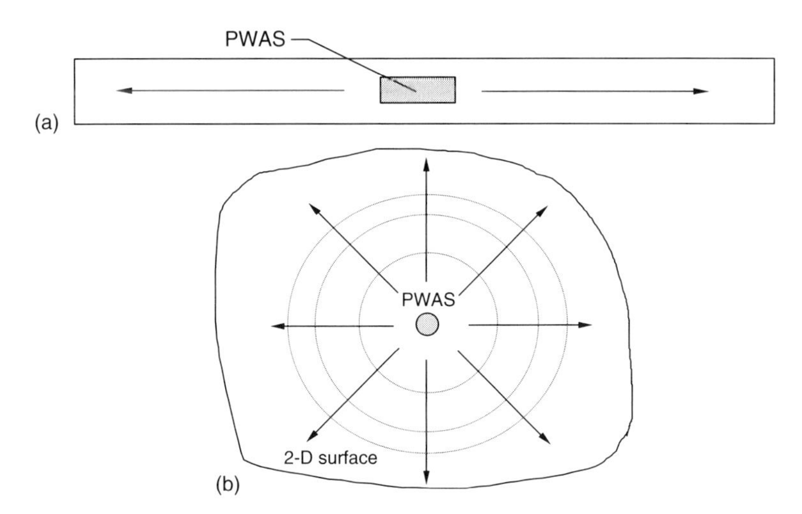

FIGURE 7.22 (a) Elastic waves generated by a PWAS in a 1-D structure; (b) circular-crested Lamb waves generated by a PWAS in a 2-D structure.

transducers act through surface tapping, applying vibrational pressure to the object's surface. PWAS, on the other hand, act through surface pinching and are strain coupled with the object surface. This imparts to PWAS a much better efficiency in transmitting and receiving ultrasonic Lamb and Rayleigh waves than conventional ultrasonic transducers. PWAS are capable of geometric tuning through matching between their characteristic direction and the half wavelength of the exited elastic wave. Rectangular shaped PWAS with high length-to-width ratio can generate unidirectional waves. Circular PWAS excite omnidirectional waves that propagate in circular wave fronts. Unidirectional and omnidirectional wave propagations are illustrated in Fig. 7.22. Omnidirectional waves can be also generated by square PWAS, although their pattern is somehow irregular in the PWAS proximity. At far enough distance ($r \gg a$), the wave front generated by square PWAS is practically identical with that generated by circular PWAS.

7.6.1 SHEAR-LAYER COUPLING BETWEEN PWAS AND STRUCTURE

The transmission of actuation and sensing between the PWAS and the structure is achieved through the adhesive layer. The adhesive layer acts as a shear layer, in which the mechanical effects are transmitted through shear effects. Figure 7.23 shows a thin-wall structure of thickness t and elastic modulus E, with a PWAS of thickness t_a and elastic modulus E_a attached to its upper surface through a bonding layer of thickness t_b and shear

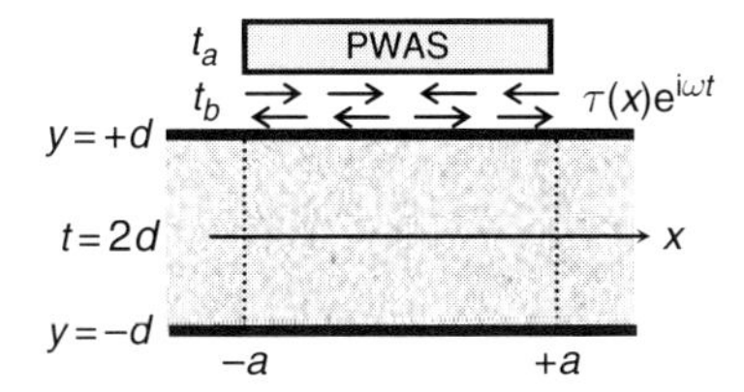

FIGURE 7.23 Interaction between the PWAS and the structure showing the bonding-layer interfacial shear stress, $\tau(X)$.

modulus G_b. The PWAS length is l_a whereas the half-length is $a = l_a/2$. In addition, the definition $d = t/2$ is used. Upon application of an electric voltage, the PWAS experiences an induced strain

$$\varepsilon_{\text{ISA}} = d_{31}\frac{V}{t_a} \tag{228}$$

The induced strain is transmitted to the structure through the bonding-layer interfacial shear stress τ. For harmonic varying excitation, the shear stress has the expression $\tau(x)e^{i\omega t}$.

The PWAS expansion is transmitted to the structure through the bonding layer, which acts predominantly in shear. Construction of the free body diagrams of the PWAS, bonding layer, and thin-wall structure over the infinitesimal length dx leads the following equilibrium equations:

$$t_a\frac{d\sigma_a}{dx} - \tau = 0 \quad \text{(PWAS)} \tag{229}$$

$$t\frac{d\sigma}{dx} + \alpha\tau = 0 \quad \text{(structure)} \tag{230}$$

The coefficient α in Eq. (230) depends on the stress, strain, and displacement distributions across the plate thickness. Under static and low-frequency dynamic conditions, one applies the usual hypothesis associated with simple axial and flexural motions, i.e., constant displacement for axial motion and linear displacement strain for flexural motion. In this case, $\alpha = 4$. For high-frequency motion, the displacement field across the plate thickness takes the more complicated forms associated with the Lamb-wave modes. However, in the present section, we will restrict our analysis to the static and low-frequency dynamic conditions. These conditions are analyzed next; as a result, the value $\alpha = 4$ in Eq. (230) will be derived. The analysis proceeds as follows. The shear stress τ applied to the upper surface is partitioned into symmetric and antisymmetric pairs, $(\tau/2, \tau/2)$ and $(\tau/2, -\tau/2)$, applied to the upper and lower surfaces, respectively (Fig. 7.24), such that

- At the upper surface, $\tau_S\big|_{y=d} + \tau_A\big|_{y=d} = \frac{\tau}{2} + \frac{\tau}{2} = \tau$
- At the lower surface, $\tau_S\big|_{y=-d} + \tau_A\big|_{y=-d} = \frac{\tau}{2} - \frac{\tau}{2} = 0.$

Under static and low-frequency dynamic conditions, the following assumptions apply: (a) For the symmetric case, uniform stress and strain distribution across the thickness is assumed. (b) For the antisymmetric case, linear stress and strain distribution across the thickness is assumed. These two cases are analyzed individually next.

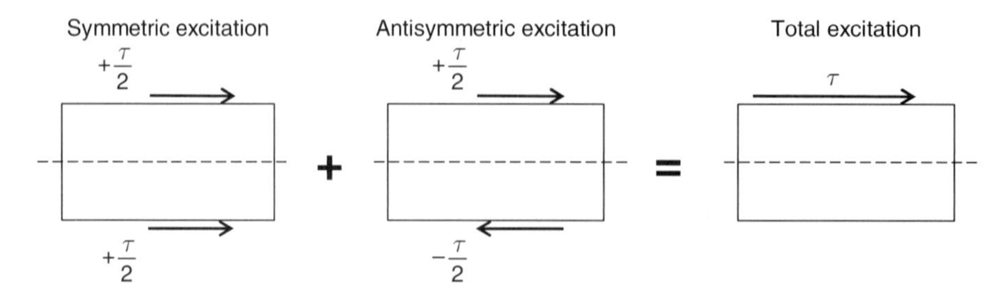

FIGURE 7.24 Symmetric and antisymmetric particle motion across the plate thickness.

7.6.1.1 Symmetric Case

In the symmetric case, the stress and strain are assumed constant across the thickness (Fig. 7.25). Because the stress is assumed uniform across the thickness, the stress distribution can be expressed as

$$\sigma(y) = \sigma_S \tag{231}$$

where σ_S is the value of the stress in the plate evaluated in the upper surface. The subscript S stands for 'symmetric'.

The stress resultants are evaluated by integration of the stress across the thickness. Because the stress distribution is symmetric, the only stress resultant is the axial force per unit width

$$N = \int_{-d}^{+d} \sigma(y)\mathrm{d}y = \int_{-d}^{+d} \sigma_S \,\mathrm{d}y = 2\,\mathrm{d}\sigma_S \tag{232}$$

The force due to the shear stresses applied to the plate surface over the length dx is

$$\mathrm{d}N_\tau = 2\frac{\tau}{2}\mathrm{d}x = \tau\,\mathrm{d}x \tag{233}$$

Hence, equilibrium of the infinitesimal element of Fig. 7.25 gives

$$\not{N} + \mathrm{d}N + \mathrm{d}N_\tau - \not{N} = 0 \tag{234}$$

Substitution of Eqs. (232) and (233) into Eq. (234) yields

$$t\frac{\mathrm{d}\sigma_S}{\mathrm{d}x} + \tau = 0 \tag{235}$$

In arriving at Eq. (235), the definition $d = t/2$ was used. Equation (235) can be rewritten in the form

$$t\frac{\mathrm{d}\sigma_S}{\mathrm{d}x} + \alpha_S\tau = 0 \tag{236}$$

where $\alpha_S = 1$.

7.6.1.2 Antisymmetric Case

In the antisymmetric case, the stress and strain are assumed to vary linearly across the thickness (Fig. 7.26). Because the stress is assumed linearly varying across the thickness, the stress distribution can be expressed as

$$\sigma(y) = \frac{y}{d}\sigma_A \tag{237}$$

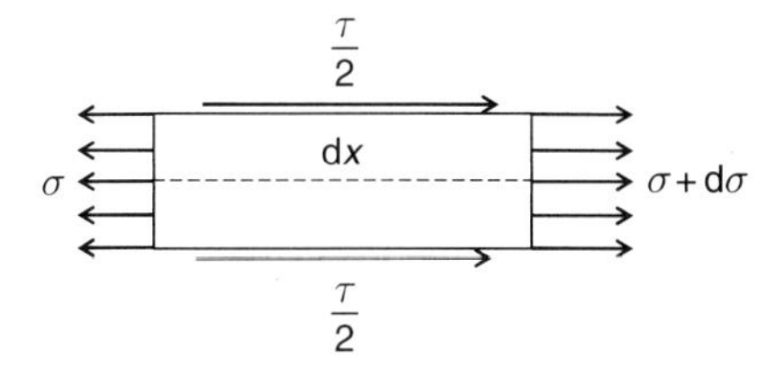

FIGURE 7.25 Symmetric stress distribution across the plate thickness.

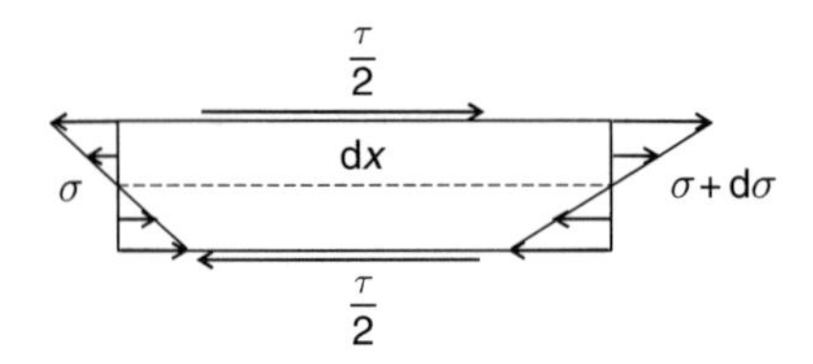

FIGURE 7.26 Antisymmetric stress distribution across the plate thickness.

where σ_A is the value of the stress in the plate evaluated in the upper surface. The subscript A stands for 'antisymmetric'. The stress resultants are evaluated by integration of the stress across the thickness. Because the stress distribution is antisymmetric, the only stress resultant is the moment per unit width

$$M = \int_{-d}^{+d} \sigma(y) y \, dy = \int_{-d}^{+d} \frac{y}{d} \sigma_A y \, dy = \sigma_A \frac{1}{d} \int_{-d}^{+d} y^2 \, dy = \sigma_A \frac{2d^2}{3} \tag{238}$$

The moment due to the shear stresses applied to the plate surface over the length dx is

$$dM_\tau = 2d \frac{\tau}{2} dx = d\tau \, dx \tag{239}$$

Hence, equilibrium of the infinitesimal element of Fig. 7.26 gives

$$\not{M} + dM + dM_\tau - \not{M} = 0 \tag{240}$$

Substitution of Eqs. (238) and (239) into Eq. (240) yields

$$\frac{2d^2}{3} d\sigma_A + d\tau \, dx = 0 \tag{241}$$

Upon simplification and using the definition $d = t/2$, we get

$$t \frac{d\sigma_A}{dx} + 3\tau = 0 \tag{242}$$

Equation (242) can be rewritten in the form

$$t \frac{d\sigma_A}{dx} + \alpha_A \tau = 0 \tag{243}$$

where $\alpha_A = 3$.

7.6.1.3 Shear-Lag Solution

The superposition $\sigma = \sigma_S + \sigma_A$ yields Eq. (230), i.e., $\alpha = \alpha_S + \alpha_A = 4$. (For Lamb-wave modes, which have a complex stress and strain distributions, the value of α will actually vary from mode to mode and will depend on the frequency-thickness product.) The strain-displacement equations in the PWAS, bonding layer, and structure are

$$\varepsilon_a = \frac{\mathrm{d}u_a}{\mathrm{d}x} \qquad \text{(PWAS)} \tag{244}$$

$$\gamma = \frac{u_a - u}{t_b} \qquad \text{(bonding layer)} \tag{245}$$

$$\varepsilon = \frac{\mathrm{d}u}{\mathrm{d}x} \qquad \text{(structure)} \tag{246}$$

whereas the stress–strain relations are

$$\sigma_a = E_a(\varepsilon_a - \varepsilon_{\mathrm{ISA}}) \qquad \text{(PWAS)} \tag{247}$$

$$\tau = G_b\gamma \qquad \text{(bonding layer)} \tag{248}$$

$$\sigma = E\varepsilon \qquad \text{(structure)} \tag{249}$$

In addition, recall the induced-strain Eq. (228), i.e., $\varepsilon_{\mathrm{ISA}} = d_{31}V/t_a$. Upon substitution, one obtains two second-order coupled differential equations in ε_a and ε, which, upon further differentiation, yield the following pair of decoupled fourth order differential equations

$$\frac{\mathrm{d}^4\varepsilon_a}{\mathrm{d}x^4} - \Gamma^2\frac{\mathrm{d}^2\varepsilon_a}{\mathrm{d}x^2} \qquad \text{(PWAS)} \tag{250}$$

$$\frac{\mathrm{d}^4\varepsilon}{\mathrm{d}x^4} - \Gamma^2\frac{\mathrm{d}^2\varepsilon}{\mathrm{d}x^2} \qquad \text{(structure)} \tag{251}$$

where

$$\Gamma^2 = \frac{G_b}{E_a}\frac{1}{t_a t_b}\frac{\alpha+\psi}{\psi} \qquad \text{(shear-lag parameter)} \tag{252}$$

and

$$\psi = \frac{Et}{E_a t_a} \tag{253}$$

Solution of Eqs. (250)–(251) is obtained in the form

$$\varepsilon_a(x) = \frac{\alpha}{\alpha+\psi}\varepsilon_{\mathrm{ISA}}\left(1 + \frac{\psi}{\alpha}\frac{\cosh\Gamma x}{\cosh\Gamma a}\right) \qquad \text{(PWAS actuation strain)} \tag{254}$$

$$\sigma_a(x) = -\frac{\psi}{\alpha+\psi}E_a\varepsilon_{\mathrm{ISA}}\left(1 - \frac{\cosh\Gamma x}{\cosh\Gamma a}\right) \qquad \text{(PWAS stress)} \tag{255}$$

$$u_a(x) = \frac{\alpha}{\alpha+\psi}\varepsilon_{\mathrm{ISA}}a\left(\frac{x}{a} + \frac{\psi}{\alpha}\frac{\sinh\Gamma x}{(\Gamma a)\cosh\Gamma a}\right) \qquad \text{(PWAS displacement)} \tag{256}$$

$$\tau(x) = \frac{t_a}{a}\frac{\psi}{\alpha+\psi}E_a\varepsilon_{\mathrm{ISA}}\left(\Gamma a\frac{\sinh\Gamma x}{\cosh\Gamma a}\right) \qquad \text{(interfacial shear stress in bonding layer)} \tag{257}$$

$$\varepsilon(x) = \frac{\alpha}{\alpha+\psi}\varepsilon_{\mathrm{ISA}}\left(1 - \frac{\cosh\Gamma x}{\cosh\Gamma a}\right) \qquad \text{(structure strain at the surface)} \tag{258}$$

$$\sigma(x) = \frac{\alpha}{\alpha+\psi} E\varepsilon_{\text{ISA}} \left(1 - \frac{\cosh \Gamma x}{\cosh \Gamma a}\right) \quad \text{(structure stress)} \tag{259}$$

$$u(x) = \frac{\alpha}{\alpha+\psi} \varepsilon_{\text{ISA}} a \left(\frac{x}{a} - \frac{\sinh \Gamma x}{(\Gamma a) \cosh \Gamma a}\right) \quad \text{(structure displacement at the surface)} \tag{260}$$

These equations apply for $|x| < a$. Outside the $|x| < a$ interval, the strain and stress variables in these equations are zero. Note that $\sigma_a \neq E_a \varepsilon_a$ because of Eq. (247). The shear-lag parameter, Γ, plays a very important role in determining the distribution of ε_a, ε, and τ along the PWAS span $(-a, a)$. The effect of the PWAS is transmitted to the structure through the interfacial shear stress of the bonding layer. A small shear stress value in the bonding layer produces a gradual transfer of strain from the PWAS to the structure, whereas a large shear stress produces a rapid transfer. Because the PWAS ends are stress free, the build up of strain takes place at the ends, and it is more rapid when the shear stress is more intense. For large values of Γa, the shear transfer process becomes concentrated toward the PWAS ends.

Example

To illustrate these equations, we considered an APC-850 PWAS with $E_a = 63\,\text{GPa}$, $t_a = 0.2\,\text{mm}$, $l_a = 7\,\text{mm}$ mounted on a thin-wall aluminum structure with $E = 70\,\text{GPa}$ and $t = 1\,\text{mm}$. (The value $t = 2\,\text{mm}$ was also considered.) The mounting is done with cyanoacrylate adhesive ($G_b = 2\,\text{GPa}$) of variable thickness, $t_b = 1\,\mu\text{m}$, $10\,\mu\text{m}$, $100\,\mu\text{m}$. The piezoelectric constant of the APC-850 material used in the PWAS is $d_{31} = -175\,\text{mm/kV}$. The applied voltage was $V = 10\,\text{V}$. Figure 7.27a presents the strain distribution in the structure and PWAS, whereas Fig. 7.27b presents the shear stress distribution. For the range of values considered here, the value of Γa varied between 16 and 58.

Examination of Fig. 7.27 reveals that the shear-lag parameter, Γa, plays a very important role in determining the distribution of ε_a, ε, and τ along the PWAS span $(-a, a)$. The effect of the PWAS is transmitted to the structure through the shear stress in the bonding layer. A small shear stress value in the bonding layer produces a gradual transfer of strain from the PWAS to the structure, whereas a large shear stress produces a rapid transfer. Because the PWAS ends are stress free, the build up of strain takes place at the ends, and it is more rapid when the shear stress is more intense. As indicated by Eq. (252), a relatively thick bonding layer produces a low Γa value, i.e., a slow transfer over the entire span of the PWAS (the '$100\,\mu\text{m}$' curves in Fig. 7.27), whereas a very thin bonding layer produces a very rapid transfer (the '$1\,\mu\text{m}$' curves in Fig. 7.27), which is confined to the ends.

Another aspect of interest is the maximum interfacial shear stress. As indicated by Fig. 7.27b, the maximum interfacial shear stress takes place at the ends. A possible concern might be that the large shear stress values at the PWAS ends would exceed the bond strength and would promote failure. A plot of the interfacial shear stress with bond thickness is presented in Fig. 7.28 for a 0.2 mm-thick PWAS under 10 V excitation. It is apparent that the maximum interfacial shear stress does not exceed 2.5 MPa, which is about an order of magnitude lower than the bonding-layer shear strength.

7.6.1.4 Pin-Force Model

It is apparent from the previous section that a relatively thick bonding layer produces a slow transfer over the entire span of the PWAS (the '$100\,\mu\text{m}$' curves in Fig. 7.27), whereas

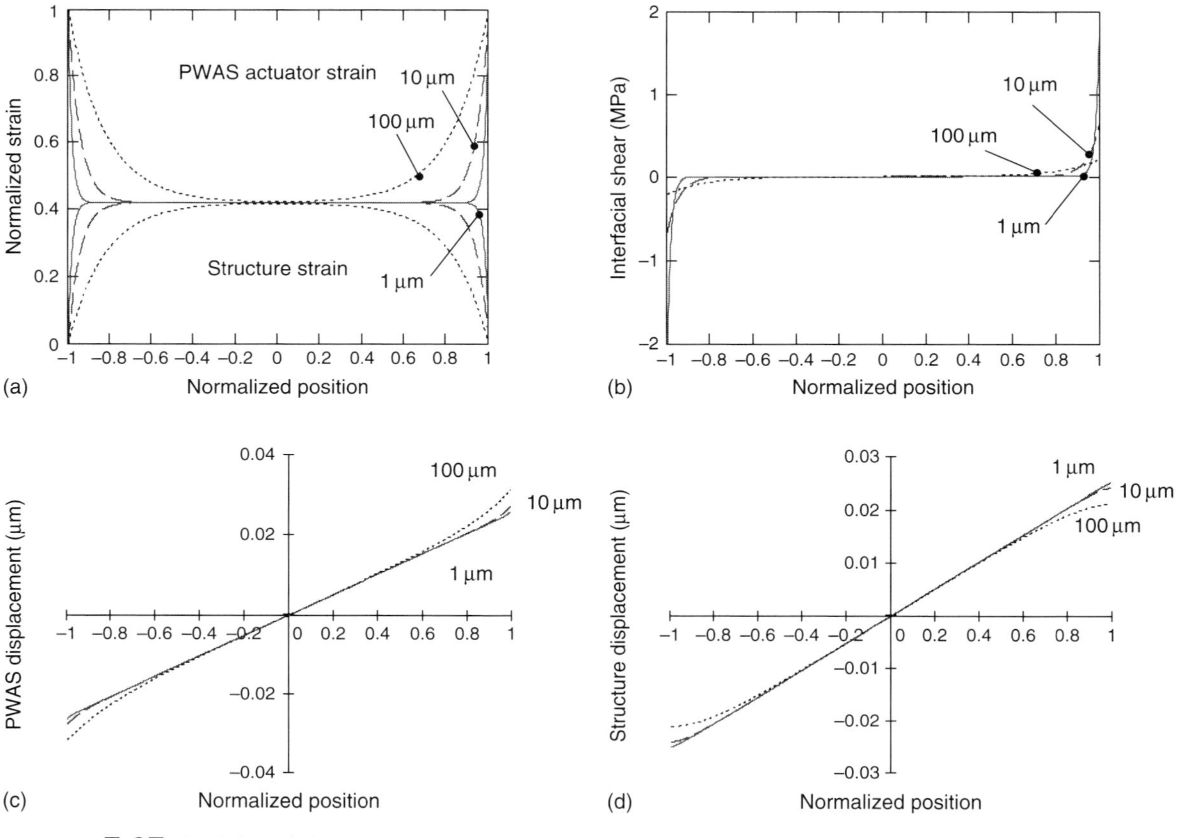

FIGURE 7.27 Variation of shear-lag transfer mechanism with bond thickness: (a) strain distribution in the PWAS and in the structure; (b) interfacial shear stress distribution; (c) displacement distribution in the PWAS; (d) displacement distribution in the structure (bond thickness $t_b = 1, 10, 100\,\mu\text{m}$).

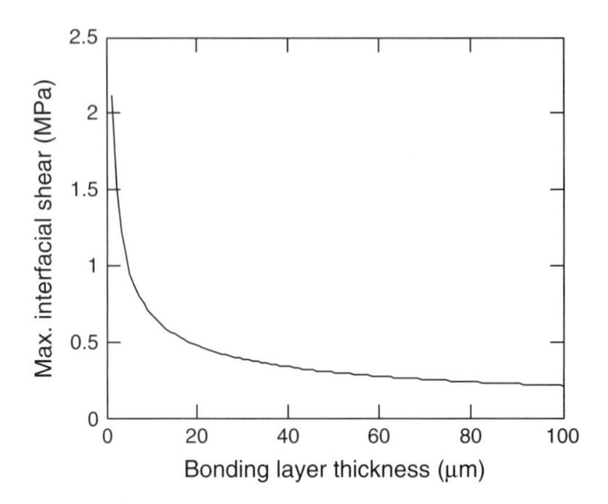

FIGURE 7.28 Variation of maximum interfacial shear stress with bond thickness ($t_b = 1 \ldots 100\,\mu\text{m}$) for a 0.2 mm thick PWAS under 10 V excitation.

a thin bonding layer produces a very rapid transfer (the '1 μm' curves in Fig. 7.27). The shear-lag analysis indicated that, as the bond thickness decreases, $\Gamma\alpha$ increases. The shear stress transfer becomes concentrated over some infinitesimal distances at the ends of the PWAS actuator. In the limit, as $\Gamma a \to \infty$, all the load transfer can be assumed to take place at the PWAS actuator ends. This leads to the concept of *ideal bonding* (also know as the *pin-force model*), in which all the load transfer takes place over an infinitesimal region at the PWAS ends, and the induced-strain action is assumed to consist of a pair of concentrated forces applied at the ends (Fig. 7.29a), i.e.,

$$\tau(x) = a\tau_a[\delta(x-a) - \delta(x+a)] \quad \text{(ideal-bonding shear stress)} \qquad (261)$$

$$F(x) = F_a[-H(x-a) + H(x+a)] \quad \text{(shear force due to pin-end forces)} \qquad (262)$$

where δ and H are the Dirac impulse function and the Heaviside step function (Fig. 7.29b), whereas

$$F_a = a\tau_a \quad \text{(pin-end forces)} \qquad (263)$$

Equation (263) represents the pin forces, $F_a = a\tau_a$, applied by the PWAS to the structure. These forces are localized at the PWAS ends (Fig. 7.29a). The pin-force model is convenient for getting simple solutions that represent a first-order of approximation to the PWAS–structure interaction. Note that this extreme situation implies that the shear

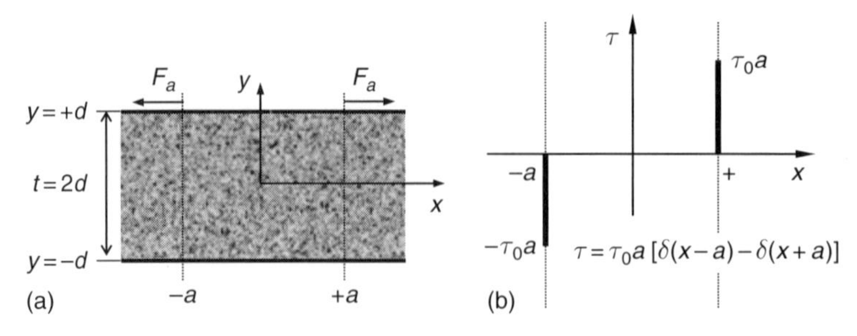

FIGURE 7.29 Pin-force model: (a) surface shear distribution; (b) direct strain induced in the structure at the upper structural surface.

stress reaches very large values over diminishing areas at the PWAS ends. Recall that the Dirac function has the localization property

$$\int f(x)\delta(x-x_0)\mathrm{d}x = f(x_0) \quad \text{(Dirac function localization property)} \tag{264}$$

We recall that the Dirac function is the derivative of the Heaviside function, i.e.,

$$\delta(x) = H'(x) \tag{265}$$

Under these assumptions, Eqs. (254)–(260) take the simple forms

$$\varepsilon_a(x) = \frac{\alpha}{\alpha+\psi}\varepsilon_{\mathrm{ISA}}\,[H(x+a)-H(x-a)] \quad \text{(PWAS actuation strain)} \tag{266}$$

$$\sigma_a(x) = -\frac{\psi}{\alpha+\psi}\varepsilon_{\mathrm{ISA}}\,[H(x+a)-H(x-a)] \quad \text{(stress in the PWAS)} \tag{267}$$

$$u_a(x) = \frac{\alpha}{\alpha+\psi}\varepsilon_{\mathrm{ISA}}x\,[H(x+a)-H(x-a)] \quad \text{(displacement in the PWAS)} \tag{268}$$

$$\tau(x) = \frac{\psi}{\alpha+\psi}t_a E_a \varepsilon_{\mathrm{ISA}}\,[-\delta(x+a)+\delta(x-a)] \quad \text{(interfacial shear stress in bonding layer)} \tag{269}$$

$$F(x) = \frac{\psi}{\alpha+\psi}t_a E_a \varepsilon_{\mathrm{ISA}}\,[-H(x+a)+H(x-a)] \quad \text{(interfacial shear force in bonding layer)} \tag{270}$$

$$\varepsilon(x) = \frac{\alpha}{\alpha+\psi}\varepsilon_{\mathrm{ISA}}\,[H(x+a)-H(x-a)] \quad \text{(structure strain at the surface)} \tag{271}$$

$$\sigma(x) = \frac{\alpha}{\alpha+\psi}E\varepsilon_{\mathrm{ISA}}\,[H(x+a)-H(x-a)] \quad \text{(structure stress at the surface)} \tag{272}$$

$$u(x) = \frac{\alpha}{\alpha+\psi}\varepsilon_{\mathrm{ISA}}x\,[H(x+a)-H(x-a)] \quad \text{(structure displacement at the surface)} \tag{273}$$

where $H(x)$ and $\delta(x)$ are the Heaviside step function and the Dirac delta function, respectively. Note that the strains in the PWAS and at the surface of the structure are equal, but the stresses are not, due to Eq. (247).

The axial force and bending moment associated with the ideal-bonding hypothesis (pin-force model) are

$$N_a = F_a \quad \text{(axial force)} \tag{274}$$

$$M_a = F_a d = F_a\frac{t}{2} \quad \text{(bending moment)} \tag{275}$$

The axial force and bending moment described by Eqs. (274) and (275) represent the excitation induced by the PWAS into the plate under the ideal-bonding hypothesis.

7.6.2 ENERGY TRANSFER BETWEEN THE PWAS AND THE STRUCTURE

The transfer of energy between the PWAS and the structure is studied for both the shear-lag model and the pin-force model.

7.6.2.1 Energy Transfer Through the Shear-Lag Model

The energy in the actuator can be evaluated as

$$
\begin{aligned}
W_a &= \int_{V_a} \frac{1}{2}\frac{\sigma_a^2}{E_a} \mathrm{d}V \\
&= \frac{\psi^2}{(\alpha+\psi)^2}\left(\frac{1}{2}E_a\varepsilon_{\mathrm{ISA}}^2\right)\int_{V_a}\frac{1}{2}\left(1-\frac{\cosh \Gamma x}{\cosh \Gamma a}\right)^2 \mathrm{d}x \quad \text{(energy retained in actuator)} \\
&= \frac{\psi^2}{(\alpha+\psi)^2}\left(\frac{1}{2}E_a\varepsilon_{\mathrm{ISA}}^2 l_a\right) I(\Gamma a) \qquad (276)
\end{aligned}
$$

where

$$
I(\Gamma a) = 1 - \frac{3}{2}\frac{\sinh \Gamma a}{\Gamma a \cosh \Gamma a} + \frac{1}{2}\frac{1}{(\cosh \Gamma a)^2} \quad \text{(bond efficiency)} \qquad (277)
$$

is a function that approaches 1 for large Γa values (Fig. 7.30). It represents a measure of the bond efficiency. For the values considered in our preliminary study, the values of $I(\Gamma a)$ were found to vary from 90 to 97%.

The term $\frac{1}{2}E_a\varepsilon_{\mathrm{ISA}}^2 l_a$ is a measure of the total induced-strain energy, whereas the term $\psi^2/(\alpha+\psi)^2$ represents how much of this energy gets stored as elastic compression in the PWAS actuator and cannot be transmitted to the structure.

The energy transmitted to the structure can be evaluated either as the elastic energy in the structure or as the work done by the interfacial shear stresses on the structural surface. The two paths are equivalent:

1. *Elastic energy in the structure*

$$
\begin{aligned}
W_1 &= \int_V \frac{1}{2}[(\sigma\varepsilon)_{\mathrm{axial}} + (\sigma\varepsilon)_{\mathrm{flexural}}]\mathrm{d}V = \frac{\alpha}{(\alpha+\psi)^2}\left(\frac{1}{2}E\varepsilon_{\mathrm{ISA}}^2\right)\int_{-a}^{a}\frac{1}{2}\left(1-\frac{\cosh \Gamma x}{\cosh \Gamma a}\right)^2 \mathrm{d}x \\
&= \frac{\alpha\psi}{(\alpha+\psi)^2}\left(\frac{1}{2}E\varepsilon_{\mathrm{ISA}}^2 l_a\right) I(\Gamma a) \qquad (278)
\end{aligned}
$$

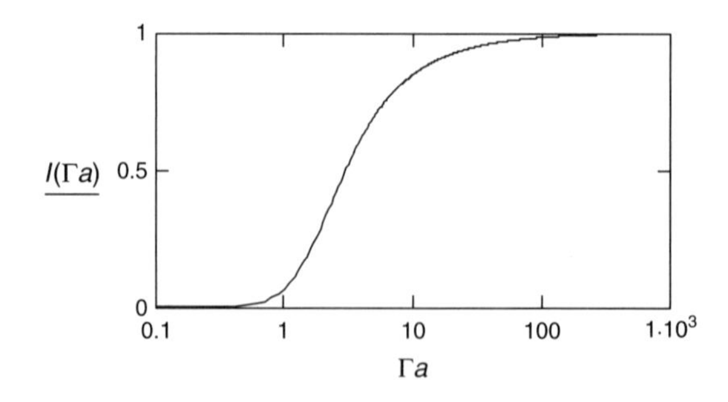

FIGURE 7.30 Bond energy efficiency variation with Γa.

2. *Work done by the shear stresses at the structural surface*

$$\begin{aligned} W_2 &= \int_{l_a} \left(\frac{1}{2} \tau u \right) \mathrm{d}x \\ &= \int_{-a}^{a} \frac{1}{2} \frac{t_a}{a} \frac{\psi}{\alpha+\psi} E_a \varepsilon_{\mathrm{ISA}} \left(\Gamma a \frac{\sinh \Gamma x}{\cosh \Gamma a} \right) \frac{\alpha}{\alpha+\psi} \varepsilon_{\mathrm{ISA}} a \left(\frac{x}{a} - \frac{\sinh \Gamma x}{(\Gamma a) \cosh \Gamma a} \right) \mathrm{d}x \\ &= \frac{\alpha \psi}{(\alpha+\psi)^2} \left(\frac{1}{2} E \varepsilon_{\mathrm{ISA}}^2 l_a \right) I(\Gamma a) \end{aligned} \tag{279}$$

These two paths of approach give the same results, and the energy transmitted to the structure can be retained as

$$W = \frac{\alpha}{(\alpha+\psi)^2} \left(\frac{1}{2} E \varepsilon_{\mathrm{ISA}}^2 l_a \right) I(\Gamma a) \quad \text{(energy transmitted to the structure)} \tag{280}$$

We note that the integral $I(\Gamma a)$ and term $\frac{1}{2} E_a \varepsilon_{\mathrm{ISA}}^2 l_a$ of Eq. (276) reappear in Eqs. (278) and (279). The term $\frac{\alpha}{(\alpha+\psi)^2}$ represents how much of the induced strain energy gets transmitted into the structure.

7.6.2.2 Energy Transfer Through the Pin-Force Model

Because the pin-force model is a constant-strain, constant-stress model, the evaluation of the energy stored in the actuator and transferred into the structure can be readily done

$$W_a = \left(\frac{1}{2} E_a \varepsilon_{\mathrm{ISA}}^2 l_a \right) \quad \text{(energy retained in the actuator)} \tag{281}$$

$$W = \frac{\alpha \psi}{(\alpha+\psi)^2} \left(\frac{1}{2} E \varepsilon_{\mathrm{ISA}}^2 l_a \right) \quad \text{(energy transferred into the structure)} \tag{282}$$

7.6.2.3 Conditions for Optimum Energy Transfer

Figure 7.31 presents a plot of the energy transferred into the structure vs. parameter ψ in nondimensional form for $\alpha = 4$. The plot was normalized by the factor $\frac{1}{2} E \varepsilon_{\mathrm{ISA}}^2 l_a$. It can be noticed that the energy transfer reaches a maximum at $\psi = 4$, i.e., as the parameter ψ equals the parameter α. Denoting the *apparent stiffness ratio*, r, by

$$r = \frac{\psi}{\alpha} = \frac{Et/\alpha}{E_a t_a} \tag{283}$$

We notice that, under the quasi-static conditions considered here, maximum energy transfer is attained at $r = 1$. This represents the well-known *stiffness matching principle* with the proviso that the stiffness of the PWAS, $E_a t_a$, needs to be matched with the *apparent stiffness* of the structure, Et/α. At the stiffness match condition, the energy in the structure and the energy in the PWAS coincide and are equal to the maximum extractable energy

$$W_{\max} = \frac{1}{4} \left(\frac{1}{2} E_a \varepsilon_{\mathrm{ISA}}^2 l_a \right) \quad \text{(maximum extractable energy)} \tag{284}$$

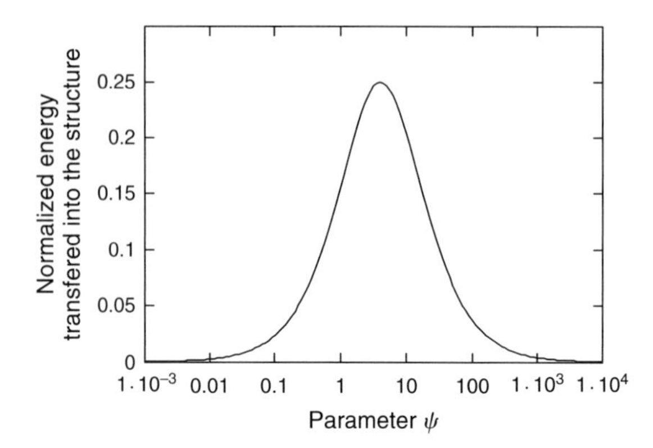

FIGURE 7.31 Energy transfer into the structure vs. parameter ψ in nondimensional form. The plot is normalized by the factor $\frac{1}{2}E\varepsilon_{\rm ISA}^2 l_a$.

The value $r = 1$ represents the *condition for maximum energy transfer* under quasi-static conditions. Under dynamic conditions, this condition will be modified by several factors:

- Because of inertia loads, the apparent structural stiffness varies with frequency
- Modal resonances will affect the apparent stiffness, which will go through minima at every resonance
- At ultrasonic frequencies, the apparent structural stiffness depends on the Lamb modes that are resonant at certain frequency
- The frequency-dependent displacement distribution across the thickness of the Lamb modes will additionally influence the apparent structural stiffness

The stiffness match principle corresponds to the *impedance match principle* commonly used in electronic circuits design in order to attain maximum power transfer conditions.

7.7 DURABILITY AND SURVIVABILITY OF PIEZOELECTRIC WAFER ACTIVE SENSORS

A number of studies have been performed to assess PWAS durability and survivability under exposure to environmental factors, large strains, and fatigue cyclic loads. Some of these results are reported. We studied the PWAS durability and survivability under temperature cycling in an oven, weather exposure in outdoor environment (sun, rain, humidity, freeze-thaw, etc.), and exposure to water and maintenance fluids (hydraulic fluids and lubrication oils). Both free PWAS and PWAS attached to an aluminum alloy plate were used. The PWAS behavior was monitored using the electromechanical (E/M) impedance method. For a free PWAS, the real part of the E/M impedance reflects its free vibration spectrum. For a bonded PWAS, the real part of the E/M impedance directly reflects the vibration spectrum of the tested specimen seen through the point-wise mechanical impedance at the point where PWAS is attached to the structure.

7.7.1 TEMPERATURE CYCLING

For temperature cyclic testing, free and bonded PWAS were placed in an oven and exposed to a temperature variation between 100°F and 175°F. The temperature cycle consisted of a slow rise from 100°F to 175°F followed by a 5-minute dwell at 175°F, then a slow descend from 175°F to 100°F followed by a 5-minute dwell at 100°F (Fig. 7.32b).

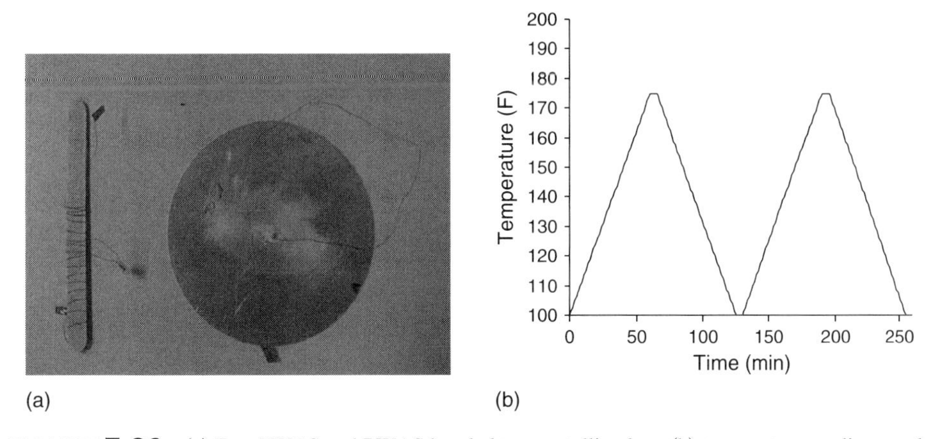

FIGURE 7.32 (a) Free PWAS and PWAS bonded to a metallic plate; (b) temperature cycling graph.

The data taken at the beginning of these experiments showed a 'settling in' effect, i.e., some amplitude reduction during the first few cycles, followed by a leveling off of the variations. In total, 1700 cycles have been performed (11 months of testing). The free PWAS survived 1700 oven cycles without significant changes in the E/M impedance spectrum (Fig. 7.33). The bonded PWAS survived only 1400 cycles in the oven without

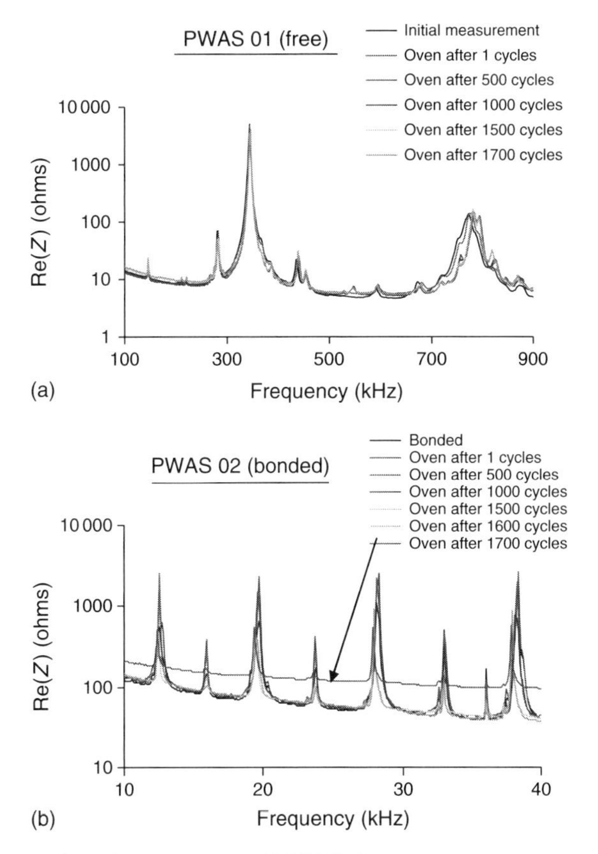

FIGURE 7.33 E/M impedance spectrum of PWAS after exposure to temperature cycling: (a) free PWAS; (b) PWAS bonded to a metallic plate.

significant changes in the E/M impedance spectrum. The spectrum taken after 1500 and 1600 cycles showed small changes. The spectrum taken after 1700 cycles showed marked changes, which were attributed to the failure of the boded PWAS. Further investigation with a Quantum 350 Scanning Acoustic Microscope[1] indicated that disbond between PWAS and substrate took place. Hence, it was concluded that the failure was not due to the piezoelectric material failure, but due to the failure of the bonded interface between the PWAS and the substrate. This failure can be attributed to the repeated differential thermal expansion between the ceramic PWAS and the metallic substrate.

7.7.2 ENVIRONMENTAL OUTDOORS EXPOSURE

Free PWAS and bonded PWAS were exposed to the outdoors environment (Fig. 7.34a) over a long time period. In this study, several adhesives and several protective coatings were examined (Table 7.11). The measured quantity was the E/M impedance spectrum. The data taken at the beginning of these experiments showed a 'settling in' effect, i.e., some amplitude reduction during the first few cycles, followed by a leveling off of the variations. After this, the E/M impedance data stayed rather constant for the duration of the test. This test has been conducted for more than 70 weeks (Fig. 7.34b). Minor repairs

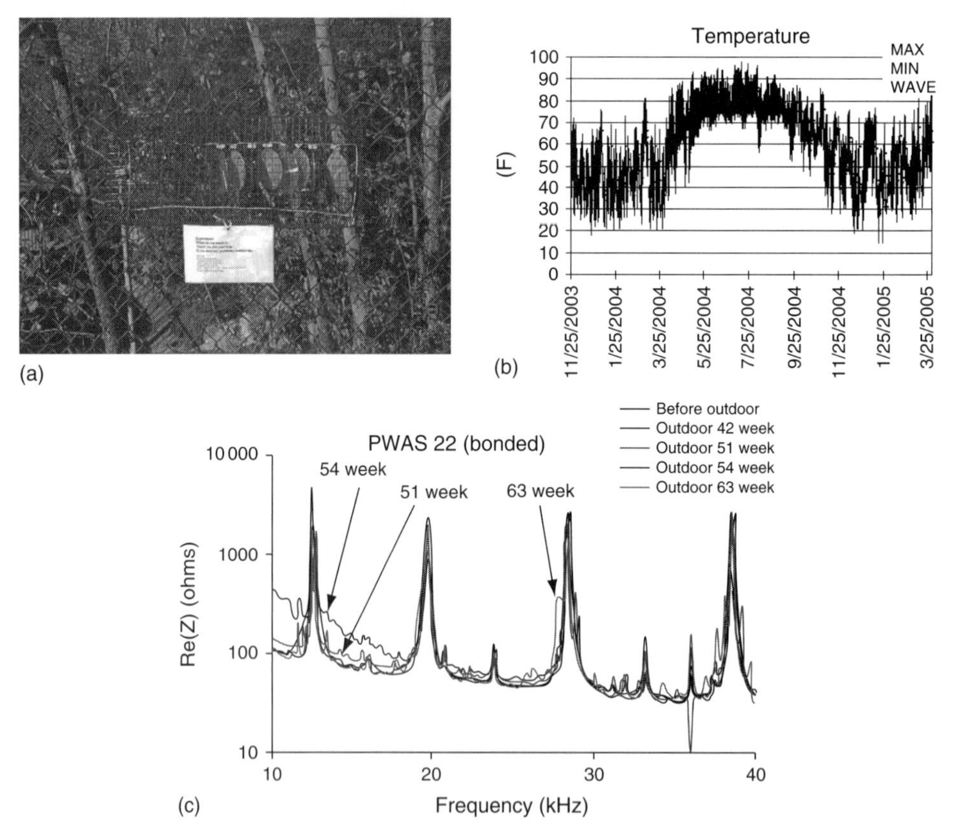

FIGURE 7.34 Environmental testing of free and bonded PWAS: (a) outdoors test fixture; (b) temperature profile; (c) E/M impedance spectrum of specimen PWAS-22.

[1] This investigation was done in collaboration with Professor Peter Nagy of the University of Cincinnati, which has this facility.

TABLE 7.11 Circular plate specimens instrumented with PWAS for environmental testing

		M-Bond 200-cynoacrylate adhesive with catalyst	M-Bond AE10-2-part 100% solid epoxy system adhesive
Protective coating	No coating	PWAS-22	PWAS-33
	M-Coat A-Polyurethane	PWAS-23	PWAS-34
	M-Coat C-Silicon	PWAS-27	PWAS-35
	M-Coat D-Acrylic	PWAS-28	PWAS-36

to the wire attachments were done. So far, no significant changes have been noticed in the PWAS E/M impedance spectrum, except for sample PWAS-22. This sample used the cyanoacrylate fast adhesive and had no protective coating. The historical evolution of the E/M impedance spectrum for this sample is shown in Fig. 7.34c: up to 42 weeks, no significant changes were recorded; at 51 weeks, small changes were observed. The main peak dropped a little and more small peaks were noticed. At 54 weeks, more changes were observed. At 63 weeks, significant changes were observed. The main peak dropped a lot and more small peaks appeared. Under optical microscope, a crack was detected in the piezoceramic PWAS. The crack was located diagonally in the right lower corner. These effects may explain the changes noted in the E/M impedance spectrum.

7.7.3 SUBMERSION EXPOSURE

The purpose of the submersion exposure tests was to determine how PWAS behaves when exposed to water and various maintenance fluids. The specimens used in this experiment were free 5-mm PWAS with two connecting wires soldered by the manufacturer. The PWAS were submerged in plastic bottles containing the fluids (Fig. 7.35). The fluids used in the submersion test were:

(0) Distilled water
(1) Saline solution
(2) Hydraulic fluid MIL-PRF-83282 Synthetic hydrocarbon
(3) Hydraulic fluid MIL-PRF-87257 Synthetic hydrocarbon
(4) Hydraulic fluid MIL-PRF-5606 Mineral
(5) Aircraft lube oil MIL-PRF-7808L Grade 3 Turbine engine synthetic
(6) Kerosene

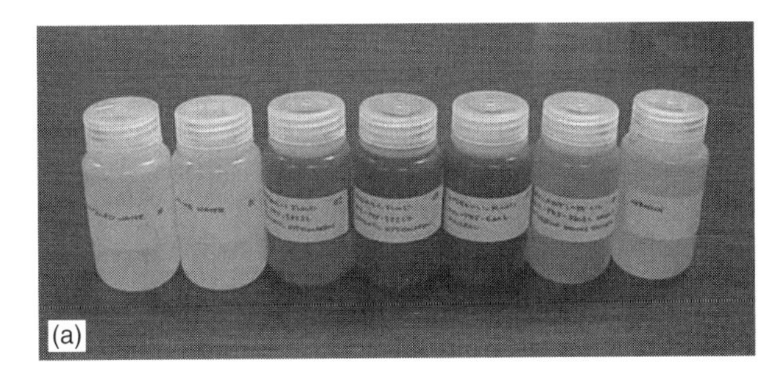

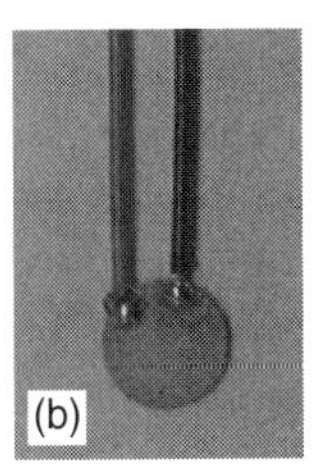

FIGURE 7.35 PWAS submersion test: (a) test containers; (b) 5-mm diameter free PWAS specimen.

This test has been conducted for more than 425 days (60 weeks). So far, no significant changes have been noticed in the PWAS E/M impedance spectrum except for the PWAS submerged in saline solution. The PWAS submerged in saline solution survived only a little over 85 days (15 weeks). The E/M impedance reading taken at 85 days exposure showed marked differences from the previous readings. The failure of this PWAS was traced to the detachment of the soldered connection. This can be attributed to the corrosive effect of the saline solution.

In a separate test, Blackshire et al. (2005) subjected PWAS bonded to an aluminum plate to electrochemical corrosion exposure using a corrosion cell and 3.5% NaCl solution. Two specimens, one unprotected, the other covered with an enamel protective coating, were used. No significant change was reported after exposure to eight corrosion cycles.

7.7.4 LARGE STRAINS AND FATIGUE CYCLIC LOADING

The behavior of PWAS transducers under large strains and under fatigue cyclic loading was studied.

In the *large strain experiments*, the specimen shown in Fig. 7.36a was used. Specimens were fabricated from 2024 T3 aluminum with a nominal thickness of 1 mm. The specimens were loaded in tension under strain control. Two specimens were used: the first one was loaded up to 5000 micro-strain, the other up to failure. Measurements were taken at 200 micro-strain intervals. The baseline impedance was recorded at zero strain and additional readings were recorded until failure of the PWAS occurred. Minimal changes occurred to the impedance signature until the value of 3000 micro-strain was exceeded. Significant changes begin to happen after 3000 micro-strain. After 6000 micro-strain, the changes in the E/M impedance were very strong (Fig. 7.37a). Eventually, the PWAS failed in tension at approximately 7200 micro-strain. The PWAS failure was in the form of a transverse crack (Fig. 7.37b).

In the *fatigue cyclic loading experiments*, the specimen shown in Fig. 7.36b was used. The specimens were fabricated from 2024 T3 aluminum with a nominal thickness of 1 mm. A 7-mm square PWAS was bonded to the specimen with M-Bond 200 cyanoacrylate adhesive. A 1-mm hole was drilled in the specimen to act as stress concentration and localize the fatigue failure. Five specimens were used (Table 7.12). The specimens were loaded in

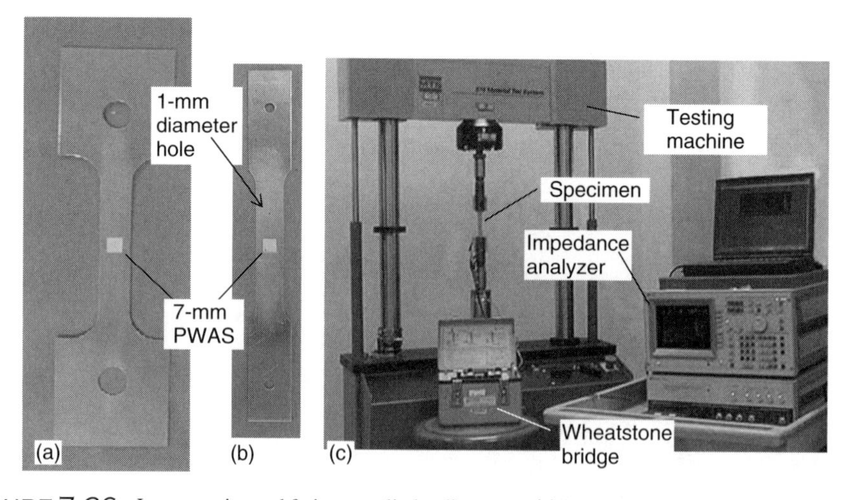

FIGURE 7.36 Large strains and fatigue cyclic loading tests: (a) large-strain test specimen; (b) fatigue cycling loading test specimen; (c) experimental setup.

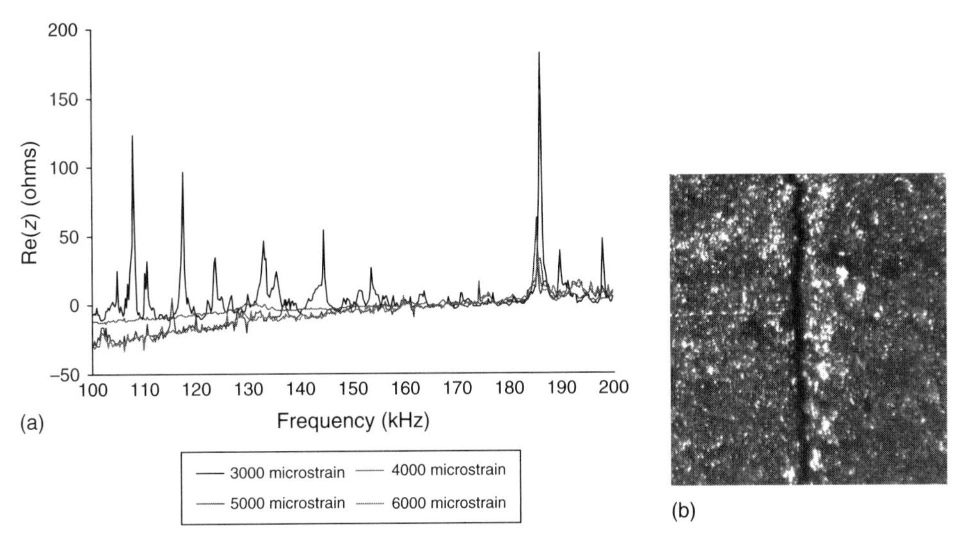

FIGURE 7.37 Large-strain tests: (a) impedance signatures up to 6000 $\mu\varepsilon$; (b) micrograph of the cracked PWAS at 7200 $\mu\varepsilon$.

TABLE 7.12 Overview of PWAS fatigue testing specimens ($R = 0.1$)

	PWAS-F1	PWAS-F2	PWAS-F3	PWAS-F4	PWAS-F5
Max load	2104 N	1560 N	1335 N	1156 N	1067 N
Min load	210 N	156 N	134 N	116 N	107 N
Mean load	1157 N	858 N	734 N	636 N	587 N
Cycles to specimen failure	178 kc	670 kc	1.3 Mc	6.25 Mc	12.2 Mc

fatigue cyclic loading with the mean loads and amplitudes adjusted such as to cause failure of the aluminum substrate at various values between 100 thousand and 10 million cycles (Fig. 7.38 and Table 7.12). The baseline impedance reading was taken at the beginning of the tests and at predetermined cyclic intervals. During the impedance measurement,

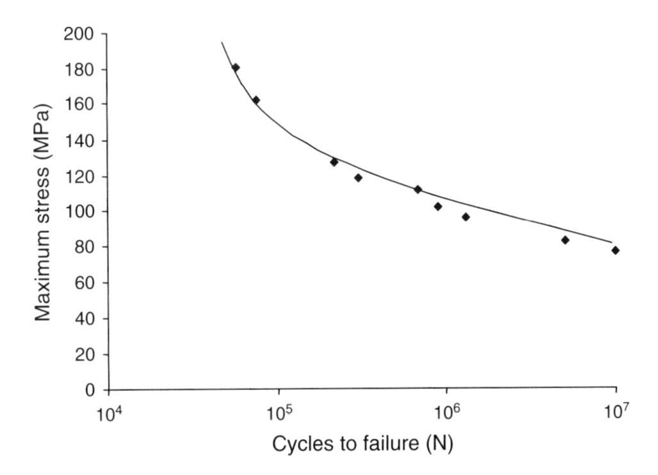

FIGURE 7.38 *S–N* curve determined during fatigue testing.

the cyclic test was stopped but the mean load was maintained. Small settle-in changes occurred in the impedance readings in the first 30 to 40 thousand cycles. Beyond this, the PWAS readings were relatively unchanged until the metallic specimen finally broke under fatigue cyclic loading. The specimen failure always occurred at the 1-mm stress-concentration hole. The PWAS survived the fatigue failure of all the metallic specimens.

7.8 SUMMARY AND CONCLUSIONS

This chapter has introduced the PWAS transducers and presented some of their sailent features. The chapter started with the analysis of PWAS resonators. Detailed analysis of the mechanical and electrical responses was presented, and the associated resonances were identified. Closed-form solutions of the E/M impedance and admittance were derived. Comparison with experimental data was presented. The use of PWAS as ultrasonic transducers was presented next. The analysis started with the investigation of the shear layer coupling between the PWAS and the support structure. Closed-form solutions were derived indicating the dependence of shear distribution on the shear layer stiffness. For very stiff shear layers (i.e., composed of high modulus thin adhesives), the shear distribution is concentrated at the PWAS ends, thus giving rise to the pin-force model. The pin-force model assumes that transmission of force between the PWAS and the structure takes place in the form of concentrated pin forces at the PWAS ends. The energy transfer between the PWAS and the structure was also analyzed, and conditions for optimum energy transfer were identified.

Durability and survivability of the PWAS transducers was examined in the last part of this chapter. Both free PWAS and PWAS attached to structural substrates were used. Exposure to temperature cycling, outdoor environments, operational fluids, large strains, and fatigue loading were considered. In most cases, the PWAS survived the tests successfully. The cases when the PWAS did not survive the tests were closely examined and possible cause of failure was discussed.

7.9 PROBLEMS AND EXERCISES

1. Consider a PWAS transducer with length $l = 7\,\text{mm}$, width $b = 1.65\,\text{mm}$, thickness $t = 0.2\,\text{mm}$, and material properties as given in Table 7.1.
   (i) Find the free capacitance, C_0, of the PWAS.
   (ii) Calculate the voltage, V, measured at PWAS terminals by an instrument with input capacitance $C_e = 1\,\text{pF}$ when the applied strain is $S_1 = -1000\,\mu\varepsilon$.
   (iii) Plot the variation of measured voltage with instrument capacitance $C_e = 0.1 \ldots 10\,\text{nF}$ when $S_1 = -1000\,\mu\varepsilon$ (use log scale for C_e axis).
   (iv) Plot variation of measured voltage with strain in the range $S_1 = 0 \ldots -1000\,\mu\varepsilon$; make a carpet plot of V vs. S_1 for various values of C_e (0.1 nF, 1 nF, 10 nF).
   (v) Extend to dynamic strain: calculate the voltage V for dynamic strain of amplitude $\hat{S}_1 = 1000\,\mu\varepsilon$ with frequency $f = 100\,\text{kHz}$ (take $C_e = 1\,\text{pF}$).
2. Consider a PWAS transducer with length $l = 7\,\text{mm}$, width $b = 1.65\,\text{mm}$, thickness $t = 0.2\,\text{mm}$, and material properties as given in Table 7.1.
   (i) Calculate the approximate value of Young modulus, Y^E, in GPa. What common material is this value close to?
   (ii) Calculate the electric field, E_3, for an applied voltage $V = 100\,\text{V}$. Express the electric field E_3 in kV/mm.

(iii) Calculate the strain S_3 for the simultaneous application of stress $T_1 = -1\,\text{MPa}$ and voltage $V = -100\,\text{V}$.
(iv) Plot on the same chart the variation of strain S_1 with voltage V for $T_1 = 0$; $-2.5, -5, -7.5, -10\,\text{MPa}$.
(v) Plot on the same chart the variation of strain S_1 with stress T_1 for $V = 0, 25, 50, 75, 100\,\text{V}$.

3. Consider a PWAS transducer with length $l = 7\,\text{mm}$, width $b = 1.65\,\text{mm}$, thickness $t = 0.2\,\text{mm}$, and material properties as given in Table 7.1.
   (i) Calculate the approximate value of Young modulus, Y^E, in GPa. What common material is this value close to?
   (ii) Calculate the electric field, E_3, for an applied voltage $V = -100\,\text{V}$. Express the electric field E_3 in kV/mm.
   (iii) Calculate the strain S_3 for the simultaneous application of stress $T_3 = -1$ MPa and voltage $V = -100$ V.
   (iv) Calculate the total displacement, ΔL, of a free stack of $N = 100$ wafers, with a 10-μm adhesive layer between wafers energized by $V = -150$ V. What is the total length of the stack, L? What is the effective induced strain $\Delta L/L$?
   (v) Plot on the same chart the variation of strain S_3 with voltage V for $T_3 = 0$; -2.5, -5, -7.5, -10 MPa.
   (vi) Plot on the same chart the variation of strain S_3 with stress T_3 for $V = -0$, $-25, -50, -75, -100\,V$
   (vii) Calculate S_2 for stress $T_3 = -1$ MPa and voltage $V = -100$ V. Compare with the calculated S_3 and discuss.

4. Consider a PWAS transducer with length $l = 7\,\text{mm}$, width $b = 1.65\,\text{mm}$, thickness $t = 0.2\,\text{mm}$, and piezoelectric material properties as given in Table 7.1. Assume internal damping ratio $\eta = 1\%$ and electric loss factor $\delta = 1\%$.
   (i) Calculate the complex compliance, $\bar{s}_{11}^E$, complex dielectric permittivity, $\bar{\varepsilon}_{33}^T$, complex coupling coefficient, $\bar{k}_{31}$, real and complex sound speeds, c and $\bar{c}$, and real and complex electrical capacitances, C and $\bar{C}$.
   (ii) Calculate the natural resonance frequencies of the first, second, and third modes of antisymmetric vibration, in kHz. Plot on the same chart the first, second, and third modes of antisymmetric vibration.
   (iii) Calculate the natural resonance frequencies of the first, second, and third modes of symmetric vibration, in kHz. Plot on the same chart the first, second, and third modes of symmetric vibration.
   (iv) Plot on separate charts the real part, the imaginary part, the amplitude, and the phase of the PWAS electromechanical admittance at 401 equally spaced points in the 10–1000 kHz frequency range. Identify on this chart some of the resonance frequencies determined earlier. State the mode type (symmetric or antisymmetric) and mode number (e.g., f_2^A for the second antisymmetric mode).
   (v) Plot on separate charts the real part, the imaginary part, the amplitude, and the phase of the PWAS complex impedance at 401 equally spaced points in the 100–1000 kHz frequency range. Identify on this chart the frequencies at which the real part of the impedance peaks. Explain how these frequencies are related, if at all, with the resonance frequencies.

8

TUNED WAVES GENERATED WITH PIEZOELECTRIC WAFER ACTIVE SENSORS

8.1 INTRODUCTION

This chapter deals with a basic aspect of the interaction between PWAS and structure during the active SHM process, i.e., the tuning between the PWAS and the Lamb waves traveling in the structure. PWAS are strain-coupled transducers that are small, lightweight, and relatively low-cost. Upon electric excitation, the PWAS transducers can generate Lamb wave in the structural material by converting the electrical energy into the acoustic energy of ultrasonic wave propagation. In the same time, the PWAS transducers are able to convert acoustic energy of the ultrasonic waves back into an electric signal. Under electric excitation, the PWAS undergo oscillatory contractions and expansions, which are transferred to the structure through the bonding layer to excite Lamb waves into the structure. In this process, several factors influence the behavior of the excited wave. We will first recall the concepts of load transferred between the PWAS actuator and structure. Then, we will model the PWAS interaction with the ultrasonic waves generated into the structure and will show that tuning opportunities exist. The tuning is especially beneficial when dealing with multimode waves, such as the Lamb waves.

The chapter adopts a gradual approach, from simple to complex. To understand the essence of the PWAS tuning mechanism, two simple cases are studied first: (a) axial waves in a bar and (b) flexural waves in a beam. It is shown that, in both cases, the tuning concept allows one to find frequencies at which the waves are strongly excited as well as frequencies at which the waves cannot be excited (i.e., they are rejected).

The chapter continues with the development of a full analytical solution for PWAS Lamb-wave tuning using the exact Lamb waves theory developed in Chapter 6. The space-domain Fourier transform is used to resolve the problem using complex contour integrals. A closed-form solution is obtained for the ideal case of perfect bonding between the PWAS and structure, which permits the use of the pin-force model for the PWAS–structure interaction. The developed general formula will show the existence of salient points, governed by the $\sin \gamma a$ function at which a particular Lamb wave mode of wavenumber γ can either be preferentially excited or be rejected by a PWAS transducer of linear dimension $2a$. Since, at any given frequency, the wavenumbers of the various Lamb wave modes existing in a plate may be widely different, the concept of preferentially tuning of certain Lamb-wave modes and rejection of others becomes apparent.

Extensive experimental validation of the PWAS Lamb-wave tuning concept is presented. Both rectangular and circular PWAS are investigated. A sweet spot for the excitation of S_0 Lamb wave mode, which is preferred for certain SHM applications, is illustrated. In matching the theory and experiments, it will be found that an effective PWAS size exists; values of the effectiveness coefficient are given in tabular form for various PWAS geometries.

Also studied are the directivity patterns associated with tuning rectangular PWAS. This aspect is very important when trying to propagate waves in selected directions, while not sending them in other directions. Full directive patterns for a 5:1 rectangular PWAS transducer are presented.

The final part of the chapter is devoted to the analysis of the PWAS Lamb-wave tuning in composite plates. This subject is considerably more difficult to study, and hence the direct approach is deemed unfeasible. Instead, the normal modes expansion technique is adopted. The theoretical foundation of this technique for the study of PWAS Lamb-wave tuning in composite plates is developed. Extensive PWAS tuning experiments are performed on a multi-layer multi-orientation carbon–fiber/epoxy composite plate using square and circular PWAS. Tuning opportunities are identified and documented. In addition, a comparison between theoretical prediction and actual measurement for the group velocity of various Lamb-wave modes in the composite plate is also presented.

8.2 STATE OF THE ART

The PWAS, under electric excitation, transfer the oscillatory contractions and expansion to the bonded layer, which transmits them to the structural surface. In this process, several factors influence the wave behavior: (i) thickness of the bonding layer; (ii) geometry of the PWAS; (iii) thickness of the structural wall and its material properties. The result of the influence of all these factors is the tuning of the PWAS with the Lamb waves traveling in the structure. We studied this phenomenon (Giurgiutiu, 2003) using the space-domain Fourier transform and plane-strain conditions; a closed-form solution was developed to describe the interaction between a PWAS and a plate structure. The closed-form solution contained harmonic functions of the product between PWAS characteristic length and the wavenumbers of various Lamb-wave modes existing in the structure. Raghavan and Cesnik (2004) extended the analysis to the case of circular PWAS transducers in polar coordinates. The resulting expression uses Bessel and Hankel functions instead of harmonic functions. Raghavan and Cesnik (2005) developed a 3-D integral solution using the double Fourier transform. The theoretical procedure is similar to that of plain-strain conditions; however, the transformation back into the space domain is quite difficult.

Giurgiutiu (2003) and Raghavan and Cesnik (2005) performed the analytical analysis for sinusoidal steady state excitation and assumed that the source is a point source. Nieuwenhuis and Neumann (2005) propose a finite element simulation to explore the operation of the wafer transducer. They modeled separately the emission and the detection processes including the effects of finite pulse width, pulse dispersion, and interaction between the PWAS element and the transmitting medium for the same point source case as studied by Giurgiutiu (2003).

Approximate thin-plate theories, such as classical plate theory (CPT), under Love-Kirchhoff kinematic assumption, and shear deformable plate theory (SDPT) or Mindlin theory have been developed for obtaining an engineering solution to a variety of problems involving the dynamic response of thin isotropic and anisotropic laminated plates. Lin and Yuan (2001) studied the interaction between an actuator and wave propagation in the host plate the Mindlin plate theory and lamination assumptions. However, the Mindlin theory is based on a system of approximate, 2-D equations of flexural and extensional motion of isotropic elastic plates (Mindlin, 1956; Mindlin and Medick, 1959; Mindlin, 1960).

Mal (1988), Mal and Lih (1992), Lih and Mal (1996) developed an exact solution to 3-D problems consisting of multilayered, angle-ply laminates of finite thickness and large lateral dimensions subjected to various types of surface loads. In these papers, the response problem was formulated using triple integral transforms (one in time and two in space), leading to an exact representation of the elastodynamic field in the transformed frequency–wavenumber domain. The inversion of the transforms required numerical evaluation of a double wavenumber integral followed by frequency inversion using the fast Fourier transform algorithm. The main computational effort in this approach involves the accurate evaluation of the double wavenumber integral. The evaluation of this integral was found to be difficult due to the presence of singularities within the integration domain and the highly oscillatory nature of the integrands at higher frequencies and large distances between the field and source points. To date, there is no algorithm available to evaluate the double wavenumber integrals that appear in the corresponding three-dimensional problems. The conventional integration schemes (e.g., Simpson, Gaussian, etc.) require millions of function evaluations, resulting in extremely slow turnaround time. Banerjee et al. (2005) proposed a scheme in which the double integral is transformed, using contour integration, into a single integral, which is then evaluated numerically using conventional integration schemes. The new integration scheme was implemented on the exact theory to obtain the time histories of the vertical surface displacement. The results at a number of field points was compared with those from FEM for propagation along different angles with respect to the fiber direction.

The interaction between transducer and structure can be also studied through the normal mode expansion technique (NME). In NME, the fields generated in the structure due to the application of the surface loading are expanded in the form of an infinite series of the normal modes of the structure itself. The velocity field of the modes must be known. It is important to define well the physical properties of the sources used to generate the waves. The source can be either a surface source or a volume source. NME technique can be used to treat waveguide excitation. The arbitrary acoustic waveguide field distribution is expanded as a superposition of waveguide modes. For complete mathematical rigor, one must prove that the field distribution of mode set is complete and orthogonal. Generally it is assumed that the set of acoustic waveguide mode functions are complete (Auld, 1990), while the orthogonality is proven. The goal is to find the amplitudes of each of the modes generated in the structure due to the application of specified surface loadings on the boundary of the structure. Moulin et al. (2000) used the NME technique for composite plate excited by a PWAS. Due to the complexity of the problem, the input, i.e., the acoustic field related to each eigenmode and the surface forcing function, was derived through the finite element method.

8.3 TUNED AXIAL WAVES EXCITED BY PWAS

General Solution

Assume a 1-D medium in which an external cause induces an actuation strain, $\varepsilon_0(x, t)$, as shown in Fig. 8.1. Such an actuation strain may be induced by surface-mounted PWAS, applied symmetrically to the top and bottom surfaces. Hence, the total strain is given by

$$\varepsilon(x, t) = \frac{\sigma(x, t)}{E} + \varepsilon_e(x, t) \tag{1}$$

where E is the Young modulus. Equation (1) can be rearranged in the form

$$\sigma = E\,(\varepsilon - \varepsilon_e) \tag{2}$$

Consider an infinitesimal element of length dx. Newton law of motion applied to the infinitesimal element yields

$$A\,d\sigma = \rho A \ddot{u}\,dx \tag{3}$$

where $\dot{u} = \partial u/\partial t$. Hence,

$$\sigma' = \rho \ddot{u} \tag{4}$$

where $u' = \partial u/\partial x$. Substitution of Eq. (2) into Eq. (4) yields

$$E\,(\varepsilon - \varepsilon_e) = \rho \ddot{u} \tag{5}$$

Recall $\varepsilon = u'$. Substitution into Eq. (5) and differentiation with respect to x yields

$$\varepsilon'' - \frac{1}{c^2}\ddot{\varepsilon} = \varepsilon_e'' \tag{6}$$

where $c^2 = E/\rho$. For harmonic excitation, we have

$$\begin{aligned} \varepsilon_e(x, t) &= \varepsilon_e(x) e^{-i\omega t} \\ \varepsilon(x, t) &= \varepsilon(x) e^{-i\omega t} \end{aligned} \tag{7}$$

Substitution of Eq. (7) into Eq. (6) gives

$$\varepsilon'' - \xi_0^2 \varepsilon = \varepsilon_e'' \tag{8}$$

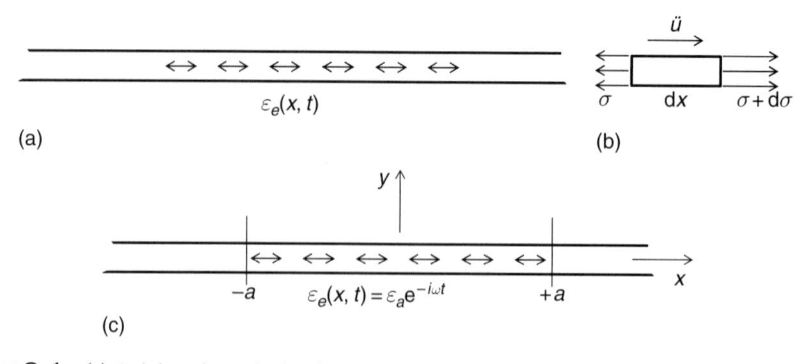

FIGURE 8.1 (a) Axial strain excitation in a 1-D medium; (b) infinitesimal element; (c) the effect of an ideally-bonded finite length PWAS.

where $\xi_0^2 = \omega^2/c^2$ is the wavenumber of axial waves in the 1-D medium. Take the space-domain Fourier transform

$$f(x) = \int_{-\infty}^{+\infty} \tilde{f}(\xi) e^{i\xi x}\, \mathrm{d}\xi$$
$$\tilde{f}(\xi) = \frac{1}{2\pi} \int_{-\infty}^{+\infty} f(x) e^{-i\xi x}\, \mathrm{d}x \tag{9}$$

The space-domain Fourier transform has the property $\tilde{f}' = i\xi \tilde{f}$. The space-domain Fourier transform of Eq. (8) is

$$-\xi^2 \tilde{\varepsilon} - \xi_0^2 \tilde{\varepsilon} = -\xi^2 \tilde{\varepsilon}_e \tag{10}$$

The solution of Eq. (10) is

$$\tilde{\varepsilon} = \frac{\xi^2}{\xi^2 - \xi_0^2} \tilde{\varepsilon}_e \tag{11}$$

Equation (11) represents the solution in the Fourier domain. Taking the inverse space-domain Fourier transform yields the solution in the space domain

$$\varepsilon(x) = \frac{1}{2\pi} \int_{-\infty}^{+\infty} \frac{\xi^2}{\xi^2 - \xi_0^2} \tilde{\varepsilon}_e(\xi) e^{i\xi x}\, \mathrm{d}\xi \tag{12}$$

Of course, the exact expression of the solution $\varepsilon(x)$ depends on the particular form of the excitation, $\varepsilon_e(x)$ and of its space-domain Fourier transform.

8.3.1 SOLUTION FOR AN IDEALLY BONDED PWAS

For an ideally bonded PWAS, the induced strain ε_a is uniform over the PWAS length. Assume the PWAS is positioned symmetrically about the origin, i.e., it stretches from $x = -a$ to $x = +a$ (Fig. 8.1c) The mathematical expression of this strain distribution is

$$\varepsilon_e(x, t) = \begin{cases} \varepsilon_a e^{-i\omega t}, & |x| < a \\ 0, & \text{otherwise} \end{cases} \tag{13}$$

The space-domain portion of Eq. (13) is the rectangular pulse function

$$\varepsilon_e(x) = \begin{cases} \varepsilon_a, & |x| < a \\ 0, & \text{otherwise} \end{cases} \tag{14}$$

The space-domain Fourier transform of Eq. (14) is the sinc function (Fig. 8.2).

$$\tilde{\varepsilon}_e = \varepsilon_a \frac{2}{\xi} \sin(\xi a) \tag{15}$$

To obtain the solution, recall Eq. (8)

$$\varepsilon'' - \xi_0^2 \varepsilon = \varepsilon_e'' \tag{16}$$

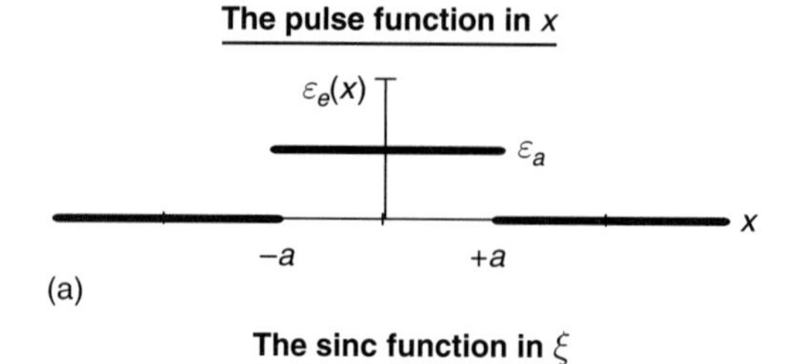

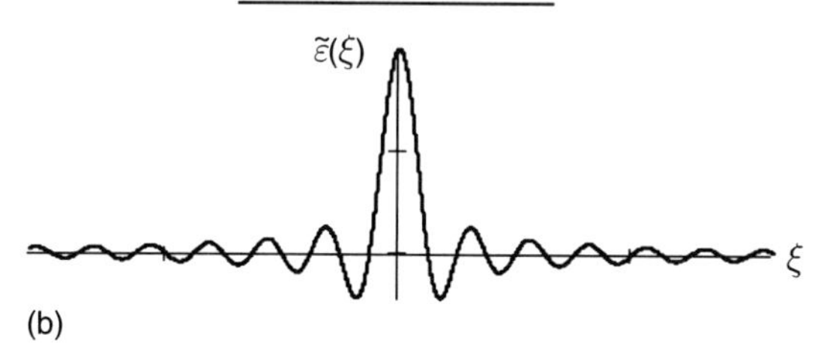

FIGURE 8.2 Strain excitation by ideally bonded PWAS: (a) the pulse function in x; (b) its Fourier transform, the sinc function in ξ.

Upon taking the space-domain Fourier transform, we obtain

$$-\xi^2\tilde{\varepsilon} - \xi_0^2\tilde{\varepsilon} = -\xi^2\varepsilon_a\frac{2}{\xi}\sin\xi a \tag{17}$$

Hence, the Fourier-domain solution is

$$\tilde{\varepsilon} = \varepsilon_a\frac{2\xi}{\xi^2 - \xi_0^2}\sin\xi a \tag{18}$$

Taking the inverse Fourier transform, we obtain the space-domain solution

$$\varepsilon(x) = \frac{2\varepsilon_a}{2\pi}\int_{-\infty}^{+\infty}\frac{\xi\sin\xi a}{\xi^2 - \xi_0^2}e^{i\xi x}\,\mathrm{d}\xi \tag{19}$$

The integral in Eq. (19) can be resolved analytically using the residues theorem and a semicircular contour C in the complex ξ domain. We note that the integrant in Eq. (19) has two poles, corresponding to the wavenumbers $-\xi_0$ and $+\xi_0$. We will resolve Eq. (19) for the forward traveling wave, which exists for $x > 0$ and generates a solution containing $i(\xi x - \omega t)$ in the exponential function. Hence, we will retain the positive pole, $+\xi_0$, inside the integration contour but exclude the negative pole, $-\xi_0$, from the integration contour. The resulting integration contour C is shown in Fig. 8.3. We note that the integration along the semicircular portion Γ of the contour C vanishes as the radius of integration becomes very large, i.e., $R \to \infty$. Therefore, the integration on the contour C resolves into the integration along the real ξ axis from $-\infty$ to $+\infty$, i.e.,

$$\oint_C = +\not{\int_\Gamma} + \int_{-\infty}^{+\infty} \tag{20}$$

According to the residue theorem, the integration around the contour C equals the sum of the residues inside the contour times a multiplicative factor $2\pi i$. Since we have

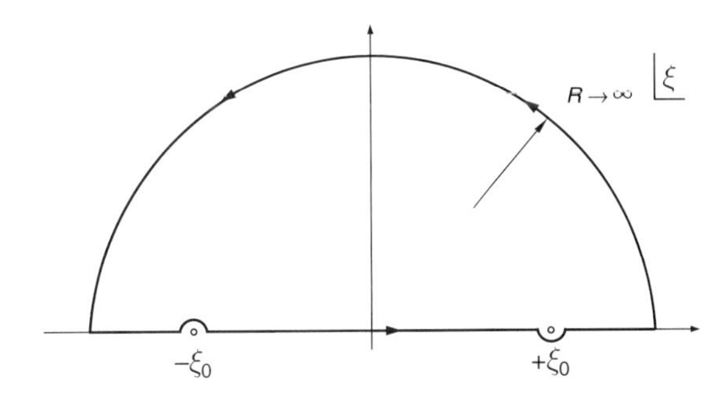

FIGURE 8.3 Contour for evaluating the strain solution under ideally bonded PWAS excitation. The residue at positive wavenumber is included, while the residue for the negative wavenumber is excluded.

retained only one pole inside the contour, the pole $+\xi_0$, the contour integral takes the expression

$$\oint_C = 2\pi i \operatorname{Res}\Big|_{\xi=\xi_0} \tag{21}$$

The residue is calculated as

$$\operatorname{Res}\left(\frac{\xi \sin \xi a}{\xi^2 - \xi_0^2} e^{i\xi x}\right) = \left(\frac{\xi \sin \xi a}{\xi + \xi_0} e^{i\xi x}\right)_{\xi=\xi_0} = \frac{\xi_0 \sin \xi_0 a}{\xi_0 + \xi_0} e^{i\xi_0 x} = \frac{1}{2} (\sin \xi_0 a)\, e^{i\xi_0 x} \tag{22}$$

Substitution of Eqs. (21)–(22) into Eq. (19) gives

$$\varepsilon(x) = \frac{2\varepsilon_a}{2\pi} 2\pi i \frac{1}{2} (\sin \xi_0 a)\, e^{i\xi_0 x} \tag{23}$$

Equation (23) yields

$$\varepsilon(x) = i\varepsilon_a \sin \xi_0 a e^{i\xi_0 x} \tag{24}$$

Equation (24) is the space-domain solution of the problem. Substitution of Eq. (24) into Eq. (7) yields the complete solution, which is the strain response in the structure due to a harmonically oscillating PWAS perfectly bonded to the structural surface. This solution has the form

$$\varepsilon(x) = i\varepsilon_a (\sin \xi_0 a)\, e^{i(\xi_0 x - \omega t)} \tag{25}$$

It is apparent that the response amplitude follows a sinusoidal variation with respect to the parameter $\xi_0 a$. A plot of this variation is presented in Fig. 8.4. Response peaks are observed at odd integer multiples of $\pi/2$, i.e., when

$$\xi_0 a = (2n-1)\frac{\pi}{2}, \quad n = 1, 2, 3, \ldots \tag{26}$$

Recalling that $\xi_0 = 2\pi/\lambda$, and that the PWAS length is $l_a = 2a$, relation (23) implies that maximum excitation will happen when the PWAS length is an odd integer multiple of the half wavelength, i.e.,

$$l_a = (2n-1)\frac{\lambda}{2}, \quad n = 1, 2, 3, \ldots \tag{27}$$

Thus, for a given wavelength, Eq. (24) provides a method to construct PWAS geometries that will optimally excite the structure with the prescribed wavelength. Of course, the

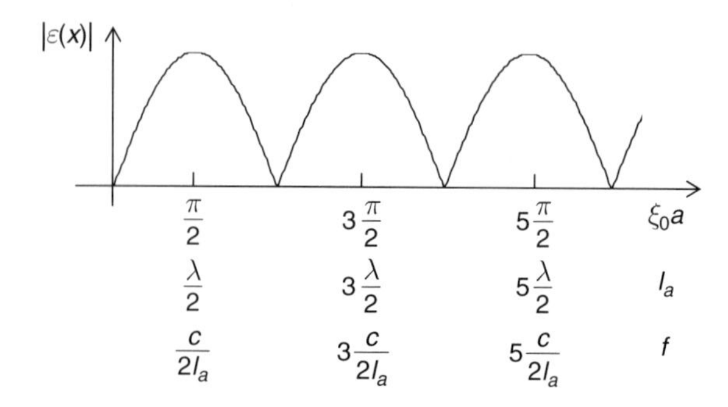

FIGURE 8.4 Optimum match between PWAS and structure for the excitation of axial waves.

first such geometry will be that in which the PWAS characteristic length is exactly half the wavelength. Higher matches are also possible, at odd integer multiples of the half wavelength.

Since the wavelength depends on frequency, $\lambda = c/f$, it is also conceivable that, for a given PWAS, certain frequencies exist at which the excitation is optimal. These optimal excitation frequencies are given by

$$f = (2n-1)\frac{c}{2l_a}, \quad n = 1, 2, 3, \ldots \tag{28}$$

8.4 TUNED FLEXURAL WAVES EXCITED BY PWAS

Consider the general equation of flexural vibrations under external moment excitation, $M_e(x, t)$,

$$EIv'''' + \rho A\ddot{v} = M_e''(x, t) \tag{29}$$

where $\dot{v} = \partial v/\partial t$ and $v' = \partial v/\partial x$. Assume harmonic variation in the time domain of the form $e^{-i\omega t}$. Hence, Eq. (29) becomes

$$EIv'''' - \omega^2 \rho Av = M_e''(x) \tag{30}$$

Divide by EI and obtain

$$v'''' - \omega^2 \frac{\rho A}{EI} v = \frac{M_e''}{EI} \tag{31}$$

The ratio M_e/EI has dimensions of curvature. We will call it the *excitation curvature*, κ_e, i.e.,

$$\kappa_e = \frac{M_e}{EI} \tag{32}$$

In addition, introduce the notation

$$\xi_F^4 = \omega^2 \frac{\rho A}{EI} = \frac{\omega^2}{\bar{a}^2}, \quad \text{where } \bar{a}^2 = \frac{EI}{\rho A} = \frac{E\frac{bd^3}{3}}{\rho(bd)} = \frac{d^2}{3}\frac{E}{\rho} \tag{33}$$

Hence, Eq. (31) becomes

$$v'''' - \xi_F^4 v = k_e'' \tag{34}$$

Take the space-domain Fourier transform

$$\begin{aligned} f(x) &= \int_{-\infty}^{+\infty} \tilde{f}(\xi) e^{i\xi x}\, \mathrm{d}\xi \\ \tilde{f}(\xi) &= \frac{1}{2\pi} \int_{-\infty}^{+\infty} f(x) e^{-i\xi x}\, \mathrm{d}x \end{aligned} \tag{35}$$

The space-domain Fourier transform has the property $\tilde{f'} = i\xi\tilde{f}$. The space-domain Fourier transform of Eq. (34) is

$$\left(\xi^4 - \xi_F^4\right) \tilde{v} = -\xi^2 \tilde{k}_e \tag{36}$$

The solution of Eq. (36) is

$$\tilde{\varepsilon} = \frac{-\xi^2}{\xi^4 - \xi_F^4} \tilde{k}_e \tag{37}$$

Equation (37) represents the solution to the problem in the Fourier domain. Taking the inverse space-domain Fourier transform yields the solution in the space domain

$$v(x) = \frac{1}{2\pi} \int_{-\infty}^{+\infty} \frac{-\xi^2}{\xi^4 - \xi_F^4} \tilde{k}_e e^{i\xi x}\, \mathrm{d}\xi \tag{38}$$

The integral in Eq. (38) can be resolved using the residues theorem and a semicircular contour C in the complex ξ domain. We note that the integrant in Eq. (38) has four poles, corresponding to the wavenumbers $+\xi_F, -\xi_F, +i\xi_F, -i\xi_F$. We will resolve Eq. (38) for the forward traveling wave, which exists for $x > 0$ and generates a solution containing $i(\xi_F x - \omega t)$ in the exponential function. Hence, we will retain the positive pole, $+\xi_F$, inside the integration contour, but exclude the negative pole, $-\xi_F$, from the integration contour. The resulting integration contour C is shown in Fig. 8.5.

This integration contour includes the real pole $+\xi_F$ and the imaginary pole $i\xi_F$. We note that the integration along the semicircular portion Γ of the contour C vanishes as

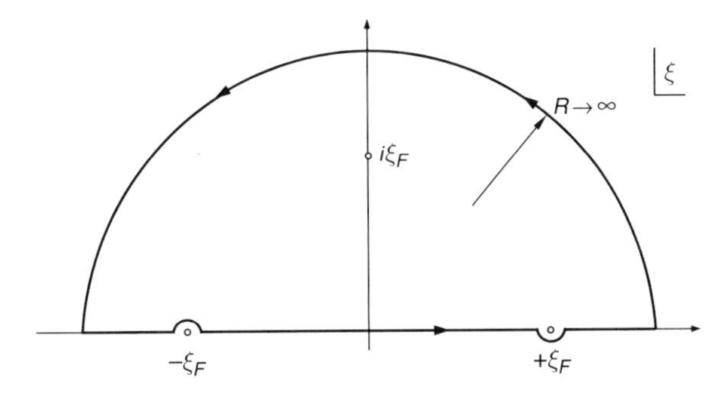

FIGURE 8.5 Contour for evaluating the flexural wave solution. The residue at positive wavenumber, $+\xi_F$, is included, while the residue for the negative wavenumber, $-\xi_F$, is excluded. Also included is the residue at the imaginary wavenumber, $+i\xi_F$.

the radius of integration becomes very large, i.e., $R \to \infty$. Therefore, the integration on the contour C resolves into the integration along the real ξ axis from $-\infty$ to $+\infty$, i.e.,

$$\oint_C = +\not{\int_\Gamma} + \int_{-\infty}^{+\infty} \tag{39}$$

According to the residue theorem, the integration around the contour C equals the sum of the residues inside the contour times a multiplicative factor $2\pi i$. Since we have retained only two poles inside the contour, $+\xi_0$ and $+i\xi_0$, the contour integral takes the expression

$$\oint_C = 2\pi i \left(\text{Res}\,|_{\xi=+\xi_0} + \text{Res}\,|_{\xi=+i\xi_0}\right) \tag{40}$$

The residue at $+\xi_F$ is calculated as

$$\text{Res}\left(\frac{\xi^2 \tilde{k}_e(\xi)}{\xi^4 - \xi_F^4} e^{i\xi x}\right)_{\xi=+\xi_F} = \left(\frac{\xi^2 \tilde{k}_e(\xi)}{4\xi^3} e^{i\xi x}\right)_{\xi=+\xi_F} = \frac{\tilde{k}_e(\xi_F)}{4\xi_F} e^{i\xi_F x} \tag{41}$$

The residue at $+i\xi_F$ is calculated as

$$\text{Res}\left(\frac{\xi^2 \tilde{k}_e(\xi)}{\xi^4 - \xi_F^4} e^{i\xi x}\right)_{\xi=+i\xi_F} = \left(\frac{\xi^2 \tilde{k}_e(\xi)}{4\xi^3} e^{i\xi x}\right)_{\xi=+i\xi_F} = \frac{\tilde{k}_e(i\xi_F)}{4i\xi_F} e^{-\xi_F x} \tag{42}$$

Substitution of Eqs. (39)–(42) into Eq. (8) gives

$$v(x) = \frac{1}{2\pi} 2\pi i \left[\frac{\tilde{k}_e(\xi_F)}{4\xi_F} e^{i\xi_F x} + \frac{\tilde{k}_e(i\xi_F)}{4i\xi_F} e^{-\xi_F x}\right] \tag{43}$$

Adding the harmonic variation in the time domain of the form $e^{-i\omega t}$ yields the complete solution

$$v(x) = i\frac{\tilde{k}_e(\xi_F)}{4\xi_F} e^{i(\xi_F x - \omega t)} + \frac{\tilde{k}_e(i\xi_F)}{4\xi_F} e^{-\xi_F x} e^{-i\omega t} \tag{44}$$

Note that the first term in Eq. (44) represents a propagating wave, while the second term does not. In fact, the second term represents a vibration that is decaying fast with x. This term represents a local vibration that does not propagate. It is called an *evanescent wave*. Thus, we will retain only the propagating wave part of Eq. (44), i.e.,

$$v(x, t) = i\frac{\tilde{k}_e(\xi_F)}{4\xi_F} e^{i(\xi_F x - \omega t)} \tag{45}$$

The imaginary number i that multiplies Eq. (45) indicates that the displacement $v(x, t)$ is in quadrature with the strain excitation, $\varepsilon_e(x, t)$.

The strain solution, ε_x, can be derived from the y-displacement solution, $v(x, t)$ as follows. First, derive the x-displacement

$$u_x = -yu'_y = -yi(i\xi_F)\frac{\tilde{\varepsilon}_e(\xi_F)}{4\xi_F} e^{i(\xi_F x - \omega t)} = y\frac{\tilde{\varepsilon}_e(\xi_F)}{4} e^{i(\xi_F x - \omega t)} \tag{46}$$

Upon space differentiation of Eq. (46), obtain the strain ε_x as

$$\varepsilon_x = u'_x = y(i\xi_F)\frac{\tilde{\varepsilon}_e(\xi_F)}{4} e^{i(\xi_F x - \omega t)} \tag{47}$$

8.4.1 SOLUTION FOR AN IDEALLY BONDED PWAS

For an ideally bonded PWAS, the excitation moment is represented by the rectangular pulse function

$$M_e(x) = \begin{cases} M_a, & |x| < a \\ 0, & \text{otherwise} \end{cases} \tag{48}$$

The space-domain Fourier transform of Eq. (48) is the sinc function (Fig. 8.6).

$$\tilde{M}_e(\xi) = M_a \frac{2}{\xi} \sin(\xi a) \tag{49}$$

Recall that $k_e = \frac{M_e}{EI}$, and denote

$$k_a = \frac{M_a}{EI} \tag{50}$$

Hence,

$$k_e(x) = \begin{cases} k_a, & |x| < a \\ 0, & \text{otherwise} \end{cases} \tag{51}$$

and

$$\tilde{k}_e(\xi) = k_a \frac{2}{\xi} \sin(\xi a) \tag{52}$$

Thus, Eq. (47) becomes

$$\varepsilon_x = y(i\xi_F) \frac{k_a \frac{2}{\xi_F} \sin \xi_F a}{4} e^{i(\xi_F x - \omega t)} = iy \frac{k_a \sin \xi_F a}{2} e^{i(\xi_F x - \omega t)} \tag{53}$$

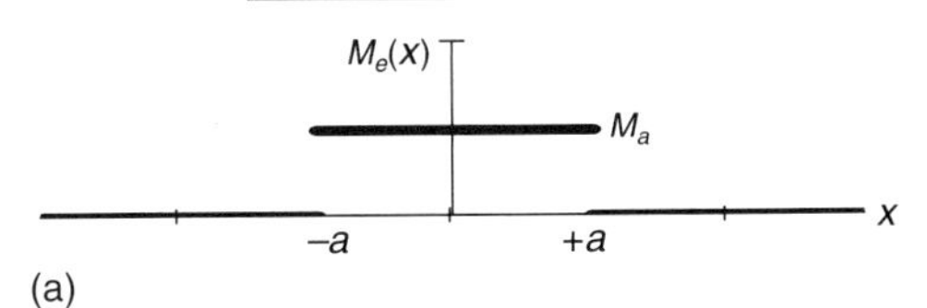

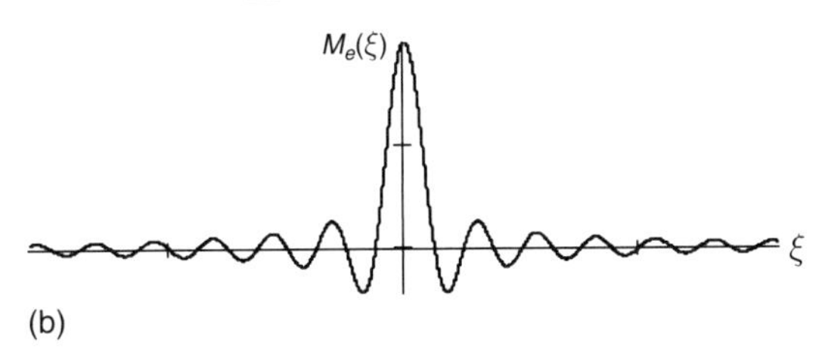

FIGURE 8.6 Moment excitation by ideally bonded PWAS: (a) the pulse function in x; (b) its Fourier transform, the sinc function in ξ.

The strain at the material surface can be obtained from Eq. (53) by taking $y = d$, i.e.,

$$\varepsilon_x\Big|_{y=d} = id\frac{k_a \sin \xi_F a}{2} e^{i(\xi_F x - \omega t)} \tag{54}$$

Recall the expression of actuation moment, M_a, in terms of the PWAS pin-force, F_a, i.e.,

$$M_a = F_a d \tag{55}$$

Substitution of Eq. (55) into Eq. (50) yields

$$k_a = \frac{F_a d}{EI} \tag{56}$$

Recall that

$$I = \frac{bd^3}{3} \quad \text{and} \quad A = b(2d) \tag{57}$$

Hence,

$$k_a = \frac{F_a d}{E\frac{bd^3}{3}} = 3\frac{2}{d}\frac{F_a}{EA} = 3\frac{2}{d}\varepsilon_a \tag{58}$$

where

$$\varepsilon_a = \frac{F_a}{EA} \tag{59}$$

Substitution of Eq. (58) into Eq. (54) yields the following expression for the strain at the material surface of the beam undergoing flexural wave excitation under ideally bonded surface PWAS

$$\varepsilon_x\Big|_{y=d} = id\frac{1}{2}3\frac{2}{d}\varepsilon_a (\sin \xi_F a)\, e^{i(\xi_F x - \omega t)} = i3\varepsilon_a (\sin \xi_F a)\, e^{i(\xi_F x - \omega t)} \tag{60}$$

It is apparent that the response amplitude follows a sinusoidal variation with respect to the parameter $\xi_F a$. A plot of this variation is presented in Fig. 8.7.

Response peaks are observed at odd integer multiples of $\pi/2$, i.e., when

$$\xi_F a = (2n - 1)\frac{\pi}{2}, \quad n = 1, 2, 3, \ldots \tag{61}$$

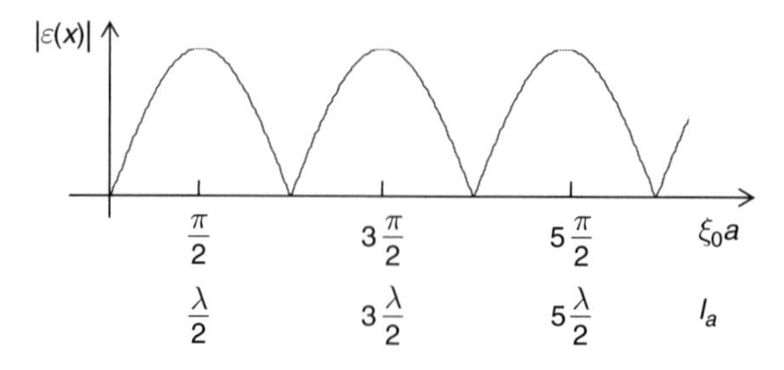

FIGURE 8.7 Optimum match between PWAS and structure for the excitation of axial waves.

Recalling that $\xi_F = 2\pi/\lambda_F$, and that the PWAS length is $l_a = 2a$, relation (23) implies that maximum excitation of flexural waves will happen when the PWAS length is an odd integer multiple of the flexural half wavelength, i.e.,

$$l_a = (2n-1)\frac{\lambda_F}{2}, \quad n = 1, 2, 3, \ldots \tag{62}$$

Thus, for a given wavelength, Eq. (62) provides a method to construct PWAS geometries that will optimally excite the structure with flexural waves of prescribed wavelength. Of course, the first such geometry will be that in which the *PWAS characteristic length is exactly half the flexural wavelength*. Higher matches are also possible. These will occur at odd integer multiples of the half wavelength.

Since the wavelength depends on frequency and wave speed, $\lambda_F = c_F/f$, it is also conceivable that, for a given PWAS, certain frequencies exist at which the excitation is optimal. These optimal excitation frequencies are given by

$$f_n = (2n-1)\frac{c_F(f_n)}{2l_a} \tag{63}$$

Equation (63) indicates that the wave speed, c_F, is also a function of frequency. This is to be expected since flexural waves are dispersive. Recall Eq. (5.141) from Chapter 5, which gives

$$c_F(\omega) = \left(\frac{Eh^2}{12\rho}\right)^{1/4}\sqrt{\omega} \tag{64}$$

Since $h = 2d$ and $\omega = 2\pi f$, Eq. (64) can be rewritten as

$$c_F(f) = \left(\frac{Ed^2}{3\rho}\right)^{1/4}(2\pi)^{1/2}f^{1/2} \tag{65}$$

Hence, Eq. (64) can be substituted into Eq. (63) to obtain a solution of the optimal excitation frequencies in terms of the structural elastic, mass, and geometric properties.

$$f_n^{\frac{1}{2}} = (2n-1)\frac{\sqrt{2\pi}}{2l_a}\left(\frac{Ed^2}{3\rho}\right)^{1/4} \tag{66}$$

or

$$f_n = (2n-1)^2\frac{\pi}{2l_a^2}\left(\frac{Ed^2}{3\rho}\right)^{1/2}, \quad n = 1, 2, 3, \ldots \tag{67}$$

8.5 TUNED LAMB WAVES EXCITED BY PWAS

In this section, we develop the theoretical foundation for selective Lamb mode excitation with PWAS transducers. The PWAS, under electric excitation, transfers the oscillatory contractions and expansion to the bonded layer, and the layer to the structural surface. In this process, several factors influence the wave behavior: thickness of the bonding layer, geometry of the PWAS, thickness, and material of the plate. Figure 8.8 presents the coupling between PWAS and two Lamb modes, S_0 and A_0. It is apparent from Fig. 8.8 that maximum coupling between the PWAS and the Lamb wave would occur when the PWAS length is an odd multiple of the half wavelength. Since different

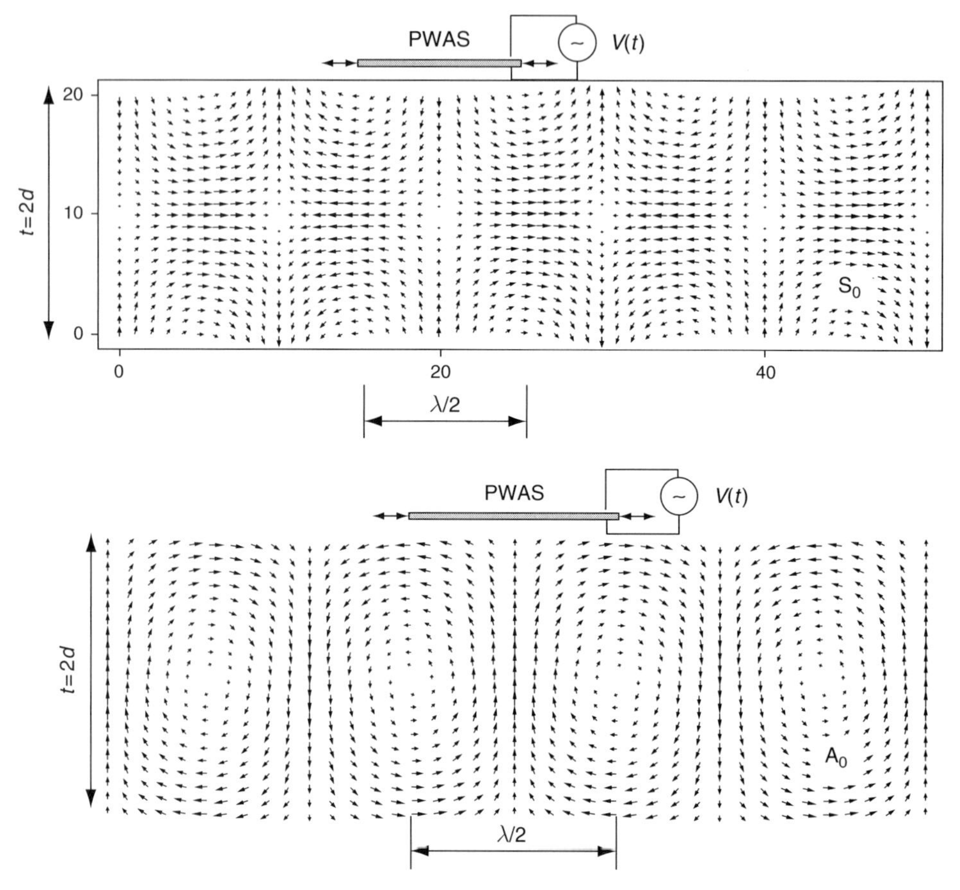

FIGURE 8.8 Typical structure of S_0 and A_0 Lamb-wave modes, and the interaction of PWAS with the Lamb wave.

Lamb modes have different wavelengths, which vary with frequency, the opportunity arises for selectively exciting various Lamb modes at various frequencies, i.e., making the PWAS tune into one or another Lamb mode.

The analysis presented in this section considers surface shear distribution resulting from PWAS action and determines the Lamb wave response of the structure. In this analysis, the choice of shear distribution is very important. As indicated in a previous section, the shear stress distribution is nonuniform, with high values at the PWAS ends. Hence, we will have to assume the shear distribution of the shear lag model corrected for frequency dependence. The solution will be developed using the space-domain Fourier transform method based on a harmonic functions kernel for straight-crested waves, and a Bessel functions kernel for circular-crested waves.

8.5.1 LAMB WAVES EXCITED BY NONUNIFORM BOUNDARY SHEAR-STRESS

A brief sketch of the proposed line of attack is presented next.

Assume harmonic shear-stress boundary excitation applied to the upper surface of the plate (Fig. 8.9), i.e.,

$$\tau_a(x, d) = \begin{cases} \tau_0 \sinh(\Gamma x), & |x| < a \\ 0, & \text{otherwise} \end{cases} \tag{68}$$

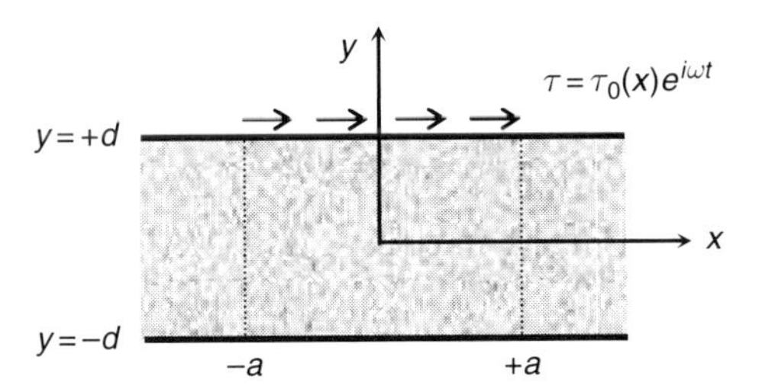

FIGURE 8.9 Plate of thickness $2d$, with a PWAS of width $2a$, under harmonic loading on the top surface.

Since the excitation is harmonic of the form $\tau_a(x)e^{-i\omega t}$, the solution is also to be assumed harmonic, and the potential functions ϕ and ψ satisfy the wave equations

$$\begin{aligned} \frac{\partial^2\phi}{\partial x^2} + \frac{\partial^2\phi}{\partial y^2} + \frac{\omega^2}{c_P^2}\phi = 0 \\ \frac{\partial^2\psi}{\partial x^2} + \frac{\partial^2\psi}{\partial y^2} + \frac{\omega^2}{c_P^2}\psi = 0 \end{aligned} \tag{69}$$

where $c_P^2 = (\lambda + 2\mu)/\rho$ and $c_S^2 = \mu/\rho$ are the longitudinal (pressure) and transverse (shear) wave speeds, λ and μ are Lamé constants, and ρ is the mass density. Recall the displacement and stress expressions

$$\begin{aligned} u_x &= \frac{\partial\phi}{\partial x} + \frac{\partial\psi}{\partial y} & \tau_{yx} &= \mu\left(2\frac{\partial^2\phi}{\partial x\,\partial y} - \frac{\partial^2\psi}{\partial x^2} + \frac{\partial^2\psi}{\partial y^2}\right) \\ u_y &= \frac{\partial\phi}{\partial y} - \frac{\partial\psi}{\partial x}, & \tau_{yy} &= \lambda\left(\frac{\partial^2\phi}{\partial x^2} + \frac{\partial^2\phi}{\partial y^2}\right) + 2\mu\left(\frac{\partial^2\phi}{\partial x^2} - \frac{\partial^2\psi}{\partial x\,\partial y}\right), \end{aligned} \quad \varepsilon_x = \frac{\partial u_x}{\partial x} \tag{70}$$

Applying the Fourier transform

$$\tilde{f}(\xi) = \int_{-\infty}^{\infty} f(\xi)e^{-i\xi x}\,\mathrm{d}x, \quad f(x) = \frac{1}{2\pi}\int_{-\infty}^{\infty} \tilde{f}(x)e^{i\xi x}\,\mathrm{d}x \tag{71}$$

yields

$$\begin{aligned} -\xi^2\tilde{\phi} + \frac{\mathrm{d}^2\tilde{\phi}}{\mathrm{d}y^2} + \frac{\omega^2}{c_P^2}\tilde{\phi} = 0 \\ -\xi^2\tilde{\psi} + \frac{\mathrm{d}^2\tilde{\psi}}{\mathrm{d}y^2} + \frac{\omega^2}{c_S^2}\tilde{\psi} = 0 \end{aligned} \tag{72}$$

$$\begin{aligned} \tilde{u}_x &= i\xi\tilde{\phi} + \frac{\mathrm{d}\tilde{\psi}}{\mathrm{d}y} & \tilde{\tau}_{yx} &= \mu\left(2i\xi\frac{\mathrm{d}\tilde{\phi}}{\mathrm{d}y} + \xi^2\tilde{\psi} + \frac{\partial^2\tilde{\psi}}{\partial y^2}\right) \\ \tilde{u}_y &= \frac{\mathrm{d}\tilde{\phi}}{\mathrm{d}y} - i\xi\tilde{\psi} & \tilde{\tau}_{yy} &= \lambda\left(-\xi^2\tilde{\phi} + \frac{\mathrm{d}^2\tilde{\phi}}{\mathrm{d}y^2}\right) + 2\mu\left(-\xi^2\tilde{\phi} - i\xi\frac{\mathrm{d}\tilde{\psi}}{\mathrm{d}y}\right) \end{aligned} \quad \tilde{\varepsilon}_x = i\xi u_x \tag{73}$$

Introduce the notations

$$p^2 = \frac{\omega^2}{c_P^2} - \xi^2, \quad q^2 = \frac{\omega^2}{c_S^2} - \xi^2 \tag{74}$$

Equation (72) becomes

$$\frac{d^2\tilde{\phi}}{dy^2} + p^2\tilde{\phi} = 0$$
$$\frac{d^2\tilde{\psi}}{dy^2} + q^2\tilde{\psi} = 0 \tag{75}$$

Equation (75) accepts the general solution

$$\tilde{\phi} = A_1 \sin py + A_2 \cos py$$
$$\tilde{\psi} = B_1 \sin qy + B_2 \cos qy \tag{76}$$

Hence, the displacements are

$$\tilde{u}_x = i\xi\,(A_1 \sin py + A_2 \cos py) + q\,(B_1 \cos qy - B_2 \sin qy)$$
$$\tilde{u}_y = p\,(A_1 \cos py - A_2 \sin py) - i\xi\,(B_1 \sin qy + B_2 \cos qy) \tag{77}$$

The four integration constants are to be found from the boundary condition. At this stage, it is advantageous to split the problem into symmetric and antisymmetric parts (Fig. 8.10).

8.5.1.1 Symmetric Solution

Assume symmetric displacements and stresses about the midplane. It can be easily observed that this amounts to

$$u_x(x,-h) = u_x(x,h) \qquad \tau_{yx}(x,-h) = -\tau_{yx}(x,h)$$
$$u_y(x,-h) = -u_y(x,h) \qquad \tau_{yy}(x,-h) = \tau_{yy}(x,h) \tag{78}$$

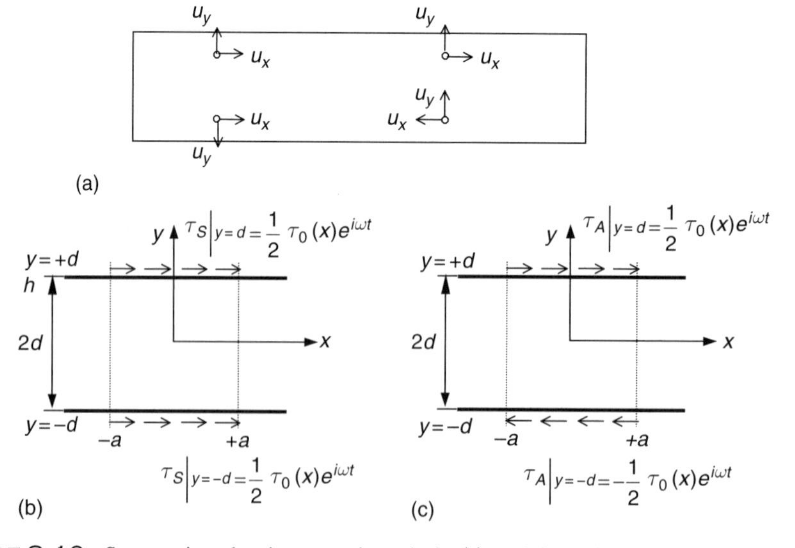

FIGURE 8.10 Symmetric and anti-symmetric analysis: (a) particle motion across the plate thickness; (b) symmetric loading and (c) anti-symmetric loading.

Note that positive shear stresses have opposite directions on the upper and lower surfaces, hence the negative sign on τ_{yx} in Eq. (78). Similar relations also apply for the Fourier transformed variables, $\tilde{u}$, $\tilde{\tau}$. Hence, the displacements and potentials for symmetric motion are

$$\begin{aligned} \tilde{u}_x &= i\xi A_2 \cos py + qB_1 \cos qy \\ \tilde{u}_y &= pA_2 \sin py - i\xi B_1 \sin qy \end{aligned} \tag{79}$$

$$\begin{aligned} \tilde{\phi} &= A_2 \cos py \\ \tilde{\psi} &= B_1 \sin qy \end{aligned} \tag{80}$$

Upon substitution in Eq. (73),

$$\tilde{\tau}_{yx} = \mu\left[-2i\xi pA_2 \sin py + (\xi^2 - q^2)B_1 \sin qy\right] \tag{81}$$

$$\tilde{\tau}_{yy} = \mu\left[(\xi^2 - q^2)A_2 \cos py + 2i\xi qB_1 \cos qy\right] \tag{82}$$

$$\tilde{\varepsilon}_x = -\xi^2 A_2 \cos py + i\xi qB_1 \cos qy$$

The symmetric boundary conditions are

$$\begin{aligned} \tau_{yx}(x, -h) &= -\tau_{yx}(x, h) = \frac{\tilde{\tau}}{2} \\ \tau_{yy}(x, -h) &= \tau_{yy}(x, h) = 0 \end{aligned} \tag{83}$$

Upon substitution, we obtain the linear system

$$\begin{aligned} -2i\xi pA_2 \sin py + (\xi^2 - q^2)B_1 \sin qy &= \frac{\tilde{\tau}}{2\mu} \\ (\xi^2 - q^2)A_2 \cos py + 2i\xi qB_1 \cos qy &= 0 \end{aligned} \tag{84}$$

Upon solution,

$$\begin{aligned} A_2 &= \frac{\tilde{\tau}}{2\mu}\frac{N_{A_2}}{D_S} \qquad N_{A_2} = 2i\xi q \cos qh, \qquad N_{B_1} = (\xi^2 - q^2)\cos ph \\ B_1 &= \frac{\tilde{\tau}}{2\mu}\frac{N_{B_1}}{D_S} \qquad D_S = (\xi^2 - q^2)^2 \cos ph \sin qh + 4\xi^2 pq \sin ph \cos qh \end{aligned} \tag{85}$$

The constants A_2 and B_1 are substituted into Eq. (79) to yield the symmetric wave solution. Upon substitution in Eq. (82), we obtain the strain wave solution at the plate's upper surface

$$\tilde{\varepsilon}_x^S = -i\frac{\tilde{\tau}}{2\mu}\frac{N_S}{D_S}, \qquad N_S = \xi q(\xi^2 + q^2)\cos ph \cos qh \tag{86}$$

8.5.1.2 Antisymmetric Solution

Assume antisymmetric displacements and stresses about the midplane, i.e.,

$$\begin{aligned} u_x(x, -h) &= -u_x(x, h) \qquad \tau_{yx}(x, -h) = \tau_{yx}(x, h) \\ u_y(x, -h) &= u_y(x, h) \qquad \tau_{yy}(x, -h) = -\tau_{yy}(x, h) \end{aligned} \tag{87}$$

Note that positive shear stresses have opposite directions on the upper and lower surfaces hence are inherently antisymmetric. Following a procedure similar to that applied to the symmetric case, we get

$$\tilde{\varepsilon}_x^A = -i\frac{\tilde{\tau}}{2\mu}\frac{N_A}{D_A}, \quad \begin{aligned} N_A &= \xi q(\xi^2+q^2)\sin ph \sin qh \\ D_A &= (\xi^2-q^2)^2 \sin ph \cos qh + 4\xi^2 pq \cos ph \sin qh \end{aligned} \tag{88}$$

8.5.1.3 Total Solution

By superposition, the total solution is the sum of the symmetric and antisymmetric solutions, i.e.,

$$\tilde{\varepsilon}_x = -i\frac{\tilde{\tau}}{2\mu}\left(\frac{N_S}{D_S}+\frac{N_A}{D_A}\right) \tag{89}$$

Applying the inverse Fourier transform and adding the harmonic time behavior yields the strain wave solution

$$\varepsilon_x(x,t) = \frac{1}{2\pi}\frac{-i}{2\mu}\int_{-\infty}^{\infty}\left(\frac{\tilde{\tau}N_S}{D_S}+\frac{\tilde{\tau}N_A}{D_A}\right)e^{i(\xi x-\omega t)}\,\mathrm{d}\xi \tag{90}$$

The integral in Eq. (90) is singular at the roots of D_S and D_A. The equations

$$\begin{aligned} D_S &= 0 \\ D_A &= 0 \end{aligned} \tag{91}$$

are exactly the Rayleigh–Lamb equations for symmetric and antisymmetric motion discussed in a previous section. They accept the simple roots

$$\begin{aligned} &\xi_0^S, \xi_1^S, \xi_2^S, \ldots \\ &\xi_0^A, \xi_1^A, \xi_2^A, \ldots \end{aligned} \tag{92}$$

The evaluation of the integral in Eq. (90) can be done by the residue theorem, using a contour consisting of a semicircle in the upper half of the complex ξ plane and the real axis (Fig. 8.11). Hence,

$$\varepsilon_x(x,t) = \frac{1}{2\mu}\sum_{\xi^S}\frac{\tilde{\tau}(\xi^S)N_S(\xi^S)}{D_S'(\xi^S)}e^{i(\xi^S x-\omega t)} + \frac{1}{2\mu}\sum_{\xi^A}\frac{\tilde{\tau}(\xi^A)N_A(\xi^A)}{D_A'(\xi^A)}e^{i(\xi^A x-\omega t)} \tag{93}$$

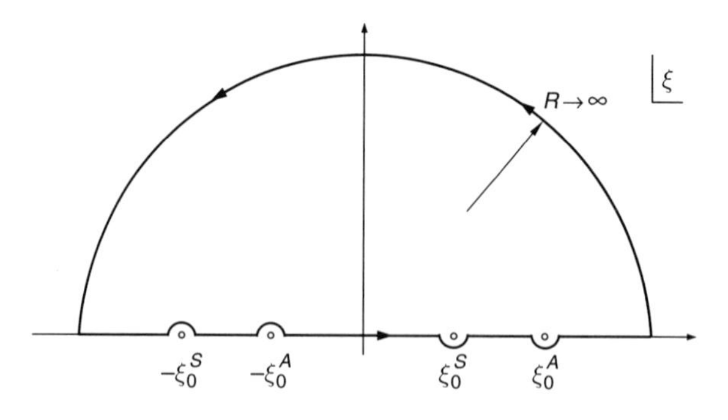

FIGURE 8.11 Contour for evaluating the surface strain under S_0 and A_0 mode. Residues at positive wavenumbers are included, and excluded for the negative wavenumbers.

where D'_S and D'_A represent the derivatives of D_S and D_A with respect to ξ. The summations in Eq. (93) spread over all the symmetric and antisymmetric roots (Lamb-wave modes) that exist for a given value of ω in a given plate. Integration with respect to x yields the displacement solution

$$u_x(x,t) = \frac{1}{2\mu}\sum_{\xi^S}\frac{1}{\xi^S}\frac{\tilde{\tau}(\xi^S)N_S(\xi^S)}{D'_S(\xi^S)}e^{i(\xi^S x-\omega t)} + \frac{1}{2\mu}\sum_{\xi^A}\frac{1}{\xi^A}\frac{\tilde{\tau}(\xi^A)N_A(\xi^A)}{D'_A(\xi^A)}e^{i(\xi^A x-\omega t)} \quad (94)$$

At low frequencies, only two Lamb-wave modes exist, S_0 and A_0, and the general solution has only two terms, i.e.,

$$\varepsilon_x(x,t) = \frac{1}{2\mu}\frac{\tilde{\tau}\left(\xi_0^S\right)N_S\left(\xi_0^S\right)}{D'_S\left(\xi_0^S\right)}e^{i\left(\xi_0^S x-\omega t\right)} + \frac{1}{2\mu}\frac{\tilde{\tau}\left(\xi_0^A\right)N_A\left(\xi_0^A\right)}{D'_A\left(\xi_0^A\right)}e^{i\left(\xi_0^A x-\omega t\right)} \quad \text{(low frequency)} \quad (95)$$

The corresponding expression for displacement is

$$u_x(x,t) = \frac{1}{2\mu}\frac{1}{i\xi_0^S}\frac{\tilde{\tau}\left(\xi_0^S\right)N_S\left(\xi_0^S\right)}{D'_S\left(\xi_0^S\right)}e^{i\left(\xi_0^S x-\omega t\right)} + \frac{1}{2\mu}\frac{1}{\xi_0^A}\frac{\tilde{\tau}\left(\xi_0^A\right)N_A\left(\xi_0^A\right)}{D'_A\left(\xi_0^A\right)}e^{i\left(\xi_0^A x-\omega t\right)} \quad \text{(low frequency)} \quad (96)$$

8.5.2 IDEAL-BONDING SOLUTION

In the case of ideal bonding, the shear stress in the bonding layer is concentrated at the ends, i.e.,

$$\tau(x) = a\tau_0\left[\delta(x-a) - \delta(x+a)\right] \quad (97)$$

The Fourier transform of Eq. (97) is

$$\tilde{\tau} = a\tau_0\left[-2i\sin\xi a\right] \quad (98)$$

Hence, Eqs. (95)–(96) become

$$\varepsilon_x(x,t) = -\frac{a\tau_0}{\mu}\sum_{\xi^S}\left(\sin\xi^S a\right)\frac{N_S(\xi^S)}{D'_S(\xi^S)}e^{i(\xi^S x-\omega t)} - \frac{a\tau_0}{\mu}\sum_{\xi^A}\left(\sin\xi^A a\right)\frac{N_A(\xi^A)}{D'_A(\xi^A)}e^{i(\xi^A x-\omega t)} \quad (99)$$

$$u_x(x,t) = -\frac{a^2\tau_0}{\mu}\sum_{\xi^S}\frac{\sin\xi^S a}{\xi^S a}\frac{N_S(\xi^S)}{D'_S(\xi^S)}e^{i(\xi^S x-\omega t)} - \frac{a^2\tau_0}{\mu}\sum_{\xi^A}\frac{\sin\xi^A a}{\xi^A a}\frac{N_A(\xi^A)}{D'_A(\xi^A)}e^{i(\xi^A x-\omega t)} \quad (100)$$

Equations (99)–(100) contain the $\sin\xi a$ function. The behavior of this function is such that it displays maxima when the PWAS length, $l_a = 2a$, equals an odd multiple of the half wavelength, and minima when it equals an even multiple. A complex pattern of such maxima and minima is involved since several Lamb modes, each with its own different wavelength, coexist at the same time. However, frequencies can be found when the response is dominated by certain modes that can be preferentially excited through *mode tuning*. An additional factor must be considered besides wavelength tuning, i.e., the mode amplitude at the top plate surface. This factor is contained in the values taken by the functions N/D'. Hence, it is conceivable that some higher modes may have little surface amplitude, while others may have larger surface amplitudes at a given frequency. Thus, two important design factors have been identified:

1. The variation of $|\sin \xi a|$ with frequency for each Lamb wave mode
2. The variation of the surface strain with frequency for each Lamb wave mode.

A plot of this solution in the 0–1000 kHz bandwidth is presented in Fig. 8.12.

The $\sin \xi a$ contained in Eq. (99) displays maxima when the PWAS length $l_a = 2a$ equals an odd multiple of the half wavelength, and minima when it equals an even multiple of the half wavelength. A complex pattern of such maxima and minima evolves since several Lamb modes coexist, each with its own different wavelength. At certain frequencies (e.g., 300 kHz in Fig. 8.12a), the amplitude of the A_0 mode goes through zero, while that of the S_0 remains strong, i.e., we have tuning of the S_0 mode, and rejection of the A_0 mode. At other frequencies, the amplitude of the S_0 mode is quite small, while that of A_0 mode is very large (e.g., 100 kHz in Fig. 8.12a). Furthermore, frequencies also exist at which both A_0 and S_0 modes are rejected, as for example the 750 kHz frequency in Fig. 8.12a.

Another important fact to be noticed in Fig. 8.12 is the difference between the strain response and the displacement response. As shown in Fig. 8.12a, the strain response of the S_0 Lamb wave remains rather constant as the frequency increases, whereas the corresponding displacement response, shown in Fig. 8.12b, decreases rapidly with frequency. This is because the strain response vs. frequency varies as the sine function $\sin \xi^S a$, whereas the displacement response varies as the sinc function, $\sin(\xi^S a/\xi^S a)$. The sinc function has a maximum at the origin, and then decreases rapidly as its argument increases. This observation indicates that the strain waves are more likely than the stress

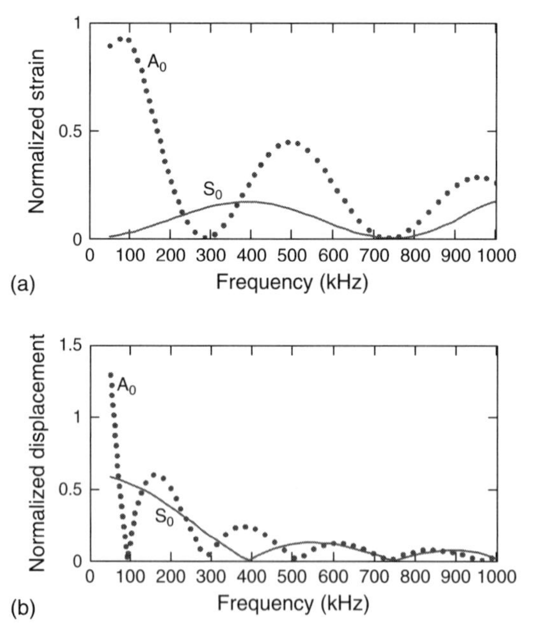

FIGURE 8.12 Predicted Lamb-wave response on a 1-mm aluminum plate under a 7-mm PWAS excitation: (a) strain response; (b) displacement response.

waves to be excited with significant amplitudes at higher amplitudes. This fact highlights the advantage of using PWAS, which are strain-coupled devices, rather than conventional ultrasonic transducers, which are displacement coupled devices. Similar observations can be also made with respect to the A_0 mode, which is also shown in Fig. 8.12. However, besides the $\sin \xi^A a$ dependency, the A_0 mode also shows a more complex dependency inherent in the $N_A(\xi^A)/D'_A(\xi^A)$ function. For this reason, the strain response peaks of the A_0 mode show a tendency to decrease with frequency, which indicates that the A_0 mode will predominantly be excitable only at lower frequencies.

8.5.3 LAMB-WAVE TUNING ANALYSIS FOR CIRCULAR PWAS

Raghavan and Cesnik (2004) extended the analysis to the case of a circular PWAS transducer (Fig. 8.13). The results for the circular transducer are similar to the case of the strip PWAS. The resulting expression of the displacement has Bessel and Hankel function instead of sine and exponential functions of the strip actuator. Subsequently, Raghavan and Cesnik (2005) developed a more general 3D integral-form solution. In this case, the equation of motion is a function of three space variables. The double Fourier transform is used. The theoretical procedure is similar to that of the 2D model; the transformation back to the space domain is quite difficult. Certain PWAS configurations, such as the rectangular wafer, the annular wafer, and the anisotropic macrofiber composite, were considered.

The results for the circular PWAS are similar to the case of the strip PWAS.

$$u_r(r,t)|_{z=d} = -\pi i \frac{a\tau_0}{\mu} e^{i\omega t} \left[\sum_{\xi^S} J_1(\xi^S a) \frac{N_S(\xi^S)}{D'_S(\xi^S)} H_1^{(2)}(\xi^S r) + \sum_{\xi^A} J_1(\xi^A a) \frac{N_A(\xi^A)}{D'_A(\xi^A)} H_1^{(2)}(\xi^A r) \right] \tag{101}$$

where J_1 is the Bessel function of the first kind and order 1, and $H_1^{(2)}$ is the second Hankel function of order 1.

The strain can be derived from the previous equation as

$$\varepsilon_r(r,t)|_{z=d} = \frac{\partial u_r\,(r,t)|_{z=d}}{\partial r} \tag{102}$$

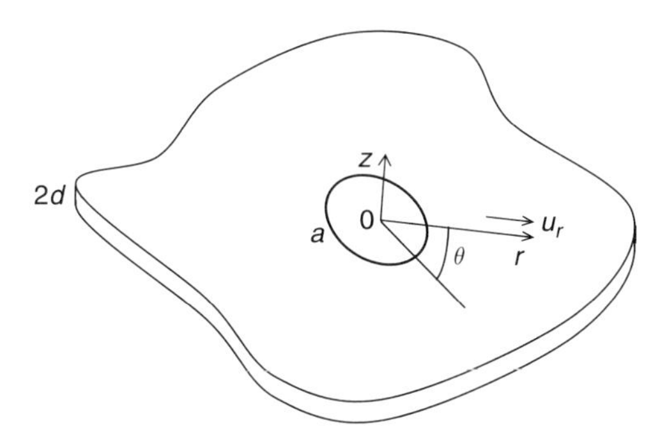

FIGURE 8.13 Infinite isotropic plate with circular surface bonded piezo-actuator and piezo-sensor.

8.6 EXPERIMENTAL VALIDATION OF PWAS LAMB-WAVE TUNING IN ISOTROPIC PLATES

Equations (99) and (102) have been coded into MATLAB program (Bottai and Giurgiutiu, 2005). The program gives the tuning curves of the structures for a given PWAS length, materials properties, material thickness, and frequency range. Experiments were conducted to validate these findings and confirm the theoretical predictions. The predicted results are compared with the experimental values in the next paragraphs.

8.6.1 INITIAL LAMB MODE TUNING RESULTS

During our early experiments, we investigated the effect of excitation frequency on the excited wave amplitude. The experimental setup is shown in Fig. 8.14.

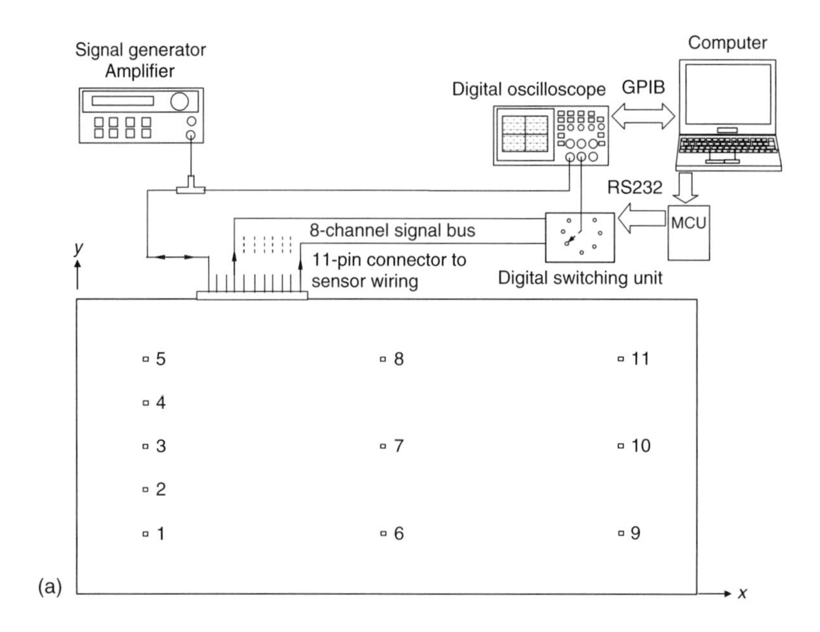

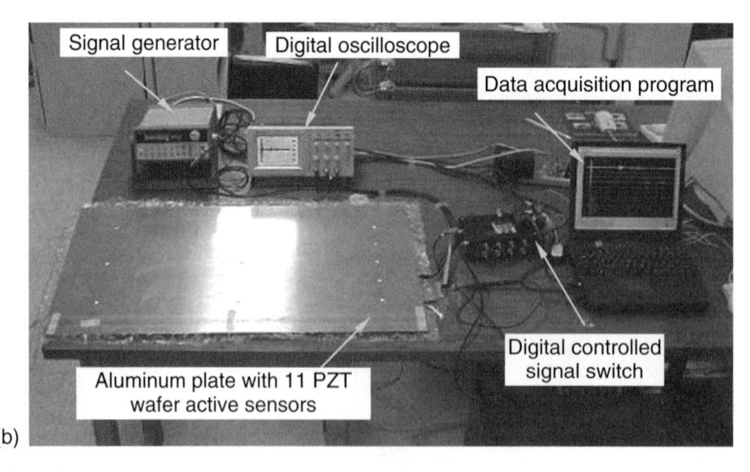

FIGURE 8.14 Experimental setup for verifying the PWAS Lamb-wave tuning: (a) schematic; (b) photo of actual setup.

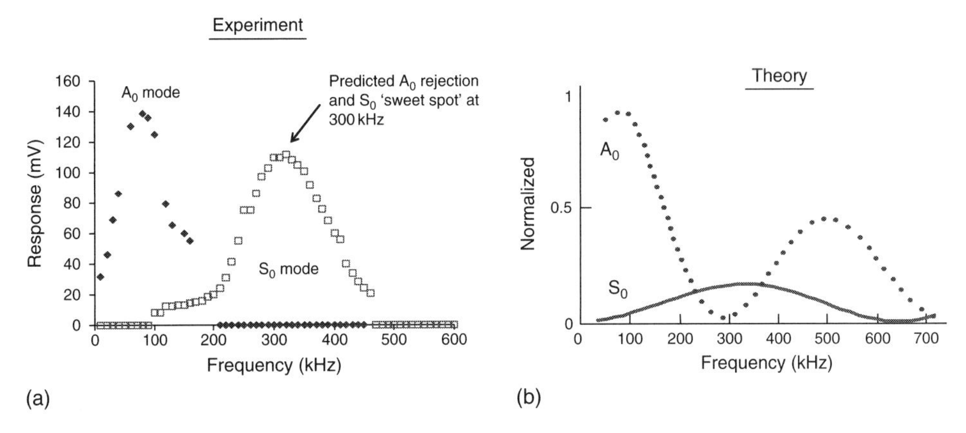

FIGURE 8.15 (a) experimental frequency tuning studies identified a maximum wave response around 300 kHz; (b) theoretical predictions are in concordance with the experimental results.

Experimental validation of these predictions is illustrated in Fig. 8.15. It was found that, at low frequencies (e.g., 10 kHz), the excitation of flexural waves was much stronger than that of axial waves. However, as frequency increased beyond 150 kHz, the excitation of flexural waves decreases, while that of axial waves increased significantly. These trends are shown in Fig. 8.15a. A 'sweet spot' for axial wave (S_0) excitation was found in the 300 to 400 kHz range. This experimental measurement confirms very well the theoretical prediction that indicates, at that frequency, a rejection of the A_0 Lamb-wave mode and a dominance of the S_0 mode (Fig. 8.15b). Systematic investigations were able to reproduce with good accuracy the group-velocity dispersion curves for the axial (S_0) and flexural (A_0) modes, and to identify optimal excitation frequencies for each Lamb wave mode. Thus, in this plate, the antisymmetric Lamb waves (A_0 mode) were best excited at around 70 kHz, while the symmetric Lamb waves (S_0 mode) were best excited at around 300 kHz. Subsequently, for the pitch-catch and pulse-echo experiments described in the following chapters, excitation at 300 kHz was adopted. This allowed us to produce S_0 Lamb waves only, which have much less dispersion at this frequency. This S_0 sweet spot is especially important for embedded PWAS ultrasonics; at this frequency, the S_0 mode has very little dispersion and can be successfully used in the pulse-echo method. These observations illustrate the mode tuning opportunities offered by PWAS excitation of Lamb waves.

8.6.2 COMPREHENSIVE STUDY OF PWAS LAMB-WAVE TUNING IN ISOTROPIC PLATES

A comprehensive study comparing these prediction formulae with experimental results was performed. Experiments were performed on large aluminum plates using square, circular, and rectangular PWAS. Frequencies up to 700 kHz were explored. Two plate thicknesses were studied, 1.07 mm and 3.15 mm. In the thinner plate, only two Lamb-wave modes, A_0 and S_0, were present in the explored frequency range. In the thicker plate, a third Lamb wave, A_1, was also present. The instrumentation was similar to that shown in Fig. 8.14, with the exception of also having a power amplifier. The signal was generated with a function generator (Hewlett Packard 33120A) and sent through a power amplifier (Krohn-Hite model 7602) to the transmitter PWAS. A data acquisition instrument (Tektronix TDS5034B) was used to measure the signal measured by the receiver PWAS. The signal used was a Hanning windowed tone burst with 3 counts. Several plates were used for these experiments:

(a) 1.07-mm thick, 1222 mm × 1063 mm large 2024-T3 aluminum alloy plate
(b) 3.175-mm thick, 505 mm × 503 mm large 6061-T8 aluminum alloy plate
(c) 3.15-mm thick, 1209 mm × 1209 mm large 2024-T3 aluminum alloy plate.

In each experiment, we used two PWAS at a distance 250 mm from one another. One PWAS was used as transmitter and the other as receiver. The frequency of the signal was swept from 10 kHz to 700 kHz in steps of 20 kHz. In this frequency range, for the 1.07 mm thickness plate, only S_0 and A_0 modes were present, while for the 3.175 mm and 3.15 mm thick plates, the S_0, A_0, and A_1 modes were present. For each frequency, the wave amplitude and the time of flight for both the symmetric S_0 mode and the antisymmetric A_0 and A_1 modes were collected. From these readings, we computed the group velocity for each mode.

8.6.3 THIN-PLATE EXPERIMENTS WITH SQUARE AND CIRCULAR PWAS

8.6.3.1 Square PWAS Tuning Results on Thin Plates

Figure 8.16 shows the results for a 7-mm square PWAS placed on a 1.07-mm 2024-T3 aluminum alloy plate. The experimental results (Fig. 8.16a) show that a rejection of the highly dispersive A_0 Lamb wave mode is observed at around 210 kHz. At this frequency, only the S_0 mode is excited, which is very beneficial for pulse-echo studies due to the low dispersion of the S_0 mode at this relatively low value of the *fd* product. On the other hand, a strong excitation of the A_0 mode is observed at around 60 kHz, at which frequency the S_0 mode is very weak.

These experimental results were numerically reproduced using Eq. (99). To obtain correct matching between the theory and experiment, we took the effective PWAS length to be 6.4 mm (Fig. 8.16b). The difference between the actual PWAS length and effective PWAS length is attributed to shear transfer/diffusion effects at the PWAS boundary. It must also be noted that the theory also predicts that the S_0 maximum would happen at 425 kHz, which is the same frequency as that of the second A_0 maximum; this prediction was also verified by the experiments.

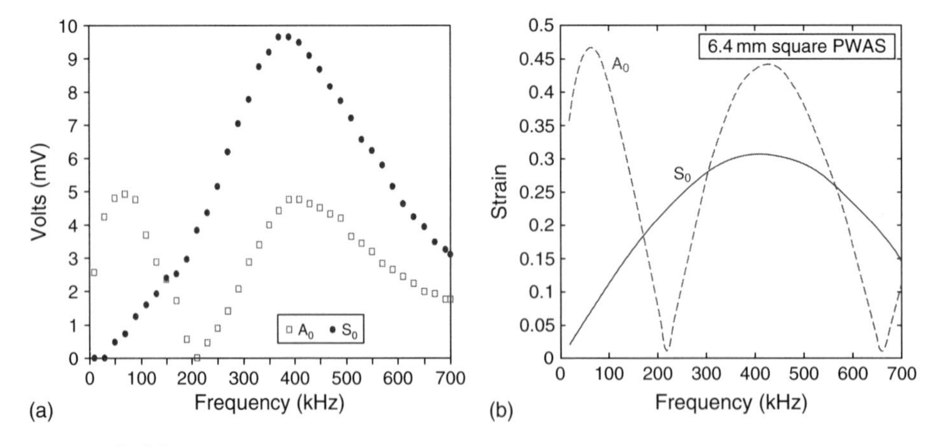

FIGURE 8.16 Lamb-wave tuning using a 7-mm square PWAS placed on 1.07-mm 2024-T3 aluminum alloy plate: (a) experimental results; (b) prediction with Eq. (99) for 6.4 mm effective PWAS length.

8.6.3.2 Round PWAS Tuning Results on Thin Plates

The results for the circular PWAS are shown in Fig. 8.17. The experimental results (Fig. 8.17a) show that a rejection of the highly dispersive A_0 Lamb-wave mode is observed at around 300 kHz. At this frequency, only the S_0 mode is excited, which is very beneficial for pulse-echo studies due to the low dispersion of the S_0 mode at this relatively low value of the *fd* product. On the other hand, a strong excitation of the A_0 mode is observed at around 50 kHz. These experimental results were reproduced using Eq. (102) with the assumption that the effective PWAS length is 6.4 mm (Fig. 8.16b). The difference between the actual PWAS length and effective PWAS length is attributed to shear transfer/diffusion effects at the PWAS boundary.

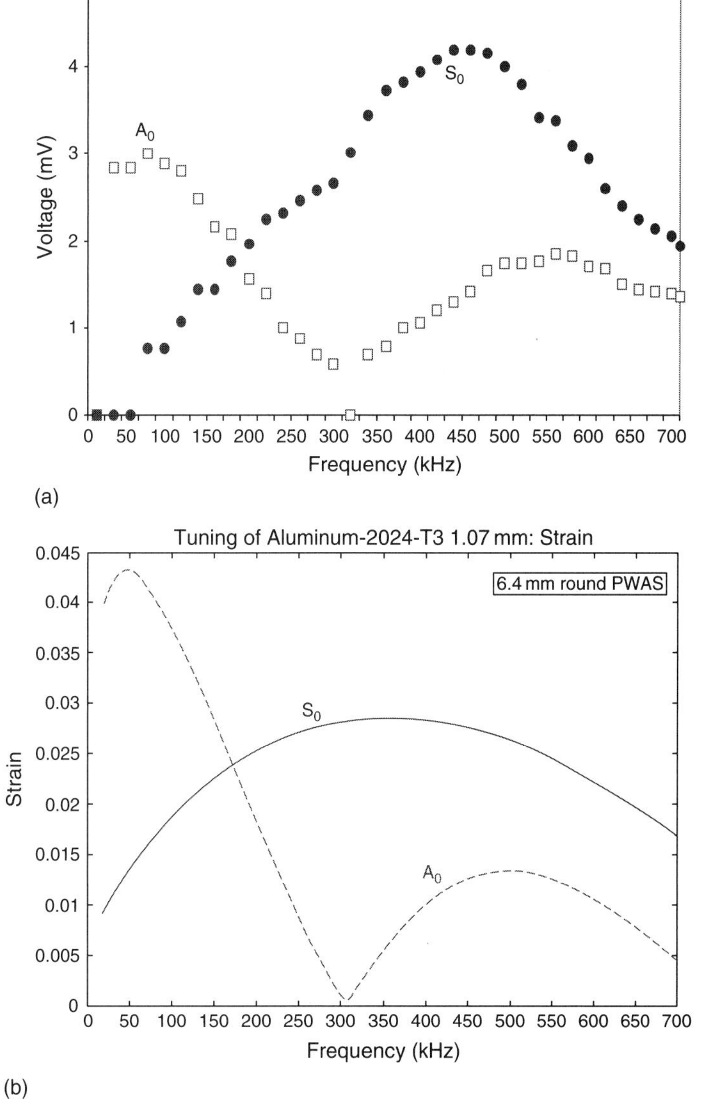

FIGURE 8.17 Lamb-wave tuning using a 7-mm round PWAS placed on 1.07-mm 2024-T3 aluminum alloy plate: (a) experimental results: (b) prediction with Eq. (102) for 6.4 mm effective PWAS length.

It is worth noticing that for the 7-mm square PWAS the S_0 'sweet spot' happens at approximately 210 kHz, whereas for the 7-mm round PWAS the S_0 'sweet spot' happens at approximately 300 kHz. On the other hand, the A_0 sweet spot happened in both cases at the same frequency of approximately 50 kHz. The difference between these two frequencies can be attributed to a number of concurrent factors such as: (a) the different assumptions that underline Eqs. (99) and (102); (b) the corner effects in the case of the rectangular PWAS. It is also remarkable that, in spite of these differences, the equivalent PWAS length is the same (6.4 mm) in both cases.

8.6.4 THICK-PLATE TUNING EXPERIMENTS WITH SQUARE AND CIRCULAR PWAS

8.6.4.1 Square PWAS Tuning Results on Thick Plates

Figure 8.18 shows the results for a 7-mm square PWAS placed on 3.15-mm 2024-T3 aluminum alloy plate. The experimental results (Fig. 8.18a) show that a partial rejection of the A_0 Lamb-wave mode is observed at around 350 kHz. At this frequency, the S_0 mode is predominantly excited. However, do the higher values of the *fd* product, the S_0 mode is already dispersive at this frequency. In addition, it is noticed that the A_0 mode is only partially rejected at this frequency, and hence it is still present in the signal. For these reasons, this S_0 sweet spot is not as effective in thick plates as it were in the thin plates described in the previous section. On the other hand, a strong excitation of the A_0 mode is observed at around 100 kHz. Again, this is accompanied by the partial excitation of the other mode, the S_0 mode. In addition, a third Lamb-wave mode, A_1, becomes excited at higher frequencies, i.e., above 500 kHz. These three observations indicate that the excitation of Lamb waves with PWAS transducers is much more complex in thick plates than in thin plates. These experimental results were reproduced using Eq. (99) with 6.4-mm effective PWAS length (Fig. 8.18b).

8.6.4.2 Round PWAS Tuning Results on Thick Plates

The results for the circular PWAS on thick plates are shown in Fig. 8.19. The experimental results (Fig. 8.19a) show that a partial rejection of the A_0 Lamb wave mode is observed at around 420 kHz. At this frequency, the S_0 mode is more strongly excited than the A_0 mode. However, for higher values of the *fd* product, the S_0 mode is already

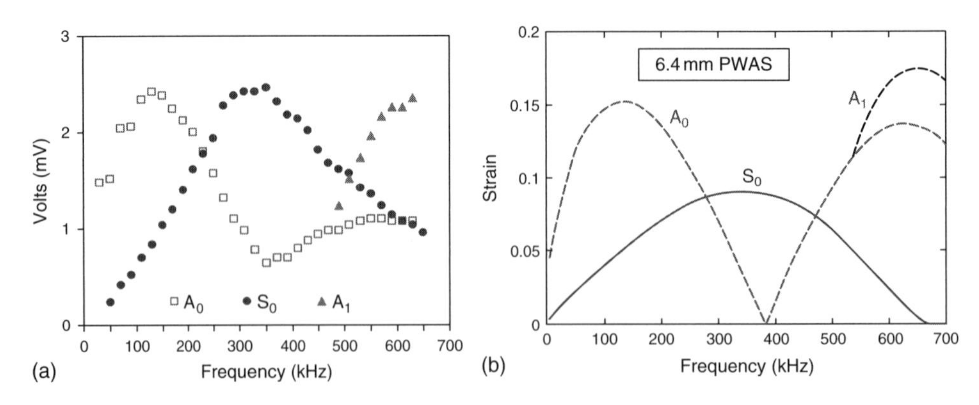

FIGURE 8.18 Lamb-wave tuning using a 7-mm square PWAS placed on 3.15-mm 2024-T3 aluminum alloy plate: (a) experimental results; (b) prediction with Eq. (99) for 6.4 mm effective PWAS length.

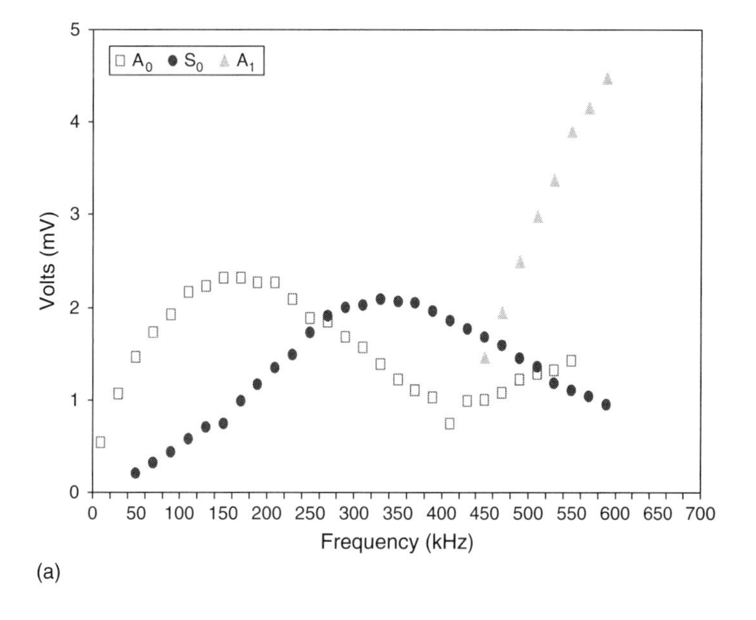

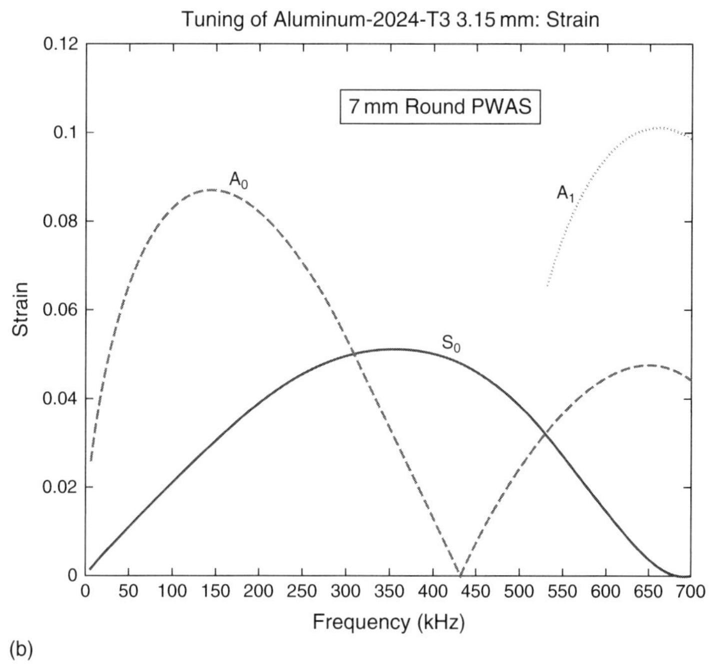

FIGURE 8.19 Lamb-wave tuning using a 7-mm round PWAS placed on 3.15-mm 2024-T3 aluminum alloy plate: (a) experimental results; (b) prediction with Eq. (102) for 7-mm effective PWAS length.

dispersive at this frequency. In addition, it is noticed that the A_0 mode, which is only partially rejected, is still present in the signal. For these reasons, this S_0 sweet spot is not as effective in thick plates as it were in the thin plates described in previous sections. On the other hand, a strong excitation of the A_0 mode is observed at around 150 kHz. Again, this is accompanied by the partial excitation of the other mode, the S_0 mode. In addition, a third Lamb-wave mode, A_1, becomes excited at higher frequencies, i.e., above

450 kHz. These three observations indicate that the excitation of Lamb waves with PWAS transducers is much more complex in thick plates than in thin plates. These experimental results were reproduced using Eq. (102) with 7-mm effective PWAS length (Fig. 8.19b).

It worth noticing that in the case of a thick plate the partial rejection of the A_0 mode happened at 350 kHz for the 7-mm square PWAS and at 420 kHz for the 7-mm round PWAS. These frequencies are much closer to each other than in the case of the thin plate. On the other hand, the A_0 sweet spot happened at 100 kHz for the square PWAS and at 150 kHz for the round PWAS. This is very different from the thin plate, where the A_0 mode sweet spot happened at about the same frequency of 50 kHz. These observations indicate that *simple scale rules do not apply* due to the highly nonlinear character of the characteristic equations governing the Lamb waves and a full analysis is required in order for accurate predictions to be achieved.

8.6.4.3 Thick Plate Effects

Both thick plates show the presence of three modes, S_0, A_0, and A_1, at frequencies between 500 and 700 kHz. It was shown previously that their velocities are close to each other and that the S_0 mode and the A_1 mode are both dispersive. The three waves are then close to each other and the superposition effect starts to be manifest, the tail of one wave interferes with the head of the next one.

Figure 8.20 shows the superposition of the three modes. At a frequency around 550 kHz in a 3.15 mm aluminum plate, the A_0 is non-dispersive while both the S_0 and the A_1 mode are dispersive (the slope of their group velocity is close to 45 degree) and hence the waves are long. The velocities of the three waves are close enough to have superposition between them. Due to the overlapping of the three waves, it is not easy to find their velocities and amplitude. This is also shown by Fig. 8.21, where two images of the three waves at two different frequencies are shown. It is interesting to notice that the mode superposition cause the formation of depressions and enhancements in wave amplitude.

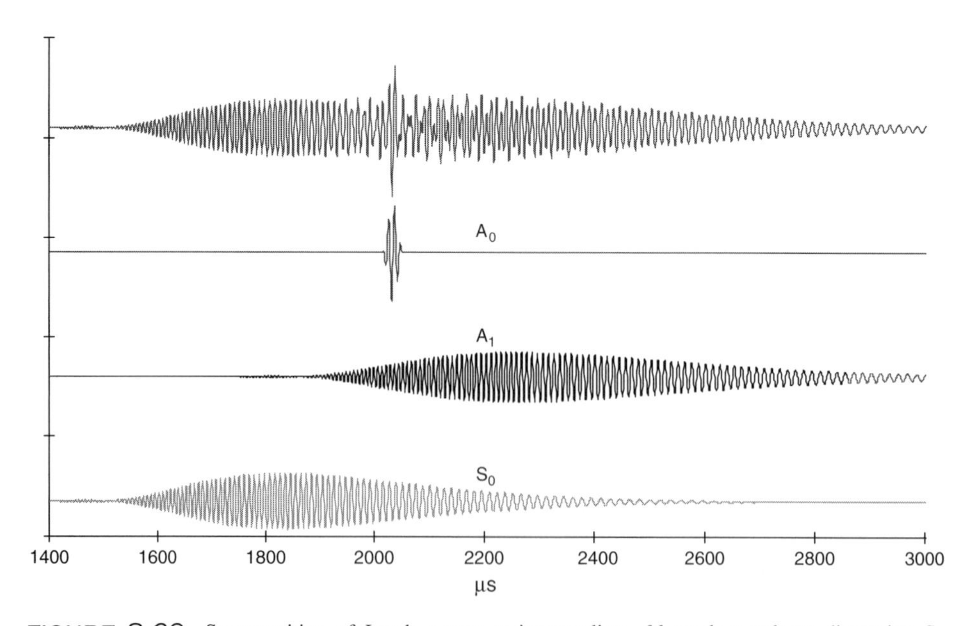

FIGURE 8.20 Superposition of Lamb waves at intermediate *fd* product values: dispersive S_0, non-dispersive A_0, dispersive A_1.

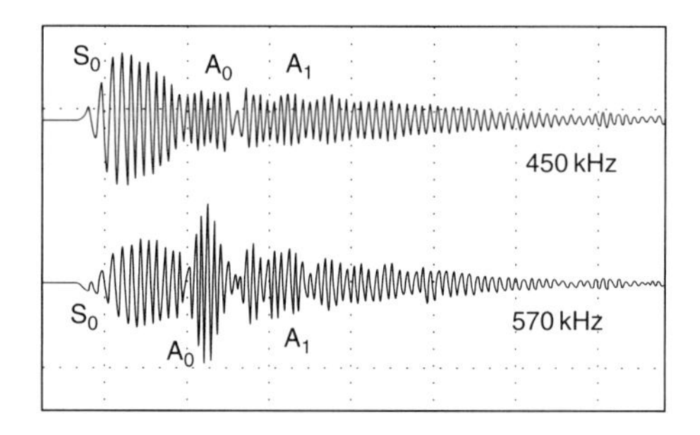

FIGURE 8.21 Wave propagation from the oscilloscope: (a) frequency 450 kHz; group velocities: S_0 4691 m/s; A_0 3167 m/s (b) Frequency 570 kHz, group velocities: S_0 3833 m/s; A_0 3149 m/s.

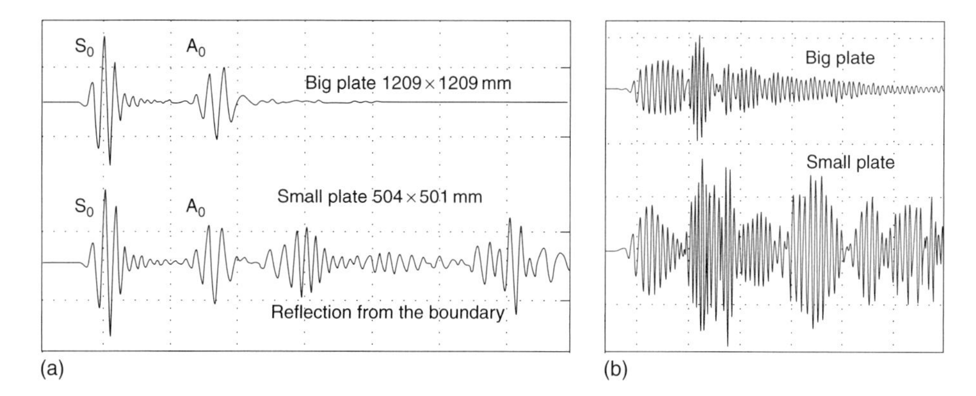

FIGURE 8.22 Waves propagation (a) 250 kHz: (b) 570 kHz.

At 450 kHz, it is still possible to determine the location and amplitude of the S_0 mode, while the superposition effect of the S_0 tail with the A_0 and A_1 modes makes it difficult to determine the location and amplitude of the two waves. At 570 kHz, the situation is clearer. Now it is possible to determine the location and amplitude of S_0 and that of the second wave. The second wave could be either the S_0 or the A_1 mode because it was not been possible to follow their movements during the change of the frequency. The third location and amplitude are difficult to determine because the tails of the two other modes superpose with the third wave.

The effects described above were even more pronounced in a small plate of size 504 mm × 501 mm. Above 450 kHz, it was difficult to locate the three waves, and the collected data seemed to be more distant from the predicted values. Moreover, the signal was disturbed by the reflection from the boundaries. Figure 8.22 compares the wave propagation of a 250 kHz tone burst in two 3.15 mm thick plates of different sizes. The boundary effects were much more pronounced in the small plate, where the reflection from the boundary was already affecting the slower A_0 mode. At 570 kHz, the superposition of the waves and the presence of the boundary reflection in the small plate made it quite difficult to determine the location and amplitude of the three modes.

8.6.5 SUMMARY OF PWAS TUNING EXPERIMENTS ON ISOTROPIC PLATES

The PWAS tuning experiments showed that it was possible to obtain selective tuning of different Lamb-wave modes in isotropic plates of various thicknesses and dimensions.

First, simulations and experiments of PWAS Lamb-wave tuning on a 1.07-mm thin aluminum plate and two kinds of PWAS geometry (square and round) were performed. The plate dimensions (1200-mm by 1200-mm) were large enough to prevent the superposition of the received signal with the reflection from the boundary. The plate thickness was small enough such that, in the examined frequency range (10–700 kHz), only the basic two Lamb-wave modes (S_0 and A_0) appeared. The results for square and round PWAS transducers were similar in behavior, but different in tuning frequencies. The measured group velocities and the tuning results were found in accordance with the predicted values.

Second, simulations and experiments with PWAS Lamb-wave tuning on a 3.15-mm thick aluminum plate and two kinds of PWAS geometry (square and round) were performed.

The size of the plate for the experiments was large enough to prevent the superposition of the received signal with the reflection from the boundary. Because of the increased plate thickness, a third Lamb-wave mode, A_1, appeared at frequencies above 450 kHz.

For frequencies below 450 kHz, where only the two basic Lamb-wave modes (S_0 and A_0) are present, the experiments were performed with the same ease as for the thin plate. The S_0 and A_0 velocity values were collected easily up to 450 kHz, showing to be in accordance with the predicted values. The tuning values showed the same concordance with the theoretical values for both square and round PWAS. For frequency above 450 kHz, the presence of the third mode (A_1) made more difficult the determination of the exact location of each of the three modes as well as their amplitudes. The trend of the curves of the three modes was, however, the same as of the predicted curves.

The tuning of square PWAS was also tested on a 3.15-mm thick plate of smaller dimensions (500-mm by 500-mm). Due to the smaller plate dimensions, reflection from the boundary appeared in these experiments. However, the experimental data for the velocities and the tuning for the S_0 and A_0 modes were in accordance with the predicted values for frequencies below 450 kHz. Above 450 kHz, both the effect of the boundaries and the effect of the third mode were not negligible and complicated the experiment. In particular, it was difficult to detect the location and amplitude of the S_0 mode.

During the experiments, it was noticed that the best concordance between the experimental data and the predicted curves was achieved for effective PWAS size smaller than the actual size. The adjustment of the real PWAS length was necessary because in the development of the theory, it was supposed that the stress induced by the PWAS was transferred to the structure at the end of the PWAS itself. This means that, in an average sense, not the entire PWAS surface is equally effective in transferring excitation to the structure. In particular, it was found that the smaller the PWAS, the greater was the relative size of the non-effective part of the PWAS transducer. Table 8.1 gives comparatively the actual PWAS size and the effective PWAS size. The percentages of the actual PWAS

TABLE 8.1 Actual and effective PWAS size for Lamb-wave tuning in isotropic plates

Actual PWAS Size	Effective PWAS Size	% of Effective PWAS	% of Noneffective PWAS
5 mm	4.5 mm	90%	10%
7 mm	6.4 mm	91.4%	8.6%
25 mm	24.8 mm	99.2%	0.8%

size that is effective in transferring excitation to the structure and the complimentary part that is not effective are also given. Also given in Table 8.1 are similar values for rectangular PWAS, which will be discussed in the next section.

8.7 DIRECTIVITY OF RECTANGULAR PWAS

PWAS can have different geometries. Until now, we considered PWAS transducers of circular or square shapes. In this section, we consider rectangular PWAS of high aspect ratio (e.g., the 5:1 aspect ratio PWAS shown in Fig. 8.23a). It will be shown that the high aspect ratio of the PWAS introduces very interesting directionally-dependent tuning effects between the Lamb waves and the PWAS transducer.

8.7.1 RECTANGULAR PWAS LAMB-WAVE TUNING STUDIES

We performed experiments to verify the wave propagation pattern of rectangular PWAS. Figure 8.23 shows the rectangular PWAS bonded to a aluminum plate 2024-T3 of dimension 1220×1064 mm. Rectangular PWAS transducers of high aspect ratio were tested to examine the directional tuning of Lamb waves. Three rectangular PWAS of 25 mm × 5 mm size, and 0.15 mm thickness (Steiner & Martin) were placed in the configuration shown in Fig. 8.23b. PWAS P1 was the transmitter and PWAS P2 and P3 were the receivers. Two sets of experiments were performed:

(i) P1 transmitting to P2
(ii) P1 transmitting to P3.

In the first case, lengthwise tuning was explored. In the second case, widthwise tuning was explored. Since the difference between length and width is quite significant (5:1 ratio), considerably different tuning behaviors were expected.

Figure 8.24 shows the result for lengthwise tuning. Since the PWAS is relatively long, the tuning frequencies are relatively low. Hence, the frequency range was restricted to 250 kHz and the frequency step was reduced such that a finer exploration can be performed. Figure 8.24a presents the experimental results. Three valleys are observed in the A_0 mode at approximately 85 kHz, 135 kHz, and 175 kHz. A peak in the A_0 mode

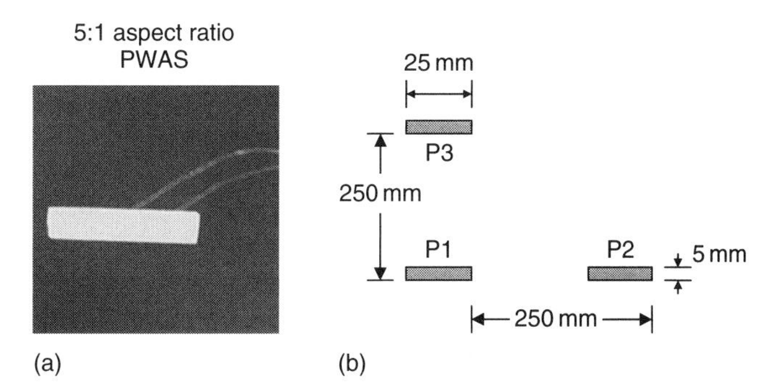

FIGURE 8.23 (a) Rectangular PWAS of aspect ratio 5:1 (25 mm × 5 mm); (b) geometric arrangement of three rectangular PWASs on a 1.07-mm 2024-T3 aluminum plate.

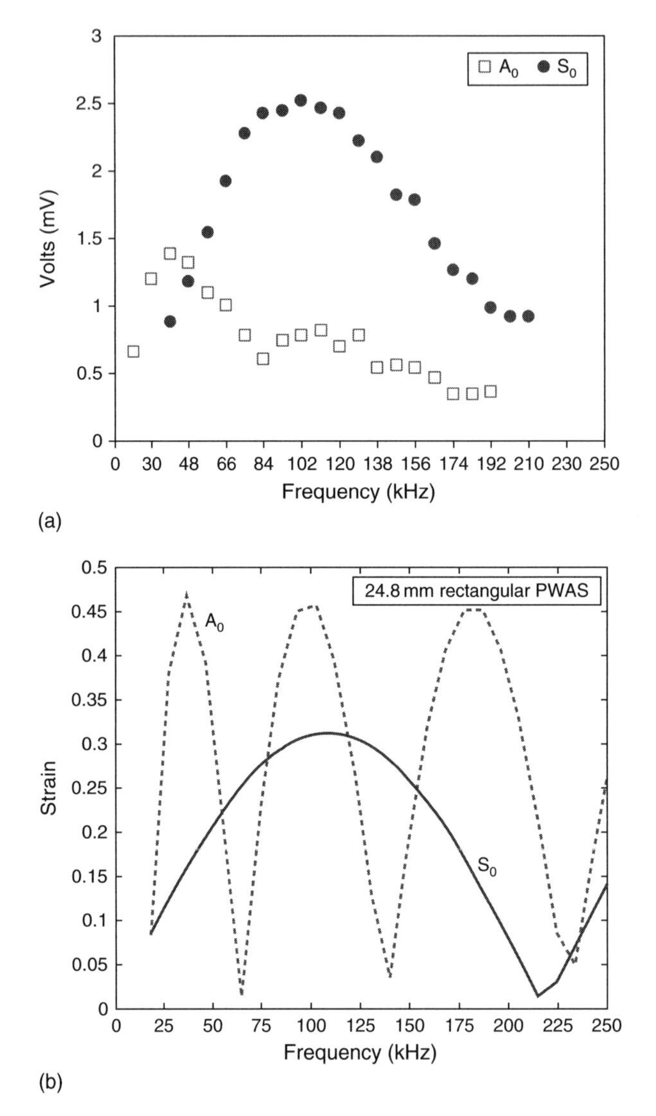

FIGURE 8.24 Lengthwise Lamb-wave tuning with a 25-mm by 5-mm rectangular PWAS placed on a 1.07-mm 2024-T3 aluminum alloy plate: (a) experimental results; (b) prediction for 24.8-mm effective length.

is observed at around 45 kHz. On the other hand, the S_0 mode seems to peak at around 110 kHz. This behavior could be reproduced relatively well using Eq. (99) with an effective PWAS length of 24.8 mm (Fig. 8.24b). The only point that could not be reproduced acceptably seems to be the third A_0 valley.

Figure 8.25 shows the results for widthwise tuning. Since the width is relatively small, the tuning frequencies are expected to be relatively high. Figure 8.25a indicates that one A_0 rejection frequency is observed at around 530 kHz. At this frequency, the S_0 mode, which is almost non-dispersive at this value of the *fd* product, is dominant. The A_0 mode, on the other hand, is found to be dominant at around 100 kHz. This behavior was reproduced quite well with Eq. (99) using an effective PWAS length of 4.5 mm.

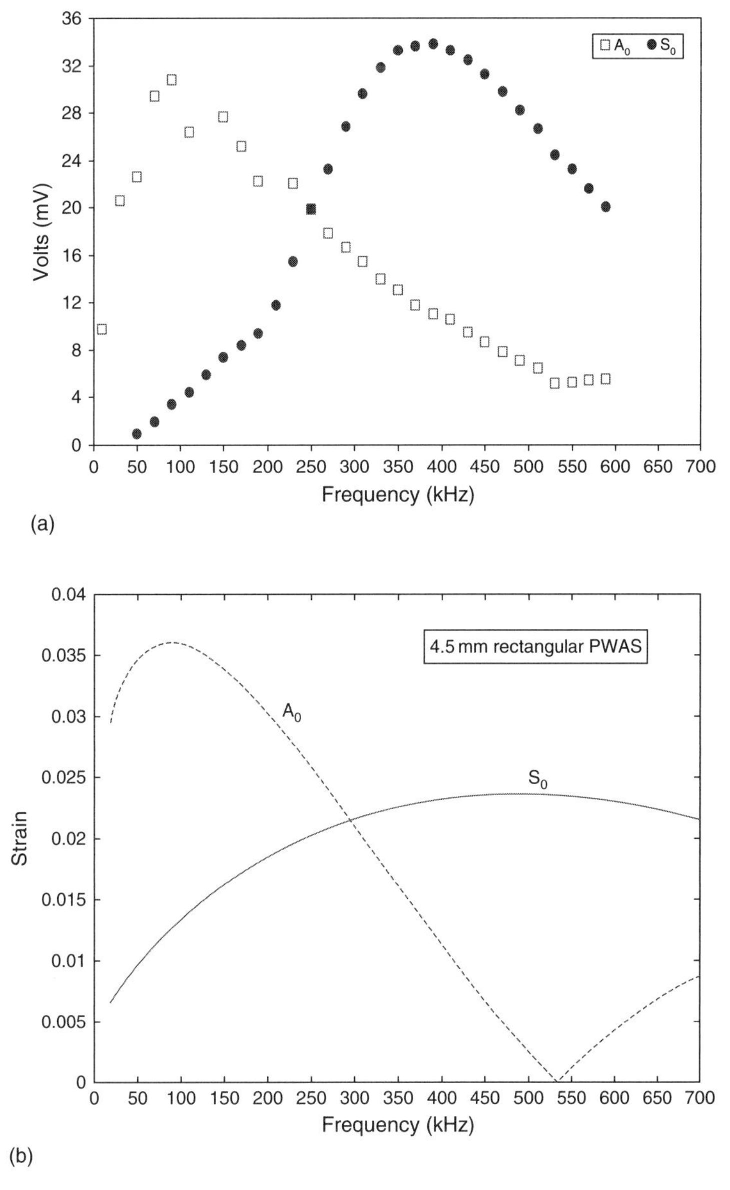

FIGURE 8.25 Widthwise Lamb-wave tuning with a 25-mm by 5-mm rectangular PWAS placed on a 1.07-mm 2024-T3 aluminum alloy plate: (a) experimental results; (b) prediction for 4.5-mm effective length.

8.7.2 DIRECTIVITY STUDIES WITH RECTANGULAR PWAS

In the previous section, we studied the PWAS Lamb-wave tuning at rectangular directions, and found that lengthwise tuning yields many tuning peaks at lower kHz frequencies, whereas widthwise tuning produces fewer peaks that are placed at higher frequency. In this section, we will explore the directivity patterns of Lamb-wave propagation from rectangular PWAS. The experimental setup is shown in Fig. 8.26. It consists of a rectangular PWAS placed in the center, and a number of six round PWAS placed along a quarter-circle arc. The rectangular PWAS was used as transmitter. Its inplane

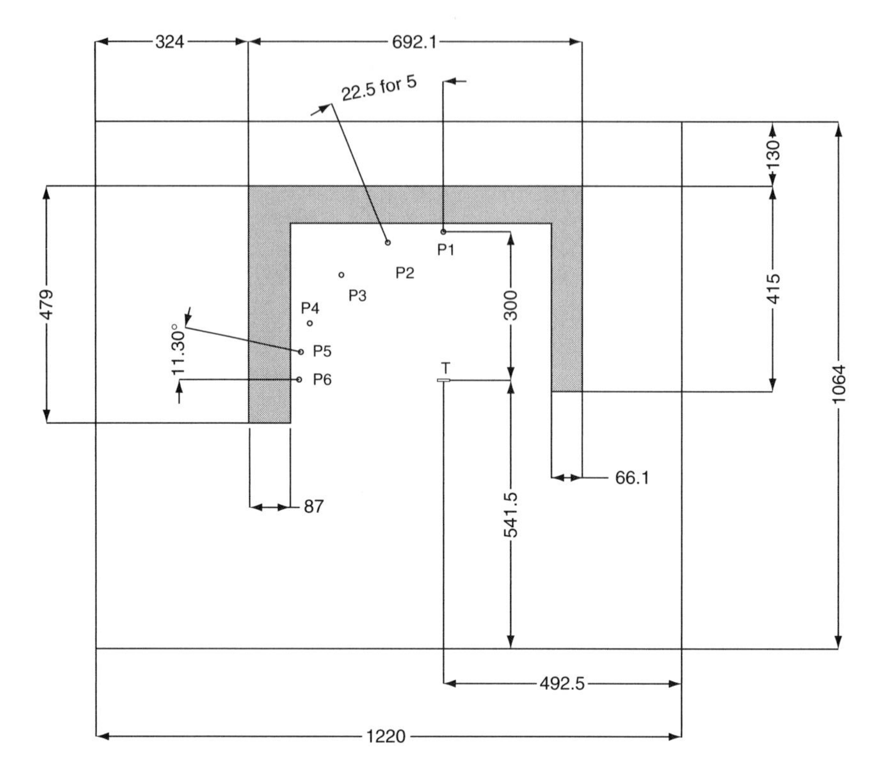

FIGURE 8.26 Experimental set up for the directivity of a rectangular PWAS.

dimensions were 25 mm × 5 mm; the thickness was 0.15 mm. The six circular PWAS transducers were used as receivers. The fact that the receiver PWAS transducers were of a circular shape ensured the nondirectional behavior needed for an unbiased observer. The circular PWAS had a diameter of 7 mm and thickness of 0.2 mm. The circular PWAS were bonded to the plate at angles of 0°, 22.5°, 45°, 67.5°, 78.3°, and 90° with respect to the vertical direction (i.e., the normal to the longest side of the rectangular PWAS transmitter). A strip of modeling clay was applied to both sides of the plate outside the experimental area in order to serve as a dam against the wave reflections from the plate boundaries. The modeling clay, which has a high viscosity, absorbed the wave energy and hence prevented the Lamb waves leaving the experimental zone and being reflected by the plate boundaries.

Figure 8.27 presents the results obtained at angular location #4. Figure 8.27a shows the experimental group velocity in comparison with the predictions obtained from Lamb wave analysis presented in previous sections. Figure 8.27b shows a plot of the signal amplitudes recorded for the various Lamb-wave modes. Similar results were obtained for the other receiver PWAS transducers. For all the receivers, the velocity detected was in agreement with the theoretical values; the amplitude of the signal was sufficiently high for good detection to take place.

After performing the measurements for all the receiver PWAS transducers, we plotted the amplitude at different frequencies as a function of the angle θ between the normal to the longest side of the PWAS. The symmetric (S_0) and antisymmetric (A_0) cases were considered. Thus, we obtained the plots presented in Figs 8.28 and 8.29. Figure 8.28 show the S_0 directivity patterns for frequency range 15–600 kHz. At low frequencies,

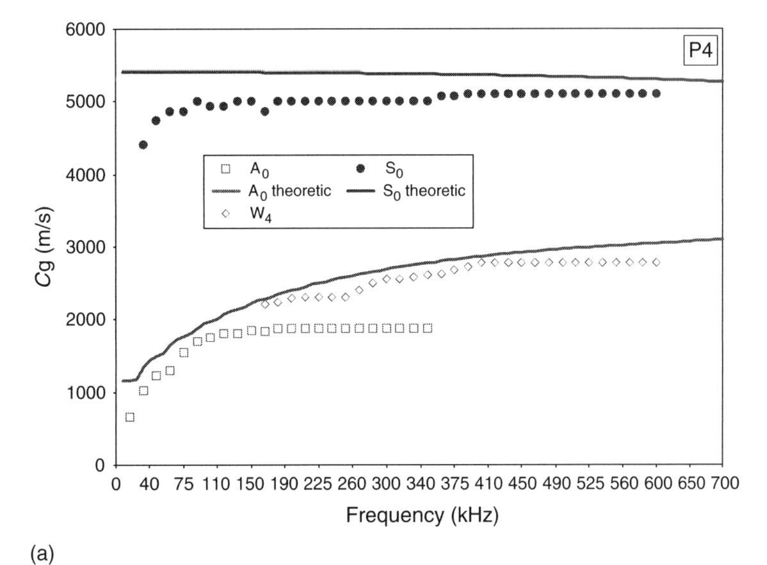

(a)

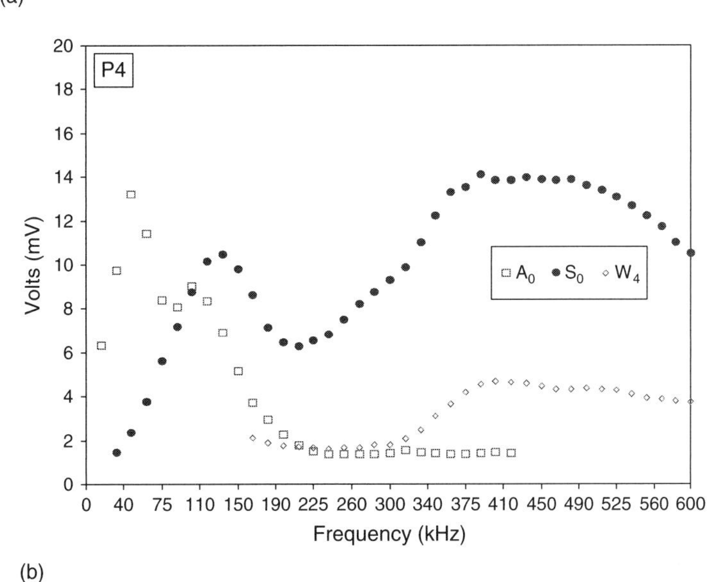

(b)

FIGURE 8.27 Experimental results for circular PWAS P4: (a) velocity; (b) signal amplitude.

the S_0 mode does not show directivity, the wave front is circular. As the frequency is beyond 105 kHz, the wave front amplitude is higher along $\theta = 0°$. In addition, it shows increasing directiveness with increasing frequency. Figure 8.29 show the A_0 directivity patterns for frequency range 15–600 kHz. The A_0 mode at low frequencies shows more directivity than at higher frequencies. In both cases, the wave propagation direction is mostly in the 0° direction when the wave amplitude is at its maximum (Figs 8.28 and 8.29).

Using the data obtained from these experiments, we verified the wavenumber dependencies on the propagation angle θ. Figure 8.30 shows that the variation of the wavenumber with frequency for the A_0 and S_0 Lamb-wave modes is virtually the same for all angular locations, and it does not depend on the angular position θ. This confirms the

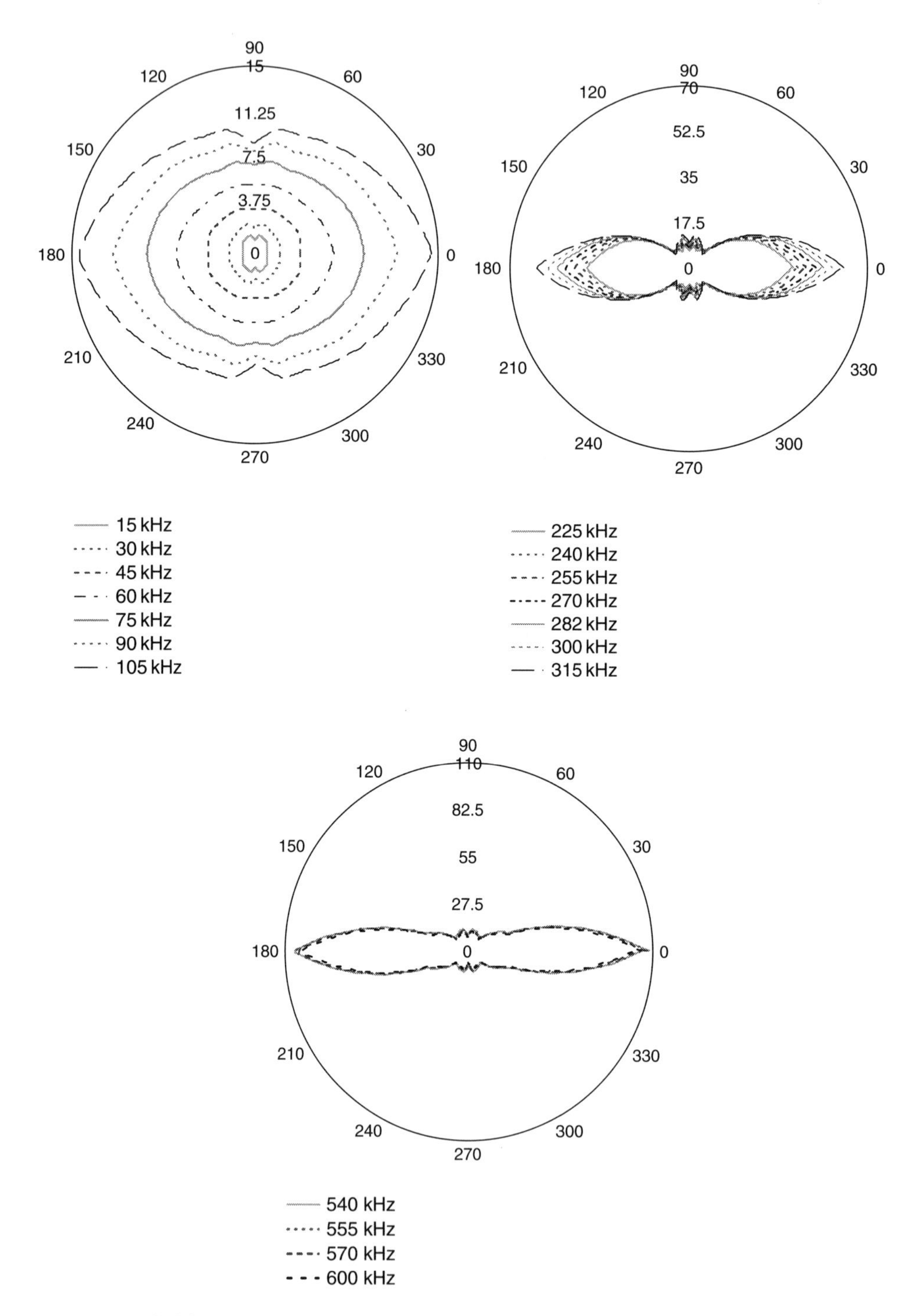

FIGURE 8.28 S_0 amplitudes at different frequencies indicating rectangular PWAS directivity.

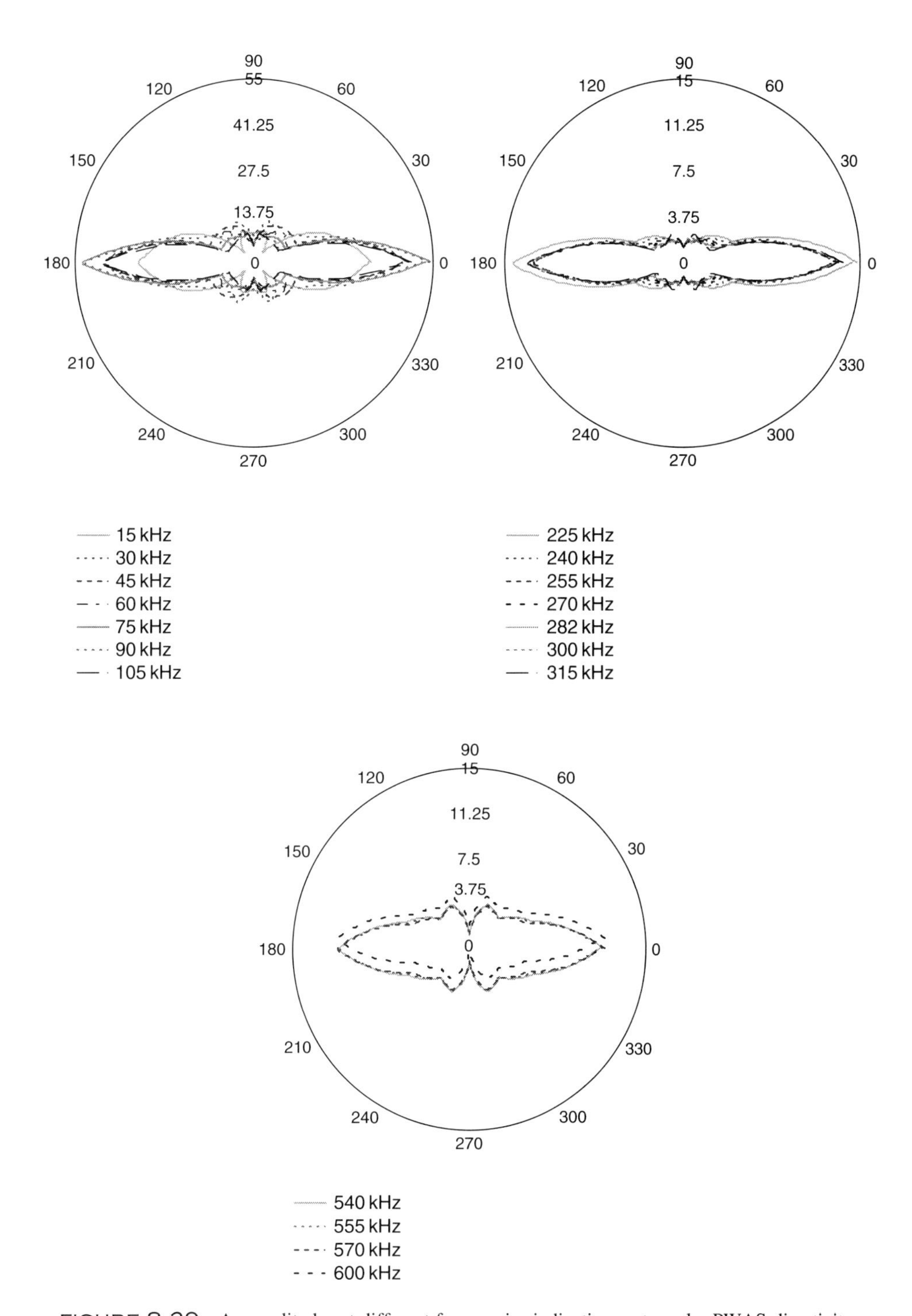

FIGURE 8.29 A_0 amplitudes at different frequencies indicating rectangular PWAS directivity.

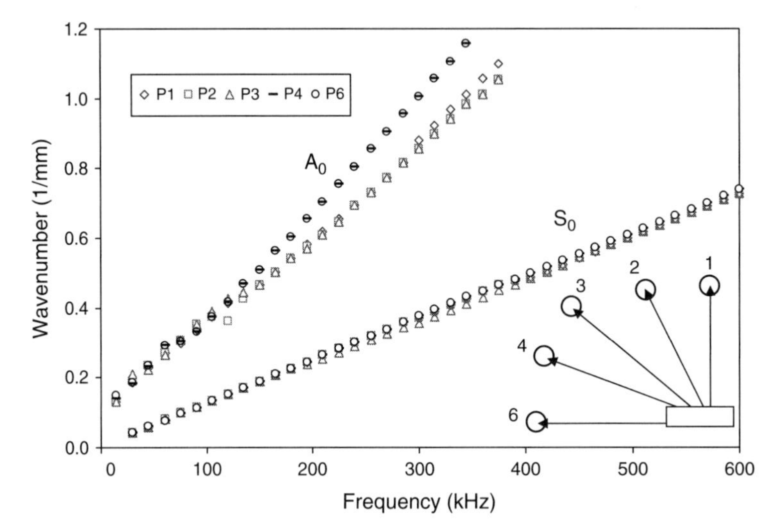

FIGURE 8.30 Wavenumber versus frequency for antisymmetric and symmetric modes at different angles.

theoretical assumption that waves propagate along all directions with the same wavenumber in spite of the transmitting PWAS transducer having a highly unbalanced aspect ratio.

From the wavenumber and frequency data, we were able to plot the amplitude tuning curves. For illustration, Fig. 8.31 shows the tuning of the S_0 mode.

8.7.3 SUMMARY OF RECTANGULAR PWAS TUNING EXPERIMENTS ON ISOTROPIC PLATES

The rectangular PWAS tuning experiments showed that it was possible to obtain tuning of different Lamb-wave modes in isotropic plates at various directions.

For experiment with rectangular PWAS transducers (P1–P2), the collection of data was more difficult than in the previous experiments. The amplitude of the signal was low and the signal had to be amplified. However, the experiments have shown results in accordance with the predicted data; in particular, it was possible to find the 'sweet spot' frequency that gives the maximum S_0 amplitude.

During the experiments, it was noticed that the best concordance between the experimental data and the predicted curves was achieved for effective PWAS size smaller than the actual size. The adjustment of the considered PWAS length was necessary because, in the development of the theory, it was supposed that the stress induced by the PWAS was transferred to the structure at the end of the PWAS itself. This means that, in an average sense, not the entire PWAS surface is equally effective in transferring excitation to the structure. The effective values for rectangular PWAS were given in Table 8.1 of the previous section. It was found that a rectangular PWAS has directionally dependent effective values. As a general rule, the larger the dimension, the smaller the percentage difference between the actual and the effective PWAS dimension. Thus, along the length dimension (25 mm), the effective size is 99.2% of the actual size, whereas along the width dimension (5 mm) the effective size is only 90% of the actual size. A possible simple model for such a behavior would be that of a fixed-size 'ineffective' zone that exists around the PWAS edges. Such a simple model would be consistent, in an average sense, with the stress-diffusion model examined in a previous chapter.

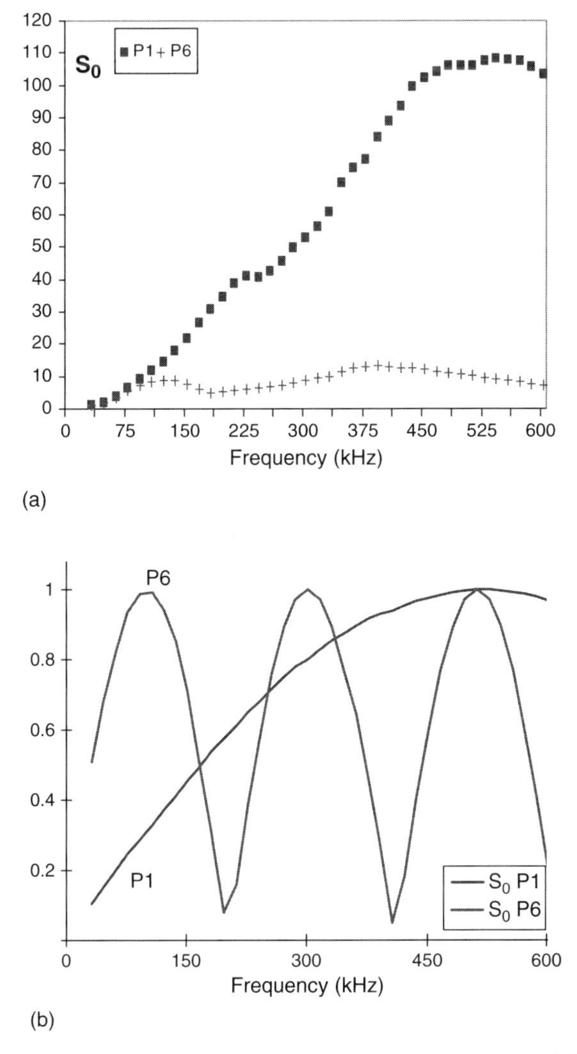

FIGURE 8.31 Rectangular PWAS tuning of S_0 Lamb wave mode at P1 and P6 directions (90° and 0°, respectively): (a) experimental values; (b) prediction.

8.8 PWAS-GUIDED WAVE TUNING IN COMPOSITE PLATES

Previous sections have discussed the PWAS-Lamb waves tuning concept for homogeneous (metallic) structures. This section will take this concept one-step further and discuss PWAS Lamb-wave tuning composite structures. The composite structures under consideration consist of a multilayer construction of orthogonal plies, each laid up at a different individual angle (ply angle). As expected, the analysis of such a multilayer inhomogeneous structure is not as simple as that of the homogenous plate structure considered in previous sections. However, this section will show that, in spite of the modeling difficulties, the PWAS-guided waves tuning concept still applies.

8.8.1 MODELING OF PWAS GUIDED-WAVES TUNING IN COMPOSITE PLATES

We have shown that the PWAS tuning on isotropic plates can be derived in closed form solution under certain simplifying assumptions. The case of tuning in composite plates is more difficult. Liu and Xi (2002) have indicated that the integral transform solution (used by us for isotropic plates) can be extended to the case of composite plates. The analytical solution of the inverse transform of the Fourier integral is solved numerically by properly addressing the infinity of the integrand along the poles on the integration axis. An alternative method, the *normal modes expansion* (NME), can be also used to determine the tuning conditions and frequencies between the PWAS and a composite plate.

The NME method has good potential for use in the determination of tuning frequency for composite plates. The value of the velocity and of the stress can be computed for each frequency and the integral in the Poynting vector can be solved numerically. The analytical solution of the inverse transform of the Fourier integral is solved numerically. The integral transform solution follows the global matrix procedure transforming in the wavenumber domain the displacement solution and then determines the global system equation for the entire laminate

$$\mathrm{AC} = \tilde{T} \tag{103}$$

where A is the global matrix for the composite plate, C contains constant vectors for the layers to be determined from the boundary conditions on the upper and lower surfaces of the plate, and $\tilde{T}$ is the Fourier transform of the external force. Solution of Eq. (103) yield the vector of constants C, and hence the displacement in the wavenumber domain for all the layers in the laminate. To obtain the displacement in the space domain, the following inverse Fourier transform integration must be solved.

$$U(z, x) = \frac{1}{2\pi} \int_{-\infty}^{\infty} \tilde{U}(z, k) e^{-ikx} \, \mathrm{d}k \tag{104}$$

Taking particular forms for the external force, it is possible to obtain the curves of the displacement due to the excitation of the PWAS. The advantage of the method developed by Liu and Xi (2002) is that the integrand has no singular point in the integration range. Solution of Eq. (104) can be done numerically because analytical solution is not possible. Moreover, even if a numerical method is used, a proper treatment is needed, as the integrand goes to infinity at the poles on the integral axis.

Rose (1999) has shown that the NME method yields tractable solutions for the generic excitation of Lamb waves in isotropic plates (Fig. 8.32). The same procedure can be used for composite plates. Below, we will review briefly the general principles of the NME method.

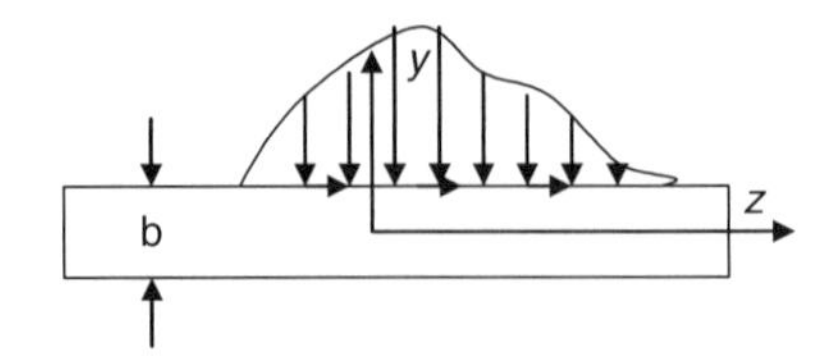

FIGURE 8.32 Plate subject to a surface traction.

8.8.1.1 Orthogonality

The proof of orthogonality requires two general acoustic field theorems:

(1) Lamb's reciprocity relation

$$\nabla \cdot (\mathbf{v}_1 \cdot \mathbf{T}_2 - \mathbf{v}_2 \cdot \mathbf{T}_1) = \mathbf{v}_2\mathbf{F}_1 - \mathbf{v}_1\mathbf{F}_2, \tag{105}$$

(2) complex reciprocity relation

$$\nabla\left(-\tilde{\mathbf{v}}_2 \cdot \mathbf{T}_1 - \mathbf{v}_1 \cdot \tilde{\mathbf{T}}_2\right) = -\frac{\delta}{\delta t}\left([\tilde{\mathbf{v}}_2 \quad \tilde{\mathbf{T}}_2]\begin{bmatrix}\rho & 0\\ 0 & \mathbf{s}^E:\end{bmatrix}\begin{bmatrix}\mathbf{v}_1\\ \mathbf{T}_1\end{bmatrix}\right) + \tilde{\mathbf{v}}_2\mathbf{F}_1 + \mathbf{v}_1\tilde{\mathbf{F}}_2 \tag{106}$$

where $\mathbf{v}_1, \mathbf{T}_1$ is a field solution driven by source $\mathbf{F}_1$, and $\mathbf{v}_2, \mathbf{T}_2$ is a second field solution driven by sources $\mathbf{F}_2$. The tilde sign above a quantity signifies complex conjugate, i.e., if $z = x + iy$, then $\tilde{z} = x - iy$. We can assume that all field quantities vary as $e^{i\omega t}$, therefore $\partial/\partial t \rightarrow i\omega$, and that the source terms are equal to zero: $\mathbf{F_1} = \mathbf{F_2} = \mathbf{0}$. Solutions '1' and '2' are free modes with propagating factors β_m and β_n respectively, i.e.,

$$\begin{aligned}\mathbf{v}_1 &= e^{-i\beta_m z}\mathbf{v}_m(x, y) \quad \text{etc.}\\ \mathbf{v}_2 &= e^{-i\beta_n z}\mathbf{v}_n(x, y) \quad \text{etc.}\end{aligned} \tag{107}$$

The structure under consideration can be a layered waveguide structure with arbitrary anisotropy and inhomogeneity. We can assume that the properties do not vary along x direction. In the case of anisotropy, this can be achieved by choosing as reference axis the principal coordinate system. Then,

$$\begin{aligned}\mathbf{v}_1 &= e^{-i\beta_m z}\mathbf{v}_m(y) \quad \text{etc.}\\ \mathbf{v}_2 &= e^{-i\beta_n z}\mathbf{v}_n(y) \quad \text{etc.}\end{aligned} \tag{108}$$

Equation (106) becomes

$$\frac{\delta}{\delta y}\left(-\tilde{\mathbf{v}}_2 \cdot \mathbf{T}_1 - \mathbf{v}_1 \cdot \tilde{\mathbf{T}}_2\right) \cdot \bar{\mathbf{y}} + \frac{\delta}{\delta z}\left(-\tilde{\mathbf{v}}_2 \cdot \mathbf{T}_1 - \mathbf{v}_1 \cdot \tilde{\mathbf{T}}_2\right) \cdot \bar{\mathbf{z}} = -i\omega\left(\rho\tilde{\mathbf{v}}_2 \cdot \mathbf{v}_1 + \tilde{\mathbf{T}}_2 : \mathbf{s}^E : \mathbf{T}_1\right) \tag{109}$$

Because $\mathbf{v}_1$ and $\mathbf{v}_2$ are orthogonal, the right-hand side of the above equation is equal to zero.

$$\frac{\delta}{\delta y}\left(-\tilde{\mathbf{v}}_2 \cdot \mathbf{T}_1 - \mathbf{v}_1 \cdot \tilde{\mathbf{T}}_2\right) \cdot \bar{\mathbf{y}} + \frac{\delta}{\delta z}\left(-\tilde{\mathbf{v}}_2 \cdot \mathbf{T}_1 - \mathbf{v}_1 \cdot \tilde{\mathbf{T}}_2\right) \cdot \bar{\mathbf{z}} = 0 \tag{110}$$

Substituting (108) in (110), we have

$$i(\beta_m - \tilde{\beta}_n)\left(-\tilde{\mathbf{v}}_n \cdot \mathbf{T}_m - \mathbf{v}_m \cdot \tilde{\mathbf{T}}_n\right) \cdot \bar{\mathbf{z}}e^{-i(\beta_m - \tilde{\beta}_n)z} = \frac{\delta}{\delta y}\left(-\tilde{\mathbf{v}}_n \cdot \mathbf{T}_m - \mathbf{v}_m \cdot \tilde{\mathbf{T}}_n\right) \cdot \bar{\mathbf{y}}e^{-i(\beta_m - \tilde{\beta}_n)z} \tag{111}$$

This equation is integrated with respect to y across the thickness. The term on the right-hand side leaves only the ( ) term; the equation becomes

$$i(\beta_m - \tilde{\beta}_n)4P_{mn} = \left(-\tilde{\mathbf{v}}_n \cdot \mathbf{T}_m - \mathbf{v}_m \cdot \tilde{\mathbf{T}}_n\right) \cdot \bar{\mathbf{y}}\Big|_{-b/2}^{b/2} \tag{112}$$

where P_{mn} is the Poynting vector given by

$$P_{mn} = \frac{1}{4}\int_{-b/2}^{b/2} \left(-\tilde{\mathbf{v}}_n \cdot \mathbf{T}_m - \mathbf{v}_m \cdot \tilde{\mathbf{T}}_n\right) \cdot \bar{\mathbf{z}}\, dy \tag{113}$$

For stress-free or rigid acoustic boundary conditions,

$$\mathbf{T} \cdot \bar{\mathbf{y}} = 0 \quad \text{or} \quad \mathbf{v} = 0 \quad \text{at} \quad y = \pm\frac{b}{2} \tag{114}$$

the right-hand side of (112) is zero, and it becomes

$$i(\beta_m - \tilde{\beta}_n)4P_{mn} = 0 \tag{115}$$

From the expression above, we derive the orthogonality relation for the waveguide modes.
The equation is satisfied for

$$\beta_m \neq \tilde{\beta}_n \quad \text{if } P_{mn} = 0 \tag{116}$$

The frequency spectrum shows that the waves occur in pairs with equal and opposite wavenumber β. For real propagating modes β_m, we get

$$P_{mm} = \text{Re}\frac{1}{2}\int_{-b/2}^{b/2} (-\tilde{\mathbf{v}}_m \cdot \mathbf{T}_m) \cdot \bar{\mathbf{z}}\, dy \tag{117}$$

P_{mn} is nonzero and represents the average power flow of the nth mode in the z direction per unit waveguide width (in the x direction) (see Fig. 8.32).

8.8.1.2 Excitation of Waveguide Modes

For the study of excitation of waveguide modes, the starting point is the complex reciprocity relation (112). In this case, we retain terms $\mathbf{F}_1$ and $\mathbf{F}_2$. Relation (110) becomes

$$\frac{\delta}{\delta y}\left(-\tilde{\mathbf{v}}_2 \cdot \mathbf{T}_1 - \mathbf{v}_1 \cdot \tilde{\mathbf{T}}_2\right) \cdot \bar{\mathbf{y}} + \frac{\delta}{\delta z}\left(-\tilde{\mathbf{v}}_2 \cdot \mathbf{T}_1 - \mathbf{v}_1 \cdot \tilde{\mathbf{T}}_2\right) \cdot \bar{\mathbf{z}} = \tilde{\mathbf{v}}_2 \mathbf{F}_1 + \mathbf{v}_1 \tilde{\mathbf{F}}_2 \tag{118}$$

The waveguide can be excited at the acoustic boundaries by volume sources, traction forces $\mathbf{T}\bar{\mathbf{y}}$, or velocity sources $\mathbf{v}$. We assume that solution '1' is of the type:

$$\begin{aligned} \mathbf{v}_1(y,z) &= \sum_m a_m(z)\mathbf{v}_m(y) \\ \mathbf{T}_1(y,z) &= \sum_m a_m(z)\mathbf{T}_m(y) \end{aligned} \tag{119}$$

We also assume that solution '2' is of the type:

$$\begin{aligned} \mathbf{v}_2(y,z) &= \mathbf{v}_n(y)e^{-i\beta_n z} \\ \mathbf{T}_2(y,z) &= \mathbf{T}_n(y)e^{-i\beta_n z} \end{aligned} \quad \text{with } \mathbf{F}_2 = \mathbf{0} \tag{120}$$

Integrating with respect to y, we obtain

$$\int_{-b/2}^{b/2} \frac{\delta}{\delta y}\left(-\tilde{\mathbf{v}}_2 \cdot \mathbf{T}_1 - \mathbf{v}_1 \cdot \tilde{\mathbf{T}}_2\right)\bar{\mathbf{y}}\, dy + \int_{-b/2}^{b/2} \frac{\delta}{\delta z}\left(-\tilde{\mathbf{v}}_2\mathbf{T}_1 - \mathbf{v}_1 \cdot \tilde{\mathbf{T}}_2\right)\bar{\mathbf{z}}\, dy = \int_{-b/2}^{b/2} \tilde{\mathbf{v}}_2\mathbf{F}_1\, dy \tag{121}$$

Substituting the solutions (119) and (120) in the integral we obtain

$$(-\tilde{\mathbf{v}}_n \cdot \mathbf{T}_1 - \mathbf{v}_1 \cdot \tilde{\mathbf{T}}_n) \cdot \overline{\mathbf{y}}\Big|_{-b/2}^{b/2} e^{-i\tilde{\beta}_n z} + \frac{\delta}{\delta z}\left(e^{-i\tilde{\beta}_n z} \sum_m a_m(z) \int_{-b/2}^{b/2} (-\tilde{\mathbf{v}}_n \cdot \mathbf{T}_m - \mathbf{v}_m \cdot \tilde{\mathbf{T}}_n) \cdot \overline{\mathbf{z}}\, \mathrm{d}y \right)$$

$$= e^{-i\tilde{\beta}_n z} \int_{-b/2}^{b/2} \tilde{\mathbf{v}}_n \cdot \mathbf{F}_1 \, \mathrm{d}y \tag{122}$$

Consider

$$P_{mn} = \mathrm{Re}\left[\frac{1}{4} \int_{-b/2}^{b/2} (-\tilde{\mathbf{v}}_n \cdot \mathbf{T}_n - \mathbf{v}_n \cdot \tilde{\mathbf{T}}_n) \cdot \overline{\mathbf{z}}\, \mathrm{d}y \right] = \mathrm{Re}\left[-\frac{1}{2} \int_{-b/2}^{b/2} \tilde{\mathbf{v}}_n \cdot \mathbf{T}_n \cdot \overline{\mathbf{z}}\, \mathrm{d}y \right] \tag{123}$$

We retain the real part because we consider only the propagating waves. The integral can be written

$$(-\tilde{\mathbf{v}}_n \cdot \mathbf{T}_1 - \mathbf{v}_1 \cdot \tilde{\mathbf{T}}_n) \cdot \overline{\mathbf{y}}\Big|_{-b/2}^{b/2} e^{-i\tilde{\beta}_n z} + \frac{\delta}{\delta z} e^{-i\tilde{\beta}_n z} \sum_m 4 a_m(z) P_{mn} = e^{-i\tilde{\beta}_n z} \int_{-b/2}^{b/2} \tilde{\mathbf{v}}_n \cdot \mathbf{F}(x, y)\, \mathrm{d}y \tag{124}$$

According to the orthogonality relation (116), the summation in (124) has only one nonzero term. Considering the propagating mode n with β_n real,

$$4P_{nn}\left(\frac{\delta}{\delta z} - i\beta_n \right) a_n(z) = (\tilde{\mathbf{v}}_n \cdot \mathbf{T} + \mathbf{v} \cdot \tilde{\mathbf{T}}_n) \cdot \overline{\mathbf{y}}\Big|_{-b/2}^{b/2} + \int_{-b/2}^{b/2} \tilde{\mathbf{v}}_n \cdot \mathbf{F}(x, y)\, \mathrm{d}y \tag{125}$$

The first term of the right-hand side is the forcing function due to the surface forces; the second term is the forcing function due to the volume sources.

In the case of a PWAS, the volume sources are zero,

$$4P_{nn}\left(\frac{\delta}{\delta z} - i\beta_n \right) a_n(z) = (\tilde{\mathbf{v}}_n(y) \cdot \mathbf{T}(y, z) + \mathbf{v}(y, z) \cdot \tilde{\mathbf{T}}_n(y)) \cdot \overline{\mathbf{y}}\Big|_{-b/2}^{b/2} \tag{126}$$

Assume that the anisotropic plate is loaded over a finite portion in the y direction on the upper surface by an infinite-width traction force in the x direction

$$\mathbf{T} \cdot \overline{\mathbf{y}} = \mathbf{t}(z) e^{i\omega t} = \left[t_y(z)\overline{\mathbf{y}} + t_z(z)\overline{\mathbf{z}} \right] e^{i\omega t}.$$

The right-hand side of Eq. (126) becomes

$$(\tilde{\mathbf{v}}_n(y) \cdot \mathbf{T}(y, z) \cdot \overline{\mathbf{y}} + \mathbf{v}(y, z) \cdot \tilde{\mathbf{T}}_n(y) \cdot \overline{\mathbf{y}})\Big|_{-b/2}^{b/2} = \tilde{\mathbf{v}}_n \left(\frac{b}{2} \right) \cdot \mathbf{t}(z)$$

The second term on the left-hand side is zero because $\mathbf{t}(z)$ is supposed to be real.

Finally, Eq. (125) becomes

$$4P_{nn}\left(\frac{\delta}{\delta z} - i\beta_n \right) a_n(z) = \tilde{\mathbf{v}}_n \left(\frac{b}{2} \right) \cdot \mathbf{t}(z)$$

This is a first-order ODE that governs the amplitudes of the general modes. Its solutions is

$$a_n(z) = \frac{e^{-i\beta_n z}}{4P_{nn}} \tilde{\mathbf{v}}_n \left(\frac{b}{2}\right) \cdot \int_c^z e^{i\beta_n \eta} \mathbf{t}(\eta)\, d\eta$$

Where c is a constant used to satisfy the boundary conditions.

Let the external tractions $\mathbf{t}$ to be nonzero only in the interval $-a \leq z \leq a$, we can write the solution as

$$a_{\pm n}(z) = \frac{e^{\mp i\beta_n z}}{4P_{nn}} \tilde{\mathbf{v}}_{\pm n} \left(\frac{b}{2}\right) \cdot \int_{-\infty}^{\infty} e^{\pm i\beta_n \eta} \mathbf{t}(\eta)\, d\eta \tag{127}$$

8.8.1.3 Ideal Bonding Solution

In the case of ideal bonding solution, the shear stress in the bonding layer is concentrated at the ends.

$$\mathbf{t}(b/2, z) = \begin{cases} a\tau_0 \left[\delta(z-a) - \delta(z+a)\right] \mathbf{z} & \text{if } |z| \leq a \\ 0 & \text{if } |z| > a \end{cases} \tag{128}$$

Substituting (128) in (127), we obtain

$$a_{\pm n}(z) = a\tau_0 \frac{\tilde{v}_{nz}(b/2)}{4P_{nn}} e^{\mp i\beta_n z} \left[\int_{-a}^{0} \delta(z-a)\, e^{+i\beta_n \eta}\, d\eta - \int_0^a \delta(z+a)\, e^{-i\beta_n \eta}\, d\eta \right] \tag{129}$$

We can write Eq. (129) as

$$a_{\pm n}(z) = BE_{\pm n} e^{\mp i\beta_n z} F^{(\pm)} \tag{130}$$

where

$$B = a\tau_0 \tag{131}$$

$$E_{\pm n} = \frac{\tilde{v}_{\pm nz}(b/2)}{4P_{nn}} \tag{132}$$

$$F^{\pm} = \int_{-a}^{0} \delta(z-a)\, e^{+i\beta_n \eta}\, d\eta - \int_0^a \delta(z+a)\, e^{-i\beta_n \eta}\, d\eta. \tag{133}$$

Function B is a constant depending on the excitation. Function $F^{\pm}$ is the Fourier integral; it can be solved to become

$$F^{\pm} = \int_{-a}^{a} \delta(\eta - a)\, e^{\pm i\beta_n \eta}\, d\eta - \int_{-a}^{a} \delta(\eta + a)\, e^{\pm i\beta_n \eta}\, d\eta = \pm 2i \sin \beta_n a$$

Substituting in (129), we obtain

$$a_{\pm n}(z) = ia\tau_0 \frac{\tilde{v}_{nz}(b/2)}{2P_{nn}} \sin \beta_n a e^{\mp i\beta_n z}$$

8.8.1.4 Nonideal Bonding Solution

Consider a plate with a bonded PWAS. The PWAS excites the plate with shear stress

$$\mathbf{t}(b/2,z)=\begin{cases} t_a\Gamma\dfrac{\psi}{\psi+\alpha}E_a\Lambda\dfrac{\sinh\Gamma z}{\cosh\Gamma a}\mathbf{z} & \text{if } |z|\le a\\ 0 & \text{if } |z|>a\end{cases} \tag{134}$$

where E_b is the elastic modulus of the structure, E_p is the elastic modulus of the PWAS, $\overline{G}=G/E_p$, G is the shear modulus of bonding layer, t is the thickness of the bonding layer, $\overline{E}=E_b/E_p$, $\theta=t/t_p$, $\psi=(E_b t_b)/(E_p t_p)$, $\Gamma^2=(\overline{G}\theta/t^2)(\psi+\alpha)/\psi$, and $\Lambda=d_{31}V/t_p$ is the induced strain. The parameter α depends on the stress distribution through the plate thickness. It comes from the integration along the thickness of the strain (or strain per y in case of antisymmetric stress). The stress distribution depends on the mode excited by the PWAS. Substituting (134) in (127), we obtain

$$a_{\pm n}(z)=-\frac{E_b\overline{G}\Lambda}{\overline{E}}\frac{1}{\Gamma\cos\Gamma}\frac{\tilde{v}_{nz}(b/2)}{4P_{nn}}e^{\mp i\beta_n z}\int_{-a}^{a}\sinh(\Gamma\eta)\,e^{\pm i\beta_n\eta}\,\mathrm{d}\eta \tag{135}$$

Γ is a constant that comes from the integration over the thickness of the stress distribution due to the wave. We can write Eq. (135) as

$$a_{\pm n}(z)=-BDE_{\pm n}e^{\mp i\beta_n z}F^{(\pm)} \tag{136}$$

where

$$B=\frac{E_b\overline{G}\Lambda}{\overline{E}} \tag{137}$$

$$E_{\pm n}=\frac{\tilde{v}_{\pm nz}(b/2)}{4P_{nn}} \tag{138}$$

$$F^{(\pm)}=\frac{\Gamma}{\cosh\Gamma a}\int_{-a}^{a}\sinh(\Gamma\eta)\,e^{\pm i\beta_n\eta}\,\mathrm{d}\eta$$

Function B depends on the geometry and material properties of the PWAS, the bond layer, and the plate structure. It also depends on the induced strain of the PWAS.

Function D depends on the geometry and material properties of the entire structure as well as on the constant α (the mode excited). Function $F^{\pm}$ is the Fourier integral; it can be solved to become

$$\begin{aligned}F^{(\pm)}&=\int_{-a}^{a}\sinh(\Gamma\eta)\,e^{\pm i\beta_n\eta}\,\mathrm{d}\eta=\frac{1}{2}\int_{-a}^{a}\left(e^{(\Gamma\pm i\beta_n)\eta}-e^{-(\Gamma\mp i\beta_n)\eta}\right)\mathrm{d}\eta\\ &=\frac{1}{2}\left[\frac{e^{(\Gamma\pm i\beta_n)\eta}}{\Gamma\pm i\beta_n}+\frac{e^{-(\Gamma\mp i\beta_n)\eta}}{\Gamma\mp i\beta_n}\right]\Bigg|_{-a}^{a}\end{aligned} \tag{139}$$

After some manipulation we obtain

$$F^{(\pm)}=\pm\frac{2i\Gamma}{(\Gamma^2-\beta_n^2)}\left[\Gamma\sin\beta_n a-\beta_n\cos\beta_n a\tanh\Gamma a\right] \tag{140}$$

For value $\Gamma a > 5$, the force becomes

$$F^{(\pm)} = \pm \frac{2i\Gamma}{(\Gamma^2 - \beta_n^2)} \left[\Gamma \sin \beta_n a - \beta_n \cos \beta_n a \right] \quad (141)$$

For higher values of Γ, we approach the ideal bonding solution.

8.8.2 PWAS TUNING EXPERIMENTS ON COMPOSITE PLATES

Pitch-catch experiments were performed in which one PWAS served as Lamb-wave transmitter and another PWAS served as receiver. The signal used in the experiments was a Hanning-windowed tone burst with 3 counts. The experimental setup is the same as that previously used for isotropic plates. The signal was generated with a function generator (Hewlett Packard 33120A) and sent through an amplifier (Krohn-Hite model 7602) to the transmitter PWAS. A data acquisition instrument (Tektronix TDS5034B) was used to measure the signal measured by the receiver PWAS. The plate used in the experiments was a quasi-isotropic composite plate $[(0/45/90/-45)_2]_S$, of A534/AF252 Uni Tape with 2.25 mm thickness and 1.236 mm × 1.236 mm size, Table 8.2 shows the material properties of each ply.

Figure 8.33 shows the layout of the experiments. The figure is focused on the central part of the composite plate. The PWAS denoted with the letter T was the transmitter while the others were receivers. The distance between the receivers and the transmitter was of 250 mm. The angle between the receivers was 22.5°. In the first experiment, we used three PWAS of square planform. The PWAS were labeled S1, S2, S3. The PWAS S2 was placed at 250 mm from S1 and S3. PWAS S1 was the transmitter while PWAS S2 was the receiver. PWAS S3 was used to check the consistency of the data. In the second experiment, we used six round PWAS transducers.

In both the experiments, the frequency of the signal was swept from 15 to 600 kHz in steps of 15 kHz. At each frequency, we collected the wave amplitude and the time of flight for the waves present in the plate.

A problem faced in the experiments was the efficacy of the ground connection. To obtain a strong signal the ground was provided by bonding a sheet of copper on the composite surface. In this way, the signal was strong and consistent during the experiments.

TABLE 8.2 Ply material properties for the quasi-isotropic plate

	A534/AF252
Density (g/cm^3)	1.54
Axial Modulus E_1 (GPa)	132
Transverse Modulus E_2 (GPa)	10.8
Poisson's ratio ν_{12}	0.24
Poisson's ratio ν_{23}	0.59
Shear Modulus G_{12} (GPa)	5.65
Shear Modulus G_{23} (GPa)	3.38
Modulus ratio E_1/E_2	12.3
Axial tensile strength X_T (MPa)	1513
Transverse tensile strength Y_T (MPa)	43.4
Strength ratio X_T/Y_T	35

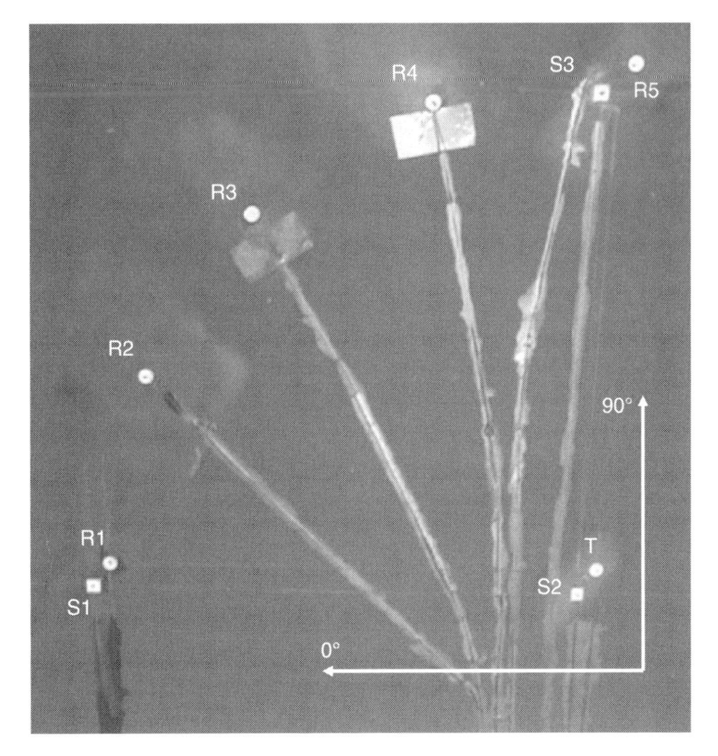

FIGURE 8.33 Experimental layout of the transmitter and receiver PWAS transducers on the quasi-isotropic composite plate.

8.8.2.1 Square PWAS Results

In this experiment we used square PWAS 7 mm long, 0.2 mm thick (American Piezo Ceramics APC-850).

Check of the bond between PWAS and host structure

To verify the consistency of the data and the quality of the bonding between the PWAS and the composite plate, we collected for some frequency the value of the amplitude of the tone burst for different kind of PWAS. As shown in Table 8.3, each transducer was at one time transmitter and at another receiver. In the 90° direction, the tone burst has the maximum amplitude at every frequency while at 0° the amplitude of the tone burst is at

TABLE 8.3 Burst amplitude measured in the composite plate at different frequencies and propagation directions

Fiber Direction	18 V	60 kHz	180 kHz	300 kHz	420 kHz
0°	S1–S2	1.93	17.64	58.18	62.7
	S2–S1	2.02	17.11	44.49	71.24
−45°	S1–S3	6.34	13.43	57.63	73.38
	S3–S1	6.64	14.17	57.58	80.98
90°	S2–S3	58.35	29.58	77.19	137.5
	S3–S2	53.54	28.18	85.9	100.7

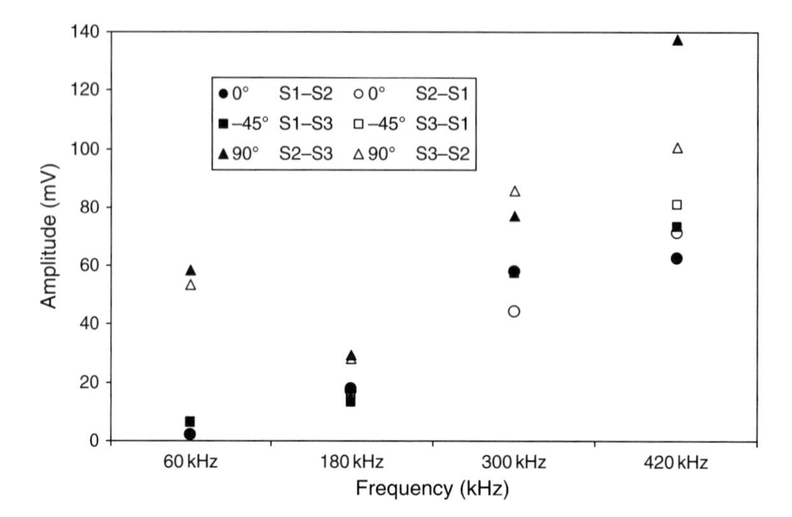

FIGURE 8.34 Signal amplitude at different frequencies in a composite plate.

its minimum. The maximum gap between the tone burst amplitude at 90° and the other directions is at the lower frequencies.

Figure 8.34 shows that the tone burst amplitude increases with the frequency in both 0° and −45° directions; for propagation along 90°, the amplitude decreases and then increases again.

S1 transmitter – S2 receiver

The layout of the experiment is shown in Fig. 8.33; transducer S1 was used as transmitter while transducer S2 was used as receiver. Figure 8.35 shows the velocities of the waves detected; as predicted by the theory, three waves were present, two Lamb-wave modes (A_0 and S_0) and one shear horizontal (SH) wave. (In Fig. 8.35 through Fig. 8.38, the SH wave is labeled as 'Wave 3'.)

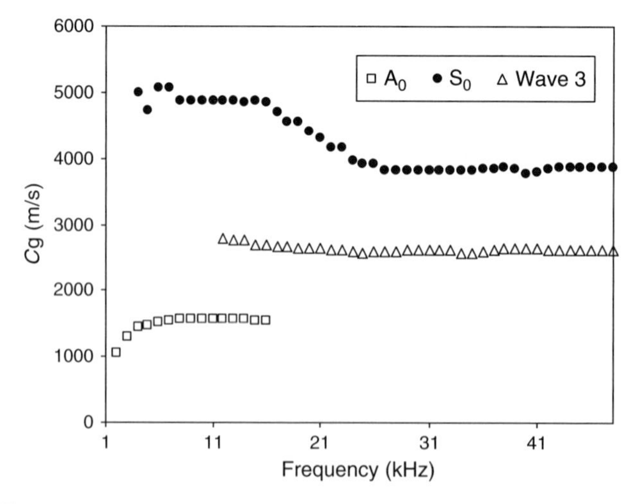

FIGURE 8.35 Group velocity dispersion curves in the composite plate ('Wave 3' denotes the SH wave).

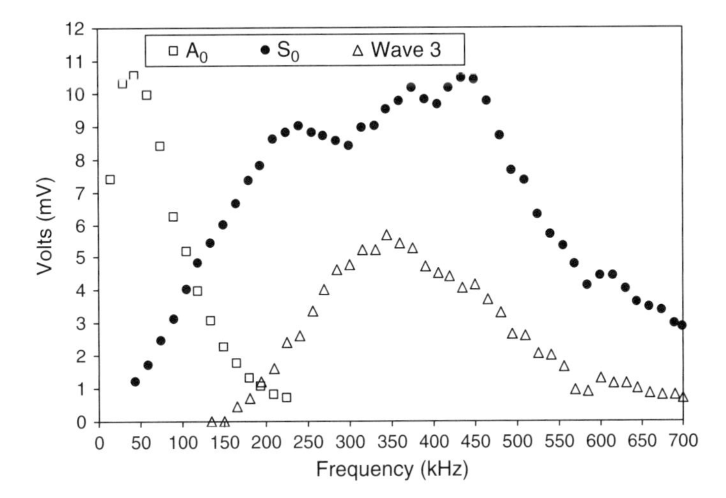

FIGURE 8.36 PWAS tuning data measured experimentally on the composite plate ('Wave 3' denotes the SH wave).

Figure 8.36 shows the amplitudes of the three waves. The A_0 mode extinguishes as soon as the SH appears. The S_0 mode has a maximum at 450 kHz and then decreases. The A_0 and S_0 modes have a slope similar to that of a metallic plate.

8.8.2.2 Round PWAS Results

Experiments with round PWAS of 7-mm diameter and 0.2-mm thickness (American Piezo Ceramics APC-850) were performed with the layout shown in Fig. 8.33. As for the square PWAS, three waves were detected. Figure 8.37 shows the waves amplitudes as detected by PWAS R1, R3, and R5, corresponding to the directions 0°, 45°, and 90° respectively. The A_0 mode reached its maximum before the S_0 mode maximum. The SH wave had amplitude one fourth of the S_0 mode amplitude.

Figure 8.38 shows the velocities of the three waves as detected by the three PWAS. The A_0 mode and the SH wave velocities are the same along all the directions while

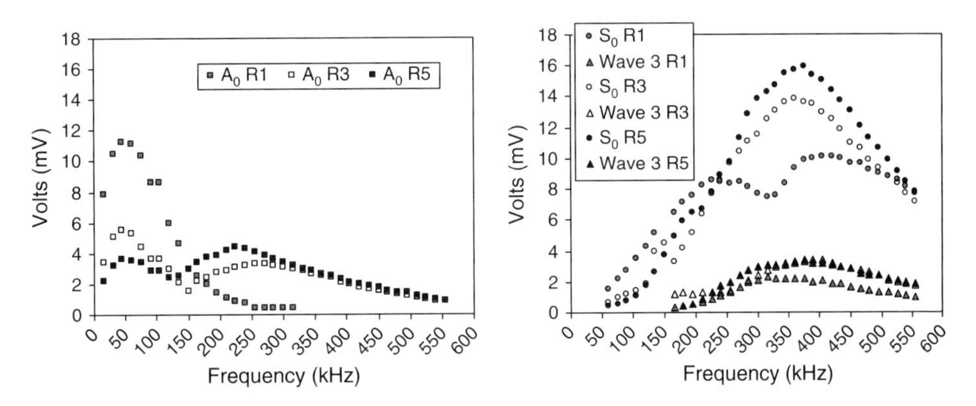

FIGURE 8.37 Wave amplitude measured with the round PWAS on the composite plate ('Wave 3' denotes the SH wave).

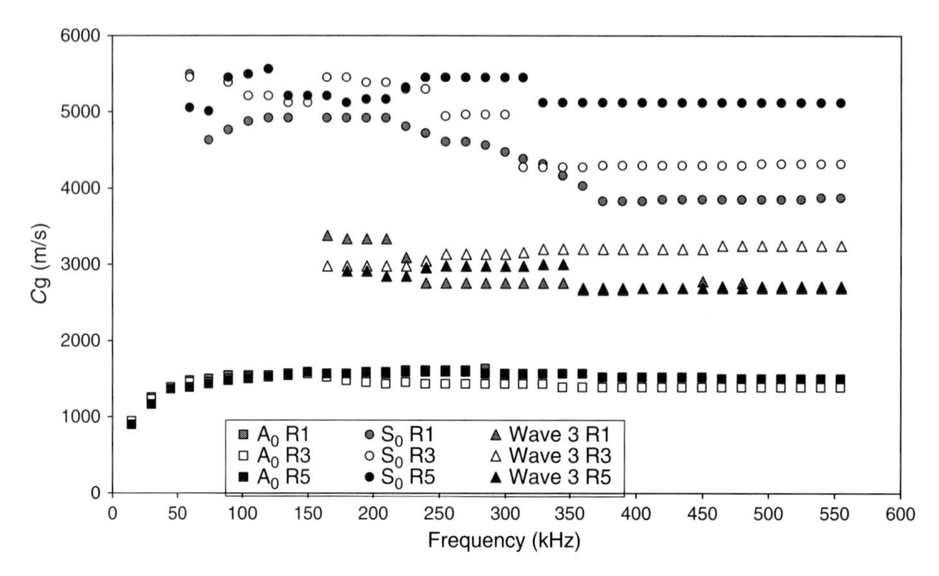

FIGURE 8.38 Group velocity dispersion curves measured with the round PWAS on the composite plate ('Wave 3' denotes the SH wave).

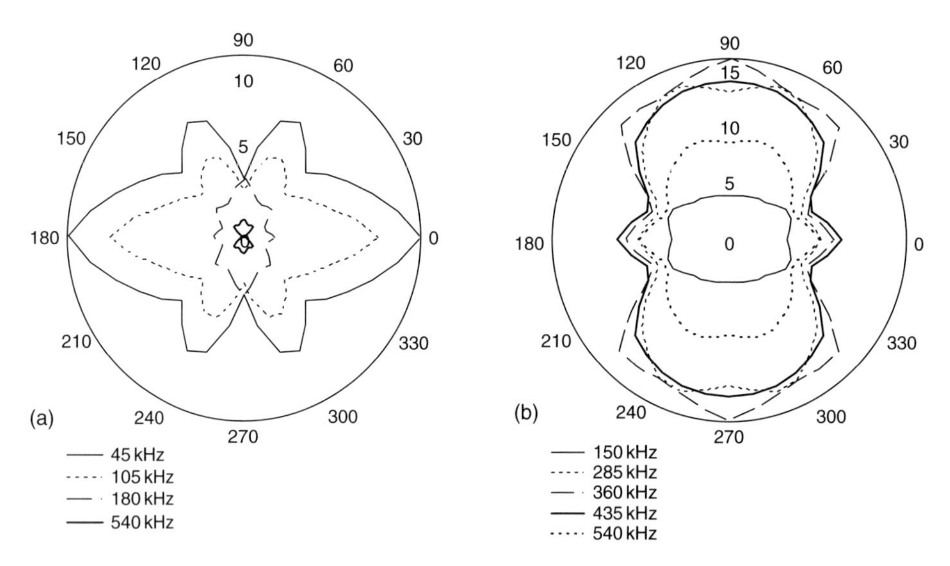

FIGURE 8.39 Wave-propagation directivity patterns for various Lamb-wave modes in a composite plate: (a) A_0 mode; (b) S_0 mode.

the S_0 mode velocity is different along different directions. In particular, the velocity is higher at 90° where the amplitude is the highest.

Figure 8.39 shows the wave-propagation directivity patterns in the composite plate. The data plotted had been interpolated by the data collected by the 5 receivers. At low frequency, the A_0 and S_0 wave amplitudes were highest along the 0° direction; at high frequencies, the direction of the highest amplitude changed to 90°. The SH wave amplitude is always higher at 90°.

8.8.2.3 Summary of PWAS Tuning Experiments on a Composite Plate

Experiments for a quasi-isotropic composite plate with two different kinds of PWAS geometry have been described. The plate dimension in the experiments was large enough to prevent the superposition of the received signal with the reflection from the boundary. The results for square and round PWAS transducers were similar. In both cases, the amplitude of the signal was high and easily detectable. The experiments clearly show the presence of tuning between PWAS and the composite material. The A_0 amplitude disappears or becomes low before the S_0 maximum. It is not possible to see the S_0 alone due to the presence of the SH mode.

8.8.3 COMPARISON OF EXPERIMENTAL RESULTS WITH NUMERICAL PREDICTIONS FOR LAMB-WAVE PROPAGATION IN COMPOSITE PLATES

We have explored how the experimental results could be compared with theoretically derived solutions. As presented in a previous chapter, our solution follows transfer matrix method for laminated composite plates described by Nayfeh (1995). The method, which is stable for low *fd* product values, allowed us to calculate the dispersion curves in composite materials of various lay-ups. From the dispersion curves output we extracted the first three modes of interest. Then, we converted the plot from wavenumber–thickness coordinates to frequency–thickness coordinates. From the phase velocity, we calculated the group-velocity dispersion curves and compared them with the experimentally measured group-velocity values. Figure 8.40 shows the theoretical and experimental group velocity data plotted on the same graph.

The experimental and theoretical data shown in Fig. 8.40 for the A_0 Lamb-wave mode agree quite well. The experimental and theoretical data for the SH and S_0 modes are in good correlation with each other for low frequencies, while at high frequencies, where the two waves are closer, it has been quite difficult to experimentally determine the wave location due to their superposition and dispersion.

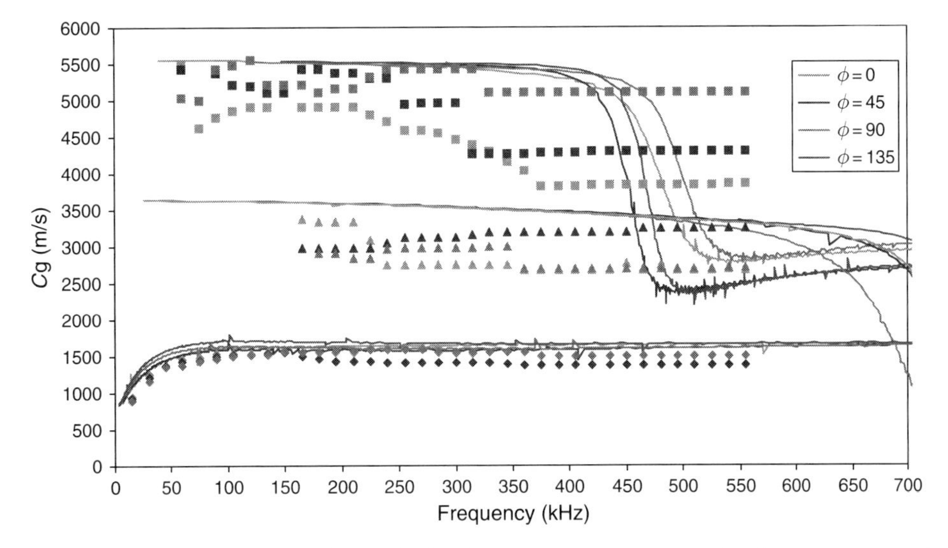

FIGURE 8.40 Comparison of theoretical and experimental group velocities for a quasi isotropic plate: solid lines = theory; dots = experiments.

8.9 SUMMARY AND CONCLUSIONS

This chapter has dealt with a basic aspect of the interaction between PWAS and structure during the active SHM process, i.e., the tuning between the PWAS and the Lamb waves traveling in the structure. PWAS are strain-coupled transducers that are small, lightweight, and relatively low-cost. Upon electric excitation, the PWAS transducers can generate Lamb waves in the structural material by converting the electrical energy into the acoustic energy of ultrasonic wave propagation. In the same time, the PWAS transducers are able to convert acoustic energy of the ultrasonic waves back into an electric signal. Under electric excitation, the PWAS undergoes oscillatory contractions and expansions, which are transferred to the structure through the bonding layer and thus excite Lamb waves into the structure. In this process, several factors influence the behavior of the excited wave. We first recalled the concepts of load transferred between the PWAS actuator and structure. Then, we modeled the PWAS interaction with the ultrasonic waves generated into the structure and showed that tuning opportunities exist. The tuning is especially beneficial when dealing with multimode waves, such as the Lamb waves.

The chapter has adopted a gradual approach, from simple to complex. To understand the essence of the PWAS tuning mechanism, two simple cases were studied first: (a) axial waves in a bar; and (b) flexural waves in a beam. It was shown that, in both cases, the tuning concept allows one to find frequencies at which the waves are strongly excited as well as frequencies at which the waves cannot be excited (i.e., they are rejected).

The chapter continued with the development of a full analytical solution for PWAS Lamb-wave tuning using the exact Lamb waves theory developed in Chapter 6. The space-domain Fourier transform was used to resolve the problem using complex contour integrals. A closed form solution was obtained for the ideal case of perfect bonding between the PWAS and structure, which permits the use of the pin-force model for the PWAS–structure interaction. The developed general formula has shown the existence of salient points, governed by the $\sin \gamma a$ function at which a particular Lamb-wave mode of wavenumber γ can be either preferentially excited or rejected by a PWAS transducer of linear dimension $2a$. Since, at any given frequency, the wavenumbers of the various Lamb-wave modes existing in a plate may be widely different, the concept of preferentially tuning of certain Lamb wave modes and rejection of others became apparent.

Extensive experimental validation of PWAS Lamb-wave tuning concept was presented. Both rectangular and circular PWAS transducers were investigated. A sweet spot for the excitation of S_0 Lamb-wave mode, which is preferential for certain SHM applications, was illustrated. In matching the theory and experiments, it was found that an effective PWAS size exists; values of the effective coefficient were given in tabular form for various PWAS geometries.

Also studied were the directivity patterns associated with tuning rectangular PWAS. This aspect is very important when trying to propagate waves in selected directions, while not sending them in other directions. Full directive patterns for a 5:1 rectangular PWAS transducer were presented.

The final part of the chapter was devoted to the analysis of PWAS Lamb-wave tuning in composite plates. This subject is considerably more difficult to study, and hence the direct approach was deemed unfeasible. Instead, the normal modes expansion technique was adopted, and the theoretical foundation of this technique for the study of PWAS Lamb-wave tuning in composite plates was developed. Extensive PWAS tuning experiments were performed on a multilayer multiorientation carbon–fiber/epoxy composite plate using square and circular PWAS transducers. Tuning opportunities were identified and documented. In addition, a comparison between theoretical prediction and actual measurement for the group velocity of various Lamb-wave modes in the composite plate was also presented.

8.10 PROBLEMS AND EXERCISES

1. Consider a PWAS of length $l = 7\,\text{mm}$, width $b = 1.65\,\text{mm}$, thickness $t = 0.2\,\text{mm}$, and material properties as given in Table 8.4. The PWAS is bonded to a 1-mm thick aluminum strip with material properties as given in Table 8.5. The PWAS length is oriented along the strip. (i) Calculate the first three frequencies at which the PWAS will tune into the axial waves propagating into the aluminum strip. (ii) Repeat the calculations considering the PWAS length oriented across the strip.
2. Consider a PWAS of length $l = 7\,\text{mm}$, width $b = 1.65\,\text{mm}$, thickness $t = 0.2\,\text{mm}$, and material properties as given in Table 8.4. The PWAS is bonded to a 1-mm thick aluminum strip with material properties as given in Table 8.5. The PWAS length is oriented along the strip. (i) Calculate the first three frequencies at which the PWAS will tune into the flexural waves propagating into the aluminum strip. (ii) Repeat the calculations considering the PWAS length oriented across the strip. (iii) Discuss the difference between these results and those of axial tuning calculated in the previous problem. Comment on the results of (i) and (ii).

TABLE 8.4 Piezoelectric wafer properties (APC-850)

Property	Symbol	Value
Compliance, in plane	s_{11}^E	$15.30 \cdot 10^{-12}\,\text{Pa}^{-1}$
Compliance, thickness wise	s_{33}^E	$17.30 \cdot 10^{-12}\,\text{Pa}^{-1}$
Dielectric constant	ε_{33}^T	$\varepsilon_{33}^T = 1750\varepsilon_0$
Thickness wise induced-strain coefficient	d_{33}	$400 \cdot 10^{-12}\,\text{m/V}$
In-plane induced-strain coefficient	d_{31}	$-175 \cdot 10^{-12}\,\text{m/V}$
Coupling factor, parallel to electric field	k_{33}	0.72
Coupling factor, transverse to electric field	k_{31}	0.36
Coupling factor, parallel to electric field, polar motion	k_p	0.63
Poisson ratio	ν	0.35
Density	ρ	$7700\,\text{kg/m}^3$
Sound speed	c	2900 m/s

Note: $\varepsilon_0 = 8.85 \times 10^{-12}\,\text{F/m}$.

TABLE 8.5 Typical material properties of aluminum and steel

	Aluminum (7075 T6)	Steel (AISI 4340 normalized)
Modulus, E	70 GPa	200 GPa
Poisson ratio	0.33	0.3
Density, ρ	$2700\,\text{kg/m}^3$	$7750\,\text{kg/m}^3$
Yield stress, Y	500 MPa	860 MPa
Cross-section radius, r	10 mm	8 mm

3. Consider a PWAS of width $b = 1.65$ mm, thickness $t = 0.2$ mm, and material properties as given in Table 8.4. The PWAS is bonded to a 1-mm thick aluminum strip with material properties as given in Table 8.5. The PWAS length is oriented along the strip. (i) Calculate the smallest PWAS length l that will tune into the flexural waves propagating into the aluminum strip at 10 kHz. (ii) Calculate the smallest PWAS length l that will reject the flexural waves propagating into the aluminum strip at 10 kHz. (iii) Calculate the PWAS length l that will tune into the axial waves propagating into the aluminum strip at 100 kHz. (iv) Calculate the PWAS length l that will reject the axial waves propagating into the aluminum strip at 100 kHz.

9

HIGH-FREQUENCY VIBRATION SHM WITH PWAS MODAL SENSORS – THE ELECTROMECHANICAL IMPEDANCE METHOD

9.1 INTRODUCTION

This chapter will present the use of PWAS transducers as high-frequency modal sensors. The use of PWAS transducers as high-frequency modal sensors has been made possible by the electromechanical (E/M) impedance method, which couples the mechanical impedance of the structural substrate with the electrical impedance measured at the PWAS transducer terminals. In this way, the mechanical resonance spectrum of the structure is reflected in a virtually identical spectrum of peaks and valleys in the real part of the E/M impedance measured at the PWAS terminals.

To understand how the PWAS modal sensor works, we will use the constrained PWAS analysis performed in Chapter 7. The constraining stiffness is actually the dynamic stiffness presented by the structure to the PWAS at the interface. In our analysis, we will perform the modal analysis of the structure and calculate the frequency-dependent dynamic stiffness as the inverse of the frequency-response function of the structure when excited by the PWAS.

The analysis will start with the consideration of a 1-D structure, i.e., a simple beam. The full vibration analysis, including both flexural and axial motions, will be performed, and the frequency-dependent dynamical stiffness presented by the structure to the PWAS will be calculated. Several small beam cases will be considered: narrow vs. wide beams; thin vs. thick beams. Subsequently, we will calculate their E/M impedance and compare it with experiments.

Next, the analysis will be extended to 2-D structures: a circular plate with a circular PWAS mounted in its center will be considered. Full analytical modeling of plate vibration, including both flexural and axial motions, will be performed. The analytical predictions will be compared with experiments conducted on a set of nominally identical specimens.

Subsequently, the E/M impedance method will be used for damage detection. First, damage detection will be tested on simple 2-D geometries. Statistical sets of circular plates with progressively increasing crack damage will be fabricated and tested. The change of the E/M impedance spectra with damage will be examined. Also examined will be the consistency of the spectra within each damage state. The spectra will be analyzed in order to determine a correlation between the spectral changes and damage; the development of a damage metric algorithm will be sought. The classification of the high-frequency spectra resulting from the high-frequency impedance measurements will be implemented with various algorithms, some based on simple overall statistics, other based on neural networks. Two approaches will be use: (a) simple statistics formula and (b) probabilistic neural networks (PNN).

Several other E/M-impedance damage-detection experiments will be performed to address various applications such as (a) damage detection in spot welds; (b) damage detection in bonded joints; (c) disbond detection in civil engineering composite overlays; (d) damage detection in aging aircraft panels.

At the end of this chapter, the whole problem will be revisited using the finite element method (FEM). Initially, we will use the conventional FEM analysis to calculate the frequency-dependent dynamic stiffness method. We will compare the FEM results with the analytical and experimental results. Then, we will use the coupled-field FEM analysis, in which the finite elements have both electrical and mechanical dof's. The results of the coupled field analysis will be compared with analytical results, conventional FEM analysis, and experiments.

9.1.1 ROOTS OF THE E/M IMPEDANCE METHOD

9.1.1.1 Mechanical Impedance Method

The mechanical impedance method has been known for some time. This method evolved in the late 1970s and early 1980s and was based on measuring the response to force excitation applied normal to a structural surface using conventional shakers and velocity transducers. It consists of exciting structural vibrations using a specialized transducer that simultaneously measures the applied normal force and the induced velocity. Lange (1978) studied the mechanical impedance method for nondestructive testing (NDT). Cawley (1984) extended the work of Lange (1978), and studied the behavior of bonded thin plates in order to identify local disbonds. He excited the vibrations of bonded plates using a specialized transducer that simultaneously measures the applied normal force and the induced velocity. Finite element analysis of the vibration of the bonded/disbonded plates was performed, and the impedance to excitation in the normal direction was predicted. The experimental work consisted of measuring the normal-direction impedance at various locations. The impedance magnitude spectrum below the anti-resonance frequency was compared with the finite element predictions, and some correlation with the presence of disbonds was attempted. Phase information was not used in the data analysis. Because these early studies, the mechanical impedance method has evolved and gained its own place among nondestructive evaluation (NDE) techniques. The mechanical impedance method is the dominant technique for detecting disbonds in laminated structures and delaminations inside composite materials up to a depth of 6 mm. Ultrasonic mechanical impedance analysis (MIA) probes and equipment are common.

The E/M impedance method differs from the mechanical impedance method on several accounts:

(a) The transducers used in the mechanical impedance method are bulky, whereas the PWAS transducers used in E/M impedance are thin and nonintrusive.
(b) The mechanical impedance transducers are not permanently attached to the structure, but have to be manually applied to various points of interest, whereas PWAS are permanently attached and can be interrogated at will.

The emerging E/M impedance method offers distinctive advantage over the mechanical impedance method. While the mechanical impedance method uses normal force excitation, the E/M impedance method uses in-plane strain excitation. The mechanical impedance transducer measures mechanical quantities (force and velocity/acceleration) to indirectly calculate the mechanical impedance, whereas the E/M impedance PWAS measures the E/M impedance directly as an electrical quantity. The effect of a PWAS affixed to the structure is to apply a local strain parallel to the surface that creates stationary elastic waves in the structure. Through the mechanical coupling between the PWAS and the host structure, on one hand, and through the electromechanical transduction inside the PWAS, on the other hand, the drive-point structural impedance is directly reflected into the effective electrical impedance as seen at the PWAS terminals.

9.1.1.2 Conventional Modal Analysis

Modal analysis and dynamic structural identification have become an intrinsic part of current engineering practice. Structural frequencies, damping, and modes shapes identified through this process are subsequently used to predict dynamic response, avoid resonances, and even monitor structural change that are indicative of damage (Harris, 1996).

Conventional modal testing (Ewins, 1984; Maia et al., 1997; Heylen et al., 1997) relies on two essential components: (a) structural excitation and (b) vibration pickups. Figure 9.1 shows that conventional structural excitation can be either through harmonic sweep using a shaker or through an impulse via an instrumented hammer. The former is more precise and can zoom in on resonant frequencies; the latter is more expedient and preferred for quick estimations. The vibration pickups can measure displacement, velocity, or acceleration. Current technologies cover miniaturized self-conditioning accelerometers (Broch, 1984) and laser velocimeters (Polytec, 2000). The accelerometers allow installation of sensor arrays that accurately and efficiently measure the mode shapes, whereas the laser offers noncontact measurements that are essential for low-mass sensitive structures. The disadvantages of accelerometers are cost, unavoidable bulkiness, and possible interference with the structural dynamics through their added mass. Laser velocimeters, on the

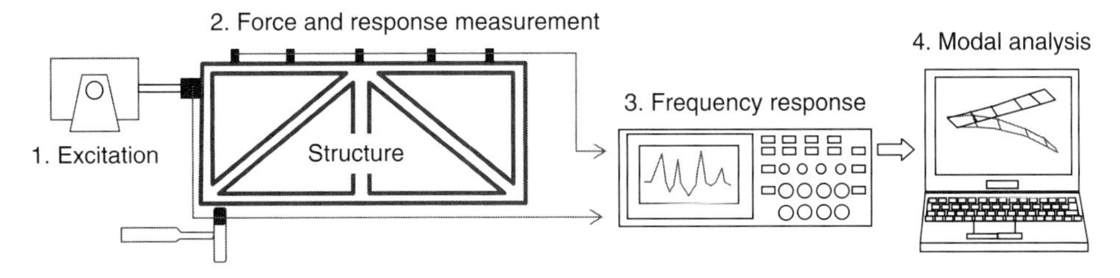

FIGURE 9.1 Schematic of conventional modal analysis structural identification experiments.

other hand, need to scan the structure to measure the mode shapes, and this significantly increases the duration of the experiments.

9.1.2 GENESIS OF THE E/M IMPEDANCE METHOD

The advent of commercially available low-cost piezoceramics has opened new opportunities for structural identification. Through their intrinsic electromechanical coupling, the piezoceramics can act as both sensors and actuators. Additionally, their frequency bandwidth is orders of magnitude larger than that of conventional shakers and even impact hammers. Small piezoelectric ceramic wafers can be permanently attached to structural surface. They could form sensor and actuator arrays that permit effective modal identification in a wide frequency band. Liang et al. (1994) performed the coupled E/M analysis of adaptive systems driven by a surface-attached piezoelectric wafer. The aim of the analysis was to determine the actuator power consumption and system energy transfer. A 1-dof analysis was performed. The electrical admittance, as measured at the terminals of a PWAS attached to the structure was derived in the form:

$$Y(\omega) = i\omega C\left(1 - \kappa_{31}^2 \frac{Z_{\text{str}}(\omega)}{Z_{\text{str}}(\omega) + Z_A(\omega)}\right) \tag{1}$$

where C is the electrical capacitance of the PWAS, $Z_{\text{str}}(\omega)$ is the 1-dof structural impedance as seen by the PWAS, and $Z_A(\omega)$ is the quasi-static impedance of the PWAS. A 1-dof numerical example was used to show that the E/M admittance response accurately reflects the system dynamic response. At coupled-system resonance, the real part of the E/M admittance was shown to have a distinct peak. However, because of the additional stiffness contributed by the PWAS, the system natural frequency shifted from 500 Hz (without PWAS) to 580 Hz (with PWAS). Experimental curve-fitting results were also presented. No modeling of the structural substrate was included, and no prediction of $Z_{\text{str}}(\omega)$ for a multi-dof structure was presented. This work was continued and extended by Sun et al. (1994) who used the half-power bandwidth method to accurately determine the natural frequency values. Mode shape extraction methodology, using multiple PWAS self and across admittances were explored. Experiments were performed on aluminum beams at frequencies up to 7 kHz. These two papers were the first to conceptualize that the E/M admittance as seen at the PWAS terminals reflects the coupled-system dynamics, and that a permanently attached PWAS could be used as structural-identification sensor. However, no theoretical modeling of the E/M impedance/admittance response for comparison with experimental data was attempted. Nor were the issues of sensor calibration, disbonding/self-diagnostics, and consistency attacked.

Subsequently, several authors reported the use of the E/M impedance method for structural health monitoring, whereby the admittance or impedance frequency spectra of pristine and damaged structures were compared. The method has been shown to be especially effective at ultrasonic frequencies, which properly capture the changes in local dynamics due to incipient structural damage. (Such changes are too small to affect the global dynamics and hence cannot be readily detected by conventional low-frequency vibration methods). The method is direct and easy to implement, the only required equipment being an electrical impedance analyzer.

9.1.3 CHALLENGES ASSOCIATED WITH MODELING THE E/M IMPEDANCE METHOD

Although relatively easy to apply experimentally, the E/M impedance method is not easy to model. A coupled-field analysis is required to capture the intricate coupling between the mechanical behavior of the structure and the electrical behavior of the PWAS such as to explain why structural damage can be detected from the analysis of the PWAS electrical impedance spectrum. Analysis methods are needed to explicitly predict the E/M admittance and impedance as it would be measured by the impedance analyzer at the PWAS terminals during the SHM process. Such a derivation is necessary to permit complete understanding of the phenomenon and to allow critical comparison with the abundant experimental results.

In response to this need, this chapter sets forth to present a step-by-step derivation of the interaction between the PWAS and the host structure, and to produce analytical expressions and numerical results for the E/M admittance and impedance seen at sensor terminal. These numerical results will be directly compared with experimental measurements. In our derivation, the limitations of the quasi-static sensor approximation adopted by previous investigators are lifted. Exact analytical expressions are being used for structural modeling of simultaneous axial and flexural vibrations. Free–free boundary conditions that can unequivocally be implemented during experimental testing (although more difficult to model) are being used.

To verify the theoretical model, experiments conducted on coupon specimens will be discussed. The E/M admittance and impedance spectra will be compared with theoretical predictions. The direct comparison between the modeled and measured E/M admittance spectra will be presented, and the capabilities and limitations of the PWAS modal sensors to detect the structural resonance frequencies from the E/M admittance response are rationally evaluated. The result of this comparison will prove that permanently attached PWAS could reliably perform structural identification. Their usefulness is especially apparent at high frequencies (ultrasonic range and beyond), where conventional vibration sensors loose their effectiveness.

The small unobtrusive PWAS used in our work do not effectively modify the structural stiffness and faithfully measure the structural dynamics. The PWAS stiffness and mass are orders of magnitude less than the structural stiffness and mass. Because the PWAS are small and unobtrusive, the dynamics of the host structure is not affected by the PWAS presence, and accurate structural identification is possible.

We start our analysis with the consideration of 1-D PWAS modal sensors that are easier to analyze whereas retaining all the important characteristics of the E/M impedance method. Closed-form solutions will be derived and verified against experiments conducted on small high-frequency metallic beams. The comparison between analysis and experiments will prove that the PWAS modal sensors can measure the high-frequency structural spectrum of these small metallic beams with remarkable fidelity. The analysis will progress to 2-D PWAS modal sensors. In our analysis, we will select circular PWAS modal sensors installed on circular plates, which allow closed-form solutions. The analytical predictions will be compared with experimental results. After these laboratory studies, a series of experiments performed on realistic specimens of various configurations will be presented.

9.2 1-D PWAS MODAL SENSORS

We start our analysis with the consideration of 1-D PWAS modal sensors that are easier to analyze whereas retaining all the important characteristics of the E/M impedance method. Closed-form solutions will be derived and verified against experiments conducted on small high-frequency metallic beams. Subsequently, a coupled-field FEM analysis, that gives even more accurate results, will be also presented.

9.2.1 ANALYSIS OF PWAS MOUNTED ON A 1-D BEAM STRUCTURE

Consider a 1-D structure with a PWAS attached to its surface (Fig. 9.2a). The PWAS has length l_a, and lies between x_a and $x_a + l_a$. Upon activation, the PWAS expands by $\varepsilon_{\text{PWAS}}$. This generates a reaction force F_{PWAS} from the beam onto the PWAS and an equal and opposite force from the PWAS onto the beam (Fig. 9.2b). This force excites the beam. At the neutral axis, the effect is felt as an axial force excitation, N_{PWAS}, and a bending moment excitation, M_{PWAS}. As the active sensor is electrically excited with a high-frequency harmonic signal, it will induce elastic waves into the beam structure. The elastic waves travel sideways into the beam structure, reflect at the beam boundaries, and set it into oscillation.

In a steady-state regime, the structure oscillates at the PWAS excitation frequency. The reaction force per unit displacement (dynamic stiffness) presented by the structure to the PWAS will depend on the internal state of the structure, on the excitation frequency, and on the boundary conditions

$$k_{\text{str}}(\omega) = \frac{\hat{F}_{\text{PWAS}}(\omega)}{\hat{u}_{\text{PWAS}}(\omega)} \tag{2}$$

where $\hat{u}_{\text{PWAS}}(\omega)$ is the displacement amplitude at frequency ω, $\hat{F}_{\text{PWAS}}(\omega)$ is the reaction force, and $k_{\text{str}}(\omega)$ is the dynamic stiffness. The symbol ^ signifies the complex amplitude of a time-varying function. Because the size of the PWAS is very small with respect to the size of the structure, Eq. (2) represents a point-wise structural stiffness.

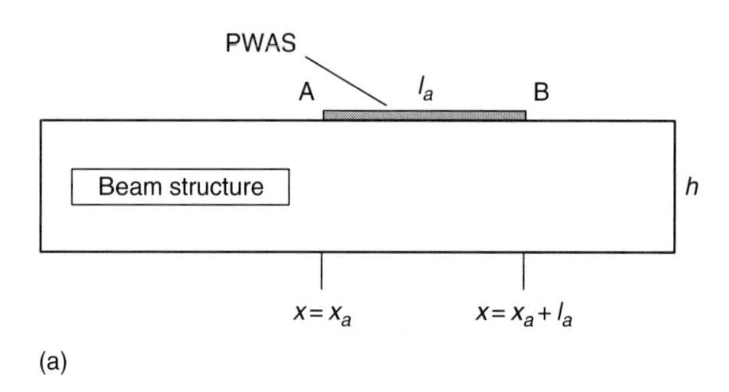

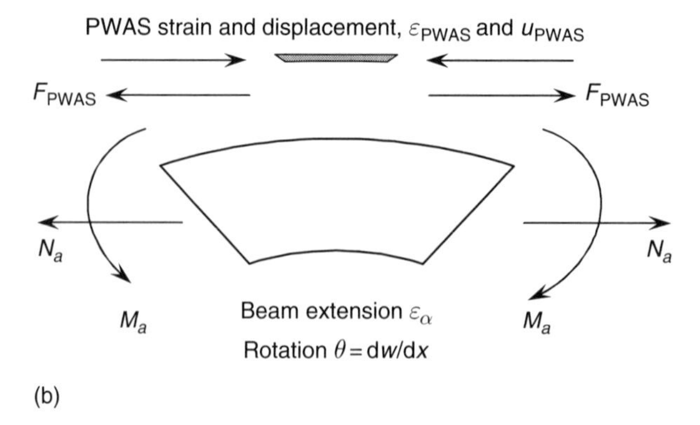

FIGURE 9.2 Interaction between a piezoelectric wafer active sensor (PWAS) and a beam-like structural substrate: (a) geometry; (b) forces and moments.

Our aim in this analysis is to understand how the E/M impedance measured at the PWAS terminals is influenced by the dynamic stiffness $k_{str}(\omega)$ presented by the structure to the PWAS. In fact, the dynamic stiffness $k_{str}(\omega)$ is *constraining* the PWAS in its oscillation. Hence, we can recall and use the constrained PWAS analysis developed in a previous chapter, which gives

$$Y = \mathrm{i}\omega \cdot C\left[1-\kappa_{31}^2\left(1-\frac{1}{\varphi\cot\varphi+r}\right)\right] \quad \text{(constrained PWAS admittance)} \tag{3}$$

$$Z = \frac{1}{\mathrm{i}\omega \cdot C}\left[1-\kappa_{31}^2\left(1-\frac{1}{\varphi\cot\varphi+r}\right)\right]^{-1} \quad \text{(constrained PWAS impedance)} \tag{4}$$

where $\varphi = \frac{1}{2}\gamma l$, γ is the wavenumber and r is the stiffness ratio. The wavenumber refers to standing waves inside the PWAS and is defined as

$$\gamma = \frac{\omega}{c} \tag{5}$$

where c is the wave speed inside the PWAS defined as

$$c^2 = \frac{1}{\rho s_{11}^E} \quad \text{(wave speed in the PWAS)} \tag{6}$$

The frequency-dependent stiffness ratio r is a function of frequency defined as the ratio

$$r(\omega) = \frac{k_{str}(\omega)}{k_{PWAS}} \tag{7}$$

where $k_{str}(\omega)$ is the frequency-dependent structural dynamic stiffness and k_{PWAS} is the PWAS static stiffness given by

$$k_{PWAS} = \frac{A_a}{s_{11}^E l_a} \tag{8}$$

9.2.2 DYNAMICS OF THE STRUCTURAL SUBSTRATE

The response of the structural substrate to the PWAS excitation is deduced from the general theory of beam vibrations developed in Chapter 3. However, the PWAS excitation departs from the typical textbook formulation because it acts as a pair of self-equilibrating axial forces and bending moments that are separated by a small finite distance, l_{PWAS}. This feature gives *gusto* to our analysis.

9.2.2.1 Definition of the Excitation Forces and Moments

The excitation forces and moments acting upon the beam structure are derived from the PWAS force, $F_{PWAS} = \hat{F}_{PWAS}e^{i\omega t}$, using the beam cross-section geometry (Fig. 9.2b),

$$M_a = F_{PWAS}\frac{h}{2}, \quad N_a = F_{PWAS} \tag{9}$$

The space-wise distribution of excitation bending moment and axial force is expressed using the Heaviside function, $H(x-x_a)$, defined as $H(x-x_a)=0$ for $x < x_a$, and $H(x-x_a)=1$ for $x_a \le x$,

$$N_e(x,t) = N_a[-H(x-x_a)+H(x-x_a-l_a)]\cdot e^{i\omega t} \tag{10}$$

$$M_e(x, t) = -M_a[-H(x - x_a) + H(x - x_a - l_a)] \cdot e^{i\omega t} \tag{11}$$

Equations (10) and (11) correspond to axial and flexural vibrations, respectively. Axial vibrations modes are usually of much larger frequency than flexural vibration modes. However, their vibrations frequencies are commensurable with those of the PWAS. Hence, in the present analysis, both axial and flexural vibrations are considered.

9.2.2.2 Axial Vibrations

The equation of motion for axial vibrations is

$$\rho A\ddot{u}(x, t) - EAu''(x, t) = N_e'(x, t) \tag{12}$$

Substitution of Eq. (10) into Eq. (12) yields

$$\rho A\ddot{u}(x, t) - EA\, u''(x, t) = \hat{N}_a[-\delta(x - x_a) + \delta(x - x_a - l_a)]\, e^{i\omega t} \tag{13}$$

where δ is Dirac's function. Assume modal expansion

$$u(x, t) = \sum_{n=0}^{\infty} C_n X_n(x) e^{i\omega t} \tag{14}$$

where $X_n(x)$ are orthonormal mode shapes, i.e., $\int X_m X_n \mathrm{d}x = \delta_{mn}$, with $\delta_{mn} = 1$ for $m = n$, and 0 otherwise. C_n are the modal amplitudes.

The mode shapes satisfy the free-vibration differential equation

$$EA\, X_n'' + \omega_n^2 \rho A X_n = 0 \tag{15}$$

Hence, multiplication by $X_n(x)$ and integration over the length of the beam yields

$$C_n = \frac{1}{\omega_n^2 - \omega^2} \cdot \frac{\hat{N}_a}{\rho A}\left[-X(x_a) + X(x_a + l_a)\right] \tag{16}$$

Thus, the axial vibration response is

$$u(x, t) = \frac{\hat{N}_a}{\rho A} \sum_{n=0}^{\infty} \frac{-X_n(x_a) + X_n(x_a + l_a)}{\omega_n^2 - \omega^2} X_n(x)\, e^{i\omega t} \tag{17}$$

9.2.2.3 Flexural Vibrations

For Euler–Bernoulli beams, the equation of motion under moment excitation is

$$\rho A\ddot{w}(x, t) + EIw''''(x, t) = -M_e''(x, t) \tag{18}$$

Substitution of Eq. (11) into Eq. (18) yields

$$\rho A\ddot{w}(x, t) + EIw''''(x, t) = \hat{M}_a[-\delta'(x - x_a) + \delta'(x - x_a - l_a)]\, e^{i\omega t} \tag{19}$$

where δ' is the first derivative of Dirac's function ($\delta' = H''$). Assume the modal expansion

$$w(x, t) = \sum_{n=N_1}^{N_2} C_n X_n(x) e^{i\omega t} \tag{20}$$

where $X_n(x)$ are the orthonormal bending mode shapes. The mode shapes satisfy the free-vibration differential equation

$$EI\,X_n'''' = \omega_n^2\,\rho A X_n \tag{21}$$

Hence, multiplication by $X_n(x)$ and integration over the length of the beam yields

$$C_n = \frac{1}{\omega_n^2 - \omega^2}\frac{\hat{M}_a}{\rho A}\int_0^l X_n(x)\,[\delta'(x-x_a) - \delta'(x-x_a-l_a)]\mathrm{d}x \tag{22}$$

Integration by parts and substitution into the modal expansion expression yields

$$C_n = -\frac{1}{(\omega_n^2 - \omega^2)}\frac{\hat{M}_a}{\rho A}[-X_n'(x_a) + X_n'(x_a + l_a)] \tag{23}$$

Hence, the flexural vibration response is

$$w(x,t) = -\frac{\hat{M}_a}{\rho A}\sum_{n=1}^{\infty}\frac{-X_n'(x_a) + X_n'(x_a + l_a)}{\omega_n^2 - \omega^2}\cdot e^{i\omega t} \tag{24}$$

9.2.2.4 Calculation of Frequency Response Function and Dynamic Structural Stiffness

To obtain the dynamic structural stiffness, $k_{\mathrm{str}}(\omega)$, presented by the structure to the PWAS, we first calculate the elongation between the two points, A and B, connected to the PWAS ends. Kinematic analysis gives the horizontal displacement of a generic point P on the beam surface in terms of the axial and flexural displacements as

$$u_P(t) = u(x) - \frac{h}{2}w'(x) \tag{25}$$

where u and w are the axial and bending displacements measured at the neutral axis.

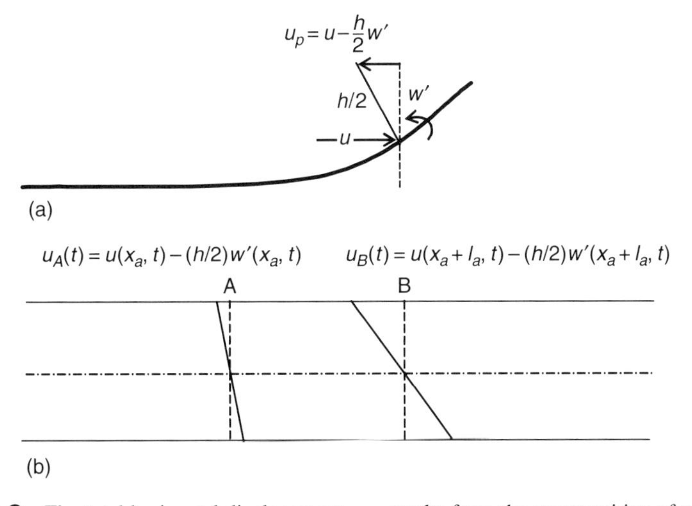

FIGURE 9.3 The total horizontal displacement, u_P, results from the superposition of axial displacement u and the rotation w'.

Letting P be A and B and then taking the difference yields

$$u_{\text{PWAS}}(t) = u_B(t) - u_A(t) = u(x_a, t) - u(x_a + l_a, t) - \frac{h}{2}\left[w'(x_a, t) - w'(x_a + l_a, t)\right] \quad (26)$$

Using Eqs. (17) and (24), Eq. (26) gives the total vibration response

$$\hat{u}_{\text{PWAS}} = \frac{\hat{F}_{\text{PWAS}}}{\rho A}\left\{\sum_{n_u} \frac{\left[U_{n_u}(x_a + l_a) - U_{n_u}(x_a)\right]^2}{\omega_{n_u}^2 - \omega^2} + \left(\frac{h}{2}\right)^2 \sum_{n_w} \frac{\left[W'_{n_w}(x_a + l_a) - W'_{n_w}(x_a)\right]^2}{\omega_{n_w}^2 - \omega^2}\right\} \quad (27)$$

where the axial and flexural vibrations frequencies and mode shapes were distinguished by the use of n_u, ω_{n_u}, $U_{n_u}(x)$ and n_w, ω_{n_w}, $W_{n_w}(x)$, respectively. Dividing Eq. (27) by $\hat{F}_{\text{PWAS}}$ yields the structural frequency response function (FRF) to the single-input single-output (SISO) excitation applied by the PWAS. This situation is similar to conventional modal testing with the proviso that the PWAS modal sensors are unobtrusive and permanently attached to the structure. For convenience, the axial FRF and flexural FRF components of the structural FRF are expressed separately, i.e.,

$$H_u(\omega) = \frac{1}{\rho A}\sum_{n_u} \frac{\left[U_{n_u}(x_a + l_a) - U_{n_u}(x_a)\right]^2}{\omega_{n_u}^2 + 2i\zeta_{n_u}\omega_{n_u}\omega - \omega^2} \quad (28)$$

$$H_w(\omega) = \frac{1}{\rho A}\left(\frac{h}{2}\right)^2 \sum_{n_w} \frac{\left[W'_{n_w}(x_a + l_a) - W'_{n_w}(x_a)\right]^2}{\omega_{n_w}^2 + 2i\zeta_{n_w}\omega_{n_w}\omega - \omega^2} \quad (29)$$

Note that modal damping, ζ, has been introduced to account for the inherent dissipative losses encountered in practical experiments. The FRF's are additive, such that the total structural FRF is simply

$$H(\omega) = H_u(\omega) + H_w(\omega) \quad (30)$$

or

$$H_w(\omega) = \frac{1}{\rho A}\left\{\sum_{n_u} \frac{\left[U_{n_u}(x_a + l_a) - U_{n_u}(x_a)\right]^2}{\omega_{n_u}^2 + 2i\zeta_{n_u}\omega_{n_u}\omega - \omega^2} + \left(\frac{h}{2}\right)^2 \sum_{n_w} \frac{\left[W'_{n_w}(x_a + l_a) - W'_{n_w}(x_a)\right]^2}{\omega_{n_u}^2 + 2i\zeta_{n_u}\omega_{n_u}\omega - \omega^2}\right\} \quad (31)$$

The SISO FRF is the same as the dynamic structural compliance seen by the PWAS modal sensor placed on the structure. The dynamic structural stiffness is the inverse of the structural compliance, i.e.,

$$k_{\text{str}}(\omega) = \frac{\hat{F}_{\text{PZT}}}{\hat{u}_{\text{PZT}}} = \rho A\left\{\sum_{n_u} \frac{\left[U_{n_u}(x_a + l_a) - U_{n_u}(x_a)\right]^2}{\omega_{n_u}^2 + 2i\zeta_{n_u}\omega_{n_u}\omega - \omega^2} + \left(\frac{h}{2}\right)^2 \sum_{n_w} \frac{\left[W'_{n_w}(x_a + l_a) - W'_{n_w}(x_a)\right]^2}{\omega_{n_w}^2 + 2i\zeta_{n_w}\omega_{n_u}\omega - \omega^2}\right\}^{-1} \quad (32)$$

For free–free beams, the axial and flexural mode shapes can be calculated with the formulae

$$U_{n_u}(x) = A_{n_u}\cos(\gamma_{n_u}x),\ A_{n_u} = \sqrt{\frac{2}{l}},\ \gamma_{n_u} = \frac{n_u\pi}{l},\ \omega_{n_u} = \gamma_{n_u}c,\ c = \sqrt{\frac{E}{\rho}},\ n_u = 1, 2, \ldots \tag{33}$$

$$W_{n_w}(x) = A_{n_w}\left[\cosh\gamma_{n_W}x + \cos\gamma_{n_w}x - \sigma_{n_w}\left(\sinh\gamma_{n_w}x + \sin\gamma_{n_w}x\right)\right] \tag{34}$$

$$\omega_{n_w} = \gamma_{n_w}^2 a,\ a = \sqrt{\frac{EI}{\rho A}},\ A_{n_w} = \frac{1}{\sqrt{\int_0^l W_{n_w}^2(x)\,\mathrm{d}x}} \tag{35}$$

Numerical values of $l \cdot \gamma_{n_w}$ and σ_{n_w} were given in Chapter 3 that dealt with vibration analysis.

9.2.3 NUMERICAL SIMULATIONS AND EXPERIMENTAL RESULTS

The analytical model was used to perform several numerical simulations that directly predict the E/M impedance and admittance signature at the PWAS terminals during structural identification. Subsequently, experiments were performed to verify these predictions. We consider a set of specimens consisting of small steel beams ($E = 200\,\text{GPa}$, $\rho = 7750\,\text{kg/m}^3$, Table 9.1) of various thickness and widths fabricated in the laboratory. The material properties are given in Table 9.1. All beams were $l = 100\,\text{mm}$ long with various widths $b_1 = 8\,\text{mm}$ (narrow beams) and $b_2 = 19.6\,\text{mm}$ (wide beams). The nominal thickness of the specimen was $h_1 = 2.59\,\text{mm}$; by gluing two specimens back-to-back, we were also able to create double thickness specimens $h_2 = 5.18\,\text{mm}$. Thus, four beam types were used (Fig. 9.4): (1) narrow-thin; (2) narrow-thick; (3) wide-thin; and (4) wide-thick.

TABLE 9.1 Typical material properties for aluminum and steel

	Aluminum (7075 T6)	Steel (AISI 4340 normalized)
Modulus (E)	70 GPa	200 GPa
Poisson ratio	0.33	0.3
Density (ρ)	$2700\,\text{kg/m}^3$	$7750\,\text{kg/m}^3$
Yield stress (Y)	500 MPa	860 MPa

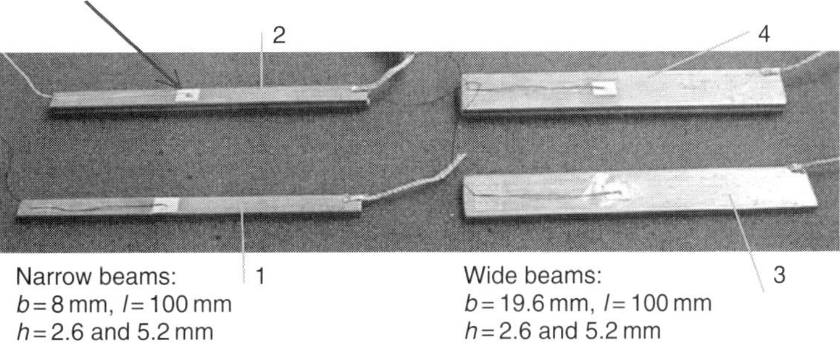

FIGURE 9.4 Experimental specimens for testing the use of PWAS modal sensors on 1-D structures.

The comparison of wide and narrow beams was aimed at identifying the width effects in the frequencies spectrum, whereas the change from double to simple thickness was aimed at simulating the effect of corrosion (for traditional structures) and disbonding/delamination on adhesively bonded and laminated structures. All specimens were instrumented with thin 7-mm square PWAS ($l_a = 6.92\,\mathrm{mm}$, $b_a = 6.91\,\mathrm{mm}$, $t_a = 0.224\,\mathrm{mm}$) bonded to the beam surface at $x_a = 40\,\mathrm{mm}$ from one end. The piezoelectric material properties of the PWAS are given in Table 9.2.

The numerical simulation was performed using the vibration analysis theory presented in the previous sections. Numerically exact expressions for the axial and flexural frequencies and mode shapes were used. Steel beams were analyzed. The damping coefficient was assumed $\zeta = 1\%$. The simulation was performed over a modal subspace that incorporates all modal frequencies in the frequency bandwidth of interest. The theoretical analysis indicates that these frequencies should be identical with the basic beam resonances, as predicted by vibration analysis.

The theoretical and experimental results are presented in Table 9.3. The 'Calc.' columns of Table 9.3 show the first 6 predicted resonances for axial and flexural vibrations.

The experimental set up is shown in Fig. 9.5. To approximate the free–free boundary conditions, the beams were supported on packing foam. Recording of the E/M impedance

TABLE 9.2 Piezoelectric properties of the PWAS material (APC-850)

Property	Symbol	Value
Compliance, in plane	s_{11}^E	$15.30 \cdot 10^{-12}\,\mathrm{Pa}^{-1}$
Compliance, thickness wise	s_{33}^E	$17.30 \cdot 10^{-12}\,\mathrm{Pa}^{-1}$
Dielectric constant	ε_{33}^T	$\varepsilon_{33}^T = 1750\varepsilon_0$
Thickness wise induced-strain coefficient	d_{33}	$400 \cdot 10^{\;12}\,\mathrm{m/V}$
In-plane induced-strain coefficient	d_{31}	$-175 \cdot 10^{-12}\,\mathrm{m/V}$
Coupling factor, parallel to electric field	k_{33}	0.72
Coupling factor, transverse to electric field	k_{31}	0.36
Coupling factor, transverse to electric field, polar motion	k_p	0.63
Poisson ratio	ν	0.35
Density	ρ	$7700\,\mathrm{kg/m^3}$
Sound speed	c	2900 m/s

Note: $\varepsilon_0 = 8.85 \times 10^{-12}\,\mathrm{F/m}$

TABLE 9.3 Theoretical and experimental results for wide and narrow beams with single and double thickness

	Beam #1 (narrow thin)			Beam #2 (narrow thick)			Beam #3 (wide thin)			Beam #4 (wide thick)		
	Calc. (kHz)	Exp (kHz)	Δ%	Calc. (kHz)	Exp (kHz)	Δ%	Calc. (kHz)	Exp (kHz)	Δ%	Calc. (kHz)	Exp (kHz)	Δ%
1	1.396	1.390	−0.4	2.847	2.812	−1.2	1.390	1.363	−1.9	2.790	2.777	−0.5
2	3.850	3.795	−1.4	7.847	7.453	−5.2	3.831	3.755	−2	7.689	7.435	−3.4
3	7.547	7.4025	−2	15.383	13.905	−10.6	7.510	7.380	−1.7	15.074	13.925	−8.2
4	12.475	12.140	−2.7		20.650		12.414	12.093	−2.6		21.825	
5	18.635	17.980	−3.6	25.430	21.787	−16.7	18.545	17.965	−3.2	24.918	22.163	−12.4
6		24.840						24.852				
7	26.035	26.317	1	26.035	26.157	0.5	26.022	26.085	0.2	25.944	26.100	0.6
		Cluster 175 kHz			Cluster 210 kHz			Cluster 35 kHz			Cluster 60 kHz	

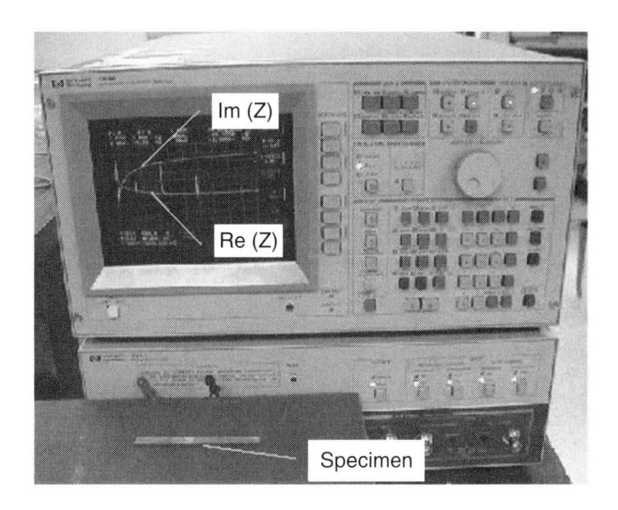

FIGURE 9.5 Experimental set up for dynamic identification of steel beams.

real part spectrum was performed in the 1–30 kHz range using an HP 4194A Impedance Analyzer. The impedance spectra are given in Fig. 9.6 through Fig. 9.9. The calculated and measured results are shown superposed. It was found that for the first four modes of single thickness beams, the predicted and measured frequency values almost superpose. For the fifth mode, there is a slight difference. The beam's natural frequencies, identified from the E/M impedance spectrum, are given in the 'Exp.' columns of Table 9.3. It should be noted that the errors are small and within the range normally accepted in experimental modal analysis. When the beam thickness was doubled, the frequencies also doubled. This is consistent with theoretical prediction. However, the error between theory and experiment seems larger for the double thickness beam, which may be caused by the compliance of the adhesive layer used in the construction of the double thickness beam. Even so, the confirmation of theoretical predictions by the experimental results is quite clear. The effect of beam width is demonstrated by comparison of 'narrow beam' and 'wide beam' columns in Table 9.3. The experimental results indicate cluster of frequencies, which move to higher frequencies as the width of the beam is reduced. We associate these clusters with width vibrations, which are not covered by the simple 1-D beam theory.

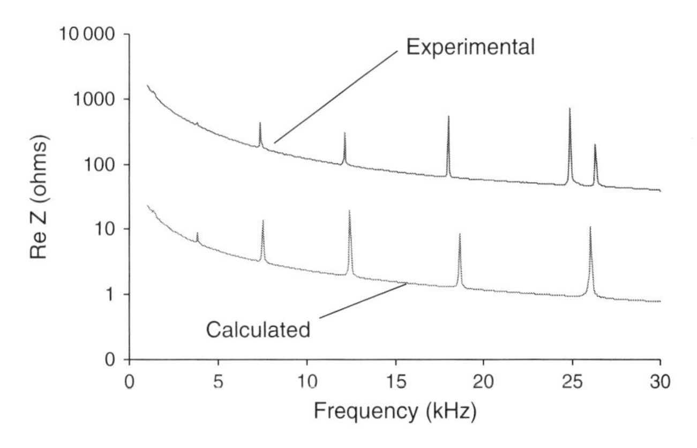

FIGURE 9.6 Experimental and calculated spectra of frequencies for single thickness narrow beam (Beam #1).

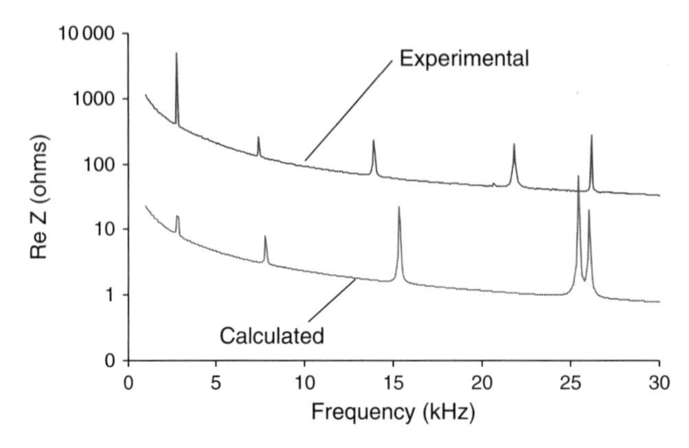

FIGURE 9.7 Experimental and calculated spectra of frequencies for double thickness narrow beam (Beam #2).

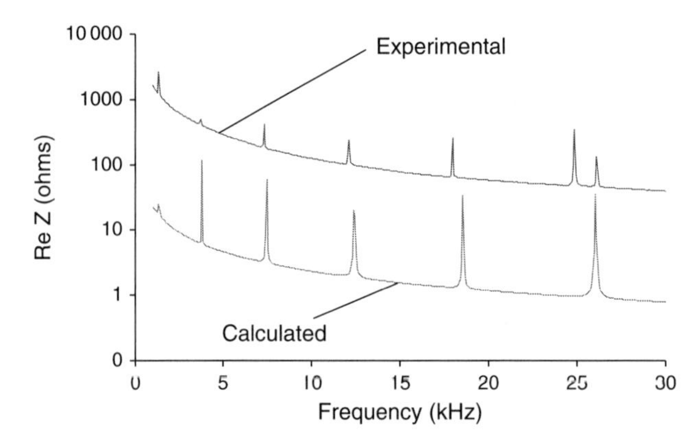

FIGURE 9.8 Experimental and calculated spectra of frequencies for single thickness wide beam (Beam #3).

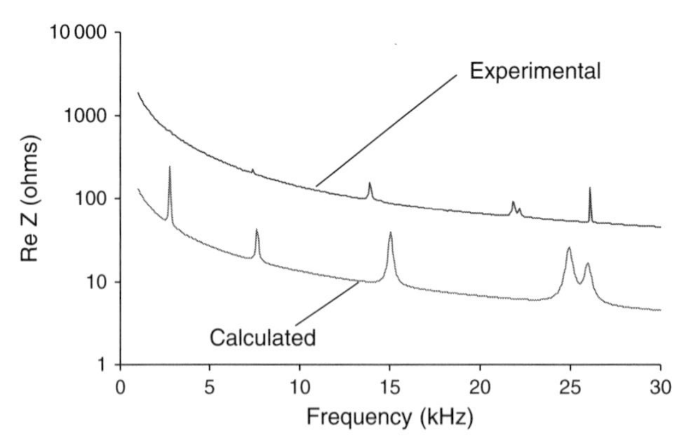

FIGURE 9.9 Experimental and calculated spectra of frequencies for double thickness wide beam (Beam #4).

The width vibrations are also influenced by the thickness, i.e., they shift to higher frequencies as the thickness is increased. This is also noticeable in Table 9.3, which shows that the lowest cluster appeared for the single thickness wide specimen and the highest cluster for double thickness narrow beam. Another important aspect that needs to be noticed is that the first axial mode of vibrations is at 25 kHz. This explains the approximately 25 kHz double frequencies observed in the single thickness beams, both narrow and wide.

These results prove that the predicted and measured results are in close agreement, within the tolerance normally expected from experimental modal analysis. The results obtained for double thickness beams are less precise because of inhomogeneity introduced by the layer of adhesive between single thickness beams.

9.2.4 COMPARISON WITH CONVENTIONAL METHODS

To highlight the advantages of the E/M impedance method over conventional modal testing, was also tried to performed the dynamic identification of these specimen types with conventional modal testing methods. A small steel beam with dimensions identical to the #1 specimen (single thickness, beam) was instrumented with two CEA-13-240UZ-120 strain gauges connected in half bridge configuration to a P-3500 strain indicator from Measurements Group, Inc. The specimen was suspended in a free–free configuration and excited with a sharp impact. The resulting signal was collected with an HP 54601B digital oscilloscope and processed numerically on a PC. Standard signal analysis algorithms (FFT) were used to extract the frequency spectrum.

The first natural frequency (1.387 kHz) was clearly displayed in the FFT spectrum. The second natural frequency (3.789 kHz) was also visible, but with a much weaker amplitude. These results are consistent with theoretical predictions and with the experimental results previously obtained with the E/M impedance method (Table 9.3). However, the impact excitation method could not excite the other higher frequencies depicted in Table 9.3; this is most probably due to the bandwidth limitations inherent in impact excitation method. We also considered the use of another conventional modal analysis method, specifically the sweep excitation method. In principle, sweep excitation would be able to reach all the natural frequencies within the sweep bandwidth. However, the application of this method to our small specimen was not found feasible because of attachment difficulties and the fact that the kHz frequency range could not be easily achieved with conventional shakers. This demonstrates that, for the type of small rigid specimens considered in our study, the E/M impedance method using PWAS modal sensors has a niche of its own that cannot be filled by conventional modal analysis methods.

9.2.5 NONINVASIVE CHARACTERISTICS OF THE PWAS MODAL SENSORS

The PWAS modal sensors used in the experiments were very small and did not significantly disturb the dynamic properties of the structure under consideration. Table 9.4 presents the mass and stiffness values for the sensor and structure. For comparison, the mass of the accelerometer is also included in Table 9.4.

TABLE 9.4 Numerical illustration of PWAS noninvasive properties

Item	Mass (g)	Percentage of structural mass	Stiffness (MN/m)	Percentage of structural stiffness
PWAS	0.082	0.5%	15	1.5%
Structure	16.4	N/A	1000	N/A
Accelerometer: 352A10, PCB Piezotronics	0.7	4.3%	N/A	N/A

The data in Table 9.4 illustrate numerically the noninvasive properties of PWAS modal. The mass and stiffness additions brought by PWAS are within the 1% range (0.5% for mass and 1.5% for stiffness). In spite of its small dimensions, the PWAS was able to adequately perform the dynamic structural identification of the test specimens, as illustrated in Fig. 9.6 through Fig. 9.9. The numerical results of this dynamic identification were given in Table 9.3. If the same identification were to be attempted with classical modal analysis, i.e., using an accelerometer and an impact hammer, the mass addition due to accelerometer would have been around 4.3%, which would obviously contaminate the results. This simple example underlines the fact that the use of PWAS modal sensors is not only advantageous but, in certain situations, irreplaceable. For small components such as found in precision machinery and computer industry, the use of PWAS could be the only practical option for in-situ structural identification.

9.2.6 PWAS SELF-DIAGNOSTICS

PWAS affixed to, or embedded into, a structure play a major role in the successful operation of a health monitoring and damage-detection system. The integrity of the sensor and the consistency of the sensor/structure interface are essential elements that can 'make or break' an experiment. The general expectation is that once the PWAS have been placed on or into the structure, they will behave consistently throughout the duration of the health-monitoring exercise. For real structures, the duration of the health-monitoring process is extensive and can span several years. It also will encompass various service conditions and several loading cases. Therefore, in-situ self-diagnostic methods are mandatory. The PWAS should be scanned periodically to determine their integrity. They should be also scanned prior to any damage-detection cycle. Self-diagnostic methods for assessing the sensor integrity are necessary to ensure that the health-monitoring process is progressing as expected. For PWAS, a sensor self-diagnostic method is readily available through the E/M impedance technique. This self-diagnosis method works as follows.

Our preliminary tests have shown that the reactive (imaginary) part of the impedance (Im Z) can be a good indication of active sensor integrity. This is justified by the fact that

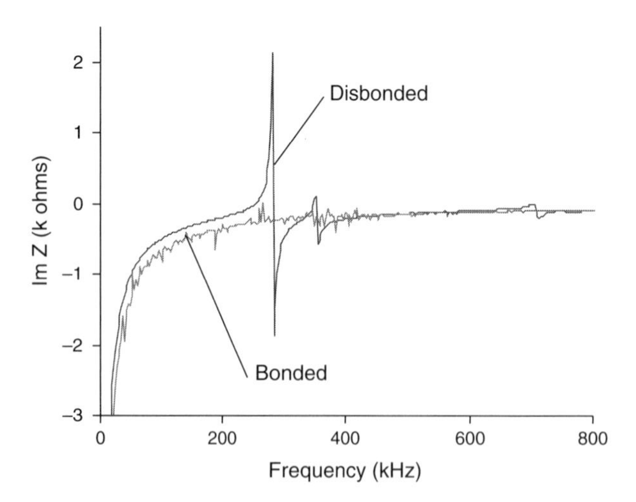

FIGURE 9.10 Piezoelectric wafer active sensor (PWAS) self-diagnostic using the imaginary part of the E/M impedance: when PWAS is disbonded, new free-vibration resonance features appear at ~267 kHz.

the PWAS is predominantly a capacitive device, and its impedance is dominated by its reactive part, $1/i\omega C$. Base-line signatures taken at the beginning of the health-monitoring process can be compared with current reading in order to identify the defective PWAS. In such comparison, the imaginary part of the PWAS complex impedance should be used. Figure 9.10 compares the Im Z spectrum of a well-bonded PWAS with that of a disbonded (free) PWAS. For the disbonded PWAS, the appearance of the free-vibration resonance and the disappearance of structural resonances constitute unambiguous features that can be used for automated PWAS self-diagnostics.

9.2.7 TYPICAL APPLICATIONS OF PWAS MODAL SENSORS

The PWAS and the associated structural dynamics identification methodology based on the E/M impedance response are ideally suited for small machinery parts that have natural frequencies in the kHz range. As an example, we considered the aircraft turbo-engine blade presented in Fig. 9.11. Two PWAS were installed, one on the blade and the other on the root. E/M impedance spectrum was recorded and natural frequencies could be easily identified (Fig. 9.12).

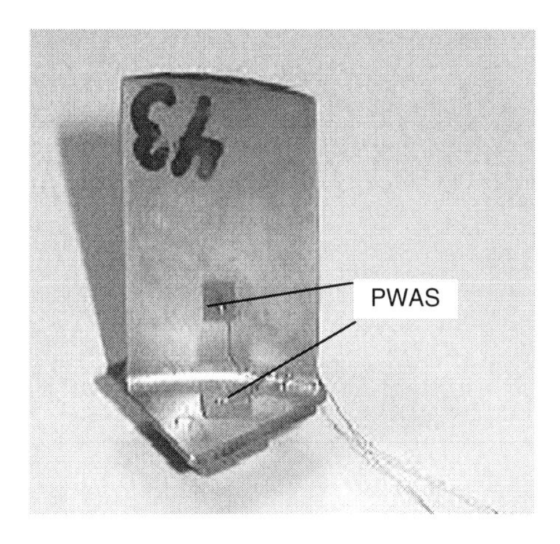

FIGURE 9.11 Aircraft turbo-engine blade equipped with PWAS.

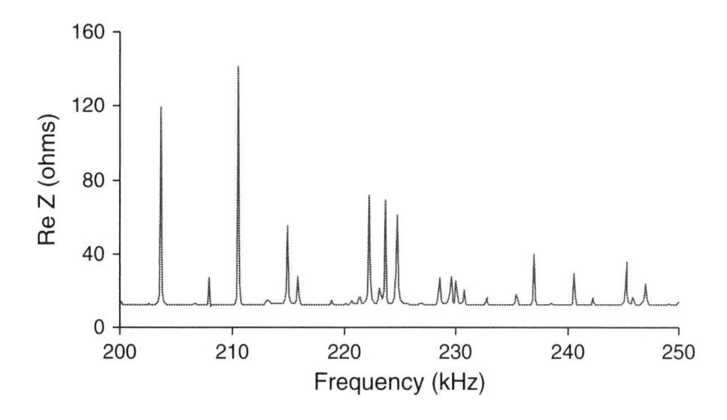

FIGURE 9.12 E/M impedance spectrum of an aircraft turbo-engine blade.

9.2.8 SECTION SUMMARY

The benefits and limitations of using embedded PWAS for structural identification at ultrasonic frequency were highlighted. An analytical model based on structural vibration theory and theory of piezoelectricity was developed and used to predict the E/M impedance response as it would be measured at the PWAS terminals. The model considers 1-D structures and accounts for both axial and flexural vibrations. Elastically constrained PWAS permanently bonded to the structure were considered. The derived mathematical expressions accounted for the dynamic response of both the sensor and the structure. Experiments were conducted on simple specimens in support of the theoretical investigation and on realistic turbine blade specimen to illustrate the method's potential. It was shown that the E/M impedance spectrum recorded by the PWAS accurately represents the mechanical response of a structure. It was further proved that the response of the structure is not modified by the presence of the PWAS, thus validating the noninvasive characteristics of the PWAS. It is shown that such sensors, of negligible mass, can be permanently applied to the structure creating a nonintrusive sensor array adequate for on-line automatic structural identification and health monitoring. The sensor calibration procedure was presented. Numerical estimation of the noninvasive properties of the proposed PWAS in comparison with conventional sensors is presented. Self-diagnostics capabilities of the proposed sensors were also investigated, and methods for automatic self-test implementation are discussed. As presented in this section, the E/M impedance method, when using just one active sensor, can only detect structural resonances. The detection of structural mode shapes is also possible but requires the simultaneous use of several sensors, their number being in direct relationship to the desired modal resolution.

A limitation of the E/M impedance technique is that it is less effective at low frequencies than at high frequencies. In our experiments, we found that below 5 kHz the resonance peaks are buried in the overall electric response. These difficulties were alleviated by narrowband tuning. The numerical results reported for the 1 to 5 kHz range were obtained with this technique. However, below 1 kHz, the E/M impedance method is simply not recommended. Above 5 kHz, the E/M impedance method is definitely superior to other experimental modal analysis and dynamic identification methods.

In view of these advantages and disadvantages, it is felt that PWAS modal sensors in conjunction with the E/M impedance method have their niche as a structural identification methodology using self-sensing, permanently attached PWAS. Because of their perceived low cost (< $10), these active sensors can also be inexpensively configured as sensor arrays. The proposed method can be a useful and reliable tool for automatic on-line structural identification in the ultrasonic frequency range. The use of PWAS modal sensors not only can be advantageous, but also, in certain situations, may be the sole investigative option. Examples include precision machinery, small but critical turbine-engine parts, computer industry components, etc.

9.3 CIRCULAR PWAS MODAL SENSORS

In this section, we intend to extend the analysis of the E/M impedance method to 2-D PWAS modal sensors. In our analysis, we will select circular PWAS modal sensors, which allow closed-form solutions. The analytical predictions will be compared with experimental results obtained on circular metallic plates instrumented with circular PWAS modal sensors.

9.3.1 INTERACTION BETWEEN A CIRCULAR PWAS AND A CIRCULAR PLATE

Assume a thin isotropic circular plate with a PWAS surface mounted at its center (Fig. 9.13). Under PWAS excitation, both axial and flexural vibrations are set in motion. The structural dynamics affects the PWAS response; it modifies the PWAS E/M impedance, i.e., the impedance measured by an impedance analyzer connected to the PWAS terminals. The purpose of this section is to model the interaction between the circular PWAS and the structure, and predict the impedance spectrum that would be measured at the PWAS terminals during the structural identification process. In our development, we will account for both the structural dynamics and the PWAS dynamics. The interaction between the circular PWAS and the structure is modeled as shown in Fig. 9.14. The structure is assumed to present to the circular PWAS an effective structural dynamic stiffness, $k_{str}(\omega)$, which includes both axial and flexural modes. The problem is formulated in terms of interaction line force, F_{PWAS}, and the corresponding displacement, u_{PWAS}, measured at the PWAS circumference. The perfect-bonding assumption implied in this model is that the surface adhesion between the PWAS and the structure can be reduced to an effective boundary interaction between the radial displacement and the line force at the PWAS circumference (similar to the pin-force model in 1-D analysis of PWAS–structure interaction).

When the PWAS is excited by an oscillatory voltage, its volume expands in phase with the voltage in accordance with the piezoelectric effect. Expansion of the PWAS mounted to the surface of the plate induces a surface reaction from the plate in the form

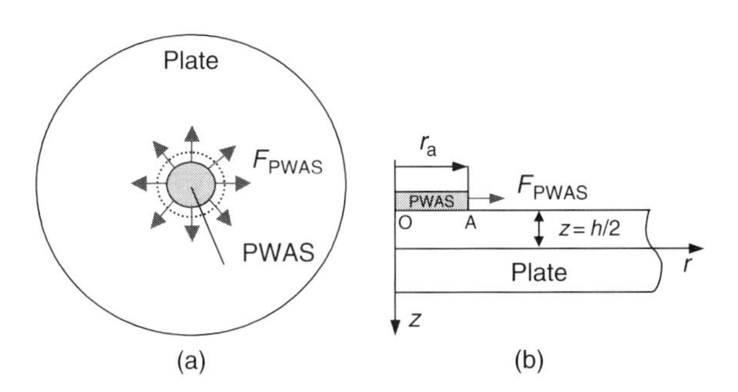

FIGURE 9.13 (a) PWAS mounted on a circular plate; (b) cross-section schematics.

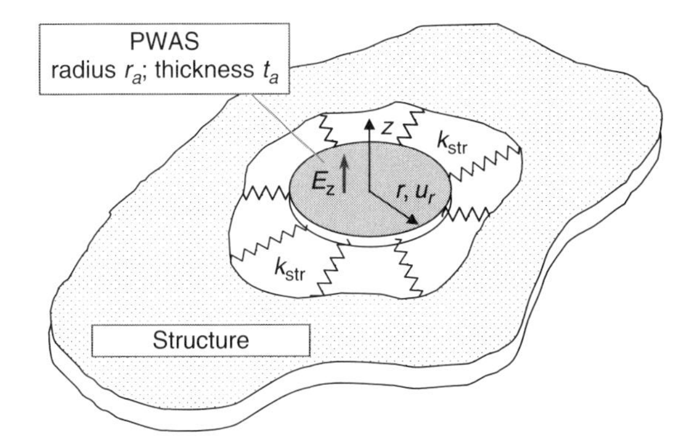

FIGURE 9.14 Circular PWAS constrained by structural stiffness, $k_{str}(\omega)$.

of the line force distributed around the PWAS circumference, $F_{\text{PWAS}}(t) = \hat{F}_{\text{PWAS}} e^{i\omega t}$. The reaction force, $F_{\text{PWAS}}(t)$, depends on the PWAS radial displacement, $u_{\text{PWAS}}(t)$, and on the frequency-dependent dynamic stiffness, $k_{\text{str}}(\omega)$, presented by the structure to the PWAS

$$F_{\text{PWAS}}(t) = k_{\text{str}}(\omega) u_{\text{PWAS}}(t) \tag{36}$$

9.3.2 MODELING OF THE CIRCULAR PLATE DYNAMICS

The equations for axisymmetric axial and flexural vibrations of circular plates are assumed in the form

$$\frac{Eh}{1-\nu^2}\left(\frac{\partial^2 u}{\partial r^2} + \frac{1}{r}\frac{\partial u}{\partial r} - \frac{u}{r^2}\right) - \rho h \frac{\partial^2 u}{\partial t^2} = -\left(\frac{\partial N_r^e}{\partial r} + \frac{N_r^e}{r}\right) \tag{37}$$

$$D\nabla^4 w + \rho h \frac{\partial^2 w}{\partial t^2} = \frac{\partial^2 M_r^e}{\partial r^2} + \frac{2}{r}\frac{\partial M_r^e}{\partial r} \tag{38}$$

where u is the in-plane displacement along the r-direction, w is the transverse displacement, h is the plate thickness, ρ is the plate density. The quantities N_r^e and M_r^e are excitation line forces and line moments acting over the whole surface of the plate. These excitation forces and moments originate in the PWAS force, $F_{\text{PWAS}}(t)$, acting at the surface of the plate at $r = r_a$. Resolving this force at the plate midplane, we get a line force and a line moment, i.e.,

$$\begin{aligned} N_a(t) &= F_{\text{PWAS}}(t) \\ M_a(t) &= \frac{h}{2} F_{\text{PWAS}}(t) \end{aligned} \tag{39}$$

Using the Heaviside step function, we write

$$\begin{aligned} N_r^e(r,t) &= N_a(t)\,[-H(r_a - r)] \\ M_r^e(r,t) &= M_a(t)\,[H(r_a - r)] \end{aligned}, \quad r \in [0, \infty) \tag{40}$$

The solutions of Eqs. (37) and (38) are sought as modal expansions

$$\begin{aligned} u(r,t) &= \left(\sum_k P_k R_k(r)\right) e^{i\omega t} \\ w(r,t) &= \left(\sum_m G_m Y_m(r)\right) e^{i\omega t} \end{aligned} \tag{41}$$

where P_k and G_m are the modal participation factors for axial and flexural vibrations, whereas $R_k(r)$ and $Y_m(r)$ are the corresponding mode shapes. For free-edge boundary conditions, the axisymmetric mode shapes $R_k(r)$ and $Y_m(r)$ are expressed in terms of Bessel functions

$$\begin{aligned} R_k(r) &= A_k J_1(\lambda_k r) \\ Y_m(r) &= A_m\,[J_0(\lambda_m r) + C_m\, I_0(\lambda_m r)] \end{aligned} \tag{42}$$

The mode shapes $R_k(r)$ and $Y_m(r)$ form orthonormal function sets that satisfy the orthonormality conditions

$$\begin{aligned}\rho h \int_0^a \int_0^{2\pi} R_k(r) R_l(r) r \, \mathrm{d}r \, \mathrm{d}\theta &= \rho h \, \pi a^2 \, \delta_{kl} \\ \rho h \int_0^a \int_0^{2\pi} Y_p(r) \cdot Y_m(r) r \, \mathrm{d}r \, \mathrm{d}\theta &= m \, \delta_{pm} = \pi a^2 \, \rho h \, \delta_{pm}\end{aligned} \tag{43}$$

where a is the outer radius of the circular plate and δ_{ij} is the Kronecker delta. Upon solution, we obtain the modal participation factors for axial and flexural vibrations.

$$\begin{aligned}P_k &= \frac{2N_a}{\rho h \cdot a^2} \frac{\left[r_a R_k(r_a) - \int_0^a R_k(r) H(r_a - r) \mathrm{d}r \right]}{\left(\omega_k^2 - 2i\varsigma_k \omega \omega_k + \omega^2 \right)} \\ G_m &= \frac{2M_a}{\rho h \cdot a^2} \frac{\left[3Y_m(r_a) + r_a \cdot Y_m'(r_a) \right]}{\left(\omega_m^2 - 2i\varsigma_m \omega \omega_m + \omega^2 \right)}\end{aligned} \tag{44}$$

where ζ_k and ζ_m are modal damping ratios. Figure 9.15 presents a plot of the axial modes, whereas Fig. 9.16 presents a plot of the flexural modes. The axial modes have in-plane

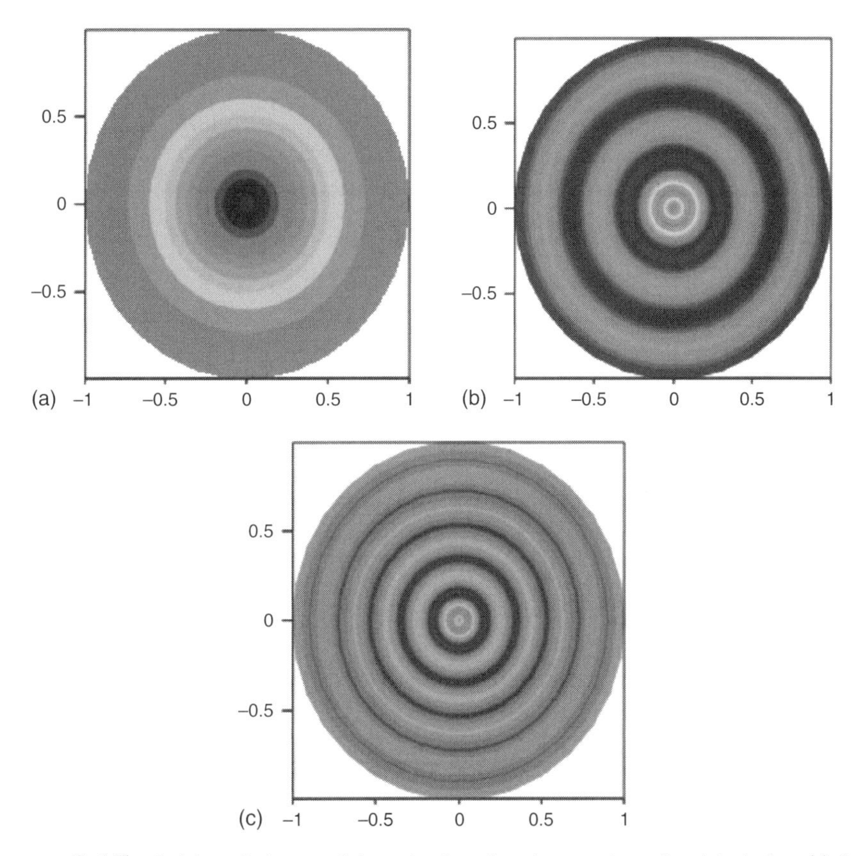

FIGURE 9.15 Axial modeshapes of free circular plate k = number of nodal circles: (a) $k = 0$; (b) $k = 5$; (c) $k = 10$.

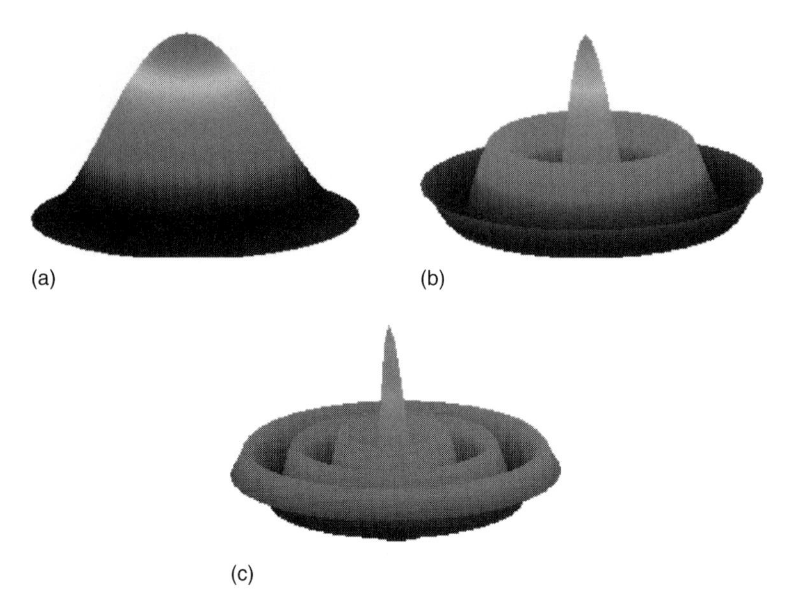

FIGURE 9.16 Flexural mode shapes of free circular plate: (a) Nodal circles $m = 1$; (b) Nodal circles $m = 4$; (c) Nodal circles $m = 7$.

motion; hence, the plot in Fig. 9.15 utilizes grey scales to represent the motion amplitude. The flexural modes have flexural motion; hence, Fig. 9.16 is a 3-D representation in which the out-of-plane flexural motion is represented along the z-axis.

9.3.3 CALCULATION OF THE EFFECTIVE STRUCTURAL STIFFNESS

We calculate the effective structural stiffness $k_{str}(\omega)$ from the structural response to PWAS excitation. The radial displacement at the rim of the PWAS can be expressed in the form

$$u_{PWAS}(r_a, t) = u(r_a, t) - \frac{h}{2} w'(r_a, t) \tag{45}$$

Note that $u(r_a, t)$ and $w(r_a, t)$ represent displacements at the plate mid-surface, whereas $u_{PWAS}(r_a, t)$ is measured at the plate upper surface (Fig. 9.13). Using Eq. (41), we write

$$u_{PWAS}(r_a, t) = \sum_k P_k R_k(r_a) e^{i\omega t} - \frac{h}{2} \sum_m G_m Y_m'(r_a) e^{i\omega t} \tag{46}$$

Substitution of Eqs. (44) into Eq. (46) yields the PWAS displacement in terms of the interface force, F_{PWAS}, and the axial and flexural dynamics of the plate. Discarding the time dependence, we write

$$\hat{u}_{PWAS}(\omega) = \frac{\hat{F}_{PWAS}(\omega)}{a^2 \rho} \left[\begin{aligned} &\frac{2}{h} \sum_k \frac{\left[r_a R_k(r_a) - \int_0^a R_k(r) H(r_a - r) \mathrm{d}r \right] R_k(r_a)}{(\omega_k^2 - 2i\varsigma_k \omega \omega_k + \omega^2)} \\ &+ \frac{h}{2} \sum_m \frac{[3Y_m(r_a) + r_a Y_m'(r_a)]\, Y_m'(r_a)}{(\omega_m^2 - 2i\varsigma_m \omega \omega_m + \omega^2)} \end{aligned} \right] \tag{47}$$

Recalling Eq. (36), we write $k_{str}(\omega) = \dfrac{\hat{F}_{PWAS}(\omega)}{\hat{u}_{PWAS}(\omega)}$, i.e.,

$$k_{str}(\omega) = a^2\rho \begin{bmatrix} \dfrac{2}{h}\sum_k \dfrac{\left[r_a R_k(r_a) - \int_0^a R_k(r)H(r_a - r)\mathrm{d}r\right] R_k(r_a)}{\left(\omega_k^2 - 2i\varsigma_k\omega\omega_k + \omega^2\right)} \\ +\dfrac{h}{2}\sum_m \dfrac{\left[3Y_m(r_a) + r_a Y_m'(r_a)\right] Y_m'(r_a)}{\left(\omega_m^2 - 2i\varsigma_m\omega\omega_m + \omega^2\right)} \end{bmatrix}^{-1} \tag{48}$$

Inversion of Eq. (48) gives the frequency response function (FRF) of the structure when subjected to PWAS excitation

$$\mathrm{FRF}_{str}(\omega) = \frac{1}{a^2\rho} \begin{bmatrix} \dfrac{2}{h}\sum_k \dfrac{\left[r_a R_k(r_a) - \int_0^a R_k(r)H(r_a - r)\mathrm{d}r\right] R_k(r_a)}{\left(\omega_k^2 - 2i\varsigma_k\omega\omega_k + \omega^2\right)} \\ +\dfrac{h}{2}\sum_m \dfrac{\left[3Y_m(r_a) + r_a Y_m'(r_a)\right] Y_m'(r_a)}{\left(\omega_m^2 - 2i\varsigma_m\omega\omega_m + \omega^2\right)} \end{bmatrix} \tag{49}$$

9.3.4 MODEL VALIDATION THROUGH NUMERICAL AND EXPERIMENTAL RESULTS

To validate the theoretical results, a series of experiments were conducted on a set of thin-gage aluminum plates. The set of specimens consisted of five identical circular plates manufactured from aircraft-grade aluminum stock ($E = 70\,\mathrm{GPa}, \rho = 2.7\,\mathrm{g/cm^3}$). The diameter of each plate was 100 mm and the thickness was approximately 0.8 mm. Each plate was instrumented at its center with a 7-mm diameter PWAS modal sensor. During the experiments, the specimens were supported on packing foam to simulate free boundary conditions. Impedance readings were taken using a HP 4194A Impedance Analyzer. The collected spectra are shown superposed in Fig. 9.17b.

Examination of Fig. 9.17b reveals that the spectra collected on the five specimens showed very little variation from specimen to specimen. Plate resonance frequencies were identified from the E/M impedance real part spectra. Table 9.5 shows the statistical data in terms of resonance frequencies and $\log_{10}$ amplitudes. It should be noted that the resonance frequencies have very little variation (1% standard deviation) whereas the $\log_{10}$ amplitudes vary more widely (1.2–3.6% standard deviation).

To validate the theory, we compared the experimental Re(Z) spectrum with the theoretical Re(Z) and the theoretical FRF spectrum. The FRF spectrum was utilized to illustrate the fact that the Re(Z) spectrum reflects the structural dynamics, i.e., the peaks of the Re(Z) spectrum coincide with the peaks the FRF peaks, which are the structural resonances. Figure 9.18a shows the FRF spectrum calculated with Eq. (49), for $r_a = 3.5\,\mathrm{mm}$, $a = 50\,\mathrm{mm}$, $h = 0.8\,\mathrm{mm}$, $\varsigma_k = 0.07\%$ and $\varsigma_m = 0.4\%$. The frequency range was 0.5–40 kHz. Six flexural resonances and one axial resonance were captured.

$$Z(\omega) = \left\{ i\omega C\left(1 - k_p^2\right) \times \left[1 + \frac{k_p^2}{1 - k_p^2} \frac{(1 + \nu_a) J_1(\varphi_a)}{\varphi_a J_0(\varphi_a) - (1 - \nu_a) J_1(\varphi_a) - \frac{a}{r_a}\chi(\omega)(1 + \nu_a) J_1(\varphi_a)}\right]\right\}^{-1} \tag{50}$$

Figure 9.18b compares the experimental and theoretical Re(Z). The theoretical Re(Z) was calculated with Eq. (50). This equation resembles the one derived in Chapter 7 with the

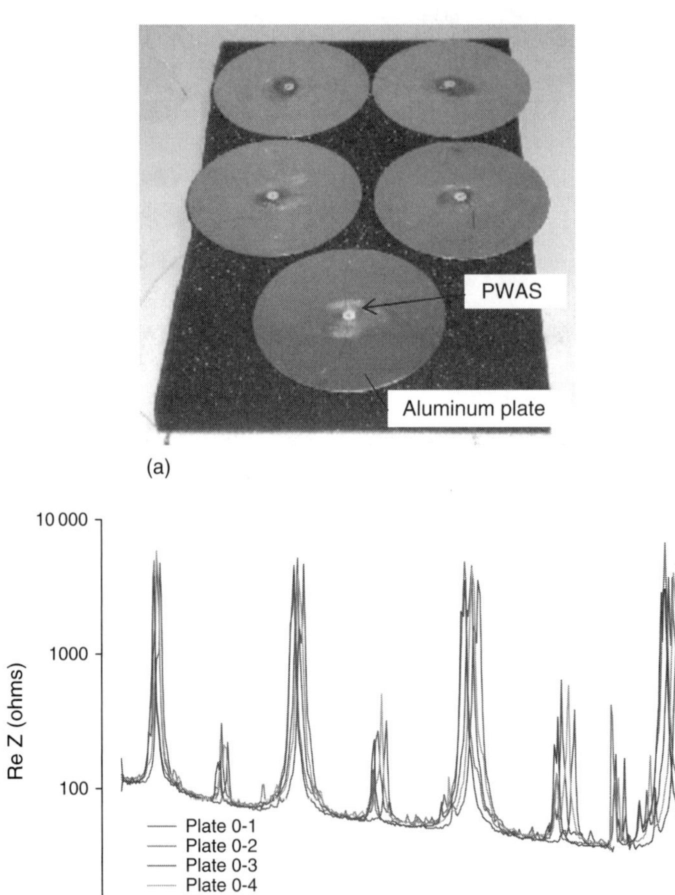

FIGURE 9.17 (a) Thin-gage aluminum plate specimens with centrally located PWAS: 100-mm diameter, 0.8-mm thickness; (b) E/M impedance spectra taken from pristine plates in the 11–40-kHz frequency band.

TABLE 9.5 Statistical summary for resonance peaks of first four axis-symmetric modes of a circular plate as measured with the PWAS using the E/M impedance method

Average Frequency (kHz)	Frequency STD (kHz) (%)	Log_{10}-Average amplitude Log_{10}-ohms	Log_{10}-Amplitude STD Log_{10}-ohms (%)
12.856	0.121 (1%)	3.680	0.069 (1.8%)
20.106	0.209 (1%)	3.650	0.046 (1.2%)
28.908	0.303 (1%)	3.615	0.064 (1.7%)
39.246	0.415 (1%)	3.651	0.132 (3.6%)

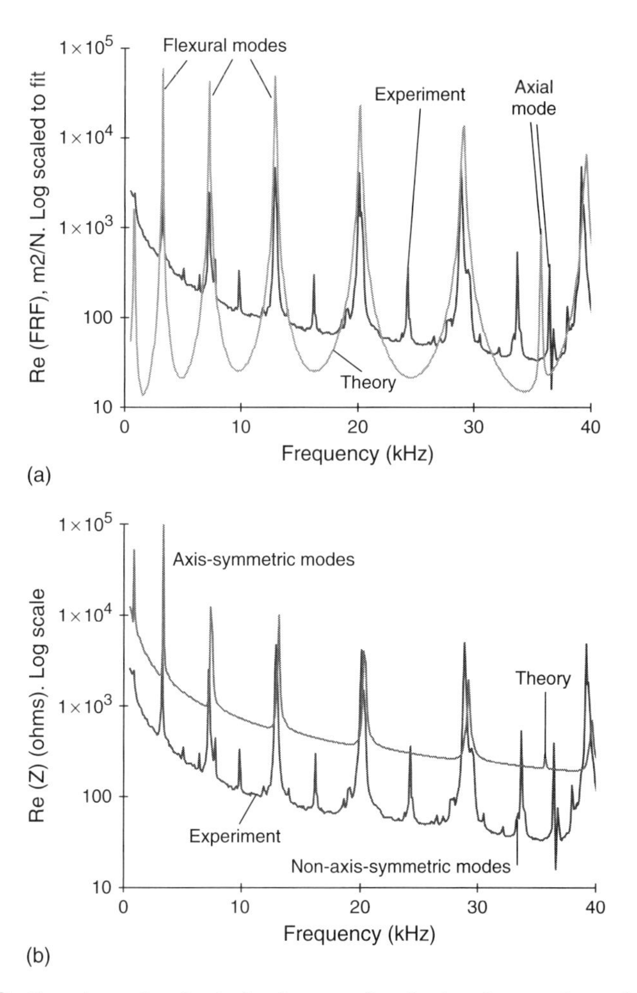

FIGURE 9.18 Experimental and calculated spectra for circular plate specimen: (a) FRF spectrum in 0.5–40 kHz frequency range; (b) E/M impedance spectrum in 0.5–40 kHz frequency range.

difference of introducing a multiplicative correction factor a/r_a, in front of the stiffness ratio $\chi(\omega)$. This correction factor was needed to account for the difference between the k_{str} distributed over the radius of the plate and the k_{PWAS} distributed over the radius of the PWAS. As seen in Fig. 9.18b, good matching between theoretical and experimental $\text{Re}(Z)$ spectra was obtained.

The numerical values of the predicted and measured frequencies are compared in Table 9.6. The match between the theoretical and the experimental resonance values was quite good. It is noted that the error is consistently very low ($<2\%$). For most modes, the matching error is less than 1%. Two exceptions are noted:

(i) The first flexural frequency has a matching error of -7.7%. This can be attributed to experimental error, because the E/M impedance does not work as well at low frequencies as it works at high frequencies. At low frequencies, the response due to mechanical resonances gets buried in the overall electric response. Hence, the peak of the first flexural mode is rather weak.

(ii) The first axial mode has a 2% error, which can be attributed to slight imperfections in the plate contour.

TABLE 9.6 Theoretical and experimental results for a circular plate with a sensor installed in the center

Frequency #	1	2	3	4	5	6	7	8
Mode	F	F	F	F	F	F	A	F
Calc. (kHz)	0.742	3.152	7.188	12.841	20.111	29.997	35.629	39.498
Exp. (kHz)	0.799	3.168	7.182	12.844	20.053	28.844	36.348	39.115
Error, Δ%	−7.708	−0.520	0.078	−0.023	0.288	0.528	1.978	0.97

Note: F, flexural mode; A, axial mode.

Thus, we can conclude that Eq. (50) permits direct comparison of experimental and theoretical E/M impedance data, which was the aim of our analysis.

Although the simulation gives a good matching with experimental results, the theoretical model presented here is limited to the analysis of purely axis-symmetrical modes. In principle, the axis-symmetric assumption is consistent with a geometry in which a circular PWAS is placed exactly at the center of a circular plate. However, if the sensor is slightly misaligned, non-axis-symmetric modes will also be excited and will appear in the spectrum. This effect is observable in Fig. 18; the low-amplitude peaks that appear at 15, 24, and 33 kHz on the experimental curves have no match on the theoretical curves. These small peaks are due to non-axis-symmetric modes that get parasitically excited because of slight misalignment in the placement of the PWAS at the center of the plate.

9.3.5 SECTION SUMMARY

A circular plate model, which accounts for sensor–structure interaction in 2-D geometry, was derived and validated through experimental testing. The model considers both the structural dynamics and the PWAS dynamics. In the structural dynamics, both the axial and the flexural vibrations were considered. The structural dynamics was incorporated into the model through the pointwise dynamic stiffness presented by the structure to the PWAS. The analytical model predicts the E/M impedance response, as it would be measured at the PWAS terminals. The real part of the E/M impedance reflects with fidelity the natural frequencies of the structure on which the PWAS is mounted. Through experimental tests, we were able to validate that the model is capable of correctly predicting the E/M impedance. We also showed that the structural frequencies can be determined directly from the E/M impedance real part. Thus, it was verified that PWAS, in conjunction with the E/M impedance, act as self-excited high-frequency modal sensors that correctly sense the structural dynamics.

9.4 DAMAGE DETECTION WITH PWAS MODAL SENSORS

This section will present the use of the high-frequency PWAS modal sensors for incipient damage detection using the E/M impedance technique. The advantage of using PWAS for damage detection resides in their very high-frequency capability, which exceeds by orders of magnitudes the frequency capability of conventional modal analysis sensors. Thus, PWAS are able to detect subtle changes in the high-frequency structural dynamics at local scales. Such local changes in the high-frequency structural dynamics are associated with the presence of incipient damage, which would not be detected by conventional modal analysis sensors operating at lower frequencies.

In this section, two types of experiments are described. The first type of experiments was performed on circular plates, which have clean and reproducible spectra, as seen in the previous section. Thus, the effect of damage on the resonance spectrum was easily observed. This first type of experiments allowed a tractable development of the damage identification algorithms. It was found that two algorithm types could be used, one based on overall statistics of the entire spectrum and the other based on the spectral features. The second type of experiments was performed on realistic specimens, which are representative of actual structures. Various specimen types were used: spot-welded joints, bonded joints, composite overlays on civil engineering structures, aircraft panel specimens, etc. These specimens had various types of damage, such as disbonds, cracks, and corrosion. The E/M impedance method was used to detect the presence of these, by comparing 'pristine' and 'damage' spectral readings. Details of these experiments are given next.

9.4.1 DAMAGE-DETECTION EXPERIMENTS ON CIRCULAR PLATES

Systematic experiments were performed on circular plates to assess the crack detection capabilities of the method. As shown schematically in Table 9.7 and Fig. 9.19, five damage groups were considered: one group consisted of pristine plates (Group 0) and four groups consisted of plates with simulated cracks placed at increasing distance from the plate edge (Group 1 through 4). Each group contained five nominally 'identical' specimens. Thus, the statistical spread within each group could also be assessed. In our study, a 10-mm circumferential slit was used to simulate an in-service crack. The simulated crack was placed at decreasing distance from the plate edge. The following radial positions of the crack from the PWAS were considered: 40, 25, 10, and 3 mm.

The experiments was conducted over three frequency bands: 10–40 kHz; 10–150 kHz, and 300–450 kHz. The data was processed by plotting the real part of the E/M impedance spectrum and by determining a damage metric to quantify the difference between spectra. The data for the 10–40-kHz band is shown in Fig. 9.19. As damage is introduced in the plate, resonant frequency shifts, peaks splitting, and the appearance of new resonances are noticed. As the damage becomes more severe, these changes become more profound. The most profound changes are noticed for Group 4. For the higher frequency bands, similar behavior was observed. Numerical values of the measured frequencies of the major resonances are given in Table 9.7.

9.4.1.1 Overall-Statistics Damage Metrics

The damage metric is a scalar quantity that results from the comparative processing of E/M impedance spectra. The damage metric should reveal the difference between spectra due to damage presence. Ideally, the damage index (DI) would be a metric, which captures only the spectral features that are directly modified by the damage presence, whereas neglecting the variations due to normal operation conditions (i.e., statistical difference within a population of specimens, and expected changes in temperature, pressure, ambient vibrations, etc.). To date, several damage metrics have been used to compare impedance spectra and assess the presence of damage. Among them, the most popular are the root mean square deviation (RMSD), the mean absolute percentage deviation (MAPD), and the

TABLE 9.7 Damage detection experiments on circular plates with PWAS and the E/M impedance method

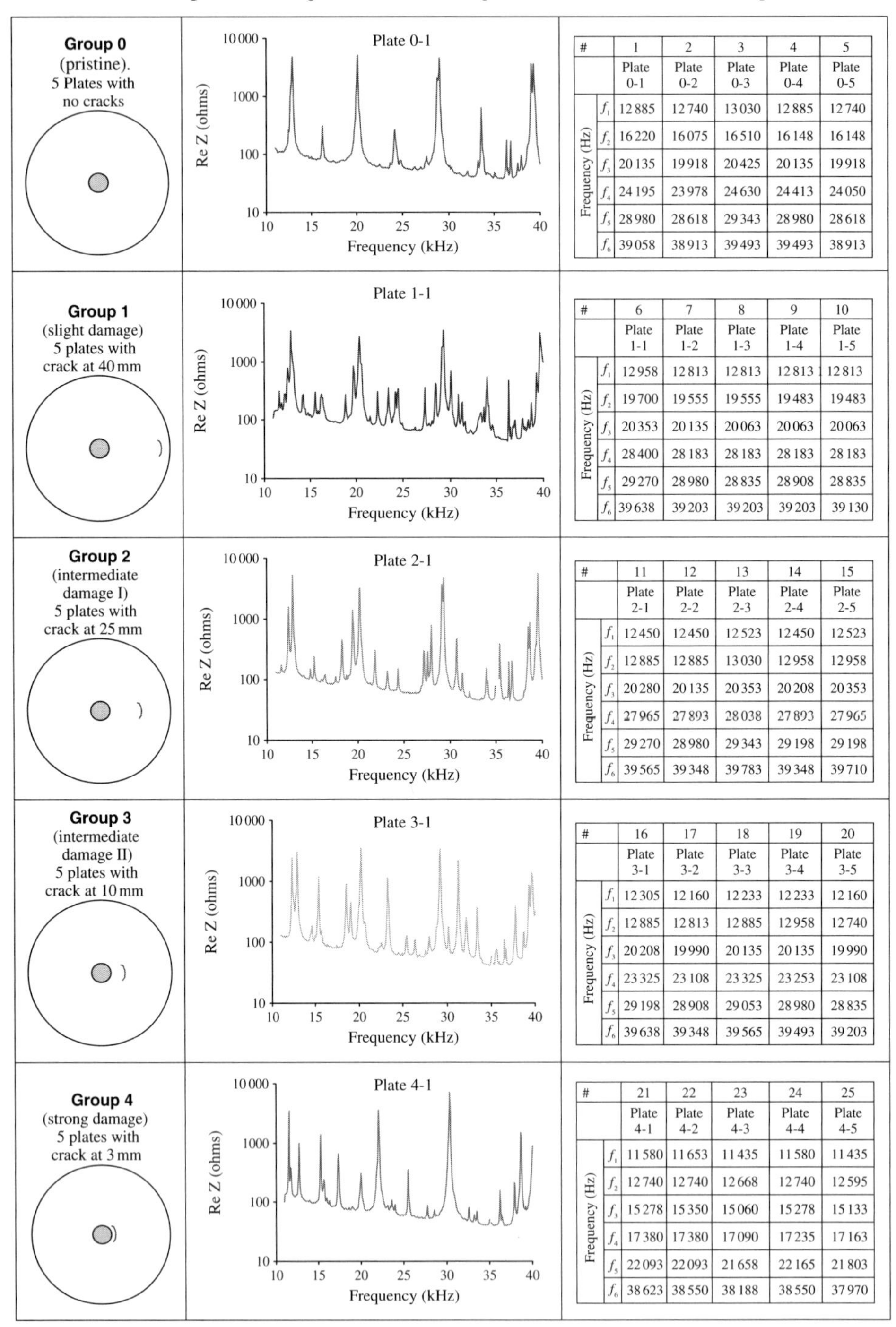

#		1	2	3	4	5
		Plate 0-1	Plate 0-2	Plate 0-3	Plate 0-4	Plate 0-5
Frequency (Hz)	f_1	12 885	12 740	13 030	12 885	12 740
	f_2	16 220	16 075	16 510	16 148	16 148
	f_3	20 135	19 918	20 425	20 135	19 918
	f_4	24 195	23 978	24 630	24 413	24 050
	f_5	28 980	28 618	29 343	28 980	28 618
	f_6	39 058	38 913	39 493	39 493	38 913

#		6	7	8	9	10
		Plate 1-1	Plate 1-2	Plate 1-3	Plate 1-4	Plate 1-5
Frequency (Hz)	f_1	12 958	12 813	12 813	12 813	12 813
	f_2	19 700	19 555	19 555	19 483	19 483
	f_3	20 353	20 135	20 063	20 063	20 063
	f_4	28 400	28 183	28 183	28 183	28 183
	f_5	29 270	28 980	28 835	28 908	28 835
	f_6	39 638	39 203	39 203	39 203	39 130

#		11	12	13	14	15
		Plate 2-1	Plate 2-2	Plate 2-3	Plate 2-4	Plate 2-5
Frequency (Hz)	f_1	12 450	12 450	12 523	12 450	12 523
	f_2	12 885	12 885	13 030	12 958	12 958
	f_3	20 280	20 135	20 353	20 208	20 353
	f_4	27 965	27 893	28 038	27 893	27 965
	f_5	29 270	28 980	29 343	29 198	29 198
	f_6	39 565	39 348	39 783	39 348	39 710

#		16	17	18	19	20
		Plate 3-1	Plate 3-2	Plate 3-3	Plate 3-4	Plate 3-5
Frequency (Hz)	f_1	12 305	12 160	12 233	12 233	12 160
	f_2	12 885	12 813	12 885	12 958	12 740
	f_3	20 208	19 990	20 135	20 135	19 990
	f_4	23 325	23 108	23 325	23 253	23 108
	f_5	29 198	28 908	29 053	28 980	28 835
	f_6	39 638	39 348	39 565	39 493	39 203

#		21	22	23	24	25
		Plate 4-1	Plate 4-2	Plate 4-3	Plate 4-4	Plate 4-5
Frequency (Hz)	f_1	11 580	11 653	11 435	11 580	11 435
	f_2	12 740	12 740	12 668	12 740	12 595
	f_3	15 278	15 350	15 060	15 278	15 133
	f_4	17 380	17 380	17 090	17 235	17 163
	f_5	22 093	22 093	21 658	22 165	21 803
	f_6	38 623	38 550	38 188	38 550	37 970

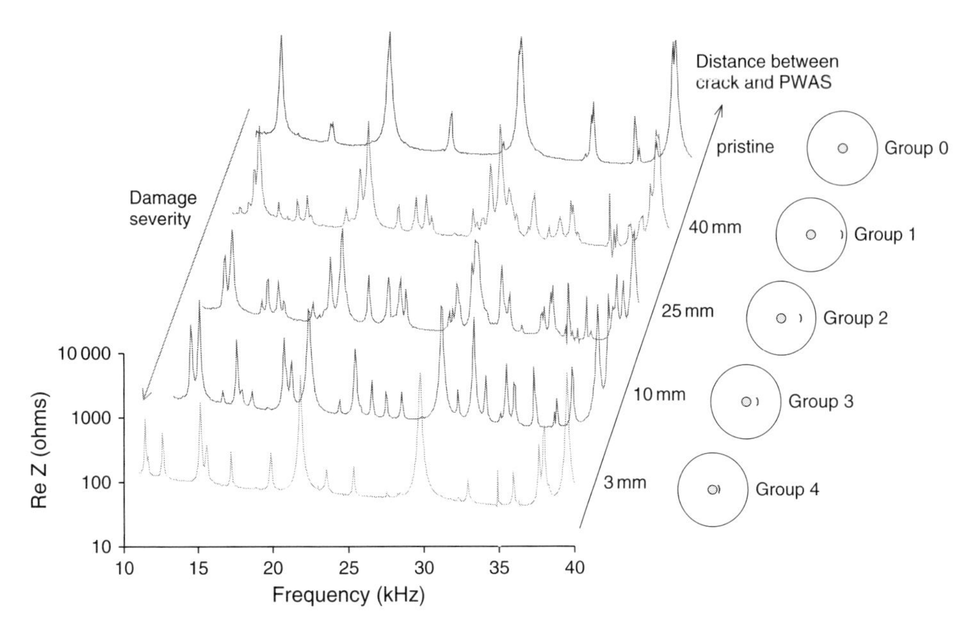

FIGURE 9.19 Dependence of the E/M impedance spectra on the damage location.

correlation coefficient deviation (CCD). The mathematical expressions for these metrics, given in terms of the impedance real part Re(Z), are as follows

$$RMSD = \sqrt{\sum_N \left[\mathrm{Re}(Z_i) - \mathrm{Re}(Z_i^0)\right]^2 / \sum_N \left[\mathrm{Re}(Z_i^0)\right]^2} \quad (51)$$

$$MAPD = \sum_N \left| \left[\mathrm{Re}(Z_i) - \mathrm{Re}(Z_i^0)\right] / \mathrm{Re}(Z_i^0) \right| \quad (52)$$

$$CCD = 1 - CC, \quad \text{where } CC = \frac{1}{\sigma_Z \sigma_{Z^0}} \sum_N \left[\mathrm{Re}(Z_i) - \mathrm{Re}(\overline{Z})\right] \cdot \left[\mathrm{Re}(Z_i^0) - \mathrm{Re}(\overline{Z}^0)\right] \quad (53)$$

where N is the number of frequencies in the spectrum and the superscript 0 signifies the pristine state of the structure. The symbols $\overline{Z}, \overline{Z}^0$ signify mean values, whereas σ_Z, σ_{Z^0} signify standard deviations.

Equations (51)–(53) yield a scalar number, which represents the relationship between the compared spectra. Thus, we expect that the resonant frequency shifts, the peaks splitting, and the appearance of new resonances that appear in the spectrum will alter the DI value and thus signal the presence of damage. The advantage of using Eqs. (51)–(53) is that the impedance spectrum does not need any pre-processing, i.e., the data obtained from the measurement equipment can be directly used to calculate the DI. In our experimental study, we used the scalar values of RMSD, MAPD, and CCD calculated with Eqs. (51)–(53).

The data-processing results for the three frequency bands (10–40 kHz, 10–150 kHz, 300–450 kHz) are summarized in Table 9.8. It seems that the CCD metric is more sensitive to the damage presence than RMSD and MAPD. However, it was also observed that for the 10–40 kHz and 10–150 kHz frequency bands the CCD variation with damage severity is not monotonic. This indicates that the choice of the frequency band may play a significant role in classification process; the frequency band with highest density of

TABLE 9.8 Overall-statistics damage metrics for various frequency bands

Frequency band	11–40 kHz				11–150 kHz				300–450 kHz			
Compared groups	0_1	0_2	0_3	0_4	0_1	0_2	0_3	0_4	0_1	0_2	0_3	0_4
RMSD %	122	116	94	108	144	161	109	118	93	96	102	107
MAPD %	107	89	102	180	241	259	170	183	189	115	142	242
CCD %	84	75	53	100	93	91	52	96	81	85	87	89

Notes: RMSD, root mean square deviation; MAPD, mean absolute percentage deviation; CCD, correlation coefficient deviation.

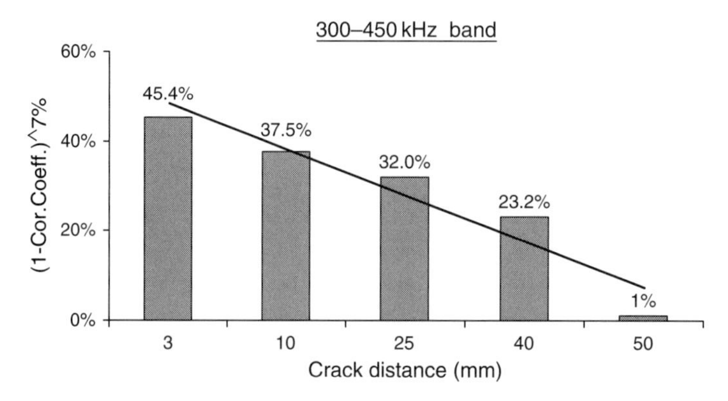

FIGURE 9.20 Monotonic variation of the CCD^7 damage metric with the crack radial position on a 50-mm radius plate in the 300–450 kHz band.

peaks is recommended. Hence, we used the frequency band 300–450 kHz for further data analysis. Figure 9.20 presents the variation of the metric CCD^7 with the crack distance from the plate center. It is apparent that as the crack is placed further away, and its influence diminishes, the value of the CCD^7 metric also diminishes. In fact, the CCD^7 damage metric tends to decrease linearly with the crack position, which may be very useful in automated damage assessment.

9.4.1.2 Probabilistic Neural Networks for Damage Identification

Probabilistic neural networks (PNN) are an efficient tool for solving classification problems. PNN are hybrid multi-layer networks that use basis functions and competitive selection concepts. In contrast to other neural network algorithms used for damage identification, PNN have a statistically derived activation function and utilize a Bayesian decision strategy for the classification problem. The kernel-based approach to probability density function (PDF) approximation is used. This method permits the construction of the PDF of any sample of data without any a priori probabilistic hypothesis. The PDF reconstruction is achieved by approximating each sample point with kernel function(s) to obtain a smooth continuous approximation of the probability distribution. In other words, the use of kernel techniques makes it possible to map a pattern space (data sample) into the feature space (classes). However, the result of such transformation should retain essential information presented in the data sample and be free of redundant information, which may contaminate the feature space. In this study, we used the resonance frequencies as a data sample to classify spectra according to the damage severity. The multivariate Gaussian kernel was chosen for PNN implementation

$$p_A(x) = \frac{1}{n \cdot (2\pi)^{d/2} \sigma^d} \sum_{i=1}^{n} \exp\left(-\frac{(x - x_{Ai})^T (x - x_{Ai})}{2\sigma^2} \right) \quad (54)$$

where i is a pattern number, x_{Ai} is ith training pattern from A category, n is total number of training patterns, d is dimensionality of measurement space, and σ is a spread parameter. Although the Gaussian kernel function was used in this work, the PNN is generally not limited to being Gaussian.

A MATLAB implementation of the PNN algorithm is shown in Fig. 9.21. PNN achieve a Bayesian decision analysis with Gaussian kernel. PNN consist of a radial-basis layer, a feedforward layer, and a competitive layer (Fig. 9.21). PNN identify class separation boundaries in the form of hyper-spheres in the input space. PNN are trained in two stages. In the first stage, through *unsupervised learning*, the training input vectors, x_n, are used to determine the basis-function parameters. Either an exact design or an adaptive design can be used. In the second stage, through *supervised learning*, the linear-layer weights are found to fit the training output values. An essential step in pattern recognition is feature extraction. Feature extraction removes irrelevant information from the data set by separating essential information from non-essential information. Feature extraction considerably reduces the problem size and dimensionality. With feature extraction, the pattern recognition is usually performed in two basic steps (Fig. 9.22). First, a number of features are extracted and placed into a features vector. Second, the features vector is used to perform the classification. In spectral analysis, the features-vector approach

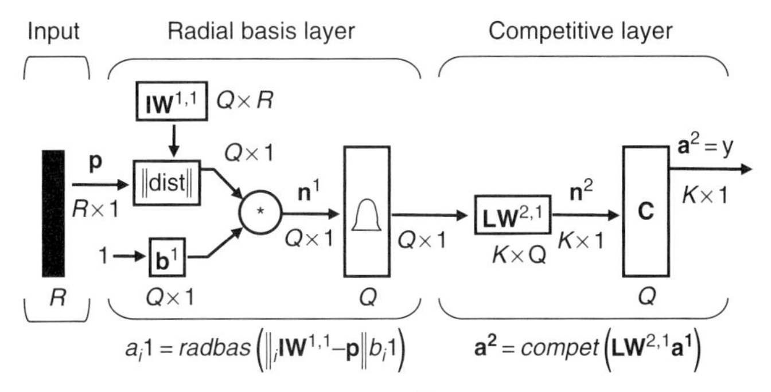

a_i1 is the ith element of $\mathbf{a}^1$ where ${}_i\mathbf{WI}^{1,1}$ is the vector made of the ith row of ${}_i\mathbf{WI}^{1,1}$

R = number of elements in input vector

Q = number or input/target pairs = number of neurons in layer 1

K = number of classes of input data = number of neurons in layer 2

FIGURE 9.21 The general architecture of a probabilistic neural net implemented in the MATLAB package (http://www.mathworks.com/access/helpdesk/help/toolbox/nnet/nnet.shtml).

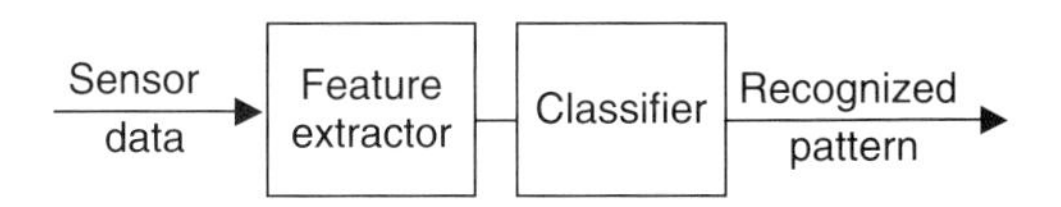

FIGURE 9.22 Schematic of a pattern recognition process utilizing features extraction.

recognizes that the number and characteristics of resonance peaks play a major role in describing the dynamic behavior of the structure.

The physics-based features extraction utilizes the deterministic relation between the peaks in the spectrum and the phenomenon of structural resonance. Hence, the spectrum is described through its essential dynamical features, i.e., resonance frequencies, resonance peaks, and modal damping coefficients. Through this process, the problem dimensionality may be reduced by at least one order of magnitude (from the 400-point spectrum to a few tens of features). However, this reduction may not be as large at very high frequencies where lightly damped structures may have very high density of resonance peaks. To account for this, an incremental approach can be used; the complexity of the features vector is gradually increased such as to balance breadth of details with computational efficiency. For example, one can consider first a reduced-order features vector containing only the resonance frequencies, and then gradually extend it to include the other features (relative amplitudes and damping factors).

9.4.1.3 Damage Detection in Circular Plates with Probabilistic Neural Network

To construct the PNN input vectors, the spectra of Table 9.7 were processed to extract the resonance frequencies for each group. Once the feature vectors are established, the classification problem can be approached in the features space. Although several features-based classification algorithms are available, in our study we explore the PNN classification algorithm. Typical results are shown in Table 9.9. These results were obtained with a features vector of size six, corresponding to the six most predominant frequencies in the 10–40 kHz spectrum. Five analysis tests were conducted: I, II, III, IV, V. In each test, one spectrum from each class was selected for training, and the other four spectra in each class were used for validation. Thus, in total, five vectors were used for training (one from each class) and twenty vectors were used for validation (four from each class). The vectors used for training are designated with letter T in Table 9.9. The vectors used for validation are designated with letter V in Table 9.9.

TABLE 9.9 Synoptic classification table for circular plates using 6-frequency feature vectors

		Group 0 (no damage)					Group 1 (r = 40 mm)					Group 2 (r = 25 mm)					Group 3 (r = 10 mm)					Group 4 (r = 3 mm)					
Plate #		1	2	3	4	5	6	7	8	9	10	11	12	13	14	15	16	17	18	19	20	21	22	23	24	25	
Test #	I	T	V	V	V	V	T	V	V	V	V	T	V	V	V	V	T	V	V	V	V	T	V	V	V	V	IN
		–	0	0	0	0	–	1	1	1	1	–	2	2	2	2	–	3	3	3	3	–	4	4	4	4	OUT
	II	V	T	V	V	V	V	T	V	V	V	V	T	V	V	V	V	T	V	V	V	V	T	V	V	V	IN
		0	–	0	0	0	1	–	1	1	1	2	–	2	2	2	3	–	3	3	3	4	–	4	4	4	OUT
	III	V	V	T	V	V	V	V	T	V	V	V	V	T	V	V	V	V	T	V	V	V	V	T	V	V	IN
		0	0	–	0	0	1	1	–	1	1	2	2	–	2	2	3	3	–	3	3	4	4	–	4	4	OUT
	IV	V	V	V	T	V	V	V	V	T	V	V	V	V	T	V	V	V	V	T	V	V	V	V	T	V	IN
		0	0	0	–	0	1	1	1	–	1	2	2	2	–	2	3	3	3	–	3	4	4	4	–	4	OUT
	V	V	V	V	V	T	V	V	V	V	T	V	V	V	V	T	V	V	V	V	T	V	V	V	V	T	IN
		0	0	0	0	–	1	1	1	1	–	2	2	2	2	–	3	3	3	3	–	4	4	4	4	–	OUT

Note: T, training vector; V, validation vector; IN, input to PNN; OUT, output from PNN.

The input to the neural network is marked with IN in Table 9.9. This input is constituted from the vectors that are presented to the neural network during the validation cycle. The output of the neural network is marked with OUT in Table 9.9. This output is constituted from the responses that the neural network gives to each input vector. The neural network can return five possible responses. These responses are represented by the numbers: 0, 1, 2, 3, 4. Each number corresponds to a group. Thus, if the neural network returns the response '3', it means that the neural network has identified the input as belonging to group 3. It is apparent from Table 9.9 that the neural network was able to correctly recognize all the presented spectra. Thus, clear distinction could be established between the spectra generated by the 'pristine' case (Group 0) and the 'damage' cases (Groups 1, 2, 3, 4). In addition, clear distinction could also be determined between the spectra of various 'damage' groups that correspond to various crack positions (Group 4, 3, 2, 1 correspond to $r = 3, 10, 25, 40\,\text{mm}$, respectively). These examples have shown that the PNN approach, in conjunction with a sufficiently large features vector, can successfully identify the damage presence and its location.

It should be noted that if the number of features in the features vector is not sufficient, the neural network might give misclassifications. This situation was met when the PNN classification was performed with a 4-frequency features vector. The four frequencies used in the features vector were the strongest resonance frequencies. In this case, good classification was attained for groups 2, 3, 4, but not for group 1. Group 1 (slight damage) was misclassified with Group 0 (pristine). This misclassification problem could not be fixed by increasing the number of training vectors. However, this problem was fixed when the size of the features vector was increased from four to six.

Further in our study, we considered an extension of the features vector size by incorporating the new resonances that appeared in the damaged plates but were not present in the pristine plates. This feature plays an important role in distinguishing healthy structures from damaged structures, especially when the damage is incipient or located away from the sensor. To achieve this, we expanded the features vector to incorporate the new resonance that appeared in the damaged plates. In order to preserve dimensionality, the pristine plates vectors were zero filled as needed. Thus, the features vector size was expanded to eleven. Then, one vector from each group was used for training, and the rest for validation. The PNN algorithm was again able to correctly classify data regardless of the choice of the training vectors. Figure 9.23 presents the percent of correctly classified data vs. number of features in the input vectors of PNN. The good classification results obtained with PNN encouraged us to use this method for damage classification in aircraft structural specimens.

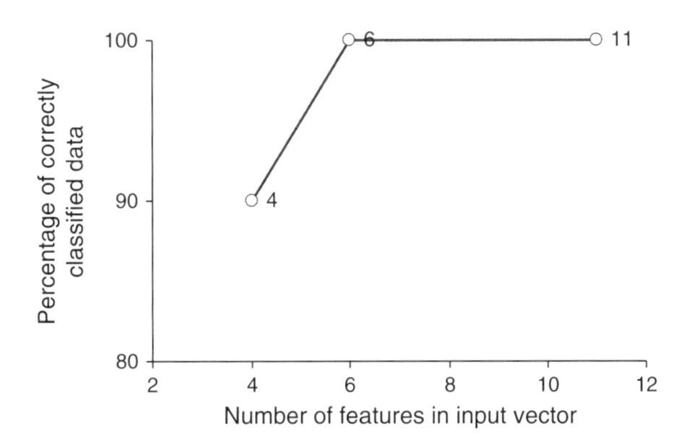

FIGURE 9.23 Number of features in input vector of PNN vs. percent of correctly classified data.

9.4.2 DETECTION OF DAMAGE PROPAGATION IN SPOT-WELDED JOINTS

The health monitoring of structural joints is a major concern. Among joining techniques, the spot welding is of great interest in a number of industries. Spot welding is the traditional assembly method for automotive bodies and has been considered in several applications in aerospace structures. The process is fast, simple, inexpensive, and therefore suitable for assembly-line mass production as used in auto industry. The resistance welding processes encompass spot welding, flash welding, percussion welding, projection welding, resistance seem welding, and others. They all share a common principle: coalescence of the faying surfaces is produced with the heat obtained from the resistance heating of the work pieces as electric current passes between the electrodes simultaneously with pressure being applied through the electrodes. The metal experiences heating, melting and solidification processes in a very short period, roughly less than one second.

In this section, we demonstrate the application of the E/M impedance technique for damage detection of a spot welded specimen. More details can be found in Giurgiutiu et al. (1999). A spot welded lap-joint shear specimen was used in this experiment (Fig. 9.24). The lap joint was constructed from dissimilar alloys, aluminum 7075-T6 and 2024-T3. This particular combination of materials chosen presents interest for the production of built-up skin-stringer structures with aerospace applications. Nominal thickness of the specimen was 2 mm (80 mil). Specimen width was 25.4 mm (1 in.) and length 167 mm (6.5 in.). The overlap length was 36 mm (1.5 in.). Spot-weld size was 9 mm (0.354 in.).

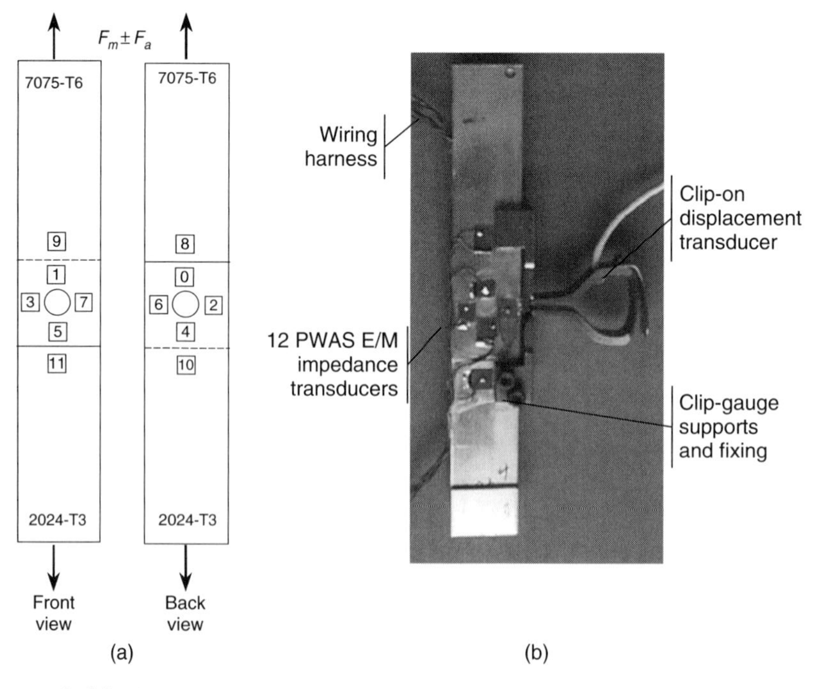

FIGURE 9.24 Spot-welded lap-joint specimen instrumented with 12 PWAS transducers: (a) schematic; (b) photograph of the actual specimen. The numbers represent the transducer stations.

9.4.2.1 Instrumentation of the Specimen

The specimen was instrumented with 12 square-shaped PWAS transducers of 6 mm (1/4 in.) size (Fig. 9.24). The PWAS transducers were manufactured in the Laboratory for Adaptive Material Systems and Structures (LAMSS), Department of Mechanical Engineering, University of South Carolina. PWAS transducer was fabricated from PZT (Lead Zirconate Titanate) single sheets supplied by Piezo Systems, Inc., Part # T107-H4ENH-602. The as-supplied PZT sheets were of dimensions 2.85 in. $\times$ 2.85 in.(72 mm $\times$ 72 mm) and had a thickness of 190 μm (7.5 mil). The sheets were cut into 6-mm strips and then into small (6 mm $\times$ 6 mm) PZT squares using proprietary methods. The small PZT squares were affixed onto 25-μm (1-mil) copper foil using specialized bonding methods. The assembled transducer was mounted onto the specimen using Micro Measurements, Inc. strain-gauge-mounting technology. The transducers were wired and numbered. Throughout the process, the electrical integrity of the transducers was measured for consistency. Rejects were dismounted and re-instrumented. Finally, support fixtures and the clip-on displacement transducer manufactured by John A. Shepic, Lakewood, Colorado, were attached. Using a Hewlett Packard 4194A Impedance Analyzer, the E/M impedance signatures of the 12 PWAS transducers affixed to the specimen were taken and stored in the PC as baseline information. The frequency range 200 to 1100 kHz was determined as best suited for this process.

9.4.2.2 Fatigue Loading and Damage Generation

The spot-welded lap-joint specimen was mounted into an MTS 810 Material Test System. Tension–tension fatigue testing at $R = 0.1$ and a max load of 2.67 kN was performed. (The typical ultimate load for the spot weld specimens is approximately 8 kN). Under this fatigue loading, the specimen fatigue life does not exceed 45 000 cycles.

In the spot-welded lap-joint specimen, fatigue cracks develop as follows: A surface crack initiates at the weld nugget/base metal interface in the 7075-T6 half of the weld specimen. Then, the surface crack grows around the periphery of the weld nugget whereas at the same time penetrates through-the-thickness the sheet thickness. After the crack penetrates the sheet, it extends in the same manner as a through crack in a center-cracked plate. At the load levels used in this study, the great majority of fatigue life is consumed before the crack penetrates the sheet thickness. In some cases, overload fracture might occur before the crack penetrates the sheet thickness. Visible light optical fractography showed the general shape of the fatigue crack in one-half of a fractured overlap-shear spot weld specimen. The initiation site is on the original faying surface of the welded specimen, and the black line separates the fatigue failure from the overload fracture.

Damage quantification and control was performed using stiffness-damage correlation principle. It has long been known that a direct correspondence exists between stiffness loss in a fatigue specimen and damage progression under repetitive (fatigue) loading. A direct relationship can be established between stiffness reduction and damage progression in materials under cyclic loading. Hence, dynamic characterization can be implemented during fatigue testing to estimate the extent of crack progression and the remaining life of the structure. Although spot-weld fatigue tests are typically used to develop S-N data, it was surmised that fatigue damage could be monitored by observing changes in specimen stiffness as a function of number of fatigue cycles. We generated controlled damage through fatigue loading and stiffness monitoring. Real-time monitoring of specimen stiffness was done using the load signal from the MTS force gauge and the displacement signal from the clip-on gauge placed across the spot weld (Fig. 9.24b). The load signal and the displacement signal are processed using a fatigue crack growth test control

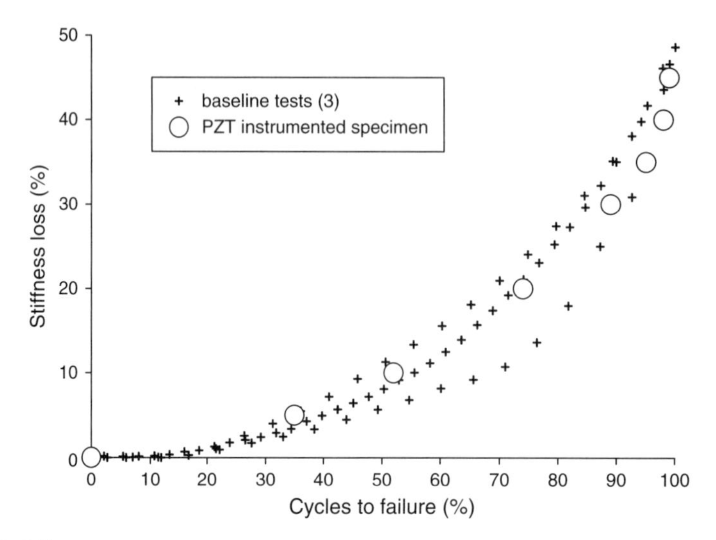

FIGURE 9.25 Graph of percentage stiffness loss vs. percentage cycles to failure for lap-shear spot weld specimens tested at $P_{max} = 2.7\,kN$ and $R = 0.1$.

and data acquisition program originally designed to monitor specimen compliance for determination of crack length in da/dN-ΔK testing. The result is a nearly continuous record of specimen stiffness as a function of fatigue life. While both spot-weld fatigue life and initial spot-weld stiffness exhibit significant scatter, a simple normalization procedure can be used to collapse all the data from tests performed under a single set of conditions. By dividing the instantaneous values of stiffness loss and fatigue cycles, normalized by the initial stiffness and by the fatigue cycles to failure, respectively, we generated '% stiffness loss' and '% cycles to failure'. After normalization, the results from several fatigue tests fell into a narrow scatter band (Fig. 9.25).

This forms the basis of the stiffness-damage correlation principle indicating that a one-to-one correspondence can be established between stiffness loss in the specimen and accumulated cycles to failure (i.e., damage). Thus, by monitoring stiffness loss, we could actually monitor and control the amount of damage accumulating in the fatigue specimen.

9.4.2.3 E/M Impedance Results

The stiffness-damage correlation principle was used to identify and control the damage progression in the spot-welded lap-joint shear specimen during the fatigue testing. The percentage stiffness loss was normalized by dividing the instantaneous values of stiffness loss and fatigue cycles, normalized by the initial stiffness and by the fatigue cycles to failure, respectively. During the test, the loading was stopped to collect health-monitoring data at predetermined damage (stiffness loss) values. This was achieved by monitoring the stiffness during fatigue cycling and stopping the experiment when the stiffness dropped to 95, 90, 80, 70, 65, 60, 55% of the initial value. These data points correspond to 5, 10, 20, 30, 35, 40, 45% stiffness loss.

The recorded data has shown two distinct groups, one from PWAS transducer 1 and the other from the rest of the transducers. PWAS transducer 1 shows significant changes during the test, and the rest showed no apparent changes. Figure 9.26a presents the superposed plots of impedance signatures of PWAS transducer 1 for increasing amounts of structural damage. Examination of Fig. 9.26a indicates that significant changes took place in the E/M impedance signature as damage progressed through the specimen.

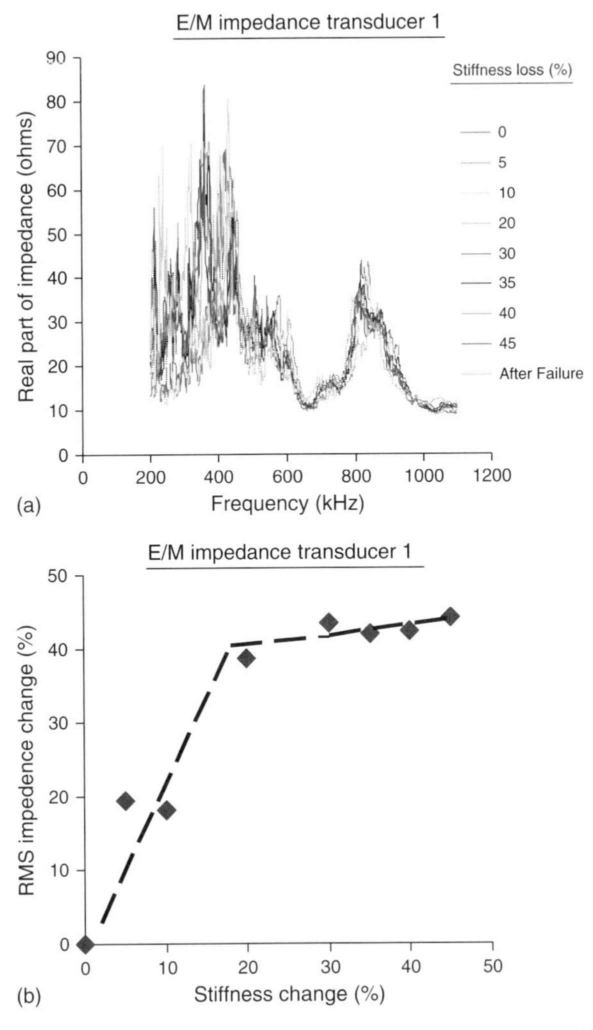

FIGURE 9.26 (a) Superposed plots of impedance signatures for increasing stiffness loss amounts, (b) correlation between RMS impedance change and specimen stiffness loss.

New frequency peaks appeared at approximately 250 and 300 kHz, whereas the peak at 400 kHz was greatly accentuated. Figure 9.26b gives a plot of the RMS impedance change versus percentage stiffness loss. The RMS impedance change calculated as:

$$\text{RMS impedance change (\%)} = \left[\frac{\sum_N \left(\operatorname{Re} Z_i - \operatorname{Re} Z_i^0\right)^2}{\sum_N \left(\operatorname{Re} Z_i^0\right)^2} \right]^{1/2} \tag{55}$$

where N is the number of sample points in the impedance signature spectrum, whereas the superscript 0 signifies the initial (baseline) state of the structure.

9.4.2.4 Explanation of the Damage Progression in the Spot-welded Joint

Failure of spot welds under tensile loading typically initiates from the heat-affected zone (HAZ) on the loading side of the spot weld. The spot-welded specimen is symmetric

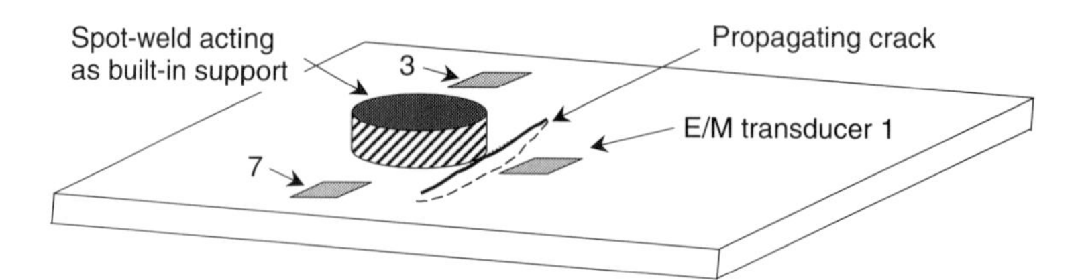

FIGURE 9.27 Representation of the bottom coupon of the spot-welded lap-shear specimen with crack propagation close to the spot weld. The crack propagation radically changes the boundary conditions in its vicinity.

with respect to the loading, and therefore, two locations have equal chance to become the crack initiation site, i.e., the region near sensor 1 on one coupon and 4 on the other. Post-test inspection of the specimen shows that a crack was formed in the HAZ near sensor 1 and advanced in the coupon next to the spot weld, as shown in Fig. 9.27. The area of influence of the boundary conditions depends on the vibration wavelength. Our tests were conducted at high frequency in the range from 200 to 1100 kHz. For a median frequency of, say, 500 kHz, the typical flexural wavelength has the value around 5 mm. Thus, the area of influence of the boundary conditions is restricted to the spot weld and crack vicinity. The only transducer present in the area of influence was PWAS 1. The rest of the transducers were outside the area of influence of the crack and did not show a distinct response.

Please also note that the RMS impedance change response of PWAS 1 shown in Fig. 9.26b flattens out after reaching the 40% mark. The explanation for this phenomenon lies in the fact that, beyond the 40% damage mark, the crack has propagated beyond the transducer proximity, and hence, its influence is not as strong as before. This explanation confirms the good localization properties of the E/M with PWAS transducers.

9.4.3 DAMAGE DETECTION IN BONDED JOINTS

The detection of damage in bonded joints is of utmost importance in many critical structures. The use of the E/M impedance for disbond detection relies on the significant change in local vibration properties of the structure at the bonded joint. When disbond takes place, the local structure transforms from a quasi monolith (the bonded state) into two distinct parts (the disbonded state). The resulting separate parts, being thinner, have a different vibration spectrum. Recall from previous chapters that the mass varies linearly with the thickness, whereas the flexural stiffness varies with the thickness cubed. Hence, the flexural vibration frequency, which varies as the square root of the stiffness to mass ratio, will vary linearly with the thickness. For example, halving the thickness will half the flexural frequency. Because the disbond under investigation is local, its extend may not exceed a few centimeters or even millimeters; the corresponding flexural frequencies are usually in the tens and hundreds of kHz range, and hence not easily detectable with conventional modal analysis equipment. However, this frequency range is well within the capabilities of the PWAS-enabled E/M impedance method. This fact makes the PWAS-enabled E/M impedance method uniquely suited for detecting incipient damage and small-scale disbonds. The following paragraphs will give a few illustrative examples of how the E/M impedance can be used to detect disbonds in several practical applications. The presentation will start with the simple example of a rectangular bonded plate element and then will evolve to the discussion of a bonded helicopter blade and a bonded space structure panel.

9.4.3.1 Bonded-Plate Rectangular Specimens

To understand the basic principles of disbond detection with the PWAS-enabled E/M impedance method, let us first consider a couple of simple experiments that illustrate the method in a simple-to-understand manner.

The *first example* uses a small rectangular specimen consisting of two identical plates bonded with a commercial adhesive (Fig. 9.28). Two small 1.55-mm thick aluminum plates (178 mm × 37 mm) were bonded using epoxy paste adhesive Hysol® EA 9309.3NA. Disbonding was simulated by producing a discontinuity in the epoxy adhesive. The discontinuity was created with a strip of Mylar® polyester film that was placed between the two plates at a mid span location. The length of the disbond was approximately 25 mm.

The specimen was instrumented with three PWAS transducers, as shown in Fig. 9.28a. The PWAS 2 was placed in the middle of the specimen above the disbonded zone. The PWAS 1 and 3 were placed above pristine zones in the adhesive layer.

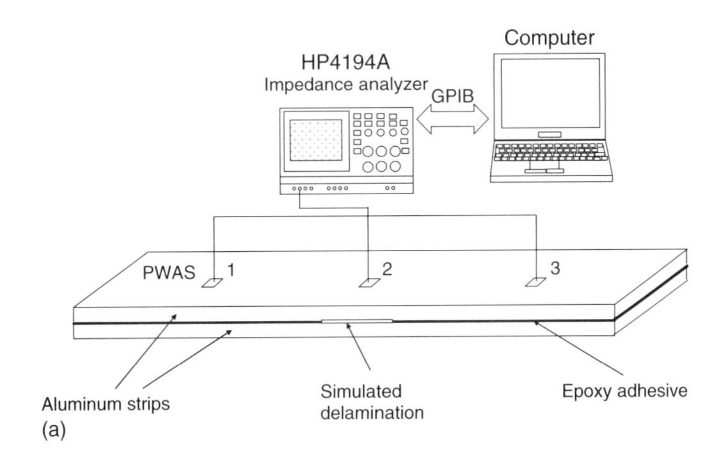

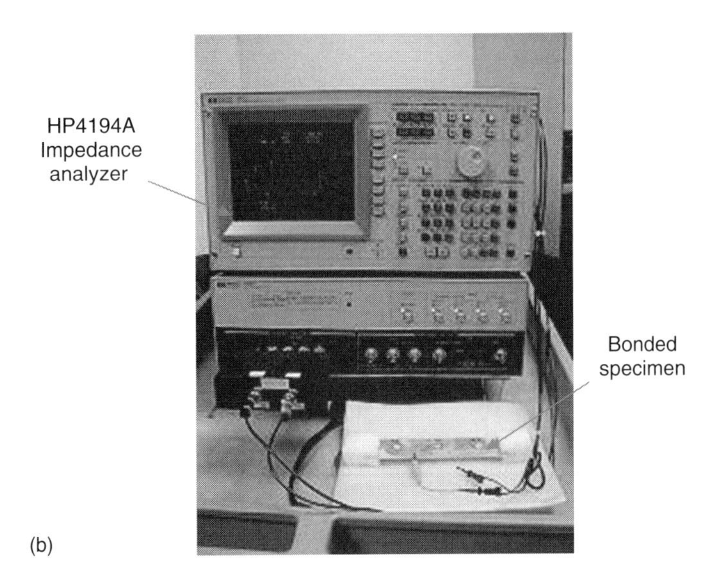

FIGURE 9.28 Instrumentation set-up for the E/M impedance method.

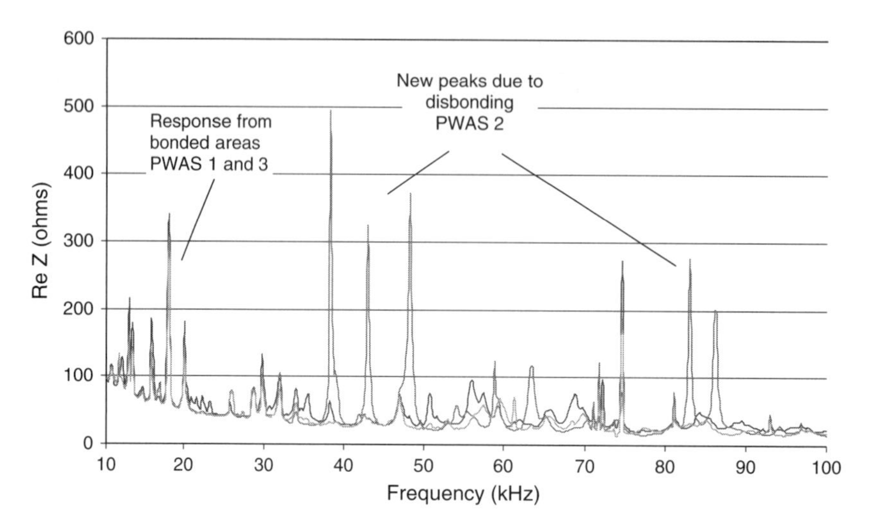

FIGURE 9.29 E/M impedance spectrum for a lap-joint specimen.

An HP4194A impedance analyzer (Fig. 9.28b) was used to measure the E/M impedance of the three PWAS transducers in the frequency range of 10 to 100 kHz. The real part of the E/M impedance spectrum recorded at the three sensors PWAS 1, PWAS 2, and PWAS 3 is presented in Fig. 9.29. It can be observed that the frequency response of PWAS 1 and 3, which are located over well-bonded areas, is very similar. In fact, as Fig. 9.29 indicates, they are actually superposed in the resonance regions. This fact indicates that the proposed method is very consistent in evaluating a good bond, and that gives consistent reading from sensor to sensor and from region to region, as long as the bond integrity status is similar.

Figure 9.29 also indicates that the spectrum of PWAS 2, which is located over the disbond, is clearly different. We see that new sharp peaks have appeared in the frequency spectrum. These new peaks represent new resonances and frequency shifts that have appeared in the spectrum because of disbonding.

The *second example* consists of an aluminum lap-joint specimen. The lap-joint specimen is more representative of actual adhesive bonding applications than the simple rectangular specimen considered in the previous example. The lap-joint specimen was fabricated from two aluminum alloy 2024-T3 strips that were partially overlapped as shown in Fig. 9.30. The two aluminum strips were 1-mm thick and 1220-mm long. One of the strip was 80-mm wide and the other was 75-mm wide. The two strips were bonded together using epoxy paste adhesive Loctite Hysol® EA 9309.3NA. The overlap of the two aluminum strips, i.e., the bonded width, was 20 mm. The specimen was instrumented with a grid of PWAS transducers as shown in Fig. 9.30b. The PWAS grid consisted of 11 columns ($i = 1, \ldots, 11$) and three rows (A, B, and C). Row A was located on the first aluminum plate, row B was located on the bond line, whereas row C was located on the second aluminum plate (C). The spacing between columns was 100 mm. The distance between rows A and B was 30 mm, whereas the distance between rows B and C was 28 mm. The PWAS that are placed on top of disbonds are B3 and B8.

An HP4194A impedance analyzer was used to measure the E/M impedance signature of the PWAS attached to the structure. Based on initial exploratory tests, the frequency range from 650 kHz to 2 MHz was selected. Measurements of the real part of the E/M impedance for several PWAS were taken. Repeated sampling of the data indicated a stable

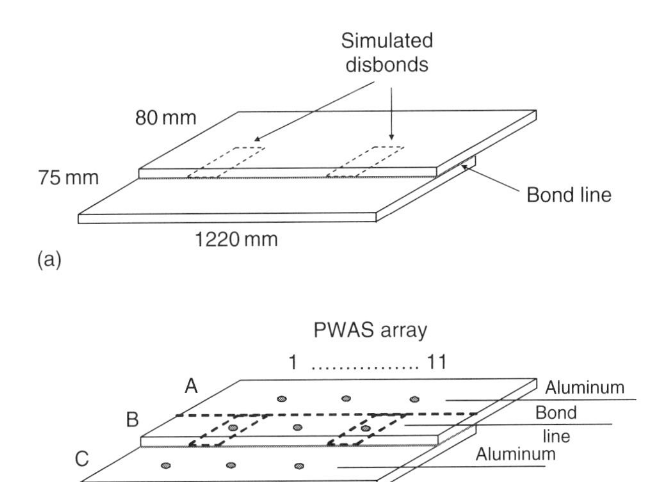

FIGURE 9.30 (a) Lap-joint specimen consisting of two aluminum strips bonded together; (b) Location of the PWAS on the lap-joint specimen.

and reproducible pattern of the impedance spectrum. The results for two measurements are presented in Fig. 9.31.

As seen in Fig. 9.31a, the resonant spectrum of a PWAS placed on the bare aluminum plate is clearly different from the resonant spectrum of PWAS placed on the bond line. This observation was consistent for all PWAS, i.e., the PWAS placed on the same row would have essentially similar E/M impedance spectrum. However, the spectra of PWAS placed on different rows were clearly different. In particular, the spectra of the PWAS placed on the bond line, whereas similar among themselves, are clearly different from the spectra of the PWAS placed on bare metal. Figure 9.31a compares the spectrum of PWAS A6 placed on bare metal with the spectrum of PWAS B6 placed on the bond line. The bare metal impedance shows a sharp peak at around 1 MHz, indicative of strong local resonance. A second peak is shown at around 1.2 MHz. This second peak is not as sharp as the first peak. A third peak is shown at around 1.5 MHz. Looking now at the spectrum of PWAS B6, which is placed on the bond line, we notice the absence of a sharp initial peak as detected on the bare metal. Instead, a rather shallow peak can be seen around the 1.05 MHz. The second peak is found around 1.55 MHz, whereas the third peak appears around 1.65 MHz. These observations indicate that the bond line structural region has higher resonance frequencies, which are attributable to the increased thickness, as well as strongly depressed peaks, which are attributable to the damping properties of the adhesive layer. The adhesive layer acts like a damper dissipating some of the vibration energy and thus reducing the peak and quality factor of the resonance peaks.

The PWAS placed on the disbond showed similar behavior. Figure 9.31b compares the E/M impedance spectrum of two adjacent PWAS placed on the bond line: PWAS B8 is placed over a disbonded region, whereas PWAS B9 is placed over a pristine bond. It is apparent that on the pristine bond (PWAS B9), the spectrum is shallow with highly damped resonance peaks, whereas the PWAS placed over a disbond (PWAS B8) has sharp resonances that can be attributed to the local release of the plate from the bond constraint.

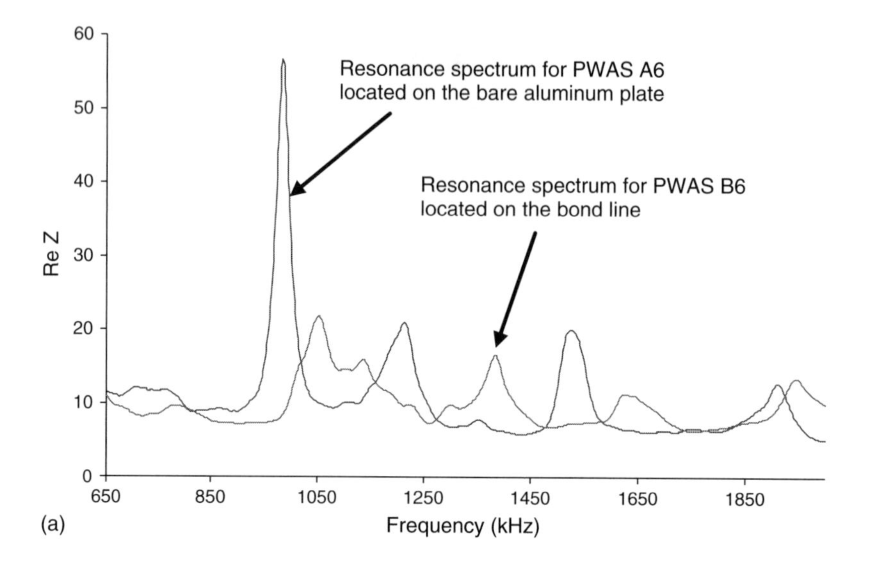

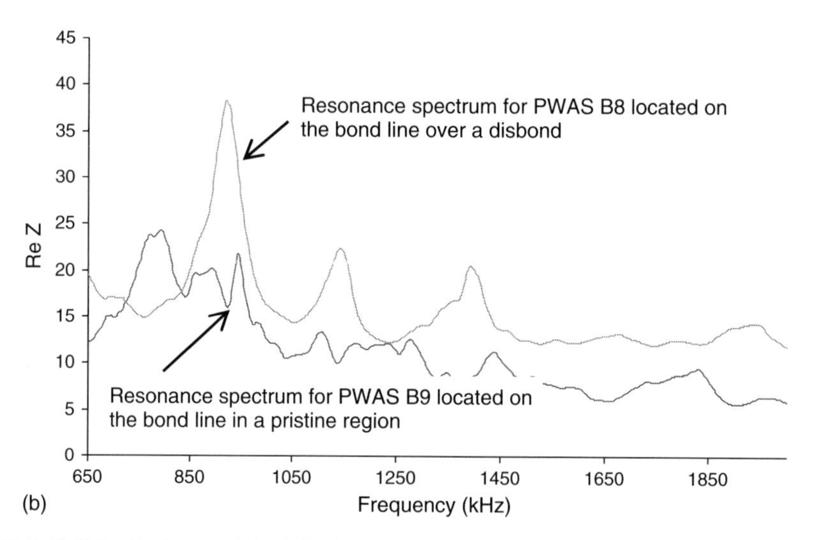

FIGURE 9.31 Real part of the E/M impedance spectrum: (a) sensor A6 (bare metal) compared with sensor B6 (bond line); (b) sensor B8 (over disbond) compared with sensor B9 (over pristine bond).

9.4.3.2 Space-structure Panels Disbond Detection with E/M Impedance Method

A metallic panel of typical space-structure construction was also tested. The specimen consists of a skin (Al 7075, 24 in. × 23.5 in. × 0.125 in.), two spars (Al 6061 I-beams, 3 in. × 2.5 in. × 0.250 in. and 24-in. long), four stiffeners (Al 6063, 1 in. × 1 in. × 0.125 in. and 18.5-in. long). The skin was attached to the two spars with several mechanical fasteners. The stiffeners were mounted on the skin with Hysol EA 9394 structural adhesive. The space-structure specimen was instrumented with several PWAS transducers, which were used to detect seeded structural damage (Fig. 9.32). Several damage-detection approaches using in-situ PWAS transducers were used. However, in this section we will only concern with the disbond-detection results. In this respect, we consider the PWAS a1, a2, and a3,

FIGURE 9.32 PWAS location on Panel 1 for the E/M impedance tests: (a) top face of Panel 1; (b) bottom face of Panel 1.

which are placed on the top stiffener, as shown in the upper part of Fig. 9.32. The PWAS a2 is place above a seeded disbond, whereas PWAS a1 and a3 are placed above pristine regions on the bonded element.

The E/M impedance method was used for disbond detection. Figure 9.33 presents the impedance spectrum for PWAS a1, a2, and a3. It can be seen that the resonant spectra of the signals from PWAS a1 and a3 located on an area with good bond are almost identical.

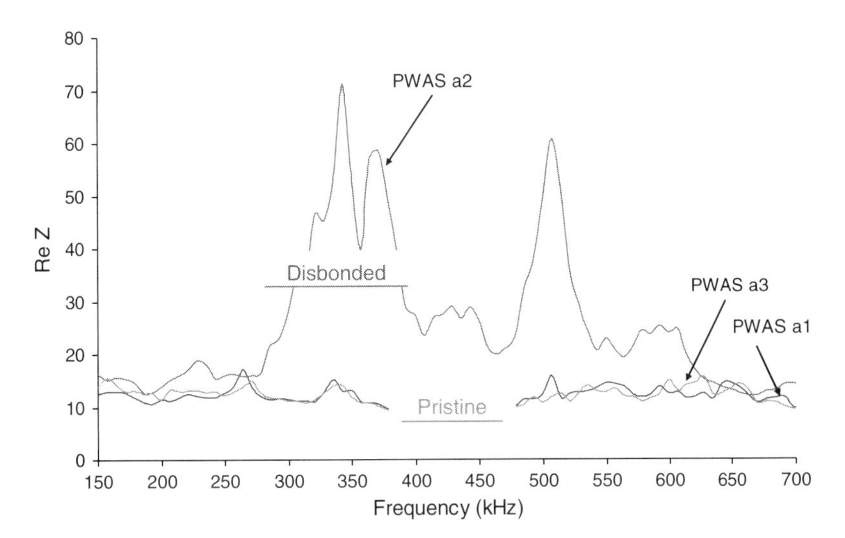

FIGURE 9.33 E/M Impedance method: resonant frequencies spectrum showing increased amplitude for the signal received at the PWAS located on the top of disbond (PWAS a2).

The resonant spectrum from PWAS a2 located on the disbond is very different showing new strong resonant peaks associated with the presence of the disbond.

9.4.3.3 Adhesively Bonded Rotor Blade Structure

Helicopter blade sections from the Apache 64H helicopter were considered. These rotor blades have a built-up construction consisting of preformed sheet-metal members adhesively bonded with high-performance structural adhesive. In-service experience with these blades has shown disbonds between the structural elements appearing because of in-flight vibrations. In our experiment, we considered a rear rotor-blade section, as shown in Fig. 9.34. The section was instrumented with several E/M impedance PWAS transducers of size 0.5 in. × 0.5 in. acting as of disbond sensors. The disbond sensors were adhesively bonded to the surface using standard strain-gauge installation procedures.

An HP4194A Impedance Analyzer was used to measure the E/M impedance signature of the disbond sensors attached to the structure. Based on initial exploratory tests, the frequency range from 100 to 750 kHz was selected. A baseline measurement of the E/M frequency response of the structure in the 'as received' condition was first recorded (Fig. 9.35, dashed line). Repeated sampling of the data indicated a stable and reproducible pattern of the impedance spectrum. Strong activity (clearly defined response peaks) was observed in the 200-kHz band. Activity of lesser amplitude also appeared in the 400-kHz and 650-kHz bands. The data was stored in PC memory as baseline signature of the structure in the 'as received' condition.

Damage was mechanically induced in the structure in the form of local disbonds. A sharp knife blade was used to induce local disbonds starting at the edge of the test section. The extent of the disbonds was about 0.5-in. spanwise, i.e., ~10% of the total bond length. The E/M impedance spectrum of the damaged structure is also shown in Fig. 9.35 (continuous line). Examination of the damage-structure E/M impedance spectra in comparison with the baseline spectra (dashed-line) reveals three important phenomena: (a) frequency shift of existing peaks; (b) increase in peak amplitudes; and (c) appearance of new peaks. The frequency shifts were toward lower frequencies. This left shift in frequency could be explained by the increase in local compliance due to disbonds. The increased impedance amplitude can be also correlated with the decrease in local damping that appears when the two faying surfaces were separated. The appearance of new peaks

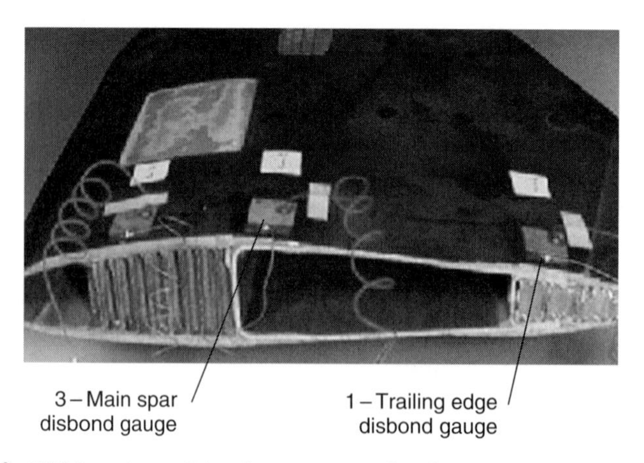

FIGURE 9.34 E/M impedance disbond sensors were placed on a rear rotor-blade section in critical areas to detect delamination between the adhesively bonded structural elements.

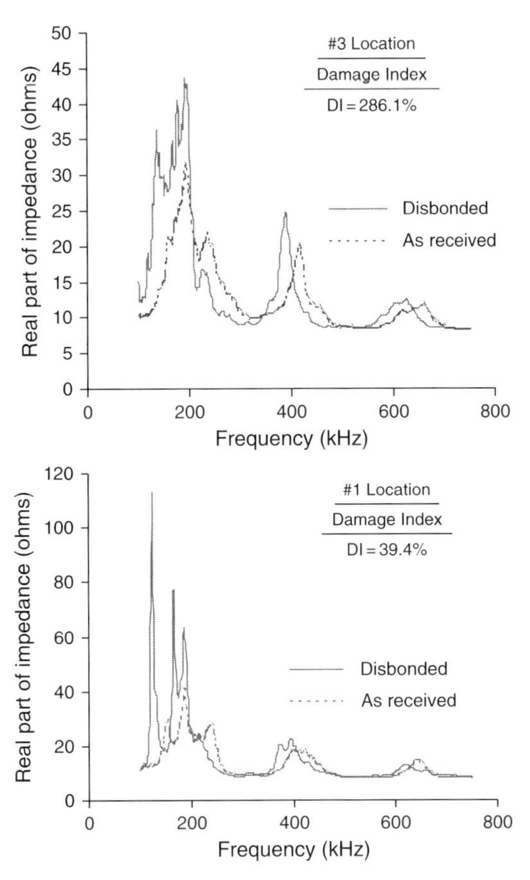

FIGURE 9.35 Comparison of the E/M impedance response curves measured for the 'as received' and 'disbonded' structure shows clear identification of the disbond: (a) spectrum of the PWAS disbond gauge at location #1; (b) spectrum of the PWAS disbond sensor at location #3.

is justified by the new local modes that are created when disbonds appear. The DI was calculated using the Euclidean norm

$$DI = \sqrt{\frac{\sum_N \left[\mathrm{Re}(Z_i) - \mathrm{Re}(Z_i^0)\right]^2}{\sum_N \left[\mathrm{Re}(Z_i^0)\right]^2}} \tag{56}$$

where N is the number of sample points in the spectrum, and the superscript 0 signifies the initial (baseline) state of the structure. The DI values are shown as text in Fig. 9.35. It can be seen that the DI has a moderate value at the location 1 (DI = 39.4%) and a higher value at the location 3 (DI = 286.1%). This difference is consistent with from the visual appearance of the E/M impedance curves. The changes in the spectrum measured at location 1 are less intense than the changes in the spectrum measured at location 3. Further work needs to be done to fully develop this E/M impedance disbond sensor. The research should concentrate on (a) Determining the sensing range of the E/M impedance disbond sensors; (b) Establishing the environmental effects on the PWAS performance (temperature and humidity); (c) Determining the correlation between PWAS size and driving voltage, on one hand, and disbond sensing range, on the other hand; and (d) Calibration of the disbond gauge in terms of minimum disbond size and minimum sensing distance as function of gauge size and driving voltage.

9.4.4 DISBOND DETECTION IN CIVIL ENGINEERING COMPOSITE OVERLAYS

This section explores the capability of PWAS transducers to perform in-situ NDE for SHM of reinforced concrete (RC) structures strengthened with fiber reinforced polymer (FRP) composite overlays. First, the principles of the disbond detection method were developed on coupon specimens consisting of concrete blocks covered with an FRP composite layer. It was found that the presence of a disbond crack drastically changes the E/M impedance spectrum measured at the PWAS terminals. It was also found that the spectra changes depend on the distance between the PWAS and the crack tip. Second, large-scale experiments were conducted on a 4.572-m long RC beam strengthened on its soffit with carbon fiber reinforced polymer (CFRP) composite overlay in the form of 51-mm wide strips. The PWAS were applied to the CFRP composite overlay at a 152-mm pitch. The beam was subject to an accelerated fatigue load regime in a three-point bending configuration up to a total of 807 415 cycles. During these fatigue tests, the CFRP overlay experienced disbonding beginning at about 500 000 cycles. The PWAS were able to detect the disbonding before they could be reliably seen by visual inspection. Good correlation between the PWAS readings and the position and extent of disbond damage was observed. These preliminary results demonstrate the potential of PWAS technology for SHM of RC structures strengthened with FRP composite overlays.

9.4.4.1 SHM of Civil Engineering Structures with Composite Overlays

The use of FRP composite overlays for structural repair and seismic strengthening applications has been studied extensively; it is now seeing its way into regular commercial applications. Composite overlays are thin sheets of fiber reinforced polymeric material adhesively bonded to conventional construction engineering materials. Fibers can be E-glass, carbon, aramid, or hybrids thereof. An FRP overlay may be applied in a wet lay-up procedure (fabrics), a dry lay-up procedure including a curing phase (prepregs), or as an adhesive application (procured laminates). Numerous experimental and analytical studies have demonstrated the ability of externally bonded FRP materials to improve the performance of reinforced concrete structural members. Improvement may take the form of enhanced load-carrying capacity, stiffness or ductility, improved performance under cyclic or repeated loading or enhanced environmental durability. Any improvement realized by an externally bonded repair system, however, is limited by the performance of the bond of the repair material to the substrate.

The degradation and loss of performance of composite overlays on concrete or masonry substrate may result from: (a) degradation of the composite overlay; (b) deterioration of the concrete substrate; and (c) loss of adhesion between the overlay and the substrate. The degradation of composite materials and the fatigue of concrete structures have been extensively researched elsewhere (see, for example, Williams, 1984, and Mallick, 1993, for composites; ACI, 1987, and Shah et al., 1995, for concrete). However, the durability of the bond between the composite and the substrate remains a critical issue. A sudden loss of bond through wide area delamination can lead to a catastrophic failure of the structure. Hence, the loss of adhesion between the composite overlay and the substrate remains a critical factor. Maintenance of good adhesion between the composite overlay and the substrate structure is of paramount importance for assuring long-term performance and for preventing failure of such structural upgrades and repairs. The loss of adhesion and performance degradation can be traced to the interface between the composite overlay and the concrete substrate. The bond degradation manifests itself in the form of disbonding cracks and delaminations. If the disbonding becomes widespread, significant loss of load-transfer capabilities can occur and structural safety may be in peril.

Bois and Hochard (2002) used the E/M impedance method to monitor damage in laminated composite structures. Applications of the E/M impedance method to the SHM of civil engineering structures were done by Ayres et al. (1996), Soh et al. (2000), Saffi and Sayyah (2001), and others. Recently, Bhalla and Soh (2003) studied the detection of earthquake-induced damage in RC structure using the E/M impedance technique, whereas Koh and Chiu (2003) performed a finite-element simulation of the disbond detection of a composite repair patch using low frequency E/M impedance and transfer function methods.

The present section presents experimental verification of the capability of PWAS to detect the disbonding of FRP composite overlays on reinforced concrete structures. Two types of experiments are presented, coupon tests and large-scale tests. The coupon tests were used to develop the disbond detection methodology based on changes in the E/M impedance spectrum in correlation with disbond cracks of various sizes. The large-scale experiments used the E/M impedance method to monitor the disbond cracks induced during the fatigue testing of a 4.572-m long RC beam strengthened on its soffit with a 51-mm wide carbon fiber (CFRP) composite strip. Good correlation between the PWAS readings and the position and extent of disbond damage was observed. These preliminary results demonstrate the potential of PWAS technology for the structural health monitoring of RC structures strengthened with FRP composite overlays.

9.4.4.2 Coupon Tests

Coupon tests were conducted to develop the disbond detection methodology based on changes in the E/M impedance spectrum in correlation with disbond cracks of various sizes. Figure 9.36 shows the coupon specimen used during these tests. The specimens consisted of a concrete substrate having an FRP composite overlay applied on its upper surface. The concrete substrate was a 51-mm × 51-mm × 178-mm concrete block fabricated in our laboratory. The FRP composite overlay consisted of glass fiber reinforced polyester (GFRP) fabricated in the laboratory from commercially available components. The composite overlay was 3.2-mm thick and 50-mm wide. During the block-casting process, two anchorage bolts (12.5-mm diameter) were inserted and set in place. These bolts attached the specimen to the test fixture and took up the reaction forces. During composite overlay fabrication, a loading hinge leaf was inserted into the left end of the

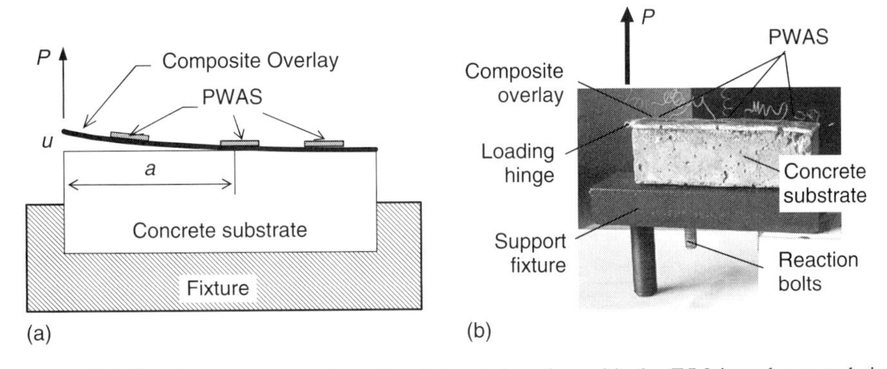

FIGURE 9.36 Coupon test specimen for disbond detection with the E/M impedance technique: (a) specimen schematic; (b) actual specimen with the PWAS disbond sensors placed at 35-mm, 90-4 mm, and 145-mm from the loading hinge.

composite overlay (Fig. 9.36b). Three PWAS were applied to the top of the composite overlay at distances of 35 mm, 90 mm, and 145 mm from the loading hinge.

The testing consisted of applying a vertical force through the loading hinge at the tip of the composite overlay and increasing it under displacement control until a crack started to propagate. The loading was performed in displacement control; hence, the start of crack propagation was accompanied by a sharp decrease in the applied force. Subsequently, the displacement was kept constant until the crack propagation process was exhausted and the crack was arrested. At this point, the crack length was recorded, measurements were taken, and the loading was resumed. With this method, we were able to grow disbond cracks of increasing lengths between the composite overlay and the concrete substrate. In total, seven cracks of increasing length were propagated (cracks 0 through 6 in Fig. 9.37). More details about the fabrication and loading of these specimens can be found in Giurgiutiu et al. (2001).

9.4.4.3 E/M Impedance Spectroscopy of the Test Coupons

Readings of the high-frequency E/M impedance of each of the three PWAS were recorded and stored for each crack length. The frequency range used during these recordings was 100–600 kHz. During data post-processing, plots of the real part of the E/M impedance were assembled. As already shown, the real part of E/M impedance, Re(Z), measured at the PWAS terminal reflects with fidelity the mechanical impedance of the structure at the PWAS location. As a crack propagates under the composite overlay, the underlying support conditions change. This change in support conditions induces a change in the resonance spectrum. Figure 9.37 shows how the E/M impedance spectrum plots of the three PWAS modify with the crack progression. The PWAS 1, which is closest to the loading end, experiences these changes first. As seen in Fig. 9.37, the pristine spectrum of PWAS 1 has two well-damped peaks, one at around 200 kHz and the other at around 390 kHz. The damping of these peaks is provided by the concrete substrate and the bonding layer. As the tip of the disbonding crack 0 reaches almost to PWAS 1, the E/M impedance changes dramatically, with the two resonance peaks amplitudes increasing very strongly and shifting to lower frequencies. The increases in the peaks amplitudes are associated with a decrease in the damping effect of the concrete substrate, which is no longer bonded to the composite in the vicinity of PWAS 1. The downward frequency shift is associated with the reduction in the local support stiffness provided by the concrete to the composite overlay. As the disbond progresses, the composite overlay tends to vibrate as a locally free plate. Because the E/M impedance test is conducted at very high frequencies, the vibration modes are highly localized, and hence very sensitive to changes in the local boundary conditions. As the cracking progressed past PWAS 1, the change in the E/M impedance spectrum becomes even stronger. This situation corresponds to crack 1 in the PWAS 1 plot of Fig. 9.37. However, once the crack tip has passed the location of the PWAS 1, no more significant changes are observed in the E/M impedance spectrum. Thus, the spectra of PWAS 1 for cracks 2 through 5 are almost identical. (No recording exists for crack 6 on PWAS 1 because the sensor was damaged.)

To quantify the damage, we used a simple DI based on the Euclidian norm between a spectrum and a baseline spectrum. The RMSD formula used was:

$$\mathrm{DI} = \sqrt{\frac{\sum_N \left[\mathrm{Re}(Z_i) - \mathrm{Re}(Z_i^0)\right]^2}{\sum_N \left[\mathrm{Re}(Z_i^0)\right]^2}} \tag{57}$$

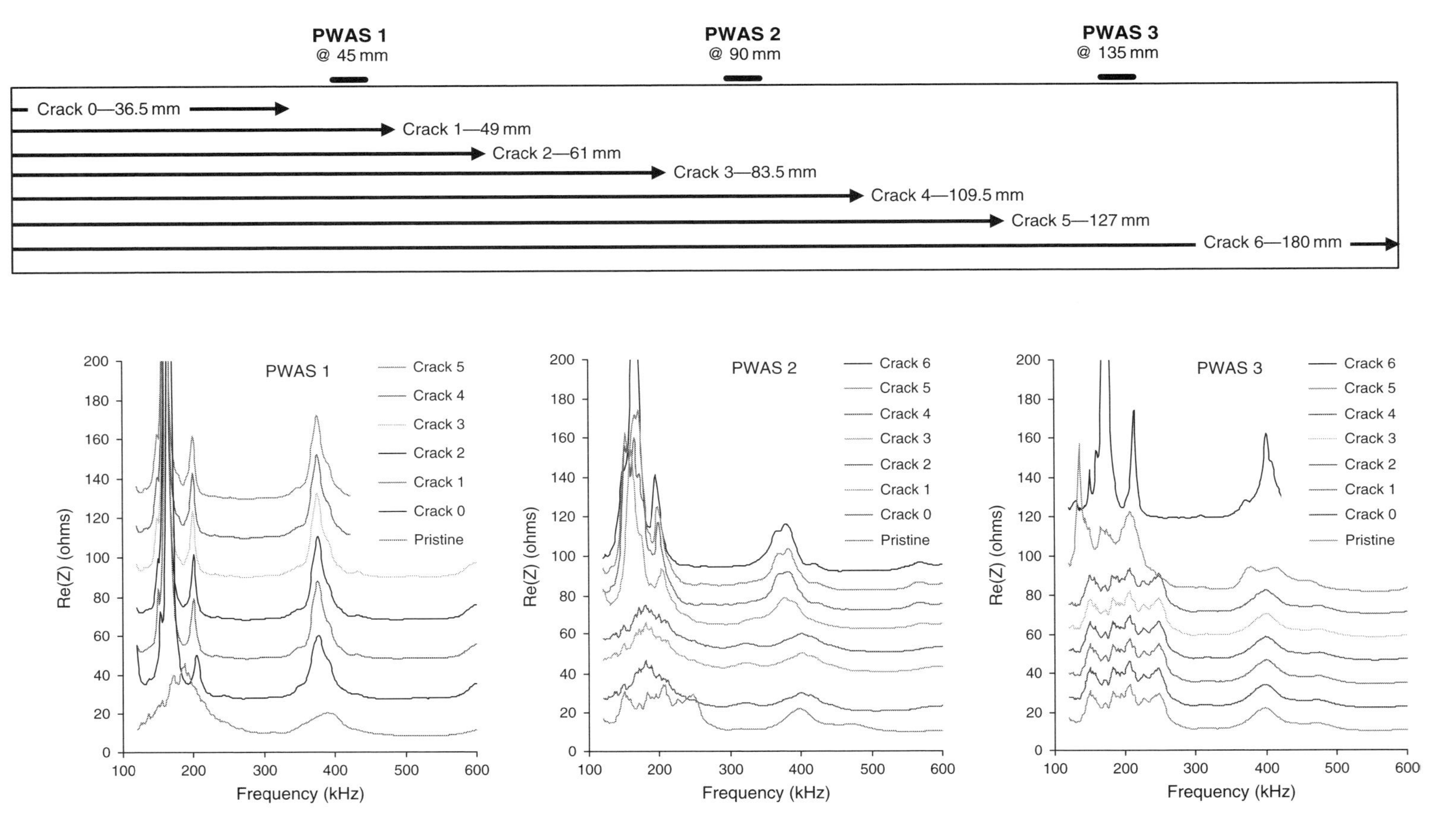

FIGURE 9.37 High-frequency E/M impedance spectra for PWAS #1 through #3 on the coupon specimen: Note that as the crack advances toward each PWAS, the E/M impedance spectrum modifies significantly (vertical shifts have been applied during plotting to allow easy examination of the spectrum shape).

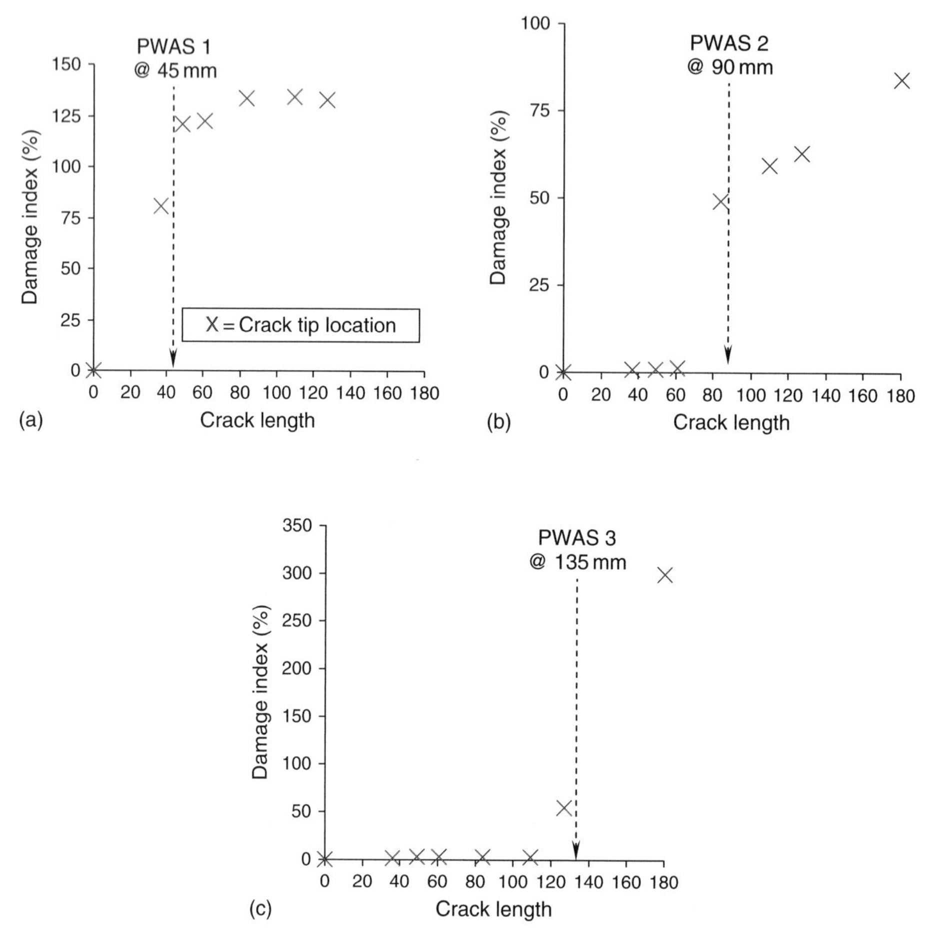

FIGURE 9.38 Progression of DI with crack length and damage location for PWAS #1 through #3. Note the dramatic change in DI as the crack tip crosses below the PWAS location.

where N is the number of points considered in the spectrum, and the superscript 0 signifies the baseline spectrum. In our work, we considered the spectrum measured on the pristine specimen before any cracks were induced to be the baseline spectrum. Figure 9.38 shows the DI plots. For PWAS #1, the DI increases rapidly at first, as the crack approaches the PWAS and crosses the PWAS location. After the crack has passed the PWAS location, the DI curve leveled off.

A similar situation is observed for the PWAS #2 and #3. Every time, large changes in the E/M impedance spectrum are observed as the crack tip approaches the PWAS location. Once the crack tip has passed the PWAS location, further changes are only marginal. For PWAS #2, these significant changes occur when the crack tip crosses the PWAS location (Fig. 9.38b). These changes are due to pronounced increases in peak amplitudes appearing at around 200 kHz and 390 kHz, simultaneous with a downward shift of the peaks (see Fig. 9.37, PWAS #2 spectrum). Similar observations can be also made for PWAS #3 (Fig. 9.38c).

These coupon tests have demonstrated that the PWAS are able to detect the disbond crack presence in their vicinity, and are insensitive to cracks in the far field. Thus, they have both crack detection and crack localization capabilities when appropriately distributed. When the cracks are far away from the PWAS, the E/M impedance spectra are practically unchanged. However, when the crack is close to the PWAS location, the detection is very strong. This detection localization property is possible only when using high-frequency vibration modes because the high-frequency vibration modes are highly localized and hence sensitive to local damage, but rather insensitive to far-field damage. PWAS allow the use of high-frequency vibration modes, something that conventional vibration methods cannot do. A damage metric expression based on the Euclidian norm and expressed as the RMSD between a baseline spectrum and the currently measured spectrum was successfully used to quantify the disbond damage intensity.

9.4.4.4 Large-Scale Tests

Large-scale tests were used to verify the disbond detection capabilities of PWAS on realistic civil engineering structures. A series of tests were conducted at the University of South Carolina to assess the strength and durability or FRP composite overlay repairs, retrofit, and rehabilitation of civil engineering structures. Part of this wide effort includes exploring sensors and methods for detecting disbonds between the FRP composite overlay and the structural concrete substrate. In particular the use of PWAS in conjunction with the E/M impedance technique to detecting disbonding was examined. Representative test results obtained during a fatigue test performed on an RC beam retrofitted with CFRP strips are presented.

The test specimen (Fig. 9.39) consisted of an RC beam with adhesively bonded preformed CFRP strips applied to its soffit. The beam was 254 mm deep, 152 mm wide; it was tested over a simple span of 4572 mm. The beam had three #4 (12.7 mm diameter) longitudinal internal steel reinforcing bars. The beam was retrofitted with a 51 mm wide by 1.4 mm thick unidirectional preformed CFRP strip (Fyfe, 2000). The CFRP was bonded to the soffit of the concrete beams using a two-part epoxy specified for the purpose and supplied by the CFRP supplier (Fyfe, 2000).

A concentrated load was cyclically applied to the beam midspan through a servo-controlled, fatigue-rated hydraulic actuator as shown in Fig. 9.39b. Sufficiently high stress values were chosen to achieve failure within a reasonable testing time. Displacement and strain readings were recorded at various locations on the beam. The beam experienced a reinforcing bar rupture between 523 and 600 kilocycles. The CFRP overlay reinforcement

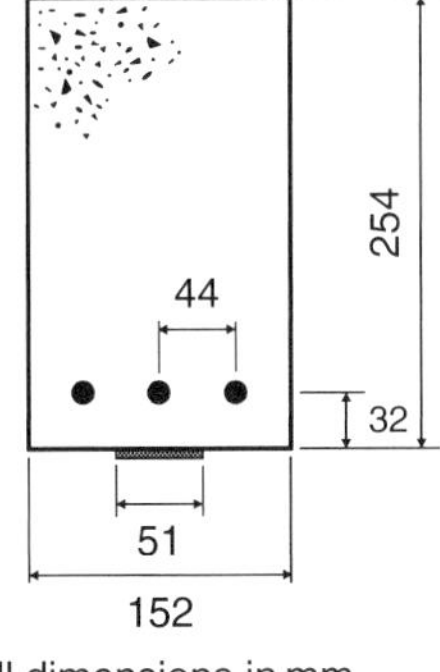

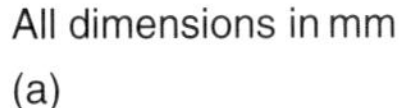

(a)

(b)

FIGURE 9.39 Reinforced concrete beam specimen setup: (a) beam details; (b) test setup.

remained intact and cycling was continued until a second reinforcing bar rupture occurred at 807 415 cycles and beam failure occurred. Post processing of strain data indicated that the strain in the CFRP composite overlay increased sharply after the rupture of the steel reinforcing bar. This increase is associated with redistribution of the load between the ruptured reinforcing steel and the composite overlay.

During the fatigue test, disbonding of the CFRP from the beam soffit was observed. Exact measurements of the extent of disbonding are virtually impossible to obtain for this type of specimen; instead, visual inspection was used to identify disbonding, as hinted by changes observed in the PWAS readings. The CFRP disbonding was initiated near the beam midspan (Fig. 9.40); it grew asymmetrically toward one beam support as fatigue loading continued. The asymmetry was biased toward the left end of the beam. The initiation of disbonding was related to the internal steel reinforcement rupture that occurred between 523 and 600 kilocycles. Visually, the first disbonds were noticed as

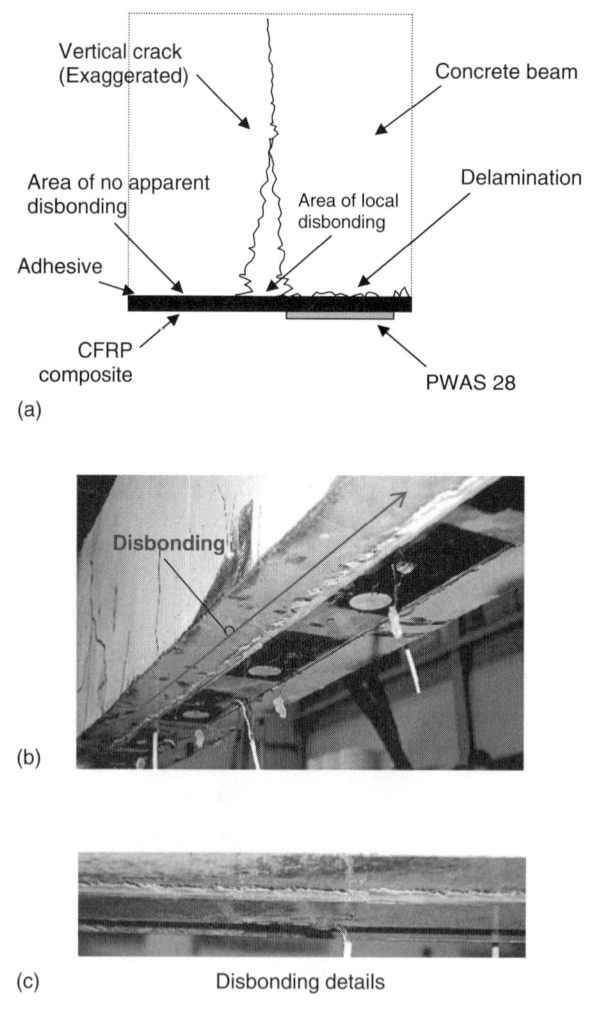

FIGURE 9.40 Disbonding of the CFRP composite overlay observed visually during the test: (a) first small disbonds appeared after rupture of internal steel reinforcement between 523 and 600 kilocycles; (b) and (c): large disbonding observed at 800 kilocycles just prior to final failure.

approximately 25-mm delaminations in the vicinity of PWAS #30 at 600 kilocycles and in the vicinity of PWAS #28 at 700 kilocycles As shown in Fig. 9.40, these disbonds were adjacent to vertical cracks in the concrete beam. As the fatigue cycles increased beyond 600 kilocycles toward the final failure at 807 415 cycles, the disbonds grew. When the rupture of the second reinforcing bar occurred at 807 415 cycles, the existing disbonds grew to such an extend that the CFRP composite overlay completely detached itself to the left end of the beam. It remained attached only on the right hand side of the beam, from its right end termination to approximately half the distance along the span.

PWAS installation and monitoring

Eighteen PWAS transducers were applied to the CFRP composite overlay after the composite overlay was attached to the RC beam soffit. The PWAS were 25-mm diameter, 0.2-mm thick, and weighed only 4 grams. The PWAS were placed along the soffit a pitch of 152-mm, symmetrical about the beam centerline (Fig. 9.41a). The PWAS installation

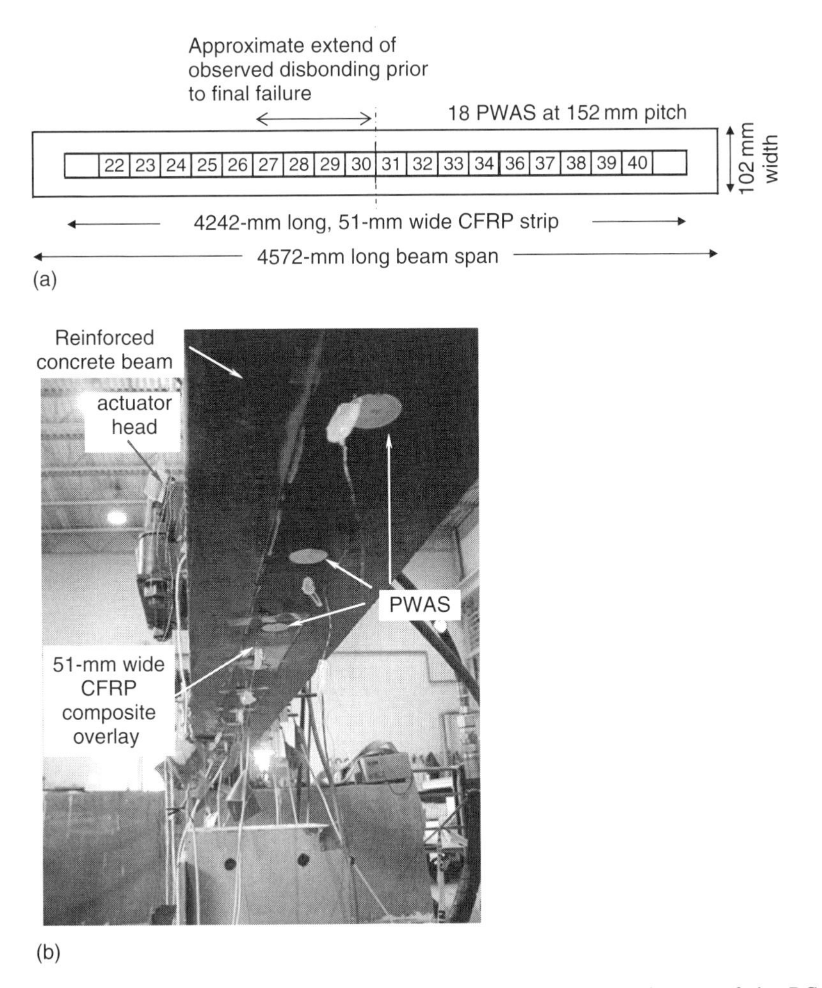

FIGURE 9.41 Installation of PWAS disbond sensors for the large-scale tests of the RC beam reinforced with a CFRP composite overlay strip: (a) location and numbering (not to scale; PWAS #35 damaged during installation, replaced with PWAS #36); (b) photo of the experimental setup showing the installation and location of the PWAS disbond sensors.

followed the procedure provided by Measurements Group, Inc. for the installation of electrical resistance strain gages. Three-meter lead wires were attached to each PWAS to allow measurement from a safe distance away from the specimen.

The E/M impedance of each PWAS was recorded with an HP4194A impedance analyzer. A custom LabVIEW program interfaced with the HP4194A was used for data collection. Baseline readings were taken with all the instrumentation in place just before the start of the loading cycles, and after the first two loading cycles ($N = 1$ and 2). Because the first loading cycle resulted in initial (expected) cracking of the beam and some settling of the specimen on the test frame, the measurements from the second cycle ($N = 2$) were retained as the baseline for future comparison.

During the fatigue test, E/M impedance readings were taken at 300, 500, 600, 700, 800 kilocycles and after the 807 kilocycles final failure. After each reading was completed, the data from each PWAS was processed and analyzed to determine if any change or shift had occurred between the previous test and the present one. If a change was detected, the specimen was inspected visually for cracking or delamination, which might correspond to the change detected by the PWAS. The purpose of the visual inspection was to develop a method of predicting, through analysis of the PWAS data, the location, direction, and rate of disbonding between the CFRP overlay and the beam substrate.

E/M impedance spectroscopy for the large-scale tests

After the fatigue test was finished, the E/M impedance data was extensively post-processed. Superposition charts of the E/M impedance spectra evolution at each sensor were produced for each PWAS. Figure 9.42 presents two typical situations. The first situation (Fig. 9.42a) is that of PWAS #28, which was located below the disbond crack. The second situation (Fig. 9.42b) is that of PWAS #33, which was not located near the disbond crack (for exact PWAS locations see Fig. 9.9). As seen in Fig. 9.42a, the PWAS #28 located above the disbond crack experienced a clear change in the E/M impedance spectrum indicative of the presence of a disbond between the FRP composite overlay and the RC beam. To the naked eye, this spectral change seems to first appear at the 600 kilocycles reading and to progress onwards. This spectral change consists of the appearance and progression of a new spectral feature at around 350 kHz. Eventually, when the disbond was fully developed subsequent to beam failure, this feature grew into a distinct peak. The associated DI calculated using Eq. (4) is presented in Fig. 9.43a. The DI curve presents three distinct sections. Below 500 kilocycles, little happens, indicative of insignificant disbond activity. However, the DI at 500 kilocycles is slightly higher than that at lower kilocycles indicating some degradation. Between 500 and 800 kilocycles, a gradual progression of the DI occurs; this can be directly correlated to the progression of the disbond crack between the composite overlay and RC beam substrate. From 800 kilocycles to the final failure at 807 kilocycles, the DI increases dramatically, indicating a complete disbond situation with no residual contact to the concrete substrate.

In contrast, one can consider the situation of the PWAS 33 that was placed in a region that experienced no disbonds. Figure 9.42b indicates that no new spectral features appeared in the E/M impedance spectrum of this PWAS. Accordingly, the associate DI (Fig. 9.43b) does not show any noticeable changes either. This situation is representative of the readings recorded on the PWAS that were placed away from disbond cracks. It indicates that the PWAS readings were not significantly affected by the large loads and the extensive load cycling applied during the fatigue test. This indirectly proves the durability and the survivability of the PWAS disbond sensors.

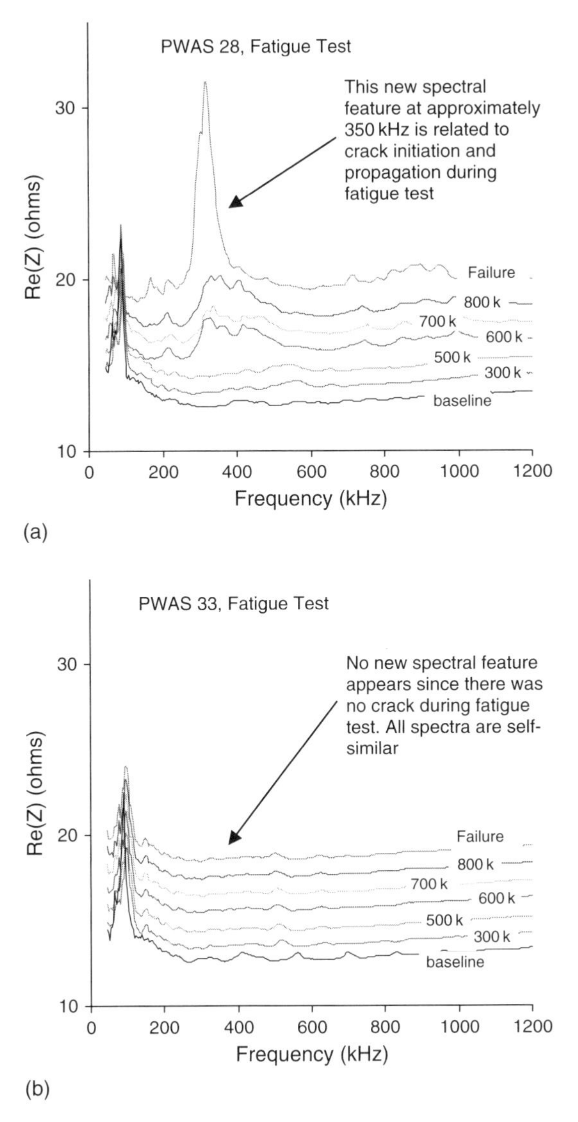

FIGURE 9.42 Experimental verification of the E/M impedance method detection of disbonds during fatigue tests on concrete beams with surface-mounted FRP composite reinforcing strips: (a) PWAS #28 near a disbond crack shows a new spectrum feature related to crack initiation and propagation; (b) PWAS #33 away from any disbond cracks does not show any change in spectral pattern. Note: for clarity, curves are shifted upwards by fixed amounts.

9.4.4.5 Section Summary

This section has presented a systematic investigation of the capabilities of PWAS to detect disbonds between a fiber-reinforced polymer (FRP) composite overlay and a concrete substrate using the E/M impedance technique. The investigation was conducted in two stages: first, coupon tests were performed to develop the disbond detection methodology based on the changes in the E/M impedance spectrum in correlation with disbond cracks of various sizes. These coupon tests have demonstrated that the PWAS are able to detect the presence of FRP disbonding in their vicinity and are insensitive to damage

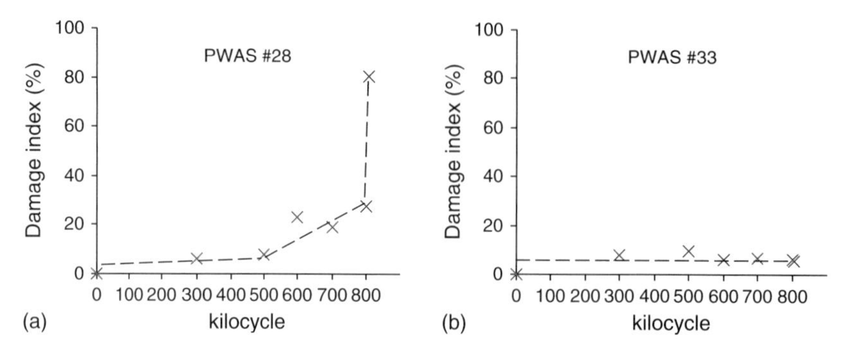

FIGURE 9.43 Progression of DI with number of kilocycles during RC beam fatigue: (a) PWAS #28 senses the presence and progression of the disbond crack through linear increase of DI between 500 and 800 kilocycle. The steep ascent between 800 kilocycles and failure corresponds to the complete wide-area disbonding of the composite overlay; (b) in contrast, PWAS #33 (which was in a beam region with no disbond crack) does not show any significant DI change.

in the far field. Thus, they have both crack detection and crack-locating capabilities. When the cracks are far away from the PWAS, the E/M impedance spectra are practically unchanged. However, when the crack is close to the PWAS location, the detection is very strong. This detection-locating property is possible only when using high-frequency vibration modes because the high-frequency vibration modes are highly localized and hence sensitive to local damage, but rather insensitive to far-field damage. A damage metric expression based on the Euclidian norm and expressed as the RMSD between a baseline spectrum and the currently measured spectrum was successfully used to quantify the disbond damage intensity.

Large-scale tests were used to verify the detection capabilities of PWAS on realistic civil engineering structures. A series of tests have been conducted at the University of South Carolina to assess the strength and durability or FRP composite overlay repairs, retrofit, and rehabilitation of civil engineering structures. Part of this wide effort includes exploring sensors and methods for detecting disbonds between the FRP composite overlay and the structural concrete substrate. In particular the examination of the use of PWAS in conjunction with the E/M impedance technique to detecting disbonding was examined. Representative test results obtained during a fatigue test performed on an RC beam retrofitted with carbon fiber-reinforced polymer are presented.

The large-scale tests proved that the PWAS could successfully detect the presence and evolution of disbond cracks that appear between the FRP composite overlay and the RC beam substrate during fatigue testing. The damage locating property of these sensors has also been verified. Good correlation between the PWAS-generated DI and the physical progression of the fatigue crack was established. In addition, it was verified that the PWAS, when applied in accordance with the methodology described in this paper, can withstand large loads and extended cycling without degradation of their sensing properties (through 807 000 cycles as demonstrated in this experiment). The technique seems also able to predict incipient damage at an early stage before being reliably observable under visual inspection.

The advantages of the PWAS technology for remotely monitoring disbond initiation and progression are apparent. However, further work is needed to establish the full understanding of this novel crack-detection technique, to map its full capabilities and possible limitations, and to ascertain its advantages and disadvantages in comparison with other disbond crack-detection methods.

9.4.5 DAMAGE DETECTION IN AGING AIRCRAFT PANELS

Realistic specimens representative of aerospace structures with aging-induced damage (cracks and corrosion) were tested. These specimens are aircraft panels with structural details typical of metallic aircraft-like structures (rivets, splices, stiffeners, etc.) The presence of such structural details complicates the structural dynamics and makes the damage detection more difficult. The specimens were made of 1 mm (0.040") thick 2024-T3 Al-clad sheet assembled with 4.2 mm (0.166") diameter countersunk rivets. Cracks were simulated with electric discharge machine (EDM) slits. In our study, we investigated crack damage and considered two specimens: pristine (Panel 0), and damaged (Panel 1). The objective of the experiment was to detect a 12.7 mm (0.5") simulated crack originating from a rivet hole (Fig. 9.44).

The panels were instrumented with eight PWAS, four on each panel (Table 9.10). On each panel, two PWAS were placed in the *medium field* (100 mm from the crack location), and two in the *near field* (10 mm from the crack location). It was anticipated that PWAS placed in a similar configuration with respect to structural details (rivets, stiffeners, etc.) would give similar E/M impedance spectra. It was also anticipated that the presence of damage would change the PWAS readings. Referring to Fig. 9.44, we observe that the PWAS S1, S2, S3, S5, S6, S7 are in pristine regions and should give similar readings, whereas S4 and S8 are in damaged regions, and should give different readings. The high frequency E/M impedance spectrum was collected for each sensor in the 200 to 550 kHz band, which shows a high density of resonance peaks. During the experiment, both aircraft panels were supported on foam to simulate free boundary conditions. The data was collected with HP 4194A impedance analyzer and then loaded into the PC through the GPIB interface.

9.4.5.1 Classification of Crack Damage in the PWAS Near-Field

The near-field PWAS were S5, S6, S7, S8. All were placed in similar structural configurations. Hence, in the absence of damage, all should give similar impedance spectra. However, S8 is close to the simulated crack originating from the rivet hole. Hence, it was anticipated that S8 should give an impedance spectrum different from that of S5, S6, S7. The change in the spectrum would be because of the presence of crack damage. Hence, we call S5, S6, S7 'pristine', and S8 'damage'. Figure 9.45a shows the superposition of the E/M impedance spectra obtained from these PWAS. Examination of these spectra reveals that sensor S8, placed next to the crack, has two distinct features that make it different from the other three spectra: (a) a higher-density of peaks; (b) an elevated dereverberated response (DR) in the 400–450 kHz range. On the other hand, the spectra of PWAS S5, S6, S7 do not show significant differences. To quantify these results, we used two methods:

1. Overall-statistics metrics of the DR
2. PNN.

Figure 9.45b shows the DR curves extracted from the Fig. 9.46a spectra. It is clear that the three DR values for the 'pristine' scenario (S5, S6, S7) are very similar. In contrast, the DR for the 'damage' scenario, S8, is clearly different. To quantify these DR differences,

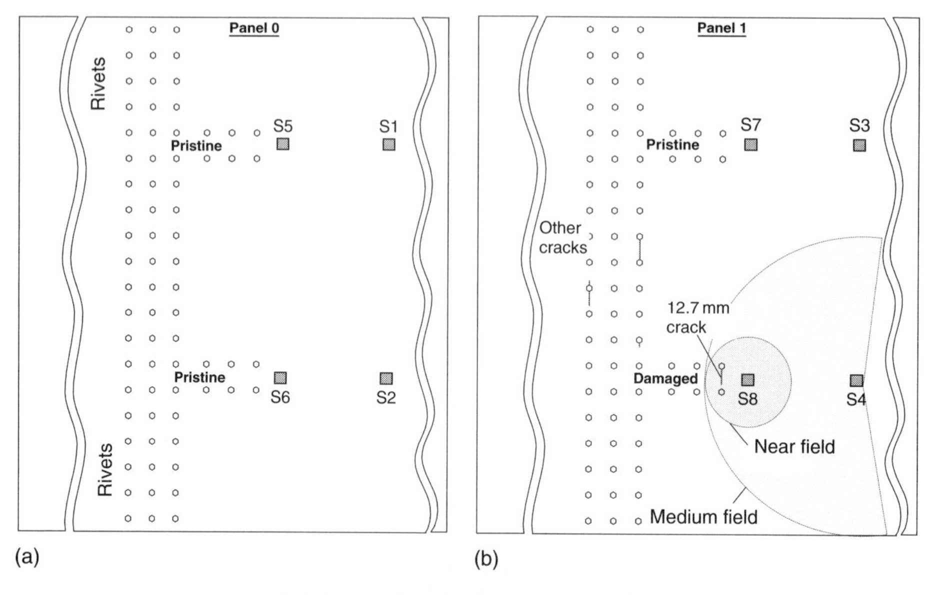

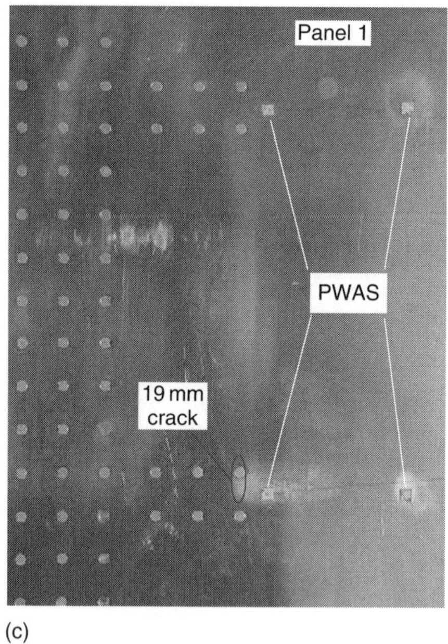

FIGURE 9.44 Schematics of the aging aircraft panel specimens and PWAS configuration: (a) panel 0, PWAS S1, S2, S5, S6; (b) panel 1, PWAS S3, S4, S7, S8; (c) actual photograph of panel 1.

TABLE 9.10 Position of PWAS on aircraft panels

	Panel 0		Panel 1	
	Pristine	Pristine	Pristine	Crack
Medium field	S1	S2	S3	S4
Near field	S5	S6	S7	S8

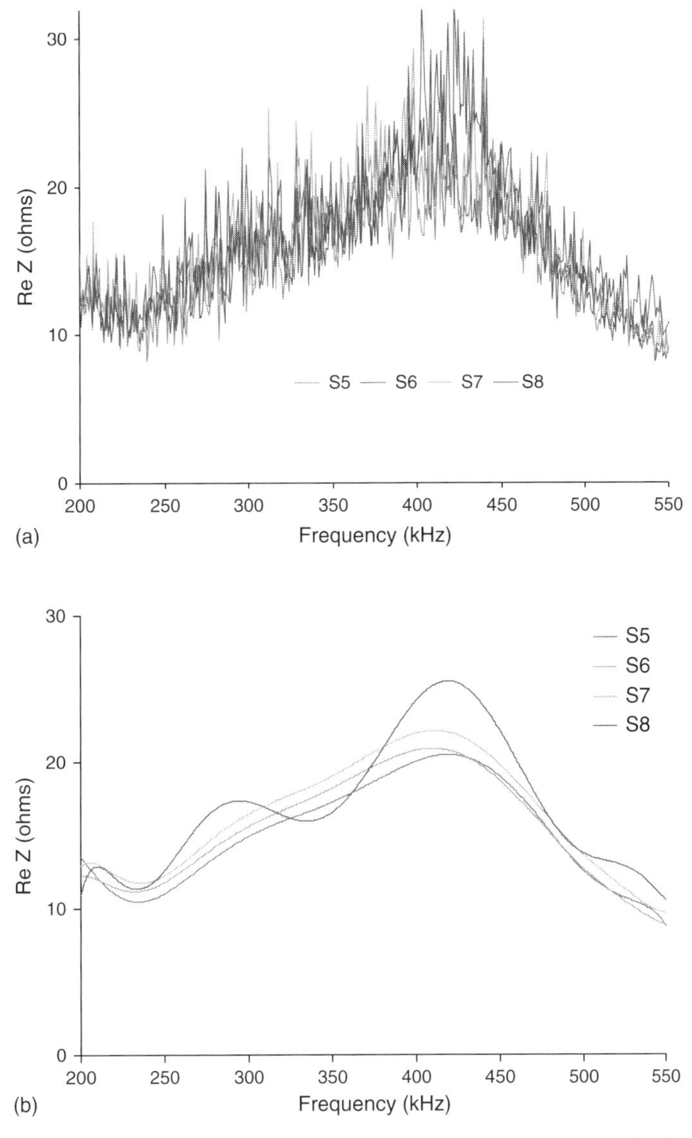

FIGURE 9.45 (a) Superposition of PWAS near-field E/M impedance spectra in the 200–550 kHz band; (b) dereverberated response.

we used the overall-statistics damage metrics defined by Eqs. (51)–(53), i.e., RMSD, MAPD, CCD. The results of this analysis are presented in Table 9.11.

Two sets of results are presented: (a) pristine vs. pristine, and (b) damage vs. pristine. The former is used to quantify the statistical differences between members of the same class, i.e., the 'pristine' PWAS S5, S6, S7. The latter is used to quantify the differences between the 'damage' sensor S8, and any of the 'pristine' PWAS, S5, S6, S7. Then, in each set, the mean value was calculated. Examination of Table 9.11 indicates that the RMSD and MAPD values for the 'damage' case are almost double that for the 'pristine' case. This indicates good damage detection capability. However, the CCD values indicate an even better detection capability, because the value for the 'damage' case is an order

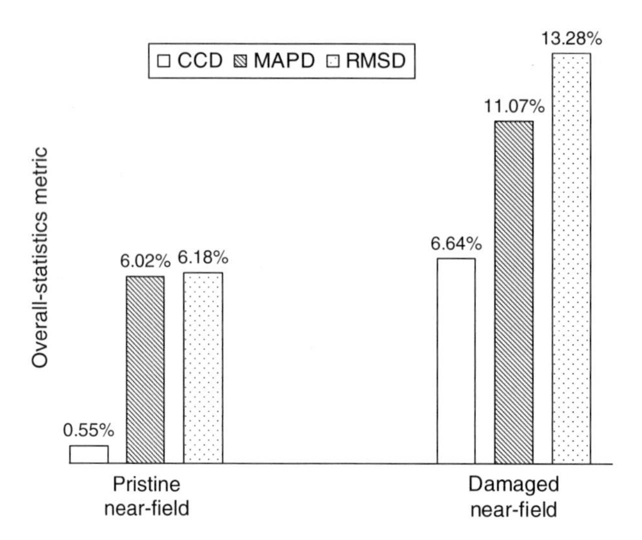

FIGURE 9.46 Overall-statistics damage metrics for comparison as calculated on the near-field spectra.

TABLE 9.11 Response of RMSD, MAPD, and CCD metrics to near-field damage (S5, S6, S7 = pristine, S8 = damage)

Class	Pristine vs. Pristine			Damage vs. Pristine		
Sensors	S5_S6	S5_S7	S6_S7	S5_S8	S6_S8	S7_S8
RMSD	4.09%	8.74%	5.71%	15.64%	14.10%	10.10%
	6.18%			13.28		
MAPD	3.75%	8.43%	5.88%	13.26%	11.89%	8.05%
	6.02%			11.07%		
CCD	0.94%	0.63%	0.07%	5.70%	7.45%	6.77%
	0.55%			6.64%		

Notes: RMSD, root mean square deviation; MAPD, mean absolute percentage deviation; CCD, correlation coefficient deviation.

of magnitude larger than for the 'pristine' case (6.64% vs. 0.55%). This confirms that CCD is potentially a very powerful damage-detection metric. The mean CCD, MAPD, RMSD values are presented graphically in Fig. 9.46. The stronger detection capability of the CCD metric is again apparent.

9.4.5.2 Classification of Crack Damage in the PWAS Medium-Field

The medium-field experiment was designed to estimate the ability of PWAS modal sensor to detect damage in a wider area. In this study, the medium field is called the area with a radius of about 100 mm where the detection of damage is still possible, but the effect of damage is not manifested as drastically as on the E/M impedance spectra in the near field. The distance between PWAS and crack for the medium-field experiment was eight times bigger than for the near-field experiment. The relative size of the near-field and medium-field of PWAS is depicted in Fig. 9.44b.

The medium field PWAS were S1, S2, S3, S4. They were located 100 mm from the first rivets in the horizontal rows of rivets. Although placed at different locations, all four PWAS were placed in similar structural situations. Hence, in the absence of damage, they should give the same spectral readings. However, S4 is not exactly in the same situation, because of the presence of the 12.7 mm simulated crack originating from the first rivet hole in Fig. 9.44. Therefore, the S4 spectrum is expected to be slightly different. Because S4 is not in the crack near field, this difference is not expected to be as large as that observed on S8 in the near-field experiments. To summarize, S1, S2, S3 are in 'pristine' situations, whereas S4 is in a 'damage' situation.

The E/M impedance spectra for S1, S2, S3, S4 are presented in Fig. 9.47. It could be noted that the spectrum of the 'damage' sensor S4 displays some higher amplitudes of some of the spectral resonances in comparison with the spectra for S1, S2, S3. On the other hand, no significant difference was observed between the S1, S2, S3 spectra, which are in the 'pristine' class. However, the changes due to damage are much slighter than those observed in the near-field experiment, and no change in the DR could be observed. In other word, the DR curves observed in the 'pristine' (S1, S2, S3) and 'damage' (S4) classes follow similar general patterns. For this reason, the analysis of the DR did not yield practical results.

To classify the medium-field spectra we used the PNN algorithm. As in the case of the circular plates, the spectral features considered in the analysis were the resonance frequencies. A feature extraction algorithm was used to obtain the features vectors. The algorithm was based on a search window and amplitude thresholds. The 48 extracted features are shown graphically in Fig. 9.48. Figure 9.48a shows the S1 results corresponding to a 'pristine' scenario, whereas Fig. 9.48b shows the S4 results corresponding to a 'damage' scenario. The resonance peaks picked by the feature extraction algorithm are marked with a cross in the data point. It can be seen that the resonance frequencies for the 'pristine' and 'damage' scenarios are different. For example, the 'pristine' scenario features several peaks above 500 kHz, whereas the 'damage' scenario does not show any peaks in this

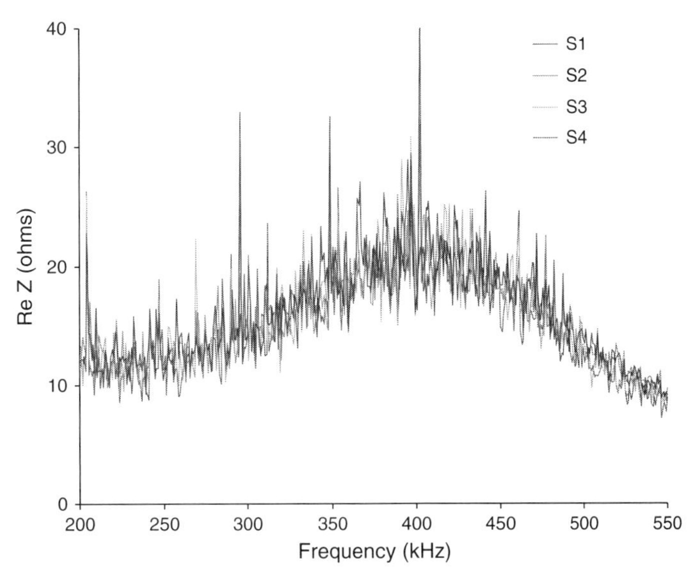

FIGURE 9.47 Superposition of PWAS medium field E/M impedance spectra in the 200–550 kHz band.

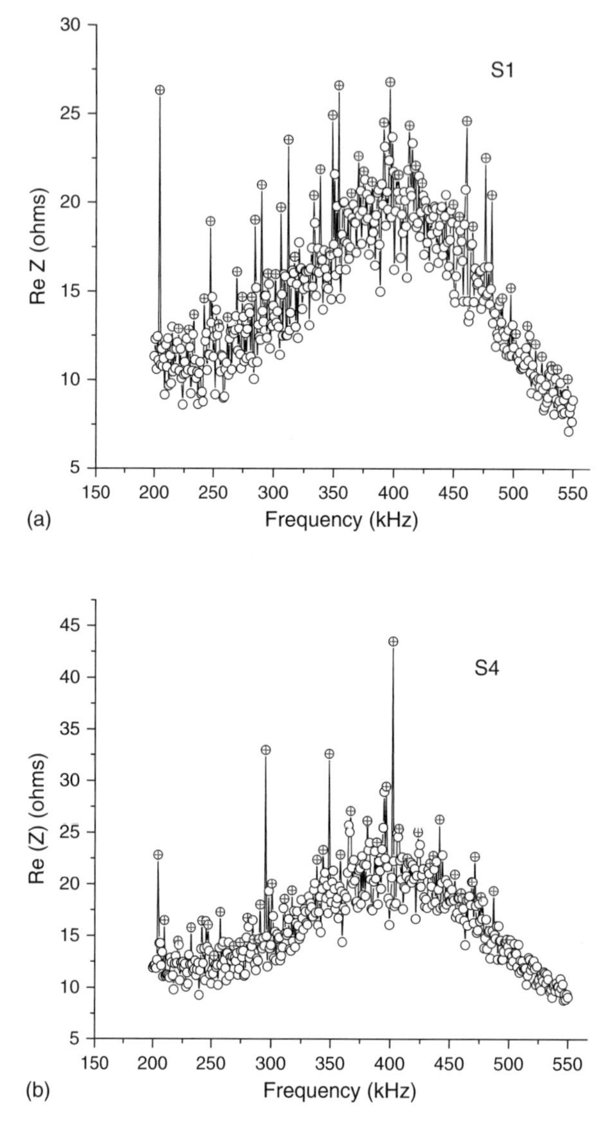

FIGURE 9.48 E/M impedance spectra for damage detection experiment in the PWAS medium field: (a) sensor S1 – 'pristine' case; (b) sensor S4 – 'damaged' case.

bandwidth. Through this process, we were able to construct four feature vectors, each 48 long, which served as inputs to the PNN. Of these, three vectors represented 'pristine' condition (S1, S2, S3) whereas the fourth vector represented a 'damage' condition (S4).

The PNN was designed to classify data into two classes: 'pristine' and 'damage'. Because we have three 'pristine' input vectors and one 'damage' input vector, we decided to use one of three 'pristine' input vectors for training and the other three vectors (two 'pristine' and one 'damage') for validation. (No training was feasible for 'damage' scenario.) This created a dichotomous situation, in which the PNN would recognize data when it belongs to the 'pristine' class and would reject it when it did not belong to the 'pristine' class. Thus, the 'damage' situation was recognized as 'non-pristine.' We emphasize that this type of classification problem often occurs in practice where the

TABLE 9.12 Damage identification in aircraft panels using probabilistic neural network classification

		Medium field				Near field				
Vector		S1	S2	S3	S4	S5	S6	S7	S8	
Test	I	T	V	V	V	T	V	V	V	IN
		–	0	0	1	–	0	0	1	OUT
	II	V	T	V	V	V	T	V	V	IN
		0	–	0	1	0	–	0	1	OUT
	III	V	V	T	V	V	V	T	V	IN
		0	0	–	1	0	0	–	1	OUT

Note: T, training; V, validation; 0, pristine; 1, damage.

response of structure in pristine condition is usually known and the response in damaged conditions is frequently unknown. The results of the PNN study indicated that the PNN was able to correctly classify data into the correspondent classes regardless of the choice of training vector for the 'pristine' class. This is indicated in Table 9.12, which presents the damage identification via PNN classification in the PWAS medium field and near field using 48-frequencies features vectors extracted from the spectra presented in Fig. 9.48. In Table 9.12, T = 'training', V = 'validation', 0 = 'pristine'; 1 = 'damage'. As shown in Table 9.12, when one pristine feature vector was used for training, the PNN was able to positively recognize the other two 'pristine' situations (0, green), and to identify the 'damage' situation (1, red). The fact that the PWAS S1 and S2 were consistently identified as 'pristine' is not surprising, because they were placed on a pristine panel. However, it is interesting to note that sensor S3, which was placed on a damaged panel, was also identified as 'pristine'. This is correct, because sensor S3, though placed on a damaged panel, is placed in a pristine location that is reasonably away from the damage location. The sensor that was more in the vicinity of the damage was sensor S4. The PNN algorithm identified the sensor S3 output as corresponding to a 'pristine' condition, and that of sensor S4 as corresponding to a 'damage' condition. This is, again, correct. This proves the detection localization capability of the PWAS-based E/M impedance method, which is only sensitive to local damage because of its high-frequency operating principle.

After successfully using the PNN method to detect damage in the medium field, we backtracked and also tried to use the PNN method for damage classification in the near field. The raw spectra obtained for the near field were processed with the features extraction algorithm using the same method as for the medium-field experiment. The PNN design, training, and validation were similar to the medium field. The PNN correctly classified data into 'pristine' and 'damage' for all training vector choices. We conclude that the classification problem for both medium field and near-field problems on aging aircraft panels can be successfully solved with the PNN algorithm.

An additional note regarding the localization property of the E/M impedance method is worth making. As seen in Fig. 9.44, Panel 1 had several other simulated cracks besides those discussed so far. However, these cracks were away from the PWAS, and hence outside their sensing range. For this reason, these other cracks did not noticeable influence the PWAS readings. The localization property of the E/M impedance method is important for finding the approximate location of the damage on the structure. Because the E/M impedance method is intended for structural health monitoring, an approximate location of the damage would be sufficient for a first alert. Further investigation of the exact damage location and of its severity would be the object of detailed NDE investigations.

Such detailed NDE investigations would be initiated once the vehicle was pulled out of service because of structural health deficiency as signaled by the SHM system.

9.4.5.3 Section Summary

The theoretical and practical aspects of the application of the E/M impedance method to the SHM of thin-wall structures using PWAS transducers were discussed. Damage-detection experiments were performed on circular plates and on aircraft panels. It was observed that the presence of damage significantly modifies the E/M impedance spectrum, which features frequency shifts, peaks splitting, and appearance of new harmonics. The rate of changes in the spectrum increases with the severity of damage. To quantify these changes and classify the spectra according to the severity of damage, two approaches were used: (a) overall-statistics damage metrics and (b) PNN. In the overall-statistics approach, RMSD, MAPD, and CCD damage metrics were considered. Through the circular plates experiments, it was found that CCD^7 damage metric is a satisfactory classifier in the high-frequency band where the resonance peaks density is high (300–450 kHz). In the PNN approach, the spectral data was first preprocessed with a features extraction algorithm. This features extraction algorithm generated the features vectors serving as input vectors to the PNN. In our investigation, we used features consisting of the numerical values of the resonance frequencies. A reduced-size features vector, containing only the four dominant resonance frequencies, permitted the PNN to classify correctly the medium and severe damage scenarios. However, the weak damage scenario gave some misclassifications. This problem was overcome by increasing the input vector size to six resonance frequencies. Furthermore, we studied the adaptive resizing of the input vector such as to accommodate the new frequencies that appeared in the spectrum when damage was present. Again, all the damage cases were properly classified.

Aging aircraft-like panels were used to study the ability of the E/M impedance method to capture incipient damage represented by a crack growing from a rivet hole. The experiments were conducted with four PWAS placed in the near field and another four placed in the medium field. Data was collected in the 200 to 550 kHz band. The overall statistics damage metrics could only be used for the near-field case, when the 'damage' spectrum showed a clear change in the DR. In this case, the CDD metric applied to the DR showed the biggest chance between the 'pristine' and the 'damage' situations. The PNN approach was able to correctly classify both near-field and medium-field spectra. The PNN inputs consisted of 48-long features vectors extracted from the spectrum with a peak-selection algorithm. The PNN had to select between four spectra (three 'pristine' and one 'damaged'). The procedure was to train the PNN on one of the 'pristine' spectra and then present the remaining three to the PNN for validation. In all cases, correct classification was obtained, and the 'damage' spectrum was recognized as 'non-pristine'. We conclude that PNN (with the appropriate choice of the spread constant) shows good potential for attacking and solving the complex classification problems associated with high-frequency spectrum-based damage detection encountered in actual applications.

The SHM tests done on aging aircraft-like specimens shows that permanently attached unobtrusive and minimally invasive PWAS, used in conjunction with E/M impedance method, can be successfully used to assess the presence of incipient damage through E/M impedance spectra classification. The E/M impedance method and the wave propagation method form complementary techniques and enabling technologies for in-situ structural health monitoring.

9.5 COUPLED-FIELD FEM ANALYSIS OF PWAS MODAL SENSORS

In previous sections, we used solely analytical models to simulate the interaction between the PWAS and the structure in order to predict the PWAS E/M impedance response during structural health monitoring. However, complicated geometries, such as encountered in real structures, cannot be modeled properly with analytical tools alone. For the analysis of complicated structural geometries, the FEM approach has gained wide-ranging acceptance in the engineering community. Could this method be also applied to predict the response of a PWAS transducer during the SHM process? The answer is *Yes*; however, the analysis is not as simple as that usually applied to simple engineering structures because it involves a coupled-field analysis in which interacting elastic, dynamic, and electrical fields have to be simultaneously analyzed. In recent years, several commercially available FEM packages have started to offer the option of coupled-field (multi-physics) elements (ANSYS, ABAQUS, etc.) In this section, we will explore how the coupled-field FEM approach can be applied to predict the E/M impedance spectrum of a PWAS interrogating a structure during the SHM process. We will start our analysis with simple models and then extended to more complicated structures.

9.5.1 COUPLED-FIELD FEM ANALYSIS OF PWAS INSTALLED ON SIMPLE STRUCTURES

To demonstrate how the coupled-field FEM approach works, we will first consider simple structures. For example, we will consider the simple example of a small metallic beam with a PWAS transducer bonded to it, as shown in our previous analysis presented at the beginning of this chapter (Fig. 9.4). Our previous analytical approach was to derive an expression for the dynamic structural stiffness $k_{str}(\omega)$ in terms of the axial and flexural modeshapes of the beam, as given in Eq. (32). Then, the function $k_{str}(\omega)$ was inserted in Eqs. (3) and (4) to predict the E/M admittance and impedance. In this section, we will first reproduce the function $k_{str}(\omega)$ numerically using the conventional FEM approach and then insert it in Eq. (3) to predict the E/M impedance. Then, we will compare the FEM results with the analytical results, as well as with experimental data. It will be shown that the conventional FEM approach gives a slightly better detection of the modal frequencies, but behaves as poorly as the analytical method in predicting the actual E/M impedance amplitude. Subsequently, we will apply the coupled-field FEM approach. This time, we will not need to use Eq. (3); rather, we will calculate the E/M impedance directly from the coupled-field FEM prediction of the behavior of PWAS voltage and current under frequency sweep. It will be shown that, through this approach, a significantly better prediction of the E/M impedance amplitude can be achieved, thus highlighting the undeniable advantage of using the coupled-field FEM approach to predict the E/M impedance spectrum of a PWAS transducer attached to a structure.

9.5.1.1 Conventional FEM Analysis of the Interaction Between a Beam Structure and a PWAS Transducer

In this section, we use the conventional FEM approach to investigate a simple beam structure having a PWAS transducer attached to its surface, as shown in Fig. 9.49. We will use FEM to calculate the dynamic structural stiffness $k_{str}(\omega)$ over a given frequency

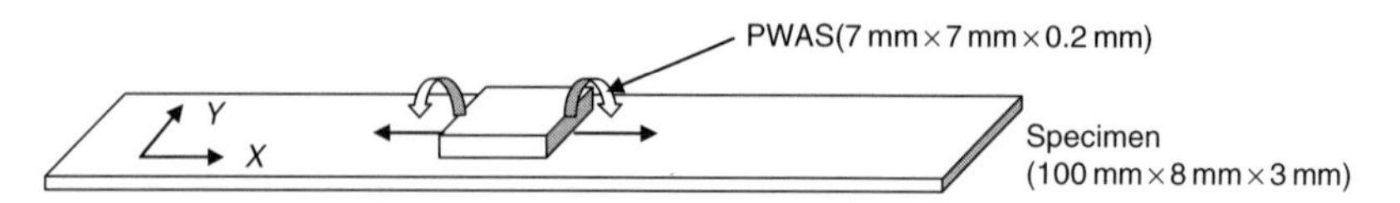

FIGURE 9.49 Modeling of the interaction between the structural beam and the PWAS transducer using a conventional FEM approach.

band. Then, we will use electromechanical coupling Eq. (3) to predict the E/M impedance at the PWAS terminals and compare it with analytical calculations and experimental data acquired with an impedance analyzer. Just like in the analytical model, we will assume the problem to be 1-D, i.e., with the force and moment interactions appropriate for the 1-D vibration of a slender beam structure aligned with the x direction.

In the 1-D situation shown in Fig. 9.49, the PWAS and the structure interact with each other through axial forces and bending moments. Ideal bonding between the PWAS and the structure is assumed; hence the PWAS–structure interaction is confined to forces acting at the PWAS ends (pin-force model). The PWAS attached to the structural surface is located offset from the beam central line; thus, surface forces at the PWAS–structure interface induce both axial forces and bending moments at the beam centerline.

The beam was modeled with ANSYS FEM program using shell elements (Fig. 9.50). To simulate the presence of the PWAS, harmonic forces and moments were applied on the structure at locations corresponding to the PWAS ends. The applied forces had unit amplitude, whereas the moments were calculated using the unit forces and the actual offset value of the PWAS location on the beam surface. The ANSYS program was run and the FRF of the structure was obtained using the harmonic analysis option. The SISO FRF is the same as the dynamic structural compliance seen by the PWAS transducer placed on the structure. The dynamic structural stiffness is the inverse of the structural compliance, i.e., the inverse of the FRF calculated with the FEM program. Thus, we obtained a numerical calculation of the dynamic structural stiffness $k_{str}(\omega)$ over the frequency band of interest.

The dynamic structural stiffness $k_{str}(\omega)$ was then used in Eq. (3) to predict the E/M impedance at the PWAS terminals. The FEM results were compared with the analytical results and experimental results, as indicated in Fig. 9.51. It clearly showed that the natural frequencies matched well with each other. In addition, the FEM approach predicts the experimentally observed double peak apparent around 27 kHz better than the analytical approach. However, none of the two methods are very good in predicting the E/M impedance amplitude in the off-resonance regions, which is way below the experimental results.

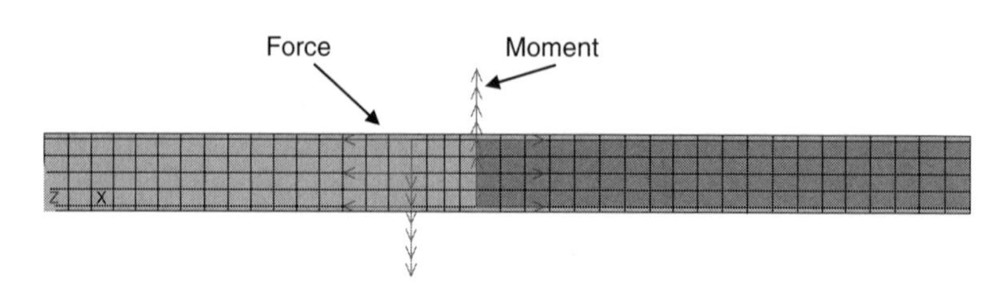

FIGURE 9.50 Meshed structure and applied loads.

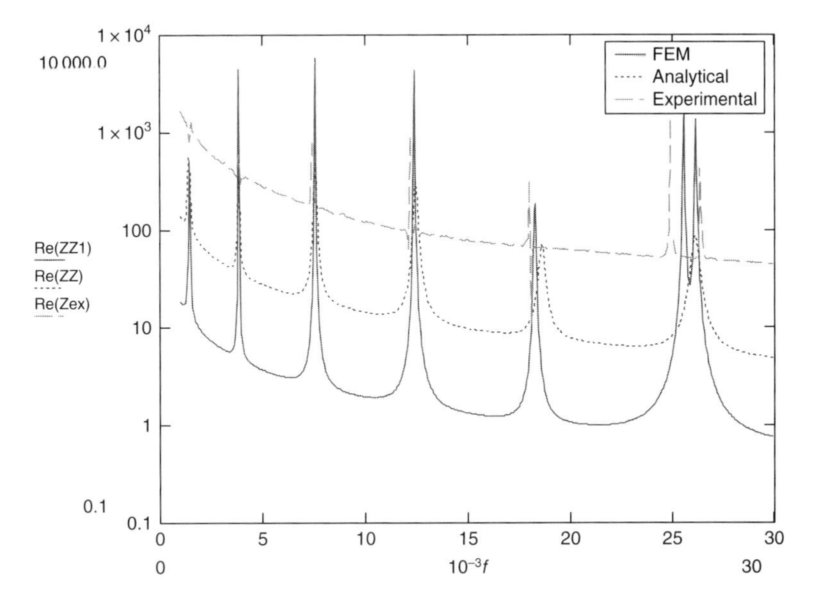

FIGURE 9.51 Comparison of real part of impedance on simple beam elements.

9.5.1.2 Coupled field 3-D Beam Structure Analysis

The 1-D analysis performed in the previous section using the conventional FEM approach has provided results that are somehow better than the analytical results; however, these results were not entirely in agreement with the experimental impedance data. This conventional FEM analysis has an important limitation: it does not include the piezoelectric properties of the PWAS in the FEM simulation and only limits itself to couple the mechanical and electrical effects through a simple formula derived under the 1-D approximation.

To overcome these shortcomings and to provide a better solution for analyzing the interaction between PWAS and structure, we will employ in this section another FEM approach, the *coupled-field FEM* approach. In the new approach, coupled-field elements are used. These piezoelectric FEM elements include both the elasto-dynamic field and the electric field in their formulation. Thus, the electric behavior is added naturally to the elastodynamic behavior and simulation is much closer to reality.

The ANSYS software provides several kinds of coupled-field elements, specifically

- PLANE13, coupled-field quadrilateral solid
- SOLID5, coupled-field brick
- SOLID98, coupled-field tetrahedron
- PLANE223, coupled-field 8-node quadrilateral
- SOLID226, coupled-field 20-node brick
- SOLID227, coupled-field 10-node tetrahedron.

The KEYOPT settings of each element activates the piezoelectric dof's, electric displacements and voltage. When we construct a structure using these elements, the coupled field analysis can be naturally implemented. Using the coupled-field FEM approach, we first analyzed the same beam structure analyzed in the previous section with conventional mechanical elements. This approach was adopted to verify that the coupled field analysis

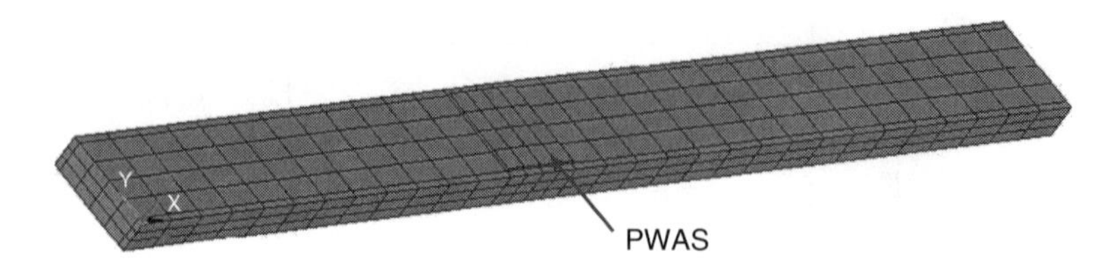

FIGURE 9.52 3-D mesh of narrow beam with PWAS.

method using coupled field elements gives credible results that are similar to the results obtained through the conventional analysis. The simulated results obtained with the coupled field analysis method were compared to previous simulation results and then with the impedance results obtained on real beams.

The PWAS transducer shown in Fig. 9.49 was modeled using SOLID5 elements with piezoelectric matrix. The PWAS was attached to the top surface of the beam as shown in Fig. 9.52. Voltage constrains was applied to the PWAS at both top and bottom surfaces representing the effect of applying a harmonic voltage to the PWAS electrodes. Harmonic analysis was performed with frequency ranging from 10 kHz to 30 kHz. The PWAS E/M impedance $Z(\omega)$ was calculated at each frequency value as the ratio V/I, where V is the voltage and I is the electric. In this calculation, the voltage V was the actual voltage potential constrain applied to the PWAS surface electrodes. The current I was calculated from the electric charge accumulated on the surface electrodes of the PWAS. The results of these calculations are presented in Fig. 9.53.

From the comparison shown in Fig. 9.53 we see that the coupled-field FEM results (denoted FEM_Coupled in Fig. 9.53) are much closer to the experimental impedance data than the conventional FEM results (denoted FEM in Fig. 9.53). These results indicate that coupled-field FEM approach can give a much better analysis of the actual interaction

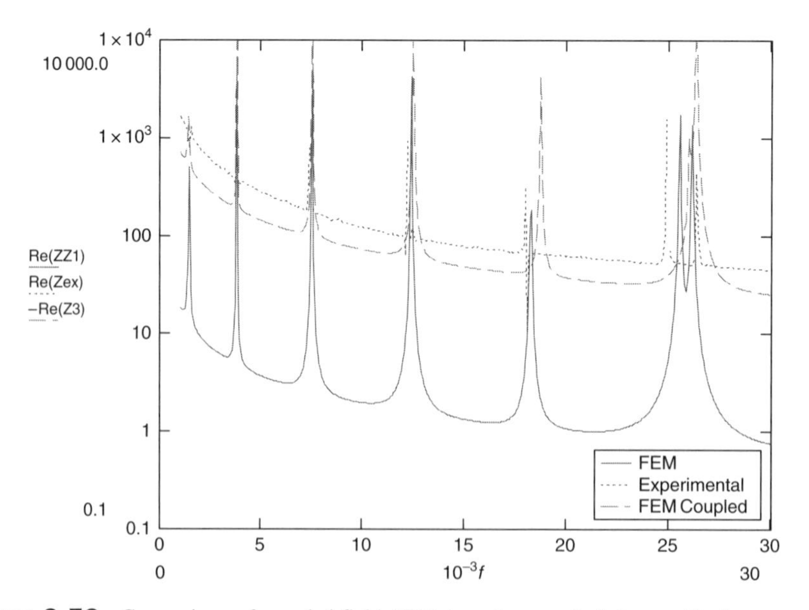

FIGURE 9.53 Comparison of coupled-field FEM impedance calculations with the conventional FEM calculations and the experimental results.

between PWAS and the structure, and can be successfully used to simulate the E/M impedance response of a structure equipped with PWAS transducers.

9.5.2 COUPLED-FIELD FEM MODELING OF TURBINE-ENGINE BLADE INSTRUMENTED WITH PWAS TRANSDUCERS

In this section, we will exemplify how the coupled-field FEM approach could be used to analyze the interaction between the PWAS transducer and a complex structure such as a titanium turbine-engine blade (Fig. 9.11). Recall that actual E/M impedance measurements of a turbine blade were experimentally obtained as presented in a previous section (Fig. 9.12). Because of the turbine-engine blade complexity, analytical methods are not able to study its frequency response and thus predict the structural dynamic stiffness and eventually the E/M impedance spectrum. However, this is possible with the coupled-field FEM approach.

To exemplify the approach, we will use a simplified model of the turbine blade that was designed and fabricated from conventional steel in the university machine shop (Fig. 9.54a). The simplified steel blade preserved the key characters of the more complex actual titanium turbine blade and thus can provide the necessary insight into how the interaction between such a complex structure and the PWAS transducer can be modeled with the coupled-field FEM approach. One of the advantages of using the simplified model is that we can fabricate several such simplified blades and introduce cracks or other damages at different positions, measure the different E/M impedance spectra, and then compare them with those resulting from the coupled-field FEM analysis. With the simplified blade model, we may also attach PWAS at different locations and study the effect of location on damage detection with the E/M impedance method.

Figure 9.54b shows the coupled-field FEM model of the simplified steel blade. As shown in the model, two PWAS are attached on the blade, one on the horizontal root surface and the other on the vertical blade surface. Following the same procedure in the narrow beam analysis, we can simulate the E/M impedance of the structure as 'felt' by the two PWAS. In performing such analysis, several issues have to be studied: (1) how exact is the modeling of the turbine blade; (2) the simulation of various damage cases of different sizes and different locations on the blade; (3) simulation of the E/M impedance changes in response to damage evaluation in the turbine blade.

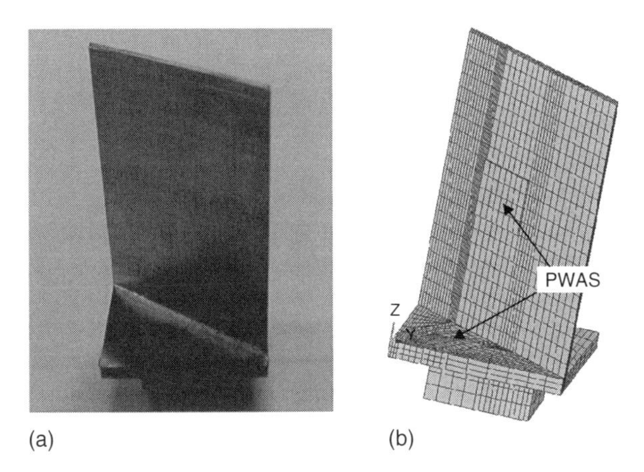

FIGURE 9.54 Simplified model of the turbine blade: (a) specimen in steel; (b) model in ANSYS.

9.6 SUMMARY AND CONCLUSIONS

This chapter has presented the use of PWAS transducers as high-frequency modal sensors. The use of PWAS transducers as high-frequency modal sensors has been made possible by the E/M impedance method, which couples the mechanical impedance of the structural substrate with the electrical impedance measured at the PWAS transducer terminals. In this way, the spectrum of the mechanical resonances of the structure is reflected in a virtually identical spectrum of peaks and valleys in the real part of the E/M impedance measured at the PWAS terminals.

To understand how the PWAS modal sensor works, we used the constrained PWAS analysis performed in Chapter 7. The constraining stiffness is actually the dynamic stiffness presented by the structure to the PWAS at their interface. In our analysis we performed the modal analysis of the structure and calculated the frequency-dependent dynamic stiffness as the inverse of the structural FRF under excitation from the PWAS. SISO conditions were considered in the calculation of the dynamic stiffness function.

The analysis started with the consideration of a 1-D structures, i.e., a simple beam. The full vibration analysis, including both flexural and axial motions, was performed, and the frequency-dependent dynamical stiffness presented by the structure to the PWAS was calculated. Several small beam cases were considered: narrow vs. wide beams; thin vs. thick beams. Subsequently, we calculated the E/M impedance and compared it with experiments performed on actual beams. Good agreement was obtained throughout.

Next, the analysis was extended to 2-D structure: a circular PWAS mounted in the center of a circular plate was considered. Full analytical modeling of plate vibration, including both flexural and axial motions, was performed. The analytical predictions were compared with experiments carefully conducted on a set of five nominally identical plates. Good agreement was obtained between the predicted and measured results. Because several plates were measured, statistical analysis could be also applied. Good consistency and repeatability, with standard deviation within 1% band was obtained.

Subsequently, the E/M impedance method was used for damage detection. First, damage detection was tested on simple 2-D geometries. Statistical sets of circular plates with progressively increasing crack damage were fabricated and tested. It was found that the E/M impedance spectra change with damage. It was also found that the spectra were consistent within each statistical set. These spectra were analyzed in order to determine a correlation between the spectral changes and damage and develop a damage metric algorithm. The classification of the high-frequency spectra resulting from the high-frequency impedance measurements was implemented with various algorithms, some based on simple overall statistics and other based on neural networks. Two approaches were used: (a) simple statistics formulae and (b) PNN. It was found that both methods give reasonable results; however, the PNN method is more predictable, although it takes more computational effort.

Several other E/M impedance damage-detection experiments were performed on various applications such as (a) damage propagation in spot welds; (b) damage detection in bonded joints; (c) disbond detection in civil engineering composite overlays; (d) damage detection in aging aircraft panels.

At the end of this chapter, the whole problem was revisited using the FEM approach. Initially, we used the conventional FEM analysis to calculate the frequency-dependent dynamic stiffness method. When this was done, we found that the FEM approach confirms the analytical results and shows only a marginal improvement on matching the experimental measurement: the resonant frequencies were well detected, but the amplitude of the E/M impedance real-part function still did not match well the measured values, especially in the off-resonance regions. Then, we did a coupled-field FEM analysis, in which we used finite elements that have both electrical and mechanical dof's. With the coupled-field analysis, the matching between prediction and experiment greatly improved. We were able to predict not only the resonance frequencies but also the amplitude the of the impedance curve in the off-resonance regions.

Another important aspect covered in this chapter is that of sensor calibration, reliability, and repeatability. These aspects are essential for the qualification of new sensor concepts. We statistically studied the intrinsic properties of the PWAS in the as-received condition (single wafers) and after adhesive attachment to the structure. Good consistency and repeatability was found in both situations. In addition, in order to prevent false readings during long-service usage of these sensors, we have explored and experimentally verified a sensor self-test procedure that is able to identify in-situ sensor disbonding through the change in the intrinsic impedance spectrum.

In summary, the E/M impedance method has proved to be a fast and reliable technique to measure the high-frequency impedance spectrum of the structure. It was demonstrated that the real part of the PWAS E/M impedance and admittance spectrum closely follows the mechanical resonance spectrum of the substrate structure. These theoretical developments were verified against experimental results, and very good agreement was obtained for 1-D and 2-D situations. Thus, PWAS transducers were identified as true modal sensors that can directly identify the mechanical resonances of the structural substrate.

Methods for structural damage detection based on the E/M impedance spectrum were examined. It was found that the E/M impedance method works best in the high kHz range. This high-frequency range is most sensitive to localized incipient damage because it can excite the high-frequency local modes. The damage detection based on the E/M impedance method utilizes the changes in high-frequency spectrum of the structure due to damage presence. Because high-frequency vibrations are energy intensive and hence localized, the damage detection with the E/M impedance method is best suited for damage detection in the sensor near field. Damage detection in the mid field is also possible, although more extensive signal processing (e.g., neural networks) is required. The E/M impedance method is insensitive to far-field damage, which presents an advantage for locating the damage position through sparse sensor networks.

Further research on the E/M impedance method should concentrate on eliminating the perturbing effects of ancillary causes that are not related to damage, but nonetheless modify the E/M impedance spectrum, such as temperature, humidity, state of sensor-to-structure bond, etc.

9.7 PROBLEMS AND EXERCISES

1. Consider a steel beam of thickness $h_1 = 2.6\,\text{mm}$, width $b_1 = 8\,\text{mm}$, length $l = 100\,\text{mm}$, and material properties given in Table 9.1. A 7-mm square PWAS ($l_a = 7\,\text{mm}$, $b_a = 7\,\text{mm}$, $t_a = 0.22\,\text{mm}$) is bonded to the beam surface at $x_a = 40\,\text{mm}$ from the left hand end. The material properties of the PWAS are given in Table 9.2. Assume 1% mechanical damping and electric loss. (i) Use the 1-D analytical expressions deduced in this chapter to calculate the admittance and impedance response in the interval 1 to 30 kHz of the PWAS attached to the structure. (ii) Plot superposed on the same chart (with appropriate scale factors) the admittance and impedance real parts and the frequency response function imaginary part. Comment on the significance of the peaks observed in these plots.

2. Consider again the data in Problem 1 above, but let either double the thickness ($h_2 = 5.2\,\text{mm}$), or wider width ($b_2 = 19.6\,\text{mm}$), or both. Use the 1-D analytical expressions deduced in this chapter to recalculate the admittance and impedance response of the PWAS in the interval 1 to 30 kHz for these other combinations of thickness and width. Discuss your results.
3. Consider a circular aluminum plate of thickness 0.8 mm, diameter 100 mm, and material properties given in Table 9.1. A 7–mm circular PWAS ($r_a = 7\,\text{mm}$, $t_a = 0.2\,\text{mm}$) is bonded in the center of the plate. The material properties of the PWAS are given in Table 9.2. Assume 1% mechanical damping and electric loss. (i) Use the 2-D analytical expressions deduced in this chapter to calculate the admittance and impedance response in the interval 1 to 40 kHz of the PWAS attached to the structure. (In these calculations use the modified Eq. 50 for impedance and its equivalent for admittance). (ii) Plot superposed on the same chart (with appropriate scale factors) the admittance and impedance real parts and the frequency response function imaginary part. Comment on the significance of the peaks observed in these plots.
4. Repeat the calculations in Problem 3 above, but exclude the axial vibration from the structural stiffness calculation. Discuss your results.
5. Repeat the calculations in Problem 3 above, but exclude the flexural vibration from the structural stiffness calculation. Discuss your results.

10

WAVE PROPAGATION SHM WITH PWAS

10.1 INTRODUCTION

This chapter deals with the use of piezoelectric wafer active sensors (PWAS) to detect structural damage using wave propagation techniques. This application of PWAS technology builds on the wealth of knowledge accumulated in the conventional nondestructive inspection (NDI), nondestructive testing (NDT), and nondestructive evaluation (NDE) fields. The main difference between our approach and conventional NDE is that our approach used permanently attached unobtrusive, minimally intrusive PWAS transducers, whereas the conventional NDE approach uses relatively large and expensive conventional ultrasonic transducers. For this reason, we see our approach as leading to an emerging new technology: *embedded ultrasonic NDE*. The embedded ultrasonic NDE will facilitate *on-demand* interrogation of the structure to determine its current state of health and predict the remaining life.

The emerging new technology of embedded ultrasonic NDE is enabled by the PWAS transducers. Because the PWAS transducers are essentially different in their operation from conventional NDE transducers (as shown in previous chapters), the development of embedded ultrasonic NDE requires the development of new ultrasonic interrogation and interpretation methods. This chapter will cover these issues and highlight new research directions that need to be explored in order to mature the embedded ultrasonic NDE concept.

After a review of the main conventional NDE ultrasonic methods, the chapter will present some simple 1-D modeling and experiments involving PWAS-generated guided waves propagating in a thin metallic strip. The PWAS transducers will be used as both transmitters and receivers. The modeling will be done with the finite element method (FEM) using (a) conventional finite elements and (b) coupled-field finite elements. In the same section, we will also present and briefly discuss pitch-catch experiments on a massive railroad track-rail specimen. In this case, due to specimen massiveness, Rayleigh surface acoustic waves will be generated, transmitted, and detected.

The next section will present and discuss 2-D wave propagation experiments performed with PWAS transducers on a thin metallic plate. The section will start with pitch-catch experiments. After presenting the experimental setup, results of the pitch-catch experiments will be discussed. Then, the section will continue with the description of pulse-echo experiments. The wave reflection signals captured by the PWAS will be subjected to multiple reflection analysis to identify how far PWAS-coupled guided waves, generated with low-power devices, will be able to travel in thin-wall structures.

The next two sections will be devoted to the PWAS-enabled embedded NDE. One section will discuss the pitch-catch embedded NDE. The other section will discuss the pulse-echo embedded NDE. The embedded pitch-catch method will be discussed in the context of detecting through-the-thickness fatigue cracks in metallic structures and delamination and disbond cracks in composite and other bonded structures. Then, the embedded pulse-echo method will be considered in metallic and composite structures. Several examples will be presented dealing with both metallic and composite specimens. The specimens under consideration will range from simple geometries to full-scale aircraft-like panels. In this way, the reader will be taken on a gradual course of understanding the principles of PWAS-based damage detection on simple geometries and then seeing how they are carried over onto complicated geometries and built-up constructions typical of practical applications.

The penultimate section of this chapter will deal with the time reversal method. This rather new technique stems from the time-reversal properties of single-mode pressure waves. In this context, the time reversal method was developed for underwater acoustics, medical treatment of kidney stones, etc. In this chapter, we will discuss the time reversal in the context of multi-mode guided Lamb waves in thin-wall structures. The time reversal method has been proposed as damage-detection technique based on the premises that nonlinearities due to damage would break down the time-reversal reconstruction of the original signal. Thus, the presence of irregularities in the time-reversal reconstruction would be indicative of damage presence. Our work will present the time-reversal theory in the case of multi-mode guided Lamb waves and will show how the time reversal process is affected by the multi-mode character of the wave field. Then, the PWAS Lamb-wave tuning concept, developed in Chapter 8, will be used to improve the time reversal process. Theoretical predictions will be compared with carefully conducted experiments.

The last subject treated in this chapter will be that of PWAS acting as passive transducers for the detection of acoustic waves generated by structural impacts (i.e., impact detection, ID) and by crack propagation (i.e., acoustic emission, AE).

10.1.1 ULTRASONIC NDE

A large number of NDE, NDT, and NDI techniques exist for identifying local damage and detecting incipient failure in critical structures. Among them, ultrasonic inspection is well established and has been used in the engineering community for several decades (Krautkramer, 1990).

Ultrasonic NDE methods rely on elastic wave propagation and reflection within the material. They try to identify the wave-field disturbances due to local damage and flaws. Ultrasonic testing involves one or more of the following measurements: time of flight (TOF) (wave transit or delay), path length, frequency, phase angle, amplitude, acoustic impedance, and angle of wave deflection (reflection and refraction). Conventional ultrasonic methods include the pulse-echo, the pitch-catch (or pulse-transmission), and the pulse-resonance techniques (Blitz and Simpson, 1996). A piezoelectric ultrasonic probe placed on the structural surface induces ultrasonic waves in the material. Good contact between the transducer and the structure is obtained by using special coupling gels. Depending on the incidence of the transducer with respect to the structural surface, the waves created in the structures may be P-waves, S-waves, or a combination of the two. P-waves are best suited for through-the-thickness detection. They are very effective in detecting anomalies along the sound path. In the pulse-echo method, defects are detected in the form of additional echoes. In the pitch-catch method, wave dispersion and attenuation due to diffused damage in the material is used as a flaw indicator.

For thickness-wise inspection, the sound path traverses only a small portion of the material volume; to achieve wide area inspection, the transducer is mechanically moved along the surface to scan and interrogate the entire structure. Ultrasonic inspection of thin wall structures (aircraft shells, storage tanks, large pipes, etc.) is a time-consuming operation that requires meticulous through-the-thickness C-scans over large areas. If done improperly, it can miss essential flaws. In addition, ultrasonic waves cannot be directly induced along the structural surface. Localized surface flaws and cracks with their plane perpendicular to the structural surface cannot be readily detected with through-the-thickness P-wave methods. One method to increase the inspection efficiency of thin-wall structures inspection is to utilize guided waves (e.g., Lamb waves) instead of the conventional P-waves. Guided waves have shown encouraging results in detecting some flaws. For the inspection of thin-wall structures, such as sheet-metal constructions, airframes, large containers, pipes, or tubes, an alternate inspection method utilizing guided Lamb waves may be more appropriate. Advanced ultrasonic techniques rely on the generation, propagation, and detection of Rayleigh, Lamb, and Love waves. These waves can be generated using conventional ultrasonic transducers and wedge couplers, provided the angle of the coupler is sufficiently large to trigger mode conversion. Alternatively, Lamb waves can be generated by comb transducers. Guided waves are being used to detect cracks, inclusions, and disbonds in metallic and composite structures. Lamb waves are appropriate for thin plate and shell structures, whereas Rayleigh waves are more useful for the detection of surface defects.

10.1.1.1 Guided Wave Methods for NDE and Damage Detection

In an infinite solid medium, elastic waves can propagate in two basic modes: pressure (P) waves and shear (S) waves. However, if the medium is bounded, wave reflections occur at the boundary and more complicated wave patterns emerge. Of particular interest are the guided waves, which remain contained in a wave-guide and can travel at large distances. Lamb waves are guided waves traveling along thin plates, whereas Rayleigh waves are guided waves constraint to the surface. Guided waves can also exist in solid and hollow cylinders, as well as in shell structures. Love waves are guided wave traveling in layered materials, whereas Stoneley waves are guided waves constrained to the material interface. The early study of guided waves is credited to Rayleigh (1887), Lamb (1917), Love (1926, 1944), and Stoneley (1924). A comprehensive analysis of Lamb waves was given by Viktorov (1967), Achenbach (1973), Graff (1975), Rose (1999), Royer and Dieulesaint (2000), and others.

Guided waves propagate along plates and shallow shells and can travel at relatively large distances with little amplitude loss. Guided waves offer the advantage of large-area coverage with a minimum of installed sensors. In recent year, a large number of papers have been published on the use of Lamb waves for NDE and damage detection (Thomson and Chimenti, 2002).

Worlton (1957, 1961) was among the first to recognize the advantages that Lamb and Rayleigh waves offer over P-waves for NDE applications. Viktorov (1967) studied the use of Rayleigh and Lamb waves for ultrasonic testing. Rose (1995) and Krautkramer (1998) reviewed the applications of portable guided wave ultrasonic devices for inspecting steam generator tubing, aging aircraft structures, and some powder-injected molded parts in manufacturing settings. Good sensitivity and range of detection were demonstrated. Recently, Rose (2002) has reiterated the inspection potential of ultrasonic guided waves for thin-wall structures. Guided Lamb waves offer an improved inspection potential over other ultrasonic methods due to the following characteristics:

- Variable mode structure and distributions
- Multi-mode character
- Sensitivity to different type of flaws
- Propagation over long distances
- Capability to follow curvature and reach hidden and/or buried part.

The potential use of guided waves for monitoring metallic aircraft structures was investigated by Alleyne and Cawley (1992) and Dalton et al. (2001). Several issues were addressed: the selection of the Lamb wave modes, the effect of skin tapering, the effect of sealants and paints, the propagation across double skin systems, and the propagation across aircraft joints. The results showed that guided waves offer good potential for structural monitoring. Long-range propagation through tapered skin presents little problem as long as dispersion is avoided. However, sealant layers can create severe damping problems. The effect of structural fasteners was not studied.

Guided Lamb waves seem to open new opportunities for cost-effective detection of damage in aircraft structures. The use of Lamb waves to detect the *corrosion* in aluminum structures using the *pitch-catch method* was explored by Chahbaz et al. (1999). An A_1 Lamb-wave mode traveling through a corroded zone ends up with a lower amplitude and longer TOF than in a pristine zone. Experiments were performed to detect corrosion around rivets and other fasteners. The pitch-catch method was also used by Grondel et al. (1999) to assess the damage progression in a riveted aircraft splice specimen during fatigue testing. The TOF changes were correlated with the appearance of microcracks and macrocracks during fatigue testing. Alleyne et al. (2001) detected corrosion in pipes with the *pulse-echo method* using the mode conversion approach. Rose (2002) outlined the inspection potential of ultrasonic guided wave for the detection of cracks, delaminations, and disbonds, and gave examples utilizing conventional angle-probe ultrasonic transducers. Light et al. (2001) studied the detection of defects in thin steel plates using ultrasonic guided Lamb waves and conventional ultrasonic equipment. De Villa et al. (2001) presented results of defect detection in thin plates using S_0 Lamb wave scanning. Conventional ultrasonic equipment consisting of wedge transmitter and bubbler receiver mounted on a scanning arm was utilized. Signal-processing methods for the determination of the arrival times and of the flaw location were explored. Flaw localization results for simulated cracks (notches) with various sizes (from 2 to 3.5 in.) and different inclinations (from 0 to 45°) were presented. Further advancements in this direction were achieved through acousto-ultrasonics (Duke, 1988).

10.1.1.2 Disbonds and Delamination Detection with Conventional Ultrasonic Methods

Ultrasonic testing of adhesively bonded joints in both aerospace and automotive applications is gaining more and more attention. Guided waves are of particular interest for the NDE of adhesively bonded joints. Guided-wave methods have considerable potential for the inspection of adherent assemblies for two reasons: they do not require direct access to the bond region and they are much more amenable to rapid scanning than are P-wave techniques. Lamb waves can be excited in one plate of a bonded assembly, propagated across the joint region, and received in the second plate of the assembly. Inspection of the joint then would be based on the differences between the signals received on one side of the assembly compared with those transmitted on the other side.

Lowe and Cawley (1994) studied the applicability of guided-wave techniques for inspection of adhesive and diffusion-bonded joints. They found that the Lamb wave techniques are sensitive to the material properties and the thickness of the adhesive layer. Rose et al. (1995) used the ultrasonic guided waves for NDE of adhesively bonded structures. They developed a double spring hopping probe (DHSP) to transmit and receive Lamb waves. This method was used to inspect a lap splice joint of a Boeing 737 aircraft. Preliminary results showed the method is capable of disbond detection. Severe corrosion was also detected using the DHSP hand-held device. Chona et al. (2003) used laser-generated Lamb waves to detect disbonds and delaminations in layered materials using a pitch-catch approach. They showed that A_0 are sensitive to damage associated with disbonds and delaminations in a multi-layer material. Lee et al. (2004) studied the problem of wave propagation in a diffusion-bonded model using spectral elements (SE) and a new local interaction simulation approach (LISA) for numerical modeling. The novelty of their work was the sensor/actuator configuration consisting of five different layers of materials with one piezoceramic element generating a thickness mode vibration. The five layers were: two piezoelectric elements used for actuating and sensing (Sonox P5), two copper layers, and in the middle a couplant layer. The experiments validated the numerical simulation, showing that the actuator/sensor configuration could operate in either S_0 or A_0 mode using an excitation frequencies of 260.5 and 100 kHz, respectively. However, the coupling layer distorts the wave propagation, due to its low impedance at the interface and low wave speed within the couplant medium.

Rose et al. (1995) used Lamb waves to detect *disbond in adhesive joints* with the pitch-catch method. In adhesive bonds, Lamb waves are able to get from one side of the bond to the other side of the bond through 'wave leakage'[1]. Two situations were considered: (a) lap splice joint disbonds detected as a loss of reception signal due to the signal no longer leaking across the splice; and (b) tear-strap disbond or corrosion observed as an increase of the reception signal due to the signal no longer leaking away in the tear strap. The use of Lamb waves for disbond detection was also reported by Singher et al. (1997), Mustafa and Chahbaz (1997), and Todd and Challis (1999). Repair patches are widely used in the aircraft industry for small repairs of the aircraft fuselage in order to extend the operational life of aging aircraft. Lamb waves can be successfully used to detect disbonds of composite repair patches (Rose et al., 1996).

Ultrasonic NDE methods for disbond detection in sandwich structures have also been investigated. Sandwich structures are characterized by high core-to-skin thickness ratio and a significant difference between the acoustic impedance of the core and that of the skin. The propagation of Lamb waves in sandwich structures have been studied as leaky Lamb waves. Rose and Agarwala (2000) studied composite sandwich plates subjected to impact damage. Experiments were conducted using conventional wedge transducers placed in a pitch-catch configuration. The guided wave test for the detection of skin to core delamination or damage detection can be simple if one chooses the correct mode and frequency: the guided-wave mode must have sufficient energy on the interface to allow energy leakage into the core. There are many examples, however, where guided wave energy remains totally in the skin, hence having the potential of sounding a false alarm. For a particular configuration, calculations and/or careful experimentation can produce the correct phase velocity and frequency values to use in the inspection. On the other hand, there can always be subtle variations in skin properties and thickness as well as in the core-to-skin bonding process. For these reasons, phase velocity-frequency tuning is needed. If one were to conduct an experiment to force energy leakage to occur during the

[1] In contrast, conventional ultrasonic P-waves cannot cross the adhesive due to impedance mismatch.

tuning process, then a received signal of about zero would indicate a good structure. If tuning did not lead to a reduction in amplitude at the receiver, one would suspect a poor bond or a damaged core. At the same time, one cannot rely on almost no signal at the receiver alone, one must tune to find points where reasonable energy would get through. If no energy can get through under any circumstance, obviously the skin itself could be excessively damaged. Regarding the skin-to-core bond, the following observations apply: (a) When the bond is good, the ultrasonic energy leaks into the core, resulting in a low amplitude signal. (b) When the bond is poor, the ultrasonic energy does not leak much into the core and the received signal has bigger amplitude.

Cost-effective and reliable damage detection is critical for the utilization of composite materials. During the late 1980s and early 1990s, work began on the application of Lamb waves to composite materials. Saravanos et al. (1994) and others have shown analytically and experimentally the possibility of detecting delamination in composite beams using Lamb waves. The sensitivity of Lamb wave propagation to fiber fracture was observed. Keilers and Chang (1995) used piezoceramics elements attached to composite beams with and without artificial delaminations to measure the beam response. Tests showed that a delamination changes the dynamic force response of the beam. Valdez (2000) performed experiments on quasi-isotropic graphite/epoxy composite specimens to detect delaminations. PWAS can be used in composite plates to form a transducer network distributed on the surface of the structure to interrogate the integrity of the plate. The A_0 mode at low frequency can be used also to interrogate a composite structure after a repair has been undertaken.

10.1.1.3 Lamb-Wave Crack Detection

Lamb-wave scattering from a damage, defect, or flaw

Lamb-wave scattering from a damage, defect, or flaw has been studied by various investigators analytically using Mindlin plate theory and numerically through finite difference method (FDM), finite element method (FEM), or boundary element method (BEM). Most of the work done to date is on circular holes or defects with a regular shape. Hinders (1996), McKeon and Hinders (1999), and Vermula and Norris (2005) studied flexural wave scattering in thin plates using Mindlin theory. For problems with more complicated geometries, the numerical methods represent the only viable approach to understanding the multiple reflections and diffractions of the ultrasonic waves within the component.

As already shown, a Lamb wave traveling in a plate has more than one mode of propagation. When the wave interacts with a structural defect it gets modified by mode conversion. Therefore, the received signal generally contains more than one mode, with the proportions of the different modes present in the wave depending on mode conversion at defects and other impedance changes. The Lamb modes are generally dispersive, which means that the shape of a propagating wave changes with distance along the propagation path. When mode conversion takes place, the geometry of the structural cross-sectional under consideration is important. If the geometry is symmetric, then only modes from the same family as the incident mode are capable to carry portions of the scattered energy; hence, for an antisymmetric-mode incident wave, the reflection will most likely be antisymmetric and the appearance of symmetric modes is very unlikely. The incident wave mode and the converted wave mode must have an inverse trend to the other, satisfying the energy conservation. This is because each of the scattered modes would share its own portion of scattered energy with the other mode through mode conversion, based on a fixed amount of incident energy.

To perform efficient damage detection, it is important to decide the mode(s) and frequency-thickness region(s) to be used. The defect sensitivity of different modes in different frequency-thickness regions is a parameter that enables the determination of the best testing regime for a particular defect type. The S_0 mode at low frequency has the characteristic of low dispersion and low leakage of energy if the plate is fluid loaded; moreover, the stresses are almost uniform through the thickness of the plate so that its sensitivity to a defect does not depend on the through-the-thickness location of the defect. Alleyne et al. (1992) and Lowe and Diligent (2002) studied the Lamb wave scattering from a notch using the FEM approach. Alleyne et al. (1992) showed the sensitivity of the S_0 mode to the depth and width of the notch. If the width is small compared with the wavelength, the transmission and reflection amplitudes are insensitive to changes in width so the ratio of notch depth to plate thickness is the controlling parameter.

The BEM approach seems a powerful tool for describing Lamb-wave mode conversion phenomena at an edge. Cho and Rose (1996) used BEM to study the mode conversion phenomena of Lamb waves at a free edge. They applied their method to the study of multi-mode reflection from a free edge in a semi-infinite steel plate. They found that S_1 is the mode that is less affected by the appearance of other modes during mode conversion. Cho (2000) studied the mode conversion in a plate with thickness variation through the hybrid BEH approach.

Grahn (2003) used a 3-D expansion approach to study the scattering problem of an incident plane S_0 Lamb wave in a plate with a circular partly through-thickness crack. He derived a linear system of equations for the expansion coefficients by projecting the different boundary conditions onto an orthogonal set of projection functions.

Lamb-wave crack detection issues in pipes, tubes, and cables

The use of guided waves for crack detection has found an immediate application in long and uniform structures such as pipes, tubing, and cables. The advantage of using guided waves is immediately apparent. Conventional across-the-thickness testing requires large pipe areas to be minutely inspected. In addition, access to the outside pipe surface requires removal of the insulation and then re-installation of the insulation after the tests. Using guided waves, a probe can be applied to the pipe at a single location, from which a considerable length of the pipe can be inspected. The insulation is only removed at the location where the probe is applied. Special probes are available for guided-wave pipe inspection. Alleyne et al. (2001) developed a ‘comb’ array probe that surrounds the pipe. Kwun et al. (1998, 2002) developed an electromagnetically coupled transducer for shear horizontal (SH) guided wave using a portable coil and a nickel foil affixed to the structure. Rose et al. (1994) studied the use of guided waves for the inspection of nuclear steam generator tubing. Reflections from circumferential cracks were obtained. Alleyne et al. (2001) used guided waves to detect circumferential cracks and corrosion in steel pipes. The properties of axisymmetric and non-axisymmetric modes were used. For circumferential cracks, it was found that the reflection coefficient is directly proportional to the circumferential extent. Thus, the reflection coefficient of a half-wall thickness notch with a circumferential extent of half a pipe diameter (16% of the pipe circumference) is approximately 5% (−26 dB). This detection sensitivity (40 dB signal-to-coherent-noise ratio) was found satisfactory for most applications. The detection of longitudinal cracks in pipes was achieved by Luo et al. (2003) using electromagnetically coupled SH (torsional) waves. Light et al. (2002) used the same method to detect through-the-thickness cracks in plates. Rizzo and Lanza di Scalea (2004) use guided waves to detect crack in tendons and suspension cables.

Lamb waves scattering from a structural defect

Because Lamb waves produce stresses throughout the plate thickness, the entire thickness of the plate is interrogated. This means that it is possible to find defects initiating at plate surface, as well as internal defects.

Lamb wave testing is complicated by the existence of at least two modes at any given frequency. It is then difficult to generate a single, pure mode, particularly above the cut off frequency-thickness value of the A_1 mode. Therefore, the received signal generally contains more than one mode, and the proportion of the different modes is modified by mode conversion phenomenon at defects and other impedance changes. The Lamb-wave modes are generally dispersive, which means that the shape of a propagating wave changes with distance along the propagation path. This makes interpretation of the signals difficult and leads to signal-to-noise ratio problems because the peak amplitude in the signal envelope decreases rapidly with distance if the dispersion is strong.

The key problem in Lamb-wave testing is the measurement of the amplitudes of the individual modes present in a multi-mode, dispersive signal. Before implementing a Lamb-wave testing technique, it is necessary to decide which mode(s) and frequency-thickness region(s) to use. Alleyne et al. (1992) studied the interaction of Lamb waves with defects in order to assess the sensitivity of different modes in different frequency-thickness regions and determine the best testing regime for a particular type of defect. An FEM numerical study of the reflection and transmission of Lamb waves in plates with and without defects was performed. The transducers under consideration were wedge transducers mounted on a steel plate. Plain-strain conditions in the xz plane were assumed. The FEM results on notches of different widths indicated that the transmission and reflection amplitudes are insensitive to changes in notch width provided the width is small compared with the wavelength. The ratio of notch depth to plate thickness $(h/2d)$ seems to be the controlling parameter.

Lowe et al. (2002) used FEM to study the reflection of S_0 Lamb-wave mode from a rectangular notch in a plate. The S_0 mode at low frequency has attractive qualities for NDE applications such as low dispersion and low leakage of energy if the plate is fluid loaded. In addition, the S_0 stresses are almost uniform across the plate thickness. Hence, the detection sensitivity does not seem to depend on the across-the-thickness location of the defect. The notch was assumed rectangular in shape (with zero width in the case of a crack), infinitely long, and normal to the direction of wave propagation. Straight-crested Lamb waves were assumed. The in-plane displacement was monitored at the mid-thickness of the plate, thus ensuring that only the symmetric S_0 propagating mode was detected, because the antisymmetric A_0 mode has zero in-plane displacement at the mid plane. The FEM simulations and the experimental measurements were in good agreement. The authors identified as an important phenomenon the interference between the reflections from the two sides of the notch, which leads to a periodic shape of the reflection as a function of notch width, exhibiting evenly spaced maxima and minima. They show that the reflection function is characterized by both the geometric ratio of the wavelength to notch width and the frequency–thickness product.

Cho and Rose (1996) used the boundary element method to study the mode conversion phenomena of Lamb waves from a free edge. The elastodynamic interior boundary value problem is formulated as a hybrid boundary integral Eq. in conjunction with the normal modes expansion (NME) technique based on the Lamb wave propagation Eq. They considered a single incident mode propagating along x_1 from boundary Γ_1. The incident displacement field of the pth mode can be expressed of amplitude α^p, the normalized displacement modal function, and the pth wavenumber k^p, i.e.,

$$u^I = \alpha^p \begin{Bmatrix} \bar{u}^p_{x_1} \\ \bar{u}^p_{x_2} \end{Bmatrix} e^{ik^p x_1} \tag{1}$$

The backscattered field induced by the pth incident wave mode could contain N independent Lamb wave normal modes determined by the incident frequency through the mode conversion Γ_2. They can be linearly superimposed on Γ_1 to express the resulting reflected fields:

$$u^R = \sum_{n=1}^{N} \beta_n \begin{Bmatrix} \bar{u}^n_{x_1} \\ \bar{u}^n_{x_2} \end{Bmatrix} e^{-ik_n x_1} \tag{2}$$

where β_n is the unknown amplitude of the nth reflected mode. The total displacement field can be defined as the superimposition of incident and reflected fields on Γ_1.

$$u = u^I + u^R \tag{3}$$

The above equation can be rewritten discretizing Γ_1 with k number of nodal points. Following the same procedure to determine the total boundary traction and substituting the total displacement field, $2k$ relations between the boundary tractions and displacements on the left-cross-sectional boundary of a waveguide Γ_1 can be defined. Once the unknown reflected amplitudes are determined, the derivation of the edge reflection coefficients of each backscattered wave mode becomes possible. The method was applied to the study of multi-mode reflection from free edge in a semi-infinite steel plate. The first case under evaluation was the mode conversion between two different modes, A_0 and A_1, taking each of them as incident mode, respectively. Only the same family modes as an incident mode are capable of carrying each portion of scattered energy from a scatterer symmetric with respect to the horizontal midplane. Therefore, there was little chance that the reflection of S_0 mode could occur in this case. The variations of the A_0 and A_1 reflection factor with frequency were studied for two cases: (a) A_0 incidence and (b) A_1 incidence. Overall, each of the reflection factor variations of the incident A_0 and the converted A_1 modes have the inverse trend to each other satisfying the energy conservation. This is because each of the scattered modes would share its own portion of scattered energy with the other mode through mode conversion, based on a fixed amount of incident energy. This behavior depends on the different mode sensitivity with respect to a different incident mode and reflector condition.

In the case of incident S_0 mode or S_1 mode, three possible scattered modes are available in this frequency range to interact with each other: S_0, S_1, and S_2. No mode conversion to any antisymmetric mode like A_0 and A_1 is possible for the reason mentioned earlier. The incident S_0 mode was found to interact mainly with S_2 rather than with S_1. The S1 reflection is stable around 0.2 ratio. The major mode conversion happens just between S_0 and S_2, and their variations represent the inverse image of the other modes to satisfy energy conservation. Unlike the other cases, the S_1 mode is nearly independent of converted mode excitations of S_0 and S_2. The incident S_1 mode is consistently dominant resulting in only minor reflections for S_0 and S_2 through the entire frequency range. For this reason, S_1 is much less affected by changes in incident frequency than other incident modes. It is apparent that S_1 is the mode that is less affected by the other modes' appearance in mode conversion.

Cho (2000) studied the mode conversion in a plate joint with thickness variation through the hybrid BEM approach. For a symmetric model, it was found that it was possible to send the mode S_0 at $fd = 2.0$ MHz-mm through a junction of structures without any significant energy loss through mode conversion, regardless of the shape of waveguide joint depicted in terms of aspect ratio s/h. Because of midplane symmetry, there is no mode conversion to antisymmetric modes. When the ratio s/h is greater than 2, the mode conversion phenomenon is almost negligible. In the case of wave scattering in a plate with nonsymmetric thickness variation and S_0 coming from the thicker plate, the mode conversion of the incident S_0 mode was found to be proportional to the slope of the inclined surface of the plate joint. The transmission of the A_0 mode increases with a decrease of the slope up to a point of $s/h = 1.8/1.0$, showing a negligible value for the case of thickness change when $s/h = 0.0/1.0$. The change of aspect ratio s/h did not have any significant influence of the mode conversion in reflected fields with incident mode S_0 at $fd = 1.0$ MHz-mm propagates in the reverse way, from a lower thickness to a higher one.

Vermula and Norris (2005) studied flexural wave scattering in thin plates using Mindlin plate theory. Mindlin plate theory includes shear-deformation and rotary-inertia effects, as in the Timoshenko beam theory. The equations of motion restrict the deformation to three degree of freedom, and they are obtained by averaging the exact equations of elasticity across the plate thickness. Hinders (1996) and McKeon and Hinders (1999) also used the Mindlin plate theory to study Lamb wave scattering from a through hole and rivets, respectively. The scattering amplitude of the S_0 Lamb wave field for the case of a plane wave incident on a hole was studied. It was found that the size of the hole affects the number of side lobes as well as the magnitude and direction of the mainlobe.

Grahn (2003) studied the scattering problem of a plane incident S_0 Lamb wave in a plate with a circular partly through-the-thickness hole. He used a 3-D approach where the wave fields in the outer part outside the hole and in the inner part beneath the hole are expanded in the possible Lamb modes and horizontally polarized shear modes. Both propagating modes and evanescent modes were included in the expansions. The expansion coefficients were obtained by utilizing the boundary conditions at the hole and the continuity conditions below the hole. A linear system of equations for the expansion coefficients was derived by projecting the different boundary conditions onto an orthogonal set of projection functions. An alternative approach for validation of the 3-D model was also studied. This approach was based on the lowest order plate theories for extensional and flexural motion. This model is only valid for very low frequencies. The two plate theories were combined and the wave fields in the different regions of the plate from the boundary conditions at the hole were obtained. For low frequencies, the agreement between the two models is good. As the frequency increases, the simplified plate model produces results that are not reliable.

In the methods discussed so far, the scattering coefficient of Lamb waves generated by the interaction of incident wave with scatterers in an plate were directly determined by using an FEM or a BEM approach in conjunction with normal mode expansion. Gunawan and Hirose (2004) adopted a different path: they proposed a mode-exciting method where all Lamb modes are simultaneously excited by appropriate boundary conditions given on both ends of a finite plate. The excited Lamb wave modes constitute a system of equations that is solved to determine the scattering coefficients for all Lamb modes. It was found that, because the scatterer is vertically symmetric, the symmetric and antisymmetric modes are uncoupled each other. The scattering coefficients ${}^1_1r^{A_0}_{A_0}$ and ${}^1_1r^{S_0}_{S_0}$ have the unit absolute values in the frequency range below the cutoff frequencies of A_1 mode ($\omega h/cT = \pi/2$) and S_1 mode ($\omega h/cT = 2.715$), respectively, where only a single-mode exists. These predictions were compared with the HBEM results obtained by Cho (2000): fairly good agreement was obtained.

Migration technique

Migration technique is a signal-processing method that originated in geophysical acoustics. In geophysical exploration, a small explosion is created on the Earth's surface; the explosion generates wave traveling in the Earth interior and on its surface. A number of geophones are used to record both the initial bang and the waves reflected from the various inhomogeneities in the Earth interior. The *migration technique* attempts to image the subsurface reflectors by moving ('migrating') the wave field recorded during the experiments to the actual spatial locations of the reflectors in the Earth's interior. The process back-propagates the recorded waves by systematically solving the wave Eq. as time-dependent boundary value problems based on Huygens' principle.

Lin and Yuan (2001, 2005) adapted the migration technique to damage detection in plates. Both finite differences simulations and experimental results obtained on an aluminum plate were presented. The antisymmetric plate waves were modeled with Mindlin plate theory, and the wave propagation solution was calculated with a finite differences algorithm. This finite differences method migrates backwards the waves diffracted by the damage to identify the damage location and extend. A linear network of sparsely spaced PWAS transducers were used to detect crack damage using the migration technique. Nine circular PWAS were mounted at 25-mm (1-inch) interval across a metallic plate and used in turn as transmitters and receivers of antisymmetric plate waves in a round-robin fashion. The damage was simulated in the form of an arc-shaped through-the-thickness slit (1.1 mm wide, 12.5 mm radius) placed at 150 mm (6-inches) from the transducers. The transmitter transducer creates a wave front that is scattered by the damage and comes back to the receiver transducers where it is recorded in digital format. In processing the experimental data, the migration time step was taken equal to the sampling interval of the A/D converter. At each time step, the received waves are migrated (back propagated) toward the damage location. Applying migration to each time section gives an image of the plate based on a single transmitter. An enhanced image is obtained by adding (stacking) the individual contributions of all the transmitters in the round-robin process. This is the *pre-stack migration* method. As an alternative, the stacking can be done first (with appropriate synchronization of the individual signals) and the migration done second. This is the *post-stack migration* method. The post-stack migration is less computationally intensive, but the pre-stack migration is more powerful in increasing the signal-to-noise ratio.

10.1.2 STRUCTURAL HEALTH MONITORING AND EMBEDDED ULTRASONIC NDE

SHM sets out to determine the health of a structure by reading a network of sensors that are permanently attached to the structure and monitored over time. SHM can be either passive or active. *Passive SHM* infers the state of the structure using passive sensors that are monitored over time and fed into a structural model. Examples of passive SHM are the monitoring of loads, stress, environmental conditions, AE from cracks, etc. Passive SHM only 'listens' to the structure but does not interact with it. Kudva et al. (1993) and Van Way et al. (1995) presented passive SHM systems utilizing an assortment of sensors such as AE transducers, resistance strain gages, fiber optics strain gages, filament crack gages, and corrosive environment sensors. Boller et al. (1999) showed that reliability of SHM systems increases when the sensors do not just 'listen' but act as both actuators and sensors. *Active SHM* uses active sensors that interrogate the structure to detect the presence of damage and to estimate its extent and intensity. One active SHM method employs active sensors, such as the PWAS transducers, which send and receive ultrasonic Lamb waves to determine the presence of cracks, delaminations, disbonds, and corrosion.

Traditional generation of guided Lamb waves has been through a wedge transducer, which impinges the plate obliquely with a tone-burst generated by a relatively large ultrasonic transducer. Snell's law ensures mode conversion at the interface; hence, a combination of pressure and shear waves are simultaneously generated into the thin plate. They constructively and destructively interfere to generate Lamb waves. Modification of the wedge angle and excitation frequency allows the selective tuning of various Lamb wave modes (Alleyne and Cawley, 1992; Rose et al., 1995). Another traditional way to selectively excite Lamb waves is through a comb transducer, in which the comb pitch is matched with the half wavelength of the targeted Lamb mode. Both the wedge and the comb probes are relatively large and expensive. If they were to be deployed in large numbers inside an aerospace structure as part of a SHM system, the cost and weight penalties would be exorbitant. Therefore, for SHM, a different type of Lamb wave transducers that are smaller, lighter, and cheaper than the conventional ultrasonic probes is required. Such transducers could be deployed into the structure as sensor networks or arrays, which would be permanently wired and interrogated at will. Thus, the opportunity for embedded ultrasonic NDE would be created.

PWAS transducers interact with the structure in much the same way that conventional NDE transducers do. For this reason, active SHM can be also viewed as *embedded ultrasonic NDE*, i.e., an NDE method that utilizes transducers that are permanently attached to the structure and can be interrogated on demand. Embedded NDE (e-NDE) is an emerging technology that will allow the transition of the conventional ultrasonic NDE methods to embedded applications in active SHM systems. Viewing active SHM from the embedded NDE viewpoint allows one to draw on the experience already developed in the NDE field and to transition it into SHM applications.

This chapter sets forth to introduce the reader to the general principles of embedded NDE and to present some of the salient achievements in its application to active SHM. The chapter starts with a brief review of guided wave theory in plate, tube, and shell structures. Then, it introduces the PWAS, which are the essential elements of embedded NDE systems. Subsequently, the chapter shows how the main NDE methods (pitch-catch and pulse-echo) are being used in embedded NDE applications. Other techniques such as the time reversal method and the migration technique are also presented. At the end, conclusions and suggestions for further work are presented.

10.2 1-D MODELING AND EXPERIMENTS

In the beginning, our attention is focused on simple 1-D wave-propagation studies that allowed a simpler analysis and foster understanding. For our studies, we selected a long and narrow thin-gage metallic strip. The strip specimen was instrumented with five pairs of PWAS. First, the PWAS instrumented specimen was FEM modeled. Then, it was subjected to experimental tests. The predicted and measured results were compared. Details of our investigation are presented next.

10.2.1 1-D SPECIMEN DESCRIPTION

The 1-D strip specimen considered in our study was made from 1.6 mm thick aircraft-grade 2024 aluminum alloy. The strip specimen was 914 mm long, 14 mm wide, and

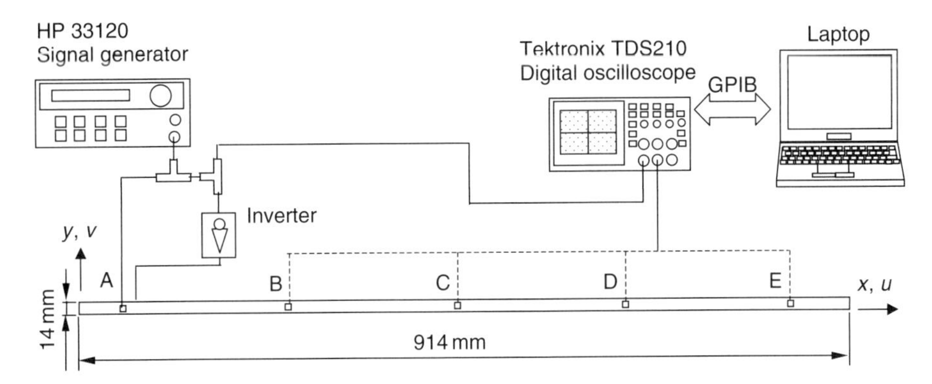

FIGURE 10.1 1-D specimen consisting of a narrow strip of 1.6 mm thick 2024 aluminum alloy, 14 mm wide, and 914 mm long. Shown are the five pairs of PWAS (A through E).

TABLE 10.1 Locations of PWAS transducers on the narrow strip specimen

Sensor number	A	B	C	D	E
x (mm)	57	257	457	657	857
y (mm)	7	7	7	7	7

1.6 mm thick. PWAS (7 mm square, 0.2 mm thick) were installed in pairs (i.e., on both sides of the strip specimen thickness) at five locations (A through E), as shown in Fig. 10.1 and Table 10.1. The double-sided installation was chosen to permit symmetric and anti-symmetric excitation. Thus, axial and flexural waves could be selectively generated.

10.2.1.1 Excitation Signal

The excitation signal considered in our studies consisted of a smoothed tone burst. The smoothed tone burst was obtained from a pure tone burst of frequency f filtered through a Hanning window. The Hanning window is described by the Eq.

$$x(t) = \frac{1}{2}\left[1 - \cos\left(\frac{2\pi t}{T_H}\right)\right], \quad t \in [0, T_H] \tag{4}$$

The number of counts, N_B, in the tone burst matches the length of the Hanning window, i.e.,

$$T_H = \frac{N_B}{f} \tag{5}$$

The tone-burst excitation was chosen in order to excite coherent single-frequency waves. This aspect is very important especially when dealing with dispersive wave types (flexure, Rayleigh, Lamb, etc.). The Hanning window smoothing was applied to reduce the excitation of side frequencies associated with the sharp transition at the start and the end of a conventional (raw) tone burst. Through these means, it was intended that the dispersion effects would be minimized and the characteristics of elastic wave propagation would be readily understood. Figure 10.2 shows a comparison of raw and smoothed 10-kHz tone bursts, as well as their Fourier transform. It is apparent that, although both

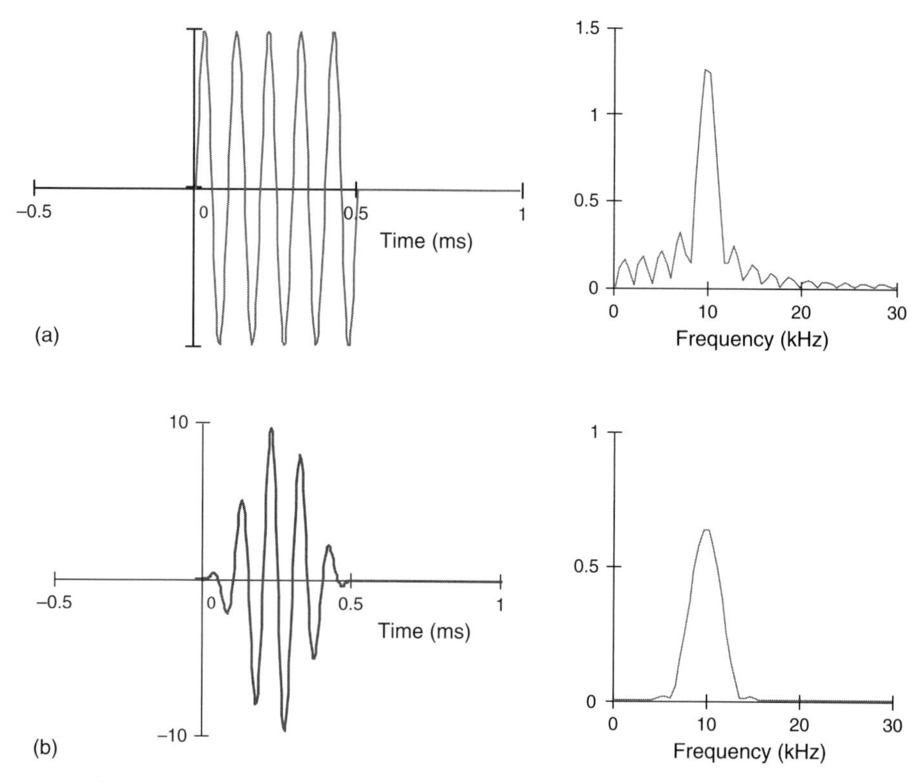

FIGURE 10.2 Example of 10-kHz, 5-count tone burst excitation: (a) raw tone burst; (b) smoothed tone burst.

tone-bursts have the same central frequency of 10 kHz, the raw tone burst (Fig. 10.2a) also excites a considerable number of side lobes, below and above the central frequency. By contrast, the smoothed tone burst (Fig. 10.2b) does not produce side lobes.

The smoothed tone burst that resulted from this process was numerically synthesized and stored in PC memory as the excitation signal. This numerically generated excitation signal was used in the finite element analysis and in the experimental investigation.

10.2.1.2 Finite Element Analysis of Wave Propagation in a Strip Specimen

Numerical simulations of the wave propagation process were performed using the commercially available finite element code, ANSYS. The narrow-strip specimen was modeled using 4-node shell elements (SHELL63). The element has six degrees of freedom at each node. The shell element displayed both bending and membrane capabilities. The discretization process used 3976 elements, arranged in rows of four across the strip specimen width (Fig. 10.3). The FEM model also permitted the simulation of structural defects, e.g., through-thickness cracks of length $2a$. Such cracks were simulated through the nodal release method. Figure 10.3 shows a simulated crack placed transverse to the strip specimen, along the center line. The crack length is controlled by the number of released nodes. To obtain better resolution for crack simulation, the discretization mesh was locally refined, as indicated in Fig. 10.3.

Two forms of the elastic wave propagation were studied: flexural waves and axial waves. To attain wave excitation, we considered creating situations in the FEM model that are equivalent to the excitation applied by the PWAS. First, we identified groups of

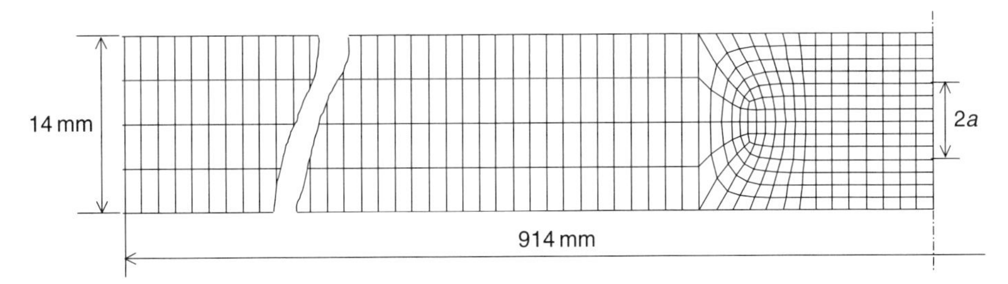

FIGURE 10.3 The mesh used in the finite-element modeling was refined around the region with simulated crack. Left hand side half of the symmetric strip specimen is shown. Crack is simulated along the specimen center line, i.e., at the right hand end of the drawing. The crack length, $2a$, is dictated by the number of released nodes.

finite elements that cover areas geometrically equivalent to that covered by the PWAS installed on the specimen (A through E in Fig. 10.1). Thus, we designated areas in the FEM model that corresponds to PWAS. To generate waves, we applied prescribed harmonic displacements to the nodes delimiting the contour of these active areas. Consistent with the physical phenomenon, the displacement applied to nodes representing opposite ends of the PWAS had to be in opposite direction. This ensures that the net effect on the structure is self-equilibrating. To generate axial waves, we applied nodal translations. For flexural waves, we applied nodal rotations. The detection of the elastic waves followed the same general principle as that applied to wave generation. The variables of interest were the differences between the displacements at the opposing ends of the PWAS, i.e., the Δu for axial waves, and $\Delta w'$ for flexural waves.

Simulation of axial waves

Figure 10.4 shows FEM simulation of axial waves in the strip specimen excited with 100-kHz 5-count Hanning-windowed axial burst at the left hand end of the strip specimen. The patterns of dilatation and contraction (in-plane motion) are shown. The wave was captured after traveling for 50 μs. Figure 10.4a gives an overall view, whereas Fig. 10.4b gives a magnified detail of the first quarter of the strip specimen. One notes that the number of peaks in the wave is greater than the burst count of 5 because both the incident wave and the wave reflected from the left hand end of the strip specimen are superposed in this wave front.

Figure 10.5 shows the wave signal that would be received at PWAS A through E. It is easily appreciated how the waves travel down the strip specimen from A to E, then

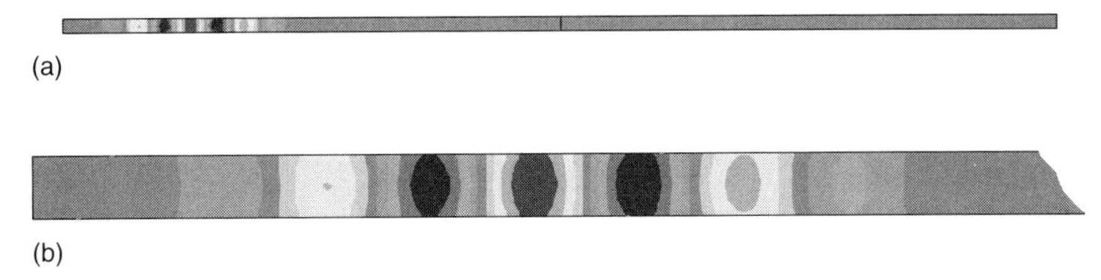

FIGURE 10.4 Finite-element simulation of axial waves in the 914 mm × 14 mm × 1.6 mm aluminum alloy strip specimen excited with 100-kHz 5-count Hanning-windowed axial burst at left hand end. The wave is captured after traveling for 50 μs: (a) overall view; (b) details of first quarter of the strip specimen. The number of peaks in the wave is greater than five because both the incident wave and the wave reflected from the left hand side end of the strip specimen are superposed in this wave front.

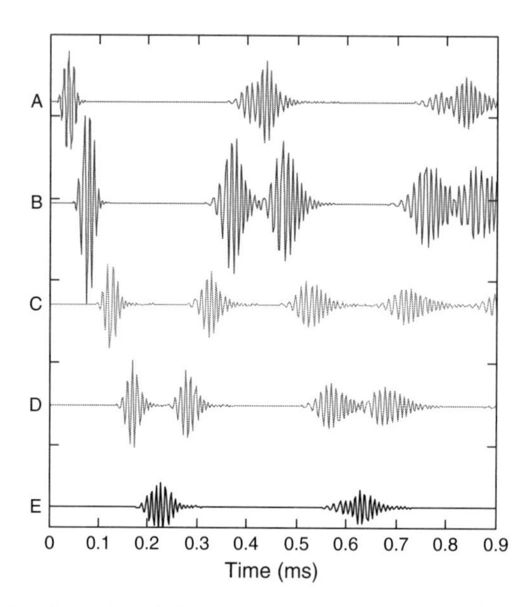

FIGURE 10.5 Finite-element simulation of the axial wave signals received at PWAS transducers A through E (100 kHz 5-count smoothed tone burst excitation at PWAS A).

reflects at the right hand end, and returns to A, followed by repetition of this pattern. One observes that, at the beginning, the five counts of the smoothed tone burst excitation can be readily identified. As the wave propagates and undergoes reflections, its coherence somehow diminishes, and some dispersion occurs. This loss of coherence may also be attributed to accumulation of numerical error.

Figure 10.6 shows FEM simulation of the pulse-echo method used for damage detection. The PWAS placed at the left hand end was used to send a 100 kHz 5-count Hanning-windowed axial burst and to receive the elastic wave responses. In a strip specimen without crack (Fig. 10.6a), the initial signal and the reflection from the right hand end appear. If an 8 mm through-the-thickness transverse crack is simulated in the center of the strip specimen, the reflection (echo) from this crack also appears (Fig. 10.6b). As the crack length increases, the amplitude of the reflection increases (Fig. 10.6c).

Simulation of flexural waves

Figure 10.7 shows FEM simulation of flexural waves in the strip specimen excited with a 100-kHz 5-count Hanning-windowed flexural burst at the left hand end. The peaks and valleys of flexural wave propagation are apparent. The wave was captured after traveling for 99.3 μs. Figure 10.7a gives an overall view, whereas Fig. 10.7b gives a magnified detail of the first quarter of the strip specimen. One notes that the number of peaks in the wave is greater than the burst count of five because of dispersion effects and of the fact that both the incident wave and the wave reflected from the left hand side end of the strip specimen are superposed in this wave front.

Figure 10.8 shows the wave signal received at PWAS A through E for a 10 kHz excitation. It is easily appreciated how the wave travels down the strip specimen from A to E, then reflects at the right hand end, and returns to A, followed by repetition of this pattern. One observes that the five counts of the smoothed tone burst excitation can be readily identified in the initial signal, but this pattern quickly deteriorates due

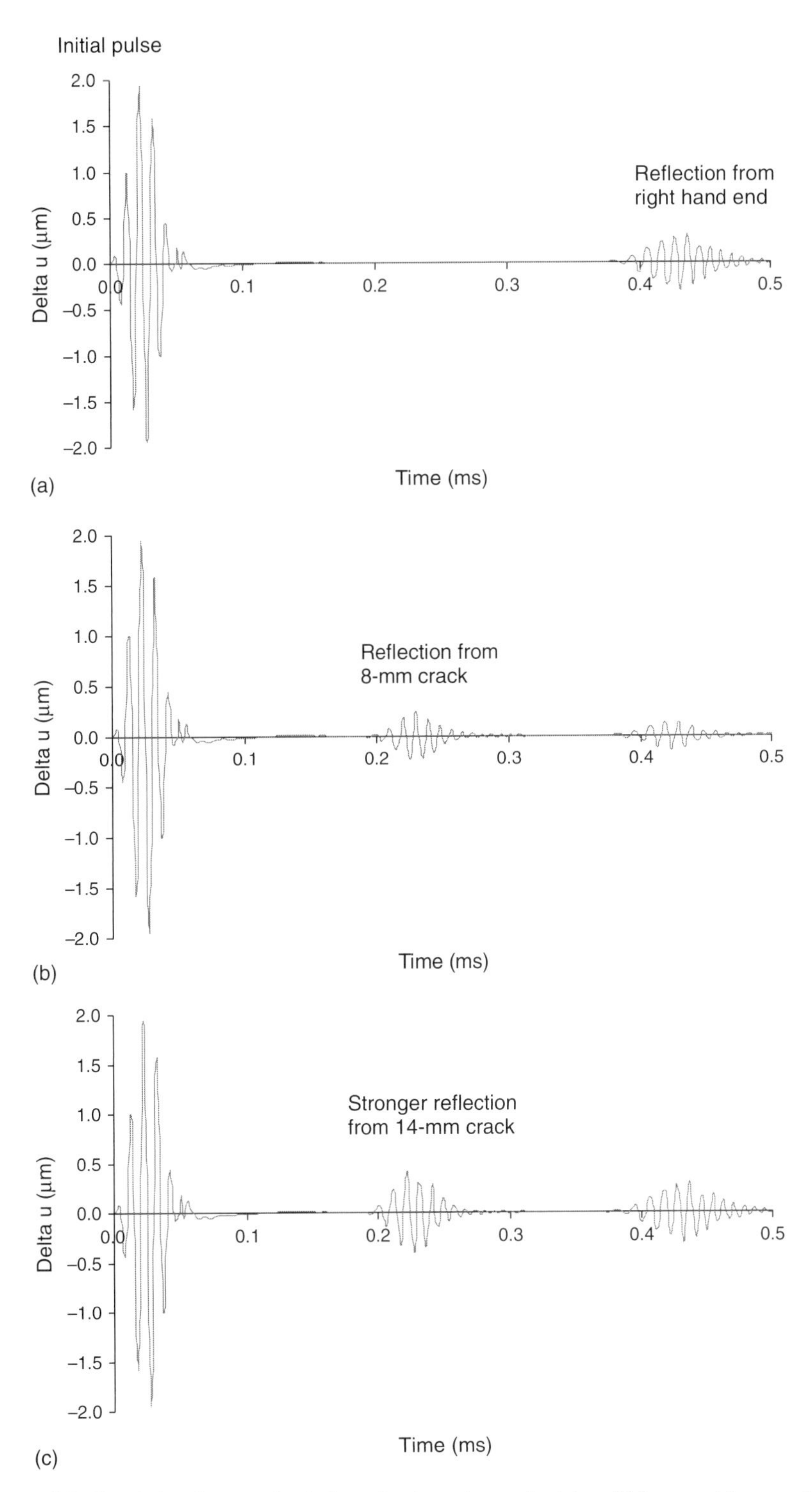

FIGURE 10.6 Finite-element simulation of pulse-echo method in a 914 mm × 14 mm × 1.6 mm aluminum alloy strip specimen using axial waves. A 100-kHz 5-count Hanning-windowed axial burst was applied at the left hand end: (a) strip specimen without crack shows only the reflection from the right hand end; (b) strip specimen with 8-mm long through-the-thickness transverse crack: shows, in addition, the reflection from crack; (c) a longer crack (14-mm) gives a stronger reflection.

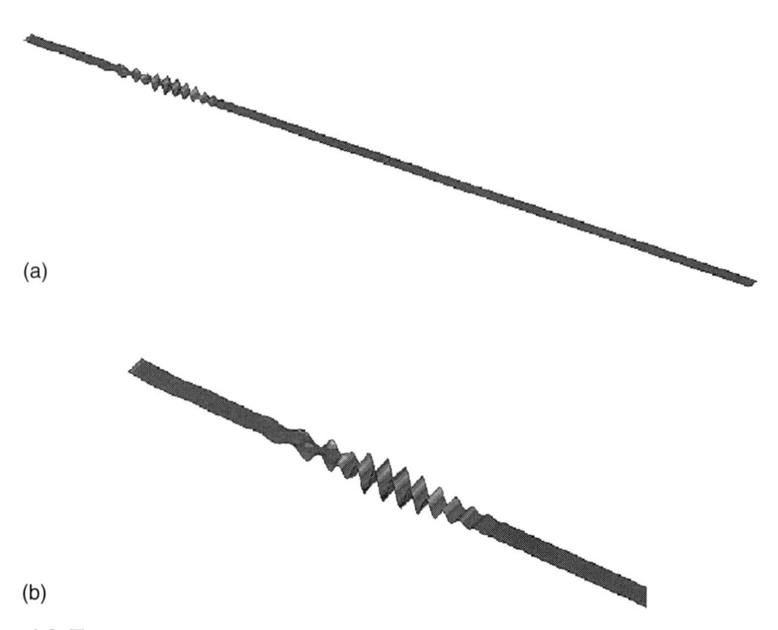

FIGURE 10.7 Finite-element simulation of flexure waves in the 914 mm × 14 mm × 1.6 mm aluminum alloy strip specimen excited with a 100-kHz 5-count Hanning-windowed burst at left hand end. The wave is captured after traveling for 99.3 μs: (a) overall view; (b) details of first quarter of the strip specimen.

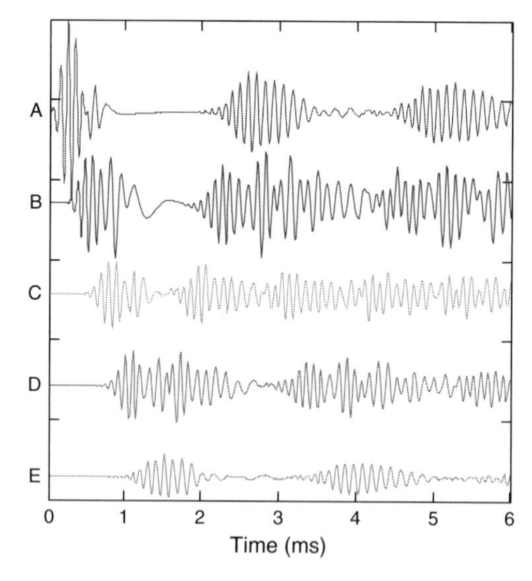

FIGURE 10.8 Finite-element simulation of flexural wave signals received at PWAS A through E (10-kHz 5-count smoothed tone burst excitation at PWAS A).

to wave dispersion. As the wave travels further and undergoes reflections, its coherence further diminishes. The loss of coherence may also be attributed to accumulation of numerical error.

Figure 10.9 shows FEM simulation of the pulse-echo method used for damage detection. The PWAS placed at the left hand end was used to send a 100 kHz 5-count

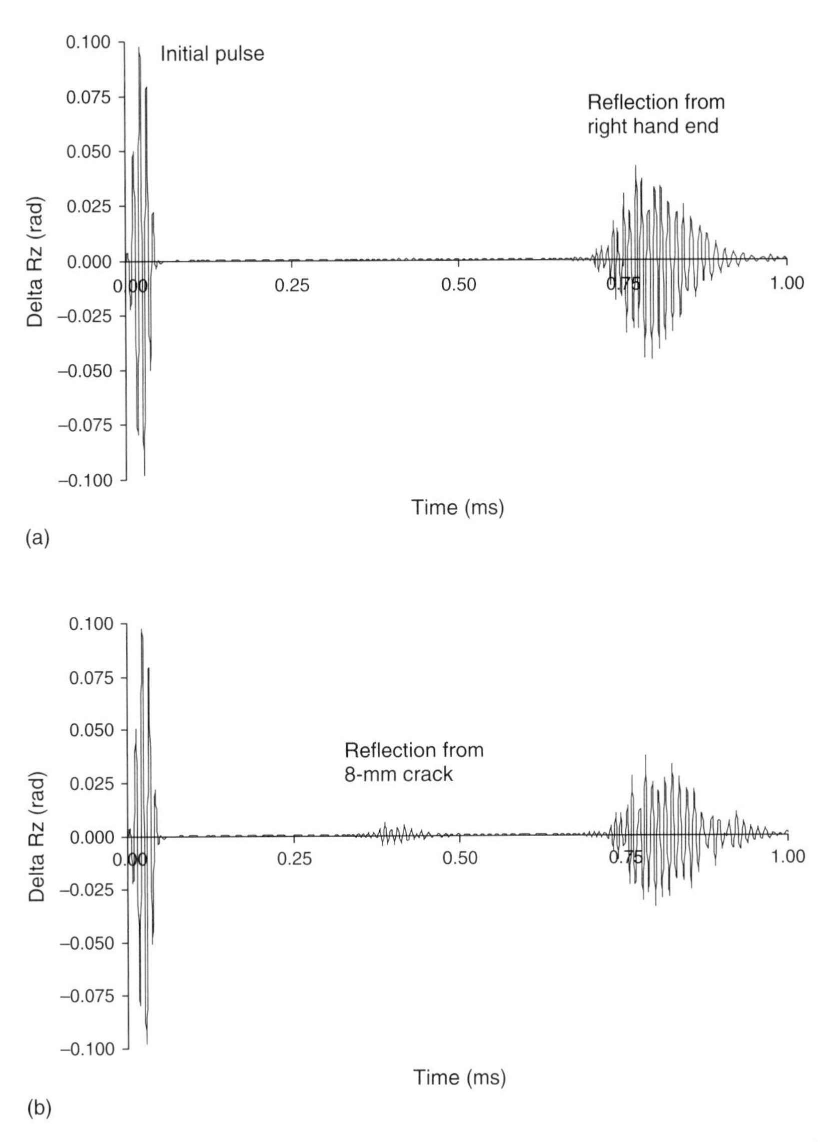

FIGURE 10.9 Finite-element simulation of pulse-echo method in a 914 mm × 14 mm × 1.6 mm aluminum alloy strip specimen using flexural waves. A 100-kHz 5-count Hanning-windowed flexural burst was applied at the left hand end: (a) strip specimen without crack shows only the reflection from the right hand end; (b) strip specimen with 8-mm long through-the-thickness transverse crack: shows, in addition, the reflection from crack; (c) a longer crack (14-mm) gives a stronger reflection.

Hanning-windowed flexural burst and to receive the elastic wave responses. In a strip specimen without crack (Fig. 10.9a), the initial signal and the reflection from the right hand end appear. If an 8-mm through-the-thickness transverse crack is simulated in the center of the strip specimen, the reflection from this crack also appears (Fig. 10.9b). As the crack length increases to 14 mm, the amplitude of the reflection increases (Fig. 10.9c).

Comparison between axial and flexural wave simulation results

The main differences between using the pulse-echo method with axial waves vs. flexural waves are revealed by the comparison of Figs. 10.6 and 10.9. Figures 10.6a and

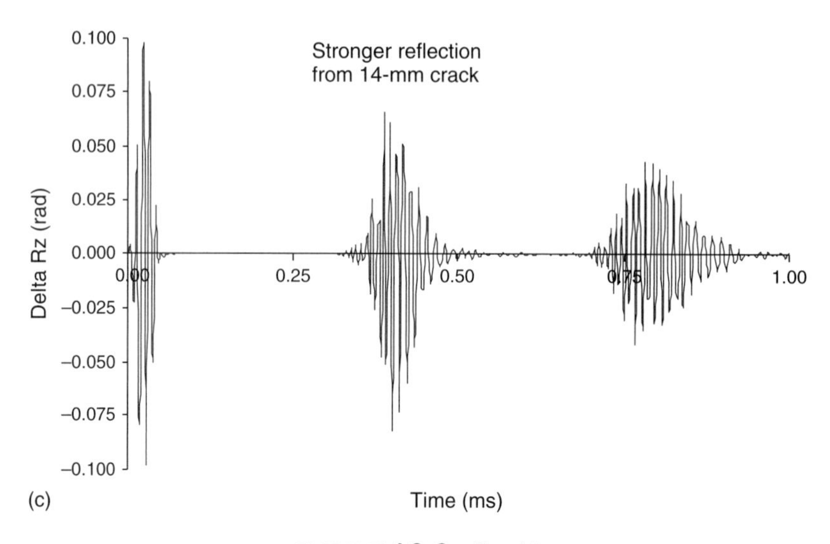

FIGURE 10.9 **Cont'd**

10.9a show that, in this thin-strip specimen, the flexural wave speed at 100 kHz is roughly half of the axial wave speed, because the same distance of 914 mm is traveled in roughly twice the time. In addition, the flexural wave echo of Fig. 10.9a shows a dispersion pattern (i.e., superposition of several waves with different frequencies), whereas the axial wave echo of Fig. 10.6a shows a much more coherent pattern. This is in agreement with the dispersion curves of modes A_0 and S_0 discussed in a previous section. At this value of the frequency-thickness product, the flexural waves, A_0, are much more dispersive than the axial waves, S_0. Figures 10.6b and 10.9b show that, *for partial cracks of the same crack size* (8 mm), *the axial wave echo is much stronger than the flexural wave echo.* Figures 10.6c and 10.9c, on the other hand, show that for a large crack (14 mm), the flexural wave echo has increased much more than the axial wave echo. These observations indicate that both the axial and the flexural waves can offer advantages, under appropriate circumstances, for damage detection, and that both should be retained for further studies. For crack-detection with the pulse-echo method, an appropriate Lamb-wave mode must be selected.

The importance of high-frequency excitation

Although the results shown above refer mainly to one frequency (100 kHz), our wave propagation simulation efforts were performed for a variety of frequencies in the range from 10 to 100 kHz. It was found that the lower-frequency limit were easier to simulate because at low frequency the wavelength is longer and spans several finite elements, and hence the distortion of each element is less severe. Consequently, at low-frequency, we could use larger elements, i.e., a coarser mesh, and less computation time. However, low-frequency waves are inappropriate for ultrasonic applications. To achieve damage detection with the pulse-echo method, the timewise length of the wave packet must be much less than the time taken for the echo to return. We observed that, at low frequencies (e.g., 10 kHz), the echo signal starts to appear before the incident signal has finished developing, thus markedly impeding the damage detection process. Hence, high excitation frequencies are important. As the frequency increases, the wavelength decreases, and hence a finer finite-element mesh is needed to capture the wave propagation process, which increases considerably the computational effort. After several trials, the compromised frequency of 100 kHz was selected. With this frequency, we were able to simulate successfully both axial and flexural wave patterns, and to identify defect-generated echoes as close as 100 mm from the source. For the detection of defects closer than 100 mm, higher frequencies are required.

10.2.2 COUPLED-FIELD FEM ANALYSIS OF WAVE PROPAGATION

In the previous section, we used the FEM approach to predict how axial and flexural waves would propagate in a narrow-strip specimen. This analysis was performed with conventional finite element formulations incorporating the elastodynamic field. Waves were generated by applying excitation displacements and rotations to the strip specimen at PWAS location. Thus, the elastodynamic wave response at the PWAS location was predicted. The piezoelectric response in the PWAS transducers was not taken into account in the analysis. Instead, we assumed that this elastodynamic wave response would be transduced into an electric response due to the piezoelectric property of the PWAS transducers.

In this section, we will do a more complete analysis by employing a coupled-field FEM analysis. In recent years, several commercially available FEM packages have started to offer the option of coupled-field (multi-physics) elements (for example, ANSYS, ABAQUS, etc.). In this section, we will explore how this coupled-field FEM approach can be applied to the case of the structurally affixed PWAS transducers used in the SHM process. Because the coupled-field elements simultaneously consider the elastodynamic and the electric fields, we will show that this type of analysis is able to predict the PWAS electric response under wave excitation in the structure, as well as the wave response in the structure under electric excitation of the PWAS.

In applying the coupled-field FEM analysis, we modeled the PWAS directly using coupled-field elements. Same strip specimen as shown in Fig. 10.1 with PWAS placed at the same locations was studied. The PWAS modeling was done with ABAQUS multi-field solid elements and attached on the conventional FEM model of the strip specimen. Part of the specimen is shown in Fig. 10.10. The bottom surface of the solid PWAS model was attached to the strip specimen surface such that the mechanical motion and forces are transferred between the two models. We excite the PWAS with electrical signals and examine the electrical response of PWAS to the wave motion in the specimen. A 5-count smoothed tone burst was used. The difference of this approach in comparison with conventional FEM approach is that the excitation signal this time was electrical voltage

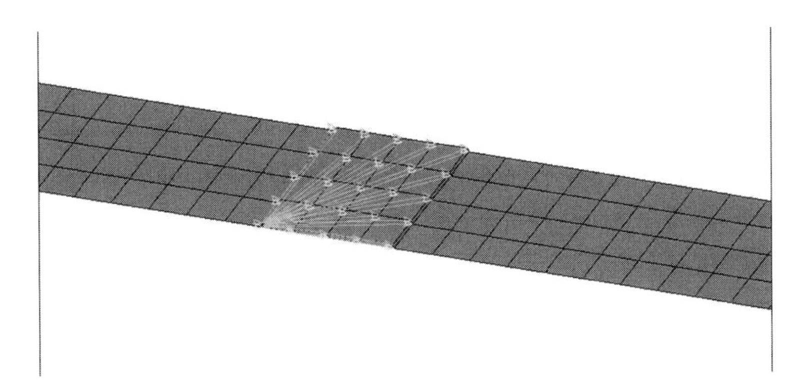

FIGURE 10.10 Meshing of the strip specimen and of the coupled field finite-element model of the PWAS transducer.

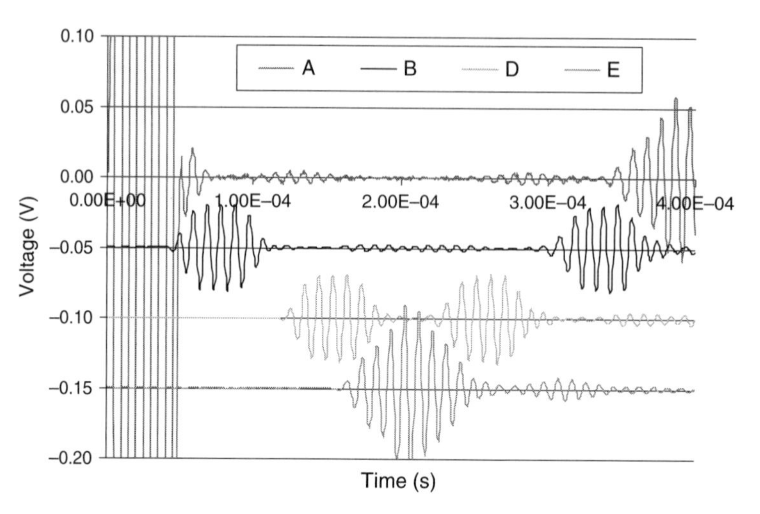

FIGURE 10.11 Finite-element simulation of pulse-echo method in narrow strip specimen using coupled-field elements.

directly fed onto the top surface electrode of the PWAS, whereas bottom electrode was electrically grounded. Thus, we are simulating the way the PWAS transducers operate in real applications.

Figure 10.11 shows the electrical signals received at PWAS A through E placed as shown in Fig. 10.1. We could observe that when the single PWAS is excited by electrical signals, the generated wave incorporates both axial mode and flexural mode, thus making the wave packet more complex than the single-mode excitation considered in previous sections.

We also simulate the effect of crack on wave propagation. As in the previous analysis using the conventional FEM approach, we put a crack at the center of the strip specimen and performed a pitch-catch and pulse-echo simulation. This time, we simulated directly the electrical signals that would be received at the various PWAS on the strip specimen (Fig. 10.12). As indicated in Fig. 10.12, the presence of the crack does affect

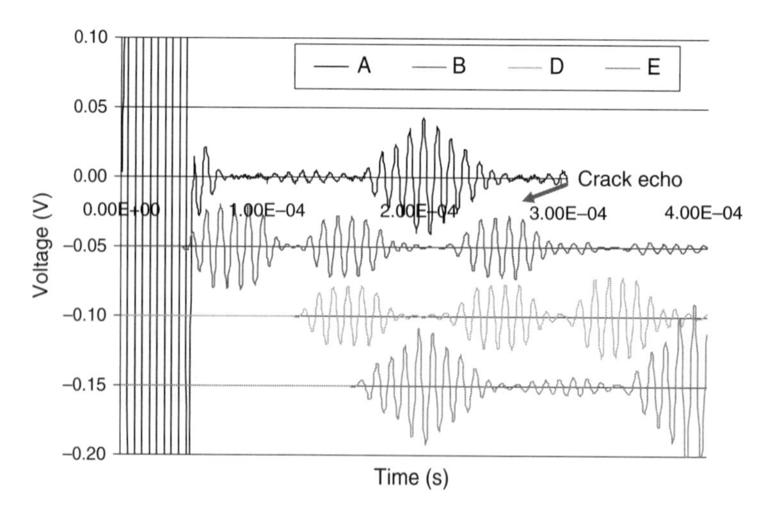

FIGURE 10.12 Coupled-field finite-element simulation of crack detection in narrow strip specimen pulse-echo method.

the wave propagation and an echo from the crack could be detected directly as electrical signals.

These simple examples have illustrated how the coupled-field FEM approach can be applied to simulate the wave propagation in a narrow metallic strip in which PWAS transducers act as transmitters and receivers of elastic waves. Our examples were kept simple in order to avoid extensive computational time. However, the method demonstrated in this simple example could be extended to predict the SHM signals in larger and more complicated specimens, as well as in full structures subjected to an SHM procedure.

10.2.3 EXPERIMENTAL INVESTIGATION OF WAVE PROPAGATION IN A STRIP SPECIMEN

The experimental setup (Fig. 10.1) consisted of a HP 33120A signal generator, a Tektronix TDS 210 digital oscilloscope, and laptop computer connected through GPIB interface. The HP 33120A signal generator was used to generate constant-amplitude tone bursts of 3–5 counts. The Tektronix TDS 210 digital oscilloscope connected to the laptop computer through the GPIB interface was used to collect data. Five PWAS pairs were placed at locations A through E as shown in Fig. 10.1. The PWAS pairs consisted of one PWAS on the top surface and another PWAS on the bottom surface of the specimen. The burst signal from the signal generator of 10 V peak-to-peak (pp) was applied directly to the PWAS. The signal was applied either to one of the two PWAS in the pair (one-sided excitation) or to both (two-sided excitation). In the latter case, both in-phase and out-of-phase excitation coupling between the upper and the lower PWAS were applied. To achieve out-of-phase coupling, the signal was split and then routed through an operational amplifier inverter circuit. Through in-phase and out-of-phase excitation, the enhancement and suppression of certain wave types (axial, flexural) could be manipulated, and thus their properties could be separately studied. One of the five pairs of PWAS (A through E) was used for transmission, whereas all PWAS were used for reception. Various combinations of transmission-reception locations were explored in a round-robin fashion.

Figure 10.13 shows signal samples captured during this investigation. On each plot, signal A represents the excitation signal applied to the PWAS at location A, whereas B through E represent the reception signals picked up by the PWAS at locations B through E. The excitation signal was 3-count constant-amplitude tone burst of 10 Vpp. The reception signals were in the 100-mVpp range.

Figure 10.13a presents the results of a 10-kHz excitation applied to the upper PWAS at location A. It can be appreciated that two wave types (flexural and axial) are simultaneously excited. The flexural wave, which travels with lower speed, is excited much stronger than the axial wave. Nevertheless, minute examination of the signal, e.g., at PWAS E, reveals that a small amplitude axial wave packet arrives well ahead of the main flexural wave packet. This figure also indicates how waves reflect at the strip specimen boundary and travel backwards and forwards through the strip specimen. Figure 10.13b shows the same phenomenon but at 100 kHz. This time, the axial waves are much better distinguishable as clear packets of smaller amplitude waves ahead of the main flexural waves. Because the axial waves are stronger at this frequency, they can be distinguished at all PWAS locations, B through E. Figure 10.13c presents the result of applying a 20-kHz symmetric excitation simultaneously to the upper and lower PWAS at location A. Under symmetric excitation, only the axial waves are generated. Their speed is high, and they undergo multiple reflections from the ends of the strip specimen. Figure 10.13d presents the result of applying a 10-kHz anti-symmetric excitation to the same PWAS pair at location A. This time, only the flexural waves are excited. They travel slower, but have

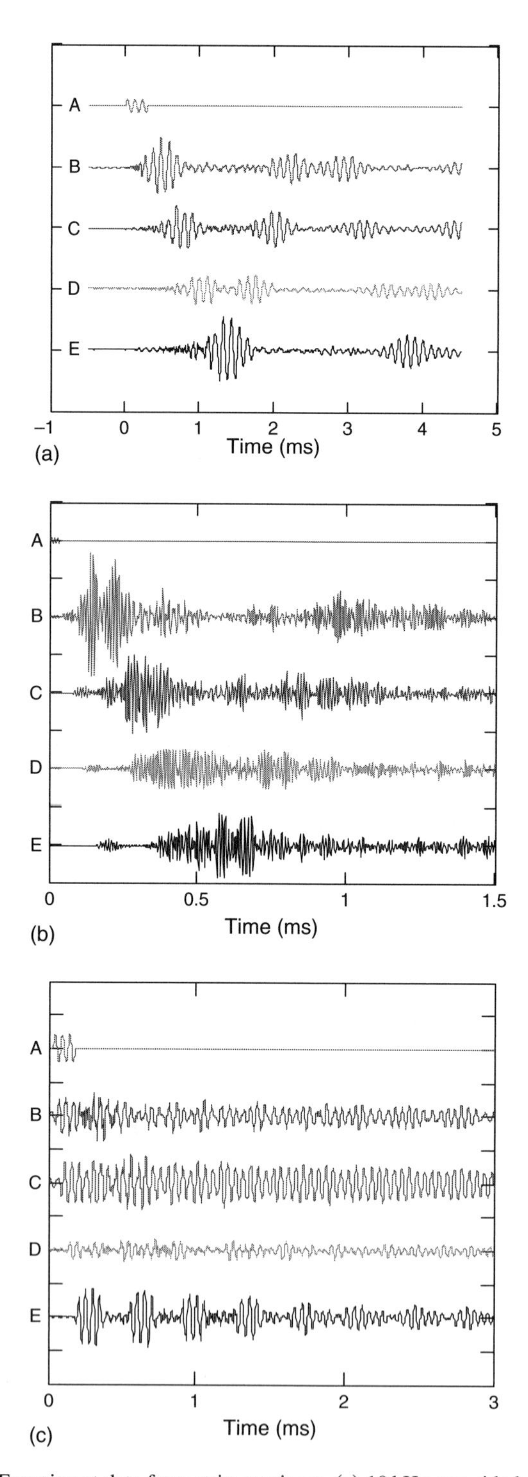

FIGURE 10.13 Experiment data from strip specimen. (a) 10 kHz one-sided excitation; (b) 100 kHz one-sided excitation; (c) 20 kHz two-sided axial excitation; (d) 10 kHz two-sided flexural excitation.

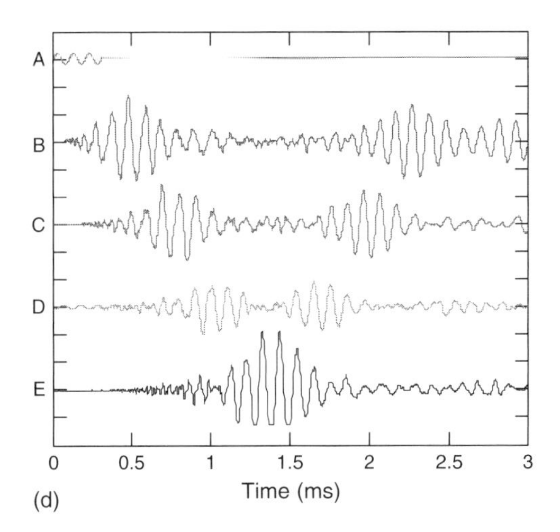

FIGURE 10.13 **Cont'd**

much larger amplitude than the axial waves. Reflections from the strip specimen-ends and their traveling back and forth are apparent. From these plots, the TOF for each wave packet could be observed, recorded, and plotted against distance. Then, the group velocity of the wave packet was calculated through linear regression.

10.2.4 PWAS RAYLEIGH WAVES GENERATION IN RAIL TRACKS

To test the feasibility of transmitting and receiving Rayleigh waves with PWAS transducers in massive structures, we conducted a small-scale demonstration experiment on a rail track specimen (Fig. 10.14). A 1.5-m ($\sim 5\,\text{ft}$) piece of rail was instrumented with five PWAS transducers. Each PWAS consisted of a 7-mm square piezoelectric wafer (0.2 mm thin) affixed to the rail surface. Ultrasonic waves in the 300-kHz band were generated at one end of the rail specimen and received at the other end, as well as at intermediate locations. Figure 10.14c shows the waves transmitted from one PWAS and received at the other two PWAS. These waves are *surface-guided waves* (Rayleigh waves) and can travel a long distance in the rail with very little attenuation. Figure 10.14c indicates that the TOF is directly proportional with distance. As shown in a previous chapter, Rayleigh wave speed can be predicted with the expression $c_R = \sqrt{G/\rho}\,(0.87 + 1.12\nu)/(1+\nu)$ where G is the shear modulus of the material, and ρ the specific mass. For carbon steels, $c_R = 2.9\,\text{km/s}$. Figure 10.14d shows the plot of distance vs. time predicted by this formula, and two experimental points corresponding to PWAS 1 and 2. The correlation between theory and experiment is very good, which gives confidence in our investigation. If cracks were present, reflections would be received. When corrosion is present, the wave-speed changes.

The PWAS Rayleigh wave concept is completely *nonintrusive*, because the ultrasonic guided waves are generated with surface mounted transducers that can be retrofitted to existing rails without any disruption to the rail traffic. In contrast, conventional axial-waves ultrasonic transducers would have to be incorporated into the rail, which is not feasible. The PWAS-based concept is *robust*, because the ultra-thin PWAS present a very low profile and can be easily protected from environmental attack. In contrast, conventional ultrasonic surface wave ultrasonic transducers are bulky and obtrusive, because they require an intermediate coupling wedge to attain pressure-to-surface mode conversion.

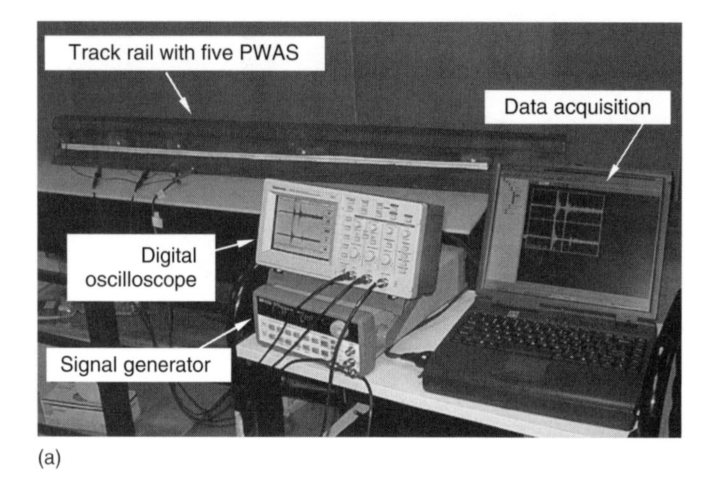

(a)

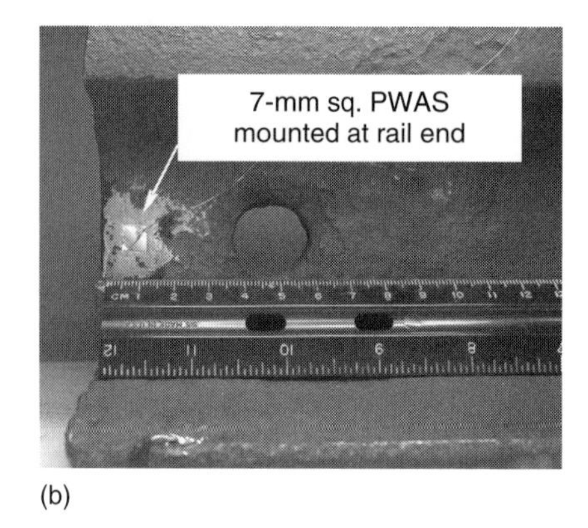

(b)

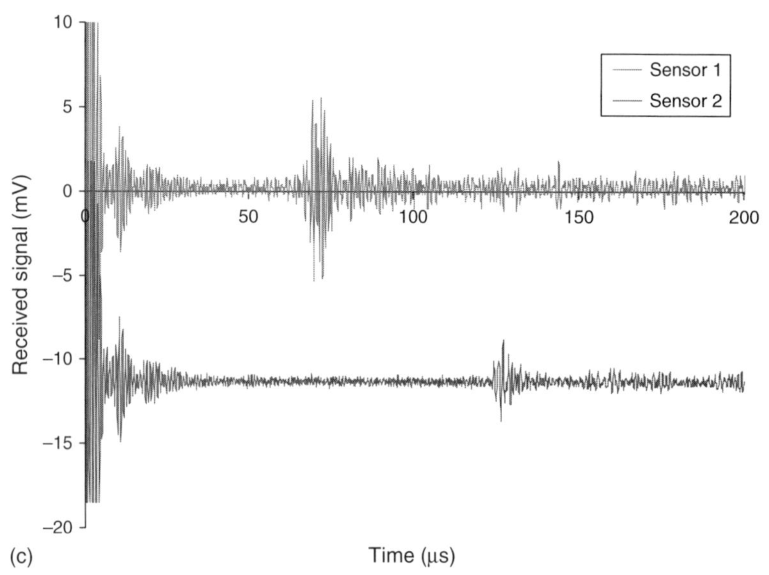

(c)

FIGURE 10.14 Transmission and reception of surface-guided Rayleigh wave with PWAS transducers in a rail track specimen: (a) track rail specimen with 5 PWAS and laboratory instrumentation; (b) details of one PWAS installation; (c) surface-guided Rayleigh waves received at 200 mm (PWAS 1) and 365 mm (PWAS 2); (d) plot of distance vs. time.

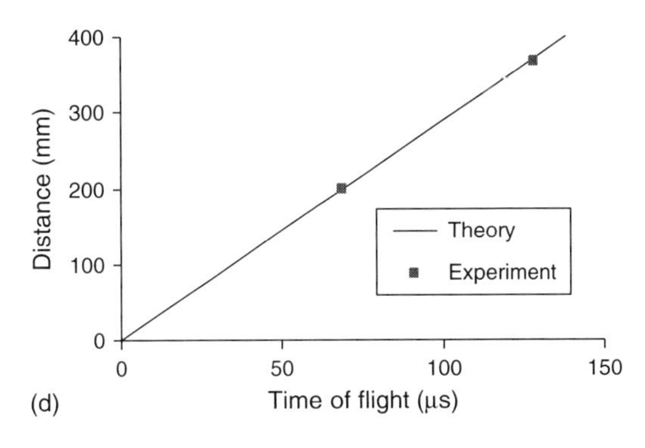

FIGURE 10.14 **Cont'd**

10.2.5 SECTION SUMMARY

The main findings of investigating PWAS-based embedded ultrasonics on 1-D specimens are:

1. The FEM modeling of wave propagation and reflection in thin strip specimens was successfully achieved with the shell elements and time integration. Both axial and flexural waves were separately studied.
2. The detection of internal cracks using the pulse-echo method with embedded ultrasonics hardware was successfully simulated at frequencies up to 100 kHz.
3. Embedded ultrasonics experiments were successfully performed on a narrow-strip specimen equipped with PWAS placed in pairs at five locations. The excitation and reception of both axial and flexural waves, separate and together, were verified.
4. Group velocity of the wave packet was successfully determined. The values were found to be within 5.6% of theoretical predictions, which is explainable in terms of manufacturer's tolerances in geometry and materials, as well as wave dispersion. The group velocity dispersion curves for flexural (A_0) and axial (S_0) Lamb wave modes were experimentally confirmed.
5. Rayleigh surface waves were generated and detected in a rail-track specimen indicating that PWAS transducers can be also successfully used on massive structures.

However, the experiments on 1-D specimens started to present considerable problems at higher frequencies. As frequency increases, the wavelength decreases, and the effect of waves bouncing from the sides of the long narrow specimen (side-edge reflections) becomes significant. These side-edge reflections considerably distort the wave propagation and mess up the signal. For these reasons, we found it necessary to move from experiments on strip specimens to experiments on large plates, in which the boundary reflections are well separated from the initial signals. As shown in the next section, large-plate experiments will give more satisfactory results and will greatly facilitate our understanding of the phenomena.

10.3 2-D PWAS WAVE PROPAGATION EXPERIMENTS

This section will present several wave propagation experiments with PWAS transducers placed on a thin metallic plate. Both pitch-catch and pulse-echo experiments will be considered.

10.3.1 2-D EXPERIMENTAL SETUP

To understand and calibrate the 2-D Lamb waves excitation and reception with PWAS transducers, a set of experiments were conducted on a thin metallic plate with $t = 1.6\,\text{mm}$. The plate was made of 2024-aluminum alloy. Its overall dimensions were $914\,\text{mm} \times 504\,\text{mm}$. The plate was instrumented with a sparse array of eleven 7-mm × 7-mm PWAS positioned on a rectangular grid (Fig. 10.15 and Table 10.2).

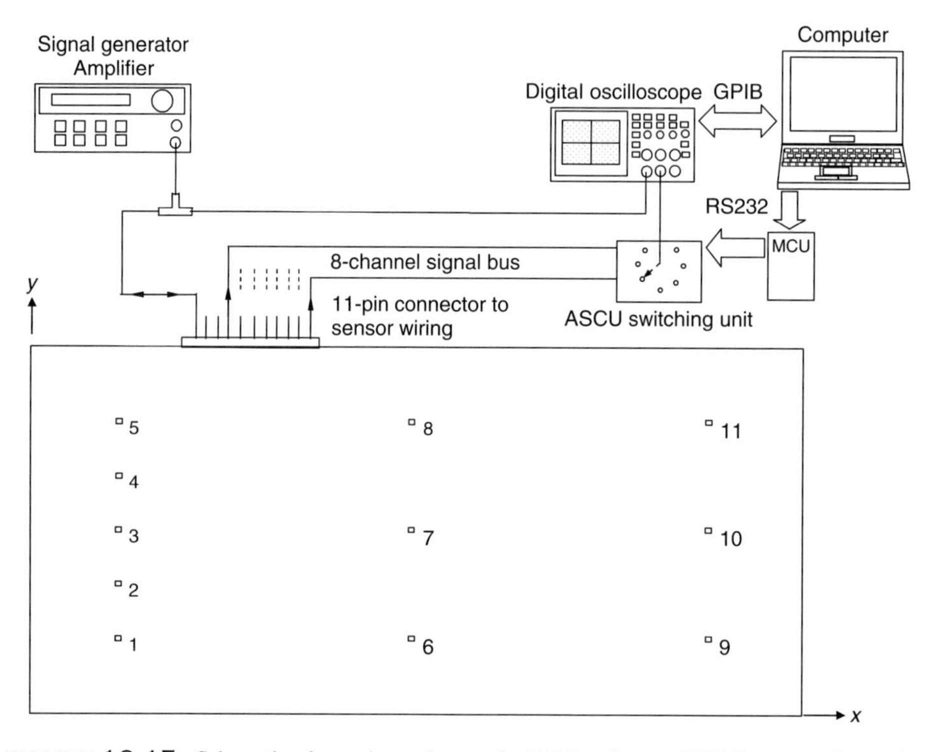

FIGURE 10.15 Schematic of experimental setup for 2-D Lamb-wave PWAS-structure interaction studies.

TABLE 10.2 Piezoelectric properties of the PWAS material (APC-850)

Property	Symbol	Value
Compliance, in plane	s_{11}^{E}	$15.30 \cdot 10^{-12}\,\text{Pa}^{-1}$
Compliance, thickness wise	s_{33}^{E}	$17.30 \cdot 10^{-12}\,\text{Pa}^{-1}$
Dielectric constant	ε_{33}^{T}	$\varepsilon_{33}^{T} = 1750\varepsilon_0$
Thickness wise induced-strain coefficient	d_{33}	$400 \cdot 10^{-12}\,\text{m/V}$
In-plane induced-strain coefficient	d_{31}	$-175 \cdot 10^{-12}\,\text{m/V}$
Coupling factor, parallel to electric field	k_{33}	0.72
Coupling factor, transverse to electric field	k_{31}	0.36
Coupling factor, transverse to electric field, polar motion	k_p	0.63
Poisson ratio	ν	0.35
Density	ρ	$7700\,\text{kg/m}^3$
Sound speed	c	2900 m/s

Note: $\varepsilon_0 = 8.85 \times 10^{-12}\,\text{F/m}$.

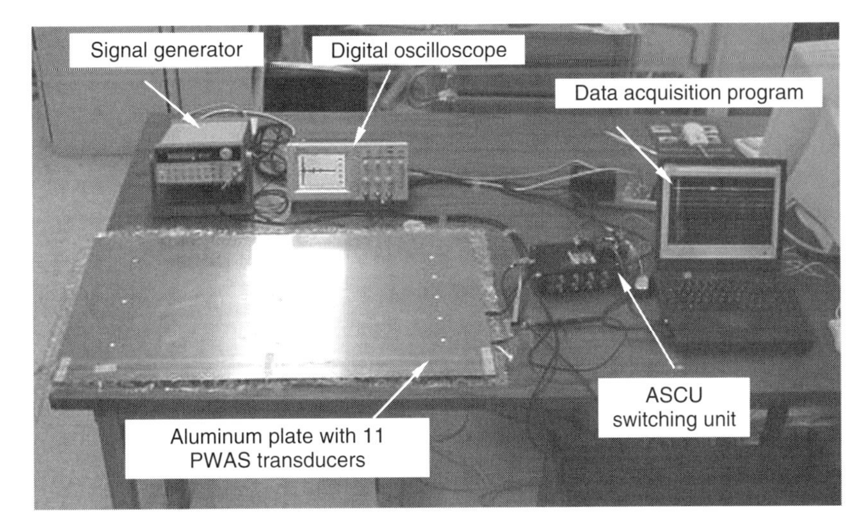

FIGURE 10.16 Experimental setup for rectangular plate wave-propagation experiment showing the plate, PWAS transducers, and instrumentation.

The PWAS were connected with thin insulated wires to a 16-channel signal bus and two 8-pin connectors (Fig. 10.16). An HP33120A arbitrary signal generator was used to generate a 10-V pp smoothed 300-kHz tone burst excitation with a 10-Hz repetition rate. The signal was sent to PWAS 11, which generated a packet of elastic waves that spread out into the entire plate. A Tektronix TDS210 two-channel digital oscilloscope, synchronized with the signal generator, was used to collect the signals captured at all PWAS. A digitally controlled switching unit and a LabVIEW data acquisition program were used. A Motorola MC68HC11 microcontroller was tested as an embedded stand-alone option.

10.3.2 2-D PITCH-CATCH PWAS EXPERIMENTS

Figure 10.17a shows a sample of the signals measured during this investigation. The first row shows the signal associated with PWAS 11 (the transmitter). One observes that all signals display the 'initial bang' packet captured through electromagnetic coupling in the cabling. The wave packets are reflection from the plate edges, and their TOF position is consistent with the distance from the PWAS to the respective edges. The other rows of signals correspond to the receptor PWAS 1 through 8. The TOF values for each PWAS receiver are consistent with the distance between the transmitter and the receiver. The consistency of the wave patterns is very good.

These raw signals were processed using a narrow-band signal correlation algorithm followed by an envelope detection method. As a result, the exact TOF for each wave packet could be precisely identified. When the TOF was plotted against radial distance between the receiving PWAS and the transmitting PWAS, a good straight-line fit (99.99% R^2 correlation) was obtained (Fig. 10.17b). The slope of this straight line is the group velocity and has the value 5.461 km/s. For the aluminum alloy used in this experiment, the theoretical group velocity for S_0 mode is 5.440 km/s. The speed detection accuracy (0.3% error) is quite good.

These systematic experiments gave conclusive results regarding the feasibility of exciting elastic waves in thin metallic plates using small inexpensive and un-obtrusive PWAS transducers. Excitation and reception of high-frequency Lamb waves was verified over a large frequency range (10 to 600 kHz). The elastic waves generated by this method

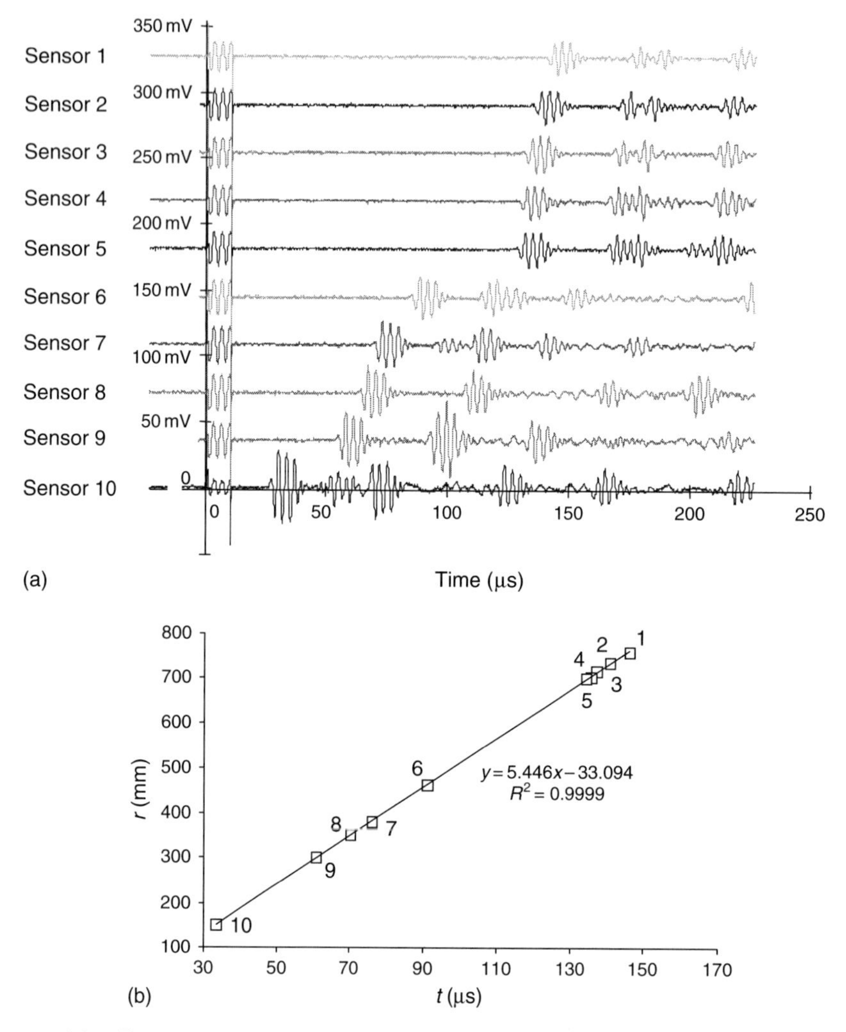

FIGURE 10.17 (a) Excitation signal and echo signals on PWAS transducers 11, and reception signals on PWAS transducers 1 through 10; (b) correlation between radial distance and time of flight.

had remarkable clarity and showed a 99.99% distance-TOF correlation. The group velocity correlated very well with the theoretical predictions.

10.3.3 2-D PULSE-ECHO PWAS EXPERIMENTAL RESULTS

The pulse-echo experiments performed on this simple specimen allowed us to understand the mechanism of Lamb wave transmission and reception with PWAS, as well as the patterns of multiple echoes resulting from multiple reflections at the plate edges. An estimation of the wave propagation range could be also achieved using multi-reflection analysis. It will be shown that this rage is beyond 2.5 m, which is quite remarkable for such a small device driven at a relatively low voltage.

10.3.3.1 Wave Reflection Signals Captured by PWAS

When PWAS were used as transmitters and receivers on the rectangular plate specimen, a pattern of multiple reflections was observed, as illustrated in Fig. 10.17a of the previous section. The next step in our analysis was to understand these reflection patterns, represented by the subsequent wave packets appearing in each signal. These packets, appearing after the first packet, represent waves that are reflected from the edges of the plate. Understanding the wave reflection patterns is essential for implementing the pulse-echo method for damage detection. In our analysis, we established a correlation between the TOF of each wave packet and the distance traveled by that particular packet (path length). The TOF determination was immediate, but the determination of the actual traveled distance was more involved. Because elastic waves propagate in a circular wave front through the 2-D plate, they reflect at all the edges and continue to travel around until fully attenuated. AutoCAD drawing software was used to assist in the analysis and calculate the actual path length of the reflected waves (Fig. 10.18).

The drawing in Fig. 10.18 shows the plate and its mirror reflections with respect to the four edges and four corners. Thus, an image containing nine adjacent plates was obtained. To identify which reflection path corresponds to a particular wave packet, we took its TOF and multiplied it by the wave speed to get a first estimate of the path length. For example, Fig. 10.18 shows that for the reflected reception at PWAS 6, the estimated distance is 645 mm. Then, a circle centered at the transmitter PWAS 11 was drawn with a radius equal to half of the estimated distance. The circle intersects with, or is very close to, one of the reflected images of the reception PWAS 6. In this way, we identified which of the reflected images of the reception PWAS 6 was the actual image to be processed. Next, the true distance (path length) between the identified PWAS image and the transmitter PWAS was determined using the 'distance' tool of the AutoCAD software. In Fig. 10.18, this true distance is 651.22 mm. In this way, the path lengths for the reflection wave packets of all PWAS were determined, and a TOF vs. path length plot could be created (Fig. 10.19). With the exception of one outlier, the TOF vs. path length plot of Fig. 10.19 shows good linearity and thus proves the consistency of our reflection analysis.

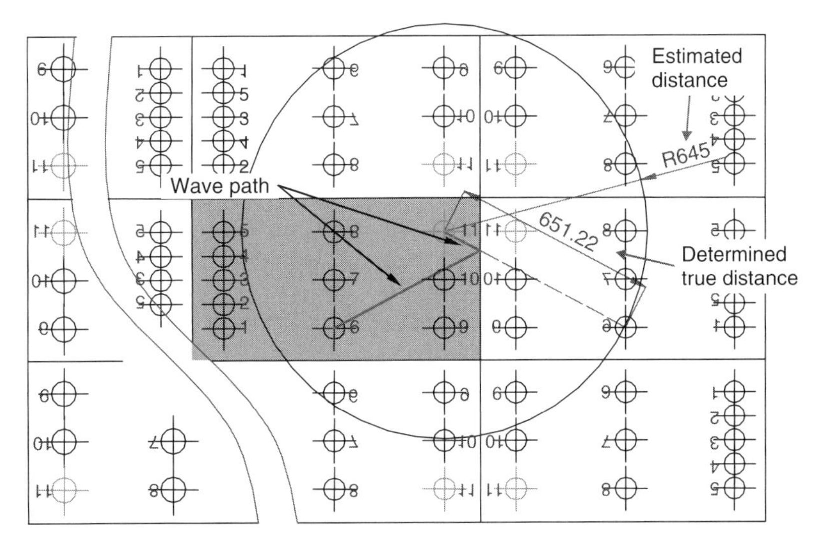

FIGURE 10.18 Method for finding the distance traveled by the reflected waves: the actual plate (shaded) is surrounded by eight mirror images to assist in the calculation of the reflection path length.

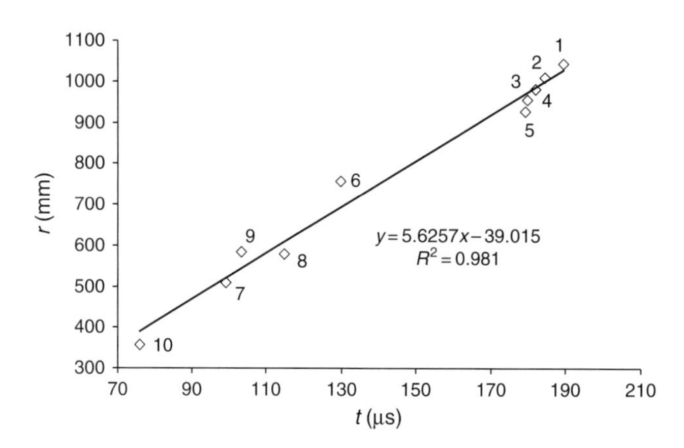

FIGURE 10.19 Correlations between path length and time of flight for the wave reflection signals (2nd wave packet) captured on PWAS transducers 1 through 10.

10.3.3.2 Pulse-Echo Reflections Analysis

The next step in our analysis was to understand the reflection patterns represented by the subsequent wave packets appearing in each signal. These packets, appearing after the first packet, represent waves that are reflected from the edges of the plate. In our analysis, we had to establish a correlation between the TOF of each wave packet and the distance traveled by that particular packet (path length). The TOF determination was immediate, but the determination of the distance actually traveled was more involved and required careful consideration of possible paths. Subsequently, we were able to perform pulse-echo analysis. After the initial bang, the transmitter PWAS 11 was also able to capture subsequent signals representing waves reflected by the plate boundaries and sent back to the transmitter. These subsequent wave packets were recorded for evaluating the pulse-echo method. Figure 10.20a shows the signal recorded on PWAS 11. This signal contains the excitation signal (initial bang) and a number of wave packets received in the pulse-echo mode. The wave generated by the initial bang undergoes multiple reflections from the plate edges, as shown in Fig. 10.20b. The values of the true path length for these reflections are given in Table 10.3. It should be noted that the path lengths for reflections R_1 and R_2 are very close. Hence, the echoes for these two reflections virtually superpose on the pulse-echo signal of Fig. 10.20a. It is also important to notice that reflection R_4 has two possible paths, R_{4a} and R_{4b}. Both paths have the same length. Hence, the echoes corresponding to these two reflection paths arrive simultaneously and form a single echo signal on Fig. 10.20a, with roughly double the intensity of the adjacent signals. Figure 10.20c shows the TOF of the echo wave packets plotted against their path lengths. The straight-line fit has a very good correlation ($R^2 = 99.99\%$). The corresponding group velocity is 5.389 km/s, i.e., within 1% of the theoretical value of 5.440 km/s.

The pulse-echo method was successfully verified using the PWAS 11 in a dual role: (i) to generate elastic waves ('initial bang') and (ii) to capture the echo signals of the waves reflected by the plate boundaries and coming back to the transmitter PWAS (Table 10.3 and Fig. 10.20). Very good correlation was obtained ($R^2 = 99.99\%$).

It is also important to notice that the multi-reflections analysis allows us to get an estimate of the range at which PWAS are able to operate. The last point plotted in Fig. 10.20c is R8. This point corresponds to a traveled distance in excess of 2000 mm. Examination of Fig. 10.20a indicates that the echo R8 is still strong and very well visible. Other echoes are still visible beyond R8. Based on these observations, we estimate that

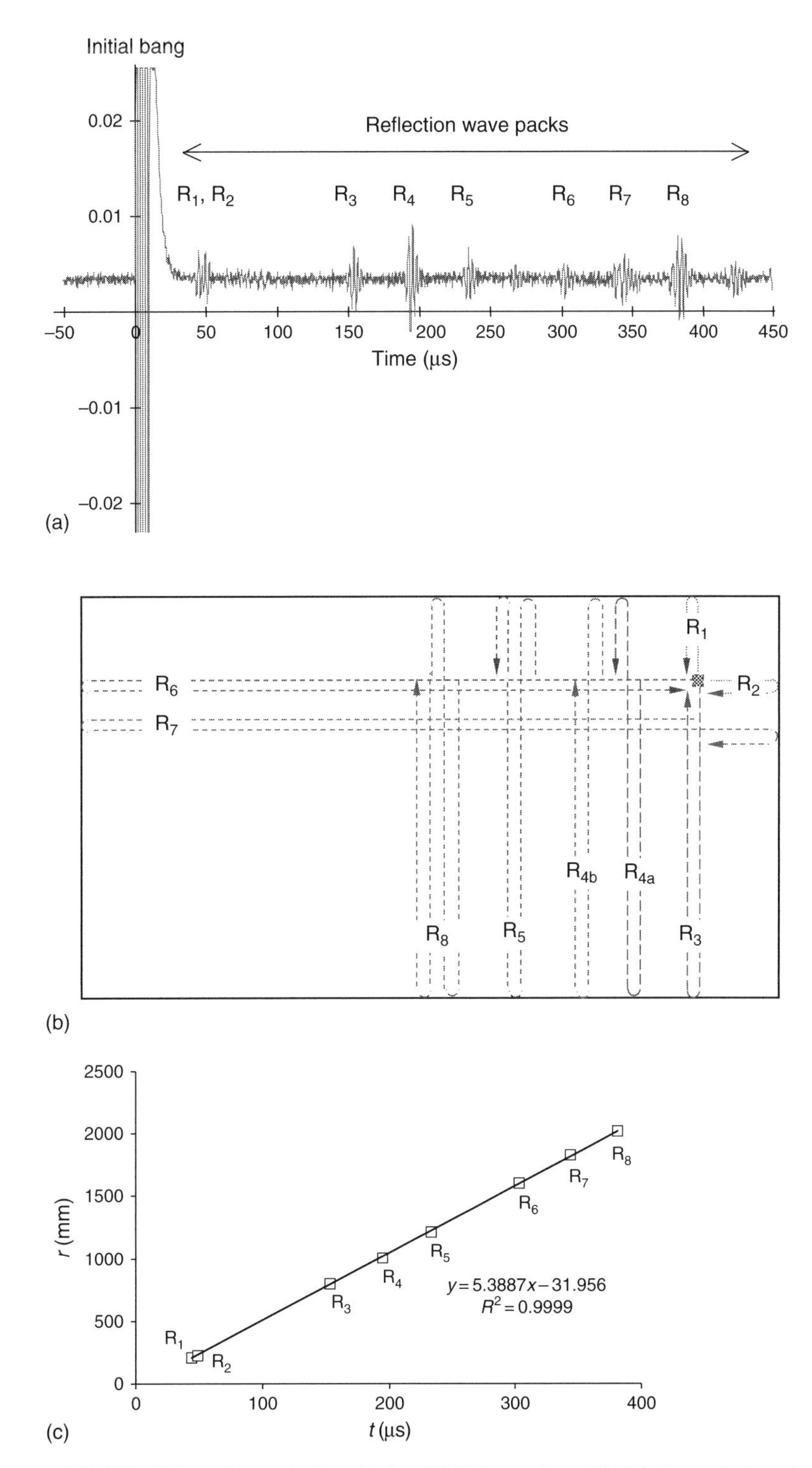

FIGURE 10.20 Pulse-echo method applied to PWAS transducer 11: (a) the excitation signal and the echo signals on PWAS 11; (b) schematic of the wave paths for each wave packet; (c) correlation between path length and time of flight ($R^2 = 99.99\%$).

TABLE 10.3 Analysis of pulse-echo signals of PWAS 11 on rectangular plate specimen

Wave packet label	R_1	R_2	R_3	R_4	R_5	R_6	R_7	R_8
Time of flight (μs)	43.8	48.8	152.8	194.4	233.2	302.8	343.2	380.8
Path length (mm)	104	114	400	504	608	800	914	1008

the range of operations of these small PWAS transducers is at least 2.5 m. This rage corresponds to a driving voltage of 10 V pp. Of course, if higher voltages were used, then the range of operation would increase even further.

10.4 PITCH-CATCH PWAS-EMBEDDED NDE

The pitch-catch method can be used to detect structural changes that take place between a transmitter transducer and a receiver transducer. The detection is performed through the examination of the guided wave amplitude, phase, dispersion, and TOF in comparison with a 'pristine' situation. Guided wave modes that are strongly influenced by small changes in the material stiffness and thickness (such as the A_0 Lamb wave) are well suited for this method. Typical applications include (a) corrosion detection in metallic structures; (b) diffused damage in composites; (c) disbond detection in adhesive joints; (d) delamination detection in layered composites, etc. Pitch-catch method can also be used to detect the presence of cracks from the wave signal diffracted by the crack.

10.4.1 THE CONCEPT OF PITCH-CATCH EMBEDDED NDE WITH PWAS

The pitch-catch method detects damage from the changes in the Lamb waves traveling through a damaged region. The method uses the transducers in pairs, one as transmitter, and the other as receiver. In the embedded pitch-catch method (Fig. 10.21), the transducers are either permanently attached to the structure or inserted between the layers of composite lay-up. Early studies in embedded pitch-catch NDE leveraged on SAW[2] technology. Culshaw and coworkers (Pierce et al., 1997, 2000; Gachagan et al., 1999) studied the use of piezoelectric transmitters and optical-fiber receivers for damage detection in composite plates using ultrasonic Lamb waves. Conventional ultrasonic transducers, interdigitated

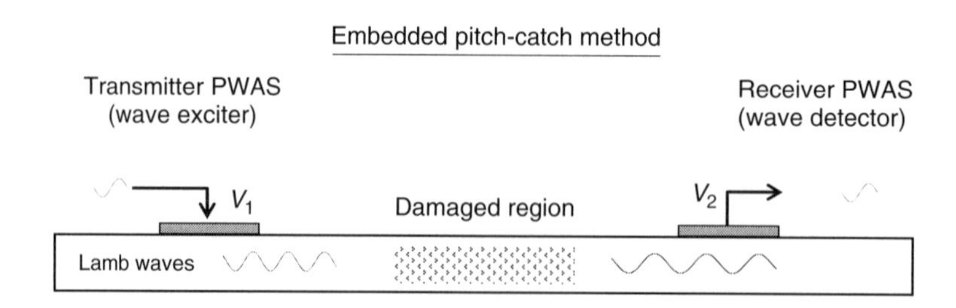

FIGURE 10.21 Embedded ultrasonics damage-detection techniques with pitch-catch method.

[2] SAW = surface acoustic wave. SAW devices have extensively been studied as delay lines in radar applications.

piezoelectric transducers, and simple PWAS were used. It was remarked that the PWAS gave good results in spite of their constructive simplicity.

Interdigitated PVDF[3] transducers were developed by Monkhouse et al. (1996, 2000) for the generation and detection of Lamb waves in embedded NDE applications. These transducers were able to generate focused directional waves in metallic specimens. However, the interdigitated transducers have a fixed operating frequency that is dictated by the interdigitated spacing. Chang and coworkers were among the first to identify the opportunity of using the piezoceramic disc wafers as both transmitters and receivers of Lamb waves. Keilers and Chang (1995) used an array of PWAS affixed to composite plates. Some of these PWAS acted as elastic waves generators, others acted as structural response detectors. The detection of damage was deduced from the differences in structural-response magnitude over the 0- to 2-kHz bandwidth. Damage detection criteria based on wave propagation were subsequently developed (Chang, 1998; Wang and Chang, 1999; Osmont et al., 2002; Wang et al., 2003; Mal et al., 2003).

10.4.2 PITCH-CATCH EMBEDDED NDE FOR CRACK DETECTION IN METALLIC STRUCTURES

Cracks in metallic structures typically run perpendicular to wall surface. A fully developed crack will cover the whole thickness (through-the-thickness crack) and will produce a tear of the metallic material. In conventional NDE, metallic-structure cracks are detected with ultrasonic or eddy current probes that have point-wise capabilities. Intensive manual scanning is required for crack detection. The aim of embedded pitch-catch NDE is to detect cracks in metallic structures using guided waves transmitted from one location and received at a different location. The analysis of the change in guided wave shape, phase, and amplitude should yield indications about the crack presence and extension.

Ihn (2003) presented an embedded pitch-catch method for the detection of cracks in metallic structures using an array of 12-mm diameter piezoceramic discs (Fig. 10.22). A 5-count 300-kHz smoothed tone burst applied to the transmitter (T) produced an omnidirectional Lamb wave into the plate. The receiver (R) detects a wave that is modified by the crack presence. The crack was grown through cyclic loading; measurements were taken at various crack length. The readings at zero crack length were taken as the baseline. The scatter wave was defined as the difference between the current received wave and the baseline wave. It was shown that the scatter amplitude increases linearly with the crack length.

Ihn (2003) applied the same embedded pitch-catch NDE method to detect multi-site crack damage in a metallic lap joint specimen (Fig. 10.23). The lap joint contained two rivet rows, each having nineteen 4.8-mm rivets. A row of 18 embedded transmitters (T) were used to send S_0 Lamb-waves to two rows of 18 embedded receivers (R). Constant amplitude cyclic tensile loading for accelerated fatigue testing was used. Multi-site crack damage was observed to develop at several locations. The test was interrupted seven times to evaluate the cracks progression (40, 65, 80, 100, 120, 140 kilocycles). Final failure occurred at 160 kilocycles. No visual inspection was possible. Crack progression was evaluated with conventional NDE methods (ultrasonic through-the-thickness scan and eddy current probes) as well as with embedded pitch-catch NDE. A damage index based on the cumulative TOF of the 420-kHz S_0 Lamb wave was used.

Figure 10.24 shows that the probability of detection (POD) of the embedded NDE pitch-catch method is comparable with that of conventional ultrasonic and eddy current

[3] PVDF = polyvinylidene difluoride, a piezoelectric polymer.

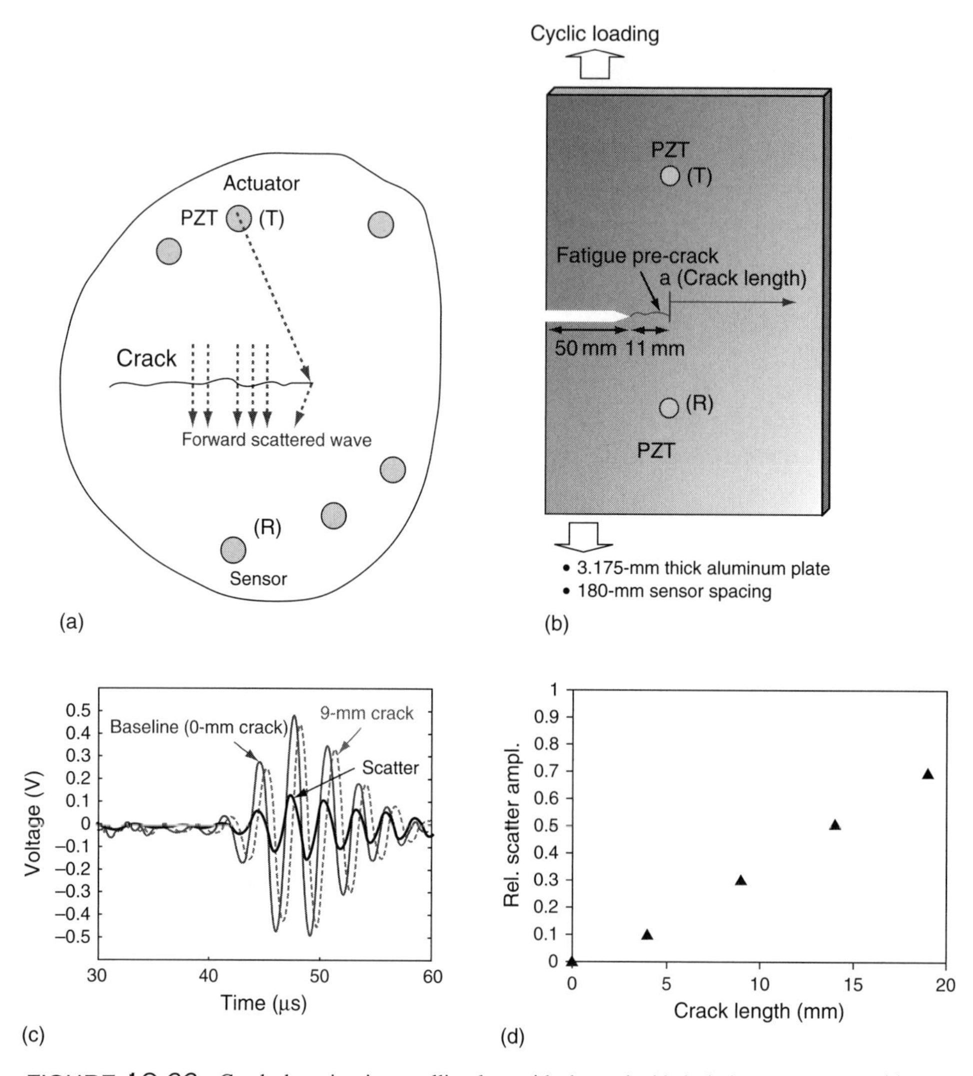

FIGURE 10.22 Crack detection in metallic plate with the embedded pitch-catch method: (a) conceptual configuration; (b) experimental setup; (c) received waves and scatter wave; (d) linear variation of scatter amplitude with crack length (Ihn, 2003).

methods for cracks larger then 5–8 mm. Circles represent embedded pitch-catch NDE; triangles represent eddy currents NDE. To obtain the POD, the different crack events monitored by conventional NDE were ranked from the largest to the smallest crack length and then compared the embedded pitch-catch method, i.e.,

$$\mathrm{POD} = \frac{\mathrm{SC}}{M + 1 - N} \tag{6}$$

where SC is the sum of the crack events recorded by the pitch-catch method; M is the total number of crack events recorded by NDE, and N is the serial event. The embedded pitch-catch NDE was also used for detecting fatigue crack growth in a metallic specimen repaired with a boron-epoxy patch. Good detection of crack larger than 4–8 mm was observed (Ihn, 2003).

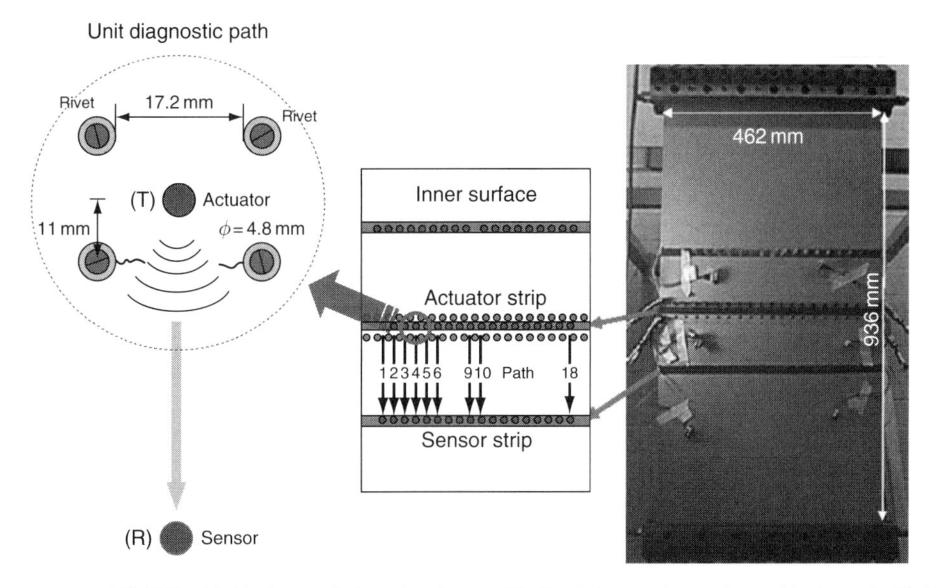

FIGURE 10.23 Multi-site crack detection in metallic lap joint specimen plate with the embedded pitch-catch method: two rows of 19 rivet holes were monitored using a row of 18 embedded transmitters (T) sending S_0 Lamb-waves to two rows of 18 embedded receivers (R) (Ihn, 2003).

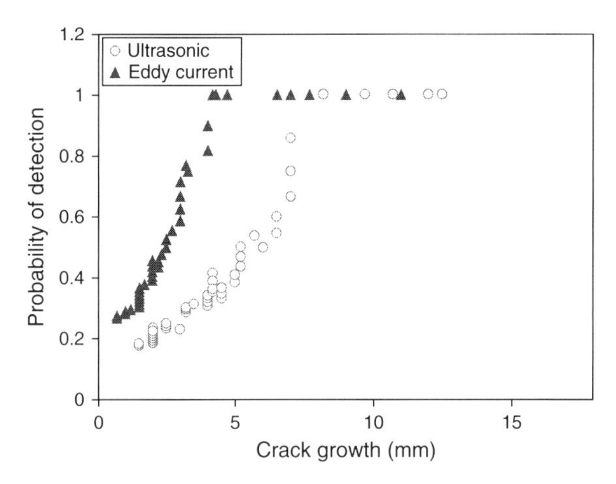

FIGURE 10.24 Probability of detection (POD) of the embedded pitch-catch in comparison with conventional ultrasonics and eddy currents (Ihn, 2003).

10.4.3 DELAMINATION DETECTION IN COMPOSITE STRUCTURES WITH EMBEDDED PITCH-CATCH NDE METHOD

Composite structures are typically resistant to through-the-thickness cracks due to the inherent crack-resistance of fiber reinforcement. However, in layered composite structures, cracks can easily propagate parallel to the wall surface, typically at the interface between layers. These cracks can be initiated by fabrication imperfections or low-velocity impact damage; subsequently, they propagated by cyclic fatigue loading. In conventional NDE, composite cracks and delaminations are detected with ultrasonic probes that can sense additional echoes due to across-the-thickness P-waves reflected by the delamination. Area coverage is achieved with surface scanning (C scans) using manual means or mechanical gantries. The aim of embedded pitch-catch NDE is to detect cracks and delaminations in composite structures using guided waves transmitted from one location and received at a different location. The disbond/delamination produces wave diffraction and mode conversion that can be analyzed and compared with the pristine signals. Analysis of the change in the guided wave shape, phase, and amplitude should yield indications about the crack presence and its extension. In addition, the sensor network built into the structure can be also used for monitoring low-velocity impact events that may be the cause of composite damage. Lamb waves have been widely used for impact damage detection in composite plates.

Chang (1998) first demonstrated that a network of piezoelectric discs could be embedded in a dielectric film, called 'SMART layer'. By emitting controlled diagnostic signals from a piezoelectric disc, the signals could be caught at neighboring piezoelectric discs. The pitch-catch technique was further used by Wang and Chang (1999) to identify impact damage in composite structures. The experimental setup consisted of an array of four piezoelectric discs connected through a selector box to either a signal generator/actuator amplifier or a sensor amplifier/DAQ card. Tone burst signals were sent from one of the piezoelectric discs and collected at all the other piezoelectric discs, in a round-robin fashion. The tone burst frequency was swept over the 40- to 150-kHz range. A_0 Lamb waves traveling in the plate were scattered by the damage. The signal interpretation consisted of analyzing the short-time Fourier transform (STFT) spectrograms to identify the TOF for each of the six paths possible between the four piezoelectric discs. The resulting TOF measurements were related to the location, size, and orientation of the damage by comparison with TOF values calculated with an analytical model. Detection results obtained with the embedded pitch-catch method compared very well with the specimen X-ray images. Similarly, Lemistre et al. (1999, 2000) used a network of transmitter-receiver piezoelectric discs (5-mm diameter, 0.1-mm thick) to detect delamination in a composite plate. A ten-cycle 365-kHz tone burst transmitted from one of the piezoelectric discs generated S_0 Lamb waves that were scattered by the damage into A_0 and SH_0 waves through diffraction and mode conversion. The received signals were analyzed with the discrete wavelet transform and compared with the pristine-plate signals. The analysis was performed using simple TOF considerations. No analytical model of the damaged plate was needed. Wavelet transform was also used by Staszewski et al. (1999), Deng et al. (1999), Su and Ye (2004), and others.

Diamanti et al. (2004) used PWAS transducers for the generation and reception of the fundamental anti-symmetric Lamb mode, A_0, in composite structures. The PWAS were arranged in a linear array in order to cover larger inspection areas. The optimal number and spacing of the transmitters was also discussed and determined. The results showed that impact damage can be successfully detected in multi-directional carbon-fiber reinforced-plastic (CFRP) laminates.

10.4.4 EMBEDDED PITCH-CATCH ULTRASONICS FOR DISBOND DETECTION

Bonded joints have a wide area of potential applications. However, the fact that the long-term durability and reliability of bonded joints is not fully understood slows down their implementation in critical applications. For this reason, the development of methods to perform reliably in situ SHM of bonded joints has received extensive attention.

Ihn (2003) used the pitch-catch technique with SMART-layer instrumentation to detect disbonds in boron-epoxy composite patch repair of cracks in metallic structures (Fig. 10.25). A SMART layer was embedded between the composite repair layers. A_0 Lamb waves, which are more sensitive to disbond and delaminations, were used.

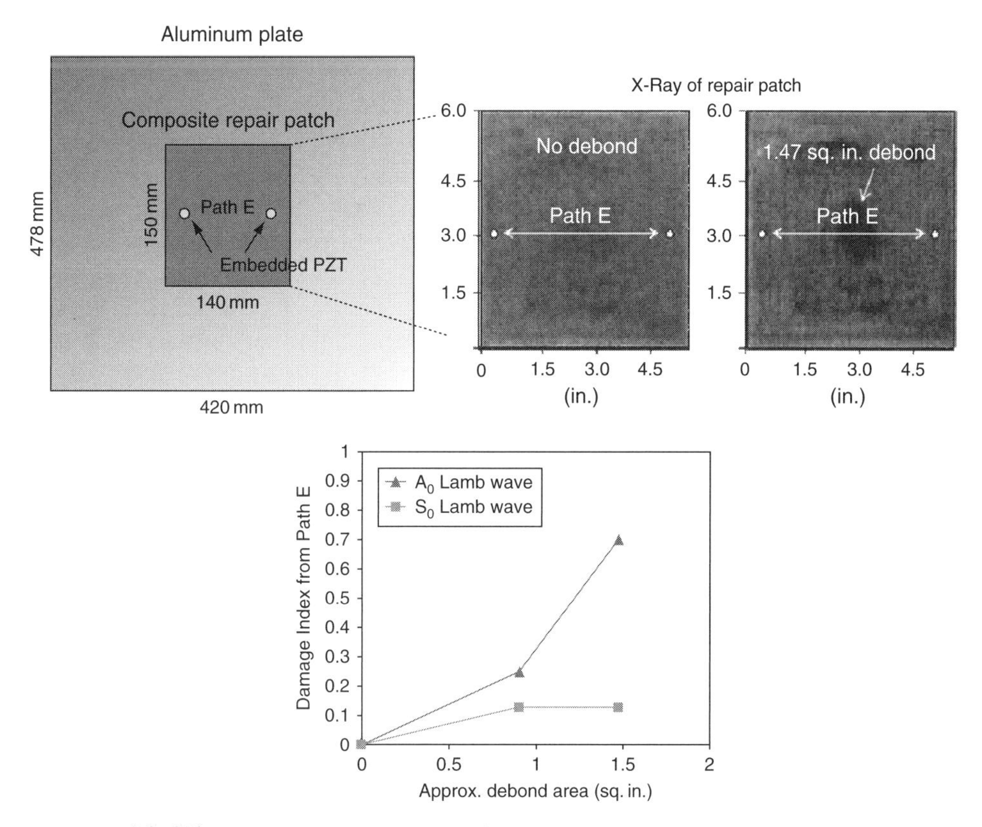

FIGURE 10.25 Embedded pitch-catch NDE for disbond detection of composite repairs of metallic structures (Ihn, 2003).

The A_0 Lamb wave frequency was 320-kHz. It was found that the damage index based on A_0 scatter energy was able to detect correctly the disbond presence.

Matt et al. (2005) considered the example of an unmanned air vehicle constructed with extensive use of adhesively bonded joints. In particular, they studied the critical adhesive connection of the composite wing skin to the wing spar. Two approaches for guided wave inspection of bonds were considered: (a) 'across the bond' in which the waves are generated in the adherent on one side of the bond and received across the bond; and (b) 'within the bond' in which the waves are both generated and detected in the bond region (Fig. 10.26a). The former approach was extensively discussed by Lanza di Scalea et al. (2001, 2004) in relation to a lap-joint of two metal plates (Fig. 10.26b), in which the guided wave leakage from one adherent into the other adherent through the bonded overlap is considered. The wave propagation is complicated by the mode conversions of the Lamb wave entering and leaving the bonded overlap. This happens because the dispersive behavior of a single plate is different from that of the multilayered overlap.

In the case of lap-shear joint, the dispersive behavior in the bonded overlap is modeled as a multilayer wave propagation problem for the three-bond states by using matrix-based methods. The mode conversion in the bonded overlap is a consequence of the different dispersive behavior of a single plate (the adherents) when compared with that of a multilayer structure (the bonded overlap). Lanza di Scalea et al. (2001, 2004) investigated three different interfaces between two aluminum plates: a fully cured epoxy layer, a poorly cured epoxy layer, and a slip interface layer. The tests were based on the measurement of the acoustic energy transferred from one adherent to the other one through the adhesive

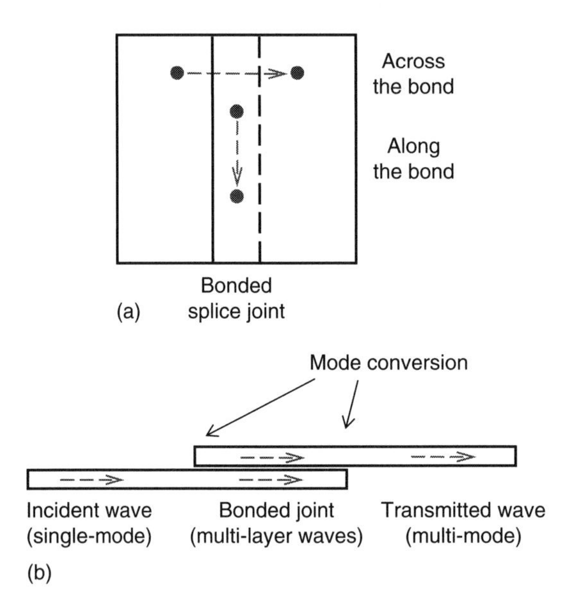

FIGURE 10.26 Embedded pitch-catch method for disbond detection: (a) 'across the bond' and 'within the bond' directions; (b) wave propagation schematics.

layer. An efficient energy transfer through the bondline is an indication of a good bond. A disbond would cause a dramatic decrease in transferred energy, whereas a poorly cured adhesive will result in less severe attenuation. The authors demonstrated that the Lamb wave modes with predominant in-plane displacements at the adherent outer edges are most sensitive to the presence of poorly cured adhesive, as the latter does not support shear-type wave propagation efficiently.

10.5 PULSE-ECHO PWAS-EMBEDDED NDE

In conventional NDE, the pulse-echo method has traditionally been used for across-the-thickness testing. For large area inspection, across-the-thickness testing requires manual or mechanical moving of the transducer over the area of interest, which is labor intensive and time consuming. It seems apparent that guided-wave pulse-echo would be more appropriate, because wide coverage could be achieved from a single location.

The embedded pulse-echo method follows the general principles of conventional Lamb-wave NDE. A PWAS transducer attached to the structure acts as both transmitter and detector of guided Lamb waves traveling in the structure. The wave sent by the PWAS is partially reflected at the crack. The echo is captured at the same PWAS acting as receiver (Fig. 10.27). For the method to be successful, it is important that a low-dispersion Lamb

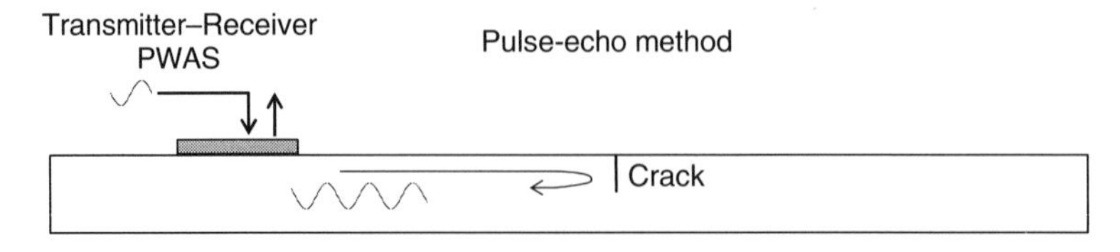

FIGURE 10.27 Principles of embedded ultrasonics damage detection with pulse-echo method.

wave be used. The selection of such a wave, e.g., the S_0 mode, is achieved through the Lamb-wave tuning methods (Giurgiutiu, 2003).

10.5.1 PWAS PULSE-ECHO CRACK DETECTION IN ALUMINUM BEAM

Moetakef et al. (1996) studied Lamb wave generation in metallic beams by piezoceramic wafers with interdigitated electrodes. This approach, bearing on the SAW technology, ensures excitation of the guided waves having the half wavelength equal to the interdigitated spacing. The transmitted signal was enhanced by phasing the excitation to individual electrodes proportional to the electrode spacing. For detection, single-electrode probes of half-wavelength width were used. Successful detection of echoes from notches was also reported.

10.5.2 PWAS PULSE-ECHO CRACK DETECTION IN COMPOSITE BEAM

Diaz Valdes and Soutis (2002) studied the detection of delaminations in a composite beam using the embedded pulse-echo method with low-frequency A_0 Lamb waves. Figure 10.28a shows the experimental setup. Rectangular PWAS (20 mm × 5 mm) were used with the length oriented along the beam axis. This ensures that low-frequency Lamb waves were predominantly excited along the beam length. Two PWAS were used, one as transmitter (actuator) and the other as receiver (sensor). A 5.5-cycle 15-kHz Hanning-smoothed tone burst was applied to the transmitter PWAS. Figure 10.28(b) shows the signal recorded in the pristine beam. The initial bang and the reflection from the end of the beam are apparent. Then, a delamination was generated in the composite beam using a scalpel blade. The size of the delamination was progressively increased, as indicated in Fig. 10.28(c). The presence of the delamination crack produced an additional echo, as shown in Fig. 10.28(d). This work was extended to 2-D composite plates by Diamanti et al. (2002).

Osmont et al. (2002) used the embedded pulse-echo method to detect damage in the foam core of a sandwich plate with glass fiber skins. The damage was simulated by a hole in the foam. Low-frequency A_0 Lamb waves (10 and 20 KHz) were used. It was shown that damage location and intensity could be deduced from the echo analysis without an

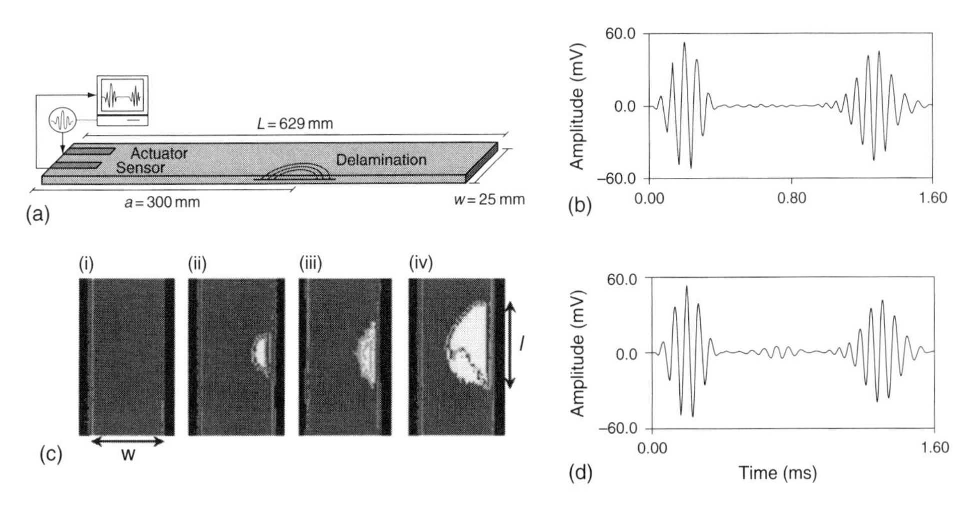

FIGURE 10.28 Detection of delaminations with the embedded pulse-echo: (a) experimental setup; (b) signals in the pristine specimen show only reflections from the beam end; (c) through-the-thickness C-scans of the specimen with conventional ultrasonics showing delamination increase; (d) signal in damaged specimen shows additional echo from the delamination crack (Diaz Valdes and Soutis, 2002).

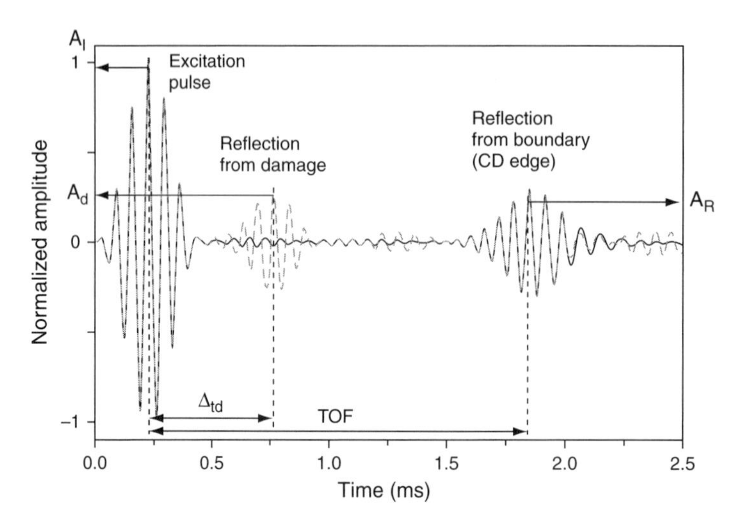

FIGURE 10.29 Comparative response of a pristine (solid line) and damaged (dashed line) carbon-fiber reinforced-plastic sandwich beam after low-velocity impact (Diamanti et al., 2004).

analytical model of the damaged plate. Barnoncel et al. (2003) used the same method to detect actual impact damages.

Diamanti et al. (2004) studied in situ disbond detection in sandwich plates with PWAS transducers. In sandwich composites, the dispersion curves at low frequencies are similar to those for free plates with a greater attenuation due to the presence of the core. S_0 mode at relatively high frequency was used for the detection of the debonding between the skin and the core and for low-velocity impact damage (Diamanti et al., 2004). Figure 10.29 shows the comparative response of a pristine and impact-damaged CFRP sandwich beam. The additional echo due to the disbond presence is clearly visible.

10.5.3 PWAS PULSE-ECHO CRACK DETECTION IN AGING AIRCRAFT-LIKE PANELS

This section discusses the use of PWAS pulse-echo method for crack detection in metallic panels of aging aircraft-like construction. For detection of through-the-thickness cracks in metallic specimens, higher frequency S_0 Lamb waves are required, such that the wavelength is sufficiently smaller than the crack length.

10.5.3.1 Aging Aircraft-like Panel Specimens

Realistic specimens representative of actual aircraft structures with aging-induced damage (cracks and corrosion) were designed and fabricated. Figure 10.30 presents a CAD drawing of these panels. The panels have a built-up construction typical of metallic aircraft structures. A lap splice joint, tear straps, and hat-shaped stringer/stiffeners are featured. The whole construction is made of 1-mm (0.040-in.) thick 2024-T3 Al-clad sheet assembled with 4.2-mm (0.166-in.) diameter countersunk rivets. Simulated cracks (EDM hairline cuts) and simulated corrosion damage (milled-out areas) were incorporated. Several specimens were constructed: (1) pristine; (2) with cracks; (3) with corrosion. The specimens were instrumented with several PWAS transducers, 7-mm square and 0.2 mm thick, made from APC-850 piezoceramic.

Wave propagation experiments were conducted on these realistic aircraft panel specimens using the PWAS pulse-echo method. The experimental apparatus was similar to

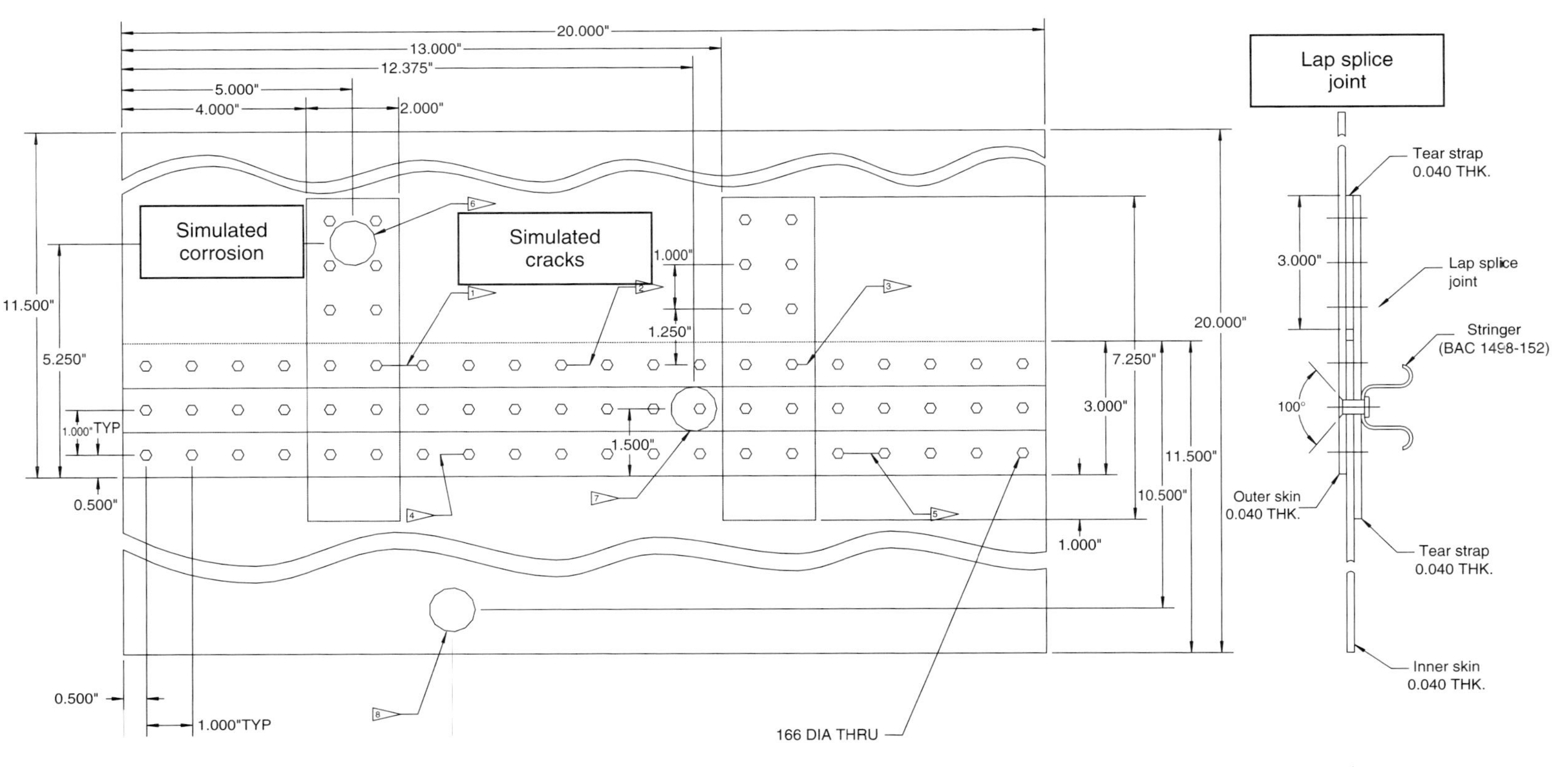

FIGURE 10.30 Aging aircraft-like lap-splice joint panel with simulated cracks (hairline slits) and corrosion (milled-out areas).

that used for the wave-propagation experiments on the rectangular specimen (Fig. 10.15). Several experiments were performed to verify the wave propagation properties and to identify the reflections due to the construction features of the panels (rivets, splice joints, etc.). We have shown earlier in this chapter that finite element simulation showed that the S_0 Lamb waves can give much better reflections from through-the-thickness cracks than the A_0 Lamb waves. This effect can be attributed to S_0 being (a) better reflected from the crack and (b) much less dispersive. The first point gives a strong signal, whereas the second ensures that the wave packet is compact and easy to interpret.

Figures 10.31 and 10.32 present a crack detection example. The challenge of this example is to detect a simulated crack emanating from a rivet hole. The detection is attempted with the PWAS pulse-echo method, by examining the PWAS signal and trying to identify the echo pattern changes that are due to the presence of the crack. The main difficulty is to distinguish between the echoes from closely placed structural features, such as the normal echoes from the rivet holes vs. the abnormal echo from the damage (crack emanating from the rivet hole). To make this distinction, we used the signal differential method. Procedural details, presented in a increasing order of complexity, are given next.

10.5.3.2 Pulse-Echo PWAS Signals Reflected by Structural Features

Figure 10.31 shows two photographs; adjacent to each photograph are the PWAS pulse-echo signals. This figure compares the situation with the lowest complexity compared with the situation of medium complexity. The situation with the lowest complexity is that in which neither the crack nor the rivet hole from which it emanates are present. Such a situation is shown in Fig. 10.31a. Only a vertical column of rivets can be seen on the far left. This vertical column of rivets, located at 200 mm from the PWAS, is indicative of the splice joint location. (In fact, the splice joint spreads outside the photo further to the left.)

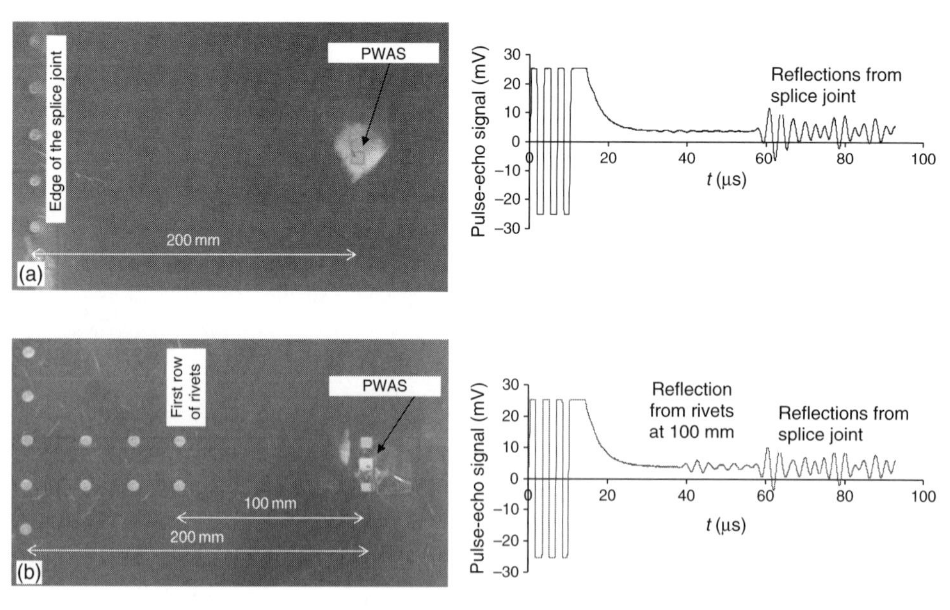

FIGURE 10.31 Effect of structural features on the pulse-echo signal: (a) the signal from a PWAS transducer placed a in a rivet-free region of a pristine panel shows only far reflection from the panel boundaries and large discontinuities, such as the splice joint at 200 mm from the PWAS transducer; (b) presence of new structural features, such as a row of rivets at 100 mm from the PWAS transducer, produces additional reflections that show up in the signal before the boundary reflections.

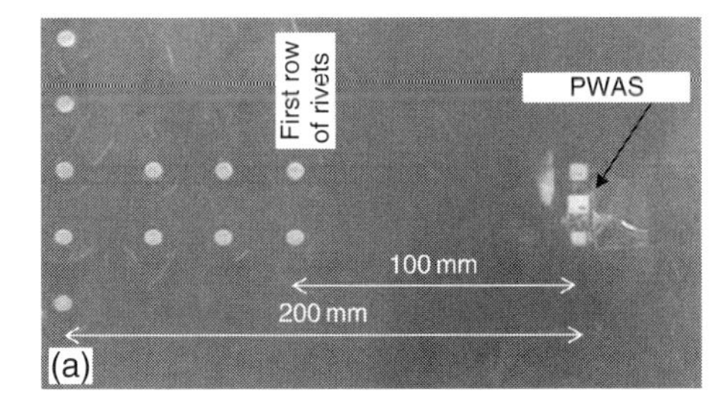

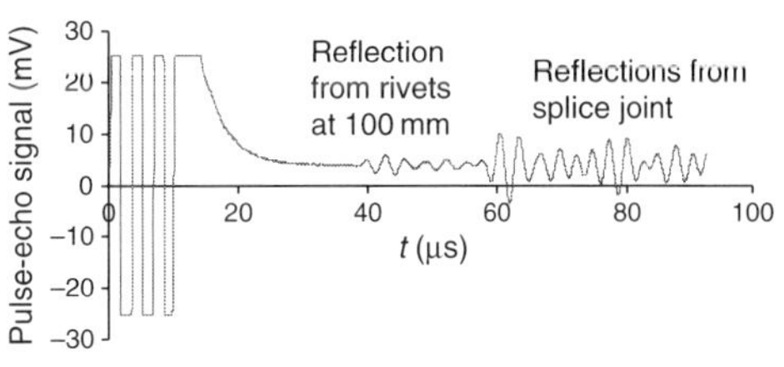

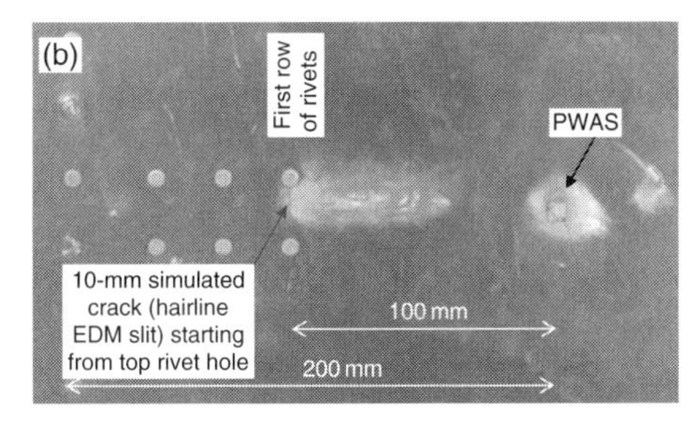

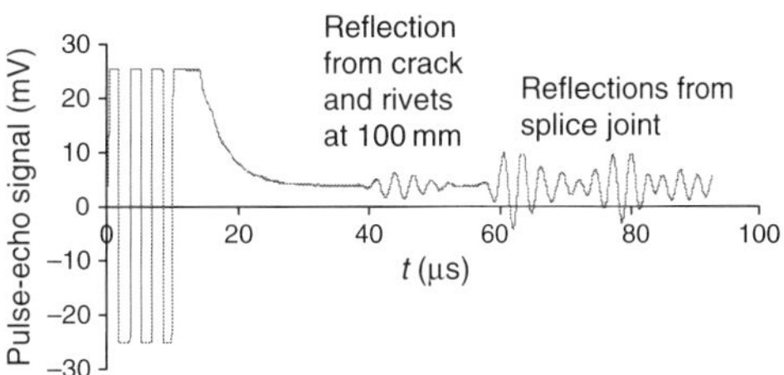

FIGURE 10.32 Effect of damage on the PWAS pulse-echo signal: (a) signal recorded in the pristine panel shows the reflections from the rivets @ 100 mm and from the panel splice joint @ 200 mm; (b) signal recorded on the damaged panel (10-mm hairline slit simulating a crack) shows a change in the reflections from the structural features @ 100 mm. However, this change is difficult to interpret due to the crack.

We see from this description that the PWAS was placed in a rivet-free region. The structural feature closest to it is the splice joint that starts at around 200 mm from the PWAS.

The PWAS pulse-echo signal is presented on the right hand right of Fig. 10.31a. The signal shows the initial bang (centered at around 5.3 μs) followed by an echo-free region. The echoes in the signal only start to arrive at approximately 60 μs. Then, a cluster of echoes start to arrive.

Figure 10.31b shows the situation with medium complexity. In addition to the vertical column of rivets in the far left, the structure has two horizontal rows of rivets that stretch from the splice joint over to the right up to 100 mm from the PWAS. The PWAS pulse-echo signal for this situation is presented on the right of Fig. 10.31b. One notices again that echoes from the structural splice start to arrive at approximately 60 μs. In addition, one notices the backscatter echoes from the rivets, which are visible at around 42 μs. The TOF relative to the initial bang middle position of 5 μs would be approximate TOF = 37 μs. This TOF is consistent with a 5.4 km/s wave speed and traveled round-trip length of 200 mm. This round-trip length of 200 mm corresponds to a distance of 100 mm from the PWAS to the reflective target. This reasoning indicates that the early echo visible at around 42 μs corresponds to reflections from the holes of the front rivets in the two horizontal rows of rivets. These front rivets are placed exactly at 100-mm distance from the PWAS.

10.5.3.3 Change in Pulse-Echo PWAS Signal due to Crack Presence

Figure 10.32 goes one-step further in increasing the complexity by adding the presence of the crack emanating from the first rivet hole in the horizontal top row of rivets. Figure 10.32a shows the same structural region as presented in Fig. 10.31b. Figure 10.32b shows a structural region identical to that shown in Fig. 10.31b but which has, in addition, a simulated crack (10-mm EDM hairline slit) starting from the first rivet in the horizontal top row. This situation represents the damaged structure. Let us now examine the changes that the presence of the crack has introduced into the pulse-echo signal. Figure 10.32b presents on the right the recorded pulse-echo signal. It shows signal features that are similar to those of Fig. 10.32a, but are somehow different at the 42 μs position. Ideally, we could compare the signal in Fig. 10.32b with that in Fig. 10.32a, and then detect the crack presence from the changes in the signal. The situation is difficult to judge because the crack and the rivet hole are at the same distance of 100 mm from the PWAS. Hence, the echo from the crack overlaps with the echoes from the adjacent rivet holes, resulting in an aggregate echo. We conclude that the detection of the crack seems particularly difficult because the echoes from the crack are mixed with the backscatter from the collocated rivets.

10.5.3.4 Signal Differential Method for Separation of Damage-related Echoes

However, this detection difficulty can be alleviated through the application of the signal differential method. This method consists in subtracting a baseline signal from the current signal. The subtraction should take into consideration both the amplitude

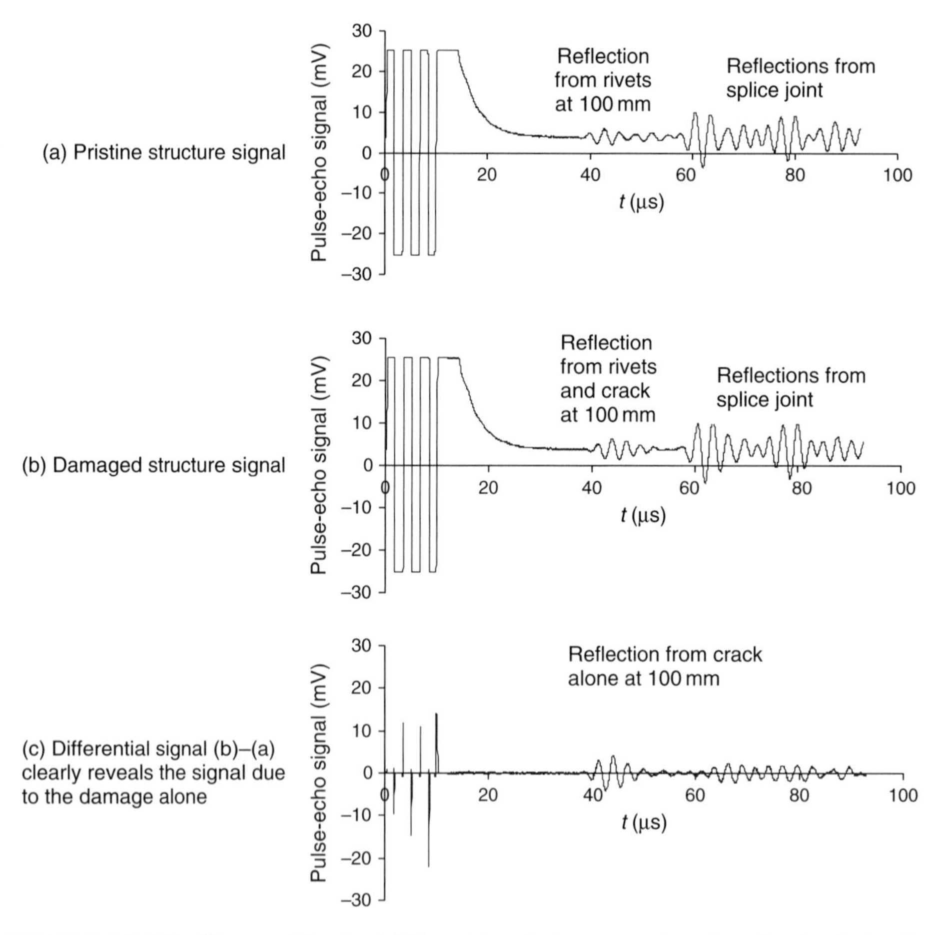

FIGURE 10.33 The use of the signal differential method to extract the pulse-echo signal related to the damage: (a) signal recorded on the pristine panel featuring the reflection from the rivets @ 100 mm and from the panel splice joint @ 200 mm; (b) signal recorded on the damaged panel featuring damage-induced modification of the 100-mm reflections; (c) signal differential method is able to reveal a strong reflection from the crack alone.

information and the phase 1 information of the signal. In our case, we will consider as baseline the signal measured on the panel without crack (Fig. 10.32a). By subtracting the baseline signal (Fig. 10.32a) from the current signal (Fig. 10.32b), the effect of the presence of the crack was readily identified.

The result of applying the signal differential method is shown in Fig. 10.33. First, let us recall the signal corresponding to the pristine structure (Fig. 10.33a). Then, let us recall the signal corresponding to the damaged structure (Fig. 10.33b). The signal from the damaged structure is not substantially different from the signal from the pristine structure. There are some differences in the echo around 42 μs, but these differences are difficult to immediately quantify. However, these difficulties are overcome by the signal differential method. The result of applying the signal differential method is given in Fig. 10.33c. This signal was obtained by subtracting the signal in Fig. 10.33a from the signal in Fig. 10.33b. The resulting signal (Fig. 10.33c) shows very clear the effect of the damage alone: it reveals a very clear and strong reflection at 42 μs, which, as indicated earlier, corresponds to a new reflector at 100 mm from the PWAS. Because this is a differential method, the reflector under discussion is something that is present in the damaged structure but was not present in the undamaged structure, i.e., the crack damaged shown at the left of Fig. 10.32b. The cleanness of the crack-detection feature and the quietness of the signal ahead of the crack-detection feature are remarkable. Thus, we conclude that this method has the potential for a clean and unambiguous detection of structural cracks.

10.6 PWAS TIME REVERSAL METHOD

The time reversal method was developed by Fink (1992) in connection with the pitch-catch method. The signal sent by the transmitter arrives at the receiver after being modified by the medium in which it travels. If the received signal is time reversed and sent back from the receiver to the transmitter, the effect of the medium is also reversed. This reversal is quite spectacular in the case of dispersive Lamb waves. In a series of remarkable experiments, Ing and Fink (1998) showed that the extensive dispersion of certain Lamb wave modes could be almost completely compensated through the time reversal method. Deutsch et al. (1997) also used the concept of reversing the reflected waves in order to focus a linear array of conventional ultrasonic transducers transmitting Rayleigh and Lamb waves. The time reversal method has proven especially useful in ultrasonic imaging of difficult media (Berryman et al., 2002).

In SHM, the time reversal method has been recently reported by several authors. Wang et al. (2003) used the time reversal method on an aluminum plate. Sohn et al. (2004) and Park et al. (2004) used the time reversal method to detect damage in composite panels. Their approach uses the assumption that 'because the time reversibility of waves is fundamentally based on the linear reciprocity of the system, this linear reciprocity and the time reversal break down if there exist any source of wave distortion due to wave scattering along the wave path. Therefore, by comparing the discrepancy between the original input signal and the reconstructed signal, damage could be detected.' This approach seems to remove the need for a historical baseline, which is inherent in other SHM methods. In addition, the method uses wavelet signal generation and processing to enhance the time reversal process in the case of multi-mode Lamb wave propagation.

Lamb-wave time reversal method is a new and tempting baseline-free damage detection technique for SHM. With this method, certain types of damage could be detected immediately without prior baseline data. However, this method is complicated by the existence of at least two Lamb modes at any given frequency and by the dispersion nature of the Lamb wave modes. The theory of Lamb wave time reversal is still under development.

This section presents a theoretical model for Lamb-wave time reversal in plate-like structure based on exact solutions of Rayleigh–Lamb wave equation. The theoretical model is first studied to predict the time reversal of single-mode and two-mode Lamb waves in plates. The validity of the proposed theoretical model is verified through experimental studies. In addition, time invariance of Lamb wave time reversal is presented.

10.6.1 THEORY OF PWAS LAMB WAVE TIME REVERSAL

10.6.1.1 Time Reversal Concept

Let us first review the time reversal concept as initially introduced by Fink (1999). The concept of *time reversal* consists of capturing an incoming wave with a receiver transducer, recording it, and then playing it backwards through the same transducer that now acts as a transmitter. Time reversal is the operation through which a signal $g(t)$ is played backwards as $g(-t)$; with an analog tape recorder, the tape is played back from the end toward the beginning; with digital recording, the last sample in the series becomes the first to be played back.

Within the range of sonic or ultrasonic frequencies where adiabatic processes dominate, the acoustic pressure field can be described by a scalar function $p(\vec{r}, t)$ that, within a heterogeneous propagation medium of density $\rho(\vec{r})$ and compressibility $\kappa(\vec{r})$, satisfies the Eq.

$$
\begin{aligned}
&(L_r + L_t)p(\vec{r}, t) = 0 \\
&L_r = \vec{\nabla} \cdot \left(\frac{1}{\rho(\vec{r})} \vec{\nabla} \right), \quad L_t = -\kappa(\vec{r})\partial_{tt}
\end{aligned}
\tag{7}
$$

Equation (7) is invariant with respect to a time-reversal process because L_t contains only second-order derivatives with respect to time (self-adjoint in time) and L_r satisfies spatial reciprocity because interchanging the source and the receiver does not alter the resulting fields (Fink, 2000).

In a non-dissipative medium, Eq. (7) guarantees that, for every burst of sound that diverges from a source, there exists (at least in theory) a set of waves that would precisely retrace the path of the sound back to the source. This remains true even if the propagation medium is inhomogeneous and presents variations of density and compressibility that may reflect, scatter, and refract the sound. If the source is point like, this allows focusing back on the source whatever the medium complexity (Fink, 2004).

The generation of this converging wave that reconstructs the initial source has been achieved through the so-called time-reversal mirrors (TRM). The TRM method involves two steps (Fink, 1999). In the first step, a source $f(t)$ emits pressure waves that propagate out and get distorted by the medium inhomogeneities. The ith transducer in the mirror array, acting as a receiver, detects the wave arriving at its location and feeds the signal to a computer. In the second step, the ith transducer acts as a transmitter and plays back the recorded signal in reverse $g(-t)$. This two-step process takes place simultaneously and synchronously in all the array transducers. In accordance with the time invariance principle of Eq. (7), the original wave is re-created, but traveling backward, retracing back to the source the passage of the original wave. Whereas traveling through the medium, the TRM wave is untangling its distortions and gets refocused at the point of original source where it arrives as $f(-t)$. Thus, after the time reversal procedure, the outgoing signal $f(t)$ is refocused back onto its original source and reconstructed in reverse as $f(-t)$. These two properties of the time reversal method have been used in many pressure-wave applications such as underwater acoustics, telecommunications, room acoustics, ultrasound medical imaging, and therapy (Fink, 2004).

10.6.1.2 Time Reversal of Lamb Waves

Because the Lamb wave's complexity and multi-modal character, the use of the time reversal concept in Lamb wave applications is relatively new and has only been explored by a few researchers. Alleyne et al. (1992) explored the time reversal method for NDE applications by transmitting a dispersive Lamb wave over a given distance and obtaining a TRM reconstruction in the form of a clean waveform with improved SNR and space resolution. Ing and Fink (1996 and 1998), Pasco et al. (2006) used TRM to focus Lamb wave energy and detect flaws or damage in plates. These studies showed that the time reversal procedure recompresses a dispersive wave, improves the spatial resolution of the testing, and makes it easy to interpret the experimental data. More recently, Lamb wave TRM has been tried for developing a baseline-free SHM technique (Kim et al., 2005; Park et al., 2004). This technique uses the reconstruction property of the time reversal procedure, i.e., an original wave is reconstructed at its source location by recording the forward wave arriving at another location and sending it back as a time-reversed signal toward the source. The crucial hypothesis of this approach is that, when damage is presented in the path between the source and the TRM, the forward wave may be mode converted, scattered, and/or reflected by the damage, and thus the TRM reconstruction procedure will break down. If the reconstructed wave is compared with the original transmission (suitable scaled), then the waveform anomalies in the reconstructed wave will be indicative of damage being present in the structure. Because the method uses in its comparison nothing else but its own signals and no information about the previous state for structure, the method is baseline-free. A typical experiment was discussed by Park et al. (2004). A steel block was placed between two surface-bonded PWAS, A and B. Without the block attached, the reconstructed wave is close in shape to the original wave. When the block was attached, the reconstructed wave differed from the original wave. Thus, the presence of damage can be detected from comparing the shape of the reconstructed wave with that of the original wave.

Although Lamb wave time-reversal technique has been attempted experimentally and shows to be effective for detecting certain types of damages for SHM, the theory of Lamb wave time reversal has not been fully studied. Wang et al. (2004) presented a theory of Lamb wave time reversal using the Mindlin plate wave theory. It predicts time reversal of the flexural plate wave, which is a satisfactory approximation of the A_0 Lamb wave mode for low and medium frequency values; however, this analysis does not consider other widely present Lamb wave modes, such as the S_0 mode, or the multi-mode Lamb waves, such as the combination mode $S_0 + A_0$. In addition, these initial studies did not include the interaction between PWAS and structure in their analysis.

This section develops an analytical model of Lamb-wave time reversal using PWAS transducers. To validate the theoretical, numerical studies and experimental tests of the time reversal of single-mode (S_0 mode or A_0 mode) and two-mode ($S_0 + A_0$ mode) Lamb waves are performed. The time invariance of the Lamb-wave time reversal process is also discussed.

10.6.1.3 Lamb Wave Excitation Signal

As already shown in previous chapters, Lamb waves are generally dispersive. After traveling a long distance, waves of different frequencies will separate from each other,

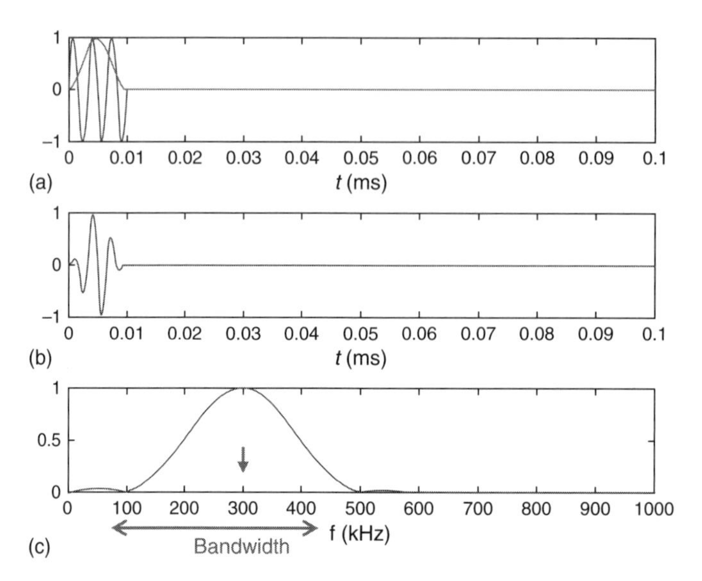

FIGURE 10.34 A 300-kHz Hanning-windowed tone burst: (a) pure-tone burst and a Hanning window; (b) Hanning-windowed tone burst; (c) magnitude spectrum of the windowed tone burst.

will distort the signal, which will become more difficult to analyze. Using input signals of limited bandwidth, one can reduce the problem of dispersion, but cannot eliminate it entirely. In our study, the Lamb waves were excited with a smoothed tone burst obtained from a pure tone of frequency f through the application of a Hanning window (Fig. 10.34a, b). The Hanning window is described by the Eq.

$$h(t) = 0.5 \cdot [1 - \cos(2\pi t/T_H)], \quad t \in [0, T_H] \tag{8}$$

The number of counts (N_B) in the tone bursts is related to the length of the Hanning window by the relation

$$T_H = N_B/f \tag{9}$$

Thus, the smoothed tone burst is governed by the Eq.

$$x(t) = h(t) \cdot \sin(2\pi f t), \quad t \in [0, T_H] \tag{10}$$

The case of a 300-kHz tone burst is illustrated in Fig. 10.34. The windowed tone burst concentrates much of the input energy at the excitation frequency of 300-kHz as indicated in its magnitude spectrum (Fig. 10.34c). In principle, this should enable us to excite the Lamb wave at the points on the dispersion curves where the group velocity is either stationary or almost stationary with respect to frequency. However, as indicated in the same Fig. 10.34c, there is also a considerable spread of the spectrum due to the sidebands of the Hanning window. An important conclusion becomes apparent: It is impossible to concentrate the energy of a finite duration input signal at a single frequency whereas using wave packets of finite length. This is due to the uncertainty principle. The same principle indicates that the signal bandwidth is inversely proportional to the time duration of the signal. Hence, for a tone-burst excitation with certain count number, N_B, the higher the frequency, the shorter the time duration, and hence the wider the main lobe bandwidth (spectral spreading). To maintain the concentration of the tone burst energy in a tight frequency band, the tone burst count number, N_B, should be increased as the frequency is increased.

10.6.2 MODELING OF PWAS LAMB WAVE TIME REVERSAL

Time reversal of Lamb wave on metallic thin plate can be modeled by the following two steps: (1) Apply tone burst excitation V_{tb} at PWAS 1 and record the forward wave V_{fd} at PWAS 2; (2) Emit the time-reversed wave V_{tr} from PWAS 2 back to PWAS 1. The return wave picked up by PWAS 1 is the reconstructed wave V_{rc}.

We constructed a time-reversal simulation program in accordance with the schematic diagram of Fig. 10.35. The simulation incorporates forward Fourier transform (FFT) and inverse Fourier transform (IFFT). The subscripts *tb*, *fd*, *tr*, and *rc* signify tone burst, forward, time reversed, and reconstructed waves, respectively. As illustrated in Fig. 10.35, the relationship between the tone burst excitation V_{tb} and the reconstructed wave V_{rc} can be easily expressed using the Fourier transform method, i.e.,

$$V_{rc}(t) = IFFT\{V_{tr}(\omega) \cdot G(\omega)\} = IFFT\{V_{tb}(-\omega) \cdot |G(\omega)|^2\} \tag{11}$$

The function $G(\omega)$ is the frequency-dependent structural transfer function that affects the wave propagation through the medium. In certain simplified cases, as illustrated below, the structural transfer function $G(\omega)$ can be obtained in analytical form, thus facilitating and expediting our simulation studies. For example, we have shown in a previous chapter that, in an isotropic plate under the excitation of an ideally bonded PWAS, the Lamb wave solution in terms of strain has the closed-form

$$\begin{aligned} \varepsilon_x(x,t)|_{y=d} = & -i\frac{a\tau_0}{\mu}\sum_{\xi^S}(\sin\xi^S a)\frac{Ns(\xi^S)}{D_S'(\xi^S)}e^{-i(\xi^S x-\omega t)} \\ & -i\frac{a\tau_0}{\mu}\sum_{\xi^A}(\sin\xi^A a)\frac{N_A(\xi^A)}{D_A'(\xi^A)}e^{-i(\xi^A x-\omega t)} \end{aligned} \tag{12}$$

10.6.2.1 Time-reversal Simulation for Two-mode Lamb Waves (A_0 and S_0)

To keep the problem analytically tractable, let us consider that wave propagation takes place at an *fd* value for which only two Lamb-wave modes exist, A_0 and S_0. In this case, the structural transfer function $G(\omega)$ can be written using the strain wave Eq. (12) as

$$G(\omega) = S(\omega)e^{-i\xi^S x} + A(\omega)e^{-i\xi^A x} \tag{13}$$

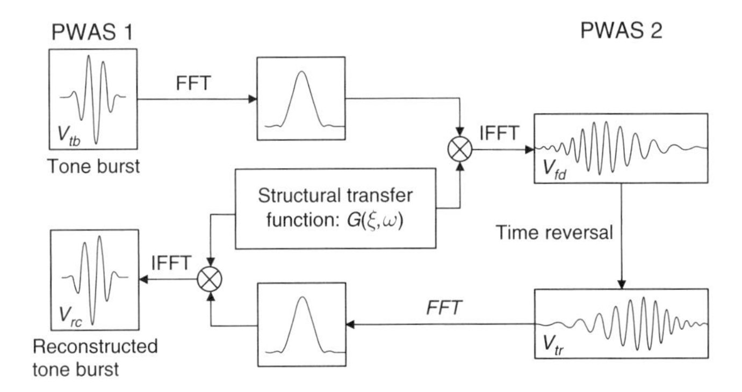

FIGURE 10.35 Lamb wave time-reversal procedure block diagram.

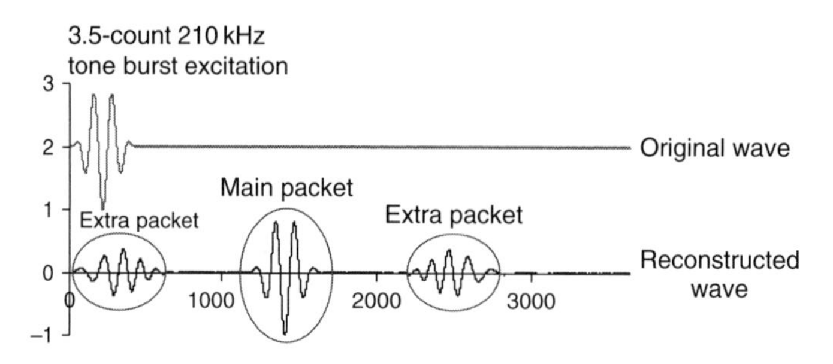

FIGURE 10.36 Reconstructed wave using 3.5-count 210-kHz tone burst excitation in simulation of two-mode Lamb wave time reversal.

where $S(\omega)$ and $A(\omega)$ are the functions corresponding to the S_0 and A_0 Lamb-wave modes, i.e.,

$$S(\omega) = -i\frac{a\tau_0}{\mu}\sin(\xi^S a)N_S(\xi^S)/D_S'(\xi^S) \tag{14}$$

$$A(\omega) = -i\frac{a\tau_0}{\mu}\sin(\xi^A a)N_A(\xi^A)/D_A'(\xi^A) \tag{15}$$

Thus,

$$|G(\omega)|^2 = |S(\omega)|^2 + |A(\omega)|^2 + S(\omega)A^*(\omega)e^{-i(\xi^S-\xi^A)x} + S^*(\omega)A(\omega)e^{i(\xi^S-\xi^A)x} \tag{16}$$

where the symbol * denotes the complex conjugate and $|S(\omega)|^2$, $|A(\omega)|^2$ represent frequency dependent real numbers. When we substitute Eq. (16) into Eq. (11), the first two terms on the R.H.S. of Eq. (16) generate together one wave packet in the reconstructed time-domain wave V_{rc}. The other two terms (the third and the fourth terms) generated two extra wave packets that will be placed symmetrically about the first packet. The locations of these two extra wave packets can be predicted using Fourier transform property of left and right time shifts.

A plot of the reconstructed wave using the time reversal procedure applied to the two-mode Lamb wave model of Eq. (13) is given in Fig. 10.36. The simulation was run for a 3.5-count 210-kHz tone burst in a 1-mm aluminum plate. Three wave packets are observed in the reconstructed wave, as predicted by Eq. (16) substituted in Eq. (11). Hence, for time reversal of a Lamb wave containing two modes (S_0 mode and A_0 mode), the reconstructed wave V_{rc} contains three wave packets. Although the input signal is not time-invariant in this case, the main wave packet in the reconstructed wave may still resemble its original tone burst excitation if $|S(\omega)|^2 + |A(\omega)|^2$ remains constant over the tone burst spectral span. This theoretical deduction explains the experimental observations of Kim et al. (2005) regarding the side packets that appear besides the main packet in the TRM-reconstructed signal.

10.6.2.2 Time-reversal Simulation for Single-mode Lamb Waves

The situation described in the previous section could be alleviated if a single-mode Lamb wave could be excited. For single-mode Lamb waves, the $G(\omega)$ would simplify considerably and only a single wave packet will appear in the TRM reconstruction. For example, let us consider that PWAS-Lamb wave mode tuning technique presented in a previous chapter is used to excite only the A_0 mode. Because of the spectral spread of the tone burst excitation, this may not be exactly experimentally possible, but an approximation in which the A_0 mode is dominant can be envisaged. If the A_0 mode is dominant, then the $G(\omega)$ function becomes

$$G(\omega) = A(\omega)e^{-i\xi^S x} \tag{17}$$

Substitution of Eq. (17) into (11) yields

$$V_{rc}(t) = IFFT\{V_{tb}(-\omega) \cdot |A(\omega)|^2\} \tag{18}$$

Equation (18) shows that the reconstructed wave V_{rc} for the case of a single-mode Lamb wave has the same phase spectrum as that of the time-reversed tone burst $V_{tb}(-t)$, whereas its magnitude spectrum is equal to that of $V_{tb}(-t)$ modified by a frequency-dependent coefficient $|A(\omega)|^2$. For narrow-band excitation, $|A(\omega)|^2$ can be assumed to be constant. Thus, Eq. (18) becomes

$$V_{rc}(t) = \text{Const} \cdot IFFT\{V_{tb}(-\omega)\} = \text{Const} \cdot V_{tb}(-t) \tag{19}$$

Equation (19) implies that the reconstructed wave V_{rc} resembles the time-reversed tone burst excitation V_{tb}. If the tone burst excitation were symmetric, i.e., $V_{tb}(t) = V_{tb}(-t)$, then the reconstructed wave V_{rc} shape would have identically the same shape as the original tone burst, and the comparison could be made directly to the original tone burst.

Thus, we have proved that the A_0 single-mode Lamb wave is exactly time reversible without the appearance of the additional side packets shown in Fig. 10.36. This assertion is expected to hold as long as the excitation is sufficiently narrow-band to keep the tuned A_0 mode dominant. It is apparent that similar conclusions can be derived for the S_0 single-mode Lamb wave, which would again be exactly time reversible as long as the excitation is sufficiently narrow-band.

10.6.3 NUMERICAL AND EXPERIMENTAL VALIDATION OF LAMB WAVE TIME REVERSAL

Several numerical and experimental tests were performed to validate the Lamb-wave time reversal theory developed in the previous section.

10.6.3.1 Experimental Setup

Figure 10.37a shows the experimental setup for the Lamb wave time reversal studies. It consists of a HP33120 function generator, a Tektronix 5430B oscilloscope, and a PC. Two specimens were used. One specimen was a 1524-mm × 1524-mm × 1-mm aluminum plate on which two 7-mm diameter round PWAS were bonded 400-mm apart (Fig. 10.37b). The other specimen was a 1060-mm × 300-mm × 3-mm aluminum plate on which two 7-mm square PWAS were bond 300-mm apart (Fig. 10.37c). The PWAS pairs were used in pitch-catch mode. Modeling clay was used on both specimens around their edges to eliminate the boundary reflection and keep the signals clean.

10.6.3.2 PWAS Mode Tuning on Specimens

PWAS mode-tuning experiments were conducted on each specimen to identify the single-mode or multi-mode Lamb-wave tuning frequencies. A Hanning-windowed tone

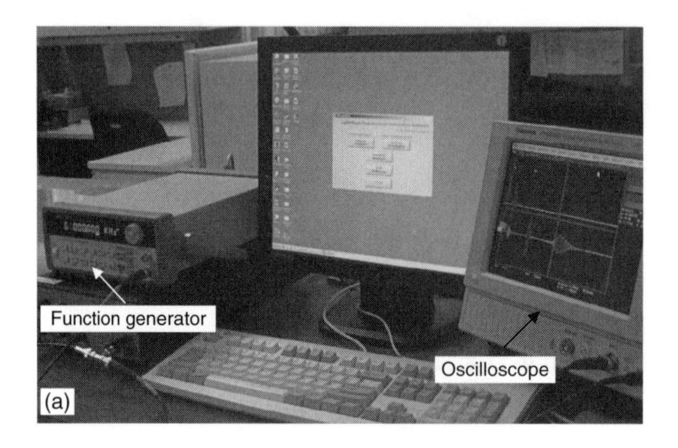

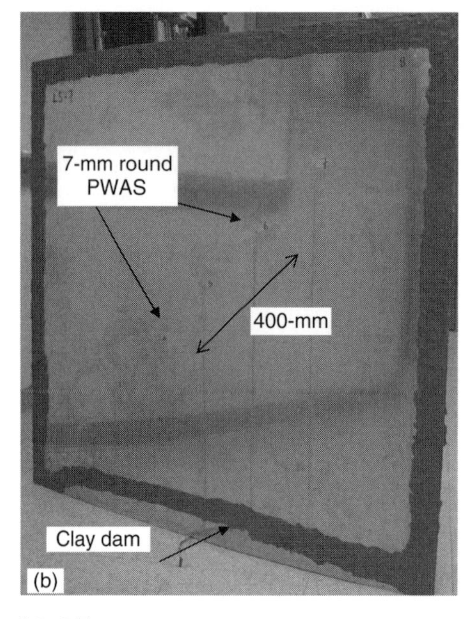

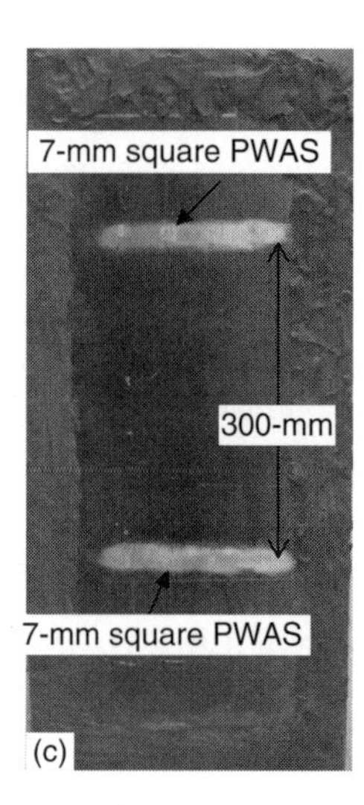

FIGURE 10.37 Time-reversal experimental setup and specimens: (a) time reversal experimental setup; (b) 1524-mm × 1524-mm × 1-mm 2024-T3 aluminum plate bonded with two 7-mm round PWAS, 400-mm apart; (c) 1060-mm × 300-mm × 3-mm 2024-T3 aluminum plate bond with two 7-mm square PWAS, 300-mm apart.

burst swept from 10 to 700-kHz in steps of 20 kHz was applied to one of the PWAS, whereas the amplitude response of the symmetric and anti-symmetric modes was recorded at the other PWAS.

Figures 10.38 and 10.39 show the PWAS mode-tuning results on the 1-mm plate with 7-mm round PWAS and on the 3-mm plate with 7-mm square PWAS, respectively. It is noted that both the normalize strain curves predicted by Eq. (12) (left) and the measured strain curves (right) follow the general pattern of a sine function, which hits zeros when the half length of the PWAS matches an odd multiple of one of the wavenumbers of the Lamb waves. For both specimens, A_0 Lamb wave mode is dominant at low frequencies. As frequency increases, the S_0 mode increases. The S_0 and A_0 modes reach similar strength around 210 kHz. As frequency further increases, the S_0 mode continues to increase overshadowing the A_0 mode. The S_0 becomes dominant at 300 kHz in the

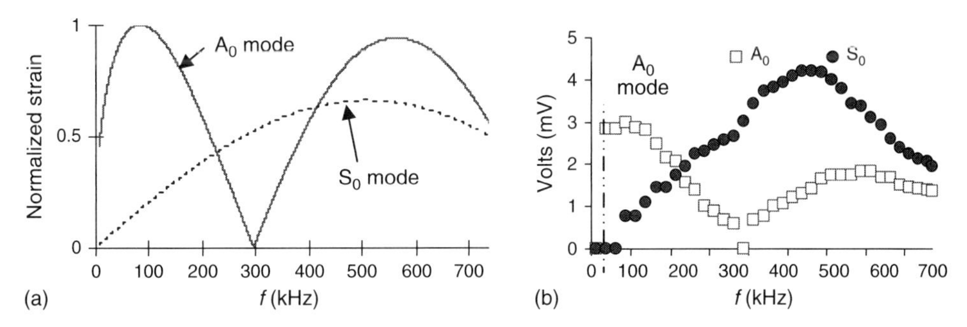

FIGURE 10.38 Lamb wave response of a 1-mm 2024-T3 aluminum plate under 7-mm round PWAS excitation: (a) normalized strain response predicted by Eq. (12); (b) experimental data in voltage units.

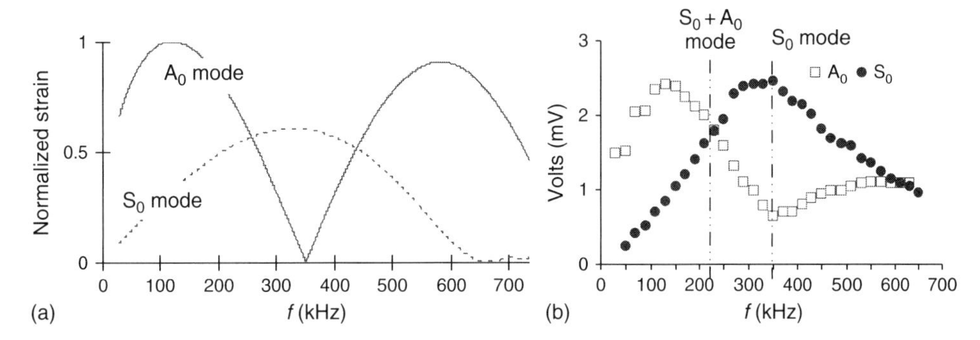

FIGURE 10.39 Lamb wave response of a 3-mm 2024-T3 aluminum plate under 7-mm square PWAS excitation: (a) normalized strain response predicted by Eq. (12); (b) experimental data.

1-mm specimen with 7-mm round PWAS. In the other specimen, having a thickness of 3 mm and PWAS of 7-mm square, the S_0 becomes dominant at around 350 kHz.

10.6.3.3 PWAS Lamb Wave Time Reversal Results

The time-reversal experiment was conducted in two main steps:

1. Forward wave generation: the function generator outputs tone burst to the PWAS transmitter to excite Lamb wave in the plate, and the PWAS receiver was connected to the oscilloscope to record the forward wave in the plate.
2. Time reversal and tone burst reconstruction: the signal from the receiver PWAS was time reversed, downloaded to the function generator volatile memory, and emitted back to the transmitter PWAS. The whole experiment was automated by using a LabVIEW program.

Numerical analysis of time reversal of the three combinations of Lamb-wave modes was performed by following the procedure presented in Fig. 10.35, where the $G(\omega)$ function is given by Eq. (13). Notice that an additional step was performed to time reverse the reconstructed wave for display comparison purpose. In addition, because we are only interested in the shape comparison of the reconstructed wave vs. the original wave, both the experimental and the numerical results were normalized appropriately.

10.6.3.4 Time Reversal of A_0 Mode Lamb Wave

Figure 10.40 shows both the normalized numerical and the experimental waves in the A_0 mode Lamb wave time reversal procedure. The Lamb wave excitation used is a 3-count tone burst with its central frequency at 36 kHz (Fig. 10.40a). Because A_0 is dominant at this frequency, the captured forward wave after propagating 400 mm mainly consists of A_0 mode wave packet, whereas S_0 mode wave packet is suppressed (Fig. 10.40b). The wave packet due to the initial bang appears sometimes in the experimental signals because of the electromagnetic (EM) coupling; this initial bang does not relate to our discussion and should be ignored. When we time reverse the forward wave and emit it back (Fig. 10.40c), the dispersed A_0 wave packet is recompressed (Fig. 10.40d). The reconstructed experimental wave resembles the time-reversed original tone burst.

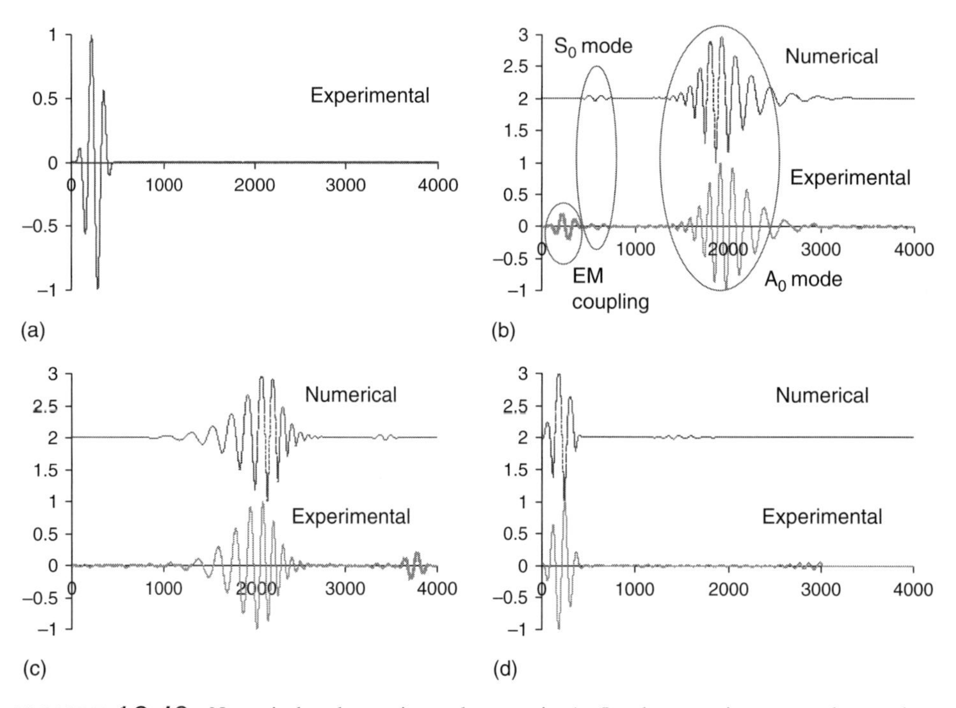

FIGURE 10.40 Numerical and experimental waves in A_0 Lamb wave time reversal procedure: (a) 3-count 36-kHz original tone burst; (b) forward wave after propagating 400 mm; (c) time reversed forward wave; (d) reconstructed wave.

10.6.3.5 Time Reversal of S_0 Mode Lamb Wave

In this experiment, a 3.5-count symmetric tone burst with its central frequency at 350 kHz (Fig. 10.41a) was used to excite S_0 mode Lamb wave in plate. As shown in Fig. 10.39, S_0 mode is maximized around 350 kHz, whereas A_0 mode is minimized. However, the 3.5-count 350-kHz tone burst excitation has a certain spectral spreading; A_0 mode Lamb wave is slightly excited and clearly observed in forward wave after propagating 300-mm in plate (Fig. 10.41b). When we time reverse the forward wave and emit it back to the source point (Fig. 10.41c), the normalized and reconstructed waves plotted in Fig. 10.41d are obtained. Although there are some residual waves, the main wave packet in the reconstructed wave resembles the original wave.

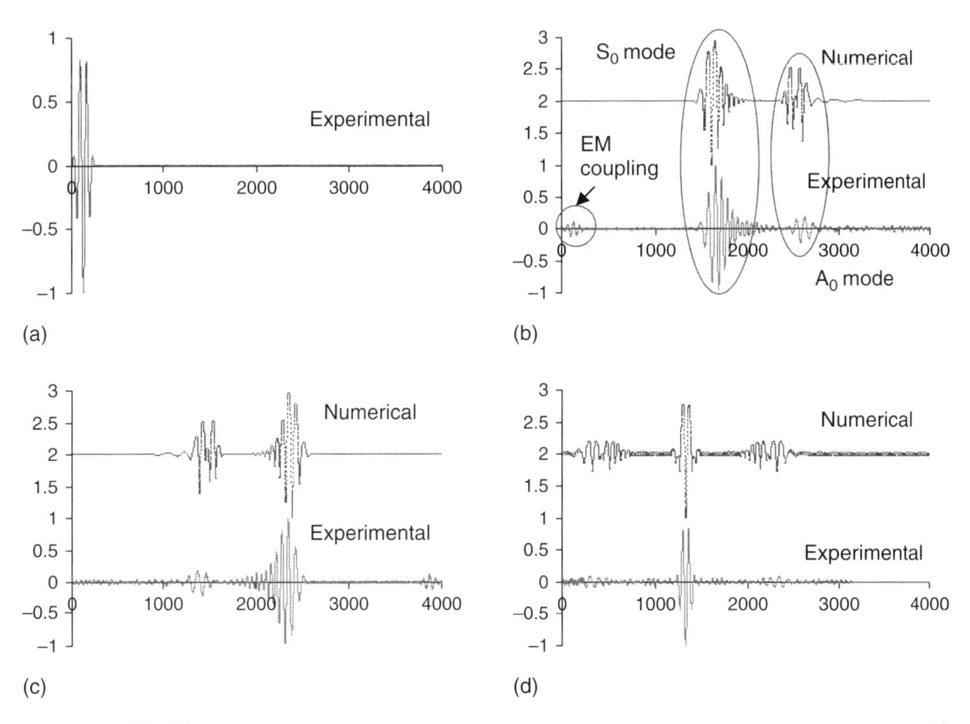

FIGURE 10.41 Numerical and experimental waves in S_0 Lamb wave time reversal procedure: (a) 3.5-count 350-kHz original tone burst; (b) forward wave after propagating 300-mm; (c) time reversed forward wave; (d) reconstructed wave.

10.6.3.6 Time Reversal of S_0+A_0 Mode Lamb Wave

A 3.5-count symmetric tone burst was tuned to 210 kHz to excite both S_0 mode and A_0 mode Lamb wave in the plate (Fig. 10.42a, b). As predicted in a previous section, three wave packets were obtained in the reconstructed wave. The first and the third wave packets are symmetric about the second one; the middle wave packet resembles its original tone burst excitation.

Figures 10.40, 10.41, and 10.42 show that the numerical and experimental waves in the time reversal procedure are very much in agreement. This indicates that the PWAS Lamb-wave time reversal theory predicts well the experimental results.

10.6.3.7 Time Invariance of Lamb-Wave Time Reversal

Let us now examine the time invariance character of the TRM reconstructed waves. If the shape of the reconstructed wave is identical to the original wave, then the procedure is time invariant. For single-mode Lamb wave time reversal, the input wave was reconstructed as shown in Figs 10.40 and 10.41. To judge time invariance, we superposed in Fig. 10.43 the original and the TRM-reconstructed waves. (The dashed line is used for the original wave, whereas the solid line is used for the TRM-reconstructed wave.) Figure 10.43a shows the A_0 results, whereas Fig. 10.43b shows the S_0 results. We notice that the TRM-reconstructed wave superposes almost identically over the original wave. This means that the TRM process for single-mode Lamb-wave excitation is time invariant. (Some small difference between the reconstructed wave and the original wave still exist possibly because of numerical error.)

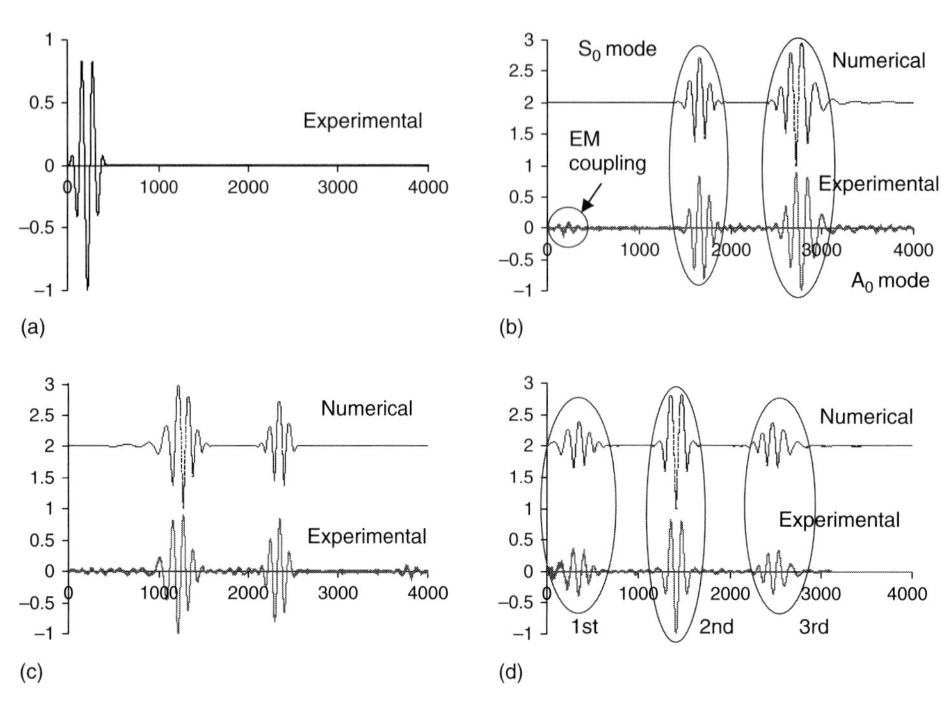

FIGURE 10.42 Numerical and experimental waves in two-mode Lamb wave time reversal procedure: (a) 3.5-count 210-kHz original tone burst; (b) forward wave after propagating 300 mm; (c) time-reversed forward wave; (d) reconstructed waves.

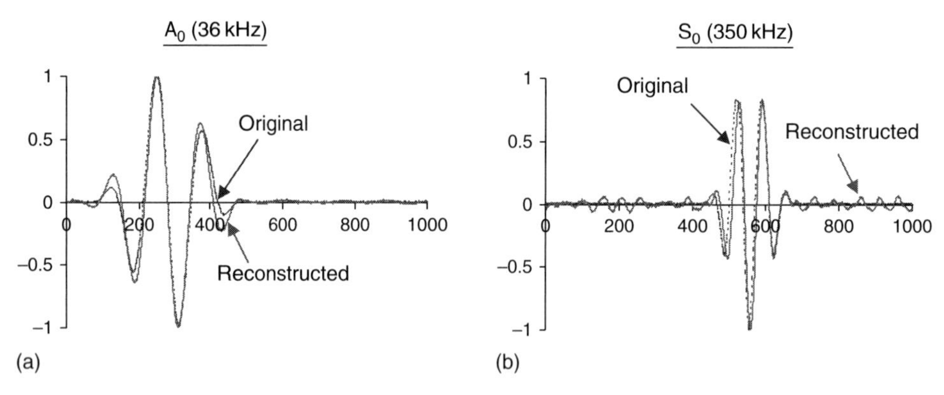

FIGURE 10.43 Superposed original tone burst and reconstructed tone burst after time reversal procedure: (a) 36 kHz, A_0 mode; (b) 350 kHz, S_0 mode.

For two-mode (210 kHz) Lamb time reversal, the reconstructed Lamb wave contains three wave packets whereas the original wave had only one packet. This indicates that, for multi-mode Lamb-wave excitation, the TRM procedure is no longer time invariant. However, the original wave is still reconstructed as the middle wave packet. Figure 10.44 shows superposed the original wave and the reconstructed middle wave. The agreement seems quite good. (The small differences that still exist between the reconstructed and the original waves may be attributed to numerical error.)

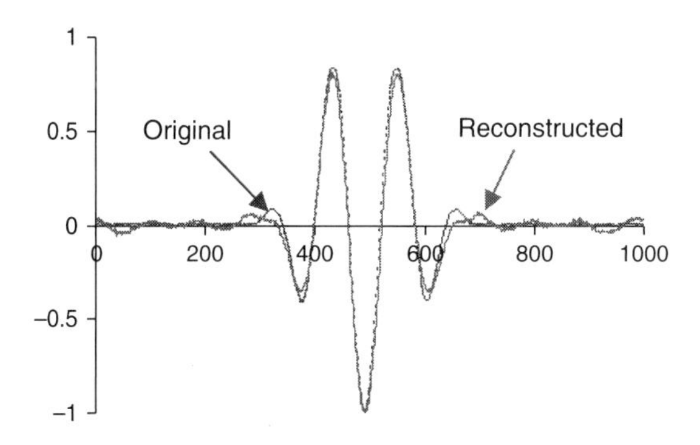

FIGURE 10.44 Superposed original tone burst and reconstructed tone burst after time reversal procedure: 210 kHz, $S_0 + A_0$ mode.

The differences between reconstructed and original waves decrease with the increase in the count number of the tone burst excitation, i.e., with the decrease of the tone burst bandwidth. This can be easily understood by considering the frequency domain behavior of the $G(\omega)$ function. As the frequency bandwidth decreases, the $G(\omega)$ function becomes almost constant. We performed a numerical study of this difference using the root mean square deviation method and calculating the similarity value as

$$Similarity(i, j) = 1 - RMSD = 1 - \sqrt{\frac{\sum_N [A_i - A_j]^2}{\sum_N (A_j)^2}} \tag{20}$$

where N is the number of points in the plot, and i, j denote two plots under comparison. This method compares the amplitude of two sets of data and uses Eq. (20) to assign a scalar value in the range from 0 to 1 (i.e., from 'not related' to 'identical').

Table 10.4 shows the similarity values for the reconstructed tone bursts in A_0 mode, S_0 mode, and $S_0 + A_0$ mode time-reversal procedures. The similarity increases with the increase of the tone burst count number. For A_0 mode time reversal, the similarity increases from 80.3% to 88.5% when the tone burst count number increases from 3 to 6. For $S_0 + A_0$ mode and S_0 mode time reversal, the similarity increases when the tone burst count number increases from 3.5 to 6.5 counts. Comparison of the similarity between the $S_0 + A_0$ mode and S_0 mode reveals that the $S_0 + A_0$ mode always possesses higher similarity than the S_0 mode for a certain count number. This happens because $S_0 + A_0$ mode in the experiment is excited at lower frequency and possesses narrower frequency span than the S_0 mode. Thus, to reconstruct better a certain Lamb wave mode using

TABLE 10.4 Similarity between reconstructed tone burst and original tone burst

Frequency (kHz)	36				210				350			
Mode	A_0				$S_0 + A_0$				S_0			
Count number	3	4	5	6	3.5	4.5	5.5	6.5	3.5	4.5	5.5	6.5
Similarity (%)	80.3	84.7	87.5	88.5	86.9	89.0	89.8	90.5	54.0	66.4	73.7	86.3

the time reversal procedure, a tone burst with lower central frequency and higher count number seems to be preferable.

10.6.4 PWAS TUNING EFFECTS ON MULTI-MODE LAMB WAVES TIME REVERSAL

In the tuning curve of Fig. 10.38, the A_0 mode and S_0 mode show similar strengths around 500-kHz. Therefore, both wave modes are simultaneously excited by a 500-kHz tone burst. The time-reversal reconstructed wave (Fig. 10.45) displays two big residual wave packets to the left and right of the reconstructed wave. In contrast, a single-mode excitation of the S_0 can be achieved at 290 kHz. When the S_0 mode is dominant, the reconstructed waveform becomes much better and has much smaller residual side packets (Fig. 10.46). However, Fig. 10.46 still shows some small residual wave packets, to the left and to the right of the main reconstructed wave. The reason for these residual wave packets is that the 16-count tone burst has a finite bandwidth, and hence a small amplitude residual A_0 mode gets excited besides the dominant S_0 mode. To eliminate the residual waves, a tone burst with increased count number, i.e., of narrower bandwidth, should be used. However, the signal will become longer in time duration and will lose its time-domain resolution. If we excite with a low frequency (say, 30 kHz), we find that the A_0 mode is dominant whereas the S_0 mode is very weak. In addition, at these lower frequencies, the bandwidth of the 16-count tone burst input signal becomes narrower in actual kHz values as compared with the high-frequency bands. Hence, the TRM reconstruction is much cleaner, as indicated in Fig. 10.47. The 30-kHz test signal yields a very good reconstruction with the time-reversal method: the A_0 mode dominates, the narrow-band

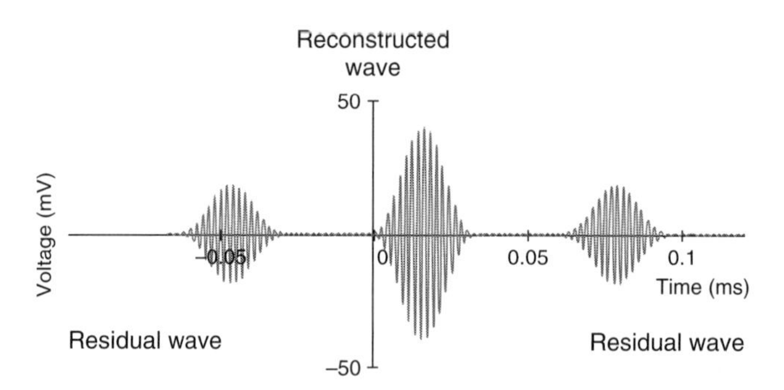

FIGURE 10.45 Untuned time-reversal reconstruction using 16-count tone burst with 500-kHz carrier frequency shows strong residual signals because multi-mode Lamb waves are present.

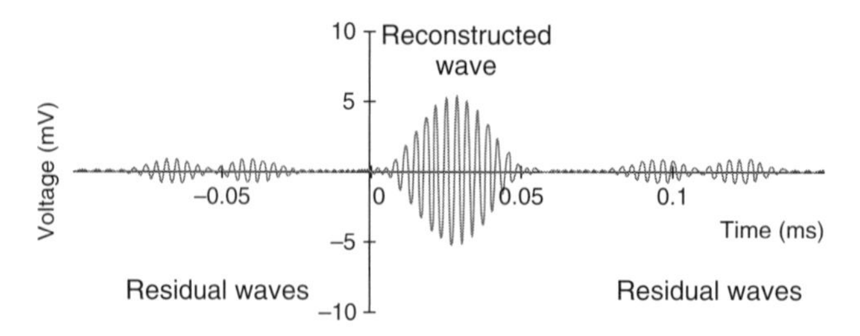

FIGURE 10.46 Time reversal with S_0 Lamb wave mode tuning shows that a reconstructed 16-count 290-kHz tone burst with carrier frequency; weak residual wave packets due to residual A_0 mode component are still present due to the side band frequencies present in the tone burst.

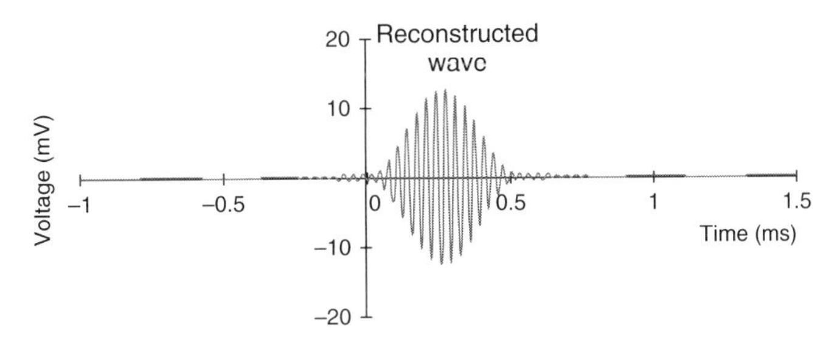

FIGURE 10.47 Time reversal with A_0 Lamb mode tuning: reconstructed input using 16-count tone burst with 30-kHz carrier frequency; no residual wave packets are present.

input signal was perfectly reconstructed by the time-reversal method, and no residual wave packets are practically observed.

Another important fact studied in our simulation was the relative amplitude between the reconstructed wave and the residual wave packets obtained during the time-reversal process at various excitation frequencies. Figure 10.48 shows the plots of the reconstructed wave packet amplitude and of the residual wave packet amplitudes over a wide frequency range (10 to 1100 kHz). As it can be seen, for an input signal with fixed number of counts (here, a 16-count tone burst), the residual wave packets amplitudes vary with input frequency. The residual reaches local minimum values and local maximum values at certain tuning frequency points. Hence, the frequency tuning technique can be used to select the optimized input frequency that will improve the reconstruction of the Lamb wave signal through the time-reversal procedure. Using this Lamb-wave mode tuning approach, we can achieve a much cleaner indication of damage presence. For example, for the PWAS size of 7 mm and plate thickness of 1 mm considered in our simulation. Figure 10.48 indicates that the 30 kHz, the 300 kHz, the 750 kHz, and probably the 1010 kHz would be optimal excitation frequencies to be used with the time-reversal damage detection procedure for this particular specimen and PWAS types.

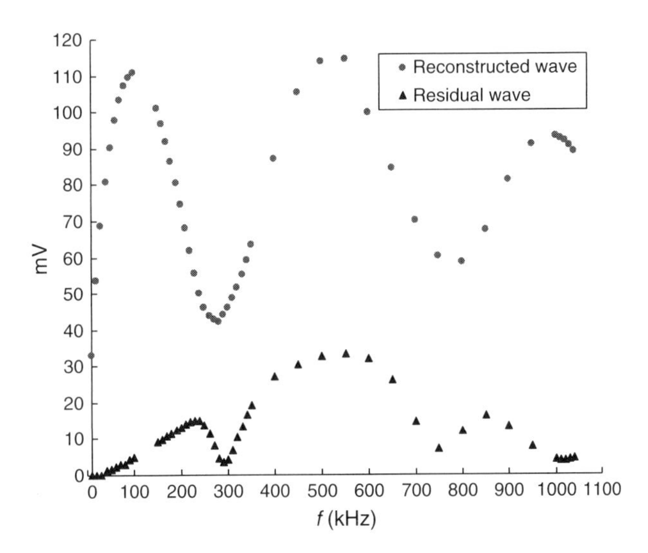

FIGURE 10.48 Reconstructed wave and residual wave in terms of their maximum amplitudes using 16-count tone burst over wide frequency range (10 to 1100 kHz).

10.6.5 SECTION SUMMARY

As a baseline free SHM technique, Lamb-wave time-reversal method has experimentally demonstrated its ability to detect certain types of damage in plate-like structure without the need of baseline data. However, unlike the time reversal of 3-D pressure waves, the time reversal of Lamb waves is complicated by the dispersion and multi-mode characteristics of the Lamb waves. Hence, the theory of time reversal of Lamb waves has not yet been fully studied. This section has presented some initial results obtained in the development of a Lamb-wave time reversal theory for PWAS SHM applications. A time reversal model based on the understanding the coupling between the PWAS transducers and the Lamb waves in the structure was developed. It was found that Lamb waves are strictly time reversible only under certain circumstances. The main conclusions are:

1. Narrow-band tone bursts of single-mode Lamb waves (e.g., S_0 mode or A_0 mode) are time reversible. Time reversibility of a single-mode Lamb-wave packet increases as the bandwidth of the tone burst excitation decreases. Single-mode Lamb wave have been obtained by using the PWAS Lamb-wave frequency tuning.
2. When time reversal is applied to two-mode Lamb wave signals, three wave packets instead of a single packet were obtained. Of the three, one is the result of true time-reversal reconstruction and resembles well the original excitation. The other two packets are unwanted signals which are artifacts of applying the time-reversal method to multi-mode signals. In other words, time reversal invariance is only valid for single-mode Lamb waves and seems to break down for signals containing more than one Lamb-wave modes.

Two sets of laboratory experiments were conducted in order to verify the prediction obtained from the theoretical model. Plates of 1-mm and 3-mm thickness, in conjunction with 7-mm round and 7-mm square PWAS were used, respectively. The results indicate that the model predicts well the experimental results. During the experiments, the tuned S_0 wave (350 kHz, Fig. 10.39b) was observed to be time reversed nicely in the 3-mm plate using 7-mm square PWAS. The tuned A_0 wave (36 kHz, Fig. 10.38b) was observed to be time reversed nicely in the 1-mm plate using 7-mm round PWAS. Outside the tuning range, where two Lamb wave modes are present (e.g., at 210 kHz, Fig. 10.39b), the time reversal generated additional side packets of Lamb waves in addition to the main wave packet.

The results presented here were done on pristine specimens in order to demonstrate the PWAS Lamb wave time reversal method. However, it is apparent that using the PWAS Lamb waves tuning technique, the single-mode Lamb wave time reversal method can identify damages in thin wall structures without prior information.

10.7 PWAS PASSIVE TRANSDUCERS OF ACOUSTIC WAVES

So far, we have seen the PWAS transducers as active transmitters and receiver of elastic waves. However, PWAS can also act as passive transducers of elastic waves. Two situations will be examined: (a) ID with PWAS transducers, and (b) AE detection with PWAS transducers.

10.7.1 IMPACT DETECTION WITH PWAS TRANSDUCERS

Impact detection (ID) with PWAS transducers in composite materials was successfully demonstrated by Wang and Chang (1999), Dupont et al. (2000), and others. In our experiments, the network of PWAS transducers of Fig. 10.16 was also used to detect low-velocity foreign-object impact on an aluminum plates. In our experiments, we used a small steel ball (0.16 g) dropped from a height of 50 mm. As shown in Fig. 10.49a, signals were collected at PWAS 1, 5, 7, 9. Figure 10.49b gives the signals recorded at PWAS 1, 5, 7, 9. The corresponding signal TOF values are $t_1 = 126\,\mu s$, $t_5 = 160\,\mu s$, $t_7 = -27\ \mu s$, $t_9 = 185\ \mu s$, relative to the oscilloscope trigger.

The distance and TOF data given in Table 10.5 can be used to calculate the impact position. Assuming the unknown impact position is (x, y), the following set of simultaneous nonlinear equations represent the relation between distance, group velocity, and TOF.

$$(x_i - x)^2 + (y_i - y)^2 = [c\,(t_i + t_0)]^2, \quad i = 1, \ldots, 4 \tag{21}$$

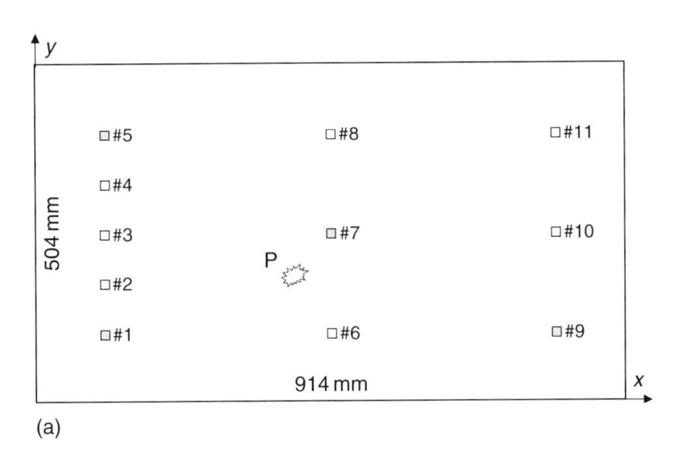

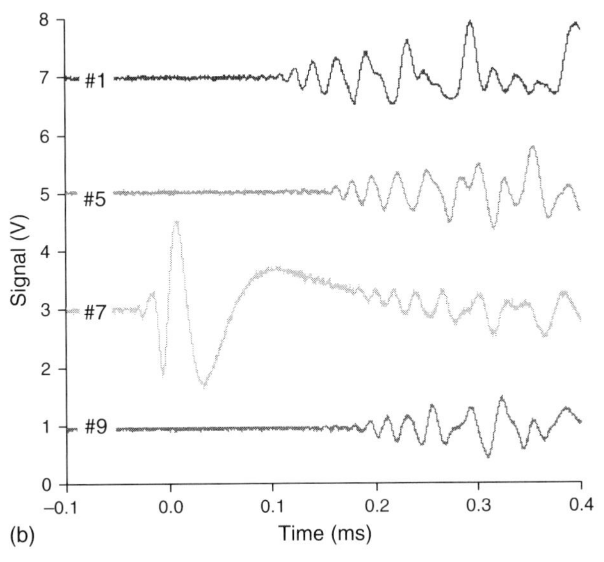

FIGURE 10.49 Impact detection experiments: (a) location of impact and PWAS transducers (b) captured ID signals with arbitrary time origin due to oscilloscope trigger. The ID event was a 0.16-g steel-ball dropped from a 50-mm height; the locations of the PWAS transducers and of the event are given in Table 10.5.

TABLE 10.5 Location of PWAS transducers by position (x, y) and radial distance, r, from the acoustic emission (AE) and impact detection (AE) events. Note that the time of flight was adjusted by 76.4 μs to account for oscilloscope trigger.

	Distance (mm)			
Sensor	x	y	r	TOF for ID (μs)
1	100	100	316	126
5	100	400	361	160
7	450	250	71	−27
9	800	100	412	185
Reconstructed ID event	402.5	189.6	N/A	
Actual AE/ID event	400	200	N/A	N/A

These equations represent a set of four nonlinear equations with four unknowns, which can be solved using error minimization routines. The unknowns are the impact location (x, y), the wave speed, c, and the trigger delay, t_0. In our studies, we tried two solution methods: (a) global error minimization and (b) individual error minimization. It was found that individual error minimization gave marginally better results, whereas the global error minimization was more robust with respect to initial guess values. The impact location determined by these calculations was $x_{\text{impact}} = 402.5\,\text{mm}$, $y_{\text{impact}} = 189.6\,\text{mm}$. These values are within 0.6 and 5.2%, respectively, of the actual impact location (400 and 200 mm).

A discussion of this method should address two important issues:

(1) the correct estimation of the arrival time
(2) the dispersive nature of the Lamb waves.

Parametric studies have revealed that the impact localization error is very susceptible to the arrival-time estimates, t_i. This is mainly due to the difficulty of estimating the wave arrival times especially when a slow rise transition is present (c.f. PWAS 5 and 9 in Fig. 10.49b). This aspect is due to the dispersive nature of the Lamb waves excited by the impact event. Because the impact is a rather broadband event (i.e., it excites a range of frequencies rather than a single frequency), the wave packet generated by the impact contains several frequency components, which travel at different wave speeds. Hence, the wave packet generated by the impact disperses rather quickly (compare 7 with, say, 9). For example, the signal from PWAS 7 (which is close to the even) seems to form a compact packet, whereas the signals from PWAS 5 and 9 (which are farther from the impact) as much more dispersed.

To alleviate these difficulties, it seems intuitive that a different TOF criterion, e.g., the energy-peak arrival time, could be used. The application of energy-peak criterion would be quite easy on the compact wave-packet signals such as 7. However, it would not be at all easy to apply this criterion on the dispersed signals such as 7 and 9. More research needs to be conducted to address the dispersion issues.

These experiments have proven that PWAS can act as passive transducers for detecting elastic wave signals generated by low-velocity impacts. The high sensitivity of the PWAS transducers is very convenient because signals of up to $\pm 1.5\,\text{V}$ were directly recorded on a digital oscilloscope without the need for any signal conditioning/pre-amplifiers. It was also shown how data processing algorithms could determine the impact location with reasonable accuracy.

10.7.2 ACOUSTIC EMISSION DETECTION WITH PWAS

Historically, AE signals have been captured with special-purpose AE sensors (e.g., Ikegami and Haugse, 2001; Mal, 2001; Dzenis and Qian, 2001), which are costly and obtrusive. Dupont et al. (2000) studied the possibility of using embedded piezoelectric wafer sensors to detect AE signals in composite materials. To illustrate the capability of PWAS to detect AE signals in metallic structures, we performed detection experiments on the rectangular plate specimen of Fig. 10.16, which was already instrumented with a network of PWAS. AE events were simulated at the location P, $x_P = 400\,\text{mm}$, $y_P = 200\,\text{mm}$ (Fig. 10.49a). Consistent with other investigators (e.g., Dzenis and Qian, 2001), the AE events were simulated by pencil lead breaks on the specimen surface (0.5 mm HB leads). Table 10.5 shows the location of the PWAS and their radial distance from the AE event.

The simulated AE signals captured at PWAS 1, 5, 7, 9 are shown in Fig. 10.50 (for display, the signals were spaced up by vertical shifts). The signal on PWAS 7, closest to the AE source ($r_7 = 71\,\text{mm}$), is the strongest. This signal displays both high-frequency and low-frequency components, corresponding to S_0 (axial) and A_0 (flexural) waves, respectively. The flexural (A_0) waves display much higher amplitudes than the axial (S_0) waves, although their travel speed is slower. The signals received at the other PWAS display similar trends, although of lower amplitudes, due to their greater distance from the AE source. These experiments have proven that the PWAS transducers are capable of detecting AE signals. The high sensitivity of these sensors is remarkable because signals of up to $\pm 0.5\,\text{V}$ were directly recorded on a digital oscilloscope without the need for any signal conditioning/pre-amplifiers.

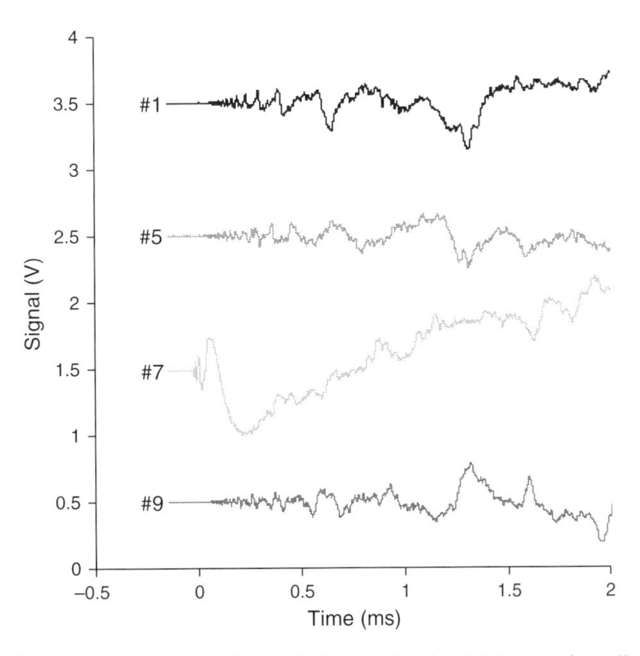

FIGURE 10.50 PWAS captured (AE) signals from a simulated AE event (coordinate locations are given in Table 10.5).

10.8 SUMMARY AND CONCLUSIONS

This chapter has presented PWAS-based wave-propagation SHM methods. This application of PWAS technology builds on the wealth of knowledge accumulated in the conventional NDI and NDE fields. The main difference between our approach and conventional NDE is that our approach used permanently-attached unobtrusive, minimally intrusive PWAS transducers. For this reason, we see our approach as leading to an emerging new technology: *embedded ultrasonic NDE (e-NDE)*. The embedded ultrasonic NDE will facilitate *on-demand* interrogation of the structure to determine its current state of health and predict the remaining life.

The emerging new technology of embedded ultrasonic NDE is enabled by the PWAS transducers. Because the PWAS transducers are essentially different in their operation from conventional NDE transducers, the development of embedded ultrasonic NDE requires the development of new ultrasonic interrogation and interpretation methods. This chapter has covered these issues and highlighted new research directions that need to be explored in order to mature the embedded ultrasonic NDE concept.

After a review of the main conventional NDE ultrasonic methods, the chapter started with the presentation of some simple 1-D modeling and experiments of PWAS-generated guided waves propagation and detection in a thin metallic strip. The FEM modeling was done with (a) conventional finite elements and (b) coupled-field finite elements. Good agreement between modeling and analysis was generally found. In the same section, we also presented and briefly discussed pitch-catch experiments on a massive railroad track-rail specimen. In this case, due to specimen massiveness, Rayleigh surface acoustic waves were generated, transmitted, and detected.

The next section presented and discussed 2-D wave propagation experiments performed with PWAS transducers on a thin metallic plate. The section started with pitch-catch experiments. After presenting the experimental setup, results of the pitch-catch experiments were shown. The pitch-catch experiments were able to reproduce very well the calculated wave-speed and group–velocity dispersion curves for several Lamb waves. Then, the section continued with the description of pulse-echo experiments. The wave reflection signals captured by the PWAS were subjected to multiple reflection analysis. It was concluded that the PWAS-coupled guided waves, generated with low-power devices, are sufficiently strong to travel at relatively large distances (in excess of 2,000 mm for a 7-mm PWAS transducer excited at 10 V pp).

The next two sections were devoted to the PWAS-enabled embedded NDE. One section discussed the pitch-catch embedded NDE. The other section discussed the pulse-echo embedded NDE. Several examples were presented dealing with both metallic and composite specimens. The specimens under considerations ranged from simple geometries to full-scale aircraft-like panels. In this way, the reader was taken to on a gradual course of understanding the principles of PWAS-based damage detection on simple geometries and then seeing how these are carried over onto complicated geometries and built-up constructions typical of practical applications. The embedded pitch-catch method was discussed in the context of detecting through-the-thickness fatigue cracks in metallic structures and delamination and disbond cracks in composite and other bonded structures. Then, the embedded pulse-echo method in metallic structures and in composites was considered.

The penultimate section of this chapter dealt with the time reversal method. This rather new technique stems from the time-reversal properties of single-mode pressure waves. In this context, the time reversal method was developed for underwater acoustic, medical treatment of kidney stones, etc. In this chapter, we discussed the time reversal in the context of multi-mode guided Lamb waves in thin-wall structures. The time reversal method has been proposed as damage-detection technique based on the premises that nonlinearities due to damage would breakdown the time-reversal reconstruction of the original signal. Thus, the presence of irregularities in time-reversal reconstruction would be indicative of damage presence. Our work presented the time-reversal theory for multi-mode guided Lamb waves and showed how the time reversal process is affected by the multi-mode character of the wave field. Then, the PWAS Lamb wave tuning concept, developed in a preceding chapter, was used to improve the time reversal process. Theoretical predictions made with our theory were compared with carefully conducted experiments.

The last subject treated in this chapter was that of PWAS acting as passive transducers for the detection of acoustic waves generated by structural impacts (ID) and by crack advancement (AE).

Although remarkable progress has been made in the development of embedded ultrasonic NDE, considerable work remains to be done. To increase the acceptance of this emerging technology, the refining of the theoretical analysis and calibration against well-planned experiments is needed. To deploy embedded NDE techniques to in-service structures, several hurdles have still to be overcome. In particular, the operational and environmental variations of the monitored structure need to be addressed. The NDE techniques presented above perform damage diagnosis through pattern comparison of a test signal with a baseline signal. This type of simple pattern comparison may be vulnerable to undesired changes of signals due to varying operational and environmental conditions of the structure such as temperature variation, humidity, etc. Considerable work still needs to be done to achieve in-service qualification of the PWAS transducers and associated hardware. Nevertheless, the PWAS-enabled embedded NDE concept has a good potential for SHM applications.

10.9 PROBLEMS AND EXERCISES

1. Consider a PWAS transducer of length $l = 7\,\text{mm}$, width $b = 1.65\,\text{mm}$, thickness $t = 0.2\,\text{mm}$, and material properties as given in Table 2. The PWAS is bonded at one end of a 1-mm thick 1000-mm long aluminum strip with $E = 70\,\text{GPa}$ and $\rho = 2.7\,\text{g/cc}$. The PWAS length is oriented along the strip. The PWAS is excited with a 3.5-counts tone burst. (i) Calculate the first frequency at which the S_0 wave propagation is dominant. (ii) Assuming that the frequency is adjusted to the value at which the S_0 wave propagation is predominant as calculated in part (i), calculate the time taken by the wave packet to travel to the other end of the strip specimen and come back. Sketch the wave pattern. (iii) Repeat part (ii) assuming that a through-the-thickness crack reflector is present at 400 mm from the PWAS. Sketch the wave pattern. (iv) Superpose the effects of (ii) and (iii) assuming that the energy is equally partitioned between the waves reflecting from the end and reflecting from the crack. Sketch the wave pattern. Discuss.
2. Consider a PWAS transducer of length $l = 7\,\text{mm}$, width $b = 1.65\,\text{mm}$, thickness $t = 0.2\,\text{mm}$, and material properties as given in Table 2. The PWAS is bonded at one end of a 1-mm thick 1000-mm long aluminum strip with $E = 70\,\text{GPa}$ and $\rho = 2.7\,\text{g/cc}$. The PWAS length is oriented along the strip. The PWAS is excited with a 3.5-counts tone burst. (i) Calculate the first frequency at which the A_0 wave propagation is dominant. (ii) Assuming that the frequency is adjusted to the value at which the A_0 wave propagation is dominant as calculated in part (i), calculate the time taken by the

wave packet to travel to the other end of the strip specimen and come back. Sketch the wave pattern. (iii) Repeat part (ii) assuming that a through-the-thickness crack reflector is present at 400 mm from the PWAS. Sketch the wave pattern. (iii) Repeat part (ii) assuming that a through-the-thickness crack reflector is present at 400 mm from the PWAS. Sketch the wave pattern. (iv) Superpose the effects of (ii) and (iii) assuming that the energy is equally partitioned between the waves reflecting from the end and reflecting from the crack. Sketch the wave pattern. Discuss.

3. (i) Explain the principles of the time reversal methods. (ii) What difficulties are encountered when the time reversal method is applied to Lamb waves and how they might be alleviated.
4. Describe how the wave propagation method and PWAS transducers can be used to detect the location of an impact in a plate. (i) Highlight the methodology. (ii) List the main difficulties to be overcome and explain how they might be alleviated.

11

In-Situ Phased Arrays with Piezoelectric Wafer Active Sensors

11.1 Introduction

Phased array principles have been widely used in radar, sonar, seismology, oceanology, and medical imaging. Phased array is a group of sensors located at distinct spatial locations in which the relative phases of the sensor signals are varied in such a way that the effective propagation pattern of the array is reinforced in a desired direction and suppressed in undesired directions. The phased array principles have allowed the development of radar and sonar systems that can scan the horizon 'electronically' without actually doing any mechanical motion. Thus the rotating dish radar (Fig. 11.1a) could be replaced by motionless phased array panels (Fig. 11.1b).

Phased arrays can act as both wave transmitters (radiators) and wave receivers (listeners). When a phased array works in transmitting mode, the relative amplitude of the signals radiated by the array in different directions determines the effective radiation pattern of the array. Thus, a phased array may be used to point toward a fixed direction, or to scan rapidly in azimuth or elevation.

The array acts as a spatial filter, attenuating all signals except those propagating from certain directions. 'Beamforming' is the name given to a wide variety of array-processing algorithms that are used to focus the array's signal-receiving or signal-transmitting abilities in a particular direction. A beam refers to the main lobe of the directivity pattern. Beamforming can apply to transmission from the array, to reception in the array, or to both. A beamforming algorithm points the array's spatial filter toward desired directions. This is similar to the dish antenna of a conventional radar swiveling to steer its beam into a desired direction; however, the phased-array beam-steering is achieved algorithmically rather than physically. Beamforming algorithm generally performs the same operations on the sensors' signals regardless of the number of sources or the character of the noise present in the wave field.

This chapter will treat the use of PWAS transducers to achieve in-situ phased arrays for active structural health monitoring. The chapter starts with a review of the phased array concepts used in conventional NDE practice. Particular attention will be paid to

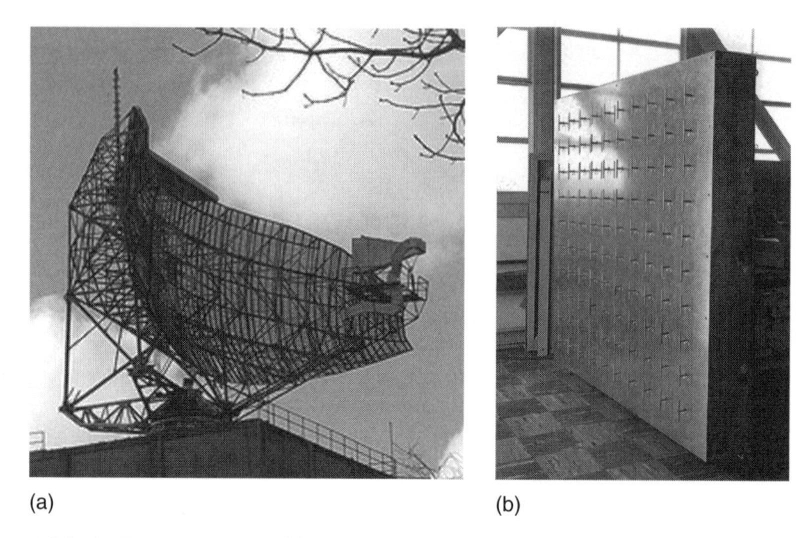

FIGURE 11.1 Radar examples: (a) rotating dish radar; (b) early phased array radar panel (Fenn et al., 2000).

the phased arrays based on the guided-wave principles. After this general overview of the state of the art, the chapter will start with the introduction of PWAS phased array principles. The simplest PWAS phased array, the linear array, will be treated first, and the concept of *embedded ultrasonics structural radar* (EUSR) will be presented. The transmitter beamforming, receiver beamforming, and phased array pulse-echo concepts will be developed from first principles. The practical implementation of the EUSR algorithm, and experimental validation and calibration of this method will be presented. The chapter will continue with an extensive experimental study of the linear PWAS arrays, such as EUSR detection studies performed on various defect types, detection experiments on curved panels, and in-situ direct imaging of crack growth during fatigue cyclic-load experiments.

The optimization of PWAS phased array beamforming is studied in some detail. The effect of r/d and d/λ ratios, of the number of elements M, and of the steering angle ϕ_0 is examined. The case of nonuniform PWAS phased arrays, in which the contribution of each element is individually weighted, is discussed. The binomial and the Dolph–Chebyshev arrays are studied in comparison with a uniform array, and tradeoffs between directivity and azimuthal coverage are identified. Actual experiments with nonuniform linear PWAS arrays are presented and comparatively discussed.

A generic PWAS phased array formulation is presented next. This generic formulation presents a generic delay-and-sum algorithm that can be particularized to either far-field (parallel rays) or near field situations. Thus the path will be opened to the study of 2-D PWAS arrays, which will cover most of the rest of this chapter. First, the effect of 2-D array configuration is studied using analytical simulation. Cross-shaped, rectangular grid, rectangular fence, circular ring, and circular grid PWAS arrays are considered. Their beamforming properties are comparatively examined. Subsequently, the implementation of a rectangular grid PWAS array into the embedded ultrasonic structural radar concept is developed in the form of the 2D-EUSR algorithm. The algorithm is tested first on simulated data, and then on actual experiments. A 4×8 rectangular PWAS array and an 8×8 square PWAS array are studied and comparatively examined. Thus, the full-circle 360° scanning properties of the 2-D PWAS phased arrays are highlighted and understood.

The last part of the chapter is devoted to the phased-array analysis using Fourier transform methods. This approach is less intuitive than the *ab initio* approach adopted so far; nevertheless, the Fourier analysis approach using the space-frequency multidimensional Fourier transform casts a clarifying light on many of the phased array properties discussed in the previous sections of this chapter. The effects of (a) the spatial sampling theorem; (b) the sampling of finite apertures; (c) the appearance of spatial aliasing and of grating lobes are exemplified and understood.

11.2 PHASED-ARRAYS IN CONVENTIONAL ULTRASONIC NDE

The phased array principle was introduced into ultrasonic testing because of its multiple advantages (Krautkramer, 1990; Rose, 1999; Ahmad et al., 2005). Some of the advantages of phased arrays over conventional ultrasonic transducers are high inspection speed, flexible data processing capability, improved resolution, and the capability of scanning without requiring mechanical movement, i.e., dynamic beam steering and focusing (Wooh and Shi, 1998). The backscattered ultrasonic signals can be analyzed and then mapped into an image. By using an ultrasonic phased array, the ultrasonic wave front can be focused in a certain point or steered in a specific direction; the inspection of a wide area can be achieved by electronically sweeping and/or refocusing without physically manipulating the transducers. Damage, which can be characterized as a local change in impedance, is diagnosed by using the echoes received from the propagating ultrasonic waves. However, in comparison to sonar and medical phased arrays, ultrasonic NDE phased arrays encounter several aspects that are of special concerns (i) the wave speed in solids is much higher than in water and tissues; (ii) the wave propagation is much more complicated, since dispersive and multi-mode waves may be present. For these reasons, the electronics required for ultrasonic phased array implementation may be rather complicated and sophisticated.

Current ultrasonic phased array technology mostly employs pressure waves generated through normal impingement on the material surface. Such phased arrays have shown clear advantages in the inspection of very thick specimens and in the sidewise inspection of thick slabs, where electronic beam scanning and focusing have produced significant improvements in the inspection efficiency.

11.2.1 PRESSURE-WAVE ULTRASONIC PHASED ARRAYS

By sequentially firing the individual elements of an array transducer at slightly different times, the ultrasonic wavefront can be focused or steered in a specific direction. Thus, inspection of a wide zone can be achieved by electronically sweeping and/or refocusing without physically manipulating the transducer.

The Krautkramer Company produces a line of phased-array transducers for the inspection of very thick specimens, and for the sidewise inspection of thick slabs, etc. as shown in Fig. 11.2 (Krautkramer, 2002). The principles of phased array ultrasonic inspection resemble the phased-array principles used in radar, sonar, seismology, oceanography, and medical imaging (Lines and Dickson, 1999). The common terminology between these application fields, such as the term ‘phased array’, shows their common ancestry.

Ultrasonic phased arrays use multiple ultrasonic elements and electronic time delays to create beams by constructive interference. They are similar to phased array radar, sonar, and other wave physics applications. However, ultrasonic applications use shorter wavelength and are subject to mode conversions and more complicated issues related to ultrasonic waves. The advantages of using a phased array of transducers for ultrasonic

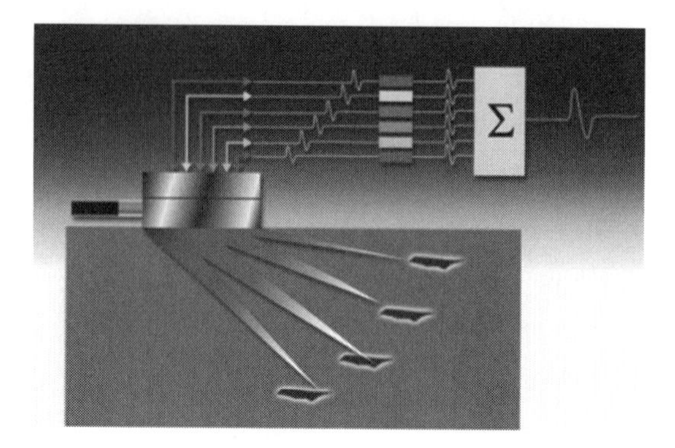

FIGURE 11.2 Conceptual representation of phased array principles (Krautkramer, 2002).

testing are multiple (Krautkramer, 1990; Rose, 1999). The arrays offer significant technical advantages for NDE inspection over conventional ultrasonics by electronically steering, scanning sweeping, and focusing the beam (Moles et al., 2005):

- They permit very rapid coverage of the inspected components
- Tailored angles of the ultrasonic transducers can be used for mapping components to maximize detection of defects
- Azimuth scanning is useful for inspections where only a minimal information is possible
- Electronic focusing permits optimizing the beam shape and size as function of the expected defect location, and consequently to optimize defect detection.

Overall, the use of phased arrays permits maximizing defect detection whereas minimizing inspection time. When using the ultrasonic phased array, besides the array itself, a typical setup includes the inspection angle, focal distance, scan pattern, etc. It can be time consuming to prepare such a setup.

11.2.2 GUIDED-WAVE ULTRASONIC PHASED ARRAYS

Rose et al. (2004) describe a high-frequency guided-wave phased-array focusing method for performing nondestructive pipe inspection with conventional phased-array transducers. This phased-array focusing method gives improved signal-to-noise ratio and better sensitivity to finding small defects in pipeline networks. Focusing is achieved in conjunction with angular beam profiling by using phased arrays and sending multiple nonaxisymmetric mode waves. The superposition process is controlled such that focusing can occur at any position inside and along the pipe. This allows very quick scanning on the entire pipe and detection of defects less than one percent cross-sectional area. The inspection could go beyond straight pipes by creating guided waves traveling beyond elbow sections.

Deutsch et al. (1998) demonstrated a phased array for the Lamb wave inspection of thin plates utilizing wedge-coupled conventional ultrasonic transducers and elaborated electronics. Steering and focusing of the Lamb waves beam were demonstrated. Potential applications of this method can be found in the wide-area ultrasonic inspection of aircraft, missiles, pressure vessels, oil tanks, pipelines, etc.

Fromme et al. (2005) designed and built a permanently attached guided ultrasonic wave array prototype to be used in the long-term monitoring of structural integrity. The array consists of a ring of 32 ultrasonic transducers equally spaced around a 70-mm diameter circle. The transducers are PZT discs of 5-mm diameter, 2-mm thickness, and a backing mass. The array was used for guided-wave transmission and reception. The excitation signal was a 5-cycle toneburst with a center frequency of 160 kHz generating the A_0 mode. The array was developed in compact form to allow fast and efficient inspection of large areas; battery operation and wireless data transfer were included.

Electromagnetic acoustic transducers (EMAT) are noncontact devices that can operate at high temperature without a coupling fluid using shear horizontal waves. MacLauchlan et al. (2004) used an EMAT phased array to conduct a real-time beam sweeping along and into a weld at over 300° F for detecting welding discontinuities. Another method for guided-waves generation is the comb transducer, which consists of a linear array of ultrasonic transducers that are fired consecutively (Alleyne et al., 2001; Rose et al., 1998; Wooh and Shi, 1999). With this method, the generated waves tend to stay in a narrow beam aligned with the array axis. Tuning of the comb transducer element spacing and the excitation frequency permits the selection of the appropriate guided-wave mode most suitable for the desired application.

Though the current NDE phased-arrays are promising, the fact that they utilize conventional ultrasonic transducers, which are bulky and expensive, make them unsuitable for SHM. Three reasons are apparent: (i) size and weight, (ii) cost, and (iii) principle of operation. Conventional ultrasonic transducers are resonant devices that consist of a piezoelectric oscillator disk, a protective layer, and a damping block. When activated, they generate high-frequency oscillations that impinge perpendicular to the contact surface. To generate an oblique incident wave, a wedge interface needs to be employed. Because of their internal complexity, conventional ultrasonic transducers are relatively bulky and expensive. It seems impractical and costly to conceive an SHM system based on conventional ultrasonic transducers permanently installed in an aircraft structure in a number sufficient to achieve the required structural coverage. However, this might be possible if a different class of transducers, which are both small and inexpensive, were available. The phased arrays discussed in this chapter employ the small and unobtrusive PWAS transducers as the constructing element to generate and receive the propagating Lamb waves. In the following sections, we will develop phased array algorithms for PWAS Lamb-wave phased arrays. We will start with 1-D PWAS phased arrays and then we will follow with 2-D phased arrays of various shapes.

11.3 1-D LINEAR PWAS PHASED ARRAYS

We will start our analysis of the PWAS phased array with the simplest configuration: we will discuss the 1-D, a.k.a. linear, phased array (Fig. 11.3). Arrays are made up of a number of elements, usually identical in size and arranged along a line, at uniform pitch. The wave pattern generated by the phased array is the result of the superposition of the waves generated by each individual element. By sequentially firing the individual elements of an array transducer at slightly different times, the ultrasonic wavefront can be focused or steered in a specific direction. Thus, inspection of a wide zone can be achieved by electronically sweeping and/or refocusing without physical manipulating the array.

Once the beam steering and focusing have been established, the detection of internal flaws is done with the pulse-echo method (Fig. 11.4). A pulse, consisting of a smooth-windowed tone-burst of duration t_p, is transmitted toward the target. The target reflects the signal and creates an echo, which is detected by the radar. By analyzing the phased-array signal in the interval $(t_p, t_p + t_0)$, one identifies the delay, τ, representing the time-of-flight (TOF) taken by the wave to travel to the target and back. Knowing TOF and wave speed allows one to precisely determine the target position.

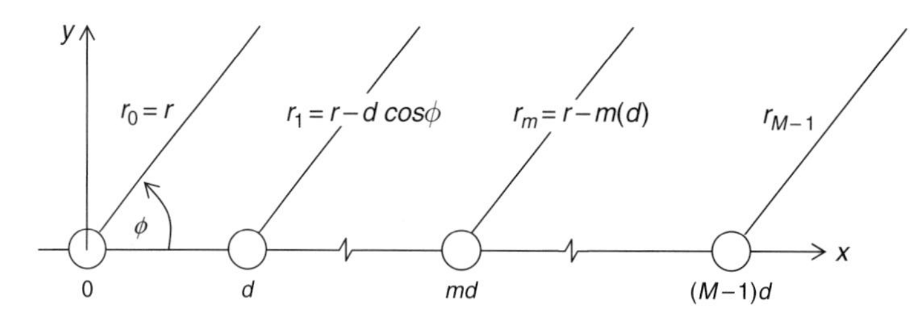

FIGURE 11.3 Uniform linear array of M omni-directional PWAS transducers spaced at pitch d.

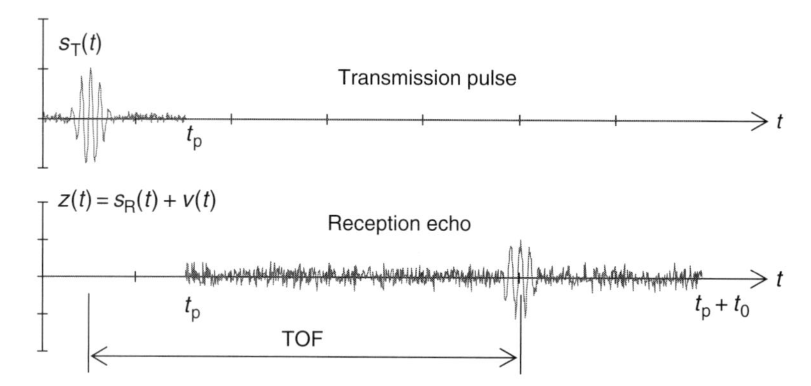

FIGURE 11.4 The pulse-echo method: (a) the transmitted pulse, $s_T(t)$; (b) the received echo, $z(t)$, consisting of the backscattered signal, $s_R(t)$, and the noise, $v(t)$. The difference between the pulse transmission and echo reception is the time-of-flight, TOF.

The single most important phenomenon that enables the use of PWAS phased arrays in conjunction with multi-modal guided waves in thin wall structures is the PWAS-Lamb wave tuning, as described in Chapter 8 of this book. The PWAS-Lamb wave tuning principle allows one to find convenient combinations of PWAS dimensions and excitation frequency that permit the preferential excitation of just one Lamb-wave mode, preferably one of minimal dispersion. In the following developments, we will assume that such tuning is possible and that a minimally dispersive Lamb wave can be tuned into. In this way, the situation depicted in Fig. 11.4 can be achieved in spite of the generally multi-modal character of the Lamb waves.

11.3.1 EMBEDDED-ULTRASONIC STRUCTURAL RADAR (EUSR)

The embedded ultrasonic structural radar (EUSR) is a concept that utilizes PWAS phased array radar principles and ultrasonic guided waves (Lamb waves) to scan large surface areas of thin-wall structures and detect cracks and corrosion (Giurgiutiu et al., 2006). In the EUSR concept, the guided Lamb waves are generated with surface-mounted PWAS that couple their in-plane motion with the in-plane particle motion of the Lamb wave as perceived at the structural surface. The guided Lamb waves stay confined inside the walls of the thin-wall structure and travel large distances with little attenuation. Thus, the target location relative to the phased array origin is described by the radial position, R, and the azimuth angle, ϕ. The EUSR algorithm works as follows. Consider a PWAS array as presented in Fig. 11.3. Each element in the PWAS array plays the role of both transmitter and receiver. A methodology is designed to change the role of each PWAS in a round-robin fashion. The responses of the structure to all the excitation signals are collected. By applying the EUSR algorithm, an appropriate delay is applied to each signal in the data set to make them all focus on a direction denoted by angle ϕ. When this angle ϕ is changed from 0 to 180 degrees, a virtual scanning beam is formed and a large area of the structure can be interrogated.

As shown in Chapter 6, Lamb waves can exist in a number of dispersive modes. However, through the PWAS-tuning approach discussed in Chapter 8, it is possible to confine the excitation to a particular Lamb wave mode of wave length $\lambda = c/F_c$, where F_c is the carrier frequency and c is the wave speed of the tuned Lamb-wave mode. The tuning is usually done such that a very-low dispersion Lamb-wave mode is selected, thus allowing us to assume quasi-constant wave speed c. Hence, the smoothed tone-burst signal generated by one PWAS is assumed of the form

$$s_T(t) = s_0(t)\cos 2\pi F_c t, \quad 0 < t < t_p \tag{1}$$

where $s_0(t)$ is a short-duration smoothing window that is applied to the carrier signal of frequency F_c between 0 and t_p. As in conventional phased array radar, we assume a uniform linear array of M PWAS, with each PWAS acting as a pointwise omni-directional transmitter and receiver. The PWAS in the array are spaced at the distance d, which is assumed to be much smaller than the distance r to a generic, far-distance point, P. Since $d << r$, the rays joining the sensors with the point P can be assimilated with a parallel fascicle, of azimuth ϕ (Fig. 11.3).

A brief explanation of the elementary beamforming concept follows. Because of the array spacing, the distance between one PWAS and the generic point P will be different from the distance between another PWAS and P. For the m-th PWAS, the distance will be shortened by $m(d\cos\phi)$. If all the PWAS are fired simultaneously, the signal from the m-th PWAS will arrive at the target P quicker by the amount

$$\delta_m(\phi) = m\frac{(d\cos\phi)}{c} \tag{2}$$

where c is the wave speed. If the PWAS are not fired simultaneously, but with some individual delays, Δ_m, $m = 0, 1, \ldots, M-1$, then the total signal received at point P will be

$$s_P(t) = \frac{1}{\sqrt{r}}\sum_{m=0}^{M-1} s_T\left(t - \frac{r}{c} + \delta_m(\phi) - \Delta_m\right) \tag{3}$$

where $1/\sqrt{r}$ represent the decrease in the wave amplitude because of the omni-directional 2-D radiation, and r/c is the delay due to the travel distance between the reference PWAS ($m = 0$) and the point P. (Wave-energy conservation, i.e., no dissipation, is assumed.)

11.3.1.1 Transmitter Beamforming

Beamforming at angle ϕ_0 with an array of M omni-directional sensors is based on the principles of constructive interference in the fascicle of parallel rays emanating from the array. The simplest way of achieving constructive interferences is to have

$$\Delta_m = \delta_m(\phi) \tag{4}$$

If Eq. (4) applies, then Eq. (3) takes the simple form

$$s_P(t) = M \cdot \frac{1}{\sqrt{r}} s_T\left(t - \frac{r}{c}\right) \tag{5}$$

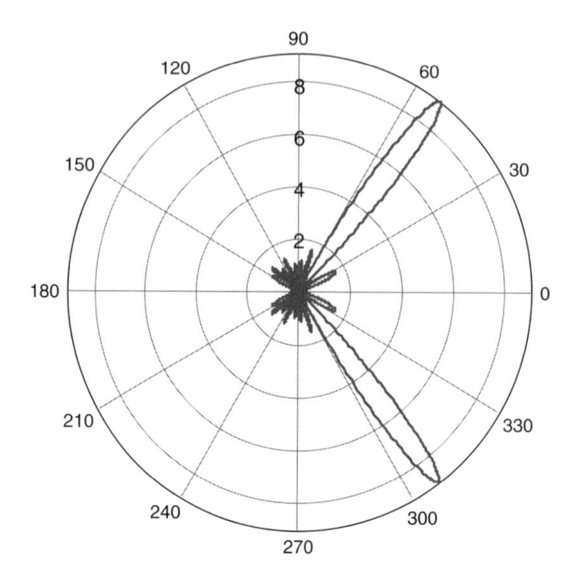

FIGURE 11.5 Calculated beamforming pattern for a 9-PWAS phased array ($\lambda/2$ spacing) with 53° target azimuth.

Equation (5) shows an M times increase in the signal strength with respect to the signal received by a single PWAS. This leads directly to the *beamforming principle*. Assume we take

$$\Delta_m = m\frac{d}{c}\cos(\phi_0) \tag{6}$$

Since Eq. (2) gives $\delta_m = m(d/c)\cos(\phi)$, then $\delta_m = \Delta_m$ will happen when $\cos(\phi) = \cos(\phi_0)$ i.e., at angles $\phi = \phi_0$ and $\phi = -\phi_0$. This indicates that constructive interference (beamforming) of the waves transmitted from the individual elements of the phased array will take place in the directions defined by the angles $\phi = \phi_0$ and $\phi = -\phi_0$. Thus, the forming of a beam at angles ϕ_0 and $-\phi_0$ is achieved through delays in the firing of the PWAS in the array as given by Eq. (6). Figure 11.5 shows the beamforming pattern for $\phi_0 = 53°$.

11.3.1.2 Receiver Beamforming

The receiver beamforming principles are reciprocal to those of the transmitter beamforming. If the point P is an omni-directional source at azimuth ϕ_0, then the signals received at the m-th sensor will arrive quicker by $m\delta_0(\phi) = m(d\cos\phi_0)/c$. Hence, we can synchronize the signals received at all the sensors by delaying them by $\Delta_m(\phi_0) = m(d/c)\cos(\phi_0)$.

11.3.1.3 Phased-array Pulse-echo

Assume that a target exists at azimuth ϕ_0 and distance R. The transmitter beamformer is sweeping the azimuth in increasing angles ϕ and receives an echo when $\phi = \phi_0$. The echo will be received on all sensors, but the signals will not be in sync. To synchronize the sensor signals, the delays $\Delta_m(\phi_0) = m(d/c)\cos(\phi_0)$ need to be applied. The process is as follows. The signal sent by the transmitter beamformer is an M times boost of the original signal

$$s_P(t) = \frac{M}{\sqrt{R}}s_T\left(t - \frac{2R}{c}\right) \tag{7}$$

At the target, the signal is backscattered with a backscatter coefficient, A. Hence, the signal received at each sensor will be

$$\frac{A \cdot M}{R} s_{\mathrm{T}}\left(t - \frac{2R}{c} + \delta_m(\phi)\right) \tag{8}$$

The receiver beamformer assembles the signals from all the sensors with the appropriate time delays, i.e.,

$$s_{\mathrm{R}}(t) = \frac{A \cdot M}{R} \sum_{m=0}^{M-1} s_{\mathrm{T}}\left(t - \frac{2R}{c} + \delta_m(\phi) - \Delta_m\right) \tag{9}$$

Constructive interference between the received signals is achieved when $\Delta_m = m(d/c)\cos(\phi_0)$. Thus, the assembled received signal will be again boosted M times with respect to the individual sensors, i.e.,

$$s_{\mathrm{R}}(t) = \frac{A \cdot M^2}{R} \sum_{m=0}^{M-1} s_{\mathrm{T}}\left(t - \frac{2R}{c}\right) \tag{10}$$

The time delay of the received signal, $s_{\mathrm{R}}(t)$, with respect to the transmit signal, $s_{\mathrm{T}}(t)$, is

$$\tau = \frac{2R}{c} \tag{11}$$

Measurement of the time delay τ observed in $s_{\mathrm{R}}(t)$ allows one to calculate the target range, $R = c\tau/2$.

11.3.1.4 Practical Implementation of the EUSR Algorithm

The practical implementation of the EUSR signal generation and collection algorithms is described next. In a round-robin fashion, one active sensor at a time is activated as the transmitter. The reflected signals are received at all the sensors. The activated sensor acts in pulse-echo mode, i.e., as both transmitter and receiver; the other sensors act as passive sensors. Thus, an $\mathrm{M} \times \mathrm{M}$ matrix of elemental signals is generated (Table 11.1). The elemental signals are assembled into synthetic beamforming responses using the synthetic beamformer algorithm as given by Eqs. (2), (3), (4), (5). The delays, Δ_j, are selected in such a way as to steer the interrogation beam at a certain angle, ϕ_0. The synthetic-beam sensor responses, $w_i(t)$, synthesized for a transmitter beam with angle ϕ_0, are assembled by the receiver beamformer into the total received signal, $s_{\mathrm{R}}(t)$, using the same delay as for the transmitter beamformer. However, to apply this method directly, one needs to know the target angle ϕ_0. Since, in general applications, the target angle is not known, we need to use an inverse approach to determine it. Hence, we write the received signal as a function of the parameter ϕ_0, using the array unit delay for the direction ϕ_0 as $\delta_0(\phi_0) = (d/c)\cos\phi_0$. (To accurately implement the time shifts when the time values fall in between the fixed values of the sampled time, we have used a spline interpolation algorithm.)

A coarse estimate of the target direction is obtained by using an azimuth sweep technique, in which the beam angle, ϕ_0, is modified until the maximum received energy is attained, i.e.,

$$\max E_{\mathrm{R}}(\phi_0), \qquad E_{\mathrm{R}}(\phi_0) = \int_{t_{\mathrm{p}}}^{t_{\mathrm{p}}+t_0} |s_{\mathrm{R}}(t, \phi_0)|^2 \, \mathrm{d}t \tag{12}$$

TABLE 11.1 $M \times M$ matrix of elemental signals generated in a round-robin activation of the active-sensor array elements

		Transmitters				Synthetic beamforming response
		T_0	T_1		T_{M-1}	
Receivers	R_0	$p_{0,0}(t)$	$p_{0,1}(t)$	...	$p_{0,M-1}(t)$	$w_0(t)$
	R_1	$p_{1,0}(t)$	$p_{1,1}(t)$	...	$p_{1,M-1}(t)$	$w_1(t)$
	R_2	$p_{2,0}(t)$	$p_{2,1}(t)$	...	$p_{2,M-1}(t)$	$w_2(t)$
	...	...	...	...	...	...
	R_{M-1}	$p_{M-1,0}(t)$	$p_{M-1,1}(t)$	...	$p_{M-1,M-1}(t)$	$w_{M-1}(t)$

After a coarse estimate of the target direction is found ϕ_0, the actual round-trip time of flight, τ_{TOF}, is calculated using an optimal estimator, e.g., the cross-correlation between the receiver and the transmitter signals

$$y(\tau) = \int_{t_p}^{t_p+t_0} s_R(t)s_T(t-\tau)\mathrm{d}t \tag{13}$$

Then, the estimated $\tau_{TOF} = 2R/c$ is obtained as the value of τ where $y(\tau)$ is maximum. Hence, the estimated target distance is

$$R_{exp} = c \cdot \frac{\tau_{TOF}}{2} \tag{14}$$

This algorithm works best for targets in the far field, for which the 'parallel-rays' assumption holds. For targets in the near and intermediate fields, a more sophisticated self-focusing algorithm, that uses triangulation principles, is used. This algorithm is an outgrowth of the passive-sensors target-localization methodologies. The self-focusing algorithm modifies the delay times used in each synthetic-beam response, $w_i(t)$. The total response is maximized by finding the focal point of individual responses, i.e., the common location of the defect that generated the echoes recorded at each sensor.

11.3.2 EUSR SYSTEM DESIGN AND EXPERIMENTAL VALIDATION

The EUSR system consists of three major modules: (a) the PWAS array; (b) the DAQ module; and (c) the signal processing module. A system diagram is shown in Fig. 11.6a. A proof-of-concept EUSR system was built in the Laboratory for Active Materials and Smart Structures (LAMSS) at the University of South Carolina to evaluate the feasibility and capability of the EUSR system.

11.3.2.1 Experimental Setup

Three specimens were used in the experiments. These specimens were 1220-mm square panels of 1-mm thick 2024-T3 Al-clad aircraft grade sheet metal stock. One of the specimens (specimen #0) was pristine and was used to obtain baseline data. The other two specimens were manufactured with simulated cracks. The cracks were placed on a line midway between the center of the plate and its upper edge (Fig. 11.6b). The cracks were 19 mm long, 0.127 mm wide. On specimen #1, the crack was placed broadside with

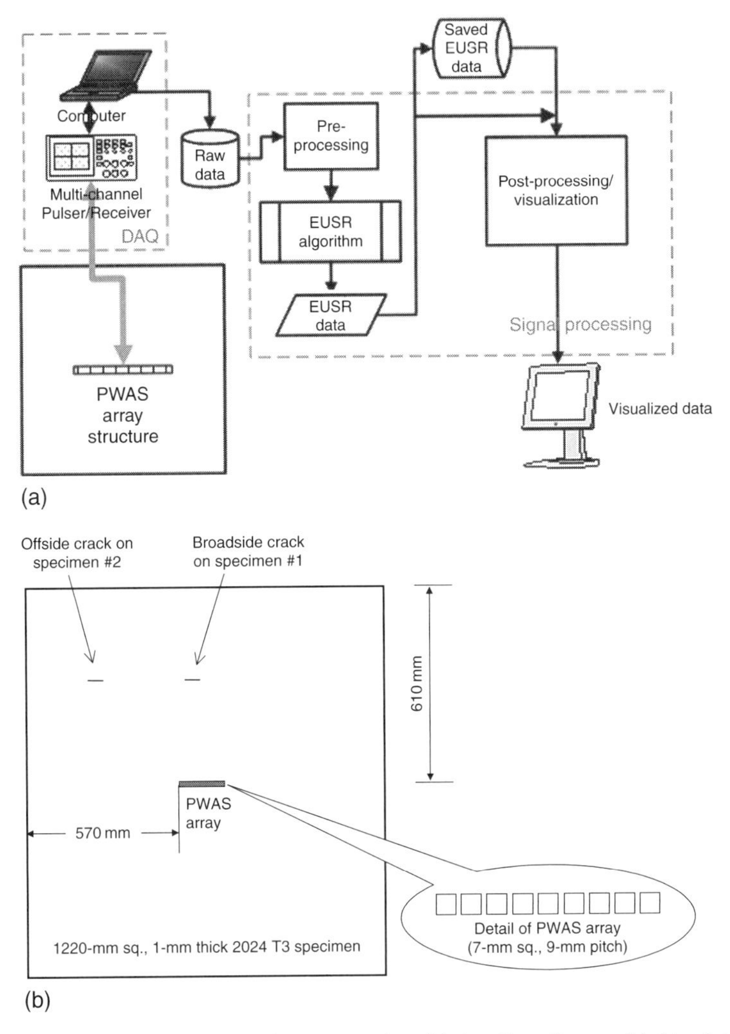

FIGURE 11.6 Proof-of-concept EUSR construction: (a) data flow diagram (b) thin plate specimens with broadside and offside cracks, and 9-element PWAS array at the center.

respect to the phase-array, at coordinates (0, 0.305 m), i.e., at $R = 305\,\text{mm}$, $\phi_0 = 90°$. On the specimen #2, the crack was placed offside with respect to the phased array, at coordinates ($-0.305\,\text{m}, 0.305\,\text{m}$), which corresponds to $R = 409\,\text{mm}$, $\phi_0 = 136.3°$ with respect to the reference point of the PWAS array. The PWAS array was constructed from nine 7-mm sq., 0.2-mm thick piezoelectric wafers (American Piezo Ceramic Inc., APC-850) placed on a straight line in the center of the plate. The sensors were spaced at pitch $d = \lambda/2$, where $\lambda = c/f$ is the wavelength of the guided wave propagating in the thin-wall structure. Since the first optimum excitation frequency for S_0 mode was 300 kHz, and the corresponding wave speed was $c = 5.440\,\text{km/s}$, the wavelength was $\lambda = 18\,\text{mm}$. Hence, the spacing in the PWAS array was selected as $d = 9\,\text{mm}$ (Fig. 11.6b).

The DAQ module consisted of an HP33120A arbitrary signal generator, a Tektronix TDS210 digital oscilloscope, and a portable PC with DAQ and GPIB interfaces. The HP33120A arbitrary signal generator was used to generate a 300 kHz Hanning-windowed tone-burst excitation with a 10 Hz repetition rate. Under the Hanning-windowed tone-burst excitation, one element in the PWAS array generated a Lamb-wave packet that spread

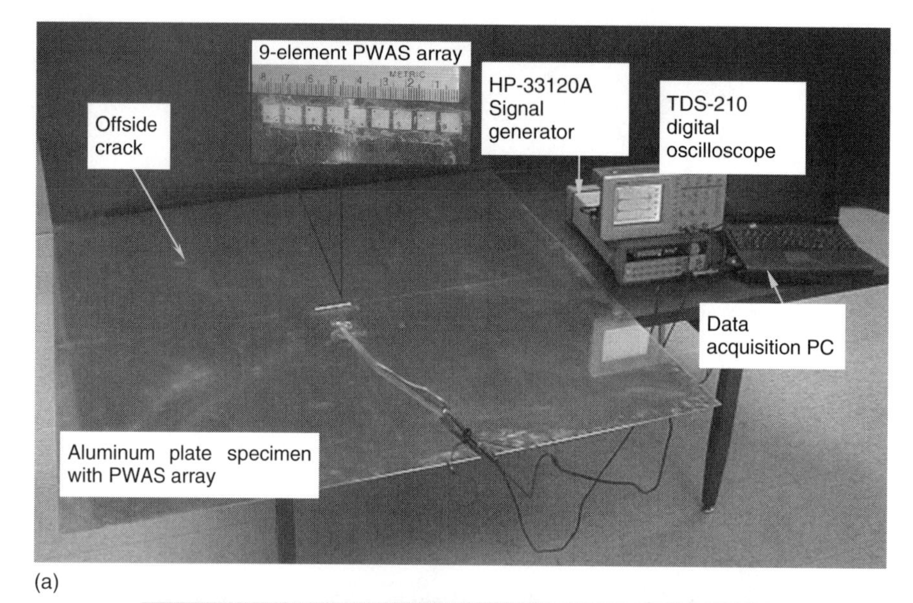

(a)

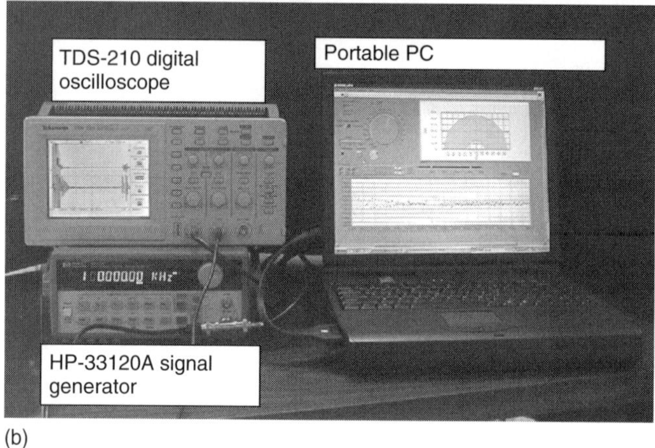

(b)

FIGURE 11.7 Experimental setup for EUSR experiment: (a) overall view showing the plate, active sensors, and instrumentation; (b) detail of the instrumentation and of the data acquisition program.

out into the entire plate in an omnidirectional pattern (circular wave front). The Tektronix TDS210 digital oscilloscope, synchronized with the signal generator, collected the response signals from the PWAS array. One of the oscilloscope channels was connected to the transmitter PWAS, whereas the other was switched among the remaining elements in the PWAS array by using a digitally controlled switching unit. A LabVIEW computer program was developed to digitally control the signal-switching, to record the data from the digital oscilloscope, and to generate the group of raw data files. Photographs of the experimental set-up are presented in Fig. 11.7.

11.3.2.2 Implementation of the EUSR Data Processing Algorithm

The signal-processing module reads the raw data files and processes them using the EUSR algorithm. Although the EUSR algorithm is not computationally intensive, the large amount of data points in each signal made this step time consuming. Hence, we elected to save the resulting EUSR data on the PC for later retrieval and post-processing.

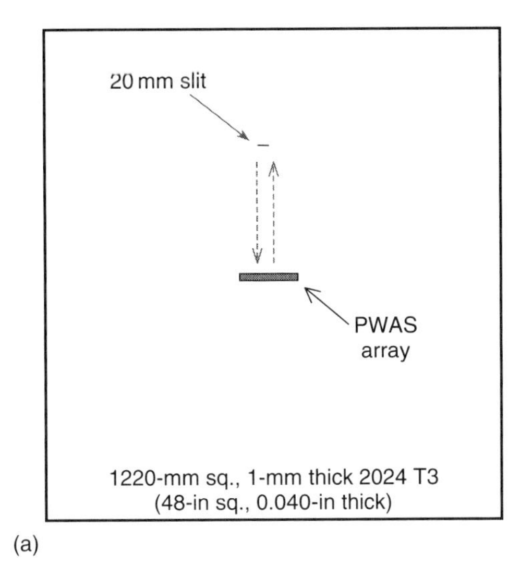

(a)

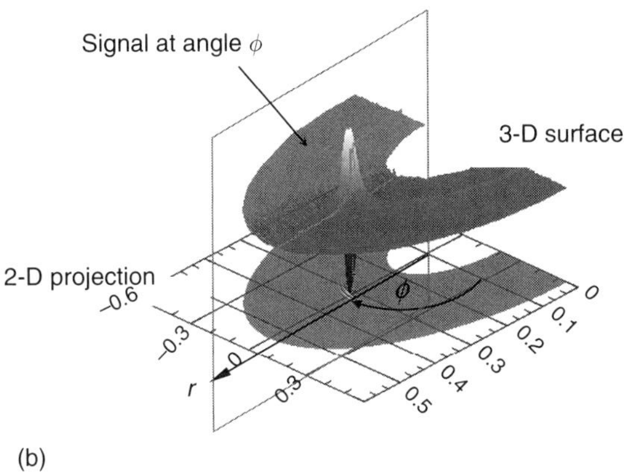

(b)

FIGURE 11.8 EUSR signal reconstruction examples: (a) schematic of the broadside crack; (b) 3-D visualization of EUSR signal reconstruction of the broadside crack.

This approach also enables other programs to access the EUSR data. Based on the EUSR algorithm, the resulting data file is a collection of signals that represent the structure response at different angles, defined by the parameter ϕ. In other words, they represent the response when the EUSR scanning beam turned at incremental angles ϕ.

After being processed, the data was transformed from the time domain to the 2-D physical domain. Knowing the Lamb wave speed c, and using $r = ct$, the EUSR signal was transformed from voltage V vs. time t to voltage V vs. distance r. The signal detected at angle ϕ was plotted on a 2-D plane at angle ϕ. Since angle ϕ was stepped from 0 to 180°, at constant increments, the plots covered a half space. These plots generate a 3-D surface, which is a direct mapping of the structure being interrogated, with the z value of the 3-D surface representing the detected signal at that (x, y) location (Fig. 11.8). If we present the z value on a color scale, then the 3-D surface is projected to the 2-D plane, and the color of each point on the plane represents the intensity of the reflections.

11.3.2.3 Experimental Results

Figure 11.9 shows the EUSR detection visualization for the broadside crack (Fig. 11.9a) and the offside crack (Fig. 11.9b) specimens. The group velocity was used to map the

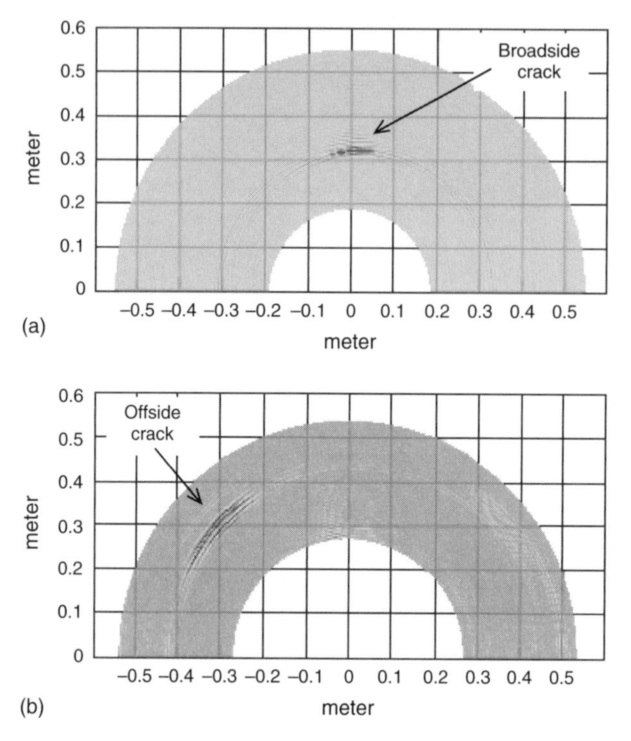

FIGURE 11.9 EUSR mapping results: (a) detection of a broadside crack ($\phi_0 = 90°$, $r = 305$ mm) (b) detection of an offside crack ($\phi_0 = 136.3°$, $r = 409$ mm).

EUSR data from the time domain to the space domain, thus the locations of the reflectors can be visually displayed. The grids represent exact mapping in meters. The shaded area represents the swept surface. The signal amplitude is presented on a color/grayscale intensity scale. The location of the crack is easily determined from the color/grayscale change. Figure 11.9a presents the results for the broadside specimen. The small area with darker color represents the high amplitude echo (reflected wave) generated when the scanning angular beam intercepted the crack. From the picture scale, we observe that this area is located at an angle of 90° and at approximately 0.3 m from the center of the plate. Careful analysis of the reconstructed signal yielded the exact $\tau_{TOF,Broadside} = 112.4\,\mu s$, corresponding to a radial position $R_{Broadside} = 305.7$ mm. This differs a mere 0.2% from the actual position of the broadside crack on this specimen ($\phi_0 = 90°$, $R = 305$ mm). The dark area in the EUSR result predicted the simulated crack quite well. Similarly, Fig. 11.9b presents the results for the offside specimen.

Figure 11.9 indicates that both the broadside crack (specimen #1) and the offside crack (specimen #2) were successfully detected. The offside crack on specimen #2 presented a detection challenge since the direct reflections from the crack were deflected away from the sensors array because of the inclination of the crack with respect to the beam axis. Only secondary backscatter signals, which are due to the discontinuity created by the crack tips in the elastic field, were recorded. These signals could be detected through the constructive interference principles contained in the EUSR algorithm. An example of the EUSR signal reconstruction for the offside crack experiment is presented in Fig. 11.9b. Recorded are residual backscatter signals arriving at the sensor array. The TOF of the crack signal was identified as $\tau_{TOF,Offside} = 151\,\mu s$. Using $c_g = 5.440$ mm/μs yields the range of the crack as $R_{Offside} = 411$ mm. The exact value is 409 mm, i.e., 0.4% error.

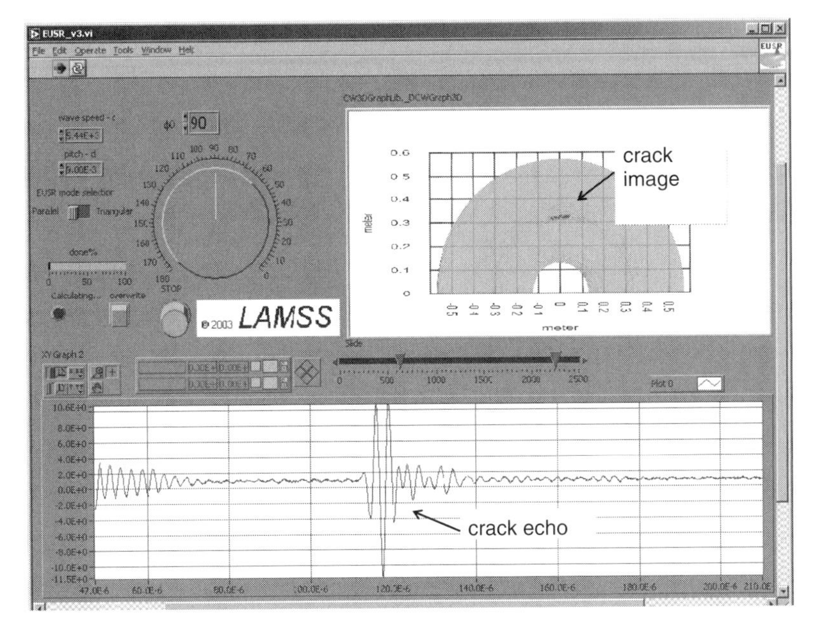

FIGURE 11.10 Graphical user interface (EUSR-GUI) front panel. The angle sweep is performed automatically to produce the structure/defect imaging picture on the right. Manual sweep of the beam angle can also be performed with the turn knob; the signal reconstructed at the particular beam angle (here, $\phi_0 = 136°$) is shown in the lower picture.

The accuracy of the EUSR method seems quite good. It is apparent that the offside crack is located just beyond the (−0.3 m, 0.3 m) coordinates, which compares very well with the actual values (−0.305 m, 0.305 m). The crack range $R_{Offside} = 411$ mm, determined from the analysis of the reconstructed echo, showed 0.4% accuracy. Both Fig. 11.8 and Fig. 11.9 prove that the detection sensitivity and accuracy of the EUSR method is good. The implementation of these concepts in a graphical user interface (GUI) is presented in Fig. 11.10. The angle sweep is performed automatically to produce the structure/defect imaging picture on the right. Manual sweep of the beam angle can be also performed with the turn knob; the signal reconstructed at the particular beam angle (here, $\phi_0 = 136°$) is shown in the lower picture. In NDE terminology, the 2-D image corresponds to a C-scan, whereas the reconstructed signal would be an A-scan.

11.3.3 SECTION SUMMARY

The delay-and-sum beamforming was implemented in PWAS phased arrays as the EUSR algorithm (Giurgiutiu et al., 2006). EUSR assumes that data from an M-element PWAS array is collected in a round-robin fashion by using one element at a time as transmitter and all the elements as receivers. With a total of M^2 data signals, EUSR conducts the beam scanning in virtual time as a signal post-processing operation. The EUSR beamforming and scanning procedure does not require complex devices or multi-channel electronic circuitry as needed by the conventional ultrasonic phased array equipment. EUSR implementation of PWAS phased arrays requires only a function generator, an oscilloscope, a switching device, and a computer. The received signals were post-processed with the embedded ultrasonics structural radar (EUSR) algorithm to obtain a direct imaging of the crack in the test plate. This image is similar to the C-scan from ultrasonic NDE, only that it was obtained from a single location using a sweeping beam of focused Lamb waves.

11.4 FURTHER EXPERIMENTS WITH LINEAR PWAS ARRAYS

11.4.1 EUSR DETECTION STUDIES

Extensive experiment were carried out to verify the detection ability of 1-D linear EUSR algorithm using the setup described in a previous section, with the only difference that the PWAS transducers consisted of eight 7-mm round wafers (Fig. 11.11)

In order to minimize the Lamb modes appearing in the wave and strengthen the signal, a frequency tuning process is established. The excitation frequency was selected as 282 kHz at which the low-dispersion S_0 mode is maximally excited. The data was collected in the round-robin pattern, resulting in a total of M^2 data sets where M is the number of elements in the array. EUSR algorithm processed the raw data and mapped the scanning beams in a 2D/3D image for further structural diagnosis.

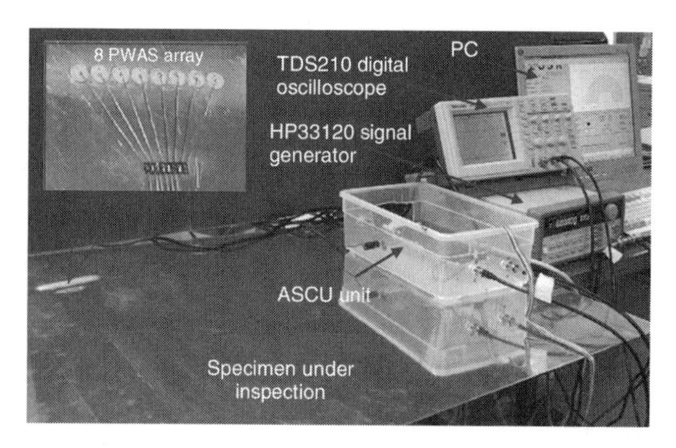

FIGURE 11.11 Experimental setup for proof-of-concept EUSR experiment with of an array of eight round PWAS transducers.

11.4.1.1 Single Broadside Pin-hole Damage

To evaluate the capability of the EUSR PWAS phased array to detect very small defects, we performed experiments in which a pin hole of increasing size was placed at 90° azimuth at 315 mm away from the array. The size of the pin hole was progressively increased from 0.5 mm to 1.0 mm, 1.57 mm, and 2.0 mm (the sizes were determined by available drill bits). Experiment results showed that EUSR was unable to detect the very small holes of size 0.5 mm and 1.0 mm. This was not surprising since a pin-hole damage can be approximately considered as a perfect wave scatterer. However, we found that the 1.57 mm hole was successfully detected by the EUSR method (Fig. 11.12b). As the

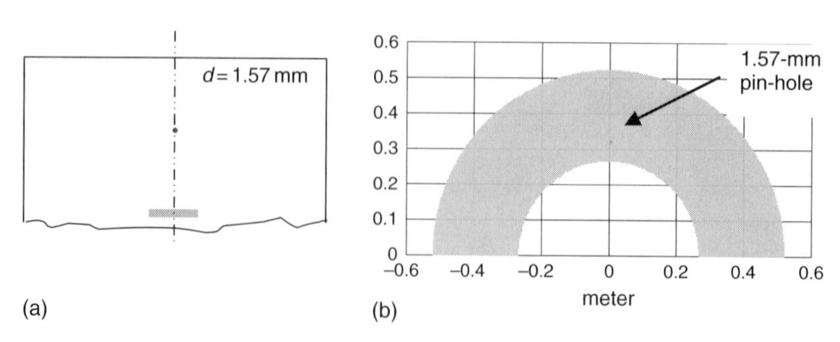

FIGURE 11.12 EUSR detection of 1.57-mm pin-hole damage: (a) specimen layout; (b) EUSR scanning image.

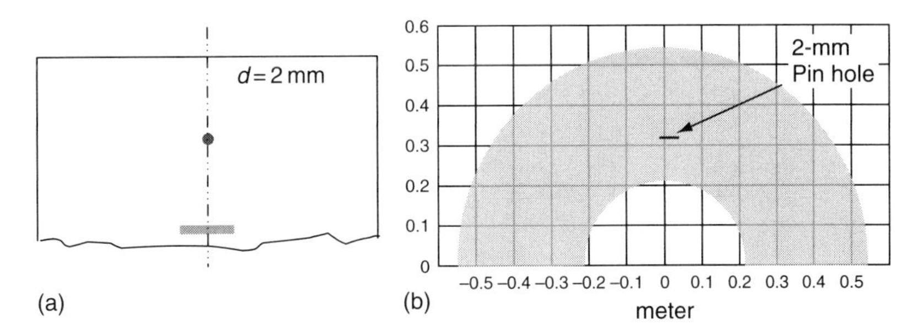

FIGURE 11.13 EUSR detection of 2-mm pin-hole damage: (a) specimen layout; (b) EUSR scanning image.

pin-hole diameter was increased, so did the detection image (Fig. 11.13b). However, for even larger holes, the increase in image size was not proportional, since backscatter is not necessarily proportional to the radius of the hole.

11.4.1.2 Horizontal Broadside Crack

A 19-mm long, 0.127-mm wide simulated crack was cut into the plate at broadside (90°) location, 305 mm away from the array. This experiment duplicated an earlier experiment; the difference was that the array used consisted of eight round PWAS of 7-mm diameter, whereas the array used in the earlier experiment consisted of nine 7-mm square PWAS. Figure 11.14b shows the EUSR scanning image for this experiment, which indicates very well both the angular location (90°) and the distance away from the array (about 0.3 m, i.e., 300 mm).

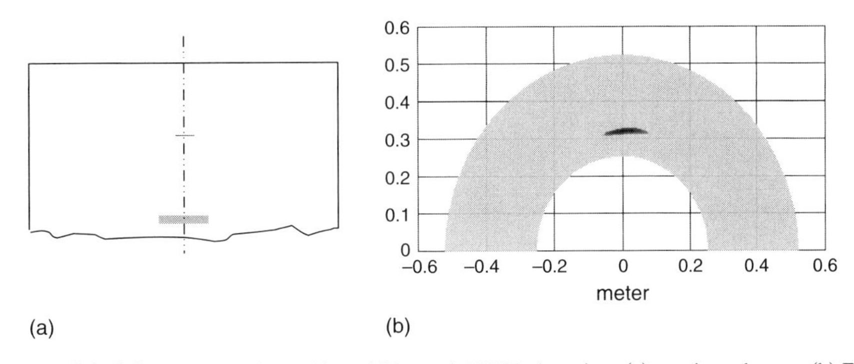

FIGURE 11.14 Single horizontal broadside crack EUSR detection: (a) specimen layout; (b) EUSR scanning image.

11.4.1.3 Sloped Broadside Crack

In this experiment, we wanted to study the influence of the crack inclination. If the crack is not parallel to the array, then its specular reflection will not be sent back to the array, but rather away from the array (Fig. 11.15a). What the array sees is only the backscatter due to the discontinuity created by the crack in the elastic field. Our crack was inclined at 30° with respect to the horizontal direction. The EUSR scanning image given in Fig. 11.15b is different from the scanning results obtained for the horizontal crack (Fig. 11.14b). The fact that the crack is inclined is somehow apparent in the EUSR image. However, the slope in the image cannot exactly indicate the 30° angle. The indication of an inclined crack is rather qualitative than quantitative. It must be added that

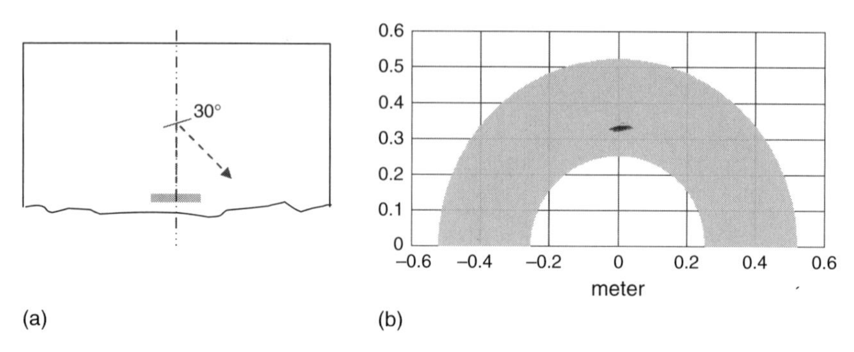

FIGURE 11.15 Single sloped broadside crack EUSR detection: (a) specimen layout; (b) EUSR scanning image.

because the specular reflection did not return to the array, the recorded signals were much weaker, which raised the need to advance signal processing, as will be described in a later chapter.

11.4.1.4 Single Horizontal Offside Crack

Figure 11.16a shows a crack at offside location with 137° azimuth and 305 mm radius. Figure 11.16b shows the corresponding EUSR scanning image. (This situation is similar to that considered in an earlier section, only that the PWAS array had eight round elements instead of nine square elements.) Different from the broadside crack situation, the orientation of this offside crack is not orthogonal to the scanning beam. Its specular reflection is not sent back to the array, but rather away from the array (Fig. 11.16a). What the array sees is only the back scattered waves due to the discontinuity created by the crack in the elastic field. Only very little energy is reflected back to be picked up by the array. However, the resulting image correctly indicates the radial distance of the array from the crack (around 300 mm); it also shows that the crack is located in the direction about 137°. This experiment verifies that by using the PWAS phased array, scanning beams in various directions can be produced.

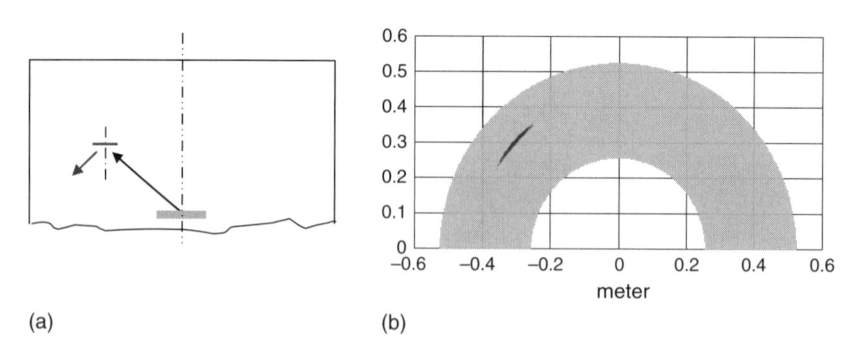

FIGURE 11.16 Single horizontal offside crack EUSR detection: (a) specimen layout; (b) EUSR scanning image.

11.4.1.5 Two Horizontal Offside Cracks

The next specimen considered in our experiments had two offside cracks placed at 67° and 117° locations (Fig. 11.17a). The resulting EUSR image is shown in Fig. 11.17b. The resulting image shows that the two cracks are both correctly imaged at locations is close

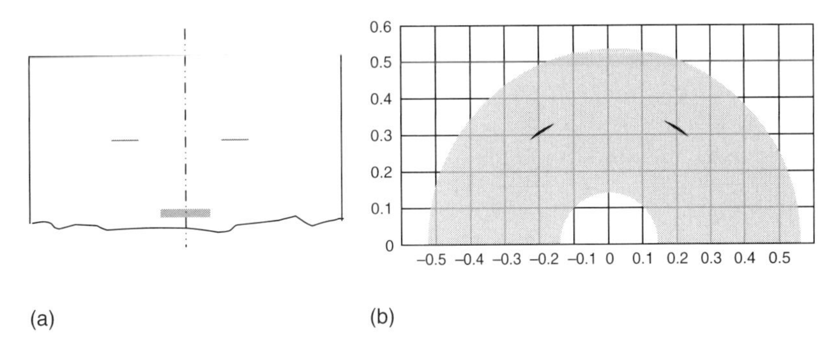

FIGURE 11.17 Two horizontal offside cracks EUSR detection: (a) specimen layout; (b) EUSR scanning image.

to their actual physical locations. This experiment verifies that the EUSR PWAS phase array method has the ability to simultaneously detect multiple defects in a structure.

11.4.1.6 Three Horizontal In-Line Cracks

An even more challenging experiment was posed by a specimen with three cracks placed at 67°, 90°, and 117° (Fig. 11.18a). This experiment is particularly challenging since the middle crack acts as a very strong reflector which sends its specular reflection back to the array, whereas the side cracks act as weak backscatters, since their specular reflections are sent away from the array. For this reason, we expected that the strong signal from the middle crack will overwhelm the weak signals from the side crack. The resulting EUSR image, shown in Fig. 11.18b, confirms our negative expectations.

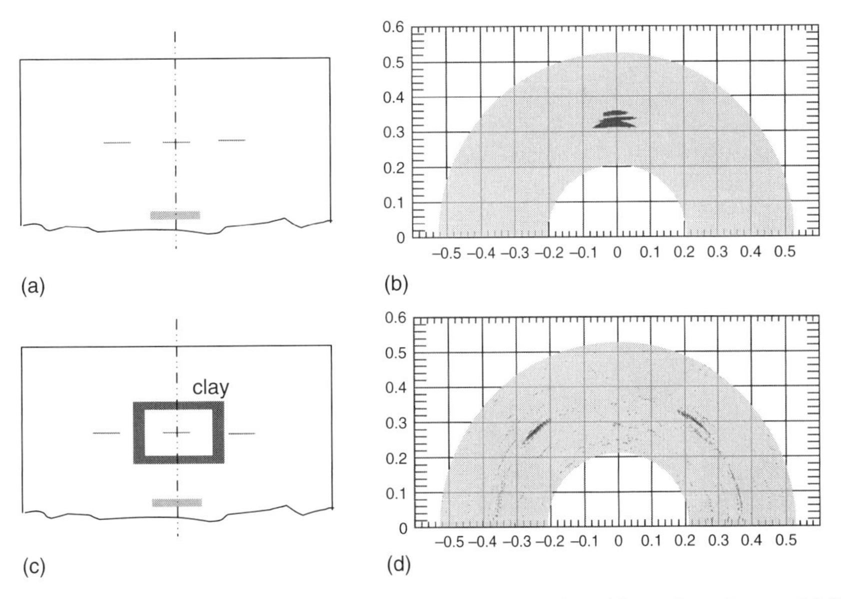

FIGURE 11.18 Three horizontal in-line cracks EUSR detection: (a) specimen layout; (b) EUSR scanning image; (c) specimen with broadside crack surrounded with clay for damping; (d) EUSR scanning image when clay was applied around the center crack.

The image presented in Fig. 11.18b only shows a single intense response at 90° azimuth representing the reflection from the middle crack placed in the broadside position; we do not see anything from the offside cracks. This confirms that the reflection from the broadside crack is so strong that the backscatters from the offside cracks are overshadowed. In order to verify this hypothesis, we applied a dam of vibration-absorbing material (modeling clay) around the broadside crack so as to isolate it (Fig. 11.18c). Hence, we were able to reveal the backscatter signals from the offside cracks, as shown in Fig. 11.18d.

After isolating the broadside crack, no reflection from it was observed any more. Figure 11.18d shows that the backscatter signals from the two offside cracks have shown up. This process confirmed our assumption that the signals from the offside cracks were submerged in the much-stronger signals reflected from the broadside crack. However, isolation of the strong reflector with modeling clay was able to reveal the presence of the other two reflectors. Thus, the offside cracks could be successfully imaged. A similar isolation result could be achieved through advanced signal processing.

11.4.2 EUSR EXPERIMENTS ON CURVED PANELS

So far, we have considered guided waves propagating in flat plates, i.e., Lamb waves. However, we have shown in Chapter 6 that guided waves also propagate in tubes and shells. The guided waves propagating in a shallow shell, which has radius of curvature much larger than the wall thickness, are very similar to the Lamb waves. Hence, it is expected that the EUSR PWAS phased array method presented in the previous section should work sufficiently well in shallow shells. To verify these theoretical hypotheses, we performed experiments on curved panels. The experimental setup was identical with that used previously in the development of the EUSR concept on the flat plate.

In order to verify the effect of curvature, the thin aluminum plate was temporarily bent into cylindrical shapes using thin wire ropes and tightening screws. The realized curvatures were related to the shortening of the chord with respect to the original flat length. After curving, the original flat length, L, became the circular arc, whereas the shortened wire rope became the chord, c. The difference between the original length, L, and the shortened length, c, was denoted by ΔL. The chord c can be calculated with the formula

$$c = 2R \sin \frac{L}{2R} \tag{15}$$

Then, the radius of curvature, R, could be calculated with the formula:

$$R = \frac{L - \Delta L}{2 \sin \frac{L}{2R}} \tag{16}$$

The values of curvature obtained with Eq. (16) are given in Table 11.2. In our experiment we went through the shortenings $\Delta L = 5, 10, 15,$ and 20 mm corresponding to curvature values $R = 3.90, 2.75, 2.24,$ and 1.95 m. These curvatures were applied in two directions. One direction was with the chord parallel to the PWAS array (direction 1), whereas the other was with the chord perpendicular to the PWAS array (direction 2). The crack imaging results of the curved panels are given in Fig. 11.19 and Fig. 11.20.

TABLE 11.2 Curvature of the specimen

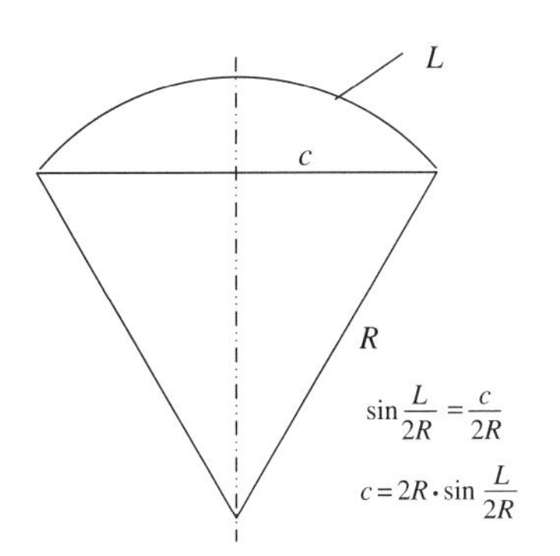

R	c	$\Delta L = L - c$
100.00E+3	1219.99	
50.00E+3	1219.97	
20.00E+3	1219.81	
10.00E+3	1219.24	
8.80E+3	1219.02	1
5.00E+3	1216.98	
3.90E+3	1215.03	5
2.75E+3	1210.02	10
2.24E+3	1204.98	15
2.00E+3	1201.17	
1.95E+3	1200.20	20
1.00E+3	1145.73	

$L = 1220$, all in mm

(a)

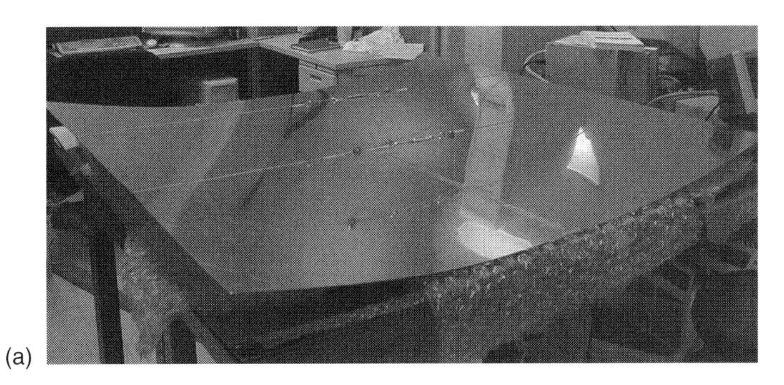

(b)

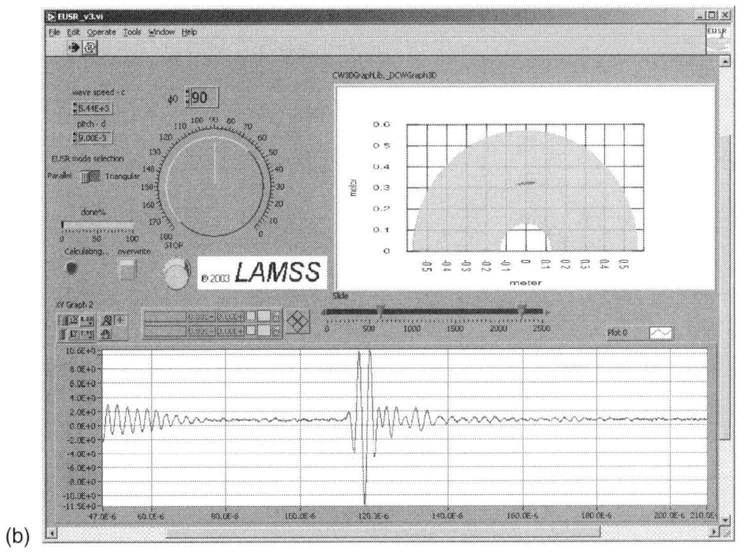

FIGURE 11.19 Crack detection at high curvature ($R = 1.95$ m) in direction 1: (a) photo of the bent specimen; (b) crack imaging and crack echo in the EUSR algorithm.

(a)

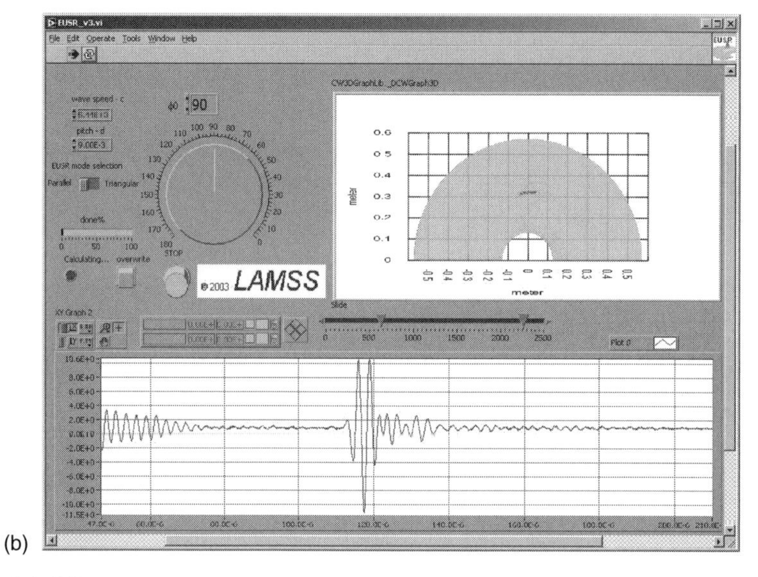

(b)

FIGURE 11.20 Crack detection at high curvature ($R = 1.95\,\text{m}$) in direction 2: (a) photo of the bent specimen; (b) crack imaging and crack echo in the EUSR algorithm.

11.4.3 IN-SITU DIRECT IMAGING OF CRACK GROWTH WITH PWAS PHASED ARRAYS

The previous section has shown how a PWAS array in conjunction with the EUSR algorithm can detect the presence of a simulated crack in a large plate using a sweeping beam of focused guided waves (S_0 Lamb waves). Though successful, these results might not be completely convincing for actual applications because of two limitations: (a) the crack was simulated, i.e., it was a machined slit, admittedly narrow, but still a slit; (b) the experiments were performed with the specimen in a controlled environment, away from vibrations and other disturbances that might be encountered in practice.

To overcome these limitations, further experiments were performed in which these restrictions were lifted. This section will present the results obtained when using a PWAS phased array to perform in-situ crack growth imaging during a fatigue test. This situation is much closer to an actual structural health monitoring situation, since the crack being monitored is an actual fatigue crack, and the data collection was performed during the fatigue experiment, i.e., in a very noisy vibrational environment. Nevertheless, the experiment was a success, and in-situ direct imaging of the crack growth was possible with a PWAS phased array permanently attached to the specimen. The imaging results were correlated with optical measurements of the crack size using penetrating liquid and a digital camera. Good consistency was obtained. Details of these experiments follow.

11.4.3.1 Design of the Experimental Specimens

The experimental specimens were designed such as to permit a controlled crack growth at an acceptable rate. For the proposed experiment, we considered a rectangular specimen with a crack in its middle. For such a specimen, the equation for the stress-intensity for mode I loading (Paris and Erdogan, 1963) is

$$K_I = \beta\sigma\sqrt{\pi a} \tag{17}$$

Where σ is the applied tensile stress, a is half of the crack length, and $\beta = K_I/K_0$. The value of the parameter β has been determined numerically for a large variety of specimen geometries and can be found in the literature (e.g., Fig. 11.3 of Chapter 1). In our case, we considered a 1 mm thick 2024-T3 aluminum specimen with dimensions 600 mm × 700 mm (24-in. × 28-in.). The specimen is loaded through a loading jig that has seventeen 16 mm holes on two rows with 50 mm pitch and 38 mm row spacing (Fig. 11.21). The presence of the loading holes imposes special requirements on the specimen design. (Since the loading jig already existed in the lab, the specimen design had to be adapted to the existing loading holes.) The loading holes weakened the specimen. Two strength concerns and one fracture concern had to be simultaneously considered with respect to the specimen-loading holes:

- The bearing strength of the hole
- The shearing strength of the 'plugs' between the holes and the sides of the specimen
- The stress concentration at the holes should not promote fatigue cracking.

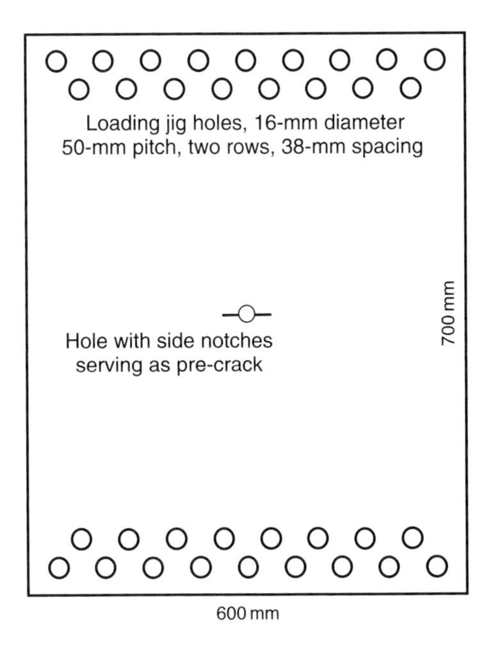

FIGURE 11.21 Schematic of specimen VG-0 for assessing the crack growth rate parameters.

In assessing these concerns, we used standard aircraft design practice guidelines for sheet metal joints (Bruhn, 1973). The cyclic fatigue load was considered varying between an upper value, F_{max}, and a lower value, F_{min}. Since the specimen was of thin sheet-metal construction susceptible to buckling under compression, only tensile loads were applied. An R-ratio of 0.1 was selected ($R = F_{min}/F_{max}$). Thus, the alternating part of the cyclic loading was $F_a = 0.45F_{max}$, whereas the mean part of the cyclic loading was $F_m = 0.55F_{max}$. The strength concerns (a) and (b) were considered to be affected only by F_{max}, whereas the stress concentration and crack propagation concern (c) was affected by both F_a and F_m. After performing the calculations, we concluded that a cyclic load of the specimen that would be safe for the bearing holes would be in the range $F_{max} = 30{,}000\,\text{kN}$, $R = 0.1$.

In the center of the specimen, we machined a hole and two notches to serve as 'pre-crack'. The length of the pre-crack was calculated such that the applied cyclic load would promote crack propagation from the pre-crack in a reasonable number of loading cycles. The cyclic load determined to be safe for the bearing holes was used to calculate the minimum pre-crack length that would promote crack propagation in the specimen. After performing the analysis, it was determined that an initial pre-crack length $2a = 50\,\text{mm}$ would be sufficient to achieve a crack intensity factor $\Delta K = 7.5\,\text{MPa}\sqrt{\text{m}}$ that will ensure an initial crack propagation at a comfortable rate according to Paris Law discussed in Chapter 1.

11.4.3.2 Experimental Results

The experiments were conducted in two stages. In the first stage, the experiments were aimed at determining the initial crack values and the loading conditions that will ensure crack nucleation and a controllable crack growth. In the second stage, the experiments were aimed at actually verifying that the crack growth can be directly imaged with the in-situ PWAS phased array using the EUSR algorithm. The first-stage experiments were aimed at giving a basis for the second-stage experiments, which were intended to utilize the crack growth parameters determined in the first stage.

Stage I Experiments: Fatigue testing without PWAS

The stage I experiments were performed on specimen VG-0; no PWAS transducers were installed (Fig. 11.21). The specimen VG-0 was made of 2024-T3 aluminum sheet with 1 mm thickness. The specimen size was 600 mm × 700 mm (24-in. × 28-in.). A notched hole was machined in the middle of the specimen. The diameter of the hole was 6.4 mm (1/4-in.) and the total pre-crack length was 53.8 mm (2.12-in.). The specimen was placed in an MTS 810 testing machine, as shown in Fig. 11.22a. Cyclic loading was applied with $F_{max} = 17{,}800\,\text{N}$ (4000 lbf) and $F_{min} = 1,780\,\text{N}$ (400 lbf), i.e., $R = 0.1$. A total of 350 kilocycles were applied at a frequency of 10 Hz. The crack length was measured periodically (every 20 kilocycles) using a microscope attached to a digital caliper. The crack-growth results are shown in Fig. 11.23 indicates that the crack growth behavior resembles Paris Law. It is significant to note that the crack growth started to accelerate after 300 kilocycles, which is consistent with Paris Law. At the end of the experiment, the crack has grown from the initial 54 mm to a final of approximately 210 mm. The corresponding stress intensity factors were $K_{initial} \cong 7.5\,\text{MPa}\sqrt{\text{m}}$ and $K_{final} \cong 15.3\,\text{MPa}\sqrt{\text{m}}$.

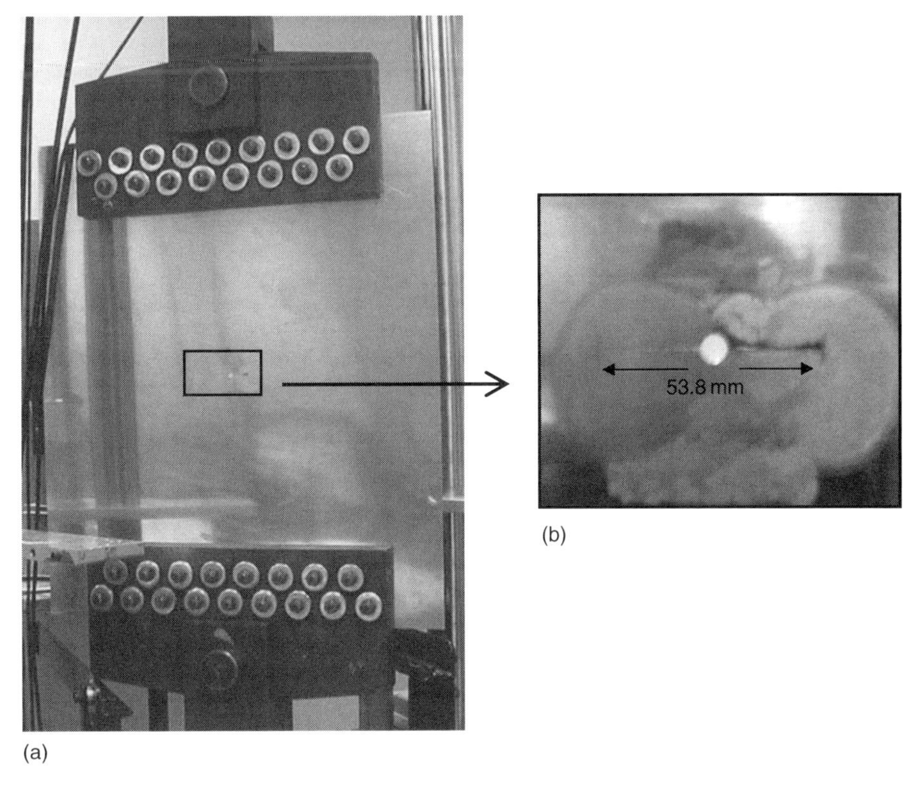

FIGURE 11.22 Experimental setup for fatigue testing without PWAS: (a) specimen VG-0 loaded in the MTS 810 testing machine; (b) hole with side notches serving as initial crack.

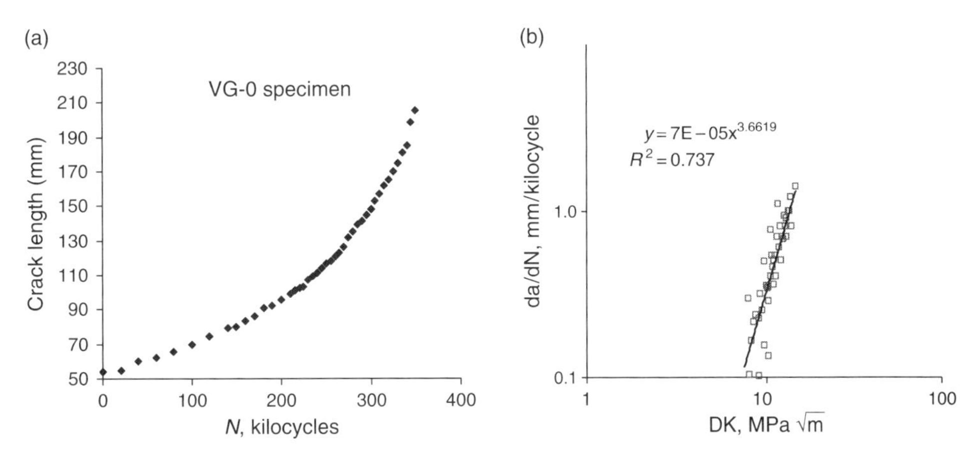

FIGURE 11.23 Crack growth results for specimen VG-0: (a) crack length vs. kilocycles; (b) Paris Law fit.

Stage II Experiments: Fatigue test with PWAS

The second specimen was dedicated to performing in-situ monitoring of fatigue crack growth using a phased array of permanently mounted PWAS transducers. This specimen was named 'VG-1' (Fig. 11.24). The specimen construction is similar to that of specimen VG-0 used for fatigue testing without PWAS, i.e., 600 mm × 700 mm 1-mm thick 2024-T3

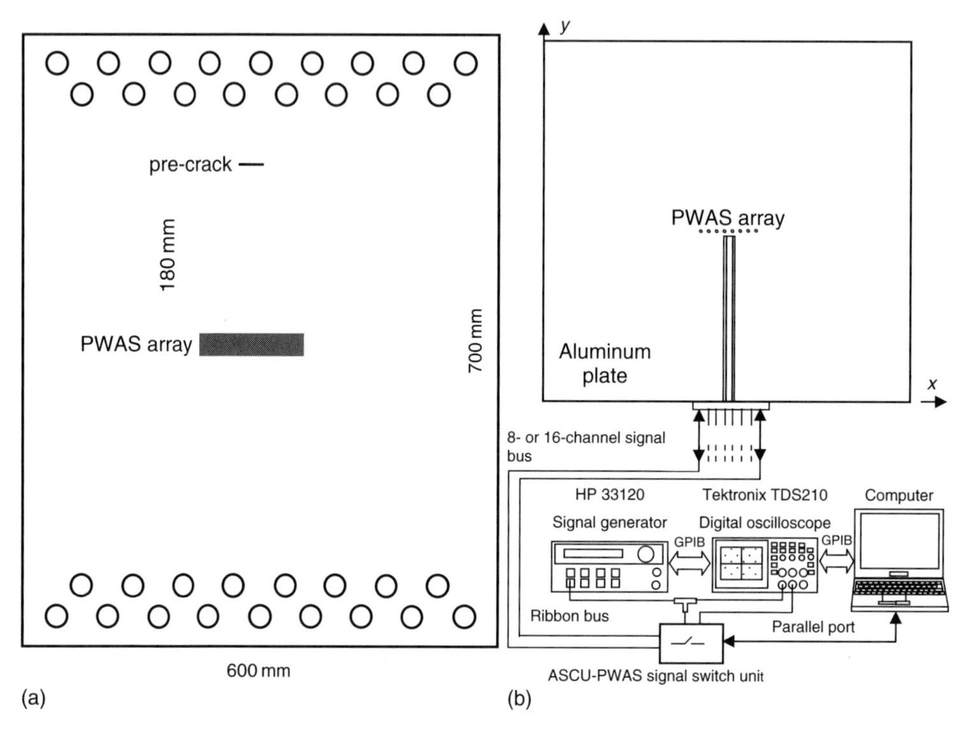

FIGURE 11.24 Schematic of experimental setup for fatigue testing with PWAS: (a) schematic of specimen VG-1 showing the installation of the PWAS array and the location of the precrack; (b) instrumentation schematics.

aluminum sheet with two rows of loading rows at each end. However, the specimen VG-1 presented the following differences from specimen VG-0:

- The pre-crack was moved from the center to one side of the specimen, approximately 180 mm north of center
- The pre-crack, made with a proprietary method, did not need a center hole and was in the form of a narrow slit, thus resembling much better an actual crack
- A 10-element PWAS array was placed in the center of the specimen.

The pre-crack was 125 μm wide and 30 mm long. The 10-PWAS array was created by slicing two rectangular PZT wafers of dimensions 25 mm × 5 mm. The wafer thickness was 200 μm. The wafer had electrodes on both sides. The procedure consisted of first bonding the rectangular wafer to the plate at 340 mm (13.4″) from the plate edge, and then slicing the attached wafer with a scoring tool. After slicing, a phased array of 10 PWAS, each approximately 5 mm (0.2-in.) square, was obtained. The PWAS elements where numbered starting with PWAS 00 on the right and ending with PWAS 09 on the left. The instrumentation setup is shown in Fig. 11.24b. It consisted of an HP 33120 signal generator, a TDS210 digital oscilloscope (serving as DAQ device), and a laptop computer.

The instrumentation was used to implement the EUSR steering beam method for direct structural imaging with guided Lamb waves using the PWAS phased array. In our case, the number of sensors used in the array was $M = 10$, i.e., from PWAS-00 through PWAS-09. Essential for the implementation of the EUSR algorithm is the round-robin collection of the M^2 array of elemental signals. The measurement procedure is performed in the following way (Fig. 11.24b): a 3-count 372 kHz tone-burst excitation signal is

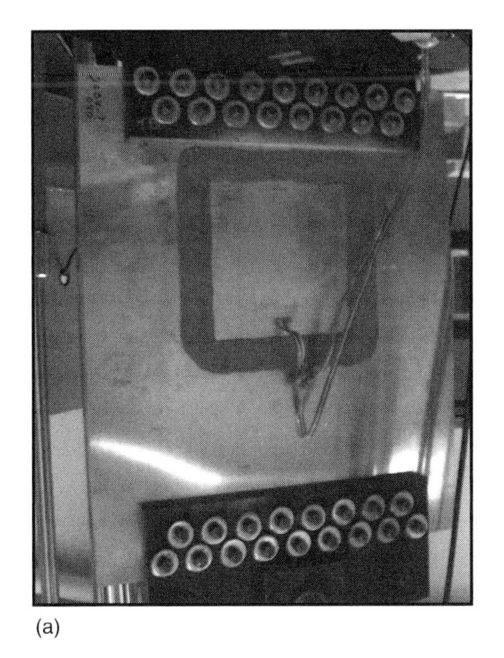

(a)

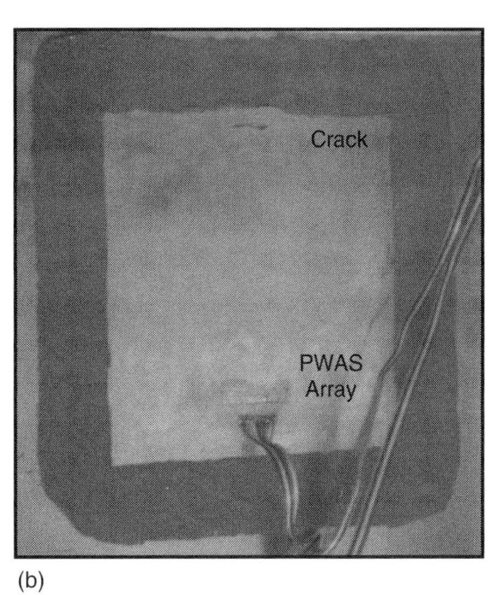

(b)

FIGURE 11.25 Experimental setup for fatigue testing with PWAS: (a) overall picture showing the specimen VG-1; (b) detail of the specimen showing the PWAS array, the crack, and the modeling clay damn.

synthesized in the function generator. (This frequency of the tone-burst signal corresponds to optimum tuning of the PWAS with the S_0 Lamb-wave mode.) The tone-burst signal is sent to one PWAS in the array where is transformed into an S_0 Lamb-wave packet. The Lamb-wave packet travels into the plate and is reflected at the plate boundary. The reflected Lamb-wave packet is received back at the PWAS array where it is converted back into an electrical signal. The signals received at each PWAS in the array (including the transmitting PWAS) are collected by a DAQ device, e.g., a digital oscilloscope. To minimize instrumentation, the collection is done on only one DAQ channel using a round-robin procedure. This generates a column of M elemental signals in the M^2 elemental-signals array. After the signal collection for one PWAS acting as exciter is finalized, the cycle is repeated for the other PWAS in a round-robin fashion. It is apparent that the round-robin data collection can be tedious if done manually. However, considerable savings can be achieved if the process is automated. To automate the round-robin data collection, we used the automatic signal collection unit ASCU-PWAS. To minimize the reflections from the loading holes and the plate boundary, a 'wave dam' made of molding clay was placed around the PWAS and crack area (Fig. 11.25).

Validation of PWAS data collection under vibratory fatigue loading

The first task in our experiment was to validate the experimental method. Two issues had to be proved:

(a) Determine if the permanently installed PWAS transducers are able to collect usable signals whereas the specimen is subjected to the intense vibrations associated with cyclic fatigue loading
(b) Determine if the PWAS transducers would survive the fatigue cycling without disbonding from the specimen.

The first issue was crucial to the experimental premises. The adverse factors to be clarified under (a) were associated with the relative weakness of the Lamb waves generated by our experimental setup in comparison with the rather large noise signals generated by the vibration of the specimen in the loading frame during the cyclic loading. Our initial tests proved that this was indeed a problem: the Lamb-wave signal was buried in the vibration noise. However, it was determined that the noise frequency (approximately, 10 to 100 Hz) was way outside the frequency band of the Lamb-wave signals (300 to 400 kHz). Hence, we were able to use a high-pass filter to reject the noise whereas maintaining the useful Lamb-wave signal. The cutoff frequency of the high-pass filter was 1 kHz. A comparison of the signal before and after applying the filter is given in Fig. 11.26. As shown in Fig. 11.26 the pulse-echo graph during fatigue cycling between PWAS 00 and PWAS 05 in the phased array with a filter taken whereas the specimen was under cyclic fatigue. Note that the reflection from the crack is clearly distinguishable. This proved that data collection can be performed on line whereas the specimen is being tested in the fatigue machine, something that, so far, had not been proven possible before.

The second concern addressed during the fatigue testing with PWAS was that of PWAS survivability. In Chapter 7, we showed that PWAS applied to dog bone fatigue specimens with stress concentrations can survive up to the specimen failure if proper bonding methods were used. Survival beyond 12 M-cycles was proved. In the present work, we used the same bonding methods: adhesive AE-10 from Measurements Group, Inc. and the corresponding surface preparation and curing procedure were used.

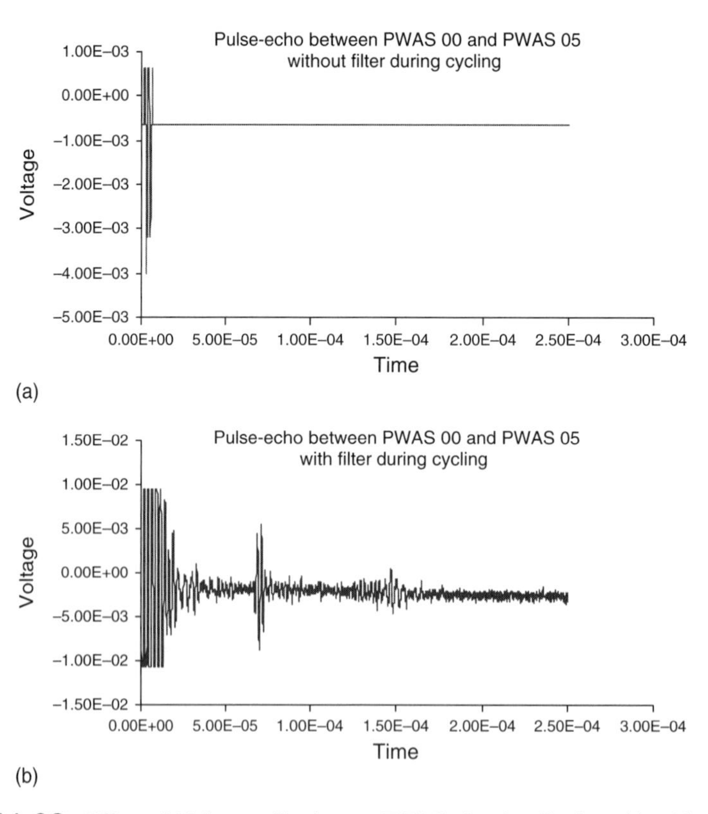

FIGURE 11.26 Effect of high-pass filtering on PWAS signal collection: (a) without filter, no Lamb-wave signals can be observed; (b) with filter, the Lamb-wave's reflection from the crack can be easily identified.

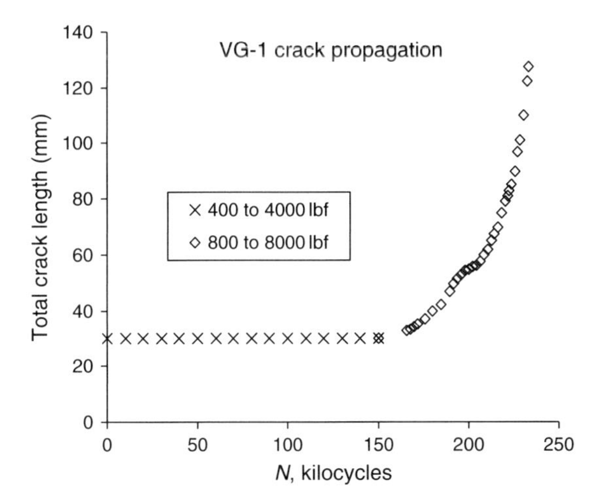

FIGURE 11.27 Crack growth history for the specimen VG-1 instrumented with a PWAS phased array.

To verify the PWAS survivability in the presence of fatigue loading, we selected in the beginning reduced cyclic loading levels such that insignificant crack growth would occur. The reduced loading was with $F_{max} = 17{,}800\,\mathrm{N}$ and $F_{min} = 1{,}780\,\mathrm{N}$, i.e., $R = 0.1$. Since the precrack size was $2a = 25\,\mathrm{mm}$, the corresponding stress intensity factor range was $\Delta K \cong 5.8\,\mathrm{MPa}\sqrt{\mathrm{m}}$. The PWAS readings were taken as follows. First, a baseline set of signals was taken. Then, the specimen was loaded into the tensile machine, and a second baseline was taken with the specimen under static load. Subsequently, the specimen was fatigued for 150 kilocycles with readings of crack size being taken every 10 kilocycles. The resulting crack-size readings are shown in the first part of Fig. 11.27. It is noticed from Fig. 11.27 that during this first part of the experiment no observable crack propagation was produced. By not propagating the crack, a comparison could be made between the baseline scans and the scans made during and after fatigue cycling. No discernable difference could be observed in the EUSR imaging. Hence, it was concluded that the adhesion method was satisfactory, and that no PWAS disbonding occurred during tests.

In-situ crack growth imaging with PWAS phased arrays during cyclic loading

The purpose of this part of the fatigue experiment was to determine if an in-situ PWAS phased array and the EUSR algorithm were able to detect and quantify the growth of the crack during fatigue loading. To achieve this, we continued the testing of specimen VG-1, but increased the load applied to the specimen to these new values, $F_{max} = 35{,}600\,\mathrm{N}$ and $F_{min} = 3{,}560\,\mathrm{N}$, i.e., $R = 0.1$, again. Crack-size readings were taken at approximately every 2 kilocycles. Under these conditions, the crack grew rapidly from 25 mm to 143 mm over a 85 kilocycles duration (Fig. 11.27). The total duration of the fatigue testing was 235 kilocycles, of which: 150 kilocycles at low load values and no crack growth; 85 kilocycles at high load values and rapid crack growth.

Recordings of the PWAS phased array ultrasonic signals were taken at selected intervals during this crack-growth process. Every reading was taken with two loading conditions: (a) under dynamic conditions, i.e., with the testing machine operating under cyclic loading; and (b) under static conditions, i.e., with the testing machine held at the mean load. As the crack growth became more rapid, the interval between two consecutive recordings also shortened. Images of the crack growth as resulting from the ultrasonic Lamb-wave PWAS phased array interrogation was reconstructed with the EUSR algorithm. A digital camera optical photograph of the actual crack was also taken, in parallel with the EUSR imaging.

The crack-growth results shown in Fig. 11.27 indicate that the crack growth behavior resembles Paris Law. It is significant to note that the crack growth started to accelerate after 220 kilocycles, which is consistent with Paris Law. At the end of the experiment, the crack had grown from the initial length of 25 mm to a final length of 143 mm. The corresponding stress intensity factors were $\Delta K_{\text{initial}} \cong 11.6\,\text{MPa}\sqrt{\text{m}}$ and $\Delta K_{\text{final}} \cong 25.3\,\text{MPa}\sqrt{\text{m}}$.

The EUSR algorithm and the associated software program were used to process the PWAS phased array data and image the crack. Figure 11.28 shows a progression of cracks sizes, as they developed in the specimen VG-1 during the fatigue testing. The upper row of images contains the optical photos taken with a digital camera. The lower row of images contains the EUSR images of the crack obtained with the PWAS phased-array method. It is apparent that the two rows of images are in good correlation.

The EUSR graphical user interface (GUI) was used to obtain an estimation of the crack size. The threshold value and the values δ and θ of the 'dial angles' are controlled from this GUI. First, an approximate position of the crack edge is obtained with the azimuth dial. If the azimuth dial is turned to an angle where the synthetic beam finds a target and gets a reflection, then the A-scan image will show a reflection echo. After a threshold value was chosen, the θ and δ angles were adjusted such that their rays touched the left and right tips (respectively) of the crack image reproduced in the EUSR GUI. The length of the crack thus estimated using the EUSR GUI can be easily found using simple geometry, i.e.,

$$2a_{EUSR} = l(\tan\delta + \tan\theta) \tag{18}$$

In our experiment, the distance between the array and the crack was $l = 180\,\text{mm}$. Hence, the EUSR-estimated crack length could be calculated at various numbers of cycles.

During data processing, it was noticed that, from 150,000 cycles to 200,000 cycles, the threshold remained the same and the EUSR image of the crack, which grew with the number of cycles, and was close to the actual crack size, as measured with optical means. However, after the crack grew to about 55 mm, the EUSR image did not grow with the same threshold and the threshold had to be changed to image the crack with dimensions

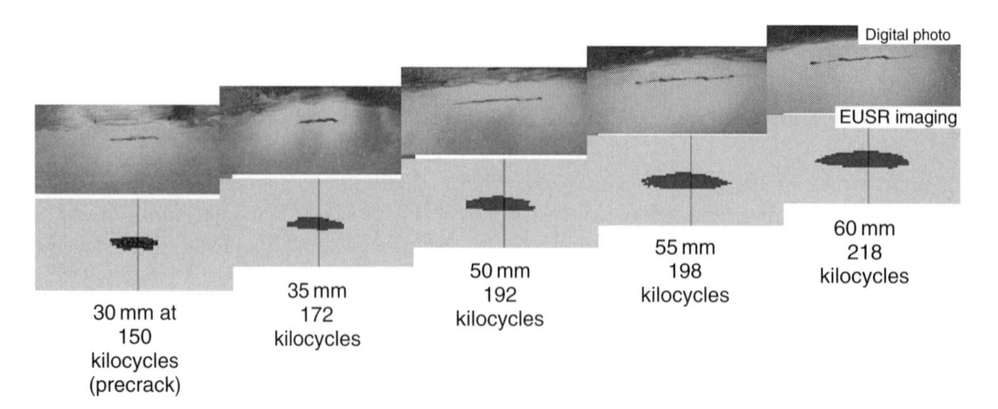

FIGURE 11.28 Comparison between images taken optically and scanned images using EUSR.

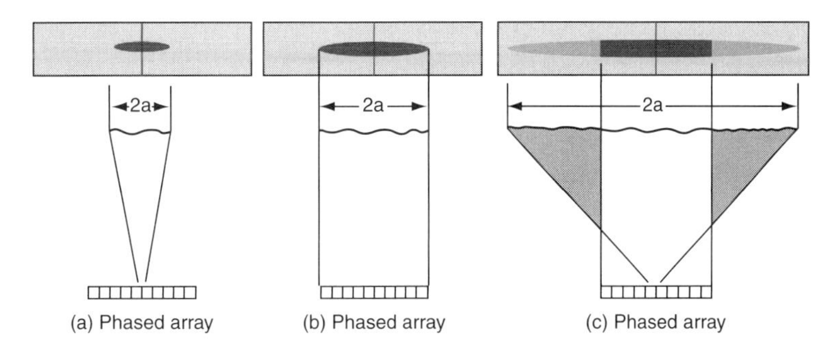

FIGURE 11.29 Aperture effects on PWAS phased array EUSR imaging of crack length: (a) if crack is less than array aperture the crack is well imaged; (b) borderline case when crack and aperture are equal in length; (c) if crack length exceeds array aperture, the EUSR image of the crack does not grow proportionally with the crack length if the threshold is kept constant. Threshold had to be lowered to see the grayed-out image of the crack.

that were comparable to the optical measurements. Several samples of the EUSR images with corresponding optical crack images are given in Fig. 11.28. The EUSR crack-length image for up to $\sim$ 200 kilocycles is close to being accurate. Beyond 200 kilocycles, the EUSR crack length image did not significantly change from the EUSR image at 200,000 cycles. This could be due to phased-array aperture effects (Fig. 11.29). The reflections of the wave from the crack are stronger when the crack is directly in front of the PWAS array. When the crack grows beyond the length of the PWAS array, the reflections are weaker and the image of the crack length cannot be seen at a high threshold because the high threshold will cover the weaker reflections (see Fig. 11.29). If the threshold is not changed when the crack length is larger than the PWAS array, the EUSR crack image will remain at the same length as the PWAS array for any crack length longer than the PWAS array. An intuitive explanation of this phenomenon is as follows: think of the PWAS array as someone looking at an object directly in front of the eyes. The person can clearly see the object directly in front of them, but cannot clearly see objects to the side or that are not directly in front. Although the objects are not clearly seen, the person is still aware of the objects. The same seems to be true with the PWAS array, it can see images that are not directly in front of it at an appropriate threshold but not well enough to quantify the accurate length.

11.4.4 SECTION SUMMARY

The PWAS phased array concept was used to achieve direct in-situ imaging of crack growth during an actual fatigue experiment. It was shown that the noise issues related to the vibrational environment can be successfully overcome and meaningful echo signals could be recorded. These signals permitted the EUSR imaging of the crack and its growth, which compared remarkably well with optical measurements made with a digital camera. Furthermore, it was proved that the permanently attached PWAS array survived the cyclic fatigue loading without disbonding or loss of functionality. Thus it was proved that permanently attached PWAS transducers arranged in a phased-array pattern can successfully monitor crack length and its propagation due to cyclic fatigue loading. Finally, it was verified that in-situ direct monitoring of crack growth can be achieved with PWAS phased-array using the EUSR algorithm.

11.5 OPTIMIZATION OF PWAS PHASED-ARRAY BEAMFORMING

In this section, we will examine methods by which the beamforming patterns of a 1-D linear array can be modified and optimized.

11.5.1 ISSUES RELATED TO PHASED ARRAY BEAMFORMING

Beamforming can be achieved by suitably applying weights and delays to the array elements to obtain the desired adaptive directional sensitivities and optimal array gains. The general principles of phased-array beamforming and signal processing were extensively presented by Johnson and Dudgeon (1993). Several other authors have also done extensive investigation in this field (Godara, 1997; Sundararaman and Adams, 2002, 2004, 2005). It seems that there are many ways to achieve the phased array beamforming, including:

(1) conventional beamforming by simple delay-and-sum algorithms;
(2) null-steering beamforming which assumed that the interference is known;
(3) optimal beamforming.

Mainlobe quality degrades when the beam is steered closer to the array, i.e., around 0° and 180°. In many applications, a 16-element array may be sufficient to ensure reasonably good directivity. Beam directivity improves with increasing d/λ values. However, if d/λ increased beyond 1/2, grating lobes start to show up; their number gets larger as d/λ increases.

Optimization of the phase array beamforming can be obtained by adjusting several parameters: (a) number of sensors, M; (b) sensor spacing, d; (c) the d/λ ratio; (d) the weights of the array. Once these parameters are fixed, the mainlobe width, grating lobes' size and location, and sidelobes' magnitude are also fixed. For the linear phased array, the objectives of beamforming optimization should include:

- minimizing the main lobe width
- eliminating grating lobes
- suppressing side lobe amplitudes.

Steering angle θ_s also affects the beamforming of a linear phased array. There is a maximum angle θ_{gr} up to which the beam can be steered without producing any detrimental grating lobes. In fact, the ratio d/λ influences both the steerability and main lobe width. When $d \ll \lambda$, grating lobes are suppressed but the directivity is rather poor, showing a broad main lobe. If $\lambda \ll d$, the directivity is superior at the intended steering angle, but the existence of additional grating lobes obfuscates the main lobe; eventually, the beam is no long steerable. A few general guidelines are as follows:

1. *Steering angle* θ_s. The width of the main lobe will change with θ_s. A grating lobe will show up within a certain θ_s range
2. *Number of elements* M. Increasing the number of elements M will improve the directivity. A sharper main lobe can be achieved by using more elements
3. *Inter element spacing* d. In most phased array designs this ratio is kept at half wavelength, i.e., $d = 0.5\ \lambda$
4. *Array aperture* $D = (M - 1)d$. If aperture length D remains constant, the main lobe width will remain unchanged as well.

The effect of increasing the number of elements M is to sharpen the main lobe; it seems that this is the safest way to improve beam steering quality. However, this effect may be associated with the corresponding increase in array aperture $D = (M - 1)d$. But we may want to keep the aperture small in order to reduce the blind zone close to the array. In this case, we may choose to sacrifice a large value of d and use many elements to fill up the aperture. In compensation, we will need to increase the carrier frequency in order to bring down the wavelength λ and keep the d/λ ratio close to 1/2. This approach, if possible, will ensure a relatively lighter and more compact design with good steering characteristics. Design strategy of linear phased array may be summarized as follows:

1. Preference is given to the larger M since it is not only improves the steering quality but also suppresses side lobe amplitudes and increases maximum steerable angle. It also guarantees there are no grating lobes present if the inter element spacing d is chosen below the critical value.
2. However, if M is increased and d is kept constant, the aperture D will also increase and thus the near-field performance of the array will degrade because of the larger blind area and reduced inspectable area.
3. The same array performance can be achieved without changing the aperture size or the near-field length by modifying the number of elements, the corresponding inter element spacing, and the carrier frequency.
4. Determination of the aperture D that meets the requirement for near-field length depends on the type of SHM inspection to be performed. Smaller array will have shorter near-field length and thus will provide an enlarged inspection area.

In the following sections we will study these effects individually in the context of PWAS phased array. The scope is to gain a better understanding of how these issues can be handled in order to achieve an optimal PWAS phased array design for SHM applications. Numerical examples will be given and discussed comparatively.

11.5.2 PARAMETERIZED BEAMFORMING FORMULAE FOR 1-D LINEAR PWAS ARRAYS

In order to conduct a parameter study and achieve phase array optimality, a parameterized beamforming formulation is first derived. A PWAS linear phased array consisting of M transducers uniformly spaced at d is considered (Fig. 11.30). The span (aperture) of the array is given by

$$D = (M - 1)d \tag{19}$$

With the coordinate system origin located in the middle of the array, the location vector of the mth element is

$$\vec{s}_m = \left[\left(m - \frac{M-1}{2}\right)d, 0\right] \tag{20}$$

And the vector $\vec{r}_m$ is

$$\vec{r}_m = \vec{r} - \vec{s}_m \tag{21}$$

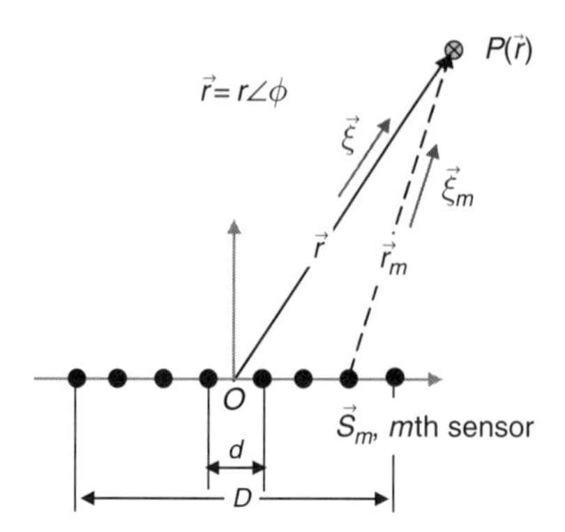

FIGURE 11.30 Schematic of an M-PWAS phased array. The coordinate origin is located in the middle of the array.

Assume a single-tone radially propagating wave from a generic PWAS source at the origin. The wave front at a point $\vec{r}$ away from the source can be expressed as

$$f(\vec{r}, t) = \frac{A}{\sqrt{|\vec{r}|}} e^{j\left(\omega t - \vec{k}\cdot\vec{r}\right)} \tag{22}$$

where $\vec{k}$ is the wavenumber, $\vec{k} = \vec{\xi}\cdot\omega/c$, and ω is wave frequency of the wave. For an M-element linear array, we apply Eq. (22) to each array element m, where $m = 0, 1, 2, \ldots, M-1$ (Fig. 11.30). The mth element is assumed at location $\vec{s}_m$, whereas the direction vector from mth element to the target is defined as $\vec{\xi}_m$. The following notations apply

$$\vec{\xi} = \frac{\vec{r}}{r}, \quad r = |\vec{r}|, \quad \vec{k} = \vec{\xi}\cdot\frac{\omega}{c} \tag{23}$$

$$\vec{r}_m = \vec{r} - \vec{s}_m, \quad r_m = |\vec{r}_m|, \quad \vec{\xi}_m = \frac{\vec{r}_m}{|\vec{r}_m|}, \quad \vec{k}_m = \vec{\xi}_m\cdot\frac{\omega}{c} \tag{24}$$

where $\vec{k}_m$ is the wavenumber of the wave propagating in the direction of $\vec{\xi}_m$. The synthetic wave front arriving at target $P(\vec{r})$, $\vec{r} = r\angle\phi$, results from the superposition of the waves generated by the M sources. If we also consider that each source can be fired with a different weight, w_m, then we get the generic expression

$$z(\vec{r}, t) = \sum_{m=0}^{M-1} w_m f(\vec{r}_m, t) = \sum_{m=0}^{M-1} w_m \cdot \frac{A}{\sqrt{r_m}} e^{\left\{j\left[\omega t - \vec{k}_m\cdot\vec{r}_m\right]\right\}} \tag{25}$$

Equation (25) can be rewritten as

$$z(\vec{r}, t) = f\left(t - \frac{r}{c}\right)\cdot\sum_{m=0}^{M-1} w_m \frac{1}{\sqrt{r_m/r}} e^{j\omega\left(\frac{r-r_m}{c}\right)} \tag{26}$$

The first multiplier represents a wave emitting from the origin and it is independent of the array elements. This wave is to be used as a reference for calculating the needed time delays for each wave transmitted from each of the phased array elements. The second multiplier, which controls the array beamforming, can be simplified by normalizing r_m by the quantity r, resulting in the beamforming factor

$$BF(w_m, M) = \frac{1}{M} \cdot \sum_{m=0}^{M-1} w_m \frac{\exp\left\{j\frac{2\pi}{\lambda}(1 - r_m)\right\}}{\sqrt{r_m}} \tag{27}$$

The scale factor $1/M$ is used to normalize the beamforming factor. By further introducing two new parameters, d/λ and r/d, the beamforming factor is rewritten as

$$BF\left(w_m, M, \frac{d}{\lambda}, \frac{r}{d}\right) = \frac{1}{M} \cdot \sum_{m=0}^{M-1} w_m \frac{\exp\left\{j2\pi\frac{d}{\lambda}\frac{r}{d}(1 - r_m)\right\}}{\sqrt{r_m}} \tag{28}$$

For the far-field case, the simplified beamforming is independent of r/d, i.e.,

$$BF\left(w_m, M, \frac{d}{\lambda}\right) = \frac{1}{M} \cdot \sum_{m=0}^{M-1} w_m \exp\left\{j2\pi\frac{d}{\lambda}\left[\left(m - \frac{M-1}{2}\right)\cos\phi\right]\right\} \tag{29}$$

The beamforming factor of Eqs. (28) and (29) has a maximum value for $\phi_0 = 90°$. This is the inherent beamforming of the linear array. The inherent beamforming for an 8-PWAS array with $w_m = 1$, $d/\lambda = 0.5$, and $r/d = 10$ is shown in Fig. 11.31 (solid line). Notice that, indeed, the maximum beam is obtained at 90°.

Now we apply 'delays' to steer the beam toward a preferred direction ϕ_0. With delays $\delta_m(\phi_0)$, the beamforming is

$$BF\left(w_m, M, \frac{d}{\lambda}, \frac{r}{d}, \phi_0\right) = \frac{1}{M} \cdot \sum_{m=0}^{M-1} w_m \frac{\exp\left\{j2\pi\frac{d}{\lambda}\frac{r}{d}(1 - r_m - \delta_m(\phi_0))\right\}}{\sqrt{r_m}} \tag{30}$$

Beamforming factor of Eq. (30) reaches its maximum in direction ϕ_0 when the delay $\delta_m(\phi_0)$ is chosen as

$$\delta_m(\phi_0) = 1 - r_m(\phi_0) = 1 - \frac{|\vec{r}(\phi_0) - \vec{s}_m|}{|\vec{r}(\phi_0)|} \tag{31}$$

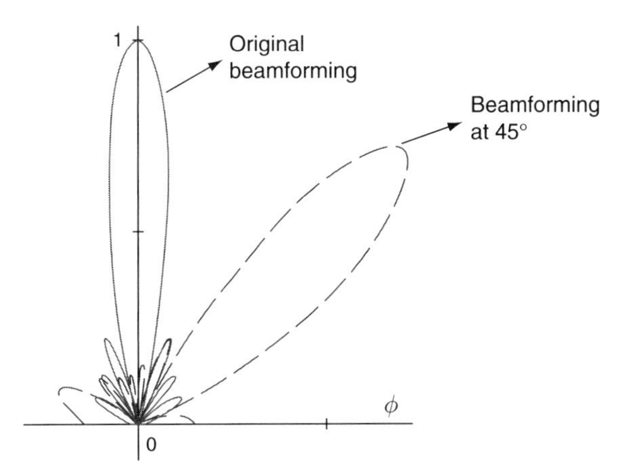

FIGURE 11.31 The original beamforming and directional beamforming at 45° of an 8-PWAS phased array with $w_m = 1$, $d/\lambda = 0.5$, $r/d = 10$.

By changing the value of ϕ_0 from 0° to 180°, we can generate a scanning beam. Simulation result of the directional beamforming at $\phi_0 = 45°$ is also shown in Fig. 11.31.

Using formula (30), we will perform several optimization studies of the PWAS phased array. The beamforming at certain direction ϕ_0 is affected by several parameters:

- spacing between neighboring PWAS elements, d;
- number of PWAS elements, M;
- steering angle, ϕ_0;
- weighting factors, w_m.

Among these parameters, the effect of spacing d is always measured by the wavelength λ. Since $\lambda = c/f$, when the wavelength λ changes with the frequency, the ratio of d/λ will also change. For the near-field triangular beamforming, there is an extra parameter needed to be considered, the ratio of r/d. Therefore, the effect of parameter d can be represented by the ratio d/λ and the ratio r/d. If all the weighting factors are the same (w_m = constant), i.e., all PWAS elements in the array are excited uniformly, this type of phased array is called uniform array. Otherwise, it is called nonuniform array. First, we will use the beamforming algorithm to explore how these parameters affect the uniform PWAS phased-array beamforming. Then, we will examine how the beamforming is modified by the weighting factors.

11.5.3 UNIFORM PWAS PHASED ARRAY STUDIES

In this section, we consider an M-element uniform PWAS array equally spaced at pitch d with equal weights, $w_m = 1$. On this uniform PWAS phased array, we consider the effect of several parameters: (i) r/d ratio; (ii) d/λ ratio; (iii) number of PWAS elements in the array, M; (iv) steering angle, ϕ_0.

11.5.3.1 Effect of r/d Ratio

The beamforming algorithm varies according to whether the reflector is located near to or far from the array. Using the concepts of antenna theory (Balanis, 2005), we define the near field as the region in which

$$0.62\sqrt{D^3/\lambda} < R_{near} \leq 2D^2/\lambda \tag{32}$$

where D is the array aperture and λ is the excitation wavelength. Far field is defined as the region where

$$R_{far} > 2D^2/\lambda \tag{33}$$

in the region $R \leq 0.62\sqrt{D^3/\lambda}$, phased array theory is no longer valid. However, other methods for damage detection, such as the PWAS electromechanical impedance technique described in Chapter 9, can be employed.

The ratio r/d determines whether a target is located in the far field or the near field of the phased array. If the target is in the near field, we have to use the exact triangular algorithm. Otherwise, we may use the parallel-ray approximation. To better quantify this

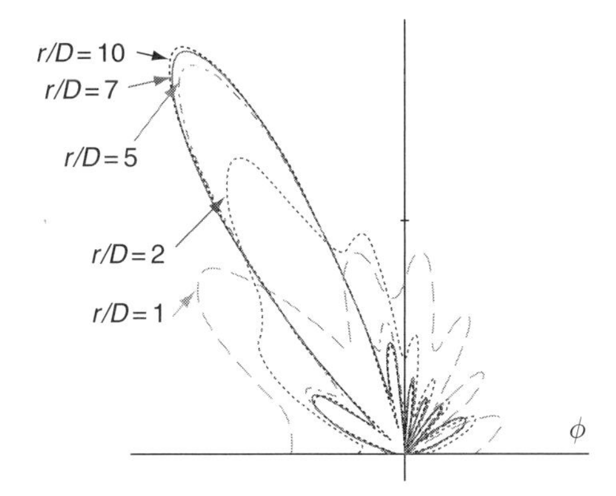

FIGURE 11.32 Beamforming at 120° of an 8-PWAS phased array spacing at 0.5λ with various r/D value.

ratio, the array span $D = (M-1)d$ is used instead of d. The definitions of Eqs. (32) and (33) can be expressed in terms of r/D as

$$0.62\sqrt{(M-1)\frac{d}{\lambda}} < \left(\frac{r}{D}\right)_{near} \leq 2(M-1)\frac{d}{\lambda} \tag{34}$$

$$\left(\frac{r}{D}\right)_{\text{far}} > 2(M-1)\frac{d}{\lambda} \tag{35}$$

These definitions show that for a particular application of an 8-PWAS array spaced at half wavelength ($d = \lambda/2$), the near field is contained in the interval $r \in (1.16D, 7D]$, whereas the far field is the outside area ($r > 7D$). Beamforming simulations with r/D value evolving from 1 to 2, 5, 7, and 10 and the PWAS array directed at 120° are shown in Fig. 11.32.

Figure 11.32 shows that very close to the array field ($r/D = 1$), directional beamforming does no longer exist (dash line). However, as the target moves away from the array (entering near field, $r/D = 2$), directional beamforming starts to take shape. The beamforming is getting better when far-field conditions are approached ($r/D = 5$). In far field, where parallel rays approximation applies, the effect of r/D vanishes. No significant difference can be noticed between $r/D = 7$ and $r/D = 10$.

11.5.3.2 Effect of d/λ Ratio

The study of the ratio d/λ shows the influence of spacing on array beamforming. Simulation results of an 8-PWAS array directed at 120° with various d/λ values are shown in Fig. 11.33. By comparing the beams in Fig. 11.33a, it can be seen that the beam width becomes smaller and smaller as d/λ increases. However, the number of sidelobes also increases. For larger d/λ value, narrow beam width (better resolution/directivity) is achieved, but many disturbing sidelobes are also presented.

For beamforming at $d/\lambda = 0.75$ and beyond (Fig. 11.33b), other strong and disturbing lobes show up at other undesired directions besides the mainlobe at the desired angle of 120°. Such lobes are called grating lobes; they are caused by spatial aliasing (Johnson and Dudgeon, 1993). Grating lobes can have magnitudes as big as the mainlobe. The grating lobes are not desired and should be avoided because they give misleading scanning results. According to the spatial sampling theorem, the spacing d between elements should be

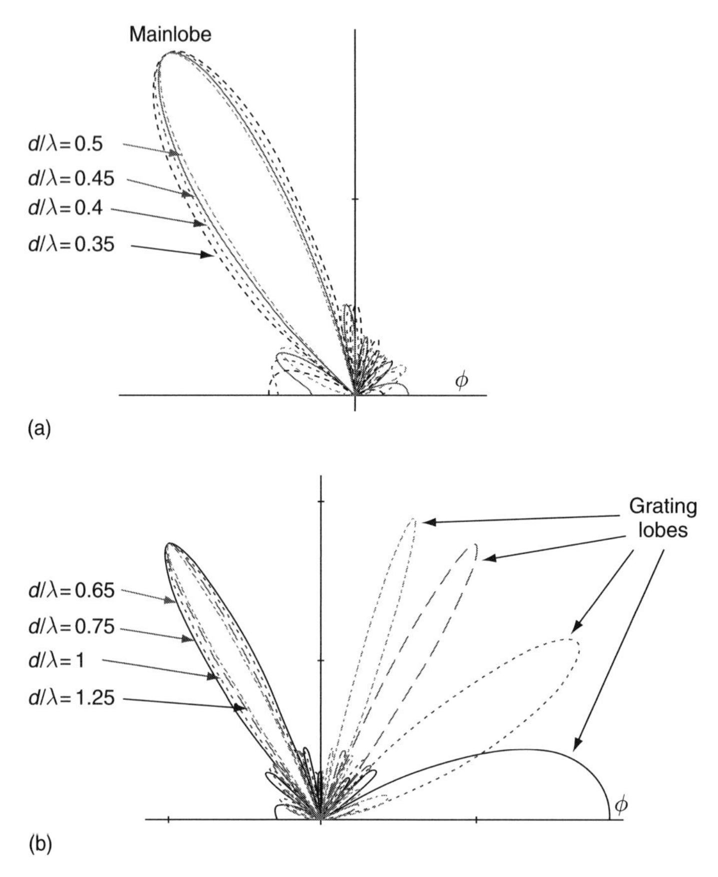

FIGURE 11.33 Beamforming at 120° of an 8-PWAS array with $r/D = 10$ using various d/λ: (a) when $d/\lambda \leq 0.5$; (b) $0.5 < d/\lambda$.

smaller or equal to the half wavelength ($d/\lambda \leq 0.5$) in order to avoid spatial aliasing. Otherwise grating lobes may appear. In practical implementation, this rule should be verified after frequency tuning. For the 300 kHz tuning frequency used in the proof-of-concept experiments, the ratio d/λ has the value $d/\lambda = 0.44$ ($d = 8\,\text{mm}$, $c = 5440\,\text{m/s}$). Therefore, no grating lobes were yet presented.

The discussion on the influence of the spacing pitch on the array beamforming indicates that larger spacing may give better directional beams despite their byproduct, larger sidelobes. However, the spacing cannot be unlimitedly increased because of the spatial sampling theorem.

11.5.3.3 Effect of Number M of PWAS in the Array

The number of elements in the array is another factor that affects the beamforming. Figure 11.34 demonstrates how beamforming is modified by different M values, e.g., $M = 8$ and $M = 16$. The comparison shows the beamforming of a 16-PWAS array having a much narrower mainlobe and slightly stronger sidelobes than that of an 8-PWAS array. Increasing the number of elements is a simple way to enhance beamforming with the small penalty of larger sidelobes. However, in practice, more elements will result in wiring issue and will be limited by the available installation space.

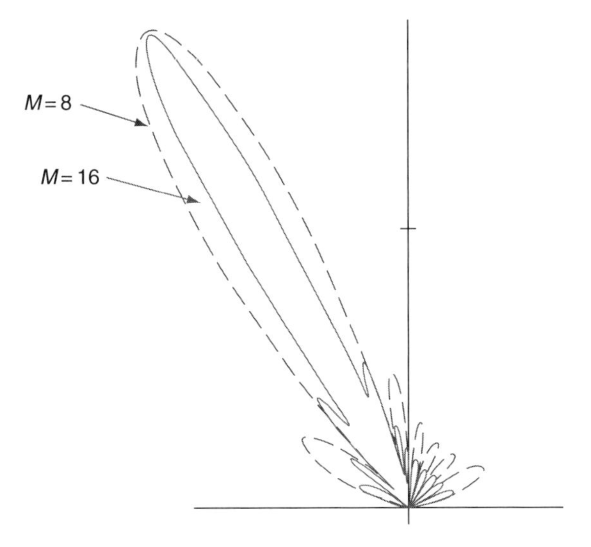

FIGURE 11.34 Beamforming at 120° of PWAS array with $r/D = 10$ and $d = 0.5\lambda$ using different M. The beam in dash line is obtained with $M = 8$ and the beam in solid line is obtained with $M = 16$.

11.5.3.4 Effect of steering angle ϕ_0

Steering angle ϕ_0 (beamforming direction) is another factor that affects the beamforming. Figure 11.35a and b shows beamforming at 0°, 30°, 60°, 90°, 120°, and 150° directions using 8-PWAS and 16-PWAS are arrays, respectively. An overall examination of Fig. 11.35 indicates that the best beamforming is achieved at $\phi_0 = 90°$, with a slender

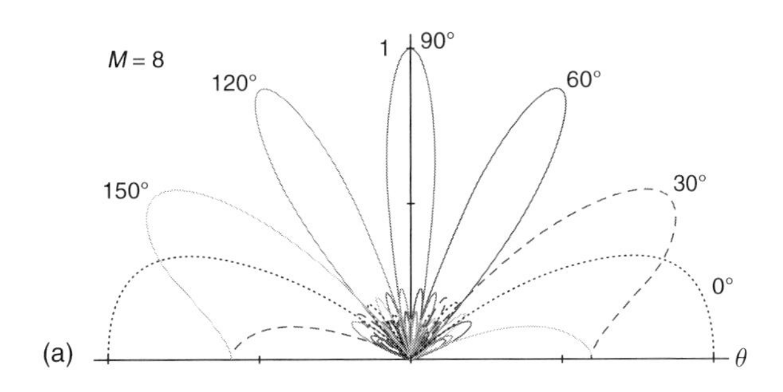

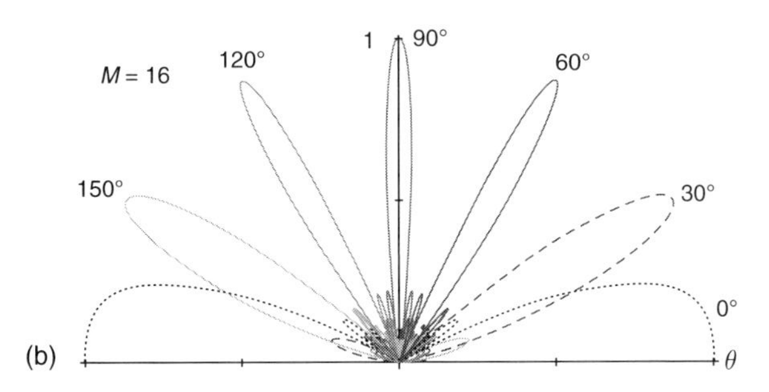

FIGURE 11.35 Beamforming at various steering angles with $r/D = 10$ and $d = 0.5\lambda$: (a) an 8-PWAS array; (b) a 16-PWAS array.

and focused beam. As the angle moves to either side of the 90° position, the beamforming worsens. At 0° and 180°, the beamforming breaks down. In fact, at 0°, the beamforming splits into two identical halves, one at 0°, the other at 180°. Such a lobe at 180° is called backlobe. As the direction increases to 30° and 60°, the back lobe shrinks and the mainlobe becomes more directional. Beamforming at 120° and 150° are symmetrical about the vertical center to that at 60° and 30°. When the mainlobe gets closer to 180° (such as 150°), the back lobe starts again increasing. The directional mainlobe is completely lost again at 180°. This observations show that a linear PWAS array does not have a complete 180° inspection range but a smaller one. The results for $M = 16$ are given in Fig. 11.35b. These results confirm the previous conclusion; however, certain improvements due to the increased number of elements are apparent. Comparing the results for $M = 8$ to those for $M = 16$, we see that an array with larger M gives a wider inspection area. No directional beam was obtained at 30° by the 8-PWAS array, but it was achieved by the 16-PWAS array.

11.5.4 NONUNIFORM PWAS PHASED ARRAY

Now we consider the effect of having different excitations for the array's elements, i.e., beamforming of a nonuniform PWAS array. The various excitations, if known, can be processed as weighting factors, w_m, i.e.,

$$BF\left(w_m, M, \frac{d}{\lambda}, \frac{r}{d}, \phi_0\right) = \frac{1}{M} \cdot \sum_{m=0}^{M-1} w_m \frac{\exp\left\{j2\pi \frac{d}{\lambda}\frac{r}{d}\left(1 - r_m - \delta_m(\phi_0)\right)\right\}}{\sqrt{r_m}} \tag{36}$$

Two widely used distributions, the binominal distribution and the Dolph–Chebyshev distribution, will be used to determine the effect of nonuniform excitation amplitudes in the array.

11.5.4.1 Binomial Array

The coefficients w_m for a binomial array can be obtained by using the binomial expansion of the expression $(1+x)^{M-1}$, i.e.,

$$(1+x)^{M-1} = 1 + (M-1)x + \frac{(M-1)(M-2)}{2!}x^2 + \frac{(M-1)(M-2)(M-3)}{3!}x^3 + \ldots \tag{37}$$

The positive coefficients of the series serve as the relative amplitude weights w_m for an M-PWAS array. Such a nonuniform array is thereby named a binomial array. For $M = 8$, the amplitude weights are {1, 7, 21, 35, 35, 21, 7, 1}. Beamforming simulation for a binomial array at different d/λ values is shown in Fig. 11.36a. Larger d/λ yields a thinner mainlobe, i.e., better resolution/directivity. Figure 11.36a also shows that the beamforming of the binomial array has no sidelobes. Actually, this is the most significant characteristic of binomial arrays. Comparison with the beamforming of an equivalent uniform array is given in Fig. 11.36b (at $d/\lambda = 0.5$). Although the binominal array has a wider mainlobe, it has no sidelobe at all which gives stronger signal suppression in undesired directions.

However, the disadvantage of using binomial array becomes apparent if we look at directional beamforming away from 90°, e.g., $\phi_0 = 45°$. Figure 11.36c shows that the binomial array had already a deteriorated beamforming at 45°, whereas the equivalent

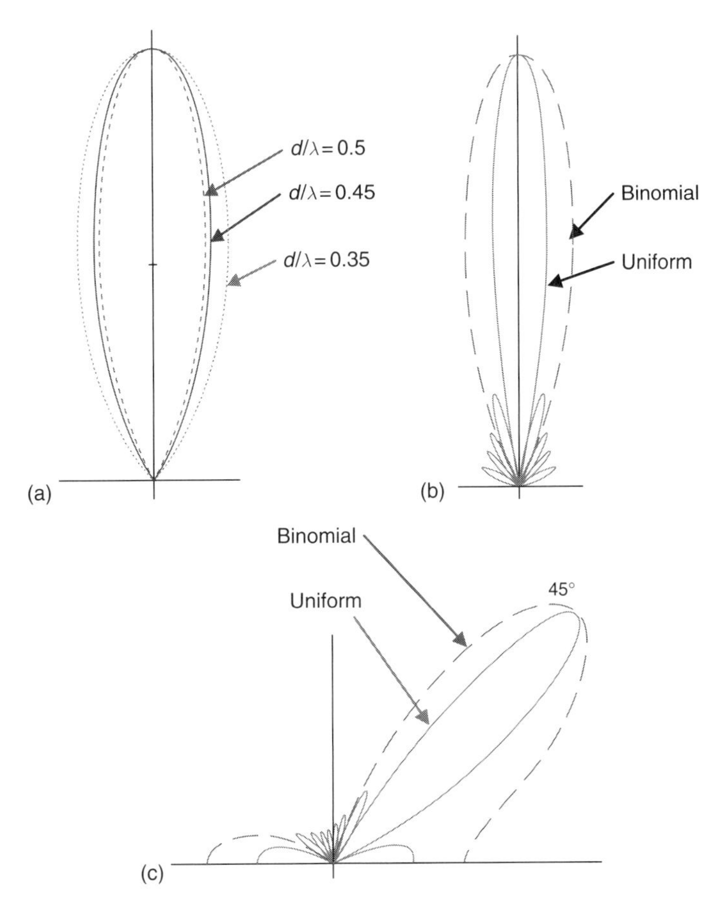

FIGURE 11.36 Beamforming of an 8-PWAS binomial array with $r/D = 10$: (a) beamforming at 90° of a binomial array with various d/λ values; (b) beamforming at 90° of a binomial array and an equivalent uniform array with $d/\lambda = 0.5$; (c) beamforming at 45° of a binomial and equivalent uniform array with $d/\lambda = 0.5$.

uniform array still has a good directional beam. In addition, the zero-sidelobe advantage of the binomial array is lost as we move away from the vertical. The binomial arrays have smaller view area than the equivalent uniform arrays and sidelobes may show up when departing from front focus.

11.5.4.2 Dolph–Chebyshev Array

Though binomial arrays have the unique property of zero sidelobes at 90°, they give larger beam width and smaller view area. Another feasible nonuniform array is the Dolph–Chebyshev array utilizing the Chebyshev distribution. By assigning a sidelobe level, i.e., the ratio of mainlobe magnitude to the first sidelobe magnitude, the Dolph–Chebyshev array coefficients can be derived (Balanis, 2005). To build an 8-PWAS Dolph–Chebyshev array, for a desired sidelobe level of 20, the normalized coefficients w_m are given as $\{0.357, 0.485, 0.706, 0.89, 1, 1, 0.89, 0.706, 0.485, 0.357\}$. Beamforming at 90° and 45° of this Dolph–Chebyshev array with $d/\lambda = 0.5$ and $r/D = 10$ is shown in Fig. 11.37a. It can be seen that, with the designed sidelobe level, the sidelobes are significantly suppressed. For beamforming of all the three arrays at 90° we see (Fig. 11.37b)

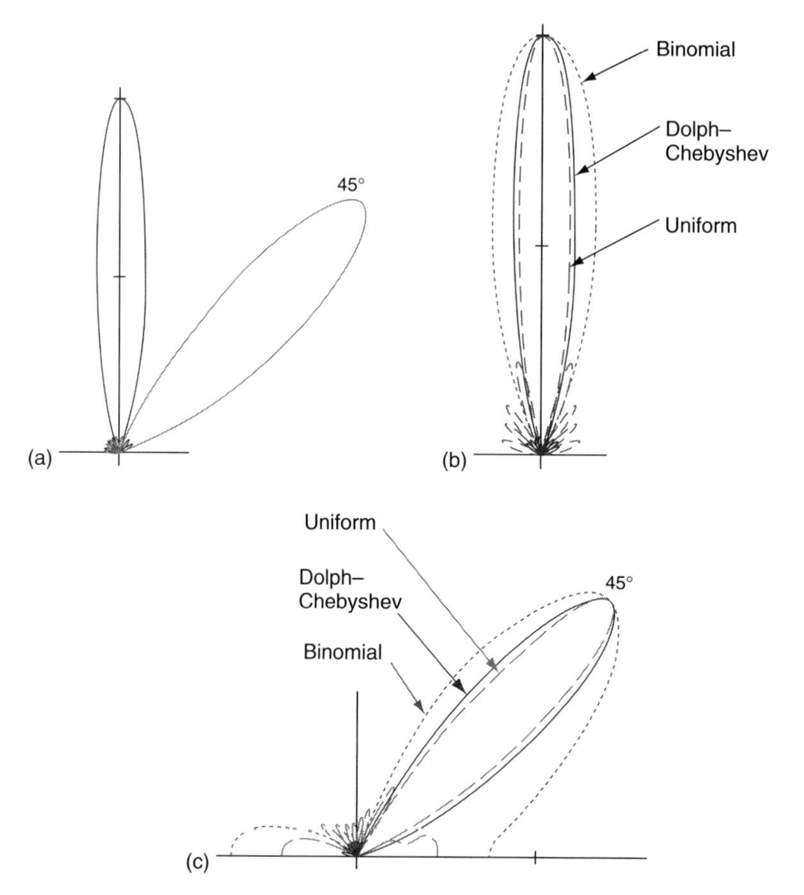

FIGURE 11.37 Beamforming of an 8-PWAS Dolph–Chebyshev array with $r/D = 10$ and $d = 0.5\lambda$: (a) beamforming at 90° and 45°; (b) beamforming at 90° of all three different arrays; (c) beamforming at 45° of all three arrays.

- considering the sidelobe level, the binomial array has the smallest sidelobes (zero sidelobe), followed by the Dolph–Chebyshev array, with the uniform array having the highest sidelobe level;
- considering the mainlobe width (directionality), the uniform array has the thinnest mainlobe and the Dolph–Chebyshev array is slightly larger, with the binominal array of largest width.

Beamforming at 45° is shown in Fig. 11.37c. At this direction, the Dolph–Chebyshev array still has low sidelobes and directional beamforming. In summary, the Dolph–Chebyshev array is a good compromise between sidelobe level and mainlobe width. While sidelobes are suppressed, its directivity at off-angles is still well maintained.

11.5.4.3 Experiments with Weighted EUSR for Nonuniform Arrays

The simulation results presented in the previous section have shown that the Dolph–Chebyshev array is expected to have smaller sidelobes in comparison to the equivalent uniform array, whereas having almost the same mainlobe width. For the binomial array, it is expected to have much smaller sidelobe level but with enlarged mainlobe width. Using the experimental data from the broadside crack specimen collected with the 8-PWAS

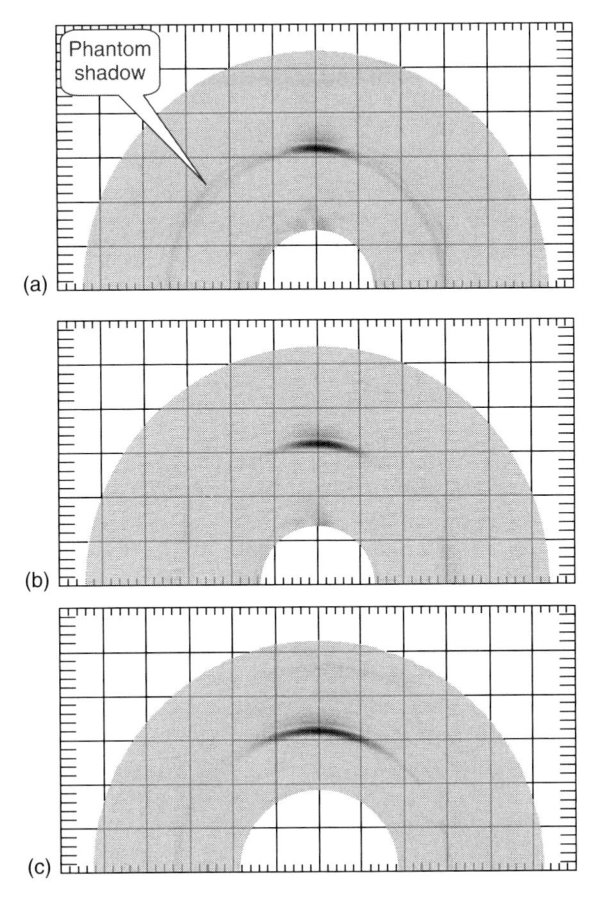

FIGURE 11.38 Crack detection using weighted EUSR algorithms on 8-PWAS nonuniform arrays: (a) image with the original uniform array; (b) image with the Dolph–Chebyshev array; (c) image with the binomial array.

uniform array, we implemented the nonuniform arrays with the weighted EUSR algorithm. Figure 11.38a gives the uniform array's EUSR image before thresholding. A phantom shadow is present as a ring. Recalling the beamforming simulation results in Fig. 11.37b, we conclude that this phantom shadow is caused by the sidelobe effect. With either the Dolph–Chebyshev array or the binomial array, the corresponding weighted EUSR algorithm should give better images (i.e., the phantom shadow is removed or reduced). The scanning results of the two nonuniform arrays are shown in Fig. 11.38b and c, respectively.

Immediately, we observe that the sidelobe phantom has been suppressed in both images, as expected. But the intensity and spread of the crack image in the binomial array image is much wider than that of the uniform array. The wider crack image further verifies that the binomial array produces the largest mainlobe width among the three arrays considered in this study.

11.5.5 SECTION SUMMARY

Beamforming properties of linear PWAS array have been investigated. It was found that several parameters can be used to manipulate the array beamforming. They are: ratio

d/λ, ratio r/d, number of elements M, steering angle ϕ_0, and weighting factors w_m. The main conclusions are:

- Larger d/λ results in thinner mainlobe width. However, the requirement of $d/\lambda \leq 0.5$ should be met to observe the spatial sampling theorem. Otherwise, grating lobes are present.
- Larger r/d ratio results in better beamforming in the near field.
- Larger M results in thinner mainlobe though the sidelobe level gets larger and the wiring may become an issue.
- Weighting can modify the beamforming, affecting the mainlobe width, the sidelobes, and inspection area.
- The effective inspection area of linear arrays is less than 180°. The actual range is affected by the number of elements and by the weights.

Experimentally, two types of nonuniform PWAS arrays have been implemented and studied: (i) the Dolph–Chebyshev array and (ii) the binomial array. This was done by using a weighted EUSR algorithm. The results coincide with the simulation results: (1) in terms of mainlobe width, uniform array is the best, followed by the Dolph–Chebyshev array, whereas the binomial array yields the largest mainlobe; (2) in terms of sidelobes, the binomial array can achieve the 'no sidelobe' effect at 90° when spaced at half wavelength, whereas the Dolph–Chebyshev array has adjustable sidelobe levels.

11.6 GENERIC PWAS PHASED-ARRAY FORMULATION

The 1-D phased array algorithm employed in many applications uses the parallel rays approximation. This assumption greatly simplifies the beamforming calculation. However, this simplifying assumption is only valid if the target is far away from the phased array. If the scanning field is not sufficiently far away, the parallel ray approximation error becomes significant, and eventually the method breaks down. The need hence exists for a generic phased-array formulation that does not rely on the parallel rays approximation, and hence can be used both in the near field and far field. Such a generic formulation will be presented in this section. In addition, the formulation will be kept sufficiently general such as not to be limited to the simple case of 1-D linear arrays. In fact, we will adopt a generic formulation in which the phased array elements could have arbitrary spatial locations.

In our derivation, we will first review the array signal processing assumptions. Then, we will develop a generic delay-and-sum formulation and then discuss its applicability in the near field and far field. After developing this generic approach, we will verify that it reduces to the simpler 1-D parallel-ray solution when a linear PWAS phased array and a target in the far field are assumed.

11.6.1 ARRAY PROCESSING CONCEPTS

Array processing is a specialized branch of signal processing that focuses on signals conveyed by propagating waves transmitted and received simultaneously by several sensors or transducers. An array processing system spatially samples the propagating field. In some applications, an array, i.e., a group of sensors located at distinct spatial locations, may be deployed to measure a propagating wave field, be it electromagnetic, acoustic, seismic, etc. The sensors serve as transducers, converting field energy to electrical energy. The output of each sensor is related to the wave field at the sensor's location by at least a conversion factor (gain) and possibly by temporal and spatial filtering. An array collects these waveforms to create a composite output signal. The goals of array processing are to combine the sensors' outputs such as to:

- Enhance the signal-to-noise ratio beyond that of a single sensor's output
- Characterize the surrounding by determining the number of sources of propagating energy, the locations of these sources, and the waveforms they are emitting
- To track the energy sources as they move in space.

11.6.1.1 Wave Propagation Assumptions

The array processing is based on previously discussed PWAS-Lamb wave propagation and tuning properties. The following assumptions are made:

1. The PWAS transducers are omnidirectional, i.e., having equal transmission and reception sensitivity in all directions. The radiation pattern of these PWAS sources is described by circular wave fronts. Far away from the source, the wave front approaches the plane-wave condition as the wave front curvature decreases. Thus, plane-wave models can be used in the far field, where the parallel-rays approximation may apply. However, the parallel-rays approximation does not hold in the near field.
2. Propagating waves are single-mode tone-burst signals that can be described by a simple function, $s(\cdot)$, in which the space–time relationship $t - \vec{\alpha} \cdot \vec{x}$ applies, with $\vec{\alpha} = \vec{k}/\omega$. When the tone-burst is not sufficiently narrow-band, dispersion may occur, and group velocity measurements will be used. If the dispersion is too large, special signal processing methods (as described in a future chapter) can be used.
3. The wave-propagation direction is equivalently represented by the wavenumber vector, $\vec{k}$, and the slowness vector, $\vec{\alpha}$, where $\vec{\alpha} = \vec{k}/\omega$.
4. The superposition principle applies. This allows several propagating waves to occur simultaneously. The constructive or destructive interference of the separate wave patterns generated by each of the array elements represents the essence of the phased array principle.
5. Common assumptions of a homogeneous, linear, lossless elastic medium are used in derivation of the phased array equations. However, the phased array principle can also be used in media that do not satisfy all these assumptions (e.g., composite materials).

11.6.1.2 Array Summing Effects

When signals are superposed in an array, signal enhancement results. The signal enhancement within an array can be explained easily. In the simplest assumption, the signal $y_m(t)$ produced at the mth sensor consists of a signal, $s(t)$, which is assumed to be more-or-less identical for all sensors, and random noise $N_m(t)$, which varies from sensor to sensor, i.e.,

$$y_m(t) = s(t) + N_m(t) \tag{38}$$

By summing the signals received by all the M sensors, one obtains

$$z(t) = \sum_{m=0}^{M-1} y_m(t) = M\, s(t) + \sum_{m=0}^{M-1} N_m(t) \tag{39}$$

Equation (39) shows that, whereas the signal, $s(t)$, is amplified M times, the random noise is also summed up. The summing up of several random noise signals usually results in noise reduction through mutual cancellation. Thus, a significant increase in the signal-to-noise ratio has been achieved. In phased arrays, the processing will be more complicated, because delays are used to create a beamforming effect. However, the basic noise reduction effect described by Eq. (39) will still apply.

11.6.2 DELAY-AND-SUM BEAMFORMING

Delay-and-sum beamforming principle lies at the foundation of phased array technology. Though more elaborate phased-array processing algorithms have been recently developed, the basic delay-and-sum approach continues to remain a widely used method because of its implicit simplicity. Essentially, the idea behind the delay-and-sum approach is very simple. When a propagating wavefront is present in the array aperture, the array elements will be excited and will generate electrical signals proportional to the excitation received from the wavefront. The individual signal outputs of the array elements are then delayed by appropriate amounts and added together. The addition results in signal reinforcement and noise reduction, similar to Eq. (39). The delays ensure that the signals from the individual array elements are properly superposed such that the addition can have a positive effect. The delays also ensure that this reinforcement effect is achieved in a preferred direction. Of course, the delays are directly related to the difference in travel time from the source to each of the array elements. For a simple 1-D linear array and a far-field source that obeys the parallel-rays approximation, the delays are calculated rather easily, as indicated by Eq. (6) given in a previous section. For a generic array and a generic field location of the source, the analysis is rather more elaborate, as described next.

11.6.2.1 Generic Delay-and-Sum Formulation

Assume, as shown in Fig. 11.39, the target located at point which is $\vec{r}_m$ away from the mth element of the phased array placed at $\vec{s}_m$. We define $\vec{\xi}$ as the unit vector pointing from the origin to $P(\vec{r})$ and $\vec{\xi}_m$ as the unit vector pointing from the mth PWAS to $P(\vec{r})$.

Consider the generic geometric arrangement of Fig. 11.39. The origin of coordinate system is defined using the PWAS phase center which equals to the quantity $\frac{1}{M}\sum \vec{s}_m$, i.e., the origin is chosen such as

$$\frac{1}{M}\sum \vec{s}_m = 0 \tag{40}$$

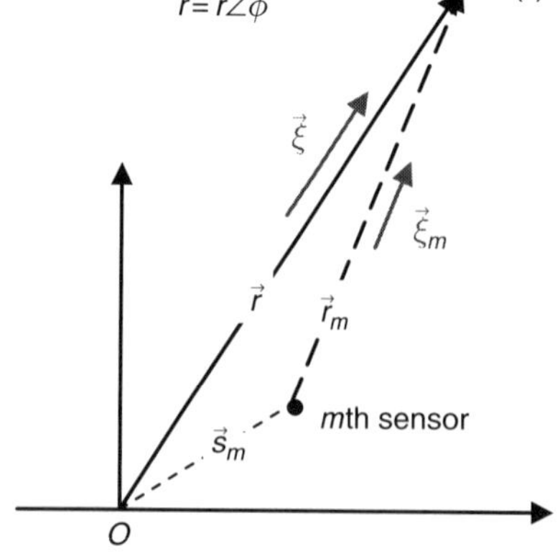

FIGURE 11.39 Geometric schematics of the mth PWAS at $\vec{s}_m$ and the reflector at $P(\vec{r})$.

The wave transmitted by the mth PWAS toward the target $P(\vec{r})$ is

$$y_m(t) = f(\vec{r}_m, t) \tag{41}$$

The delay-and-sum beamforming consists of two steps:

1. Apply a delay Δ_m and an optional weighting factor w_m to the output of the mth PWAS
2. Sum up the output signals of all the M PWAS.

This processing can be expressed as

$$z(t) = \sum_{m=0}^{M-1} w_m y_m(t - \Delta_m) \tag{42}$$

The delay Δ_m can be adjusted to focus the array's output beam on a particular propagating direction $\vec{\xi}_0$, whereas w_m is the weighting factors for enhancing the beam shape and reducing sidelobe levels. By using the delay-and-sum method, the wave field generated at $P(\vec{r}, \phi)$ by M-PWAS array locating at $\vec{s}_m$, $m = 0, \ldots, M-1$ can be found.

11.6.2.2 Far Field and Near Field

The beamforming algorithm varies according to whether the reflector is located near to or far from the array. Using the concepts of antenna theory (Balanis, 2005), we define the near field as the region in which

$$0.62\sqrt{D^3/\lambda} < R_{near} \leq 2D^2/\lambda \tag{43}$$

where D is the array aperture and λ is the excitation wavelength. Far field is defined as the region where

$$R_{far} > 2D^2/\lambda \tag{44}$$

In the inner region $R \leq 0.62\sqrt{D^3/\lambda}$, phased array theory is no longer valid. However, other methods for damage detection, such as the PWAS E/M impedance technique described in Chapter 9, can be employed.

If the reflector $P(\vec{r})$ is close to the array, i.e., in near field (Fig. 11.40a), the propagating wavefront is curved (circular wavefront) and the wave propagating directions are dependent on the location of each array element. For this situation, wave propagation direction varies from element to element and different individual unit vectors $\vec{\xi}_m$, $m = 1, 2, \ldots$ need to be assigned to each array element (Fig. 11.40a).

However, if the reflector $P(\vec{r})$ is far from the array (i.e., in the far field to the array), the propagating wave field generated by the array approaches a plane wave (Fig. 11.40b). Hence, the propagation directions of the waves emitted from each of the array elements become approximately parallel, i.e.,

$$\vec{\xi}_0 \approx \vec{\xi}_1 \ldots \approx \vec{\xi}_{M-1} \approx \vec{\xi} \tag{45}$$

For the far-field situation, the propagating direction is $\vec{\xi}$ for all the plane waves, regardless of the location of the sensors.

In the following sections, we will first deduce the beamforming formulation for arbitrary PWAS phased array using the full wave traveling paths and then simplify it to the far field using the parallel-ray approximation.

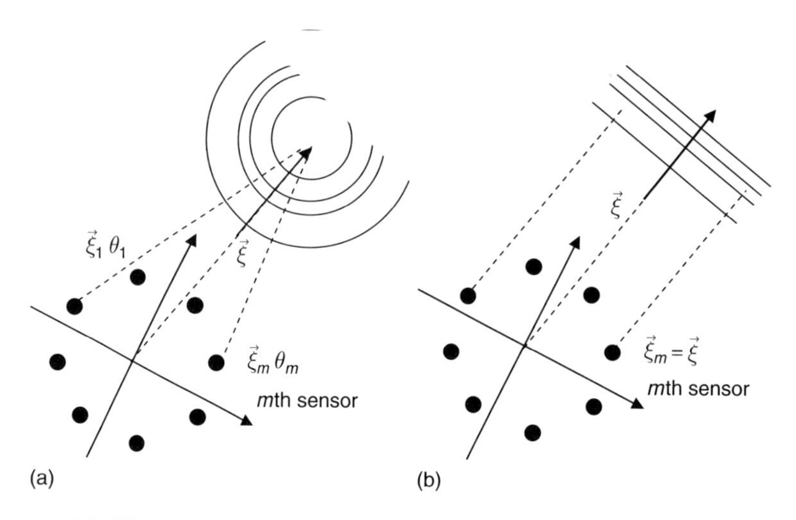

FIGURE 11.40 Beamforming in near and far field to the array: (a) near field; (b) far field.

11.6.3 GENERIC BEAMFORMING FORMULATION FOR PWAS PHASED ARRAY

Assume that tuning between the PWAS and the Lamb waves in the structure has been achieved such that a low dispersion Lamb wave of wave speed c is dominant (details of such a tuning process were given in Chapter 7). Hence, the wave front at a point located $|\vec{x}|$ away from the PWAS source can be expressed as

$$f(\vec{x}, t) = \frac{A}{\sqrt{|\vec{x}|}} e^{i\left(\omega t - \vec{k}\cdot\vec{x}\right)} \tag{46}$$

where $\vec{k} = \vec{\xi} \cdot \omega/c$ is the wavenumber and ω is the angular frequency of the wave front. To use the exact traveling paths for each element, the unit vectors for each of the directions from the array elements to the target $P(\vec{r})$ have to be defined. We call these unit vectors $\vec{\xi}_m$, $m = 0, 1, \ldots, M-1$. Referring to the geometric schematic of Fig. 11.39a, in which $\vec{s}_m$ is the position vector of the mth element in the array, we define the following notations

$$\vec{\xi} = \frac{\vec{r}}{r}, \quad r = |\vec{r}|, \quad \vec{k} = \frac{\omega}{c}\vec{\xi} \tag{47}$$

$$\vec{r}_m = \vec{r} - \vec{s}_m, \quad r_m = |\vec{r}_m|, \quad \vec{\xi}_m = \frac{\vec{r}_m}{|\vec{r}_m|}, \quad \vec{k}_m = \frac{\omega}{c}\vec{\xi}_m \tag{48}$$

where $\vec{k}$ and $\vec{k}_m$ are the wavenumbers for waves propagating in the directions of $\vec{\xi}$ and $\vec{\xi}_m$, respectively. Using these notations, the wavefront coming from the mth element toward the target $P(\vec{r}, \phi)$ can be written as

$$f(\vec{r}_m, t) = \frac{A}{\sqrt{r_m}} e^{\left\{i\left[\omega t - \vec{k}_m\cdot\vec{r}_m\right]\right\}} \tag{49}$$

The combined wavefront arriving at $P(\vec{r}, \phi)$ through the superposition of the effects of all the M elements is

$$z(\vec{r}, t) = \sum_{m=0}^{M-1} f(\vec{r}_m, t) = \sum_{m=0}^{M-1} \frac{A}{\sqrt{r_m}} e^{\{i[\omega t - \vec{k}_m \cdot \vec{r}_m]\}} \tag{50}$$

We will call the signal in Eq. (50) the *synthetic wave front* of the phased array. The generic Eq. (50) will be now used to study the near-field and far-field conditions. For near-field conditions, we will examine the exact traveling path, a.k.a. the *triangular algorithm*. For far-field conditions, we will study the *parallel-ray approximation*. Both conditions will still maintain a generic arbitrary phased array arrangement. In the end, we will show that the parallel-ray approximation reduces to the 1-D linear array algorithm when the generic array is reduced to a equally spaced linear array.

11.6.3.1 Near Field: Exact Traveling Path Analysis (Triangular Algorithm)

For the generic situation, exact traveling wave paths are used in the beamforming formulation. Using Eqs. (47) and (48), we rewrite Eq. (49) as

$$f(\vec{r}_m, t) = f\left(\vec{r}, t - \frac{r}{c}\right) \frac{1}{\sqrt{r_m/r}} e^{i\omega \frac{r - r_m}{c}} \tag{51}$$

Similarly, the synthetic signal $z(\vec{r}, t)$ defined by Eq. (50) becomes

$$z(\vec{r}, t) = f\left(t - \frac{r}{c}\right) \sum_{m=0}^{M-1} w_m \frac{1}{\sqrt{r_m/r}} e^{j\omega \frac{1-(r_m/r)}{c}} \tag{52}$$

Normalize the quantity r_m by the quantity r, and define the normalized vectors

$$\tilde{r}_m = r_m/r, \quad m = 0, 1, \ldots, M-1 \tag{53}$$

Hence, Eq. (52) becomes

$$z(\vec{r}, t) = f\left(t - \frac{r}{c}\right) \sum_{m=0}^{M-1} w_m \frac{1}{\sqrt{\tilde{r}_m}} e^{j\omega\left(\frac{1-\tilde{r}_m}{c}\right)} \tag{54}$$

Equation (54) is made up from the multiplication of two terms. The first term, $f(t - r/c)$, is a function that does not depend on the locations and weights of the phased array elements and on the location of target. This term represents the individual wave signal that would be produced by a single PWAS element placed at the origin. We will leave this term alone. The second term, however, depends on the locations and weights of the phased array elements and on the location of target. Hence, this term will change if we change the array configuration. It will also change if the target changes. We will call this second term the *beamforming factor* (denoted as *BF*). In accordance with Eq. (54), the *BF* function is given by

$$BF(w_m, \tilde{r}_m) = \sum_{m=0}^{M-1} w_m \frac{1}{\sqrt{\tilde{r}_m}} e^{j\omega\left(\frac{1-\tilde{r}_m}{c}\right)} \tag{55}$$

To steer the output wave front $z(\vec{r}, t)$ in a certain direction ϕ_0, i.e., to reinforce the waves propagating in a desired direction in comparison with the waves propagating in other directions, we introduce the delays $\Delta_m(\phi_0)$ and write the beamforming factor as

$$BF(w_m, \tilde{r}_m) = \sum_{m=0}^{M-1} w_m \frac{1}{\sqrt{\tilde{r}_m}} e^{j\omega\left(\frac{1-\tilde{r}_m}{c} - \Delta_m(\phi_0)\right)} \tag{56}$$

Equation (56) shows that a maximum of the beamforming factor $BF(w_m, \tilde{r}_m)$ can be achieved if one can make the exponentials equal to one, which will be achieved when their exponents are zero, i.e.,

$$\exp\{(1-\tilde{r}_m)/c - \Delta_m(\phi_0)\} = 1 \quad \text{when} \quad (1-\tilde{r}_m)/c - \Delta_m(\phi_0) = 0 \tag{57}$$

To achieve this, we will have to apply to each element of the PWAS phased array the delay

$$\Delta_m(\phi_0) = \frac{1-\tilde{r}_m}{c} \tag{58}$$

When the delays of Eq. (58) are used, the beamforming factor corresponding to the particular direction ϕ_0 reaches a maximum and takes the value

$$BF(w_m, \tilde{r}_m, \phi_0) = \sum_{m=0}^{M-1} w_m \frac{1}{\sqrt{\tilde{r}_m}} \tag{59}$$

Equation (59) shows that further manipulation of the value of the beamforming function at the desired direction ϕ_0 can be achieved by adjusting the weighting factors w_m. One way of using this effect is to try to compensate the effect caused by the PWAS elements being placed at different locations, i.e., by defining

$$w_m = \sqrt{\tilde{r}_m} \tag{60}$$

If Eq. (60) is used, then the value of the beamforming factor in the desired direction ϕ_0 reaches the optimum value M, i.e.,

$$BF(w_m, \vec{r}, \vec{s}_m, \phi_0) = \sum_{m=0}^{M-1} 1 = M \tag{61}$$

Substituting Eq. (61) into (54) yields the synthetic wave of the M-PWAS array at P is under optimum conditions as

$$z(\vec{r}, t) = M f\left(t - \frac{r}{c}\right) \tag{62}$$

We see that the synthetic wave $z(\vec{r}, t)$ has become reinforced M times relative to the individual wave signal that would be produced by a single PWAS element placed at the origin.

The process described above indicates that, with proper delays and weights, the phased array can be made to generate beamforming in a certain direction, ϕ_0, and at a certain location, r. As different from the simplified parallel-ray algorithm, the exact algorithm presented here is able to focus both azimuthally through the angle, ϕ_0, and radially through the presumed target location, r. For this reason, it does not depend on the commonly used parallel-ray approximation and hence can be used in the near field where the parallel algorithm fails. However, the implementation of this exact algorithm is more elaborate and requires more computational time.

11.6.3.2 Far Field: Parallel-Ray Approximation (Parallel Algorithm)

In the previous section, we deduced a generic beamforming formula for PWAS phased arrays that is exact but requires intensive computation. However, if the target is in the far field and the parallel-rays assumption applies, the algorithm can be simplified and the computational time can be reduced. If the target is located far away, we can assume that the rays emanating from the array elements toward the target are approximately parallel. Hence the $\vec{\xi}_m$ unit vectors become approximately equal and the formulae become independent of the relative location of the phased array elements. This implies that we assume that the propagation directions of the waves emanating from the individual elements are considered to be parallel to each other, i.e.,

$$\vec{\xi}_m \approx \vec{\xi}, \qquad m = 0, 1, \ldots, M-1 \tag{63}$$

Under these conditions, Eqs. (47) and (48) become

$$\vec{k}_m \approx \vec{\xi} \cdot \frac{\omega}{c} = \vec{k} \quad \text{and} \quad \sqrt{r_m} \approx \sqrt{r} \tag{64}$$

Furthermore, Eq. (49) becomes

$$f(\vec{r}_m, t) \approx f\left(\vec{r}, t - \frac{r}{c}\right) e^{i\frac{\omega}{c}\vec{\xi}\cdot\vec{s}_m} \tag{65}$$

To generate beamforming, we apply delays $\Delta_m(\phi_0)$ and weights w_m. Thus, the beamforming factor becomes

$$BF(w_m, \vec{s}_m) = \sum_{m=0}^{M-1} w_m e^{j\frac{\omega}{c}\left(\vec{\xi}\cdot\vec{s}_m - \Delta_m(\phi_0)\right)} \tag{66}$$

To achieve beamforming in a certain direction ϕ_0, one chooses the delay such as to maximize the complex exponential functions, i.e.,

$$\Delta_m(\phi_0) = \frac{\vec{\xi}\cdot\vec{s}_m}{c} \tag{67}$$

If we also choose the weights as unity, i.e., $w_m = 1$, then the beamforming factor will take the value M, and the synthetic signal $z(\vec{r}, t)$ will become M times reinforced with respect to the individual reference signal $f(\vec{r}, t)$.

As noted, the delay and weight Eq. (67) holds only when the far-field assumption is valid. However, the generic equations derived in Section 11.6.3.1 do not have this limitation and can be used for any situation, regardless of the target position.

11.7 2-D PLANAR PWAS PHASED ARRAY STUDIES

In the previous section, we have explored the capabilities of 1-D linear arrays of PWAS transducers and have shown that they can successfully detect the presence of cracks by performing large area scanning and imaging with guided Lamb waves. However, we have also encountered some of major shortcomings of 1-D linear arrays, such as

(a) the degradation of beamforming properties at angles close to 0° and 180°
(b) the half-plane mirror effect, which does not allow discriminating between a target placed above the array and a target placed below the array (Fig. 11.41).

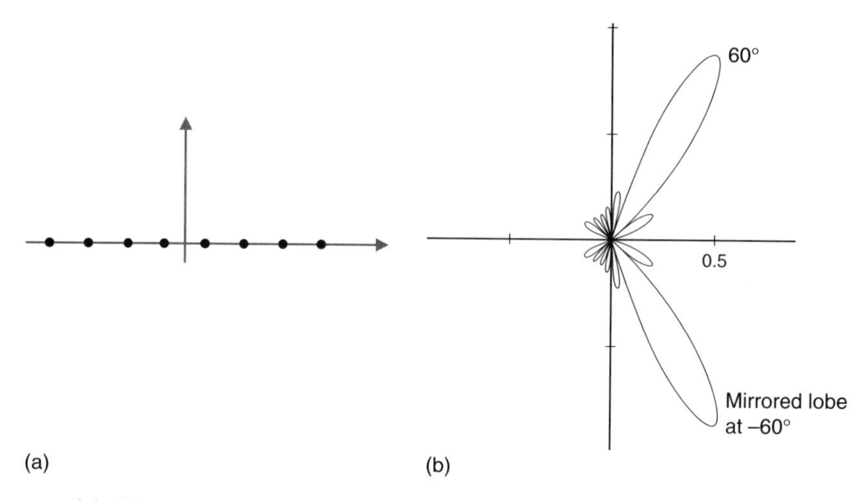

FIGURE 11.41 1-D linear array: (a) 1-D linear array along a straight line; (b) beamforming of an 8-PWAS linear array showing beamforming symmetric about the array itself.

We have also seen that various array optimization techniques can be employed to alleviate the first issue. However, the second issue is an inherent basic property of the 1-D linear array and cannot be overcome through array optimization. Because of its inherent geometrical limitation, the 1-D linear array beamforming is naturally symmetric about the array itself, as illustrated in Fig. 11.41b. The image in Fig. 11.41b shows a mainlobe directed at 60°, and also a mirrored lobe shows up symmetrically at −60°. This inherent mirror symmetry results in the inspection area being limited to, at most, 180°.

11.7.1 MOTIVATION AND BACKGROUND

The way to overcome the above-described issues inherent in the 1-D linear arrays is to adopt a 2-D array design configuration. Rather than placing the array elements along a straight line, we will explore positioning them along a cross shape, a rectangular fence, a circular ring, a rectangular grid, a circular grid, etc. In this way, we will achieve a 2-D planar array. Planar arrays not only provide more control and optimization parameters for improving the array beamforming and array performance, but are also more versatile and able to provide much finer beamforming with lower sidelobes. Most importantly, the 2-D planar arrays are able to perform full-range 360° scanning and create a complete image of the specimen subjected to structural health monitoring.

The concept of 2-D planar array is not new. In fact, it has been extensively researched in the context of radar and sonar phased arrays (Balanis, 2005). More recently, they have also been explored for embedded NDE applications. For example, Wilcox (2003) explored two types of circular array using electromagnetic acoustic transducers (EMAT). Type I consisted of a fully populated disc, whereas type II contained only a single ring. Simulation results with S_0 Lamb waves in a 5 mm thick aluminum plate and an ideal reflector located at 3 o'clock position with respect to the array indicated that the type I array gives reasonably good results, whereas the type II array had many large sidelobes. However, the use of a deconvolution method was predicted to be able to remove the sidelobes such that the type II array would perform as well as the type I array. Subsequently, the type II array was selected for experimental implementation. The experimental set-up consisted of two concentric ring arrays (one for transmission, the other for reception) mounted on a metallic plate. The array was used to detect a simulated defect consisting of a small steel disk bonded to the plate. The results were compared with the simulation studies. Though the simulation studies indicated that the deconvolution method would be effective in removing the sidelobes, the experimental results showed that its actual effectiveness is rather limited and many sidelobes are still present. Further experiments with this EMAT circular ring array were performed by Fromme et al. (2005). The circular ring array was permanently attached to a 5-mm thick aluminum plate of size 2.45 m by 1.25 m. The damage was simulated by a through-the-thickness hole of 15-mm radius. Further results on using this EMAT phased array for in-situ damage detection have been reported by Wilcox et al. (2005).

In the following sections, we will first perform simulation of the beamforming patterns of several 2-D planar phased arrays (cross shaped, rectangular fence, circular ring, rectangular grid, circular grid, etc.) in order to examine their beamforming patterns and evaluate their potential effectiveness for SHM applications. The analytical simulation will be performed with the general formulation discussed in a previous section, under the assumption that the damage to be detected lies in the far field, and hence the parallel-rays approximation applies. All simulations will assume a 1-mm thick aluminum plate and 7-mm round PWAS elements tuned to obtain a quasi non-dispersive S_0 Lamb-wave mode with a propagating speed $c = 5440\,\mathrm{m/s}$. Subsequently, we will use these simulations to select an array type which is most appropriate for implementation. The selected array type will be constructed in several variants. Experiments performed on actual specimens will be presented and the experimental results will be discussed. At the end, a short summary of the main theoretical and experimental results obtained with 2-D planar phased arrays using PWAS-coupled guided waves for structural health monitoring will be given.

11.7.2 CROSS-SHAPED ARRAY

The cross-shaped array layout is illustrated in Fig. 11.42a. A cross-shaped array can be considered as two perpendicular linear arrays. Cartesian coordinates are used. Array elements are distributed along x and y directions, spaced at d_x and d_y, respectively. For a cross-shaped array having M PWAS in the x direction and N PWAS in the y direction,

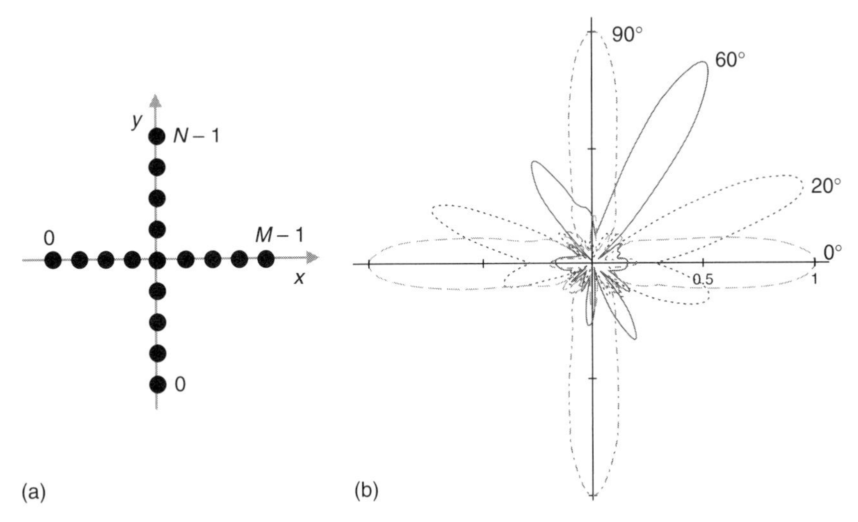

FIGURE 11.42 Cross-shaped array design: (a) schematic of a cross-shaped array with M elements in x direction and N elements in y direction; (b) beamforming at various angles with $d_x = d_y = 0.5\lambda$, $r/d = 10$, $M = N = 9$.

the positions of the mth element (in the x direction) and the nth element (in the y direction) are determined as

$$m\text{th element} : \left\{ \left(m - \frac{M-1}{2} \right) d_x, 0 \right\} \tag{68}$$

$$n\text{th element} : \left\{ 0, \left(n - \frac{N-1}{2} \right) d_y \right\} \tag{69}$$

Beamforming results at various directions using cross-shaped phased array having nine PWAS elements in each direction ($M = N = 9$) are shown in Fig. 11.42b. For 20° and 60°, unique directional beams are obtained at desired directions, though with significant sidelobes. But for 0° and 90°, duplicated beams in the opposite direction are also present (180° for beamforming at 0°, and 270° for beamforming at 90°). Therefore, when using cross-shaped arrays, the 360° full-range beamforming is only conditionally attained. For the directions along the array main axes, such as 0°, 90°, 180°, 270°, duplicate beams will always appear.

11.7.3 RECTANGULAR GRID ARRAY

Rectangular grid arrays are constructed by aligning linear arrays, as illustrated in Fig. 11.43a. Cartesian coordinates are used. Array elements are distributed in the x and y directions, spaced at d_x and d_y, respectively. For an $M \times N$ rectangular PWAS array, the coordinates of the (m, n)th element are

$$\left\{ \left(m - \frac{M-1}{2} \right) d_x, \left(n - \frac{N-1}{2} \right) d_y \right\} \tag{70}$$

Beamforming at various directions using an 8×8 PWAS array is shown in Fig. 11.43b. For this array, unique directional beams at all desired directions, 0°, 20°, 80°, 150°, 260°, 330°, have been successfully obtained, with low sidelobes. The simulation results show that a rectangular array is a good candidate for 360° full-range damage detection.

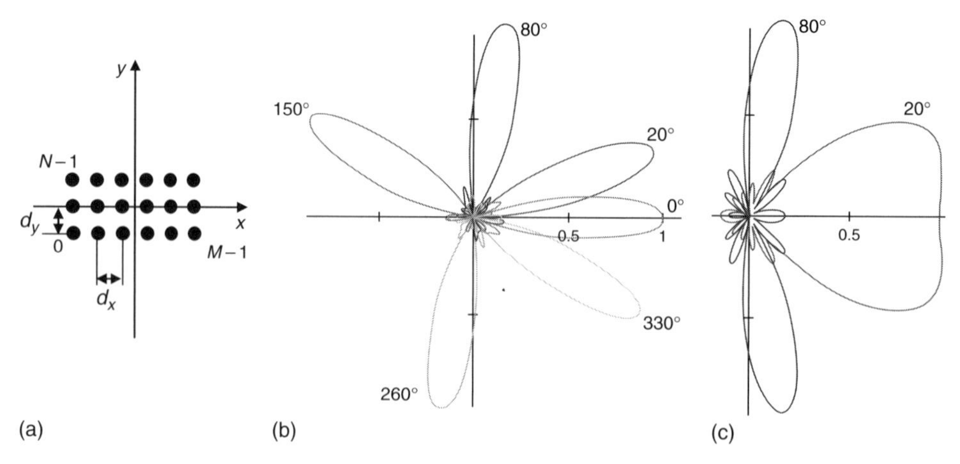

FIGURE 11.43 Rectangular grid array design: (a) general schematic of an $M \times N$ rectangular array spacing at d_x and d_y; (b) beamforming at various angles with $d_x = d_y = 0.5\lambda$, $r/d = 10$, $M = N = 8$; (c) beamforming at 20° and 80° for $N = 1$, with scanning range reduced to 180° (two main lobes symmetrical about the horizontal axis appear, since this has become a 1-D array).

A rectangular array will reduce to a 1-D linear array when either M or N decreases to one. Beamforming at 20° and 80° with $N = 1$ is given in Fig. 11.43c. The resulting lobes are symmetric about 0°/180° (x direction), as expected.

11.7.4 RECTANGULAR FENCE ARRAY

Rectangular grid arrays have good directionality in the 360° full range. However, to construct a rectangular grid array, a large number of PWAS have to be used to cover all the grid nodes, which will results in the complexity of wiring and data collection. A substitute candidate for the rectangular grid array is the rectangular fence array, where only the elements at the grid periphery are used (Fig. 11.44a). For a rectangular fence array with M elements in x direction and N elements in y direction, the coordinates of the (m, n)th element are

$$\begin{aligned}
&\text{Side (1)} \quad \left(\left(-\frac{M-1}{2}\right)d_x, \left(n-\frac{N-1}{2}\right)d_y\right)\\
&\text{Side (2)} \quad \left(\left(m-\frac{M-1}{2}\right)d_x, \left(N-\frac{N-1}{2}\right)d_y\right)\\
&\text{Side (3)} \quad \left(\left(m-\frac{M-1}{2}\right)d_x, \left(-\frac{N-1}{2}\right)d_y\right)\\
&\text{Side (4)} \quad \left(\left(M-\frac{M-1}{2}\right)d_x, \left(n-\frac{N-1}{2}\right)d_y\right)
\end{aligned} \tag{71}$$

The beamforming pattern of a rectangular fence array having 8-PWAS elements on each side (28 elements total) is given in Fig. 11.44b. It can be seen that good directional beamforming within the 360° full range has been obtained. When comparing with the equivalent 8×8 rectangular array, one notices that the rectangular fence array maintains a good mainlobe but has more severe sidelobes.

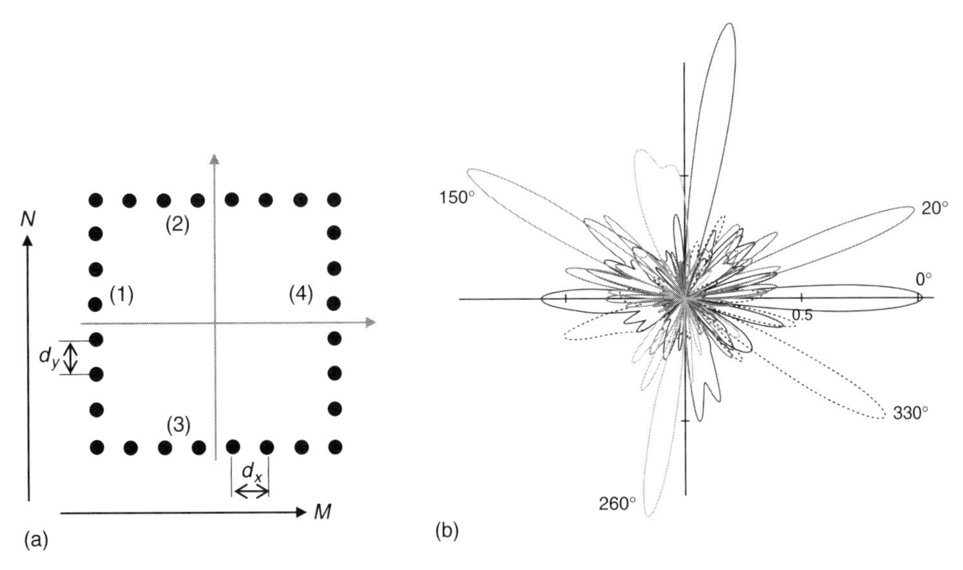

FIGURE 11.44 2-D rectangular fence array design: (a) schematic of a rectangular ring array spacing at d_x and d_y; (b) beamforming at various directions for such an array with $d_x = d_y = 0.5\lambda$, $r/d = 10$, $M = N = 8$.

11.7.5 CIRCULAR RING ARRAY

The circular ring array is obtained by placing the PWAS elements uniformly along a circle (Fig. 11.45a). Using polar coordinates, the coordinates of the mth element in an M-PWAS ring array is

$$(x, y) = (R\cos\theta_m, R\sin\theta_m) \tag{72}$$

where

$$R = \frac{Md}{2\pi}, \qquad \theta_m = m \cdot \frac{2\pi}{M} \tag{73}$$

Beamforming patterns of a circular ring phased array consisting of 64 PWAS elements spaced at $d = 0.5\lambda$ is shown in Fig. 11.45b. When comparing these results with those obtained for the equivalent 64-PWAS rectangular grid array (Fig. 11.43b), one notices that beamforming of the circular ring array gives a finer mainlobe (better directivity) but much stronger sidelobes. In addition, one notices that the ring array has a diameter of about 163 mm whereas the rectangular grid array has a diagonal of 80 mm, i.e., the equivalent rectangular grid is much more compact. Figure 11.45 also shows that the circular ring array reduces to a classic dipole when for $M = 2$ (Fig. 11.45c).

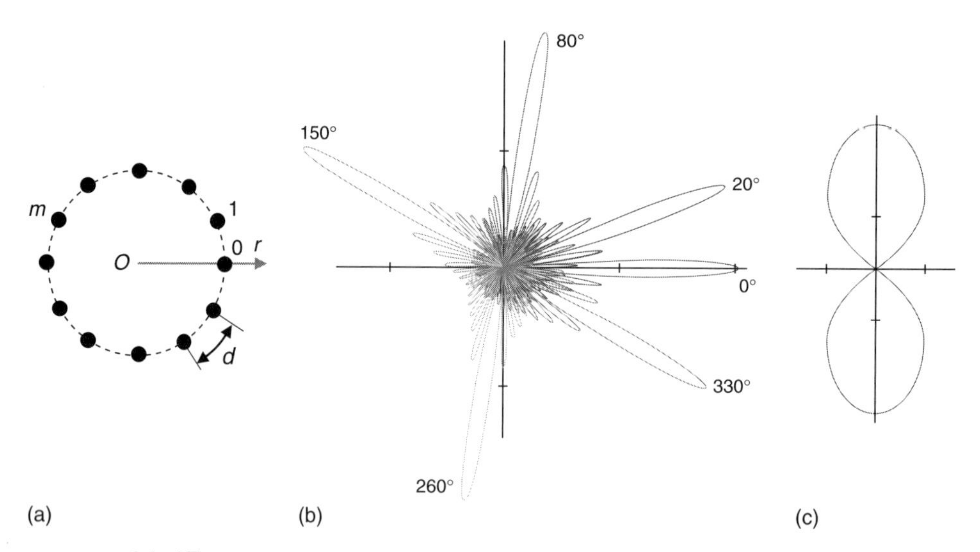

FIGURE 11.45 2-D circular ring array design: (a) schematic of the circular ring array; (b) beamforming at various angles with $d = 0.5\lambda$, $r/d = 10$, $M = 64$; (c) beamforming when $M = 2$ when the circular ring array is reduced to a classic dipole.

11.7.6 CIRCULAR GRID ARRAY

A circular grid array can be constructed by assembling a group of circular ring arrays with a common center (Fig. 11.46a). The spacing between neighboring rings is defined as d_r (radial spacing) whereas the spacing between neighboring elements in each ring is defined as d_c (circumferential spacing). The numbering of the elements will start from

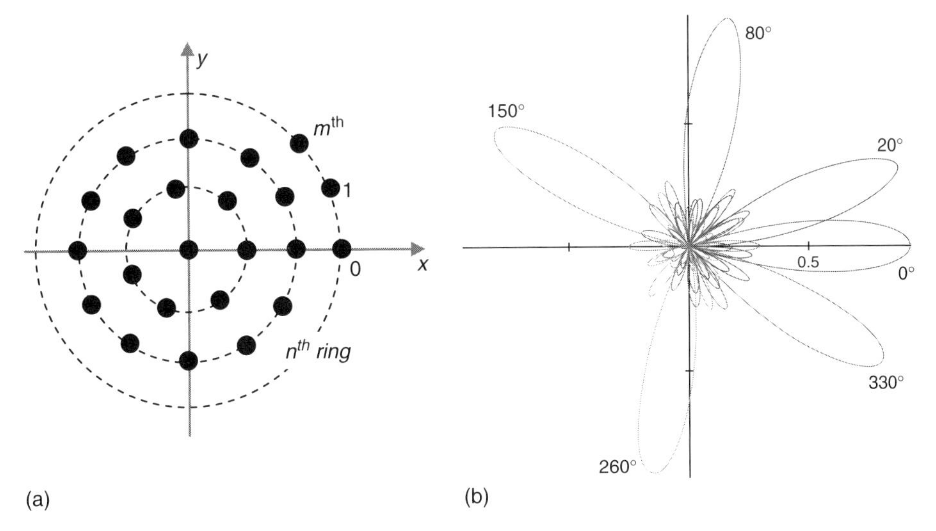

FIGURE 11.46 Sixty-six-element circular grid array: (a) layout of the circular disk array; (b) beamforming within 0°~360° range at different angles with $N = 5$ (five-ring array), $d_c/\lambda = d_r/\lambda = 0.5$, $d_r/r = 1/10$.

the most internal ring and move outwards. At each ring, it starts from $+x$ direction (i.e., 0°). The radius R_n of nth ring is simply

$$R_n = n\,d_r, \quad n = 0, 1, 2, \ldots, (N-1) \tag{74}$$

The number of elements M_n in the nth ring can be determined by rounding up to the closest integer the value $2\pi n d_r/d_c$, i.e.,

$$M_n = \left\lceil \frac{2\pi n d_r}{d_c} \right\rceil, \quad n = 0, 1, 2, \ldots, (N-1) \tag{75}$$

where the ceiling function $\lceil\ \rceil$ is defined as

$$\lceil x \rceil = \inf\{n \in \mathbb{Z} | x \leq n\} \tag{76}$$

The beamforming pattern at various directions of a five-ring concentric circular array is shown in Fig. 11.46b. The number of elements in each ring is {1, 7, 13, 19, 26}, requiring a total of 66 PWAS. When comparing these results with the results obtained for the corresponding 8×8 rectangular grid array (Fig. 11.43b), one notices that both arrays have good directionality within the 360° full range and have similar beamforming performance. However, the circular array seems to yield a little higher sidelobe levels.

11.7.7 SECTION SUMMARY

The 2-D planar array studies presented in this section indicate that, of all the analyzed arrays, the cross-shaped array has the poorest performance, because it presents 'blind spots' at 0°, 90°, 180°, 270° where it cannot distinguish between front and back because of a very strong back lobe. Therefore, we discard this array as a potential candidate for implementation. The other four array types studied in this section (rectangular grid, rectangular fence, circular ring, circular grid) have both relative benefits and relative shortcomings; hence, the choice for implementation between these four array types will depend to a large extent on the specialist's preferences and available manufacturing facilities. In the experimental studies to be presented in the coming sections, we will focus on rectangular grid arrays of various M/N ratios.

11.8 THE 2-D EMBEDDED ULTRASONIC STRUCTURAL RADAR (2D-EUSR)

In this section, we will describe the 2-D embedded ultrasonics structural radar concept that was developed from the original EUSR concept that uses 1-D PWAS arrays. We achieved the 2-D EUSR using 2-D planar PWAS arrays (Giurgiutiu and Yu, 2006). Though the results presented here are specific to rectangular grid PWAS phased arrays, they can be easily adapted to other array configurations (rectangular fence, circular ring, circular grid). The benchmark crack detection situations to be discussed in this section are presented in Fig. 11.47.

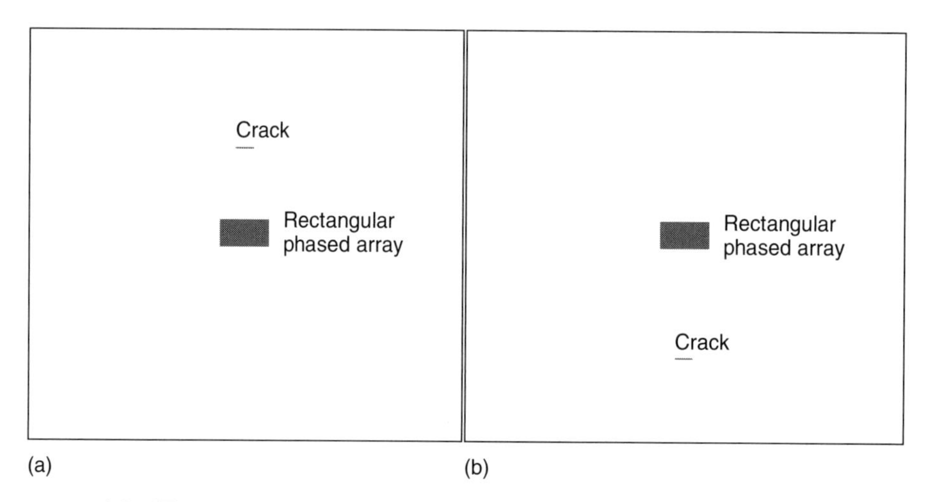

FIGURE 11.47 Benchmark crack-detection scenarios used to develop and test the 2D-EUSR algorithm: (a) crack at 90°; (b) crack at 270°.

11.8.1 EUSR ALGORITHM FOR 2-D $N \times M$ RECTANGULAR CONFIGURATION ARRAY

As different from the 1-D implementation of the EUSR algorithm, the implementation of a 2-D analysis requires that we define the 'front' and the 'back' of the 2-D array. That is to say, we need to establish numbering rule for setting up the firing order in the array. The numbering used in 1-D situation started from left to right (such that the front side is determined). Extending to the 2-D situation, the rule of numbering still starts from left to right and then moves from the back to front. By this means, the front side of the array can be determined. If the 0° direction (or front side of the array) is determined, the numbering of the $N \times M$ array will start from the left-bottom corner and increase from left to right (as the arrow indicates). Figure 11.48a illustrates this concept for the case of an 8×8 array.

The 2-D rectangular grid array can be treated as a number of N linear arrays arranged one after the other, which is equivalent to an $N \times M$ matrix. There are M elements in

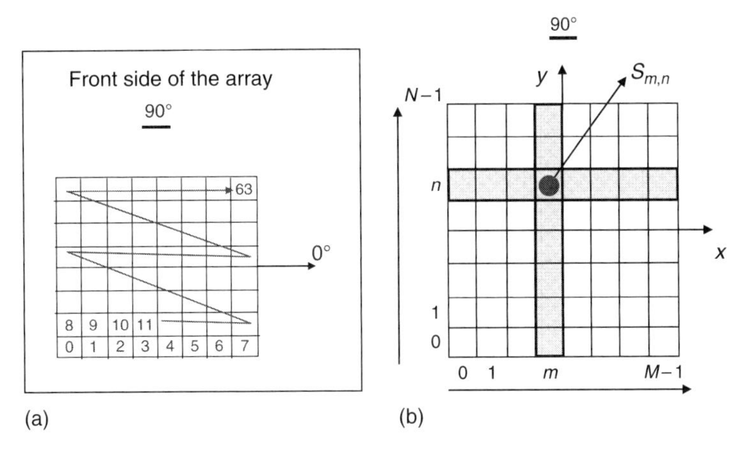

FIGURE 11.48 2-D $N \times M$ rectangular PWAS array indexing: (a) array numbering starting from the left bottom, determining the front side of the array; (b) details of the numbering and resulting element coordinates.

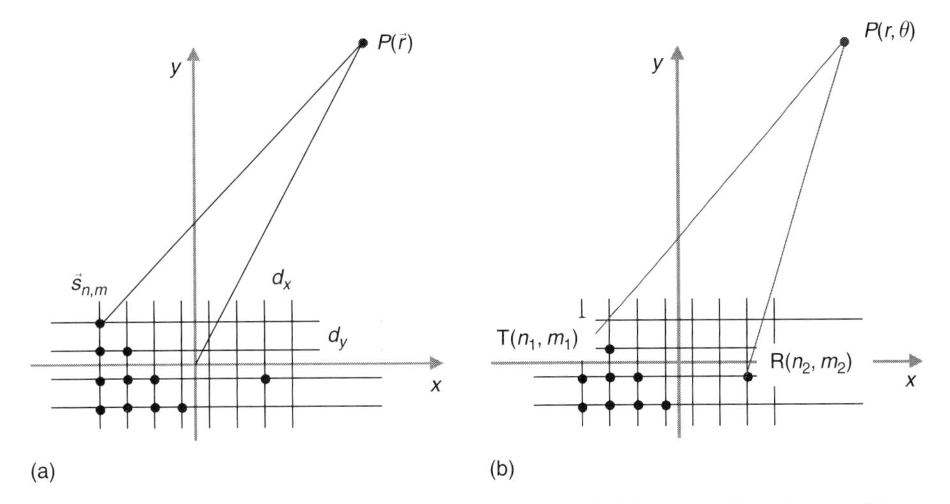

FIGURE 11.49 Simulation data: (a) geometric schematic of a 2-D rectangular grid array; (b) transmitter and receiver localization.

each row and N rows in the array, i.e., a matrix with N rows and M columns. To identify each element, both column (x direction) and row (y direction) information is needed. As before, we set the coordinate system with the origin in the phase center. Hence, an element located at (n, m) in the array, i.e., at the nth row and mth column (Fig. 11.49), will have the x and y coordinates given by

$$[x, y] = \left[\left(m - \frac{M-1}{2}\right) d_x, \left(n - \frac{N-1}{2}\right) d_y\right] \tag{77}$$

where d_x and d_y are the element spacing values in the x and y directions, respectively. For simplicity, we will set them equal in numerical examples, i.e., $d_x = d_y = d$. Hence, the location vector $\vec{s}_{n,m}$ of the (n, m) element will be given by

$$\vec{s}_{n,m} = \left(m - \frac{M-1}{2}\right) d_x \vec{i} + \left(n - \frac{N-1}{2}\right) d_y \vec{j} \tag{78}$$

The geometric schematic of the array elements and of the vector directions to the target $P(\vec{r})$ is shown in Fig. 11.49a.

Once the data is available, EUSR reads it. Since EUSR is based on the pulse-echo method, there needs to be a distinction between transmitters and receivers. Index (n_1, m_1) and index (n_2, m_2) then are assigned to a transmitter and a receiver, respectively. As stated in the original EUSR algorithm, one PWAS is selected to be transmitter whereas all the rest PWAS serves as the receivers in a round-robin fashion. By adding all these signals together with appropriate delays, the signal will be reinforced. The delaying and adding procedure at a direction ϕ_0 for both transmission and reception can be expressed as

$$z(\phi_0) = \sum_{\mathrm{T}} \sum_{\mathrm{R}} f(t - \Delta_{T,\phi_0} - \Delta_{R,\phi_0}) \tag{79}$$

Using indexing, Eq. (80) becomes

$$z(\phi_0) = \sum_{n_1=0}^{N-1} \sum_{m_1=0}^{M-1} \left\{ \sum_{n_2=0}^{N-1} \sum_{m_2=0}^{M-1} f(t - \Delta_{(n_1,m_1),\phi_0} - \Delta_{(n_2,m_2),\phi_0}) \right\} \tag{80}$$

Therefore, for the 2-D rectangular array, the 2D-EUSR algorithm parameters should also include the number of rows and columns and both x and y spacing information.

11.8.2 DEVELOPMENT OF SIMULATION DATA FOR TESTING THE 2D-EUSR ALGORITHM

Before applying the 2D-EUSR algorithm to actual experimental data, it was considered wise and prudent to verify its effectiveness on simulated data. For this purpose, we developed a set of simulated pulse-echo wave signals that would be representative of the benchmark case-study presented in Fig. 11.47.

According to the general theory of wave propagation and reflection, the wave front of a pure-tone single-mode wave at distance R away from a source is

$$f(r = R) = \frac{1}{\sqrt{R}} f(t - \delta(R, c)) \tag{81}$$

In the Eq. (81), the $1/\sqrt{R}$ factor represents the wave decay when propagating in 2-D circular wavefronts; the wave amplitude decays inversely proportional to the root square of the traveling distance, R, such that the total energy in the wavefront remains constant. The delay $\delta(R, c)$ is an intrinsic time delay depending on the traveling distance R and the wave propagating speed c. Using Eq. (81), we simulated the signals that would be received whereas using a 4×8 rectangular grid array to detect an ideal reflector located at position (r, θ) as shown in Fig. 11.49b. When using the ideal reflector, we assume that the energy arriving at the reflector from the transmitter is entirely sent back toward the array. This situation may not be entirely encountered in practice, where cracks are actually directional specular reflectors; however, this situation is sufficiently good for our study, which focuses on time delays calculation and manipulation. The excitation frequency f, wave propagation speed c, the sampling interval Δt, and the number of data points N_{data} are user-defined input. The excitation signal employed in this simulation is the same 3-count toneburst used for 1-D array studies. Since peak-to-peak value will be used for the time-of-flight detection, a left shift of t_0 is applied to the toneburst so that the transmitting signal starts at its peak value.

Also given in the simulation are the following parameters: spacing d (determined by the ratio d/λ), number of elements in x direction (M columns), and elements in

y direction (N rows). These parameters describe the specific array configuration. The reflector (simulated defect) is located at $P(r, \theta)$. During the wave propagation process, the signal will first decay over the distance traveled from transmitter element in the array to reflector, and then decay again over the distance traveled from reflector to the receiver element in the array. The transmitter element has position (n_1, m_1) in the array, whereas the receiver element has position (n_2, m_2). Since the coordinates of the elements (transmitters or receivers) are given by their positions in the array matrix, the distances can be formulated as follows (all values have been normalized with regard to the distance r):

$$\text{Reflector position} \quad \vec{R} = \cos\theta\vec{i} + \sin\theta\vec{j} \tag{82}$$

$$\text{Array element position } \vec{s}(n, m) = \left(n - \frac{M-1}{2}\right) d\vec{i} + \left(m - \frac{N-1}{2}\right) d\vec{j} \tag{83}$$

The distance between the array element and the reflector

$$\text{dist}(n, m) = \left|\vec{R} - \vec{S}(n, m)\right| \tag{84}$$

Hence, the time delay is

$$\Delta(n, m) = \frac{\text{dist}(n, m)}{c} \tag{85}$$

For a signal sent from the transmitter $\text{T}(m_1, n_1)$ and arriving at receiver $\text{R}(m_2, n_2)$, the total delay is

$$\Delta = \Delta(n_1, m_1) + \Delta(n_2, m_2) \tag{86}$$

Considering that a reflector can simultaneously also serve as a transmitter, we calculate the signal received back at the array as

$$R(t) = \begin{cases} T(t) + \dfrac{1}{\sqrt{\text{dist}(n_1, m_1)}} \cdot \dfrac{1}{\sqrt{\text{dist}(n_2, m_2)}} \cdot T(t - \Delta) & \text{if } m_1 = m_2 \text{ and } n_1 = n_2 \\ \dfrac{1}{\sqrt{\text{dist}(n_1, m_1)}} \cdot \dfrac{1}{\sqrt{\text{dist}(n_2, m_2)}} \cdot T(t - \Delta) & \text{otherwise} \end{cases} \tag{87}$$

where $T(t)$ is the transmitter signal.

Using Eqs. (81) and (87), we generated a full set of signals using the round-robin transmitter–receiver pair method of the EUSR algorithm. The simulated data set is saved in the same format as the format prepared for the experimental data to be collected later. In this way, we ensure that a direct utilization of the simulated data signals in the EUSR processing is possible.

11.8.3 GENERATION OF SYMMETRIC CRACK DATA THROUGH INDEX MANIPULATION

We have shown in Fig. 11.48 that the ordering of the array elements has to be judiciously chosen so as to ensure the proper orientation of the array with respect to the target. In this context, Fig. 11.48 has indicated how an array must be numbered such that a crack target placed north of the array (i.e., at the 90° location) will be properly identified. We will now explain how the same principle can be advantageously used to

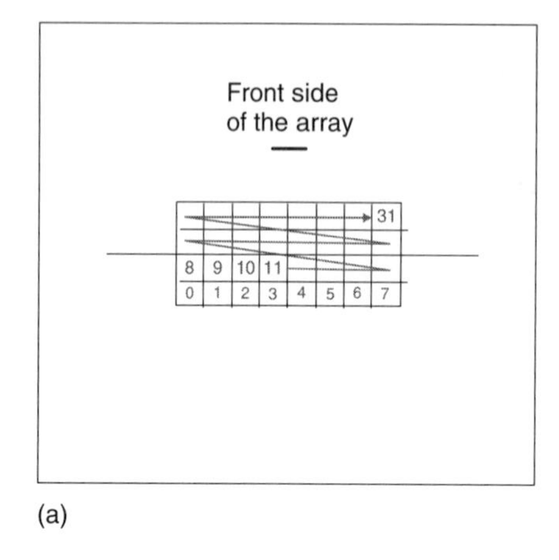

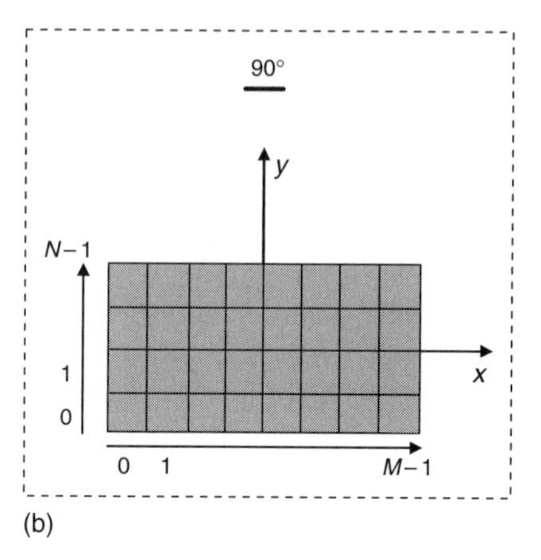

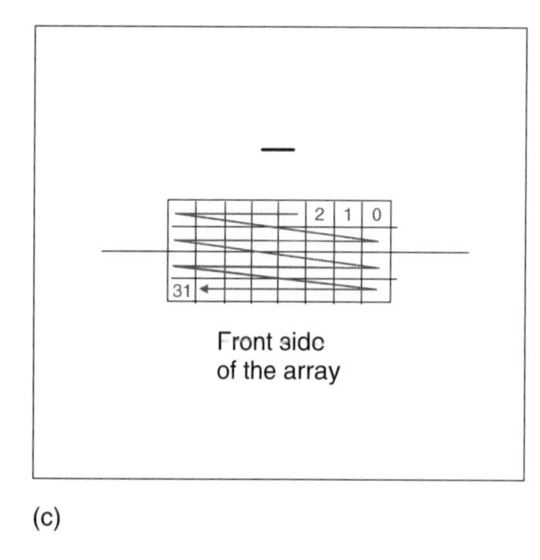

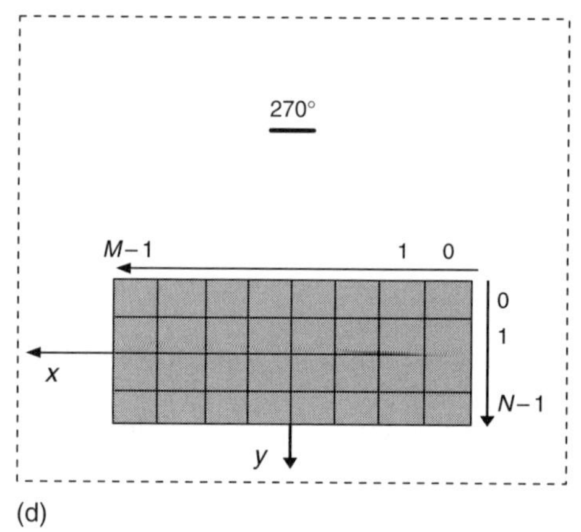

FIGURE 11.50 Numbering of the array elements such as to generate two test data sets from a single data collection: (a) array numbering method #1, placing the crack target in front of the array; (b) numbering rule for method #1; (c) array numbering method #2, placing the crack target behind the array; (d) array numbering for method #2.

generate a new data set representing a target placed South of the array (270° location), i.e., symmetric with respect to the array x-axis.

This method is illustrated in Fig. 11.50 and consists of establishing a numbering rule for the order in which excitation is passed along the array elements. First, recall that the original numbering in the 1-D situation used to start from left to right, such that the front side of a linear array can be always determined. However, since the 1-D linear array algorithms could not distinguish between the front and rear of the array, this numbering convention was of no immediate consequence in the case of 1-D linear arrays. Extending now to the situation of 2-D rectangular grid arrays, we will first maintain the rule of numbering from left to right and then will extend in the second dimension from the back to front of the array. This will generate the numbering method #1, as illustrated in Fig. 11.50a,b. This method results in a 90°-positioned crack target, i.e., in front of the array.

The second way of numbering the array elements (method #2) is to start from the back of the array, proceed right-to-left, and then front-to-back. This method #2 is illustrated in Fig. 11.50c,d. This method presents the crack as being positioned in the rear of the array, i.e., at the 270° position. By re-indexing the simulated signals in accordance with method #2, we were able to generate a set of simulated data representing a crack positioned in the rear of the array. However, this second set of data was obtained without any investment in additional numerical simulation, but simply through re-indexing. The two different data sets can now be used to demonstrate that the 2D-EUSR is able to correctly differentiate between a 90° and 270° placed target, something that the 1-D linear arrays were unable to do.

11.8.4 THE TESTING OF 2D-EUSR ALGORITHM ON SIMULATED DATA

Before studying the results of testing the 2D-EUSR on simulated data, it is appropriate to recall the results of beamforming studies performed in a previous section. In this respect, it was illustrated that if a rectangular grid array has an equal number of rows and columns (i.e., $N = M$) then the beamforming pattern is very nice all around the 360° azimuth, with small side and back lobes. But what happens if $N \neq M$? If $N \neq M$, then the symmetry is broken and quite significant backlobe effects start to appear. Such a situation is illustrated in Fig. 11.51a for a rectangular grid array in which $N = (1/2)M(N = 4, M = 8)$. We notice that, for this array configuration, the beamforming has significant backlobes, as illustrated for the case of 90° in Fig. 11.51b which shows a backlobe amplitude that is almost 1/3 of the mainlobe amplitude. By symmetry, we infer that the same 1:3 ratio will also apply at 270°. Careful examination of Fig. 11.51a reveals that, in fact, strong backlobes would be present throughout the 360° range. This indicates that, although the rectangular array

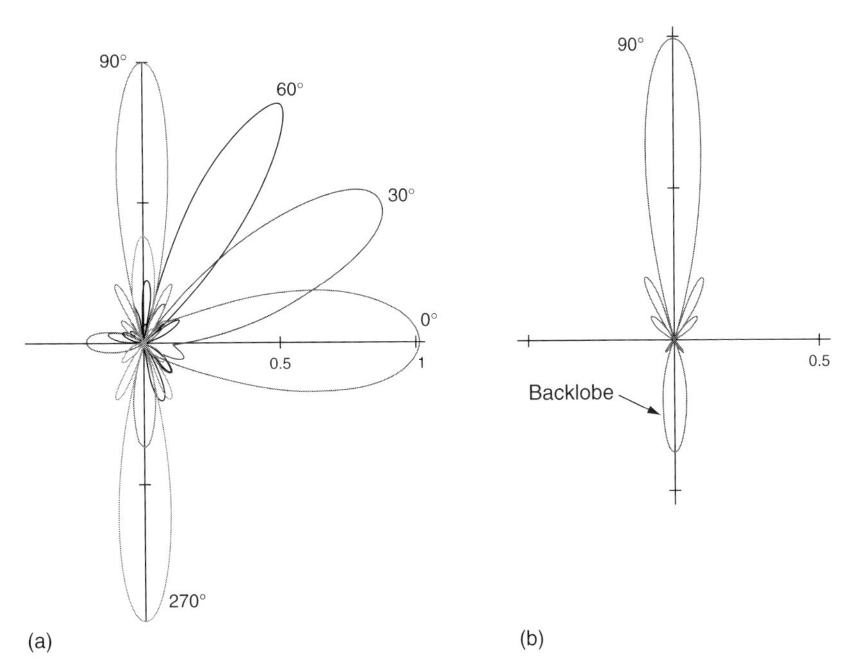

FIGURE 11.51 Beamforming of the 2-D 4 × 8 rectangular array: (a) theoretical beamforming at various directions; (b) beamforming at 90° revealing the presence of a strong backlobe.

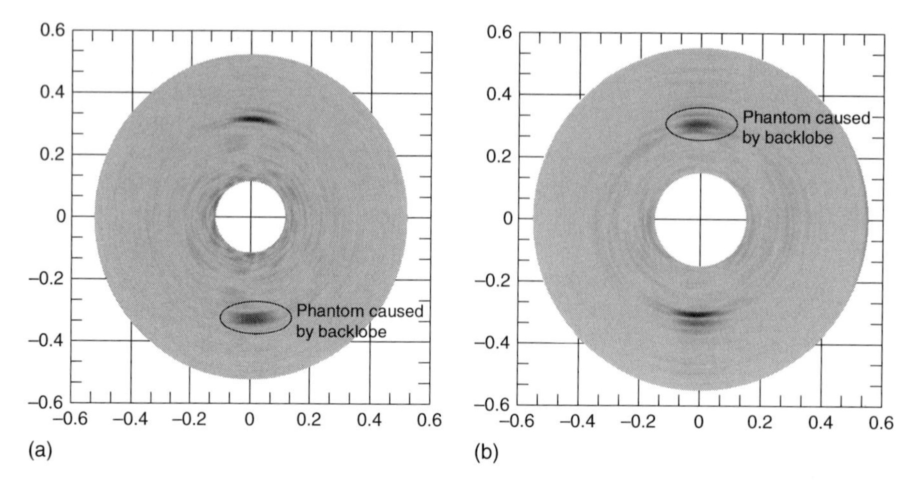

FIGURE 11.52 2-D 4 × 8 rectangular array EUSR image in the original form: (a) 90° crack image; (b) 270° crack image.

delivers good mainlobe throughout the 360° range, the fact that it has $N \neq M$ results in nonnegligible backlobes.

Now, we turn our attention to the results of applying the 2D-EUSR algorithm to the 2-D simulated data discussed in the previous section. These results are shown in Fig. 11.52 in the form of reconstructed 360° scan images. Figure 11.52a represents a crack target placed in front of the array. It can be seen that the strongest detection is located at the 90° position, where the actual crack exists. In the same time, a phantom image (false detection) is present at the 270° position, i.e., symmetric with respect to the horizontal axis of the array. This phantom is due to the backlobe effect. Figure 11.52b shows the same effect in the case of the actual target being placed at 270°. In this case, the phantom due to the backlobe effect appears at 90°.

However, since the main lobe is still much larger than the backlobe, the phantom effect due to the backlobe can be removed through thresholding. By setting a threshold value above the backlobe level but below the main lobe value, we can clean up the image and remove the phantom (the values under the threshold are simply forced to

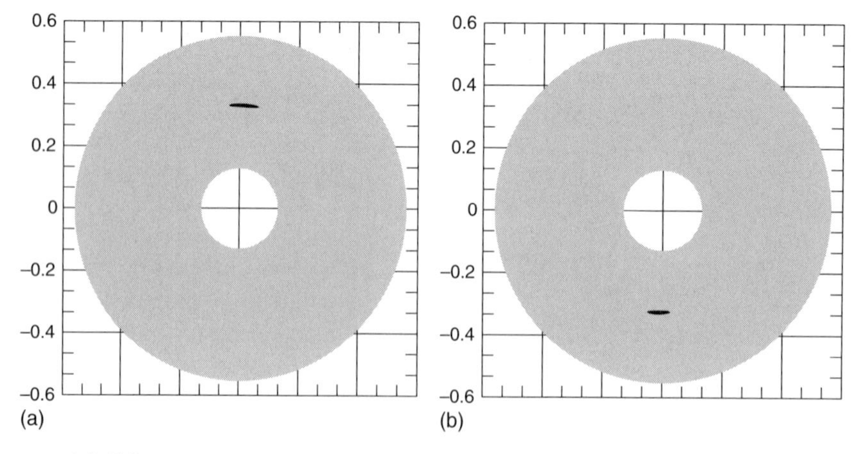

FIGURE 11.53 2-D 4 × 8 rectangular array EUSR image after thresholding: (a) 90° crack image; (b) 270° crack image.

zeroes). After applying thresholding, the phantom detection is eliminated, as shown in Fig. 11.53a; now the single crack existing in our simulated specimen is very clearly identified. Similar results are obtained in removing the 90° phantom present in the 270° data of Fig. 11.52b. After applying the thresholding, the image is cleaned and correctly indicates the single crack located at 270° (Fig. 11.53b).

11.9 DAMAGE DETECTION EXPERIMENTS USING RECTANGULAR PWAS ARRAY

To verify the damage detection ability of the 2D-EUSR algorithm, two rectangular PWAS arrays (a 4×8 array and an 8×8 array) were used. The 4×8 PWAS array was used to detect a crack in an aluminum plate placed in front of the array and parallel to it. Such a crack would act as a specular reflector and send the echo signal directly back to the array.

The 8×8 PWAS array was used to detect a more difficult situation. The crack was again placed in front of the array, but its orientation was inclined at 45°. This inclined crack sends its specular reflection signal away from the array and thus makes the detection more difficult. What the 8×8 array detects is the secondary signals resulting from scatter at the crack ends.

11.9.1 DAMAGE DETECTION EXPERIMENTS USING A 4 x 8 PWAS ARRAY

A rectangular 4×8 phased array, with a total of 32 PWAS elements, was installed in the center of a 1-mm thick 2024 T3 aluminum alloy plate. The phased array was fabricated from a large 0.2-mm thick PZT plate that was diced into an 4×8 pattern. Each element in the array was a 7-mm square PWAS with an element spacing $d = 8\,\text{mm}$. The aluminum plate was of square shape with a nominal dimension of 1220 mm (4-ft). The damage to be detected was a 20-mm through-the-thickness simulated crack (0.125-mm slit). The crack was placed in front of the array at 90° position (Fig. 11.54a). The experimental set-up (Fig. 11.54b) consisted of an HP33120 function generator for sending the excitation signals and a TDS210 digital oscilloscope for collecting the received signals. A laptop personal computer (PC) was used via the GPIB interface to control the function generator and collect data from the digital oscilloscope. A computer-controlled automatic signal collection unit (ASCU) was used for switching the excitation to the transmitter element and collecting the signal from the receiver elements of the array. The ASCU device was designed and fabricated in the Laboratory for Adaptive Materials and Smart Structures (LAMSS) of the University of South Carolina. The excitation signal was a 3-count toneburst with carrier frequency of 285 kHz. The carrier frequency of 285 kHz was determined through a tuning process aimed at maximizing the quasi-nondispersive S_0 Lamb waves and minimizing the highly dispersive A_0 Lamb waves.

Different from the traditional phased arrays which require complicated equipment for sending all phased signals simultaneously, the 2D-EUSR method collects the data using very simple equipment and a round-robin pattern. At each time, one transmitter in the array sends out the excitation whereas all the elements in the array serve as receivers and pick up the signal reflections. In a round-robin fashion, all the elements in the array serve in turn as transmitters. A round-robin pattern was implemented such that all combinations of transmission–reception between the array elements were explored. At any time in this process, one PWAS element acts as transmitter and sends out an excitation signal whereas all PWAS elements serve as receivers to collect the reflection signals. The result

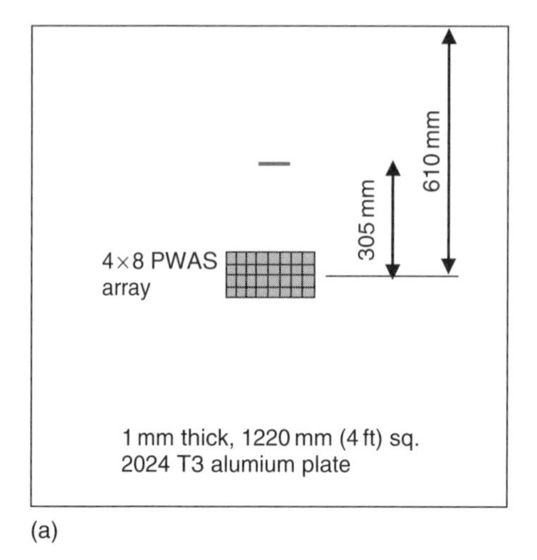

(a)

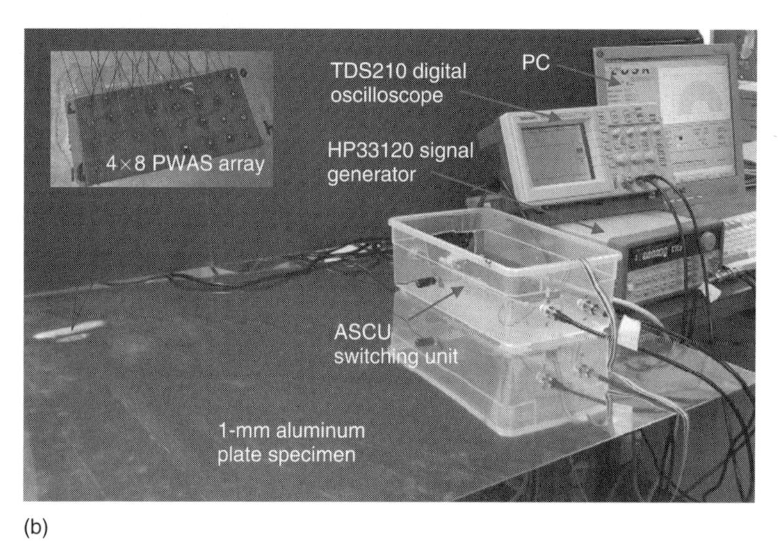

(b)

FIGURE 11.54 The 360° full range crack detection using a 4×8 rectangular PWAS array: (a) specimen layout (sizes are not proportional; crack and array are exaggerated for illustration); (b) laboratory experiment setup.

of the data collection was a matrix of $(4 \times 8)^2 = 32 \times 32 = 1024$ elemental signals that was stored in the PC mass storage device. The data collection followed the indexing rule described in a previous section. The data was recorded in a single Excel data file. After array description is entered in the ASCU interface (i.e., the value of N and M signifying the number of rows and the number of elements in each row), the ASCU system scans the transmitter and receiver elements in the order $n = 0, \ldots, N-1$ and $m = 0, \ldots, M-1$. The Excel data sheet is organized as continuous blocks for each transmitter; within each transmitter block, the receiver signals are arranged in continuous columns. Subsequently, the 2D-EUSR algorithm was applied to the matrix of stored signals in post-processing. The 2D-EUSR algorithm applies the phased-array algorithm to obtain a full-field 360° scanning image of the specimen. The 2D-EUSR algorithm realizes the phased-array process in virtual time by applying delays to the elemental signals stored in the PC memory. Through this process of virtual signal phasing, the 2D-EUSR algorithm performs the automatic steering of the interrogation beam through the full 360° range and constructs the image of the scanned specimen. This image is equivalent to a C-scan, only that it is performed with Lamb waves guided along the plate and not with pressure waves traveling across the plate thickness. On this image, the structural damage appears clearly as a darker area (color images are also available on demand). In addition, the 2D-EUSR algorithm presents an A-scan (signal vs. time image) in a direction that can be selected at will.

11.9.2 DISCUSSION OF 2-D EUSR IMAGING RESULTS WITH A 4 × 8 PWAS ARRAY

In this section we will discuss the results of 2-D EUSR imaging using experimental data. The discussion will be made in comparison with the results obtained with simulated data. The purpose of this comparison is to understand which image artifacts are due to experimental error and which are intrinsic to the phase array method.

Figure 11.55 presents the images obtained for the detection of a 90° crack with a 4 × 8 PWAS array in two situations: (a) simulation data, and (b) experimental data. It is apparent that the experimental data (Fig. 11.55b) is much more noisy than the simulation data (Fig. 11.55a); this fact is not surprising, and it is to be expected. However, both the theoretical and experimental data present a phantom detection positioned at the mirror location (i.e., 270°). This phantom is an artifact of the 4 × 8 array. This mirror phantom is due to backlobe inherent in a rectangular array with unequal number of elements in the x and y directions, as illustrated by the beamforming studies previously presented in Fig. 11.51.

Figure 11.56 presents the images obtained for the detection of a 270° crack with a 4 × 8 PWAS array in two situations: (a) simulation data; and (b) experimental data. It is apparent that the experimental data (Fig. 11.56b) is much more noisy than the simulation data (Fig. 11.56a); this fact is not surprising, and it is to be expected. However, both the theoretical and experimental data present a phantom detection positioned at the mirror location (i.e., 90°). This phantom is an artifact of the 4 × 8 array. This mirror phantom is

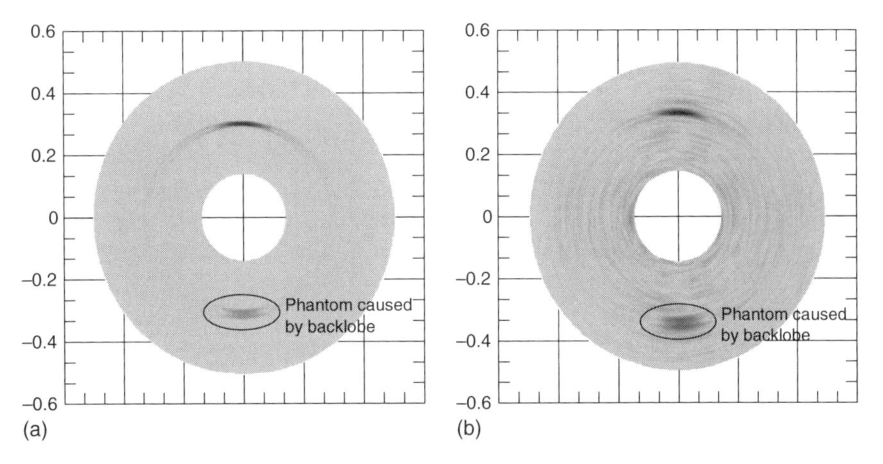

FIGURE 11.55 2-D EUSR damage detection of a single broadside defect at 90° using a 4 × 8 rectangular array: (a) simulation data for an ideal broadside reflector; (b) experimental data for a broadside crack.

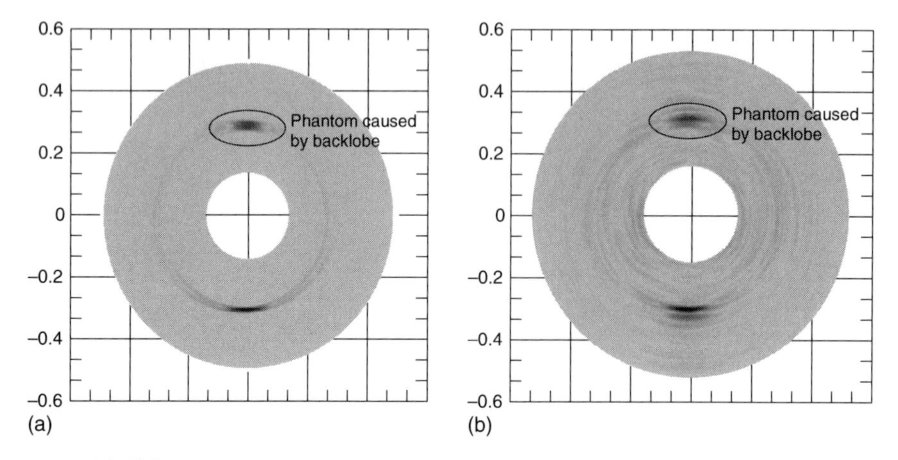

FIGURE 11.56 2-D EUSR damage detection of a single broadside defect at 270° using a 4 × 8 rectangular array: (a) simulation data for an ideal broadside reflector; (b) experimental data for a broadside crack.

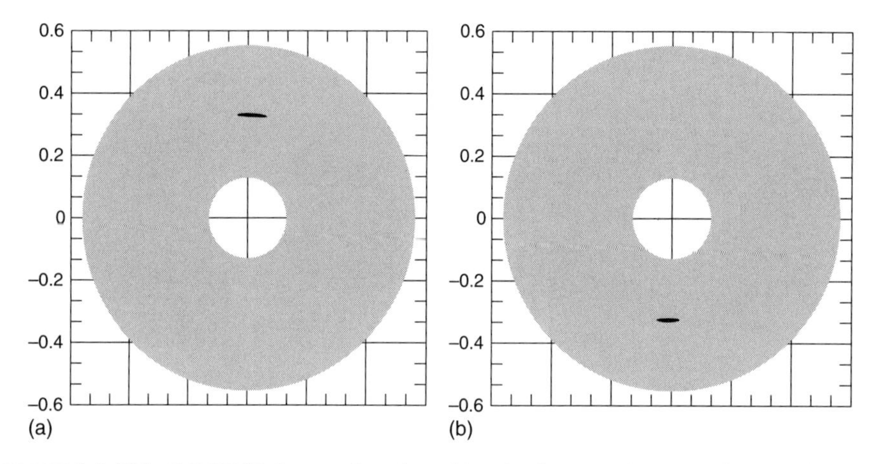

FIGURE 11.57 2-D EUSR damage detection with a 4 × 8 rectangular array after applying thresholding: (a) single crack at 90°; (b) single crack at 270°.

due to backlobe inherent in a rectangular array with unequal number of elements in the x and y directions (Fig. 11.51).

However, the images shown in Figs 11.55 and 11.56 can be cleaned up by applying thresholding. After applying thresholding, the phantoms caused by the backlobe effect are eliminated and the location of the crack damage is correctly indicated at 90° and 270°, respectively (Fig. 11.57a,b).

11.9.3 DETECTION OF OFFSIDE CRACKS WITH A 4 x 8 PWAS ARRAY

The capability of a 4 × 8 PWAS array to detect offside crack was investigated using a specimen layout similar to the one used with the linear PWAS array. The crack is assumed to be located at 120°. The results of the initial analysis is presented in Fig. 11.58a. It is apparent that the crack is correctly imaged, but a disturbing phantom placed symmetrically about the horizontal axis is also present. The phantom is an artifact

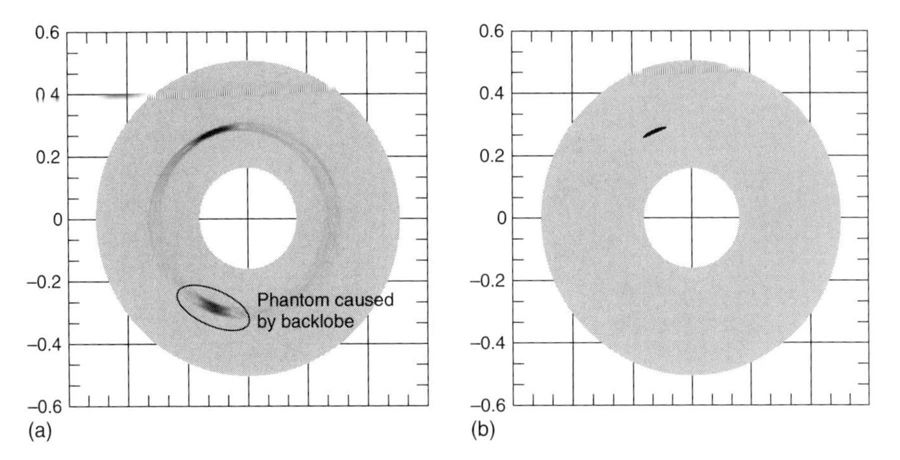

FIGURE 11.58 2-D damage detection of simulated single crack at 120°: (a) raw processing; (b) after thresholding.

of the 4 × 8 PWAS array. The phantom can be successfully removed by thresholding, as indicated in Fig. 11.58b.

11.9.4 DAMAGE DETECTION WITH AN 8 x 8 PWAS ARRAY

We have seen in the previous section that considerable problems have come from the fact that an unbalanced rectangular array has a pronounce backlobe as indicated in Fig. 11.51. This backlobe was successfully removed through thresholding. However, we should also seek other more fundamental ways to deal with this problem. A natural idea is to use a balanced rectangular array, i.e., a rectangular array in which the number of elements in rows and columns are equal. In this scope, we explored the 8 × 8 PWAS array.

11.9.4.1 Theoretical Beamforming of an 8 x 8 PWAS Array

The theoretical beamforming of an 8 × 8 PWAS array is shown in Fig. 11.59a. The array is assumed of to have the same element spacing as the previously discussed 4 × 8 array. If we compare Fig. 11.59 with Fig. 11.51 corresponding to the 4 × 8 array, we notice that indeed, as expected, the backlobes have decreased considerably. This decrease is especially apparent for the 90° direction: comparing Fig. 11.59b with the corresponding Fig. 11.51b, one notes that the backlobe reduction is at least of 3 times. This indicates that the square 8 × 8 array offers substantial reduction in the backlobe effect in comparison with the 4 × 8 array. To verify this fact, we performed 2D-EUSR analysis using simulated signals corresponding to damage located at 90°. The results are shown in Fig. 11.60. When comparing Fig. 11.60 with Fig. 11.52a, we notice that the raw image is greatly improvement and hardly needs any retouching.

11.9.4.2 Experimental Damage Detection with an 8 x 8 PWAS Array

To confirm the encouraging theoretical results presented in the previous section, we performed an experiment with an 8 × 8 PWAS array. The 8 × 8 array was constructed using 7-mm round PWAS elements (Fig. 11.61b). The specimen under investigation was again 1-mm thick 2024 T3 aluminum alloy plate. The damage to be detected was a simulated 20-mm crack placed at 90°. However, the simulated crack had an inclination of

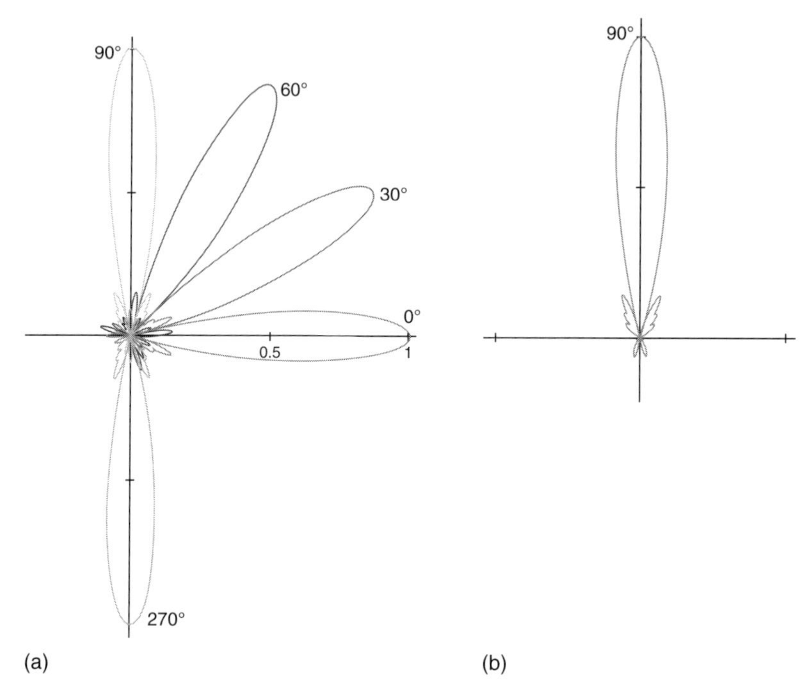

FIGURE 11.59 Beamforming of the 2-D 8 × 8 rectangular array: (a) theoretical beamforming at various directions; (b) beamforming at 90°.

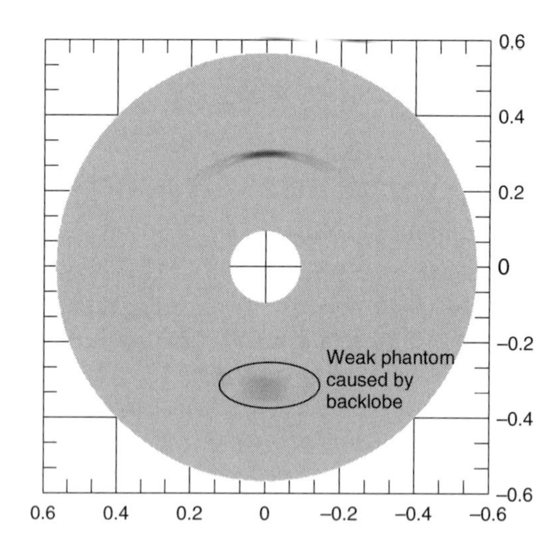

FIGURE 11.60 2D-EUSR imaging of a 90° damage with an 8 × 8 PWAS array using simulation data indicates strong damage detection and very weak mirror phantom shadow.

45° with respect to the horizontal axis (Fig. 11.61a). This inclination of the crack made the problem an order of magnitude more difficult, because the specular reflection from the crack will not return to the array, but will be deflected away. The excitation frequency was set to 300 kHz, which corresponded to the frequency tuning of circular 7-mm PWAS elements.

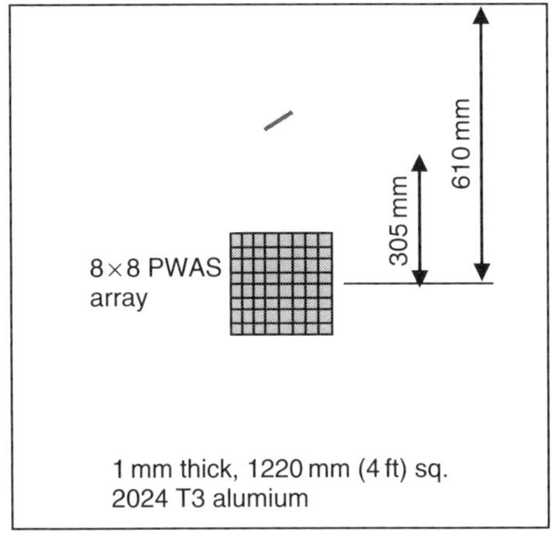

(a)

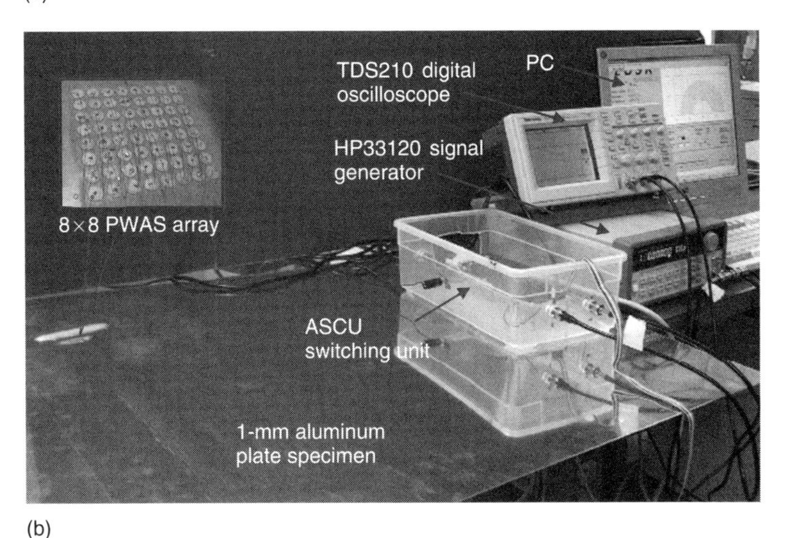

(b)

FIGURE 11.61 A 360° full-range crack detection using an 8 × 8 rectangular PWAS array: (a) specimen layout (sizes are not proportional; crack and array are exaggerated for illustration); (b) laboratory experimental setup.

11.9.4.3 EUSR Mapped Images

2D-EUSR scanning results and damage imaging obtained with this 8 × 8 PWAS array are given in Fig. 11.62. The original image resulting from direct 2D-EUSR imaging is shown in Fig. 11.62a. One observers that the crack located at 90° to the array is well indicated. It is noticed that the indication is unique within the 360° range with no phantom image at other directions. However, the image is nosier than the image obtained with the 4 × 8 array (Fig. 11.55b). This noisier image is attributed to the fact that the 45° crack used in this experiment gives an order of magnitude smaller echo than the horizontal crack considered in the 4 × 8 experiment. The echo from the horizontal crack came from a specular total reflection, whereas the echo from the 45° crack considered here comes only from diffraction backscatter effects. In order to remove the noise, we applied thresholding; as a result we obtained the much cleaner image of Fig. 11.62b, in which the background noise was removed. Another effect that may be attributed to the fact that

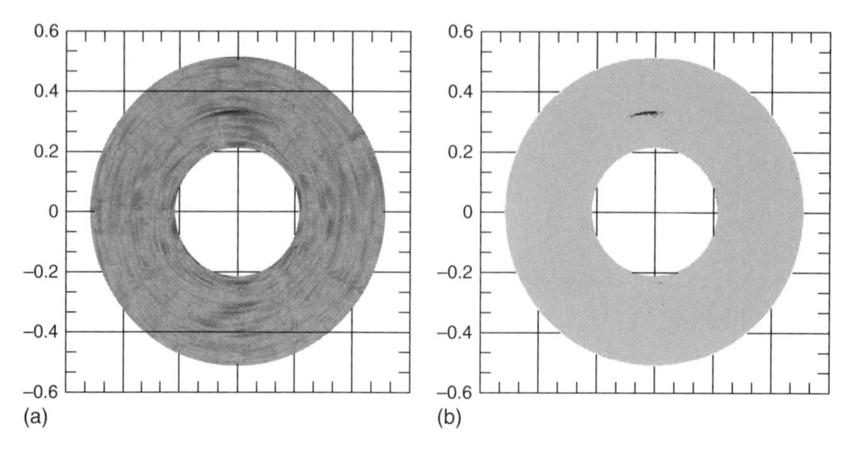

FIGURE 11.62 The 2-D 8 × 8 PWAS array EUSR image: (a) original EUSR image and A-scan at 95°; (b) enhanced EUSR image using post-processing.

the crack is inclined is that the strongest reflection seems to reside at 95°, slightly to the left of the crack central position.

11.10 PHASED ARRAY ANALYSIS USING FOURIER TRANSFORM METHODS

So far, our analysis of the PWAS phased arrays has been performed in the time domain by applying simple delay-and-sum principles to a spatially-defined array of transmitters–receivers. However, further insight into the function of phased arrays can be gained via Fourier analysis. Using Fourier analysis, one gains a better understanding of the fundamental principles of the phased-array method. In this section, we will recall some fundamental results of the space–time Fourier transform and apply them to the study of wave propagation through spatial apertures. This will be expanded to the study of sampled wave signals propagating through sampled spatial apertures, which are representative of phased array geometries.

11.10.1 SPATIAL-FREQUENCY ANALYSIS

11.10.1.1 Fourier Transform

A short review of the Fourier transform principles is given next; further details of the Fourier transform definitions and properties are given in Appendix A.

Consider a signal $x(t)$ defined on $(-\infty, +\infty)$. Its Fourier transform is defined as

$$X(\omega) = \int_{-\infty}^{+\infty} x(t) \cdot e^{-j\omega t} \mathrm{d}t \tag{88}$$

The inverse Fourier transform is defined as

$$x(t) = \frac{1}{2\pi} \int_{-\infty}^{+\infty} X(\xi) e^{j\xi t} \mathrm{d}\xi \tag{89}$$

A rectangular pulse $p_T(t)$ as shown in Fig. 11.63a is defined in time domain as

$$p_T(t) = \begin{cases} 1, & -T/2 \le t \le T/2 \\ 0, & \text{otherwise} \end{cases}$$

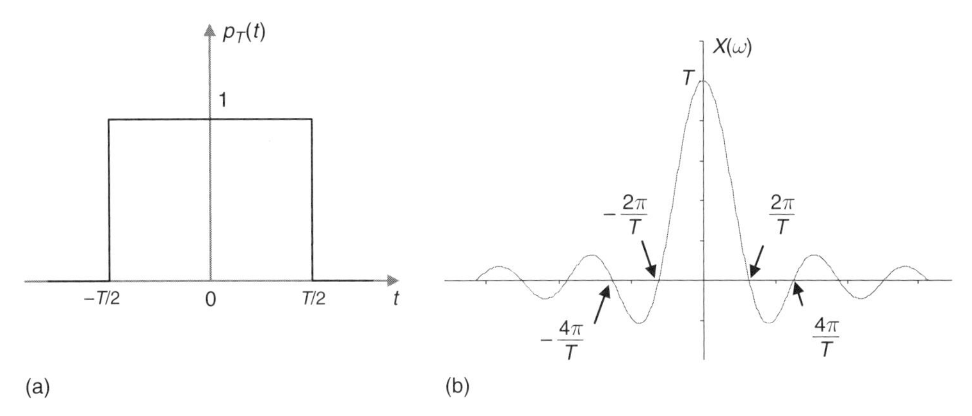

FIGURE 11.63 A rectangular pulse and its Fourier transform: (a) a pulse of duration T seconds; (b) the Fourier transform of the pulse.

Its Fourier transform is the sinc function, i.e., $\text{sinc}(x) = \frac{\sin \pi x}{\pi x}$, i.e.,

$$X(\omega) = T \cdot \text{sinc}\frac{\omega}{2\pi/T} \tag{90}$$

The spectrum of this signal is plotted in Fig. 11.63b. Notice that the sinc(x) function is equal to 1 at the point $x = 0$. The function sinc(x) crosses zero at $x = \pm 1, \pm 2, \ldots$. Therefore, $X(\omega = 0) = T$ and $X(\omega)$ crosses zero when $\omega = \pm 2\pi/T, \pm 4\pi/T, \ldots$

For a sampled signal (i.e., the discretization of a continuous signal), we have the sampled version of signal $x(t)$ as $x_s(n\Delta t)$ where Δt is the sampling interval and $n = 0, 1, \ldots, N$, with N being the number of sample points. The sampled signal can be represented using a summation of impulses of the type $p(t) = \sum_{n=-\infty}^{\infty} \delta(t - n\Delta t)$, i.e.,

$$x_s(t) = x(t)p(t) = \sum_{n=-\infty}^{\infty} x(t)\delta(t - n\Delta t) = \sum_{n=-\infty}^{\infty} x(n\Delta t)\delta(t - n\Delta t) \tag{91}$$

We define $\omega_s = 2\pi/\Delta t$ as the angular sampling frequency. As shown in Appendix A, the impulse train $p(t)$ can be represented using Fourier series as $p(t) = \sum_{n=-\infty}^{\infty} c_k e^{jn\omega_s t} = \sum_{n=-\infty}^{\infty} \frac{1}{\Delta t} e^{jn\omega_s t}$. Thus,

$$x_s(t) = x(t)p(t) = \sum_{n=-\infty}^{\infty} \frac{1}{\Delta t} x(t) e^{jk\omega_s} \tag{92}$$

Using the property of Fourier transform about multiplication with a complex exponential in time domain (see Appendix A), the Fourier transform of a sampled signal is

$$X_s(\omega) = \frac{1}{\Delta t} \sum_{n=-\infty}^{\infty} X(\omega - n\omega_s) \tag{93}$$

11.10.1.2 Multidimensional Fourier Transform: Wavenumber–Frequency Domain

The multidimensional Fourier transform is briefly reviewed. Consider a signal $f(\vec{r}) = f(r_1, r_2, \ldots, r_N)$ defined in an N-dimensional space. Each variable r_i extends from $-\infty$ to ∞. Apply Fourier transform to each dimension by assigning a harmonic variation $e^{j\xi_i r_i}$. This process yields

$$F(\vec{\xi}) = \int_{\vec{r}} f(\vec{r}) e^{-j\vec{\xi}\cdot\vec{r}} \mathrm{d}\vec{r} \tag{94}$$

where $\vec{\xi} \cdot \vec{r} = \xi_1 r_1 + \xi_2 r_2 + \ldots + \xi_N r_N$, $e^{-j(\xi_1 r_1 + \xi_2 r_2 + \ldots + \xi_N r_N)} = e^{-j\vec{\xi}\cdot\vec{r}}$, and $\mathrm{d}\vec{r} = \mathrm{d}r_1 \mathrm{d}r_2, \ldots, \mathrm{d}r_N$. Therefore, the Fourier transform and inverse Fourier transform of an N-dimensional signal are

$$F(\vec{\xi}) = \int \ldots \int_{-\infty}^{+\infty} f(\vec{r}) e^{-j\vec{\xi}\cdot\vec{r}} \mathrm{d}\vec{r} \tag{95}$$

$$f(\vec{r}) = \frac{1}{(2\pi)^N} \int \ldots \int_{-\infty}^{+\infty} F(\vec{\xi}) e^{j\vec{\xi}\cdot\vec{r}} \mathrm{d}\vec{\xi} \tag{96}$$

If the signal is a ***space–time*** signal, i.e., $f(\vec{r}, t)$ where $\vec{r}$ denotes the spatial dimensions and t denotes the time, then $\vec{k}$ denotes the transformed variable corresponding to the spatial variable $\vec{r}$ and ω denotes the transformed variable corresponding to the time variable t. Usually $\vec{k}$ is named ***wavenumber*** or ***spatial frequency***, whereas ω is usually known as ***frequency***, or ***temporal frequency***. The combination $(\vec{k}, \omega)$ is called the *wavenumber–frequency domain*. The spatial–temporal Fourier transform and its inverse, the wavenumber–frequency Fourier transform, are

$$F(\vec{k}, \omega) = \int_{\vec{r}} \int_{t} f(\vec{r}, t) e^{-j(\omega t - \vec{k}\cdot\vec{r})} \mathrm{d}t \mathrm{d}\vec{r} \tag{97}$$

$$f(\vec{r}, t) = \frac{1}{(2\pi)^4} \int_{\vec{k}} \int_{\omega} F(\vec{k}, \omega) e^{j(\omega t - \vec{\xi}\cdot\vec{r})} \mathrm{d}\omega \mathrm{d}\vec{k} \tag{98}$$

The fundamental harmonic function in this transform is $e^{j(\omega t - \vec{k}\cdot\vec{r})}$. This is a ***plane wave***. For a fixed time t_0 and a constant phase θ_0, the surface $\vec{k} \cdot \vec{r} = \omega t_0 - \theta_0$ is called a ***wave front***. Furthermore, we define the unit vector in the direction of $\vec{k}$ as

$$\vec{\xi} = \frac{\vec{k}}{|\vec{k}|} \tag{99}$$

The unit vector $\vec{\xi}$ is normal to the wave front and called is as the ***direction of wave propagation***. Parallel wave front crests are separated by a wavelength distance λ given by

$$\lambda = \frac{2\pi}{|\vec{k}|} \tag{100}$$

11.10.1.3 Space–Frequency Transform of a Single-mode Plane Wave

First, we consider a single-tone single-mode plane wave

$$f(\vec{r}, t) = e^{j(\omega_0 t - \vec{k}_0\cdot\vec{r})} \tag{101}$$

where ω_0 is the temporal frequency (which is related to the temporal period, T_0, by the relation $\omega_0 = 2\pi/T_0$) and $\vec{k}_0$ is the wavenumber vector. The wavelength is $\lambda_0 = 2\pi/k_0$.

The spatial–frequency transform of this signal is

$$
\begin{aligned}
F(\vec{k},\omega) &= \int_{\vec{r}}\int_{t} e^{j(\omega_0 t-\vec{k}_0\vec{r})}e^{-j(\omega t-\vec{k}\vec{r})}\mathrm{d}\vec{r}\mathrm{d}t \\
&= \int_{\vec{r}}\int_{t} e^{\{-j(\omega-\omega_0)t+j(\vec{k}-\vec{k}_0)\vec{r}\}}\mathrm{d}\vec{r}\mathrm{d}t \\
&= \int_{\vec{r}} e^{j(\vec{k}-\vec{k}_0)\vec{r}}\mathrm{d}\vec{r}\int_{t} e^{-j(\omega-\omega_0)t}\mathrm{d}t \\
&= \int_{\vec{r}} e^{-j\vec{k}_0\vec{r}}e^{j\vec{k}\vec{r}}\mathrm{d}\vec{r}\int_{t} e^{j\omega_0 t}e^{-j\omega t}\mathrm{d}t
\end{aligned}
\tag{102}
$$

Using the common Fourier transform pair $e^{j\omega_0 t} \leftrightarrow \delta(\omega-\omega_0)$, Eq. (102) generates

$$
F(\vec{k},\omega) = \delta(\vec{k}-\vec{k}_0)\delta(\omega-\omega_0) \tag{103}
$$

Second, we consider a wave propagating in a particular direction $\vec{\xi}_0$, i.e.,

$$
f(\vec{r},t) = f\left(t-\vec{\xi}_0\frac{\vec{r}}{c}\right) \tag{104}
$$

Define the slowness vector $\vec{\alpha}_0$ of the propagating wave as

$$
\vec{\alpha}_0 = \frac{\vec{\xi}_0}{c} \tag{105}
$$

Hence, Eq. (104) becomes

$$
f(\vec{r},t) = f(t-\vec{\alpha}_0\vec{r}) \tag{106}
$$

Similarly, the wavenumber–frequency spectrum of this wave can be obtained

$$
\begin{aligned}
F(\vec{k},\omega) &= \int_{\vec{r}}\int_{t} e^{j\omega_0(t-\vec{\alpha}_0\cdot\vec{r})}e^{-j(\omega t-\vec{k}\vec{r})}\mathrm{d}\vec{r}\mathrm{d}t \\
&= \int_{\vec{r}}\int_{t} e^{j(\vec{k}-\omega_0\vec{\alpha}_0)\vec{r}}e^{-j(\omega-\omega_0)t}\mathrm{d}\vec{r}\mathrm{d}t \\
&= \delta(\omega-\omega_0)\delta(\vec{k}-\omega_0\vec{\alpha}_0)
\end{aligned}
\tag{107}
$$

In Eq. (107), the first multiplier $\delta(\omega-\omega_0)$ is the Fourier transform of the original temporal signal $f(t)$, i.e., $f(t) \leftrightarrow F(\omega)$, $F(\omega) = \delta(\omega-\omega_0)$. Therefore Eq. (107) can be written as

$$
F(\vec{k},\omega) = F(\omega)\delta(\vec{k}-\omega_0\vec{\alpha}_0) \tag{108}
$$

Equation (108) shows that the propagating wave $f(t-\vec{a}_0\vec{r})$ contains energy only along the direction $\vec{k} = \omega_0\vec{a}_0$ in the wavenumber–frequency space[1].

[1] The delta function $\delta(\xi)$, takes values 1 for $\xi = 0$, i.e., $\delta(0) = 1$, and 0 otherwise, i.e., $\delta(\xi) = 0$, $\xi \neq 0$. Hence $\delta(\vec{k}-\omega_0\vec{\alpha}_0) = 1$ when $\vec{k} = \omega_0\vec{\alpha}_0$

11.10.1.4 Filtering in the Wavenumber–frequency Space

In the signal processing field, a filter is a device that rejects unwanted frequency components (i.e., the frequencies that fall into the stopband) whereas keeping others (i.e., the frequencies in the passband). Extending this concept to space-time domain, we may implement spatial temporal filtering to retain signal components at certain temporal frequency, or those propagating in certain directions whereas filtering out noise and/or signal components in other directions. Extending this concept to the wavenumber–frequency domain, we consider an input signal $F(\vec{k}, \omega)$ multiplied by a spatial-temporal filter having the wavenumber–frequency response $H(\vec{k}, \omega)$. The process generates the modified output wavenumber–frequency spectrum $Z(\vec{k}, \omega)$, i.e.,

$$Z(\vec{k}, \omega) = H(\vec{k}, \omega)F(\vec{k}, \omega) \tag{109}$$

A schematic of this process is shown in Fig. 11.64. The output spectrum $Z(\vec{k}, \omega)$ corresponds to a modified space–time signal $z(\vec{x}, t)$, which can be reconstructed using the inverse Fourier transformation formulae. This spatial-temporal filter concept includes the direction-finding array signal processing discussed in this chapter.

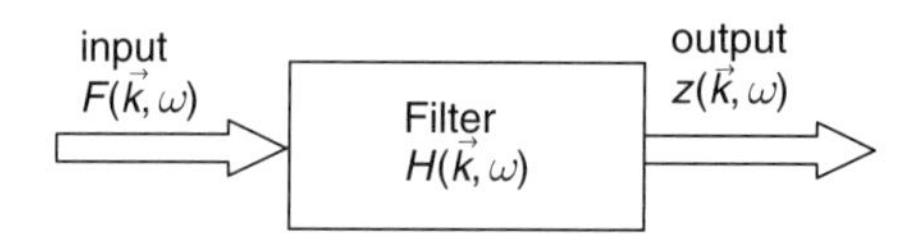

FIGURE 11.64 Spatial–temporal filtering using a wavenumber–frequency filter $H(\vec{k}, \omega)$.

11.10.2 APERTURES AND ARRAYS

Propagating waves may vary in space and time. Some wave sensors are designed to gather energy propagating from specific directions, thus having a given spatial extent. The energy is spatially integrated and the sensor is focused in particular directions. This type of sensors are said to be *directional*. In contrast, sensors that sample the wave field from all directions are *omnidirectional*. Sensors that gather signal energy over finite spatial areas are called *aperture* sensors. A sensor *array* consists of a group of sensors, directional or not, that are combined to produce a single output. Sensor arrays can be treated as sampled or discretized apertures. We start with the discussion of the aperture, which is the first step in understanding the Fourier description of the phased array method.

11.10.2.1 Continuous Finite Apertures

The effects of a finite aperture on a spatial signal $f(\vec{x}, t)$ are represented through the *aperture function* $w(\vec{x})$. The aperture function has two main properties:

(1) the size and shape of the aperture;
(2) within the aperture, the relative weight of the wave field resulting from the aperture.

The aperture weighting function can implement the condition in which the sensor is weighted within the aperture region and zeroed outside. In terms of signal processing, the aperture weighting function serves as a filter to eliminate signals outside a field of interest. A continuous finite aperture in the spatial domain is similar to a continuous finite duration impulse signal. For example, consider an aperture that covers the region $|\vec{x}| \leq R$. The shape and size of the aperture are known, i.e., a circle of radius R. The aperture (weighting) function for this sensor can be represented as simply

$$w(\vec{x}) = \begin{cases} 1, & |\vec{x}| \leq R \\ 0, & \text{otherwise} \end{cases} \tag{110}$$

If we observe a field through a finite aperture, then the sensor's output becomes

$$z(\vec{x}, t) = w(\vec{x}) f(\vec{x}, t) \tag{111}$$

Corresponding relationship in space–time domain can be expressed using convolution[2] in the frequency domain, i.e.,

$$Z(\vec{k}, \omega) = C\, W(\vec{k}) * F(\vec{k}, \omega) \tag{112}$$

Where C is a constant. The functions $Z(\vec{k}, \omega)$, $W(\vec{k})$, and $F(\vec{k}, \omega)$ are given by the corresponding space–time Fourier transforms, i.e.,

$$W(\vec{k}) = \int_{\vec{x}} w(\vec{x}) e^{j\vec{k}\vec{x}} \mathrm{d}\vec{x} \tag{113}$$

$$F(\vec{k}, \omega) = \int_{\vec{x}} \int_{t} f(\vec{x}, t) e^{-j(\omega t - \vec{k}\vec{x})} \mathrm{d}\vec{x} \tag{114}$$

Substituting Eqs. (113) and (114) into (112) gives

$$Z(\vec{k}, \omega) = \int_{\vec{l}} W(\vec{l}) * F(\vec{k} - \vec{l}, \omega) \mathrm{d}\vec{l} \tag{115}$$

First consider a single plane wave $s(t)$ propagating in a particular direction with the slowness vector $\vec{\alpha}_0$, i.e., $f(\vec{r}, t) = s(t - \vec{\alpha}_0 \cdot \vec{r})$. Using Eq. (108), the wavenumber–frequency spectrum of this propagating field is

$$F(\vec{k}, \omega) = S(\omega) \delta(\vec{k} - \omega \vec{\alpha}_0) \tag{116}$$

where $S(\omega)$ is the Fourier transform of the source signal $s(t)$ and $\delta(\vec{k})$ is the spatial impulse function. Substitution of Eq. (116) into (115) yields

$$\begin{aligned} Z(\vec{k}, \omega) &= \int_{\vec{l}} W(\vec{l}) * S(\omega) \delta(\vec{k} - \vec{l} - \omega \vec{\alpha}_0) \mathrm{d}\vec{l} \\ &= S(\omega) \int_{\vec{l}} W(\vec{l}) * \delta(\vec{k} - \omega \vec{\alpha}_0 - \vec{l}) \mathrm{d}\vec{l} \\ &= S(\omega) W(\vec{k} - \omega \vec{\alpha}_0) \end{aligned} \tag{117}$$

Equation (117) shows that when $\vec{k} = \omega \vec{\alpha}_0$, we have $Z(\omega \vec{\alpha}_0, \omega) = S(\omega) W(\vec{0})$. Along this particular path in the wavenumber–frequency space, the output spectrum equals the

[2] Convolution property: multiplication in the time domain is convolution in the frequency domain: $x(t)y(t) = \frac{1}{2\pi} X(\omega) * Y(\omega)$. See Appendix A for details.

signal spectrum amplified by a constant value $W(\vec{0})$, the value of $W(\cdot)$ at the origin. In this case, all the information from the original signal spectrum $S(\omega)$ is passed through the aperture output. For other values of $\vec{k}$, the spectrum $S(\omega)$ will be multiplied by a frequency dependent gain $W(\vec{k}-\omega\vec{\alpha}_0)$. This frequency dependent gain will distort the relative strengths and phases of the frequency components in the original signal spectrum, effectively filtering it.

If the original signal is a superposition of various plane waves, i.e., $f(\vec{x},t) = \sum_i s_i(t-\vec{\alpha}_{0,i}\cdot\vec{x})$, then its space–time spectrum will be $F(\vec{k},\omega) = \sum_i S_i(\omega)\delta(\vec{k}-\omega\vec{\alpha}_{0,i})$. After applying the aperture function, the output spectrum will be $Z(\vec{k},\omega) = \sum_i S_i(\omega)W(\vec{k}-\omega\vec{\alpha}_{0,i})$. When $\vec{k}=\omega\vec{\alpha}_{0,j}$ where j is the index corresponding to one of the propagating signals, the output spectrum can be rewritten as

$$Z(\omega\vec{\alpha}_{0,j},\omega) = S_j(\omega)W(\vec{0}) + \sum\nolimits_{i,i\neq j} S_i(\omega)W\left\{\omega(\vec{\alpha}_{0,j}-\vec{\alpha}_{0,i})\right\} \tag{118}$$

From Eq. (118) we see that if the aperture smoothing function spectrum $W(\cdot)$ can be designed in such a way that $W\left\{\omega(\vec{\alpha}_{0,j}-\vec{\alpha}_{0,i})\right\}$ is small compared to $W(\vec{0})$, then the aperture will act as a ***spatial filter*** that passes signals propagating in the direction determined by the slowness vector $\vec{\alpha}_{0,j}$ whereas rejecting the others.

11.10.2.2 1-D Linear Aperture

Now, let us consider the simplest finite continuous aperture, the 1-D linear aperture, which is nonzero only along a finite length D in the x-axis direction (Fig. 11.65a). Since this aperture lies only along the x axis, its aperture function depends only on component k_x. The simplified 1-D aperture function is

$$w(x) = \begin{cases} 1, & |x| \leq D/2 \\ 0, & \text{otherwise} \end{cases} \tag{119}$$

We see that the aperture is sensitive only to the variable x of the field. Recall Eq. (90) that gives the Fourier transform of a rectangular pulse of time duration T in terms of the sinc

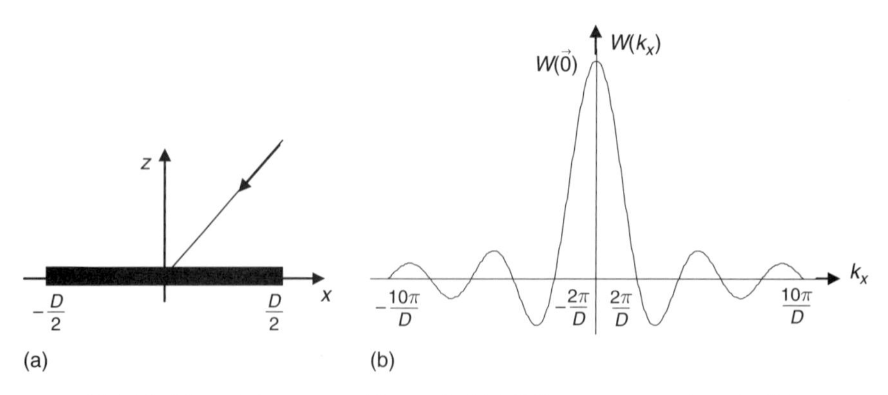

FIGURE 11.65 Linear 1-D aperture with span D: (a) 1-D linear aperture schematic; (b) wavenumber spectrum of the aperture function.

function. Similarly, the space Fourier transform of a linear aperture can be represented using the sinc function

$$W(k_x) = D\,\mathrm{sinc}\left(\frac{k_x}{2\pi/D}\right) \tag{120}$$

or

$$W(k_x) = D\frac{\sin \pi \cdot \dfrac{k_x}{2\pi/D}}{\pi \cdot \dfrac{k_x}{2\pi/D}} = \frac{\sin(k_x D/2)}{k_x/2} \tag{121}$$

Equation (121) shows that the aperture function spectrum depends only on the wavenumber component k_x and the maximum aperture length D. The spectrum is shown in Fig. 11.65b.

According to the properties of the sinc function, some characteristics of the 1-D linear aperture can be concluded as follows:

1. **Main lobe**: At the point $k_x = 0$, the aperture function reaches its maximum value $W(0)$, $W(0) = D$. This is called the height of the main lobe. We see that the mainlobe height is completely determined by the size of the aperture.
2. **Sidelobe**: The aperture function has infinite number of sidelobes with decreasing amplitude. At the points $k_x = \pm 2\pi/D, \pm 4\pi/D, \ldots$ the aperture function crosses zero.
3. **SLL**: The ratio of the height of the largest sidelobe to the height of the mainlobe is called sidelobe level (SLL), measuring an aperture's ability to reject unwanted signals and focus on particular propagating waves. By differentiating the aperture function, the location and the height of the largest sidelobe can be determined. For 1-D linear aperture of span D, the first largest sidelobe occurs at about $k_x \approx 8.9868/D$ or $2.86\pi/D$. The value of $|W(k_x)|$ at this point is approximately $0.2172D$, corresponding to SLL $= 0.2172$. We see this ratio is a constant, independent of the aperture length D, which means changing the size of the aperture will not improve the aperture's filtering ability.
4. **Resolution**: As we stated before, the aperture function can be interpreted as a spatial filter with the mainlobe as the passband and the sidelobes as the stopband. Spatial resolution is the minimum spatial measurement with which two plane waves can be distinguished. According to the classical definition of resolution given by the Rayleigh criterion[3], the resolution equals the value of the smallest wavenumber where the aperture function equals zero. Spectrum in Fig. 11.65b shows that it happens when $k_x = \pm 2\pi/D$, which is half of the mainlobe width $4\pi/D$. Therefore, the mainlobe width determines the aperture's ability to separate propagating waves. Suppose there are two plane waves passing an aperture, the resulting spectrum is

$$Z(\vec{k}_x, \omega) = S_1(\omega)W(\vec{k}_x - \omega\vec{\alpha}_{0,1}) + S_2(\omega)W(\vec{k}_x - \omega\vec{\alpha}_{0,2}) \tag{122}$$

Equation (122) indicates that each plane wave causes an aperture function to appear in the wavenumber–frequency spectrum. According to the 1-D linear aperture properties, the aperture function is zero when $k_x = \pm 2\pi/D, \pm 4\pi/D, \pm 6\pi/D, \ldots$, and the mainlobe width is $4\pi/D$.

[3] Two incoherent plane waves that propagate in two slightly different directions are resolved if the mainlobe peak of one aperture function falls on the first zero of the other aperture function.

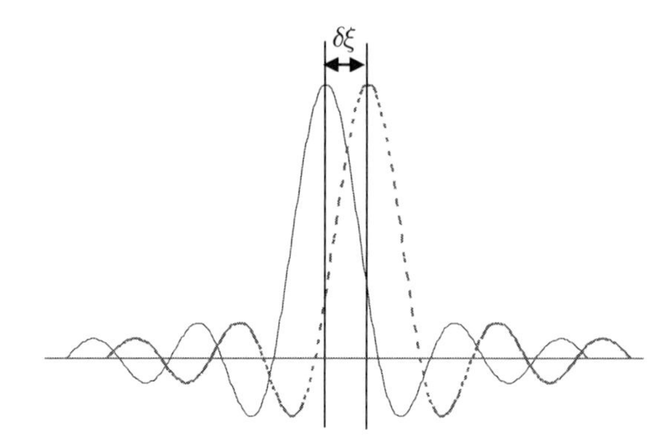

FIGURE 11.66 Aperture resolution in terms of distinguishing two propagating waves.

For a propagating plane wave, the relationship between wavenumber $\vec{k}$ and wavelength λ is $\vec{k} = 2\pi\vec{\xi}/\lambda$, where $\vec{\xi}$ is the direction vector of $\vec{k}$. By finite differences approximation, we have

$$\Delta\vec{k} = \frac{2\pi\Delta\vec{\xi}}{\lambda}$$

Since $|k_x| = 2\pi/D$, we will have

$$\Delta\xi = \frac{\lambda}{D} \tag{123}$$

Equation (123) shows that the directional resolution $\Delta\xi$, i.e., the spacing between two neighboring aperture functions in the spectrum (Fig. 11.66), depends on both the wave length and the aperture span.

11.10.2.3 Spatial Sampling

In contrast to an aperture, an array consists of a group of individual sensors that spatially sample the surroundings. Each sensor could be an independent aperture or an omnidirectional sensor. In terms of receiving, it can also be treated as a group of individual sensors that spatially sample the wave field of the source.

If a signal $f(x, t)$ (which is a function of both spatial dimension and time) is spatially sampled at the **spatial sampling interval** d along the x axis, a sequence of temporal signals can be obtained as $\{y_m(t_0)\} = \{f(md, t_0)\}$, $-\infty < m < +\infty$, (t_0 is a particular time). The spatial sampling function is

$$\mathrm{P}(x) = \sum_{m=-\infty}^{+\infty} \delta(x - md) \tag{124}$$

The sampled signal is obtained by multiplying the original signal with the sampling function

$$f_s(x, t_0) = f(x, t_0)\mathrm{P}(x) = \sum_{m=-\infty}^{+\infty} f(md, t_0)\delta(x - md) \tag{125}$$

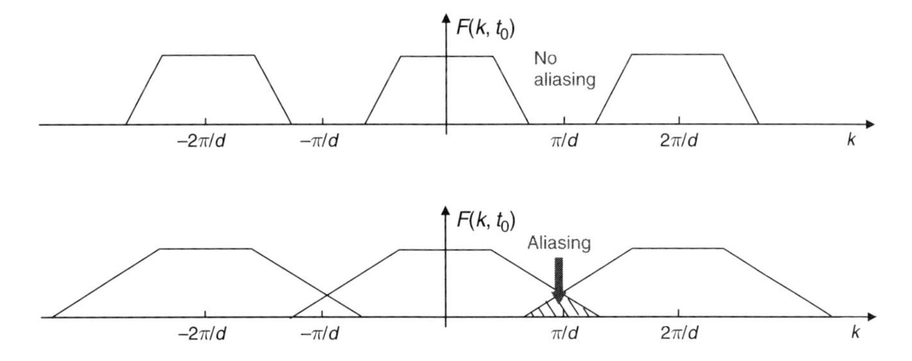

FIGURE 11.67 Aliasing occurs if the spatial signal is not spatially bandlimited below k_{NQ}.

Similar to the sampling of temporal signal, the spatial sampling should assure that the original signal $f(x, t_0)$ can be correctly reconstructed from a series of sampled points $\{y_m(t_0)\}$. Using Shannon sampling theorem, the space-time signal $f(x, t_0)$ should have no wavenumber components above the Nyquist wavenumber k_{NQ} with $k_{\mathrm{NQ}} = \pi/d$. If the original signal is wavenumber bandlimited to the wavenumber bandwidth k_{B}, the spatial sampling interval d must meet the requirement of $d \leq \pi/k_{\mathrm{B}}$ (since $k_{\mathrm{B}} \leq k_{\mathrm{NQ}}$). Under this condition, the original signal can be considered well represented by its spatial samples, i.e.,

$$f(x, t_0) \approx f_{\mathrm{s}}(x, t_0) = \sum_{m=-\infty}^{+\infty} f(md, t_0)\delta(x - md) \tag{126}$$

Using Fourier transform properties, the Fourier transform of the spatially sampled signal is

$$F_{\mathrm{s}}(k, t_0) = \frac{1}{d} \sum_{m=-\infty}^{+\infty} F(k - mk_{\mathrm{s}}, t_0) \tag{127}$$

Where $F(k, t_0)$ is the spatial Fourier transform of the original signal and k_{s} is the sampling spatial frequency, $k_{\mathrm{s}} = 2\pi/d$. Equation (127) indicates that the wavenumber spectrum of the sampled signal is periodic with wavenumber period k_{s}. However, this is not a problem as long as the original signal was wavenumber bandlimited as indicated above.

If the signal $f(x, t_0)$ is not spatially bandlimited to below k_{NQ}, since the spatial spectrum $F_{\mathrm{s}}(k, t_0)$ of the sampled signal is periodic at multiples of k_{s}, then overlapping, or ***aliasing***, occurs (Fig. 11.67). The spatial sampling theorem can be summarized as follows: if a continuous spatial signal is bandlimited to a frequency below k_{NQ}, then it can be periodically sampled without loss of information as long as the sampling spacing d is smaller than or equal to π/k_{B}.

11.10.2.4 Arrays and the Sampled Finite Continuous Apertures

An array is composed of individual sensors which sample the wave field at discrete spatial locations. We will analyze the wave field of a propagating wave $f(x, t)$ measured by a 1-D linear array. This array is constructed by spatially sampling a 1-D linear finite aperture, resulting in M equally spaced point-wise sensors separated by d spacing along the x axis. By definition, the wavenumber frequency spectrum of the original signal $f(x, t)$ is

$$F(k, \omega) = \int_{-\infty}^{+\infty} \int_{-\infty}^{+\infty} f(x, t) \exp\{-j(\omega t - kx)\} \mathrm{d}x\mathrm{d}t \tag{128}$$

After the wave field is sampled by the array at every distance d, the resulting wavenumber frequency spectrum of the sampled signal $\{y_m(t)\} = \{f(md, t)\}$ is

$$F_s(k, \omega) = \int_{-\infty}^{+\infty} \sum_{m=-\infty}^{+\infty} y_m(t) \exp\{-j(\omega t - kx)\} \mathrm{d}t \tag{129}$$

Using Eq. (127), the spectrum of the spatially sampled signal becomes

$$F_s(k, \omega) = \frac{1}{d} \sum_{m=-\infty}^{+\infty} F\left(k - m\frac{2\pi}{d}, \omega\right) \tag{130}$$

Recalling Eq. (111), the output signal $z(x, t)$ equals to the product of $w(x)$ and $f(x, t)$. Using the convolution theory, we obtain the output wavenumber–frequency spectrum

$$Z(k, \omega) = F(k, \omega)W(k) = \frac{d}{2\pi} \int_{-\pi/d}^{\pi/d} F(k^*, \omega)W(k - k^*)\mathrm{d}k^* \tag{131}$$

where $W(k)$ is the wavenumber–frequency spectrum of the sampled aperture function $\{w_m\}$, i.e.,

$$W(k) - \sum_m w_m e^{jkmd} \tag{132}$$

The function $\{w_m\}$ is called ***discrete aperture function***. Substituting Eq. (130) into (131), we can express Eq. (131) as

$$Z(k, \omega) = \frac{1}{2\pi} \int_{-\pi/d}^{\pi/d} \left[\sum_{m=-\infty}^{+\infty} F\left(k^* - m\frac{2\pi}{d}\right)\right] W(k - k^*)\mathrm{d}k^* \tag{133}$$

Hence, $Z(k, \omega)$ is a window-smoothing result of the spectrum $F(k, \omega)$ with $W(k)$ as the smoothing window. When the spacing d becomes too large for the spatial bandwidth, the aliasing effect will occur.

11.10.2.5 Properties of 1-D Linear Array of M Elements

For a 1-D linear array consisting of M elements, the aperture function $w_m(x)$ is

$$w_m(x) = \begin{cases} 1, & |m| \leq M_{1/2} \\ 0, & |m| > M_{1/2} \end{cases} \tag{134}$$

$M_{1/2}$ equals $M/2$ when M is even and $(M - 1)/2$ when M is odd. The spectrum in the spatial field is

$$W(k) = \sum_{|m| \leq M_{1/2}} 1 \cdot e^{jkmd} = \frac{\sin \frac{1}{2}Mkd}{\sin \frac{1}{2}kd} \tag{135}$$

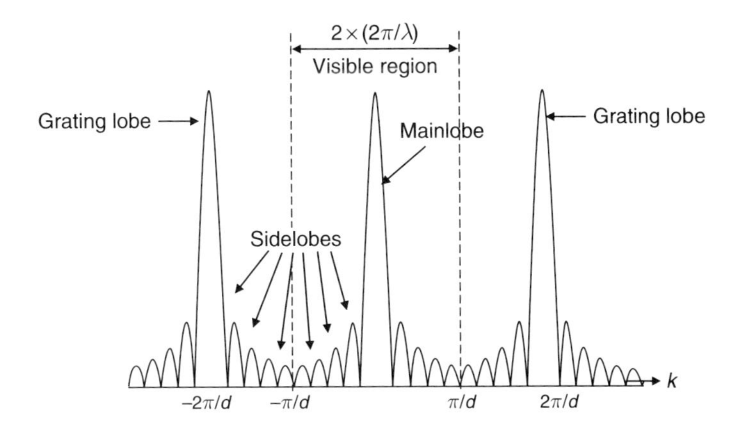

FIGURE 11.68 Aperture function magnitude spectrum of a linear array.

To obtain Eq. (135), we used the formulae

$$\sum_{n=0}^{n-1} r^n = \frac{1-r^N}{1-r} \quad \text{and} \quad \sum_{n=0}^{N-1} e^{inx} = \frac{\sin(\frac{1}{2}Nx)}{\sin(\frac{1}{2}x)} e^{ix(N-1)/2} \tag{136}$$

to get

$$\sum_{n=-N}^{N} e^{inx} = \frac{\sin(\frac{1}{2}Nx)}{\sin(\frac{1}{2}x)} \tag{137}$$

A plot of the $W(k)$ magnitude $|W(k)|$ is given in Fig. 11.68. As shown in Fig. 11.68, $W(k)$ repeats the aperture function of a continuous aperture every $2\pi/d$. Comparing Fig. 11.68 to Fig. 11.65b, we notice that the aperture function for a linear array is a periodic function of variable k with period $2\pi/d$, whereas for a finite continuous aperture it is not.

Observing Fig. 11.68, we see that, for each period, there is one mainlobe and a lot of sidelobes. The mainlobes are located at $m(2\pi/d)$, $m = 0, \pm 1, \pm 2, \ldots$ and have a height $W(0)$. The function $W(k)$ crosses zero at positive x-axis when $k = 2\pi/Md$. Therefore the mainlobe width equals $4\pi/Md$. We see that the mainlobe width decreases when M or d increase. Recalling the resolution analysis in Section 11.10.2.2, we can see that the resolution improves when M or d increases.

To avoid aliasing, the signal wavenumber k has to be less than or equal to π/d. Since $k = \omega/c$, we can state the relationship between d and ω for avoiding aliasing in the form

$$d \leq \frac{\pi c}{\omega} \tag{138}$$

In terms of wavelength λ, ($\lambda = c/f$ and $\omega = 2\pi f$), we get the relationship between the sensor spacing d and wavelength λ as

$$d \leq \frac{\lambda}{2} \tag{139}$$

Hence, the spatial aliasing is avoided as long as the sensor spacing is less than or equal to a half wavelength. This shows that though increasing the spacing d gives thinner mainlobe width and higher resolution, the spacing d cannot be enlarged without limitation since d larger than $\lambda/2$ will result in aliasing.

11.10.2.6 Grating Lobes

In previous section, we found that the wavenumber–frequency spectrum of a linear array is $2\pi/d$ periodic. As shown in Fig. 11.68, the mainlobe is repeated at the locations of any integer multiple of $2\pi/d$, e.g., $\pm 2\pi/d$. These lobes are often called ***grating lobes***. Grating lobes have the same amplitude as the mainlobe. As a result, the signals propagating at the spatial frequencies of grating lobes cannot be distinguished from the signal propagating at the mainlobe spatial frequency.

Similar to the 1-D finite continuous aperture of Section 11.10.2.2, we now consider a linear array consisting of M sensors pitched at distance d along the x-axis (Fig. 11.69). The array aperture function is $W(k_x)$ where $k_x = k_0 \cdot \cos(\phi_0) = \frac{2\pi}{\lambda_0} \cos \phi_0$. As shown, $W(k_x)$ is $2\pi/d$ periodic and grating lobes occur at the locations of integer multiples of the period, $\pm 2\pi/d, \pm 4\pi/d$, etc. For an incident wave of angle ϕ_0, we want to find the angle ϕ_{gr} corresponding to the first grating lobe locating at $2\pi/d$.

Denote

$$k_{gr} = \frac{2\pi}{d} = -\frac{2\pi}{\lambda_0} \cos \phi_0 \quad \rightarrow \quad \cos \phi_{gr} = -\frac{d}{\lambda_0}$$

Since the amplitudes of a harmonic function stay within the range $[-1, 1]$, we get the physical restriction

$$\frac{d}{\lambda_0} \leq 1 \tag{140}$$

For the upper limit when $d = \lambda$, the corresponding incident angle ϕ_{gr} is 180°, i.e., along the opposite x-axis direction. The incident angle for the mainlobe is 90° along the z-axis direction. These means that the signals coming in the x *and* z directions yield the same spatial response, i.e., the signals from these two directions are indistinguishable.

Since $k_x = 2\pi \cos \phi_0 / \lambda_0$ and $|\cos \phi_0| \leq 1$, we have $k_x \leq \pm 2\pi/\lambda_0$. This region is called the ***visible region*** for a given wavelength λ_0 with the real incident angle ranging from $-90°$ to $90°$. Recalling the no-aliasing condition, the wavelength should be $\lambda \geq 2d$. If $d < \lambda < 2d$, the spatial field is undersampled and aliasing occurs, although without grating lobes.

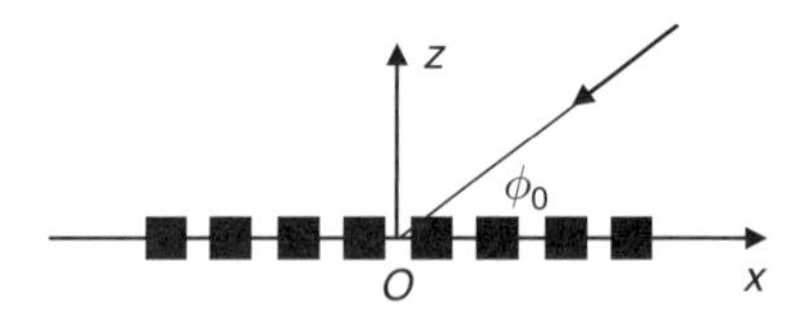

FIGURE 11.69 An 1-D linear array of M sensors separated by d with an incident wave at the direction ϕ_0.

11.11 SUMMARY AND CONCLUSIONS

This chapter has dealt with the use of PWAS transducers to achieve in-situ phased arrays for active structural health monitoring. The chapter started with a review of the phased array concepts used in conventional NDE practice. Particular attention was paid to the phased arrays based on the guided-wave principles. After this general review of the state of the art, the chapter started with the introduction of PWAS phased array principles. The simplest PWAS phased-array, the linear array, was treated first, and the concept of *embedded ultrasonics structural radar* (EUSR) was presented. The transmitter beamforming, receiver beamforming, and phased-array pulse-echo concepts were developed from first principles. The practical implementation of the EUSR algorithm, and experimental validation and calibration of this method were presented. The chapter continued with an extensive experimental studies of the linear PWAS arrays, such as EUSR detection studies performed on various defect types, detection experiments on curved panel, and in-situ direct imaging of crack growth during a fatigue cyclic loading experiment.

The optimization of PWAS phased array beamforming was studied in detail. The effect of r/d and d/λ ratios, of the number of elements M, and of the steering angle ϕ_0 was examined. The case of nonuniform PWAS phased arrays, in which the contribution of each element is individually weighted, was discussed. The binomial and the Dolph–Chebyshev arrays were studied in comparison with an uniform array, and tradeoffs between directivity and azimuthal coverage were identified. Actual experiments with nonuniform linear PWAS arrays were presented and comparatively discussed.

A generic PWAS phased array formulation was discussed next. This generic formulation presented a generic delay-and-sum algorithm that can be particularized to either far-field (parallel rays) or near-field situations. Thus the path was opened to the study of 2-D PWAS arrays, which covered most of the rest of the chapter. First, the 2-D array configuration was studied using analytical simulation. Cross-shaped, rectangular grid, rectangular fence, circular ring, and circular grid PWAS arrays were considered. Their beamforming properties were comparatively examined. Subsequently, the implementation of a rectangular grid PWAS array into the embedded ultrasonic structural radar concept was developed in the form of the 2D-EUSR algorithm. The algorithm was tested first on simulated data, and then on actual experiments. A 4×8 rectangular PWAS array, and an 8×8 square PWAS array were studied and comparatively examined. Thus, the full-circle 360° scanning properties of the 2-D PWAS phased arrays were demonstrated and understood.

The last part of the chapter was devoted to the phased-array analysis using Fourier transform methods. This approach is less intuitive than the ab initio approach adopted in the rest of the chapter. Nevertheless, the Fourier analysis approach using the space-frequency multi-dimensional Fourier transform casts a clarifying light on many of the phased array properties discussed in the previous sections of this chapter. The effects of the spatial sampling theorem, the sampling of finite apertures, and the appearance of spatial aliasing and of grating lobes were exemplified and understood.

11.12 PROBLEMS AND EXERCISES

1. What is the most important phenomenon that enables the successful use of PWAS phased arrays in conjunction with multi-modal guided waves in thin-wall structures?
2. (i) Using the straight-wavefront assumption, calculate the optimum phased-array pitch for a PWAS phased-array made up of 10-mm PWAS ideally bonded to a 1-mm thick aluminum plate ($E = 72.4\,\text{GPa}$, $\rho = 2.78\,\text{g/cc}$, $\nu = 0.33$). (ii) At what frequency would this PWAS array operate?
3. Outline the main advantages and disadvantages of the usage of 1-D and 2-D PWAS phased arrays.

4. Assume a PWAS phased array tuned to a 300-kHz S0 Lamb wave in a 1 mm thick aluminum plate ($E = 72.4$ GPA, $\rho = 2.78\,\text{g/cc}$, $\nu = 0.33$). The array contains eight PWAS elements placed linearly at 7-mm pitch. The PWAS phased array is used to detect far-field damage. Calculate the array delay times required to steer the array beam in the 40° direction
5. What are grating lobes and how do they appear?
6. Describe the effect of spatial aliasing and give a numerical example.

12

SIGNAL PROCESSING AND PATTERN RECOGNITION FOR PWAS-BASED STRUCTURAL HEALTH MONITORING

12.1 INTRODUCTION

This chapter examines two aspects of considerable importance to the structural health monitoring (SHM) process: signal processing and damage identification/pattern recognition algorithms. The SHM community is constantly exposed to various aspects of signal processing, spectra collection, data processing and analysis, pattern recognition, and decision making. This chapter will review some of these aspects. Both the state- of-the-art and general principles, on one hand, and specific examples, on the other hand, will be presented. However, an exhaustive presentation of these vast topics will not be attempted, since it would be beyond the scope and possibilities of a single chapter.

The signal-processing subject is described in terms of two major algorithmic paths: the *short-term Fourier transform* (STFT) and the *wavelet transform* (WT). Both these approaches fall into the larger class of *time-frequency analysis*. Their use is to gain a better understanding of the signal spectrum time-wise behavior and, based on this, extract that part (or bandwidth) of the signal that is of relevance to the SHM process. In many instances, the signal processing can be used as a pre- or post-processing action that enhances the efficiency of SHM techniques presented in previous chapters.

This damage identification/pattern recognition process will be discussed in the context of the analysis of spectrum features (e.g. resonance frequencies, resonance peaks). The existence of adequate algorithms would allow one to classify the spectral data into classes according to the damage state of the structure. An elementary classification problem would consist of distinguishing between 'damaged' and 'pristine' structures. More advanced techniques would allow classifying the structure's health by assessing the progress of damage based on its severity and identify the damage location. Specific examples will be presented in the context of the electromechanical (E/M) impedance method, which permits the direct determination of structural modal spectrum at high frequencies. The high-frequency modal spectra determined with the E/M impedance methods are more susceptible to incipient local damage than the relatively low-frequency spectra determined with the conventional modal analysis techniques. We will explore several issues related to the problem of damage identification during the SHM process, such as:

- Develop damage metrics for active sensors structural health monitoring
- Test and calibrate these damage metrics using controlled experiments
- Study the scale-up issues associated with the transition of the new damage metrics to full-scale structural health monitoring projects.

However, this chapter could not be exhaustive and all encompassing, since to do so would require a lot more space than available in this book. The reader with further interest in studying this subject may want to refer to specialized literature. An electronic keyword search would yield a wealth of information. The following keywords (listed in alphabetical order) are suggested: *Bayesian networks*; *damage diagnosis and remaining life prognosis*; *damage severity method*; *fault diagnostic schemes*; *fuzzy logic*; *genetic algorithms*; *hierarchical diagnosis and prognosis*; *hybrid methods*; *independent component analysis*; *inverse interpolation method*; *knowledge-based reasoning*; *machine learning and artificial intelligence*; *matched damage processing*; *matched field processing*; *matched processing method*; *neural networks*; *novelty detection method*; *probabilistic methods*; *statistical methods*; and *support vector machine method*, etc.

12.2 FROM FOURIER TRANSFORM TO SHORT-TIME FOURIER TRANSFORM

The classical Fourier transform (FT) determines the frequency contents (spectrum) of a stationary signal by comparing it with an infinite number of sine and cosine functions of different frequencies. As shown in Appendix A, the mathematical expression of the direct Fourier transform (FT) of the inverse Fourier transform (IFT) and of the inverse Fourier transform (IFT) are, respectively,

$$X(\omega) = \int_{-\infty}^{+\infty} x(t)e^{-j\omega t}\mathrm{d}t \tag{1}$$

$$x(t) = \frac{1}{2\pi}\int_{-\infty}^{+\infty} X(\omega)e^{j\omega t}\mathrm{d}\omega \tag{2}$$

The Fourier transform spectrum allows us to determine which frequencies exist in a stationary signal whose statistical parameters are constant over time. However, for nonstationary signals, it fails to tell how the frequencies evolve with time. The basic concept of short-time Fourier transform is directly developed from the classic Fourier transform. It breaks up the nonstationary signal into small segments (assuming the signal is stationary or quasi-stationary over the segments), and then applies Fourier transform to each segment to ascertain the frequencies that exist in that segment (Fig. 12.1). The totality of such spectra indicates how the spectrum is varying in time (Cohen, 1995).

The segmentation requires a windowing technique. Some details of this method are given next.

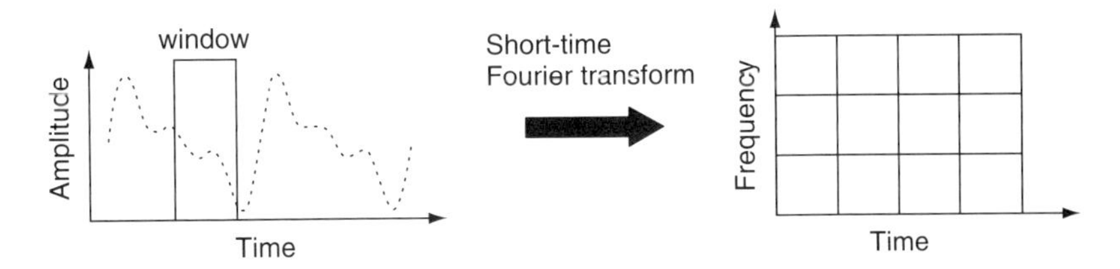

FIGURE 12.1 Schematic of windowing process and short-time Fourier transform.

12.2.1 SHORT-TIME FOURIER TRANSFORM AND THE SPECTROGRAM

Suppose we have a signal $s(t)$. To study the signal properties at time t_0, a windowing processing is applied. The window, indicated by a function $w(t)$, is centered at t_0. The windowing process will produce a modified signal

$$s_w(t, t_0) = s(t)w(t - t_0) \tag{3}$$

This modified signal is a function of the fixed time, t_0, and the varying time t.

The corresponding Fourier transform of the signal $s_w(t)$ is

$$S_w(t_0, \omega) = \frac{1}{2\pi}\int e^{-j\omega t} s_w(t, t_0)\mathrm{d}t = \frac{1}{2\pi}\int e^{-j\omega t} s(t)w(t - t_0)\mathrm{d}t \tag{4}$$

The energy density spectrum at time t_0 is

$$P_{\mathrm{ST}}(t_0, \omega) = |S_w(\omega)|^2 = |\frac{1}{2\pi}\int e^{-j\omega t} s(t)w(t - t_0)\mathrm{d}t|^2 \tag{5}$$

For each different time, we get a different spectrum; the totality of these spectra is the time–frequency distribution P_{ST}, called ***spectrogram***. Equation (4) is called ***short-time Fourier transform*** (STFT) since only the signal around the interested time t_0 is analyzed.

12.2.2 UNCERTAINTY PRINCIPLE IN SHORT-TIME FOURIER TRANSFORM PROPERTIES

As we have just mentioned, the short-time Fourier transform cuts the original signal into piecewise stationary segments and then performs the Fourier transform analysis to obtain time localization. However, the signal cannot be cut finer and finer since short-duration signals have inherently large frequency bandwidths and the spectra of such short-duration signals have little to do with the properties of the original signal due to the uncertainty principle applied to the small time intervals. The uncertainty principle states that the product of the standard deviations in time and in frequency (i.e., time and frequency resolutions) has a limited value (Cohen, 1995). The decrease (increase) in frequency resolution results in an increase (decrease) in time resolution and vice versa. Mathematically, this principle can be expressed as

$$\Delta\omega \cdot \Delta t \geq \frac{1}{2} \tag{6}$$

Normalize the signal segment in Eq. (3) as

$$\overline{x}_w(t) = \frac{x_w(t)}{\sqrt{\frac{1}{T}\int_T |x_w(t)|^2\mathrm{d}t}} = \frac{x(t)w(t - t_0)}{\sqrt{\frac{1}{T}\int_T |x(t)w(t - t_0)|^2\mathrm{d}t}} \tag{7}$$

where T is the window length. The Fourier transform of the signal segment is

$$\overline{X}_w(\omega) = \frac{1}{2\pi}\int e^{-j\omega t}\overline{x}_w(t)\mathrm{d}t \tag{8}$$

The function $\overline{X}_w(\omega)$ gives an indication of the spectral content at the time t_0. The mean time and duration for $\overline{x}_w(t)$ are

$$\langle\hat{t}\rangle = \int t\cdot|\overline{x}_w(t)|^2\,\mathrm{d}t \tag{9}$$

$$T_{t_0}^2 = \int(t-\langle\hat{t}\rangle)^2\cdot|\overline{x}_w(t)|^2\,\mathrm{d}t \tag{10}$$

Similarly, the mean frequency and bandwidth are

$$\langle\hat{\omega}\rangle = \int\omega\cdot\left|\overline{X}_w(\omega)\right|^2\mathrm{d}\omega \tag{11}$$

$$B_{t_0}^2 = \int(\omega-\langle\hat{\omega}\rangle)^2\cdot\left|\overline{X}_w(\omega)\right|^2\mathrm{d}\omega \tag{12}$$

Substituting Eqs. (10) and (12) into the uncertainty principle of Eq. (6) yields

$$B_{t_0}\cdot T_{t_0}\geq\frac{1}{2} \tag{13}$$

This is the uncertainty principle for STFT. It is a function of time t_0, the signal $x(t)$, and the window function $w(t)$. Note that this uncertainty principle shown in Eq. (13) places limits only on the STFT procedure but has no constraints on the original signal. The uncertainty principle of the original signal does not change because we modify it by windowing (Cohen, 1995).

If we cut the signal into too small parts, they will lose the connection to the original signal and therefore cannot correctly indicate the original signal's properties. Therefore, as we narrow the time window ($T_t \to 0$), Eq. (13) implies that we should expect $B_t \to \infty$, that is, for infinitely short duration signals the bandwidth becomes infinite.

12.2.3 WINDOW FUNCTIONS

Since the window function is the means of chopping up the signal, the STFT of the signal is directly affected by the properties of the window function. The choice of an adequate window function is essential for a meaningful STFT usage. By definition, the window function is a function that is zero-valued outside of a chosen interval. For instance, a function that is constant inside the interval and zero elsewhere is called a rectangular window (Fig. 12.2a). When a signal is multiplied by a window function, the product is also zero-valued outside the interval. That is to say, all that is left of the signal is what is 'viewed' through the window. Window functions are widely used in spectral analysis and filter design.

The rectangular window is very easy to implement and has excellent resolution characteristics for signals of comparable strength, but it is a poor choice for signals of disparate amplitudes. This characteristic is sometimes described as 'low dynamic range'. Due to the abrupt change at the ends of the rectangular window, its spectrum has severe leakage (right side of Fig. 12.2a). Several other often used high or moderate resolution

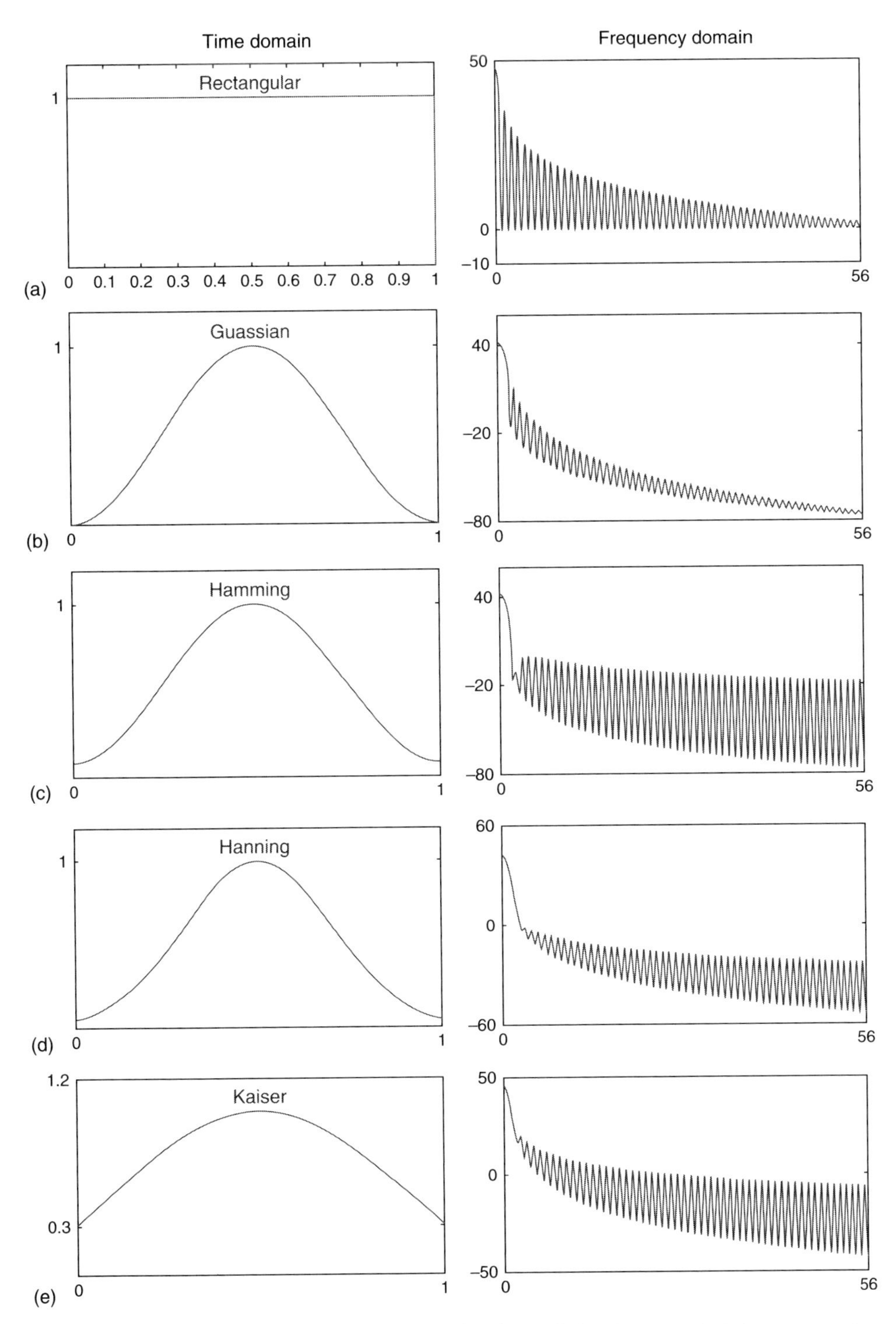

FIGURE 12.2 Several often used window functions of duration τ and their corresponding Fourier spectra: (a) a rectangular window and its Fourier spectrum; (b) a Gaussian window and its Fourier spectrum; (c) a Hamming window and its Fourier spectrum; (d) a Hanning window and its Fourier spectrum; (e) a Kaiser window and its Fourier spectrum.

windows and their corresponding Fourier spectra are also presented in Fig. 12.2 for a window length of $\tau = 1$. Their mathematical expressions are listed below

- Rectangular window $w(x) = 1,\ 0 \le t \le \tau$
- Gaussian window $w(x) = e^{-\frac{1}{2}\left(\frac{x-\frac{\tau}{2}}{\sigma\frac{\tau}{2}}\right)^2}$, with $\sigma \le 0.5,\ 0 \le t \le \tau$
- Hamming window $w(x) = 0.53836 - 0.46164\cos\left(\frac{2\pi x}{\tau}\right),\ 0 \le t \le \tau$
- Hanning window $w(x) = 0.5\left(1 - \cos\left(2\pi\frac{x}{\tau}\right)\right),\ 0 \le t \le \tau$
- Kaiser window $w(x) = I_0\left(\pi\alpha\sqrt{1-\left(\frac{2x}{\tau-1}\right)^2}\right),\ 0 \le t \le \tau$

In summary, by multiplying the original signal $x(t)$ with a window function $w(t)$ that peaks around t_0 and falls off rapidly over a limited time duration τ, the signal $[t_0 - \tau,\ t_0 + \tau]$ is emphasized inside the interval, whereas the signal outside this interval is suppressed.

12.2.4 TIME-FREQUENCY ANALYSIS BY SHORT-TIME FOURIER TRANSFORM

STFT has a direct connection to the Fourier transform, making it easy to apply and understand. It gives the time–frequency distribution of the signal by shifting a fixed–size window through the signal and taking the Fourier transform of the windowed portions. When using STFT analysis, there are two parameters affecting the piecewise stationary property of the segments:

- When cutting the original signal, we need to make sure that the segmental signals are stationary or quasi-stationary so that Fourier transform is valid over the intervals. On one side, the window should be short to ensure that the local quasi-stationary condition is satisfied within the window. On the other hand, however, if the window were too small, the uncertainty principle of Eq. (6) would generate unrealistic high-frequency content. Hence, how should one determine the window size (time interval) τ?
- When moving from one window to the next window, we should make sure that the step is not too big such that an important part of the signal is omitted, but also we should avoid making the step too small, which would result in a wide-band frequency spectrum, as mentioned by the uncertainty principle. Hence, how should one determine the window spacing (step) $d\tau$?

The STFT analysis implemented in MATLAB was used to run some comparative cases. It allowed us to select the window type (rectangular, Hanning, Hamming, or Kaiser), window size, and the step size. The window size is controlled as

$$\text{window size} = \frac{\text{length of input signal}}{\text{usercontrol}} \tag{14}$$

where the *usercontrol* is an integer selected by the user to divide the length of the input signal. Step size is how much the window moves forward in terms of data points. It is controlled as

$$\text{step} = \frac{\text{windowsize}}{\text{usercontrol}} = \frac{\text{length of input signal}}{(\text{usercontrol})^2} \tag{15}$$

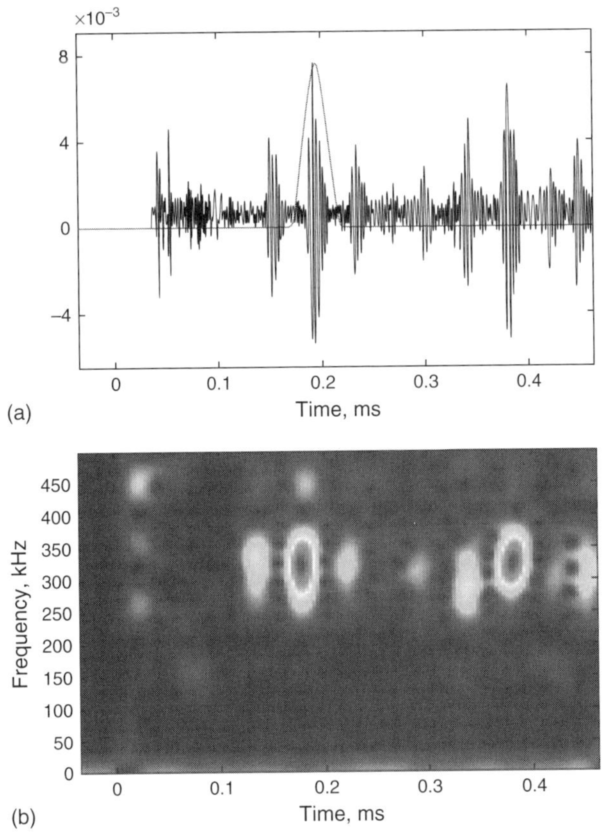

FIGURE 12.3 STFT analysis on a pulse-echo signal containing many echoes: (a) pulse-echo signal and an analyzing Hanning window with $\tau = 0.0512$ ms; (b) STFT spectrogram using the Hanning window with $\tau = 0.0512$ ms, moving at $d\tau = 0.0032$ ms.

A simple MATLAB program was developed to demonstrate how the window size τ and moving step $d\tau$ affect the STFT spectrogram by analyzing a typical pulse-echo signal collected on our thin plate specimen as shown in Fig. 12.3a. The signal is excited at 333 kHz. The spectrogram shown in Fig. 12.3b is obtained by using a Hanning window of $\tau = 0.0512$ ms, moving at $\tau = 0.0032$ ms

Figure 12.4 illustrates the effect of the STFT parameters, such as window type, window length, τ, and step $d\tau$. The spectrum in Fig. 12.4a is obtained by using a rectangular window. Observing along the frequency dimension, the rectangular window spectrum obviously has lower frequency resolution due to the leakage introduced by sidelobes.

The spectrogram in Fig. 12.4b is obtained by using the same window as in Fig. 12.3b but at larger moving interval $\tau = 0.0128$ ms. It can be seen that both spectra correctly show dark spots representing the wave packets along the time scale while correctly indicating the frequencies, which are focused on the excitation frequency around 330 kHz. However, comparing the two spectrograms, it is easy to notice that the spectrogram with larger moving step (Fig. 12.4b) is not as smooth as the one using smaller moving step (Fig. 12.3b). That is to say, the spectrogram in Fig. 12.4b has severe discontinuity. The spectrogram in Fig. 12.4b also clearly indicates other weaker wave packets around the two strongest packets close to 0.2 ms and 0.4 ms, respectively.

Figure 12.4c is obtained by using a larger window size ($\tau = 0.1024$) while moving at the same pace as in Fig. 12.3b. The spectrogram for a larger window shows better

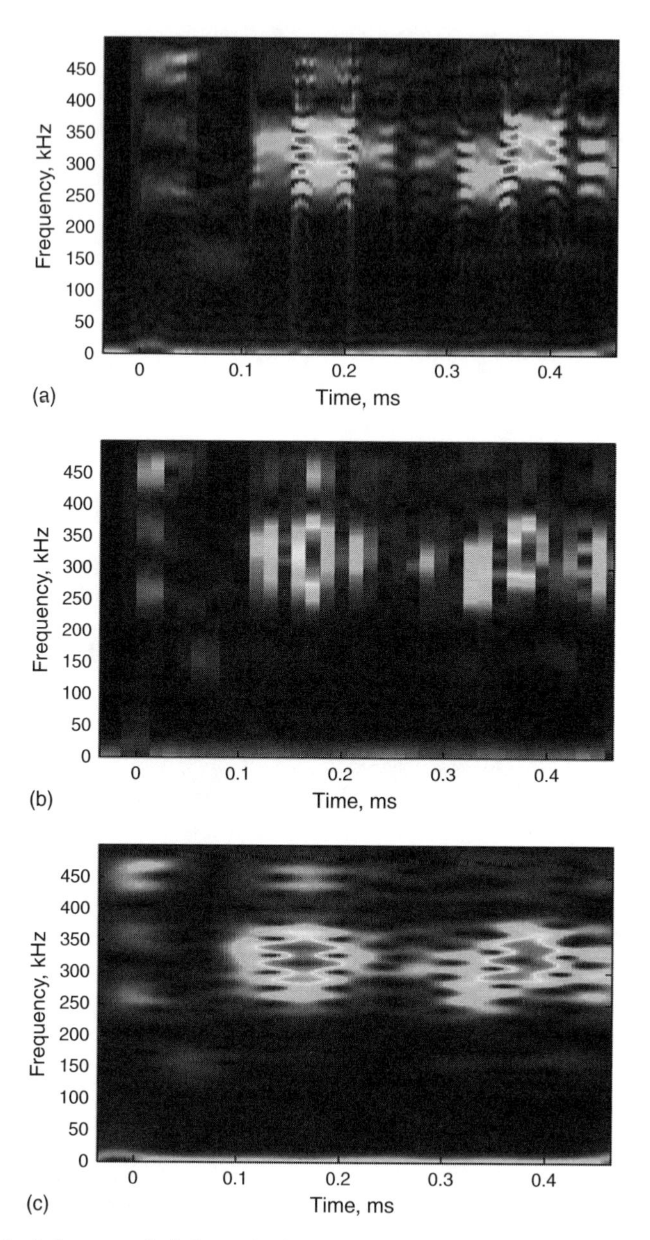

FIGURE 12.4 Influences of window selection on STFT spectrogram: (a) STFT spectrogram using a rectangular window with $\tau = 0.0512\,\text{ms}$, moving at $d\tau = 0.0032\,\text{ms}$; (b) STFT spectrogram using the Hanning window with $\tau = 0.0512\,\text{ms}$, moving at $d\tau = 0.0128\,\text{ms}$; (c) STFT spectrogram using the Hanning window with $\tau = 0.1024\,\text{ms}$, moving at $d\tau = 0.0032\,\text{ms}$.

frequency resolution (the frequency width of the strong spots are smaller compared to those in Fig. 12.4b) but larger width along the time scale. In the spectrogram, the time information of the original signal cannot be correctly revealed. Not only does it have broader time resolution, it also loses the information about the smaller wave packets.

Hence, we concluded that the window type, window size, and window moving step would affect the performance of the STFT. With smaller window size and moving step, more local details about the signal can be revealed. However, when the window size

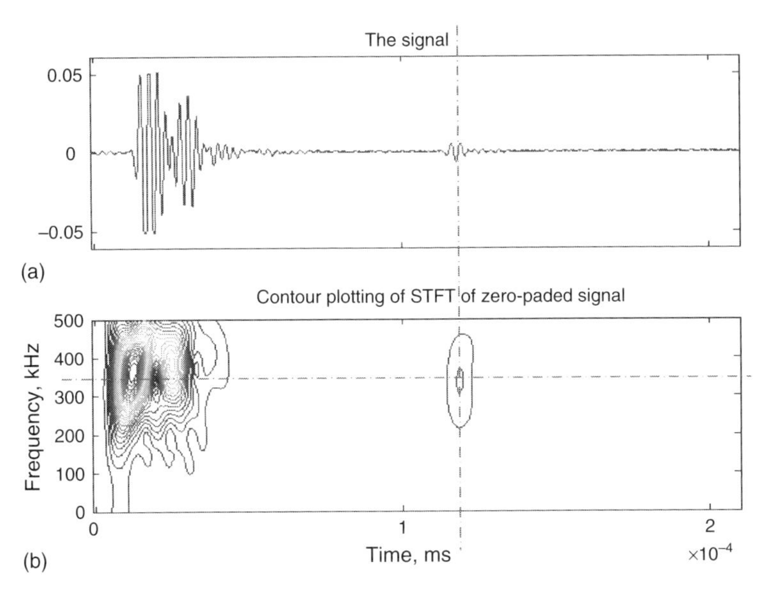

FIGURE 12.5 Short-time Fourier transform analysis: (a) original signal; (b) STFT contour plot.

becomes too small (for example, several sample points per window); a broad frequency spectrum will result and longer calculation time will be needed.

Figure 12.5 presents the results of the STFT analysis using a Hanning window and enhanced resolution. In order to enhance the frequency resolution, the original signal was preprocessed by zero padding. Default value of *usercontrol* is 16. Figure 12.5b gives the STFT spectrogram using contour plots. The magnitude of each point in the time–frequency plane is assigned to a certain contour line. It can be found that in the time–frequency plane, the strongest response mostly appears at the frequency around 330 kHz and close to the time when wave packets in Fig. 12.11a reach their peaks (local maxima). This further confirms that STFT is a useful method for the analysis of nonstationary signal due to its ability to indicate both time and frequency information.

12.2.5 THE SHORT-FREQUENCY TIME FOURIER TRANSFORM

The short-frequency time Fourier transform is the frequency domain dual of the short-time Fourier transform. We have used the STFT method to analyze the frequency properties of a signal around a time instant t_0. Conversely, we may want to study the time properties at a particular frequency (this is of important interest for dispersive guided wave applications). This can be implemented by applying a window function $H(\omega)$ in the frequency domain to the Fourier spectrum $W(\omega)$ of the signal and then using the inverse Fourier transform to obtain the corresponding time signal $x_\omega(t)$. To be focused on the frequency of interest, the window $H(\omega)$ should be narrow, or equivalently, the corresponding $h(t)$ should be broad (Cohen, 1995).

12.3 WAVELET ANALYSIS

In the previous section, we have presented the analysis of nonstationary signals using the Short-time Fourier transform (STFT) to obtain the time–frequency spectrogram. Although STFT was able to reveal how the signal's properties evolve with time, the

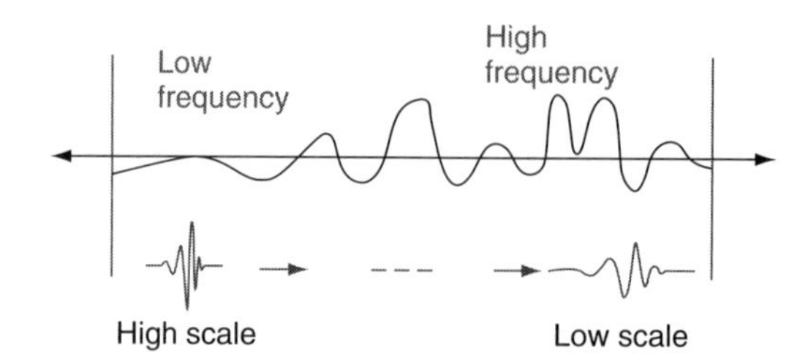

FIGURE 12.6 Wavelet transform using adjustable windows.

equal-time intervals (fixed window size) of the STFT method proved a drawback. The STFT windowing process results in a trade-off between the time and frequency resolutions and therefore accuracy cannot be simultaneously obtained in both time and frequency domains. The presence of short-duration high-frequency signal bursts occur are hard to detect, whereas in our pulse-echo damage detection, burst signals are often used. In this chapter, another time–frequency method, the wavelet transform, will be used as an alternative to overcome the disadvantages of the STFT. It is well known that the wavelets can keep track of time and frequency information, zooming in on short bursts or zooming out to detect long, slow oscillations, as demonstrated in Fig. 12.6.

As a new area in mathematics, wavelet transforms have many different applications. The wavelet transforms come in three different forms, the continuous wavelet transform, the discrete wavelet transform, and the wavelet packets (Forbes et al., 2000). Continuous wavelet transform (CWT) is a form of wavelet transform used extensively for time–frequency analysis, whereas the discrete wavelet transform (DWT) is widely used for image compression, sub-band coding, and time series analysis such as denoising. Wavelet packets can be used to look at the frequency changes in a signal over time; however, for some signals, this analysis may give very rough results. Our work was focused on the possible use of the CWT and DWT in support of signal processing for structural health monitoring with PWAS transducers.

12.3.1 TIME FREQUENCY ANALYSIS USING CONTINUOUS WAVELET TRANSFORM

12.3.1.1 Mathematical Development of Continuous Wavelet Transform

CWT is similar to the STFT, except that it uses a different functional basis. We recall that FT and STFT use a basis of an infinite number of sine and cosine functions of different frequencies. In contrast, CWT uses a functional basis consisting of dilated and shifted versions of a single basis function called ***mother wavelet***. An illustration of these concepts is given in Fig. 12.7.

The CWT of signal $x(t)$ by using mother wavelet $\psi(t)$ is

$$WT(a,\tau) = \frac{1}{\sqrt{|a|}} \int x(t)\psi^*\left(\frac{t-\tau}{a}\right)\mathrm{d}t \tag{16}$$

Where $\psi(t)$ is called the ***mother wavelet***. The name 'wavelet' implies a small wave packet, limited to finite time span. This is in contrast with the sine and cosine functions that uniformly span the whole time axis. a is the ***scaling*** (or dilation), and τ is the ***translation*** (or time shift) of the wavelet with respect to the signal. The factor $1/\sqrt{|a|}$ is introduced for energy normalization at different scales. If the mother wavelet $\psi(t)$

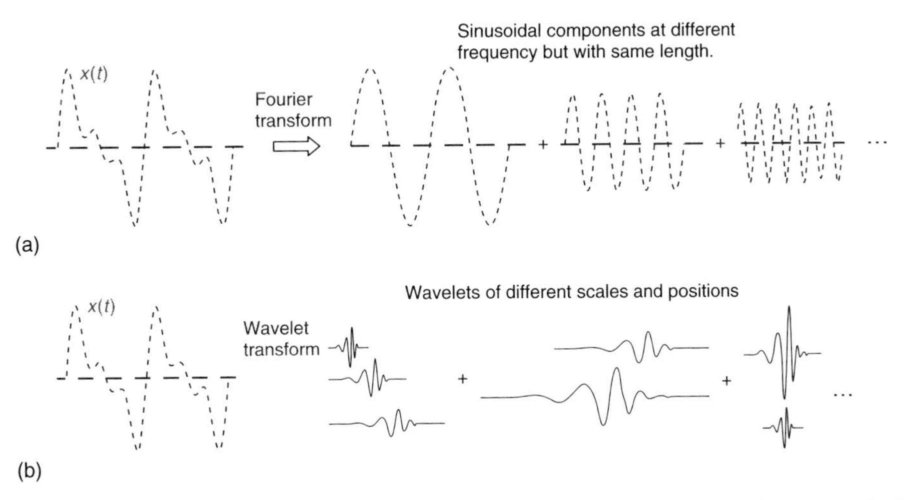

FIGURE 12.7 Analyzing a signal based on Fourier transform and wavelet transform respectively: (a) using Fourier transform with fixed window length; (b) using wavelet transform with adjustable window length.

is considered as the impulse response of a bandpass filter, then Eq. (16) allows us to understand the CWT through the bandpass analysis (Phillips, 2003):

- The CWT is the inner product or cross correlation of the signal $x(t)$ with the scaled and shifted wavelet $\psi\left[(t-\tau)/a\right]/\sqrt{a}$. This cross correlation is a measure of the similarity between signal and the scaled and shifted wavelet.
- For a fixed scale a, the CWT is the convolution of the signal $x(t)$ with the time reversed wavelet $\psi\left[-t/a\right]/\sqrt{a}$. That is to say, the CWT is the output obtained when the signal is fed to the filter with an impulse response $\psi\left[-t/a\right]/\sqrt{a}$. Therefore, Eq. (16) becomes a convolution between the signal and the scaled time reversed wavelet (containing $-t$)

$$WT(a,\tau)=\frac{1}{\sqrt{|a|}}x(t)*\psi^*\left(-\frac{t}{a}\right) \tag{17}$$

- As scale a increases, the center frequency and the bandwidth of the bandpass filter will increase (Mertins, 1999).

The resulting time–frequency representation of the magnitude squared, $|CWT(a,\tau)|^2$, is named ***scalogram***, represented as

$$\text{Scalogram:}\qquad |WT(a,\tau)|^2=\frac{1}{|a|^2}\left|\int x(t)\psi^*\left(\frac{t-\tau}{a}\right)\mathrm{d}t\right|^2 \tag{18}$$

12.3.1.2 Mother Wavelets

For wavelet analysis, we need first to determine the mother wavelet to be used. Many functions can be used as mother wavelet. Actually, if a function $f(t)$ is continuous, has null moments, decreases quickly toward 0 when t moves to $+/-$ infinity, or is null outside a segment of τ, then it can be a candidate for being a mother wavelet. In other words, the square integrable function $\psi(t)$ can be a mother wavelet if it satisfies the ***admissibility condition*** (Valens, 2004).

$$\int \frac{|\Psi(\omega)|^2}{|\omega|} \mathrm{d}\omega < +\infty \tag{19}$$

where $\Psi(\omega)$ stands for the Fourier transform of $\psi(t)$. The admissibility condition implies that the Fourier transform of $\psi(t)$ vanishes at the zero frequency, i.e.,

$$|\Psi(\omega)|^2_{\omega=0} = 0 \tag{20}$$

This means that wavelets must have a bandpass-like spectrum.

The property represented by Eq. (20) also means that the average value of the wavelet in the time domain must be zero

$$\int \psi(t)\mathrm{d}t = 0 \tag{21}$$

Therefore, such a wavelet must be oscillatory. In other words, $\psi(t)$ must resemble a ***wave packet***. (This is how the name 'wavelet' was initially introduced.) Generally, wavelets are localized waves; outside their time location, they quickly drop to zero. (By contrast, the sinusoids used in Fourier transform keep oscillating over the whole time domain.)

Another additional condition on the wavelet is the ***regularity conditions***, which makes the wavelet transform decrease quickly with decreasing scale a. The regularity condition states that the wavelet function should have some smoothness and concentration in both time and frequency domain. The explanation of regularity is quite complex; a complete explanation can be arrived at by using the concept of vanishing moments (Poularikas, 1996).

In our application, the signals to be analyzed are general nonstationary signals and the commonly used wavelets can be used for analysis. We have selected the analyzing wavelet from the wavelet functions available in the MATLAB simulation tool.

Similar to the rectangular window used in STFT, the simplest mother wavelet is a square wave named ***Haar wavelet*** (Fig. 12.8a). However, due to its abrupt change at the ends of the wavelet, it does not approximate continuous signals very well.

The ***Morlet wavelet*** is the most commonly used wavelet for time-scale analysis (Mertins, 1999). It is a modulated Gaussian function well localized in both time and frequency domain. Therefore, its resulting wavelet transform coefficients can correctly reflect the behavior of the signal at each scale and translation. Figure 12.9 shows a Morlet wavelet and its corresponding spectrum. We will use this Morlet wavelet for the CWT analysis presented in the following sections. Other often used mother wavelets also listed in Fig. 12.8 are Daubechies, Coiflet, and Symlet (MATLAB, 2005).

12.3.1.3 The Time–Frequency Resolution of CWT

Different from STFT, the time–frequency resolution of CWT depends on the scale. CWT will modify the length of the wavelets at different scale (frequency) to analyze adaptively signals with localized information. Nevertheless, the uncertainty principle of Eq. (6) still applies. At low scale (high frequency), the time resolution is good, but the frequency resolution is bad due to the fact that the wavelet is well localized in time but poorly in frequency. Similarly, at high scale (low frequencies), the frequency resolution is good, but the time resolution is poor, which means a signal with rapid changes in the time domain will be present at high frequency range in frequency domain. With such adjustable 'windows,' CWT is expected to give a better time–frequency representation than STFT.

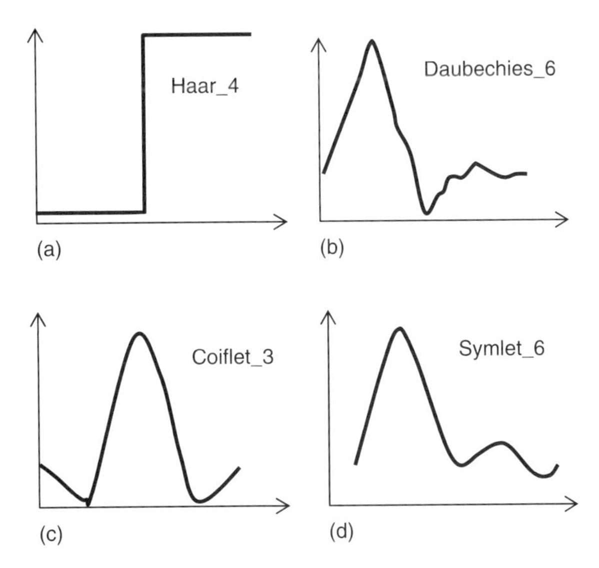

FIGURE 12.8 Several often used wavelets: (a) Haar wavelet; (b) Daubechies wavelet; (c) Coiflet wavelet; (d) Symlet wavelet.

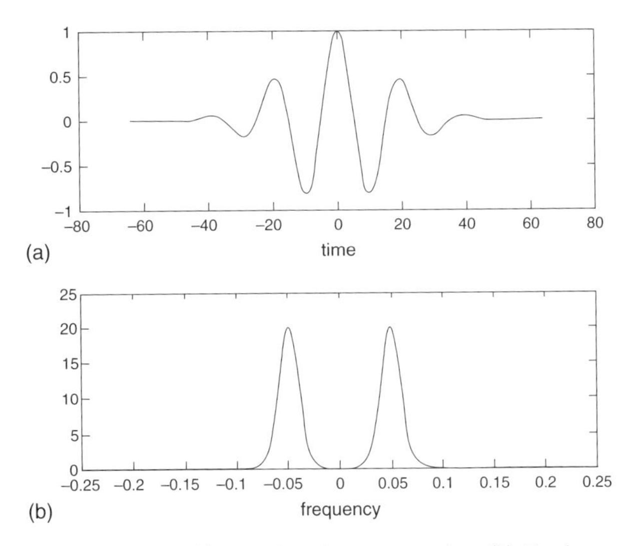

FIGURE 12.9 Morlet wavelet: (a) time domain representation; (b) Fourier magnitude frequency spectrum.

12.3.1.4 Implementation of the CWT and Comparison with STFT

Continuous wavelet transform constitutes an improvement over STFT for processing reflected nonstationary signal and identifying the echoes representing defects. The CWT implemented in MATLAB produces a spectrum of time-scale vs. amplitude called ***scalogram***. However, the scalogram cannot be used for direct time–frequency analysis. Instead, we used the relation between scale and frequency to generate a time–frequency spectrum. The scale–frequency relation is

$$f = \frac{\text{center frequency of the wavelet}}{\text{scale} \times \text{sampling interval}} \tag{22}$$

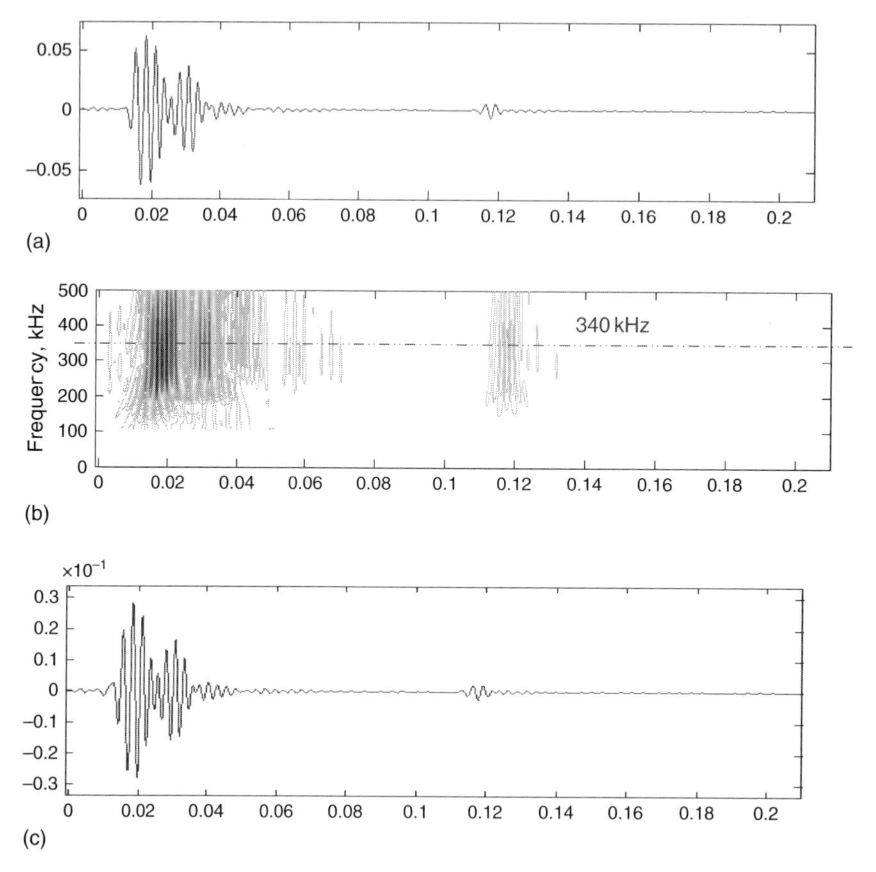

FIGURE 12.10 Continuous wavelet transform on received signal: (a) original signal; (b) CWT spectrum in scale; (c) CWT filtering of the 343 kHz frequency component.

We developed an extensive graphical user interface (GUI) to compare signals and various processing methods. GUI user controls included the selection of the mother wavelet and the scale range. An example of the CWT analysis performed on a signal received by sensor 8 in the array when sensor 1 is triggered is shown in Fig. 12.10. Contour plots were used. The contours were coded such that the numerical values are mapped to different closed loops and the density of these loops represents the magnitude of the area. Areas with high-density contours have stronger amplitudes and areas with low density contours have weaker amplitudes. The time–frequency coordinates provide a more conventional way of interpreting the data. It shows that most of the energy of our signal is concentrated on the frequency around 340 kHz. Also, the small echo at time about 0.13 ms is observed at contour plot scale of 23, which corresponds to a frequency around 340 kHz.

For a particular frequency of interest, the coefficients at corresponding scale (frequency) can be also retrieved, known as the ***CWT filtering***. Figure 12.10c shows the filtered 343 kHz frequency of the excitation level.

To understand the advantages of CWT over STFT, let us recall the STFT analysis of the wave burst signal already presented in Fig. 12.5. The recalled image is presented in Fig. 12.11.

Hence, we see that a common characteristic of STFT and CWT is that both gives two-dimensional spectra for time–frequency analysis. Yet, STFT and CWT are different in several aspects. The basic difference is that wavelets use a size-adjustable window

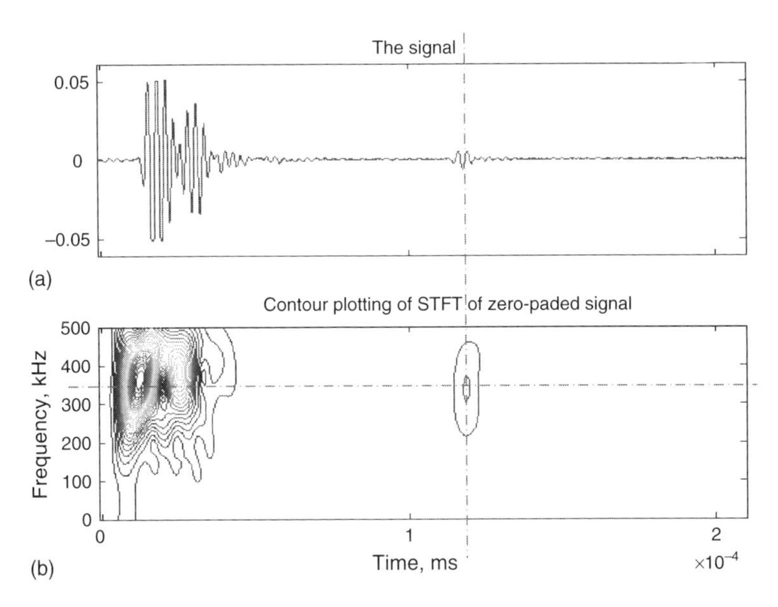

FIGURE 12.11 Short-time Fourier transform analysis: (a) original signal; (b) STFT contour plot.

that is more advantageous than the fixed window used by STFT (recall that STFT processes the signal with a sliding constant-length window). In CWT, the window length is variable. In a local area where the signal has high-frequency content, e.g., the first wave packet in Fig. 12.10a, the window will be shorter, whereas where it has a low frequency, e.g., the small echo in the middle of Fig. 12.10a, the window will be longer. In contrast, STFT cannot get the best results for signals having large frequency changes. In CWT, the window length is adjusted by the CWT algorithm according to the local frequency scale. In addition, CWT can easily extract the coefficients at a certain scale that approximately corresponds to the frequency of interest. This is useful for monitoring frequency components that are important for assessing the structural state.

In summary, the common characteristic of STFT and WT is that both methods are 2-D time–frequency or time–scale spectrums. The difference is that to get time–frequency spectrum, STFT processes signals with a sliding window yet of constant length along the time axis. In details, a window of constant length is centered at some position in the time axis and used to weight the original signal. Then the general Fourier transform is applied on the weighted signal. Next, the center of window will move to next time instant and repeat the processing so that both time information, from the center of the windows, and the frequency information can be obtained. However, for the CWT method, the difference comes from the window length. The length of each window can be adjusted based on the wavelet scale value. A time–scale spectrum rather than time–frequency spectrum is obtained in CWT.

Another difference is that STFT is based on Fourier transform. Fourier analysis consists of breaking up a signal into uniform sinusoids of various frequencies extending to $+/-$ infinity in time. In contrast, wavelet analysis breaks up a signal into a series of wavelets that are shifted and scaled. The wavelets are limited in both time and frequency.

CWT has the advantage that it can adjust the window length according to the need of real signals. Therefore, detailed information (high frequency components) can be obtained with a narrow window and general information (low frequency components) with a large window. However, the difficulty of CWT comes from the wavelets. It is hard to pick out the correct wavelets for a specific target signal. In addition, since wavelet analysis is more sophisticated, its application requires more knowledge and experience. In many instances, STFT may be the more accessible way to achieve quickly the 2-D time–frequency analysis.

12.3.2 MULTIRESOLUTION ANALYSIS (MRA)

Theoretically, the discrete wavelet transform (DWT) provides a tool for decomposing signals into elementary building blocks, which are called wavelets.

While the CWT uses arbitrary scales and arbitrary wavelets, DWT decomposes the signal into mutually orthogonal set of wavelets. The orthogonal requirement on DWT is the main difference from the CWT. The wavelets used in DWT are constructed from a scaling function that must be orthogonal to its discrete translations. This is much like Fourier transform representing signals in terms of elementary periodic waves, yet with the difference being that wavelets are aperiodic.

Since DWT expansion is localized in both time and frequency domains due to the reason that wavelets are built out of dilates and shifts of mother wavelet, DWT is considered especially useful for analysis of nonstationary or aperiodic signals.

Using the mother wavelet $\psi(t)$, a set of time-scaled and time-shifted wavelets can be obtained, called ***baby wavelets*** and defined as

$$\Psi_{m,n}(t) = 2^{m/2}\Psi(2^m t - n) \tag{23}$$

For these baby wavelets, a set of values $C(1/2^m, n/2^m)$ are

$$C(1/2^m, n/2^m) = \langle x(t), \Psi_{m,n}(t) \rangle \tag{24}$$

Where $\langle f(t), g(t) \rangle = \int f(t)g(t)\mathrm{d}t$ is the inner product of $f(t)$ and $g(t)$.

In the case of an orthogonal wavelet, an analysis formula called discrete wavelet transform is defined

$$\text{DWT:} \quad c_{m,n} = \int_{-\infty}^{+\infty} x(t)\Psi_{m,n}(t)\mathrm{d}t \tag{25}$$

In addition, the synthesis of the signal is

$$x(t) = \sum_m \sum_n c_{m,n}\Psi_{m,n}(t) \tag{26}$$

Based on DWT, the multiresolution analysis is derived. Define the space W_m to be set of all signal $x(t)$ which can be synthesized from the baby wavelets $\Psi_{m,n}(t)$, $-\infty < n < +\infty$. These spaces are orthogonal to each other and any signal $x(t)$ can be synthesized as

$$x(t) = \sum_{m=-\infty}^{+\infty} x_m(t) \tag{27}$$

$$x_m(t) = \sum_{n=-\infty}^{+\infty} c_{m,n}\Psi_{m,n}(t) \tag{28}$$

Another way to express MRA is to define V_m to be the set of all signals $x(t)$ which can be synthesized from all the baby wavelets $\Psi_{k,n}(t)$, where $k < m$ and $-\infty < k < +\infty$. Therefore,

$$x(t) = \sum_{k=-\infty}^{m-1} \sum_n c_{k,n}\Psi_{k,n}(t) \tag{29}$$

The spaces V_m are nested inside each other, i.e., $\{0\} \subset \ldots \subset V_{-2} \subset V_{-1} \subset V_0 \subset \ldots L^2$. As m goes to infinity, V_m enlarges to become all energy signals (L^2) and as m goes to negative infinite, V_m shrinks down to zero. It is clear from the definitions that every signal in V_{m+1} is a sum of a signal in V_m and W_m since

$$x(t) = \sum_{k=-\infty}^{m} \sum_n c_{k,n} \Psi_{k,n}(t) = \sum_{k=-\infty}^{m-1} \sum_n c_{k,n} \Psi_{k,n}(t) + \sum_n c_{k,n} \Psi_{k,n}(t) \tag{30}$$

Hence,

$$V_{m+1} = V_m + W_m \tag{31}$$

Equation (31) shows that the space W_m is the difference in the subspace sense between adjacent spaces V_m and V_{m+1}. The spaces W_m and V_m can be visualized as shown in Fig. 12.12.

Hence, the multiresolution analysis refers to the signal processing in the nested sequence of subspaces. To decompose a signal $x(t)$ in space V_0, we write the following nested expressions

$$\begin{aligned} V_0 &= V_{-1} + W_{-1} \\ &= (V_{-2} + W_{-2}) + W_{-1} \\ &= [(V_{-3} + W_{-3}) + W_{-2}] + W_{-1} \\ &= \{[(V_{-4} + W_{-4}) + W_{-3}] + W_{-2}\} + W_{-1} \end{aligned} \tag{32}$$

Then, the various decompositions of $x(t)$ are

$$\begin{aligned} x(t) &= A_1(t) + D_1(t) \\ &= (A_2(t) + D_2(t)) + D_1(t) \\ &= [(A_3(t) + D_3(t)) + D_2(t)] + D_1(t) \\ &= \{[(A_4(t) + D_4(t)) + D_3(t)] + D_2(t)\} + D_1(t)_1 \end{aligned} \tag{33}$$

Where $D_i(t)$ is in W_{-i}, called the ***details*** at level i and $A_i(t)$ is in V_{-i}, called the ***approximation*** at level i.

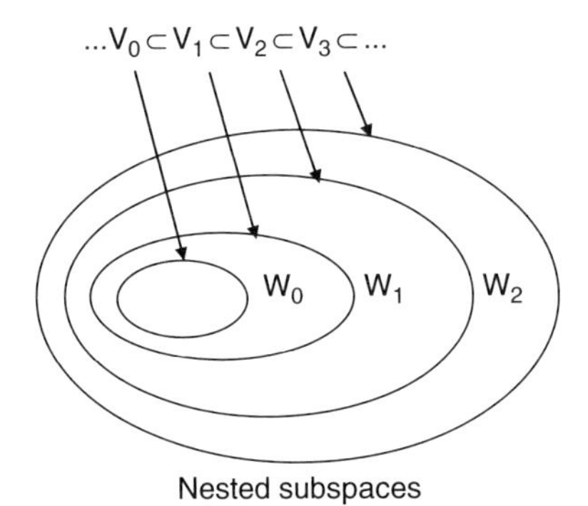

FIGURE 12.12 Nested subspaces formed by W_m and V_m.

According to ***the 2-scale property of multiresolution*** which states that, for a signal $x(t)$ in the space V_m, if and only if $x(2t)$ is in the next space V_{m+1}, i.e.,

$$\Psi_{m,n}(2t) = \frac{1}{\sqrt{2}}\Psi_{m+1,n}(t) \tag{34}$$

The wavelet $\Psi(t)$ has a scaling function $\phi(t)$ that produces the multiresolution subspaces V_m. The baby scaling functions are defined as

$$\phi_{m,n}(t) = \sqrt{2^m}\phi(2^m t - n) \tag{35}$$

The scaling function has a property called the ***2-scale equation*** and it gives rise to a filter with coefficients $h_0(t)$ such that

$$\phi(t) = \sum_k h_0(k)\sqrt{2}\phi(2t - k) \tag{36}$$

Similarly, for the wavelet, by using another filter $h_1(t)$, the two-scale equation is

$$\phi(t) = \sum_k h_0(k)\sqrt{2}\phi(2t - k) \tag{37}$$

12.3.2.1 Discrete Wavelet Transform (DWT)

As mentioned in the previous MRA, the discrete wavelet transform (DWT) is a representation of a signal $x(t)$ using the orthonormal basis consisting of a countable infinite set of wavelets (Schniter, 2005). Denoting the wavelet basis as $\psi_{m,n}(t)$, the DWT is

$$x(t) = \sum_{m=-\infty}^{+\infty} \sum_{n=-\infty}^{+\infty} c_{m,n}\Psi_{m,n}(t) \tag{38}$$

The DWT coefficient $c_{m,n}$ is given by

$$c_{m,n} = \int_{-\infty}^{+\infty} x(t)\Psi^*_{m,n}(t)\mathrm{d}t \tag{39}$$

Here wavelets are orthonormal functions obtained by shifting and dilating a mother wavelet $\psi(\mathrm{t})$

$$\Psi_{m,n}(t) = 2^{-m/2}\Psi(2^{-m}t - n) \tag{40}$$

As m increases, the wavelets stretch by a factor of two and as n increases, the wavelets shift to the right. By this means, wavelet $\psi(\mathrm{t})_{m,\mathrm{n}}$ is constructed such that it describes a particular level of details in the signal $x(t)$. As m becomes smaller, the wavelet becomes finer 'grained' and the level of details increases. Therefore, the DWT can give a multiresolution description of a signal, which is very useful in analyzing practical signals.

Using the scaling function $\phi(t)$, the definition of DWT can be represented in another way. We intend to approximate a signal $x(t)$ closely by projecting it into subspace $V_{J,m}$ using the basis $\phi_{J,l}(t)$:

$$cA_0(l) = \int_{-\infty}^{+\infty} x(t)\phi_{J,l}(t)\mathrm{d}t \tag{41}$$

The projection coefficients, $cA_0(n)$, is the approximation of the original signal $x(t)$. The signal $x(t)$ can be approximately recovered as

$$x(t) \approx \sum_l cA_0(l)\phi_{J,l}(t) \tag{42}$$

After the approximation, the signal is now in the subspace V_J. It can be decomposed using the subspaces V_{J-k} and W_{J-k} together with their bases $\phi_{J-k,n}(t)$ and $\Psi_{J-k,n}(t)$. When the scale gets larger, the index $J-k$ gets more negative. Assuming $k=1$ yields

$$V_J = W_{J-1} + V_{J-1} \tag{43}$$

Using the bases $\phi_{J-1,n}(t)$ and $\Psi_{J-1,n}(t)$, the signal $x(t)$ can be represented as

$$\begin{aligned} x(t) &= \sum_l cA_0(l)\phi_{J,l}(t) = \sum_n cA_1(n)\phi_{J-1,n}(t) + \sum_n cD_1(n)\phi_{J-1,n}(t) \\ &= A_1(t) + D_1(t) \end{aligned} \tag{44}$$

$A_1(t)$ and $D_1(t)$ are the approximation and detail at level 1 respectively. The approximation $A_1(t)$ can be further approximated in the same way as Eq. (44). This is the basis of DWT decomposition.

12.3.2.2 Practical Implementation of DWT Algorithm

In wavelet analysis, we use the bases $\phi_{m,n}(t)$ and $\Psi_{m,n}(t)$ to decompose a signal into approximations and details. The approximations are the high-scale, low-frequency components of the signal, which could be used to recover the clean signal. The details are the low-scale, high-frequency components, which could be used to recover the noise. In signal processing, the calculation of DWT is implemented by passing it through a series of filters, called ***filter banks***.

The first step is to pass the signal through two different filters (Fig. 12.13):

- The signal is first passed through a lowpass filter with impulse response $g[n]$.
- Simultaneously the signal is also decomposed using a highpass filter $h[n]$.
- The outputs consist of the detail coefficients from the highpass filter and approximation coefficients from the lowpass filter.

Since in $g[n]$, half the frequencies of the signal $x[n]$ have been removed, half the signal samples can be discarded with Nyquist rule. Therefore, the filter outputs are ***downsampled*** by a factor of 2 (Fig. 12.14):

$$y_L[n] = \sum_{k=-\infty}^{+\infty} x[k]\cdot g[2n-k] \tag{45}$$

$$y_H[n] = \sum_{k=-\infty}^{+\infty} x[k]\cdot h[2n-k] \tag{46}$$

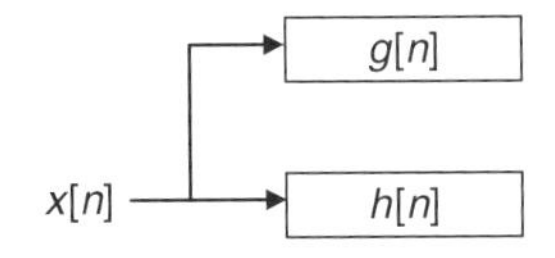

FIGURE 12.13 Using lowpass and highpass filters for DWT decomposition.

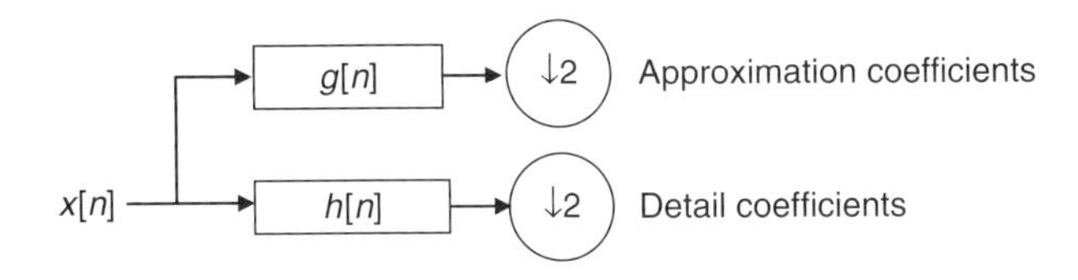

FIGURE 12.14 Downsampling process for DWT decomposition.

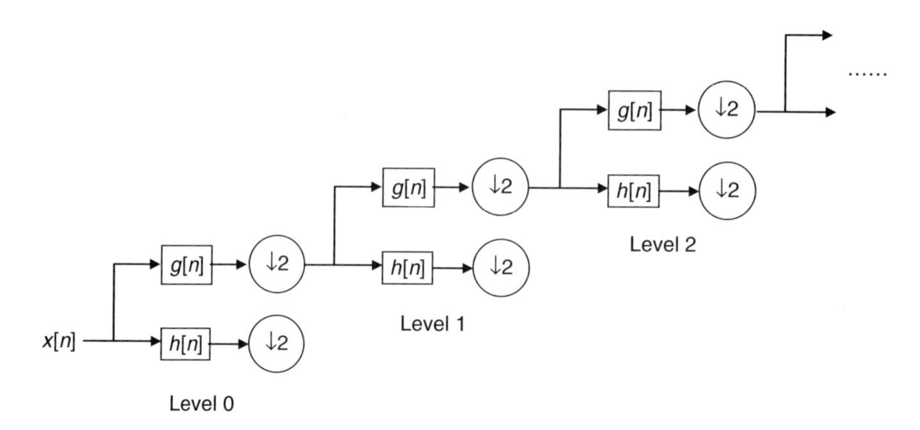

FIGURE 12.15 DWT decomposition implemented as a set of filter banks.

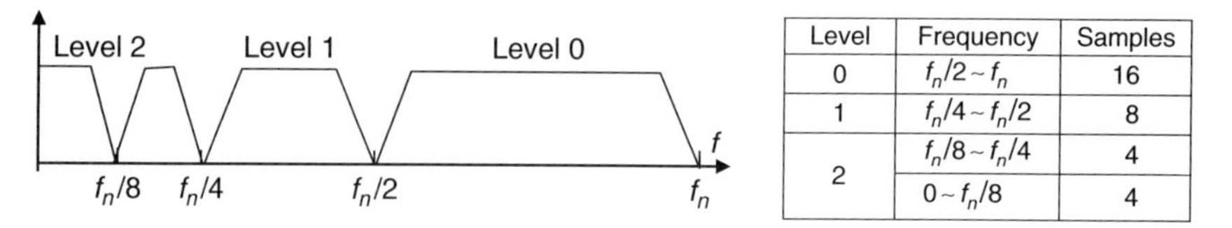

Level	Frequency	Samples
0	$f_n/2 \sim f_n$	16
1	$f_n/4 \sim f_n/2$	8
2	$f_n/8 \sim f_n/4$	4
	$0 \sim f_n/8$	4

FIGURE 12.16 Frequency ranges resulted by DWT decomposition.

Notice that the decomposition process has halved the time resolution since only half of each filter output characteristics is kept. The halved time resolution also results in a doubled frequency resolution thereby.

The decomposition will be repeated to further increase the frequency resolution. The approximation coefficients are continually decomposed with high- and low-pass filters and then downsampled. This is represented as a binary tree with nodes representing a subspace with different time-frequency localization, known as the filter bank (Fig. 12.15). At each levels of the filter bank, the signal is decomposed into low and high frequencies. Due to the decomposition processes, the signal with a scale of 2^n, where n is the number of level, will generate a frequency domain representation as shown in the Fig. 12.16 (given a signal with 16 samples, frequency range $0 \sim f_n$, and 3-level decomposition).

12.3.3 DENOISING USING DIGITAL FILTERS AND DWT

The signals collected during nondestructive testing experiments are always contaminated by the environmental or measuring disturbances, known as the noise. For most engineering signals, the low-frequency content is the most important part that engineers are interested in, whereas the high-frequency content, on the other hand, forms the noise. However, for high-frequency guided waves used in NDE/SHM, extra attention has to be paid to separate the useful high frequency component from the noise. For denoising applications, we first used the traditional digital filters; then we tried the DWT denoising method.

12.3.3.1 Digital Filters

Digital filters are widely used in signal processing to remove or to keep certain parts of the signal. Digital filters are uniquely characterized by their frequency responses $H(\omega)$ in the frequency domain, which is the discrete time Fourier transform of the time response $h(t)$ (Vaseghi, 1999). Based on different types of $h(t)$ design, digital filters are categorized into two groups, known as ***finite-duration impulse response*** (***FIR***) filters, for which the impulse response is nonzero for only a finite number of samples, and ***infinite-duration impulse response*** (***IIR***), for which the impulse response has an infinite number of nonzero samples. IIR filters are also known as the feedback filters since their filter coefficients include feedback terms in a difference equation. Compared to FIR filters, IIR filters can achieve the desired design specifications with a relatively lower order so that fewer unknown parameters need to be computed and stored, which might lead to a lower design and implementation complexity.

When designing a filter for a specific application, we are always expecting a highly frequency selective filter with sharp cutoff edges, or short transition bands. However, ideal sharp edges correspond to mathematical discontinuities and cannot be realized in practice. Therefore the ideal filter design looks for an implementable filter whose frequency response $H(\omega)$ best approximates the specified ideal magnitude and phase responses. For most filter designs, there are tradeoffs among competing parameters.

12.3.3.2 Denoising Using Digital Filters

The simplest filter is an ideal filter with zero phase. Four commonly used ideal frequency responses are (1) lowpass filter; (2) highpass filter; (3) bandpass filter; (4) bandstop filter, as shown in Fig. 12.17.

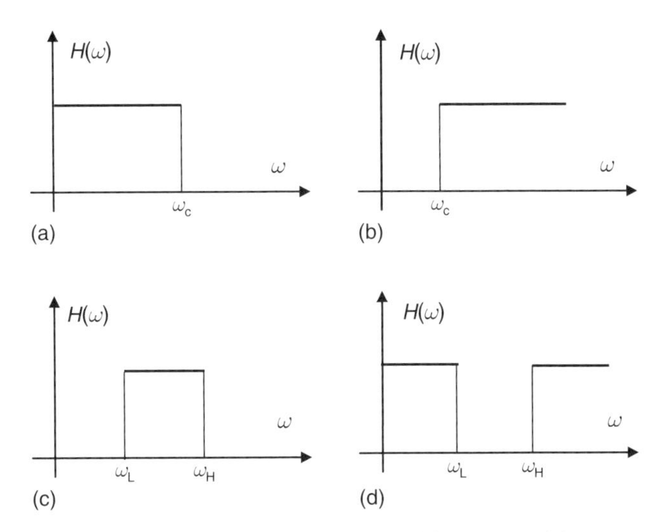

FIGURE 12.17 Common ideal digital filter types. (a) lowpass; (b) highpass; (c) bandpass; (d) bandstop.

The corresponding frequency responses can be expressed as

$$\text{Lowpass} \qquad H(\omega) = \begin{cases} 1, & \omega < \omega_c \\ 0, & \omega \geq \omega_c \end{cases} \tag{47}$$

$$\text{Highpass} \qquad H(\omega) = \begin{cases} 0, & \omega < \omega_c \\ 1, & \omega \geq \omega_c \end{cases} \tag{48}$$

$$\text{Bandpass} \qquad H(\omega) = \begin{cases} 1, & \omega_L < \omega < \omega_H \\ 0, & \text{Otherwise} \end{cases} \tag{49}$$

$$\text{Bandstop} \qquad H(\omega) = \begin{cases} 1, & \omega < \omega_L \text{ or } \omega > \omega_H \\ 0, & \text{Otherwise} \end{cases} \tag{50}$$

In actual implementations, the ideal filters need to be approximated with a realizable shape. That is to say, the sharp cutoff edges need to be replaced with ***transition bands*** in which the designed frequency response would change smoothly from one band to another. Therefore, the cutoff edges are replaced with nonzero width transition bands located around the ideal cutoff edges. An example of an implementable lowpass filter is shown in Fig. 12.18.

Here, ω_p is the passband cutoff frequency and ω_s is the stopband cutoff frequency. The resulting transition band has a width of $\Delta\omega = \omega_s - \omega_p$. In the transition band, the frequency responses is desired to change smoothly, i.e., without fluctuations or overshoots. This requirement may be satisfied by a design with constraints on such transition bands. As shown in the Fig. 12.18, δ_p is the passband ripple and the maximum allowable error in the passband, whereas δ_s is the stopband ripple and the maximum allowable error in the stopband.

In digital filters, the filter order N is a variable to be used in the filter design for achieving optimal filtering performance. The realizable filter can be found by optimizing the width of the transition bands or reducing the passband and/or stopband error, etc.

$$H(\omega) = \frac{B(\omega)}{A(\omega)} = e^{-j\omega N} \frac{\sum_{k=0}^{M} b_k e^{-j\omega k}}{\sum_{k=0}^{N} a_k e^{-j\omega k}} \tag{51}$$

Equation (51) is the frequency response $H(\omega)$ of an IIR filter at order N. The order N of the IIR filter determines the number of previous output samples that need to be stored and then fed back to compute the current output samples.

In addition, many practical filter designs are specified in terms of constraints on the magnitude responses and no constraints on the phase response. Then, the objective of filter design is to find a functional filter (either IIR or FIR) whose magnitude frequency response $H(\omega)$ approximates the given specified design constraints. There are four classical filter functions available in most simulation software: (1) Butterworth, (2) Chebyshev I, (3) Chebyshev II, (4) Elliptic. Each filter has its own characteristics in terms of performance as listed in Table 12.1.

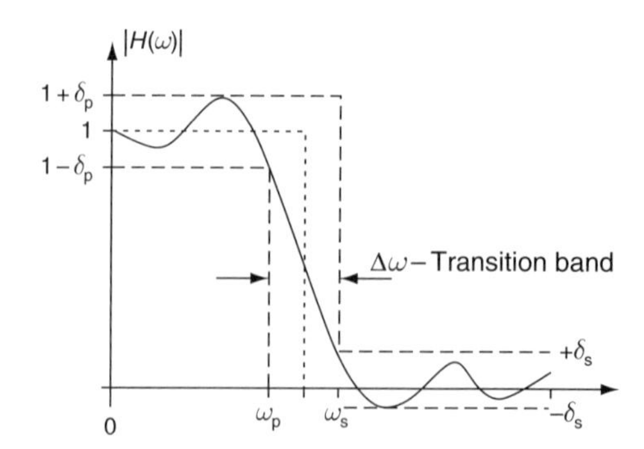

FIGURE 12.18 Frequency response of an implementable lowpass filter.

TABLE 12.1 Typical commonly used digital filters

Filter type	MATLAB function	Short description	Illustration
Butterworth	Butter	It is characterized by a magnitude response that is maximally flat in the passband and monotonic overall.	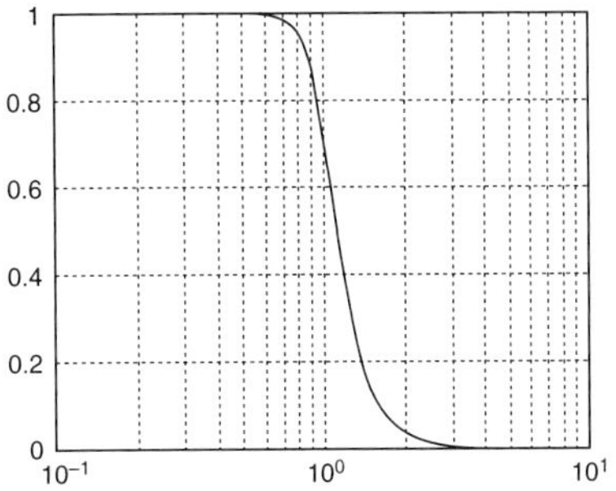
Chebyshev I	Cheby1	It has equi-ripples in the passband and monotonic in the stopband. It rolls off faster than type II, but at the expense of greater deviation from unity in the passband.	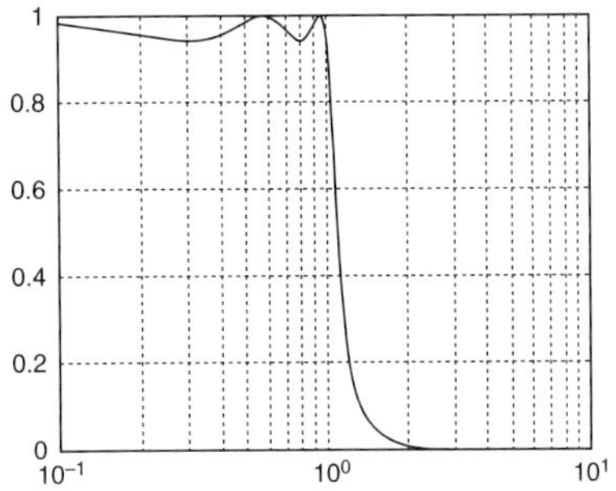
Chebyshev II	Cheby2	It is monotonic in the passband and equi-ripple in the stopband. Type II does not roll off as fast as type I, but it is free of passband ripple.	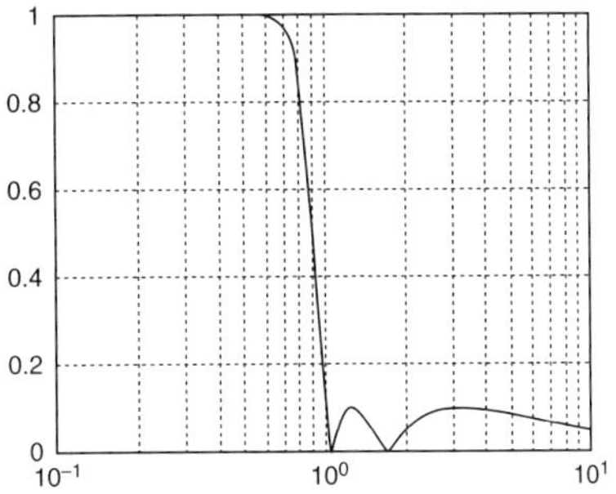
Elliptic	ellip	It offers steeper roll-off than Butterworth or Chebyshev, but is equi-ripple in both pass and stop bands. Elliptic filters can meet given specification with lowest filter order of any type.	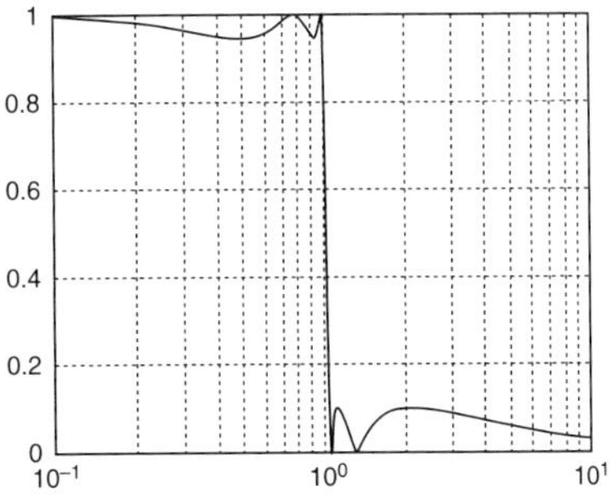

Denoising studies were performed by using a digital lowpass filter to remove the high frequency noise. The Chebyshev II filter was used because it has a monotone behavior in the low-pass band and a relatively short transition band. Other parameters such as the filtering mode, cutoff frequency, and filter order were also optimally adjusted.

To compare the denoising performance of various methods, we used as data the statistical denoising examples of Demirli and Saniie (2001). We simulated an ideal wave-packet signal with a carrier frequency of 2 MHz, sampling it at 100 MHz over 2 μs duration. Gaussian white noise (GWN) was added at certain signal-to-noise ratio (SNR) values. The original clean signal is modeled as

$$x(t) = \beta e^{-\alpha(t-\tau)^2} \cos\left[2\pi f\,(t-\tau)+\phi\right] \tag{52}$$

The signal parameters include the bandwidth factor α, the arriving time τ, the carrier frequency f_c, phase ϕ, and amplitude β. The energy of the signal can be formulated simply as

$$E_x = \frac{\beta^2}{2}\sqrt{\frac{\pi}{2\alpha}} \tag{53}$$

The SNR is determined explicitly in the case of GWN with variance σ_v^2 as

$$\mathrm{SNR} = 10\log\left(\frac{E_x}{\sigma_v^2}\right) \tag{54}$$

Hence, given SNR, the variance of the given GWN can be found as

$$\sigma_v^2 = \frac{E_x}{10^{\mathrm{SNR}/10}} \tag{55}$$

Using Eq. (55), a zero mean GWN at certain SNR can be determined and added to the original signal to construct the noise-contaminated signal.

$$\mathrm{GWN}(t) = \sigma\cdot N(0,1) = \sqrt{\frac{\beta^2}{2}\cdot\frac{\sqrt{\frac{\pi}{2\alpha}}}{10^{\mathrm{SNR}/10}}}\cdot N(0,1) \tag{56}$$

Where $N(0, 1)$ is random number observed standard distribution. An original signal with noise is simulated with $\alpha = 25$, $\tau = 1.07\,\mu s$, $\phi = 0.87\,\mathrm{rad}$, $\beta = 1$, and $\mathrm{SNR} = 5$. It is assembled using Eqs. (52) and (56), i.e.,

$$x_{\mathrm{Noise}}(t) = x(t) + GWN(t) \tag{57}$$

Signal $x_{\mathrm{Noise}}(t)$ is shown in Fig. 12.19a.

The denoising was done by using a Chebyshev type II bandpass filter with order 6 and cutoff frequencies at 3 MHz and 8 MHz (Fig. 12.19b). Result is shown in Fig. 12.19c.

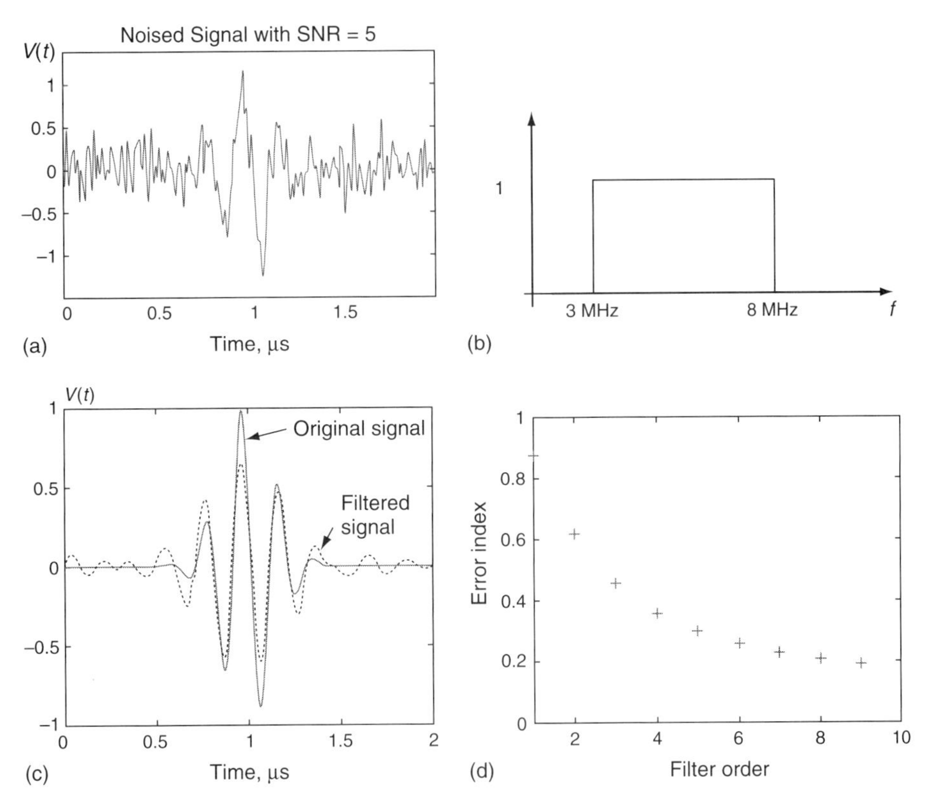

FIGURE 12.19 Denoising with filtering: (a) original signal at SNR = 5; (b) ideal Chebyshev II filter with cutoff frequency at 3 MHz and 8 MHz; (c) filtered signal (dot line) at order 8; (d) error index.

We see that the filtered signal has big deviation from the real signal. Part of the signal centered in the middle was lost, whereas large disturbances appeared as side lobes. To explore how the filter order affects the denoising, an error index was used, defined as

$$\varepsilon = \frac{\sum |fx_i - x_i|}{\sum |x_i|} \tag{58}$$

Where fx is the filtered signal. Error index is plotted at various filter orders (Fig. 12.19d). It shows with larger filter order, the error index gets decreased. So higher filter order is desired for better denoising. However, considering the practical feasibility of implementation, the higher the filter order, the harder is to build such a filter. It is not good to increase the order unlimitedly.

Though digital filtering is relatively easier to implement, it does not always provide sufficient accuracy and flexibility. In the following sections, we will investigate other denoising approaches based on the wavelet transform.

12.3.3.3 Denoising Using DWT

Though digital filtering is a classical method widely used to denoise, the filtered signal obtained by using a digital filter of order 8 (Fig. 12.19b) is far from 'clean' compared to the ideal signal. As we know, multilevel discrete wavelet transform (DWT) decomposition is one of the popular methods used to reduce noise. For example, Fig. 12.20 illustrates the

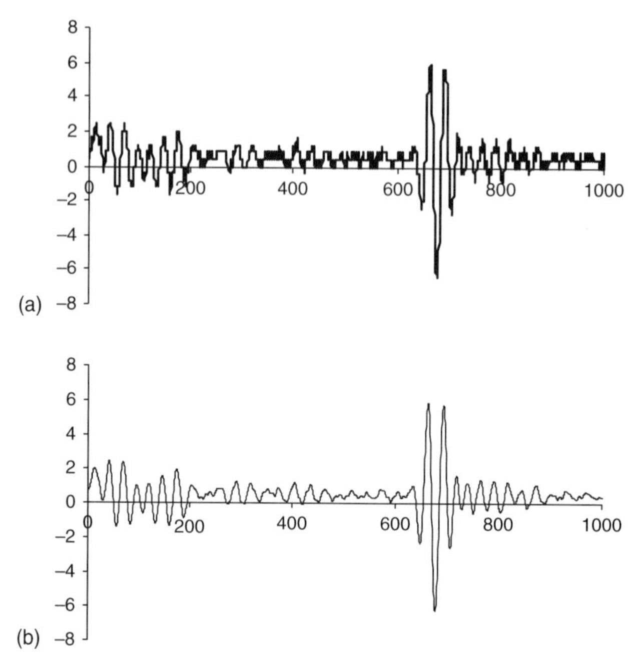

FIGURE 12.20 Discrete wavelet transform for denoising: (a) original signal; (b) clean signal after DWT denoising.

DWT denoising of an experimental signal captured by us during EUSR tests. To further understand the use of DWT denoising, we will use again the noisy signals constructed with Eq. (57) and will denoise them with the DWT method.

The DWT denoising efficiency was found to be affected by two factors: SNR and sampling frequency. The higher the SNR and/or the sampling frequency are, the more efficient the DWT denoising is. The results of DWT denoising for signals at various SNR values are presented in Fig. 12.21. We have found that at high SNR (i.e., less noise content in the signal), the DWT worked well and eliminated most of the noises. As SNR decreased, the quality of DWT denoising also decreased. At this situation, densification processing is necessary for improved denoising result.

DWT denoising has been employed to remove noise and improve precision of experimental signals. Fig. 12.21a shows a part of the original signal captured in our experiments. Fig. 12.21b shows the signal cleaned through DWT denoising using Morlet wavelet. It is seen that the high frequency local disturbance has been removed from the original signal and the curve has become much smoother.

The contaminated signal previously processed by the digital filter was used again, as presented in Fig. 12.22a. Since the sampling rate was 100 MHz, only 200 data points were recorded during the 2-μs duration. This was found to be insufficient for proper analysis. To overcome this, we increased the number of points (and hence the original sampling frequency) by inserting extra points of zero amplitude. The number of points inserted between each pair of original points was m. Thus, the total number of points was increased from N to $N^* = N + m \cdot (N - 1)$. We call m the densification factor. Figure 12.22 shows the comparisons of signal denoising with DWT at different levels and with different densification factors.

The original signal had SNR = 5. Figure 12.22b shows the DWT denoised signal with a densification factor $m = 8$ superposed on the ideal signal. Fig 12.22c also shows the

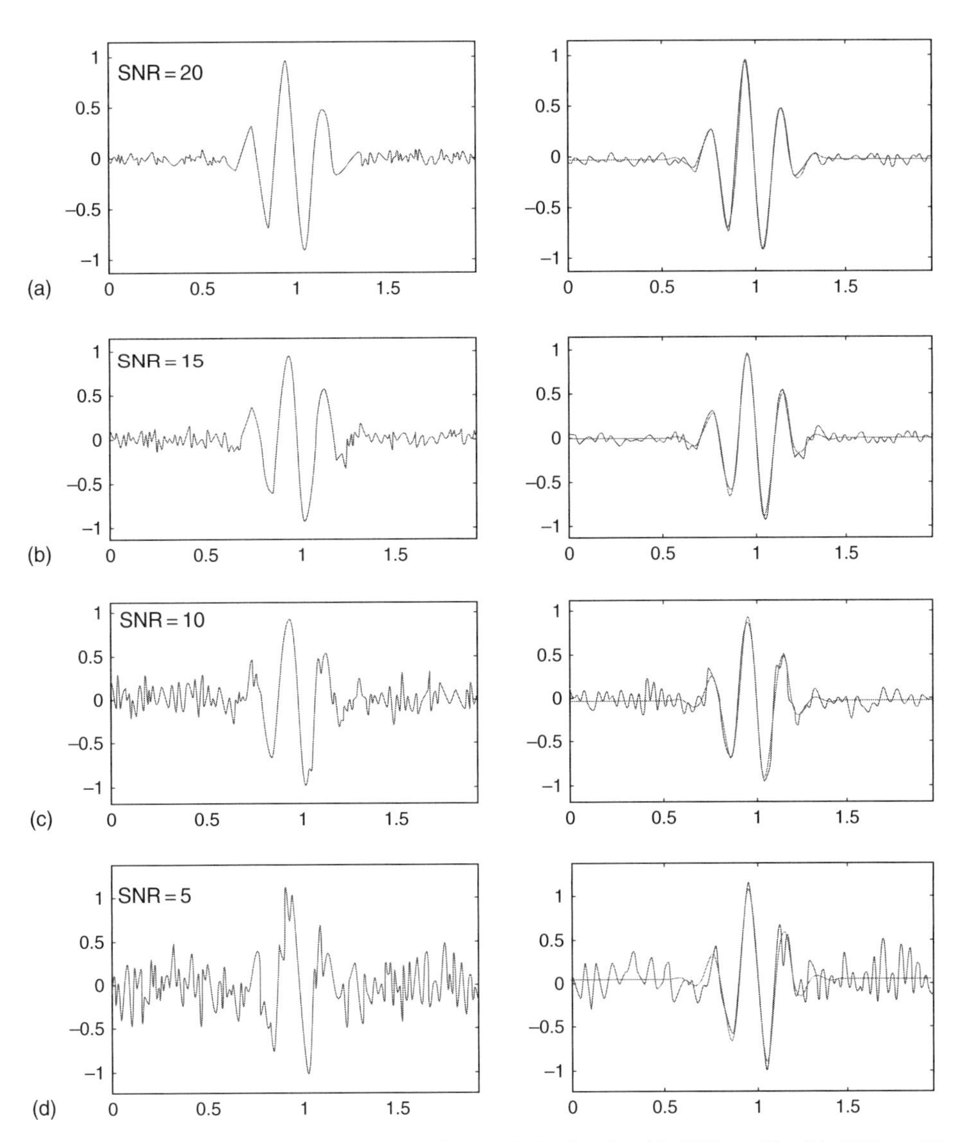

FIGURE 12.21 DWT denoising at different SNR levels: (a) SNR = 20; (b) SNR = 15; (c) SNR = 10; (d) SNR = 5.

DWT-denoised signal with a higher densification factor, $m = 50$, superposed on the ideal signal. Note that, for the plot in Fig. 12.22c, the difference between ideal signal and denoised signal is imperceptible. Figure 12.22d plots the error index (logarithm scale) for densification of $m = 8$ and $m = 50$, respectively. Another effect noticed in Fig. 12.22d is that using a higher densification factor gives a smaller error. It is also noticed from Fig. 12.22d that increasing the DWT level reduces the error in a log-lin manner up to a certain DWT level. However, beyond that, the denoising process breaks down, and a drastic increase in error is experienced. Thus, a critical DWT level exists at which the denoising is optimal. The value of this critical DWT level depends on the densification factor, m.

DWT denoising has been employed to remove noise and improve precision of experimental signals. Figure 12.23a shows an original signal captured in our experiments.

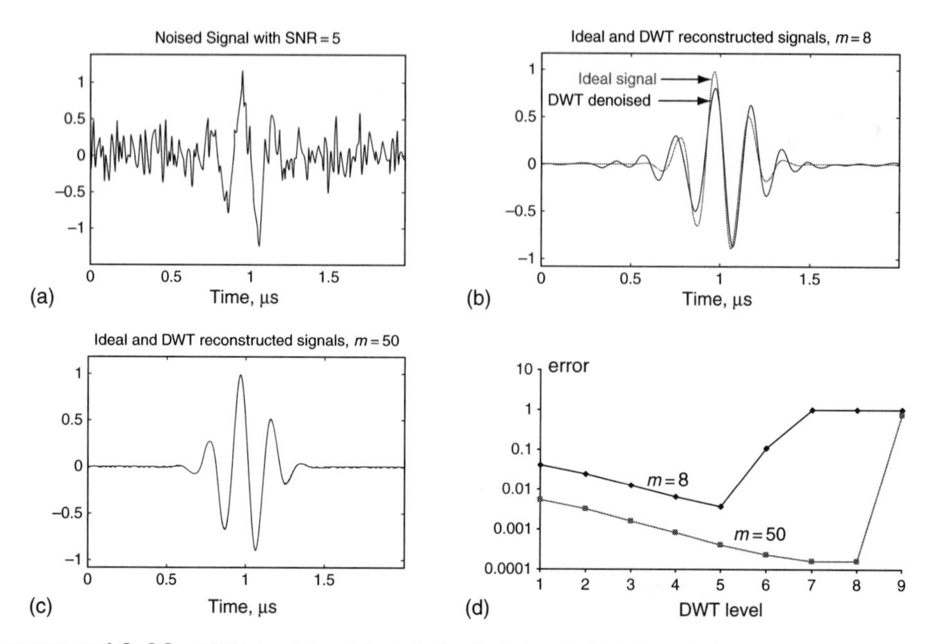

FIGURE 12.22 DWT denoising: (a) original noised signal with SNR = 5; (b) low quality denoising with $m = 8$, level = 1; (c) high quality denoising with $m = 50$, level = 5; (d) variations of error index at different level of $m = 8$ and $m = 50$.

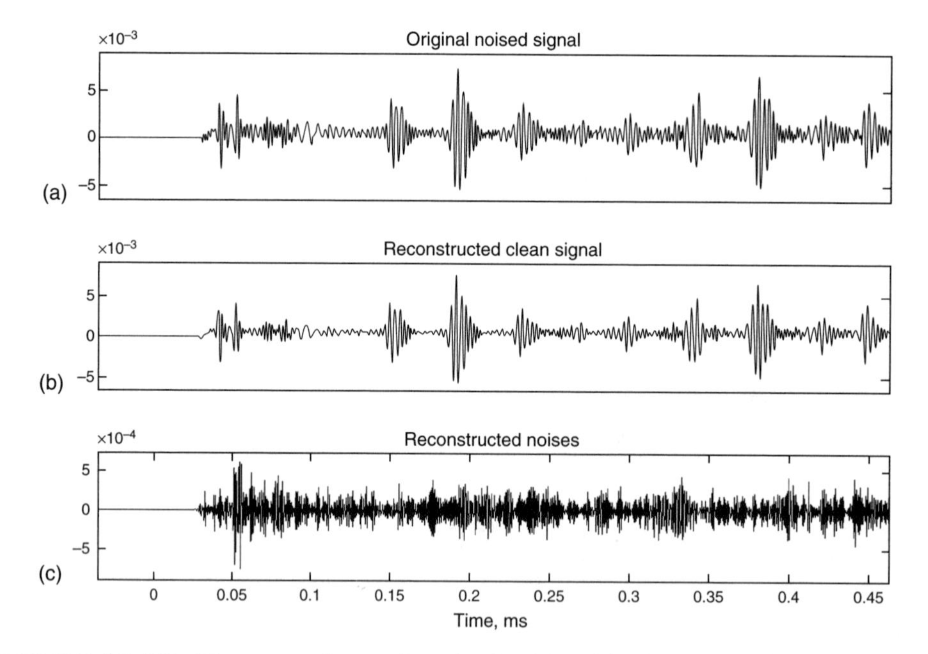

FIGURE 12.23 Discrete wavelet transform for denoising: (a) original signal (with initial bang removed); (b) clean signal after DWT denoising; (c) removed noise component.

Figure 12.23b gives the signal cleaned through DWT denoising using Morlet wavelet and Fig. 12.23c is the removed noise component. It is seen that the high frequency local disturbance has been removed from the original signal and the curve has become much smoother.

12.4 STATE OF THE ART DAMAGE IDENTIFICATION AND PATTERN RECOGNITION FOR STRUCTURAL HEALTH MONITORING

According to Fritzen and Bohle (1999), structural damage, such as fatigue cracks in metals or delamination of layer composite structures, causes characteristic local changes of stiffness, damping, and/or mass. Corresponding to these changes, shifts of the dynamic characteristics (frequencies, modeshapes, modal damping) occur. The deviation between the pristine dynamic properties and the damaged dynamic properties can be used to detect damage, and diagnose its location and extent. For example, Morita et al. (2001) used a damage–frequency correlation matrix trained on the first five structural frequencies of a five-story steel frame to identify damage presence and location from changes in frequencies and modeshapes. As presented by Farrar et al. (2001), the implementation issues for an integrated structural health monitoring system are: (a) hardware implementation of an active micro electromechanical system consisting of excitation, sensors, microprocessor, and wireless modem integrated pack; and (b) signal processing to identify damage. The latter involves a feature-extraction data-compression process, which identifies damage-sensitive properties from the measured vibration response. Data compression into small-size feature vectors is necessary to keep the problem size within manageable limits. In addition, realistic damage scenarios are always accompanied by nonlinear and/or nonstationary structural response, for which adequate analysis methods are required.

Damage identification algorithms are vital for practical implementation of SHM system. Damage identification algorithms can classify experimental data into several classes (i.e., pristine, lightly damaged, and severely damaged) depending on the damage severity and/or location. For example, when a researcher compares the frequency spectra of pristine and damaged structures, the difference between spectra is obvious in many cases, although, the researcher does not think about the classification algorithm, which classifies spectra into pristine and damaged scenarios. For successful solution of a classification problem, the deviation of inherent parameters of the structure due to temperature, humidity, or integrity changes as reflected in the measured data should be accounted for. According to Doebling et al. (1996), structural damage, such as fatigue cracks in metals or delamination of layer composite structures, causes characteristic local changes of stiffness, damping, and/or mass. Corresponding to these changes, shifts of the dynamic characteristics (frequencies, modeshapes, modal damping) occur. Fritzen and Bohle (1999) studied parameter selection strategies for modal-analysis–based damage detection using the experimental data collected during the controlled experiments on the I-40 bridge over the Rio Grande, New Mexico. Modal analysis measurements in the range 2–12 Hz were collected at 26 accelerometer locations. An initial cut (600 mm long by 10 mm wide) was progressively expanded until it virtually cut through a girder. However, positive detection was only possible when the cut was almost through the girder, highlighting the difficulty of detecting incipient damage with low-frequency modal analysis.

Deviation between the pristine dynamic properties and the damaged dynamic properties can be used to detect damage and diagnose its location and extent. For example, Morita et al. (2001) used a flexibility method and damage–frequency correlation matrix obtained for the first five structural frequencies of a five-story steel frame to identify damage presence and location from changes in frequencies and modeshapes. Hu and Fukunaga (2001) studied structural damage identification using piezoelectric sensors. The piezoelectric sensors were used as segment-wise curvature sensors. The resulting frequency response, modeled with a finite element analysis, was used to detect the presence and location of damage.

As presented by Farrar et al. (2001), the implementation issues for an integrated structural health monitoring system are:

- Hardware implementation of an active micro electromechanical system consisting of excitation, sensors, microprocessor, and wireless modem integrated pack.
- Signal processing to identify damage.

The latter involves a feature-extraction data-compression process, which identifies damage-sensitive properties from the measured vibration response.

The feature-extraction technique presented by Farrar et al. (2001) implements statistical algorithms that analyze distributions in the extracted features in an effort to determine the damage state of the structure. Statistical modeling was used to quantify changes in data features due to the presence of damage or variations in operational and environmental conditions. Three types of algorithms were suggested: group classification, regression analysis, and outlier detection. Data compression into small-size features vectors was necessary to keep the problem size within manageable limits. In addition, realistic damage scenarios were always accompanied by nonlinear and/or nonstationary structural response, for which adequate analysis methods are required. Todoroki et al. (1999) developed a health monitoring system schematic in which the probability of damage was estimated using statistical methods applied to strain gauge data continuously collected from the sensors. Hickinbotham and Austin (2000) studied a novelty detection for flight data from aircraft strain gauge measurements using the frequency of occurrence method. Using a Gaussian mixture model, they were able to identify abnormal sensor data. However, a significant percentage of normal flights were misclassified as abnormal. Sohn and Farrar (2000) studied statistical process control and projection techniques for damage detection. Principal component analysis with linear and quadratic projections was used to compress the features vectors using the Bayesian classification method. Auto regression models were used to fit the measured time series.

Ho and Ewins (1999) performed a numerical evaluation of a damage index. Using a finite element model, a damage indicator plot was developed from the modeshapes curvature. The model was verified against perturbations related to the measurement noise, the modeshape spatial resolution, and the location of damage. It was found that these methods can detect damage; however, situations in which damage presence was not detected were also identified. Lecce et al. (2001) used piezoelectric wafer active sensors for damage detection in aircraft structures. The FRF of a 9×9 sensor array was used to identify damage. Global damage index expressions based on the absolute value of the difference between the pristine and damaged spectra in the 0 to 5000 Hz range were considered. Ying et al. (2001) used stochastic optimal coupling control to study the seismic response mitigation of adjacent high-rise structures. The response evaluation criterion was calculated as the relative difference between the root mean square deviation (RMSD) of the response and the RMSD of the target. Ni et al. (2001) studied the viability of active control of cable-stayed bridges. The authors used the stochastic seismic response analysis and a control algorithm aimed at minimizing RMSD displacement response.

Extensive modal-analysis studies have focused on using automatic pattern recognition based on neural networks (NN) algorithms for structural damage identification. Worden et al. (2000) reviewed the use of multilayer perceptron (MLP) and radial basis function (RBF) NN and genetic algorithms for modeling the time response of a nonlinear Duffing oscillator with hysteretic response. Zang and Imregun (2000) used damage detection with a neural network that recognizes specific patterns in a compressed FRF. The compression of the FRF data was achieved with a principal component analysis algorithm.

Chan et al. (1999) used neural network novelty filtering to detect anomalies in bridge cables using the measured frequency of the cables. An auto associative neural network, which is a multilayer feed forward perceptron NN was used. This network was trained to reproduce, at the output layer, the patterns presented at the input layer. However, since the patterns are passed through hidden layers, which have fewer nodes than the input layer (bottleneck layers), the network is forced to learn just the significant prevailing features of the patterns. The novelty index was developed based on Euclidean norm and used as a threshold to identify the anomalous data. Ni et al. (2001) studied the application of adaptive PNN (probabilistic neural network) to suspension bridge damage detection. The PNN was designed to implement the Bayesian decision analysis with a Parzan window estimator into the artificial neural network. Sohn et al. (2001) studied the novelty detection under changing environmental conditions to identify structural damage in tested structures. An auto-associative neural network with three hidden layers was used. The novelty index was defined as the Euclidean distance between the target outputs and the neural network outputs. The method was demonstrated on a simplified analytical model representing a computer hard disk storage device. Sundareshan et al. (2001) studied a neural system for structural health monitoring using embedded piezoceramic elements. The authors attempted to hardwire the artificial neurons into the monitored structure using piezoceramic fibers and transducer–bus interface modules. Liu et al. (1999) used a multilayer feedforward neural network to perform structural damage identification. Time-domain parameter estimation was used, and the natural frequencies were identified as the 'most reliable indicators of damage'.

Development of suitable damage metrics and damage identification algorithms remains an open question. The damage index is a scalar quantity that serves as a metric of the damage present in the structure.

This short literature review reveals that there are three major concepts currently being pursued in the classification of data into categories during the health monitoring process:

1. Direct statistical approach (group classification, regression analysis, and outlier detection)
2. Overall statistics approach (RMSD, MAPD, and CC)
3. Neural networks (RBF, PNN, bottleneck) using spectral features (frequency, amplitude, and damping factor, etc.).

These approaches are available for implementation in the E/M impedance structural health monitoring. In this research, all the above-mentioned approaches were studied.

12.4.1 DAMAGE IDENTIFICATION AND DAMAGE METRIC ALGORITHMS

Development of suitable damage metrics and damage identification algorithms remains an open question in the practical application of structural health monitoring methods. The **damage index** is a scalar quantity that serves as a metric for the damage present in the structure. Let us discuss, for example, the use of the E/M impedance technique for SHM. Ideally, the damage index should be able to evaluate the E/M impedance spectrum and indicate damage presence, location, and severity. Sun et al. (1995) used a damage index based on the **root mean square deviation** (RMSD) change of the E/M impedance real part spectrum. The damage index compares the amplitudes of the two spectra (damaged vs. pristine) and assigns a scalar value based on the formula:

$$\mathrm{RMSD} = \sqrt{\frac{\sum_N \left[\mathrm{Re}(Z_i) - \mathrm{Re}(Z_i^0)\right]^2}{\sum_N \left[\mathrm{Re}(Z_i^0)\right]^2}} \tag{59}$$

where N is the number of sample points in the spectrum, and the superscript 0 signifies the pristine (base-line) state of the structure. Though simple and extensively used, the RMSD metric has an inherent problem: perturbing effects unrelated to damage (e.g., temperature variation) shift up and down the spectrum, and directly affect the damage index value. Compensation of such effects is not straightforward, and may not even be possible. The use of other damage metrics, base on alternative statistical formulae (**absolute percentage deviation**, the **covariance**, the **correlation coefficient**, etc.), did not alleviate the problem (Monaco et al., 2000; Leece et al., 2001; Tseng and Naidu, 2002).

Improved E/M-impedance damage index algorithms, which are less sensitive to noise and environmental effects, are being currently sought. Quinn et al. (1999) developed an E/M impedance damage-index algorithm based on the differences of the piecewise integration of the frequency response curve between the damaged and undamaged cases. In addition, improved characterization of the structure was sought by the separation of transverse and longitudinal outputs through directionally attached piezoelectric sensors. Lopes and Inman (1999) and Lopes et al. (1999) studied neural network techniques for damage identification, localization, and quantification based on the high-frequency E/M impedance spectra. During analytical simulation, a three-level normalization scheme was applied to the E/M impedance spectrum based on the resonance frequencies. First, the sensitivity of certain resonance frequencies to the location of the damage was identified. Second, the frequency change with damage amplitude at each location was calculated. Third, the normalized percentage frequency change for each damage severity was computed. One-layer and two-layer neural networks were constructed and successfully trained on analytical models with simulated damage. However, when applied to actual experiments (a 4-bay bolted structure and a 3-bay screw-connected space frame), the neural network approach had to revert to overall statistics such as: (i) the area between damaged and undamaged impedance curves; (ii) the RMSD of each curve; and (iii) the correlation coefficient between damaged and undamaged curves. Though not entirely successful, these studies indicated that a better damage metric might be achieved via a features-based pattern-recognition approach.

12.4.2 PATTERN-RECOGNITION METHODS FOR DAMAGE IDENTIFICATION

Automatic pattern recognition (Gonzales, 1978) plays a preeminent role in artificial intelligence research (Winston, 1993). Early successes obtained with automatic chemical classification (e.g., the DENDRAL mass spectrogram analyzer, Lindsay et al., 1980) and medical inference (e.g., the MYCIN bacterial diagnosis expert system, Shotliffe and Buchanan, 1975) spurred research interest in this field.

Figure 12.24 presents the typical steps of a pattern-recognition process: A sensor transforms the physical state into a measured signal. The preprocessor removes unnecessary or corrupting information from the signal. The feature extractor computes the features required for classification. The classifier yields a decision regarding the class membership of the pattern being analyzed, i.e., achieves the pattern-recognition goal. Farrar et al. (2001) pointed out that statistical modeling for feature discrimination uses algorithms that analyze distributions in the features spaces in an effort to determine the

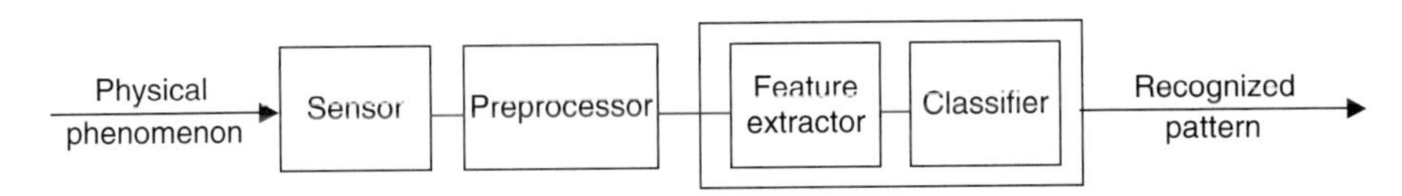

FIGURE 12.24 Schematic of a pattern-recognition process.

damage state of the structure. Statistical modeling is used to quantify when changes in data features can be considered significant. In addition, such algorithms can be used to discriminate between changes caused by damage and changes caused by varying operational and environmental conditions. Statistical modeling algorithms fall into three general categories. (1) group classification, (2) regression analysis, and (3) outlier detection. The appropriate algorithm will depend on the ability to perform supervised and unsupervised learning. The latter involves statistical process control, principal component analysis, and linear or quadratic discriminants. Building on statistical and adaptive signal processing (Kay, 1988, 1993; Manolakis et al., 2000), recent automatic pattern-recognition research has focused on using artificial neural networks (Nigrin, 1993; Bishop, 1995; Mehrotra et al., 1997; Looney, 1997; Omidvar and Dayhoff, 1998; Cowell et al., 1999; Principe et al., 2000). Currently, artificial neural network algorithms are widely available as software packages (Hagan et al., 1996; Demuth and Beale, 2000).

12.5 NEURAL NETWORKS

In this section, the utilization of neural networks (NN) for categorization of structural state into several classes according to damage intensity and location is discussed.

Neural networks are biologically inspired artificial intelligence representations that mimic the functionality of the nervous system. Biological neurons consist of **synapses**, **dendrites**, **axons**, and **cell bodies**. Specific to the biological neuron is that they **fire** (i.e., send signals) only when **activated**. Similarly, artificial neurons are set on by **activation** or **transfer function** (a.k.a. **threshold function**). The activation function $g(\cdot)$ can be discontinuous (e.g., symmetrized Heaviside step function, a.k.a. **hardlimit**) or continuous (e.g., logistic sigmoid, a.k.a. **logsig**). Linear transfer functions are also used. In an artificial neural net, each simulated neuron is viewed as a node connected to other nodes (neurons) through **links**. Each link is associated with a **weight**. The weight determines the nature and strength of one neuron's influence on another. Neural nets are organized in layers: **input layer**, **output layer**, and **hidden layers**. (The output of the neurons in hidden layers is not observable from the outside world.) Artificial neural nets consisting of **multipliers**, **adders**, and **transfer functions** can be implemented as software algorithms and/or hardwired electronic devices.

12.5.1 INTRODUCTION TO NEURAL NETWORKS

Figure 12.25 shows a schematic of a biological neuron, which consist of synapses, dendrites, axons, and the cell body. The biological neuron sends signals only when activated. Similarly, artificial neurons are set on by activation or transfer function, $f(n)$ (Fig. 12.25b). The activation function $f(n)$ can be discontinuous (e.g., symmetrized Heaviside step function, a.k.a. hardlimit) or continuous (e.g., logsig). Linear transfer functions are also used. In an artificial neural net, each simulated neuron is viewed as a node connected to other nodes (neurons) through links associated to weights. By

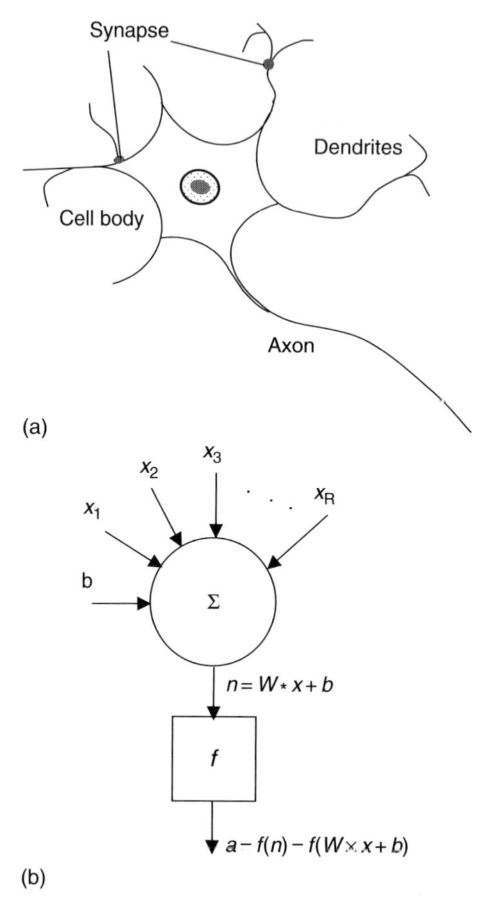

FIGURE 12.25 The neural networks concept: (a) the biological neuron; (b) the mathematical representation.

analogy to a biological neuron where the weights represent the strength of the synapse, the weights used in an artificial neuron determine the nature and strength of one neuron's influence on another. The output of the neuron represents the signal on the axon. Neural networks are organized in layers (input layer, output layer, and hidden layers). For practical applications, artificial neural networks are implemented as software algorithms and/or hardwired electronic devices. Some typical examples of neural networks are:

- Feedforward
- Competitive
- Probabilistic neural networks

Feedforward NN compute the output directly from the input, in one pass. No feedback is involved. Feedforward NN can recognize regularity in data and can serve as pattern identifiers. A typical feedforward NN is the perceptron. The mathematical representation of the perceptron is

$$f(n) = Heaviside\,(Wx + b) \tag{60}$$

where $f(n)$ is the output transfer function.

The perceptron scalar output equals 1 when a pattern has been identified and 0 otherwise. A generalized perceptron uses nonlinear transfer functions, $f(n)$, i.e., basis functions. Interconnected perceptrons form perceptron layers. Several perceptron layers lead to a multilayer perceptron (feedforward) NN. In pattern recognition, multilayer feedforward NN can be trained to identify class separation boundaries in the form of hyper-planes in the multidimensional input space.

Competitive neural networks are characterized by two properties:

(1) They compute some measure of the distance between stored prototype patterns and the input pattern
(2) They perform a competition between the output neurons to determine which neuron represents the prototype pattern closest to the input.

Probabilistic neural networks (PNN) are hybrid multilayer networks that use basis functions and competitive selection concepts. Probabilistic NN utilize a Bayesian decision analysis with Gaussian kernel (Parzan window) and consist of an input layer, a radial-basis layer, and a competitive layer. The radial-basis functions depend on the distance between the input vectors and some prototype vectors, representative of the various input classes. In the testing phase, the PNN classifies each input vector by determining the closest correspondent class previously assigned during the training process. These characteristics of PNN make it an appropriate choice for classification of the features vectors obtained from the reduction of E/M impedance spectra. A detailed discussion of PNN is given next.

12.5.2 TYPICAL NEURAL NETS

Some typical neural nets are: (a) **feedforward**, (b) **recurrent**; (c) **competitive**; (d) **probabilistic**. These will be briefly discussed next.

Feedforward neural nets compute the output directly from the input, in one pass. No feedback is involved. Feed-forward nets can recognize regularity in data; they can serve as pattern identifiers. A typical feedforward neural net is the **perceptron**. The perceptron scalar output equals 1 when a pattern has been identified, and 0 otherwise. The mathematical representation of the perceptron is $y(\mathbf{x}) = g\left(\sum_{i=0}^{N} w_i x_i\right) = g(\mathbf{w}^T\mathbf{x})$, where $g(\cdot)$ is the output transfer function, and x_0 is always set to 1 ($x_0 = 1$) such that w_0 plays the role of bias (Fig. 12.26a). A generalized perceptron (Fig. 12.26b) uses nonlinear transfer functions, $\phi_j(\cdot)$, a.k.a. **basis functions**, which are placed between the input and the adder: $y(\mathbf{x}) = g\left(\sum_{i=0}^{N} w_i \phi_i(\mathbf{x})\right) = g\left(\mathbf{w}^T \phi(\mathbf{x})\right)$. Interconnected perceptrons form **perceptron layers**. Several perceptron layers lead to a **multilayer perceptron** (feedforward) neural net. In pattern recognition, multilayer feedforward NN can be trained to identify **class separation boundaries** in the form of **hyper-planes** in the multidimensional input space (Fig. 12.27a).

Recurrent neural nets have feedback connections from output to the input. They can learn to identify a pattern. A typical recurrent network is the **Hopfield network**, which can be easily hardwired using resistors, capacitors, and feedback connections. They have been used for associative memories, and optimization problems. The output of a Hopfield network is the very pattern that was identified.

Competitive neural nets are characterized by two properties: (a) they compute some measure of the distance between stored prototype patterns and the input pattern; (b) they perform a competition between the output neurons to determine which neuron represents

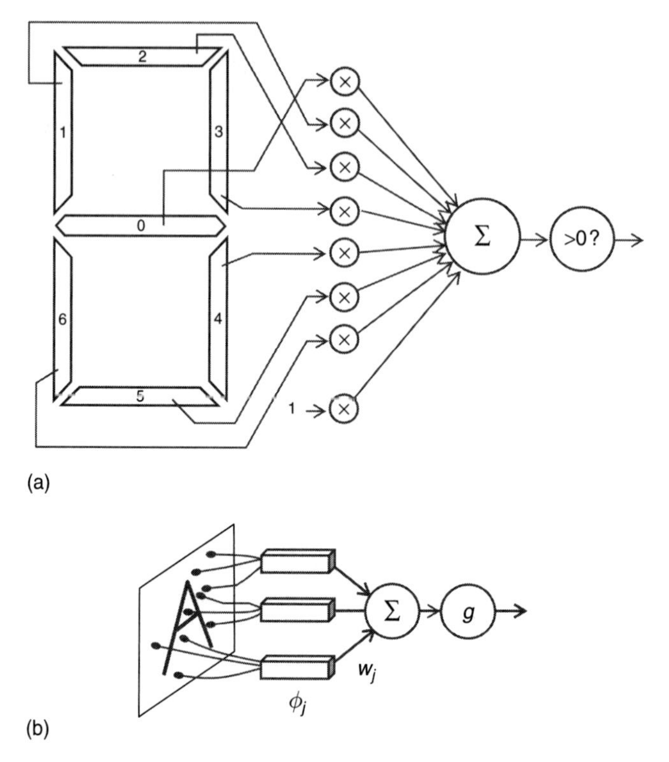

FIGURE 12.26 (a) simple perceptron capable of identifying any and only one of the 10 decimal digits written in a 7-segment display format; (b) character-identification perceptron.

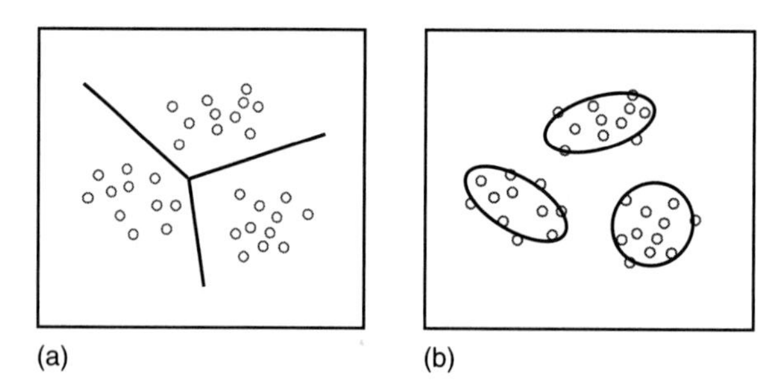

FIGURE 12.27 Separation of features vectors into class clusters: (a) a multilayer perceptron yields straight boundaries (hyperplanes); (b) radial-basis NN yields curved enclosures for each class.

the prototype pattern closest to the input. **Adaptive competitive neural nets** adjust the prototype patterns as new inputs are applied to the neural net. Thus, an adaptive competitive neural net learns to cluster inputs into different input classes. A typical competitive network is the two-layer **Hamming network**: The first layer is a **feedforward network**, whereas the second layer is a **recurrent network** that performs the neuron competition and determines a winner. The output of a Hamming network is a vector containing a '1' at a row position corresponding to the identified class.

Probabilistic neural nets are hybrid multilayer networks that use basis functions and competitive selection concepts. Probabilistic NN achieve a Bayesian decision analysis with Gaussian kernel (Parzan window). A probabilistic NN consists of a **radial-basis layer**, a **feedforward layer**, and a **competitive layer**. The radial basis functions depend on the distance between the input vectors and some **prototype vectors**, μ_j, representative of the various input classes.

$$y_k(\mathbf{x}) = \sum\nolimits_{m=0}^{M} w_{km}\phi_m(\mathbf{x}), \quad \phi_m(\mathbf{x}) = \exp\left(-\frac{\|\mathbf{x}-\mu_m\|^2}{2\sigma_m^2}\right), \phi_0 = 1 \tag{61}$$

NN identify class separation boundaries in the form of **hyper-spheres** in the input space (Fig. 12.27b).

12.5.3 TRAINING OF NEURAL NETS

Multi-layer feedforward neural nets are trained through **backpropagation**, which is a generalized learning rule for neural nets with nonlinear differentiable transfer functions. The method is called *backpropagation* because it computes first the changes to the weights in the final layer, and then reuses much of the same computation to calculate the changes to the weights in the penultimate layer, and so on until it reaches *back* to the initial layer. Standard backpropagation uses the gradient descent algorithm; other optimization techniques such as Newton-Raphson, conjugate gradients, etc. have also been used.

The training through backpropagation is **iterative**, and can take several presentations of the training set. During training, the weights are optimized by minimization of a suitable error function, e.g., the sum-of-squared-errors $E = \frac{1}{2}\sum_n \sum_k \left\{y_k(\mathbf{x}^n; w_{kj}) - t_k^n\right\}^2$. The backpropagation training has an **exit strategy**, i.e., the training is stopped when the output error falls below given **convergence thresholds**. The duration of learning depends considerably on the **rate parameter** (a.k.a. regularization parameter), which controls the changes made to the neural network weights after each training cycle. The rate parameter controls the step size in the gradient descent; hence, if the step size is large, non-linear effects become significant, and strongly affect the validity of the gradient calculation formulae. If the training is too long, the network becomes too specialized on the training data and looses generalization, i.e., has unsatisfactory performance on validation data. A method for improving generalization is **early stopping** (Fig. 12.28).

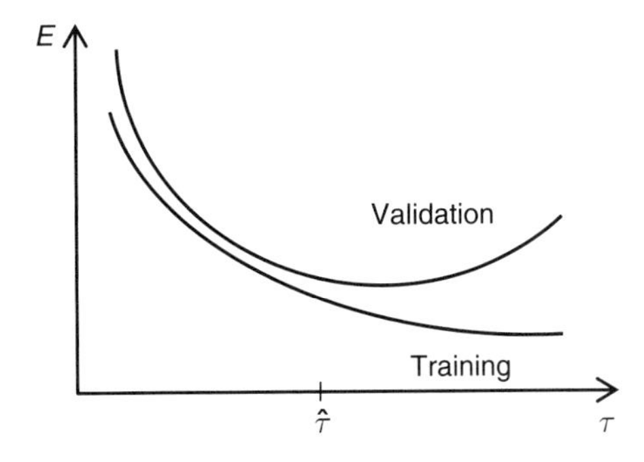

FIGURE 12.28 Early stopping improves generalization performance.

Probabilistic neural nets are trained in two stages:

The first stage consists of **unsupervised learning**. In the first stage, the training input vectors, $\mathbf{x}^n$, are used to determine the basis-function parameters, μ_j and σ_j. Either an **exact design** or an **adaptive design** can be used. The exact design sets the prototype vectors equal to the input training vectors, $\mu_j = \mathbf{x}^n$, $j = n$. The adaptive design clusters the input vectors and determines prototype vectors that are representative of several input vectors. One way of implementing the adaptive design is through incremental addition of neurons. Radial-basis neurons are added until the sum-squared error falls beneath an error goal, or a predetermined number of neurons is reached. At each iteration, if an input vector results in lowering the network error, then it is used to create a new radial-basis neuron. The radial spread parameters, σ_j, are adjusted to achieve optimal coverage of the input space.

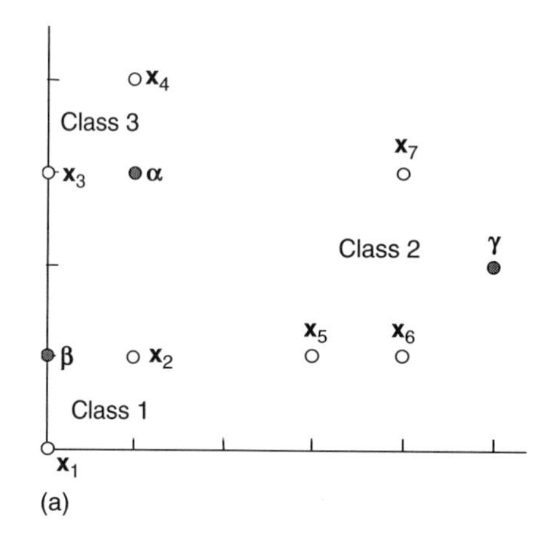

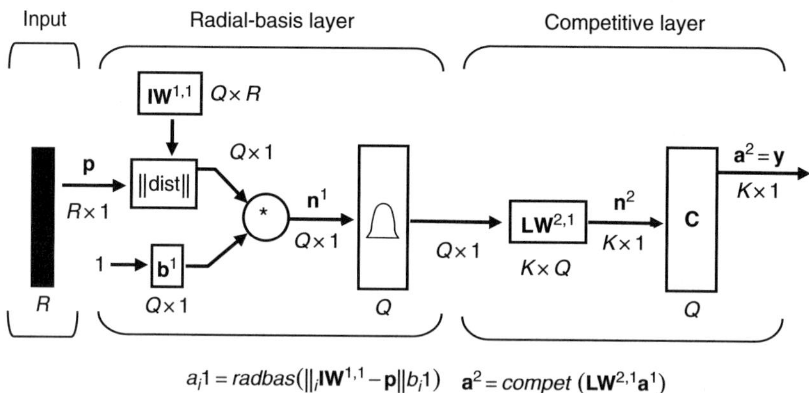

a_i1 = is the i-th element of $\mathbf{a}^1$ where $_i\mathbf{IW}^{1,1}$ is the vector made of the i-th row of $_i\mathbf{IW}^{1,1}$
R = number of elements in input vector
Q = number or inpu5/target pairs = number of neurons in layer 1
K = number of classes of input data = number of neurons in layer 2

(b)

FIGURE 12.29 Classification using probabilistic neural networks (PNN): (a) $\mathbf{x}_1, \mathbf{x}_2, \mathbf{x}_3, \mathbf{x}_4, \mathbf{x}_5, \mathbf{x}_6, \mathbf{x}_7$ are the training vectors defining classes 1, 2, and 3 and $\boldsymbol{\alpha}, \boldsymbol{\beta}, \boldsymbol{\gamma}$ are the new vectors that have been classified; (b) the general architecture of the probabilistic neural net.

The second stage consists of **supervised learning**. In the second stage, the basis functions are kept fixed (matrix Φ), and the linear-layer weights, w_{nj}, are found to fit the training output values, $\mathbf{t}^n$, using backpropagation. The formal solution for the weight is given by: $\mathbf{W} = \left(\Phi^{\dagger}\mathbf{T}\right)^T$ where $\Phi^{\dagger}$ denotes the pseudo-inverse of Φ. In practice, this is solved using the singular value decomposition to avoid problems due to the possible ill-conditioning of matrix Φ.

Figure 12.29a presents a simple probabilistic network classification example (MATLAB, 2000). The training vectors $\mathbf{x}_1$, $\mathbf{x}_2$, $\mathbf{x}_3$, $\mathbf{x}_4$, $\mathbf{x}_5$, $\mathbf{x}_6$, $\mathbf{x}_7$, were used to train the neural net and define the classes 1, 2, and 3. Then, the new vectors α, β, γ, were used to perform validation of the neural net. The new vectors were successfully assigned. Figure 12.29b reproduces the general architecture of the probabilistic neural network.

Basis function optimization: One of the main advantages of the radial-basis neural net in comparison with the multilayer perceptron is the possibility of choosing a suitable design for the hidden neurons without having to perform a full nonlinear optimization of the NN. Two design directions are available: (a) selecting an appropriate number of basis functions and (b) selecting appropriate values of the basis function parameters. The basis function parameters should be chosen to form a representation of the probability density of the input data. This leads to an unsupervised procedure for optimizing the basis function parameters which depends only on the training input data and which ignores any training target information. The basis function centers, μ_j, can be regarded as *prototypes* of the training input vectors.

12.5.4 NEURAL NETS FOR STRUCTURAL DAMAGE IDENTIFICATION

Extensive modal-analysis studies have focused on using an automatic pattern recognition method for structural damage identification. Worden (2000) reviewed the use of multilayer perceptron (MLP) and radial basis function (RBF) neural networks (NN) and genetic algorithms for modeling the time response of a nonlinear Duffing oscillator with hysteretic response. Worden et al. (2000) studied two structural health-monitoring methods, the first performing novelty detection in the low-frequency (0–250 Hz) frequency response function (FRF) transmissibility, the second extracting frequencies and modeshapes and quantify a damage indicator. Farrar et al. (1999), Ying et al. (2001), and Jin et al. (2001) considered the statistical pattern recognition method for vibration-based structural health monitoring using stochastical analysis. Zang and Imregun (2000) used damage detection with a neural network that recognizes specific patterns in a compressed FRF. The compression of the FRF data was achieved with a principal component analysis (PCA) algorithm. Chan et al. (1999) used neural network novelty filtering to detect anomaly in bridge cables using the measured frequency of the cables. Ni et al. (2001) studied the application of adaptive probabilistic neural network (PNN) to suspension bridge damage detection. Sohn et al. (2001) studied the novelty detection under changing environmental conditions. Liu et al. (1999), Lloyd amd Wang (1999), Loh and Huang (1999), Todoroki et al. (1999), and Sohn and Farrar (2000) studied the statistical and neural network processing of time-domain signals. Chang et al. (1999), Ho and Ewins (1999), and Hu and Fukunaga (2001) followed the model-based approach to structural damage detection. Sundareshan et al. (2001) attempted to hardwire the artificial neurons into the health-monitored structure using piezoceramic fibers and transducer–bus interface modules. Fritzen and Bohle (1999) studied parameter selection strategies for modal-analysis–based damage detection using the experimental data collected during the controlled experiments on the I-40 bridge over the Rio Grande, New Mexico. Modal analysis measurements in the range 2–12 Hz were collected at 26 accelerometer locations. An initial cut (600 mm long by 10 mm wide) was progressively expanded until it completely cut through the bridge. However, positive detection was possible only when the cut was almost through the frame, highlighting the difficulty of detecting incipient damage with low-frequency modal analysis.

12.5.5 PROBABILISTIC NEURAL NETWORKS

Probabilistic neural networks (PNN) were first proposed by Specht (1990) as an efficient tool for solving classification problems. In contrast to other neural network algorithms used for damage identification (e.g., Chan et al., 1999), PNN has statistically derived activation function and utilizes Bayesian decision strategy for the classification problem. This unique feature allows the decision boundary to approach asymptotically the Bayes optimal decision surface under certain easily met conditions (Specht, 1990). The accuracy of the decision boundaries depends on the accuracy with which the underlying probability density functions are estimated.

The Bayesian strategy implies that the classification procedure is performed in a way that minimizes the probability of misclassification. During the structural health monitoring process, several categories usually should be considered. Each category in a multi-category classification problem corresponds to the particular state of the structure under examination and represents a degree of damage. The task is to classify the structural health based on the set of measurements (features) represented in a features vector. Denoting possible categories by $\theta_A, \theta_B, \ldots, \theta_N$, and the p-dimensional features vector by $x = [x_1, x_2, \ldots, x_N]^T$ it is possible to formulate the Bayes decision rule in the following way:

$$\mathrm{d}(x) \in \theta_A \quad \text{if} \quad h_A l_A p_A(x) > h_K l_K p_K(x) \quad \text{for all} \quad K \neq A \tag{62}$$

where $\mathrm{d}(x)$ is the decision on the test vector x. The variables h_A and $h_K = 1 - h_A$ are the a priori probabilities of the categories θ_A and θ_K. The variable l_A is the loss associated with misclassifying when $\theta \in \theta_A$, and l_K is the loss associated with misclassifying $d(x) \notin \theta_K$ when $\theta \in \theta_K$. The functions $p_A(x)$ and $p_K(x)$ are the probability density functions (PDF) for categories θ_A and θ_K, respectively.

The decision surface boundary is defined by the ratio of a priori probabilities and losses of one category and the a priori probabilities and losses of another category.

$$p_A(x) = C_K p_K(x) \text{ for all } K \neq A \text{ where } C_K = h_K l_K / h_A l_A \tag{63}$$

The ratio of a priori probabilities can be estimated based on the previous damage identification results. For example, it may occur that from 100 data vectors acquired during the SHM process only 5 represented the damaged case (θ_A). In this scenario, the a priori probabilities are $h_A = 0.05$, $h_K = 0.95$, and the ratio of a priori probabilities would be 19. It can be shown (Specht, 1990) that, in PNN, the ratio of a priori probabilities is accounted for by introducing the ratio of the number of training patterns from each category:

$$C_K^{\mathrm{PNN}} = \frac{h_K l_K}{h_A l_A} \cdot \frac{n_A}{n_K} \tag{64}$$

where n_K is the number of training patterns from category K, and n_A is the number of training patterns from category A.

In order to achieve the realistic scenario discussed above, one should use 19 training patterns for the pristine case and only one training pattern for the damaged case.

To simplify multicategorical situations, it can be assumed that a priori probabilities are equal for all categories involved in the study. The ratio of losses in (63) and (64) can be determined only from the decision significance. Thus, evaluation of the ratio of losses is subjective. Further in this work, the losses are considered unbiased and, thus, the constant C_K will equal unity for all categories ($C_K \equiv 1$). Thus, the Bayes decision rule (62) will be simplified to:

$$d(x) \in \theta_A \text{ if } h_A p_A(x) > h_K p_K(x) \; \forall K \neq A \tag{65}$$

PNN utilizes a kernel-based approach to probability density function (PDF) approximation. This was introduced by Parzen (1962) and allows one to construct a PDF of any sample of data without any a priori probabilistic hypothesis (Rasson and Lissoir, 1998). The reconstruction of the probability density is achieved by approximating each sample point with kernel function(s) to obtain a smooth continuous approximation of the probability distribution (Dunlea et al., 2001). In other words, using the kernel technique it is possible to map a pattern space (data sample) into the feature space (classes). However, the result of such transformation should retain essential information presented in the data sample and be free of redundant information, which may contaminate the feature space.

The multivariate Gaussian kernel is usually used for probabilistic neural network implementation. In the original Specht formulation (Specht, 1990), this kernel was expressed as

$$p_A(x) = \frac{1}{n \cdot (2\pi)^{d/2} \sigma^d} \sum_{i=1}^{n} \exp\left(-\frac{(x - x_{Ai})^T (x - x_{Ai})}{2\sigma^2} \right) \tag{66}$$

where i is a pattern number, x_{Ai} is the ith training pattern from A category, n is the total number of training patterns, d is a dimensionality of measurement space, and σ is a spread parameter.

Although the Gaussian kernel function was used in this work, its form is generally not limited to being Gaussian. Burrascano et al. (2001) considered damage identification using an eddy current nondestructive technique and utilized the PNN with different kernel types.

The architecture of a PNN is shown in Fig. 12.30. The neural network is organized for classification of input patterns x into several categories or classes. The input layer consists of distribution units, which transfer the same input values to all the pattern units in the second (pattern) layer. The pattern layer computes distances from the input features vector x to the training input vector x_{Ai}. The result of this operation indicates the degree of proximity between the input and the target vectors. Subsequently, it is multiplied, element-by-element, by the bias b and fed into the nonlinear activation function. The activation function used for neurons in the pattern layer is of the same form as the Gaussian kernel in Eq. (66), namely

$$p_A(x) = \exp\left(- \|x - x_{Ai}\|^2 \cdot b^2 \right) \tag{67}$$

In Eq. (67) the bias b is responsible for the sensitivity of the neuron and is inversely proportional to the spread (smoothing) parameter. The spread parameter determines the width of the area in the input space to which each neuron responds (Demuth and Beale, 1998). In general, networks with relatively large spread parameters are more sensitive because the function (67) is wider and can cover a larger data set. However, too large spread may generate misclassification problems. The output of the pattern layer is a vector

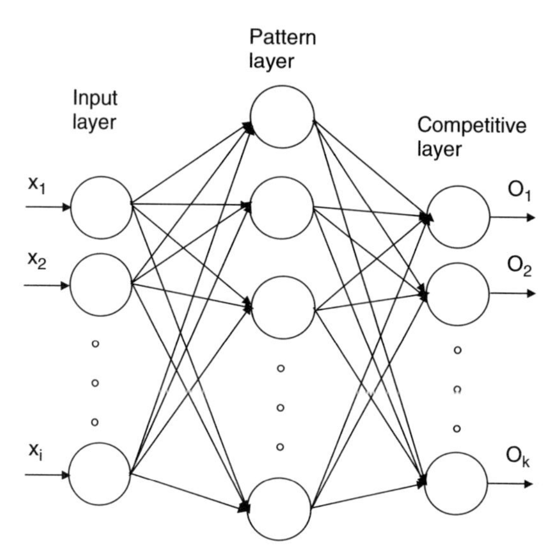

FIGURE 12.30 Probabilistic neural network showing the input layer, the pattern layer, and the output (competitive) layer.

of probabilities estimated with Eq. (67). Neurons with high probability represent a particular category (class), whereas neurons with near-zero probability will not be admitted. The competitive layer is used to perform this task. The competitive layer outputs a one (1) for the class with maximum probability and zeroes (0) for the other classes. Thus, the network classifies a certain input vector into a particular class based on the maximum probability value.

The literature review has revealed that several issues need to be addressed when performing the implementation of a PNN to a particular classification problem:

- Choice of input data
- Normalization of input data
- Choice of training data
- Choice of spread constant

These issues are briefly addressed next.

12.5.5.1 Choice of Input Data

The NN should be able to distinguish the healthy and the damaged scenarios using the input data vectors. The most widely used technique is based on the frequency domain data. The frequency spectrum of a healthy structure has a certain distribution of resonance frequencies associated with the corresponding vibration modes. When damage appears, this initial distribution is changed, and (or) new resonance frequencies, which were absent in the original spectrum, may appear. The location of damage is assumed to be in direct relationship with the degree of change from the original spectral distribution and (or) magnitude and number of newly appeared harmonics. Although this technique is direct and in most cases efficient, it may also have some drawbacks. For very small damage, the shift of the natural frequencies in the spectrum, especially at low order modes, is small (Chrondos and Dimarogonas, 1980). If no new resonances appear, it is difficult to distinguish between the weak-damage case and the pristine case. Another problem may be the possible ambiguity arising when measuring the deviation of frequencies from the original values. Depending on its nature, a damage may cause the downward or upward shift of the original (pristine) natural frequency. It can be shown that Euclidean distance metrics are not capable of distinguishing between these two cases. Choosing of other parameters to be used as input data for the NN may solve this problem (Atalla, 1996) although this would require the preprocessing of the frequency spectra. However, in damage identification procedures, it is desirable that the input data is accessed directly with minimum preprocessing, and that the data stored in the features vectors allows monitoring of damage progress or location. Direct use of frequency domain data meets this criterion, despite the noted limitations.

12.5.5.2 Normalization of Input Data

The spread of magnitude values inside the input features vector may have a side effect on PNN performance, which computes the distance between the training and the validation vectors. During classification, PNN favors the elements of the training and validation vectors with large magnitude variation and disregards those with small variation. To ensure good network performance, it is desirable that the input data be normalized on the [0, 1] interval. Attala, (1996) suggested the following linear transformation to ensure that each entry of the input vector falls within a prescribed bound defined by the minimum and maximum values of the correspondent feature:

$$x_i^{\text{new}} = \frac{x_i^{\text{old}} - \min(x_i^{\text{old}})}{\max(x_i^{\text{old}}) - \min(x_i^{\text{old}})} \tag{68}$$

where i is a pattern number. For example, for five input vectors, which contain frequency domain data, the pattern number i will represent the first frequency in all five vectors. In other words, it is possible to think of i as a row number in the input space. When high accuracy is desirable over certain frequency ranges, Eq. (68) can be modified by introducing a scaling factor, r_i, in front of the fraction. In many cases, the normalization of input data helps to improve the classification procedure, although the decision on its implementation always rests with the user and is based on the nature of the classification problem and the structure of the input data. In this work, as it will be shown shortly, the PNN is able to classify successfully the damage scenarios in circular plate specimens without normalization of the frequency domain input data.

12.5.5.3 Choice of Training Data

The PNN used in this study utilizes Gaussian kernel centered on the training data. The training data should represent the most common scenario for the particular classification category. It is advised that, for each class, the input vectors with the widest spread of pattern values are used in the training set. This ensures coverage of problematic cases, which might be incorrectly classified otherwise. In the best situation, the training set resembles the probability distribution of the structural parameters corresponding to the particular classes. To guarantee the compliance with the Bayes decision rule (65), the size of the training set is usually chosen according to the values of a priori probabilities (64). The number of features in the features vector plays a significant role in the success of the classification process. It will be shown that the larger the size of the input features-space (training and validation vectors), the better the classification results. In one particular case, when the training and validation vectors consisted of four frequency values, the PNN was able successfully to classify patterns into three categories according to the damage location, but failed to distinguish between the pristine structure and the structure with weak damage. This problem was overcome by taking six frequencies in the features input space. Subsequently, the PNN successfully assigned the proper categories to each damage scenario.

12.5.5.4 Choice of Spread Constant

It is possible to think of a spread parameter as being proportional to the standard deviation parameter of PDF. Attala (1996) suggested a method that adjusts the proportionality constant using the gradient descent algorithm in order to minimize the sum square error of the network over the validation set. However, the choice of the spread constant significantly depends on the input data and where the decision boundary is defined. For $\sigma \to \infty$, the decision boundary approaches a hyperplane, whereas for $\sigma \to 0$ it represents the nearest neighbor classifier. A very large value of spread parameter would cause the estimated density to be Gaussian regardless of the true underlying distribution (Specht, 1990). In this study, the spread constant was chosen heuristically. It was noticed that the difference of spread constant in several orders of magnitude did not noticeable modify the PNN performance. However, the limitations for the values of this parameter were observed. Further discussion on this subject can be found in Wasserman (1993).

12.6 FEATURES EXTRACTORS

Feature extraction is an important stage in reducing the problem dimensionality and separating essential from nonessential information. As an illustration, consider the E/M impedance spectrum reproduced in Fig. 12.33b. The spectrum contains 400 data points collected at 0.075 kHz intervals in the 10 to 40 kHz range. Two spectra are presented: pristine and damaged. The two spectra are obviously different. However, quantifying the difference is neither obvious nor immediate. In a featureless approach, one would attempt to directly compare the two 400-long strings raw input vectors of numbers (raw input vectors) associated to each spectrum. Such a flat approach would lead to a large-dimensional problem, and the associate computational difficulty. Hence, we will use a features-vector approach.

The features-vector approach recognizes that the number and characteristics of resonance peaks play a major role in describing the dynamic behavior of the structure. Hence, they form **essential features** of the impedance spectrum. Distinction between a ‘pristine’ and a ‘damaged’ structure can be based on these features. For example, in the 10–40 kHz band, the pristine-plate spectrum of Fig. 12.33b displays four dominant resonance peaks (12.8, 20.1, 28.9, and 39.2 kHz), and three minor peaks (15.9, 24.5, 34.1 kHz). By considering these features, the problem dimensionality can be reduced by almost 2 orders of magnitude, from 400-points to less than 10 features. For the damaged plate, also shown in Fig. 12.33b, we distinguish six dominant peaks (11.6, 12.7, 15.2, 22.0, 30.0, 39.1 kHz) and six smaller ones. Though the number of features has increased, the problem size is still at one order of magnitude less than in a featureless approach. Similar results are observable in the 10–150 kHz and 300–450 kHz bands. The use of high frequencies will ensure that our damage metric will detect incipient damage at an early stage.

12.6.1 FEATURES EXTRACTION ALGORITHM

The **features extraction algorithm** utilizes a rectangular search window that transverses the spectrum from left to right. A peak is found inside the search window when the difference in height between the local data maximum inside the window and the data values at both ends of the window is less than the height of the window. The height of the window is specified as a percentage of the total amplitude of the data in the range (the amplitude is defined as the difference between the maximum and the minimum of the data). The width of the window is specified as a percentage of the total number of points in the data range. Generally, the smaller the height and the width values, the more peaks are likely to be identified. However, too many peaks of low amplitude may result in an unnecessarily long and noisy features vector. A threshold identifier, based on the energy content of each peak, will be used to prune the features vector. Training of the threshold identifier will be part of the development of the feature extraction algorithm. *The features vector will contain the resonance frequencies identified in the spectrum.* We did not introduce the amplitudes in the features vector because, as indicated in Section C.1.3.2, the spectrum amplitudes are prone to random and bias error, and vary with environmental conditions (temperature, etc.).

12.6.2 ADAPTIVE RESIZING OF THE FEATURES VECTOR

It is apparent from Fig. 12.33b that the number of features in the 'pristine' spectrum is different than the number of features in the 'damaged' spectrum. As damage takes place, two phenomena can occur: (a) the values of the existing features can shift; (b) new features, associated with new vibration modes and local resonances, may appear. For example, for the dominant peaks in Fig. 12.33b, we observe that the features vector for the 'pristine' spectrum has dimension $N_{\text{pristine}} = 4$, whereas the features vector for the 'damaged' spectrum has dimension $N_{\text{damaged}} = 6$. Induction heuristics approach will be used to add and subtract features to the feature space, i.e., to enlarge or reduce its dimensionality. The zero-fill method will be used to upgrade the lower-dimension features vector to higher-dimension features vectors. Thus, the 'pristine' features vector will be updated to $N^*_{\text{pristine}} = 6$, with the new cells being filled with zero values. After the features vector is updated with new frequencies, the cells are reordered to form an increasing-frequency sequence.

12.6.3 ADAPTIVE DAMAGE DETECTOR AND CLASSIFIER

The **adaptive damage detector and classifier** algorithm will process the features vector using a neural network based on competitive probabilistic classification and radial-basis functions. Radial-basis neural net and multilayer perceptrons play similar roles in pattern recognition since both provide techniques for approximating arbitrary non-linear functional mappings between multidimensional spaces. In both cases, the mappings are expressed in terms of parameterized compositions of functions of single variables. However, the particular structure of the two neural nets is very different (Bishop, 1995):

1. The activation of a multilayer perceptron is constant on parallel $(d-1)$-dimensional hyperplanes, whereas the activation of the basis function is constant on concentric $(d-1)$-dimensional hyperspheres.
2. The training process of the multilayer perceptron is highly nonlinear with problems of local minima, or nearly flat regions in the error function, which lead to slow convergence even with advanced optimization strategies. By contrast, the radial-basis neural net learning is very fast.

3. A multilayer perceptron has many layers of weights, and a complex pattern of connectivity. By contrast, a radial-basis NN has a simple architecture consisting of two layers of weights, in which the first layer contains the parameters of the basis functions, and the second layer generates the output vectors by forming linear combinations of the activations of the basis functions.
4. For a multilayer perceptron, all the parameters are determined through a single global training strategy involving supervised training. By contrast, a radial-basis NN is trained in two stages: (a) first, the basis functions are determined by unsupervised training techniques using the input data alone; (b) second, the second-layer weights are found by fast linear supervised training.

Following Ni et al. (2001), an adaptive probabilistic neural network (PNN) will be utilized in the damage detection algorithm. The probabilistic neural network will be developed to sort the input vectors into 'pristine' and 'damaged', using a **multi-level** structural health monitoring classification, signifying various degrees of damage intensity (e.g., 'just damaged', 'damaged but safe', 'severely damaged', etc.). The radial-basis PNN characteristic of being capable of both **supervised** and **unsupervised learning** will be used as follows:

- During **unsupervised learning**, the natural classification of the input features vectors (resonance frequencies) will be detected to establish the radial-basis function footprints.
- During **supervised learning**, structural health monitoring classification labels will be assigned to the natural footprints identified during unsupervised learning.

To promote adaptability, the PNN will be designed with separate smoothing parameters for features vector component. The value of each smoothing parameter will be optimized such as to become almost a measure of that variable's importance. A genetic algorithm (GA) will be used to optimize the smoothing parameters of the adaptive PNN. The GA will test several waiting schemes until it finds the one that gives the best classification of the training data. A novelty index (anomaly metric) will be developed by passing the training vectors again through the network after the network had been trained. The novelty index is used as a threshold to identify anomalous data during the NN usage. The novelty index will be defined as the Euclidean distance between the target outputs and the neural network output.

12.7 CASE STUDY: E/M IMPEDANCE SPECTRUM FOR CIRCULAR PLATES OF VARIOUS DAMAGE LEVELS

12.7.1 TYPICAL E/M IMPEDANCE SPECTRA FOR CIRCULAR PLATES WITH CRACK DAMAGE

Typical E/M impedance data takes the form of frequency domain plots of the E/M impedance real part, which accurately represents the local structural impedance at the sensor location. In this study, the HP 4194A Impedance Analyzer was utilized. The analyzer performs impedance measurements with 401 data points along the frequency axis. Thus, the complete data set contains three columns with 401 data points each, corresponding to (a) frequency, (b) real part of the impedance Re (Z), and (c) imaginary part of the impedance Im (Z). The real part of E/M impedance, Re (Z), is the best illustrator of information on local structural dynamics since the resonance frequencies and amplitudes are easily distinguished and the frequency shifts are immediately noticed. We performed damage detection experiments on five statistical groups of specimens with increasing amounts of damage (Fig. 12.31). Each group contained five identical 0.8-mm thick and 100-mm diameter aluminum circular plates. A 7-mm diameter, 0.2-mm thick PWAS was placed in the center of each plate. A 10-mm simulated crack was produced in each specimen in the form of a narrow through-the-thickness slit cut with a water-jet machining. The slit was incrementally brought closer to the plate center (Fig. 12.31a).

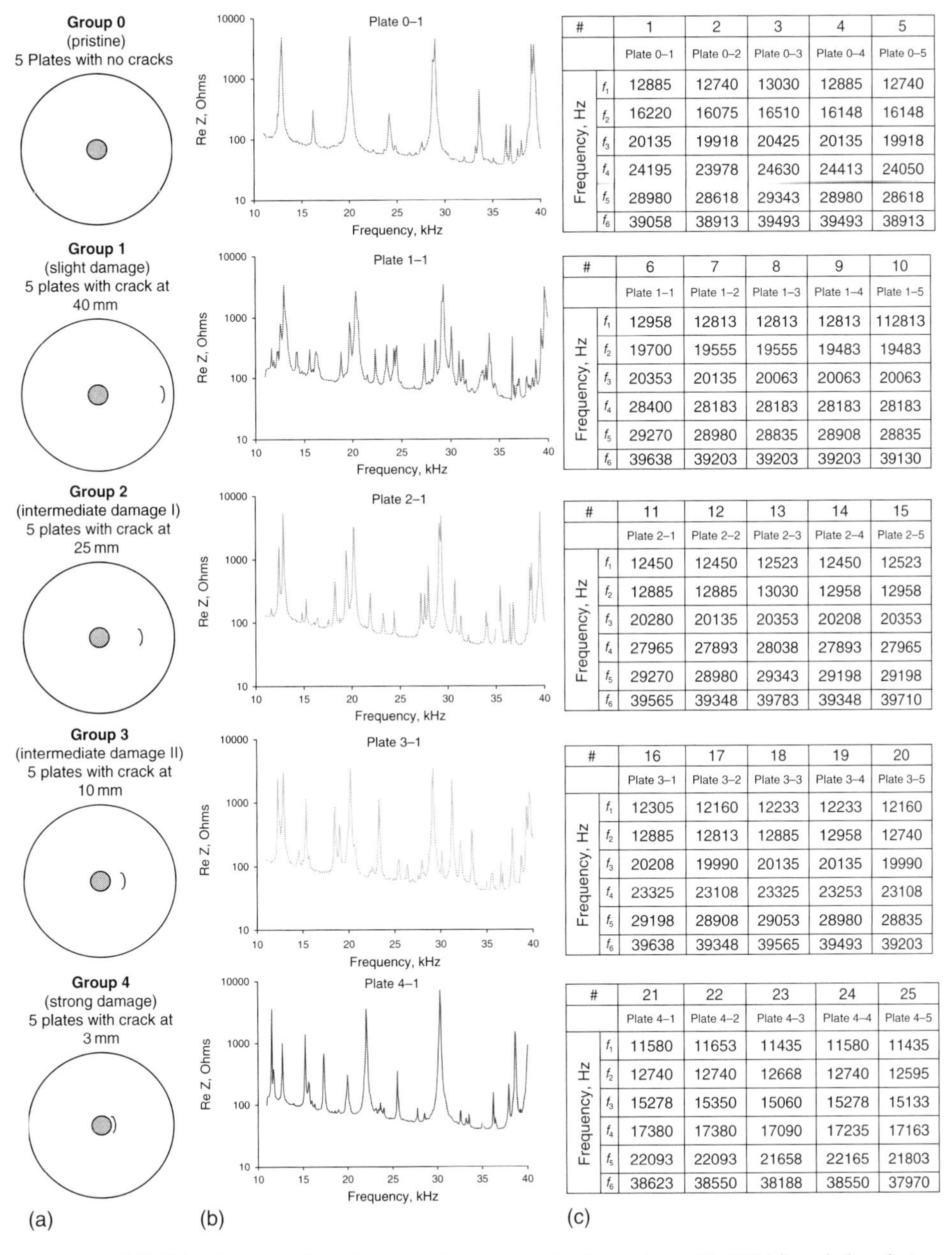

#		1	2	3	4	5
		Plate 0–1	Plate 0–2	Plate 0–3	Plate 0–4	Plate 0–5
Frequency, Hz	f_1	12885	12740	13030	12885	12740
	f_2	16220	16075	16510	16148	16148
	f_3	20135	19918	20425	20135	19918
	f_4	24195	23978	24630	24413	24050
	f_5	28980	28618	29343	28980	28618
	f_6	39058	38913	39493	39493	38913

#		6	7	8	9	10
		Plate 1–1	Plate 1–2	Plate 1–3	Plate 1–4	Plate 1–5
Frequency, Hz	f_1	12958	12813	12813	12813	112813
	f_2	19700	19555	19555	19483	19483
	f_3	20353	20135	20063	20063	20063
	f_4	28400	28183	28183	28183	28183
	f_5	29270	28980	28835	28908	28835
	f_6	39638	39203	39203	39203	39130

#		11	12	13	14	15
		Plate 2–1	Plate 2–2	Plate 2–3	Plate 2–4	Plate 2–5
Frequency, Hz	f_1	12450	12450	12523	12450	12523
	f_2	12885	12885	13030	12958	12958
	f_3	20280	20135	20353	20208	20353
	f_4	27965	27893	28038	27893	27965
	f_5	29270	28980	29343	29198	29198
	f_6	39565	39348	39783	39348	39710

#		16	17	18	19	20
		Plate 3–1	Plate 3–2	Plate 3–3	Plate 3–4	Plate 3–5
Frequency, Hz	f_1	12305	12160	12233	12233	12160
	f_2	12885	12813	12885	12958	12740
	f_3	20208	19990	20135	20135	19990
	f_4	23325	23108	23325	23253	23108
	f_5	29198	28908	29053	28980	28835
	f_6	39638	39348	39565	39493	39203

#		21	22	23	24	25
		Plate 4–1	Plate 4–2	Plate 4–3	Plate 4–4	Plate 4–5
Frequency, Hz	f_1	11580	11653	11435	11580	11435
	f_2	12740	12740	12668	12740	12595
	f_3	15278	15350	15060	15278	15133
	f_4	17380	17380	17090	17235	17163
	f_5	22093	22093	21658	22165	21803
	f_6	38623	38550	38188	38550	37970

FIGURE 12.31 Damage detection experiments on circular plates with PWAS and the electro-mechanical (E/M) impedance method: (a) schematic of damage; (b) E/M impedance spectra; (c) frequencies for each specimen

Each group of five plates represented a different state of damage. Group 0 represented the pristine condition.

12.7.1.1 High-Frequency Structural Impedance Data

The advantage of the E/M impedance method is its high frequency capability. Thus, it can measure with ease the high-order harmonics, which are more sensitive to local damage than the low-order harmonics. Figure 12.32 compares spectra taken in three high-frequency bands: 10–40 kHz, 10–150 kHz, and 300–450 kHz. The high-frequency modes

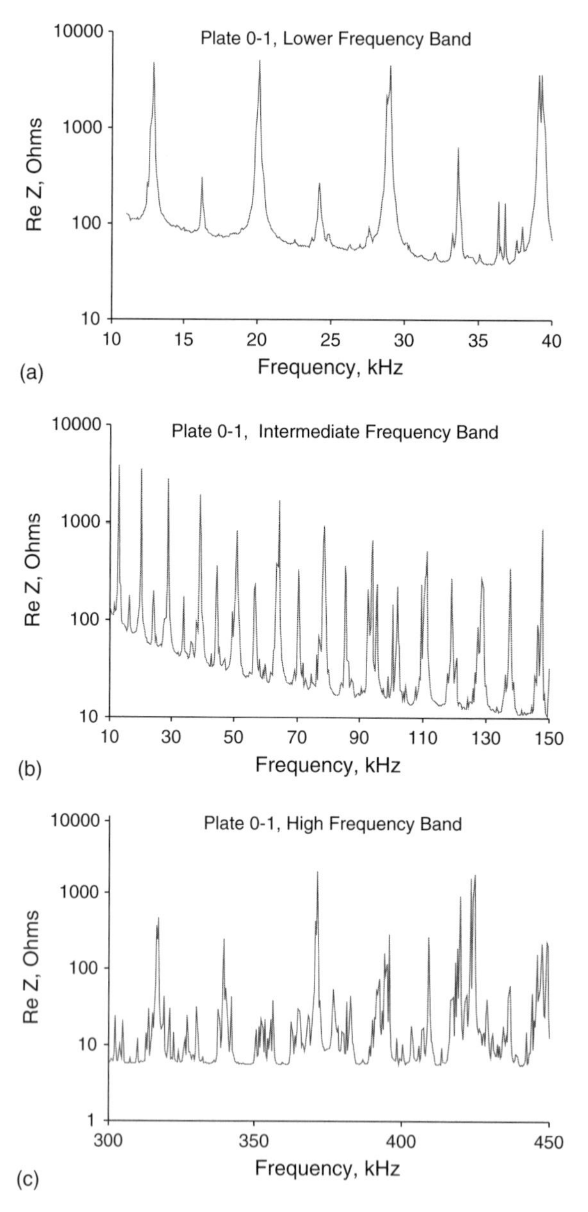

FIGURE 12.32 Modal density increases with the frequency band (E/M impedance measurements on Group 0, Plate1): (a) 10–40 kHz, 4 major peaks and 4 minor peaks; (b) 10–150 kHz, 18 major peaks, 8 minor peaks; (c) 300–450 kHz, 8 major peaks with multiple crests, numerous minor peaks.

are very well excited. As expected, the modal density increases with the frequency band, which enriches the information contents of the data. However, during data collection over the high frequency band, a researcher should carefully consider the size of frequency interval to ensure the good resolution of the measurements. Poor resolution of the spectrum may hide important harmonics.

12.7.1.2 The Effect of Damage Location on E/M Impedance Spectra

An important issue to be addressed for the practical application of the E/M impedance method is the sensitivity of the technique to the presence of damage. It is known that the impedance spectrum is sensitive to the state of structural integrity, but it is still not clear how far the method can detect damage and what size of the damage will noticeably alter the impedance signature. Thus, the task was to quantify the effect of structural damage on the E/M impedance spectrum. This can be done by considering similar-shaped specimens with the damage placed gradually at various distances from the sensor. In the damage detection experiment (Fig. 12.31), five statistical groups of specimens with increasing amounts of damage were tested. Each group contained five identical 0.8-mm thick and 100-mm diameter aluminum circular plates. A 7-mm diameter, 0.2-mm thick PZT-wafer active sensor was placed in the center of each plate. A 10-mm simulated crack was produced in each specimen in the form of a narrow through-the-thickness slit cut with a water jet. The slit was incrementally brought closer to the plate center (Fig. 12.31a). Each group of five plates represented a different state of damage. Group 0 represented the pristine condition. The distance between a sensor and a slit decreases as the group number increases. Within each group, the damage was 'identically' replicated to all five group members. Thus, the method's repeatability and reproducibility was tested. During the experiments, the specimens were supported on commercial foam to simulate free boundary condition around the plate's circumference. High-frequency structural impedance readings were taken in the three frequency bands: 10–40 kHz; 10–150 kHz; and 300–450 kHz. Clear separation between modes was noticed in the 10–40 kHz frequency band.

Figure 12.33 compares the 'pristine' and the 'damaged' spectra, for two damage locations. When the damage is remote, the frequency spectrum is only slightly modified:

- Small shifts take place in the existing resonances
- New low-intensity resonances appear.

When the damage is close to the sensor, these modifications are intense:

- Large shifts take place in the existing resonances
- Strong new resonances appear
- Some peaks split.

A similar situation was observed for higher-frequency bands. We conclude that, in general, as damage progresses, two phenomena happen: (a) the resonance frequencies change; and (b) new resonance frequencies appear.

Figure 12.33 compares 'pristine' and 'damaged' spectra, for various damage intensities. Minor damage (Fig. 12.33a) slightly modifies the frequency spectrum by: (i) small shifts in the existing resonances and (ii) new low-intensity resonances. Whereas intense damage (Fig. 12.33b) drastically modifies the frequency spectrum by: (i) large shifts in the existing resonances; (ii) strong new resonances and (iii) peak splits.

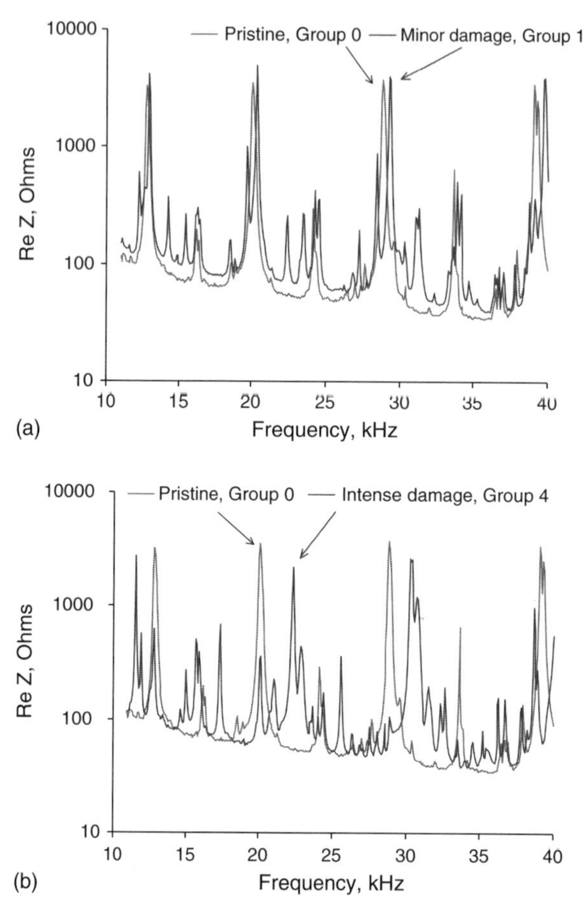

FIGURE 12.33 E/M impedance results in the 10–40 kHz band: (a) minor damage (Group 1 vs. Group 0); (b) intense damage (Group 4 vs. Group 0).

12.7.1.3 Inherent Statistical Variation in the E/M Impedance Data

Knowledge of the statistical variation in nominally identical situations is of great importance for the assessment and calibration of any health monitoring method. During the experiments, particular attention was given to producing nominally 'identical' specimens. However, slight variations were unavoidable, giving rise to random error in the E/M impedance data. Consider the circular plates damage experiment described in a previous chapter. Figure 12.31 illustrates the experimental specimens and E/M impedance spectra. The points of resonances in the 11–40 kHz frequency band were chosen for statistical data processing. As shown in Table 12.2, the resonant frequencies fall in a very tight interval (1% STD), indicating good experimental reproducibility. However, the resonance amplitudes show a wider spread (10–30% STD). Similar results were obtained for the other damage scenarios. Figure 12.34 illustrates statistical effects in Group 0 (pristine) in the lower-frequency band (10–40 kHz). Average and standard deviation (STD) values for the four dominant resonances are also given. It is noted that the resonant frequencies fall in a very tight interval (1% STD), indicating good experimental reproducibility. However, the resonance amplitudes show a wider spread (10–30% STD). Similar results were obtained for the other damage cases.

TABLE 12.2 Statistical summary for resonance peaks of four axisymmetric modes of a circular plate as measured with the piezoelectric active sensor using the E/M impedance method

Statistical Summary for Circular Plates in Group 0 – Pristine			
Average frequency, kHz	Frequency STD, kHz (%)	$\log_{10}$ – Average amplitude, $\log_{10}$ – Ohms	$\log_{10}$ – Amplitude STD, $\log_{10}$ – Ohms (%)
12.856	0.121 (1%)	3.680	0.069 (1.8%)
20.106	0.209 (1%)	3.650	0.046 (1.2%)
28.908	0.303 (1%)	3.615	0.064 (1.7%)
39.246	0.415 (1%)	3.651	0.132 (3.6%)

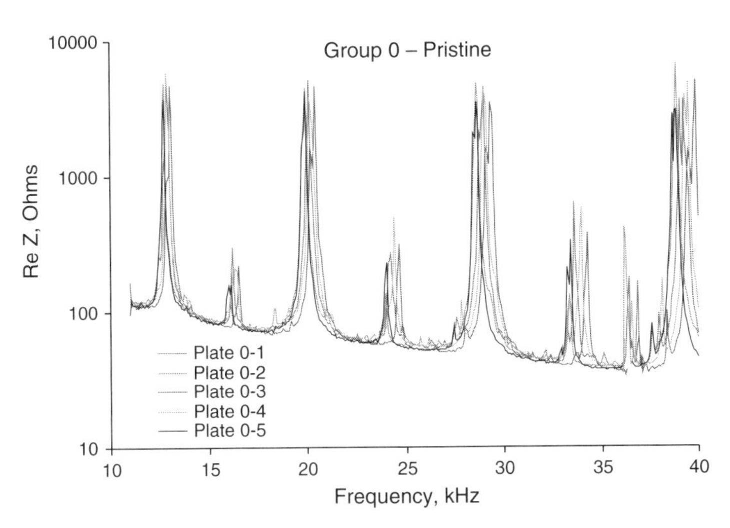

FIGURE 12.34 Statistical effects on impedance signal: superposed resonance spectra of five 'identical' Group 0 (pristine) specimens.

12.7.1.4 Effect of Temperature on the E/M Impedance Spectrum

Bias error in the E/M impedance data is usually due to the change in an active-sensor's piezoelectric properties with temperature. Ayres et al. (1996) and Pardo de Vera and Guemes (1997), among others, have shown that changes in the environmental temperature induce shifts in the E/M impedance spectrum without essentially changing its appearance. In addition, Park et al. (1999), and Berman and Quattrone (1999) observed that the effect of temperature revealed itself in shifting resonance frequencies of the impedance spectra. Other effects, such as humidity and freeze–thaw Cycles, may also be important and need be investigated. In this chapter, the temperature-effect was neglected since the measurements were always conducted in the laboratory at the room temperature.

12.7.2 FEATURE-VECTOR DATA REDUCTION

In many cases, it is desirable to reduce the complete data set to a small number of spectral features, which uniquely represent the current structural condition. This leads to an important stage of feature extraction, i.e., reduction of problem dimensionality and separation of the essential information from the nonessential information. In spectral analysis, the features-vector approach recognizes that the number and characteristics of resonance peaks play a major role in describing the dynamic behavior of the structure. By considering the essential features of the spectrum, i.e. the peaks at resonance frequencies, meaningful information corresponding to the particular state of structural health can be extracted. This information can be organized in a features vector where each feature corresponds to a certain number of resonance frequencies. As an illustration, consider the E/M impedance spectra reproduced in Fig. 12.31. Each spectrum contains 401 data points positioned at 0.075 kHz intervals in the 10 to 40 kHz range. Five spectra are presented: one pristine and four with various damage location. Although the difference between spectra is obvious, quantification of this difference is neither obvious nor immediate. In a featureless approach, one would attempt to directly compare the two 401-long strings of numbers (raw input vectors) associated with each spectrum. Such a flat consideration would lead to a large-dimensional problem, and the associate computational difficulty.

Based on the features vector approach, it is possible to distinguish between a 'pristine' and a 'damaged' structure by considering the number and the position of the resonance peaks. For example, in the 10–40 kHz band, the pristine-plate spectrum of Fig. 12.31 displays four dominant resonance peaks (12.8, 20.1, 28.9, and 39.2 kHz), and four minor peaks (15.9, 24.5, 34.1, 36.35 kHz). By considering these features, the dimensionality of the problem can be reduced by almost 2 orders of magnitude, from 400-points to less than 10 features. The details for these features vectors are given in Fig. 12.31c. For the damaged plate, also shown in Fig. 12.31, the six dominant peaks (11.6, 12.7, 15.2, 17.4, 22.0, 38.6 kHz) can be distinguished. Though the number of features has increased, the problem size is still at one order of magnitude less than in a featureless approach. Similar results are observable in the 10–150 kHz and 300–450 kHz bands. The use of high frequencies ensures that the damage metric detects incipient damage at an early stage. Besides the resonance frequency, f_r, and amplitude at resonance, A_r, other candidates for spectral features may be the damping factor and the position of the spectral baseline. However, it was found that f_r and A_r serve as the most reliable indicators of structural damage.

12.7.2.1 Features Extraction Algorithm

The features extraction algorithm used in this study utilizes a rectangular search window that transverses the spectrum from left to right. A peak is found inside the search window when the difference in height between the local data maximum inside the window and the data values at both ends of the window is less than the height of the window. The height of the window is specified as a percentage of the total amplitude of the data in the range (the amplitude is defined as the difference between the maximum and the minimum of the data). The width of the window is specified as a percentage of the total number of points in the data range. Generally, the smaller the height and the width values, the more peaks are likely to be identified. However, too many peaks of low amplitude may result in an unnecessarily long and noisy features vector. A threshold identifier, based on the energy content of each peak, is used to prune the features vector. As the result of the data extraction, the features vector contains the resonance frequencies identified in the spectrum.

12.7.2.2 Adaptive Resizing of the Features Vector

It is apparent from Fig. 12.31 that the number of features in the 'pristine' spectrum is different than the number of features in the 'damaged' spectra. As damage takes place, two phenomena may occur: (a) the values of the existing features may change; (b) the new features, associated with the new vibration modes and local resonances, may appear. For example, considering the dominant peaks in Fig. 12.31, one observes that the features vector for the 'pristine' spectrum (plate 0_1) has dimension $N_{pristine} = 8$, whereas the features vector for the 'damaged' spectrum (plate 4_1) has dimension $N_{damaged} = 11$.

Induction heuristics approach was used to add and subtract features to the feature space, i.e., to enlarge or reduce its dimensionality. The zero-fill method was used to upgrade the lower-dimension features vector to a higher-dimension features vector. Thus, the 'pristine' features vector was updated to $N^*_{\text{pristine}} = 11$, with the new cells being filled with zero values. After the features vector is updated with new frequencies, the cells are reordered to form an increasing-frequency sequence.

12.7.3 STATISTICAL ANALYSIS OF E/M IMPEDANCE SPECTRA

12.7.3.1 Overall Statistics Damage Metrics

The damage index is a scalar quantity that results from the processing of two impedance spectra and reveals the difference between them. Theoretically, the best damage index would be the metric that captures features of the spectra directly modified by the damage presence, and neglects the variations due to normal conditions (i.e., statistical difference within one population of specimens or normal deviation of temperature, pressure, level of vibrations etc.). Unfortunately, this goal is not easy, if possible, to achieve. To date, several damage metrics are widely used to compare impedance spectra and assess the damage presence. Among them, the most popular are: the root mean square deviation (RMSD), the mean absolute percentage deviation (MAPD), the covariance (Cov), and the correlation coefficient deviation (CCD). The mathematical expressions for RMSD, MAPD, Cov, and CCD can be expressed as follows

$$RMSD = \sqrt{\frac{\sum_N \left[\mathrm{Re}(Z_i) - \mathrm{Re}(Z_i^0)\right]^2}{\sum_N \left[\mathrm{Re}(Z_i^0)\right]^2}} \tag{69}$$

$$MAPD = \sum\nolimits_N \left|\left[\mathrm{Re}(Z_i) - \mathrm{Re}(Z_i^0)\right]/\mathrm{Re}(Z_i^0)\right| \tag{70}$$

$$Cov = \sum\nolimits_N \left[\mathrm{Re}(Z_i) - \mathrm{Re}(\overline{Z})\right] \cdot \left[\mathrm{Re}(Z_i^0) - \mathrm{Re}(\overline{Z}^0)\right] \tag{71}$$

$$CCD = 1 - \frac{1}{\sigma_Z \sigma_{Z^0}} \sum\nolimits_N \left[\mathrm{Re}(Z_i) - \mathrm{Re}(\overline{Z})\right] \cdot \left[\mathrm{Re}(Z_i^0) - \mathrm{Re}(\overline{Z}^0)\right] \tag{72}$$

where N is the number of sample points in the spectrum (here, $N = 401$), and the superscript 0 signifies the pristine (base-line) state of the structure. $\overline{Z}$, $\overline{Z}^0$ are the mean values and σ_Z, σ_{Z^0} are the standard deviations for the current and the baseline spectra.

Equations (70)–(72) result in scalar numbers, which represent the relationship between the compared spectra. Thus it is expected that the variations in Re(Z) and appearance of new harmonics may alter the scalar value of a damage index identifying the damage. The advantage of using Eqs. (70)–(72) is that the input impedance spectra does not need any pre-processing, i.e., the data obtained from measurement equipment can be directly used to calculate the damage index.

On the other hand, the frequency shift of resonance peaks may not have an obvious effect on the overall statistics damage index. In this case, the choice of the appropriate frequency band becomes a critical issue. Consider a scenario where one of the resonance peaks falls in the upper or lower limits of the fixed frequency band. Due to statistical variation between specimens in a population, the resonance peak corresponding to this modeshape may not be present in the spectrum for another specimen because it falls out of the chosen frequency band. This effect may noticeably affect the value of damage index when the density of resonance peaks is low, i.e., only few peaks are present in the spectrum. For this reason, it has been suggested that the overall statistic metric be applied for frequency bands with a high density of resonance peaks. Another phenomenon, which is not captured adequately with overall statistics formulas, is the shift of the baseline in the impedance spectrum. This shift usually occurs due to variations in installation of the sensors or under normal operational conditions. Although unrelated to damage, the baseline shift alters the values of damage index, which then needs a proper adjustment.

In the experimental study, several overall-statistics damage metrics were used to classify the different groups of specimens presented in Chapter 9, Section 9.4. The experimental specimens were described in depth in the previous section. The calculations were performed for the following overall-statistics damage metrics: root mean square deviation (RMSD); mean absolute percentage deviation (MAPD); covariance (Cov); correlation coefficient deviation (CCD). The covariance damage metric was disregarded in further analysis in favor of the correlation coefficient deviation. The results of the calculations are summarized in Table 12.3. The correlation coefficient deviation was found to be the best metric of damage presence. Table 12.3 presents correlation coefficient deviation for three frequency bands: 10–40 kHz, 10–150 kHz, 300–450 kHz. It was observed that for 10–40 kHz and 10–150 kHz frequency ranges distribution of variation of correlation coefficient is not uniform, although it was expected to decrease as the crack moves away from the sensor. During investigation, it was found that the choice of frequency band for data analysis plays a significant role in this process. It is recommended that the researcher choose the frequency band where the highest density of amplitude peaks is observed. Following this approach, the frequency band 300–450 kHz was chosen for further data analysis. The variation of correlation coefficient, $(CCD)^7$, was studied for 4 damage scenarios. The results are presented in Fig. 12.35. The $(CCD)^7$ damage metric tends to decrease linearly as the crack moves away from the sensor.

TABLE 12.3 Table overall-statistics damage metrics for various frequency ranges

Frequency range	11–40 kHz				11–150 kHz				300–450 kHz			
Groups	0_1	0_2	0_3	0_4	0_1	0_2	0_3	0_4	0_1	0_2	0_3	0_4
RMSD%	122	116	94	108	144	161	109	118	93	96	102	107
MAPD%	107	89	102	180	241	259	170	183	189	115	142	242
CCD%	84	75	53	100	93	91	52	96	81	85	87	89

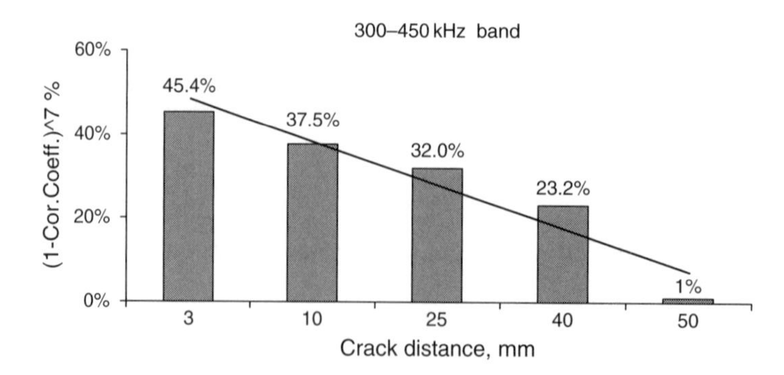

FIGURE 12.35 Damage metric variation with the distance between the crack and the sensor in the 300–450 kHz frequency band.

From the results of the overall-statistics procedure for classification of damage location in circular plates, the following conclusions can be drawn:

- The crack presence significantly modifies the real part of the E/M impedance spectrum. The amplitude variations are well captured by the overall statistic metrics.
- The difference between 'pristine' and 'damaged' spectra decreases as the distance between the sensor and the crack increases. The relationship between spectral classes and distances is more prominent in the high-frequency band, which has a high density of resonance peaks.
- To obtain consistent results during the health monitoring process, the appropriate frequency band (usually in the hundreds of kHz) and the appropriate damage metric must be used.

In general, overall-statistics approach may be a good choice for SHM systems where the precision of the damage metric is less important than the ability of the method to detect the damage at an early stage. This approach is quick and easy to apply, because no additional pre-processing of the measurement data is needed. However, the results of SHM process using the overall-statistics approach are not always straightforward, and the researcher should pay particular attention to the frequency band of investigation, and to the changes in impedance spectra due to environmental and other factors encountered during the normal operation of the structure.

12.7.3.2 Damage Detection Using Features-Based Statistics

In previous sections, the features of the E/M impedance spectrum, which may be used for damage identification, were discussed. It was also mentioned that these features should be sensitive to the changes in the local structural dynamics produced by the damage. Each feature, say frequency, falls into a certain statistical distribution, which can be determined by considering a statistical sample of the data. In such a sample, the numerical values of the frequencies for the particular harmonic of the spectrum will have a statistical spread and information on the type of statistical distribution can be extracted. Knowing the probability distribution, which is illustrated with probability density functions, the classification problem can be attacked. In the elementary case of structural damage identification when the data is available for two scenarios (the 'pristine' and the 'damaged'), there is a need to assign a criterion for classification and, accordingly, determine to which of the corresponding two classes the data belongs. In other words, this statistical problem is reduced to testing the 'pristine' hypothesis against the 'damaged' hypothesis. For the situations when the structural damage grows in time, or is discretely distributed, a similar approach can be employed to classify data in the several categories depending on damage size or location.

The use of the E/M impedance method allows direct identification of local structural dynamics by analyzing the impedance spectrum. The spectrum gives information on structural resonances. Therefore, its parameters can be used to solve the classification problem. For simplicity, consider the case where the data is classified into the 'pristine' and the 'damaged' classes. To reduce the number of variables, consider only the resonance frequencies as the feature variables for damage identification. Without loss of generality, it is appropriate to assume that the probability distribution of a particular resonance frequency measured on several specimens is normal (Gaussian). In

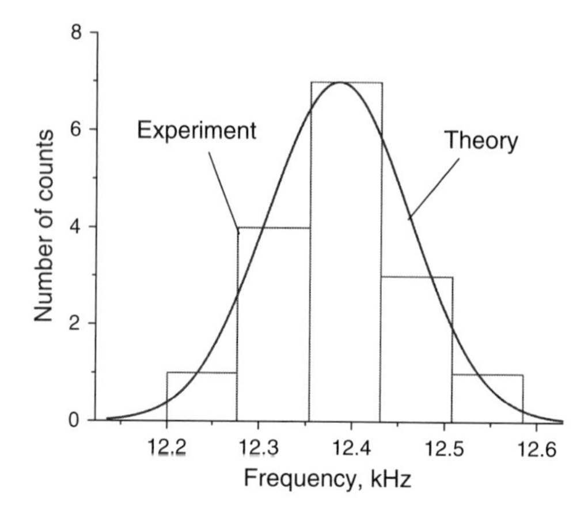

FIGURE 12.36 Statistical distribution for the third harmonic of the 16 circular plates.

support of this assumption, the statistical distribution of the third resonance frequency of 16 'identical' circular plates was studied. Figure 12.36 presents the histogram derived from the experimental data. The theoretical normal distribution curve was also superimposed on this graph. Since the sample histogram closely fits the normal distribution, we conclude that the population of resonance frequencies has a normal distribution.

12.7.3.3 The Use of *t*-Test for Damage Identification

In testing statistical hypotheses, the null and alternative hypothesis situations need to be assigned. During the health monitoring process, the structure is assumed to be in the 'pristine' condition unless strong evidence is found to contradict this assumption (Devore, 1999). Thus, it is convenient to denote the 'pristine' condition as the null hypothesis (H_0) and the 'damaged' as the alternative hypothesis (H_a). A test on the hypotheses should be performed to decide whether the 'pristine' hypothesis should be rejected.

In this study, the set of specimens presented in Figure 12.31 was used. The set of specimens consists of five groups of plates with various damage location. For hypothesis testing, consider two damage cases: (a) when the crack is far away from the center of the plate (weak damage, group 1), and (b) when the crack is very close to the center of the plate (strong damage, group 4). The pristine group, which corresponds to the null hypothesis H_0, is group 0. Consider a strong damage situation when the crack is close to the center of the plate, and denote the means as μ_0 (null) and μ (alternative). There are three possible choices of the alternative hypothesis: $\mu > \mu_0$; $\mu \neq \mu_0$; $\mu < \mu_0$; It is known that natural frequencies of structures tend to shift downward when the presence of damage results in reduction of stiffness. However, the amount of this shift depends on the severity of damage and on the location of damage relative to the nodes of the mode shape. Hence, the relationship is nonlinear (Chrondos and Dimarogonas, 1980). Based on these facts, we assume that it is appropriate to test the null hypothesis against the alternative hypothesis when $\mu < \mu_0$, i.e., $\mu_0 - \mu > 0$.

The *t*-test was performed on the data obtained for the third resonance frequency for group 0 and group 4. The data, which represents two independent random samples

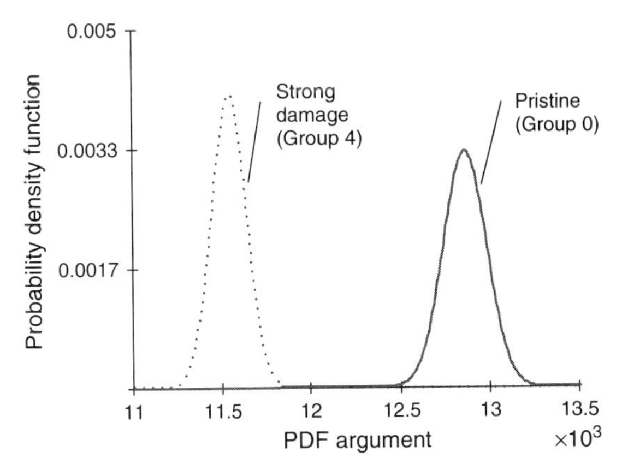

FIGURE 12.37 Distribution of probability density function for Group 0 (pristine) and Group 4 (strong damage); investigated spectral feature was the third resonance frequency, f_3.

selected from independent normal populations, is shown in Fig. 12.37. The t-statistic is calculated according to

$$t = \frac{(\overline{X}_1 - \overline{X}_2)}{\sqrt{\left(\frac{s_1^2}{n_1} + \frac{s_2^2}{n_2}\right)}} \tag{73}$$

where $\overline{X}_1$, $\overline{X}_2$ and s_1^2, s_2^2 are the mean and the variances of samples of sizes n_1 and n_2.

The t-statistics is used to determine the p-value for each hypothesis testing. P-value is the area of the upper tail of the normal distribution curve controlled with the calculated value of t (Devore, 1999). Based on the p-value calculated during the two-sample t-test, the null hypothesis can be either accepted or rejected. Assume that the type-I error, α, does not exceed 0.05. The two-sample t-test gives $t = 18.96$, and the mean difference = 1319 Hz. Hence, we obtain $p = 0.00$ (Table 12.4). Since $p < \alpha$, the null hypothesis is rejected at the level α which means that $\mu < \mu_0$ and hence the structure is damaged. Thus, using the third resonance frequency as a structural health feature, clear separation between the 'strong damage' and the 'pristine' conditions is possible.

Another extreme scenario to be considered is that in which the damage is far away from the center of the plate, and, hence, its effect on the plate specimen is difficult to distinguish. This situation is found when comparing groups 0 and 1. The data for this case is given also in Table 12.4. Assume type-I error, α, at the same level of 0.05. The two-sample t-test for group 0 and group 1 gives: $t = 0.23$, the mean difference = 14 Hz, and $p = 0.414$. In this case $p > \alpha$ and hence the null hypothesis cannot be rejected. Figure 12.38 illustrates

TABLE 12.4 Two-sample t-test

	f_3 Hz		f_4 Hz		f_5 Hz		f_6 Hz	
Groups	Average	p-value	Average	p-value	Average	p-value	Average	p-value
0	12856	N/A	20106	N/A	28907	N/A	39246	N/A
1	12841	0.414	20135	0.601	28965	0.637	39275	0.553
2	12479	0.001	19468	0.001	27950	0.001	38506	0.010
3	12218	0.000	18482	0.000	28994	0.707	37578	0.001
4	11536	0.000	15219	0.000	30009	1.000	38376	0.003

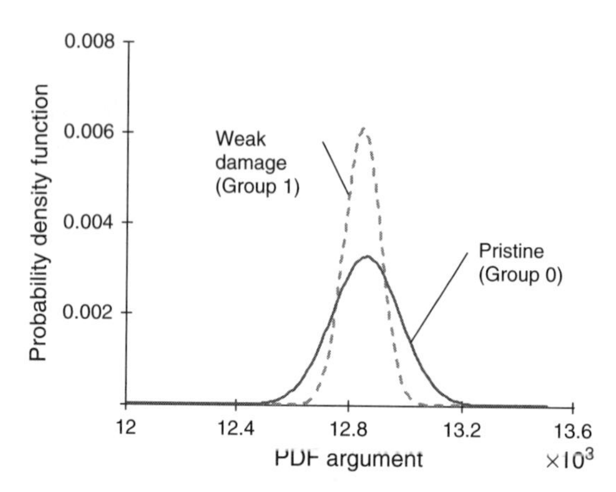

FIGURE 12.38 Distribution of probability density functions for Group 0 (pristine) and Group 1 (weak damage); investigated spectral feature was the third resonance frequency, f_3.

this situation. Thus, using the 3rd frequency as structural health feature it is impossible to distinguish between the 'weak damage' and the 'pristine' damage conditions. Although this result shows a limitation, it is consistent with the physical explanation that the weak damage contributes little to structural dynamics and sometimes may not be distinguishable from the pristine case within the practical statistical spread of certain frequencies. For example, Fig. 12.38 shows that a remote crack introduces very little variation in frequency values and is hard to identify.

The procedure illustrated above for the f_3 frequency as a structural health monitoring feature can be repeated with the other frequencies taken as structural health features. It was noticed that, depending on the damage location, some resonance frequencies shift significantly, but some do not show considerable shifts. This situation is difficult to account for in decision making for a multiple number of frequencies. Thus, to obtain reliable results from the SHM process, the cumulative effect of all the frequency shifts should be considered. Kernel-based statistical analysis has been developed to accurately classify data into several classes according to their relevance. The implementation of Kernel analysis in automatic learning algorithms based on neural networks (NN) opens a broad horizon for the analysis of large data sets. In such applications, the machine solves the classification problem according to the Bayes decision rule.

12.7.4 NEURAL NETS DAMAGE DETECTION IN CIRCULAR PLATE SPECIMENS USING THE E/M IMPEDANCE DATA

In this section we examine the implementation of neural nets approach to damage detection with the E/M impedance method using the laboratory data collected on circular plate specimens. A schematic of the probabilistic neural net used for this example was shown in Fig. 12.29b.

The case of damage identification in circular plates discussed in previous sections is used to test the ability of PNN to classify the data into several categories relative to the damage location. Following Fig. 12.31, five classes are assigned: 0 – no damage (pristine); 1 – damage is located 40 mm from the sensor (weak damage); 2 – damage is located 25 mm apart from the sensor; 3 – damage is located 10 mm apart from the sensor; and 4 – damage is located 3 mm from the sensor (strong damage). Examples of the E/M impedance spectra obtained for each situation are shown in Fig. 12.31b. The frequency domain data is chosen to solve the classification problem. At the beginning, a small number of frequencies was used, namely the four dominant frequencies of the pristine plate. This studies the variation in behavior of the four dominant frequencies of the pristine plate with the appearance of damage. Table 12.5 shows the input data matrix for this classification problem. The network used for this case consists of four inputs, five neurons in the pattern layer, and five neurons in the output layer. The number of inputs corresponds to the number of frequencies in an input features vector, the pattern layer is formed according to the number of input/target pairs, and the number of output neurons represents the categories in which the input data supposed to be classified.

The features vector used for the training and validation with the correspondent classification results are presented in Table 12.6. The results of the four tests are given in Table 12.6 where T and V denote the testing and validation vectors correspondingly. It can be seen that the PNN is able to classify successfully the input data into classes that represent strong and intermediate damage. This means that variation of the natural frequencies for those cases is large enough to separate classes. However, when the damage is remote, the PNN produces inconsistent results. For example, in Test I the 'weak damage' data (group 1) was misclassified as the 'pristine' (group 0) four times, whereas the 'pristine' data (group 0) was misclassified as the 'weak damage' (group 1) two times. As the number of training vectors was increased, the incidence of misclassification diminished (e.g. Test II). However, even for a maximum number of the training data, misclassification error between group 0 and group 1 could not be avoided. At the next step, the number of training vectors in misclassified groups was increased. Though this was taken to the extreme in test 4, the correct results were not obtained. It is worth noting that the same situation was observed while solving the classification problem with the statistical analysis discussed in the previous section.

Some of the natural frequencies do not noticeably deviate for the cases of the pristine structure and the structure with remote damage. The obvious way of improvement of such a situation is to increase the number of natural frequencies in the input features vectors and see if their distribution gives better variance from class to class. This is the scope of an example given next.

The number of frequencies in the input vectors was increased to six. This led to six inputs in the PNN. It is expected that due to additional information, the PNN will distinguish the most difficult classification cases, i.e., the weak damage class (group 1), from the pristine class (group 0). Two additional rows were added into the input data matrix as is shown in Table 12.7. The synoptic results are given in Table 12.8, which indicates that, regardless of training vector, all input data was correctly classified into the five classes. In the discussed examples, only the deviation of natural frequencies from the original value corresponding to the healthy structure was considered. The appearance of new harmonics was not introduced yet. Nevertheless, this feature plays an important role in distinguishing between healthy and damaged structures, especially for the cases when the damage is incipient or located far away from the sensing module. It is possible to account for the appearance of new harmonics by using the opposite procedure: introduce zero values for frequencies in the features vector of a healthy structure where these harmonics are not present. Thus, the features vector can be expanded to the desirable length. As an example, 11 frequencies were used in the input features vectors to create a network. The input features vectors and classification results are presented in Table 12.9 and Table 12.10. Similar to the previous situation, the PNN was able to classify correctly data regardless of the choice of training vector. Summarizing, promising results were obtained for the damage classification using PNN and the E/M impedance method. As a general remark, it is recommended to have an adequately large number of frequencies in the features vector to ensure good performance of PNN.

TABLE 12.5 Input data matrix for PNN classification of circular plates: the 4-resonance frequencies study (all values in Hz)

1	2	3	4	5	6	7	8	9	10	11	12	13	14	15
Group 0 (pristine)					Group 1 (weak damage)					Group 2				
12885	12740	13030	12885	12740	12958	12813	12813	12813	12813	12450	12450	12523	12450	12523
20135	19918	20425	20135	19918	20353	20135	20063	20063	20063	19483	19410	19483	19410	19555
28980	28618	29343	28980	28618	29270	28980	28835	28908	28835	27965	27893	28038	27893	27965
39058	38913	39855	39493	38913	39638	39203	39203	39203	39130	38695	38333	38550	38405	38550

16	17	18	19	20	21	22	23	24	25
Group 3					Group 4 (strong damage)				
12305	12160	12233	12233	12160	11580	11653	11435	11580	11435
18395	18395	18540	18540	18395	15278	15350	15060	15278	15133
29198	28908	29053	28980	28835	38623	38550	38188	38550	37970
37753	37463	37680	37535	37463	40000	39928	39203	40000	39493

TABLE 12.6 Synoptic classification table for circular plates: the 4-resonance frequencies case

Test	1	2	3	4	5	6	7	8	9	10	11	12	13	14	15	16	17	18	19	20	21	22	23	24	25	
	Group 0 (pristine)					Group 1					Group 2					Group 3					Group 4 (strong damage)					
I	T	V	V	V	V	T	V	V	V	V	T	V	V	V	V	T	V	V	V	V	T	V	V	V	V	IN
	X	1	2	2	1	X	1	1	1	1	X	3	3	3	3	X	4	4	4	4	X	5	5	5	5	OUT
II	T	T	V	V	V	T	T	V	V	V	T	V	V	V	V	T	V	V	V	V	T	V	V	V	V	IN
	X	X	2	2	1	X	X	2	2	2	X	3	3	3	3	X	4	4	4	4	X	5	5	5	5	OUT
III	T	T	T	V	V	T	T	V	V	V	T	V	V	V	V	T	V	V	V	V	T	V	V	V	V	IN
	X	X	X	2	1	X	X	2	2	2	X	3	3	3	3	X	4	4	4	4	X	5	5	5	5	OUT
IV	T	T	T	V	T	T	T	V	V	V	T	V	V	V	V	T	V	V	V	V	T	V	V	V	V	IN
	X	X	X	2	X	X	X	2	2	2	X	3	3	3	3	X	4	4	4	4	X	5	5	5	5	OUT

1–25 – plate number
1–5 – correspondent class number
T – training vector
V – validation vector
X – vector absent for classification
IN – input vectors of PNN
OUT – output result of PNN

TABLE 12.7 Input data matrix for PNN classification of circular plates: the 6-resonance frequencies study (all values in Hz)

1	2	3	4	5	6	7	8	9	10	11	12	13	14	15
Group 0 (pristine)					Group 1					Group 2				
12885	12740	13030	12885	12740	12958	12813	12813	12813	12813	12450	12450	12523	12450	12523
16220	16075	16510	16148	16148	19700	19555	19555	19483	19483	12885	12885	13030	12958	12958
20135	19918	20425	20135	19918	20353	20135	20063	20063	20063	20280	20135	20353	20208	20353
24195	23978	24630	24413	24050	28400	28183	28183	28183	28183	27965	27893	28038	27893	27965
28980	28618	29343	28980	28618	29270	28980	28835	28908	28835	29270	28980	29343	29198	29198
39058	38913	39493	39493	38913	39638	39203	39203	39203	39130	39565	39348	39783	39348	39710

16	17	18	19	20	21	22	23	24	25
Group 3					Group 4 (strong damage)				
12305	12160	12233	12233	12160	11580	11653	11435	11580	11435
12885	12813	12885	12958	12740	12740	12740	12668	12740	12595
20208	19990	20135	20135	19990	15278	15350	15060	15278	15133
23325	23108	23325	23253	23108	17380	17380	17090	17235	17163
29198	28908	29053	28980	28835	22093	22093	21658	22165	21803
39638	39348	39565	39493	39203	38623	38550	38188	38550	37970

TABLE 12.8 Synoptic classification table for circular plates: 6-resonance frequencies case

Test	1	2	3	4	5	6	7	8	9	10	11	12	13	14	15	16	17	18	19	20	21	22	23	24	25	
	Group 0 (pristine)					Group 1					Group 2					Group 3					Group 4 (strong damage)					
I	T	V	V	V	V	T	V	V	V	V	T	V	V	V	V	T	V	V	V	V	T	V	V	V	V	IN
	X	1	1	1	1	X	2	2	2	2	X	3	3	3	3	X	4	4	4	4	X	5	5	5	5	OUT
II	V	T	V	V	V	V	T	V	V	V	V	T	V	V	V	V	T	V	V	V	V	T	V	V	V	IN
	1	X	1	1	1	2	X	2	2	2	3	X	3	3	3	4	X	4	4	4	5	X	5	5	5	OUT
III	V	V	T	V	V	V	V	T	V	V	V	V	T	V	V	V	V	T	V	V	V	V	T	V	V	IN
	1	1	X	1	1	2	2	X	2	2	3	3	X	3	3	4	4	X	4	4	5	5	X	5	5	OUT
IV	V	V	V	T	V	V	V	V	T	V	V	V	V	T	V	V	V	V	T	V	V	V	V	T	V	IN
	1	1	1	X	1	2	2	2	X	2	3	3	3	X	3	4	4	4	X	4	5	5	5	X	5	OUT
V	V	V	V	V	T	V	V	V	V	T	V	V	V	V	T	V	V	V	V	T	V	V	V	V	T	IN
	1	1	1	1	X	2	2	2	2	X	3	3	3	3	X	4	4	4	4	X	5	5	5	5	X	OUT

1–25 – plate number
1–5 – correspondent class number
T – training vector
V – validation vector
X – vector absent for classification
IN – input vectors of PNN
OUT – output result of PNN

TABLE 12.9 Input data matrix for PNN classification of the circular plates: the 11-resonance frequencies case (all values in Hz)

1	2	3	4	5	6	7	8	9	10	11	12	13	14	15
Group 0					Group 1					Group 2				
12885	12740	13030	12885	12740	12958	12813	12813	12813	12813	12450	12450	12523	12450	12523
16220	16075	16510	16148	16148	19700	19555	19555	19483	19483	12885	12885	13030	12958	12958
20135	19918	20425	20135	19918	20353	20135	20063	20063	20063	19483	19410	19483	19410	19555
24195	23978	24630	24413	24050	28400	28183	28183	28183	28183	20280	20135	20353	20208	20353
28980	28618	29343	28980	28618	29270	28980	28835	28908	28835	27965	27893	28038	27893	27965
33620	33330	34273	33983	33403	39638	39203	39203	39203	39130	29270	28980	29343	29198	29198
36375	36158	36593	36520	36158	0	0	0	0	0	30720	30575	30793	30648	30938
39058	38913	39493	39493	38913	0	0	0	0	0	35215	35143	35288	35143	35288
0	0	0	0	0	0	0	0	0	0	38695	33333	38550	38405	38550
0	0	0	0	0	0	0	0	0	0	39565	39348	39783	39348	39710
0	0	0	0	0	0	0	0	0	0	0	0	0	0	0

16	17	18	19	20	21	22	23	24	25
Group 3					Group 4				
12305	12160	12233	12233	12160	11580	11653	11435	11580	11435
12885	12813	12885	12958	12740	12740	12740	12668	12740	12595
15350	15205	15423	15423	15205	15278	15350	15060	15278	15133
18540	18395	18540	18540	18395	17380	17380	17090	17235	17163
20208	19990	20135	20135	19990	22093	22093	21658	22165	21803
23325	23108	23325	23253	23108	30285	30140	29560	30285	29778
29198	28908	29053	28980	28835	38623	38550	38188	38550	37970
31228	30938	31010	31083	31010	0	0	0	0	0
33403	34925	33258	33258	33113	0	0	0	0	0
37535	37463	37680	37535	37463	0	0	0	0	0
39638	39348	39565	39493	39203	0	0	0	0	0

TABLE 12.10 Synoptic table for classification of circular plates: 11-resonance frequencies case

Test	1	2	3	4	5	6	7	8	9	10	11	12	13	14	15	16	17	18	19	20	21	22	23	24	25	
	Group 0 (pristine)					Group 1					Group 2					Group 3					Group 4 (strong damage)					
I	T	V	V	V	V	T	V	V	V	V	T	V	V	V	V	T	V	V	V	V	T	V	V	V	V	IN
	X	1	1	1	1	X	2	2	2	2	X	3	3	3	3	X	4	4	4	4	X	5	5	5	5	OUT
II	V	T	V	V	V	V	T	V	V	V	V	T	V	V	V	V	T	V	V	V	V	T	V	V	V	IN
	1	X	1	1	1	2	X	2	2	2	3	X	3	3	3	4	X	4	4	4	5	X	5	5	5	OUT
III	V	V	T	V	V	V	V	T	V	V	V	V	T	V	V	V	V	T	V	V	V	V	T	V	V	IN
	1	1	X	1	1	2	2	X	2	2	3	3	X	3	3	4	4	X	4	4	5	5	X	5	5	OUT
IV	V	V	V	T	V	V	V	V	T	V	V	V	V	T	V	V	V	V	T	V	V	V	V	T	V	IN
	1	1	1	X	1	2	2	2	X	2	3	3	3	X	3	4	4	4	X	4	5	5	5	X	5	OUT
V	V	V	V	V	T	V	V	V	V	T	V	V	V	V	T	V	V	V	V	T	V	V	V	V	T	IN
	1	1	1	1	X	2	2	2	2	x	3	3	3	3	X	4	4	4	4	X	5	5	5	5	X	OUT

1–25 – plate number
1–5 – correspondent class number
T – training vector
V – validation vector
X – vector absent for classification
IN – input vectors of PNN
OUT – output result of PNN

12.7.5 PROPOSED NEW ALGORITHM FOR PWAS DAMAGE DETECTION WITH THE E/M IMPEDANCE METHOD

A two-tier damage metric algorithm using neural nets framework is needed. This will significantly impact the advancement of knowledge in the field and will assist the practical implementation and industrial development of structural health monitoring technology. The focus of research should be on the development, calibration, and testing of a damage metric algorithm for incipient damage detection during active-sensor structural health monitoring:

- The damage-detection paradigm is based on the presumption that the presence of incipient damage will significantly modify the high-frequency impedance spectrum of the monitored structure in the proximity of the sensor.
- The proposed new algorithm adopts the pattern-recognition approach and constructs a **features extractor** and an **anomaly detector-classifier** to compare a given spectrum against a baseline ('pristine') spectrum in order to answer, with sufficient certainty, two main questions: (a) *Is the damage present?* and (b) *How severe is the damage?*
- The algorithmic architecture to be used will be the neural nets framework.

The proposed new algorithm is developed in three major phases: (A) **development** and **training** of a damage-metric algorithm using existing data; (B) **validation** of the damage-metric algorithm using validation data selected from the existing data set; (C) **testing** of the damage-metric algorithms using new data.

The *development and training phase* is performed using active-sensor already existing spectral data. This phase will produce a **damage-metric algorithm** consisting of two distinct components that sequentially process the data:

- *Features extractor*: This algorithm will perform **data-compression** to reduce the input dimensionality by extracting the relevant features contained in the active-sensor spectral signals. The features will be stored in the features vector, which is part of a **features space**. The features selected for this algorithm will take advantage of the natural connection between the appearance of structural damage and the changes in the high-frequency dynamics of the structure. Thus, the 400-points structural spectrum collected by the active sensor will be collapsed into a small number of resonance-related features. An important characteristic of the feature extractor developed in this phase of the algorithm will be its ability to perform recursive extension of the features-space dimensionality. This capability is needed for coping with new resonances that appear in the spectrum as damage occurs and progresses throughout the structure.
- *Anomaly detector and classifier*: This algorithmic component will process the features vector using a NN anomaly-detection algorithm based on competitive probabilistic classification and radial-basis functions. The anomaly detector will sort the input vectors into 'pristine' and 'damaged' (dichotomous classification). The classifier will further sort the data into multilevel structural health monitoring classes, signifying various degrees of damage intensity (e.g., 'pristine'; 'damaged but safe'; 'severely damaged', etc.). An important characteristic of the radial-basis NN is its ability to perform both supervised and unsupervised learning:
  (1) **Unsupervised learning** will detect the natural footprint patterns existing in the input vectors. This step will set up the parameters of the radial-basis functions.
  (2) **Supervised learning** will be used to assign classification labels to the natural footprints identified during unsupervised learning.

The *validation phase* of the algorithm will verify that the damage-metric algorithm is able to identify correctly a set of input data similar to the one used during training. To this purpose, the existing data set will be split into a **training set** and a **validation set**. The validation set will be set aside and the damage-metric algorithm will be developed using only the training set. Then, the validation set will be used to perform the validation procedure and assess the algorithm accuracy with respect to the validation data.

The *testing phase* of the algorithm development will verify and confirm the damage-metric algorithm against new data called the **test set**. The test set will be generated using simulation, laboratory experiments on typical specimens, and field tests on actual structures. The testing phase will achieve an evaluation of the overall performance of the damage-metric algorithm and of its noise-rejection capabilities.

12.8 SUMMARY AND CONCLUSIONS

This chapter has examined two aspects of considerable importance to the structural health monitoring (SHM) process: signal processing and damage identification/pattern recognition algorithms. The SHM community is constantly exposed to various aspects of signal processing, spectra collection, processing, analysis and decision-making. This chapter has reviewed some of these aspects. Both the state-of-the-art and general principles, on one hand, and specific examples, on the other hand, have been presented. However, an exhaustive presentation of these vast topics could not be attempted, since it would be beyond the scope and possibilities of a single chapter.

The signal processing subject was described in terms of two major algorithmic paths: the *short-term Fourier transform* (STFT) and the *wavelet transform* (WT). Both these approaches fall into the larger class of *time-frequency analysis*. They are used to better understand the time-wise behavior of the signal spectrum and extract that part (or bandwidth) of the signal that is of relevance to the SHM process. In many instances, the signal processing can be used as a pre- or post-processing action that enhances the SHM techniques presented in previous chapters.

This damage identification/pattern recognition process has been discussed in the context of analyzing the spectral features (e.g., resonance frequencies, resonance peaks, etc.). An elementary classification problem that distinguished between 'damaged' and 'pristine' structures was considered. As more advanced classification technique that allows classifying the structural health by damage progression based on damage severity and location was considered. Specific examples were presented in the context of the electromechanical (E/M) impedance method, which permits the direct determination of structural modal spectrum at very high structural frequencies. The high-frequency modal spectrum determined with the E/M impedance method is more susceptible to incipient local damage than the relatively low-frequency spectra determined with the conventional modal analysis techniques. Several issues related to the problem of damage identification during the SHM process were explored, such as:

(i) Develop damage metrics for active-sensor SHM
(ii) Test and calibrate these damage metrics using controlled experiments
(iii) Study the scale-up issues associated with the transition of the new damage metrics to full-scale structural health monitoring projects.

However, this chapter could not be exhaustive and all-encompassing, since to do so would require a lot more space than available in this book. The reader with further interest in studying this subject may want to refer to specialized literature. An electronic keyword search using the keywords mentioned in the first section of this chapter would yield a wealth of information.

12.9 PROBLEMS AND EXERCISES

1. What are the main differences between the conventional Fourier transform and the short-time Fourier transform?
2. What are the main differences between the short-time Fourier transform and the wavelet transform?
3. What are the main differences between the continuous wavelet transform and the discrete wavelet transform?
4. What are the main differences between multilayer feedforward neural networks and probabilistic neural networks?

APPENDIX A

MATHEMATICAL PREREQUISITES

A.1 FOURIER ANALYSIS

A fundamental concept in the study of signals is the frequency content of a signal. For a large class of signals, the frequency content can be generated by decomposing the signal into frequency components given by sinusoids. Following is the basics about Fourier transforming.

A.1.1 FOURIER SERIES OF PERIODIC SIGNAL

For a periodic signal $x(t)$

$$x(t) = a\sin(\omega_0 t) \tag{1}$$

with $T = 2\pi/\omega_0$, the Fourier expansion expressed as a series is available

$$x(t) = \sum_{k=-\infty}^{+\infty} C_k e^{jk\omega_0 t}, \quad -\infty < t < \infty \tag{2}$$

where

$$\begin{cases} C_0 = \dfrac{1}{T}\displaystyle\int_T x(t)\mathrm{d}t \\ C_k = \dfrac{1}{T}\displaystyle\int_T x(t)e^{-jk\omega_0 t}\,\mathrm{d}t, \quad k \neq 0 \end{cases} \tag{3}$$

A.1.2 FOURIER TRANSFORM OF APERIODIC SIGNAL

An aperiodic signal $x(t)$ has the Fourier transform

$$X(\omega) = \int_{-\infty}^{+\infty} x(t)e^{-j\omega t}\,\mathrm{d}t, \quad -\infty < \omega < +\infty \tag{4}$$

With

$$\begin{cases} X(\omega) = 2\int_0^{+\infty} x(t)\cos\omega t\,dt, & x(t) \text{ is even} \\ X(\omega) = -2j\int_0^{+\infty} x(t)\sin\omega t\,dt, & x(t) \text{ is odd} \end{cases} \tag{5}$$

The generalized Fourier transform is defined starting from the Fourier transform of a periodic signal as following. For a periodic signal $x(t)$, its Fourier expansion is

$$x(t) = \sum_{k=-\infty}^{+\infty} C_k e^{jk\omega_0 t} \tag{6}$$

Using common Fourier transform pair

$$e^{j\omega_0 t} \leftrightarrow 2\pi\delta(\omega - \omega_0) \tag{7}$$

Its Fourier transform, or say the generalized Fourier transform of this periodic signal, is

$$X(\omega) = \sum_{k=-\infty}^{+\infty} C_k \cdot 2\pi\delta(\omega - k\omega_0) = 2\pi \sum_{k=-\infty}^{+\infty} C_k\delta(\omega - k\omega_0) \tag{8}$$

$\delta(\omega - k\omega_0)$ is a train of impulses located at $\omega = k\omega_0$, $k = 0, \pm 1, \pm 2, \ldots$

A.1.3 DISCRETE TIME FOURIER TRANSFORM

The Fourier transform of a discrete-time signal $x[n](n = 0, 1, 2, \ldots)$ is called discrete time Fourier transform (DTFT) and has the definition

$$X(\Omega) = \sum_{n=-\infty}^{+\infty} x[n]e^{-j\Omega n} \tag{9}$$

$X(\Omega)$ has the property of ***periodicity***. It is completely determined by computing $X(\Omega)$ only over any 2π interval, for example, the interval $0 \leq \Omega \leq 2\pi$.

A.1.3.1 *N*-Point Discrete Fourier Transform

If $X(\Omega)$ is the DTFT of a discrete-time signal $x[n]$, the N-point discrete Fourier transform (DFT) X_k of the signal $x[n]$ can be defined in two ways, given $x[n] = 0$ for all $n < 0$ and $n \geq N$.

A.1.3.2 Definition 1: From DTFT to DFT

The DTFT of the N-point discrete signal $x[n]$ is

$$X(\Omega) = \sum_{n=-\infty}^{+\infty} x[n]e^{-j\Omega n} = \sum_{n=0}^{N-1} x[n]e^{-j\Omega n} \tag{10}$$

Set $\Omega = k(2\pi/N)$, $k = 0, 1, 2, \ldots, N-1$. Then

$$X_k = X\left(\Omega = k\frac{2\pi}{N}\right) \tag{11}$$

Therefore the N-point DFT can be considered as the sampled DTFT spectrum at $\Omega = k(2\pi/N)$ in the frequency domain.

A.1.3.3 Definition 2: Direct Definition

The definition for N-point DFT of a discrete time signal $x[n]$ is

$$X_k = \sum_{n=0}^{N-1} x[n] e^{-jk\frac{2\pi}{N}n}, \quad k = 0, 1, 2, \ldots, N-1 \tag{12}$$

where X_k is a function of the discrete variable k. DFT is completely specified by the N values $X_0, X_1, \ldots, X_{N-1}$ since $X(\Omega)$ is periodic.

A.1.4 DISCRETE FOURIER TRANSFORM ANALYSIS OF PULSE AND SINUSOID SIGNALS

In this section, we will start by finding the DTFT of a rectangular pulse train. Based on this conclusion, we will further understand how to obtain DTFT of other signals.

A.1.4.1 DTFT of Rectangular Signal

We have a rectangular signal $p[n]$ as shown in Fig. A.1

$$p[n] = \begin{cases} 1, & -q \le n \le q \\ 0, & \text{otherwise} \end{cases} \tag{13}$$

DTFT $P(\Omega)$ of the rectangular signal is

$$P(\Omega) = \frac{\sin\left(q + \frac{1}{2}\right)\Omega}{\sin\frac{\Omega}{2}} \tag{14}$$

Equation (14) can be proved as following. Since $p[n]$ is an even function, according to the Fourier transform property of the even functions, DTFT of $p[n]$, $P(\Omega)$, is a real-valued function

$$P(\Omega) = \sum_{n=-q}^{+q} e^{-j\Omega n} = \sum_{n=-q}^{+q} (e^{-j\Omega})^n \tag{15}$$

Using the expansion property

$$\sum_{a}^{b} r^n = \frac{r^a - r^{b+1}}{1 - r} \tag{16}$$

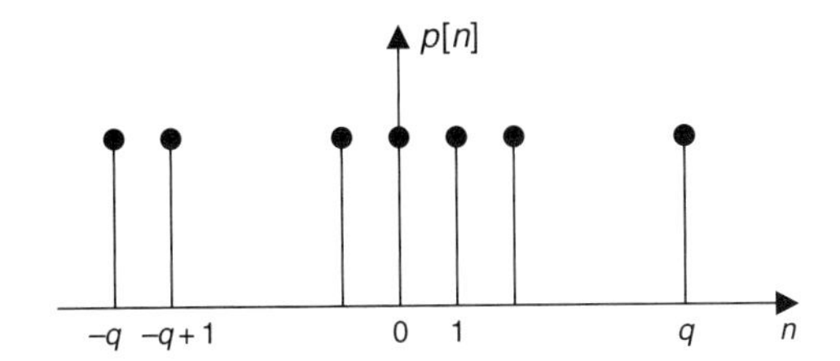

FIGURE A.1 Rectangular pulses in the time domain.

Equation (15) becomes

$$P(\Omega) = \sum_{n=-q}^{+q} (e^{-j\Omega})^n = \frac{(e^{-j\Omega})^{-q} - (e^{-j\Omega})^{q+1}}{1 - e^{-j\Omega}} \tag{17}$$

$$P(\Omega) = \frac{e^{j\Omega(q+\frac{1}{2})} - e^{-j\Omega(q+\frac{1}{2})}}{e^{j\frac{\Omega}{2}} - e^{-j\frac{\Omega}{2}}} \tag{18}$$

$$P(\Omega) = \frac{\sin(q+\frac{1}{2})\Omega}{\sin\frac{\Omega}{2}} \tag{19}$$

A.1.4.2 DTFT and DFT of q-point Shifted Rectangular Signal $x[n] = p[n-q]$

The q-point shifted rectangular signal as shown in Fig. A.2 is

$$x[n] = p[n-q] = \begin{cases} 1, & 0 \le n \le 2q \\ 0, & \text{otherwise} \end{cases} \tag{20}$$

Its DTFT and DFT results are

$$\text{DTFT} \qquad X(\Omega) = \frac{\sin(q+\frac{1}{2})\Omega}{\sin\Omega/2} e^{-jqn} \tag{21}$$

$$\text{DFT} \qquad |X_k| = \begin{cases} 2q+1, & k=0 \\ 0, & \text{otherwise} \end{cases} \tag{22}$$

The results are proved as following. Using the shifting property of Fourier transform, the DTFT of the q-point shifted rectangular signal is obtained from Eq. (19):

$$X(\Omega) = \frac{\sin(q+\frac{1}{2})\Omega}{\sin\Omega/2} e^{-jqn} \tag{23}$$

Set $N = 2q-1$. Then the DFT of $x[n]$ is

$$X_k = X\left(\Omega = k\frac{2\pi}{N}\right) \tag{24}$$

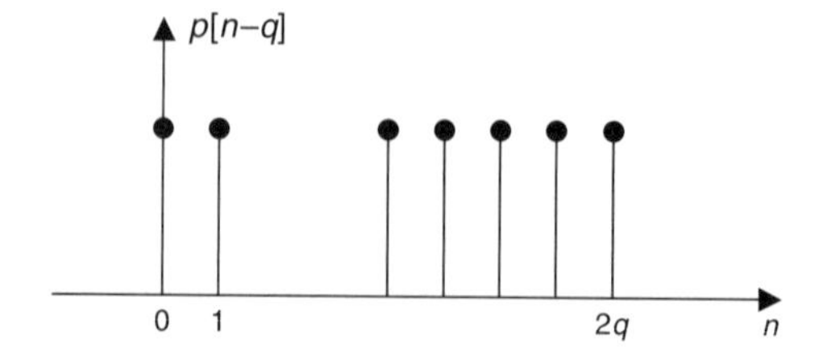

FIGURE A.2 q-point shifted pulse train $x[n] = p[n-q]$.

And

$$|X_k| = \left|X\left(\Omega = k\frac{2\pi}{N}\right)\right|$$

$$|X_k| = \left|\frac{\sin(q+\frac{1}{2})\cdot k\frac{2\pi}{N}}{\sin\frac{k\frac{2\pi}{N}}{2}}\right|\cdot\left|e^{-jq\Omega}\right|$$

$$|X_k| = \left|\frac{\sin k\pi}{\sin\left(\frac{\pi}{2q+1}k\right)}\right|, \quad k = 0, 1, 2\ldots 2q \tag{25}$$

Case I: $k = 0$

$$|X_k| = \lim_{k\to 0}\frac{\sin k\pi}{\sin\left(\frac{\pi}{2q+1}k\right)} = \lim_{k\to 0}\frac{\pi\cos k\pi}{\frac{\pi}{2q+1}\cos\left(\frac{\pi}{2q+1}k\right)} = 2q+1 \tag{26}$$

Case II: $k \neq 0$

$$|X_k| = 0 \tag{27}$$

A.1.4.3 DTFT of Discrete Cosine Signal

A discrete cosine signal of a frequency $\omega_0 (0 \leq \omega_0 \leq 2\pi)$ can be defined as

$$x[n] = \cos n\omega_0 \cdot p[n-1] = \begin{cases}\cos n\omega_0, & 0 \leq n \leq 2q\\ 0, & \text{otherwise}\end{cases} \tag{28}$$

The DTFT of an infinite discrete time cosine signal can be found as

$$\text{DTFT}(\cos n\omega_0) = \sum_{r=-\infty}^{+\infty}\pi[\delta(\Omega+\omega_0-2\pi r)+\delta(\Omega-\omega_0-2\pi r)] \tag{29}$$

Using Fourier transform property of production and DTFT of q-point shifted rectangular signal shown in Eq. (21), the DTFT of the limited-length cosine is

$$\text{DTFT}(x[n]) = X(\Omega) = \frac{1}{2\pi}\int_{-\pi}^{\pi}P(\Omega-\lambda)\pi[\delta(\lambda+\omega_0)+\delta(\lambda-\omega_0)]\,\mathrm{d}\lambda \tag{30}$$

$$X(\Omega) = \frac{1}{2\pi}\int_{-\pi}^{\pi}P(\Omega-\lambda)\pi\delta(\lambda+\omega_0)\mathrm{d}\lambda + \frac{1}{2\pi}\int_{-\pi}^{\pi}P(\Omega-\lambda)\pi\delta(\lambda-\omega_0)\mathrm{d}\lambda \tag{31}$$

Set $t = \Omega - \lambda$, then $\mathrm{d}t = -\mathrm{d}\lambda$. Also from the property of delta function, we know $\int_{-\infty}^{+\infty}p(x)\delta(x-x_0)\mathrm{d}x = p(x_0)$. Hence Eq. (31) becomes

$$X(\Omega) = \frac{1}{2}[P(\Omega+\omega_0)+P(\Omega-\omega_0)] \tag{32}$$

where

$$P(\Omega) = \frac{\sin(q+\frac{1}{2})\Omega}{\sin\frac{\Omega}{2}}e^{-jqn} \tag{33}$$

A.1.4.4 *N*-point DFT of Discrete Cosine Signal $x[n]$

Let $N = 2q+1$, i.e., $q = (N-1)/2$. Using the DTFT described in Eq. (32), the DFT is

$$X_k = X\left(\Omega = k\frac{2\pi}{N}\right) = \frac{1}{2}\left[P\left(k\frac{2\pi}{N}+\omega_0\right)+P\left(k\frac{2\pi}{N}-\omega_0\right)\right] \quad (34)$$

and

$$P\left(\Omega = k\frac{2\pi}{2q+1}\pm\omega_0\right) = \frac{\sin\left(q+\frac{1}{2}\right)\left(k\frac{2\pi}{2q+1}\pm\omega_0\right)}{\sin\frac{1}{2}\left(k\frac{2\pi}{2q+1}\pm\omega_0\right)}e^{-jq\left(k\frac{2\pi}{2q+1}\pm\omega_0\right)} \quad (35)$$

$k = 0, 1, 2, \ldots, N-1$. Assume that for $k = r$

$$\frac{2\pi r}{2q+1} = \omega_0 \quad (36)$$

Substitute Eq. (36) into Eq. (35)

$$P\left(\Omega = k\frac{2\pi}{2q+1}\pm\frac{2\pi r}{2q+1}\right) = \frac{\sin\left(q+\frac{1}{2}\right)\left(k\frac{2\pi}{2q+1}\pm\frac{2\pi r}{2q+1}\right)}{\sin\frac{1}{2}\left(k\frac{2\pi}{2q+1}\pm\frac{2\pi r}{2q+1}\right)}e^{-jq\left(k\frac{2\pi}{2q+1}\pm\frac{2\pi r}{2q+1}\right)} \quad (37)$$

then

$$P\left(\Omega = k\frac{2\pi}{2q+1}\pm\frac{2\pi r}{2q+1}\right) = \frac{\sin\pi(k\pm r)}{\sin\frac{\pi(k\pm r)}{2q+1}}e^{-jq\frac{2\pi(k\pm r)}{2q+1}} \quad (38)$$

$$P\left(\Omega = k\frac{2\pi}{2q+1}\pm\frac{2\pi r}{2q+1}\right) = \begin{cases} 2q+1, & k = N-r \\ 0, & \text{otherwise} \end{cases} \quad (39)$$

or

$$P\left(\Omega = k\frac{2\pi}{2q+1}\pm\omega_0\right) = \begin{cases} 2q+1, & k = N-r \\ 0, & \text{otherwise} \end{cases} \quad (40)$$

Therefore, the DFT coefficients can be determined as

$$X_k = \begin{cases} q+\frac{1}{2}\left(\frac{N}{2}\right), & k = N-r, r \\ 0, & \text{otherwise} \end{cases} \quad (41)$$

where r can be determined from Eq. (36)

$$r = \frac{2q+1}{2\pi}\omega_0 = \frac{N}{2\pi}\omega_0 \quad (42)$$

Here we have to be sure that r is an integer. From Eq. (42), it is obvious that the value of r is determined by the angular frequency ω_0 of the cosine signal. When the real value of r is not an integer, that is to say, r will fall into the interval between two neighboring integers, a phenomenon called *leakage* occurs. The leakage phenomenon will be investigated in detail in the following section.

A.1.5 INVESTIGATION OF DFT SPECTRUM LEAKAGE

For signals like $\cos\omega_0 t$ or $\sin\omega_0 t$, the only nonzero frequency component in the spectrum within the range of $[0, f_s]$ (f_s, the sampling frequency) should be at the point $f = f_0$, with $f_0 = \omega_0/2\pi$. However, most of the time we cannot see such an ideal spectrum. Rather, the spectrum will spread out about the point f_0. This is the phenomenon called as *leakage*. We will present some examples to further explain leakage and investigate the reason for it. All the examples use cosine signals with unit magnitude.

A.1.5.1 Case Study of the Leakage Phenomenon

Example 1

Consider an N-point discrete time signal $x[n] = \cos\omega_0 n$, $n = 0, 1, 2, \ldots, N-1$. Take $N = 21$.

Therefore, the variable q takes the value

$$q = \frac{N-1}{2} = 10 \tag{43}$$

Case 1: $\omega_0 = 10\pi/21$

In this case, r is found to be $r = 5$. Hence, it is an integer. Theoretically, there should not be any leakage. Recalling Eq. (41), we can obtain

$$X_r = q + \frac{1}{2} = 10.5 \tag{44}$$

For $k = r = 5$ and $k = N - q = 16$. A MATLAB program was used to verify the result, as shown in Fig. A.3a.

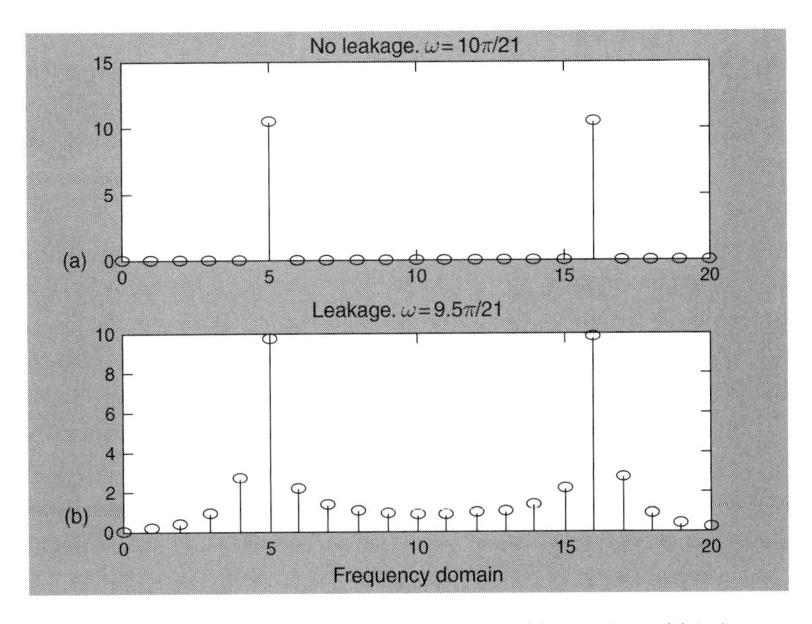

FIGURE A.3 Leakage under certain conditions: (a) no leakage; (b) leakage.

Case 2: $\omega_0 = 9.5\pi/21$

In this case, $r = 4.75$. It is not an integer. So leakage should occur in the DFT spectrum. Results are shown in Fig. A.3b. In the plotting, we can see that in the leakage spectrum, the energy is no longer focused on the point $r = 5$, which corresponds to the fundamental frequency. The energy spreads out about the point.

Example 2

In practical application, DFT is calculated using the fast Fourier transform (FFT) algorithm. Application of FFT algorithm requires that the number of samples, N, should be an integer which is the power of 2. In Example 2, signal used in Example 1 is defined with new N values at $N = 64$, 128, 256, 512. Two cases will be discussed to see how the number N affects the leakage.

According to Eq. (42), considering now N is an even number as the power of 2, in order to obtain an integer r, we set

$$r = \frac{N}{4} \tag{45}$$

Therefore, the angular frequency can be found as

$$\omega_0 = \frac{r}{\frac{N}{2\pi}} = \frac{\pi}{2} \tag{46}$$

For $\omega_0 = \pi/2$, theoretically there should be no leakage. To make the leakage occur, we simply modify the r as $r = (N-1)/4$. The corresponding angular frequency becomes

$$\omega_0 - \frac{N-1}{N} \cdot \frac{\pi}{2} \tag{47}$$

Simulation testing results are listed in Table A.1.

The spectrum energy is calculated using the formula

$$E = \sum_{k=0}^{N-1} X_k^2 \tag{48}$$

TABLE A.1 Parameter analysis for leakage and non-leakage DFT at various N

N	Non-leakage				Leakage				Amplitude error
	$\omega_0 = \frac{\pi}{2}$	$r = \frac{N}{4}$	$X_r = \frac{N}{2}$	Energy	$\omega_0 = \frac{N-1}{N}\frac{\pi}{2}$	$r = \frac{N-1}{4}$	$X_r = \frac{N}{2}$	Energy	
64	$\frac{\pi}{2}$	32	32	2048	$0.9844\frac{\pi}{2}$	15.75	29.164	2080	8.86%
128	$\frac{\pi}{2}$	64	64	8192	$0.9922\frac{\pi}{2}$	31.75	59.974	8256	6.29%
256	$\frac{\pi}{2}$	128	128	32768	$0.9961\frac{\pi}{2}$	63.75	115.594	32896	9.69%
512	$\frac{\pi}{2}$	256	256	131072	$0.9980\frac{\pi}{2}$	127.75	230.835	131330	9.83%

TABLE A.2 Spectrum energy of leakage and non-leakage spectrum at various N

		Non-leakage spectrum energy	Leakage spectrum energy	Energy error
N	64	2048	2080	1.5625%
	128	8192	8256	0.7813%
	256	32768	32896	0.3906%
	512	131072	131330	0.1968%

Simulation results are listed in Table A.2. We found that the energy in leakage spectrum was slightly larger than the energy in non-leakage spectrum. With increasing N value, the difference decreases.

Spectrums of the two cases with N at different values were obtained and plotted in Fig. A.4. It shows for the leakage situations, with larger N, the leakage gets less spread out, or, the energy is more concentrated around the fundamental frequency. However, the error in the amplitude is still significant (8.86% for $N = 64$ while 9.83% for $N = 512$).

Example 3

Consider a continuous time signal $x(t) = \cos \omega_0 t$.

The continuous signal needs to be sampled such that computer-processing can be done accurately. For example, consider the pure-tone signal presented in Fig. A.5. The signal duration is $t = 8$ s. For an N-point sampling, the sampling frequency is given by

$$f_s = \frac{1}{T_s} = \frac{1}{\frac{t}{N}} = \frac{N}{t} \tag{49}$$

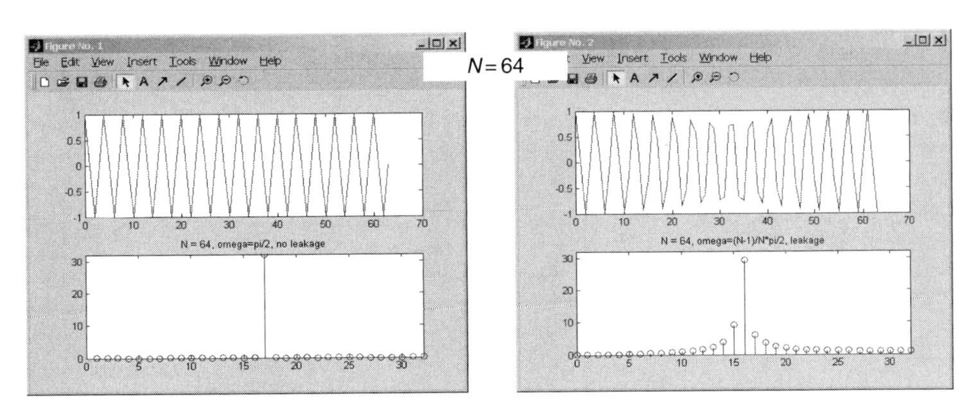

FIGURE A.4 Leakage phenomenon at different sample counts.

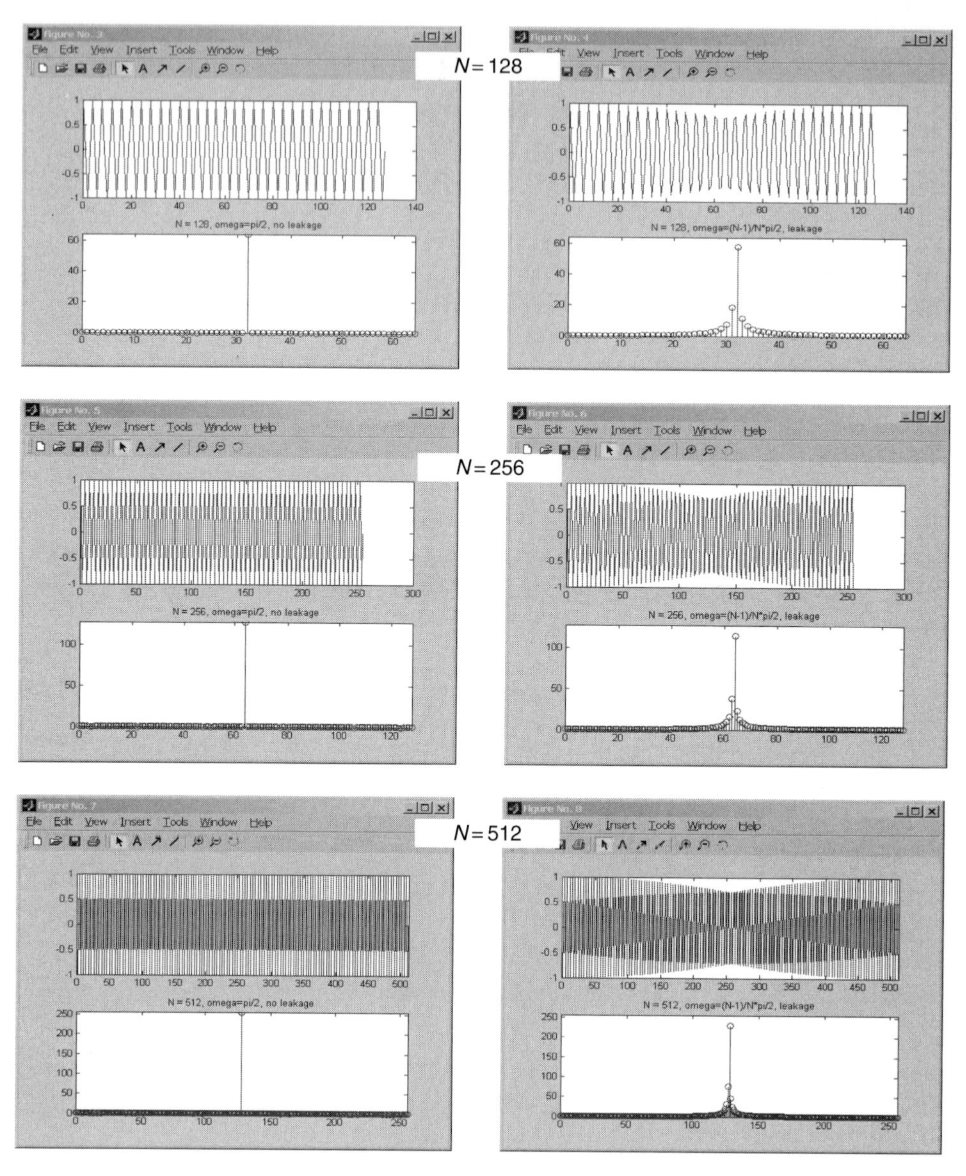

FIGURE A.4 Cont'd

Example 3 is aimed on finding out how the number of samples N, i.e., the sampling method, affects the DFT spectrum of a continuous signal. Also, two cases are discussed:

1. $\omega_0 = \frac{\pi}{2}$, no leakage
2. $\omega_0 = \frac{N-1}{N} \cdot \frac{\pi}{2}$, leakage.

The spectrum plots for sampled $x(t)$ at various sampling rates (various numbers of sample points, N) are shown in Fig. A.5. It shows that higher sampling rates leads to leakage being compensated and the numerical spectrum approximating better the ideal non-leakage spectrum.

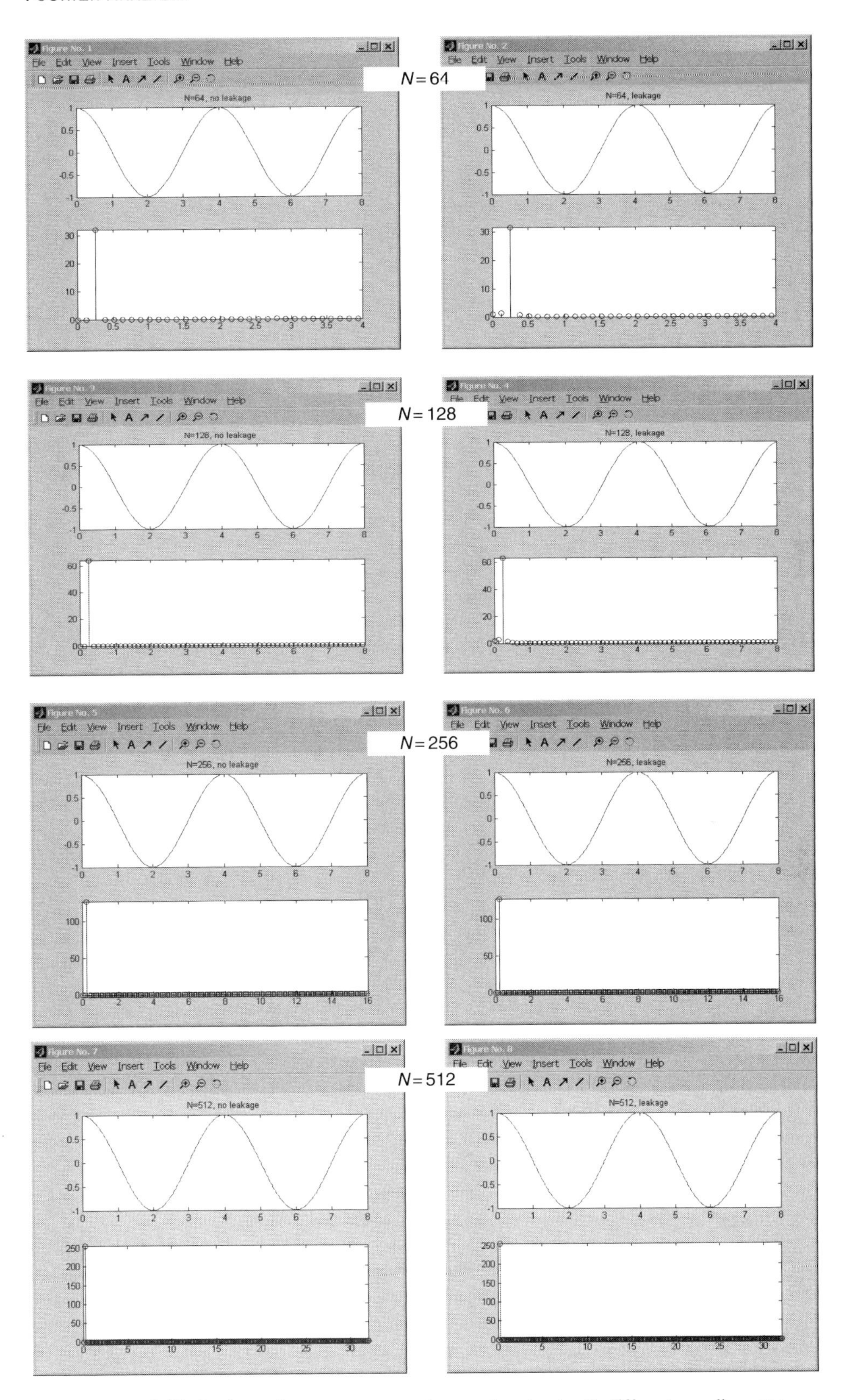

FIGURE A.5 Leakage phenomenon on continuous sine signal with different sampling rates.

A.1.5.2 Discussion

From the examples above, we can conclude that when the number of sample points, N, is sufficiently large, the energy leakage is very small. Actually, in the continuous-time signal example, when N equals 256 and 512, there is already no visible leakage. Hence, in the DFT processing, more sample points (higher sampling rate) can increase the spectrum precision.

However, this precision is gained at the expense of time. According to DFT algorithm, the operating time depends on the number of samples N. During the MATLAB simulation, it has been found that the programs run very slowly for higher sample numbers. It is essential to make a compromise between precision and efficiency.

A.2 SAMPLING THEORY

Under certain conditions, a continuous-time signal can be completely represented by and recoverable from a set of samples which are recorded at points equally spaced in time (Oppenheim, 2002). This is referred to as the sampling theorem, which constructs a bridge between the continuous-time signals and discrete-time signals.

A.2.1 IMPULSE-TRAIN SAMPLING

In order to develop the sampling theorem, a convenient way to represent the sampling of a continuous-time signal at regular intervals is needed. A useful way to do this is with a periodic impulse train multiplied by the continuous-time signal $x(t)$ which is to be sampled. This is known as the ***impulse-train sampling***, as depicted in the Fig. A.6.

The interval ΔT is known as the *sampling period* and the fundamental frequency of the impulse train $p(t)$ is $\omega_s = 2\pi/\Delta T$, known as the *sampling frequency*. In the time domain, the sampled points $x_s(t)$ shown in Fig. A.6b can be represented as

$$x_s(t) = x(t)p(t) \tag{50}$$

where

$$p(t) = \sum_{n=-\infty}^{+\infty} \delta(t - n \cdot \Delta T) \tag{51}$$

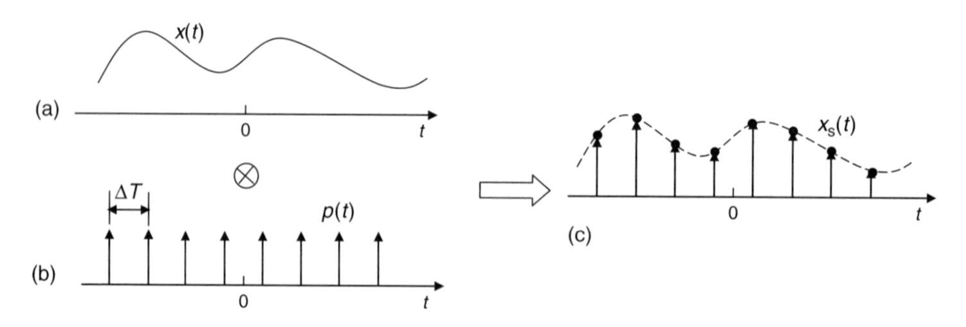

FIGURE A.6 Pulse train sampling with regular interval ΔT: (a) continuous-time signal $x(t)$; (b) impulse train $p(t)$ with uniform interval ΔT; (c) sampled signal $x_s(t)$.

Equation (50) can be further expanded as

$$x_s(t) = \sum_{n=-\infty}^{+\infty} x(n\Delta T)\delta(t - n \cdot \Delta T) \tag{52}$$

Note that the function $p(t)$ is a periodic function with period ΔT. Hence, it can be expanded in a Fourier series about the frequency $\omega_s = 2\pi/\Delta T$, i.e.,

$$p(t) = \sum_{n=-\infty}^{+\infty} c_n e^{jn\omega_s t} \tag{53}$$

The Fourier coefficients c_n are computed as

$$\begin{aligned} c_n &= \int_{-\Delta t/2}^{\Delta t/2} p(t)e^{-jn\omega_s t}\mathrm{d}t = \int_{-\Delta t/2}^{\Delta t/2} \sum_m \delta(t - m\Delta T)e^{-jn\omega_s t}\mathrm{d}t \\ &= \frac{1}{\Delta T}\int_{-\Delta t/2}^{\Delta t/2} \delta(t)e^{-jn\omega_s t}\mathrm{d}t = \frac{1}{\Delta T}\left\{e^{-jn\omega_s t}\right\}_{t=0} = \frac{1}{\Delta T} \end{aligned} \tag{54}$$

Note that, in Eq. (54), the function $p(t)$ represented by Eq. (51) was simplified as

$$p(t) \equiv \delta(t) \quad \text{for } t \in \left(\frac{-\Delta T}{2}, \frac{\Delta T}{2}\right) \tag{55}$$

Since all the other δ functions fall outside the $\left(\frac{-\Delta T}{2}, \frac{\Delta T}{2}\right)$ intervals.

Therefore the Fourier series of the impulse train with a period of ΔT in Eq. (51) is

$$p(t) = \sum_{n=-\infty}^{+\infty} \delta(t - n \cdot \Delta T) = \sum_{n=-\infty}^{\infty} c_k e^{jn\omega_s t} = \sum_{n=-\infty}^{\infty} \frac{1}{\Delta T} e^{jn\omega_s t} \tag{56}$$

Substitute Eq. (56) into Eq. (52) to obtain

$$x_s(t) = x(t)p(t) = \sum_{n=-\infty}^{\infty} \frac{1}{\Delta T} x(t)e^{jk\omega_s} \tag{57}$$

Recall the Fourier transform property that if a time signal $x(t)$ is multiplied with a complex exponential in the time domain and if the Fourier transform pair is $x(t) \leftrightarrow X(\omega)$, then we have $x(t)e^{j\omega_0 t} \leftrightarrow X(\omega - \omega_0)$, with ω_0 being a real number. Hence, the Fourier transform of a sampled signal represented by Eq. (57) is

$$X_s(\omega) = \frac{1}{\Delta T} \sum_{n=-\infty}^{\infty} X(\omega - n\omega_s) \tag{58}$$

where $X(\omega)$ if the Fourier transform of the original continuous-time signal, i.e., $x(t) \leftrightarrow X(\omega)$.

From Eq. (58) we see that the Fourier transform $X_s(\omega)$ of the sampled continuous-time signal is a periodic function of ω consisting of a superposition of shifted and scaled $X(\omega)$ at the points $\omega = n\omega_s(-\infty < n < \infty)$.

A.2.2 THE SAMPLING THEOREM

From Eq. (58) we see that for a band-limited signal $x(t)$, its Fourier transform $X(\omega)$ is zero about a certain frequency $B(X(\omega) = 0$ if $\omega \geq B)$. If the value B meets the condition

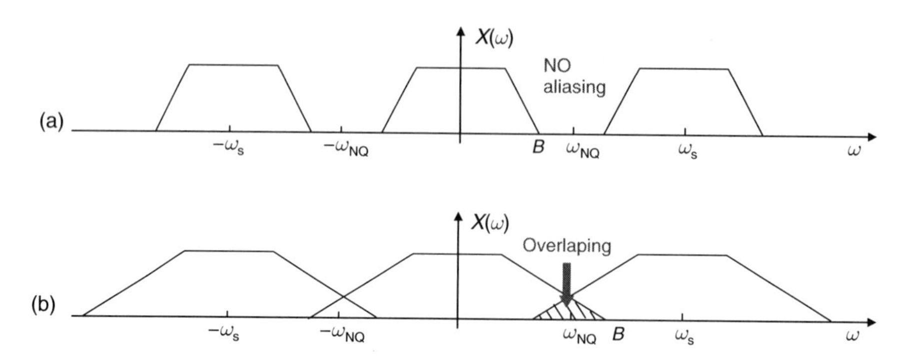

FIGURE A.7 Sampling theorem and aliasing effect. (a) Fourier transform $X(\omega)$ of a band-limited signal with $X(\omega) \leq B$ and $B \leq \omega_{NQ}$, $\omega_{NQ} = \omega_s$; (b) overlapping occurs when $B > \omega_{NQ}$, causing aliasing.

that $B \leq \omega_{NQ}$, $\omega_{NQ} = \omega_s$ as shown in Fig. A.7a, there is no overlapping. At this situation, if the sampled signal $x_s(t)$ is applied to an ideal lowpass filter with a cutoff frequency B, the only component of $X_s(\omega)$ that passes the filter is $X(\omega)$. Hence the output of the filter is equal to $x(t)$, which shows that the original signal $x(t)$ can be completely and exactly reconstructed from the sampled signal $x_s(t)$. By this result, a signal with bandwidth B can be reconstructed completely and exactly from the sampled signal $x_s(t)$ if the sampling frequency ω_s is chosen to be greater than or equal to $2B(\omega_s \geq 2B)$. This result is well known as the *sampling theorem*. The upper limit of the bandwidth is called Nyquist frequency, where $\omega_{NQ} = B$, and the corresponding minimum sampling frequency is $\omega_s = 2B$.

However, if the above situation $\omega_s \geq 2B$ is not satisfied, as shown in Fig. A.7b $B > \omega_{NQ}$, the $X(\omega)$ replicas overlap in the frequency domain. As a result of overlapping of frequency components, it is impossible to reconstruct the original signal $x(t)$ completely and exactly by using a lowpass filter to process the sampled signal. The output of a lowpass filter with a cutoff frequency B will contain high frequency components of $x(t)$ transposed to low frequency components of $x(t)$, as the shade area shown in Fig. A.7b. This phenomenon is called *aliasing*. Aliasing will result in a distorted version of the original signal $x(t)$.

A.3 CONVOLUTION

Convolution is a mathematical way of combining two signals to form a third signal and is of significant importance to linear time-invariant discrete-time systems. In a linear time-invariant system, convolution relates three signals of interest: the input signal, the output signal, and the system impulse response (Smith, 1999). The input signal is decomposed into simple additive components, then each of these components is passed through a linear system, and the resulting output components are synthesized. If impulse decomposition is used, such a decomposition procedure can be described by the mathematical operation, convolution.

A.3.1 DISCRETE TIME SIGNAL CONVOLUTION

Two important terms are used widely is discrete time signal convolution, delta function known as $\delta[n]$ and impulse response $h[n]$. The delta function $\delta[n]$ is a normalized impulse, only having a sample value of one while all the others being zeros as shown in Fig. A.8a. The delta function is also called the unit impulse.

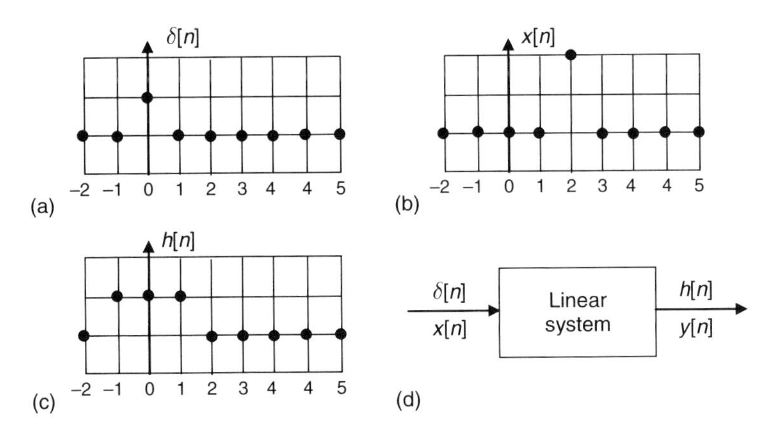

FIGURE A.8 Discrete time signal convolution: (a) delta function; (b) input impulse $x[n]$; (c) impulse response; (d) the input, output, and impulse response of a linear system.

The *impulse response* is the output of a system when $\delta[n]$ is the input, usually given the symbol $h[n]$. If two systems are different in any way, they will have different impulse responses.

Any impulse can be represented as a *shifted* and *scaled* $\delta[n]$. Consider the signal shown in Fig. A.8b. It is composed of all zeros except the sample point at $n = 2$, which has a scale of 2. We can see it is the same as $\delta[n]$: (1) shifted to the right by 2; (2) increased by 2. In equation form, $x[n]$ can be represented as

$$x[n] = 2\delta[n-2] \tag{59}$$

If for a linear system, its impulse response $h[n]$ is like the signal shown in Fig. A.8c, what will be the output when $x[n]$ is the input? We say, if the properties of homogeneity and shift invariance are used, input $x[n] = 2\delta[n-2]$ will result in a shifted and scaled impulse response by the same amounts as how $x[n]$ is modified from $\delta[n]$. That is to say, the output $y[n]$ will be $2h[n-2]$. It can be seen that if the system's impulse response is given, the output of the system is also known.

The mathematical equation of discrete convolution can be put as

$$y[n] = x[n] * h[n] \tag{60}$$

Using the concept of using pulse decomposition to obtain output, Eq. (60) can be

$$y[i] = \sum_{j}^{M-1} h[j]x[i-j] \tag{61}$$

A.3.2 CONTINUOUS TIME CONVOLUTION

Given two continuous time signals $x(t)$ and $y(t)$, the convolution is defined by

$$x(t) * y(t) = \int_{-\infty}^{+\infty} x(\tau)y(t-\tau)\mathrm{d}\tau \tag{62}$$

If both $x(t)$ and $y(t)$ are zero for all $t < 0$, Eq. (62) becomes

$$x(t) * y(t) = \begin{cases} 0, & t < 0 \\ \int_0^t x(\tau)y(t-\tau)\mathrm{d}\tau, & t \geq 0 \end{cases} \tag{63}$$

A.3.3 PROPERTIES OF CONVOLUTION

The properties of convolution can be summarized and demonstrated by the discrete signal convolution, listed in Table A.3.

Given two continuous signals $x(t)$ and $y(t)$ with Fourier transform $X(\omega)$ and $Y(\omega)$, the Fourier transform of the convolution $x(t)^* y(t)$ is equal to the product $X(\omega)Y(\omega)$, resulting in a relationship

$$x(t) * y(t) \leftrightarrow X(\omega)Y(\omega) \tag{64}$$

The pair shown in Eq. (64) is very useful for practical implementation of the convolution in the time domain. The convolution can be calculated by first finding out the Fourier transform of each signal and obtaining the product of their Fourier transform, and then using inverse Fourier transform to find out the convolution result in the time domain.

Conversely, for the convolution of $X(\omega)^* Y(\omega)$ in the frequency domain, then

$$x(t)y(t) \leftrightarrow \frac{1}{2\pi}[X(\omega) * Y(\omega)] \tag{65}$$

Equation (65) indicates that multiplication in the time domain corresponds to convolution in the Fourier transform domain.

TABLE A.3 Properties of convolution

Mathematical properties	Commutative property: $a[n]*b[n]=b[n]*a[n]$
	Associative property: $(a[n]*b[n])*c[n]=a[n]*(b[n]*c[n])$
	Distributive property: $a[n]*b[n]+a[n]*c[n]=a[n]*(b[n]+c[n])$
Delta function	$x[n]*\delta[n]=x[n]$

A.4 HILBERT TRANSFORM

The envelope of a family of curves or surfaces is a curve or surface that is tangent to every member of the family. Envelope extracts the amplitude of a periodic signal. It can be used to simplify the process of detecting the time of arrival for the wave packets in our EUSR system. In the EUSR, the envelope of the signal is extracted by applying Hilbert transform to the cross-correlation signal.

The Hilbert transform is defined as

$$H(x(t)) = -\frac{1}{\pi}\int_{-\infty}^{+\infty}\frac{x(\tau)}{t-\tau}\mathrm{d}t \tag{66}$$

It can also be defined in terms of convolution theory

$$H(x(t)) = x(t) * \frac{1}{\pi t} \tag{67}$$

Hilbert transform is often used to construct a complex signal

$$\tilde{x}(t) = \tilde{x}_{\mathrm{Re}}(t) + \mathrm{i}\cdot\tilde{x}_{\mathrm{Im}}(t) \tag{68}$$

where

$$\tilde{x}_{\mathrm{Re}}(t) = x(t)$$
$$\tilde{x}_{\mathrm{Im}}(t) = H(x(t)) \tag{69}$$

The real part of the constructed signal, $\tilde{x}_{\mathrm{Re}}(t)$, is the original data $x(t)$, while the imaginary part $\tilde{x}_{\mathrm{Im}}(t)$ is the Hilbert transform of $x(t)$. Considering the convolution theory, the Hilbert transform (imaginary part) is a version of the original signal (real part) $x(t)$ after a 90° phase shift.

Suppose we have a harmonic signal $u_c(t) = A\cos(2\pi f_c t + \phi_0)$, f_c is the carrying frequency and ϕ_0 is the initial phase. We will have the Hilbert transform according to the Hilbert transform property that $H(\cos(\omega t)) = \sin(\omega t)$

$$H(u_c(t)) = A\sin(2\pi f_c t + \phi_0) = A\cos\left(2\pi f_c t + \phi_0 - \frac{\pi}{2}\right) \tag{70}$$

Thus, the Hilbert-transformed signal has the same amplitude and frequency content as the original signal and includes phase information that depends on the phase of the original signal. The magnitude of each complex value $\tilde{x}(t)$ has same the amplitude as the original signal.

Therefore, we can say that the magnitude of the analytical signal $\tilde{x}(t)$ is the envelope of the original signal. Just by observing the envelope signal, the wave packages can be easily recognized.

One aspect of Hilbert transform for envelope detection is that the result may differ from the theoretical envelope if the frequency composition is relatively complicated. For example, we have two simple signals showing in Fig. A.9a. The 3-count tone burst has a carry frequency of 300 kHz and the sinusoid has a frequency of 1500 kHz. The curves in Fig. A.9b are (i) the synthetic signal by simply adding the two basic signals and (ii) the 'envelope' signal obtained with the Hilbert transform method. We see that

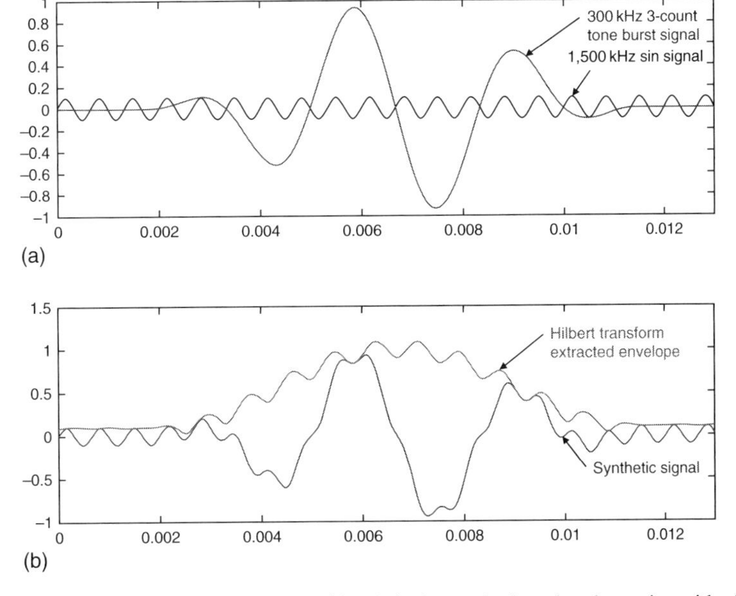

FIGURE A.9 Hilbert transform example: (a) original constitutive signals: a sinusoid of frequency 1500 kHz and a 3-count tone-burst of 300 kHz; (b) the synthetic signal ($A_1 \neq A_2, \phi1 \neq \phi2$) and its envelope extracted by Hilbert transform.

the 'envelope' is far from being the smooth envelope that we expected to see, thus illustrating one of the shortcomings of the Hilbert-transform envelope extraction approach. These simulation results can be verified analytically using two properties of the Hilbert transform, as follows.

- The often used Hilbert transform pair: $H(A\cos\varphi) = A\sin(\varphi)$
- Hilbert transform linearity property: $H(x(t)+y(t)) = H(x(t)) + H(y(t))$.

Assume we have a signal consisting of two frequencies, $u(t) = A_1\cos\varphi_1 + A_2\cos\varphi_2$, where $\varphi_1 = \varphi_1(t)$ and $\varphi_2 = \varphi_2(t)$.

The Hilbert transform $H(u(t))$ and the constructing analytical signal $\tilde{u}(t)$ are

$$H\{u(t)\} = A_1\sin\varphi_1 + A_2\sin\varphi_2 \tag{71}$$

$$\tilde{u}(t) = A_1\cos\varphi_1 + A_2\cos\varphi_2 + j(A_1\sin\varphi_1 + A_2\sin\varphi_2) \tag{72}$$

To get the amplitudes $|u(t)|$ and $|\tilde{u}(t)|$, several situations are possible:

Situation 1: $\varphi_1 = \varphi_2$

This means we actually have a single frequency composition. We have $|u(t)| = A_1 + A_2$ and $|\tilde{u}(t)| = \sqrt{A_1^2 + A_2^2 + 2A_1A_2} = A_1 + A_2$. Therefore, $|u(t)| = |\tilde{u}(t)|$, it is verified.

Situation 2: $\varphi_1 \neq \varphi_2$ but $A_1 = A_2$

Now $u(t) = A\cdot 2\cos\left(\frac{\varphi_1+\varphi_2}{2}\right)\cos\left(\frac{\varphi_1-\varphi_2}{2}\right)$ and the amplitude $|u(t)| = 2A$. The amplitude of $\tilde{u}(t)$ is

$$\begin{aligned}
|\tilde{u}(t)| &= A\sqrt{(\cos\varphi_1 + \cos\varphi_2)^2 + (\sin\varphi_1 + \sin\varphi_2)^2} \\
&= A\sqrt{\left(2\cos\left(\frac{\varphi_1+\varphi_2}{2}\right)\cos\left(\frac{\varphi_1-\varphi_2}{2}\right)\right)^2 + \left(2\sin\left(\frac{\varphi_1+\varphi_2}{2}\right)\cos\left(\frac{\varphi_1-\varphi_2}{2}\right)\right)^2} \\
&= 2A\sqrt{\left(\cos^2\left(\frac{\varphi_1+\varphi_2}{2}\right) + \sin^2\left(\frac{\varphi_1+\varphi_2}{2}\right)\right)\cos^2\left(\frac{\varphi_1+\varphi_2}{2}\right)} \\
&= 2A\cdot\left|\cos\left(\frac{\varphi_1+\varphi_2}{2}\right)\right|
\end{aligned}$$

If $\left|\cos\left(\frac{\varphi_1+\varphi_2}{2}\right)\right| = 1$, $|u(t)| = |\tilde{u}(t)|$; otherwise $|u(t)| \neq |\tilde{u}(t)|$.

Situation 3: $\varphi_1 \neq \varphi_2$ and $A_1 \neq A_2$, $|u(t)| = |A_1\cos\varphi_1 + A_2\cos\varphi_2|$

$$\begin{aligned}
|\tilde{u}(t)| &= \sqrt{(A_1\cos\varphi_1 + A_2\cos\varphi_2)^2 + (A_1\sin\varphi_1 + A_2\sin\varphi_2)^2} \\
&= \sqrt{A_1^2\cos^2\varphi_1 + A_2^2\cos^2\varphi_2 + 2A_1A_2\cos\varphi_1\cos\varphi_2} \\
&\quad + \sqrt{A_1^2\sin^2\varphi_1 + A_2^2\sin^2\varphi_2 + 2A_1A_2\sin\varphi_1\sin\varphi_2} \\
&= \sqrt{A_1^2 + A_2^2 + 2A_1A_2(\cos\varphi_1\cos\varphi_2 + \sin\varphi_1\sin\varphi_2)} \\
&= \sqrt{A_1^2 + A_2^2 + 2A_1A_2\cos(\varphi_1 - \varphi_2)}
\end{aligned}$$

For this case, it is difficult to tell $|u(t)| = |\tilde{u}(t)|$ or $|u(t)| \neq |\tilde{u}(t)|$ directly, but there are some conditions that $|u(t)| \neq |\tilde{u}(t)|$, e.g., when $\varphi_1 = 90°, \varphi_2 = 0°$, $|u(t)| = |A_2|$, $|\tilde{u}(t)| = \sqrt{A_1^2 + A_2^2}$. Clearly, we can see that $|u(t)| \neq |\tilde{u}(t)|$. So we can conclude that $u(t)$ does not always equal to $\tilde{u}(t)$. Example of this situation is demonstrated in Fig. A.9.

Hence, we can conclude, when the signal has contains more than one frequency, the Hilbert transform for envelope extraction may have aliasing.

A.5 CORRELATION METHOD

Cross correlation is used to detect similarities in two signals. The cross correlation R_{xy} of two signals $x(t)$ and $y(t)$ is defined by

$$R_{xy}(t) = \int_{-\infty}^{+\infty} x(\tau)y(\tau - t)\mathrm{d}\tau \tag{73}$$

For discrete-time signals, Eq. (73) can be expressed as:

$$R_{xy}[m] = \frac{1}{N}\sum_{n=0}^{N-1} x[n]y[n - m] \tag{74}$$

To illustrate how the method works, we use a 3-count sinusoidal signal (tone-burst). It is generated at the transmitting PWAS using a sinusoidal carrier signal with period $T = 1\,\mu\text{s}$ starting at $t_0 = 0\,\mu\text{s}$ and oscillating for three cycles. This signal arrives at the receiving sensor at $t_1 = 600\,\mu\text{s}$ (Fig. A.10a). The first signal is the transmitted signal; the second signal, below it, is the receiving signal. The cross-correlation coefficient $R_{xy}(t)$ is shown at the bottom; it has a peak right at $t_1 = 600\,\mu\text{s}$.

The process of generating the cross-correlation coefficients can be explained as follows: the input signal slides along the time axis of the correlated signal with a small step while the similarity of the overlapped part of two signals is compared. Before the tail of the input signal hits it, the two signals are not related and the corresponding cross-correlated coefficients are zero. When the input signal arrives at $t_1 = 600\,\mu\text{s}$, the two signals completely overlap, i.e., they match perfectly. At this moment, the cross-correlation coefficient $R_{xy}(t)$ also reaches the maximum value (Fig. A.10b). After the input signal moves away from the best match point $t_1 = 600\,\mu\text{s}$, the similarity between the overlapped parts becomes smaller again. Also the cross-correlation coefficients decrease. When the input signal totally moves away the compared signal, the overlapping is none and the coefficients become zeros.

Figure A.10b illustrates the cross-correlation results of window-smoothed tone-burst signals. Assume the time-of-flight (TOF) of the received signal is still $t_1 = 600\,\mu\text{s}$. As explained before, the received signal is compared with the transmitted signal by sliding through the time axis. When the transmitted signal moves forward, the resulting coefficients $R_{xy}(t)$ change from 0 to small values, increasing, and then hitting the peak, and decreasing till they become zero again. From bottom cross-correlation coefficients curve *iii*), we see a peak showing up at $t_1 = 600\,\mu\text{s}$, which is exactly the arrival time of the received signal.

An extra bonus while using the cross-correlation method is that, for real signals that have unavoidable noise, the method can reduce the annoyance of noise since the noise is not related to the signals and not auto-related either. We employed an example signal from the proof-of-concept experiment of the EUSR system collected by PWAS. The excitation

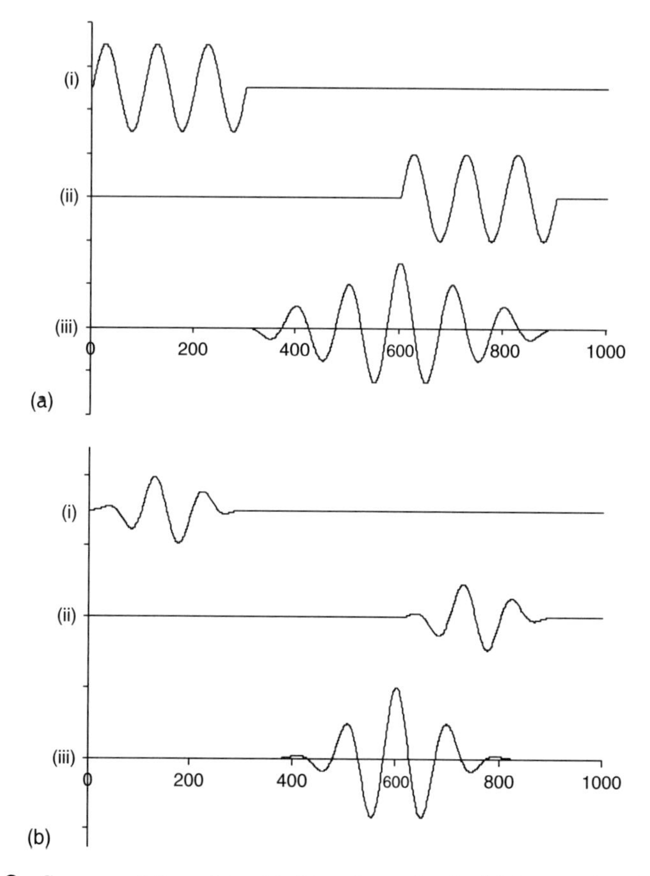

FIGURE A.10 Cross-correlation of simulated tone-burst signals: (a) 3-count sine signal; (b) 3-count Hanning-window smoothed signal. *Note*: (i) transmitted signal $x(t)$; (ii) received signal $y(t)$; (iii) cross-correlation of $x(t)$ and $y(t)$.

signal was a window-smoothed 3-count tone-burst of 340 kHz carrier frequency. This signal was used as the baseline signal in the cross-correlation method, shown as Fig. A.11a. The received signal, shown in Fig. A.11b, carried significant noise, which hampered the detection of the time of flight of the echo reflected by the crack. Cross correlation

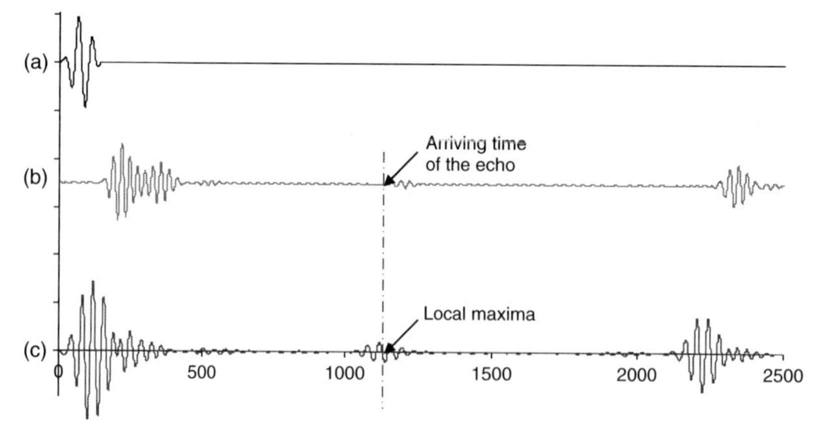

FIGURE A.11 Cross correlation of PWAS signals: (a) window-smoothed tone-burst baseline; (b) PWAS received signal; (c) cross correlation of the two signals.

generated the coefficient curve in Fig. A.11c. Note that it has several local maxima, corresponding to the arrival time of each wave packet. The noise may contaminate the original signal, but it will not significantly affect the cross-correlation result. By applying some peak detection method, we can pick out those peaks and, therefore, determine the arrival time of each wave packet.

A.6 TIME AVERAGED PRODUCT OF TWO HARMONIC VARIABLES

Assume two time-dependent harmonic variables

$$
\begin{aligned}
V(t) &= \hat{V}\cos(\omega t + \phi) \\
I(t) &= \hat{I}\cos \omega t
\end{aligned}
\tag{75}
$$

Using complex notations, we write

$$
\begin{aligned}
V(t) &= \hat{V}e^{i(\omega t + \phi)} \\
I(t) &= \hat{I}e^{i\omega t}
\end{aligned}
\tag{76}
$$

Recalling Euler identity $e^{i\alpha} = \cos\alpha + \mathrm{i}\sin\alpha$, we understand that the formulation (76) implies the convention of using only the real part of the complex exponential, since,

$$
\begin{aligned}
V(t) &= \hat{V}\cos(\omega t + \phi) = \mathrm{Re}(\hat{V}e^{i(\omega t + \phi)}) \\
I(t) &= \hat{I}\cos \omega t = \mathrm{Re}(\hat{I}e^{i\omega t})
\end{aligned}
\tag{77}
$$

The *product* of the two variables defined by Eq. (75) is given by

$$
P(t) = V(t)\cdot I(t) = \hat{V}\cos(\omega t + \phi)\cdot \hat{I}\cos \omega t \tag{78}
$$

Using standard trigonometric identities (Spiegel, 1968), we express Eq. (78) in the form

$$
\begin{aligned}
P(t) &= \hat{V}\cos(\omega t + \phi)\cdot \hat{I}\cos \omega t = \frac{1}{2}\hat{V}\hat{I}(\cos\phi + \cos(2\omega t + \phi)) \\
&= \frac{1}{2}\hat{V}\hat{I}\cos\phi + \frac{1}{2}\hat{V}\hat{I}\cos(2\omega t + \phi)
\end{aligned}
\tag{79}
$$

The function $P(t)$ defined by Eq. (79) is a harmonic function that contains a constant part, $\frac{1}{2}\hat{V}\hat{I}\cos\phi$, and a harmonic part $\frac{1}{2}\hat{V}\hat{I}\cos(2\omega t + \phi)$ that oscillates with twice the frequency of the original variables given by Eq. (75).

The *time-averaged product* of the two variables defined by Eq. (75) is given by the time average of the function $P(t)$, i.e.,

$$
< P > = \frac{1}{T}\int_0^T P(t)\mathrm{d}t \tag{80}
$$

where T is the period given by $T = 2\pi/\omega$. Using Eq. (79), we write Eq. (80) as

$$
< P > = \frac{1}{T}\int_0^T P(t)\mathrm{d}t = \frac{1}{T}\int_0^T \frac{1}{T}\hat{V}\hat{I}\cos\phi\,\mathrm{d}t + \frac{1}{T}\int_0^T \frac{1}{T}\hat{V}\hat{I}\cos(2\omega t + \phi)\mathrm{d}t \tag{81}
$$

Upon rearrangement, we write Eq. (81) as

$$< P > = \frac{1}{2}\hat{V}\hat{I}\cos\phi\left(\frac{1}{T}\int_0^T \mathrm{d}t\right) + \frac{1}{2}\hat{V}\hat{I}\left(\frac{1}{T}\int_0^T \cos(2\omega t + \phi)\mathrm{d}t\right) \tag{82}$$

It is apparent that the first integral in Eq. (81) equals 1, whereas the second integral equals zero (Spiegel, 1968). Hence, the time-averaged product of the two harmonic variables defined by Eq. (75) is given by

$$< P > = \frac{1}{2}\hat{V}\hat{I}\cos\phi \tag{83}$$

The result of Eq. (83) can be also expressed in terms of the complex notations (76). Note that

$$\mathrm{Re}\left(e^{i(\omega t+\phi)}e^{-i\omega t}\right) = \mathrm{Re}\left(e^{i\phi}\right) = \cos\phi \tag{84}$$

Using Eq. (84) and the definition of the conjugate $\bar{z}$ of a complex number z, i.e.,

$$\bar{z} = x - \mathrm{i}y, \quad \text{where } z = x + \mathrm{i}y \tag{85}$$

we can write

$$\hat{V}e^{i(\omega t+\phi)}\bar{\hat{I}e^{-i\omega t}} = V\bar{I} \tag{86}$$

Substituting Eqs (84) and (86) into Eq. (83) yields $< P > = \mathrm{Re}\left(\frac{1}{2}V\bar{I}\right)$. Since $\cos(-\alpha) = \cos(\alpha)$, we also have $< P > = \mathrm{Re}\left(\frac{1}{2}\bar{V}I\right)$. Hence, we can state the general assertion that the *time-averaged product of the two harmonic variables $I(t)$ and $V(t)$ is given by*

$$< P > = < VI > = \mathrm{Re}\left(\frac{1}{2}V\bar{I}\right) = \mathrm{Re}\left(\frac{1}{2}\bar{V}I\right) \tag{87}$$

Equation (87) indicates that *the time-averaged product of two harmonic variables is one half the product of one variable times the conjugate of the other.* If the harmonic variables are both time and space dependant, e.g.,

$$\begin{aligned} V(t) &= \hat{V}e^{i(\omega t - \gamma x + \phi)} \\ I(t) &= \hat{I}e^{i(\omega t - \gamma x)} \end{aligned} \tag{88}$$

then Eq. (87) still applies. The proof is as follows:

$$P(x,t) = V(x,t)\cdot I(x,t) = \hat{V}\cos(\omega t - \gamma x + \phi)\cdot\hat{I}\cos(\omega t - \gamma x) \tag{89}$$

Using standard trigonometric identities (Spiegel, 1968), we express Eq. (89) in the form

$$\begin{aligned} P(x,t) &= \hat{V}\cos(\omega t - \gamma x + \phi)\cdot\hat{I}\cos(\omega t - \gamma x) = \frac{1}{2}\hat{V}\hat{I}(\cos\phi + \cos(2\omega t - 2\gamma x + \phi)) \\ &= \frac{1}{2}\hat{V}\hat{I}\cos\phi + \frac{1}{2}\hat{V}\hat{I}\cos(2\omega t - 2\gamma x + \phi) \end{aligned} \tag{90}$$

The time average of Eq. (90) is obtained by integration over the time $T = 2\pi/\omega$ followed by division by T, i.e.,

$$< P > = \frac{1}{T}\int_0^T P(x,t)\mathrm{d}t = \frac{1}{T}\int_0^T \left[\frac{1}{2}\hat{V}\hat{I}\cos\phi + \frac{1}{2}\hat{V}\hat{I}\cos(2\omega t - 2\gamma x + \phi)\right]\mathrm{d}t = \frac{1}{2}\hat{V}\hat{I}\cos\phi \tag{91}$$

On the other hand,

$$\mathrm{Re}V\tilde{I} = \mathrm{Re}\left[\hat{V}e^{i(\omega t-\gamma x+\phi)}\hat{I}e^{-i(\omega t-\gamma x)}\right] = \hat{V}\hat{I}\mathrm{Re}e^{i\phi} = \hat{V}\hat{I}\cos\phi \tag{92}$$

Comparison of Eqs (91) and (92) proves Eq. (87).

A.7 HARMONIC AND BESSEL FUNCTIONS

A.7.1 ORTHOGONAL PROPERTIES OF HARMONIC FUNCTIONS

$$\begin{aligned}
\int_0^{2\pi} \sin px \sin qx\,\mathrm{d}x &= \int_0^{2\pi} \frac{1}{2}[\cos(p-q)x - \cos(p+q)x]\mathrm{d}x \\
&= \frac{1}{2}\int_0^{2\pi} \cos(p-q)x\,\mathrm{d}x - \frac{1}{2}\int_0^{2\pi}\cos(p+q)x \\
&= \begin{cases} \frac{1}{2}\frac{1}{p-q}\sin(p-q)x|_0^{2\pi} - \frac{1}{2}\frac{1}{p+q}\sin(p+q)x|_0^{2\pi} = 0, & \text{for } p \neq q \\ \frac{1}{2}2\pi - \frac{1}{2}\frac{1}{2p}\sin(p+q)x|_0^{2\pi} = \pi, & \text{for } p = q \end{cases} \\
&= \pi\delta_{pq}
\end{aligned} \tag{93}$$

where the symbol δ_{pq} is the Kronecker delta ($\delta_{pq} = 1$ for $p = q$, and $\delta_{pq} = 0$ for $p \neq q$, p and q integers). A similar argument also applies for nonintegers, i.e.,

$$\int_0^l \sin\gamma_p x \cos\gamma_q x\,\mathrm{d}x = \frac{1}{2}\int_0^l \left[\cos(\gamma_p - \gamma_q)x - \cos(\gamma_p + \gamma_q)x\right]\mathrm{d}x = \pi\delta_{pq} \tag{94}$$

$$\int_0^l \cos\gamma_p x \cos\gamma_q x\,\mathrm{d}x = \frac{1}{2}\int_0^l \left[\cos(\gamma_p - \gamma_q)x - \cos(\gamma_p + \gamma_q)x\right]\mathrm{d}x = \pi\delta_{pq} \tag{95}$$

A.7.2 BESSEL FUNCTIONS OF FIRST AND SECOND KIND

The *Bessel equation* is an ordinary differential equation of the form

$$x^2\frac{\mathrm{d}^2y}{\mathrm{d}z^2} + x\frac{\mathrm{d}y}{\mathrm{d}x} + (x^2 - v^2)y = 0 \tag{96}$$

where $v \geq 0$ is a given real number. In the general case when v is not an integer, the solution of Eq. (96) is of the form

$$y(x) = C_1 J_v(x) + C_2 J_{-v}(x) \tag{97}$$

where the function $J_v(x)$ is the *Bessel function of first kind* and order v. The function $J_v(x)$ does not have a closed-form solution, but it is expressed in terms of a convergent series.

For details, the reader is referred to standard books (e.g., Abramowitz and Stegun, 1964; Bowman, 1958; Watson, 1995). The Bessel functions are available as built-in functions in most mathematical software packages (e.g., MathCAD, MATLAB, etc.) In the case when v is not an integer, the functions $J_v(x)$ and $J_{-v}(x)$ are independent and Eq. (97) is a general solution of Eq. (96).

If the constant v is an integer, i.e., $v = n$, than $J_{-n}(x) = (-1)^n J_n(x)$, which means that $J_n(x)$ and $J_{-n}(x)$ are no longer independent and Eq. (97) is no longer a general solution. To resolve this situation, a new function, the *Bessel function of second kind*, $Y_n(x)$, is introduced using the definition

$$Y_n(x) = \lim_{v \to n} Y_v(x), \quad \text{where } Y_v(x) = \frac{1}{\sin v\pi}[J_v(x)\cos v\pi - J_{-v}(x)] \tag{98}$$

The function defined by Eq. (98) is also known as *Neumann function* as well as *Weber function*. It can be shown that $Y_{-n}(x) = (-1)^n Y_n(x)$. The general solution of Eq. (96) for all values of v is written in the form

$$y(x) = AJ_v(x) + BY_v(x) \tag{99}$$

A.7.3 PROPERTIES OF THE BESSEL FUNCTIONS

Bessel functions of different orders are related by the following formulae

$$J_{-n}(x) = (-1)^n J_n(x); \quad I_{-n}(x) = I_n(x) \tag{100}$$

$$J_0'(x) = -J_1(x) \tag{101}$$

$$J_1'(x) = J_0(x) - \frac{1}{x}J_1(x) \tag{102}$$

$$J_2'(x) = \left(1 - \frac{4}{x^2}\right)J_1(x) + \frac{2}{x}J_0(x) \tag{103}$$

$$\frac{d}{dx}[x^v J_v(x)] = x^v J_{v-1}(x) \tag{104}$$

$$\frac{d}{dx}[x^{-v} J_v(x)] = -x^{-v} J_{v+1}(x) \tag{105}$$

$$J_v(x) = \frac{x}{2v}J_{v-1}(x) + \frac{x}{2v}J_{v+1}(x) \tag{106}$$

$$J_v'(x) = J_{v-1}(x) - \frac{v}{x}J_v(x) \tag{107}$$

$$J_v'(x) = -J_{v+1}(x) + \frac{v}{x}J_v(x) \tag{108}$$

$$J_v'(x) = \frac{1}{2}J_{v-1}(x) - \frac{1}{2}J_{v+1}(x) \tag{109}$$

$$J_{1/2}(x) = \sqrt{\frac{2}{\pi x}}\sin x \tag{110}$$

$$J_{-1/2}(x) = \sqrt{\frac{2}{\pi x}}\cos x \tag{111}$$

A.7.4 MODIFIED BESSEL FUNCTIONS

The substitution of ix for x in Eq. (96) leads to the *modified Bessel equation*, i.e.,

$$x^2\frac{d^2y}{dz^2}+x\frac{dy}{dx}-(x^2+v^2)y=0 \tag{112}$$

The general solution of Eq. (112) is expressed in the form

$$y(x)=CI_v(x)+DK_v(x) \tag{113}$$

where C and D are arbitrary constants, whereas $I_v(x)$ and $K_v(x)$ are the *modified Bessel functions of the first and second kind* defined as

For v a integer n, it can be shown that

$$I_{-n}(x)=I_n(x) \tag{114}$$

$$I_0'(x)=I_1(x) \tag{115}$$

The modified Bessel functions are also known as *hyperbolic Bessel functions*.

A.7.5 HANKEL FUNCTIONS

In many applications, there is a practical need for solutions of Bessel equation (96) that are complex for real values of x. Such solutions can be expressed in terms of the *Hankel functions* given by

$$\begin{aligned}H_v^{(1)}(x)&=J_v(x)+iY_v(x)\\ H_v^{(2)}(x)&=J_v(x)-iY_v(x)\end{aligned} \tag{116}$$

The functions defined by Eq. (116) are known as *the first and second Hankel functions of order* v. They are also known as the Bessel functions of third kind and order v. The general solution of Eq. (96) expressed in terms of the Hankel functions takes the form

$$y(x)=C_1H_v^{(1)}(x)+C_2H_v^{(2)}(x) \tag{117}$$

Hankel functions are particularly useful for studying the behavior of the solution in the far field as $|x|\to\infty$. But both Hankel functions are singular at the origin.

A.7.6 ASYMPTOTIC BEHAVIOR OF BESSEL FUNCTIONS

It can be shown that, for large x,

$$J_v(x)\approx\sqrt{\frac{2}{\pi x}}\cos\left(x-\frac{1}{2}v\pi-\frac{1}{4}\pi\right) \tag{118}$$

$$J_{2n}(x)\approx(-1)^n\frac{1}{\sqrt{\pi x}}(\cos x+\sin x) \tag{119}$$

$$J_{2n+1}(x)\approx(-1)^{n+1}\frac{1}{\sqrt{\pi x}}(\cos x-\sin x) \tag{120}$$

Equations (118) through (120) indicate that, at large x, the Bessel functions behave like harmonic functions of decaying amplitude. Of particular interest is the behavior of the Bessel functions of order 0 and 1, i.e.,

$$\begin{aligned} J_0(x) &\approx \frac{1}{\sqrt{\pi x}}(\cos x + \sin x) \\ J_1(x) &\approx -\frac{1}{\sqrt{\pi x}}(\cos x - \sin x) \end{aligned} \tag{121}$$

A.7.7 ORTHOGONAL PROPERTIES OF BESSEL FUNCTIONS

Similar to harmonic functions, Bessel functions also show orthogonal properties (Kreyszig, 1999). For example, for the Bessel functions of the first kind, one writes (Kreyszig, 1999)

$$\int_0^R xJ_n(\lambda_{mn}x)J_n(\lambda_{kn}x)\mathrm{d}x = 0, \quad k \neq m \tag{122}$$

where λ_{mn} and λ_{kn} are roots of the characteristic equation

$$J_n(\lambda R) = 0 \tag{123}$$

Let $\alpha_{1n} < \alpha_{2n} < \alpha_{3n} \ldots$ denote the positive roots of $J_n(\alpha) = 0$. Then, Eq. (123) holds for

$$\lambda_{mn} = \frac{\alpha_{mn}}{R}, \quad m = 1, 2, 3, \ldots \tag{124}$$

Examination of Eq. (122) reveals that the orthogonality of the Bessel function is calculated with respect to the weight function $p(x) = x$.

A.7.8 NORMALITY PROPERTIES OF BESSEL FUNCTIONS

According to Eq. (122), Bessel functions are orthogonal, i.e., the weighted integral of the product of two dissimilar Bessel functions is zero. But what happens if Eq. (122) is applied to two identical Bessel functions? In this case, we get the norm of the Bessel function, i.e.,

$$\|J_n(\lambda_{nm}x)\|^2 = \int_0^R xJ_n^2(\lambda_{mn}x)\mathrm{d}x, \quad m = 1, 2, 3, \ldots \tag{125}$$

It can be shown (Kreyszig, 1999) that Eq. (125) simplifies to give

$$\int_0^R xJ_n^2(\lambda_{mn}x)\,\mathrm{d}x = \frac{R^2}{2}J_{n+1}^2(\lambda_{nm}R), \quad m = 1, 2, 3, \ldots \tag{126}$$

Therefore, the *norm of a Bessel function* is given by

$$\|J_n(\lambda_{nm}x)\| = \frac{R}{\sqrt{2}}J_{n+1}(\lambda_{nm}R), \quad m = 1, 2, 3, \ldots \tag{127}$$

A.7.9 FOURIER–BESSEL SERIES

Since the Bessel functions form an orthogonal set, any given function $f(x)$ can be expanded in terms of this set, i.e., we can write

$$f(x) = a_1 J_n(\lambda_{1n} x) + a_2 J_n(\lambda_{2n} x) + \cdots \tag{128}$$

The coefficients a_m are given by the expression

$$a_m = \frac{2}{R^2 J_{n+1}^2(\alpha_{nm})} \int_0^R x J_n(\lambda_{mn} x) f(x) \mathrm{d}x \tag{129}$$

The proof of Eq. (129) comes from multiplying the series expansion (128) by the weight function $xJ_n(\lambda_{mn}x)$, integrating over the interval $(0, R)$ and using the orthogonality relation (122) and the normality relation (126).

APPENDIX B

ELASTICITY NOTATIONS AND EQUATIONS

B.1 BASIC NOTATIONS

$$\frac{\partial}{\partial x}(\;) = (\;)' \quad \text{and} \quad \frac{\partial}{\partial t}(\;) = \dot{(\;)} \tag{1}$$

$$\delta_{ij} = \begin{cases} 1 & \text{if } i = j \\ 0 & \text{otherwise} \end{cases} \quad \text{(Kronecker delta)} \tag{2}$$

B.1.1 BASIC CARTESIAN VECTOR NOTATIONS AND DIFFERENTIAL OPERATORS

$$\vec{u} = u_x\vec{e}_x + u_y\vec{e}_y + u_z\vec{e}_z \quad \text{(displacement vector)} \tag{3}$$

$$\vec{\nabla} = \frac{\partial}{\partial x}\vec{e}_x + \frac{\partial}{\partial y}\vec{e}_y + \frac{\partial}{\partial z}\vec{e}_z \quad \text{(del operator)} \tag{4}$$

$$\text{grad } \phi = \vec{\nabla}\phi = \frac{\partial \phi}{\partial x}\vec{e}_x + \frac{\partial \phi}{\partial y}\vec{e}_y + \frac{\partial \phi}{\partial z}\vec{e}_z \quad \text{(gradient of scalar function } \phi\text{)} \tag{5}$$

$$\text{div } \vec{A} = \vec{\nabla} \cdot \vec{A} = \frac{\partial}{\partial x}A_x + \frac{\partial}{\partial y}A_y + \frac{\partial}{\partial z}A_z \quad \text{(divergence of vector function } \vec{A}\text{)} \tag{6}$$

$$\text{curl } \vec{A} = \vec{\nabla} \times \vec{A} = \left(\frac{\partial A_z}{\partial y} - \frac{\partial A_y}{\partial z}\right)\vec{e}_x + \left(\frac{\partial A_x}{\partial z} - \frac{\partial A_z}{\partial x}\right)\vec{e}_y + \left(\frac{\partial A_y}{\partial x} - \frac{\partial A_x}{\partial z}\right)\vec{e}_z \quad \text{(curl of vector function } \vec{A}\text{)} \tag{7}$$

$$\nabla^2 = \vec{\nabla} \cdot \vec{\nabla} = \frac{\partial^2}{\partial x^2} + \frac{\partial^2}{\partial y^2} + \frac{\partial^2}{\partial z^2} \quad \text{(Laplace operator)} \tag{8}$$

$$\nabla^2\phi = \frac{\partial^2\phi}{\partial x^2} + \frac{\partial^2\phi}{\partial y^2} + \frac{\partial^2\phi}{\partial z^2} \quad \text{(Laplacian of scalar function } \phi) \tag{9}$$

$$\nabla^4 = \nabla^2\nabla^2 = \frac{\partial^4}{\partial x^4} + \frac{\partial^4}{\partial y^4} + \frac{\partial^4}{\partial z^4} + 2\frac{\partial^4}{\partial x^2\partial y^2} + 2\frac{\partial^4}{\partial y^2\partial z^2} + 2\frac{\partial^4}{\partial z^2\partial x^2} \quad \text{(biharmonic operator)} \tag{10}$$

where the vectors $\vec{e}_x$, $\vec{e}_y$, $\vec{e}_z$ are unit vectors in the direction of axes x, y, z, respectively.

B.1.2 BASIC TENSOR NOTATIONS

$$(\;)_{ii} = (\;)_{11} + (\;)_{22} + (\;)_{33} \quad \text{(Einstein implied summation)} \tag{11}$$

$$(\;)_{i,j} = \frac{\partial(\;)_i}{\partial x_j} \quad \text{(differentiation shorthand)} \tag{12}$$

B.2 3-D STRAIN–DISPLACEMENT RELATIONS

B.2.1 CARTESIAN NOTATIONS

$$\varepsilon_{xx} = \frac{\partial u_x}{\partial x} \qquad \varepsilon_{xy} = \frac{1}{2}\left(\frac{\partial u_x}{\partial y} + \frac{\partial u_y}{\partial x}\right)$$
$$\varepsilon_{yy} = \frac{\partial u_y}{\partial y} \qquad \varepsilon_{yz} = \frac{1}{2}\left(\frac{\partial u_y}{\partial z} + \frac{\partial u_z}{\partial y}\right) \tag{13}$$
$$\varepsilon_{zz} = \frac{\partial u_z}{\partial z} \qquad \varepsilon_{zx} = \frac{1}{2}\left(\frac{\partial u_z}{\partial x} + \frac{\partial u_x}{\partial z}\right)$$

Symmetry relations apply, i.e., $\varepsilon_{yx} = \varepsilon_{xy}$, $\varepsilon_{zy} = \varepsilon_{yz}$, $\varepsilon_{xz} = \varepsilon_{zx}$

B.2.2 TENSOR NOTATIONS

$$\varepsilon_{ij} = \frac{1}{2}\left(\frac{\partial u_i}{\partial x_j} + \frac{\partial u_j}{\partial x_i}\right) \quad i, j = 1, 2, 3 \tag{14}$$

Using the differentiation shorthand (12), we write Eq. (14) in the form

$$\varepsilon_{ij} = \frac{1}{2}(u_{i,j} + u_{j,i}) \quad i, j = 1, 2, 3 \tag{15}$$

In expanded form:

$$\varepsilon_{11} = \frac{1}{2}\left(\frac{\partial u_1}{\partial x_1} + \frac{\partial u_2}{\partial x_2}\right) = \frac{\partial u_1}{\partial x_1} \qquad \varepsilon_{12} = \frac{1}{2}\left(\frac{\partial u_1}{\partial x_2} + \frac{\partial u_2}{\partial x_1}\right)$$
$$\varepsilon_{22} = \frac{1}{2}\left(\frac{\partial u_2}{\partial x_2} + \frac{\partial u_2}{\partial x_2}\right) = \frac{\partial u_2}{\partial x_2} \qquad \varepsilon_{23} = \frac{1}{2}\left(\frac{\partial u_2}{\partial x_3} + \frac{\partial u_3}{\partial x_2}\right) \tag{16}$$
$$\varepsilon_{33} = \frac{1}{2}\left(\frac{\partial u_3}{\partial x_3} + \frac{\partial u_3}{\partial x_3}\right) = \frac{\partial u_3}{\partial x_3} \qquad \varepsilon_{31} = \frac{1}{2}\left(\frac{\partial u_3}{\partial z} + \frac{\partial u_1}{\partial x_3}\right)$$

B.3 DILATATION AND ROTATION

B.3.1 DILATATION

Dilatation, Δ, is a measure of the *volumetric strain*, ε_v, i.e., associated with an uniform expansion of the elastic medium. Several equivalent expressions of the dilatation are as follows.

$$\Delta = \varepsilon_{xx} + \varepsilon_{yy} + \varepsilon_{zz} \tag{17}$$

$$\Delta = \frac{\partial u_x}{\partial x} + \frac{\partial u_y}{\partial y} + \frac{\partial u_z}{\partial z} \tag{18}$$

$$\Delta = \vec{\nabla} \cdot \vec{u} \tag{19}$$

$$\Delta = \varepsilon_{kk} = \varepsilon_{11} + \varepsilon_{22} + \varepsilon_{33} \quad k = 1, 2, 3 \tag{20}$$

$$\Delta = \frac{\partial u_k}{\partial x_k} = \frac{\partial u_1}{\partial x_1} + \frac{\partial u_2}{\partial y_2} + \frac{\partial u_3}{\partial z_3} \quad k = 1, 2, 3 \tag{21}$$

$$\Delta = u_{k,k} = u_{1,1} + u_{2,2} + u_{3,3} \quad k = 1, 2, 3 \tag{22}$$

B.3.2 ROTATION

Cartesian notations

$$\omega_{xy} = \frac{1}{2}\left(\frac{\partial u_x}{\partial y} - \frac{\partial u_y}{\partial x}\right)$$
$$\omega_{yz} = \frac{1}{2}\left(\frac{\partial u_y}{\partial z} - \frac{\partial u_z}{\partial y}\right) \tag{23}$$
$$\omega_{zx} = \frac{1}{2}\left(\frac{\partial u_z}{\partial x} - \frac{\partial u_x}{\partial z}\right)$$

Tensor notations

$$\omega_{ij} = \frac{1}{2}\left(\frac{\partial u_i}{\partial x_j} - \frac{\partial u_j}{\partial x_i}\right) \quad i, j = 1, 2, 3 \tag{24}$$

or

$$\omega_{ij} = \frac{1}{2}(u_{i,j} - u_{j,i}) \tag{25}$$

Vector notations

$$\vec{\omega} = \frac{1}{2}\vec{\nabla} \times \vec{u} \tag{26}$$

Using the expansion rule for the vector product we write

$$\vec{\omega} = \frac{1}{2}\vec{\nabla} \times \vec{u} = \frac{1}{2}\begin{vmatrix} \vec{e}_x & \vec{e}_y & \vec{e}_z \\ \frac{\partial}{\partial x} & \frac{\partial}{\partial y} & \frac{\partial}{\partial z} \\ u_x & u_y & u_z \end{vmatrix} = \frac{1}{2}\left(\frac{\partial u_z}{\partial y} - \frac{\partial u_y}{\partial z}\right)\vec{e}_x + \frac{1}{2}\left(\frac{\partial u_x}{\partial z} - \frac{\partial u_z}{\partial x}\right)\vec{e}_y$$
$$+\frac{1}{2}\left(\frac{\partial u_x}{\partial z} - \frac{\partial u_z}{\partial x}\right)\vec{e}_z \tag{27}$$

The following equivalent notations are used interchangeably:

$$\vec{\omega} = \omega_x \vec{e}_x + \omega_y \vec{e}_y + \omega_z \vec{e}_z \tag{28}$$

and

$$\vec{\omega} = -\omega_{yz} \vec{e}_x - \omega_{zx} \vec{e}_y - \omega_{xy} \vec{e}_z \tag{29}$$

where

$$\begin{aligned} \omega_x &= -\omega_{yz} = \frac{1}{2}\left(\frac{\partial u_z}{\partial y} - \frac{\partial u_y}{\partial z}\right) \\ \omega_y &= -\omega_{zx} = \frac{1}{2}\left(\frac{\partial u_x}{\partial z} - \frac{\partial u_z}{\partial x}\right) \\ \omega_z &= -\omega_{xy} = \frac{1}{2}\left(\frac{\partial u_y}{\partial x} - \frac{\partial u_x}{\partial y}\right) \end{aligned} \tag{30}$$

B.4 3-D STRESS–STRAIN RELATIONS IN ENGINEERING CONSTANTS

The elasticity relation in *compliance formulation* can be expressed as

$$\begin{aligned} \varepsilon_{xx} &= \frac{1}{E}\sigma_{xx} + \frac{-v}{E}\sigma_{yy} + \frac{-v}{E}\sigma_{zz} & & \varepsilon_{xy} = \frac{1}{2G}\sigma_{xy} \\ \varepsilon_{yy} &= \frac{-v}{E}\sigma_{xx} + \frac{1}{E}\sigma_{yy} + \frac{-v}{E}\sigma_{zz} & \text{and} \quad & \varepsilon_{yz} = \frac{1}{2G}\sigma_{yz} \\ \varepsilon_{zz} &= \frac{-v}{E}\sigma_{xx} + \frac{-v}{E}\sigma_{yy} + \frac{1}{E}\sigma_{zz} & & \varepsilon_{zx} = \frac{1}{2G}\sigma_{zx} \end{aligned} \tag{31}$$

where E is *Young elastic modulus*, v is *Poisson ratio*, and G is the *shear modulus* given by

$$G = \frac{1}{2(1+v)}E \tag{32}$$

Inversion of Eq. (31) yields the *stiffness formulation* of the constitutive relation

$$\begin{aligned} \sigma_{xx} &= \frac{(1-v)}{(1+v)(1-2v)}E\varepsilon_{xx} + \frac{v}{(1+v)(1-2v)}E\varepsilon_{yy} + \frac{v}{(1+v)(1-2v)}E\varepsilon_{zz} & & \sigma_{xy} = 2G\varepsilon_{xy} \\ \sigma_{yy} &= \frac{v}{(1+v)(1-2v)}E\varepsilon_{xx} + \frac{(1-v)}{(1+v)(1-2v)}E\varepsilon_{yy} + \frac{v}{(1+v)(1-2v)}E\varepsilon_{zz} & \text{and} \quad & \sigma_{yz} = 2G\varepsilon_{yz} \\ \sigma_{zz} &= \frac{v}{(1+v)(1-2v)}E\varepsilon_{xx} + \frac{v}{(1+v)(1-2v)}E\varepsilon_{yy} + \frac{(1-v)}{(1+v)(1-2v)}E\varepsilon_{zz} & & \sigma_{zx} = 2G\varepsilon_{zx} \end{aligned} \tag{33}$$

B.5 3-D STRESS–STRAIN RELATIONS IN LAME CONSTANTS

Lame constants are defined in relation with Eq. (33) as:

$$\lambda = \frac{v}{(1+v)(1-2v)}E$$
$$\mu = G = \frac{1}{2(1+v)}E \tag{34}$$

Conversely, we can relate the engineering elasticity constants, E and G, to the Lame constants, i.e.,

$$E = \frac{\mu(3\lambda+2\mu)}{\lambda+\mu} \quad \text{(Young modulus)} \tag{35}$$

$$G = \mu \quad \text{(Shear modulus)} \tag{36}$$

$$B = \frac{3\lambda+2\mu}{3} \quad \text{(Bulk modulus)} \tag{37}$$

$$v = \frac{\lambda}{2(\lambda+\mu)} \quad \text{(Poisson ratio)} \tag{38}$$

Substitution of Eq. (34) into Eq. (33) yields the 3-D stress–strain relations in Lame constants, as detailed in the next subsection.

B.5.1 CARTESIAN NOTATIONS

$$\begin{aligned}\sigma_{xx} &= (\lambda+2\mu)\varepsilon_{xx}+\lambda\varepsilon_{yy}+\lambda\varepsilon_{zz} \\ \sigma_{yy} &= \lambda\varepsilon_{xx}+(\lambda+2\mu)\varepsilon_{yy}+\lambda\varepsilon_{zz} \\ \sigma_{zz} &= \lambda\varepsilon_{xx}+\lambda\varepsilon_{yy}+(\lambda+2\mu)\varepsilon_{zz}\end{aligned} \quad \text{or} \quad \begin{aligned}\sigma_{xx} &= \lambda\Delta+2\mu\varepsilon_{xx} \\ \sigma_{yy} &= \lambda\Delta+2\mu\varepsilon_{yy} \\ \sigma_{zz} &= \lambda\Delta+2\mu\varepsilon_{zz}\end{aligned} \quad \text{and} \quad \begin{aligned}\sigma_{xy} &= 2\mu\varepsilon_{xy} \\ \sigma_{yz} &= 2\mu\varepsilon_{yz} \\ \sigma_{zx} &= 2\mu\varepsilon_{zx}\end{aligned} \tag{39}$$

where Δ is the dilatation defined by Eq. (17).

B.5.2 TENSOR NOTATIONS

$$\sigma_{ij} = \lambda\delta_{ij}\varepsilon_{kk}+2\mu\varepsilon_{ij} \quad i,j=1,2,3 \tag{40}$$

where

$$\varepsilon_{kk} = \varepsilon_{11}+\varepsilon_{22}+\varepsilon_{33} \quad \text{(Einstein implied summation)} \tag{41}$$

and δ_{ij} is the Kronecker delta defined by Eq. (2). Equation (40) can also be written in the form

$$\sigma_{ij} = \lambda\Delta+\mu\varepsilon_{ij} \tag{42}$$

where Δ is the dilatation defined in Eq. (17).

B.6 3-D STRESS–DISPLACEMENT RELATIONS

B.6.1 CARTESIAN NOTATIONS

Substitution of Eqs (13) into Eqs (39) yields

$$\begin{aligned}
&\sigma_{xx} = (\lambda+2\mu)\varepsilon_{xx}+\lambda\varepsilon_{yy}+\lambda\varepsilon_{zz} && \sigma_{xx} = \lambda\Delta+2\mu\varepsilon_{xx} && \sigma_{xy} = \mu\left(\frac{\partial u_x}{\partial y}+\frac{\partial u_y}{\partial x}\right)\\
&\sigma_{yy} = \lambda\varepsilon_{xx}+(\lambda+2\mu)\varepsilon_{yy}+\lambda\varepsilon_{zz} \quad \text{or} && \sigma_{yy}= \lambda\Delta+2\mu\varepsilon_{yy}, \quad \text{and} && \sigma_{yz} = \mu\left(\frac{\partial u_y}{\partial z}+\frac{\partial u_z}{\partial y}\right)\\
&\sigma_{zz} = \lambda\varepsilon_{xx}+\lambda\varepsilon_{yy}+(\lambda+2\mu)\varepsilon_{zz} && \sigma_{zz} = \lambda\Delta+2\mu\varepsilon_{zz} && \sigma_{zx} = \mu\left(\frac{\partial u_z}{\partial x}+\frac{\partial u_x}{\partial z}\right)
\end{aligned} \tag{43}$$

B.6.2 TENSOR NOTATIONS

Substitution of Eqs (15) into Eqs (40) yields

$$\sigma_{ij} = \lambda\delta_{ij}\varepsilon_{k,k}+2\mu\, u_{ij} \quad i, j, k = 1, 2, 3 \tag{44}$$

where Δ is the dilatation defined in Eq. (20). Equation (44) can also be written in the form

$$\sigma_{ij} = \lambda\Delta+\mu(u_{i,j}+u_{j,i}) \tag{45}$$

where Δ is the dilatation defined in Eq. (22).

B.7 3-D EQUATIONS OF MOTION

B.7.1 CARTESIAN NOTATIONS

$$\begin{aligned}
\frac{\partial\sigma_{xx}}{\partial x}+\frac{\partial\sigma_{xy}}{\partial y}+\frac{\partial\sigma_{xz}}{\partial z} &= \rho\frac{\partial^2 u_x}{\partial t^2}\\
\frac{\partial\sigma_{yx}}{\partial x}+\frac{\partial\sigma_{yy}}{\partial y}+\frac{\partial\sigma_{yz}}{\partial z} &= \rho\frac{\partial^2 u_y}{\partial t^2}\\
\frac{\partial\sigma_{zx}}{\partial x}+\frac{\partial\sigma_{zy}}{\partial y}+\frac{\partial\sigma_{zz}}{\partial z} &= \rho\frac{\partial^2 u_z}{\partial t^2}
\end{aligned} \tag{46}$$

where ρ is the mass density. Symmetry relations apply, i.e., $\sigma_{yx} = \sigma_{xy}$, $\sigma_{zy} = \sigma_{yz}$, $\sigma_{xz} = \sigma_{zx}$

B.7.2 TENSOR NOTATIONS

$$\frac{\partial\sigma_{ij}}{\partial x_j} = \rho\ddot{u}_i \quad i, j = 1, 2, 3 \tag{47}$$

We can compact Eq. (47) even further using the differentiation shorthand rule (12), i.e.,

$$\sigma_{ij,j} = \rho\ddot{u}_i \quad i, j = 1, 2, 3 \tag{48}$$

B.8 TRACTIONS

B.8.1 CARTESIAN NOTATIONS

Consider a cut surface of external normal

$$\vec{n} = n_x\vec{e}_x + n_y\vec{e}_y + n_z\vec{e}_z \quad \text{(external normal)} \tag{49}$$

Then, the traction vector acting on the cut surface of normal $\vec{n}$ is given by

$$\vec{t} = t_x\vec{e}_x + t_y\vec{e}_y + t_z\vec{e}_z \quad \text{(traction vector)} \tag{50}$$

The components of the traction vector $\vec{t}$ are related to the stresses $\sigma_{xx}, \sigma_{yy}, \sigma_{zz}, \sigma_{yz}, \sigma_{zx}, \sigma_{xy}$, by the equation

$$\begin{aligned} t_x &= \sigma_{xx}n_x + \sigma_{xy}n_y + \sigma_{xz}n_z \\ t_y &= \sigma_{yx}n_x + \sigma_{yy}n_y + \sigma_{yz}n_z \\ t_z &= \sigma_{zx}n_x + \sigma_{zy}n_y + \sigma_{zz}n_z \end{aligned} \tag{51}$$

B.8.2 TENSOR NOTATIONS

$$t_i = \sigma_{ij}n_j \quad i, j = 1, 2, 3 \tag{52}$$

B.9 3-D GOVERNING EQUATIONS–NAVIER EQUATIONS

B.9.1 CARTESIAN NOTATIONS

$$\begin{aligned} (\lambda+\mu)\left(\frac{\partial^2 u_x}{\partial x^2} + \frac{\partial^2 u_y}{\partial x \partial y} + \frac{\partial^2 u_z}{\partial x \partial z}\right) + \mu\left(\frac{\partial^2 u_x}{\partial x^2} + \frac{\partial^2 u_x}{\partial y^2} + \frac{\partial^2 u_x}{\partial z^2}\right) &= \rho \ddot{u}_x \\ (\lambda+\mu)\left(\frac{\partial^2 u_x}{\partial y \partial x} + \frac{\partial^2 u_y}{\partial y^2} + \frac{\partial^2 u_z}{\partial y \partial z}\right) + \mu\left(\frac{\partial^2 u_y}{\partial x^2} + \frac{\partial^2 u_y}{\partial y^2} + \frac{\partial^2 u_y}{\partial z^2}\right) &= \rho \ddot{u}_y \\ (\lambda+\mu)\left(\frac{\partial^2 u_x}{\partial z \partial x} + \frac{\partial^2 u_y}{\partial z \partial y} + \frac{\partial^2 u_z}{\partial z^2}\right) + \mu\left(\frac{\partial^2 u_z}{\partial x^2} + \frac{\partial^2 u_z}{\partial y^2} + \frac{\partial^2 u_z}{\partial z^2}\right) &= \rho \ddot{u}_z \end{aligned} \tag{53}$$

B.9.2 VECTOR NOTATIONS

$$(\lambda+\mu)\vec{\nabla}(\vec{\nabla}\cdot\vec{u}) + \mu\nabla^2\vec{u} = \rho\ddot{\vec{u}} \tag{54}$$

where $\vec{\nabla}$ and ∇^2 are given by Eqs (4) and (8), respectively. Equivalent forms of the Navier Eq. (54) are:

$$(\lambda+\mu)\vec{\nabla}\Delta + \mu\nabla^2\vec{u} = \rho\ddot{\vec{u}} \tag{55}$$

$$(\lambda+\mu)\vec{\nabla}(\vec{\nabla}\cdot\vec{u}) + \mu\nabla^2\vec{u} = \rho\ddot{\vec{u}} \tag{56}$$

$$(\lambda+2\mu)\vec{\nabla}(\vec{\nabla}\cdot\vec{u}) - 2\mu\vec{\nabla}\times\vec{\omega} = \rho\ddot{\vec{u}} \tag{57}$$

$$(\lambda+2\mu)\vec{\nabla}\Delta - 2\mu\vec{\nabla}\times\vec{\omega} = \rho\ddot{\vec{u}} \tag{58}$$

B.9.3 TENSOR NOTATIONS

$$(\lambda+\mu)u_{j,ji}+\mu u_{i,jj}=\rho\ddot{u}_i \quad i,j=1,2,3 \tag{59}$$

B.10 2-D ELASTICITY

B.10.1 PLANE-STRESS CONDITIONS

Plane stress conditions are used in the analysis of thin-walled plates and shell. The plate surface is assumed free; hence, the z direction stress is assumed zero ($\sigma_{zz}=0$). Also zero are the surface shear stresses $\sigma_{yz}=0$ and $\sigma_{zx}=0$. Hence, the 3-D elasticity relations reduce to

$$\begin{aligned}
\varepsilon_{xx}&=\frac{1}{E}\sigma_{xx}+\frac{-v}{E}\sigma_{yy}\\
\varepsilon_{yy}&=\frac{-v}{E}\sigma_{xx}+\frac{1}{E}\sigma_{yy} \quad \text{and} \quad \varepsilon_{xy}=\frac{1}{2G}\sigma\\
\varepsilon_{zz}&=\frac{-v}{E}\sigma_{xx}+\frac{-v}{E}\sigma_{yy}
\end{aligned} \tag{60}$$

In our analysis, we are only interested in the strains ε_{xx}, ε_{yy}, ε_{xy}. Solution of Eq. (60) yields

$$\begin{aligned}
\sigma_{xx}&=\frac{E}{1-v^2}\left(\varepsilon_{xx}+v\varepsilon_{yy}\right)\\
&\qquad\qquad \sigma_{xy}=2G\varepsilon_{xy}\\
\sigma_{yy}&=\frac{E}{1-v^2}\left(v\varepsilon_{xx}+\varepsilon_{yy}\right)
\end{aligned} \tag{61}$$

B.10.2 PLANE-STRAIN CONDITIONS

Plane-strain conditions are used in the analysis of structures very thick in the z direction. The z direction strain is assumed zero ($\varepsilon_{zz}=0$). Also zero are the shear strains $\varepsilon_{yz}=0$ and $\varepsilon_{zx}=0$. Hence, the 3-D elasticity relations (33) that express stresses in terms of strains are reduced to

$$\begin{aligned}
\sigma_{xx}&=\frac{E}{(1+v)(1-2v)}\left[(1-v)\varepsilon_{xx}+v\varepsilon_{yy}\right]\\
\sigma_{yy}&=\frac{E}{(1+v)(1-2v)}\left[v\varepsilon_{xx}+(1-v)\varepsilon_{yy}\right] \quad \text{and} \quad \sigma_{xy}=2G\varepsilon_{xy}\\
\sigma_{zz}&=\frac{vE}{(1+v)(1-2v)}\left[\varepsilon_{xx}+\varepsilon_{yy}\right]
\end{aligned} \tag{62}$$

Solution of Eq. (62) yields

$$\begin{aligned}
\varepsilon_{xx}&=\frac{1+v}{E}\left[(1-v)\,\sigma_{xx}-v\sigma_{yy}\right]\\
&\qquad\qquad \text{and} \quad \varepsilon_{xy}=\frac{1}{2G}\sigma_{xy}\\
\varepsilon_{yy}&=\frac{1+v}{E}\left[-v\sigma_{xx}+(1-v)\,\sigma_{yy}\right]
\end{aligned} \tag{63}$$

B.11 POLAR COORDINATES

Assume polar coordinates (r, θ), and corresponding displacements (u_r, u_θ). Plane-stress conditions are assumed, i.e., the z direction stresses are assumed identical zero.

B.11.1 COORDINATES TRANSFORMATION

$$\begin{cases} x = r\cos\theta \\ y = r\sin\theta \end{cases} \qquad \begin{cases} r = \sqrt{x^2+y^2} \\ \theta = \arg(x, y) \end{cases} \tag{64}$$

B.11.2 BASIC DIFFERENTIAL OPERATIONS AND OPERATORS

$$\vec{u} = u_r\vec{e}_r + u_\theta\vec{e}_\theta \quad \text{(displacement vector)} \tag{65}$$

$$\operatorname{grad}\phi = \vec{\nabla}\phi = \frac{\partial\phi}{\partial r}\vec{e}_r + \frac{1}{r}\frac{\partial\phi}{\partial\theta}\vec{e}_\theta \quad \text{(gradient of a scalar function)} \tag{66}$$

$$\operatorname{div}\vec{A} = \vec{\nabla}\cdot\vec{A} = \frac{1}{r}\frac{\partial}{\partial r}(rA_r) + \frac{1}{r}\frac{\partial}{\partial\theta}A_\theta \quad \text{(divergence of a vector function)} \tag{67}$$

$$\nabla^2 = \frac{\partial^2}{\partial r^2} + \frac{1}{r}\frac{\partial}{\partial r} + \frac{1}{r^2}\frac{\partial^2}{\partial\theta^2} \quad \text{(Laplace operator)} \tag{68}$$

$$\nabla^2\phi = \frac{\partial^2\phi}{\partial r^2} + \frac{1}{r}\frac{\partial\phi}{\partial r} + \frac{1}{r^2}\frac{\partial^2\phi}{\partial\theta^2} \quad \text{(Laplacian of scalar function } \phi) \tag{69}$$

$$\begin{aligned} \nabla^4 = \nabla^2\nabla^2 &= \left(\frac{\partial^2}{\partial r^2} + \frac{1}{r}\frac{\partial}{\partial r} + \frac{1}{r^2}\frac{\partial^2}{\partial\theta^2}\right) \\ &\times \left(\frac{\partial^2}{\partial r^2} + \frac{1}{r}\frac{\partial}{\partial r} + \frac{1}{r^2}\frac{\partial^2}{\partial\theta^2}\right) \quad \text{(biharmonic operator)} \end{aligned} \tag{70}$$

where the vectors $\vec{e}_r$, $\vec{e}_\theta$ are unit vectors in the direction of axes r, θ, respectively.

Note that, in polar coordinates, the shorthand notations $\vec{\nabla}\phi$, $\vec{\nabla}\cdot\vec{A}$, $\nabla^2\phi$, $\nabla^2\nabla^2\phi$, still apply, but the actual formulae are not intuitively immediate.

B.11.3 STRAIN–DISPLACEMENT RELATIONS

$$\begin{aligned} \varepsilon_r &= \frac{\partial u_r}{\partial r} \\ \varepsilon_\theta &= \frac{1}{r}\frac{\partial u_\theta}{\partial\theta} + \frac{u_r}{r} \\ \varepsilon_{r\theta} &= \frac{1}{r}\frac{\partial u_r}{\partial\theta} + \frac{\partial u_\theta}{\partial r} - \frac{u_\theta}{r} \end{aligned} \tag{71}$$

B.11.4 STRESS–STRAIN RELATION

$$\begin{aligned} \varepsilon_r &= \frac{1}{E}\sigma_r + \frac{-\nu}{E}\sigma_\theta \\ \varepsilon_\theta &= \frac{-\nu}{E}\sigma_r + \frac{1}{E}\sigma_\theta \end{aligned} \qquad \varepsilon_{r\theta} = \frac{1}{G}\sigma_{r\theta} \tag{72}$$

$$\begin{aligned} \sigma_r &= \frac{E}{1-\nu^2}(\varepsilon_r + \nu\varepsilon_\theta) \\ \sigma_\theta &= \frac{E}{1-\nu^2}(\nu\varepsilon_r + \varepsilon_\theta) \end{aligned} \qquad \sigma_{r\theta} = G\varepsilon_{r\theta} \tag{73}$$

$$\sigma_r = \frac{E}{1-\nu^2}\left(\frac{\partial u_r}{\partial r} + \nu\frac{u_r}{r} + \nu\frac{1}{r}\frac{\partial u_\theta}{\partial \theta}\right)$$
$$\sigma_\theta = \frac{E}{1-\nu^2}\left(\nu\frac{\partial u_r}{\partial r} + \frac{u_r}{r} + \frac{1}{r}\frac{\partial u_\theta}{\partial \theta}\right) \tag{74}$$
$$\sigma_\theta = G\left(\frac{1}{r}\frac{\partial u_r}{\partial \theta} + \frac{\partial u_\theta}{\partial r} - \frac{u_\theta}{r}\right)$$

B.12 CYLINDRICAL COORDINATES

A cylındrıcal coordinate system consists of three independent variables, r, θ, z and corresponding displacements u_r, u_θ, u_z.

B.12.1 COORDINATES TRANSFORMATION

The relation to Cartesian coordinates is given by

$$\begin{cases} x = r\cos\theta \\ y = r\sin\theta \\ z \neq 0 \end{cases} \qquad \begin{cases} r = \sqrt{x^2+y^2} \\ \theta = \arg(x, y) \\ z \neq 0 \end{cases} \tag{75}$$

B.12.2 BASIC DIFFERENTIAL OPERATIONS AND OPERATORS

$$\vec{u} = u_r\vec{e}_r + u_\theta\vec{e}_\theta + u_z\vec{e}_z \quad \text{(displacement vector)} \tag{76}$$

$$\text{grad } \phi = \vec{\nabla}\phi = \frac{\partial\phi}{\partial r}\vec{e}_r + \frac{1}{r}\frac{\partial\phi}{\partial\theta}\vec{e}_\theta + \frac{\partial\phi}{\partial z}\vec{e}_z \quad \text{(gradient of a scalar function)} \tag{77}$$

$$\text{div } \vec{A} = \vec{\nabla}\cdot\vec{A} = \frac{1}{r}\frac{\partial}{\partial r}(rA_r) + \frac{1}{r}\frac{\partial}{\partial\theta}A_\theta + \frac{\partial}{\partial z}A_z \quad \text{(divergence of a vector function)} \tag{78}$$

$$\text{curl } \vec{A} = \vec{\nabla}\times\vec{A} = \left(\frac{1}{r}\frac{\partial A_z}{\partial\theta} - \frac{\partial A_\theta}{\partial z}\right)\vec{e}_r + \left(\frac{\partial A_r}{\partial z} - \frac{\partial A_z}{\partial r}\right)\vec{e}_\theta + \frac{1}{r}\left(\frac{\partial(rA_\theta)}{\partial r} - \frac{\partial A_r}{\partial\theta}\right)\vec{e}_z \quad \text{(curl of vector function } \vec{A}\text{)} \tag{79}$$

$$\nabla^2 = \frac{\partial^2}{\partial r^2} + \frac{1}{r}\frac{\partial}{\partial r} + \frac{1}{r^2}\frac{\partial^2}{\partial\theta^2} + \frac{\partial^2}{\partial z^2} \quad \text{(Laplacian)} \tag{80}$$

$$\nabla^2\phi = \frac{\partial^2\phi}{\partial r^2} + \frac{1}{r}\frac{\partial\phi}{\partial r} + \frac{1}{r^2}\frac{\partial^2\phi}{\partial\theta^2} + \frac{\partial^2\phi}{\partial z^2} \quad \text{(Laplacian of scalar function } \phi\text{)} \tag{81}$$

$$\nabla^4 = \nabla^2\nabla^2 = \left(\frac{\partial^2}{\partial r^2} + \frac{1}{r}\frac{\partial}{\partial r} + \frac{1}{r^2}\frac{\partial^2}{\partial\theta^2} + \frac{\partial^2}{\partial z^2}\right) \times \left(\frac{\partial^2}{\partial r^2} + \frac{1}{r}\frac{\partial}{\partial r} + \frac{1}{r^2}\frac{\partial^2}{\partial\theta^2} + \frac{\partial^2}{\partial z^2}\right) \quad \text{(biharmonic operator)} \tag{82}$$

where the vectors $\vec{e}_r$, $\vec{e}_\theta$, $\vec{e}_z$ are unit vectors in the direction of axes r, ϕ, z, respectively.

Note that, in cylindrical coordinates, the shorthand notations $\vec{\nabla}\phi$, $\vec{\nabla}\cdot\vec{A}$, $\vec{\nabla}\times\vec{A}$, $\nabla^2\phi$, $\nabla^2\nabla^2\phi$, still apply, but the actual formulae are not intuitively immediate.

B.12.3 STRAIN–DISPLACEMENT RELATIONS

$$\varepsilon_{rr} = \frac{\partial u_r}{\partial r} \qquad \varepsilon_{r\theta} = \frac{1}{r}\frac{\partial u_r}{\partial \theta} + \frac{\partial u_\theta}{\partial r} - \frac{u_\theta}{r}$$

$$\varepsilon_{\theta\theta} = \frac{1}{r}\frac{\partial u_\theta}{\partial \theta} + \frac{u_r}{r} \quad \text{and} \quad \varepsilon_{rz} = \frac{1}{2}\left(\frac{\partial u_z}{\partial r} + \frac{\partial u_r}{\partial z}\right) \tag{83}$$

$$\varepsilon_{zz} = \frac{\partial u_z}{\partial z} \qquad \varepsilon_{\theta z} = \frac{1}{2}\left(\frac{\partial u_\theta}{\partial z} + \frac{1}{r}\frac{\partial u_z}{\partial \theta}\right)$$

B.12.4 EQUATIONS OF MOTION

$$\frac{\partial \sigma_{rr}}{\partial r} + \frac{1}{r}\frac{\partial \sigma_{r\theta}}{\partial \theta} + \frac{\partial \sigma_{rz}}{\partial z} + \frac{\sigma_{rr} - \sigma_{\theta\theta}}{r} = \rho \ddot{u}_r$$

$$\frac{\partial \sigma_{r\theta}}{\partial r} + \frac{1}{r}\frac{\partial \sigma_{\theta\theta}}{\partial \theta} + \frac{\partial \sigma_{\theta z}}{\partial z} + \frac{2}{r}\sigma_{r\theta} = \rho \ddot{u}_\theta \tag{84}$$

$$\frac{\partial \sigma_{rz}}{\partial r} + \frac{1}{r}\frac{\partial \sigma_{\theta z}}{\partial \theta} + \frac{\partial \sigma_{zz}}{\partial z} + \frac{1}{r}\sigma_{rz} = \rho \ddot{u}_z$$

B.12.5 STRESS–STRAIN RELATION

$$\varepsilon_{rr} = \frac{1}{E}\sigma_{rr} + \frac{-v}{E}\sigma_{\theta\theta} + \frac{-v}{E}\sigma_{zz} \qquad \varepsilon_{r\theta} = \frac{1}{2G}\sigma_{r\theta}$$

$$\varepsilon_{\theta\theta} = \frac{-v}{E}\sigma_{rr} + \frac{1}{E}\sigma_{\theta\theta} + \frac{-v}{E}\sigma_{zz} \quad \text{and} \quad \varepsilon_{\theta z} = \frac{1}{2G}\sigma_{\theta z} \tag{85}$$

$$\varepsilon_{zz} = \frac{-v}{E}\sigma_{rr} + \frac{-v}{E}\sigma_{\theta\theta} + \frac{1}{E}\sigma_{zz} \qquad \varepsilon_{zr} = \frac{1}{2G}\sigma_{zr}$$

and

$$\sigma_{rr} = \frac{(1-v)}{(1+v)(1-2v)}E\varepsilon_{rr} + \frac{v}{(1+v)(1-2v)}E\varepsilon_{\theta\theta} + \frac{v}{(1+v)(1-2v)}E\varepsilon_{zz} \qquad \sigma_{r\theta} = 2G\varepsilon_{r\theta}$$

$$\sigma_{\theta\theta} = \frac{v}{(1+v)(1-2v)}E\varepsilon_{rr} + \frac{(1-v)}{(1+v)(1-2v)}E\varepsilon_{\theta\theta} + \frac{v}{(1+v)(1-2v)}E\varepsilon_{zz} \quad \text{and} \quad \sigma_{\theta z} = 2G\varepsilon_{\theta z}$$

$$\sigma_{zz} = \frac{v}{(1+v)(1-2v)}E\varepsilon_{rr} + \frac{v}{(1+v)(1-2v)}E\varepsilon_{\theta\theta} + \frac{(1-v)}{(1+v)(1-2v)}E\varepsilon_{zz} \qquad \sigma_{zr} = 2G\varepsilon_{zr} \tag{86}$$

as well as

$$\sigma_{rr} = (\lambda + 2\mu)\varepsilon_{rr} + \lambda\varepsilon_{\theta\theta} + \lambda\varepsilon_{zz} \qquad \sigma_{rr} = \lambda\Delta + 2\mu\varepsilon_{rr} \qquad \sigma_{r\theta} = 2\mu\varepsilon_{r\theta}$$

$$\sigma_{\theta\theta} = \lambda\varepsilon_{rr} + (\lambda + 2\mu)\varepsilon_{\theta\theta} + \lambda\varepsilon_{zz} \quad \text{or} \quad \sigma_{\theta\theta} = \lambda\Delta + 2\mu\varepsilon_{\theta\theta} \quad \text{and} \quad \sigma_{\theta z} = 2\mu\varepsilon_{\theta z} \tag{87}$$

$$\sigma_{zz} = \lambda\varepsilon_{rr} + \lambda\varepsilon_{\theta\theta} + (\lambda + 2\mu)\varepsilon_{zz} \qquad \sigma_{zz} = \lambda\Delta + 2\mu\varepsilon_{zz} \qquad \sigma_{zr} = 2\mu\varepsilon_{zr}$$

B.13 SPHERICAL COORDINATES

A spherical coordinate system consists of three independent variables, r, θ, ϕ and corresponding displacements u_r, u_θ, u_ϕ.

B.13.1 COORDINATES TRANSFORMATION

The relation to Cartesian coordinates is given by

$$\begin{cases} x = r\sin\theta\cos\phi \\ y = r\sin\theta\sin\phi \\ z = r\cos\theta \end{cases} \qquad \begin{cases} r = \sqrt{x^2+y^2+z^2} \\ \phi = \arg(x, y) \\ \theta = \cos^{-1}(z/r), \ 0 \le \theta \le \pi \end{cases} \tag{88}$$

B.13.2 BASIC DIFFERENTIAL OPERATIONS AND OPERATORS

$$\vec{u} = u_r\vec{e}_r + u_\phi\vec{e}_\phi + u_\theta\vec{e}_\theta \quad \text{(displacement vector)} \tag{89}$$

$$\operatorname{grad}\Phi = \vec{\nabla}\Phi = \frac{\partial\Phi}{\partial r}\vec{e}_r + \frac{1}{r}\frac{\partial\Phi}{\partial\theta}\vec{e}_\theta + \frac{1}{r\sin\theta}\frac{\partial\Phi}{\partial\phi}\vec{e}_\phi \quad \text{(gradient of a scalar function)} \tag{90}$$

$$\operatorname{div}\vec{A} = \vec{\nabla}\cdot\vec{A} = \frac{1}{r^2}\frac{\partial}{\partial r}(r^2A_r) + \frac{1}{r\sin\theta}\frac{\partial}{\partial\theta}(A_\theta\sin\theta) + \frac{1}{r\sin\theta}\frac{\partial}{\partial\phi}A_\phi$$

$$\text{(divergence of a vector function)} \tag{91}$$

$$\operatorname{curl}\vec{A} = \vec{\nabla}\times\vec{A} = \frac{1}{r\sin\theta}\left(\frac{\partial}{\partial\theta}(A_\phi\sin\theta) - \frac{\partial A_\theta}{\partial\phi}\right)\vec{e}_r + \left(\frac{1}{r\sin\theta}\frac{\partial A_r}{\partial\phi} - \frac{1}{r}\frac{\partial}{\partial r}rA_\phi\right)\vec{e}_\theta$$

$$+\frac{1}{r}\left(\frac{\partial}{\partial r}(rA_\theta) - \frac{\partial A_r}{\partial r}\right)\vec{e}_\phi \quad \text{(curl of vector function } \vec{A}) \tag{92}$$

$$\nabla^2\Phi = \frac{1}{r^2}\frac{\partial}{\partial r}\left(r^2\frac{\partial\Phi}{\partial r}\right) + \frac{1}{r^2\sin\theta}\frac{\partial}{\partial\theta}\left(\sin\theta\frac{\partial\Phi}{\partial\theta}\right) + \frac{1}{r^2\sin^2\theta}\frac{\partial^2\Phi}{\partial\phi^2}$$

$$\text{(Laplacian of scalar function } \phi) \tag{93}$$

$$\nabla^4\Phi = \nabla^2\nabla^2\Phi \ \text{(biharmonic operator)} \tag{94}$$

where the vectors $\vec{e}_r$, $\vec{e}_\theta$, $\vec{e}_\phi$ are unit vectors in the direction of axes r, θ, ϕ, respectively.

Note that, in spherical coordinates, the shorthand notations $\vec{\nabla}\Phi$, $\vec{\nabla}\cdot\vec{A}$, $\vec{\nabla}\times\vec{A}$, $\nabla^2\Phi$ still apply, but the actual formulae are not intuitively immediate.

B.13.3 STRAIN–DISPLACEMENT RELATIONS

$$\varepsilon_{rr} = \frac{\partial u_r}{\partial r} \qquad \varepsilon_{r\phi} = \frac{1}{2}\left(\frac{1}{r\sin\theta}\frac{\partial u_r}{\partial\phi} + \frac{\partial u_\phi}{\partial r} - \frac{u_\phi}{r}\right)$$

$$\varepsilon_{\theta\theta} = \frac{1}{r}\frac{\partial u_\theta}{\partial\theta} + \frac{u_r}{r} \quad \text{and} \quad \varepsilon_{r\theta} = \frac{1}{2}\left(\frac{1}{r}\frac{\partial u_r}{\partial\theta} + \frac{\partial u_\theta}{\partial r} - \frac{u_\theta}{r}\right)$$

$$\varepsilon_{\phi\phi} = \frac{1}{r\sin\theta}\frac{\partial u_\phi}{\partial\phi} + \frac{u_r}{r} + u_\phi\frac{\cot\theta}{r} \qquad \varepsilon_{\theta\phi} = \frac{1}{2}\left(\frac{1}{r}\frac{\partial u_\phi}{\partial\theta} - \frac{u_\phi\cot\theta}{r} + \frac{1}{r\sin\theta}\frac{\partial u_\phi}{\partial\phi}\right) \tag{95}$$

B.13.4 EQUATIONS OF MOTION

$$\begin{aligned}
&\frac{\partial \sigma_{rr}}{\partial r} + \frac{1}{r\sin\theta}\frac{\partial \sigma_{r\phi}}{\partial \phi} + \frac{1}{r}\frac{\partial \sigma_{r\theta}}{\partial \theta} + \frac{2\sigma_{rr} - \sigma_{\phi\phi} - \sigma_{\theta\theta} + \sigma_{r\theta}\cot\theta}{r} = \rho \ddot{u}_r \\
&\frac{\partial \sigma_{r\theta}}{\partial r} + \frac{1}{r\sin\theta}\frac{\partial \sigma_{\phi\phi}}{\partial \phi} + \frac{1}{r}\frac{\partial \sigma_{\theta\phi}}{\partial \theta} + \frac{3\sigma_{r\phi} + 2\sigma_{\phi\theta}\cot\theta}{r} = \rho \ddot{u}_\phi \\
&\frac{\partial \sigma_{r\theta}}{\partial r} + \frac{1}{r\sin\theta}\frac{\partial \sigma_{\phi\theta}}{\partial \phi} + \frac{1}{r}\frac{\partial \sigma_{\theta\theta}}{\partial \theta} + \frac{3\sigma_{r\phi} + (\sigma_{\theta\theta} - \sigma_{\phi\phi})\cot\theta}{r} = \rho \ddot{u}_\theta
\end{aligned} \tag{96}$$

B.13.5 STRESS–STRAIN RELATION

$$\begin{aligned}
\varepsilon_{rr} &= \frac{1}{E}\sigma_{rr} + \frac{-\nu}{E}\sigma_{\theta\theta} + \frac{-\nu}{E}\sigma_{\phi\phi} & & \varepsilon_{r\theta} = \frac{1}{2G}\sigma_{r\theta} \\
\varepsilon_{\theta\theta} &= \frac{-\nu}{E}\sigma_{rr} + \frac{1}{E}\sigma_{\theta\theta} + \frac{-\nu}{E}\sigma_{\phi\phi} & \text{and} \quad & \varepsilon_{\phi r} = \frac{1}{2G}\sigma_{\phi r} \\
\varepsilon_{\phi\phi} &= \frac{-\nu}{E}\sigma_{rr} + \frac{-\nu}{E}\sigma_{\theta\theta} + \frac{1}{E}\sigma_{\phi\phi} & & \varepsilon_{\theta\phi} = \frac{1}{2G}\sigma_{\theta\phi}
\end{aligned} \tag{97}$$

and

$$\begin{aligned}
\sigma_{rr} &= \frac{(1-\nu)}{(1+\nu)(1-2\nu)}E\varepsilon_{rr} + \frac{\nu}{(1+\nu)(1-2\nu)}E\varepsilon_{\theta\theta} \\
&\quad + \frac{\nu}{(1+\nu)(1-2\nu)}E\varepsilon_{\phi\phi} & & \sigma_{r\theta} = 2G\varepsilon_{r\theta} \\
\sigma_{\theta\theta} &= \frac{\nu}{(1+\nu)(1-2\nu)}E\varepsilon_{rr} + \frac{(1-\nu)}{(1+\nu)(1-2\nu)}E\varepsilon_{\theta\theta} \\
&\quad + \frac{\nu}{(1+\nu)(1-2\nu)}E\varepsilon_{\phi\phi} & \text{and} \quad & \sigma_{\phi r} = 2G\varepsilon_{\phi r} \\
\sigma_{\phi\phi} &= \frac{\nu}{(1+\nu)(1-2\nu)}E\varepsilon_{rr} + \frac{\nu}{(1+\nu)(1-2\nu)}E\varepsilon_{\theta\theta} & & \sigma_{\theta\phi} = 2G\varepsilon_{\theta\phi} \\
&\quad + \frac{(1-\nu)}{(1+\nu)(1-2\nu)}E\varepsilon_{\phi\phi}
\end{aligned} \tag{98}$$

as well as

$$\begin{aligned}
\sigma_{rr} &= (\lambda+2\mu)\varepsilon_{rr} + \lambda\varepsilon_{\theta\theta} + \lambda\varepsilon_{\phi\phi} & & \sigma_{rr} = \lambda\Delta + 2\mu\varepsilon_{rr} & & \sigma_{r\theta} = 2\mu\varepsilon_{r\theta} \\
\sigma_{\theta\theta} &= \lambda\varepsilon_{rr} + (\lambda+2\mu)\varepsilon_{\theta\theta} + \lambda\varepsilon_{\phi\phi} & \text{or} \quad & \sigma_{\theta\theta} = \lambda\Delta + 2\mu\varepsilon_{\theta\theta} & \text{and} \quad & \sigma_{\phi r} = 2\mu\varepsilon_{\phi r} \\
\sigma_{\phi\phi} &= \lambda\varepsilon_{rr} + \lambda\varepsilon_{\theta\theta} + (\lambda+2\mu)\varepsilon_{\phi\phi} & & \sigma_{\phi\phi} = \lambda\Delta + 2\mu\varepsilon_{\phi\phi} & & \sigma_{\theta\phi} = 2\mu\varepsilon_{\theta\phi}
\end{aligned} \tag{99}$$

BIBLIOGRAPHY

Abramowitz, M. and Stegun, I. (1964) *Handbook of Mathematical Functions with Formulas, Graphs, and Mathematical Tables*. Dover Publications, Inc.

Achenbach, J.D. (1973) *Wave Propagation in Elastic Solids*. Elsevier.

ACI (1987) *Considerations for Design of Concrete Structures Subjected to Fatigue Loading*. American Concrete Institute.

Ahmad, R., Kundu, T., and Placko, D. (2005) Modeling of phased array transducers. *J. Acoust. Soc. Am.*, **117**, 1762–1776.

Airy, J.R. (1911) The vibrations of circular plates and their relation to Bessel functions. *Proc. Phys. Soc. (London)*, **23**, 225–232.

Alleyne, D.N. and Cawley, P. (May, 1992a) The interaction of Lamb waves with defects. *IEEE Trans. Ultrasonics. Ferroelectrics and Frequency Control*, **39**(3).

Alleyne, D.N. and Cawley, P. (1992b) Optimization of Lamb wave inspection techniques. *NDTE Int.*, **25**(1), 11–22.

Alleyne, D., Pialucha, T., and Cawley, P. (1992) A signal regeneration technique for long-range propagation of dispersive Lamb waves. *Ultrasonics*, **31**(3), 201–204.

Alleyne, D.N., Pavlakovic, B., Lowe, M.J.S., and Cawley, P. (2001) Rapid, long range inspection of chemical plant pipework using guided waves. *Rev. Progress in QNDE*, **20**, 180–187.

ANSI/IEEE Std. 176. (1987) *IEEE Standard on Piezoelectricity*. New York: The Institute of Electrical and Electronics Engineers, Inc.

Atalla, M.J. (1996) Model updating using neural network. Ph.D. Dissertation, Virginia Polytechnic Institute and State University, April 1996.

Auld, B.A. (1990) Acoustic fields and waves in solids. Vols. 1 and 2. John Wiley & Son.

Ayres, T., Chaudhry, Z., and Rogers, C. (1996) Localized health monitoring of civil infrastructure via piezoelectric actuator/sensor patches. *Proceedings, SPIE's 1996 Symposium on Smart Structures and Integrated Systems, SPIE* Vol. 2719, pp. 123–131.

Balanis, C.A. (2005) *Antenna Theory Analysis and Design*. Hoboken, NJ: John Wiley & Sons, Inc.

Banerjee, S., Prosser, W., and Mal, A. (January, 2005) Calculation of the response of a composite plate to localized dynamic surface loads using a new wave number integral method. *Transaction of the ASME*, **72**.

Banks, H.T., Smith, R.C., and Wang, Y. (1996) Smart material structures: Modeling, estimation and control. Masson, Paris: John Wiley & Sons.

Barnoncel, D., Osmont, D., and Dupont, M. (2003) Health monitoring of sandwich plates with real impact damages using PZT devices. In *Structural Health Monitoring 2003* (F.-K. Chang, ed.) DEStech Pub, pp. 871–878.

Berman, J. and Quattrone, R. (1999) Characterization of impedance-based piezosensor subject to combined hygro-thermal-mechanical load. Final Project Report, US Army Construction Engineering Research Laboratory.

Berryman, J., Borcea, L., Papanicolaou, G., and Tsogka, C. (2002) Statistically stable ultrasonic imaging in random media. *J. Acoust. Soc. Am.*, **112**, 1509–1522.

Bhalla, S. and Soh, C.K. (October, 2003) Structural impedance based damage diagnosis by piezo-transducers. *Earthquake Eng. Struct. Dyn.*, **32**(12), 1897–1916.

Bishop, C.M. (1995) *Neural Networks for Pattern Recognition*. Clarendon Press, ISBN 0-19-853849-9.

Blackshire, J.L., Giurgiutiu, V., Cooney, A., and Doane, J. (2005) Characterization of sensor performance and durability for structural health monitoring. *SPIE* Vol. 5770, paper # 5770-08.

Blevins, R.D. (1979) *Formulas for Natural Frequency and Mode Shape*. Litton Educational Publishing Inc.

Blitz, J. and Simpson, G. (1996) *Ultrasonic Methods of Non-Destructive Testing*. Chapman & Hall.

Bois, C. and Hochard, C. (2002) Measurement and modeling for the monitoring of damaged laminate composite structures. *1st European Workshop on Structural Health Monitoring*, 10–12 July 2002, Paris, France, pp. 425–432.

Boller, C., Biemans, C., Staszewski, W. et al. (1999) Structural damage monitoring based on an actuator-sensor system. *Proceedings of SPIE Smart Structures and Integrated Systems Conference*, 1–4 March 1999, Newport, CA.

Bottai, G. and Giurgiutiu, V. (2005) Simulation of the Lamb wave interaction between piezoelectric wafer active sensors and host structure. *SPIE* Vol. 5765, paper # 5765–29.

Bowman, F. (1958) *Introduction to Bessel Functions*. New York: Dover Pub. Inc.

Broch, J.T. (1984) *Mechanical Vibration and Shock Measurements*. Brüel & Kjær.

Bruhn, E.F. (1973) *Analysis and Design of Flight Vehicle Structures*. Jacobs Publishing, Inc.

Burrascano, P., Cardelli, E., Faba, A. et al. (2001) Application of probabilistic neural networks to eddy current non destructive test problems. *EANN 2001 Conference*, 16–18 July 2001, Cagliari, Italy.

Cawley, P. (1984) The impedance method for non-destructive inspection. *NDT Int.*, **17**(2), 59–65.

Chahbaz, A., Gauthier, J., Brassard, M., and Hay, D.R. (1999) Ultrasonic technique for hidden corrosion detection in aircraft wing skin. *Proceedings of the 3rd Joint Conference on Aging Aircraft, 1999.*

Chan, T.H.T., Ni, Y.Q., and Ko, J.M. (1999) Neural network novelty filtering for anomaly detection of Tsing Ma bridge cables. *Proceedings of the 2nd International Workshop on Structural Health Monitoring*, 8–10 September 1999, Stanford University, Stanford, CA, USA, pp. 430–439.

Chang, F.-K. (1995) Built-in damage diagnostics for composite structures. *Proceedings of the 10th International Conference on Composite Structures (ICCM-10)*, 14–18 August 1995, Whistler, BC, Canada, Vol. 5, pp. 283–289.

Chang, F.-K. (1998) Manufacturing and design of built-in diagnostics for composite structures. *52nd Meeting of the Society for Machinery Failure Prevention Technology*, March 30–April 3 1998, Virginia Beach, VA.

Chang, F.-K. (2001) Structural health monitoring: Aerospace assessment. *Aero Mat 2001, 12th ASM Annual Advanced Aerospace Materials and Processes Conference*, 12–13 June 2001, Long Beach, CA.

Chang, T.Y.P., Chang, C.C., and Xu, Y.G. (1999) Updating structural parameters: An adaptive neural network approach. *Proceedings of the 2nd International Workshop on Structural Health Monitoring*, 8–10 September 1999, Stanford University, Stanford, CA, USA, pp. 379–389.

Chaudhry, Z., Sun, F.P., and Rogers, C.A. (1994) Health monitoring of space structures using impedance measurements. *Fifth International Conference on Adaptive Structures*, 5–7 December 1994, Sendai, Japan, pp. 584–591.

Cho, Y. (May, 2000) Estimation of ultrasonic guided wave mode conversion in a plate with thickness variation. *IEEE Trans. Ultrasonics. Ferroelectrics and Frequency Control*, **47**(3).

Cho, Y. and Rose, J.L. (April, 1996) A boundary element solution for a mode conversion study on the edge reflection of Lamb waves. *J. Acoust. Soc. Am.*

Chona, R., Suh, C.S., and Rabroker, G.A. (2003) Characterizing defects in multi-layer materials using guided ultrasonic waves. *Optics and Lasers in Engineering*, **40**, 371–378.

Collins, J.A. (1993) *Failure of Materials in Mechanical Design*. John Wiley & Sons.

Colwell, R.C. and Hardy, H.C. (December, 1937) The frequencies and nodal systems of circular plates. *Phil. Mag.*, Ser. 7, **24**(165), 1041–1055.

Cowell, R.G., Dawid, A.P., Lauritzen, S.L., and Spiegelhalter, D.J. (1999) *Probabilistic Networks and Expert Systems*. Springer-Verlag.

Crawley, E.A. and deLuis, J. (1987) Use of piezoelectric actuators as elements of intelligent structures. *AIAA J.*, **25**(10), 1375–1385.

Crawley, E.F. and Anderson, E.H. (January, 1990) Detailed models of piezoceramic actuation of beams. *J. Intell. Mater. Syst. Struct.*, **1**(1), 4–25.

Cruse, T.A. (September, 2004) Future directions in NDE and structural health monitoring. *J. Nondestr. Eval.*, **23**(3), 77–79, Springer Verlag.

Culshaw, B., Pierce, S.G., and Staszekski, W.J. (1998) Condition monitoring in composite materials: An integrated system approach. *Proceedings of the Institute of Mechanical Engineers*, Vol. 212, Part I, pp. 189–202.

Dalton, R.P., Cawley, P., and Lowe, M.J. (2001) The potential of guided waves for monitoring large areas of metallic aircraft structures. *J. Nondestr. Eval.*, **20**, 29–45.

Demuth, H. and Beale, M. (1998) Neural network toolbox for use with MATLAB®. Version 3, Mathworks, Inc.

Demuth, H. and Beale, M. (2000) *MATLAB Neural Network Tool Box*. Natick, MA: The Math Works, Inc.

Deng, X., Wang, Q., and Giurgiutiu, V. (1999) Structural health monitoring using active sensors and wavelet transforms. *SPIE's 6th Annual International Symposium on Smart Structures and Materials*, Newport Beach.

Deutsch, W.A.K., Cheng, A., and Achenbach, J.D. (1997) Self-focusing of Rayleigh waves and Lamb waves with a linear phased array. *Res. Nondestr. Eval.*, **9**(2), 81–95.

Deutsch, W.A.K., Deutsch, K., Cheng, A., and Achenbach, J.D. (December, 1998) Defect detection with Rayleigh and Lamb waves generated by a self-focusing phased array. *NDT.net*, **3**(12), 1–6.

De Villa, F., Roldan, E., Tirano, C. et al. (2001) Defect detection in thin plates using S_0 Lamb wave scanning. *SPIE* Vol. 4335, pp. 121–130.

Devore, J.L. (1999) *Probability and Statistics for Engineering and the Sciences*. 5th ed. Duxbury, 2000.

Diamanti, K., Hodgkinson, J.M., and Soutis, C. (2002) Damage detection of composite laminates using PZT generated Lamb waves. *1st European Workshop on Structural Health Monitoring*, 10–12 July 2002, Paris, France, pp. 398–405.

Diamanti, K., Soutis, C., and Hodgkinson, J.M. (2004) Lamb waves for the non-destructive inspection of monolithic and sandwich composite beams. *Composites, Part A*, Vol. 36, 2005, 189–195, Elsevier Pub.

Diaz Valdes, S.H. and Soutis, C. (May, 2002) Real-time nondestructive evaluation of fiber composite laminates using low-frequency Lamb waves. *J. Acoust. Soc. Am.*, **111**(5), 2026–2033.

Duke, J.C. Jr. (1988) *Acousto-Ultrasonics – Theory and Applications*. Plenum Press.

Dunlea, S., Moriarty, P., and Fegan, D.J. (2001) Selection of TeV γ-rays using the Kernel multivariate technique. *Proceedings of ICRC*.

Dupont, M., Osmont, R., Gouyon, R., and Balageas, D.L. (2000) Permanent monitoring of damage impacts by a piezoelectric sensor based integrated system. In *Structural Health Monitoring 2000* (F.-K. Chang, ed.) Technomic, pp. 561–570.

Dzenis, Y. and Qian, J. (2001) Analysis of microdamage evolution histories in composites. *7th ASME NDE Topical Conference*, April 23–25, San Antonio, TX.

Ewins, D.J. (1984) *Modal Test: Theory and Practice*. Research Studies Press Ltd., Letchworth, Hertfortshire, England.

Farrar, C.R., Duffey, T.A., Doebling, S.W., and Nix, D.A. (1999) A statistical pattern recognition paradigm for vibration-based structural health monitoring. *Proceedings of the 2nd International Workshop on Structural Health Monitoring*, 8–10 September, Stanford University, Stanford, CA, USA, pp. 764–773.

Farrar, R.C., Sohn, H., Fugate, M.L., and Czarnecki, J.J. (2001) Statistical process control and projection techniques for damage detection. *Proceedings of the SPIE's Conference on Advanced Nondestructive Evaluation for Structural and Biological Health Monitoring*, 4–8 March 2001, Newport Beach, CA, USA, Vol. 4335, pp. 1–19.

Fenn, A., Temme, D.H., Delaney, W.P., and Courtney, W.E. (2000) The development of phased-array radar technology. *Lincoln Lab. J.*, **12**(2), 321–340.

Fink, M. (1992) Time reversal of ultrasonic fields – Part I: Basic principles. *IEEE Trans. Ultrasonics. Ferroelectrics and Frequency Control*, **39**(5), 555–566.

Fink, M. (1999) Time-reversed acoustics. *Sci. Am.*, **281**(5), 91–97.

Fink, M., Cessareau, D., Derode, A., Prada, C., Roux, P., Tanter, M., Thomas, J.-L., Wu, F. (2000) Time reversed acoustics. *Rep. Prog. Phys*, **63**, 1933–1995.

Fink, M., Montaldo, G., Tanter, M. (2004) Time reversed acoustics. *2004 IEEE Ultrasonics Symposium*, pp. 850–859.

Fitzgerald, A.E., Higginbotham, D.E., and Grabel, A. (1967) *Basic Electrical Engineering*. New York: McGraw-Hill.

Fritzen, C.P. and Bohle, K. (1999) Parameter selection strategies in model-based damage detection. *Proceedings of the 2nd International Workshop on Structural Health Monitoring*, 8–10 September 1999, Stanford University, Stanford, CA, USA, pp. 901–911..

Fromme, P., Wilcox, P., Lowe, M., and Cawley, P. (2005) A guided ultrasonic waves array for structural integrity monitoring. *Quant. Nondestr. Eval.*, **24**, 1780–1787.

Fuller, C.R., Snyder, S.D., Hansen, C.H., and Silcox, R.J. (1990) *Active Control of Interior Noise in Model Aircraft Fuselages Using Piezoceramic Actuators*. American Institute of Aeronautics and Astronautics Paper # 90-3922.

Fyfe Company LLC. (2000) *Fyfe UC Composite Laminate Strip System*. Material Specification Sheet.

Gachagan, A., Hayward, G., McNab, A. et al. (1999) Generation and reception of ultrasonic guided waves in composite plates using comformable Piezoelectric transmitters and optical-fiber detectors. *IEEE Trans. Ultrasonics. Ferroelectrics and Frequency Control*, **46**(1), 72–81.

Gazis, D.C. (May, 1959) Three dimensional investigation of the propagation of waves in hollow circular cylinders. *J. Acoust. Soc. Am.*, **31**(5), 568–578.

Giurgiutiu, V. and Rogers, C.A. (1997) Electro-Mechanical (E/M) impedance method for structural health monitoring and non-destructive evaluation. *International Workshop on Structural Health Monitoring*, 18–20 September 1997, Stanford University, CA, pp. 433–444.

Giurgiutiu, V., Reynolds, A., and Rogers, C.A. (October, 1999) Experimental investigation of E/M impedance health monitoring of spot-welded structural joints. *J. Intell. Mater. Syst. Struct.*, **10**, 802–812.

Giurgiutiu, V. and Zagrai, A.N. (December, 2000) Characterization of Piezoelectric wafer active sensors. *J. Intell. Mater. Syst. Struct.*, **11**, 959–976.

Giurgiutiu, V., Lyons, J., Petrou, M. et al. (2001) Fracture mechanics testing of the bond between composite overlays and concrete substrate. *J. Adhe. Sci. Technol.*, **15**(11), 1351–1371, The Netherlands: VSP International Science Pub.

Giurgiutiu, V. and Zagrai, A.N. (January, 2002) Embedded self-sensing Piezoelectric active sensors for online structural identification. *Transactions of ASME, J. Vibration and Acoustics*, **124**, 116–125.

Giurgiutiu, V. (2003) Lamb wave generation with Piezoelectric wafer active sensors for structural health monitoring. *SPIE* Vol. 5056, paper # 5056-17.

Giurgiutiu, V. (2005) Tuned Lamb-wave excitation and detection with Piezoelectric wafer active sensors for structural health monitoring. *J. Intell. Mater. Syst. Struct.*, **16**(4), 291–306.

Giurgiutiu, V., Bao, J., and Zagarai, A.N. (2006) Structural health monitoring system utilizing guided Lamb waves embedded ultrasonic structural radar. *US Patent, Patent No. US 6,996, 480 B2*, 7 February 2006.

Giurgiutiu, V. and Yu, L. (2006) Omnidirectional 2-D Embedded-Ultrasonics Structural Radar (EUSR) with Piezoelectric Wafer Active Sensors (PWAS) Phased Arrays for Wide-Area Damage Detection/SHM/NDE. USC-IPMO, Disclosure ID No. 00597, July 2006.

Graff, K.F. (1975) *Wave Motion in Elastic Solids*. Mineola, NY: Oxford University Press.

Grahn, T. (2003) Lamb wave scattering from a circular partly through-thickness hole in a plate. *Wave Motion* Vol. 37, No. 1, January 2003, pp. 63–80.

Grondel, S., Moulin, E., and Delebarre, C. (1999) Lamb wave assessment of fatigue damage in aluminum plates. *Proceedings of the 3rd Joint Conference on Aging Aircraft*.

Gunawan, A. and Hirose, S. (March, 2004) Mode-exciting method for Lamb wave-scattering analysis. *J. Acous. Soc. Am.*, **115**(3).

Hagan, M.T., Demuth, H.B., and Beale, M. (1996) *Neural Network Design.* PWS Publishing Co., International Thompson Publishing, ISBN 0-534-94332-2.

Harris, C.M. (1996) Shock and vibration handbook. USA: McGraw-Hill.

Hartman, A. and Schijve, J. (1970) The effects of environment and load frequency on the crack propagation law for macro fatigue crack growth in aluminum alloys. *Eng. Fract. Mech.*, **1**, 615.

Haskell, N.A. (1953) Dispersion of surface waves on multilayer media. *Bull. Seismological Soc. Am.*, **43**, 17–34.

Herakovich, C.T. (1998) *Mechanics of Fibrous Composites.* John Wiley & Sons, Inc.

Heylen, W., Lammens, S., and Sas, P. (1997) Modal analysis theory and testing. Heverrlee, Belgium: Katholieke Universiteit Leuven.

Hickinbotham, S.J. and Austin, J. (2000) Novelty detection for flight data from airframe strain gauges. *Proceedings of the European COST F3 Conference on System Identification & Structural Health Monitoring*, Madrid, Spain, pp. 773–780.

Hinders, M.K. (1996) Lamb waves scattering from rivets. *Quant. Nondestr. Eval.*, **15**.

Ho, Y.Q. and Ewins, D.J. (1999) Numerical evaluation of damage index. *Proceedings of the 2nd International Workshop on Structural Health Monitoring*, 8–10 September 1999, Stanford University, Stanford, CA, USA, pp. 995–1009.

Hu, N. and Fukunaga, H. (2001) Structural damage identification using Piezoelectric sensors. *Proceedings of the SPIE's Conference on Advanced Nondestructive Evaluation for Structural and Biological Health Monitoring*, 4–8 March 2001, Newport Beach, CA, USA, Vol. 4335, pp. 371–382.

Iguchi, S. (1953) Dei Eigenschwingungen and Klangfiguren der vierseitig freien rechteckigen Platte. *Ingr.-Arch.*, Bd. 21, ser. 303, Heft 5-6, 1953, pp. 304–322.

Ihn, J.-B. and Chang, F.-K. (2002b) Multi-crack growth monitoring at riveted lap joints using Piezoelectric patches. *Proceedings of SPIE's 7th Annual International Symposium on NDE for Health Monitoring and Diagnostics*, 17–21 March 2002, San Diego, CA, Vol. 4702, pp. 29–40.

Ihn, J.-B. (2003) Built-in diagnostics for monitoring fatigue crack growth in aircraft structures. PhD Dissertation, Stanford University, Department of Aeronautics and Astronautics.

Ikeda, T. (1996) *Fundamentals of Piezoelectricity.* Oxford University Press.

Ikegami, R. and Haugse, E.D. (2001) Structural health management for aging aircraft. *SPIE* Vol. 4332, pp. 60–67.

Ing, R. and Fink, M. (1996) Time recompression of dispersive Lamb waves using a time reversal mirror – Application to flaw detection in thin plates. *IEEE Ultrasonics Symposium*, **1**, 659–663.

Ing, R.K. and Fink, M. (1998) Time-reversed Lamb waves. *IEEE Trans. Ultrasonics. Ferroelectrics and Frequency Control*, **45**(4), pp. 1032–1043.

Itao, K. and Crandall, S.H. (1979) Natural modes and natural frequencies of uniform, circular, free-edge plates. *J. Appl. Mech.*, **46**, 448–453.

Jin, S., Livingston, R.A., and Morzougui, D. (2001) Stochastic system invariant spectrum analysis applied to smart systems in highway bridges. *Proceedings of the SPIE's Conference on Smart Systems for Bridges, Structures, and Highways*, 4–8 March 2001, Newport Beach, CA, USA, Vol. 4330, pp. 301–312.

Johnson, D.H. and Dudgeon, D.E. (1993) *Array Signal Processing: Concepts and Techniques.* Upper Saddle River, NJ: PTR Prentice Hall.

Johnson, D.E., Hilburn, J.L., Johnson, J.R., and Scott, P.D. (1995) *Basic Electric Circuit Analysis.* 5th ed. New Jersey: Prentice Hall.

Kausel, E. (1986) Wave propagation in anisotropic layered media. *Int. J. Numer. Methods Eng.*, **23**, 1567–1578.

Kay, S.M. (1988) *Modern Spectral Estimation: Theory and Application.* Prentice Hall, ISBN 0-13-598582-X.

Kay, S.M. (1993) *Fundamentals of Statistical Signal Processing: Estimation Theory.* Prentice Hall, ISBN 0-13-345711-7.

Keilers, C.H. and Chang, F.-K. (September, 1995) Identifying delamination in composite beams using built-in Piezoelectrics: Part I Experiments and analysis, Part II An identification method. *J. Intell. Mater. Syst. Struct.*, **6**, 649–672.

Kim, S., Sohn, H., Greve, D., and Oppenheim, I. (2005) Application of a time reversal process for baseline-free monitoring of a bridge steel girder. *International Workshop on Structural Health Monitoring*, 15–17 September, Stanford, CA, 2005.

Kirchhoff, G.R. (1850) Uber die Schwingungen Einer Kriesformigen Elastischen Scheibe. *Poggendorffs Annalen*, **81**, 258–264.

Knopoff, L. (1964) A matrix method for elastic waves problems. *Bull. Seismological Soc. Am.*, **43**, 431–438.

Knowles, J.K. (September, 1997) *Linear Vector Spaces and Cartesian Tensors*. New York: Oxford University Press, Inc.

Koh, Y.L. and Chiu, W.K. (August, 2003) Numerical study of detection of disbond growth under a composite repair patch. *Smart Mater. Struct.*, **12**(4), 633–641.

Krautkramer, J. and Krautkramer, H (1990) *Ultrasonic Testing of Materials*. Springer-Verlag.

Krautkramer (June, 1998) Emerging technology – guided wave ultrasonics. *NDTnet*, **3**(6), http://www.ndt.net/article/0698/kk_gw/kk_gw.htm.

Krautkramer (2002) *Products Catalog*, http://www.geinspectiontechnologies.com/solutions/Testing Machines/Ultrasonics/phasedarray.html.

Kreyszig, E. (1999) *Advanced Engineering Mathematics*. 8th ed. John Wiley & Sons.

Kudva, J.N., Marandis, C. and Gentry, J. (1993) Smart structures concepts for aircraft structural health monitoring. *SPIE* Vol. 1917, pp. 964–971.

Kwun, H. and Bartels, K.A. (1998) Magnetostrictive sensors technology and its applications. *Ultrasonics*, **36**, 171–178.

Kwun, H., Light, G.M., Kim, S.Y. et al. (2002) Permanently installable, active guided-wave sensor for structural health monitoring. *First European Workshop on Structural Health Monitoring*. DEStech Pub., pp. 390–397.

Lakshmanan, K.A. and Pines, D.J. (June, 1997) Modeling damage in composite rotorcraft flexbeams using wave mechanics. *J. Smart Mater. Struct.*, **6**(3), 383–392, Bristol, England: Institute of Physics.

Lamb, H. (1917) On waves in an elastic plate. *Proc. R. Soc. Lond. A.*, **93**, 114.

Lange, I. (1978) Characteristics of the impedance method of inspection and of impedance inspection transducers. *Sov. J. NDT*, 958–966.

Lanza di Scalea, F., Bonomo, M., and Tuezzo, D. (January, 2001) Ultrasonic guided wave inspection of bonded lap joints: Noncontact method and photoelastic visualization. *Res. Nondestr. Eval.*, **13**(3), 153–171.

Lanza di Scalea, F., Rizzo, P., and Marzani, A. (January, 2004) Propagation of ultrasonic guided waves in lap-shear adhesive joints: Case of incident a0 Lamb wave. *J. Intell. Mater.*

Lecce, L., Viscardi, M., and Zumpano, G. (2001) Multifunctional system for noise control and damage detection on a typical aeronautical structure. *SPIE* Vol. 4327, 210–212.

Lee, B.C., Palacz, M., Krawczuk, M. et al. (September, 2004) Wave propagation in a sensor/actuator diffusion bond model. *J. Sound and Vibration*, **276**(3–5), 671–687.

Leissa, A. (1969) *Vibration of Plates*. NASA SP-160, US Government Printing Office.

Lemistre, M., Gouyon, R., Kaczmarek, H., and Balageas, D. (1999) Damage localization in composite plates using wavelet transform processing on Lamb wave signals. *2nd International Workshop of Structural Health Monitoring*, 8–10 September 1999, Stanford University, pp. 861–870.

Lemistre, M., Osmont, D., and Balageas, D. (2000) Active health system based on wavelet transform analysis of diffracted Lamb waves. *SPIE* Vol. 4073, pp. 194–202.

Lester, H.C. and Lefebvre, S. (July, 1993) Piezoelectric actuator models for active sound and vibration control of cylinders. *J. Smart Mater. Struct.*, **4**, 295–306.

Liang, C., Sun, F.P., and Rogers, C.A. (January, 1994) Coupled electro-mechanical analysis of adaptive material system-determination of the actuator power consumption and system energy transfer. *J. Intell. Mater. Syst. Struct.*, **5**, 12–20.

Lindner, D.K. (1999) *Introduction to Signals and Systems*. New York: McGraw-Hill.

Light, G.M., Minachi, A., and Spinks, R.L. (2001) Development of a guided wave technique to detect defects in thin steel plates prior to stamping. *Proceeding of the 7th ASME NDE Topical Conference. NDE*-Vol. 20, pp. 163–165.

Light, G.M., Kwun, H., Kim, S.Y., and Spinks, R.L. (2002) Magnetostrictive sensor technology for monitoring bondline quality and defect growth under adhesively bonded patches on simulated wing structure. *6th Joint FAA/DOD/NASA Conference on Aging Aircraft*, 16–19 September 2002, San Francisco, CA.

Lih, S.S. and Mal, A.K. (1996) Response of multilayered composite laminates to dynamic surface loads. *Composites*, Part B, **27B**, 633–641.

Lin, X. and Yuan, F.G. (2001) Diagnostic Lamb waves in an integrated piezoelectric sensor/actuator plate: Analytical and experimental studies. *Smart Mater. Struct.*, **10**, 907–913.

Lin, X. and Yuan, F.G. (December, 2005) Experimental study of applying migration technique in structural health monitoring. *Struct. Health Monit. – An Int. J.*, **4**, 341–353.

Lindsay, R., Buchanan, B.G., Feigenbaum, E.A., and Lederberg, J. (1980) *Applications of Artificial Intelligence for Chemical Inference: The DENDRAL Project*. New York: McGraw-Hill.

Lines, D. and Dickson, K. (1999) Optimization of high-frequency array technology for lap-joint inspection. *Proceedings of the 3rd Joint Conference on Aging Aircraft*.

Liu, G.R. and Xi, Z.C. (2002) *Elastic Waves in Anisotropic Laminates*. Boca Raton, FL: CRC Press.

Liu, P., Sana, S., and Rao, V.S. (1999) Structural damage identification using time-domain parameter estimation techniques. *Proceedings of the 2nd International Workshop on Structural Health Monitoring*, 8–10 September 1999, Stanford University, Stanford, CA, USA, pp. 812–820.

Lloyd, G.M. and Wang, M.L. (1999) Neural network novelty filtering for anomaly detection of Tsing Ma bridge cables. *Proceedings of the 2nd International Workshop on Structural Health Monitoring*, 8–10 September 1999, Stanford University, Stanford, CA, USA, pp. 713–722.

Loh, C.H. and Huang, C.C. (1999) Damage identification of multi-story steel frames using neural networks. *Proceedings of the 2nd International Workshop on Structural Health Monitoring*, 8–10 September 1999, Stanford University, Stanford, CA, USA, pp. 390–399.

Looney, C.G. (1997) *Pattern Recognition Using Neural Networks*. Oxford University Press, ISBN 0-19-507920-5.

Lopes, V. Jr., Park, G., Cudney, H., and Inman, D.J. (1999) Smart structures health monitoring using artificial neural network. *Proceedings of the 2nd International Workshop of Structural Health Monitoring*, 8–10 September 1999, Stanford University, pp. 976–985.

Lopes, V. Jr. and Inman, D.J. (1999) Active damage detection using smart materials. *Proceeding of the Eurodiname'99 Dynamic Problems in Mechanics and Mechatronics*, 11–16 June 1999, Guzburg, Germany.

Love, A.E.H. (1926) *Some Problems of Geodynamics*. Cambridge University Press.

Love, A.E. H. (1944) *A Treatise on the Mathematical Theory of Elasticity*. New York: Dover Pub.

Lowe, M.J.S. and Cawley, P. (1994) The applicability of plate wave techniques for the inspection of adhesive and diffusion bonded joints. *J. Nondestr. Eval.*, **13**, 185–199.

Lowe, M.J.S. (July, 1995) Matrix technique for modeling ultrasonic waves in multilayered media. *IEEE Trans. Ultrasonics. Ferroelectrics and Frequency Control*.

Lowe, M.J.S. and Diligent, O. (January, 2002) Low-frequency reflection characteristics of the S0 Lamb wave from a rectangular notch in a plate. *J. Acoust. Soc. Am.*

Luo, W., Rose, J.L., and Kwun, H. (2003) A two dimensional model for crack sizing in pipes. *Rev. Prog. Quant. Nondestr. Eval.*, **23**, 187–192, AIP Pub.

Maia, N., Silva, J., He, J. et al. (1997) *Theoretical and Experimental Modal Analysis*. Research Studies Press Ltd.

Mal, A.K. (1988) Wave propagation in layered composite laminates under periodic surface loads. *Wave Motion*, **10**, 257–266.

Mal, A.K. and Lih, S.S. (1992) Elastodynamic response of a unidirectional composite laminate to concentrated surface loads, Parts I & II. *ASME J. Appl. Mech.*, **55**, 878–892.

Mal, A.K. (2001) NDE for health monitoring of aircraft and aerospace structures. *Proceeding of the 7th ASME NDE Topical Conference. NDE*-Vol. 20, pp. 149–155.

Mal, A.K., Ricci, F., Gibson, S., and Banerjee, S. (2003) Damage detection in structures from vibration and wave propagation method. *SPIE* Vol. 5047, pp. 202–210.

Mallick, P.K. (1993) *Fiber Reinforced Composites – Materials, Manufacturing, and Design.* New York: Marcel Dekker.

Manolakis, D.G., Ingle, V.K., and Kogon, S.M. (2000) *Statistical and Adaptive Signal Processing.* McGraw Hill, ISBN 0-07-040051-2.

Matt, H., Bartoli, I., and Lanza di Scalea, F. (October, 2005) Ultrasonic guided wave monitoring of composite wing skin-to-spar bonded joints in aerospace structures. *J. Intell. Mater Syst. Struct.*

McKeon, J.C.P. and Hinders, M.K. (1999) Lamb waves scattering from a through hole. *J. Sound and Vibration*, **224**(5), 843–862.

McLachlan, N. (1948) *Bessel Functions for Engineers.* London: Oxford University Press.

McMillan, J.L. and Pelloux, R.M.N. (1967) Fatigue crack propagation under program and random loads. *ASTM STP-415*, p. 505.

Measurements Group Inc. (1992) Student manual for strain gage technology. Bulletin 309D.

Mehrotra, K., Mohan, C.K., and Ranka, S. (1997) *Elements of Artificial Neural Networks.* MIT Press, ISBN 0-262-13328-8.

Meirovitch, L. (1986) *Elements of Vibration Analysis.* 2nd ed. New York: McGraw-Hill.

Meitzler, A.H. (April, 1961) Mode coupling occurring in the propagation of elastic pulses in wires. *J. Acoust. Soc. Am.*, **33**(4), 435–445.

Mindlin, R.D., Schacknow, A., and Deresiewicz, H. (1956) Flexural vibrations of rectangular plates. *J. Appl. Mech.*, **23**, 430, New York.

Mindlin, R.D. and Medick, M.A. (December, 1959) Extensional vibration of elastic plates. *J. Appl. Mech.*

Mindlin, R.D. (1960) Waves and vibrations in isotropic elastic plates. In *Structural Mechanics* (J.N. Goodier and N.J. Hoff, eds) New York: Pergamon Press.

Moetakef, M.A., Joshi, S.P., and Lawrence, K.L. (1996) Elastic wave generation by Piezoceramic patches. *AIAA J.*, **34**(10), 2110–2117.

Monaco, E., Franco, F., and Lecce, L. (2001) Experimental and numerical activities on damage detection using magnetostrictive actuators and statistical analysis. *J. Intell. Mater. Syst. Struct.*, **11**(7), 567–578.

Monkhouse, R.S.C., Wilcox, P.D., and Cawley, P. (1996) Flexible interdigital PVDF Lamb wave transducers for the development of smart structures. *Rev. Prog. Nondestr. Eval.*, **16A**, 877–884.

Monkhouse, R.S.C., Wilcox, P.D., Lowe, M.J.S. et al. (2000) The rapid monitoring of structures using interdigital Lamb wave transducers. *Smart Mater. Struct.*, **9**, 304–309.

Morita, K., Teshigawa, M., Isoda, H. et al. (2001) Damage detection tests of five-story frame with simulated damages. *Proceedings of the SPIE's Conference on Advanced Nondestructive Evaluation for Structural and Biological Health Monitoring*, 4–8 March 2001, Newport Beach, CA, USA, Vol. 4335, pp. 106–114.

Moulin, E., Assaad, J., and Delebarre, C. (January, 2000) Modeling of Lamb waves generated by integrated transducers in composite plates using a coupled fine element-normal modes expansion method. *J. Acoust. Soc. Am.*, **107**(1).

Mustafa, V. and Chahbaz, A. (March, 1997) Imaging of disbond in adhesive joints with Lamb waves. *NDT.net*, **2**(3) http://www.ndt.net/article/tektren2/tektren2.htm.

Nayfeh, A.H. (April, 1991) The general problem of elastic wave propagation in multilayered anisotropic media. *J. Acous. Soc. Am.*, **89**(4), Pt. 1.

Nayfeh, A.H. (1995) *Wave Propagation in Layered Anisotropic Media with Application to Composites.* Elsevier.

Ni, Y.Q., Jiang, S.F., and Ko, J.M. (2001) Application of adaptive probabilistic neural networks to damage detection of Tsing Ma suspension bridge. *Proceedings of the SPIE's Conference on Health Monitoring and Management of Civil Infrastructure System*, 4–8 March 2001, Newport Beach, CA, USA, Vol. 4337, pp. 347–356.

Nieuwenhuis, J.H. and Neumann, J.J. (November, 2005) Generation and detection of guided waves using PZT wafer transducers. *IEEE Trans. Ultrasonics. Ferroelectrics and Frequency Control*, **52**(11).

Nigrin, A. (1993) *Neural Networks for Pattern Recognition.* MIT Press, ISBN 0-262-14054-3.

Omidvar, O. and Dayhoff, J. (Eds) (1998) *Neural Networks and Pattern Recognition.* Academic Press, ISBN 0-12-526420-8.

Osmont, D., Dupont, M., Gouyon, R. et al. (2000) Damage and damaging impact monitoring by PZT sensors-based HUMS. *SPIE* Vol. 3986, pp. 85–92.

Osmont, D., Barnoncel, D., Devillers, D., and Dupont, M. (2002) Health monitoring of sandwich plates based on the analysis of the interaction of Lamb waves with damages. In *Structural Health Monitoring* (D.L. Balageas, ed.) DEStech Pub, pp. 336–343.

Pardo de Vera, C. and Guemes, J.A. (1997) Embedded self-sensing Piezoelectrics for damage detection. *Proceedings of the International Workshop on Structural Health Monitoring*, 18–20 September 1997, Stanford University, CA, pp. 445–455.

Paris, P.C. and Erdogan, F. (1963) A critical analysis of crack propagation laws. *J. Basic Eng. ASME Transactions, Ser. D*, **85**(4), 528–534.

Park, G., Kabeya, K., Cudney, H., and Inman, D.J. (1999) Impedance-based structural health monitoring for temperature varying applications. *JSME Int. J.*, **42**(2), 249–258.

Park, G., and Inman, D.J. (2001) Impedance-based structural health monitoring, monograph. In *Nondestructive Testing and Evaluation Methods for Infrastructure Condition Assessment* (S.C. Woo, ed.) New York, NY: Kluwer Academic Publishers.

Park, H.W., Sohn, H., Law, K.H., and Farrar, C.R. (2004) Time reversal active sensing for health monitoring of a composite plate. *J. Sound and Vibration*, http://www.sciencedirect.com/science/journal/0022460X.

Parzen, E. (1962) On estimation of a probability density function and mode. Ann. Math. Stat., **33**, pp. 1065–1076.

Pasco, Y., Pinsonnault, J., et al. (2006) Time reversal method for damage detection of cracked plates in the medium frequency range: The case of wavelength-size cracks. *SPIE's 13th International Symposium on Smart Structures and Materials and 11th International Symposium on NDE for Health Monitoring and Diagnostics*, 26 February–2 March, 2006, San Diego, CA, paper #6176.

Pierce, S., Staszewski, W., Gachagan, A. et al. (1997) Ultrasonic condition monitoring of composite structures using a low profile acoustic source and an embedded optical fiber sensor. *SPIE* Vol. 3041, pp. 437–448.

Pierce, S., Culshaw, B., Manson, G. et al. (2000) The application of ultrasonic Lamb wave techniques to the evaluation of advanced composite structures. *SPIE* Vol. 3986, pp. 93–103.

Poisson, S.-D. (1829) Memoires de l'Academie royale des Sciences de l'Institut de France. Tome VIII.

Polytec PI, Inc. (2000), www.polytecpi.com.

Principe, J.C., Euliano, N.R., and Lefebvre, W.C. (2000) *Neural and Adaptive Systems: Fundamentals Through Simulation*. ISBN 0-471-35164-9.

Quinn, P., Palacios, L., Carman, G., and Speyer, J. (1999) Health monitoring of structures using directional piezoelectrics. *ASME Mechanics and Materials Conference*, 27–30 June 1999, Blacksburg, VA.

Raghavan, A. and Cesnik, C.E.S. (July, 2004) Modeling of piezoelectric-based Lamb-wave generation and sensing for structural health monitoring. *Smart Struct. Mater.*, 419–430.

Raghavan, A. and Cesnik, C.E.S. (2005) Piezoelectric-actuator excited-wave field solution for guided-wave structural health monitoring. *SPIE* Vol. 5765, pp. 313–323.

Rasson, J.P. and Lissoir, S. (1998) Symbolic Kernel discriminant analysis. NTTS'98, *International Seminar on New techniques and Technologies for Statistics*.

Rayleigh, J.W.S. (August, 1873) On the nodal lines of a square plate. *Phil Mag.*, Ser. 4, **46**(304), 166–171.

Rayleigh, J.W.S. (1887) On waves propagated along the plane surface of an elastic solid. *Proc. London Math. Soc.*, **17**, 4–11.

Rich, T.P. and Cartwright, D.J. (Eds) *Case Studies of Fracture Mechanics*. Report AMMRC MS77-5, US Army Material Development and Readiness Command, Alexandria, VA, 1977.

Ritz, W. (1909) Theorie der Transversalschwingungeneiner quadratischen Platte mitfrien Rdndern. *Ann. d. Physik*, **28**, pp. 737–786.

Rizzo, P. and Lanza di Scalia, F. (2004) Discrete wavelet transform to improve guided-wave-based health monitoring of tendons and cables. *SPIE* Vol. 5391, pp. 523–532.

Rokhlin, S.I. and Wang, L. (June, 2002) Stable recursive algorithm for elastic wave propagation in layered anisotropic media: Stiffness matrix method. *J. Acoust. Soc. Am.*

Rose, J.L., Ditri, J.J., Pilarski, A. et al. (1994) A guided wave inspection technique for nuclear steam generator tubing. *NDT&E Int.*, **27**, 307–310.

Rose, J.L. (1995) Recent advances in guided wave NDE. *1995 IEEE Ultrasonics Symposium*, pp. 761–770.

Rose, J.L., Rajana, K.M., and Hansch, M.K.T. (1995) Ultrasonic guided waves for NDE of adhesively bonded structures. *J. Adhes.*, **50**, 71–82.

Rose, J.L., Rajana, K.M., and Barnisher, J.N. (1996) Guided waves for composite patch repair of aging aircraft. *Rev. Prog. Quant. Nondestr. Eval.*, **15**, 1291–1298, Plenum Press.

Rose, J.L. (1999) *Ultrasonic Waves in Solid Media.* New York: Cambridge University Press.

Rose, J.L., Hay, T., and Agarwala, V.S. (2000) Skin to honeycomb core delamination detection with guided waves. *4th Joint DOD/FAA/NASA Conf. on Aging Aircraft.*

Rose, J.L. (August, 2002) A baseline and vision of ultrasonic guided wave inspection potential. *ASME J. Pressure Vessel Technol.* – Special Issue on Nondestructive Characterization of Structural Materials, **124**(3), 273–282.

Royer, D. and Dieulesaint, E. (2000) *Elastic Waves in Solids*. Springer.

Saffi, M. and Sayyah, T. (2001) Health monitoring of concrete structures strengthened with advanced composite materials using smart piezoelectric material. *Smart Mater. Struct.*, **11**, 317–329.

Saravanos, D.A., Birman, V., and Hopkins, D.A. (1994) Detection of delaminations in composite beams using piezoelectric sensors. *Proceedings of the 35th AIAA Structures, Structural Dynamics and Materials Conference.*

Shah, S.P., Swartz, S.E., and Ouyang, C. (1995) *Fracture Mechanics of Concrete.* John Wiley & Sons.

Shotliffe, E.H. and Buchanan, B.G. (1975) *MYCIN: Computer-Based Medical Consultations.* New York: Elsevier.

Sicard, R., Goyette, J., and Zellouf, D. (2002) A numerical dispersion compensation technique for time recompression of Lamb wave signals. *Ultrasonics*, **40**(1–8), 727–732.

Silk, M.G. and Bainton, K.F. (January, 1979) The propagation in metal tubing of ultrasonic wave modes equivalent to Lamb waves. *Ultrasonics*, 11–19.

Singher, L., Segal, Y., and Shamir, J. (1997) Interaction of a guided wave with a nonuniform adhesion bond. *Ultrasonics*, **35**, 385–391.

Soh, C.K., Tseng, K.K.-H., Bhalla, S., and Gupta, A. (2000) Performance of smart piezoelectric patches in health monitoring of RC bridge. *Smart Mater. Struct.*, **9**(4), 533–542.

Sohn, H., and Farrar, R.C. (2000) Statistical process control and projection techniques for damage detection. *Proceedings of the European COST F3 Conference on System Identification & Structural Health Monitoring*, Madrid, Spain, pp. 105–114.

Sohn, H., Worden, K., and Farrar, C.R.A. (2001) Novelty detection under changing environmental conditions. *Proceedings of the SPIE's Conference on Smart Systems for Bridges, Structures, and Highways*, 4–8 March 2001, Newport Beach, CA, USA, Vol. 4330, pp. 108–127.

Sohn, H., Wait, J.R., Park, G., and Farrar, C.R. (2004a) Multi-scale structural health monitoring for composite structures. *2nd European Workshop on Structural Health Monitoring*, 7–9 July 2004, Munich, Germany, pp. 721–729.

Specht, D.F. (1990) Probabilistic neural networks. *Neural Networks*, **3**, 109–118.

Spiegel, M.R. (1968) *Mathematical Handbook of Formulas and Tables.* Schaum's Outline Series, New York: Mc Graw-Hill.

Staszewski, W.J., Pierce, S.G., Worden, K., and Culshaw, B. (1999) Cross-wavelet analysis for Lamb wave damage detection in composite materials using optical fibre. *Key Eng. Mater.*, **167–168**, 373–380.

Stoneley, R. (1924) Elastic waves at the surface of separation of two solids. *Proc. R. Soc. London, Ser. A*, **106**, 416–428.

Su, Z. and Ye, L. (March, 2004) Fundamental Lamb mode-based delamination detection for CF/CP composite laminates using distributed piezoelectrics. *Struct. Health Monit. – An Int. J.*, **3**(1), 43–68.

Sun, F.P., Liang, C., and Rogers, C.A. (1994) Experimental modal testing using piezoceramic patches as collocated sensors-actuators. *Proceeding of the 1994 SEM Spring Conference & Exhibits*, 6–8 June 1994, Baltimore, MI.

Sun, F.P., Chaudhry, Z., Rogers, C.A., and Majmundar, M. (1995) Automated real-time structure health monitoring via signature pattern recognition. *Proceedings, SPIE North American Conference on Smart Structures and Materials*, 26 February–3 March, 1995, San Diego, CA, Vol. 2443, pp. 236–247.

Sundareshan, M.J., Schulz, M.J., Ghoshal, A., and Martin, W.N. (2001) A neural system for structural health monitoring. *SPIE* Vol. 4328, pp. 130–141.

Tang, B., Henneke, E.G. II (March, 1989) Long wavelength approximation for Lamb wave characterization of composite laminates. *Res. Nondestr. Eval.*, **1**(1), 51–64.

Thomson, W.T. (1950) Transmission of elastic waves through a stratified solid medium. *J. Appl. Phys.*, **21**, 89–93.

Thomson, D.O. and Chimenti, D.E. (Eds) (2002) Review of Progress in Quantitative Nondestructive Evaluation, Chapter 2C 'Guided Waves' and Chapter 7 'NDE Applications'. AIP Conference Proceedings Vol. 615.

Todd, P.D. and Challis, R.E. (January, 1999) Quantitative classification of adhesive bondline dimensions using Lamb waves and artificial neural networks. *IEEE Trans. Ultrasonics. Ferroelectrics and Frequency Control*, **46**(1), 167–181.

Todoroki, A., Shimamura, Y., and Inada, T. (1999) Plug and monitor system via ethernet with distributed sensors and CCD cameras. *Proceedings of the 2nd International Workshop on Structural Health Monitoring*, 8–10 September 1999, Stanford University, Stanford, CA, USA, pp. 571–580.

Tseng, K.-H., Soh, C.K., and Naidu, A.S.K. (June, 2002) Non-parametric damage detection and characterization using smart piezoceramic material. *Smart Mater. Struct.*, **11**(3), 317–329.

Tzou, H.S. and Tseng, C.I. (1990) Distributed piezoelectric sensor/actuator design for dynamic measurement/control of distributed parametric systems: A piezoelectric finite element approach. *J. Sound and Vibration*, **138**, pp. 17–34.

Valdez, S.H.D. (September, 2000) Structural integrity monitoring of CFRP laminates using piezoelectric devices. PhD Thesis, Imperial College of Science, Technology and Medicine.

Van Way, C.B., Kudva, J.N., Schoess, J.N. et al. (1995) Aircraft structural health monitoring system development – overview of the air force/navy smart metallic structures program. *SPIE* Vol. 2443, pp. 277–284.

Vermula C. and Norris, A.N. (2005) Flexural wave propagation and scattering on thin plates using Mindlin theory *Wave Motion*, **26**, I-12, June, 2005.

Viktorov, I.A. (1967) *Rayleigh and Lamb Waves – Physical Theory and Applications*. New York: Plenum Press.

Wang, L. and Rokhlin, S.I. (2001) Stable reformulation of transfer matrix method for wave propagation in layered anisotropic media. *Ultrasonics*, **39**, 413–424.

Wang, C.S. and Chang, F.-K. (1999) Built-in diagnostics for impact damage identification of composite structures. In *Structural Health Monitoring 2000* (F.-K. Chang, ed.) Technomic, pp. 612–621.

Wang, C.H., Rose, J.T., and Chang, F.-K. (2003) A computerized time-reversal method for structural health monitoring. *SPIE* Vol. 5046, paper #5046-7, pp. 48–58.

Wang, C.H., Rose, J.T., and Chang, F.K. (2004) A synthetic time-reversal imaging method for structural health monitoring. *Smart Mater. Struct.*, **13**(2), 415–423.

Wang, X., Lu, G., and Guillow, S.R. (January, 2002) Stress wave propagation in orthoptropic laminated thick-walled spherical shells. *Int. J. Solids Struct.*

Wasserman, P.D. (1993) Advanced methods in neural computing. Van Nostrand Reinhold.

Watson, G.N. (1995) *A Treatise on the Theory of Bessel Functions*. 2nd ed. New York: Cambridge University Press.

Wilcox, P. (June, 2003a) Omni-directional guided wave transducer arrays for the rapid inspection of large areas of plate structures. *IEEE Trans. Ultrasonics. Ferroelectrics and Frequency Control*, **50**(6), 699.

Wilcox, P. (2003b) A rapid signal processing technique to remove the effect of dispersion from guided wave signals. *Ultrasonics*, **31**(3), 201–204.

Wilcox, P., Lowe, M., and Cawley, P. (April, 2005) Omnidirectional guided wave inspection of large metallic plate structures using an EMAT array. *IEEE Trans. Ultrasonics. Ferroelectrics and Frequency Control*, **52**(4), 653–665.

Williams, J.G. (1984) *Fracture Mechanics of Polymers*. Ellis Harwood Ltd.

Winston, P.H. (1993) *Artificial Intelligence*. Addison-Wesley, ISBN 0-201-53377-4.

Wooh, S.C. and Shi, Y. (1998) Influence of phased array element size on beam steering behavior. *Ultrasonics*, **36** (1998) 737–749.

Wooh, S.C. and Shi, Y. (1999) A simulation study of the beam steering characteristics for linear phased arrays. *J. Nondestr. Eval.*, **18**(2).

Worden, K. (2000) Nonlinearity in structural dynamics: The last ten years. *Proceedings of the European COST F3 Conference on System Identification & Structural Health Monitoring*, Madrid, Spain, pp. 29–51.

Worlton, D.C. (1957) Ultrasonic testing with Lamb waves. *Non-Destructive Test.*, **15**, 218–222.

Worlton, D.C. (1961) Experimental confirmation of Lamb waves at megacycles frequencies. *J. Appl. Phy.*, **32**, 967–971.

Ying, Z.G., Ni, Y.Q., and Ko, J.M. (2001) Seismic response mitigation of adjacent high-rise structures via stochastic optimal coupling-control. *SPIE* Vol. 4330, pp. 289–300.

Zagrai, A. and Giurgiutiu, V. (2001) Electro-mechanical impedance method for crack detection in thin plates. *J. Intell. Mater. Syst. Struct.*, **12**(10), 709–718.

Zang, C. and Imregun, M. (2000) Structural damage detection via principal component analysis and artificial neural networks. *Proceedings of the European COST F3 Conference on System Identification & Structural Health Monitoring*, Madrid, Spain, pp. 157–167.

Zemenek, J. (1972) An experimental and theoretical investigation of elastic wave propagation in a cylinder. *J. Acoust. Soc. Am.*, **51**(1), 265–283, Pt. 2.

Zhongqing, S. and Ye L. (February, 2005) Lamb wave propagation-based damage identification for quasi-isotropic CF/EP composite laminates using artificial neural algorithm: Part II – Implementation and validation. *J. Intell. Mater. Syst. Struct.*, **16**.

INDEX

G

I

O